P9-EEH-147

Diversity Amid Globalization

WORLD REGIONS, ENVIRONMENT, DEVELOPMENT

Fifth Edition

Les Rowntree
University of California, Berkeley

Martin Lewis
Stanford University

Marie Price
George Washington University

William Wyckoff
Montana State University

Prentice Hall

Boston Columbus Indianapolis New York San Francisco Upper Saddle River
Amsterdam Cape Town Dubai London Madrid Milan Munich Paris Montréal Toronto
Delhi Mexico City São Paulo Sydney Hong Kong Seoul Singapore Taipei Tokyo

Geography Editor: Christian Botting
Marketing Manager: Maureen McLaughlin
Project Editor: Anton Yakovlev
Assistant Editor: Kristen Sanchez
Editorial Assistant: Christina Ferraro
Marketing Assistant: Nicola Houston
Managing Editor, Geosciences and Chemistry: Gina M. Cheselka
Project Manager, Production: Maureen Pancza
Design Director: Mark Stuart Ong
Interior and Cover Design: Wanda Espana/Wee Design Group
Art Project Managers: Connie Long and Ronda Whitson
Senior Manufacturing & Operations Manager: Nick Sklitsis
Operations Specialist: Maura Zaldivar
Media Producer: Tim Hainley
Senior Media Production Supervisor: Liz Winer
Associate Media Project Manager: David Chavez
Photo Research Manager: Elaine Soares
Photo Researchers: Kristin Piljay and Caroline Commins
Composition/Full Service: Element LLC
Production Editor, Full Service: Suganya Karrupasamy
Copy Editor: Marcia Youngman
Illustrations: Spatial Graphics, Kevin Lear

Cover Photograph: Landscape with grain terraces, Highlands, Madagascar, Africa; Alamy

Credits and acknowledgments borrowed from other sources and reproduced, with permission, in this textbook appear on the appropriate page within the text or on p. 691.

Library of Congress Cataloging-in-Publication Data

Diversity amid globalization : world regions, environment, development / Les Rowntree ...[et al.]. — 5th ed.
p. cm.
Includes bibliographical references and index.
ISBN-13: 978-0-321-71448-0
ISBN-10: 0-321-71448-2
1. Geography. I. Rowntree, Lester, 1938-
G128.D58 2012
910—dc22

2010044253

Printed in the United States
10 9 8 7 6 5 4 3 2 1

Prentice Hall
is an imprint of

www.pearsonhighered.com

ISBN 0-321-71448-2/978-0-321-71448-0

BRIEF CONTENTS

ABOUT OUR SUSTAINABILITY INITIATIVES

This book is carefully crafted to minimize environmental impact. The materials used to manufacture this book originated from sources committed to responsible forestry practices. The paper is FSC® certified. The binding, cover, and paper come from facilities that minimize waste, energy consumption, and the use of harmful chemicals.

Pearson closes the loop by recycling every out-of-date text returned to our warehouse. We pulp the books, and the pulp is used to produce items such as paper coffee cups and shopping bags. In addition, Pearson aims to become the first climate neutral educational publishing company.

The future holds great promise for reducing our impact on Earth's environment, and Pearson is proud to be leading the way. We strive to publish the best books with the most up-to-date and accurate content, and to do so in ways that minimize our impact on Earth.

MIX
Paper from
responsible sources
FSC® C011387

Prentice Hall
is an imprint of

CONTENTS

PREFACE

Diversity Amid Globalization is an issues-oriented textbook for college and university world regional geography classes that explicitly recognizes the geographic changes accompanying globalization. With this focus we join the many who argue that globalization is the most fundamental reorganization of the planet's socioeconomic, cultural, and geopolitical structure since the Industrial Revolution; this premise provides the point of departure for this book. As geographers we think it essential for students to understand two interactive tensions. First, they need to appreciate and critique the consequences of converging environmental, cultural, political, and economic systems through globalization. Second, they need to also deepen their understanding of the creation and persistence of geographic diversity and difference. Consequently, the interaction and tension between these opposing forces of homogenization and diversification forms a current running throughout the book's chapters and is reflected in our title, *Diversity Amid Globalization*.

NEW TO THE FIFTH EDITION

Numerous updates and changes have been made to the fifth edition of *Diversity Amid Globalization* to increase and enhance its efficacy as a world regional geography textbook.

- **Expanded, Integrated Treatment of Globalization**, including a revised introductory discussion in chapter 1, and new introductory paragraphs within the regional chapter-opening spreads that briefly frame how the forces of globalization are affecting the region.
- ***Exploring Global Connections*** are new sidebar essays in each regional chapter that discuss globalization in terms of transregional links—about how different regions are connected.
- ***Environment and Politics*** are new sidebar essays in each regional chapter that explore the complicated interaction in a globalized world between some aspect of the environment (global warming, rainforest destruction, sustainability, infrastructural construction, historic preservation, etc) and the politics and policies enabling or inhibiting pertinent social action.
- **New *Development Issues* maps** in each regional chapter highlight some measure of economic and social development, encouraging students to appreciate the many ways to define and measure development in different regional settings.
- **Updated *Population* maps** now contain more effective density and urban size categories that better represent the reality of the contemporary world.
- **The most current information on global warming** is integrated, including the policies and programs enacted by different countries and regions to address climate change, emission reduction, and energy usage. Updated material is introduced initially in chapter 2, and subsequently integrated through the different regional chapters in dedicated climate change sections.
- **Current data and information** are integrated into all text, tables, and maps.
- **MasteringGeography™** Used by over one million science students, the Mastering platform is the most effective and widely used online tutorial, homework, and assessment system for the sciences. Now available with *Diversity Amid Globalization,* **MasteringGeography™** offers:
  - Assignable activities that include: MapMaster™ interactive maps, geography videos from Television for the Environment's global *Life* and *Earth Report* series, *Encounter World Regional Geography* Google Earth multimedia, Thinking Spatially and Data Analysis.
  - Additional end-of-chapter questions
  - Test Bank questions and Reading Quizzes
  - A student Study Area with MapMaster interactive maps, geography videos, *In the News* RSS feeds, web links, flashcard glossary, quizzes, and more.
  - An eText version of *Diversity Amid Globalization*.

CHAPTER ORGANIZATION

As is true of most other regional geography textbooks, *Diversity Amid Globalization* is organized to describe and explain the major world regions of Asia, Europe, Africa, the Americas, and so on. Our 12 regional chapters, however, depart somewhat from traditional world regional textbooks. Instead of filling these regional chapters with descriptions of individual countries, we place most of that material on the MasteringGeography™ Study Area. This leaves us free to develop five important themes as the structure for each regional chapter. We begin with Environmental Geography by discussing the physical geography of each region, as well as current environmental issues. Next, we assess the region's Population and Settlement geography, in which demography, land use, and settlement (including cities) are discussed. Third is a section on Cultural Coherence and Diversity that examines the geography of language and religion and also explores current cultural tensions resulting from the interplay of globalization and diversity. Following that is a thematic section on each region's Geopolitical Framework that examines the political geography of the region, including micro-regionalism, separatism, ethnic conflicts, global terrorism, and supranational organizations. We then conclude each regional chapter with a section on Economic and Social Development in which we analyze each region's economic framework; this section also examines social development issues, a topic that includes regional and global gender issues.

The regional framework follows two substantive introductory chapters that provide the conceptual and theoretical framework of human and physical geography necessary to understand the contemporary globalized world. In the first chapter students are introduced to the notion of globalization and are asked to ponder the costs and benefits of the globalization process, a critical perspective that is becoming increasingly necessary in the contentious global arena. Following this the geographical foundation for each of the five thematic sections is examined. This discussion draws heavily on the major concepts fundamental to an introductory university geography course. The second chapter, "The Changing Global Environment," as mentioned, presents the themes and concepts of global physical geography, including landforms and geology, climate, global warming, hydrology, biogeography, and food production.

CHAPTER FEATURES

Within each regional chapter several unique features complement the thematic pedagogy of our approach:

- **Comparable Maps** Of the many maps in each regional chapter, 8 are constructed on the same theme and with similar data so that readers can easily draw comparisons between different regions. Thus, in every regional chapter readers will find maps of physical geography, climate, environmental issues, population density, language, religion, development issues, and geopolitical issues of the region.
- **Other Maps** In addition, each regional chapter also has a double handful of other maps illustrating such major themes as urban growth, ethnic tensions, social development, regional development, and linkages to the global economy. As well, the fifth edition presents 30 maps of specific subregions. These maps are linked to a specific theme (or themes) within each chapter and are explicitly discussed in the text. These new subregion maps help the text provide a stronger emphasis on the local scale and how global processes affect places
- **Comparable Regional Data Sets** Again, to facilitate comparison between regions, as well as to provide important insight into the characteristics of each region, each chapter contains two tables. The first provides population data of all sorts, including population density, level of urbanization, total fertility rates, proportion of the population under 15 and over 65 years of age, and net migration rates for each country within the region. The second table presents economic and social development data for each country including GNI per capita, GDP growth, life expectancy, percent of the population living on less than $2 per day, infant mortality rates, and the UN gender equity index.
- **Sidebar Essays** Within each chapter there is an array of sidebar essays that complement text material.
  - **Cityscapes** These sidebars combine text, maps, and photos to convey a sense of place for a major city within each regional chapter and speak to the fact that our globalized world is now an urban one.
  - **People on the Move** sidebars capture the human geography behind contemporary migration as people migrate—legally and not so legally—responding to the varied currents of globalization.
  - **Global to Local** sidebars highlight how global trends affect local areas and local people. The sidebars foreground the fact that people often respond to homogenizing global forces very differently, and this variation in responses gives unique character to local people and places.
  - **Geographic Tools** sidebars discuss the various tools used by geographers to understand the world. Examples are GIS, aerial photography, field surveys, repeat photography, and so on. This content also links these tools and skills to employment possibilities in geography and allied fields.
  - **Environment and Politics** sidebars reflect that politics and policy shape the global landscape in many different ways, and these new sidebars focus on those relationships with case studies from the region.
  - **Exploring Global Connections** sidebars explore how activities in different world regions are linked so that students understand that in a globalized world regions are neither isolated nor discrete entitiies.
- **Review and Research Questions** There are two sets of review and research questions at the end of each regional chapter. The first set helps readers review basic terminology and concepts. The second set, "Thinking Geographically," asks readers to draw together more abstract themes and problems by doing further research on an issue. Resources for answering these critical thinking exercises are found in the chapter bibliographies, in the *Instructor Resource Manual,* and on MasteringGeography.
- **Regional Novels and Films** Students can learn much about a world region through novels and films with strong local content, thus we include a list of selected works at the end of each regional chapter that we believe complements the understanding of a region's people and places.

ACKNOWLEDGMENTS

We have many people to thank for their help in the conceptualization, writing, rewriting, and production of *Diversity Amid Globalization*. First, we'd like to thank the thousands of students in our world regional geography classes who have inspired us with their energy, engagement, and curiosity; challenged us with their critical insights; and demanded a textbook that better meets their need to understand the diverse people and places of our dynamic world.

Next, we are deeply indebted to many professional geographers and educators for their assistance, advice, inspiration, encouragement, and constructive criticism as we labored through the different stages of this book. Among the many who provided invaluable comments on various drafts and editions of *Diversity Amid Globalization* are:

Gilian Acheson, *Southern Illinois University–Edwardsville*
Joy Adams, *Humboldt State University*
Dan Arreola, *Arizona State University*
Bernard BakamaNume, *Texas A&M University*
Brad Baltensperger, *Michigan Technological University*
Laurence Becker, *Oregon State University*
Max Beavers, *Samford University*
James Bell, *University of Colorado*
William H. Berentsen, *University of Connecticut*
Kevin Blake, *Kansas State University*
Karl Byrand, *University of Wisconsin–Sheboygan County*
Michelle Calvarese, *California State University, Fresno*
Craig Campbell, *Youngstown State University*
Elizabeth Chacko, *George Washington University*
Philip Chaney, *Auburn University*
David B. Cole, *University of Northern Colorado*
Malcolm Comeaux, *Arizona State University*
Jonathan C. Comer, *Oklahoma State University*
Catherine Cooper, *George Washington University*
Jeremy Crampton, *George Mason University*
Kevin Curtin, *University of Texas at Dallas*
James Curtis, *California State University–Long Beach*
Dydia DeLyser, *Louisiana State University*
Francis H. Dillon, *George Mason University*
Jason Dittmer, *Georgia Southern University*
Jerome Dobson, *University of Kansas*
Caroline Doherty, *Northern Arizona University*
Vernon Domingo, *Bridgewater State College*
Roy Doyon, *Ball State University*
Jane Ehemann, *Shippensburg University*
Dean Fairbanks, *California State University–Chico*
Chuck Fahrer, *Georgia College and State University*
Caitie Finlayson, *Florida State University*
Doug Fuller, *George Washington University*
Gary Gaile, *University of Colorado*
Sherry Goddicksen, *California State University–Fullerton*
Sarah Goggin, *Cypress College*
Reuel Hanks, *Oklahoma State University*
Steven Hoelscher, *University of Texas, Austin*
Erick Howenstine, *Northeastern Illinois University*
Peter J. Hugil, *Texas A&M University*
Eva Humbeck, *Arizona State University*
Ryan S. Kelly, *University of Kentucky*
Richard H. Kesel, *Louisiana State University*
Rob Kremer, *Front Range Community College*
Robert C. Larson, *Indiana State University*
Alan A. Lew, *Northern Arizona University*
Catherine Lockwood, *Chadron State College*
Max Lu, *Kansas State University*
Luke Marzen, *Auburn University*
Kent Matthewson, *Louisiana State University*
James Miller, *Clemson University*
Bob Mings, *Arizona State University*
Wendy Mitteager, *SUNY Oneonta*
Sherry D. Morea-Oakes, *University of Colorado, Denver*
Anne E. Mosher, *Syracuse University*
Tim Oakes, *University of Colorado*
Nancy Obermeyer, *Indiana State University*
Karl Offen, *University of Oklahoma*
Thomas Orf, *Las Positas College*
Kefa Otiso, *Bowling Green State University*
Joseph Palis, *University of North Carolina*
Jean Palmer-Moloney, *Hartwick College*
Bimal K. Paul, *Kansas State University*
Michael P. Peterson, *University of Nebraska–Omaha*
Richard Pillsbury, *Georgia State University*
Brandon Plewe, *Brigham Young University*
Jess Porter, *University of Arkansas at Little Rock*
Patricia Price, *Florida International University*
Erik Prout, *Texas A&M University*
David Rain, *United States Census Bureau*
Rhonda Reagan, *Blinn College*
Craig S. Revels, *Portland State University*
Scott M. Robeson, *Indiana State University*
Pamela Riddick, *University of Memphis*
Paul A. Rollinson, *Southwest Missouri State University*
Yda Schreuder, *University of Delaware*
Kathy Schroeder, *Appalachian State University*
Kay L. Scott, *University of Central Florida*
Duncan Shaeffer, *Arizona State University*
Dimitrii Sidorov, *California State University–Long Beach*
Susan C. Slowey, *Blinn College*
Andrew Sluyter, *Louisiana State University*
Christa Smith, *Clemson University*
Joseph Spinelli, *Bowling Green State University*
William Strong, *University of Northern Alabama*
Philip W. Suckling, *University of Northern Iowa*
Curtis Thomson, *University of Idaho*
Suzanne Traub-Metlay, *Front Range Community College*
James Tyner, *Kent State University*
Nina Veregge, *University of Colorado*
Gerald R. Webster, *University of Alabama*
Keith Yearman, *College of DuPage*

Emily Young, *University of Arizona*
Bin Zhon, *Southern Illinois University–Edwardsville*
Henry J. Zintambia, *Illinois State University*

In addition, we wish to thank the many publishing professionals who have been involved with this project, for it is a privilege working with you. We thank Paul F. Corey, President of Pearson's Science division, for his early and continued support for this book project; Geography editor and good friend Christian Botting, for his professional guidance, enduring patience, and high standards; Project Manager Anton Yakovlev, for his daily attention to production matters and his diplomatic interaction with four demanding (and sometimes cranky) authors; Editorial Assistant Christina Ferraro for gracefully taking care of the many, many incidental tasks connected to this project; Marketing Manager Maureen McLaughlin for her sales and promotion work; Production Project Manager Maureen Pancza, *Element* Production Editor Suganya Karrupasamy and *Element* VP of Production Cindy Miller for somehow turning thousands of pages of manuscript into a finished book; *Spatial Graphics* Production Manager and Cartographer Kevin Lear for his outstanding work on our maps, and, finally, photo researchers Kristin Piljay and Caroline Commins for their fine work finding photographs for this new edition of *Diversity Amid Globalization*.

Last, the authors want to thank that special group of friends and family who were there when we needed you most—early in the morning and late at night; in foreign countries and familiar places; when we were on the verge of crying yet needing to laugh; for your love, patience, companionship, inspiration, solace, understanding, and enthusiasm: Eugene Adogia, Elizabeth Chacko, Meg Conkey, Rob Crandall, Karen Wigen, and Linda, Tom, and Katie Wyckoff. Words cannot thank you enough.

Les Rowntree
Martin Lewis
Marie Price
William Wyckoff

THE TEACHING AND LEARNING PACKAGE

In addition to the text itself, the authors and publisher have worked with a number of talented people to produce an excellent instructional package. This package includes traditional supplements as well as new online multimedia and assessment.

MasteringGeography™ with Pearson eText

Used by over one million science students, the Mastering platform is the most effective and widely used online tutorial, homework, and assessment system for the sciences. MasteringGeography™ offers:

- Assignable activities that include: MapMaster interactive maps, geography videos, *Encounter World Regional Geography* Google Earth multimedia, Thinking Spatially and Data Analysis.
- Additional end-of-chapter questions
- Test Bank questons and Reading Quizzes
- Student study area with MapMaster interactive maps, videos, *In the News* RSS feeds, web links, glossary flash cards, quizzes, and more.
- Pearson eText for *Diversity Amid Globalization*, Fifth Edition, which gives students access to the text whenever and wherever they can access the Internet, and includes powerful interactive and customization functions. www.masteringgeography.com

Instructor

- ***Instructor Resource Manual* (online download only)** Includes Learning Objectives, detailed Chapter Outlines, Key Terms, For Thought and Discussion topics to generate classroom motivation, and Exercise Activities that link to features in the students' *Mapping Workbook* and the MapMaster Interactive Maps. www.pearsonhighered.com/irc
- **TestGen® Computerized Test Bank (online download only)** TestGen® is a computerized test generator that lets instructors view and edit *Test Bank* questions, transfer questions to tests, and print the test in a variety of customized formats. This *Test Bank* includes approximately 1000 multiple choice, true/false, and short answer/essay questions. New to this edition are questions that focus on figures and maps from the text. Questions are correlated against the revised National Geography Standards and Bloom's Taxonomy to help instructors better map the assessments against both broad and specific teaching and learning objectives. The *Test Bank* is also available in Microsoft Word®, and is importable into Blackboard and WebCT. www.pearsonhighered.com/irc
- **Blackboard Test Bank (download only)** The Blackboard Test Bank provides *Test Bank* questions for import into Blackboard Learning System. www.pearsonhighered.com/irc
- **Instructor Resource Center on DVD (0-321-72310-4)** Everything instructors need where they want it. The Pearson Prentice Hall *Instructor Resource Center on DVD* helps make instructors more effective by saving them time and effort. All digital resources can be found in one well-organized, easy-to-access place. The IRC on DVD includes:
  - All textbook images as JPEGs, PDFs, and PowerPoint® presentations
  - Pre-authored Lecture Outline PowerPoint presentations, which outline the concepts of each chapter with embedded art and can be customized to fit instructors' lecture requirements
  - CRS "Clicker" Questions in PowerPoint format, correlated against the National Geography Standards and Bloom's Taxonomy
  - The TestGen software, Test Bank questions, and answers for both Macs and PCs, correlated against the National Geography Standards and Bloom's Taxonomy
  - Electronic files of the *Instructor Resource Manual* and *Test Bank*
  - Approximately 100 Geography Video Clips

  This Instructor Resource content is also available completely online via the Instructor Resources section of www.mygeoscienceplace.com and www.pearsonhighered.com/irc.
- **Television for the Environment *Earth Report* Geography Videos on DVD (0-32-166298-9)** This three-DVD set is designed to help students visualize how human decisions and behavior have affected the environment, and how individuals are taking steps toward recovery. With topics ranging from the poor land management promoting the devastation of river systems in Central America to the struggles for electricity in China and Africa, these 13 videos from Television for the Environment's global *Earth Report* series recognize the efforts of individuals around the world to unite and protect the planet.
- **Television for the Environment *Life* World Regional Geography Videos on DVD (0-13-159348-X)** From the Television for the Environment's global *Life* series this two-DVD set brings globalization and the developing world to the attention of any world regional geography course. These 10 full-length video programs highlight matters such as the growing number of homeless children in Russia, the lives of immigrants living in the United States trying to aid family still living in their native countries, and the European conflict between commercial interests and environmental concerns.

- **Television for the Environment *Life* Human Geography Videos on DVD (0-13-241656-5)** This three-DVD set is designed to enhance any human geography course. These DVDs include 14 full-length video programs from Television for the Environment's global *Life* series, covering a wide array of issues affecting people and places in the contemporary world, including the serious health risks of pregnant women in Bangladesh, the social inequalities of the "untouchables" in the Hindu caste system, and Ghana's struggle to compete in a global market.
- **Aspiring Academics: A Resource Book for Graduate Students and Early Career Faculty (0-13-604891-9)** Drawing on several years of research, this set of essays is designed to help graduate students and early career faculty begin their careers in geography and related social and environmental sciences. This teaching aid stresses the interdependence of teaching, research, and service—and the importance of achieving a healthy balance in professional and personal life—in faculty work, and does not view it as a collection of unrelated tasks. Each chapter provides accessible, forward-looking advice on topics that often cause the most stress in the first years of a college or university appointment.
- **Teaching College Geography: A Practical Guide for Graduate Students and Early Career Faculty (0-13-605447-1)** Provides a starting point for becoming an effective geography teacher from the very first day of class. Divided in two parts, the first set of chapters addresses "nuts-and-bolts" teaching issues in the context of the new technologies, student demographics, and institutional expectations that are the hallmarks of higher education in the 21st century. The second part explores other important issues: effective teaching in the field; supporting critical thinking with GIS and mapping technologies; engaging learners in large geography classes; and promoting awareness of international perspectives and geographic issues.
- **AAG Community Portal for Aspiring Academics & Teaching College Geography** This website is intended to support community-based professional development in geography and related disciplines. Here you will find activities providing extended treatment of the topics covered in both books. The activities can be used in workshops, graduate seminars, brown bags, and mentoring programs offered on campus or within an academic department. You can also use the discussion boards and contributions tool to share advice and materials with others. www.pearsonhighered.com/aag/

Student

- **MasteringGeography™ Student Access Code Card (0-321-72052-0)** This is the student access code card for MasteringGeography™.
- **Mapping Workbook (0-321-72312-0)** This workbook, which can be used in conjunction with either the main text or an atlas, features political and physical outline maps of every global region, printed in black and white. These maps, along with the names of the regional key locations and physical features, are the basis for identification exercises. Conceptual exercises are included to further exam students' comprehension of the Key Points presented in the main text chapters. These exercises review both the physical and human environment of the regions. An answer key for the mapping workbook exercises is also available.
- **Study Guide (0-321-75091-8)** The *Study Guide* includes additional learning objectives, a complete chapter outline, critical thinking exercises, problems and short essay work using actual figures from the text, and a self-test with an answer key in the back.
- **Encounter World Regional Geography Workbook & Website (0-321-68175-4)** *Encounter World Regional Geography* provides rich, interactive explorations of world regional geography concepts through Google Earth™ explorations. Students explore the globe through the themes of Environment, Population, Culture, Geopolitics, and Economy and Development, answering multiple choice and short answer questions. All chapter explorations are available in print format as well as online quizzes, accommodating different classroom needs. All worksheets are accompanied with corresponding Google Earth™ media files, available for download from www.mygeoscienceplace.com.
- **Encounter Earth: Interactive Geoscience Explorations Workbook & Website (0-321-58129-6)** *Encounter Earth* gives students a way to visualize key topics in their introductory geoscience courses. Each exploration consists of a worksheet, available in the workbook and as a PDF file, and a Google Earth™ KMZ file, containing placemarks, overlays, and annotations referred to in the worksheets. All chapter explorations are available in print format as well as online quizzes, accommodating different classroom needs. All worksheets are accompanied with corresponding Google Earth™ media files, available for download from www.mygeoscienceplace.com.
- **Goode's World Atlas 22nd Edition (0-321-65200-2)** Goode's World Atlas has been the world's premiere educational atlas since 1923, and for good reason. It features over 250 pages of maps, from definitive physical and political maps to important thematic maps that illustrate the spatial aspects of many important topics. The 22nd edition includes 160 pages of new, digitally produced reference maps, as well as new thematic maps on global climate change, sea level rise, CO^2 emissions, polar ice fluctuations, deforestation, extreme weather events, infectious diseases, water resources, and energy production.
- **Dire Predictions (0-13-604435-2)** Periodic reports from the Intergovernmental Panel on Climate Change (IPCC) evaluate the risk of climate change brought on by humans. But the sheer volume of scientific data remains inscrutable to the general public, particularly to those who may still question the validity of climate change. In just over 200 pages, this practical text presents and expands upon the essential findings of the 4th Assessment Report in a visually stunning and undeniably powerful way to the lay reader. Scientific findings that provide validity to the implications of climate change are presented in clear-cut graphic elements, striking images, and understandable analogies.

ABOUT THE AUTHORS

Les Rowntree is a Visiting Scholar at the University of California, Berkeley, where he researches and writes about environmental issues. This career change comes after three decades of teaching both Geography and Environmental Studies at San Jose State University in California. As an environmental geographer, Dr. Rowntree's interests focus on international environmental issues, biodiversity conservation, and human-caused global change. He sees world regional geography as a way to engage and inform students by giving them the conceptual tools needed to critically assess global issues. Dr. Rowntree has done research in Iceland, Alaska, Morocco, Mexico, Australia, and Europe, as well as in his native California. Current writing projects include a book on the natural history of California's coast, as well as textbooks in geography and environmental science.

Martin Lewis is a Senior Lecturer in History at Stanford University. He has conducted extensive research on environmental geography in the Philippines and on the intellectual history of global geography. His publications include *Wagering the Land: Ritual, Capital, and Environmental Degradation in the Cordillera of Northern Luzon, 1900–1986* (1992), and, with Karen Wigen, *The Myth of Continents: A Critique of Metageography* (1997). Dr. Lewis has traveled extensively in East, South, and Southeast Asia. His current research focuses on the geographical dimensions of globalization. In April 2009 Dr. Lewis was recognized by *Time* magazine, as a favorite lecturer.

Marie Price is a Professor of Geography and International Affairs at George Washington University. A Latin American specialist, Marie has conducted research in Belize, Mexico, Venezuela, Cuba, and Bolivia. She has also traveled widely throughout Latin America and Sub-Saharan Africa. Her studies have explored human migration, natural resource use, environmental conservation, and regional development. She is a non-resident fellow of the Migration Policy Institute, a non-partisan think tank that focuses on immigration. Dr. Price brings to *Diversity Amid Globalization* a special interest in regions as dynamic spatial constructs that are shaped over time through both global and local forces. Her publications include the co-edited book, *Migrants to the Metropolis: The Rise of Immigrant Gateway Cities* (2008, Syracuse University Press) and numerous academic articles and book chapters.

William Wyckoff is a geographer in the Department of Earth Sciences at Montana State University specializing in the cultural and historical geography of North America. He has written and co-edited several books on North American settlement geography, including *The Developer's Frontier: The Making of the Western New York Landscape* (1988), *The Mountainous West: Explorations in Historical Geography* (1995) (with Lary M. Dilsaver), *Creating Colorado: The Making of a Western American Landscape 1860–1940* (1999), and *On the Road Again: Montana's Changing Landscape* (2006). In 2003 he received Montana State's Cox Family Fund for Excellence Faculty Award for Teaching and Scholarship. A World Regional Geography instructor for 26 years, Dr. Wyckoff emphasizes in the classroom the connections between the everyday lives of his students and the larger global geographies that surround them and increasingly shape their future.

Linking people and places through globalization

From population to environment, economic development to cultural diversity, the process of globalization is dramatically changing our world. Using a consistent thematic structure to examine world regions, ***Diversity Amid Globalization: World Regions, Environment, Development,* Fifth Edition** explores the diverse geography of the world, and the dynamic causes and effects of globalization.

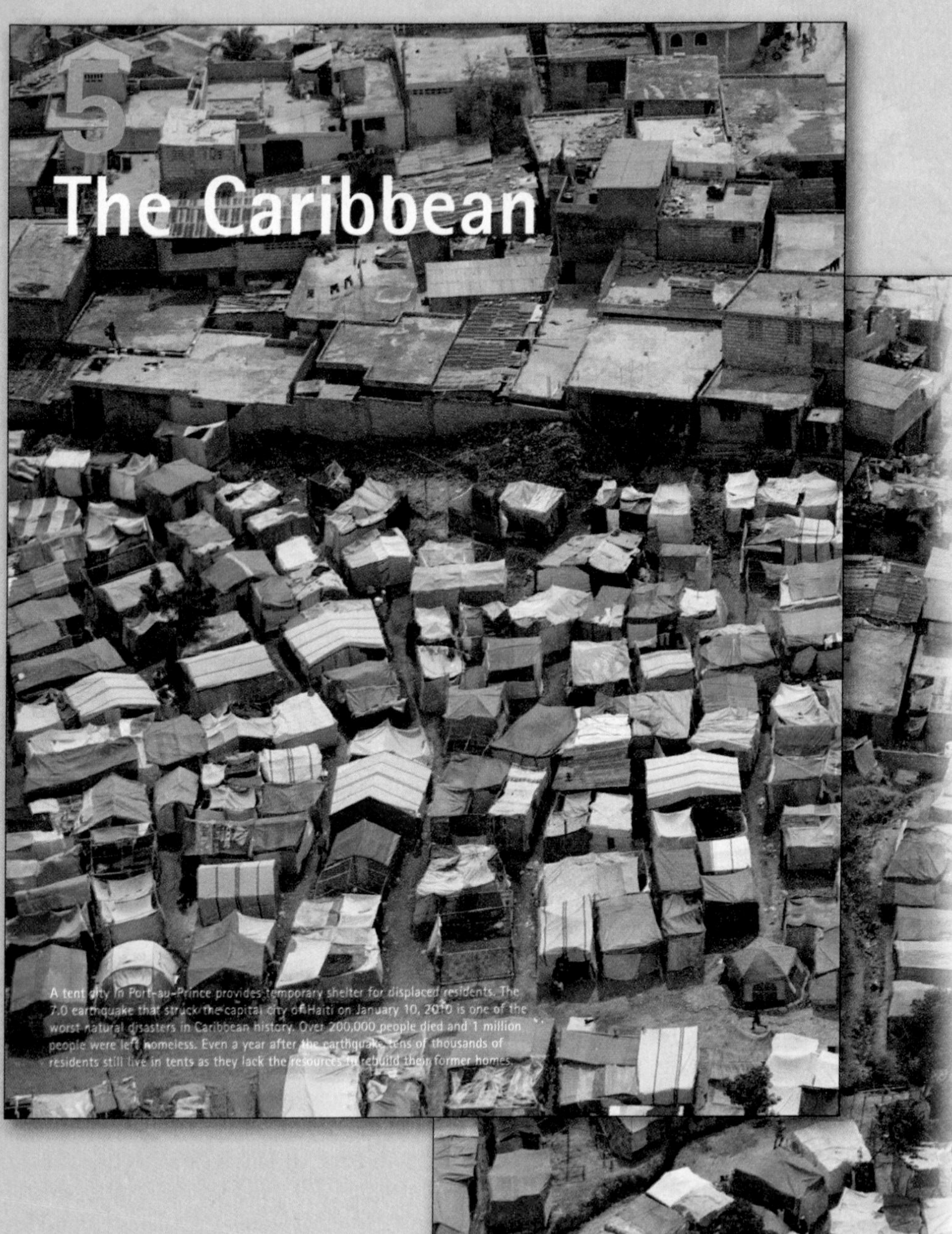

A tent city in Port-au-Prince provides temporary shelter for displaced residents. The 7.0 earthquake that struck the capital city of Haiti on January 10, 2010 is one of the worst natural disasters in Caribbean history. Over 200,000 people died and 1 million people were left homeless. Even a year after the earthquake, tens of thousands of residents still live in tents as they lack the resources to rebuild their former homes.

DIVERSITY AMID GLOBALIZATION

Named for some of its former inhabitants—the Carib Indians—the modern Caribbean is a culturally complex and economically peripheral world region. Settled by various colonial powers, the residents of this region are mostly a mix of African, European, and South Asian peoples. Greatly impacted by the global economy but with relatively little influence on it, the livelihoods of Caribbean peoples are filled with uncertainty.

ENVIRONMENTAL GEOGRAPHY
Climate change threatens the Caribbean, with the potential for stronger and more frequent hurricanes, loss of territory due to sea level rising, and destruction of coral reefs.

POPULATION AND SETTLEMENT
The earthquake that leveled Port-au-Prince, Haiti, in 2010 showed the vulnerability of this large but very poor Caribbean city. Throughout the region, large numbers of people have emigrated in search of economic opportunity and are sending back millions of dollars in remittances.

CULTURAL COHERENCE AND DIVERSITY
Creolization, the blending of African, European, and even Amerindian elements, has resulted in many unique Caribbean expressions of culture, such as *rara, reggae,* and steel drum bands.

GEOPOLITICAL FRAMEWORK
The first area of the Americas to be extensively explored and colonized by Europeans, the region has seen many rival European claims and, since the early 20th century, has experienced strong U.S. influence.

ECONOMIC AND SOCIAL DEVELOPMENT
Environmental, locational, and economic factors make tourism a vital component of this region's economy, particularly in Puerto Rico, Cuba, the Dominican Republic, and the Bahamas.

179

The changing face of globalization

Today, the world is experiencing the most fundamental reorganization of its socioeconomic, cultural, and geopolitical structure since the Industrial Revolution. ***Diversity Amid Globalization*** focuses on these changes by examining both the homogenizing and diversifying forces inherent to the globalization processes.

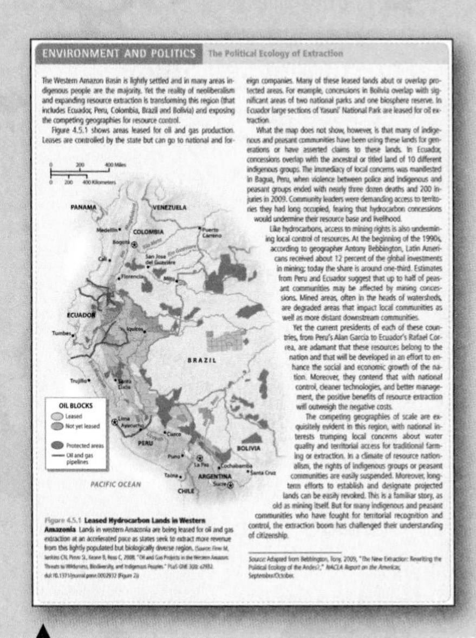

ENVIRONMENT AND POLITICS The Political Ecology of Extraction

The Western Amazon Basin is lightly settled and in many areas indigenous people are the majority. Yet the reality of neoliberalism and expanding resource extraction is transforming this region (that includes Ecuador, Peru, Colombia, Brazil and Bolivia) and exposing the competing geographies for resource control.

Figure 4.5.1 shows areas leased for oil and gas production. Leases are controlled by the state but can go to national and foreign companies. Many of these leased lands abut or overlap protected areas. For example, concessions in Bolivia overlap with significant areas of two national parks and one biosphere reserve. In Ecuador large sections of Yasuní National Park are leased for oil extraction.

What the map does not show, however, is that many of indigenous and peasant communities have been using these lands for generations or have asserted claims to these lands. In Ecuador, concessions overlap with the ancestral or titled land of 10 different indigenous groups. The immediacy of local concerns was manifested in Bagua, Peru, when violence between police and indigenous and peasant groups ended with nearly three dozen deaths and 200 injuries in 2009. Community leaders were demanding access to territories they had long occupied, fearing that hydrocarbon concessions would undermine their resource base and livelihood.

Like hydrocarbons, access to mining rights is also undermining local control of resources. At the beginning of the 1990s, according to geographer Antony Bebbington, Latin Americans received about 12 percent of the global investments in mining; today the share is around one-third. Estimates from Peru and Ecuador suggest that up to half of peasant communities may be affected by mining concessions. Mined areas, often in the heads of watersheds, are degraded areas that impact local communities as well as more distant downstream communities.

Yet the current presidents of each of these countries, from Peru's Alan Garcia to Ecuador's Rafael Correa, are adamant that these resources belong to the nation and that will be developed in an effort to enhance the social and economic growth of the nation. Moreover, they contend that with national control, cleaner technologies, and better management, the positive benefits of resource extraction will outweigh the negative costs.

The competing geographies of scale are exquisitely evident in this region, with national interests trumping local concerns about water quality and territorial access for traditional farming or extraction. In a climate of resource nationalism, the rights of indigenous groups or peasant communities are easily suspended. Moreover, long-term efforts to establish and designate projected lands can be easily revoked. This is a familiar story, as old as mining itself. But for many indigenous and peasant communities who have fought for territorial recognition and control, the extraction boom has challenged their understanding of citizenship.

Figure 4.5.1 **Leased Hydrocarbon Lands in Western Amazonia** Lands in western Amazonia are being leased for oil and gas extraction at an accelerated pace as states seek to extract more revenue from this lightly populated but biologically diverse region. (Source: Finer M, Jenkins CN, Pimm SL, Keane B, Ross C, 2008, "Oil and Gas Projects in the Western Amazon: Threats to Wilderness, Biodiversity, and Indigenous Peoples." PLoS ONE 3(8): e2932. doi:10.1371/journal.pone.0002932 (Figure 2))

Source: Adapted from Bebbington, Tony, 2009, "The New Extraction: Rewriting the Political Ecology of the Andes?," *NACLA Report on the Americas*, September/October.

▲

NEW! Environment and Politics
These new sidebar essays examine the complicated interaction between the environment, politics, and policies that enable (or inhibit) social action in a globalizing world.

Figure 2.15 **Methane Emissions from Arctic Wetlands** About 10 percent of global methane emissions are released from the Arctic tundra. Although a natural process, these methane emissions have increased with global warming. Some scientists are concerned that the tundra may reach a tipping point where the amount of methane emissions increases dramatically, resulting in a sudden change to the global climate.

Figure 8.11 **Glacial Retreat in the Austrian Alps** The blue sign in the foreground marks where the snout of the Pasterze glacier was in 1985, and illustrates how far the glacier has retreated in the last 25 years, shrinkage that is probably a result of global warming. This glacier is the largest in Austria and has its origin on Austria's highest mountain, the Grossglockner.

▲

Integrated Coverage of Global Warming and Climate Change
Each regional chapter has a section on the projected effects of global warming, as well as actions being taken (or not) for mitigating climate change.

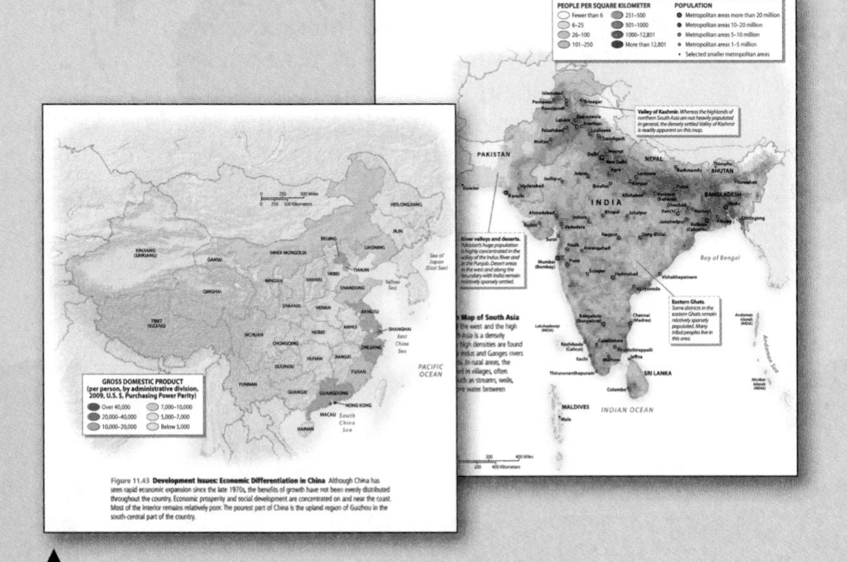

Figure 11.43 **Development Issues: Economic Differentiation in China** Although China has seen rapid economic expansion since the late 1970s, the benefits of growth have not been evenly distributed throughout the country. Economic prosperity and social development are concentrated on and near the coast. Most of the interior remains relatively poor. The poorest part of China is the upland region of Guizhou in the south-central part of the country.

▲

Consistent, Compelling Thematic Cartography
Eight thematic maps in each regional chapter empower readers to make critical inter-regional comparisons of physical geography, environmental issues, population and settlement, language, religion, geopolitics, and development issues. The **new Development Issues maps** highlight a specific measure of economic and social development, encouraging students to appreciate the many ways to define development in different regions.

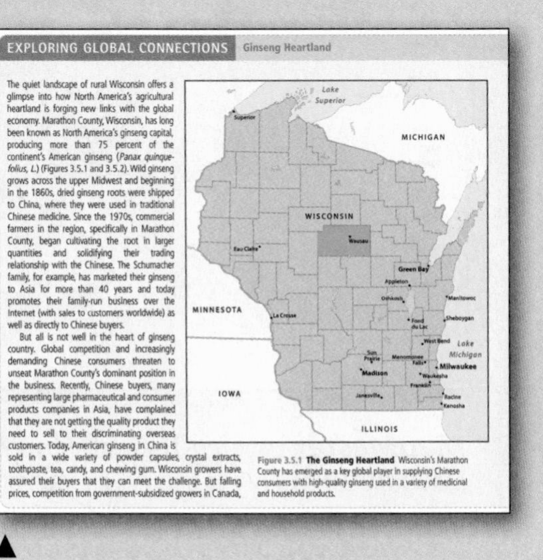

EXPLORING GLOBAL CONNECTIONS Ginseng Heartland

The quiet landscape of rural Wisconsin offers a glimpse into how North America's agricultural heartland is forging new links with the global economy. Marathon County, Wisconsin, has long been known as North America's ginseng capital, producing more than 75 percent of the continent's American ginseng (*Panax quinquefolius, L.*) (Figures 3.5.1 and 3.5.2). Wild ginseng grows across the upper Midwest and beginning in the 1860s, dried ginseng roots were shipped to China, where they were used in traditional Chinese medicine. Since the 1970s, commercial farmers in the region, specifically in Marathon County, began cultivating the root in larger quantities and solidifying their trading relationship with the Chinese. The Schumacher family, for example, has marketed their ginseng to Asia for more than 40 years and today promotes their family-run business over the Internet (with sales to customers worldwide) as well as directly to Chinese buyers.

But all is not well in the heart of ginseng country. Global competition and increasingly demanding Chinese consumers threaten to unseat Marathon County's dominant position in the business. Recently, Chinese buyers, many representing large pharmaceutical and consumer products companies in Asia, have complained that they are not getting the quality product they need to sell to their discriminating overseas customers. Today, American ginseng in China is sold in a wide variety of powder capsules, crystal extracts, toothpaste, tea, candy, and chewing gum. Wisconsin growers have assured their buyers that they can meet the challenge. But falling prices, competition from government-subsidized growers in Canada,

Figure 3.5.1 **The Ginseng Heartland** Wisconsin's Marathon County has emerged as a key global player in supplying Chinese consumers with high-quality ginseng used in a variety of medicinal and household products.

▲

NEW! Exploring Global Connections
These new informative sidebar essays discuss globalization in terms of transregional links, illustrating how different regions are interconnected.

www.masteringgeography.com

Immerse your students in the study of world regions and globalization with MasteringGeography. Used by over one million students, the Mastering platform is the most effective and widely used online tutorial, homework, and assessment system in the sciences.

Assignable Content:

- MapMaster™ Place Name Interactive Map activities
- MapMaster Layered Thematic Interactive Map activities
- Geography videos with multiple-choice quizzes
- *Encounter World Regional Geography* Google Earth™ activities
- End-of-chapter questions
- Reading Quizzes
- Thinking Spatially & Data Analysis activities
- Test Bank Questions

For Student Self Study:

- Pearson eText
- Geography Videos
- MapMaster Place Name and Layered Thematic Interactive Maps
- Outline maps
- Quizzes
- Country Profiles
- *In the News* RSS feeds
- Web links
- Key Terms Study Tools (Glossary, Key Terms, and Flash Cards)
- Authors' Blog at www.geocurrents.info

MapMaster™

MapMaster is a powerful tool that helps students practice, improve, and master map reading, spatial reasoning, geographic literacy, and critical-thinking skills. All of the interactive map activities are designed to be modular and flexible. When combined with associated quizzes correlated against the U.S. National Geography Standards, MapMaster can be used for assigned homework, pre- and post-testing, and formal assessment by instructors, as well as for self-study by students.

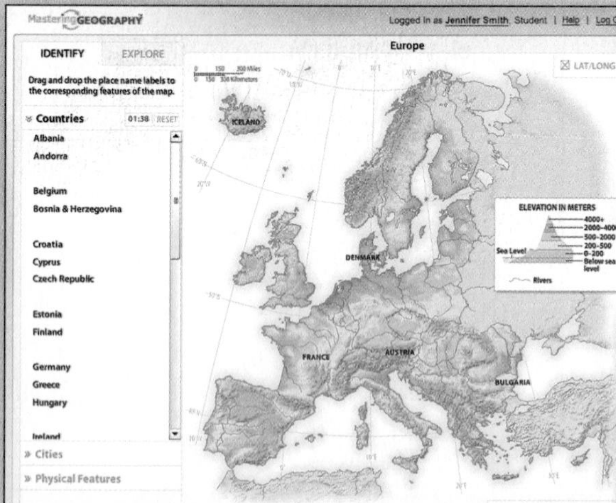

◀ ***MapMaster* Place Name Interactive Map Activities** have students identify place names of political and physical features at regional and global scales, as well as explore select country data from the CIA World Factbook. Multiple-choice quizzes for the place name labeling activities and country data sets offer instructors flexible opportunities for summative assessment and pre- and post-testing.

***MapMaster* Layered Thematic Interactive Map Activities** act as a mini-GIS tool, allowing students to layer various thematic maps to analyze spatial patterns and data at regional and global scales. Multiple-choice and short-answer quizzes are organized around themes of Physical Environment, Population, Culture, Geopolitics, and Economy, giving instructors flexible, modular options for assessing student learning. ▶

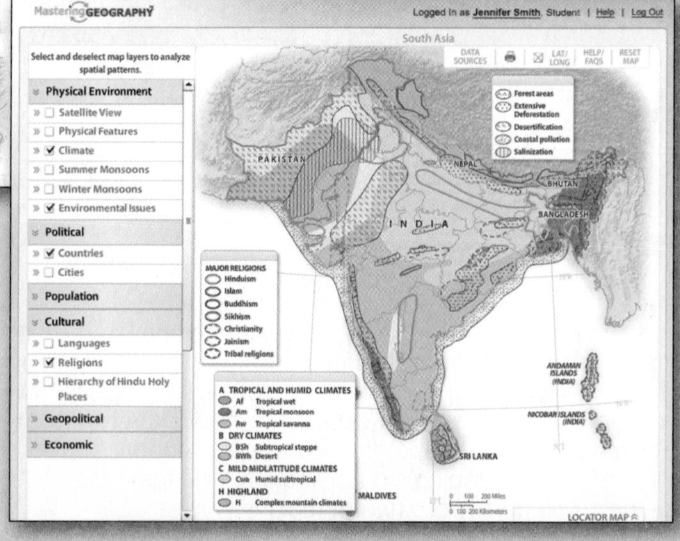

world regions and globalization

◀ ***Encounter World Regional Geography* with Google Earth™**
Students use the dynamic features of Google Earth to visualize and explore world regions and answer questions related to the themes of Environment, Population, Culture, Geopolitics, and Economy & Development. All Explorations include corresponding Google Earth KMZ media files, and questions include hints and specific wrong-answer feedback to help coach students towards mastery of the concepts.

Geography Videos ▶
Short videos from Television for the Environment's *Life* and *Earth Report* global series explore a range of topics related to world regions, globalization, physical geography, and climate change. Examples of topics include the growing number of homeless children in Russia, the serious health risks of pregnant women in Bangladesh, and the struggles for electricity in China and Africa. Accompanying video quizzes include multiple choice and short-answer questions with hints and specific wrong-answer feedback for each of the episodes.

Thinking Spatially & Data Analysis
These activities help students develop spatial reasoning and critical thinking skills by identifying and labeling features from maps, illustrations, photos, graphs, and charts. Students then examine related data sets, answering multiple-choice questions and increasingly higher order conceptual short-answer questions, both of which include hints and specific wrong-answer feedback.

Motivating students to learn and explore

With easy-to-customize, easy-to-assign, and automatically graded assessments, instructors can maximize class time and motivate students to learn outside of class and arrive prepared for lecture.

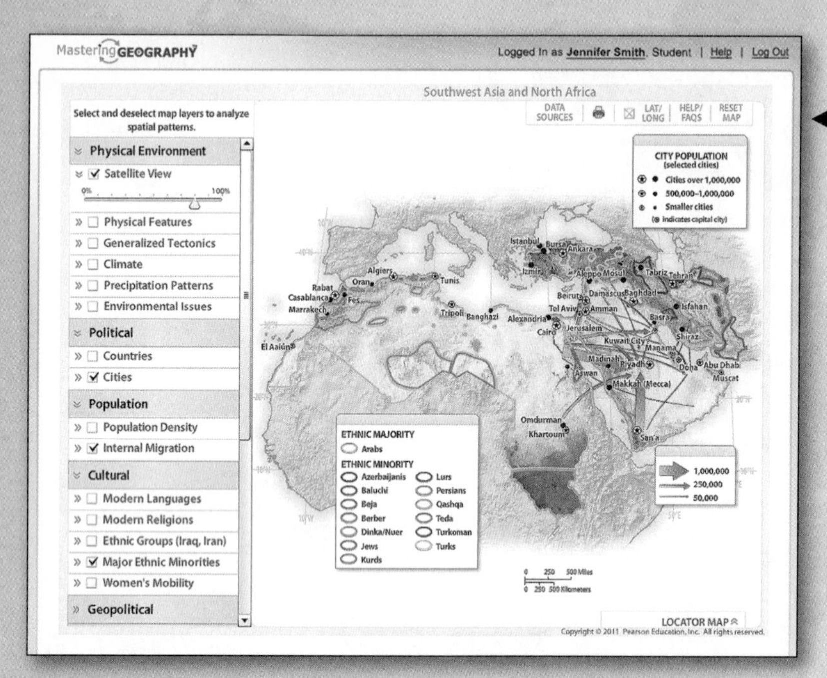

◄ MasteringGeography offers:

- Assignable activities, including MapMaster™ interactive maps, geography videos, *Encounter World Regional Geography* Google Earth™ multimedia, Thinking Spatially & Data Analysis
- End-of-chapter questions
- Test Bank questions
- Student study area with MapMaster interactive maps, videos, RSS feeds, web links, quizzes, flashcard glossary, and more
- Optional Pearson eText

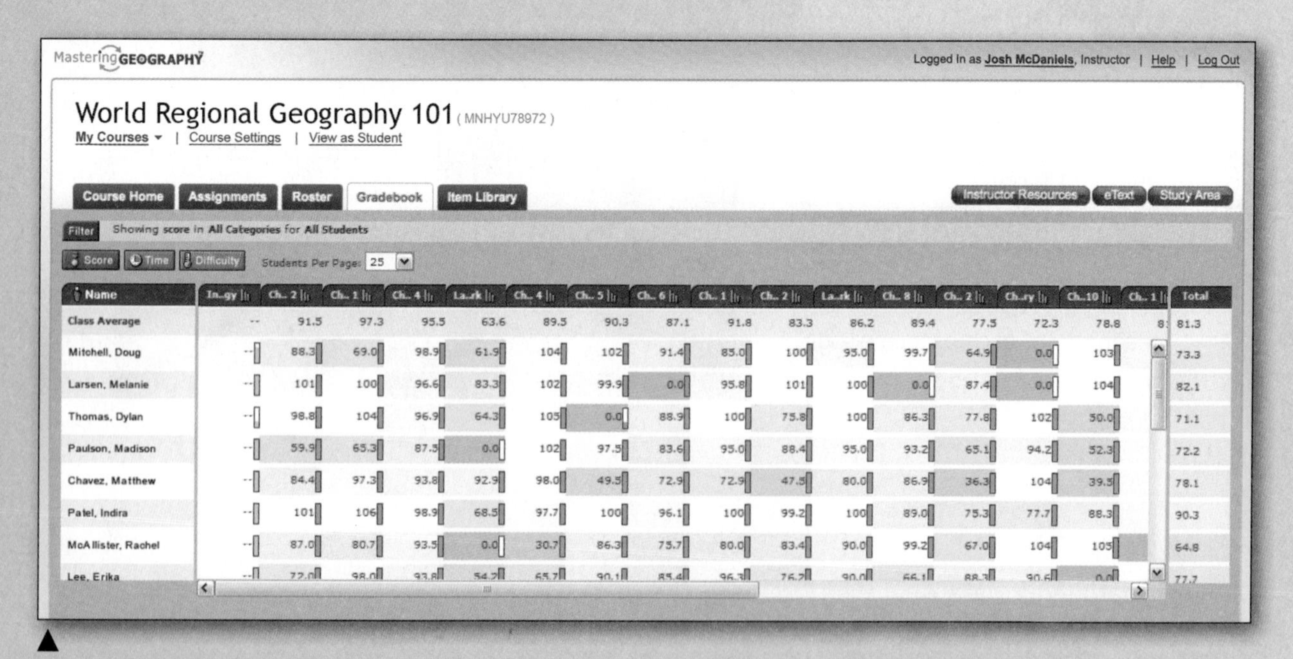

▲

MasteringGeography Gradebook

Provides instructors with quick results and easy-to-interpret insights into student performance. Every assignment is automatically graded, with shades of red highlighting struggling students and challenging assignments.

geography beyond the classroom

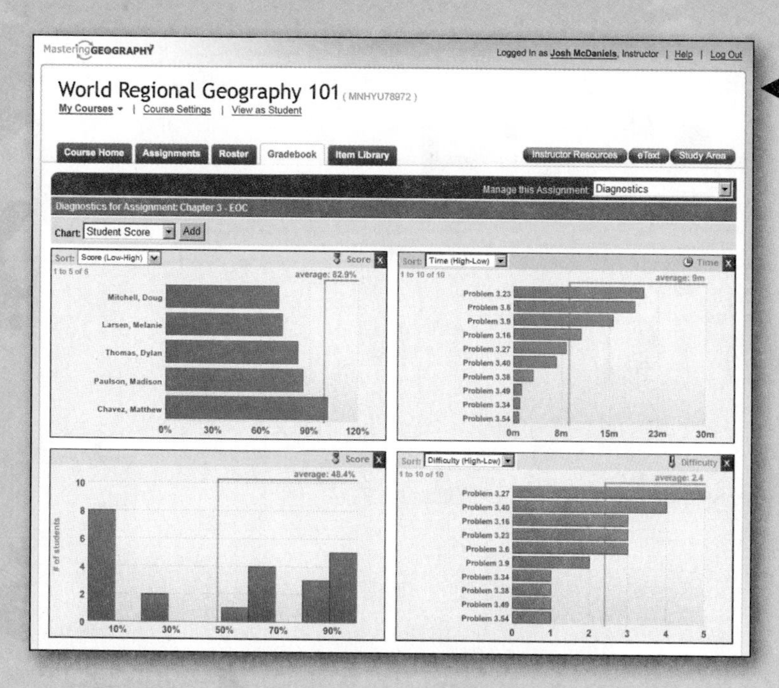

Gradebook Diagnostics

Enables an instructor to see class performance as a whole to fine-tune lectures before class.

Continuously Improving Content

MasteringGeography offers a dynamic pool of assignable content that improves with student usage. Detailed analysis of student performance statistics—including time spent, answers submitted, solutions requested, and hints used—ensures the highest quality content.

1. **We conduct a thorough analysis** of each problem by reviewing student performance data that has been generated by real students.
2. **We make enhancements** to improve the clarity and accuracy of content, answer choices, and instructions for each problem.
3. **We repeat the process.** This ongoing, **two-way learning process** helps students learn as students help the system improve.

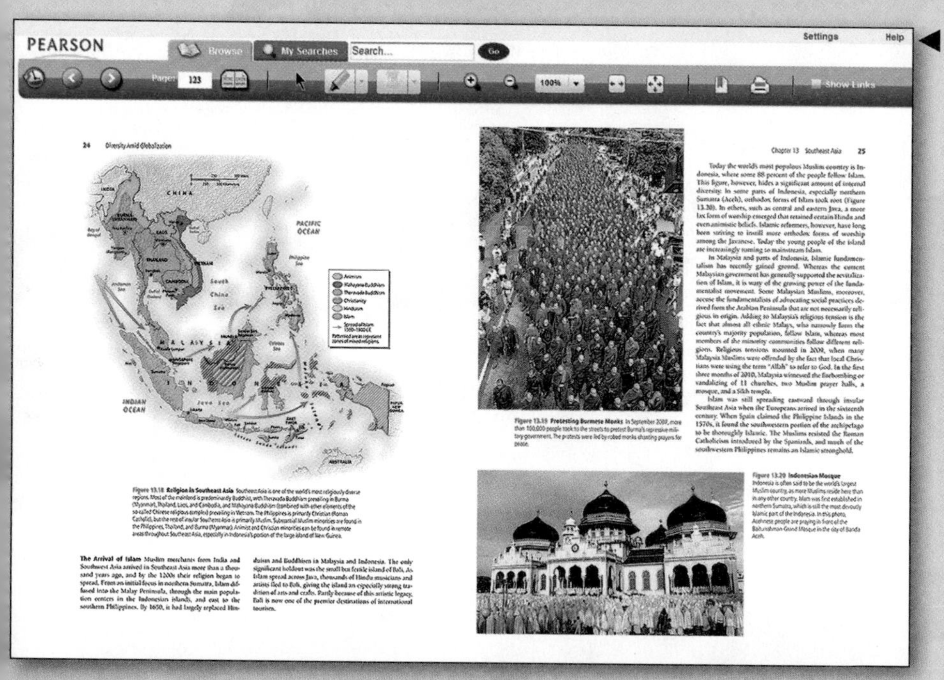

Pearson eText

Search, annotate, zoom, highlight, and more—all electronically with Pearson eText.

Pearson eText gives students access to the text whenever and wherever they can access the Internet. The eText pages look exactly like the printed text, and include powerful interactive and customization functions.

1 Diversity Amid Globalization

In many ways Singapore and its harbor are icons of globalization. Not only is the city itself a focal point of international finance, but the Port of Singapore is the world's busiest transshipment harbor.

1

Some of the most important challenges facing the world in the 21st century are associated with **globalization**, the increasing interconnectedness of people and places through converging economic, political, and cultural activities. Once-distant regions and cultures are now increasingly linked through commerce, communications, and travel. Although earlier forms of globalization existed, especially during Europe's colonial period, the current degree of planetary integration is stronger than ever. In fact, many observers argue that contemporary globalization is the most fundamental reorganization of the world's socioeconomic structure since the Industrial Revolution. While few dispute the widespread changes brought about by globalization, the topic is controversial because not everyone agrees on whether the benefits outweigh the costs.

Although economic activities may be the main force behind globalization, the consequences affect all aspects of land and life: Cultural patterns, political arrangements, and social development are all undergoing profound change. Because natural resources are now global commodities, the planet's physical environment is also affected by globalization. Financial decisions made thousands of miles away alter local ecosystems, and the cumulative effects of these far-ranging activities often have negative consequences for the world's climates, oceans, waterways, and forests.

These immense, widespread global changes make understanding our contemporary world a challenging yet necessary task. Although globalization cuts across many academic disciplines, world regional geography is a fundamental starting point because of its focus on regions, environment, geopolitics, culture, and economic and social development. This book seeks to impart knowledge of globalization by carefully outlining the basic patterns of world geography and showing how they are being reorganized by global interconnections (Figure 1.1).

CONVERGING CURRENTS OF GLOBALIZATION

Most scholars agree that the major component of globalization is the economic reorganization of the world. Although different forms of a world economy have existed for centuries, a well-integrated and truly global economy is primarily the product of the past several decades. The attributes of this system, while familiar, bear repeating:

- Global communication systems that link all regions and most people on the planet instantaneously (Figure 1.2)
- Transportation systems capable of moving goods quickly by air, sea, and land
- Transnational business strategies that have created global corporations more powerful than many sovereign nations
- New and more flexible forms of capital accumulation and international financial institutions that make 24-hour trading possible
- Global agreements that promote free trade
- Market economies and private enterprises that have replaced state-controlled economies and services
- An abundance of planetary goods and services that have arisen to fulfill consumer demand, real or imagined (Figure 1.3)
- Economic disparities between rich and poor regions and countries that drive people to migrate, both legally and illegally, in search of a better life
- An army of international workers, managers, and executives who give this powerful economic force a human dimension

As a result of this global reorganization, economic growth in some areas of the world has been unprecedented

Figure 1.1 Global Communications The effects of globalization are everywhere, even in remote villages in developing countries. Here, in a small village in southwestern India, a rural family earns a few dollars a week by renting out viewing time on its globally linked television set.

Figure 1.2 Global-to-Local Connections A controversial aspect of globalization is the outsourcing of jobs from developed to developing countries. One example is the relocation of customer call centers away from Europe and North America to India where educated English-speaking employees work for relatively low wages. This photo shows a call center in Bangalore, India.

during recent decades. International corporations, along with their managers and executives, have amassed vast amounts of wealth, profiting from the new opportunities unleashed by globalization. Not everyone, however, has gained from economic globalization, nor have all world regions shared equally in the benefits. While globalization is often touted as benefiting everyone through trickle-down economics, there is mounting evidence that this trickling process is neither happening in all places nor for all peoples. Additionally, as was demonstrated in the global recession of 2008–2010, economic interconnectivity can also increase economic vulnerability; this is illustrated by the precipitous decline in Hawaii's tourist trade as the economies of both Japan and the United States went flat at the same time.

Globalization and Cultural Change

Economic changes also trigger cultural changes. The spread of a global consumer culture that masks local diversity often accompanies globalization, frequently creating deep and serious social tensions between traditional cultures and new, external globalizing currents. Global TV, movies, Facebook, Twitter, and videos promote Western values and culture that are imitated by millions throughout the world. But reactions to westernization, can take sometimes violent forms, such as that from radical Islamism.

Fast-food franchises are changing—some would say corrupting—traditional diets, with the explosive growth of McDonald's, Burger King, and KFC outlets in the world's cities. While these changes may seem harmless to North Americans because of their familiarity, they are not just expressions of the deep cultural changes the world is experiencing through globalization but also generally unhealthy and environmentally destructive. The expansion of the cattle industry, for example, a result of the new global demand for beef, is doing serious environmental damage to tropical rain forests.

Although the media give much attention to the rapid spread of Western consumer culture, nonmaterial culture is also becoming more dispersed and homogenized through globalization. Language is an obvious example. Many a Western tourist in Russia or Thailand has been startled by locals speaking an English made up largely of Hollywood movie phrases. But far more than speech is involved as social values also are dispersed globally. Changing expectations about human rights, the role of women in society, and the intervention of nongovernmental organizations are also expressions of globalization that may have far-reaching effects on cultural change.

It would be a mistake, however, to view cultural globalization as a one-way flow that spreads from the United States

Figure 1.3 Global Shopping Malls Once a fixture only of suburban North America, the shopping mall is now found throughout the world. This mall is in downtown Kunming, the capital city of Yunnan province, China.

and Europe into the corners of the world. In actuality, when forms of U.S. popular culture spread abroad, they are typically melded with local cultural traditions in a process known as *hybridization*. The resulting cultural "hybridities," such as hip hop and rap music or Asian food, can themselves resonate across the planet, adding yet another layer to globalization.

In addition, ideas and forms from the rest of the world are also having a great impact on U.S. culture. The growing internationalization of American food, the multiple languages spoken in the United States, and the spread of Japanese comic book culture among U.S. children are all examples of globalization's effects within the United States (Figure 1.4).

Globalization and Geopolitics

Globalization also has important geopolitical components. To many, an essential dimension of globalization is that it is not restricted by territorial or national boundaries. For example, the creation of the United Nations following World War II was a step toward creating an international governmental structure in which all nations could find representation. The simultaneous emergence of the Soviet Union as a military and political superpower at that time led to a rigid division into Cold War blocs that slowed further geopolitical integration. But with the peaceful end of the Cold War in the late 1980s and early 1990s, the former communist countries of eastern Europe and the Soviet Union were opened almost immediately to global trade and cultural exchange that has changed those countries immensely.

Further, there is a strong argument that globalization—almost by definition—has weakened the political power of individual states by strengthening the power of regional economic and political organizations such as the European Union or the World Trade Organization. In some world regions, a weakening of traditional state power has resulted in stronger local and separatist movements as illustrated by the turmoil on Russia's southern borders or the plethora of separatist organizations in Europe.

Figure 1.4 Global Culture in the United States The multilingual welcome offered by a public library in Montgomery County, Maryland, not only illustrates the many different languages spoken by people in the suburbs of Washington, DC, but also reminds us that expressions of globalization are found throughout North America.

Environmental Concerns

As mentioned, the expansion of a globalized economy is creating and intensifying environmental problems throughout the world. **Transnational firms**, which do global business through international subsidiaries, disrupt local ecosystems with their incessant search for natural resources and manufacturing sites. Landscapes and resources previously used by only small groups of local peoples are now thought of as global commodities to be exploited and traded on the world marketplace. As a result, native peoples are often deprived of their traditional resource base and displaced into marginal environments. On a larger scale, economic globalization is aggravating worldwide environmental problems such as global warming, air pollution, water pollution, and deforestation. And yet it is only through global cooperation, such as the UN treaties on biodiversity protection or global warming, that these problems can be addressed. These topics are discussed further in Chapter 2.

Social Dimensions

Globalization has a clear demographic dimension. Although international migration is not new, increasing numbers of people from all parts of the world are crossing national boundaries, legally and illegally, temporarily and permanently (Figure 1.5). Migration from Latin America and Asia has drastically changed the demographic configuration of the United States, and migration from Africa and Asia has transformed western Europe. Countries such as Japan and South Korea that have long been perceived as ethnically homogeneous now have substantial immigrant populations. Even a number of relatively poor countries, such as Nigeria and the Ivory Coast, encounter large numbers of immigrants coming from even poorer countries, such as Burkina Faso. Although international migration is still curtailed by the laws of every country—much more so, in fact, than the

Figure 1.5 International Migration Workers from southern India dig a hole to install a street sign in Dubai, United Arab Emirates. This Persian Gulf emirate is experiencing a massive construction boom as it shifts from an oil-based economy to an economy based on real estate, tourism, and international finance. As a result, temporary migrant workers from India and Pakistan constitute much of the labor force.

movement of goods or capital—it is still rapidly mounting, propelled by the uneven economic development associated with globalization.

Finally, there also is a significant criminal element to contemporary globalization, including terrorism (discussed later in this chapter), drugs, pornography, slavery, and prostitution. Illegal narcotics, for example, are definitely a global commodity (Figure 1.6). Some of the most remote parts of the world, such as the mountains of northern Burma, are thoroughly integrated into the circuits of global exchange through the production of opium and, therefore, into the world heroin trade. Even many areas that do not directly

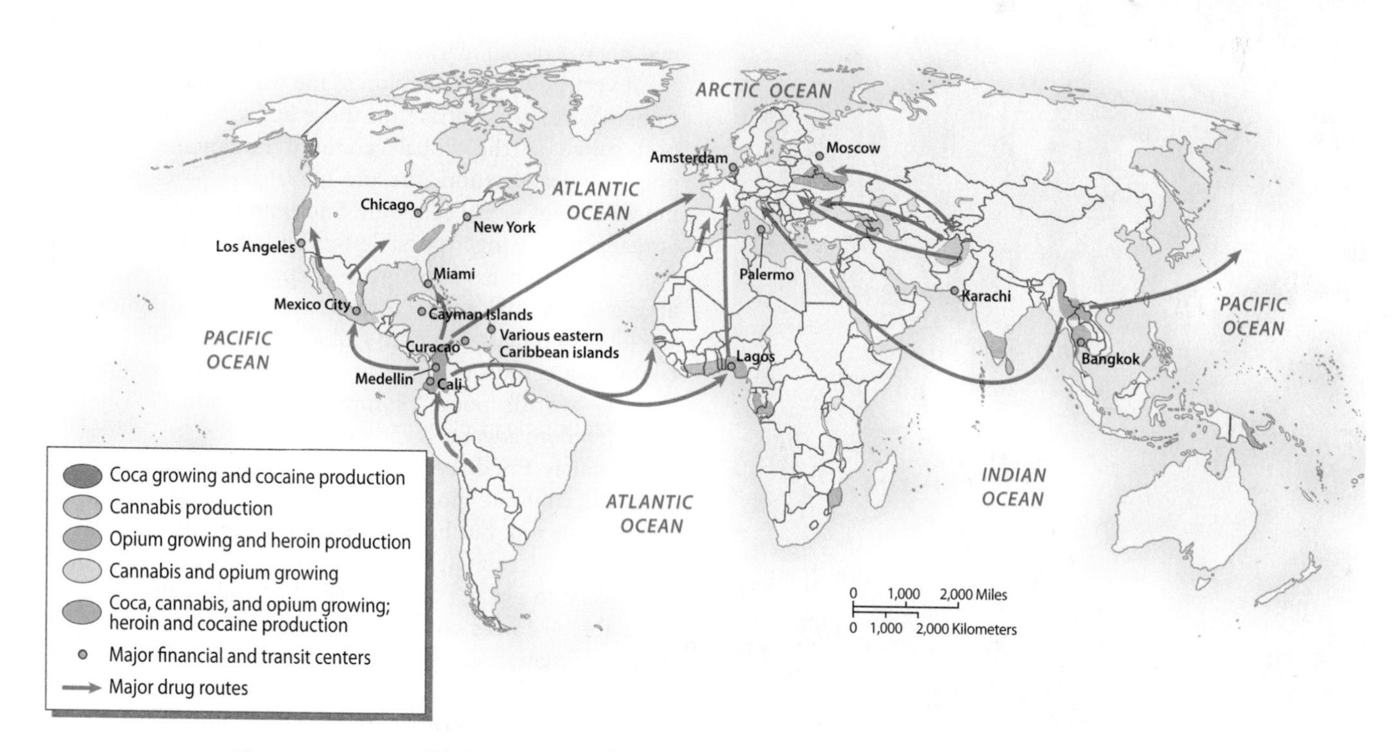

Figure 1.6 The Global Drug Trade The cultivation, processing, and transshipment of coca (cocaine), opium (heroin), and cannabis (marijuana) are global issues. The most important cultivation centers are Colombia, Mexico, Afghanistan, and northern Southeast Asia, and the major drug financing centers are located mostly in the Caribbean, the United States, and Europe. In addition, Nigeria and Russia also play significant roles in the global transshipment of illegal drugs.

produce drugs are involved in their global sale and transshipment. Nigerians often occupy prominent positions in the international drug trade, as do members of the Russian mafia. Many Caribbean countries have seen their economies become reoriented to drug transshipments and the laundering of drug money. Prostitution, pornography, and gambling have also emerged as highly profitable global businesses. Over the past decades, for example, parts of eastern Europe have become major sources of both pornography and prostitution, finding a lucrative but morally questionable niche in the new global economy.

ADVOCATES AND CRITICS OF GLOBALIZATION

Globalization, especially in its economic form, is one of today's most contentious issues (Figure 1.7). Supporters believe that it results in a greater economic efficiency that will eventually result in rising prosperity for the entire world. In contrast, critics think that globalization will largely benefit those who are already prosperous, leaving most of the world poorer than before as the rich and powerful exploit the less fortunate.

Economic globalization is generally applauded by corporate leaders and economists, and it has substantial support among the leaders of both major political parties in the United States. Beyond North America, moderate and conservative politicians in most countries generally support free trade and other aspects of economic globalization. Opposition to economic globalization is widespread in the labor and environmental movements, as well as among many student groups worldwide. Hostility toward globalization is sometimes deeply felt, as massive protests at World Bank and World Trade Organization meetings have made obvious.

Figure 1.7 Protests Against Globalization Meetings of international groups such as the World Trade Organization, World Bank, and International Monetary Fund (IMF) commonly draw large numbers of protesters against globalization. This demonstration took place at a Washington, DC, meeting of the World Bank and IMF.

The Pro-globalization Stance

Advocates of globalization argue that globalization is a logical and inevitable expression of contemporary international capitalism that will benefit all nations and all peoples. Economic globalization can work wonders, they contend, by enhancing competition, allowing the flow of capital to poor areas, and encouraging the spread of beneficial new technologies and ideas. As countries reduce their barriers to trade, inefficient local industries will be forced to become more efficient in order to compete with the new flood of imports, thereby enhancing overall national productivity. Those that cannot adjust will most likely go out of business, making the global market place more efficient.

Every country and region of the world, moreover, ought to be able to concentrate on those activities for which it is best suited in the global economy. Enhancing such geographic specialization, the pro-globalizers argue, creates a more efficient world economy. Such economic restructuring is made increasingly possible by the free flow of capital to those areas that have the greatest opportunities. By making access to capital more readily available throughout the world, economists contend, globalization should eventually result in a certain global **economic convergence**, implying that the world's poorer countries will gradually catch up with the more advanced economies.

Thomas Friedman, one of the most influential advocates of economic globalization, argues that the world has not only shrunk but has also become economically "flat" so that financial capital, goods, and services can flow freely from place to place. For example, the need to attract capital from abroad forces countries to adopt new economic policies. Friedman describes the great power of the global "electronic herd" of bond traders, currency speculators, and fund managers who either direct money to or withhold it from developing economies, resulting in economic winners and losers (Figure 1.8).

To the committed pro-globalizer, even the global spread of **sweatshops**—crude factories in which workers sew clothing, assemble sneakers, and perform other labor-intensive

Figure 1.8 The "Electronic Herd" One facet of economic globalization is the rapid movement of capital as bond traders, currency speculators, and fund managers direct money in or out of the economies of developing countries. In this picture, traders and clerks work on the Eurodollar Futures floor of the Chicago Mercantile Exchange.

tasks for extremely low wages—is to be applauded (Figure 1.9). People go to work in sweatshops because the alternatives in the local economy are even worse. Once countries achieve full employment through sweatshops, the argument goes, they can gradually improve conditions and move into more sophisticated, better-paying industries. Proponents of economic globalization also commonly argue that multinational firms based in North America or Europe often offer better pay and safer working conditions than do local firms and thus contribute to worker well-being. Extreme pro-globalizers go so far as to assert that poor countries ought to take advantage of their generally lax environmental laws to attract highly polluting industries from the wealthy countries because acquiring such industries would enhance their overall economic positions.

The pro-globalizers also strongly support the large multinational organizations that facilitate the flow of goods and capital across international boundaries. Three such organizations are particularly important: the World Bank, the International Monetary Fund (IMF), and the World Trade Organization (WTO). The primary function of the World Bank is to make loans to poor countries so that they can invest in infrastructure and build more modern economic foundations. The IMF is concerned with making short-term loans to countries that are in financial difficulty—those having trouble, for example, making interest payments on the loans that they had previously taken. The WTO, a much smaller organization than the other two, works to reduce trade barriers between countries to enhance economic globalization. It also tries to mediate between countries and trading blocs that are engaged in trade disputes (Figure 1.10).

To support their claims, pro-globalizers argue that countries that have been highly open to the global economy have generally had much more economic success than those that have isolated themselves by seeking self-sufficiency. The world's most isolated countries, Burma (Myanmar) and North Korea, have become economic disasters, with little growth and rampant poverty, whereas those that have opened themselves to global forces, such as Singapore and Thailand, in the same period have seen rapid growth and substantial reductions in poverty.

Critics of Globalization

Virtually all of the claims of the pro-globalizers are strongly contradicted by the anti-globalizers. Opponents often begin by arguing that globalization is not a "natural" process. Instead, it is the product of an explicit economic policy promoted by free-trade advocates, capitalist countries (mainly the United States, but also Japan and the countries of Europe), financial interests, international investors, and multinational firms. Opponents also point out that the processes of globalization today are more pervasive than those of the historical past, even during the period of European colonialism. Thus, while global economic and political linkages have been around for centuries, their contemporary expression is unprecedented.

Because the globalization of the world economy appears to be creating greater inequity between rich and poor, the "trickle-down" model of developmental benefits for all people in all regions has yet to be validated. On a global scale, the richest 20 percent of the world's people consume 86 percent of the world's resources, whereas the poorest 80 percent use only 14 percent. The growing inequality of this age of globalization is apparent on both global and national scales.

Figure 1.9 Global Sweatshops One of the most debated aspects of economic globalization is the crude factories, or sweatshops, in which workers sew clothing, assemble sneakers, stitch together soccer balls, and perform other labor-intensive work for low wages. This is a Chinese-owned factory in Nicaragua.

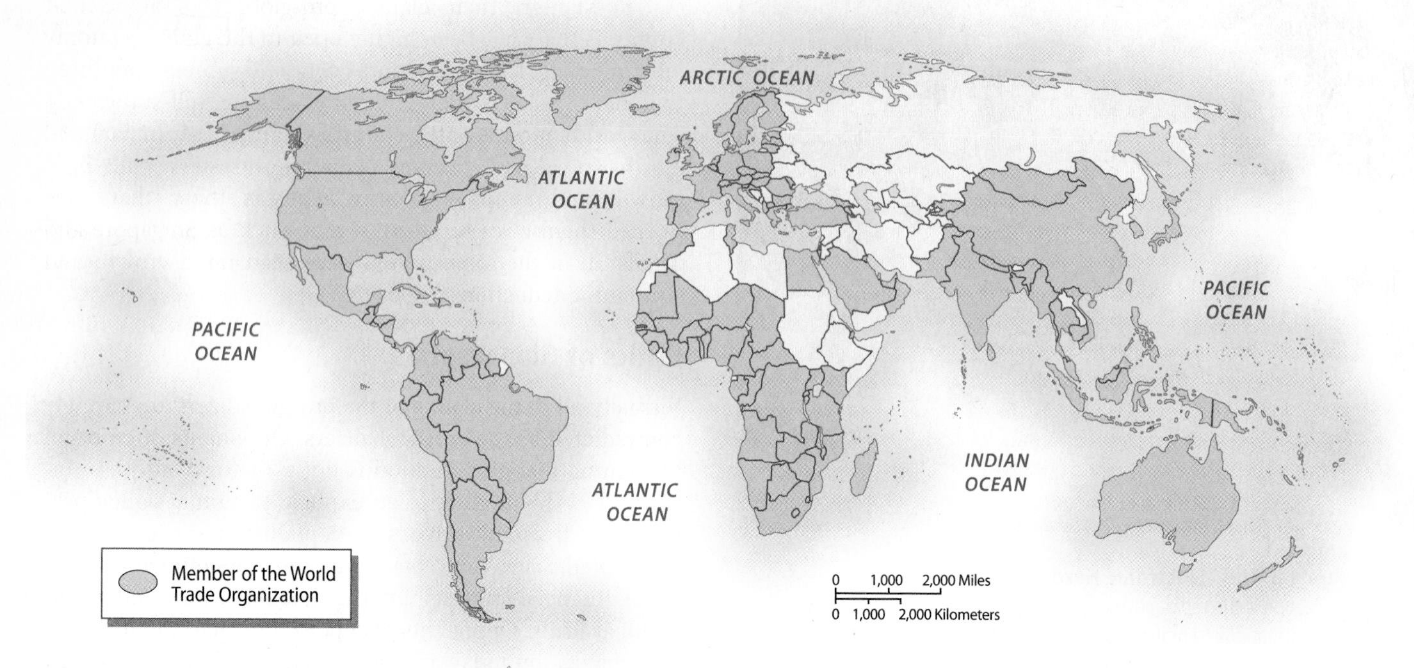

Figure 1.10 World Trade Organization One of the most powerful institutions of economic globalization is the World Trade Organization (WTO), which was created in 1995 to oversee trade agreements, encourage open markets, enforce trade rules, and settle disputes. The WTO currently consists of 153 member countries. In addition to these member countries, more than 30 states have "observer status," including Russia, Iran, and Iraq.

Globally, the wealthiest countries have grown much richer over the past two decades, while the poorest have become more impoverished. Nationally, even in developed countries such as the United States, the wealthiest 10 percent of the population have reaped almost all of the gains that globalization has offered; the poorest 10 percent have either seen their income decline in recent decades as wages have remained static or they have actually lost their jobs to outsourcing.

Opponents also contend that globalization promotes free-market, export-oriented economies at the expense of localized, sustainable activities. World forests, for example, are increasingly cut for export timber rather than serving local needs. As part of their economic structural adjustment package, the World Bank and the IMF encourage developing countries to expand their resource exports so they will have more hard currency to make payments on their foreign debts. This strategy, however, usually leads to overexploitation of local resources. Opponents also note that the IMF often requires developing countries to adopt programs of fiscal austerity that often entail substantial reductions in public spending for education, health, and food subsidies. By adopting such policies, critics warn, poor countries will end up with even more impoverished populations than before.

Anti-globalizers also dispute the evidence on national development offered by the pro-globalizers. Highly successful developing countries such as South Korea, Taiwan, and Malaysia, they argue, have indeed been engaged with the world market, but they have generally done so on their own terms rather than those of the IMF and other advocates of full-fledged economic globalization. These countries have actually protected many of their domestic industries from foreign competition and have, at various times, controlled the flow of capital.

Furthermore, anti-globalizers contend that the "free-market" economic model commonly promoted for developing countries is not the one that Western industrial countries used for their own economic development. In Germany, France, and even to some extent the United States, governments historically have played a strong role in directing investment, managing trade, and subsidizing chosen sectors of the economy.

Those who challenge globalization also worry that the entire system—with its instantaneous transfers of vast sums of money over nearly the entire world on a daily basis—is inherently unstable. The noted critic John Gray, for example, argues that the same "electronic herd" that Thomas Friedman applauds is a dangerous force because it is susceptible to "stampedes." International managers of capital tend to panic when they think their funds are at risk; when they do so, the entire intricately linked global financial system can quickly become destabilized, leading to a crisis of global proportions. The rapid downturn of the global economy in late 2008 seems to support that assertion (Figure 1.11).

Even when the "herd" spots opportunity, trouble may still ensue. As vast sums of money flow into a developing country,

Figure 1.11 Global Economic Recession Unused shipping containers pile up on parking lots and vacant space near a residential area in northwest Hong Kong because of the downturn in China's export economy during the global recession of 2009. During the height of the economic downturn, about 25 percent of the world's container ships were idled.

they may create a speculatively inflated **bubble economy** that cannot be sustained. Such a bubble economy emerged in Thailand and many other parts of Southeast Asia in the mid-1990s. Analysts have also used the concept of bubble economy to explain the tragic collapse of the Icelandic and Irish economies in 2009 (Figure 1.12).

A Middle Position

A number of experts, not surprisingly, argue that both the anti-globalization and the pro-globalization stances are exaggerated. Those in the middle ground tend to argue that economic globalization is indeed unavoidable; even the anti-globalization movement, they point out, is made possible by the globalizing power of the Internet and is, therefore, itself an expression of globalization. They further contend that while globalization holds both promises and pitfalls, it can be managed, at both the national and international levels, to reduce economic inequalities and protect the natural environment. These experts stress the need for strong yet efficient national governments, supported by international institutions (such as the UN, World Bank, and IMF) and globalized networks of environmental, labor, and human rights groups.

Unquestionably, globalization is one of the most important issues of the day—and certainly one of the most complicated. While this book does not pretend to resolve the controversy, it does encourage readers to reflect on these critical points as they apply to different world regions.

Figure 1.12 Economic Turmoil in Iceland Protesters burn an effigy of Iceland's prime minister during a demonstration against the government's handling of the 2009 economic crisis. Given the relatively small size of Iceland's economy, the collapse of its bubble economy was the largest downturn in percentage terms suffered by any country in history.

DIVERSITY IN A GLOBALIZING WORLD

As globalization increases, many observers foresee a world far more uniform and homogeneous than today's. The optimists among them imagine a universal global culture uniting all humankind into a single community untroubled by war, ethnic strife, or resource shortage—a global utopia of sorts.

A more common view, however, is that the world is becoming blandly homogeneous as different places, peoples, and environments lose their distinctive character and become indistinguishable from their neighbors. While diversity may be difficult for a society to live with, it also may be dangerous to live without. Nationality, ethnicity, cultural distinctiveness—all are the legitimate legacy of humanity. If this diversity is blurred, denied, or repressed through global homogenization, humanity loses one of its defining traits.

But even if globalization is generating a certain degree of homogenization, the world is still a highly diverse place (Figure 1.13). One still finds marked differences in culture (language, religion, architecture, foods, and many other attributes of daily life), economy, and politics—as well as in the physical environment. Such diversity is so vast that it cannot readily be extinguished, even by the most powerful forces of globalization. In fact, globalization often provokes a strong reaction on the part of local people, making them all the more determined to maintain what is distinctive about their way of life. Thus, globalization is understandable only if one also examines the diversity that continues to characterize the world and, perhaps most importantly, the tension between these two forces—the homogenization of globalization and the reaction against it in terms of protecting cultural and political diversity. Unfortunately, this tension sometimes takes negative and violent forms, as illustrated today by radical Islam and tribal warfare in Nigeria and Sudan (Figure 1.14).

The politics of diversity also demands increasing attention as we try to understand worldwide tensions over terrorism, ethnic separateness, regional autonomy, and political

Figure 1.14 Tribal Warfare Rebels from the Sudan Liberation Movement (SLM) declared war against the Sudanese government in the province of Darfur. The two groups that originated the rebellion have now morphed into dozens of warring factions, complicating the search for peace amid a profound humanitarian crisis.

Figure 1.13 Local Cultures This family in Ghana reminds us that although few places are beyond the reach of globalization, many unique local landscapes, economies, and cultures still exist. For example, most of the belongings of the family in this photograph are local in origin and come from long-standing cultural tradition.

independence. Groups of people throughout the world seek self-rule of territory they can call their own. Today, most wars are fought *within* countries, not *between* them. As a result, our interest in geographic diversity takes many forms and goes far beyond simply celebrating traditional cultures and unique places. People have many ways of making a living throughout the world, and it is important to recognize this fact as the globalized economy becomes increasingly focused on mass-produced retail goods. Furthermore, a stark reality of today's economic landscape is unevenness: While some people and places prosper, others suffer from unrelenting poverty. This, unfortunately, also is a form of diversity amid globalization.

In summary, globalization can be defined as the increasing interconnectedness of people and places through converging processes of economic, political, and cultural change. Globalization is pervasive and unrelenting and is a dominant feature of the contemporary world. Furthermore, its benefits are uneven. Although economic restructuring is a prime cause, globalization is not simply the growth and expansion of international trade. It is also represented by converging and homogenizing forces that many people resist. Thus, an equally important theme is how geographic diversity comes in conflict with globalization—how globalization in different places and at different times is suppressed, renegotiated, hybridized, protected, preserved, or extinguished. Because of the importance of this theme, globalization and diversity should be examined as inseparable—often in conflict but at other times complementary.

GEOGRAPHY MATTERS: Environments, Regions, Landscapes

Geography is one of the most fundamental sciences, a discipline awakened and informed by a long-standing human curiosity about our surroundings and the world environment. The term *geography* has its roots in the Greek words for "describing the Earth," and this discipline has been advanced since classical times by all cultures and civilizations. Along with the inherent satisfaction of knowing about different environments also come practical benefits through exploration, resource use, world commerce, and travel. In some ways, geography can be compared to history: While historians describe and explain what has happened over time, geographers describe and explain Earth's spatial dimensions, and how the world differs from place to place (see "Geographic Tools: From Satellites to Shovels").

Given the broad scope of geography, it is no surprise that geographers have many different conceptual approaches to investigating the world. At the most basic level, geography can be broken into two complementary pursuits, *physical and human geography*. Physical geography examines climate, landforms, soils, vegetation, and hydrology, while human geography concentrates on the spatial analysis of economic, social, and cultural systems.

A physical geographer, for example, studying the Amazon Basin of Brazil might be interested primarily in the ecological diversity of the tropical rain forest or the ways the destruction of that dense vegetation changes the local climate and hydrology. The human geographer, in turn, would focus on the social and economic factors explaining the migration of settlers into the rain forest, or the tensions and conflicts over land and resources between these migrants and indigenous peoples.

Another basic division is that between focusing on a specific topic or theme contrasted with analyzing a place or a region. The former approach is referred to as *thematic* or *systematic geography*, while the latter is called *regional geography*. These two perspectives are complementary and by no means mutually exclusive. This textbook, for example, draws upon a regional scheme for its overall architecture, dividing the globe into 12 separate world regions, yet it organizes each chapter thematically, examining the topics of environment, population and settlement, cultural differentiation, geopolitics, and economic development in a systematic way. In doing so, each chapter combines four kinds of geography: physical, human, thematic, and regional geography.

A fundamental component of geographic inquiry is the examination of the basic yet highly complex relationship between humans and their environment. Though this analysis may take many different forms, three overarching questions form the foundation:

- Where and under what conditions does nature control, constrain, or disrupt human activities, and what kinds of societies are most vulnerable to these environmental influences (Figure 1.15)?
- How have human activities affected natural systems such as the world's climate, vegetation, rivers, and oceans, and what are the consequences of these actions?
- Is there any evidence that humans can live in balance with nature so that human use of natural resources might be sustainable over a longer period of time, or is environmental damage an inevitable consequence of human settlement?

While there are no simple answers to these three questions, the study of world regional geography offers valuable insights from different cultural and regional perspectives.

Areal Differentiation and Integration

Geography is considered the spatial science, charged with the study of Earth's space (or surface area), and the uniqueness of places as well as the similarities between them. One component of that responsibility is describing and explaining the differences that distinguish one piece of the world from another. The geographical term for this is **areal differentiation** (*areal* means "pertaining to area"). Why is one

GEOGRAPHIC TOOLS From Satellites to Shovels

Geographers use a wide array of tools—from satellites to shovels—to explain the world and solve environmental problems (Figure 1.1.1). While some geographers get their information from aerial photos and satellite images, others work at ground level, digging holes to sample soils, mapping trees and plants, or even interviewing people in their homes and at the mall. While geographers' toolboxes and methods may differ, the goals are the same: a better understanding of global environments and the people who inhabit them.

Many of our tools are commonplace, like maps, compasses, handheld GPS (global positioning system) devices, or computers. Other tools, though, are highly specialized, such as multi-spectral satellite images of vegetation, computer models of refugee migration, or soil-boring tools to capture prehistoric pollen samples. As readers of this book about the world, you might be interested in how geographers collect and analyze information about different places and different issues. Toward that goal, each chapter includes a sidebar called *Geographic Tools* that illustrates how specific tools and methods are used to expand our understanding of the world. The complete chapter list follows:

Figure 1.1.1 Satellites to Shovels A graduate student gathers a soil sample for her master's thesis on the ecology of native plants in coastal California.

- Chapter 2: Reconstructing Climates of the Past
- Chapter 3: Counting North Americans: The Censuses of 2010 and 2011
- Chapter 4: Participatory Mapping in the Slums of Rio de Janeiro
- Chapter 5: Tracking Hurricanes
- Chapter 6: Monitoring Land Cover and Conservation Changes in Niger
- Chapter 7: Measuring the Urban Impact of Islamic Banking
- Chapter 8: Landscape Restoration in England
- Chapter 9: Mapping Russian Forests
- Chapter 10: Remote Sensing, GIS, and the Tragedy of the Aral Sea
- Chapter 11: The East Asian Gazetteer and the Reconstruction of Historical Geography
- Chapter 12: Survey Work on Contraceptive Use in West Bengal
- Chapter 13: Statistical Comparisons and the Assessment of Philippine Deforestation
- Chapter 14: Fire Ecology and Management in Australia

Many colleges and universities offer courses in the use of these tools, particularly GIS (geographic information systems), remote sensing (interpreting aerial photographs and satellite imagery), and computer-aided cartography (mapmaking). Such tools can provide students with technical skills that are much sought after in today's job market. If you are interested, ask your world regional geography professor about those kinds of opportunities at your college.

Figure 1.15 Natural Hazards and Globalization The eruption of Iceland's Eyjafjallajokull volcano in April 2010 brought chaos to world commerce by closing down air traffic in Europe and the North Atlantic. Along with its disruption to air passengers, the effects rippled through the world economy in thousands of ways. Farmers in Kenya, for example, could not get their fresh fruits and vegetables to European markets.

Figure 1.16 Areal Differentiation In this satellite photo the California–Mexico border is apparent because of the different sizes of agricultural fields. The larger red pattern in the upper left is the irrigated fields of the United States, while in the bottom right the much smaller and less irrigated fields of Mexico have a different landscape signature when seen from space. Explaining such differences is a central focus of geography.

part of Earth humid and lush while another just a few hundred kilometers away is an arid desert (Figure 1.16)?

Geographers also are interested in the connections between different places and how they are linked. This concern is one of **areal integration**, which is the study of how places interact with each other. An example is the analysis of how and why the economies of Singapore and the United States are closely intertwined even though the two countries are situated in entirely different physical, cultural, and political environments. The questions of areal integration, obviously, are becoming increasingly important because of the varied global linkages that are inherent to globalization.

Regions: Formal and Functional

The human intellect seems driven to make sense of the universe by lumping phenomena together into categories of similarity. Biology has its taxa of living objects, history its eras and periods of time, geology its epochs of Earth history. Geography, too, organizes information about the world by compressing it into units of similarity, which are called **regions**.

Sometimes the unifying threads of a region are physical, such as climate and vegetation, resulting in a regional designation like the *Sahara Desert* or the *Amazonian rain forest*. Other times the threads are more complex, combining economic and cultural traits, as in the use of the term *Corn Belt* for parts of the central United States. Human beings commonly compress large amounts of information into stereotypes; in a way, a geographic region is just that—a spatial stereotype for a portion of Earth that has some special signature or characteristic that sets it apart from other places (Figure 1.17).

Some caution is advised, however. First, no region is homogeneous throughout its area; thus even though there may be a single characteristic that unites an area, there will also be diversity and differences within it. Second, regional borders are fuzzy. Except for political borders that separate countries, states, or provinces, regional boundaries are rarely as abrupt as they appear on a map. Just as mountain ranges usually rise gradually from lowland areas or deserts transition gradually into grasslands, so do manufacturing regions thin out toward the periphery, or linguistic regions blend gradually with each other. Put differently, while those traits used to define a geographic region are most explicit and clear in the core area, the defining traits usually weaken and grow less apparent moving away from the core to the boundary.

The Cultural Landscape: Space into Place

Human beings transform space into distinct places that are unique and heavily loaded with meaning and symbolism. This diverse fabric of *placefulness* is of great interest to geographers because it tells us much about the human condition as it varies throughout the world. Places can tell us how humans interact with nature and among themselves; where there are tensions and where there is peace; where people are rich and where they are poor.

A common tool for the analysis of place is the concept of the **cultural landscape**, which is, simply stated, the visible, material expression of human settlement, past and present. Thus the cultural landscape is the tangible expression of the human habitat. It visually reflects the most basic human needs—shelter, food, work. Additionally, the cultural landscape acts to bring people together (or keep them apart), because it is a marker of cultural values, attitudes, and symbols. As cultures vary greatly around the world, so do cultural landscapes (Figure 1.18).

Most of the French countryside, for example, with its modestly sized agricultural fields enclosed by hedgerows and small stone houses clustered together in villages, looks very different from a typical farm landscape of the Midwestern United States, where fields cover several square miles and people are separated from each other by large distances. From the contrasting look of these two landscapes, we can learn much about the social and economic systems underlying the two different societies.

Increasingly, however, we see the uniqueness of places being eroded by the homogeneous landscapes of globalization—shopping malls, fast-food outlets, business towers, theme parks, industrial complexes. Understanding the forces behind the spread of these landscapes is important because they tell us much about the expansion of global economies and cultures. While a modern shopping mall in Hanoi, Vietnam, may seem familiar to someone from North America, this new landscape represents yet another component of globalized world culture that has been implanted into a once remote and distinctive city.

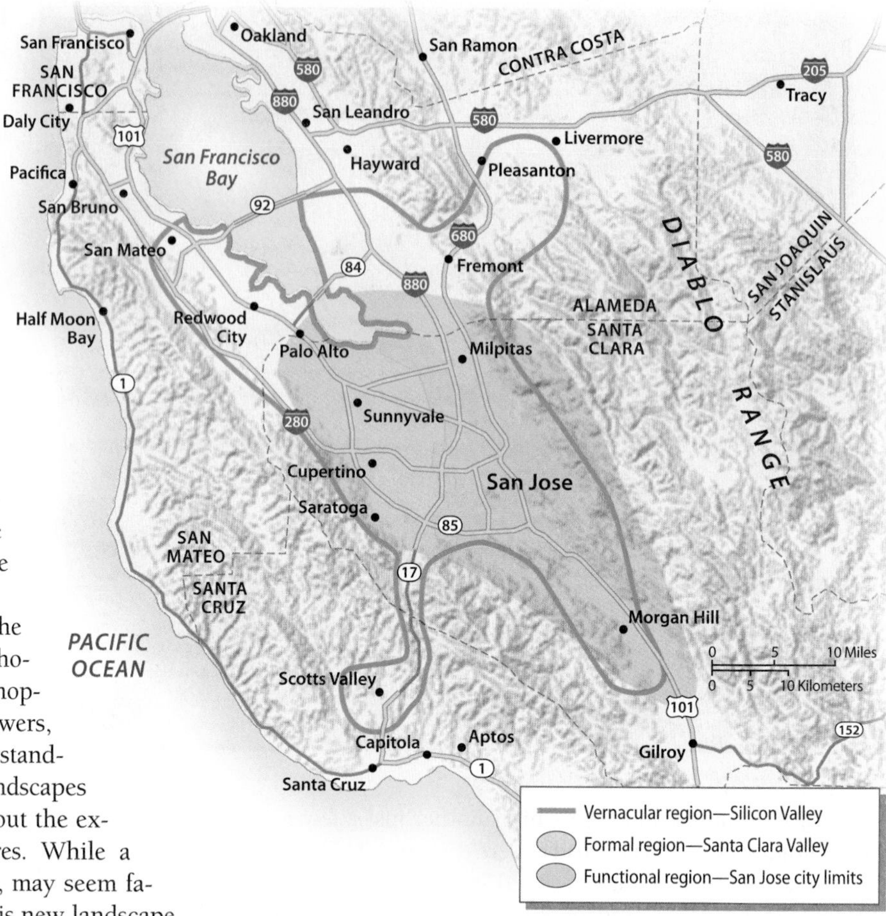

Figure 1.17 Geographic Regions This map illustrates three different kinds of geographic regions: vernacular, formal, and functional. A vernacular region is an abstraction that has indistinct cognitive borders, as shown by the general outline of what the public considers to be Silicon Valley. Other vernacular regions would be "The Midwest," "The Deep South," "The Pacific Northwest," and so on. In contrast, formal regions have distinct boundaries, such as that for the Santa Clara Valley, defined as the lowland between two bordering mountain ranges. Last, a functional region is based on a certain activity or organizational structure, such as the civic government of San Jose, delimited by its legal city limits.

Scales: Global to Local

There is a sense of scale to all systematic inquiry, whatever the discipline. In biology some scientists study the smaller units of cells, genes, or molecules, while others take a larger

Figure 1.18 The Cultural Landscape Despite globalization the world's landscapes still show great diversity, as shown by this village and its surrounding rice terraces on the island of Luzon, Philippines. Geographers use the cultural landscape concept to better understand how people interact with their environment.

view, analyzing plants, animals, or ecosystems. Similarly, some historians may focus on a specific individual at a specific point in time, while others take a broader view of international events over many centuries.

Geographers also work at different scales. While one may concentrate on analysis of a local landscape—perhaps a single village in southern China—another will focus on the broader regional picture, examining all of southern China (Figure 1.19). Others do research on a still larger global scale, perhaps studying emerging trade networks between southern India's center of information technology in Bangalore and North America's Silicon Valley, or investigating how India's monsoon might be connected to and affected by the Pacific Ocean's El Niño. But even though geographers may be working at different scales, they never lose sight of the interactivity and connectivity between local, regional, and global scales, of the ways that the village in southern India might be linked to world trade patterns, or how the late arrival of the monsoon (delayed, perhaps, by El Niño in the Pacific Ocean) could affect agriculture and food supplies in different parts of India (see "Global to Local: How the World Is Connected").

GLOBAL TO LOCAL How the World Is Connected

As noted in the text, the ability to move between different scales—global, regional, local—is critical for understanding contemporary world regional geography because of the way globalization links all peoples and places. Few villages today, however remote, are unconnected to the modern world. Global economies draw upon local crops and resources, and, conversely, fluctuations in world commodity prices affect the well-being of local people (Figure 1.2.1). Global TV and videos introduce foreign styles, ideas, mannerisms, and expectations into small towns and remote settlements in formerly isolated parts of the world, and global tourism brings strangers (some welcome, some not) to far-flung places, often corrupting and compromising the very uniqueness the tourists had sought.

In this book, we use the phrase *global to local* to capture the range of scales necessary for placing regions, places, landscapes, and people in their globalized context. Because our premise is that no place in today's world is completely isolated from the larger currents of a globalizing society, we need this perspective to examine the complicated linkages connecting people and places with the larger, dynamic world. These points are elaborated upon in the *Global to Local* sidebars in each regional chapter.

Chapter 2: The Globalization of Bushmeat and Animal Poaching
Chapter 3: Alaska's Climate Refugees
Chapter 4: Can Eco-Friendly Fair Trade Coffee Promote Biodiversity and Rural Development?
Chapter 5: Caribbean Carnival Around the World
Chapter 6: Roses to Europe, Thanks to Kenya
Chapter 7: How Morocco Went Hollywood in Ouarzazate
Chapter 8: European Hip Hop and Hyperlocality
Chapter 9: The Tale of Two Siberias
Chapter 10: The Mongolian Cashmere Industry Readjusts
Chapter 11: The Korean Wave
Chapter 12: The Sialkot Industrial Complex
Chapter 13: Poipet and Southeast Asia's Other "Sin Cities"
Chapter 14: Nauru and the Mixed Benefits of Globalization

Figure 1.2.1 Global to Local A farmer in Guilin, China, uses a laptop to check local and global grain prices before taking his crop to market.

Figure 1.19 Geographic Scales Geographers work at many different scales, from large regional or even global analyses, to micro-studies of one specific environment or landscape. This photo is of Zhujiajiao village in China's Shanghai province.

THEMES AND ISSUES IN WORLD REGIONAL GEOGRAPHY

Following two introductory chapters, this book adopts a regional perspective, grouping all of Earth's countries into a framework of 12 world regions (Figure 1.20). We begin with a region familiar to most of our readers, North America, and then move to Latin America, the Caribbean, Africa, the Middle East, Europe, Russia, and the different regions of Asia, before concluding with Australia and Oceania. Each of the 12 regional chapters employs the same five-part thematic structure—environmental geography, population and settlement, cultural coherence and diversity, geopolitical framework, and economic and social development. The concepts and data central to each theme are discussed in the following sections.

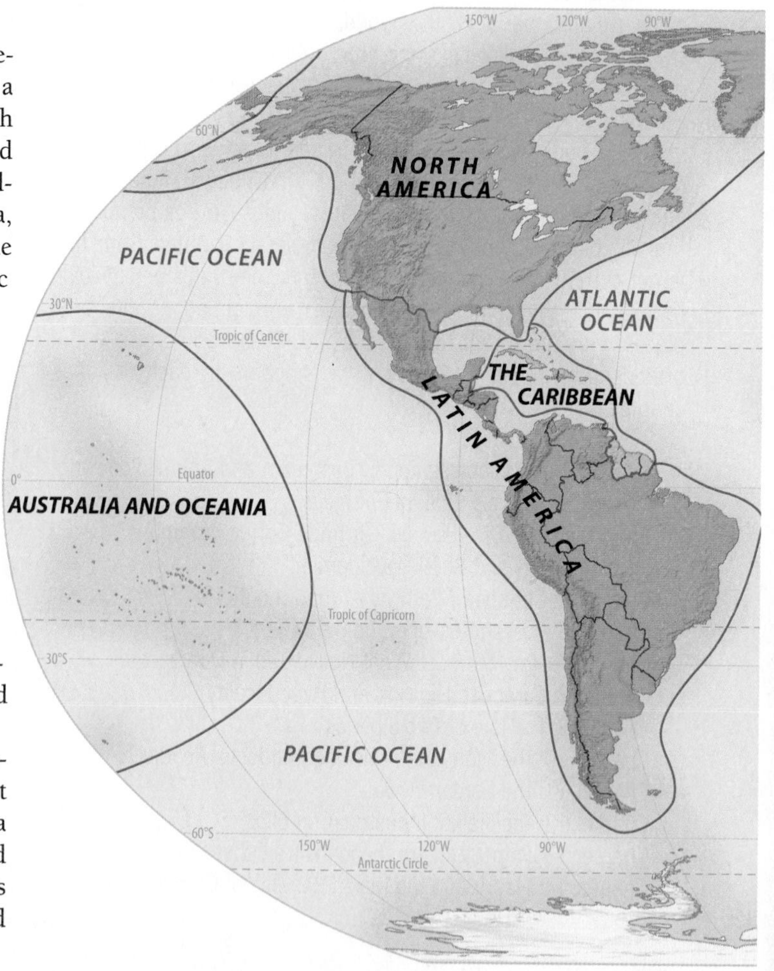

Environmental Geography

Chapter 2 provides background on global environmental geography, outlining the global environmental elements fundamental to human settlement—climate, geology, hydrology, and vegetation, along with a discussion of the linkages between environment, globalization, and world food problems.

As well, in the regional chapters, the Environmental Geography section discusses the environmental issues relevant to each world region, topics such as global warming, sea level rise, acid rain, tropical rainforest destruction, and wildlife conservation. These environmental issues sections are not simply a list of problems but also discuss plans and policies developed to resolve those issues.

Population and Settlement: People on the Land

The human population is far larger than it has ever been, and there is considerable debate about whether continued growth will benefit or harm the human condition. This concern is not only about the total population of the world but also about the geographic patterns of human settlement. As the world map shows, while some parts of the world are densely populated, other places remain almost empty (Figure 1.21).

With nearly 7 billion humans on Earth, we currently add about 131 million people each year, at a rate of 15,000 births per hour. About 90 percent of this population growth takes place in the world regions of Africa, South and East Asia, and Latin America. Because of rapid growth in developing countries and emerging economies, perplexing questions dominate discussions of many global issues. Can these countries absorb the rapidly increasing population and still achieve the necessary economic and social development to ensure some level of well-being and stability for their populations? What role, if any, should the developed countries of North America, Europe, and East Asia play in helping developing countries with their population problems? While population is a complex and contentious topic, several points help focus the issues:

- Very different rates of population growth are found in different regions of the world. While some countries are growing rapidly, others have no natural growth at all. Instead, any population growth comes from in-migration. India is an example of the former; Italy of the latter.
- Population planning takes many forms, from the fairly rigid one- or two-child policies of China and other Asian countries to the "more children, please" programs of no-growth western European countries (Figure 1.22).
- Not all attention should be focused on natural growth because migration is increasingly the root cause of population change in the globalized world. Most international migration is driven by a desire for a better life in the richer regions of the developed world. But there are also millions of migrants who are refugees from civil strife, persecution, and environmental disasters.
- The greatest migration in human history is now going on as people move from rural to urban environments. In 2009 a landmark was reached when population experts estimated that more than half the world's population now lived in towns and cities. As with natural population growth, the developing countries of Africa, Latin America, and Asia are experiencing the most rapid changes in urbanization.

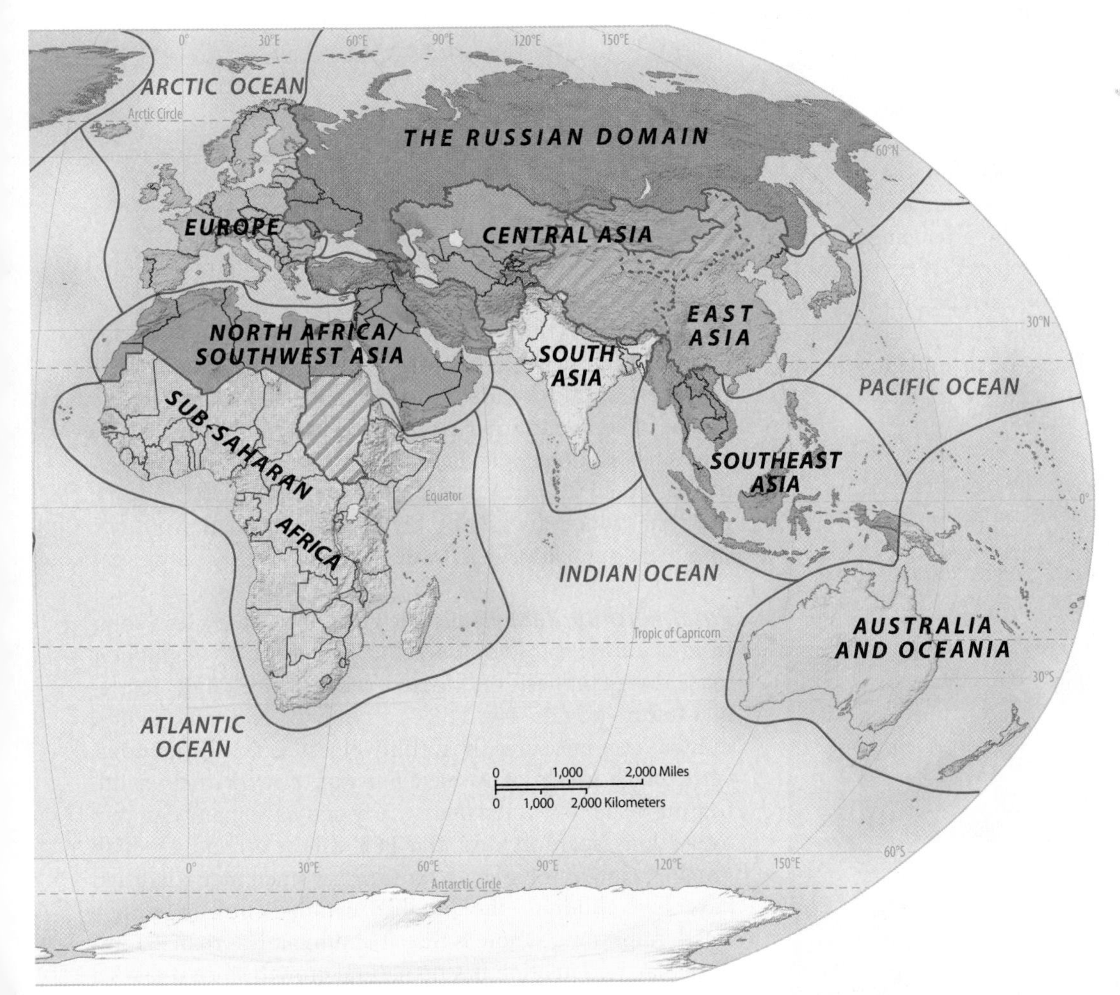

Figure 1.20 World Regions These regions are the basis for the 12 regional chapters in this book. Countries or areas within countries that are treated in more than one chapter are designated on the map with a striped pattern. For example, western China is discussed in both Chapter 10, on Central Asia, and Chapter 11, on East Asia. Also, three countries on the South American continent are discussed as part of the Caribbean region because of their close cultural similarities to the island region.

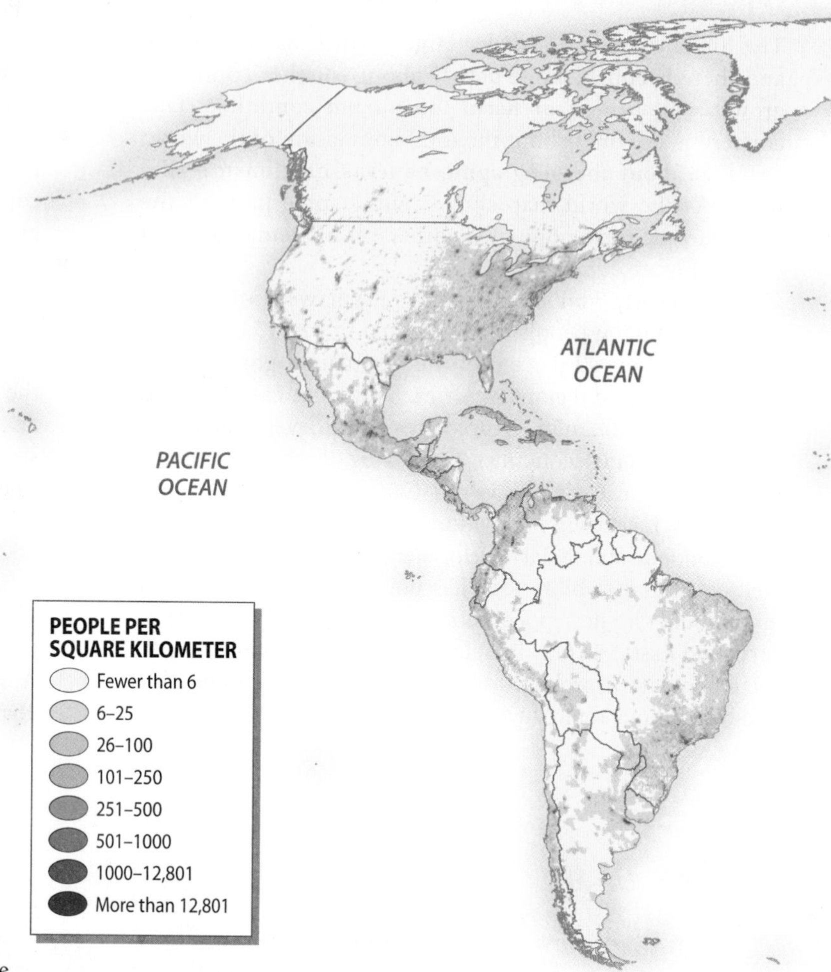

Figure 1.21 World Population This map shows differing population densities in the areas of the world. East Asia stands out as the most populated region, with high densities in Japan, Korea, and eastern China. The second-most populated region is South Asia, dominated by India, which is second only to China in population. In North Africa and Southwest Asia, population clusters are often linked to the availability of water for irrigated agriculture, as is apparent with the population cluster along the Nile River. Higher population densities in Europe, North America, and other countries are usually associated with large cities, their extensive suburbs, and nearby economic activities.

Population Growth and Change Because of the centrality of population growth, each regional chapter in this book includes a table of population data for the countries within that region (Table 1.1). Although at first glance the statistics in these tables might seem daunting, this information is crucial to understanding the population geography of the regions.

Natural Population Increase A common starting point for measuring demographic change is the rate of natural increase (RNI), which depicts as a percentage the annual growth rate for a country or region. This statistic is produced by subtracting the number of deaths in a given year from the births. Important to remember is that gains or losses through migration are not considered in the RNI.

Further, instead of using raw numbers for a large population, demographers divide the gross numbers of births or deaths by the total population, thereby producing a figure per 1,000 of the population. This is referred to as the *crude birthrate* or the *crude death rate*. For example, in 2010 the crude birthrate for the whole world was 20 per 1,000, with a crude death rate of 8 per 1,000. Thus, the natural growth rate was 12 per 1,000. Converting that figure to a percentage allows us to express it as the RNI; therefore the rate of natural increase for the world in 2010 was 1.2 percent per year.

Because birthrates vary greatly amongst peoples and cultures (and between countries and regions of the world), rates of natural increase also vary greatly. In Africa, for example, several countries have crude rates of more than 40 births per 1,000 people. Because their death rates are generally less than 20 per 1,000, RNIs are greater than 3 percent per year, which are among the highest population growth numbers found anywhere in the world.

Figure 1.22 Family Planning Policies Many countries in the developing world have concluded that unrestrained population growth may keep them from realizing their development goals, and therefore have put family planning policies in place that promote birth control. This poster in Vietnam urges smaller families.

Total Fertility Rate While the crude birthrate gives some insight into current conditions in a country, demographers place more emphasis on the total fertility rate (TFR) to predict future growth. The TFR is a synthetic and hypothetical number that measures the fertility of a statistically fictitious, yet average group of women moving through their childbearing years. If women marry early and have many children over a long span of years, the TFR will be a relatively high number. Conversely, if data show that women marry late and have few children, the number will be correspondingly small. Important to note is that any number less than 2.1 implies that a population has no natural growth since it takes a

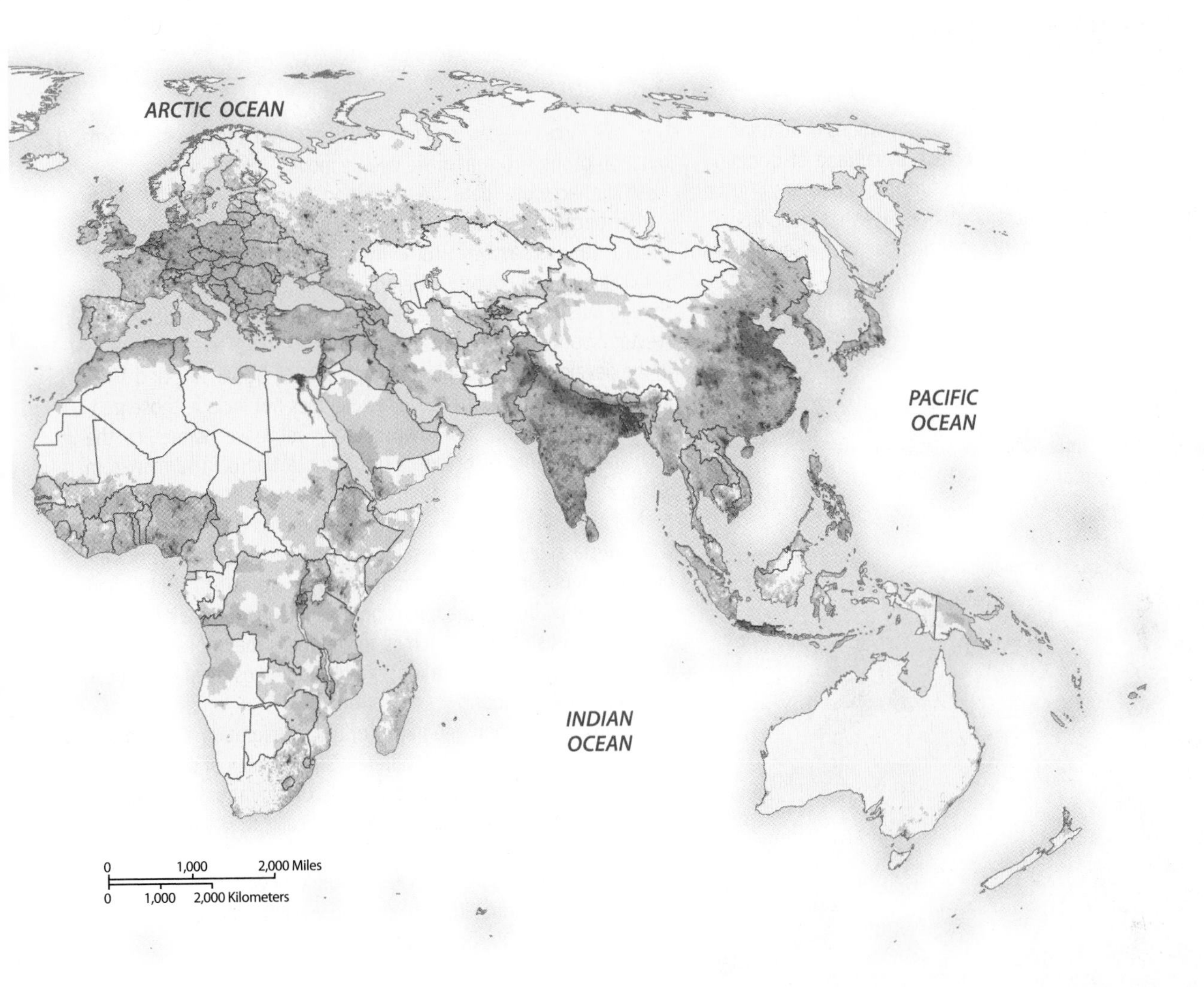

TABLE 1.1 Population Indicators of the World's Ten Largest Countries

Country	Population (millions) 2010	Population Density (per square kilometer)	Rate of Natural Increase (RNI)	Total Fertility Rate	Percent Urban	Percent <15	Percent >65	Net Migration (Rate per 1000) 2005–10[a]
China	1,338.1	140	0.5	1.5	47	18	8	–0.3
India	1,188.8	362	1.5	2.6	29	32	5	–0.2
United States	309.6	32	0.6	2.0	79	20	13	3.3
Indonesia	235.5	124	1.4	2.4	43	28	6	–0.6
Brazil	193.3	23	1.0	2.0	84	27	7	–0.2
Pakistan	184.8	232	2.3	4.0	35	38	4	–1.6
Bangladesh	164.4	1,142	1.5	2.4	25	32	4	–0.7
Nigeria	158.3	171	2.4	5.7	47	43	3	–0.4
Russia	141.9	8	–0.2	1.5	73	15	13	0.4
Japan	127.4	337	–0.0	1.4	86	13	23	0.2

[a]*Net Migration Rate from the United Nations, Population Division, World Population Prospects: The 2008 Revision Population Database.*
Source: Population Reference Bureau, World Population Data Sheet, 2010.

PEOPLE ON THE MOVE The Geographies of Global Migration

Globalization has led to one of the largest migrations in human history as people leave their homes to take advantage of booming economies and better living conditions in other countries (Figure 1.3.1). Either desperate conditions at home—what geographers call "push forces"—or the lure of promising pastures elsewhere—the "pull forces"—can lead to migration. Often, though, it's a combination of the two that make people move.

While accurate data on migration is notoriously difficult to gather because people often migrate without documentation, the United Nations says that currently at least 190 million people are legally living in a country different than where they were born. Although most of those people have moved to cities in the developed western world, there is also considerable migration to magnet cities in the developing world, places such as Delhi, Dubai, Rio, and Mexico City. Furthermore, that UN statistic does not include those people who stay within their own country, yet leave home, like the highland peasants of Bolivia moving to La Paz or a farmer from Iowa cashing in to take a job in Atlanta.

Nor does that statistic count the estimated 35 million people who are classified as refugees, those who have fled violence or natural disasters. Or the untold number who are not in established refugee camps but, instead, making do somehow on their own. Although the push and pull forces for refugees may not have been caused directly by globalization, often, if one digs deep enough, a connection can be found.

Because the human geography of global migration wears many different faces and tells many different stories, each of our chapters sheds light on this complicated process of migration through our *People on the Move* sidebars. More specifically, they are:

Chapter 3: Skilled Immigrants Fuel Big City Growth
Chapter 4: National Identity and Japanese Brazilians
Chapter 5: The Diaspora Drifts Back
Chapter 6: The New African Americans
Chapter 7: Jordanian Bedouins and the Badia Project
Chapter 8: Smuggling Illegal Immigrants into Europe
Chapter 9: Chinese Immigrants in Russia's Far East
Chapter 10: Migration In and Out of Kazakhstan
Chapter 11: Rural Migrants Flock to China's Cities
Chapter 12: Punjabis Leaving Agriculture and Leaving India
Chapter 13: The Rohingyas Seek Sanctuary
Chapter 14: Does Australia Have a Population Problem?

Figure 1.3.1 People on the Move A boat carrying 179 illegal immigrants who had hoped to reach Europe is stopped by the Spanish Coast Guard before it reached Spain's Canary Islands.

minimum of two children to replace their parents. From population data collected in the past decade, the current TFR for the world is 2.5. While that is the average for the whole world, the variability between regions is striking. To illustrate, the current TFR for Africa is 4.7, whereas in no-growth Europe it is only 1.6 (Figure 1.23).

Young and Old Populations One of the best indicators of the momentum (or lack of) for continued population growth is the youthfulness of a population because these data show the proportion of a population about to enter their prime reproductive years. The common statistic for this measure is the *percentage of a population under age 15*. As a global average, 27 percent of the population is younger than age 15, but in fast-growing Africa, that figure is 41 percent, with several African countries approaching 50 percent. This suggests strongly that rapid population growth in Africa will continue for at least another generation, despite the tragedy of the HIV/AIDS epidemic. In contrast, Europe has only 16 percent of its population under 15, and North America 20 percent.

The other end of the age spectrum is also important, and it is measured by the *percentage of a population over age 65*. That figure is useful for inferring the needs of a society to provide social services for its senior citizens and pensioners. Japan and most European countries, for example, have a relatively high proportion of people over age 65. As a result, concerns are raised about whether there will be enough wage earners to support the social security needs of a large elderly population.

Figure 1.23 Fertility and Mortality Birthrates and death rates vary widely around the world. Fertility rates result from an array of variables, including state family-planning programs and the level of a woman's education. This family is in Bangladesh, a country that is working hard to reduce its birthrate through education programs.

The structure of a population, which includes the percentage of young and old, is presented graphically as a **population pyramid.** This graph plots the percentage of all different age groups along a vertical axis that divides the population into male and female (Figure 1.24). The large percentage of young people in a fast-growing population provides a wide base and the small percentage of elderly a narrow tip, thereby giving the graph its pyramidal shape. Older populations, in contrast, with fewer young people and an aging population, give the graph a very different shape, with a narrow base and broader upper segment.

Life Expectancy Another demographic indicator that contains information about health and well-being in a society is *life expectancy,* which is the average length of a life expected at the birth of a typical male or female in a specific country. Because a large number of social factors such as health services, nutrition, and sanitation influence life expectancy, these data are often used as an indicator of the level of social development in a country. Because this book uses life expectancy as a social indicator, life expectancy data are found in the economic and social development tables, not with other population statistics.

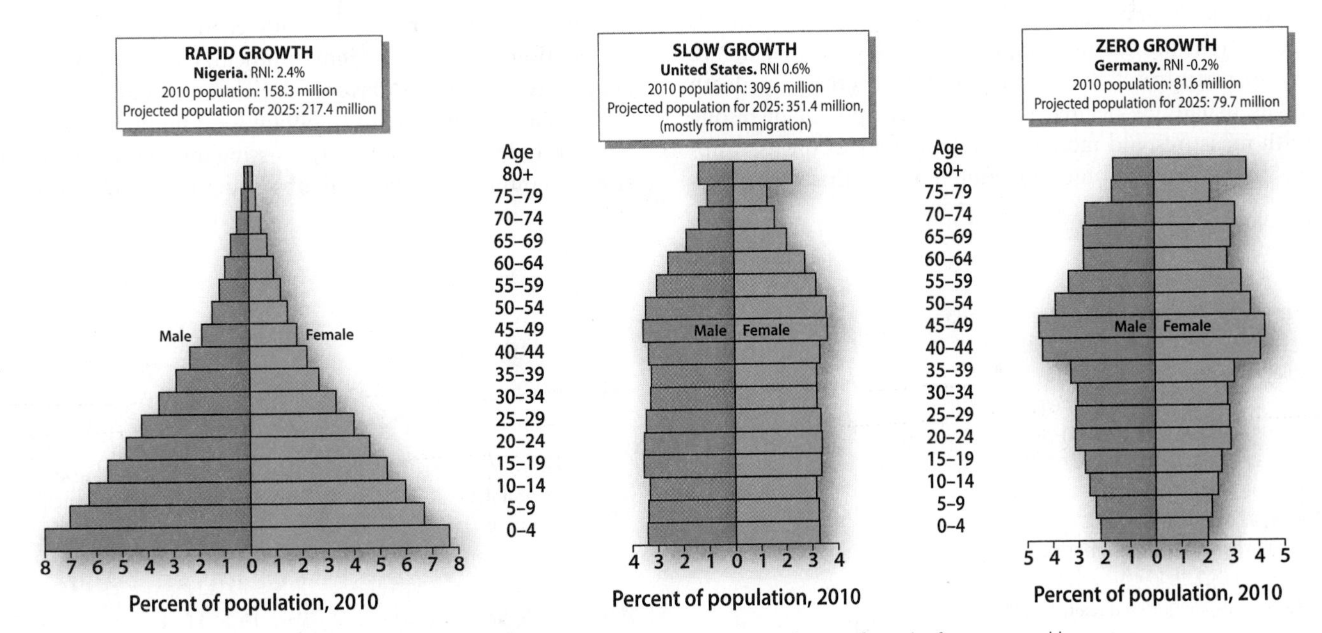

Figure 1.24 Population Pyramids The term *population pyramid* comes from the form assumed by a rapidly growing country such as Nigeria, when data for age and sex are plotted graphically as a percentage of the total population. The broad base illustrates the high percentage of young people in the country's population, which indicates that rapid growth will probably continue for at least another generation. This pyramidal shape contrasts with the narrow bases of the slow- and negative-growth countries, the United States and Germany, which have fewer people in the childbearing years.

Not surprisingly because social conditions vary widely around the world, so do life expectancy figures. In general, though, life expectancy has been increasing over the decades, implying that the conditions supporting life and longevity are improving. To illustrate, in 1975, the average life expectancy figure for the world was 58 years, whereas today it is 69. In Sub-Saharan Africa, however, life expectancy has changed very little over the last 30 years because of the HIV/AIDS epidemic. As a result, the life expectancy for the region is the same (52) as it was in 1975. In Russia, life expectancy has fallen in the past two decades, probably because of the deterioration of social services accompanying economic restructuring in the post-Soviet era.

The Demographic Transition The historical record shows that population growth rates commonly change over time. More specifically, in Europe, North America, and Japan, growth declined as countries became increasingly industrialized and urbanized. From this historical data, demographers generated the **demographic transition model**, a four-stage conceptualization that tracks changes in birthrates and death rates through time as a population urbanizes (Figure 1.25).

In the demographic transition model, Stage 1 is characterized by both high birthrates and death rates, leading to a very slow rate of natural increase. Historically, this stage is associated with Europe's preindustrial period, a period that also predated common public health measures such as sewage treatment, the scientific understanding of disease transmission, and the most fundamental aspects of modern medicine. Not surprisingly, death rates were high and life expectancy was short. Tragically, these conditions are still found today in some parts of the world.

In Stage 2, death rates fall dramatically while birthrates remain high, producing a rapid rise in the RNI. Again, in both historical and modern times, this decrease in death rates is usually associated with the development of public health measures and modern medicine. One of the assumptions of the demographic transition model is that these services become increasingly available only after some degree of economic development and urbanization takes place.

However, even as death rates fall, it takes time for people to respond with lower birthrates, which begin in Stage 3. This, then, is the transitional stage in which people apparently become aware of the advantages of smaller families in an urban and industrial setting.

In Stage 4, a very low RNI results from a combination of low birthrates and very low death rates. As a result, there is very little—if any—natural population increase. Today, the United States, Japan, and most European countries are clearly in Stage 4 of the demographic transition model.

Migration Patterns Never before in human history have so many people been on the move. Today, more than 190 million people live outside the country of their birth and thus are officially designated as migrants by international agencies. Much of this international migration is directly linked to the new globalized economy because half of the migrants live either in the developed world or in developing countries with vibrant industrial, mining, or petroleum extraction economies. In the oil-rich countries of Kuwait and Saudi Arabia, for example, the labor force is composed primarily of foreign migrants. In total numbers, fully one-third of the world's migrants live in seven industrial countries: Japan, Germany, France, Canada, the United States, Italy, and the United Kingdom. Moreover, most of these migrants have moved to cities; in fact, 20 percent of migrants live in just 20 world cities. Further, because industrial countries usually have very low birthrates, immigration accounts for a large proportion of their population growth. For example, about one-third of the annual growth in the United States is due to in-migration.

But not all migrants move for economic reasons. War, persecution, famine, and environmental destruction cause people to flee to safe havens elsewhere. Accurate data on refugees are often difficult to obtain for a number of reasons (such as individuals not legally crossing international boundaries or countries deliberately obscuring the number for po-

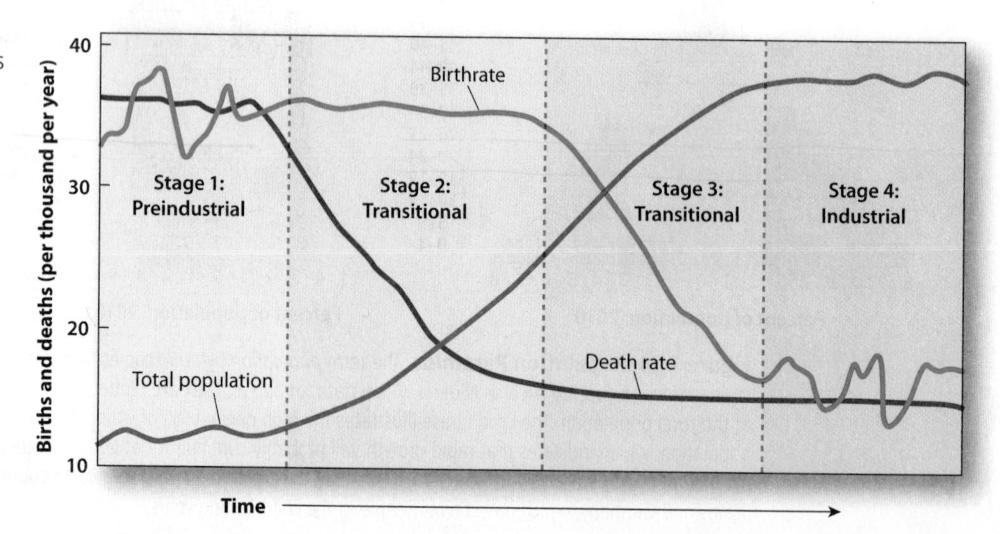

Figure 1.25 Demographic Transition As a country goes through industrialization, its population moves through the four stages in this diagram, referred to as the *demographic transition*. In Stage 1, population growth is low because high birthrates are offset by high death rates. Rapid growth takes place in Stage 2, as death rates decline. Stage 3 is characterized by a decline in birthrates. The transition ends with low growth once again in Stage 4, resulting from a relative balance between low birthrates and death rates. With a large number of developed countries now showing no natural growth, demographers talk of adding a 5th stage to the traditional demographic transition model, one where there is negative natural growth.

Figure 1.26 Refugees About 35 million people are refugees from ethnic warfare and civil strife. Most of these refugees are in Africa and western Asia. These women wait in line for water at a refugee camp in southern Sudan.

litical reasons), but UN officials estimate that some 35 million people should be considered refugees. More than half of these are in Africa and western Asia (Figure 1.26).

Push and Pull Forces While its causes may be complicated, the migration process can be understood through three interactive concepts. First, *push forces*, such as civil strife, environmental degradation, or unemployment, drive people from their homelands. Second, *pull* forces, such as better economic opportunity or health services, attract migrants to certain locations, within or beyond their national boundaries. Connecting the two are the *informational networks* of families, friends, and sometimes labor contractors who provide information on the mechanics of migration, transportation details, and housing and job opportunities.

Net Migration Rates The amount of immigration (in-migration) and emigration (out-migration) is measured by the **net migration rate**, a statistic that depicts whether more people are entering or leaving a country. A positive figure means the population is growing because of in-migration, whereas a negative number means more people are leaving than arriving. As with other demographic indicators, the net migration rate is provided for the number of migrants per 1,000 of a base population. To illustrate, the net migration rate for the United States is 3 per 1,000 people, whereas Canada, which receives even more immigrants than the United States, has a net migration rate of 8. In contrast, Mexico, the source of many migrants to North America, has a net migration rate of −3.

Some of the highest net migration rates are found in countries that depend heavily on migrants for their labor force, such as the United Arab Emirates, with a net migration rate of 16, and Kuwait at 8. Countries with the highest negative migration rates are Marshall Islands −23, Federated Micronesia −15, and a number of the Caribbean island states, with net migration rates around −5.

Settlement Geography Data on **population density**, which is the average number of people per area unit (square mile or square kilometer), conveys important information about settlement in a specific country. In Table 1.1 one sees the striking difference between the high population density of India contrasted to the much lower figure for the United States. Flying over these two countries and looking down at the settlement patterns would explain this contrast (Figure 1.27). While much of the United States is covered by farms covering hundreds of acres, with houses and barns several miles from their neighbors, the landscape of India is made up of small villages, distanced from each other by only a mile or so. This results in a population density three times higher in India than in the United States. Japan has a similar settlement pattern to India, even though it is primarily an urban, industrial country. Bangladesh, one of the most densely settled countries in the world, must squeeze its large and rapidly growing population into a limited amount of dry land built by the delta of two large rivers, the Ganges and the Brahmaputra.

Because population densities differ considerably between rural and urban areas, the national figure can be a bit misleading. Many of the world's cities, for example, have densities of more than 30,000 people per square mile (10,344 per square kilometer), with the central areas of Mumbai (Bombay) and Shanghai easily twice as dense because of the prevalence of high-rise apartment buildings. In contrast, most North American cities have densities of fewer than 10,000 people per square mile (3,861 per square kilometer)

Figure 1.27 Contrasting Settlement Density: U.S. and India In the U.S. the population density is 32 per square kilometer, contrasted with 356 per square kilometer in India. Besides the difference in the size of each country's population, another factor in the contrasting densities is the difference in settlement patterns. In the United States people commonly settle on dispersed farms on large acreage, as illustrated by the landscape in Iowa, whereas as in India there is dense settlement in both towns and rural landscapes.

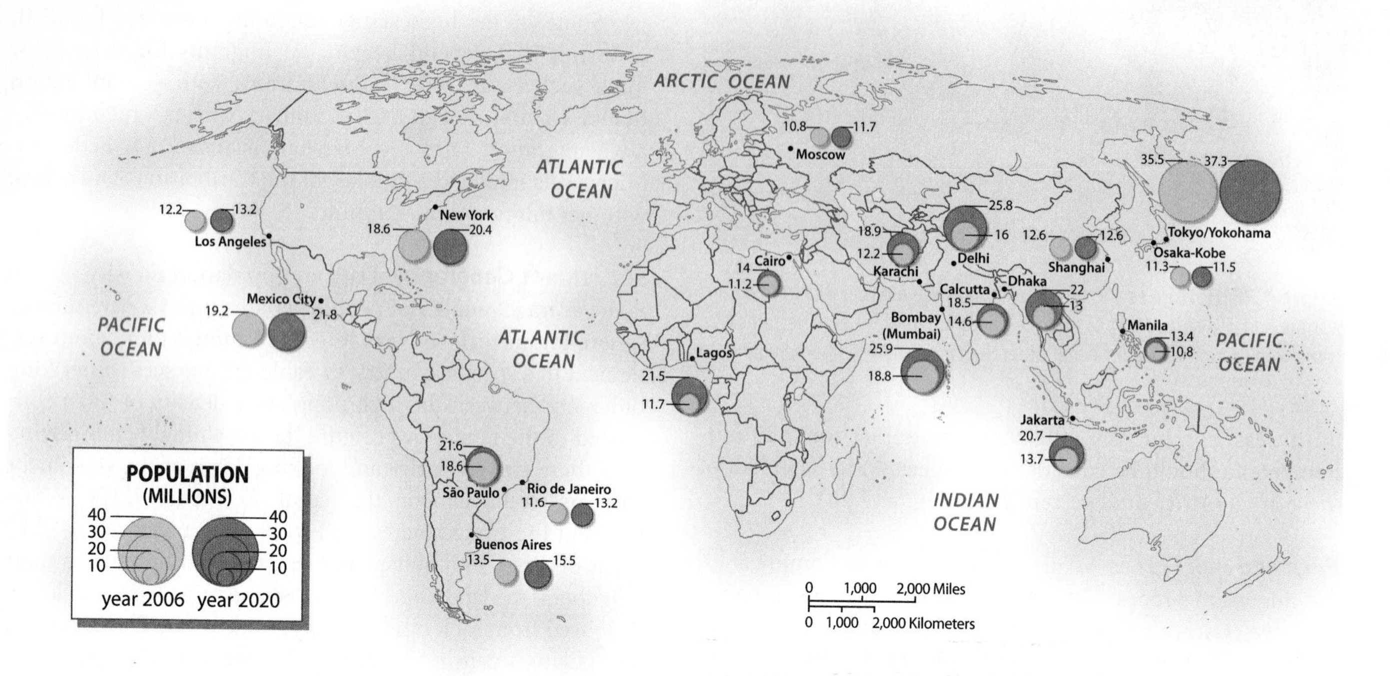

Figure 1.28 Growth of World Cities This map shows the 20 largest cities in the world, along with their projected growth by the year 2020. Note that the greatest city growth in the next decade will be in the developing world. In contrast, there is little growth forecast for large cities like Tokyo or New York in the developed world.

because of the cultural preference for single-family dwellings on individual urban lots.

An Urban World Cities are the focal points of the contemporary, globalizing world, the fast-paced centers of deep and widespread economic, political, and cultural change. Because of this vitality and the options cities offer to impoverished and uprooted rural peoples, they are also magnets for migration. The scale and rate of growth of some world cities is absolutely staggering. Estimates are that between natural growth and in-migration, Mumbai (Bombay), India, will add over 7 million people by the year 2020, which, assuming growth is constant throughout the period (which is questionable), would mean the urban area adds over 10,000 new people each week. The same projections would have Lagos, Nigeria, which has the highest annual growth, adding almost 15,000 per week (Figure 1.28).

Based upon data on the **urbanized population**, which is the percentage of a country's population living in cities, as mentioned, at least half the world's population now lives in cities. Further, demographers predict that the world will be 60 percent urbanized by the year 2025.

Tables in this book's regional chapters include data on the urbanization rate for each country. To illustrate, more than 80 percent of the populations of Europe, Japan, Australia, and the United States live in cities. Generally speaking, most countries with such high rates of urbanization are also highly industrialized because manufacturing tends to cluster around urban centers. In contrast, the urbanized rate for developing countries is usually less than 50 percent, with figures closer to 40 percent not uncommon. Urbanization figures also show where there is high potential for urban migration. If the urbanized population is relatively small, as in Zimbabwe (Africa), where only some 37 percent of the population lives in cities, the probability of high rates of urban migration in the next decades is great.

Conceptualizing the City Given the important role of cities in world regional geography, it seems appropriate to offer a brief conceptual framework for examining urban settlements.

- Cities are interdependent with other cities and towns, both regionally and globally, and are linked in an urban system of interactive settlements. As some cities expand or come to specialize in a specific function, such as banking or manufacturing, the relationships and interactions within that urban system may change. The largest city within the city system may have **urban primacy**, a term used to describe a city that is disproportionately large and dominates economic, political, and cultural activities within the country. In many countries, one large primate city usually stands out in that country's urban system. Examples include Bangkok, Mexico City, Cairo, Paris, and Buenos Aires.
- Although cities are highly complicated settlements, they often share traits of **urban structure**, a term that refers to the distribution and patterning of land use within the city. Land use is linked to specific functions, such as finance, governmental activities, housing, retailing, and industry. Almost all large cities, for example, contain a *central business district* (CBD) in which banks and business headquarters are located. Because these functions

Figure 1.29 Urban Form This photo of downtown Caracas, Venezuela, shows that city form expresses both local and global influences. In the foreground are low-income "ranchitos" where recent migrants to the city live, whereas in the background are the high-rise buildings of the financial center that are linked to global trade.

are not randomly distributed, but instead follow well-understood location principles, it is possible to generate models of urban structure that convey the commonalties between cities within a specific region. An example is the model of Latin American cities discussed in Chapter 4.

- Closely related to urban structure is the landscape of the city, the **urban form** or physical arrangement of buildings, streets, parks, and architecture that gives each city its unique sense of place. For example, Cairo, Egypt, clearly looks and feels very different from New Delhi, India, or Caracas, Venezuela (Figure 1.29). The reasons for the range of urban landscapes include differences in histories, economies, planning policies, and cultural values (see "Cityscapes: Exploring the World's Urban Places").
- Because urban migration has taken place so rapidly in developing countries, many cities have reached a state of **overurbanization**, in which the urban population grows more quickly than the provision of necessary support services such as housing, transportation, waste disposal, and water supply. Unfortunately, this is a fairly common situation in the developing world. The hundreds of thousands of "street people" of Kolkata (Calcutta), India, are legendary, although the phenomenon of mass homelessness is by no means restricted to South Asia.
- Overurbanization often results in **squatter settlements**, illegal developments of makeshift housing on land neither owned nor rented by their inhabitants. Such settlements often are built on steep hillsides or on river floodplains that expose the occupants to the dangers of landslide and floods. Squatter settlements also are often found in the open space of public parks or along roadways, where they are regularly destroyed by government authorities, usually to be quickly rebuilt by migrants who have no other alternatives (Figure 1.30). Even for those who have found more permanent housing, water and sewer connections are often rudimentary at best. Water service can be sporadic and restricted to a few hours a day, and open sewers are a disturbingly com-

Figure 1.30 Squatter Settlements Because of the massive migration of people to world cities, adequate housing for the rapidly growing population becomes a daunting problem. Often, migrant housing needs are filled with illegal squatter settlements, such as this one in New Delhi, India.

CITYSCAPES Exploring the World's Urban Places

More than half of the people in the world live in cities. From North America to Asia, Africa to Australia, we have become an urban species, clustering in cities that continue to expand upward and outward at unprecedented rates of growth. Although many countries have long been urbanized, until this decade most of the world's population still lived in the countryside. But all that is changing rapidly, with profound implications for the new urban dwellers, as well as for the larger global environment.

What are these cities like and what do people do there? How are they changing as globalization aggravates tensions between the traditional and modern worlds (Figure 1.4.1)?

Although the broad currents of urbanization are discussed throughout this book, we focus on a specific town or city within each regional chapter to convey a particular sense of that place, of what it looks and feels like, and what issues the people of that city face. These *Cityscapes* sidebars describe an array of cities, large and small, rich and poor (but usually a combination of both), and offer a series of regional snapshots that illustrate how these special localities are growing and changing. More specifically, they are:

Chapter 3: Miami
Chapter 4: Panama City—Latin America's Aspiring Crossroads
Chapter 5: The Caribbean Metropolis of Santo Domingo
Chapter 6: Africa's Golden and Global City
Chapter 7: Dubai's Rocky Road to Urbanization
Chapter 8: Berlin Reinvents Itself
Chapter 9: St. Petersburg
Chapter 10: The Storied City of Samarkand
Chapter 11: The Varied Neighborhoods of Tokyo
Chapter 12: Life on the Streets of Mumbai (Bombay)
Chapter 13: Metro Manila, Primate City of the Philippines
Chapter 14: Apia, Capital of Samoa

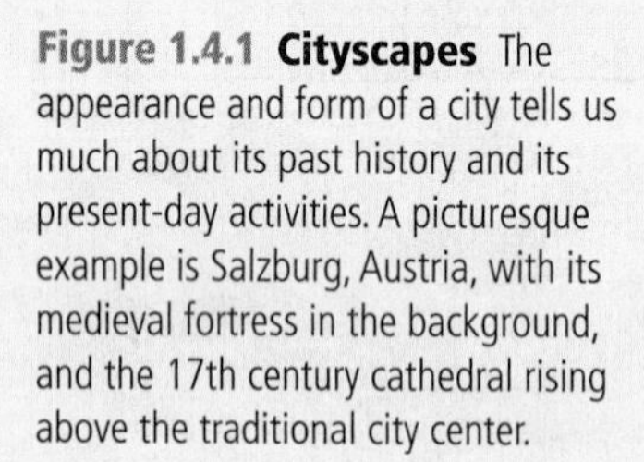

Figure 1.4.1 Cityscapes The appearance and form of a city tells us much about its past history and its present-day activities. A picturesque example is Salzburg, Austria, with its medieval fortress in the background, and the 17th century cathedral rising above the traditional city center.

mon feature of many cities. The World Bank estimated recently that 60 percent of the people in less-developed cities lived in homes that are not connected to sewers.

Cultural Coherence and Diversity: The Geography of Tradition and Change

Social scientists like to say that culture is the weaving binding together the world's diverse social fabric. If true, one glance at the daily news suggests this complex global tapestry is unraveling with widespread cultural tension and conflict. As noted earlier, with the recent rise of global communication systems (satellite TV, films, videos, etc.), stereotypical Western culture is spreading at a rapid pace, and while some cultures accept these new cultural influences willingly, others resist this new form of cultural imperialism through protests, censorship, even terrorism.

The geography of cultural cohesion and diversity, then, entails an examination of tradition and change, of language and religion, of group belonging and identity, and of the complex and varied currents that underlie 21st-century ethnic factionalism and separatism (Figure 1.31).

Culture in a Globalizing World Given the diversity of cultures around the world, coupled with the dynamic changes associated with globalization, traditional definitions of cul-

Figure 1.31 Ethnic Tensions Unfortunately, much contemporary cultural change is characterized by violence among different ethnic groups. In this photo, a commuter in the Indian state of Gujarat passes by a car burned during rioting between Muslims and Hindus.

ture must be stretched somewhat to provide a viable conceptual framework. A basic definition provides a starting point. **Culture** is learned, not innate, and is shared behavior held in common by a group of people, empowering them with what could be called a "way of life."

In addition, culture has both abstract and material dimensions: speech, religion, ideology, livelihood, and value systems, but also technology, housing, foods, and music. These varied expressions of culture are relevant to the study of world regional geography because they tell us much about the way people interact with their environment, each other, and the larger world. Finally, especially given the widespread influences of globalization, it is best to think of culture as dynamic rather than static. That is, culture is a process, not a condition, something that is constantly adapting to new circumstances. As a result, there are always tensions between the conservative, traditional elements of a culture and the newer forces promoting change (Figure 1.32).

(a)

(b)

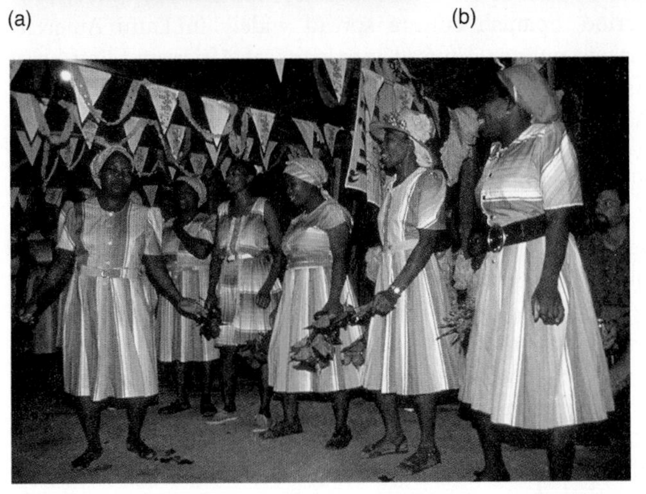

(c)

(d)

Figure 1.32 Folk, Ethnic, Popular, and Global Culture A broad spectrum of cultural groups characterizes the world's contemporary cultural geography. Representing global culture (a) is a Korean woman in downtown Washington, DC, with a mobile phone and laptop computer. In Amsterdam, the Netherlands, two visitors (b) wear the universal clothes of popular culture, while at the opposite end of the spectrum are the ethnic cultures of (c) Garifuna women in Honduras and (d) men in Peru performing the Chonguinada dance.

ENVIRONMENT AND POLITICS Controversies and Complexities

Because the environment is fundamental to our existence and humans have many different opinions and values about its usage, environmental issues are almost always controversial, whether the topic is territorial rights, resource usage, or simply preserving and protecting nature (Figure 1.5.1). Who gets to use water, cut trees, till the land, or control the territory? And with controversy comes politics, at all scales and in all forms, as stakeholders air their differences and local, national, or international governmental representatives mediate and seek solutions.

Because of the extended reach of international commerce, economic globalization has aggravated many environmental issues. What were once local environmental controversies have often been expanded and recast in the global arena. In the regional chapters of this book we have taken a closer look at the interactions between *Environment and Politics* in sidebars that examined a variety of issues.

Figure 1.5.1 Environment and Politics Local, regional, national, and international politics reshape the global environment in many different ways. Here, in Nogales, Mexico, is the other side of the border fence built by the United States to restrict illegal entry into the United States from Mexico

When Cultures Collide Cultural change often takes place within the context of international tensions. Sometimes one cultural system will replace another; at other times, resistance will stave off change. More commonly, however, a newer, hybrid form of culture results in an amalgamation of two cultural traditions. Historically, colonialism was the most important perpetuator of these cultural collisions; today, globalization in its varied forms can be thought of as the major vehicle of cultural tensions and change.

Cultural imperialism is the active promotion of one cultural system at the expense of another. Although there are still many expressions of cultural imperialism today, the most severe examples occurred in the colonial period when European cultures spread worldwide, often overwhelming, eroding, and even replacing indigenous cultures. During this period, Spanish culture spread widely in Latin America; French culture diffused into parts of Africa and Southeast Asia; and British culture overwhelmed South and Southwest Asia. New languages were mandated, new education systems were implanted, and new administrative institutions replaced the old. Foreign dress styles, diets, gestures, and organizations were added to existing cultural systems. Many vestiges of colonial culture are still evident today. In India, the makeover was so complete that many are fond of saying, with only slight exaggeration, that "the last true Englishman will be an Indian."

Today's cultural imperialism is seldom linked to an explicit colonizing force but more often comes as a fellow traveler with economic globalization. Though many expressions of cultural imperialism carry a Western (even U.S.) tone—

such as McDonald's, MTV, Marlboro cigarettes, and the use of English as the dominant language of the Internet—these facets result more from a search for new consumer markets than from deliberate efforts to spread modern U.S. culture throughout the world.

The reaction against cultural imperialism is **cultural nationalism**, which is the process of protecting and defending a cultural system against diluting or offensive cultural expressions, while at the same time actively promoting national and local cultural values. Often cultural nationalism takes the form of explicit legislation or official censorship that simply outlaws unwanted cultural traits. Examples of legislated cultural nationalism are common. France has long fought the Anglicization of its language by banning "Franglais," the use of English words such as "Le week-end," in official French. More recently, France has sought to protect its national music and film industries by legislating that radio DJs play a certain percentage of French songs and artists each broadcast day (40 percent currently). In addition, many Muslim countries censor Western cultural influences by restricting and censoring international TV, an element they consider the source of many undesirable cultural influences. Most Asian countries as well are increasingly protective of their cultural values, and many are demanding changes to tone down the sexual content of MTV and other international TV networks.

As noted, the most common product of cultural collision is the blending of forces to form a new, synergistic form of culture, called **cultural syncretism or hybridization** (Figure 1.33). To characterize India's culture as British, for example, would be to grossly oversimplify and exaggerate England's colonial influence. Instead, Indians have adapted many British traits to their own circumstances, infusing them with their own meanings. India's use of English, for example, has produced a unique form of "Indlish" that often befuddles visitors to South Asia. Nor should we forget that India has added many words to our English vocabulary—*khaki, pajamas, veranda,* and *bungalow,* among others. Clearly, both the Anglo and Indian cultures have been changed by the British colonial presence in South Asia.

Figure 1.33 Cultural Hybridity The hybridization of culture is clearly visible in this photo of South Asians (India vs. Pakistan) playing cricket, which was, before the colonial period, a uniquely English sport. Note, too, that this traditional sport now carries the logos and advertisements of global culture.

Language and Culture in Global Context Language and culture are so intertwined that in the minds of many language is the characteristic that best differentiates and defines cultural groups (Figure 1.34). Furthermore, because language is the primary means for communication, it obviously folds together many other aspects of cultural identity such as politics, religion, commerce, and customs. In addition, language is fundamental to cultural cohesiveness. It not only brings people together but also sets them apart; it can be an important component of national or ethnic identity and a means for creating and maintaining boundaries of regional identity.

Because most languages have common historical (and even prehistorical) roots, linguists have grouped the thousands of languages found throughout the world into a handful of language families. This is simply a first-order grouping of languages into large units, based on common ancestral speech. For example, about half of the world's people speak languages of the Indo-European family, a large group that includes not only European languages such as English and Spanish but also Hindi and Bengali, dominant languages of South Asia.

Within language families are smaller units that also give clues to the common history and geography of peoples and cultures. *Language branches and groups* (also called *subfamilies*) are closely related subsets within a language family in which there are usually shared and similar sounds, words, and grammar. Well known are the similarities between German and English and between French and Spanish. Because of their similarities, these languages are placed into the same linguistic groupings.

Additionally, individual languages often have very distinctive forms associated with specific regions, which are called *dialects*. Although dialects of the same language have their own unique pronunciation and grammar (think of the distinctive differences, for example, between British, North American, and Australian English), they are—sometimes with considerable effort—mutually intelligible.

When people from different cultural groups cannot communicate directly in their native languages, they often agree on a third language to serve as a common tongue, a **lingua franca**. Swahili has long served that purpose between the many tribal languages of eastern Africa, and French was historically the lingua franca of international politics and diplomacy. Today, English is increasingly the

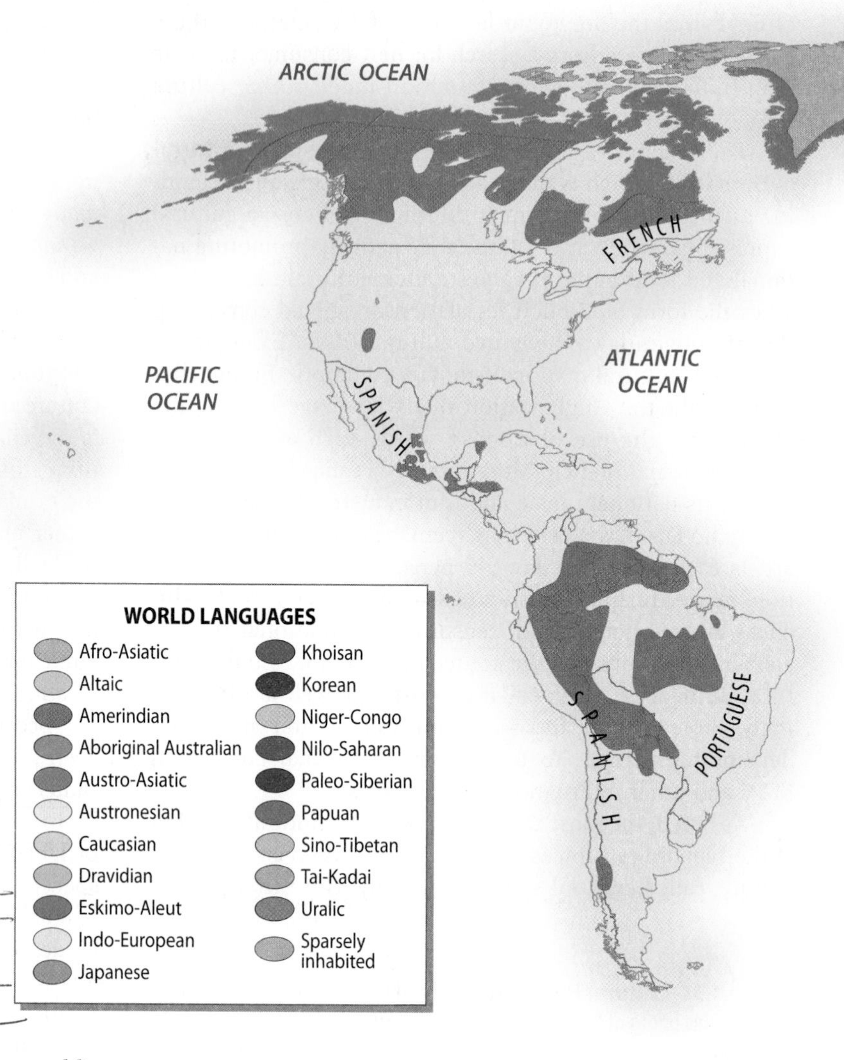

Figure 1.34 World Languages Most languages of the world belong to a handful of major language families. About 50 percent of the world's population speaks a language belonging to the Indo-European language family, which includes languages common to Europe, but also major languages in South Asia, such as Hindi. They are in the same family because of their linguistic similarities. The next largest family is the Sino-Tibetan family, which includes languages spoken in China, the world's most populous country.

common language of international communications (Figure 1.35).

A Geography of World Religions Another extremely important defining trait of cultural groups is religion (Figure 1.36). Indeed, in this era of a totalizing global culture, religion is becoming increasingly important in defining cultural identity. Recent ethnic violence and unrest in far-flung places such as the Balkans, Iraq, and Indonesia illustrate the point.

Universalizing religions, such as Christianity, Islam, and Buddhism, attempt to appeal to all peoples, regardless of location or culture; these religions usually have a proselytizing or missionary program that actively seeks new converts. In contrast are the **ethnic religions**, which remain identified closely with a specific ethnic, tribal, or national group. Judaism and Hinduism, for example, are usually regarded as ethnic religions because they normally do not actively seek new converts. Instead, people are born into ethnic religions.

Christianity, a universalizing religion, is the world's largest religion in both areal extent and number of adherents. Although fragmented into separate branches and churches, Christianity as a whole has 2.1 billion adherents, or about one-third of the world's population. The largest number of Christians can be found in Europe, Africa, Latin America, and North America. Islam, which has spread from its origins on the Arabian Peninsula as far east as Indonesia and the Philippines, has about 1.3 billion members.

While not as severely fragmented as Christianity, Islam should not be thought of as a homogeneous religion because it is also split into separate groups. The two major branches are Shiite Islam, which constitutes about 11 percent of the total Islamic population and represents a majority in Iran and southern Iraq, and the more dominant Sunni Islam that is found from the Arab-speaking lands of North Africa to Indonesia. Both of these forms of Islam are experiencing fundamentalist revivals in which proponents are interested in maintaining purity of faith distanced from Western influences.

Judaism, the parent religion of Christianity, is also closely related to Islam. Although tensions are often high between Jews and Muslims, these two religions, along with Christianity, actually share historical and theological roots in the Hebrew prophets and leaders. Judaism now numbers about 14 million adherents, having lost perhaps one-third of its total population due to the systematic extermination of Jews by the Nazis during World War II.

Hinduism, which is closely linked to India, has about 900 million adherents. Outsiders often regard Hinduism as polytheistic because Hindus worship many deities. Most Hindus argue, however, that all of their faith's gods are merely representations of different aspects of a single divine, cosmic unity. Historically, Hinduism is linked to the caste system, with its segregation of peoples based on ancestry and occupation. Today, however, because India's democratic government is committed to reducing the social distinctions between castes, the connections between religion and caste are now much less explicit than in the past.

Buddhism, which originated as a reform movement within Hinduism 2,500 years ago, is widespread in Asia, ex-

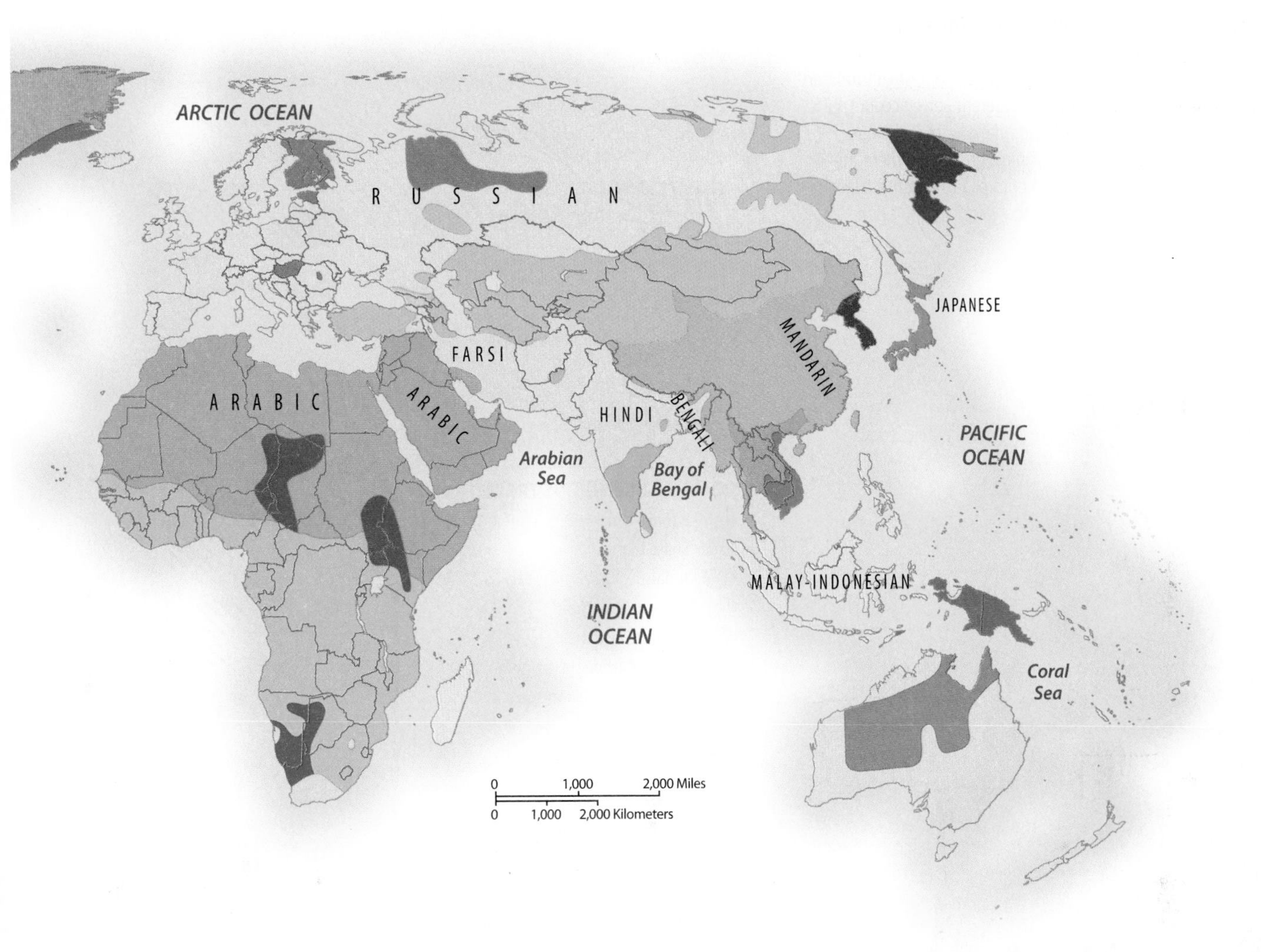

tending from Sri Lanka to Japan and from Mongolia to Vietnam (Figure 1.37). In its spread, Buddhism came to coexist with other faiths in certain areas making it difficult to accurately estimate the number of its adherents. Estimates of the total Buddhist population range from 350 million to 900 million people.

Finally, in some parts of the world, religious practice has declined significantly, giving way to **secularization**, in which people consider themselves either nonreligious or outright atheistic. Though secularization is difficult to measure, social scientists estimate that about 1.1 billion people fit into this category worldwide. Perhaps the best example of secularization comes from the former communist lands of Russia and eastern Europe, where, historically, there was overt hostility between government and church. Since the demise of Soviet communism in the 1990s, however, many of these countries have experienced modest religious revivals.

Figure 1.35 English as Global Language A bilingual traffic sign in Dubai, United Arab Emirates, is a reminder that English has become a universal form of communication for transportation, business, and science.

Figure 1.36 Major Religious Traditions This map shows the major religions throughout the world. For most people, religious tradition is a major component of cultural and ethnic identity. While Christians of different sorts account for about 34 percent of the world's population, this religious tradition is highly fragmented. Within Christianity, there are about twice as many Roman Catholics as Protestants. Islam accounts for about 20 percent of the world's population; Hindus make up about 14 percent.

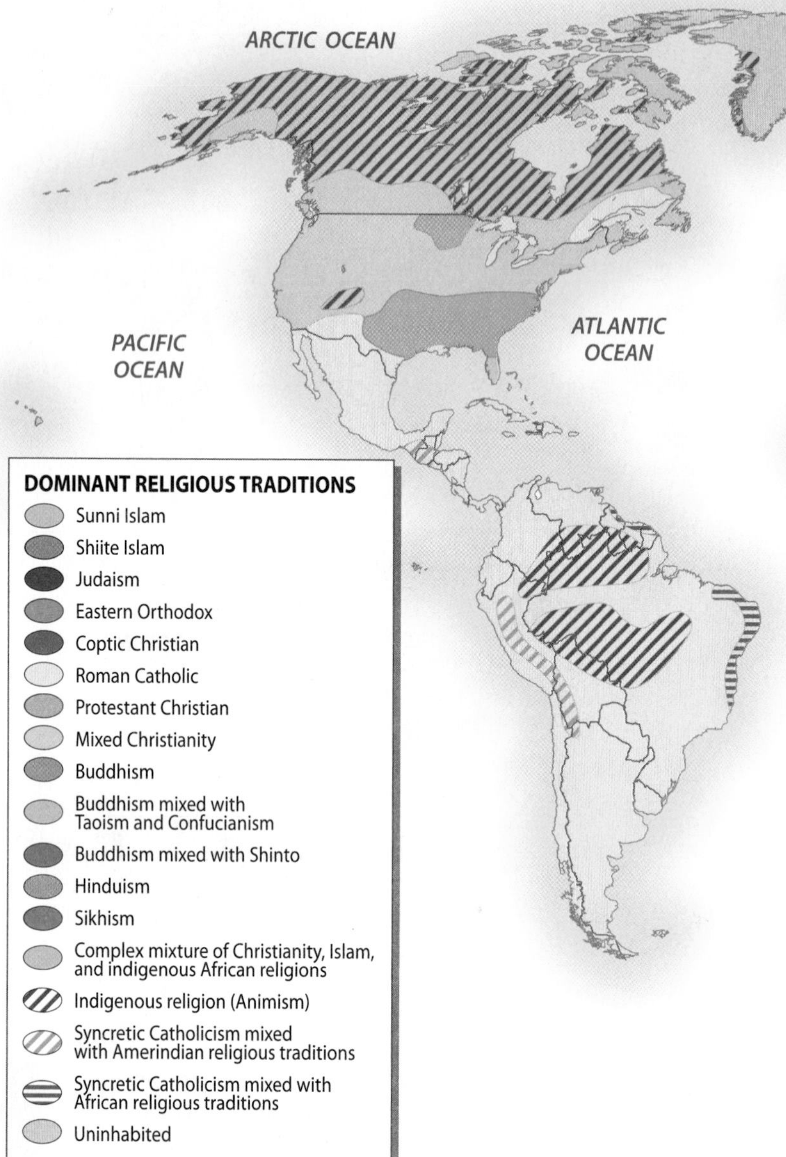

Figure 1.37 Religious Landscapes The varied expressions of a culture's religion commonly appear in the landscape—the mosques and minarets of Islam, the churches and cathedrals of Christianity, the synagogues of Judaism, the shrines of Hinduism, or the temples and statues of Buddhism. An example is this statue of the sea deity *Kwun Yum* at the Tin Hau Temple in Repulse Bay, Hong Kong.

Secularization has also grown more pronounced recently in western Europe. Although historically and to some extent still culturally a Roman Catholic country, France possibly has more people (mostly migrants) attending Muslim mosques on Fridays than it has attending Christian churches on Sundays. Japan and the other countries of East Asia are also noted for their high degree of secularization.

Geopolitical Framework: Unity and Fragmentation

The term *geopolitics* is used to describe and explain the close link between geography and politics. More specifically, geopolitics focuses on the interactivity between power, territory, and space at all scales, from the local to the global. Un-

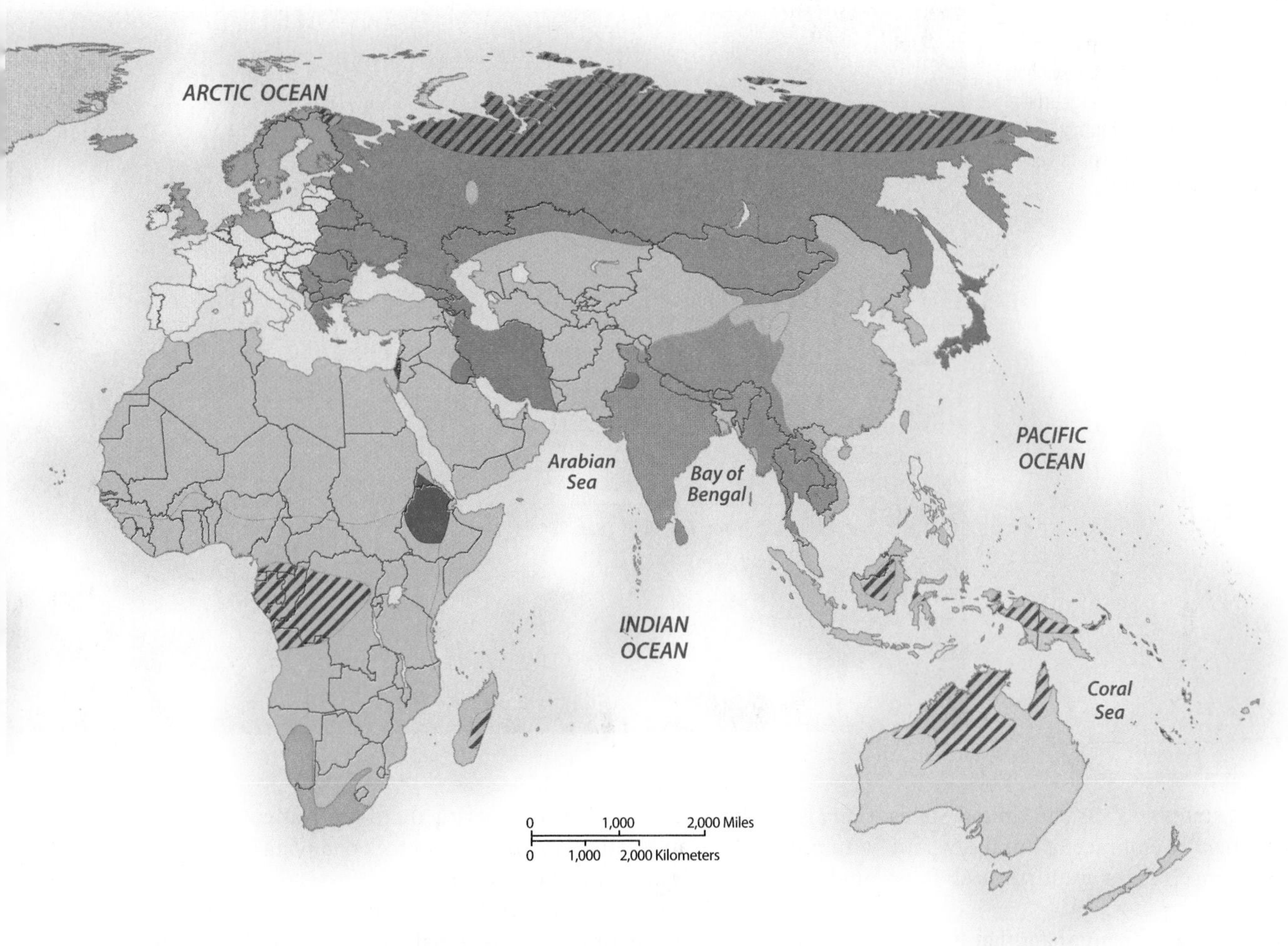

questionably, one of the world's characteristics of the last several decades has been the speed, scope, and character of political change in various regions of the world, thus discussions of geopolitics are central to world regional geography.

With the demise of the Soviet Union in 1991 came opportunities for self-determination and independence in eastern Europe and Central Asia, resulting in fundamental changes to economic, political, and even cultural alignments (Figure 1.38). Religious freedom helped drive national identities in some new Central Asian republics, while eastern Europe was primarily concerned with new economic and political links to western Europe. Russia itself still wavers perilously between different geopolitical pathways. All of these topics are discussed further in chapters 8, 9, and 10.

While international conflicts remain a nagging concern of diplomats and governments, more common are strife and tension within—rather than between—nation-states. Civil unrest, tribal tensions, terrorism, and religious factionalism have created a new fabric and scale of political tension.

Nation-States and Beyond A traditional reference point for examining geopolitics is the concept of the nation-state. The term *state* refers to a political entity with territorial boundaries recognized by other countries and internally governed by an organizational structure, whereas *nation* refers to a large group of people who share numerous socio-cultural elements such as language, religion, tradition, and identity. A **nation-state**, then, is a relatively homogeneous cultural group occupying its own fully independent political territory. Historical France and England are often used as the type examples of nation-states, while contemporary countries such as Albania, Egypt, Bangladesh, Japan, and the two Koreas qualify as areas where there is close overlap between nation and state.

Overall, however, globalization has weakened the concept of the nation-state. Today, in fact, most of the world's 200 or so political entities are questionable fits with the traditional definition. International migration, for example, has led to large populations of immigrants within countries, people who do not necessarily share the national culture of the majority. In England, large numbers of South Asians form their own communities, speak their own languages, have their own religions, and dress to their own standards. Similarly, France is home to a mosaic of peoples from its former colonial lands in Africa and Asia. In North America Canada and the United States also have large immigrant populations who are legal citizens of the political state, yet who have either changed the very nature of the national culture by their presence or live on its margins.

Figure 1.38 The End of the Cold War With the fall of Soviet communism, many nations rushed to remove the symbols of their former governments, to forget the past and make room for a new future. In this photo taken in 1990, a monument to Lenin is toppled in Bucharest, Romania.

The fact that states such as the United States, Canada, and Germany officially embrace cultural diversity with policies declaring themselves as multicultural states underscores these changes.

As well, there are many states that have groups of people who seek autonomy from the central government with the right to govern themselves. This is the case with the French-speaking people of Quebec province in Canada, as well as the Catalonians and Basques of Spain, to name just a few separatist groups (Figure 1.39). A closer look at France, Italy, and the United Kingdom would reveal more separatist groups. Even in North America there are Native American groups and Native Hawaiians who seek complete autonomy from the U.S. or Canadian governments.

Not to be overlooked is how the rise of giant multinational firms and regional organizations have eclipsed the power of political units. This is certainly the case for the 27 member states of the European Union, a topic discussed in more detail in Chapter 8.

Finally, there are cultural groups who lack political voice and representation due to the way political borders have been drawn. In Slovakia, to illustrate, a group who trace their cul-

Figure 1.39 Ethnic Separatism A major aspect of contemporary geopolitics is the way ethnic groups are demanding recognition, autonomy, and often independence from larger political units. These Basque women in southwestern France are protesting the outlawing of a Basque youth group French and Spanish authorities suspect of aiding Basque terrorists.

tural heritage to adjacent Hungary agitate for redrawing the political border between the two states. In Southwest Asia, the Kurdish people have long been considered a nation without a state because they are divided by the political borders of Turkey, Syria, Iraq, and Iran (Figure 1.40).

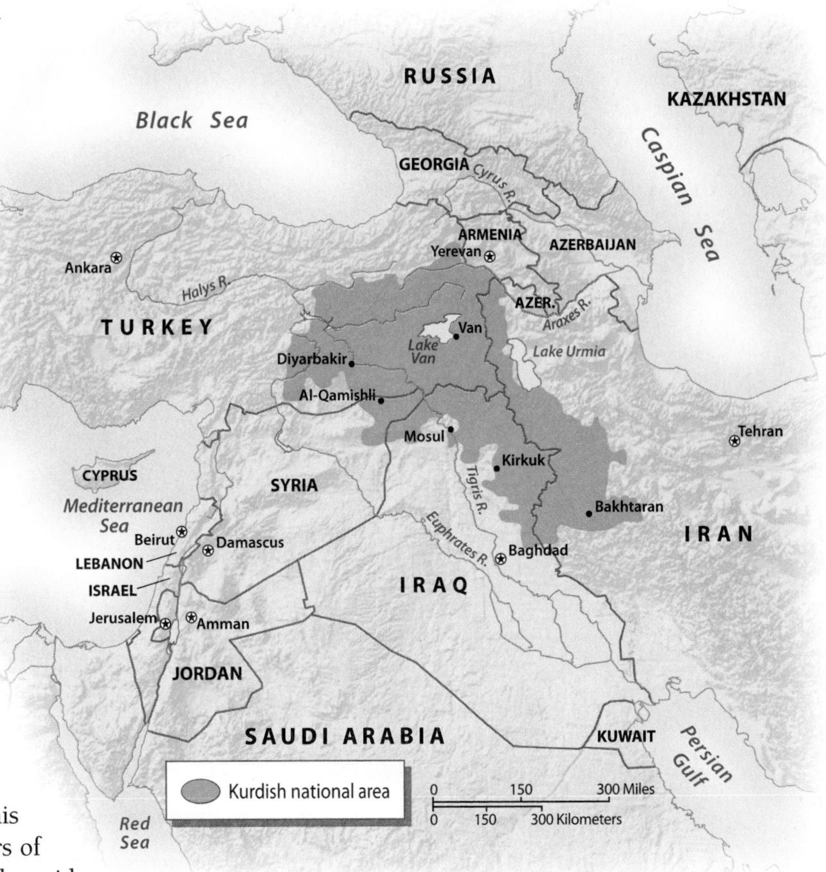

Figure 1.40 A Nation Without a State Not all nations or large cultural groups control their own political territories and thus are without a state. As this map shows, the Kurdish people of Southwest Asia occupy a larger cultural territory that is in four different political states—Turkey, Iraq, Syria, and Iran. As a result of this political fragmentation, the Kurds are considered a minority in each of these four countries.

Colonialism and Decolonialization One of the overarching themes in world geopolitics is the waxing and waning of European colonial power over much of the world. **Colonialism** refers to the formal establishment of rule over a foreign population. A colony has no independent standing in the world community but instead is seen only as an appendage of the colonial power. Generally speaking, the main period of colonialization by European states was from 1500 through the mid-1900s. (Figure 1.41).

Decolonialization refers to the process of a colony's gaining (or regaining) control over its own territory and establishing a separate, independent government. As was the case with the Revolutionary War in the United States, this process often begins as a violent struggle. As wars of independence became increasingly common in the mid-20th century, some colonial powers recognized the inevitable and began working toward peaceful disengagement from their colonies. In the late 1950s and early 1960s, for example, Britain and France granted independence to most of their former African colonies, often after periods of warfare or civil unrest. This process was nearly completed in 1997, when Hong Kong was peacefully restored to China by the United Kingdom.

But decades and even centuries of colonial rule are not easily erased; thus, the influence of colonialism is still commonly found in the new nations' governments, education, agriculture, and economies. While some countries may enjoy special status with, and receive continued aid from, their former colonial masters, others remain disadvantaged because of a reduced resource base exploited by colonial powers. In fact some scholars regard the continuing economic ties between certain imperial powers and their former colonies as a form of exploitative neocolonialism. On the other hand, some remaining colonies such as the Dutch Antilles in the Caribbean have made it clear that they have no wish for independence, finding both economic and political advantages in continued dependency. Because the consequences of colonialism differ greatly from place to place, the final accounting of its effects is far from complete (Figure 1.42).

Global Terrorism and Insurgency As mentioned earlier, challenges to a centralized political state or authority have long been part of global geopolitics as rebellious and separatist groups seek independence, autonomy, and territorial control. Armed conflict in the form of insurgency has also been part of this process as illustrated by the American and Mexican Revolutions that successfully fought against European colonial powers. Terrorism, which can be defined as violence directed at nonmilitary targets, has also been common, albeit to a far lesser degree than today.

Further, until the September 2001 terrorist attacks on the United States by Al Qaeda, terrorism was usually directed at specific local targets and committed by insurgents with focused goals. The Irish Republican Army (IRA) bombings in Great Britain and Basque terrorism in Spain are illustrations. The attacks on the World Trade Center and the Pentagon (as well as the thwarted attack on the Capitol), however, went well beyond conventional geopolitics as a small group of religious extremists attacked the symbols of Western culture, finance, and power. Experts assume the goal of the Al Qaeda terrorists was less to disrupt world commerce and politics and more to make a statement about the strength of their own convictions and power. Regardless of

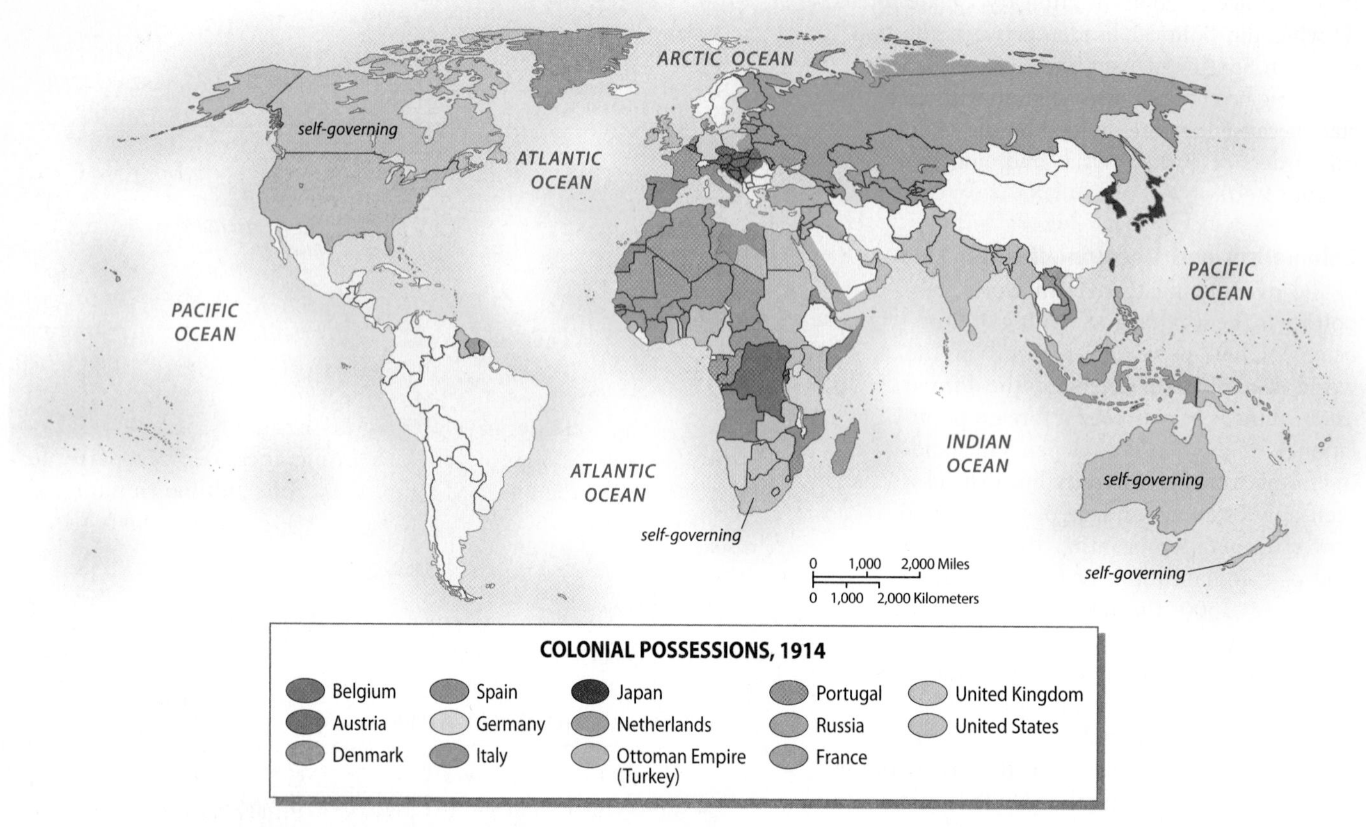

Figure 1.41 The Colonial World, 1914 This world map shows the extent of colonial power and territory just prior to World War I. At that time, most of Africa was under colonial control, as were Southwest Asia, South Asia, and Southeast Asia. Australia and Canada were very closely aligned with England. Also note that in Asia, Japan controlled colonial territory in Korea and northeastern China, which was known as Manchuria at that time.

motives, those acts of terrorism underscore the need to expand our conceptualization of the linkages between globalization and geopolitics (Figure 1.43).

More to the point, many experts argue that global terrorism is both a product of and a reaction to globalization. Unlike earlier geopolitical conflicts, the geography of global terrorism is not defined by a war between well-established political states. Instead, the Al Qaeda terrorists appear to belong to a web of small, well-organized cells located in many different countries. These cells are linked in a decentralized network that provides guidance, financing, and political autonomy and that has used the tools and means of globalization to its advantage. Members communicate instantaneously via mobile phones and the Internet. Transnational members travel between countries quickly and frequently. The network's activities are financed through a complicated array of holding companies and subsidiaries that traffic in a range of goods, including honey, diamonds, and opium. To recruit members and political support, the network feeds on the unrest and inequities (real and imagined) resulting from economic globalization. The network's terrorist acts then target symbols of those modern global values and activities it opposes. Even though the 9/11 attacks focused terror on the United States, the casualties and resultant damage were international, as citizens from more than 80 countries were killed in the World Trade Center tragedy.

Although Al Qaeda is the most visible and possibly the most threatening of contemporary global terrorist groups, the U.S. State Department names 45 groups on its list of foreign terrorist organizations. While most of these are clustered in North Africa and Southwest and Central Asia, this list also includes insurgent groups in South America, Africa, and Asia.

The military responses to global terrorism and insurgency involve several components ranging from neutralizing terrorist activities, known as counterterrorism, to counterinsurgency, a more complicated, multifaceted strategy that combines military warfare with social and political service activities designed to win over the local population and deprive insurgents of a political base. These counterinsurgency activities include, first, clearing and then holding territory held by insurgents, followed by building schools, medical clinics, and a viable (and legal) economy. Often these nonmilitary activities are referred to as nation-building since the goal is to replace the separatist insurgency with a viable social, economic, and political fabric more complementary to the larger geopolitical state. This, of course, is the current strategy employed by the United States in Iraq and Afghanistan.

Figure 1.42 Colonial Vestiges in Vietnam The red star flag of communist Vietnam flies in front of the Hotel de Ville in Ho Chi Minh City (formerly Saigon). This juxtaposition of the contemporary government's symbol and an artifact of the French colonial period captures the process of decolonialization and independence.

Economic and Social Development: The Geography of Wealth and Poverty

The pace of global economic change and development has accelerated dramatically in the past several decades, first ascending rapidly, then, in late 2008, dropping precipitously as the world fell into an economic recession (Figure 1.44). While the linkages between these economic fluctuations and globalization are complex, the overarching question remains: Do the positive changes of economic globalization outweigh the negative? Answers vary and are often elusive and incomplete; however, a starting point in world regional geography is to link economic change to social development.

Economic development is commonly accepted as desirable because it generally brings increased prosperity to people, regions, and nations. By conventional thinking, at least, this usually translates into social improvements such as better health care, improved education systems, and more progressive labor practices. However, one of the most troubling expressions of recent economic growth has been the geographic unevenness of prosperity and social improvement. While some regions prosper, others languish and, in fact, fall farther behind more developed countries. As a result, the gap between rich and poor regions has increased over the past several decades, and this economic and social unevenness has, unfortunately, become one signature of globalization. Today, according to the World Bank, about half the people in the world live on less than $2 a day, which is the commonly accepted definition of poverty (Figure 1.45).

These inequities are problematic because of their inseparable interaction with political, environmental, and social issues. For example, political instability and civil strife within a nation are often driven by the economic disparity between a poor periphery and an affluent, industrial core, between the haves and have-nots, causing instability throughout the country, and influencing international economic interac-

Figure 1.43 Global Terrorism The September 11, 2001 attacks by Al Qaeda terrorists on the World Trade Center, the Pentagon, and the thwarted plan to destroy the U.S. Capitol building resulted in more than 3000 deaths.

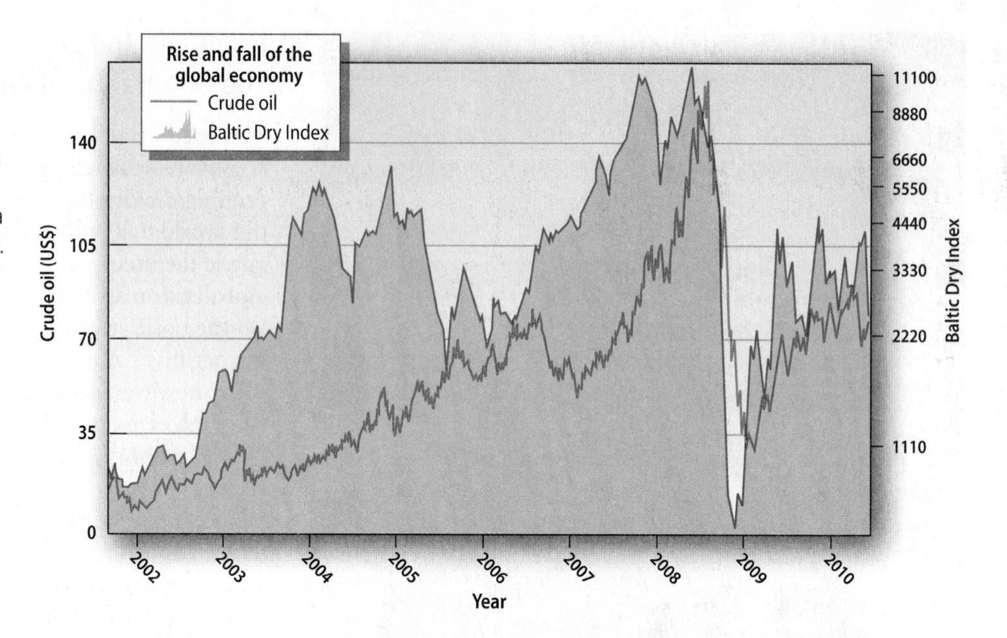

Figure 1.44 The 2009 Global Recession Two indicators of the global economic activity, the price of crude oil and the Baltic Dry Index (BDI), which is linked to world shipping activity, reflect the expansion and, more recently, the sudden contraction of the global economy in 2009. The last data points show a slow recovery as of mid-2010.

tions. The concept of a developed and affluent economic core and a less-well off periphery has also been applied at a larger global scale in development studies.

More- and Less-Developed Countries Until the 20th century, economic development was centered in North America, Japan, and Europe while most of the rest of the world remained gripped in poverty. This uneven distribution of economic power led scholars to devise a **core–periphery model** of the world. According to this scheme, the United States, Canada, western Europe, and Japan constituted the global economic core of the north, whereas most of the areas to the south made up a less-developed periphery. Although oversimplified, this core–periphery dichotomy does contain some truth. All the G8 countries—the exclusive club of the world's major industrial nations, made up of the United States, Canada, France, England, Germany, Italy, Japan, and Russia—are located in the Northern Hemisphere. In addition, many critics postulate that the developed countries achieved their wealth primarily by exploiting the poorer countries of the southern periphery, historically through colonial relationships and today with economic imperialism.

Figure 1.45 Living on Less than $2 a Day More than half the world's people live on less than $2 a day, which is the UN's definition of poverty. In South Asia, where these workers are making bricks, about three-quarters of the population is considered impoverished.

As a result, much is made today of "north–south tensions," a phrase implying that the rich and powerful countries of the Northern Hemisphere are still at odds with the poor and less, powerful countries of the south. Over the past several decades, however, the global economy has grown much more complicated. A few former colonies of the south, most notably Singapore, have become very wealthy. In addition, a few northern countries, namely Russia, have experienced economic decline in recent decades. Further, the developed countries of Australia and New Zealand never fit into the north–south division because of their Southern Hemisphere location. For these reasons, many global experts conclude that the designation *north–south* is now outdated and, thus, should be avoided.

The *third world* is another term often used erroneously as a synonym for the developing world. This phrase implies a low level of economic development, unstable political organizations, and a rudimentary social infrastructure. Historically, however, the term came from the Cold War vocabulary used to describe countries that were independent and neither allied with the capitalist and democratic first world or the communist second world superpowers of the Soviet Union and China, thus had little to do with levels of economic development. Today, because the Soviet Union no longer exists and China has changed its economic orientation, the term *third world* has lost its original political meaning. In this book, therefore, we avoid that term and instead use relational terms that capture the complex spec-

EXPLORING GLOBAL CONNECTIONS A Closer Look at Globalization

Globalization comes in many shapes and forms as it connects far-flung people and places. While many of these interactions are expected and common knowledge, such as the global reach of multinational corporations, other global connections are more surprising (Figure 1.6.1). Who would expect to find Australian firefighters dowsing California wildfires, or Russians investing in Thailand's coastal resorts, or Bronx hip hop music becoming the favored voice of Europe's youth?

Indeed, global connections are complex and ubiquitous—so much so that an understanding of the many different shapes, forms, and scales of these interactions is a key component of the study of global geography. To complement that study, each chapter contains an *Exploring Global Connections* sidebar that examines a wide variety of topics. More specifically they are:

Chapter 3: Ginseng Heartland
Chapter 4: The Growth of European and Asian Investment in Latin America
Chapter 5: Cuba's Medical Diplomacy
Chapter 6: China's Investment in Africa
Chapter 7: Sunny Side Up: North Africa's Desertec Initiative
Chapter 8: A New Cold War over Heating Europe
Chapter 9: From Russia with Love: The Megatons to Megawatts Program
Chapter 10: Foreign Military Bases in Central Asia
Chapter 11: Taiwan Seeks Diplomatic Recognition
Chapter 12: The Tata Group's Global Network
Chapter 13: ACFTA: the ASEAN–China Free Trade Area
Chapter 14: China Buys Australia

Figure 1.6.1 Exploring Global Connections By definition globalization means there are new and different interactions between far-flung places around the world, and examining those relationships is an important part of world regional geography.

If there's one symbol of modern globalization that is understood by all, it is the shipping container, an invention of the 1970s that changed the way goods are transported on both land and sea.

trum of economic and social development—*more-developed country (MDC)* and *less-developed country (LDC)*. The global pattern of more- and less-developed countries can be inferred from a map of gross national income (Figure 1.46), one of several indicators used to assess development and economic wealth.

Indicators of Economic Development The terms *development* and *growth* are often used interchangeably when referring to international economic activities. There is, however, value in keeping them separate. *Development* has both qualitative and quantitative dimensions. Common dictionary definitions use phrases such as "expanding or realizing potential" and "bringing gradually to a fuller or better state." When we talk about economic development, then, we usually imply structural changes, such as a shift from agricultural to manufacturing activity that also involves changes in the allocation of labor, capital, and technology. Along with these changes are assumed improvements in standard of living, education, and political organization. The structural changes experienced by Southeast Asian countries such as Thailand and Malaysia in the past several decades illustrate this process.

Growth, in contrast, is simply the increase in the size of a system. The agricultural or industrial output of a country may grow, as it has for India in the past decade, and this growth may—or may not—have positive implications for development. Many growing economies, in fact, have actually experienced increased poverty with economic expansion. When something grows, it gets bigger; when it develops, it

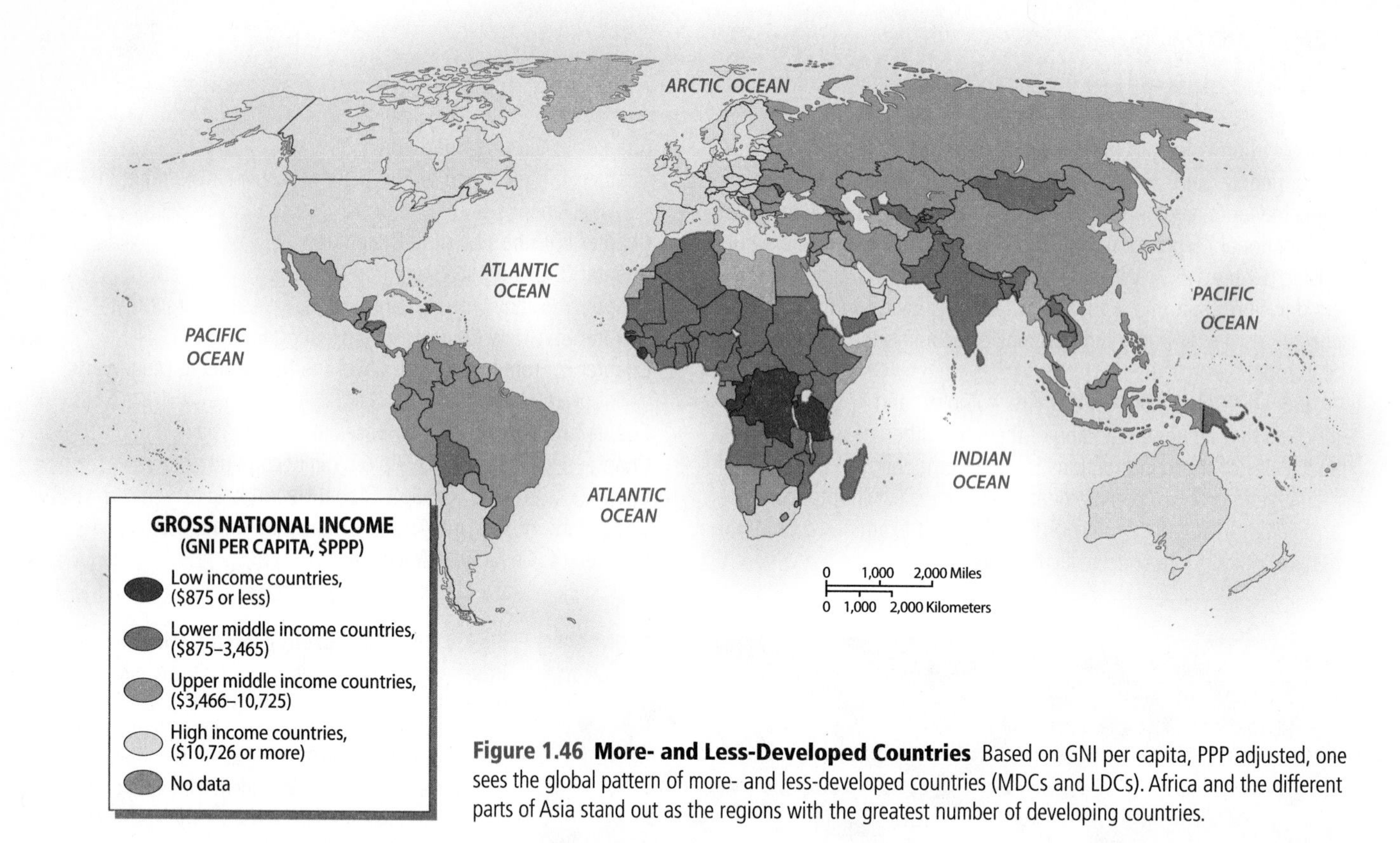

Figure 1.46 More- and Less-Developed Countries Based on GNI per capita, PPP adjusted, one sees the global pattern of more- and less-developed countries (MDCs and LDCs). Africa and the different parts of Asia stand out as the regions with the greatest number of developing countries.

improves. Critics of the world economy are often heard to say that we need less growth and more development.

In this book, each of the regional chapters includes a table of both economic and development indicators (Table 1.2). A few introductory comments are necessary to explain these data.

Gross Domestic Product and Income The traditional measure of the size of a country's economy is the value of all final goods and services produced within its borders, called the **gross domestic product (GDP)**. When combined with net income from abroad, this domestic income constitutes a country's **gross national income (GNI)** (formerly referred to

TABLE 1.2 Development Indicators for the World's Ten Largest Countries

Country	GNI per capita, PPP 2008	GDP Average Annual % Growth 2000–08	Human Development Index (2007)[a]	Percent Population Living Below $2 a Day	Life Expectancy (2010)[b]	Under Age 5 Mortality Rate (1990)	Under Age 5 Mortality Rate (2008)	Gender Equity (2008)[c]
China	6,010	10.4	0.772	36.3	74	46	21	103
India	2,930	7.9	0.612	75.6	64	116	69	92
United States	48,430	2.4	0.956		78	11	8	100
Indonesia	3,590	5.2	0.734	60.0	71	86	41	98
Brazil	10,070	3.6	0.813	12.7	73	56	22	102
Pakistan	2,590	5.4	0.572	60.3	66	130	89	80
Bangladesh	1,450	5.8	0.543	81.3	66	149	54	106
Nigeria	1,980	2.5	0.511	83.9	47	230	186	85
Russia	15,440	6.7	0.817	<2	68	27	13	98
Japan	35,190	1.1	0.960		83	6	4	100

[a]*United Nations, Human Development Report, 2009.*

[b]*Population Reference Bureau, World Population Data Sheet, 2010.*

[c]*Gender Equity—Ratio of female to male enrollments in primary and secondary school. Numbers below 100 have more males in primary/secondary school, numbers above 100 have more females in primary/secondary schools.*

Source: World Bank, World Development Indicators, 2010.

Figure 1.47 PPP and Local Currencies The concept of purchasing power parity (PPP) is used to generate a sense of the true cost of living for different countries so that economic data can be adjusted for local conditions based upon a common market basket of goods. This outdoor produce market is in Chichicastenango, Guatemala.

as gross national product [GNP]). Although the term is commonly used, GNI is an incomplete and sometimes misleading economic indicator because it ignores nonmarket economic activity such as bartering or household work, and also because it does not take into account ecological degradation or depletion of natural resources. For example, if a country were to clear-cut its forest, an activity that could severely limit future growth if forest resources were in short supply, this cutting would nevertheless increase the GNI for that particular year. Further, diverting educational funds to purchase military weapons might also increase a country's GNI in the short run, but the economy would likely suffer in the future because of its less-well-educated population. In other words, GNI is a snapshot of a country's economy for a specific period, not an infallible indicator of continued vitality or social well-being.

Because GNI data vary widely between countries and are commonly expressed in mind-boggling numbers such as billions and trillions of dollars, a better comparison is to divide GNI by the country's population, thereby generating a **gross national income (GNI) per capita** figure. This way, one can compare large and small economies in terms of how they may (or may not) be benefiting the population. For example, the annual GNI for the United States is over $9 trillion. Dividing that figure by the population of 309 million results in a GNI per capita of around $48,000. Japan has a GNI about half the size of the United States; however, because that country has a much smaller population of 127 million, its unadjusted GNI per capita is about $39,000; thus, one could conclude that the two economies are almost comparable.

An important qualification to these GNI per capita data is the concept of adjustment through **purchasing power parity (PPP),** an adjustment that takes into account the strength or weakness of local currencies (Figure 1.47). When not adjusted by PPP, GNI data are based on the market exchange rate for a country's national currency as compared to the U.S. dollar. As a result, the GNI data might be inflated or undervalued, depending on the strength or weakness of that currency. If the Japanese yen were to fall overnight against the dollar as a result of currency speculation, Japan's GNI would correspondingly drop, despite the fact that Japan had experienced no real decline in economic output. Because of these possible distortions, the PPP adjustment provides a more accurate sense of the local cost of living. To illustrate, when Japan's GNI per capita is adjusted for PPP, which takes out the inflationary factor, the figure is $35,000, which is lower than the unadjusted GNI per capita figure.

Economic Growth Rates A country's rate of economic growth is measured by the average annual growth of its GDP over a 5-year period, a statistic called *GDP average annual percent growth*. The average growth rate for developing countries such as China, India, and Nigeria is considerably higher than those of the developed countries of the United States and Japan (see Table 1.2). This difference in economic growth rates is expected, given that developing countries are just that—developing; thus one would expect a higher annual growth rate than from a mature, developed economy like that of the United States.

Indicators of Social Development Although economic growth is a major component of development, equally important are the conditions and quality of human life. As noted earlier, the standard assumption is that economic development and growth will spill over into the social infrastructure, leading to improvements in public health, gender equity, and education. Unfortunately, even the briefest glance at the world reveals that poverty, disease, illiteracy, and gender inequity are still widespread, despite a growing global economy. Even in China, which has experienced unprecedented economic growth in recent decades, half the population is still impoverished; in Pakistan, two-thirds live below the poverty line of $2 per day, and in some African countries, that figure approaches 90 percent.

However, there are hints of improvement in some developing countries. For example, the percentage of those living in deep poverty, which the UN defines as living on less $1 per day, has fallen from 28 percent in 1990 to just under 20 percent today. Life expectancy, too, has increased in those countries, rising from 60 in 1990 to 65.

The Human Development Index For the past three decades, the United Nations has tracked social development in the world's countries through the **Human Development Index (HDI),** which combines data on life expectancy, literacy, educational attainment, gender equity, and income (Figure 1.48). In a December 2009 analysis, the 182 countries that provided data to the UN are ranked from high to low, with Norway achieving the highest score, Australia in second place, and Iceland third (despite, somehow, its economic meltdown of 2008) At the lowest end of the HDI are a handful of African countries, including Central African Republic, Sierra Leone, with Niger in last place. Noteworthy is that second to last place is occupied by Afghanistan, a

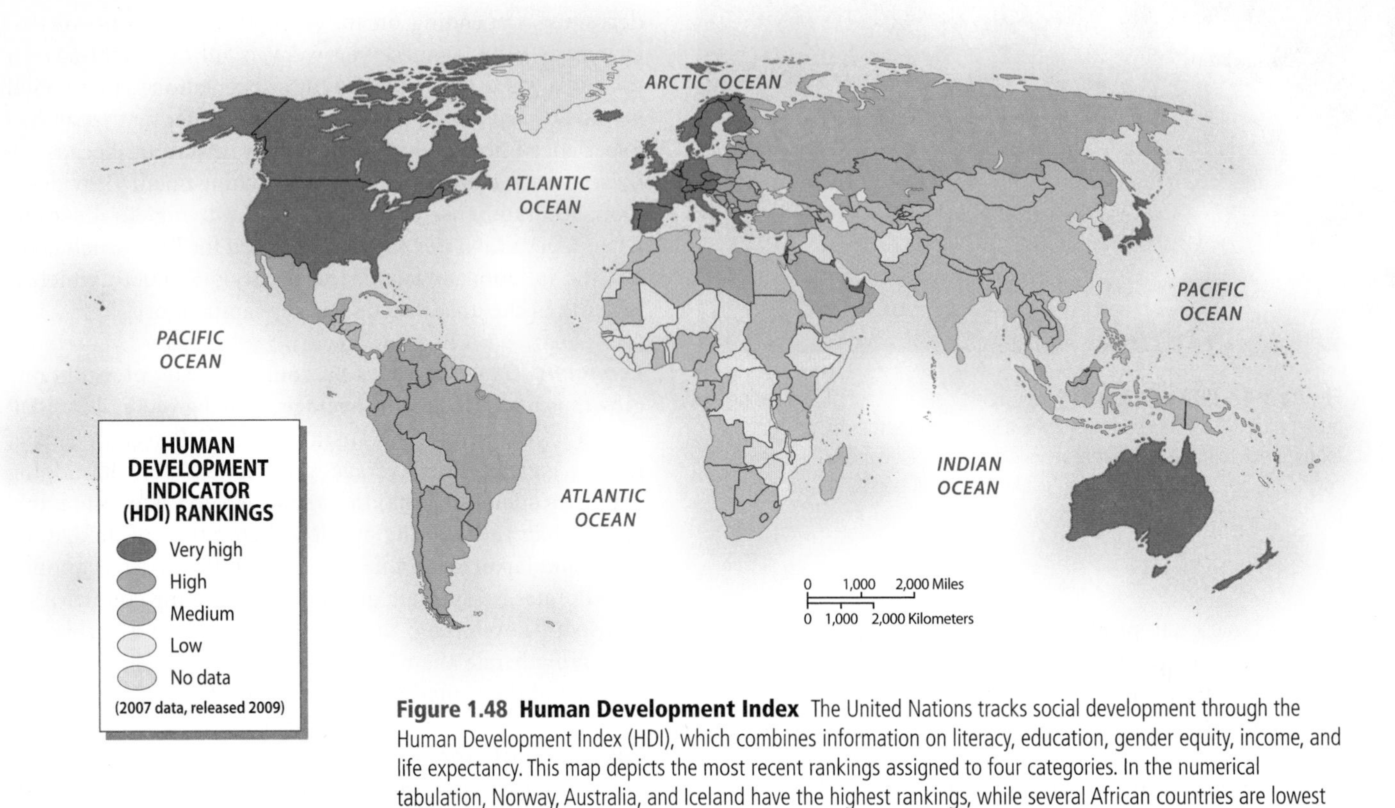

Figure 1.48 Human Development Index The United Nations tracks social development through the Human Development Index (HDI), which combines information on literacy, education, gender equity, income, and life expectancy. This map depicts the most recent rankings assigned to four categories. In the numerical tabulation, Norway, Australia, and Iceland have the highest rankings, while several African countries are lowest in the scale.

country where U.S. economic and military investment is considerable.

Although the HDI can be criticized for using national data that overlook the diversity of development within a country, overall, the HDI conveys a reasonably accurate sense of a country's human and social development; thus, we include country-by-country data in social development tables throughout this book.

Poverty and Infant Mortality The international definition of *poverty* is living on less than $2 per day. *Deep poverty* is defined as existing on less than $1 per day. Granted, the cost of living varies greatly around the world, nevertheless the United Nations has found these definitions work well for measuring poverty and its associated social conditions. While poverty data are usually presented at the national level, the UN and other agencies are attempting to compile data at a local scale in order to better understand—and, hopefully, improve—the economic landscape within a country. The patterns of poverty in the African country of Madagascar provide an instructive example (Figure 1.49).

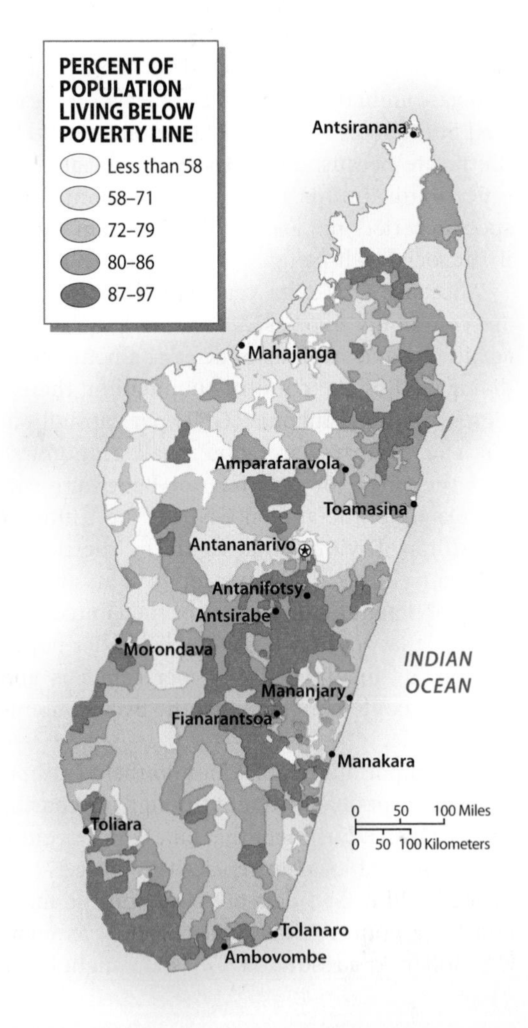

Figure 1.49 The Landscape of Poverty As is true of most other countries, the distribution of poverty within Madagascar is uneven, with clusters of abject poverty contrasting with regions that are less poor. This map shows that the highest rates of poverty are in the central highlands, where the country's population is concentrated and the density is highest. In contrast, the northern and eastern lowlands are less poor. *(Map adapted from Where the Poor Are: An Atlas of Poverty, Center for International Earth Science Information Network, 2006)*

Figure 1.50 Child Mortality The mortality rate of children under the age of 5 is an important indicator of social conditions such as health services, food availability, and public sanitation. In this photo kindergarten children in Jakarta, Indonesia, receive an oral polio vaccine.

Under age five mortality depicts the number of children who die per 1,000 humans within that age bracket, and is another widely used indicator of social conditions. Aside from the tragedy of infant death, child mortality also reflects the wider conditions of a society, namely, the availability of food, health services, and public sanitation. If those factors are lacking, children under age five suffer most, therefore their death rate is taken as an indication of whether a country has the necessary social infrastructure to sustain life (Figure 1.50). In the social development tables throughout this book, child mortality data are given for two points in time, 1990 and 2008, to indicate whether the social structure improved over that period. Although most countries have seen improvement over the past decades, disturbing differences in child mortality rates still exist on a global level.

Gender Equity The ratio of male to female students enrolled in primary and secondary schools is a measurement of *gender equity*. If the gender equity ratio is below 100, then more males are enrolled in schools than females. Conversely, a ratio over 100 means more females are enrolled than males in primary and secondary schools. Used by the UN as a measure of equality in education, this statistic is linked closely to literacy rates, which in turn are linked to social development. The assumption is that if females are not enrolled in school, a high rate of female illiteracy results in negative social development.

Although sex discrimination takes other forms, such as the preference for male children in India and China or the barriers erected against women in business and politics, the United Nations is currently focusing on correcting the gender inequity in education (Figure 1.51). Specifically, the UN has made neutralizing educational disparity one of its Millennium Development Goals. Gender equity now has higher priority because increasing women's literacy has positive outcomes that complement social development. For example, some studies indicate a strong correlation between lower birthrates and female literacy, probably because literate women are more open to family planning measures.

Besides gender equity, there are many other linkages between economic and social development. This introductory chapter has discussed only a few and on a general scale; the regional chapters of this book examine other issues in more detail.

Figure 1.51 Women and Literacy Gender inequities in education lead to higher rates of illiteracy for women. However, when there is gender equity in education, female literacy has a number of positive outcomes in a society. For example, educated women have a higher participation rate in family planning, which usually results in lower birthrates.

Summary

- As the world becomes increasingly interdependent, globalization is driving a fundamental reorganization of cultures and economies through instantaneous global communication, the growth of transnational firms, and the spread of Western consumer habits.
- Because globalization involves both positive and negative changes, it is a controversial and contentious topic. While proponents argue that economic globalization will benefit everyone as regional activities become more efficient in the face of global competition, opponents counter this argument with evidence that only the rich are benefiting and that poorer countries are falling farther behind. The truth, however, is more complicated, as is globalization itself, thus unwise to generalize it as either positive or negative.
- In most regions of the developing world, population and settlement issues revolve around four issues: rapid population growth, family planning (or its absence), migration to new centers of economic activity (both within and outside the region), and rapid urbanization.
- A major theme in cultural geography is the tension between the forces of global cultural homogenization and the countercurrents of local cultural and ethnic identity. Throughout the world, small groups are setting themselves apart from larger national cultures with renewed interest in ethnic traits, languages, religion, territory, and shared histories. Not merely a matter of colorful folklore and local customs, this cultural diversity is translated into geopolitics and linked to agendas of regional autonomy or separatism.
- One issue—terrorism—tops the list of global geopolitical issues and can be triggered by local groups with limited agendas (such as Basque autonomy), as well as larger, more complex terrorist networks (namely Al Qaeda) with wider, more abstract goals. Besides terrorism, there are many other kinds of geopolitical tensions, including ethnic strife, border disputes, and movements for regional autonomy and independence.
- The theme of economic and social development is dominated by one issue—the increasing disparity between rich and poor, between countries and regions that already have wealth and are getting even richer through globalization and those that remain impoverished. Often, blatant inequities in social development, education, health care, and working conditions accompany these disparities in wealth.

Key Terms

areal differentiation *(page 11)*
areal integration *(page 13)*
bubble economy *(page 9)*
colonialism *(page 35)*
core-periphery model *(page 38)*
cultural imperialism *(page 28)*
cultural landscape *(page 13)*
cultural nationalism *(page 29)*
cultural syncretism or hybridization *(page 29)*
culture *(page 27)*
decolonialization *(page 35)*
demographic transition model *(page 22)*
economic convergence *(page 6)*
ethnic religion *(page 30)*
globalization *(page 2)*
gross domestic product (GDP) *(page 40)*
gross national income (GNI) *(page 40)*
gross national income (GNI) per capita *(page 41)*
lingua franca *(page 29)*
nation-state *(page 33)*
net migration rate *(page 23)*
overurbanization *(page 25)*
population pyramid *(page 21)*
purchasing power parity (PPP) *(page 41)*
rate of natural increase (RNI) *(page 18)*
region *(page 13)*
secularization *(page 31)*
squatter settlement *(page 25)*
sweatshops *(page 6)*
total fertility rate (TFR) *(page 18)*
transnational firm *(page 4)*
universalizing religion *(page 30)*
urban form *(page 25)*
urban primacy *(page 24)*
urban structure *(page 24)*
urbanized population *(page 24)*

Review Questions

1. What is globalization? What are the different components or attributes? Give examples of each.
2. Summarize the arguments used by proponents and opponents of globalization.
3. Why is scale important to geographic analysis?
4. Define three different kinds of geographic regions. Give examples of each.
5. How do you calculate the RNI for a country? How and why does RNI vary amongst different countries?
5. What are the four stages of the demographic transition? Explain how they differ from each other, and what the outcome is in terms of RNI.
6. What is *overurbanization* and why is it happening? What are the consequences?
7. What is a workable definition of *culture* in this age of globalization?
8. Define and give examples of *culture hybridity.*
9. What is the global distribution of Islam? What are the two major divisions within Islam? Give examples of countries where each group is dominant.
10. Define GNI. What are its shortcomings as a measure of development?
11. What are the best indicators of social development in Less Developed Countries (LDCs)?

Thinking Geographically

1. Select an economic, political, or cultural activity in your city and discuss how it has been influenced by globalization.
2. Choose a specific country or region of the world and examine the benefits and liabilities that globalization has posed for that country or region. Remember to look at different facets of globalization, such as the environment and cultural cohesion and conflict, as well as the economic effects on different segments of the population.
3. Drawing on information in current newspapers and magazines and on TV and the Internet, apply the concepts of cultural imperialism, nationalism, and syncretism to a region or place experiencing cultural tensions.
4. Select an African country with a colonial past and, first, trace out its pathway of decolonization, and then, second, describe and analyze its contemporary relations with its former colonial master.
5. Using the tables of social indicators in the regional chapters of this book, identify traits shared by countries in which there is a high percentage of female illiteracy. What general conclusions do you reach based on your inquiry?

Bibliography

Boo, Katherine. 2004. "Letter from India. The Best Job in Town: The Americanization of Chennai." *The New Yorker,* July 5, 2004, 54–69.

Friedman, Thomas L. 2009 *Hot, Flat, and Crowded 2.0: Why We Need a Green Revolution—And How It Can Renew America.* New York: Farrar, Straus and Giroux.

Friedman, Thomas L. 2006. *The World is Flat: A Brief History of the Twenty-first Century.* New York: Farrar, Straus and Giroux.

Gilpin, Robert. 2000. *The Challenge of Global Capitalism: The World Economy in the 21st Century*. Princeton, NJ: Princeton University Press.

Lemann, Nicholas. 2010. "Terrorism Studies: Social Scientists Do Counterinsurgency." *The New Yorker,* April 26, 2010, 73–77.

McFalls, Joseph, Jr. 2007. *Population: A Lively Introduction*. 5th ed. Washington, DC: Population Reference Bureau.

Schaeffer, Robert K. 2009. *Understanding Globalization: The Social Consequences of Political, Economic, and Environmental Change.* 4th ed. Lanham, MD: Rowman & Littlefield.

Stiglitz, Joseph E. 2006. *Making Globalization Work.* New York: W.W. Norton.

The World Bank. 2007. *The World Bank Atlas*. Washington, DC: International Bank for Reconstruction and Development/The World Bank.

Log in to www.masteringgeography.com for MapMaster™ interactive maps, geography videos, RSS feeds, flashcards, web links, an eText version of *Diversity Amid Globalization,* and self-study quizzes to enhance your study of Diversity Amid Globalization.

MapMaster™ presents 13 Place Name and 13 Layered Thematic interactive maps to help students practice and master their geographic literacy, spatial reasoning, and critical thinking skills.

2 The Changing Global Environment

The Athabasca Glacier in Canada's Jasper National Park has retreated about a mile over the last century because of global warming, and is currently receding at between 70 and 100 feet (20–30 meters) per year.

The human imprint is everywhere on Earth, from the highest mountains to the deepest ocean depths; from dry deserts to lush tropical forests; from frozen Arctic ice caps to the atmospheric layers we breathe. Hundreds of spent oxygen canisters clutter the heights of Mt. Everest, for example, and castoff plastic water bottles litter coastlines and oceans worldwide (Figure 2.1). As deserts bloom in North America with irrigated cotton fields, tropical forests in Brazil are laid waste by logging to create pastures for cattle.

Although many changes to the global environment are intentional and have improved the quality of human life, other changes are inadvertent and have proven harmful to both humans and the environment. Because of the importance of these issues, a study of the changing global environment is central to the study of world regional geography since environmental issues are deeply intertwined with globalization and diversity. The destruction of tropical rain forests, for example, is a response to international demand for wood products, beef, soybeans, and biofuels. Similarly, global warming is closely linked to world industry, commerce, and consumption.

Figure 2.1 The Human Imprint Trash dumped at sea litters an isolated beach in the Outer Hebrides islands of the Atlantic, attesting to the fact that there's probably not an environment on Earth untouched and unaffected by human activities.

GEOLOGY AND HUMAN SETTLEMENT: A Restless Earth

Geology shapes the fundamental form of Earth's surface by giving a distinctive character to world landscapes through the physical fabric of mountains, hills, valleys, and plains. The geologic environment is also critical to a wide spectrum of human activities and concerns; the relationship between soil fertility and agriculture or the distribution of mineral resources are examples. Additionally, the geologic environment presents humans with challenges and hazards in the form of devastating earthquakes, unstable landscapes, and explosive volcanoes. Clearly, a basic understanding of the physical processes that shape Earth's surface is crucial to comprehending human settlement in different parts of the world.

Plate Tectonics

The starting point for understanding the dynamic geology of Earth is **plate tectonics,** a geophysical theory that postulates that Earth is made up of a large number of geologic plates that move very slowly across its surface. This theory explains and describes both the inner workings of our planet as well as many surface landscape features. Additionally, plate tectonic theory also explains the world distribution of hazardous earthquakes and volcanoes.

Earth's interior is separated into three major zones with very different physical characteristics. Those three parts are the core, the mantle, and the outer crust. Tectonic plate theory is built on the assumption that a significant heat exchange takes place deep within Earth due to the cooling of the inner core. As a result, molten material is circulated through the mantle area (Figure 2.2).

This exchange of heat occurs in numerous **convection cells,** large areas of very slow-moving molten rock within Earth. Much like a stove provides heat for boiling water in a pan, radioactive decay deep within Earth's core drives these convection cells. As the molten material reaches Earth's surface, it cools and becomes denser, causing the material to sink back into the mantle. Although geophysicists assume that convection cells circulate molten material at different rates, generally speaking that movement is only a few inches or centimeters per year, a rate roughly similar to the growth of human fingernails.

In turn, these interior convection cells drag **tectonic plates,** which are huge, continent-size blocks of rock, in different directions across Earth's surface. On top of the tectonic plates sit assemblages of continents and the ocean

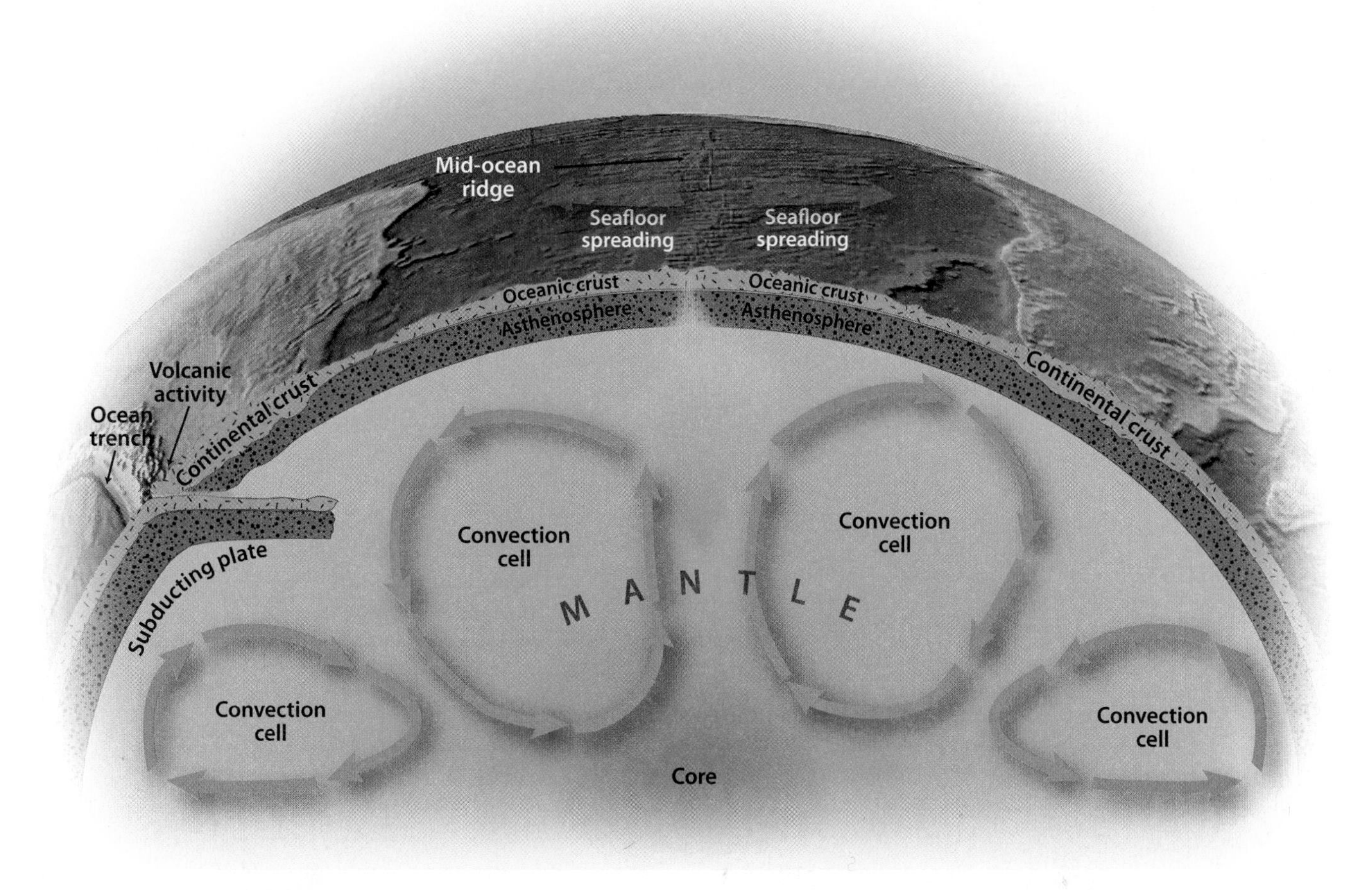

Figure 2.2 Plate Tectonics According to plate tectonics theory, large convection cells circulate molten rock in different directions within Earth's mantle. Near the crust the slow movement of the cells drags tectonic plates away from the mid-oceanic ridges, resulting in the collision of plates in convergent plate boundaries. While the rate of movement differs between convection cells, in general it is only a few inches (or centimeters) per year.

basins. The map of world tectonic plates (Figure 2.3) shows that these plates vary significantly in size. It is important to note that the continents are not identical to the tectonic plates; in fact, rarely is there a close match between the two. More often continents straddle several tectonic plates. For example in western North America, California lies atop two different tectonic plates, the Pacific Plate and the North American Plate, on what is called a **convergent plate boundary**. This means that the two plates are converging, or being forced together by convection cells deep within Earth. The notorious San Andreas Fault, which runs north and south through coastal California, is the actual plate boundary, and this explains why the large cities of Los Angeles and San Francisco are vulnerable to destructive earthquakes (Figure 2.4).

As well, major mountain ranges have resulted from the tectonic convergence or collision of these two plates. The volcanic Cascade Range of Oregon and Washington and the massive Sierra Nevada of California are two examples. Similarly, in Latin America, the Andes Mountains are also products of two colliding plates, the eastward-moving Nazca Plate and the westward-moving South American Plate.

Often in these collision zones one tectonic plate dives below another, creating a **subduction zone** characterized by deep trenches where the ocean floor has been pulled downward by tectonic movement. Subduction zones exist off the western coast of South America, the northwest coast of Oregon, Washington, and Alaska, and also near the Philippines, where the Mariana Trench forms the world's deepest ocean depths at 35,000 feet (10,700 meters). These subduction zones are also the location for Earth's most powerful earthquakes, as illustrated by the 8.8 magnitude Chile quake of 2009 and the 9.2 magnitude Alaska quake of 1964.

In other parts of the world, tectonic plates move in opposite directions, forming a **divergent plate boundary**. As plates diverge, magma usually flows from Earth's interior, creating mountain ranges with active volcanoes. In the North Atlantic, Iceland is formed on a divergent plate boundary that bisects the Atlantic Ocean. In other places,

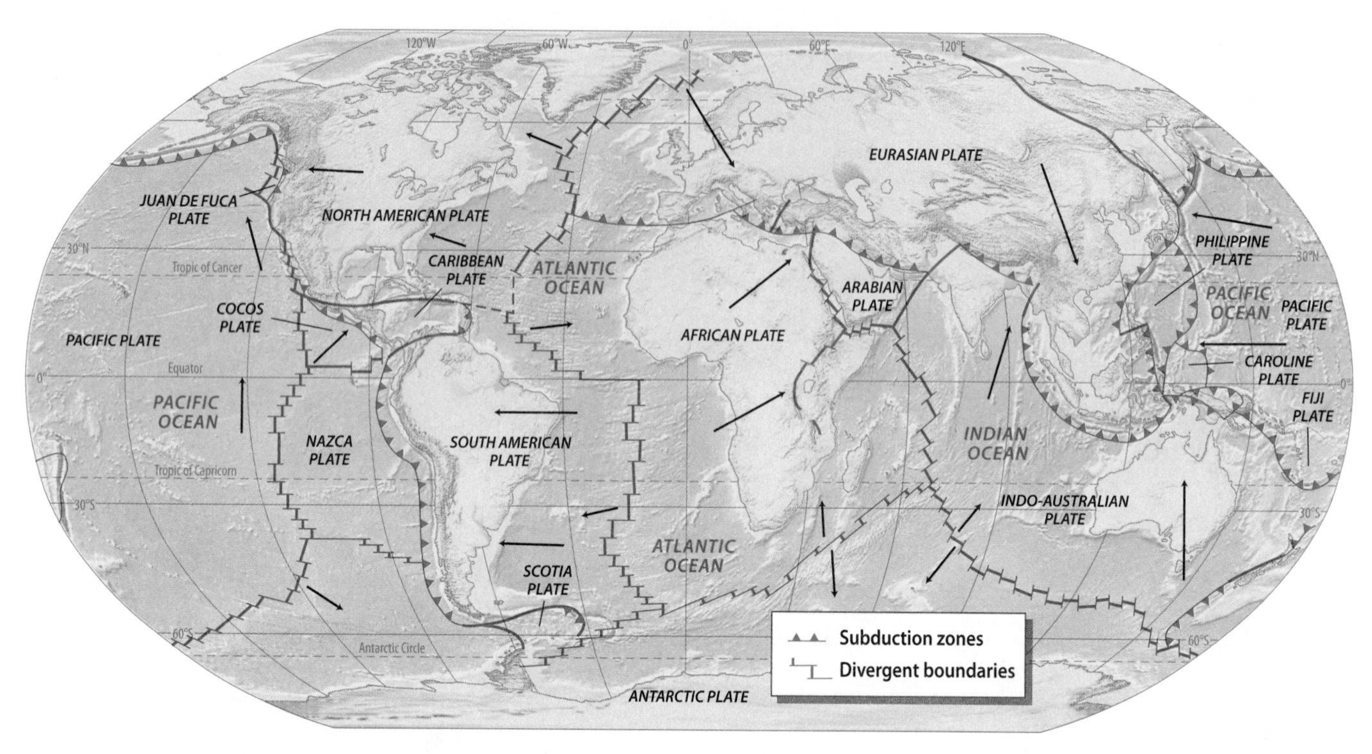

Figure 2.3 Global Pattern of Tectonic Plates Similar to the fractures of a jigsaw puzzle, Earth's tectonic plates vary in size and shape. Where plates collide, earthquakes and volcanoes are common, as are folded and faulted mountain ranges. In other areas, where one plate dives below another, deep oceanic trenches result, as is the case off the coast of South America and in the western Pacific Ocean.

divergent boundaries form deep depressions, or **rift valleys**, such as that occupied by the Red Sea between northern Africa and Saudi Arabia. To the west, in Africa, a splinter of this boundary has created the extensive African rift, a part of which is the famous Olduvai Gorge, where tectonic activity has preserved the earliest traces of our human ancestors in volcanic soils.

Geologic evidence suggests that some 250 million years ago all the world's plates were tightly consolidated into a supercontinent centered on present-day Africa. Through time, this large area, called *Pangaea*, was broken up as convection cells moved the tectonic plates apart. A hint of this former continent can be seen in the jigsaw puzzle fit of South America to Africa, and of North America to Europe.

Although tectonic plate theory explains many of the world's large mountain ranges, it does not account for all highlands. Although both the Himalaya and Alpine mountain ranges were created by the colliding forces of tectonic plates, there are also many mountain ranges far removed from tectonic boundaries. In North America, the Rocky Mountains serve as an illustration, as do the Ural Mountains

Figure 2.4 The San Andreas Fault Zone in California A string of lakes and reservoirs traces the notorious San Andreas Fault in coastal California. This view is looking north, with San Francisco and the Golden Gate visible in the upper right. When combined with other faults in the area, the Bay area has 63% chance of suffering a major earthquake in the next 30 years.

Figure 2.5 The World's Mountains Many, but not all, of the mountainous areas of the world are created by the collision of tectonic plates, examples being the Alps of Europe, the Andes of South America, and the Himalayas of South Asia. However, some significant mountains, such as the Rockies of North America, are relatively far from current tectonic boundaries, illustrating that tectonic stresses and strains are transmitted long distances in Earth's crust.

in the center of Russia (Figure 2.5). This reminds us that local geologic forces also play an important role in shaping the landscape.

Geologic Hazards: Earthquakes and Volcanoes

Although floods and hurricanes usually take a higher human toll than geologic hazards, nonetheless earthquakes and volcanoes can have a major impact on human settlement and activities. In January 2010, for example, over 230,000 people died from a magnitude 7.0 earthquake in Haiti, and although deaths were much lower from a stronger quake (8.8 magnitude) in Chile several weeks later, millions became homeless and the country's economy was in shambles (Figure 2.6). The vastly different effects of those two quakes underscore the fact that vulnerability to geologic hazards differs considerably around the world.

Figure 2.6 Earthquake Damage in Haiti In January 2010 Haiti was devastated by a magnitude 7.0 earthquake that killed an estimated 230,000 people, injured a similar amount, and left a million Haitians homeless.

The most vulnerable societies are those where urbanization has taken place rapidly and recently, creating situations where much of the population dwells in substandard housing. Tragically, even if reasonable building codes and construction standards exist, enforcement is often lax, leading to shoddy, hazardous buildings. Another variable is the ability of a society to respond immediately after an earthquake with medical care, search and rescue operations, and the provision of food and shelter for its population. Since these activity resources are often a function of the relative wealth of a country, poorer countries are generally more susceptible to disaster than richer ones. At a global scale, this means that billions of people in the fast growing cities of the Caribbean, Southwest, South, Southeast, and East Asia—all of whom live in active seismic zones—are vulnerable to a geologic disaster (Figure 2.7).

However, this does not mean that North Americans are immune to geologic disaster. Mentioned earlier is the vulnerability of San Francisco and Los Angeles to a large earthquake. Even with high building standards and rigorous code enforcement, estimates are that 10 to 30 thousand deaths will result when one of those cities experiences a major earthquake. A similarly high toll can be expected from a major quake in other vulnerable cities, namely Portland, Seattle, and Salt Lake City. Even Midwestern and East Coast cities are vulnerable, although earthquakes are rare, yet the regional geology of those areas is capable of producing a catastrophic quake.

In addition to earthquakes, volcanic eruptions are also found along most tectonic plate boundaries and can also cause major destruction. In 1985, for example, a volcanic eruption resulted in 23,000 deaths in Colombia, South America. In some cases eruptions can be predicted days in advance, which usually provides enough time for evacuation. During the 1991 eruption of Mt. Pinatubo in the Philippines, 60,000 people were evacuated, although 800 did die in the disaster (Figure 2.8). Because of their

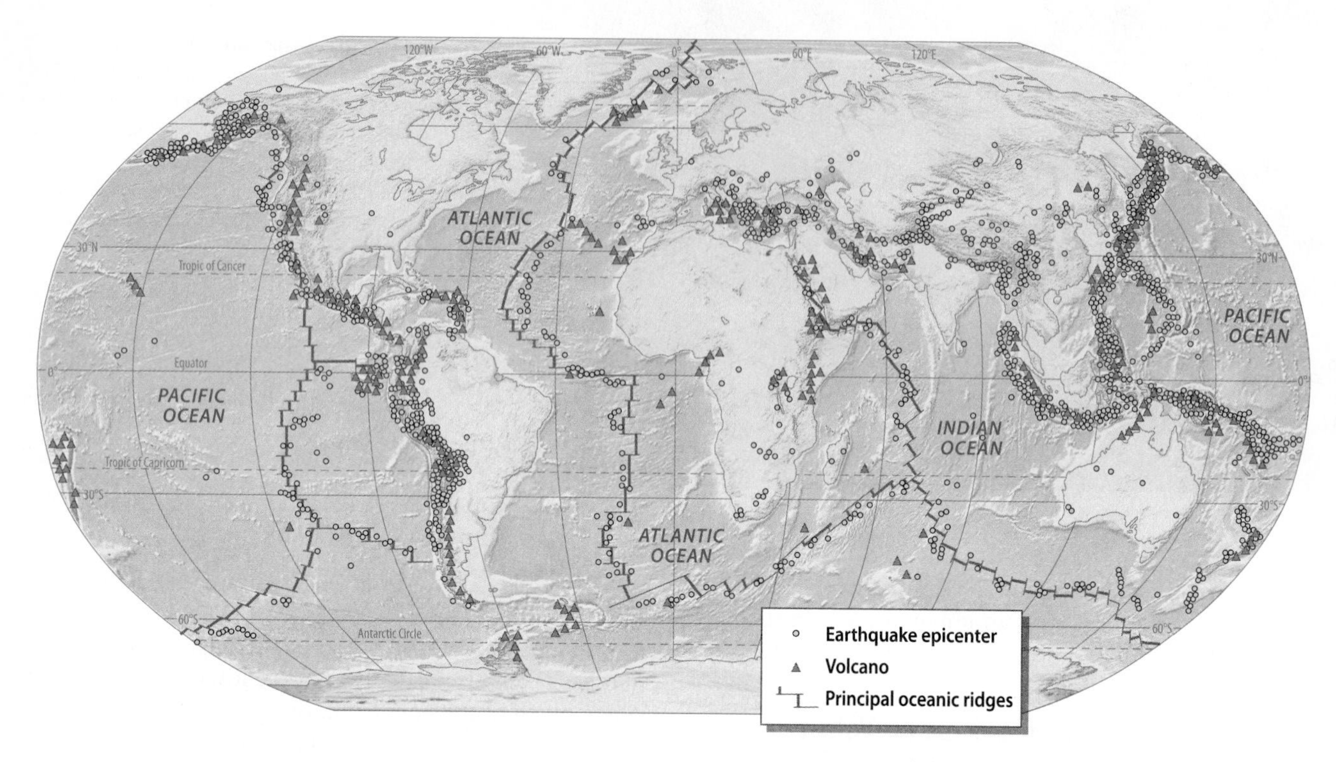

Figure 2.7 Global Earthquakes and Volcanoes The distribution of earthquakes and volcanoes is closely associated with tectonic plate boundaries. The circum-Pacific zones of activity, from the western Americas (both North and South) to East Asia is particularly active and is often referred to as the Pacific Rim of Fire. Another dangerous zone runs east-west, from northern South Asia to Mediterranean Europe. Between these two active earthquakes, billions of the world's population are at risk from destructive earthquakes.

predictability, the loss of life from volcanoes is generally a fraction of that from earthquakes. In the 20th century, an estimated 75,000 people were killed by volcanic eruptions while approximately 1.5 million died in earthquakes.

Unlike earthquakes, volcanoes also provide some benefits to people. In Iceland, New Zealand, and Italy, geothermal activity produces energy to heat houses and power factories. In other parts of the world, such as the islands of Indonesia, volcanic ash has enriched soil fertility for food crops. Additionally, local economies benefit from tourists attracted by scenic volcanoes in such places as Hawaii, Japan, and the Pacific Northwest.

Figure 2.8 Volcanic Eruptions Active volcanoes are also associated with plate tectonic borders throughout the world. Unlike earthquakes, there are usually a string of precursors before volcanic eruptions, allowing time for humans to evacuate the danger zone. This dramatic photo is from the 1991 eruption of Mt. Pinatubo in the Philippines.

GLOBAL CLIMATES: A Troublesome Forecast

Human settlement and food production are closely linked to patterns of local weather and climate. Where it is dry, such as in the arid parts of Southwest Asia, life and landscape differ considerably from the wet tropical areas of Southeast Asia. This is because people in different parts of the world adapt to weather and climate in widely varying ways, depending on their culture, economy, and technology. Although some desert areas of California are covered with high-value irrigated agriculture that produces vegetables for the global marketplace year around, most of the world's arid regions support very little agriculture and barely participate in global commerce. Moreover, when drought hits one portion of the world, such as Russia's grain belt or Africa's Sahel, the socioeconomic repercussions are felt throughout the world. As a result, climate links us together in our globalized economy, providing opportunities for some, hardships for others, yet challenges for all in the struggle to supply food and shelter.

Additionally, one of the most pressing problems facing the world today is that human activities are changing Earth's climate through global warming. Just what the future will bring is not entirely clear. However, even if the forecast is uncertain, there is little question that many forms of life—including humans—will face difficulties adjusting to the changes brought about by global warming (Figure 2.9).

Figure 2.9 Polar Bears Threatened by Global Warming Ice floes are an important part of the polar bear's habitat. Because Arctic ice is melting more rapidly now due to global warming, the 20,000 bears living in the wild face an uncertain future. Recently, polar bears have been reported drowning trying to swim long distances between ice sheets.

Climatic Controls

Although weather and climate differ tremendously around the world, an accepted set of atmospheric processes control and influence meteorological conditions. More specifically, there are five main factors that explain differences in global weather and climate—solar energy, latitude, interaction between land and water, world pressure systems, and global wind patterns.

Solar Energy Both the surface of Earth and the atmosphere immediately above it are heated by energy from the Sun as our planet revolves around that large star. Solar energy is one of the most important variables producing different world climates; it explains, for example, the great differences between tropical climates near the equator and the cold climates closer to the poles.

Most incoming solar energy, or **insolation**, is absorbed by Earth's land and water surfaces. These surfaces, in turn, heat the lower atmosphere through the process of reradiation. This reradiated energy is trapped by clouds and water moisture in the air adjacent to Earth, providing a warm envelope that makes life possible on our planet. Because there is some similarity between this process and the way a garden greenhouse traps sunlight to make that structure's interior warmer than the outside, this natural process of atmospheric heating is known as the **greenhouse effect** (Figure 2.10). Were it not for this process, Earth would be far too cold for human habitation. More specifically, without the greenhouse effect our climate would be much like that on Mars.

Latitude Because of the curvature of Earth, the highest amounts of insolation are received in the equatorial region, the area between the equator and 23.5° north and south latitude. Poleward of this equatorial region, Earth receives less insolation. As a result, not only are the tropics much warmer than the middle or high latitudes, but there is also a buildup of heat energy that must be redistributed through other processes, namely global wind systems, ocean currents, and even massive tropical storms, typhoons and hurricanes (Figure 2.11).

Interaction between Land and Water Because land and water differ in their abilities to absorb and reradiate insolation, the global arrangement of oceans and continents is a major influence on world climates. Land areas heat and cool faster than bodies of water, which explains why temperature extremes such as hot summers and cold winters are always found away from coasts. Conversely, because water bodies retain solar heat longer than land areas, oceanic or maritime climates usually have moderate temperatures without the seasonal extremes found inland. The climatic characteristics of land and water differ so much that geographers use the term **continentality** to describe inland climates with hot summers and cold, snowy winters such as those found in interior North America and Russia. In contrast, **maritime climates**, or those close to the ocean, generally have cool, cloudy summers (Oregon, Washington, and British Columbia are good examples), with winters that are cold but lack the extreme subfreezing temperatures of interior locations.

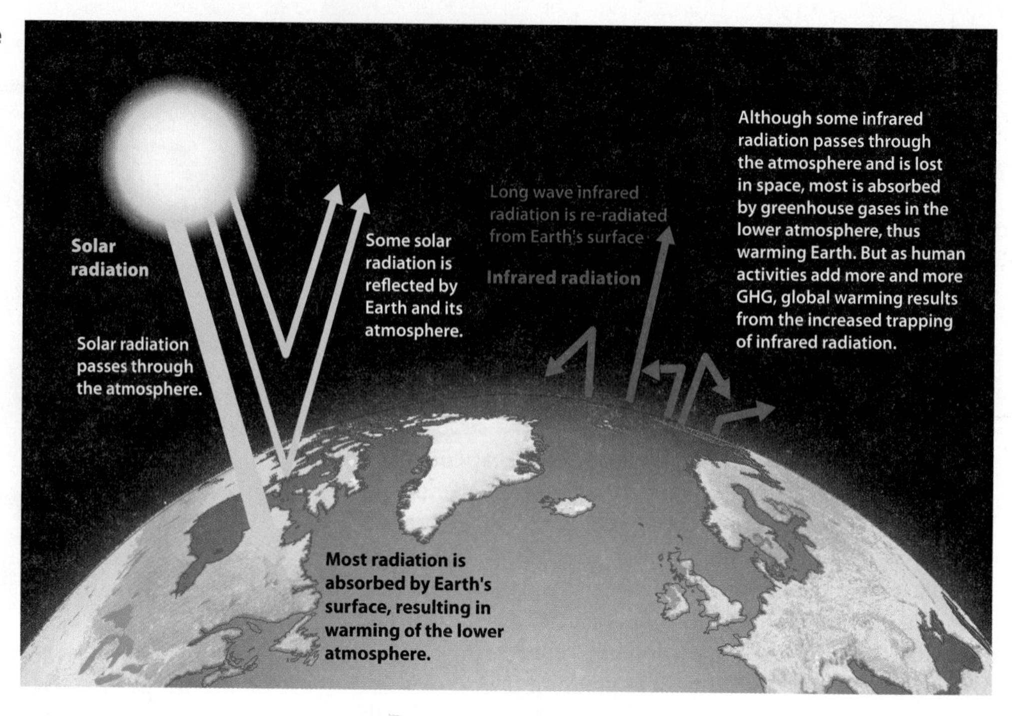

Figure 2.10 Solar Energy and the Greenhouse Effect The greenhouse effect is the trapping of solar radiation in the lower atmosphere, resulting in a warm envelope surrounding Earth.

Global Pressure Systems The uneven heating of Earth due to latitudinal differences and the arrangement of oceans and continents produces a regular pattern of high- and low-pressure cells that, in turn, drive the world's wind and storm systems (Figure 2.12). To illustrate, the interaction between high- and low-pressure systems in the North Pacific produces the storms that are driven onto the North American continent in both winter and summer. Similar processes in the North Atlantic produce winter and summer weather for Europe.

Farther south in the subtropical zones, large oceanic cells of high pressure cause different conditions. These high-pressure cells expand during the warm summer months because of the subsidence of warm air from the equatorial regions. As they enlarge, the cells produce the warm, rainless summers of the Mediterranean area of Europe and California. In the equatorial zone itself, summer weather spawns the strong tropical storms known as typhoons in Asia and hurricanes in North America and the Caribbean. The map of global pressure systems shows that this same

Figure 2.11 Hurricanes and Typhoons Strong tropical storms result from the differential heating of Earth's surface and, while often damaging to human settlement, are a natural process of atmospheric physics. This photo was taken in Baja, Mexico during Hurricane Jimena in September 2009.

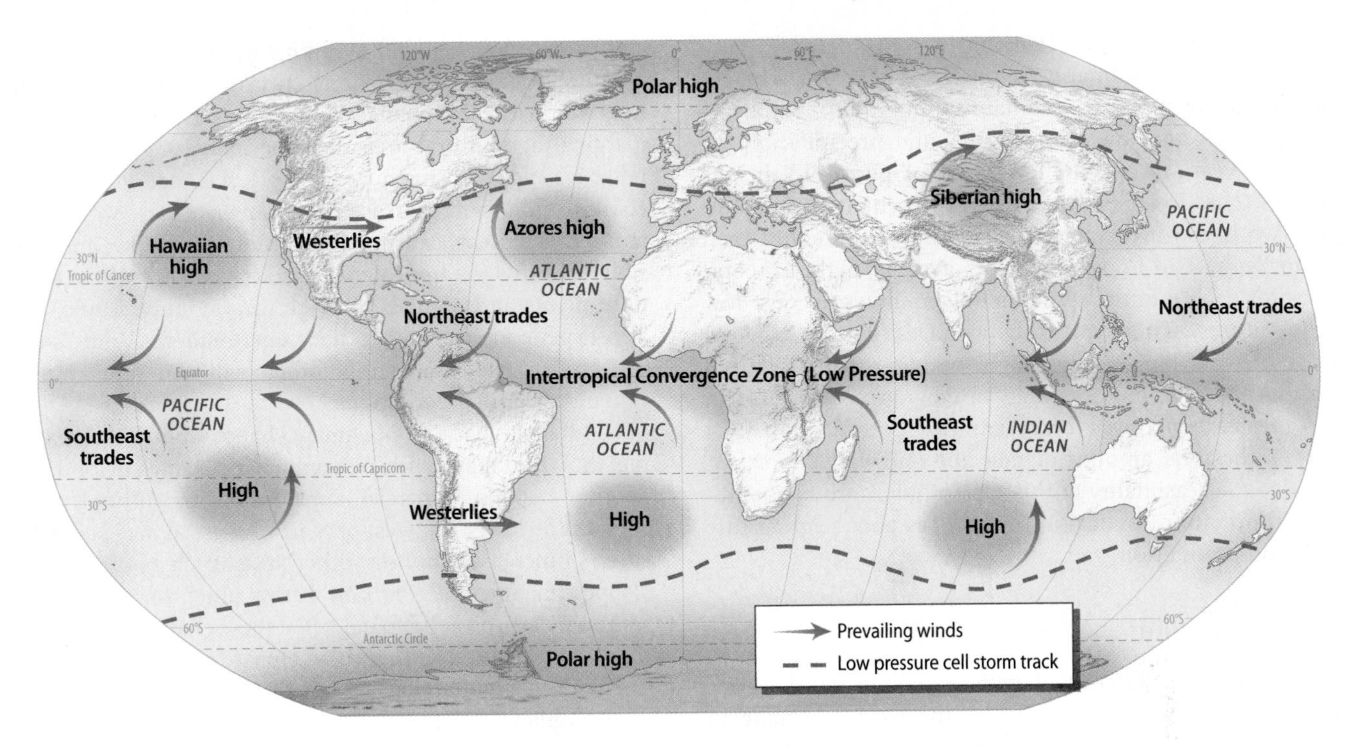

Figure 2.12 Global Pressure Systems The global pattern of high- and low- pressure systems results from the unequal heating of Earth's surface. The Intertropical Convergence Zone (ITCZ) is where the pressure systems and winds of the Northern and Southern hemispheres meet; poleward of the ITCZ, westerly winds move storm systems through the middle latitudes.

alternating pattern of high- and low-pressure systems is also found in the Southern Hemisphere.

Global Wind Patterns The pressure systems also produce global wind systems. It is important to remember that air flows from high pressure to low (just as water flows from high elevations to low), thus, winds flow away from high-pressure and into low-pressure cells. This explains the monsoon in India, for example, which arrives in June as moisture-laden air masses flow from the warm Indian Ocean over land into the low-pressure area above northern India and Tibet. In the winter the opposite is true: As high pressure builds over these same areas, winds flow outward from cold Tibet and the snowy Himalayas toward the low pressure over the warm Indian Ocean.

World Climate Regions

The interaction of the meteorological processes just discussed produces the world's weather and climate. Before going further, though, it is important to note the difference between these two terms. *Weather* is the short-term, day-to-day expression of atmospheric processes: Our weather can be rainy, cloudy, sunny, hot, windy, calm, or stormy all within a short time period. This ever-changing weather is measured at regular intervals each day, often hourly. Data are then compiled on temperature, pressure, precipitation, humidity, and so on; and, over a period of time, statistical averages from these daily observations provide a quantitative picture of what are statistically common conditions. From these data, a sense of a regional *climate* is generated. Usually at least 30 years of daily weather data are required before climatologists and geographers construct a picture of an area's climate. In summary, weather is the short-term expression of highly variable meteorological processes, whereas climate is the long-term, average conditions. Or as weather pundits like to say, "climate is what you expect; weather is what you get" (see "Geographic Tools: Reconstructing Climates of the Past").

Where similar conditions prevail over a larger area, boundaries are drawn cartographically around that area, which is then called a **climate region**. Knowing the climate type for a specific part of the world not only conveys a clear sense of average rainfall and temperatures, but also allows larger inferences about human activities and settlement. If an area is categorized as desert, then we infer that rainfall is so limited that any agricultural activities require irrigation. In contrast, if we see an area characterized by the climatic designation for tropical monsoon, we know there are warm temperatures and adequate amounts of rainfall for agriculture.

A standard scheme of climate types is used throughout this book, and each regional chapter contains a map showing the different climate regions in some detail. Figure 2.13 shows these climatic regions for the world. The regional climate maps also contain **climographs**, graphs of average high and low temperatures and precipitation for an entire year at a specific location. In Figure 2.13 the climate of Cape Town, South Africa, is presented as an example. Two lines for temperature data are presented on each climograph: The upper one plots average high temperatures for each month, while the lower shows average low temperatures. These monthly averages convey a good sense of what typical days might be like during different seasons. Besides temperatures, climographs also contain bar graphs depicting average monthly precipitation. Not only is the total amount of rain and snowfall important, but also the seasonality of this precipitation provides valuable information for making inferences about agriculture and food production.

Global Warming

Human activities connected with economic development and industrialization are changing the world's climate in ways that will have significant consequences for all living organisms, be they plants, animals, or human. More specifically, **anthropogenic**, or human-caused, pollution of the lower atmosphere is increasing the natural greenhouse effect so that worldwide **global warming**—an increase in the temperature of Earth's atmosphere—is taking place. This warming, in turn, may change rainfall patterns so that some marginal agricultural areas become increasingly arid; melt polar ice caps, causing a rise in sea levels that will threaten coastal settlement; increase the number of heat waves that strike urban areas, causing death from heat stroke; and possibly lead to a greater intensity in tropical storms. In addition, plants and animals, both terrestrial and oceanic, will also experience changes to their environment. While some will probably become extinct, others will have to migrate to find more familiar conditions.

Moreover, because of climate change from global warming, the world may experience a dramatic change in food production regions. Whereas some prime agricultural areas, such as the lower Midwest of North America, may suffer because of increased dryness, other areas, such as the Russian steppes, may actually become better suited for agriculture. Although it is too early to tell when these changes may occur, there is little question that these climate changes will have a dramatic effect on human settlement, world trade, and global food supplies.

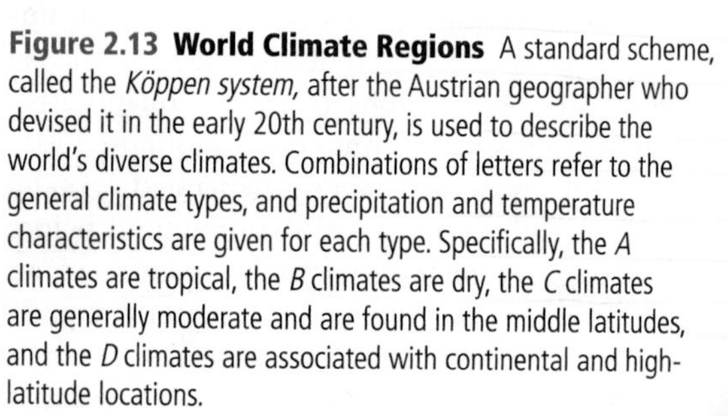

Figure 2.13 World Climate Regions A standard scheme, called the *Köppen system,* after the Austrian geographer who devised it in the early 20th century, is used to describe the world's diverse climates. Combinations of letters refer to the general climate types, and precipitation and temperature characteristics are given for each type. Specifically, the *A* climates are tropical, the *B* climates are dry, the *C* climates are generally moderate and are found in the middle latitudes, and the *D* climates are associated with continental and high-latitude locations.

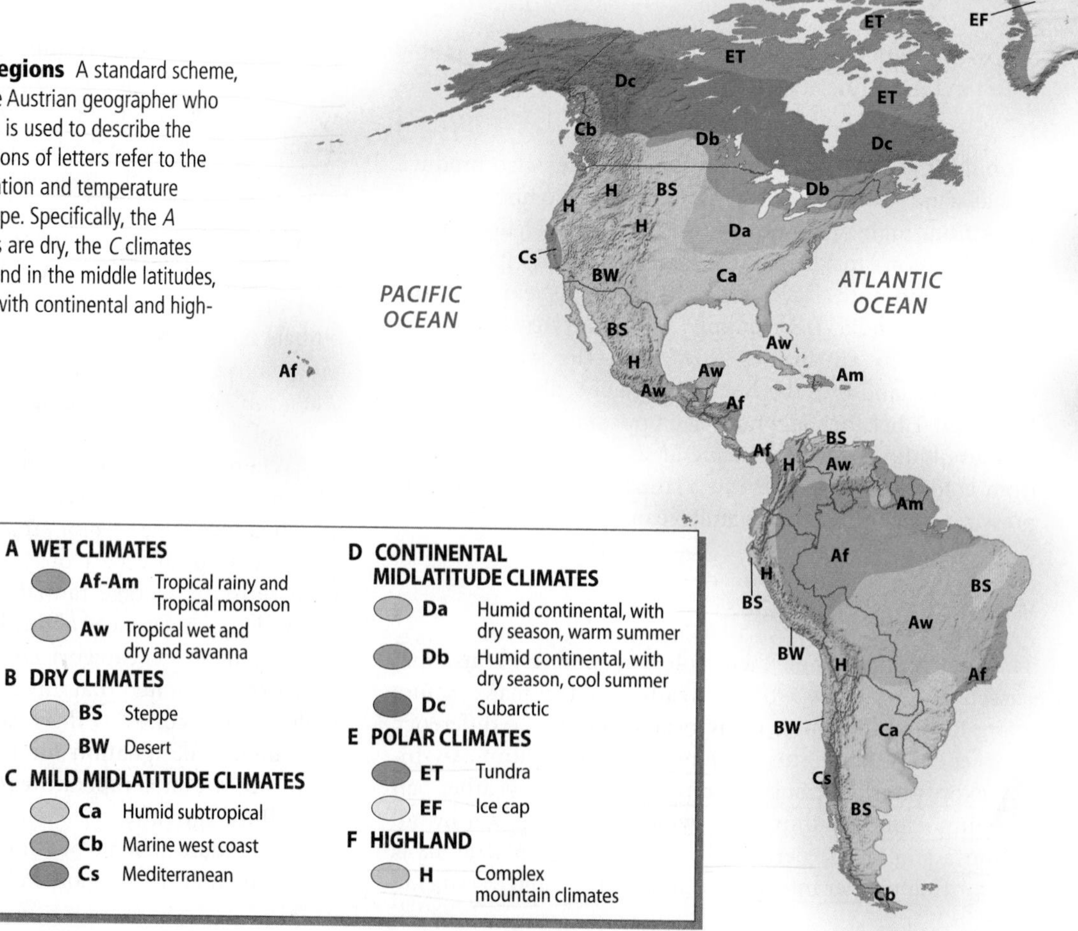

Causes of Global Warming As noted, the natural greenhouse effect provides us with a warm atmospheric envelope that supports human life. This warmth comes from the trapping of incoming and outgoing (or re-radiated) solar radiation by an array of natural greenhouse gases in the atmospheric layer closest to Earth. Although natural greenhouse gases have varied somewhat over long periods of geologic time, they were relatively stable until recently. However, within the past 130 years, coincidental with the Industrial Revolution in Europe and North America, the composition and amount of those greenhouse gases have changed dramatically, primarily from the burning of fossil fuels. As a result, the average temperature for Earth has increased markedly over the past century (Figure 2.14).

Four major greenhouse gases account for the bulk of this change:

1. *Carbon dioxide (CO_2)* accounts for more than half of the human-generated greenhouse gases. The increase in atmospheric CO_2 is primarily a result of burning fossil fuels such as coal and petroleum. To illustrate, in 1860 atmospheric CO_2 was measured at 280 parts per million (ppm). Today it is around 390 ppm, and it is projected to exceed 450 ppm by 2050, assuming that fossil fuels remain a major source for the world's energy needs.
2. *Chlorofluorocarbons (CFCs)* make up nearly 25 percent of the human-generated greenhouse gases and come mainly from widespread use of aerosol sprays and refrigeration (including air conditioning). Although CFCs have been banned in North America and most of Europe, they are still increasing in the atmosphere at the rate of 4 percent each year. These gases are highly stable and reside in the atmosphere for a long time, perhaps as long as 100 years. As a result, their role in global warming is highly significant. Recent research has shown that a molecule of CFC absorbs 1,000 times more infrared radiation from Earth than a molecule of CO_2.
3. *Methane (CH_4)* has increased 151 percent since 1750 as a result of vegetation burning associated with rainforest clearing, anaerobic activity in flooded rice fields, cattle and sheep effluent, and leakage from pipelines and refineries connected with natural gas production

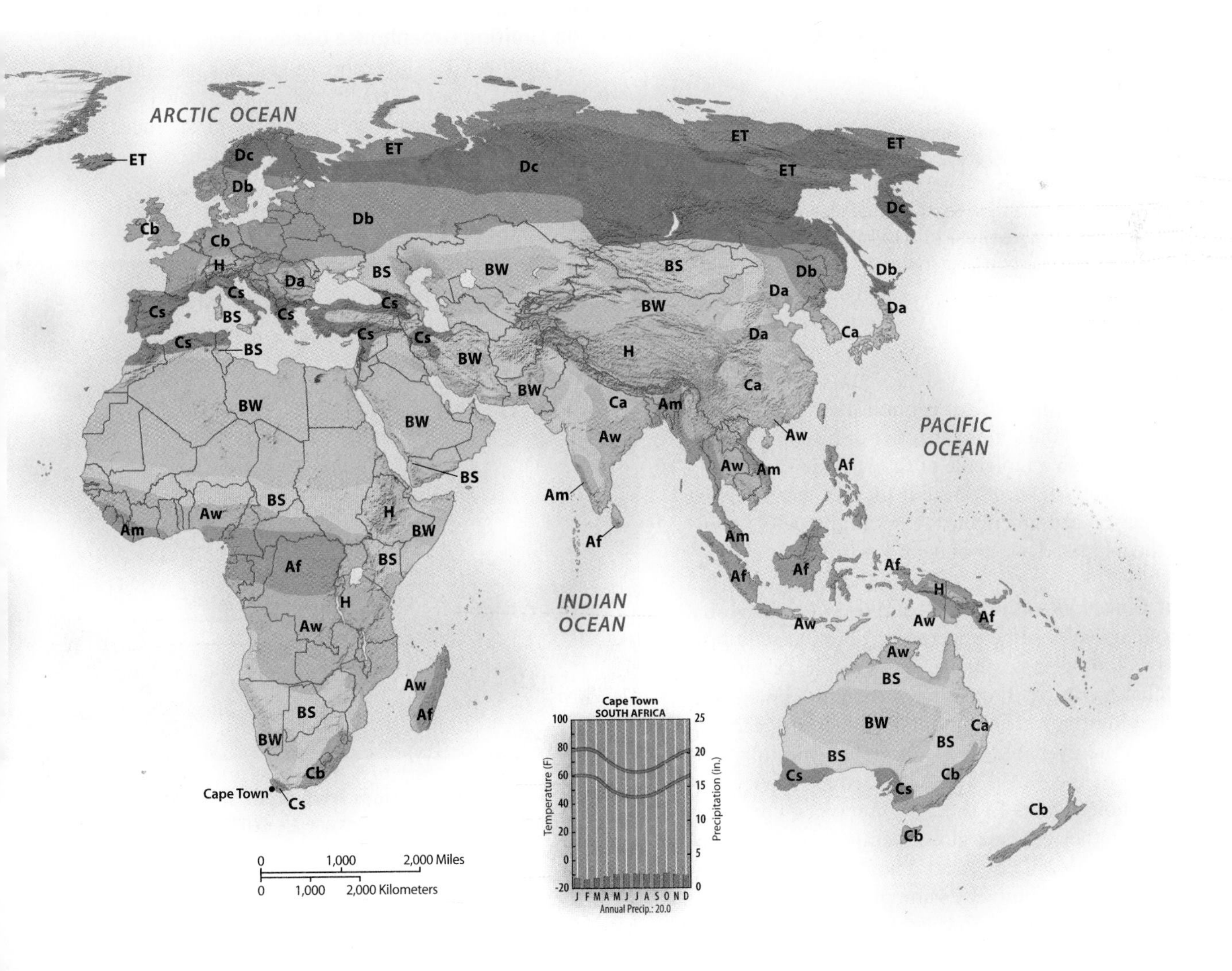

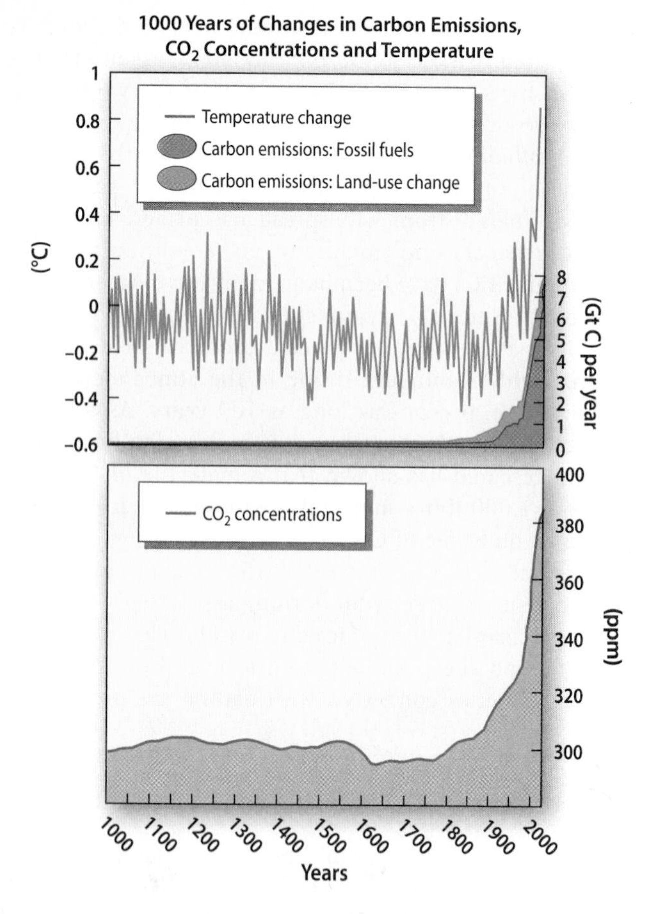

Figure 2.14 Increase in CO_2 and Temperature These two graphs show the relationship between the rapid increase of CO_2 in the atmosphere and the associated rise in average annual temperature for the world. The graphs go back 1,000 years and show both CO_2 and temperature to have been relatively stable until the recent industrial period, when the burning of fossil fuels (coal and oil) began on a large scale.

(Figure 2.15). Currently, CH_4 accounts for about 16 percent of anthropogenic greenhouse gases, and each molecule is 23 times more effective at trapping atmospheric heat than a molecule of CO_2 (Figure 2.15).

4. *Nitrous oxide (N_2O)* is responsible for just over 5 percent of human-caused greenhouse gases and results mainly from the increased use of industrial fertilizers.

Effects of Global Warming The complexity of the global climate system leaves some uncertainty about exactly how the world's climate may change as a result of human-caused global warming. Increasingly, though, climate scientists using high-powered computer models are reaching consensus on the probable effects of global warming. Unless countries of the world drastically reduce their emission of greenhouse gases in the next few years, computer models predict that average global temperatures will increase 2 to 4°F (1 to 2°C) by 2030. This may not seem dramatic at first but it is about the same magnitude of change as the cooling that caused the ice age glaciers to cover much of Europe and North America 30,000 years ago. More problematic is that this temperature increase is projected to double by 2100.

As mentioned, this climate change could cause a shift in major agricultural areas. For example, the Wheat Belt in the United States might receive less rainfall and become warmer and drier, endangering grain production as we know it today. While more northern countries, such as Canada and Russia, might experience a longer growing season because of global warming, the soils in these two areas are not nearly as fertile as in the United States. As a result, food experts are predicting a decrease in the world's grain production by 2030. In addition, the southern areas of the United States and the Mediterranean region of Europe can expect a warmer and drier climate that will demand even more irrigation for crops.

Warmer global temperatures will also cause sea levels to rise as oceans warm and polar ice sheets and mountain glaciers melt. Currently, projections are for an increase of at least 3.3 feet (1 meter) by 2100. Although this may not seem like much, even the smallest increase will endanger low-lying island nations throughout the world as well as coastal areas in Europe, Asia, and North America (Figure 2.16).

Globalization and Global Warming: The International Debate on Limiting Greenhouse Gases The international debate on limiting GHG (greenhouse gas) has passed through two phases in the last several decades and is now is entering a third, a stage that many experts say is the most crucial since

Figure 2.15 Methane Emissions from Arctic Wetlands About 10 percent of global methane emissions are released from the Arctic tundra. Although a natural process, these methane emissions have increased with global warming. Some scientists are concerned that the tundra may reach a tipping point where the amount of methane emissions increases dramatically, resulting in a sudden change to the global climate.

Figure 2.16 Sea Level Rise from Global Warming Predictions are that global warming will cause Earth's oceans to rise by at least 39 inches (1 meter) by the end of the 21st century because of the melting of ice caps and glaciers, coupled with the thermal expansion of warming ocean waters. Some settlements, such as Venice, Italy, will suffer greatly from sea level rise. Currently, Venice is flooded about four times a year from natural high tides. but with the predicted sea level rise, this flooding may take place at least 30 times a year.

emission reduction actions taken in the next 10 years will determine the extent of global warming for the rest of the century.

The first phase began at the UN-sponsored 1992 Earth Summit in Rio de Janeiro when 167 countries signed an agreement to voluntarily limit their greenhouse gas emissions. However, because none of the signatories reached their emission reduction target, a more formal second phase began with the 1997 meeting in Kyoto, Japan. At that meeting the 30 western industrialized countries that were responsible for most GHG emissions agreed to reduce their emissions back to 1990 levels by the year 2012. Unlike the Rio agreement, which was voluntary, the Kyoto Protocol had the force of international law once it was ratified. But ratification came neither quickly nor easily. One reason was that the United States, which at that time was the largest emitter of greenhouse gas pollution, never ratified the agreement because of political resistance based on concerns that emission regulation would harm the U.S. economy. Additionally, Congress was uncomfortable with the fact that India and China, as developing countries, were not subject to any kind of emissions restrictions even though they had substantial and increasing emissions (Table 2.1).

Despite resistance from the U.S. government, the Kyoto Protocol became international law in early 2005 when Russia ratified the agreement. But even with the political change from presidential administrations that opposed Kyoto (Bush) to one supporting it (Obama), U.S. congressional opposition remained strong. There were also problems with Kyoto at a global scale, namely because most signatories were not on track to meet their 2012 emission reduction targets. As well, when China, which, to repeat, was not bound by Kyoto, surpassed the United States as the largest GHG emitter in 2008, it became increasingly clear that the 1997 agreement had to be fundamentally revised.

Thus, the third phase of international emission reduction negotiations has begun with a series of meetings to con-

TABLE 2.1 The World's Major CO_2 Polluters

Country	Total CO_2 Emissions in MMT[a] (2008)[b]	Percent Change (1998–2008)	Percent of World Total, CO_2 Emissions	Per Capita Emissions of CO_2, Metric Tons
China	6,533	65	21	4.9
United States	5,832	4	19	19.1
Russia	1,729	18	6	12.3
India	1,494	64	5	1.3
Japan	1,214	9	4	9.5
Germany	828	–5	3	10.0
World	*30,377*	32	*58% of world total*	4.5

[a]*Million metric tons.*
[b]*The most current and compatible international data from the Energy Information Administration (EIA) are for 2008.*

GEOGRAPHIC TOOLS Reconstructing Climates of the Past

Geographers, along with scientists from many other fields, are deeply involved in studying global warming. This research involves not only analyzing modern-day weather records for signs of climate change but also reconstructing the climates of past centuries to provide a longer term view of Earth's weather and climate fluctuations. Because accurate weather instruments have been used only for the last 200 years, a number of tools are necessary to probe into the climate of past centuries, using what are called *climate proxy tools*. These methods have now become refined to the point where they provide a fairly accurate record of global climates going back at least 15,000 years.

Scientists use three main tools to reconstruct past climates: tree-ring analysis, or *dendrochronology*; pollen analysis, or *palynology*; and ice-core isotope analysis from the world's glaciers and ice caps. The first two are discussed here.

Figure 2.1.1 Tree Rings and Past Climates Annual growth rings in trees give clues to past climates by showing years when growth was greater or lesser than normal as indicated by the width of the annual tree ring. In this photo there are several years where narrow tree rings may indicate a drought that inhibited tree growth.

The annual growth rings of a tree give clues about the growing conditions in a given year. Because trees are sensitive to moisture, a wider growth ring suggests a wetter year, whereas narrower rings imply drier years. Additionally, yearly growth rings also give clues about temperature, with wider growth rings resulting from more moderate temperatures and narrow rings suggesting a harsher (usually colder) year. Rather than cutting down trees to examine the rings, scientists instead extract from the trunk thin cores that contain growth rings showing the complete history of the tree's growth. After these rings are measured precisely under a microscope, they are subjected to a number of statistical tests to assure validity in replicating past climatic conditions.

One obstacle to the use of tree rings for constructing past climates is that most trees live only for several hundred years. So how do scientists use tree rings to go back thousands of years? Logs found in prehistoric dwellings, or dead trees found in bogs or underwater in lakes, can be cored and tree ring chronologies constructed for the period when the tree was living. When cobbled together from various sources, tree-ring chronologies can provide a good sense of the climate back 5,000 years.

The pollen given off by plants, shrubs, and trees is also useful to reconstruct growing conditions hundreds and thousands of years ago. Plants, shrubs, and trees produce pollen as part of their reproductive process, and commonly the fine-grained pollen material is carried by the wind to other plants. Pollen also collects on lake surfaces, and eventually it sinks to the lake bottom where it collects in the mud and is preserved for thousands of years.

Palynologists, then, seek out these prehistoric lakes, many of which are now bogs, and take samples from the lake bottom to retrieve the ancient pollen. Using high-powered microscopes, they then can identify the pollen and link it to specific plant, shrub, and tree species, thus reconstructing the vegetation that covered the landscape thousands of years ago and, by inference, the climate of that period.

Given the need to decipher past climates to better understand global warming, geography students who become acquainted with the methods and tools of climate reconstruction can find rewarding careers with research labs and government agencies doing this kind of work.

struct a new international agreement. While the content of a new agreement is far from clear, many believe (at least in late 2010) it will build on the structure of the Copenhagen Accord, a working document signed by 188 countries at the UN-sponsored Copenhagen meeting in December 2009. The major points of this document are:

- An agreement by the signatories that global warming is an urgent problem, and that they commit themselves to action based upon ". . . common but differentiated responsibilities and respective capabilities . . . and on the basis of equity." This wording sets the scene for a more flexible document than the Kyoto Protocol, and one more sensitive to achieving equity between traditional western industrial powers and emerging economies such as China and India.
- Further, the Copenhagen signatories agreed action is urgently needed for adapting to the effects of global warm-

ing, particularly in the least developed countries, small island states, and Africa. Unlike Kyoto, which was primarily concerned with reducing emissions, Copenhagen recognizes that global warming is already causing hardships for many countries because of more frequent drought, sea level rise, and forest degradation, and that action must be taken soon to help those countries. Equally important is an agreement that western developed countries will provide financial resources to support adaptation.

- In what is referred to as the "portfolio approach," instead of imposing emission regulations from the top down by a UN agency, the Copenhagen Accord recognizes the importance of local, regional, and national policies for emission reduction. In the absence of a national program in the United States, state programs, such as California's emission reduction plan, have become increasingly important
- A very significant component of the Copenhagen Accord is the emphasis placed on reducing GHG emissions from deforestation and forest degradation, particularly from tropical areas where large amounts of CO_2 are emitted during rain-forest cutting. Additionally, developed countries commit themselves to financial resources to help reduce deforestation in emerging economies, a measure that could have profound implications for preserving tropical rain forests.

Despite the progress made with the Copenhagen Accord (and many environmental groups were profoundly disappointed that not more was achieved), significant stumbling blocks remain. Three stand out:

First, China and India have now become significant contributors to global GHG pollution, and yet currently both those countries emphatically reject the notion that they must adhere to an international emission reduction plan because of fear it will injure their growing economies (Figure 2.17). Here the question of equity becomes central, for both China and India still consider themselves developing countries, different in structure, character, and history than the developed economies of the United States, Europe, Russia, and Japan, and therefore think they should be treated differently than those western countries who created the global warming problem by their unrestricted GHG emissions during the 19th and 20th centuries.

Second, the issue of just how much financial aid developed countries will contribute to less developed countries as aid for adaptation to global warming is highly controversial, not simply because of the dollar and cents cost, but also because some developed countries see this foreign aid as a form of scapegoating by developing countries—accusing them of using global warming as an excuse to extort money from richer countries when the causes of their problems are failures within their own governments.

Last, questions are raised as to whether the United Nations is the appropriate agency through which to forge a global warming agreement. Why not, ask some, form a working group of the major polluters (China, United States, Russia, India, Europe) and discuss solutions at that smaller scale instead of involving over 200 UN members? The counterargument, of course, is that global warming affects all countries, large and small, polluter or not, and therefore a global solution is necessary.

WATER: A Scarce Global Resource

Water is central to human life and our supporting activities, agriculture, industry, transportation, even recreation. And yet water is unevenly distributed around the world, being plentiful in some areas and distressingly scarce in others.

The Global Water Budget

A world map shows that more than 70 percent of the surface area of the world is covered by oceans. As a result, 97 percent of the total global water supply is saltwater and only 3 percent is freshwater. Of the minuscule amount of freshwater on Earth, almost 70 percent is locked up in polar ice caps and mountain

Figure 2.17 China's Energy Needs and Global Warming Emissions Several years ago China surpassed the United States as the largest emitter of greenhouse gases. A major component of this global warming pollution is that China is also the world's largest consumer of coal, most of which fuels the country's growing network of power plants that are needed to supply electricity for the country. Although China has some so-called "clean" coal-fired power plants that reduce emissions, most power generating plants are not. In fact, many lack even the most basic pollution control devices. As a result some experts predict China's greenhouse gas emissions will actually double in the next two decades, a worrisome prospect that could have significant effects on the global climate.

glaciers. Furthermore, groundwater supplies account for almost 30 percent of the world's freshwater, which leaves less than 1 percent accessible from surface rivers and lakes.

Another way to conceptualize this limited amount of freshwater is to think of the total global water supply as 100 liters, or 26 gallons. Of that amount, only 3 liters (0.8 gallon) would be freshwater; and of that small supply, only a mere 0.003 liters, or about half a teaspoon, would be available to humans.

Water planners use the concept of **water stress** to map where water problems exist and, as well, to predict where future problems will occur (Figure 2.18). Water stress data are generated by calculating the amount of freshwater available in relation to current and future population amounts. Africa stands out as a region of high water stress, a region where hydrologists predict that three-quarters of Africa's population will experience water shortages by the year 2025. Other problem areas will be northern China, India, much of Southwest Asia, Mexico, the western U.S., and even parts of Russia. Although global warming may actually increase rainfall in some parts of the world, scientists forecast that in general climate change will aggravate global water problems. Four areas of concern occupy water planners: scarcity, sanitation, access, and management.

Water scarcity Currently, about half the world's population lives in areas where water shortages are common, and, as the population in these areas increases, these water problems will become more acute. Additionally, since 70 percent of the world's freshwater usage is for agriculture, food production will probably decline as water becomes increasingly scarce.

Water sanitation Where clean water is not available, people use polluted water for their needs, resulting in a high rate of sickness and death. As a general statement, polluted water sources are the greatest source of illness and death worldwide. Not surprisingly, the largest source of water pollution is lack of sewage treatment facilities.

Water access By definition, when a resource is scarce, access is problematic, and these hardships take many forms. Women and children, for example, often bear the burden of providing water for family use, and when this means walking long distances to pumps and wells, and then waiting in long lines to draw water, their daily time budget is severely curtailed for other activities such as school or work (Figure 2.19). Given the amount of human labor involved in providing water for crops, it is not surprising that some studies have shown that in certain areas people expend as many calories of energy irrigating their crops as they gain from the food itself.

Ironically, some recent international efforts to increase people's access to clean water have gone astray and, instead, have actually aggravated access problems. Historically, domestic water supplies have been public resources organized and regulated, either informally by common consent, or more formally as public utilities, resulting in free or low

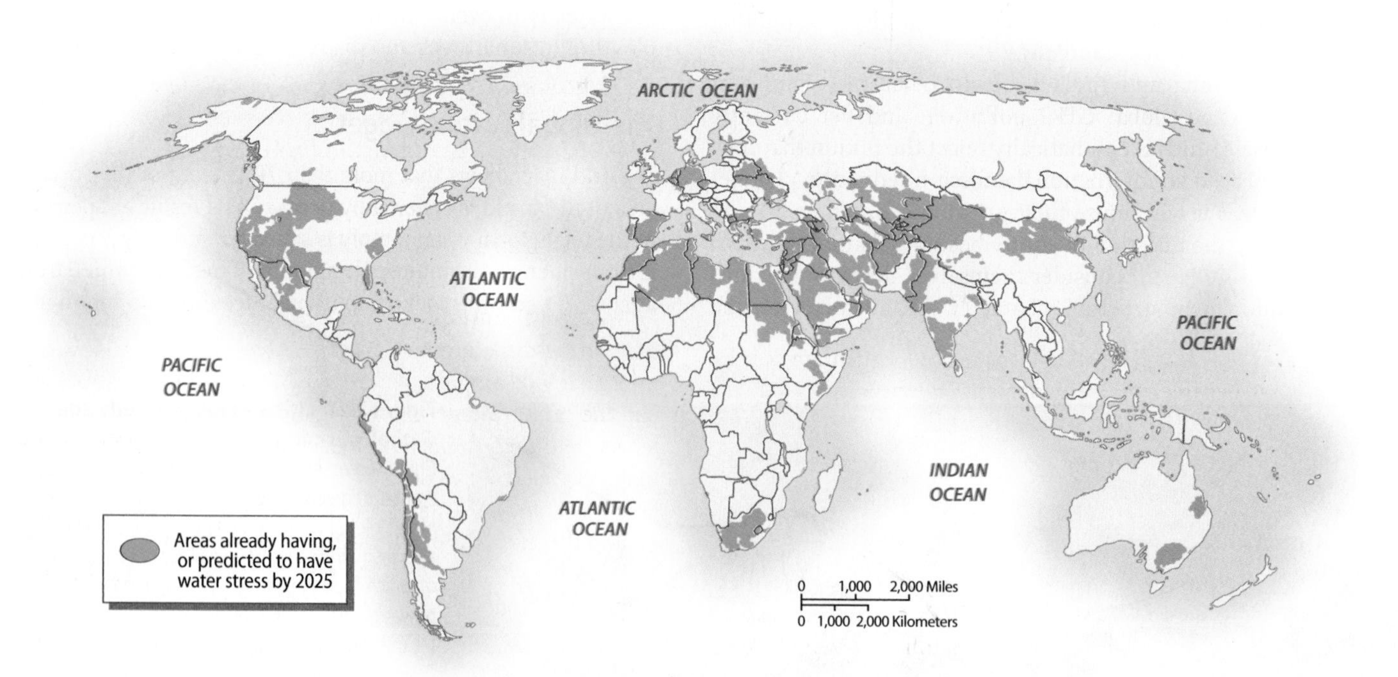

Figure 2.18 Global Water Stress This map shows those parts of the world forecast by the UN to have serious water shortages by the year 2025. In many of the areas, water stress is already evident, namely in northern Africa and Southeast Asia. Global warming will aggravate the water situation as temperatures increase. Africa, in particular, will suffer greatly; projections are that fully 75 percent of its population will face water shortages by 2025.

Figure 2.19 Women and Water In many parts of the world women and young girls spend much of their day providing water for their homes and villages. For young girls this task often interferes with their schooling. These women and girls are fetching water in the Punjab region of India.

cost water. In the recent decades, however, the World Bank and the International Monetary Fund have promoted the privatization of water systems as a condition for providing loans and economic aid to developing countries. While the agency's goals have been laudable, the means have been controversial because, commonly, the international engineering firms that have upgraded rudimentary water systems by installing modern water treatment and delivery technology usually have increased the costs of water delivery to recoup their investment. Although the people may now have access to cleaner and more reliable water, in many cases the price is higher than they can afford, forcing them to either do without or go to other unreliable and polluted sources.

In Cochabamba, Bolivia, the privatization of the water system resulted in a 35 percent increase in water costs. In response, the people rebelled and rioted with demonstrations that became tragically violent. Eventually the water system was returned to public control. Reportedly, however, still today half the city's population is without reliable water.

Water resource management The governmental regulation and allocation of water is usually rife with conflict, often resulting in inefficient uses of this scarce resource as political bodies vie with each to monopolize water supplies. In the western United States, the conflict between or within states is legendary as demonstrated by California internal water politics, or between California and Arizona. In other parts of the arid and semiarid world, where rivers start and end in different independent entities, such as the Nile or the Tigris Rivers, water and war, unfortunately, come close to being synonymous.

HUMAN IMPACTS ON PLANTS AND ANIMALS: The Globalization of Nature

One aspect of Earth's uniqueness compared to other planets is the rich diversity of plants and animals covering its continents. Geographers and biologists think of the cloak of vegetation as the "green glue" that binds together life, land, and atmosphere. Humans are very much a part of this interaction. Not only are we evolutionary products of a specific **bioregion**, an assemblage of local plants and animals (most probably the tropical savanna of Africa), but our human prehistory also included the domestication of certain plants and animals. From this process has come agriculture. Further, humans have changed the natural pattern of plants and animals dramatically by plowing grasslands, burning woodlands, cutting forests, and hunting animals.

However, the pace and magnitude of these changes have accelerated in recent decades, and have led to a crisis in the biological world as ecosystems are devastated and entire species exterminated. Many of these problems can be explained by the globalization of nature and of local ecologies. Until the last half century, tropical forests were primarily homes for small populations of indigenous peoples who made modest demands on local environments for their sustenance and subsistence. But today these same tropical forests are capital for multinational corporations that clear-cut forests for international trade in wood products or search out plants and animals to meet the needs of far-removed populations. For example, Japanese lumber companies cut South American rain forests; German pharmaceutical corporations harvest medicinal plants in Africa; poachers kill North American bears and then sell their gallbladders on the Asian black market as elixirs and sexual stimulants. Unfortunately, the list of destructive human interactions with nature is a long one (see "Global to Local: The Globalization of Bushmeat and Animal Poaching").

Biomes and Bioregions Biome is the biogeographic term used to describe a grouping of the world's flora and fauna into a large ecological province or region; in this book we use the terms **biome** and bioregion interchangeably. These terms are closely connected with climate regions because the major characteristics of a climate region—temperature, precipitation, and seasonality—are also the major factors influencing the distribution of natural vegetation and animals. A brief overview of the most important bioregions follows (Figure 2.20).

Tropical Forests and Savannas

Tropical forests are found in the equatorial climate zones of high average annual temperatures, long days of sunlight throughout the year, and heavy amounts of rainfall. This bioregion covers about 7 percent of the world's land area (roughly the size of the contiguous United States) in Central

Figure 2.20 **World Bioregions** Although global vegetation has been greatly modified by clearing land for agriculture, settlements, and cutting forests for lumber and paper pulp, there is still a recognizable pattern to the world's bioregions, ranging from tropical forests to Arctic tundra. Each bioregion has its own array of ecosystems, containing plants, animals, and insects.

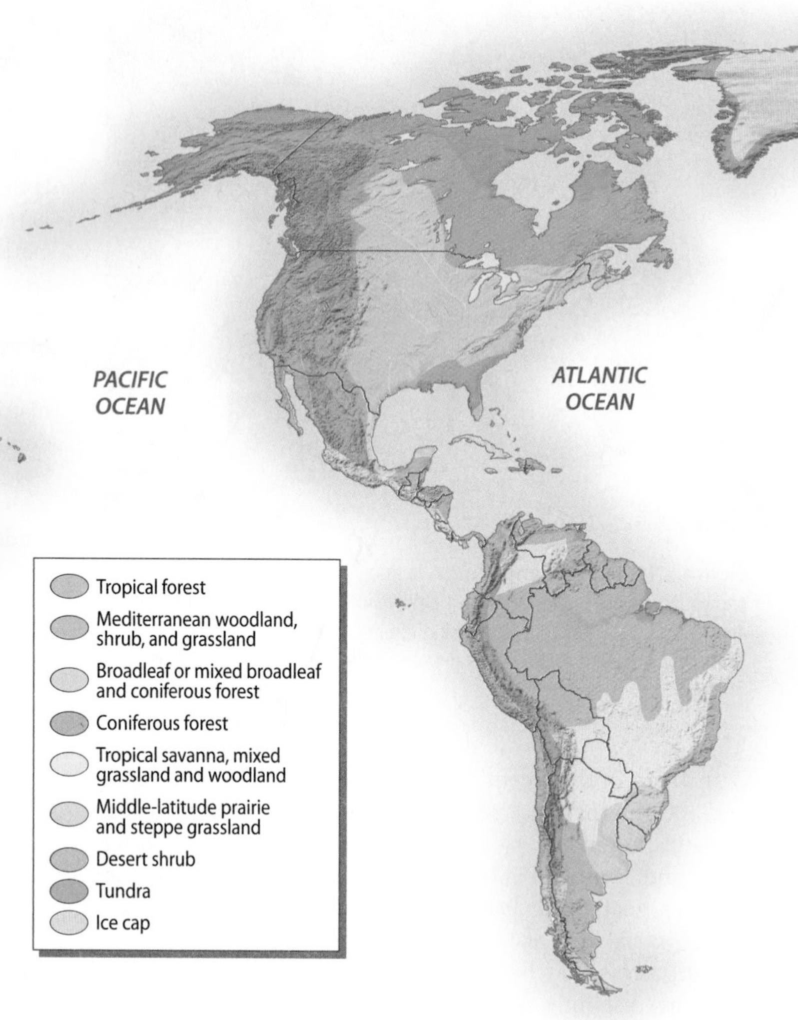

and South America, Sub-Saharan Africa, Southeast Asia, Australia, and on many tropical Pacific islands. More than half of the known plant and animal species live in the tropical forest bioregion, making it the most diverse of all biomes.

The dense tropical forest vegetation is usually arrayed in three distinct levels that are adapted to decreasing amounts of sunlight from the canopy to the forest floor (Figure 2.21). The tallest trees, around 200 feet (61 meters) high, receive open sunlight; the middle level (around 100 feet, or 31 meters) gets filtered sunlight; the third level is the forest floor, where plants can survive with very little direct sunlight. Even though much organic material accumulates on the forest floor in the form of falling leaves, tropical forest soils tend to be very low in stored nutrients. The nutrients are stored instead in the living plants. As a result, tropical forest soils are not well suited for intensive agriculture.

Figure 2.21 **Tropical Rain Forest** As fragile as they are diverse, tropical rain-forest environments feature a complex, multilayered canopy of vegetation. Plants on the forest floor are well adapted to receiving very little direct sunlight.

Deforestation in the Tropics

Tropical forests are being devastated at an unprecedented rate, creating a crisis that tests our political, economic, and ethical systems as the world searches for solutions to this pressing problem. Although deforestation rates differ from region to region, each year an area of tropical forest about the size of Wisconsin or Pennsylvania is denuded. Spatially, almost half of this activity is in the Amazon Basin of South America. However, deforestation appears to be occurring faster in Southeast Asia, where some estimates suggest that logging is occurring three times faster than in the Amazon. If these estimates are accurate, Southeast Asia might be completely stripped of forests within 15 years (Figure 2.22).

An important side effect of tropical deforestation is the release of CO_2 into the atmosphere. Current estimates suggest that fully 20 percent of all human-caused greenhouse gas emissions result from cutting and burning tropical forests.

Behind this widespread cutting of tropical forests lies the recent globalization of commerce in international wood products. Reportedly, Japan was the first to globalize its timber business by reaching far beyond its national boundaries to purchase timber in North and South America and South-

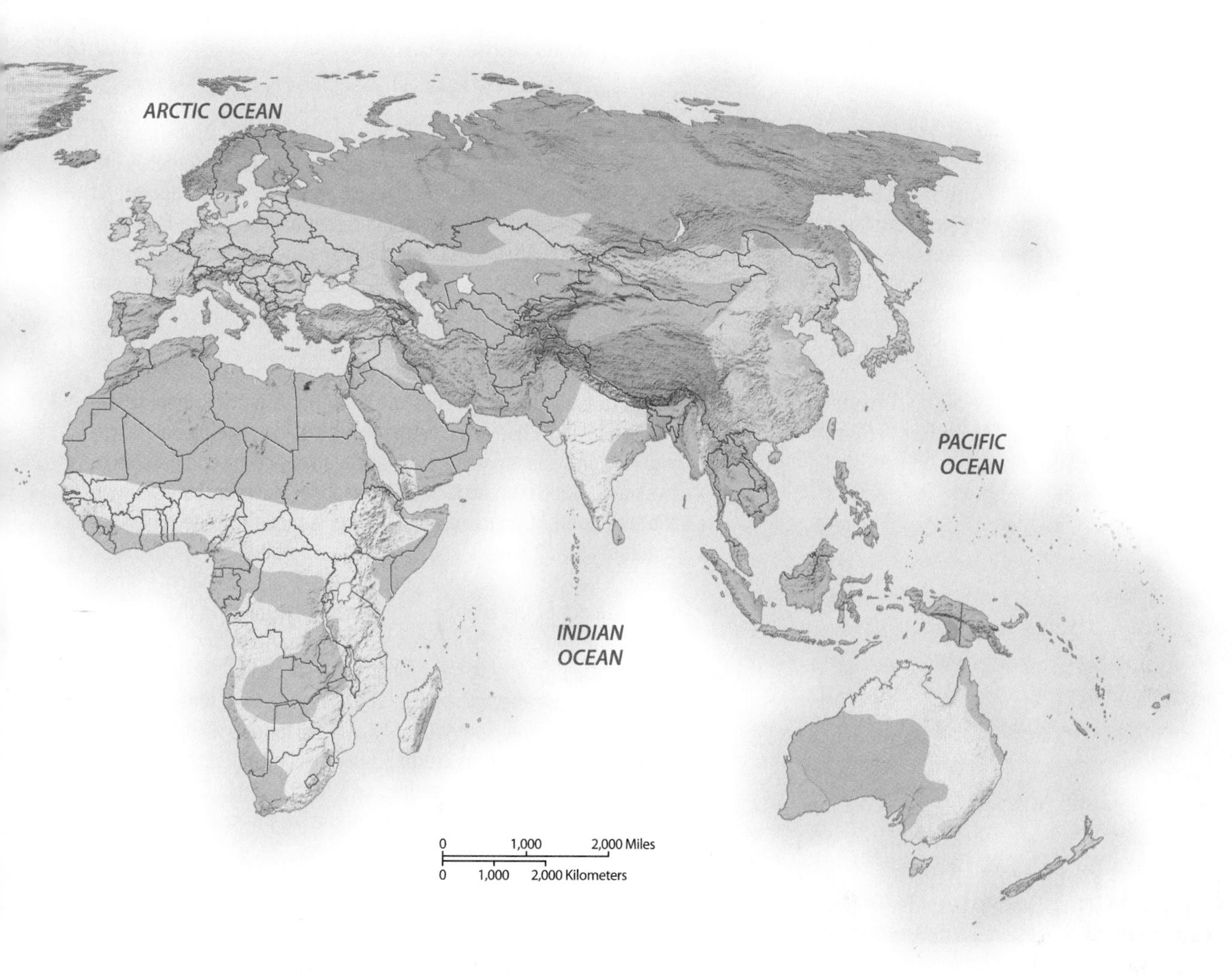

east Asia. This was a two pronged strategy for meeting the increased demand for wood products in Japan while at the same time protecting its own rather limited forests for recreational use. Currently, about one-half of all tropical forest timber is destined for Japan. Unfortunately, much of it is used for throwaway items such as chopsticks and newspapers.

Another factor contributing to the rapid destruction of tropical rain forests is the world's seemingly insatiable appetite for beef. Cattle species originally bred to survive the hot weather of India are now raised on grassland pastures created by cutting tropical forests throughout the tropical world. Unfortunately, because tropical forest soils are poor in nutrients, cattle ranching is not a sustainable activity in these new grassland areas. After a few years, soil nutrients become exhausted, making it necessary to move ranching activities into newly cleared forestland. As a result, more forest is cut, more pasture is created, more soil is destroyed, and the process goes on at the expense of forestlands.

A third factor behind tropical forest destruction is that these forest areas are often the last settlement frontiers for the rapidly growing population of the developing world. Brazil has used much of its interior Amazon rain forest for settlement to ease population pressure along its densely settled eastern coast, where rural lands are controlled by powerful landowners. Brazil had a choice: Either address the troublesome issue of land reform and break up the coastal estates or open up the interior rain forests. It chose the path of least resistance, allowing settlers to clear and homestead

Figure 2.22 Tropical Forest Destruction There are many ill effects resulting from tropical rain-forest destruction including loss of plants and animals, the loss of homeland for native peoples, and the release of large amounts of global warming greenhouse gases. This photo is of a new cacao plantation on Borneo Island, East Malaysia.

GLOBAL TO LOCAL The Globalization of Bushmeat and Animal Poaching

Today, as a result of globalization, plants and animals in one part of the world are commonly threatened by the desires and decisions of people thousands of miles away. A middle-aged Beijing gentleman's sex life, for example, may determine whether an endangered African rhino lives or dies; Europe's ravenous appetite for fish endangers Africa's treasured lions, leopards, and great apes; and Internet Websites and chat rooms facilitate the poaching of endangered species.

In Africa, the market for wild animal meat—"bushmeat" in local parlance—is a major conservation issue. Justin Brashares, a University of California, Berkeley conservation biologist, has revealed a close connection between international fishing fleets along Africa's western coast and the increased killing of bushmeat on land. The relationship is simple: When West Africans cannot find affordable fish in local markets, they turn to bushmeat for their protein supply. And this problem has worsened with the recent appearance of European fishing fleets off Africa's coastline. More specifically, the European fish catch off West Africa has increased 20-fold in the past several decades, resulting in fewer fish for local fishermen, thus leading to a higher kill rate of terrestrial bushmeat as great apes, lions, leopards, elephants, and hippos are poached from national parks and game reserves to fulfill the food needs of West Africa's growing population.

Figure 2.2.1 Bushmeat in Africa This bushmeat market in Gabon displays a collection of bush pigs, duikers (small antelope) and monkeys. Wildlife conservation agencies have found the illegal killing and marketing of bushmeat has increased recently as the local population looks for alternatives to seafood resources depleted by European fishing fleets.

Also taking a heavy toll on the world's wildlife are global poaching networks that constitute a black-market activity comparable to the trade in illegal narcotics and weapons. Much of the activity in the trafficking of poached animal parts is thought to be run by international drug cartels and is estimated to be worth around $20 billion a year. In Brazil, for example, more than 20 million animals are taken out of the country illegally each year, making that country one of the top sources for smuggled fauna. According to government officials, poaching has brought 208 species to the brink of extinction. A blue macaw, one of Brazil's most threatened species, with only 200 left in South America, will sell for over $50,000 in Europe. Further, a toucan is worth $7,000 in the United States, and a collection of rain-forest butterflies sells for $3,000 in China.

As well, the Internet poses "one of the biggest challenges facing CITES (the Convention on International Trade in Endangered Species)," said Paul Todd of the International Fund for Animal Welfare (IFAW) in March 2010. The IFAW found that thousands of endangered species are sold on auction sites, classified ads, and chat rooms, mostly in the United States and Europe, but also in China, Russia, and Australia. Much of what is traded is illegal African elephant ivory, but the group also found exotic birds, pelts from polar bears and leopards, and tiger bones and bear bladders that are used as sexual stimulants.

the tropical forestlands of the interior. Many other countries are doing the same, looking at the vast tracts of tropical rain forest as a safety valve of sorts, a landscape that can be used to temporarily deflect the pressures of land hunger by opening up land for migration and settlement.

Deserts and Grasslands

Large areas of arid and semiarid climate lie poleward of the tropics, and here are found the world's extensive deserts and grasslands. Fully one-third of Earth's land area qualifies as true desert, with annual rainfall of less than 10 inches (25 centimeters). In areas receiving more rainfall, grassy plants appear, often forming a lush cover during the wet season. In North America, the midsection of both Canada and the United States is covered by grassland known as **prairie**, which is characterized by thick, long grasses. In other parts of the world, such as Central Asia, Russia, and Southwest Asia, shorter, less dense grasslands form the **steppe**.

The boundary between desert and grassland has always fluctuated naturally because of changes in climate. During wet periods, grasslands might expand, only to contract once again during drier decades. The transition zone between the two is a precarious environment for humans, as the United States learned during the 1930s, when the semiarid grasslands of the western prairie lands turned into the notorious "Dust Bowl." At that time, thousands of farmers watched their fields devastated by wind erosion and drought—a disaster that led to an exodus from these once-productive lands.

Farming marginal lands may actually worsen the situation, leading to **desertification**, or the spread of desertlike conditions into semiarid areas (Figure 2.23). This has happened on a large scale throughout the world—in Africa, Australia, and South Asia, to name just a few regions. In fact, in the past several decades, an area estimated to be about the size of Brazil has become desertified through poor cropping practices, overgrazing, and the buildup of salt in soils from irrigation. In northern China, an area the size of Denmark became desertified between 1950 and 1980, with the expansion of farming into marginal lands. Historically this region averaged 3 sandstorms a year; today 25 such storms are common each year, with the choking sand and dust traveling eastward into China's large cities.

Temperate Forests

The large tracts of forests found in middle and high latitudes are called *temperate forests*. Their vegetation is different from the low-latitude forests found in the equatorial regions. In temperate forests, two major tree types dominate. One type is softwood coniferous evergreen trees, such as pine, spruce, and fir, which are found in higher elevations and higher latitudes. The second category comprises deciduous trees, which drop their leaves during the winter. Examples are elm, maple, and beech. Because these trees are hardwood, hence harder to mill, they are generally less favored by the timber industry than softwood species.

In North America, conifers dominate the mountainous West, Alaska, and Canada's western mountains, while deciduous trees are found on the Eastern Seaboard of the United States north to New England. There the two tree types intermix before giving way to the softwood forests of Maine and the Maritime provinces of eastern Canada.

In the coniferous forests of western North America, the struggle between timber harvesting and environmental concerns remains controversial. Whereas timber interests argue that increased cutting is the only way to meet high demand for lumber and other wood products, environmental concerns over the protection of habitat for endangered species (the northern spotted owl, for example) have caused the government to make large tracts of forest off-limits to commercial logging (Figure 2.24). Further complicating the future of western forests are global market forces. Many Japanese and Chinese timber firms pay premium prices for logs cut from U.S. and Canadian forests, outbidding domestic firms for these scarce resources. Because these trees are often cut from public lands, this facet of globalization raises an interesting question about the appropriate use of public forests that are maintained by tax money.

In Europe, hardwoods were the natural tree cover in western countries such as France and Great Britain before these once-extensive forests were cleared away for agriculture. In the higher latitudes of Norway and Sweden, coniferous species prevail in the remaining forests. These conifers also populate extensive forests across Germany and Eastern Europe, as well as through Russia into Siberia, creating an almost unbroken landscape of dense forests.

The Siberian forest is a resource that could become a major source of income for financially strapped Russia. Some argue that if the Siberian forests are put on the market for global trade, it will reduce logging pressure on North America's western forests, which, in turn, could make it easier to enact and enforce comprehensive environmental protection in the United States. On the other hand, cutting this large forest would release huge amounts of CO_2 and methane into the atmosphere, aggravating global warming. This example illustrates once again how the multifaceted forces of globalization are intertwined with the fate of local ecosystems.

FOOD RESOURCES: Environment, Diversity, and Globalization

If the human population continues to grow at expected rates, food production must double by 2025 just to provide each person in the world with a basic subsistence diet. Every minute of

Figure 2.23 Desertification Climatic fluctuations and human misuse combine to produce desertification in certain localities, where marginal lands are overcropped or grazed heavily, resulting in expansion of nearby deserts. Globally, many rangelands remain threatened by desertification.

Figure 2.24 Clear-Cut Forest Commercial logging in Washington's Olympic Peninsula has dramatically reshaped this landscape. Throughout the Pacific Northwest, environmental lobbies have successfully restricted logging to protect habitat for endangered species and recreation.

each day, 258 people are born who need food; during that same minute, about 10 acres of cropland are lost because of environmental problems such as soil erosion and desertification. Furthermore, many scientists argue that the interaction between global warming, water problems, and food scarcity will be the defining issues of the next decade (Figure 2.25).

The Green Revolutions

The world's population doubled during the last 40 years of the 20th century and, remarkably, during that same period, global food production also doubled to keep pace with this population explosion. This increase in food production came primarily from the expansion of intensive industrial agriculture into those areas that historically produced only subsistence crops.

More specifically, since 1950, the increases in global food production have come from changes associated with the **Green Revolution**, innovations that involved new agricultural techniques using genetically altered seeds coupled with high inputs of chemical fertilizers and pesticides. The first stage of the Green Revolution combined three processes: first, the change from traditional mixed crops to monocrops, or single fields, of genetically altered, high-yield rice, wheat, and corn seeds; second, intensive applications of water, fertilizers, and pesticides; and, third, additional increases in the intensity of agriculture by reduction in the fallow, or field-resting time, between seasonal crops.

Since the 1970s, a second stage of the Green Revolution has evolved. This phase emphasizes new strains of fast-growing wheat and rice specifically bred for tropical and subtropical climates (Figure 2.26). When combined with irrigation, fertilizers, and pesticides, these new varieties allow farmers to grow two or even three crops a year on a parcel that previously supported only one. Using these methods, India actually doubled its food production between 1970 and 1992.

However, some argue that the practices associated with the Green Revolution carry high environmental and social costs. Because these crops draw heavily on fossil fuels, there has been a 400 percent increase in the agricultural use of fossil fuels during the past several decades. As a result, Green Revolution agriculture now consumes almost 10 percent of the world's annual oil output. Earlier, cheap oil prices facilitated much of this increased agricultural usage of fossil fuel; but today, higher oil prices raise questions about the sustainability of these revolutionary methods. More to the point, the question is whether these higher energy costs will raise the price of food beyond what people in developing countries can afford.

The environmental costs of the Green Revolution have also been significant, resulting in damage to habitat and wildlife from diversion of natural rivers and streams for irrigation, pollution of rivers and water sources by pesticides and chemical fertilizers applied in heavy amounts to fields, and increased regional air pollution from factories and chemical plants that produce these agricultural chemicals.

There is also evidence that the Green Revolution can bring high social costs and disruption to some areas. Because the financial costs to farmers participating in the Green Revo-

Figure 2.25 Food Supplies and Global Warming This farmer in southwest China gathers water to irrigate his crops during one of the worst droughts in China's history. Most predictions are that droughts like this will become increasingly common as global warming worsens, impacting global food supplies to the point of causing considerable hardship for large populations in Africa and Asia.

Figure 2.26 The Green Revolution in India Because of its large and expanding population, India is a region where food supplies may be a problem in the near future. The country doubled its food production between 1970 and 1992, primarily due to the expansion of rice and wheat yields. Nonetheless, the future of food security in this large country is uncertain. In this photo, a woman is planting rice seedlings in the Rishi Valley of Andhra Pradesh state.

lution are higher than with traditional farming, successful farmers must have access to capital and bank loans to purchase hybrid seeds, fertilizers, pesticides, and even new machinery. For those with high social standing, a family support system, or good credit, the rewards can be great; but for those without access to loans or family support, the toll can be heavy, for traditional farmers cannot compete against those raising Green Revolution crops in the regional marketplace.

In some wheat-growing areas of India, the social and economic distance between the well-off Green Revolution farmers and poor traditional farmers has become a major source of economic, social, and political tension. An area where once all shared a common plight has now become a highly stratified society of rich and poor. These social costs can be condoned only because of the pressing need for food in this rapidly growing country.

Problems and Projections

Even though agriculture has been able to keep up with population growth in the past decades, few experts are confident that this pace will continue. Although the regional chapters of this book include fuller discussion of food and agriculture issues, four general points offer a framework for understanding the world's food security issues:

- While overall food production remains an important global issue, it is in fact local and regional problems that often keep people from obtaining needed food supplies. To many experts, the issue is not so much global food production as the widespread poverty and civil unrest at local levels that keep people from growing, buying, or receiving adequate food.
- Political problems are more commonly responsible for food shortages and famines than are natural events such as drought and flooding. Distribution of world food supplies is often highly politicized, starting at the global level and continuing down to the local. Usually food aid goes to political friends and allies, while those allied with enemies go without.
- Globalization is causing dietary preferences to change worldwide, and the implications of these changes could be profound. Currently two-thirds of the world population is primarily vegetarian, eating only small portions of meat since it is usually beyond their means. However, because of recent economic booms in some developing countries, an increasing number of people in these countries are now regularly eating meat, having moved up the food chain from a subsistence diet based on local resources to a global diet containing a greater amount of meat (Figure 2.27). In many cases this change in diet comes not just from new economic prowess but also from changing cultural tastes and values, as people are exposed to new products through economic and cultural globalization.

  There are, however, constraints on the world's ability to supply these new tastes because of the agricultural and environmental support system needed to produce and market meat. According to food experts, the global food production system could sustain only half the current world population if everyone ate the same meat-rich diet of North America, Europe, and Japan because this diet contains three or four times as much meat as the traditional diet of people in developing countries.
- Most food supply experts agree that the two world regions of greatest concern are Africa and South Asia. Until 1970, Africa was self-sufficient in food, but since that time there have been serious disruptions and even breakdowns in the food supply system. These problems stem from rapid population increase and civil disruption from

Figure 2.27 Cattle Ranching in Brazil Growing global demands for meat are reshaping the world's agricultural landscapes. Cattle ranching in western Brazil's rain forest reflects these changing market conditions, and the practice is also introducing new environmental problems.

tribal warfare. As a result, one of every four people faces food shortages in Sub-Saharan Africa. (Chapter 6 contains more detail on these issues). South Asia's future is also problematic. The United Nations predicts that currently almost 200 million people in South Asia suffer from chronic undernourishment. (Further discussion of these problems is found in Chapter 12.)

There is some good news in this otherwise bleak picture. Because population growth rates are generally declining in the industrializing areas of East Asia and food production there is still increasing, the United Nations predicts that the percentage of undernourished people in that region will actually drop by 5 percent in the next 10 years. Similar gains were made in Latin America because of lower population growth and higher agricultural production. World food experts report that in 2010 the proportion of chronically hungry in South America fell to about 6 percent, half the 1990 rate.

Summary

- Global environmental change is driven by human activities. While some changes are intentional and are beneficial for humankind, such as the irrigation of arid lands to increase the world's food supplies, other environmental changes have resulted in unintentional negative consequences. Global warming from fossil fuel consumption is an example.
- Globalization is both a help and a hindrance to world environmental problems. As a positive force, globalization is central to sharing information and increasing public awareness of environmental issues. In addition, many would argue that globalization facilitates a new willingness of countries to work together under the umbrella of international agreements to resolve environmental problems. Such cooperation has led to international treaties on ocean pollution, global warming, and protection of wildlife species.
- The arrangement of tectonic plates on Earth is not only responsible for diverse global landscapes, but also for earthquake and volcanic hazards that threaten the well-being of billions of humans, particularly in the large cities of developing countries in Asia and South America.
- Climate change and global warming resulting from pollution of the atmosphere by greenhouse gases is a by-product of industrialization, both past and present. A major facet of this problem is the burning of fossil fuels—petroleum and coal—that releases CO_2 into the atmosphere. Historically, the developed countries of Europe and North America were the major greenhouse gas producers; today, however, the developing economies of China and India have become major polluters as well.
- Plants and animals throughout the world face an extinction crisis because of habitat destruction from human activities, economic exploitation, and blatantly illegal poaching activities. Tropical forests are a focus of these problems since they contain the most plant and animal species of any bioregion, yet are threatened by numerous forces including the world's demands for wood products, cattle ranching for global meat-eaters, and resettlement of a country's own people.
- Even though the world's population is growing more slowly than in the past, and food supplies are increasing, there are still serious issues of food security for at least half the world's population. A raft of causes lie behind these problems, including governmental policies emphasizing export crops at the expense of crops fulfilling local food needs; international free trade policies and local protectionism that manipulates food markets; and political turmoil and favoritism that hinders effective food distribution within countries.

Key Terms

anthropogenic *(page 56)*
biome *(page 63)*
bioregion *(page 63)*
climate region *(page 55)*
climograph *(page 56)*
continentality *(page 53)*
convection cells *(page 48)*
convergent plate boundary *(page 49)*
desertification *(page 66)*
divergent plate boundary *(page 49)*
global warming *(page 56)*
greenhouse effect *(page 53)*
Green Revolution *(page 68)*
insolation *(page 53)*
maritime climate *(page 53)*
plate tectonics *(page 48)*
prairie *(page 66)*
rift valley *(page 50)*
steppe *(page 66)*
subduction zone *(page 49)*
tectonic plates *(page 48)*
water stress *(page 62)*

Review Questions

1. Explain the role of convection cells in tectonic plate theory.
2. Describe several kinds of tectonic plate boundaries, along with the landforms and landscapes usually associated with these boundaries.
3. Describe the natural greenhouse effect.
4. How do continental and maritime climates differ? What causes this difference?
5. Which major human activities cause global climate change and warming? More specifically, how have greenhouse gases changed because of these activities?

6. Describe the ecological characteristics of a tropical rain forest.
7. What are the different causes of tropical forest deforestation? How are they linked to globalization?
8. What is *desertification*? Why, and where, is it a problem?
9. Describe the characteristics of the Green Revolution.
10. What are the reasons behind food shortages in most parts of the world?

Thinking Geographically

1. What are the most threatening natural hazards in your region—earthquakes, tornadoes, hurricanes, floods, drought? Do research through local agencies or the public library to learn about disaster preparedness plans for your community.
2. Which climate region do you live in? What are the major weather problems faced by people in your area? How do they adjust to these problems?
3. Visit some of the many Internet websites that relate to global warming. Once you have an overview of the different positions, concentrate on one or two of the most contentious issues, such as the debate within the United States between environmentalists and business interests, or the difference in opinion between developed and developing countries. Discuss the different stakeholders and vested interests holding contrasting views.
4. How has the vegetation in your area been changed by human activities in the past 100 years? Have these changes led to the extinction of any plants or animals or placed them on the endangered list? If so, what is being done to protect them or restore their habitat?
5. Study the globalization of wood products by acquainting yourself with the source areas for different items, such as building lumber, paper, furniture, or other items found in a local import store. More specifically, does the lumber used in construction in your area come from Canada or the United States? Which items in local stores come from tropical forests?
6. Conduct a detailed analysis of the food issues in a foreign country of your choice and answer the following questions. In general, has food supply kept up with population growth? Has the country suffered recently from food shortages or famine? If so, were these shortages the result of natural causes, such as drought or floods, or distribution problems resulting from civil disruption? Which segments of the population have food security, and which segments have recurring problems obtaining adequate food? How is this inequity best explained? Finally, have food preferences or diets changed recently? If so, how and why?

Bibliography

Ecological Society of America. 2007. *Frontiers in Ecology and the Environment.* Special Issue: Ecology in an Era of Globalization. May, 5(4).

Gleick, Peter, et al. 2008 *The World's Water: The Biennial Report on Freshwater Resources, 2008–2009.* Washington, DC: Island Press.

IPCC. 2007. *Climate Change 2007: The Scientific Basis. A Report of Working Group 1 of the Intergovernmental Panel on Climate Change.* New York: United Nations Environmental Programme (UNEP).

IPCC. 2007. *Climate Change 2007: Synthesis Report. Summary for Policymakers. A Report of Working Group 1 of the Intergovernmental Panel on Climate Change.* New York: United Nations Environmental Programme (UNEP).

Mann, Michael, and Lee Krump. 2009. *Dire Predictions: Understanding Global Warming.* Upper Saddle River, N.J.: Prentice Hall.

Paarlberg, Robert. 2010. *Food Politics: What Everyone Needs to Know.* New York: Oxford University Press.

Rosegrant, Mark, et al. 2003. "Will the World Run Dry? Global Water and Food Security." *Environment* 45(7), 24–36.

Shiva, Vandana. 2000. *Stolen Harvest: The Hijacking of the Global Food Supply.* Cambridge, MA: South End Press.

Williams, Michael. 2003. *Deforesting the Earth: From Prehistory to Global Crisis.* Chicago: University of Chicago Press.

Log in to www.masteringgeography.com for MapMaster™ interactive maps, geography videos, RSS feeds, flashcards, web links, an eText version of *Diversity Amid Globalization,* and self-study quizzes to enhance your study of The Changing Global Environment.

MapMaster™ presents 13 Place Name and 13 Layered Thematic interactive maps to help students practice and master their geographic literacy, spatial reasoning, and critical thinking skills.

3
North America

The border between the United States and Mexico has become increasingly contentious amid questions over immigration policy, growing drug-related violence, and security concerns.

DIVERSITY AMID GLOBALIZATION

North America plays a pivotal role in economic globalization, both as a leading driver of global change as well as a growing participant in an increasingly interconnected international economy. The region's role as a destination for varied global immigrants has also made it one of the world's most diverse cultural settings, home to a relatively new amalgam of people that have arrived there from every corner of the earth.

ENVIRONMENTAL GEOGRAPHY

Fewer ski days in southern Quebec, widespread drought in the U.S. Southwest, and sea-level rises in many vulnerable coastal communities may be only a few of the many consequences of global warming within the North American region.

POPULATION AND SETTLEMENT

Sprawling suburbs characterize the expanding peripheries of hundreds of North American cities, creating a multitude of challenges including lengthier commutes for urban workers and the loss of prime agricultural lands along the edge of metropolitan areas.

CULTURAL COHERENCE AND DIVERSITY

Cultural pluralism remains strong in North America. Currently, there are about 38 million immigrants living in the United States alone. The tremendous growth in Hispanic and Asian immigrants since 1970 has fundamentally reshaped the region's cultural geography.

GEOPOLITICAL FRAMEWORK

Cultural pluralism continues to shape political geographies within the region. American immigration policy remains hotly contested in the United States and persisting regional and native rights issues confront Canadians.

ECONOMIC AND SOCIAL DEVELOPMENT

The geographical impacts of the recent economic crisis in North America were very uneven. Particularly hard-hit were portions of the Sun Belt in Arizona, Nevada, and California where speculative home construction between 2000 and 2007 was replaced thereafter by a period of increasing foreclosures, falling prices, and rising unemployment.

North America has been fundamentally refashioned by globalization. Stroll down any busy street in Toronto, Tucson, or Toledo, and count the ways in which international products, foods, culture, and economic connections shape the everyday scene. From farmers to software engineers, most North Americans are employed in occupations that are either directly or indirectly linked to the global economy. That close relationship was dramatically illustrated in the global economic downturn in 2008–2009 as many North American workers saw their jobs disappear. Similarly, overseas businesses were hurt by declining demand from North American consumers.

Sizable foreign-born populations in each nation also provide direct links to every part of the world. Tourism brings in millions of additional foreign visitors and billions of dollars that are spent everywhere from Las Vegas to Disney World. In more subtle ways, North Americans embrace globalization in their everyday lives. They consume ethnic foods, tune in to televised international sporting events, enjoy the sounds of salsa and Senegalese music, surf the Internet from one continent to the next, and invest their pensions in global mutual funds.

Globalization also is a two-way street, and North American capital, culture, and power are ubiquitous. By any measure of multinational corporate investment and global trade, the region plays a dominant role that far outweighs its population of 340 million residents. North American consumer goods, information technology, and investment capital circle the globe. In addition, North American foods and popular culture are diffusing globally at a rapid pace. North American music, cinema, and fashion have also spread rapidly around the world. Soaring downtown skylines from Mumbai to Beijing increasingly resemble their North American counterparts.

The North American region encompasses the United States and Canada, a culturally diverse and resource-rich region that has seen unparalleled human modification and economic development over the past two centuries (Figure 3.1). The result is one of the world's most affluent regions, where two highly urbanized, mobile, and rapidly globalizing societies have the highest rates of resource consumption on Earth. Indeed, the region superbly exemplifies a **postindustrial economy** that is shaped by modern technology, by innovative financial and information services, and by a popular culture that dominates both North America and the world beyond.

Politically, North America is home to the United States, the last remaining global superpower. Such status brings the country onto center stage in times of global tensions, whether they are in the Middle East, South Asia, or West Africa. In addition, North America's largest metropolitan area of New York City (22 million people) is home to the United Nations and other global political and financial institutions.

North of the United States, Canada is the other political unit within the region. While slightly larger in area than the United States (3.83 million square miles [9.97 million square kilometers] versus 3.68 million square miles [9.36 million square kilometers]), Canada's population is only about 10 percent that of the United States.

The United States and Canada are commonly referred to as "North America," but that regional terminology can sometimes be confusing. In some geography textbooks the region is called "Anglo America" because of its close and abiding connections with Britain and its Anglo-Saxon cultural traditions. The increasingly visible cultural diversity of the region, however, has discouraged the widespread use of the term in more recent years. While more culturally neutral, the term *North America* also has its problems. As a physical feature, the North American continent commonly includes Mexico, Central America, and often the Caribbean. Culturally, however, the United States–Mexico border seems a better dividing line, although the growing Hispanic presence in the southwestern United States, as well as ever-closer economic links across the border, make problematic even that regional division. While the future may find Mexico even more intimately tied to its northern neighbors, our coverage of the "North American" region concentrates on Canada and the United States, two of the world's largest and most affluent nation-states.

Widespread abundance and affluence characterize North America. The region is extraordinarily rich in natural resources, such as navigable waterways, good farmland, fossil fuels, and industrial metals. The modern scene displays the results of combining that rich resource base with the acquisitive nature of its European colonizers and an accelerating pace of technological innovation. Indeed, contemporary North America displays both the bounty and the price of the development process. On one hand, the region shares the benefits of modern agriculture, globally competitive industries, excellent transport and communications infrastructures, and two of the most highly urbanized societies in the world. The cost of abundance, however, has been high: European settlers all but eliminated native populations, and later populations logged forests, converted grasslands into farms, eroded precious soils, threatened numerous species with extinction, diverted great rivers, and often wasted natural resources. Today, although home to only about 5 percent of the world's population, the region consumes about 25 percent of the world's commercial energy budget and produces carbon dioxide emissions at a per person rate more than 15 times that of India.

Nevertheless, economic growth has vastly improved the standard of living for many North Americans, who enjoy high rates of consumption and varied urban amenities that are the envy of the less-developed world. Satellite dishes, sushi, and shopping malls are within easy reach of most North American residents. Amid this material abundance, however, there are persisting disparities in income and in the quality of life. Poor rural and inner-city populations still struggle to match the affluence of their wealthier neighbors, and poverty has been slow to disappear, despite the unprecedented economic growth of the second half of the 20th century.

North America's unique cultural character also defines the region. The cultural characteristics that hold this region together include a common process of colonization, a heritage of Anglo dominance, and a shared set of civic beliefs in repre-

Figure 3.1 North America North America plays a pivotal role in globalization. The region also contains one of the world's most highly urbanized and culturally diverse populations. With 340 million people and extensive economic development, North America is also one of the largest consumers of natural resources on the planet.

sentative democracy and individual freedom. But the history of the region has also juxtaposed Native Americans, Europeans, Africans, and Asians in fresh ways, and the results are two societies unlike any other (Figure 3.2). Adding to the mix is a popular culture that today exerts a powerful homogenizing influence on North American society.

ENVIRONMENTAL GEOGRAPHY: A Threatened Land of Plenty

North America's physical and human geographies are enormously diverse and intricately linked. Hurricane Katrina's tumultuous arrival along the Gulf coast in August 2005 exemplifies the complexities and significance of those interconnections. Three intertwined variables came into play to explain Katrina's long-lasting impact within the region. First, the storm itself was large and powerful, sweeping inland across southern Louisiana and coastal Mississippi with winds of more than 120 miles per hour. The U.S. Census Bureau estimates that almost 10 million Americans in the region experienced hurricane force winds on the morning the storm came ashore.

Second, the built environment across the region was extensive, with many population centers vulnerable to the effects of a large hurricane. New Orleans was by far the largest urban area in the path of the storm. Unfortunately, for generations, city planners and developers downplayed the hazards of its low-lying, bowl-like setting. Aging levees were breached near Lake Pontchartrain and along several commercial canals, putting 80 percent of the city under water (Figure 3.3). To the east, populated coastal areas of southern Mississippi were also devastated by the storm.

Third, the region's social geography made evacuating urban areas difficult in the face of the storm. Poverty rates across the region are almost double the national norm. Some of the worst-hit areas of New Orleans were more than 80 percent black. The city's poor and elderly residents were its least mobile, forced to ride out the storm in their flooded neighborhoods or in the overcrowded Superdome. Added to this were the problems of coordinating a viable emergency response to the disaster. The resulting deaths, damage, and slow economic recovery were a powerful reminder of how the costs and impacts of a "natural" environmental disaster are inevitably wed to a region's cultural, social, and economic characteristics. More than five years later, New Orleans is still struggling to recover economically from the disaster. Thousands of homes remain empty, the city's public transport system has been painfully slow to recover, problems with chronic illness and depression have grown significantly, and poverty remains widespread. In addition, city planners and federal engineers continue to debate the environmental wisdom of resettling the lowest-lying portions of the city. Inevitably, other storms will come calling on this vulnerable portion of the Gulf Coast.

The Costs of Human Modification

Katrina's story is a reminder that North Americans have modified their physical setting in many ways. Processes of globalization and accelerated urban and economic growth have transformed North America's landforms, soils, vegetation, and climate. Indeed, problems such as acid rain, nuclear waste storage, groundwater depletion, and toxic chemical spills are all manifestations of a way of life unimaginable only a century ago (Figure 3.4).

Figure 3.2 Seattle's Cultural Landscape Practitioners of Falun Dafa, a Chinese meditation practice, begin their day in Seattle's culturally diverse International District just southeast of downtown.

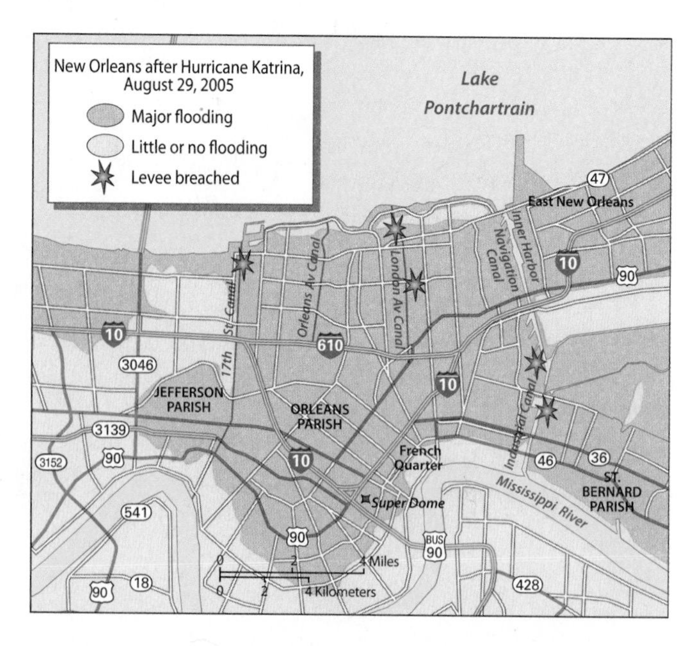

Figure 3.3 New Orleans after Hurricane Katrina Hurricane Katrina put much of New Orleans under water and inflicted particular devastation upon the city's poor, black population. (James M. Rubenstein, *The Cultural Landscape*, 8th ed., © 2008. Reprinted with permission of Pearson Prentice Hall)

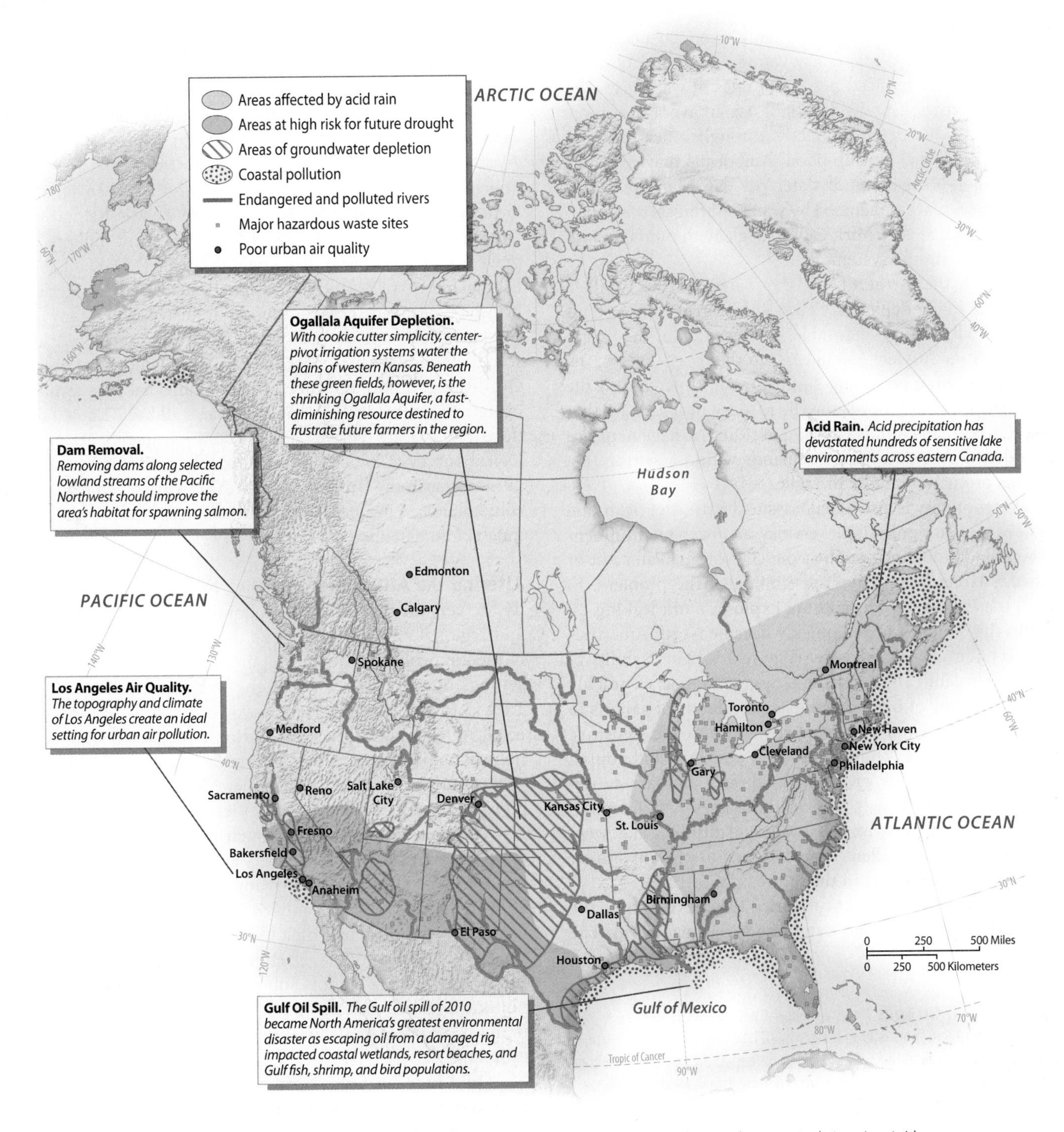

Figure 3.4 Environmental Issues in North America Many environmental issues threaten North America. Acid rain damage is widespread in regions downwind from industrial source areas. Elsewhere, widespread water pollution, cities with high levels of air pollution, and zones of accelerating groundwater depletion pose health dangers and economic costs to residents of the region. Since 1970, however, both Americans and Canadians have become increasingly responsive to the dangers posed by these environmental challenges.

Transforming Soils and Vegetation The arrival of Europeans on the North American continent impacted the region's flora and fauna as countless new species were introduced, including wheat, cattle, and horses. As the number of settlers increased, forest cover was removed from millions of acres. Grasslands were plowed under and replaced with grain and forage crops not native to the region. Widespread soil erosion was increased by unsustainable cropping and ranching practices, and many areas of the Great Plains and South suffered lasting damage.

Managing Water The United States also consume huge amounts of water. While conservation efforts and technology have slightly reduced per capita rates of water use over the past 25 years, city dwellers still use an average of more than 175 gallons daily. Indirectly, through other forms of food and industrial consumption, Americans may consume more than 1,400 gallons of water per day. Many places in North America are threatened by water shortages. Metropolitan areas such as New York City struggle with outdated municipal water supply systems.

Beneath the Great Plains, the waters of the Ogallala Aquifer are being depleted. Central-pivot irrigation systems are steadily lowering water tables across much of the region by as much as 100 feet (30 meters) in the past 50 years, the costs of pumping are rising steadily, and 50 percent of the area's irrigated land may see wells run dry by 2020. Farther west, California's complex system of water management is a reminder of that state's large demands within a setting that remains prone to periodic drought.

Water quality is also a major issue. North Americans are exposed to water pollution every day, and even environmental laws and guidelines, such as the U.S. Clean Water Act or Canada's Green Plan, have not eliminated the problem. In 2010, the Deepwater Horizon rig explosion and leaking oil well in the Gulf of Mexico quickly became North America's greatest environmental disaster (Figure 3.5). The inability of British Petroleum (BP) and the U.S. government to quickly control the flow of oil allowed tens of millions of gallons to escape into the Gulf, damaging sea and bird life, sensitive coastal ecosystems, and the region's resource and tourist economies. The full environmental and human consequences of the spill will take years to assess. New regulations and restrictions on deepwater oil drilling that were imposed after the accident were designed to prevent similar disasters from occurring in the future.

Elsewhere, North America's varied fisheries illustrate the complexities of continental water-resource management. The case of the Asian carp suggests how unpredictable the process can be. Imported from East Asia by southern catfish farmers in the 1970s, the carp removed algae and suspended matter from ponds in the lower Mississippi Valley. But many carp escaped their pens during large floods in the early 1990s. Twenty years later, carp have migrated northward and are now knocking on the door of the Great Lakes as they discover artificial ship canals which link the lakes with the Mississippi system. In 2010, the Army Corps of Engineers spent millions to prevent such an invasion that could decimate the Great Lakes fishing industry. Elsewhere, eastern Canada's cod fisheries, British Columbia's coastal salmon industry, and many of Alaska's freshwater and saltwater fishing grounds have all seen their annual catches decline dramatically in recent years. Overfishing, deadly infections from commercial fish-farming operations, and global climate shifts have all been blamed for the declines. Data released in 2009 by the United States Geological Survey (U.S.G.S.) are also troubling. Fish sampled in 291 streams in the United States revealed widespread mercury contamination in the nation's freshwater fish population, much of it related to coal-burning power plants.

Figure 3.5 Gulf Oil Spill North Americans witnessed the area's greatest environmental disaster in history in 2010 as millions of barrels of oil from a leaking well in the Gulf of Mexico damaged the region's sensitive ecosystems, fishing industry, and tourism economy. This view shows oil washing up on the beach at Gulf Shores, Alabama.

Altering the Atmosphere Indeed, North Americans modify the very air they breathe; in doing so, they change local and regional climates as well as the chemical composition of the atmosphere. The development associated with urban settings often produces nighttime temperatures some 9 to 14°F (5 to 8°C) warmer than nearby rural areas. At the local level, industries, utilities, and automobiles contribute carbon monoxide, sulfur, nitrogen oxides, hydrocarbons, and particulates to the urban atmosphere. While some of the region's worst offenders are U.S. cities such as Houston and Los Angeles, Canadian cities such as Toronto, Hamilton, and Edmonton also experience significant problems of air quality. In 2009, the U.S. Environmental Protection Agency (E.P.A.) issued a report identifying almost 600 urban American neighborhoods in which elevated cancer rates were associated with air pollution. High-impact localities such as Los Angeles or the Illinois suburbs of St. Louis might have pollution-related cancer rates that are 500 times higher than for rural, clean-air localities in settings such as central Montana.

On a broader scale, North America is plagued by **acid rain**, industrially produced sulfur dioxide and nitrogen oxides in the atmosphere that damage forests, poison lakes, and kill fish. Many acid rain producers are located in the Midwest and southern Ontario, where industrial plants, power-generating facilities, and motor vehicles contribute atmospheric pollution. Prevailing winds transport the pollutants and deposit damaging acid rain and snow across the Ohio Valley, Appalachia, the northeastern United States, and eastern Canada (see Figure 3.4).

Air pollution is also going global, freely drifting into Canada and the United States on prevailing winds. One recent estimate suggests at least 30 percent of the region's ozone comes from beyond its borders. Both China and Mexico are major contributors of airborne pollutants.

The Price of Affluence

Globalization has brought many benefits to North America, but with the accompanying urbanization, industrialization, and heightened consumption, the region is also paying an environmental price for its affluence. Energy consumption within the region, for example, remains extremely high, imposing a growing list of environmental and economic costs both within North America and beyond. North Americans consume energy at a per capita rate almost double that of the Japanese and more than 16 times that of India's population. For all its economic growth and material wealth, the 20th century in North America will also be remembered for its toxic waste dumps, frequently unbreathable air, and wildlands lost to development.

Still, many environmental initiatives in the United States and Canada have addressed local and regional problems (see "Environment and Politics: Deconstructing a Landscape: Dam Removal in the United States"). For example, the improved water quality of the Great Lakes over the past 30 years is an achievement to which both nations contributed and from which both benefit. Conservation practices have greatly reduced water- and wind-related soil erosion in North America, although experts estimate that one-third of U.S. and one-fifth of Canadian cropland are still at high-to-severe risk of future erosion. Tougher air quality standards have also selectively reduced emissions in many North American cities. Similarly, the U.S. Superfund program (begun in 1980) and Canada's Environmental Protection Act (CEPA—begun in 1988)

ENVIRONMENT AND POLITICS — Deconstructing a Landscape: Dam Removal in the United States

For most of its history, the United States government has supported policies to increasingly control the flow of the nation's streams and rivers. To that end, more than 75,000 dams have been constructed in the country's history through agencies such as the U.S. Army Corps of Engineers. They support a myriad of flood control, irrigation, and hydroelectric projects.

Since the 1970s, however, new environmental and political initiatives have encouraged the reversal of these long-standing policies. The 1973 passage of the Endangered Species Act mandated that federal agencies preserve threatened plants and animals along critical river habitat. The legislation became the political genesis of the dam removal movement. A varied political coalition of environmentalists, fishermen, and Native American groups have increasingly pressured governmental agencies to remove dams that many argue have outlived their usefulness and do more harm than good. Other groups have emerged as dam supporters, including boating enthusiasts, nearby farmers, and consumers of hydroelectric power.

By 2009, American Rivers (an environmental organization) reported that more than 200 dams have been removed nationwide. Pennsylvania's Saucon Park Dam, for example, was targeted for removal in the upper drainage of the Lehigh River. Built in the 1920s to create a reservoir for recreation, the aging dam has recently depleted fish stocks, contributed to stream bank erosion, and raised fears about catastrophic failure and flooding. The dam's successful removal has restored fish habitat for brown trout and brook trout populations and spurred the area's recreation economy.

In the Olympic Mountains of western Washington State, a pair of dams along the Elwha River (built in 1913 and 1929) is also targeted by the dam removal movement (Figure 3.1.1). Supporters argue that the value of the restored salmon habitat for 400,000 fish will far exceed the modest value of the hydroelectric power generated by the dams. While local political resistance has stalled the effort for years, recent federal stimulus money has jumpstarted the project and the dams may be fully removed by 2013 or 2014.

Figure 3.1.1 Dam along the Elwha River, Washington This dam along the Elwha River in western Washington's Olympic Mountains is scheduled for removal by 2014, helping to restore the watershed's salmon fisheries and expand the area's recreational opportunities.

An even larger initiative was approved along California's Klamath River in 2009, preparing the way for the removal of four dams owned by PacifiCorp, an electric utility. The removal is scheduled for completion by 2020 and will restore more than 300 miles of prime West-coast salmon habitat, aid nearby tribal communities, and stimulate the region's fishing and tourist economy.

These modest successes, dam removal advocates argue, may pave the way for larger initiatives. The Colorado River's Glen Canyon Dam is often cited as the biggest target, but given both the political and logistical challenges involved in such a mammoth effort, nothing is likely to happen in the near future. Still, the deconstruction of America's dams may be just beginning and it marks a policy shift destined to fundamentally rework North America's river systems and in the process restore streams to some of their former environmental health.

Figure 3.6 **Tehachapi Wind Farm, California** Joshua trees and spinning turbines mingle in the high desert at the Tehachapi Wind Farm west of Mojave. The area's 5,000 turbines generate enough electricity to serve the residential needs of 350,000 Californians annually. Wind energy represents a growing percentage of North America's energy budget.

have significantly cleaned up hundreds of North America's toxic waste sites.

Perhaps most importantly, there is increasing support in North America for green industries and technologies. The growing popularity of **sustainable agriculture** exemplifies the trend, where organic farming principles, a limited use of chemicals, and an integrated plan of crop and livestock management combine to offer both producers and consumers environmentally friendly alternatives. By 2006, for example, 4 percent of U.S. apple and lettuce production was certified organic by the United States Department of Agriculture.

In the United States alone, Americans invested almost $30 billion in alternative energy sources in 2006 and recent court rulings and political initiatives have encouraged more environmentally friendly innovation (Figure 3.6). While fossil fuels will continue to dominate U.S. energy consumption in the early 21st century, the growing technological and economic appeal of **renewable energy sources**, such as hydroelectric, solar, wind, geothermal are likely to fundamentally rework North America's economic geography in coming years as policymakers, industrial innovators and consumers are all attracted to their enduring availability and potentially lower environmental costs (Figure 3.7). In addition, the U.S. Recovery Act, passed in 2009 to stimulate economic activity, earmarked approximately $80 billion to renewable energy initiatives. Domestic production of renewable energy is projected to double between 2009 and 2012.

A Diverse Physical Setting

North America's complex landscape is dominated by vast interior lowlands bordered by more mountainous topography in the western portion of the region (see Figure 3.1). In the eastern United States, extensive coastal plains stretch

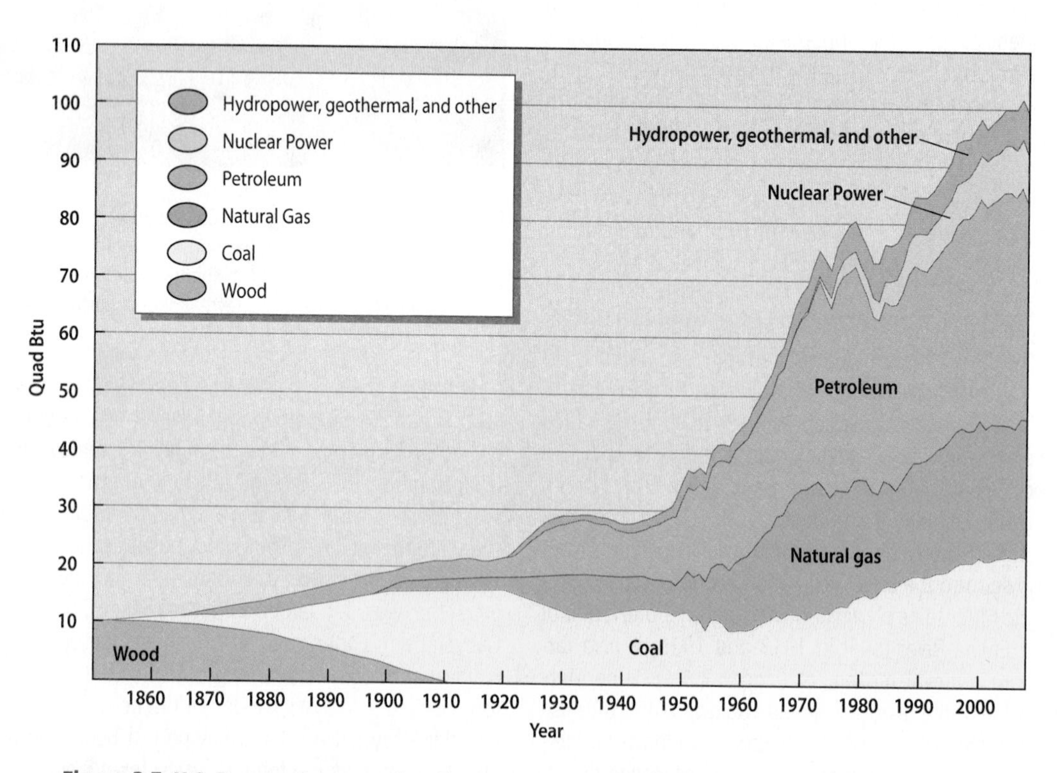

Figure 3.7 **U.S. Energy Consumption** The growing popularity of fossil fuels is evident in U.S. energy consumption during the late 19th century as coal, oil, and then natural gas supplanted wood consumption. Nuclear power and an assortment of renewable energy sources are poised to play a larger role during the 21st century. (James M. Rubenstein, *Contemporary Human Geography*, 2010. Reprinted with permission of Pearson Prentice Hall)

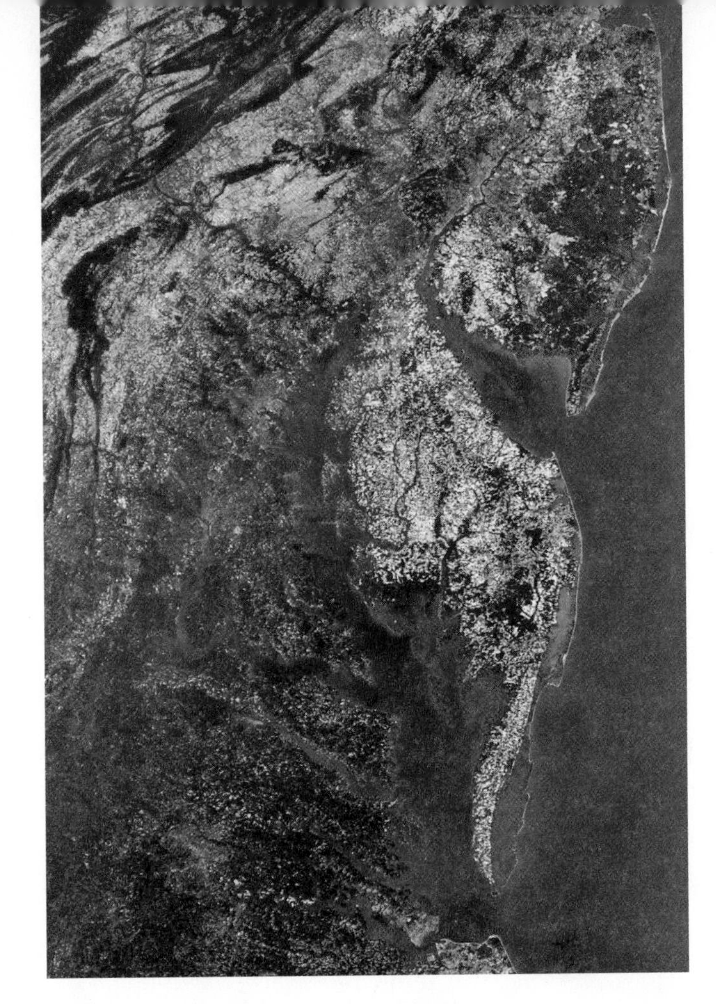

Figure 3.8 **Satellite Image of Chesapeake Bay** This view of the Mid-Atlantic coast reveals the intricate shoreline of Chesapeake Bay (lower center). The coast sector is characterized by drowned river valleys, barrier islands, and sandy beaches. The Piedmont zone and Appalachian Highlands appear to the northwest.

from southern New York to Texas and include a sizable portion of the lower Mississippi Valley. The Atlantic coastline is complex and is made up of drowned river valleys, bays, swamps, and low barrier islands (Figure 3.8). The nearby Piedmont, which is the transition zone between nearly flat lowlands and steep mountain slopes, consists of rolling hills and low mountains that are much older and less easily eroded than the lowlands. West and north of the Piedmont are the Appalachian Highlands, an internally complex zone of higher and rougher country reaching altitudes from 3,000 to 6,000 feet (915 to 1,829 meters). Far to the southwest, Missouri's Ozark Mountains and the Ouachita Plateau of northern Arkansas resemble portions of the southern Appalachians. Much of the North American interior is a vast lowland extending east–west from the Ohio River valley to the Great Plains, and north–south from west central Canada to the coastal lowlands near the Gulf of Mexico. Glacial forces, particularly north of the Ohio and Missouri rivers, have actively carved and reshaped the landscapes of this lowland zone.

In the West, mountain-building (including large earthquakes and volcanic eruptions), alpine glaciation, and erosion produce a regional topography quite unlike that of eastern North America. The Rocky Mountains reach more than 10,000 feet (3,048 meters) in height and stretch from Alaska's Brooks Range to northern New Mexico's Sangre de Cristo Mountains (Figure 3.9). West of the Rockies, the Colorado Plateau is characterized by highly colorful sedimentary rock eroded into spectacular buttes and mesas. Nevada's sparsely settled basin and range country features north–south-trending mountain ranges alternating with structural basins with no outlet to the sea. North America's western border is marked by the mountainous and rain-drenched coasts of southeast Alaska and British Columbia; the Coast Ranges of Washington, Oregon, and California; the lowlands of the Puget Sound (Washington), Willamette Valley (Oregon), and Central Valley (California); and the complex uplifts of the Cascade Range and Sierra Nevada.

Patterns of Climate and Vegetation

North America's climates and vegetation are highly diverse, mainly as a response to the region's size, latitudinal range,

Figure 3.9 **Rocky Mountains** Montana's Glacier National Park reveals the characteristic signatures of alpine glaciation that are found in many portions of the Rocky Mountain region, both in the United States and in Canada.

and varied terrain (Figure 3.10). Much of North America south of the Great Lakes is characterized by a long growing season, 30 to 60 inches (76.2 to 152.4 centimeters) of precipitation annually, and a deciduous broadleaf forest (later cut down and replaced by crops). From the Great Lakes north, the coniferous evergreen or **boreal forest** dominates the continental interior. Near Hudson Bay and across harsher northern tracts, trees give way to **tundra**, a mixture of low shrubs, grasses, and flowering herbs that grow briefly in the short growing seasons of the high latitudes.

Drier continental climates found from west Texas to Alberta feature large seasonal ranges in temperature and unpredictable precipitation that averages between 10 and 30 inches (25.4 and 76.2 centimeters) annually. The soils of much of this region are fertile, and originally supported **prairie** vegetation dominated by tall grasslands in the East and by short grasses and scrub vegetation in the West. Western North American climates and vegetation are greatly complicated by the region's many mountain ranges. The Rocky Mountains and the intermontane interior experience the typical seasonal variations of the middle latitudes, but patterns of climate and vegetation are greatly modified by the effects of topography. Farther west, marine west coast climates dominate north of San Francisco, while a dry summer Mediterranean climate occurs across central and southern California.

Global Warming in North America

Global warming has already profoundly reshaped North America and accelerating rates of human-influenced climate change appear destined to complicate the situation further. High latitude and alpine environments are particularly vulnerable to global warming. Changes in arctic temperatures, sea ice, and sea levels have increased coastal erosion, impacted migrating whale and polar bear populations, and stressed traditional ways of life for Eskimo and Inuit populations (see "Global to Local: Alaska's Climate Refugees"). In western North American mountains, expanding mountain

GLOBAL TO LOCAL Alaska's Climate Refugees

For most residents of North America, global warming is an occasional news item and a topic of muddled political and scientific debate. But for a growing number of native peoples in Alaska, the local, tangible consequences of global warming are literally lapping at the front door. In fact, a 2009 federal government report identified 31 Alaskan villages that face "imminent threats" to their very survival. A dozen settlements have already begun a process of relocation to localities deemed less vulnerable to the vagaries of warming in arctic and subarctic settings that have witnessed some of the most dramatic short-term environmental shifts on the planet. These migrants may be early examples of what a recent Environmental Justice Foundation report called "climate refugees," a growing army of perhaps 150 million people worldwide who may soon find themselves forced to move in a variety of vulnerable environmental settings. Ironically, many of these refugees are indigenous peoples in marginal hinterlands who have contributed almost nothing to greenhouse gas emissions.

Figure 3.2.1 Coastal Village in Western Alaska The village of Shishmaref reveals the costs of global warming in many vulnerable high-latitude environments.

For the native Alaskans, the local impacts of global warming have come from multiple directions, threatening their economic viability and cultural survival. Western Alaska's coastal Eskimo villages of Shishmaref (population 600) and Newtok (population 300) illustrate the daunting challenges that residents face. One major problem is that the permafrost (permanently frozen ground) is thawing, causing homes to shift and buckle and increasing the vulnerability of roads and structures that are found near unstable coastal and beach environments. The result is a landscape filled with mud and makeshift boardwalks (Figure 3.2.1). Making matters worse, seasonal sea ice that once protected coastal villages from heightened wave erosion is melting earlier in the summer and freezing later in the fall. The changing environmental conditions have also disrupted seal, polar bear, and caribou populations, which in turn have reduced the size and dependability of traditional hunts. These transformations are neither hypothetical nor subtle. Villagers are literally witnessing the disappearance of their homes as neighborhoods sink into the sea.

While several local residents attended the Copenhagen climate conference in 2009, most villagers do not hold out much hope for immediate relief. They point out that the same federal government that originally forced them into a sedentary coastal lifestyle (traditionally, most were nomadic peoples) now resists helping them relocate from their doomed villages or to address the larger issues of global warming. Their identities now rooted in these places, these 21st century climate refugees have become early casualties in a land where the costs of global warming have arrived suddenly and with drastic consequences.

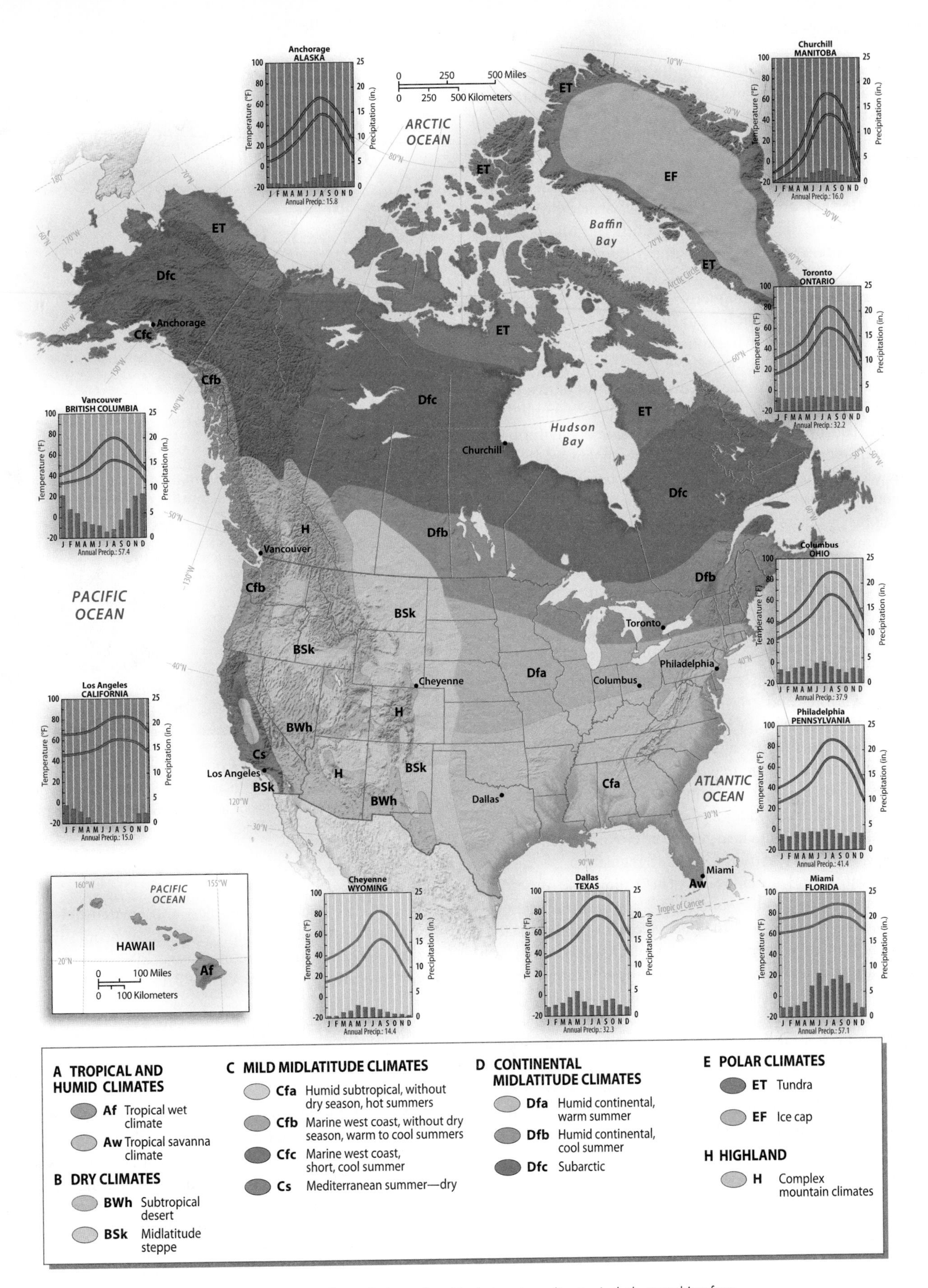

Figure 3.10 Climate Map of North America North American climates include everything from tropical savanna (Aw) to tundra (ET) environments. Most of the region's best farmland and densest settlements lie in the mild (C) or continental (D) midlatitude climate zones.

pine beetle populations are rapidly infesting lodgepole and whitebark pine forests in the northern Rocky Mountains. A 2009 study of 76 stands of old-growth forest throughout western North America confirmed that tree mortality rates have dramatically risen in the past 15 years and higher temperatures were cited as the primary culprit. Elsewhere, Mexican butterflies are breeding in Texas, armadillos (earlier natives of Texas) are marching into southern Illinois, and exotic fish species are migrating north into the Chesapeake Bay and offshore Southern California.

The cumulative long-term consequences of climate change for North Americans are enormous. Many coastal localities, especially low-lying zones in the arctic, along the East Coast, and in the Gulf of Mexico, will be increasingly vulnerable to rising sea levels. Basic redistributions of native plants, animals, and crops are already underway. For example, high-latitude distributions of tundra and permafrost environments may shift more than 300 miles (480 km) poleward by 2050, dramatically impacting wildlife populations and human settlements in these regions. Many of North America's spectacular alpine glaciers are rapidly disappearing, often with important implications for downstream fisheries and population centers (Figure 3.11). For northern U.S. and Canadian farmers, longer growing seasons may open up new possibilities for agriculture, but weeds and plant pathogens will migrate north with the crops. The Great Lakes region may become wetter by the end of the century. But both hotter and drier conditions are likely in the Southwest, portending more western wildfires and growing resource conflicts in that rapidly growing region of the United States (see Figure 3.4).

In the United States, a growing federal political response to global climate change may produce more dramatic policy responses. For example, a ruling in 2009 by the E.P.A. declared that carbon dioxide (as a greenhouse gas) posed a potential danger to human health and the environment, opening the way for more restrictive federal regulation of utilities, power plants, and automobiles. The ruling also raises the prospect of a politically contentious **cap-and-trade energy policy** in the United States in which overall carbon emissions would be limited or *capped* by a central authority within a system that would allow polluters to buy and sell carbon credits (or *allowances*) in an emissions-trading marketplace. Over time, the total cap on carbon emissions is lowered, polluters are encouraged to utilize lower-cost alternatives, and successful participants that reduce their emissions can profit from selling their unused credits in the marketplace. Such a system can also go global if multiple participants agree on emissions standards and long-term goals.

Figure 3.11 Glaciers in Retreat With prospects for further global warming, the outlook is bleak for many of North America's alpine glaciers. This pair of images records change at Grinnell Glacier, in Montana's Glacier National Park, between 1940 and 2006.

POPULATION AND SETTLEMENT: Reshaping a Continental Landscape

The North American landscape is the product of four centuries of extraordinary human change. During that period, Europeans, Africans, and Asians arrived in the region, disrupted Native American peoples, and created new patterns of human settlement. Today, 340 million people live in the region, and they are some of the world's most affluent and highly mobile populations (Table 3.1).

Modern Spatial and Demographic Patterns

Metropolitan clusters dominate North America's population geography, producing strikingly uneven patterns of settlement across the region (Figure 3.12). In Canada, about 90 percent of the population is found within 100 miles (160 km) of the U.S. border. Within this broad region, Canada's "Main Street" corridor contains most of that nation's urban population, led by the cities of Toronto (5.5 million) and Montreal (3.8 million). The federal capital of Ottawa (1.2 million) and the industrial center of Hamilton (700,000) are also within the Canadian urban corridor. **Megalopolis**, the largest settlement agglomeration in the United States, includes the Washington, DC/Baltimore area (8.3 million), Philadelphia (6.4 million), New York City (22 million), and

TABLE 3.1 Population Indicators

Country	Population (millions) 2010	Population Density (per square kilometer)	Rate of Natural Increase (RNI)	Total Fertility Rate	Percent Urban	Percent <15	Percent >65	Net Migration (Rate per 1000) 2005–10[a]
Canada	34.1	3	0.4	1.7	80	17	14	6.3
United States	309.6	32	0.6	2.0	79	20	13	3.3

[a]*Net Migration Rate from the United Nations, Population Division, World Population Prospects: The 2008 Revision Population Database.*

Source: Population Reference Bureau, World Population Data Sheet, 2010.

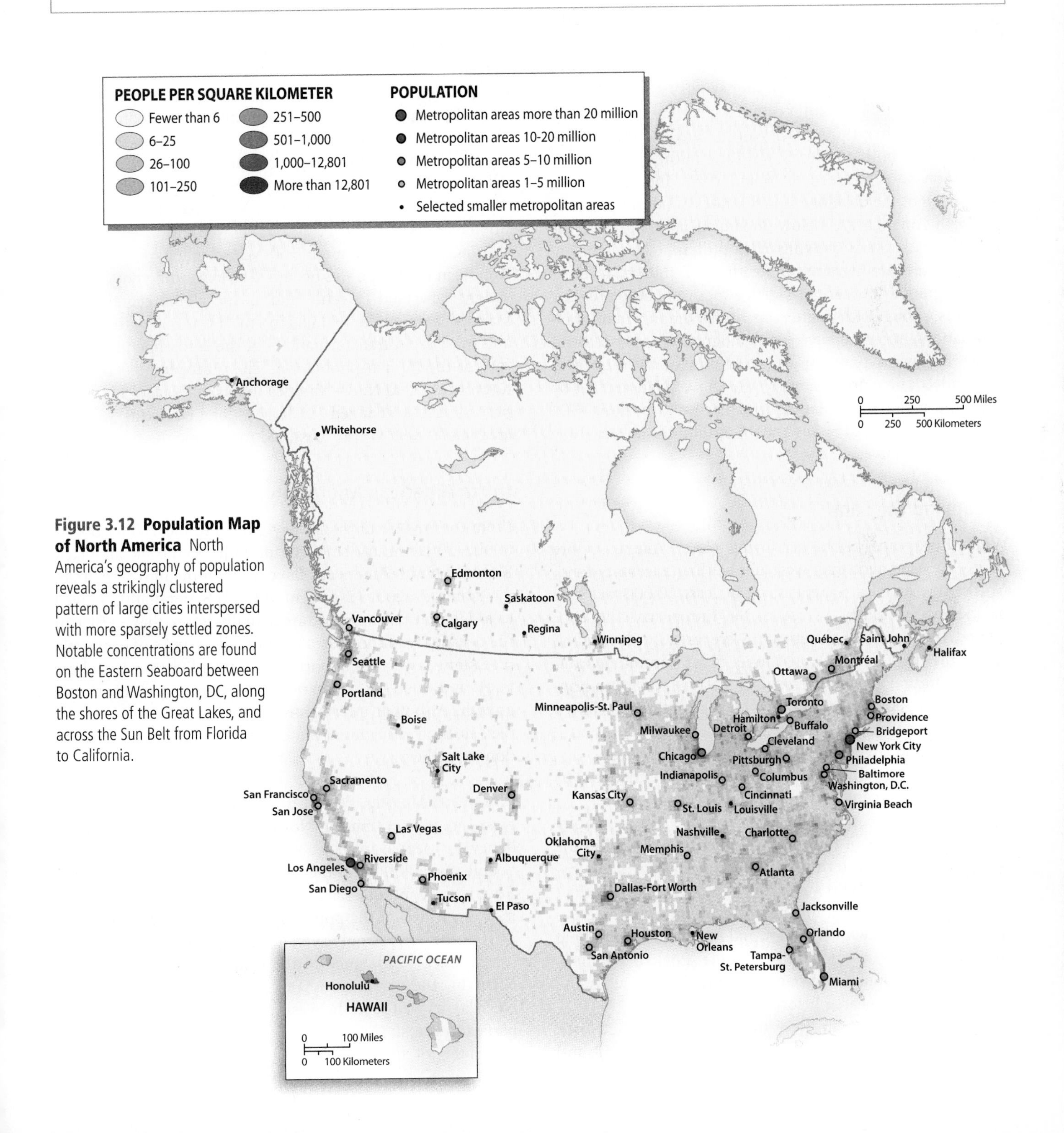

Figure 3.12 Population Map of North America North America's geography of population reveals a strikingly clustered pattern of large cities interspersed with more sparsely settled zones. Notable concentrations are found on the Eastern Seaboard between Boston and Washington, DC, along the shores of the Great Lakes, and across the Sun Belt from Florida to California.

the greater Boston metropolitan area (7.5 million). Beyond these two national core areas, other sprawling urban centers cluster around the southern Great Lakes, various parts of the South, and along the Pacific Coast.

North America's population has increased greatly since the beginning of European colonization. Before 1900, high rates of natural increase produced large families. In addition, waves of foreign immigration swelled settlement, a pattern that continues today. In Canada, a population of fewer than 300,000 Indians and Europeans in the 1760s grew to an impressive 3.2 million a century later. For the United States, a late colonial (1770) total of around 2.5 million increased more than 10-fold to more than 30 million by 1860. Both countries saw even higher rates of immigration in the late 19th and 20th centuries, although birthrates gradually fell after 1900. After World War II, birthrates rose once again in both countries, resulting in the "baby boom" generation born between 1946 and 1965. Today, however, as in much of the developed world, rates of natural increase in North America are below 1 percent annually, and the overall population is growing older. Still, the region continues to attract immigrants: more than 37 million foreign-born migrants now live in North America. These growing numbers, along with higher birthrates among immigrant populations, recently have led demographic experts to increase long-term population projections for the 21st century. Indeed, new predictions by the United Nations that by 2050 the region's population will reach 448 million (404 million in the United States and 44 million in Canada) may prove conservative.

Occupying the Land

When Europeans began occupying North America more than 400 years ago, they were not settling an empty land. North America was populated for at least 12,000 years by peoples as culturally diverse as the Europeans who conquered them. Native Americans were broadly distributed across the region and adapted to its many natural environments. Cultural geographers estimate Native American populations in 1500 CE at 3.2 million for the continental United States and another 1.2 million for Canada, Alaska, Hawaii, and Greenland. In many areas, European diseases and disruptions reduced these Native American populations by more than 90 percent as contacts increased.

The first stage of a dramatic new settlement geography began with a series of European colonies, mostly within the coastal regions of eastern North America (Figure 3.13). Established between 1600 and 1750, these regionally distinct societies were anchored on the north by the French settlement of the St. Lawrence Valley and extended south along the Atlantic Coast, including separate English colonies. Scattered developments along the Gulf Coast and in the Southwest also appeared before 1750.

The second stage in the Europeanization of the North American landscape took place between 1750 and 1850, and it was highlighted by settlement of much of the better agricultural land within the eastern half of the continent. Pioneers surged across the Appalachians, following the American Revolution (1776) and a series of Indian conflicts. They found much of the Interior Lowlands region almost ideal for agricultural settlement. Much of southern Ontario, or Upper Canada, was also opened to widespread development after 1791.

The third stage in North America's settlement expansion picked up speed after 1850 and continued until just after 1910. During this period, most of the region's remaining agricultural lands were settled by a mix of native-born and immigrant farmers. Farmers were challenged and sometimes defeated by drought, mountainous terrain, and short northern growing seasons. In the American West, settlers were attracted by opportunities in California, the Oregon country, Mormon Utah, and the Great Plains. In Canada, thousands occupied southern portions of Manitoba, Saskatchewan, and Alberta. Gold and silver discoveries led to initial development in areas such as Colorado, Montana, and British Columbia's Fraser Valley.

Incredibly, in a mere 160 years, much of the North American landscape was occupied as expanding populations sought new land to settle and as the global economy demanded resources to fuel its growth. It was one of the largest and most rapid transformations of the landscape in the history of the human population. This European-led advance forever reshaped North America in its own image, and in the process it also changed the larger globe in lasting ways by creating a "New World" destined to reshape the Old.

North American Migration

From the mythic days of Davy Crockett and Daniel Boone to the 20th-century sojourns of John Steinbeck and Jack Kerouac, North Americans have been on the move. In 2008, for example, about 12 percent of the U.S. population moved to a different residence, a rate actually lower than earlier in the decade. Migration rates tend to fall in times of economic recession (fewer job opportunities, inability to sell real estate) and increase in periods of more rapid economic growth. Although interregional population flows are complex in both the United States and Canada, several trends dominate the picture.

Westward-Moving Populations The most persistent regional migration trend in North America has been the tendency for people to move west. Indeed, for two centuries people have followed the setting sun, and many North Americans continue that pattern to the present. By 1990, more than half of the population of the United States lived west of the Mississippi River, a dramatic shift from colonial times. Since 1990, some of the fastest-growing states have been in the American West (including Arizona, and Nevada), as well as in the western Canadian provinces of Alberta and British Columbia.

Much of the above-average growth in the Mountain States since 2000 was fueled by new job creation in high-

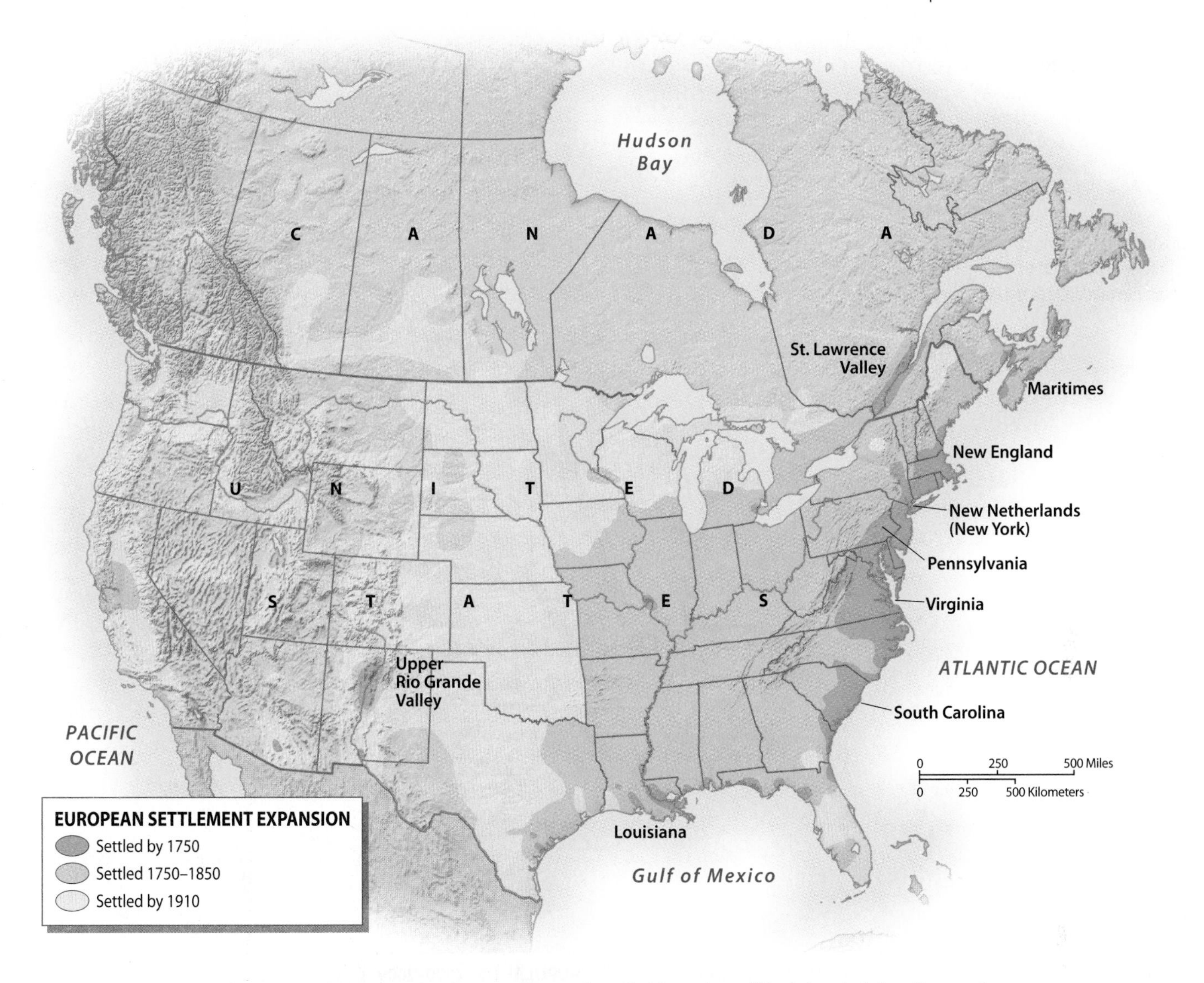

Figure 3.13 European Settlement Expansion Sizable portions of North America's East Coast and the St. Lawrence Valley were occupied by Europeans before 1750. The most remarkable surge of settlement occurred during the next century as Europeans opened vast areas of land and dramatically disrupted Native American populations.

technology industries and services, as well as by the region's scenic, recreational, and retirement amenities (Figure 3.14). Many of these new migrants include out-bound Californians. The sustained move to the Interior West has many implications: the demand for water in the arid region continues to grow; the mix of natives and recent migrants creates cultural tensions; and the area's growing political and economic power is redefining its traditionally peripheral role in national affairs. New census data in 2010, for example, suggest the region will gain power in the House of Representatives and play a larger role in national elections.

Recently, however, the economic slowdown hit the region hard. The contraction of the construction and leisure-time industries in settings such as Las Vegas and Phoenix slowed growth in these metropolitan areas to their lowest rates in decades and home values in some neighborhoods declined by more than 50 percent. Even so, overall population growth within the Mountain West still outpaced national averages between 2006 and 2009.

Black Exodus from the South African Americans have also generated distinctive patterns of interregional migration. Most blacks remained economically tied to the rural South after the Civil War. Conditions changed, however, in the early 20th century. Many African Americans migrated because of declining demands for labor in the agricultural South and growing industrial opportunities in the North and West. Migrants ended up in cities where jobs were located. Boston, New York, Philadelphia, Detroit, Chicago, Los Angeles, and Oakland became key destinations for southern blacks. Since 1970, however, more blacks have moved from North to South. Sun Belt jobs and federal civil rights guarantees now attract

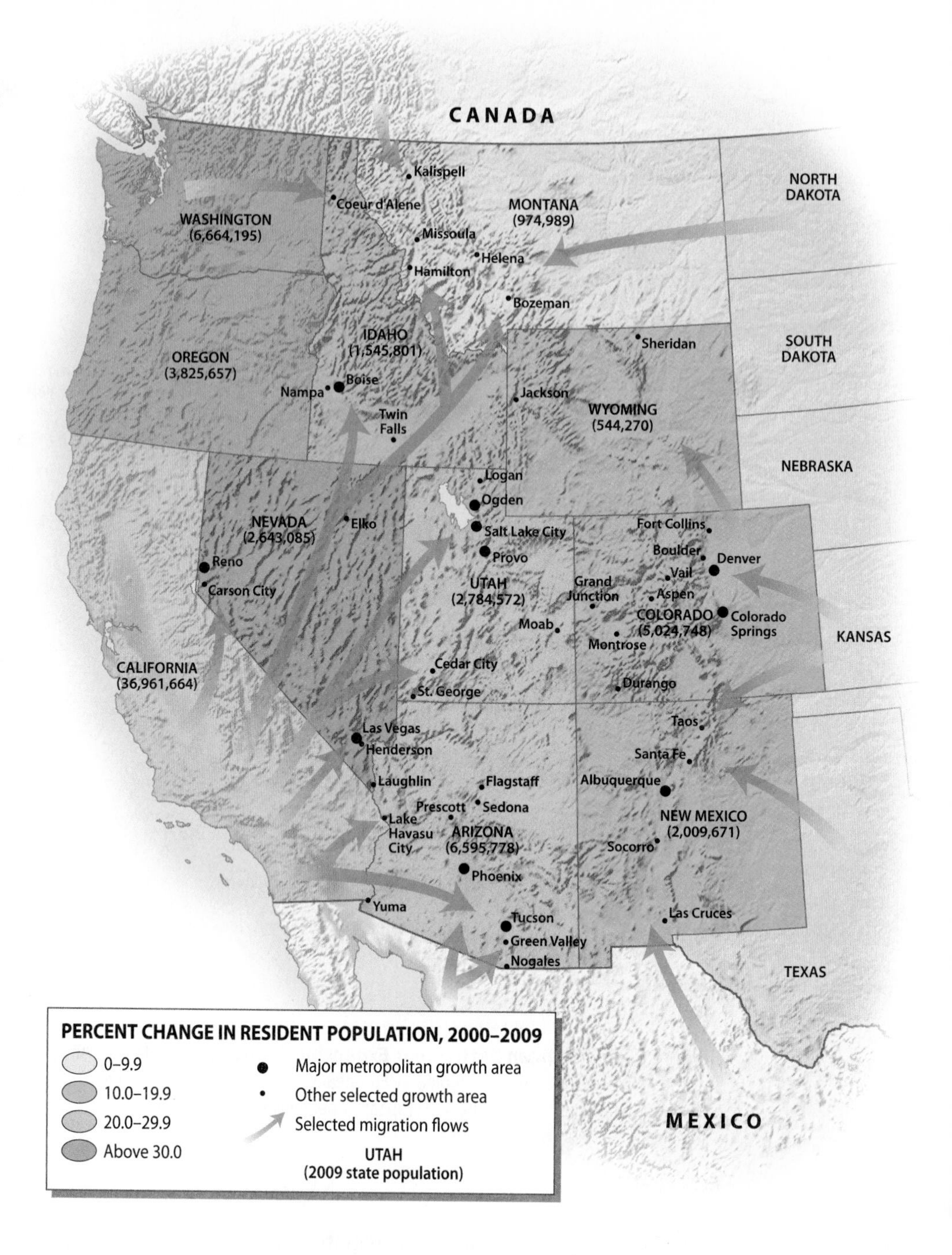

Figure 3.14 Intermountain West Growth, 2000–2009 Nevada (32 percent growth) and Arizona (29 percent growth) were the two fastest-growing states between 2000 and 2009. Many people have flocked to the region's larger cities to find employment, often in growing high-technology industries. Other amenity-bound migrants and retirees are attracted to the region's smaller towns and recreational areas. (U.S. Census)

many northern urban blacks to growing southern cities. The net result is still a major change from 1900: At the beginning of the century, more than 90 percent of African Americans lived in the South, while today only about half of the nation's 41 million blacks reside within the region.

Rural-to-Urban Migration Another continuing trend in North American migration has taken people from the country to the city. Two centuries ago, only 5 percent of North Americans lived in urban areas (cities of more than 2,500 people), whereas today more than 75 percent of the North American population is urban. Shifting economic opportunities account for much of the transformation: As mechanization on the farm reduced the demand for labor, many young people left for new employment opportunities in the city.

Growth of the Sun Belt South Twentieth-century moves to the American South are clearly related to other dominant trends in North American migration, yet the pattern deserves closer inspection. Particularly after 1970, southern states from the Carolinas to Texas grew much more rapidly than states in the Northeast and Midwest. Since 2000, the South has remained a key regional destination for domestic migrants within the United States, adding people even more rapidly than the West. Florida, Texas, Georgia, and North Carolina have recently experienced sizable population gains, with migrants heading for job-rich suburbs as well as high-amenity retirement locations. In 2008, for example, Raleigh, North Carolina, and Austin, Texas, were the nation's fastest-growing metropolitan areas, replacing previous high-fliers such as Las Vegas and Phoenix. Factors that have contributed to the South's growth are its buoyant economy, modest living costs, adoption of air conditioning, attractive recreational opportunities, and appeal to snow-weary retirees. Movements have been selective, however; many rural agricultural and mountain counties within the South have seen few new residents, while amenity-rich coastal settings and job-generating metropolitan areas have witnessed spectacular growth. Dallas–Fort Worth's bustling metropolitan area (6.7 million), for example, is now larger than that of Philadelphia (6.4 million) (Figure 3.15).

Nonmetropolitan Growth During the 1970s, certain areas in North America beyond its large cities began to see significant population gains, including many rural settings that had previously lost population. Selectively, that pattern of **nonmetropolitan growth** in which people leave large cities and move to smaller towns and rural areas continues today. Some participants in the process are part of the growing retiree population in both Canada and the United States, but a substantial number are younger, so-called *lifestyle migrants*. They find or create employment in affordable smaller cities and rural settings that are rich in amenities and often removed from the perceived problems of urban America. In fact, recent nonmetropolitan population growth has exceeded metropolitan growth in most western states. Other smaller communities outside the West, such as Mason City, Iowa; Mankato, Minnesota; and Traverse City, Michigan, also are seen as desirable destinations for migrants interested in downsizing from their metropolitan roots.

Settlement Geographies: the Decentralized Metropolis

North America's settlement landscape reflects the population movements, shifting regional economic fortunes, and technological innovations of the last century. The ways in which settlements are organized on the land—the actual appearance of cities, suburbs, and farms—as well as the very ways in which North Americans socially construct their communities have changed greatly in the past century. Today's cloverleaf interchanges, sprawling suburbs, outlet malls, and theme parks would have struck most 1900-era residents as utterly extraordinary.

Settlement landscapes of North American cities boldly display the consequences of **urban decentralization**, in which metropolitan areas sprawl in all directions and suburbs take on many of the characteristics of traditional downtowns. Although both Canadian and U.S. cities have experienced decentralization, the impact is particularly profound in the United States, where inner-city problems, poor public transportation, widespread automobile ownership, and fewer regional-scale planning initiatives have encouraged many middle-class urban residents to move beyond the central city. Even beyond North America, observers note a globalization of urban sprawl: many Asian, European, and Latin American cities are taking on attributes of their North American counterparts as they experience similar technological and economic shifts. Indeed, much as they have in

Figure 3.15 Downtown Dallas, Texas Rapid job creation has transformed Sun Belt cities such as Dallas. Healthy growth in office space, specialty retailing, and entertainment districts has fueled downtown Dallas's expansion and reshaped the look of the central-city skyline.

Seattle and Albuquerque, suburban Wal-Marts, semiconductor industrial parks, and shopping malls may become increasingly familiar sights on the peripheries of Kuala Lumpur or Mexico City.

Historical Evolution of the City in the United States Changing transportation technologies decisively shaped the evolution of the city in the United States (Figure 3.16). The pedestrian/horsecar city (pre-1888) was compact, essentially limiting urban growth to a 3- or 4-mile-diameter ring around downtown. The invention of the electric trolley in 1888 expanded the urbanized landscape farther into new "streetcar suburbs," often 5 or 10 miles from the city center. A star-shaped urban pattern resulted, with growth extending outward along and near the streetcar lines. The biggest technological revolution came after 1920 with the widespread adoption of the automobile. The automobile city (1920–1945) promoted the growth of middle-class suburbs beyond the reach of the streetcar and added even more distant settlement in the surrounding countryside. Following World War II, growth in the outer city (1945 to the present) promoted more decentralized settlement along commuter routes as built-up areas appeared 40 to 60 miles from downtown.

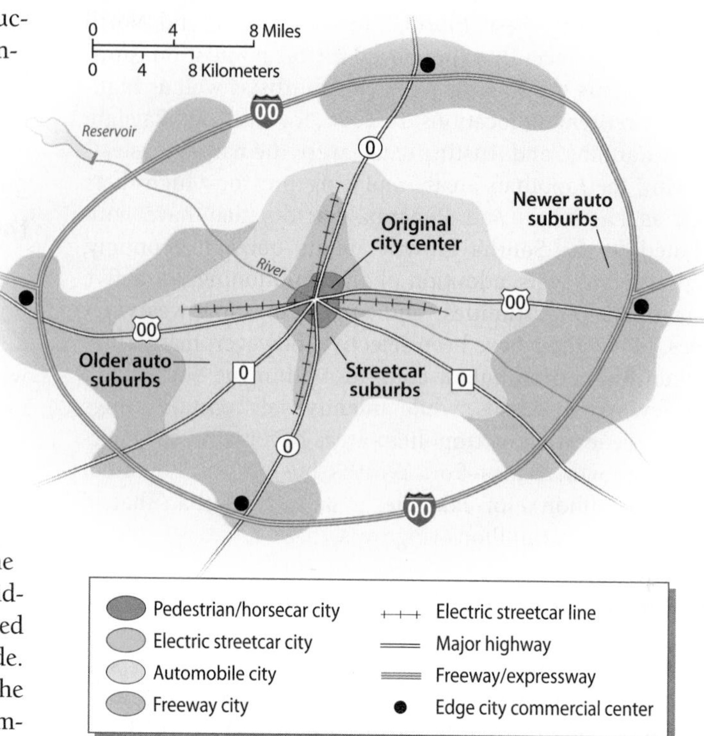

Figure 3.16 Growth of the American City Many U.S. cities became increasingly decentralized as they moved through eras dominated by the pedestrian/horsecar, electric streetcar, automobile, and freeway. Each era left a distinctive mark on metropolitan America, including the recent growth of edge cities on the urban periphery.

Urban decentralization also reconfigured land-use patterns in the city, producing metropolitan areas today that are strikingly different from their counterparts of the early 20th century. In the city of the early 20th century, idealized in the **concentric zone model**, urban land uses were neatly organized in rings around a highly focused central business district (CBD) that contained much of the city's retailing and office functions. Residential districts beyond the CBD were added as the city expanded, with higher-income groups seeking more desirable locations on the outside edge of the urbanized area.

Today's **urban realms model** recognizes these new suburbs characterized by a mix of peripheral retailing, industrial parks, office complexes, and entertainment facilities. These areas of activity, often called "edge cities," have fewer functional connections with the central city than they have with other suburban centers. For most residents of an edge city, jobs, friends, and entertainment are located in other suburbs, rather than in the old downtown. Tysons Corner, Virginia, located west of Washington, DC, is an excellent example of the edge-city landscape on the expanding periphery of a North American metropolis (Figure 3.17).

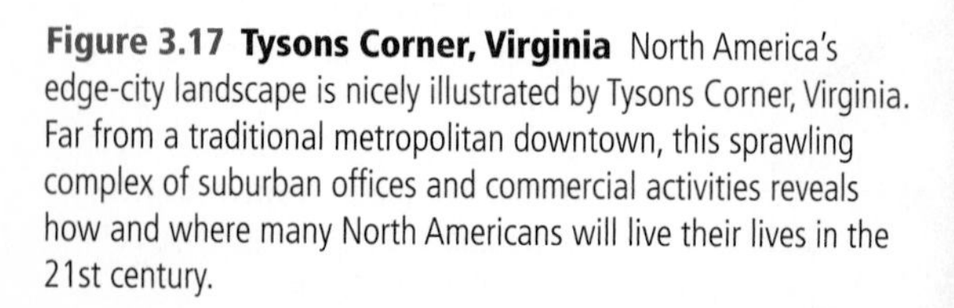

Figure 3.17 Tysons Corner, Virginia North America's edge-city landscape is nicely illustrated by Tysons Corner, Virginia. Far from a traditional metropolitan downtown, this sprawling complex of suburban offices and commercial activities reveals how and where many North Americans will live their lives in the 21st century.

The Consequences of Sprawl The rapid evolution of the North American city continues to transform the urban landscape and those who live in it (see "Geographic Tools: Counting North Americans: The Censuses of 2010 and 2011"). As suburbanization accelerated in the 1960s and 1970s, many inner cities, especially in the Northeast and Midwest, suffered losses in population, increased levels of crime and social disruption, and a shrinking tax base that often brought them to the brink of bankruptcy. Today, inner-city poverty rates average almost three times those of nearby suburbs. Unemployment rates remain above the national average. Central cities in the United States are also places of racial tension, the product of decades of discrimination, segregation, and poverty. Although the number of middle-class African Americans and Hispanics is growing, many exit the central city for the suburbs, further isolating the urban underclass that remains behind. In Detroit, for example, almost 90 percent of inner-city residents are black and the city's poverty rate is almost triple the national norm.

Amid these challenges, inner-city landscapes are also experiencing a selective renaissance. Referred to as **gentrification**, the process involves the displacement of lower-income residents of central-city neighborhoods with higher-income residents, the rehabilitation of deteriorated inner-city landscapes, and the construction of new shopping complexes, entertainment attractions, or convention centers in selected downtown locations. The older and more architecturally diverse housing of the central city is also a draw, serving as specialty shops and restaurants for a cosmopolitan urban clientele and offering residential opportunities for upscale singles who wish to live near downtown. Seattle's Pioneer Square, Toronto's Yorkville district, and Baltimore's Harborplace exemplify how such public and private investments shape the central city.

Many city planners and developers involved in such efforts are advocates of **new urbanism**, an urban design movement stressing higher density, mixed-use, pedestrian-scaled neighborhoods where residents might be able to walk to work, school, and local entertainment. Pittsburgh's recent urban renaissance offers an affordable housing market, an older highly skilled workforce, and mixed-use neighborhoods such as the SouthSide Works development, a 34-acre assortment of residences, offices, and stores created on the site of an old steel plant on the Monongahela River (Figure 3.18).

Figure 3.18 Pittsburgh's SouthSide Works Neighborhood New investments have transformed Pittsburgh's SouthSide Works neighborhood into an upscale office, shopping, and entertainment district.

The suburbs are also changing. Construction of new corporate office centers, fashion malls, and industrial facilities has created true "suburban downtowns" in suburbs that are no longer simply bedroom communities for central-city workers. Indeed, such localities have been growing players in the continent's globalization process. Many of North America's key internationally connected corporate offices (IBM, Microsoft), industrial facilities (Boeing, Cisco, Oracle), and entertainment complexes (Disneyland, Walt Disney World, and the Las Vegas Strip) are now in such settings, and they are intimately tied to global information, technology, capital, and migration flows.

The edge-city lifestyle has also transformed both Canada and the United States into a continent of suburban commuters in which people live in one suburb and work in another. The average daily commute now exceeds 30 miles (48 km) per person in cities such as Atlanta, Georgia, and Birmingham, Alabama. In 2007, almost 90 percent of Americans commuted to work by automobile and less than 1 in 20 used public transportation.

The outward trajectory of low-density suburbs may be slowing. Indeed, one interesting consequence of the global recession of the late 2000s is that North America's low-density suburban fringe stopped growing in many localities and older central city settings witnessed a population resurgence as people found it harder to move and as speculative edge-city real estate ventures floundered. Longer-term, however, North Americans are likely to continue their outward drift, creating a vast suburban periphery where the boundary between country and city blurs in a mosaic of clustered housing developments, shopping complexes, and remaining open space (Figure 3.19).

Figure 3.19 Life on the Urban Periphery New homes mingle with rolling grasslands in suburban Douglas County, Colorado, miles south of Denver's Central Business District, which is barely visible on the northern horizon. Today, Douglas County is a mix of upscale residential developments, commercial nodes, and open space.

GEOGRAPHIC TOOLS Counting North Americans: The Censuses of 2010 and 2011

Every 10 years, hundreds of thousands of census takers canvass the North American continent, gathering basic information on household size, age, and a variety of other social and economic characteristics. For the United States, April 1, 2010 is the big day and for the Canadians the census is taken on May 10, 2011. Thereafter, census workers fan out and follow up on households that did not mail in the form or that may have never received one in the first place.

For the Canadians, the Constitutional Act of 1867 mandated the count, typically done in years ending in 1. Since then, the census gathering has grown to include a number of additional economic, housing-related, and social surveys, including more data gathered during intervening years. In Canada, census questions must be approved by the federal cabinet and are distributed through an agency known as Statistics Canada (www.statcan.gc.ca). Both a short and long form is used to gather a variety of information from every Canadian household.

In the United States, the Census Bureau (www.census.gov) compiles and analyzes census data and has done so since 1790. In 2010, only a single short, ten-question form was used in the U.S. census. In the recent census, millions of bilingual forms were also utilized for the first time, in an effort to increase the response rate in Latino communities across the country. The longer form, once used to harvest more detailed information from one out of every six households, has been replaced by an ongoing initiative known as the American Community Survey (ACS). The ACS now collects and updates information by sampling the U.S. population continually throughout the decade.

In both countries, census data are utilized to adjust election districts and to apportion tax dollars more efficiently. In the United States, for example, more than $400 billion is annually allocated according to census data to help fund services such as schools, hospitals, job-training centers, senior centers, and infrastructure improvement. Businesses also use the census as North America's largest market-research survey. It helps restaurants and retailers plan locations for new stores and can suggest what products and services might do well in particular neighborhoods. In addition, the censuses are one of the most widely used tools of spatial analysis by North American human geographers interested in solving a variety of demographic, social, and economic problems. Information gathered in 2010 and 2011 will be used as benchmark data for thousands of experts examining a myriad of urban, economic, and political issues across the continent.

Figure 3.3.1 **Fall Creek Place, Indianapolis** This renovated inner-city neighborhood features a pleasing mix of century-old homes and new traditional-style houses designed to offer residents affordable housing.

For instance, planners and private developers in the city of Indianapolis relied heavily on census data as they tackled the problem of providing more affordable housing to inner-city residents (Figure 3.3.1). The spatial patterns are apparent when one compares census tract data showing the percentage of renter households (Figure 3.3.2) with the distribution of median household income (Figure 3.3.3). In many low-income inner-city tracts in Indianapolis, rental households make up more than 75 percent of the total, while on

Settlement Geographies: Rural North America

Rural North American landscapes trace their origins to early European settlement. Over time, these immigrants from Europe showed a clear preference for a dispersed rural settlement pattern as they created new farms on the North American landscape. In portions of the United States settled after 1785, the federal government surveyed and sold much of the rural landscape. Surveys were organized around the simple, rectangular pattern of the federal government's township-and-range survey system, which offered a convenient method of dividing and selling the public domain in 6-mile-square townships (Figure 3.20). Canada developed a similar system of regular surveys that stamped much of southern Ontario and the western provinces with a strikingly rectilinear character.

Commercial farming and technological changes further transformed the settlement landscape. Railroads opened corridors of development, provided access to markets for commercial crops, and helped to establish towns. By 1900, several transcontinental lines spanned North America, radically transforming the farm economy and the pace of rural life. After 1920, however, even greater change accompanied the arrival of the automobile, farm mechanization, and better rural road networks. The need for farm labor declined with mechanization, and many smaller market centers became unnecessary as farmers equipped with automobiles and trucks could travel farther and faster to larger, more diverse towns.

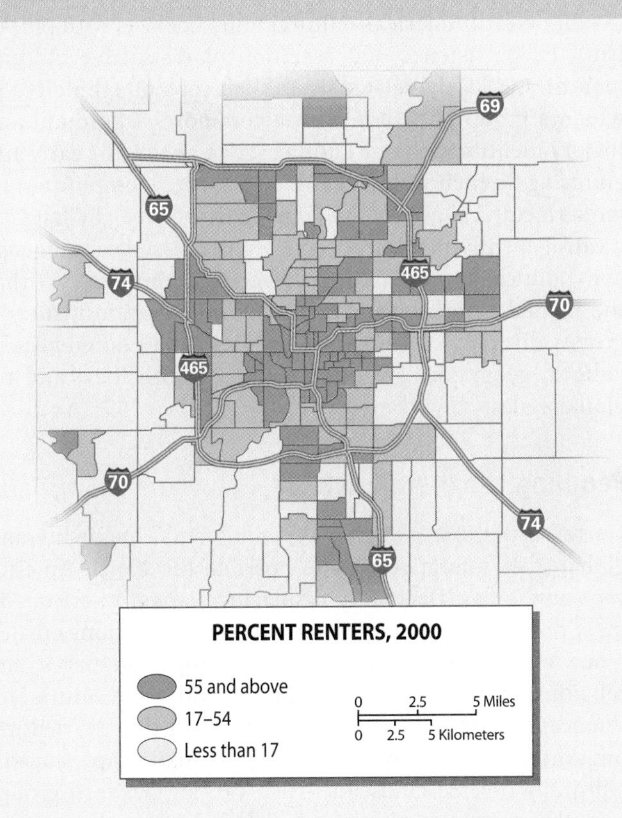

Figure 3.3.2 Indianapolis, Percentage of Renter Households, 2000 Older housing stock near the center of Indianapolis has often been converted into rental units. City planners and developers are trying to create more single-family housing opportunities in such localities. (Reprinted from Rubenstein, 2005, *An Introduction to Human Geography*, 8th ed., Upper Saddle River, NJ: Prentice Hall)

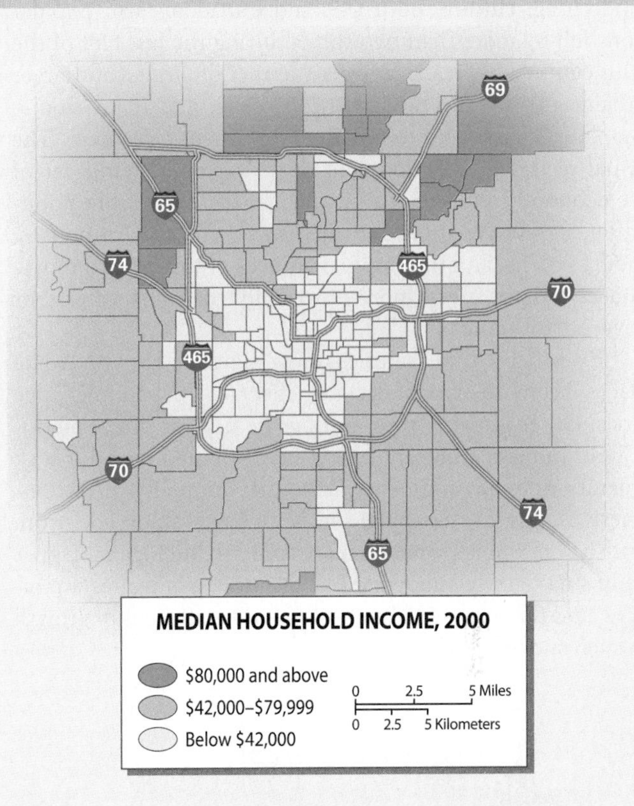

Figure 3.3.3 Indianapolis, Median Household Income, 2000 Higher-income residents tend to cluster toward the newer housing units, larger lot sizes, and expanded employment and entertainment opportunities of the urban periphery. In Indianapolis, the city's wealthy north and northeast sectors contrast dramatically with census precincts near the center of the city. (Reprinted from Rubenstein, 2005, *An Introduction to Human Geography*, 8th ed., Upper Saddle River, NJ: Prentice Hall)

the edge of the city wealthy residents enjoy high rates of home ownership. The geographical patterns helped identify the older and poorer Fall Creek Place neighborhood (on the city's northern downtown fringe) as a focus of urban revitalization efforts. The newest census, once fully analyzed, can help reveal how effective those revitalization efforts have been and signal more generally how well Indianapolis and other North American localities have fared in the first decade of the 21st century.

Figure 3.20 Iowa Settlement Patterns The regular rectangular look of this Iowa town and the nearby rural setting reveals the North American penchant for simplicity and efficiency. In the United States, the township-and-range survey system stamped such predictable patterns across vast portions of the North American interior.

Today, many areas of rural North America face population declines as they adjust to the changing conditions of modern agriculture. Both U.S. and Canadian farm populations fell by more than two-thirds during the last half of the 20th century. Typically, a fewer number of farms (but larger in acreage) dot the modern rural scene, and many young people leave the land to obtain employment elsewhere. The visual record of abandonment offers a painful reminder of the economic and social adjustments that come from population losses. Weed-choked driveways, empty farmhouses, roofless barns, and the empty marquees of small-town movie houses tell the story more powerfully than any census or government report.

Elsewhere, rural settings show signs of growth. Some places begin to experience the effects of expanding edge cities. Other growing rural settings lie beyond direct metropolitan influence but are seeing new populations who seek amenity-rich environments removed from city pressures. These trends are shaping the settlement landscape from British Columbia's Vancouver Island to Michigan's Upper Peninsula. Newly subdivided land, numerous real estate offices, and the presence of espresso bars are all signs of growth in such surroundings.

CULTURAL COHERENCE AND DIVERSITY: Shifting Patterns of Pluralism

North America's cultural geography is both globally dominant and internally pluralistic. On one hand, history and technology have produced a contemporary North American cultural force that is second to none in the world. Many people outside the United States speak of cultural imperialism when they describe the increasing global dominance of American popular culture, which they often see as threatening the vitality of other cultural values. Yet North America is also a mosaic of different peoples who retain part of their traditional cultural identities and celebrate their pluralistic roots.

The Roots of a Cultural Identity

Powerful forces formed a common dominant culture within North America. While both the United States (1776) and Canada (1867) became independent from Great Britain, the two countries remained closely tied to their Anglo roots. Key Anglo legal and social institutions solidified the common set of core values that many North Americans shared with Britain and, eventually, with one another. Traditional Anglo beliefs emphasized representative government, separation of church and state, liberal individualism, privacy, pragmatism, and social mobility. From those shared foundations, particularly within the United States, consumer culture blossomed after 1920, producing a common set of experiences oriented around convenience, consumption, and the mass media.

But North America's cultural unity coexists with pluralism, the persistence and assertion of distinctive cultural identities. Closely related is the concept of **ethnicity**, in which a group of people with a common background and history identify with one another. For Canada, the early and enduring French colonization of Quebec complicates its modern cultural geography. Canadians face the challenge of creating a truly bicultural society where issues of language and political representation are central concerns. Within the United States, given its unique immigration history, a greater diversity of ethnic groups exists, and differences in cultural geography are often found at both local and regional scales.

Peopling North America

North America is a region of immigrants. Quite literally, global-scale migrations made possible the North America we know today. Decisively displacing Native Americans in most portions of the region, immigrant populations created a new cultural geography of ethnic groups, languages, and religions. Early migrants often had considerable cultural influence, despite minuscule numbers. Over time, varied immigrant groups and their changing destinations produced a culturally diverse landscape. Also varying between groups was the pace and degree of **cultural assimilation**, the process in which immigrants were absorbed by the larger host society.

Migration to the United States In the United States, variations in the number and source regions of migrants produced five distinctive chapters in the country's history (Figure 3.21). In Phase 1 (prior to 1820), English and African influences dominated. Enslaved slaves, mostly from West Africa, contributed additional cultural influences in the South. Northwest Europe served as the main source region of immigrants between 1820 and 1870 (Phase 2). The emphasis, however, shifted away from English migrants. Instead, Irish and Germans dominated the flow and provided more cultural variety.

As Figure 3.21 shows, immigration reached a much higher peak around 1900, when almost 1 million foreigners entered the United States *annually*. During Phase 3 (1870–1920), the majority of immigrants were southern and eastern Europeans. Political strife and poor economies in Europe existed during this period. News of available land and expanding industrialization in the United States offered an escape from such difficult conditions. By 1910, almost 14 percent of the nation was foreign-born. Very few of these immigrants, however, targeted the job-poor U.S. South, creating a cultural divergence that still exists.

Between 1920 and 1970 (Phase 4), more immigrants came from neighboring Canada and Latin America, but overall totals fell sharply, a function of more restrictive fed-

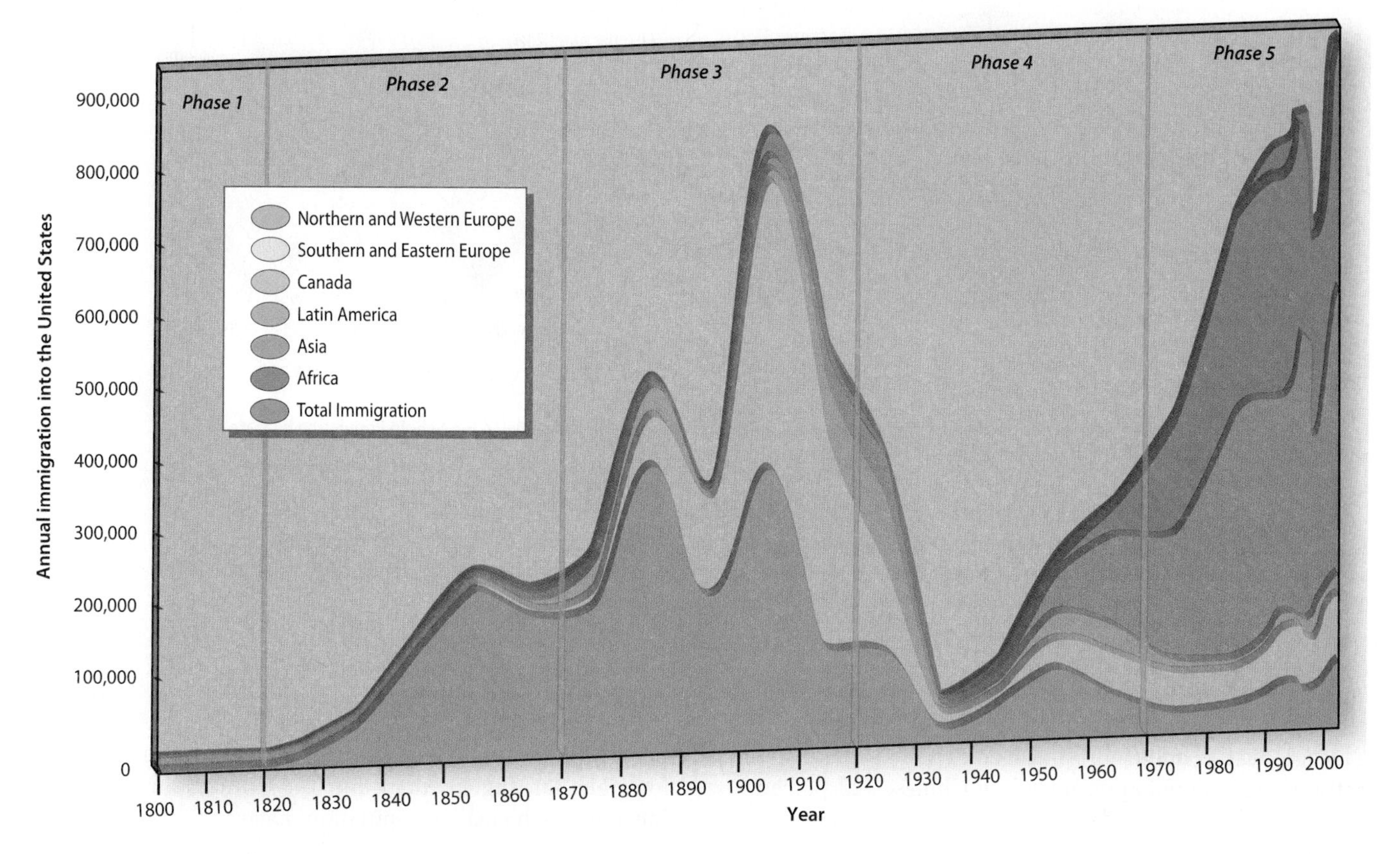

Figure 3.21 U.S. Immigration, by Year and Group Annual immigration rates peaked around 1900, declined in the early 20th century, and then surged again, particularly since 1970. The source areas of these migrants have also shifted. Note the decreased role of Europeans currently versus the growing importance of Asians and Latin Americans. (Modified from Rubenstein, 2005, *An Introduction to Human Geography*, 8th ed., Upper Saddle River, NJ: Prentice Hall)

eral immigration policies (the Quota Act of 1921 and the National Origins Act of 1924), the Great Depression, and the disruption caused by World War II.

Since 1970 (Phase 5), the region has witnessed a sharp reversal in numbers, and now annual arrivals surpass those of the early 20th century (see Figure 3.21). Most legal migrants since 1970 originated in Latin America or Asia. The current surge was made possible by economic and political instability abroad, a growing postwar American economy, and a loosening of immigration laws (the Immigration Acts of 1965 and 1990 and the Immigration Reform and Control Act of 1986). Illegal immigration also rose after 1970, although several federal laws passed in 1996 greatly increased the number of U.S. border patrol agents and made it more difficult for immigrants, both legal and illegal, to receive federal welfare benefits. Today, about 11–12 million unauthorized immigrants live in the United States. Illegal immigration into the United States actually fell modestly between 2007 and 2010, partly a function of uncertain job prospects in the construction and service sectors of the economy.

An estimated 12 million Mexican-born residents (or about 11 percent of Mexico's population) now live in the United States. Mexicans make up about 64 percent of the nation's Hispanic population, but those numbers are changing. In the next 25 years, most of the projected increase in the U.S. Hispanic population will be fueled by births within the country versus new immigrants. While half of U.S. Hispanics live in California or Texas, they are increasingly moving to other areas (Figure 3.22). Recently, states such as Wisconsin, Georgia, Kansas, and Arkansas have witnessed dramatic increases in their immigrant Hispanic populations, a trend likely to continue in the early 21st century. The cultural implications of these Hispanic migrations are profound: today the United States is the fifth largest Spanish-speaking nation on Earth.

In percentage terms, migrants from Asia constitute the fastest-growing immigrant group, and various Asian ethnicities, both native and foreign-born, account for 5 percent of the U.S. population. Chinese is now the third most spoken language in the United States (behind English and Spanish). California remains a key entry point for new migrants and is home to more than one-third of the nation's overall Asian population (see Figure 3.22). Asian migrants often move to large cities such as Los Angeles, San Francisco, and New

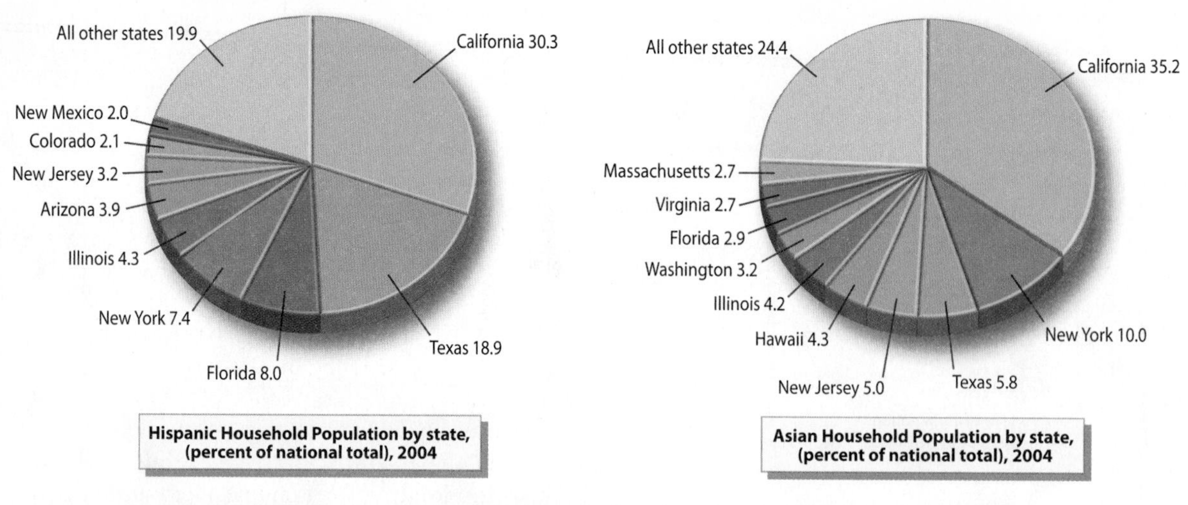

Figure 3.22 Distribution of U.S. Hispanic and Asian Household Populations, by State, 2004 California and Texas claim about half of the nation's Hispanic population, but a growing number are locating elsewhere. California alone is still home to more than one-third of the country's Asian population. (U.S. Census, American Community Survey Reports, issued February 2007)

York City. Beyond these key gateway cities, Asians are also moving to growing communities in Washington, DC, Chicago, Seattle, and Houston. Diverse cultures make up the nation's Asian population, including Chinese (23 percent), Indian (19 percent), Filipino (18 percent), Vietnamese (11 percent), and Korean (10 percent).

The future cultural geography of the United States will be dramatically redefined by these recent immigration patterns (Figure 3.23). Indeed, the increasing ethnic diversity of the country is simply one more manifestation of the globalization process and the powerful pull of North America's economy and political stability. By 2050, Asians may total about 10 percent of the U.S. population, and almost one American in three will be Hispanic. Indeed, it is likely that the U.S. non-Hispanic white population will achieve minority status by that date.

Finney County, Kansas, suggests how U.S. communities are changing. Set deep in the heartland, far from any international border or port city, this southeastern Kansas county entered "majority-minority" status in 2008, meaning its proportion of non-Hispanic whites is now under 50 percent. Garden City, the largest community in the county, has long been home to Hispanic populations who have worked in the agricultural sector, especially in the area's growing meatpacking industry. But recently, thousands of Somali and Southeast Asian immigrants have been added to the mix, drawn to the area's steady work and modest cost of living. The result is a complex cultural mosaic that suggests how immigration is now reworking the cultural geography of every corner of the country.

The Canadian Pattern The peopling of Canada included early French arrivals that concentrated in the St. Lawrence Valley. After 1765 many migrants came from Britain, Ireland, and the United States. Canada then experienced the same surge and reorientation in migration flows seen in the United States around 1900. Between 1900 and 1920, more than 3 million foreigners ventured to Canada, an immigration rate far higher than for the United States, given Canada's much smaller population. Eastern Europeans, Ital-

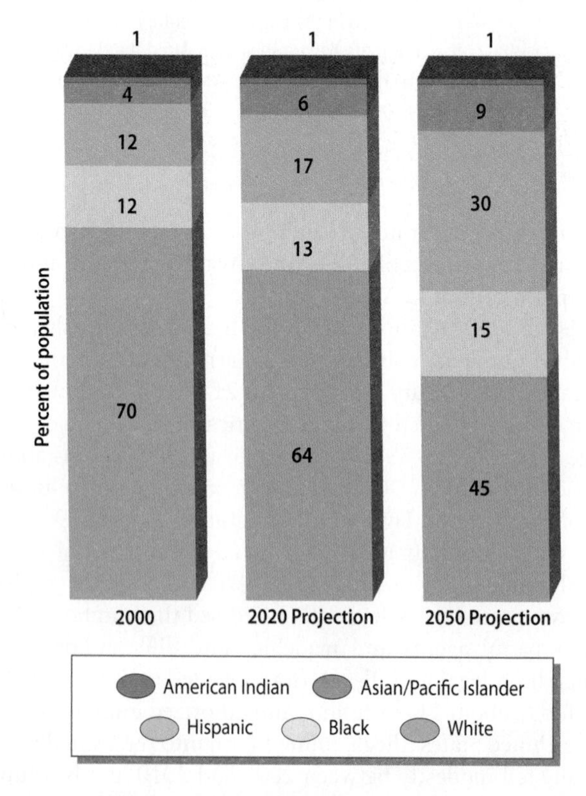

Figure 3.23 Projected U.S. Ethnic Composition, 2000 to 2050 By the middle of the 21st century, almost one in three Americans will be Hispanic, and non-Hispanic whites will achieve minority status amid an increasingly diverse U.S. population. (Modified from the U.S. Census, Census of Population)

ians, Ukrainians, and Russians were the most important nationalities in these later movements. Today, about 60 percent of Canada's recent immigrants are Asians and its 19 percent foreign-born population is among the highest in the developed world. In Toronto, the city's 44 percent foreign-born population reveals a slight bias toward European backgrounds. On Canada's west coast, Vancouver (38 percent foreign-born) has been a key destination for Asian immigrants, particularly Chinese (Figure 3.24). The city's reputation as a culturally diverse, cosmopolitan locality was further enhanced in 2010 when it was ranked as "the world's most livable city" and when it proved to be an attractive host for the Winter Olympic Games. Thousands of global visitors enjoyed the city and billions of viewers took in the setting on television.

Culture and Place in North America

Cultural and ethnic identity is often strongly tied to place. North America's cultural diversity is expressed geographically in two ways. First, similar people congregate near one another and derive meaning from the territories they occupy in common. Second, culture marks the visible scene: the everyday landscape is filled with the artifacts, habits, language, and values of different groups. Boston's Italian North End simply looks and smells different from nearby Chinatown, and rural French Quebec is a world away from a Hopi village in Arizona (Figure 3.25).

Persisting Cultural Homelands French Canadian Quebec superbly exemplifies the cultural homeland: it is a culturally distinctive nucleus of settlement in a well-defined geographical area, and its ethnicity has survived over time, stamping the cultural landscape with an enduring personality. Overall, about 23 percent of Canadians are French, but more than 80 percent of the population of Quebec speaks French, and language remains the cultural glue that unites the homeland. Indeed, policies adopted after 1976 strengthened the French language within the province by requiring French instruction in the schools and by mandating national bilingual programming by the Canadian Broadcasting Corporation (CBC). Ironically, many Quebecois feel that the greatest cultural threat may come not from Anglo-Canadians but rather from recent immigrants to the province. Southern Europeans or Asians in Montreal, for example, show little desire to learn French, preferring instead to put their children in English-speaking private schools.

Another well-defined cultural homeland is the Hispanic Borderlands (see Figure 3.25). It is similar in geographical magnitude to French Canadian Quebec, significantly larger in total population, but not specifically linked to a single political entity such as a state or province. Historical roots of the homeland are deep, extending back to the 16th century, when Spaniards opened the region to the European world. The homeland's historical core is in northern New Mexico, including Santa Fe and much of the surrounding rural hinterlands. A rich legacy of Spanish place-names, earth-toned

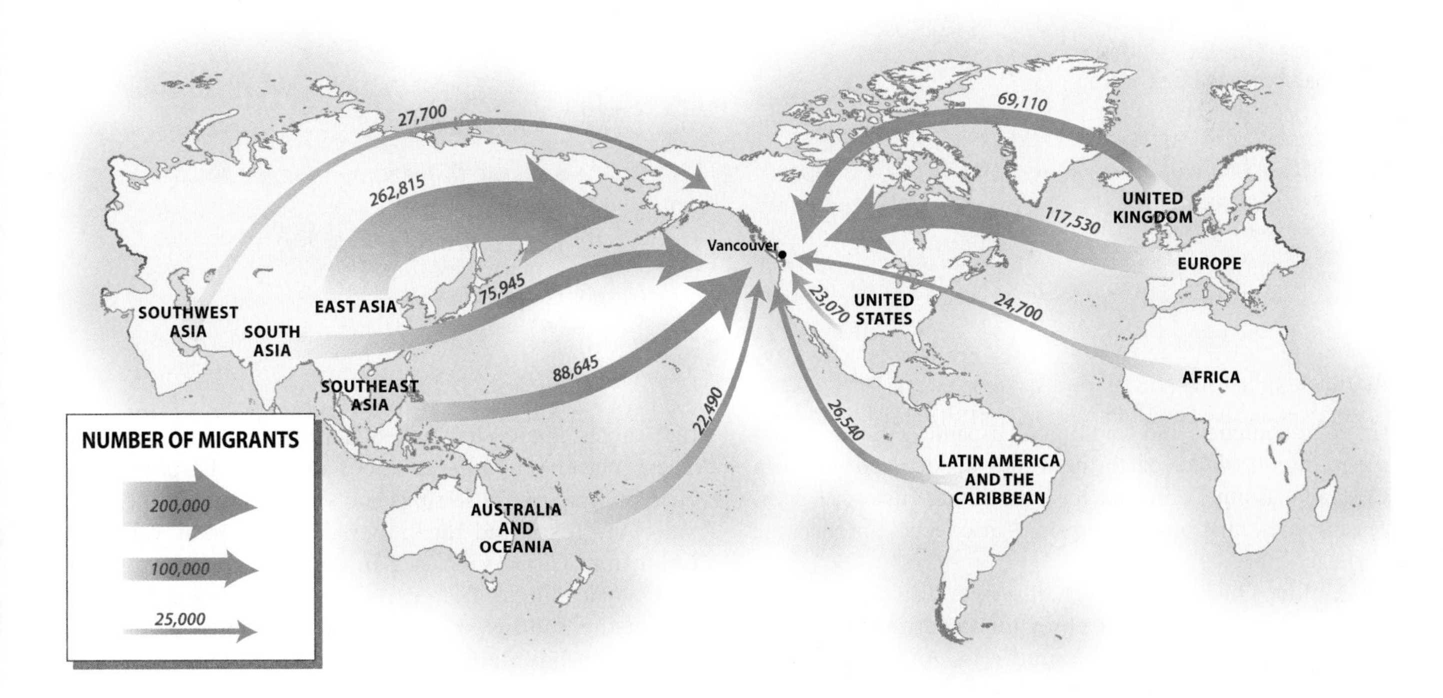

Figure 3.24 Vancouver's Immigrant Population, by Place of Birth, 2001 Vancouver is home to many Europeans, particularly from Great Britain, although most of the city's immigrants have come from East, Southeast, and South Asia. About equal numbers of immigrant residents have moved to Vancouver from the United States, Latin America/Caribbean, Africa, and Australia/Oceania. (Data from *Statistics Canada*, 2001 Census)

Figure 3.25 Selected Cultural Regions of North America From northern Canada's Nunavut to the Southwest's Hispanic Borderlands, different North American cultural groups strongly identify with traditional local and regional homelands. Such patterns stamp the modern landscape with an enduring personality. The map displays a sampling of these regions across North America. (Modified from Jordan, Domosh, and Rowntree, 1998, *The Human Mosaic*, Upper Saddle River, NJ: Prentice Hall)

Catholic churches, and traditional Hispanic settlements dot the rolling highlands of northern New Mexico and southern Colorado. From California to Texas, other historical sites, place-names, missions, and presidios also reflect the Hispanic heritage.

Unlike Quebec, however, massive 20th-century migrations from Latin America brought an entirely new wave of Hispanic settlement to the Southwest. About 45 million Hispanics now live in the United States, with more than half in California, Texas, New Mexico, and Arizona combined. Indeed, by 2015, Hispanics will likely outnumber non-Hispanic whites in California. Within the homeland, Hispanics have created a distinctive Borderlands culture that mixes many elements of Latin and North America. These newer migrants augment the rural Hispanic presence in agricultural settings such as the lower Rio Grande Valley in Texas and the Imperial and Central Valleys in California. Cities such as San Antonio and Los Angeles also play leading roles in expressing the Hispanic presence within the Southwest. Regionally distinctive Latin foods and music add internal cultural variety to the region. New York City, Chicago, and Miami serve as key points of Hispanic cultural influence beyond the homeland (see "Cityscapes: Miami").

African Americans also retain a cultural homeland, but it has diminished in intensity because of out-migration (see

Figure 3.25). Reaching from coastal Virginia and North Carolina to East Texas, the Black Belt is a zone of African American population remaining from the cotton South, when a vast majority of American blacks resided within the region. Today, although many blacks have left for cities, dozens of rural counties in the region still have large black majorities. Blacks account for more than one-quarter of the populations of Mississippi (37 percent), Louisiana (33 percent), South Carolina (29 percent), Georgia (29 percent), and Alabama (26 percent). More broadly, the South is home to many black folk traditions, including music such as black spirituals and the blues, which have now become popular far beyond their rural origins. Regrettably, even though the rural neighborhoods of the black homeland differ greatly in density and appearance from African American urban neighborhoods in the North, poverty plagues both types of communities.

A second rural homeland in the South is Acadiana, a zone of persisting Cajun culture in southwestern Louisiana (see Figure 3.25). This homeland was created in the 18th century when French settlers were expelled from eastern Canada (an area known as "Acadia") and relocated to Louisiana. Nationally popularized today through their food and music, the Cajuns have a lasting attachment to the bayous and swamps of southern Louisiana.

Native American populations are also strongly tied to their homelands. Indeed, many native peoples maintain intimate relationships with their surroundings, weaving elements of the natural environment together with their material and spiritual lives. About 5.2 million Indians, Inuits, and Aleuts live in North America, and they claim allegiance to more than 1,100 tribal bands. Particularly in the American West and the Canadian and Alaskan North, native peoples control sizable reservations, including the Navajo Reservation in the Southwest, as well as self-governing Nunavut in the Canadian North (see Figure 3.25). Although these homelands preserve traditional ties to the land, they are also settings for pervasive poverty, health problems, and increasing cultural tensions (Figure 3.26). Within the United States, many Native American groups have taken advantage of the special legal status of their reservations and have built gambling casinos and tourist facilities that bring in much-needed capital, but also challenge traditional lifeways.

Figure 3.26 Native American Poverty Navajo youngsters enjoy a game of basketball on the Navajo Reservation in northeast Arizona. Poor housing, low incomes, and persistent unemployment plague many Native American settings across the rural West.

A Mosaic of Ethnic Neighborhoods North America's cultural mosaic is also enlivened by smaller-scale ethnic signatures that shape both rural and urban landscapes. For example, distinctive rural communities that range from Amish settlements in Pennsylvania to Ukrainian neighborhoods in southern Saskatchewan add cultural variety. When much of the agricultural interior was settled, immigrants often established close-knit communities. Among others, German, Scandinavian, Slavic, Dutch, and Finnish neighborhoods took shape, held together by common origins, languages, and religions. Although many of these ties weakened over time, rural landscapes of Wisconsin, Minnesota, the Dakotas, and the Canadian prairies still display a number of these cultural imprints. Folk architecture, distinctive settlement patterns, ethnic place-names, and the simple elegance of rural churches selectively survive as signatures of cultural diversity upon the visible scene of rural North America.

Ethnic neighborhoods also enrich the urban landscape and reflect both global-scale and internal North American migration patterns. Complex social and economic processes are clearly at work. Employment opportunities historically fueled population growth in North American cities, but the cultural makeup of the incoming labor force varied, depending on the timing of the economic expansion and the relative accessibility of an urban area to different cultural groups. The ethnic geography of Los Angeles exemplifies the interplay of the economic and cultural forces at work (Figure 3.27). Since most of its economic expansion took place during the 20th century, its ethnic geography reflects the movements of more recent migrants. African American communities on the city's south side (Compton and Inglewood) represent the legacy of black population movements out of the South. Hispanic (East Los Angeles) and Asian (Alhambra and Monterey Park) neighborhoods are a reminder that about 40 percent of the city's population is foreign-born.

Particularly in the United States, ethnic concentrations of nonwhite populations increased in many cities during the 20th century as whites exited for the perceived safety of the suburbs. In terms of central-city population, African Americans make up more than 60 percent of Atlanta, while Los Angeles is now more than 40 percent Hispanic (almost 5 million Hispanics reside in Los Angeles County, greater than the

CITYSCAPES Miami

Miami represents the quintessentially diverse North American city of the 21st century (Figure 3.4.1). With a blossoming regional metropolitan population of more than 5.5 million people (including the Fort Lauderdale area), Miami is one of the largest cities in the South. It features a unique, subtropical urban landscape, rich ethnic diversity, and an economy closely wed to nearby Latin America.

In some ways, Miami has always been defined by its exotic weather, by the warm Atlantic breezes that rustle through its palm trees, and by the ever-changing clouds that drift above its urban skyline, beach resorts, and nearby bays. In 1896 it was the warm weather (average January high temperature of 76°F (24°C)) and the region's prospects as a great winter resort that enticed business executive Henry Flagler to build a railroad into South Florida. Many northeastern urbanites, eager for winter warmth, came south and, particularly after World War II, a growing number retired there. Added to this mix was an expanding collection of Caribbean and Latin American migrants who trickled, then flooded into Miami as its regional economy grew. Much of the city's initial surge in Latin population came with Castro's takeover of Cuba in 1959. Between 1960 and 1980, hundreds of thousands of refugees left Cuba and went to Miami's Cuban community. Many brought capital and entrepreneurial skills, establishing hundreds of

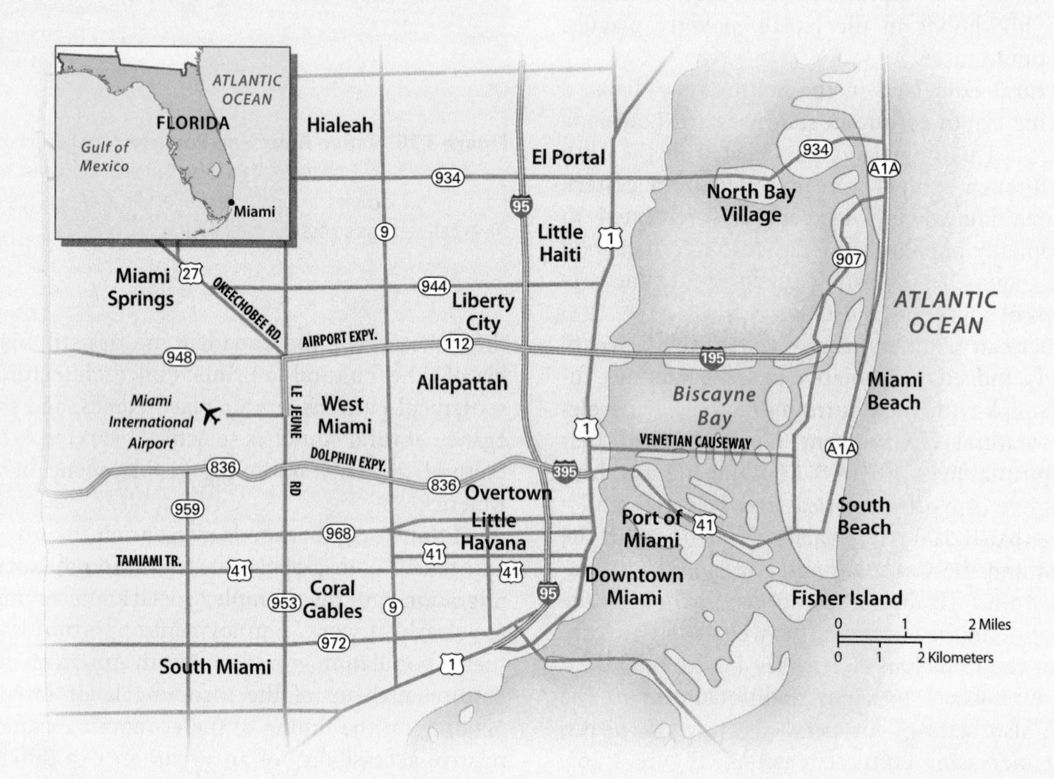

Figure 3.4.1 Miami With a regional metropolitan population of more than 5.5 million, Miami displays an extraordinary diversity of cultural landscapes that reflect both its exotic character in the American South as well as its intimate links to Latin America.

population of Costa Rica). Ethnic concentrations are also growing in the suburbs of some U.S. cities: Southern California's Monterey Park has been called the "first suburban Chinatown," and growing numbers of middle-class African Americans and Hispanics shape suburban neighborhoods in metropolitan settings such as Atlanta and San Antonio.

Patterns of North American Religion

Many religious traditions also shape North America's human geography. Reflecting its colonial roots, Protestantism is dominant within the United States, accounting for about 60 percent of the population. In some settings, hybrid American religions sprang from broadly Protestant roots. By far the most successful are the Latter-day Saints (Mormons), regionally concentrated in Utah and Idaho, and claiming more than 5 million North American members. While many traditional Catholic neighborhoods have lost population in the urban Northeast, Catholic numbers are growing in the West and South, reflecting both domestic migration patterns and higher rates of Hispanic immigration and births. Strong Protestant religious traditions en-

Cuban businesses in the city. By the late 1970s, Latin America's economic growth demanded closer ties with the United States. What better place than Miami to serve as a gateway?

Walk through Miami's bustling airport today (45 percent of all airport arrivals are international passengers) and you are immediately aware of the unique role this city plays in linking the economic and cultural worlds of Latin America with those of North America. Spanish-speaking travelers and international flight announcements dominate airport conversation as planes arrive from and depart to cities throughout Central and South America. As a result, modern Miami has become the leading North American corporate and trade center for Latin America. Indeed, some have termed it the "capital of the Americas." Today, Miami handles a huge volume of trade with Latin America and the Caribbean. Airplane connections, money flows, and even the illegal drug trade demonstrate Miami's centrality. Newspapers such as the *Colombian Post, Venezuela Al Dia,* and the *Diario Las Americas* keep local residents and Latin visitors in touch with news from the home country.

Miami's neighborhoods reflect its unique history and ethnic richness. Southwest of bustling downtown, the culturally diverse neighborhood of Coral Gables contains Spanish-style homes and affluent shopping and entertainment centers. But on the city's north side, the largely African American Liberty City and Overtown communities struggle with urban poverty and drug problems. Elsewhere, Caribbean influences shape the sights, sounds, and smells of Little Haiti and Allapattah (Little Dominican Republic). The city's largest Latin community is Cuban and traditionally is centered around Little Havana southwest of downtown. Along Little Havana's Calle Ocho (8th Street), the smell of roast pork mingles with cigar smoke and nearby Maximo Gomez Park is dotted with Cuban checkers and domino players (Figure 3.4.2). In nearby Miami Beach, a revitalized art deco–style commercial district features a mix of hotels and restaurants that cater to older Jewish retirees as well as to a younger generation of partygoers bound for the trendy South Beach nightclub scene (Figure 3.4.3). Elsewhere, many older buildings have been razed, replaced by high-rise luxury condominiums.

Figure 3.4.2 Little Havana Many of Miami's older Cuban residents cluster along the Calle Ocho on the city's west side and enjoy spending time socializing at Maximo Gomez Park.

Figure 3.4.3 Miami Beach Miami Beach's art deco district has achieved nationwide fame. Many of the resort town's older hotels, commercial businesses, and apartment houses have received a colorful facelift in the past 25 years.

liven many African American communities, both in the rural South and in dozens of urban settings where churches serve as critical centers of community identity and solidarity (Figure 3.28).

Within Canada, almost 40 percent of the population is Protestant, with the United Church of Canada and the Anglican Church claiming the largest numbers of followers. Roman Catholicism is important in regions that received large numbers of Catholic immigrants. French Canadian Quebec is a bastion of Catholic tradition and makes Canada's population (43 percent) distinctly more Catholic than that of the United States (24 percent).

Millions of other North Americans practice religions outside of the Protestant and Catholic traditions or are unaffiliated with traditional religions. Orthodox Christians congregate in the urban Northeast, where many Greek, Russian, and Serbian Orthodox communities were established between 1890 and 1920. The telltale domes of Ukrainian Orthodox churches still dot the Canadian prairies of Alberta, Saskatchewan, and Manitoba. More than 7 million Jews live in North America, concentrated in East and West

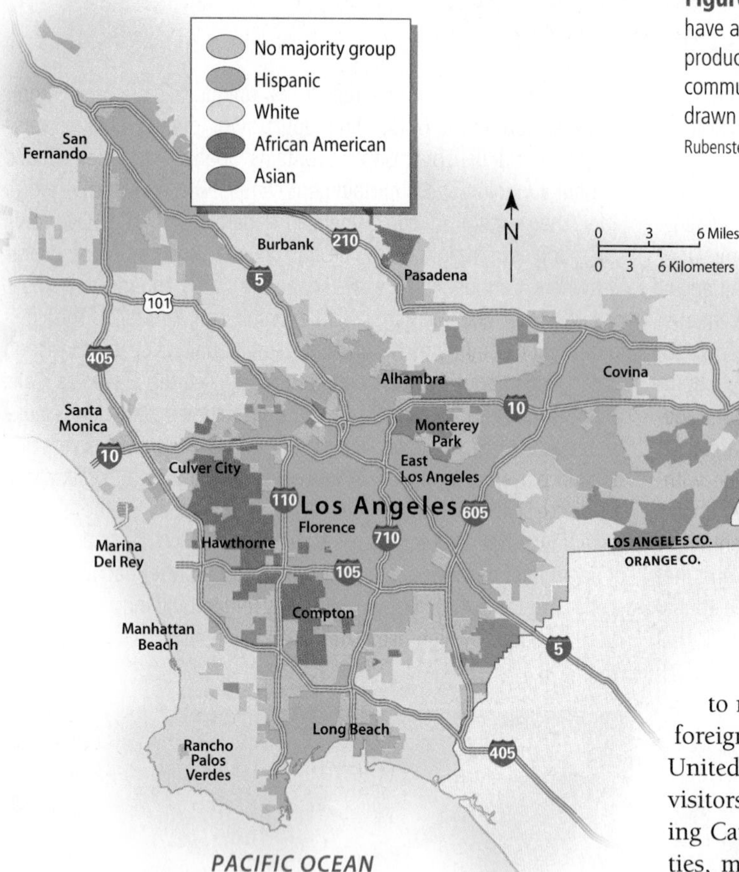

Figure 3.27 Ethnicity in Los Angeles Varied economic opportunities have attracted a wide variety of migrants to North American cities, producing varying ethnic mosaics of distinctive neighborhoods and communities. In Los Angeles, several cycles of economic expansion have drawn a diverse collection of residents from around the globe. (Reprinted from Rubenstein, 2008, *An Introduction to Human Geography*, Upper Saddle River, NJ: Prentice Hall)

Coast cities. In the United States, the rapidly growing organization known as the Nation of Islam also has a strong urban orientation, reflecting its appeal to many economically dispossessed African Americans. Many other Muslims (6 million), Buddhists (1 million), and Hindus (1 million) also live in the United States. While only about 8 percent of people in the United States classify themselves as nonbelievers, a recent survey showed that 30 percent of the population claimed to have a largely secular lifestyle in which religion was rarely practiced.

The Globalization of American Culture

North America's culture is becoming more global at the same time that global cultures are becoming more North American (influenced particularly by the United States). But the process of cultural globalization is becoming more complex. No longer can we think of simple flows of foreign influences into North America or the juggernaut of U.S. cultural dominance invading every traditional corner of the globe. In the 21st century, the story of cultural globalization will increasingly feature a mix of influences that flow in many directions at once, and that feature new hybrid cultural creations.

North Americans: Living Globally

More than ever before, North Americans in their everyday lives are exposed to people from beyond the region. With more than 37 million foreign-born migrants living across the region, diverse global influences are free to mingle in new ways. In addition, tens of thousands of foreigners arrive daily, mostly as tourists. In 2008, the United States recorded more than 50 million international visitors arriving in the country, many arriving from neighboring Canada and Mexico. At American colleges and universities, more than 670,000 international students (57 percent from Asia) add a global flavor to the ordinary curriculum. Canada experiences a similar saturation of such global influences, with the U.S. presence particularly dominant.

Globalization presents cultural challenges for North Americans. In the United States, one key issue revolves around the English language, which some have described as the "social glue" that holds the nation together. Since 1980, the continuing flood of non-English-speaking immigrants into the country has sharpened the debate over the role English should play in American culture. Many in the United States argue that English should be the country's only officially recognized language in order to compel immigrants to learn it and thus speed their assimilation into the host culture. Some immigrants suggest they need to maintain their traditional languages both to function within their ethnic communities and to preserve their cultural heritage.

Figure 3.28 African American Church, Oakland, California Oakland's First African Methodist Episcopal Church has been a Bay Area cultural institution for more than a century. It remains a focus of the community today.

At the same time, the evidence suggests that North America's immigrants are learning English more rapidly than ever before, seeing it as a powerful tool to accelerate their economic opportunities, both in the United States and Canada. A 2006 poll of Hispanic residents revealed that almost 60 percent of Hispanics agreed that "immigrants have to speak English to say that they are part of American society." The growing popularity of **Spanglish**, a hybrid combination of English and Spanish spoken by Hispanic Americans, also illustrates the complexities of North American globalization. Spanglish includes interesting hybrids such as *chatear*, which means to have an online conversation.

North Americans are going global in other ways. By 2009, about 75 percent of North Americans regularly used the Internet, opening the door for far-reaching journeys in cyberspace. North Americans also travel much more widely than ever before. Residents of the United States take more than 27 million overseas trips annually (75 percent for leisure travel, 25 percent for business travel). Within North America, the popularity of ethnic restaurants has peppered the region with a bewildering variety of Cuban, Ethiopian, Basque, and Pakistani eateries. Chinese and Italian foods dominate the taste buds of all North Americans, with Mexican food a rapidly growing choice in the United States. The growing affinity for foreign beverages mirrors the pattern; imported beer sales in the United States expanded much more rapidly than domestic sales between 2000 and 2008. American consumers drink more than 800 million gallons of imported beer annually (Figure 3.29). In another example of globalization, Heineken, famed for its fine Dutch brew, now owns the rights to resell in the United States both the Tecate and Dos Equis brands, two popular Mexican beers. Americans also have increased their consumption of foreign red wines and have rapidly Europeanized their coffee-drinking habits. In fashion, *Gucci*, *Armani*, and *Benetton* are household words for millions who keep their eyes on European styles. Although British pop music has been an accepted part of North American culture for five decades, the beat of German techno bands, Gaelic instrumentals, and Latin rhythms also has become an increasingly seamless part of daily life within the region. Indeed, from acupuncture and massage therapy to soccer and New Age religions, North Americans are tirelessly borrowing, adapting, and absorbing the larger world around them.

The Global Diffusion of U.S. Culture In a parallel fashion, U.S. culture has forever changed the lives of billions of people beyond the region. Although the economic and military power of the United States was notable by 1900, it was not until after World War II that the country's popular culture reshaped global human geographies in fundamental ways. The Marshall Plan and Peace Corps initiatives exemplified the growing presence of the United States on the world stage even as European colonialism waned. Rapid improvements in global transportation and information technologies, much of them engineered in the United States, also brought the world

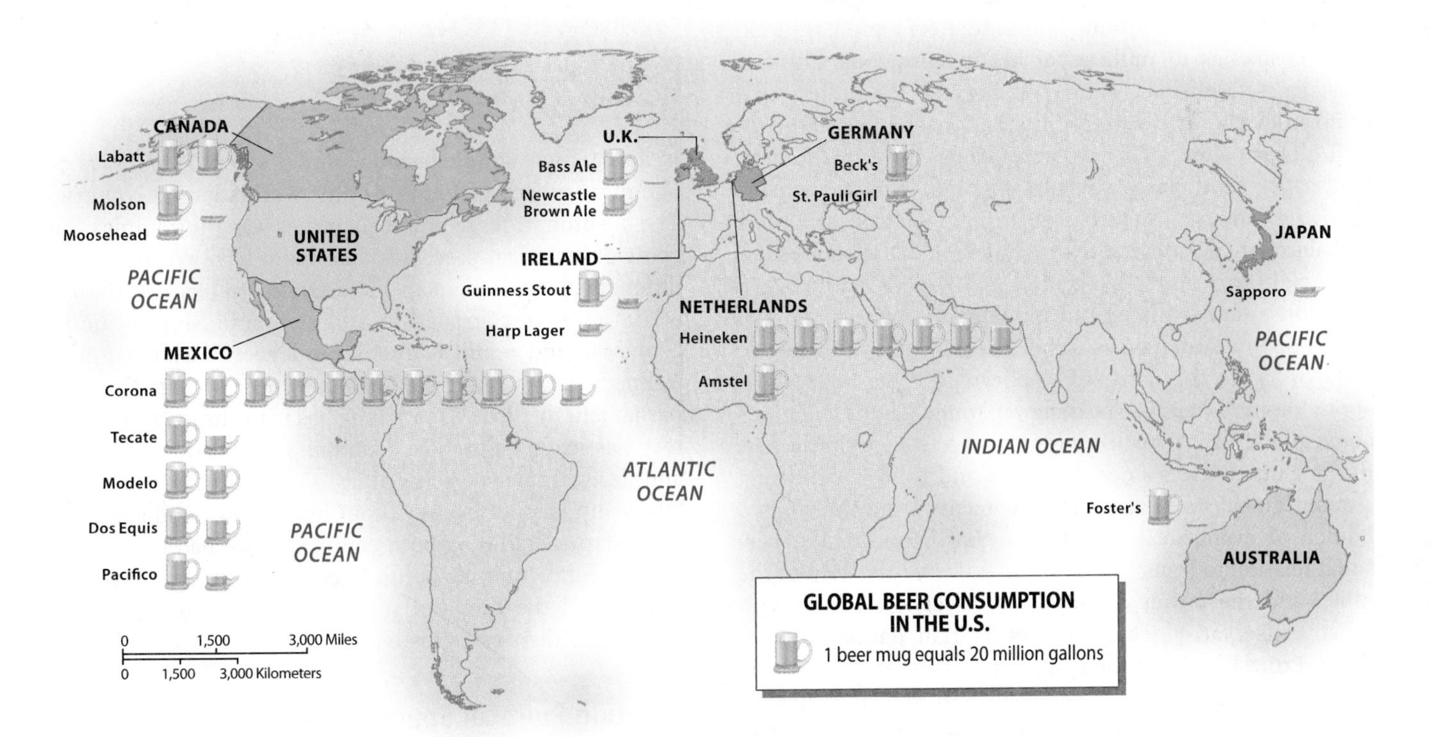

Figure 3.29 Annual Beer Imports to the United States, 2002 Whether they are aware of it or not, North Americans are increasingly eating and drinking globally. Rising beer imports, including many upscale foreign labels, exemplify the pattern. The nation's beer drinkers know no bounds to their thirsts, drawing diversely from Asian, Australian, European, and Latin American producers. (Data from *Modern Brewery Age*, 2003)

more surely under the region's spell. Perhaps most critical was the marriage between growing global demands for consumer goods and the rise of the multinational corporation, which was superbly structured to meet and cultivate those needs.

The results of these connections are not a simple Americanization of traditional cultures or a single, synthesized global culture shaped in the U.S. image. Still, millions of people, particularly the young, are strongly attracted by the North American emphasis on individualism, consumption, youth, and mobility. The wide popularity of English-language teaching programs in places from China to Cuba is testimony to the cultural power of the United States. Indeed, English is already an official language in more than 60 countries around the world. Global information flows illustrate the country's cultural influence on the world. Companies such as Time Warner and Walt Disney increasingly dominate multiple entertainment media. U.S. book and magazine publishing is thriving, and there is an expanding international appetite for everything from technical journals to romance novels and science fiction. The United States also dominates television; as television sets, cable networks, and satellite dishes spread, so do U.S. sitcoms, CNN, and MTV. The virtual world of the Internet and of social networking sites such as Facebook also reveals a strong North American presence that is increasingly shaping global cultural change. Movie screens tell a similar story.

The United States shapes the popular cultural landscape of every corner of the globe. Global corporate advertising, distribution networks, and mass consumption bring Cokes and Big Macs to Moscow and Beijing, golf courses to Thai jungles, Mickey and Minnie Mouse to Tokyo and Paris, and Avon cosmetics to millions of beauty-conscious Chinese. Western-style business suits have become the professional uniform of choice, while T-shirts and jeans offer standardized global comfort on days away from work. In the built landscape, central-city skylines become indistinguishable from one another, suburban apartment blocks take on a global sameness, and one airport hotel looks the same as another eight time zones away.

North American cultural influences are also reshaping the globe in unusual ways. While many people are aware of the worldwide popularity of blue jeans and Big Macs, what about divine healing, speaking in tongues, and religious conversion? Evangelical Protestantism has been on the rise in the United States since the 1970s and is now witnessing surging global popularity. The Pentecostal movement is a branch of evangelical Christianity that emphasizes a personal relation to Jesus Christ, the literal interpretation of the Bible, and the power of religious conversion. Along with other conservative, charismatic Christian denominations, these Protestant revivalists represent one of the world's fastest-growing religious movements and they are a unique aspect of North American culture that is reworking the map of religion in large areas of Africa and Latin America.

Evangelicals now make up sizable minorities in countries such as Kenya, Guatemala, and Brazil, particularly appealing to young, upwardly mobile populations. Often this Pentecostal message reflects a blend of Western Christianity and local, traditional religious practices. In Africa, for example, the movements can blend ancestor worship with evangelical Christianity and feature frenzied testimonials of conversion in churches such as Nigeria's Mountain of Fire and Miracles. In Guatemala, Pentecostal religions have grown explosively since 1990 and services take place all over Guatemala City, from luxury hotels to abandoned buildings. Also blossoming in the country are dozens of charismatic Christian radio and television stations, all proclaiming loudly the arrival of a new religious fervor. The result is a popular hybrid cultural phenomenon that fuses conservative North American Protestantism with the changing lives of 21st-century Latin Americans.

But U.S. cultural influences have not gone unchallenged. Canadian government agencies routinely chastise their radio, television, and film industries for letting in too much U.S. cultural influence. As an antidote to their overbearing southern neighbor, the government requires certain levels of Canadian content in much of the media programming. The French also criticize U.S. dominance in such media as the Internet. Public subsidies to France's Centre National de la Cinematographie are designed to foster filmmaking for a national audience deluged with English-language productions. Elsewhere, Iran has banned satellite dishes and many U.S. films, although illegal copies of top box-office hits often find their way through national borders.

GEOPOLITICAL FRAMEWORK: Patterns of Dominance and Division

In disarmingly simple fashion, North America's political geography brings together two of the world's largest countries. The creation of these political entities was the complex outcome of historical processes that might have created quite a different North American map. Once established, the two countries have coexisted in a close relationship of mutual economic and political interdependence. President John F. Kennedy summarized the links in a speech to the Canadian parliament in 1962: "Geography has made us neighbors, history has made us friends, economics has made us partners, and necessity has made us allies." That cozy continental relationship has not been without its tensions, and some persist today. In addition, both countries have had to deal with fundamental internal political complexities that have not only tested the limits of their federal structures, but also challenged their very existence as states.

Creating Political Space

The United States and Canada sprang from very different political roots. The United States broke cleanly and violently from England. The American Revolution fostered a powerful sense of nationalism that sped the process of spatial expansion and contributed toward the creation of a continental, indeed global, power. By contrast, Canada was a country of convenience, born from a peaceful, incremental separation from Britain and then assembled as a collection of distinctive regional societies that only gradually acknowledged their common political destiny.

Uniting the States Turning back the clock to the early 18th century reveals a political geography very much in the making. Beyond scattered frontiers of European settlement lay vast domains of Native American–controlled political space. Although boundaries were not formally surveyed or mapped, native peoples carved up the continent in an elaborate geography of homelands, allied territories, and enemy terrain. The creation of the United States replaced this continental division of political space with another.

With amazing rapidity, European nations and then the United States imposed their own political boundaries across the region. The 13 English colonies, sensing their common destiny after 1750, finally united two decades later and clashed violently with their colonial parent. By the 1790s, the young nation's political claims had reached the Mississippi River; the new republic was busily coercing land cessions from native peoples; and the Ordinance of 1787 had provided a template for western territory and state formation that served as the model of expansion for the next century. Soon the Louisiana Purchase (1803) nearly doubled the country's size, creating a new political domain that was as vast as it was unexplored. By mid-century, Texas had been annexed; treaties with Britain had secured the Pacific Northwest; and an aggressive war with Mexico had captured much of the Southwest. The territorial acquisition of Alaska (1867) and Hawaii (1898) eventually rounded out the present political domain of 50 states.

Assembling the Provinces Canada was created under quite different circumstances. The modern pattern of provinces was assembled in a slow and uncertain fashion. After the American Revolution, England's remaining territorial claims in the region came under the control of colonial administrators in British North America. The Quebec Act of 1774 allowed for continued French settlement in the St. Lawrence Valley and provided the initial template for governing the region. Soon, however, Anglo settlers near Lakes Ontario and Erie pressed for more local colonial representation. The result was the Constitutional Act of 1791, which divided the colony into Upper Canada (Ontario) and Lower Canada (Quebec). Frustration with that system led to the Act of Union in 1840, thereby reuniting the two Canadas. In 1867 the British North America Act united the provinces of Ontario, Quebec, Nova Scotia, and New Brunswick in an independent Canadian Confederation. The peaceful separation from the mother country also guaranteed to Quebec special legal and cultural privileges that set the stage for some of the modern power struggles within the country.

Once created, the Canadian Confederation grew in piecemeal fashion, more out of geographical convenience than from any compelling nationalism at work to unite the northern portion of the continent. Within a decade, the Northwest Territories (1870), Manitoba (1870), British Columbia (1871), and Prince Edward Island (1873) joined Canada, and the continental dimensions of the country took shape. Soon, the Yukon Territory (1898) separated from the Northwest Territories; Alberta and Saskatchewan gained provincial status (1905); and Manitoba, Ontario, and Quebec were enlarged (1912) north to Hudson Bay. Newfoundland finally joined in 1949. The addition of Nunavut Territory (1999), carved from the Northwest Territories, represents the latest change in Canada's political geography.

Continental Neighbors

Geopolitical relationships between Canada and the United States have always been close: their common 5,525-mile (8,900-kilometer) boundary requires both nations to pay close attention to one another. During the 20th century, the two countries lived largely in political harmony with one another. The Great Lakes, in particular, have been a setting for remarkable continental cooperation (Figure 3.30). In 1909, the Boundary Waters Treaty created the International Joint Commission, an early step in the common regulation of

Figure 3.30 Satellite Image of the Great Lakes North America's Great Lakes region features one of the most environmentally complex political boundaries in the world. Both Canada and the United States share responsibility (at a variety of local, state/provincial, federal levels) for managing the ecological health of the five Great Lakes shown (W to E: Superior, Michigan, Huron, Erie, and Ontario).

cross-boundary issues involving Great Lakes water resources, transportation, and environmental quality. The St. Lawrence Seaway (1959) opened the Great Lakes region to better global trade connections. With the signing of the Great Lakes Water Quality Agreement (1972) and the U.S.-Canada Air Quality Agreement (1991), the two nations have joined more formally in cleaning up Great Lakes pollution and in reducing acid rain in eastern North America. In 2009, environmental agreements were updated, opening the way to new cooperative cleanup efforts between the two countries.

The two nations are also key trading partners. The United States receives about 85 percent of Canada's exports and supplies more than half of its imports. Conversely, Canada is the United States' most important trading partner, accounting for roughly 20 percent of its exports and imports. One landmark agreement reached in 1989 was the signing of the bilateral (two-way) Free Trade Agreement (FTA). Five years later, the larger **North American Free Trade Agreement (NAFTA)** extended the alliance to Mexico. Paralleling the success of the European Union (EU), NAFTA has forged the world's largest trading bloc, including more than 400 million consumers and a huge free-trade zone that stretches from beyond the Arctic Circle to Latin America. But NAFTA has also posed challenges for in the region: U.S. jobs have been lost to Mexico and Mexico's economy has become increasingly dependent on demand from the United States, creating problems in times of declining consumption in North America.

Political conflicts still divide North Americans (Figure 3.31). Environmental issues produce cross-border tensions, especially when environmental degradation in one nation affects the other. For example, Montana's North Flathead River originates in British Columbia, where Canadian logging and coal mining operations periodically threaten fisheries and recreational lands south of the border.

More generally, tighter U.S. regulations since 2009 have made it more difficult to cross the border in either direction. Reflecting security concerns in the United States, the regulations demand that persons crossing the border present a passport or other approved form of identification just as they would on the border with Mexico. Many people in Canada cite the new rules as potentially harmful to tourism between the two countries.

Agricultural and natural resource competition also causes occasional controversy between the two neighbors. The appearance of mad cow disease in Canadian livestock curtailed exports to the United States and elsewhere. Furthermore, when the disease appeared in U.S. cattle in 2003, Canadian sources were suspected, raising tensions between the two nations. Problems have periodically developed when Canadian wheat and potato growers were accused of dumping their products into U.S. markets, thus depressing prices and profits for U.S. farmers. Similar issues have arisen in the logging industry, although a 2006 agreement signed between the two countries has lessened tensions over that issue. In the far north, the two countries disagree on the maritime boundary between Yukon Territory and the state of Alaska. In addition, the United States does not agree with the assertion that a potential Northwest Passage opening across a more ice-free Arctic Ocean would essentially be within Canada's territorial waters.

The Legacy of Federalism

The United States and Canada are **federal states** in that both nations allocate considerable political power to units of government beneath the national level. Other nations, such as France, have traditionally been **unitary states**, in which power is centralized at the national level. Federalism leaves many political decisions to local and regional governments and often allows distinctive cultural and political groups to be recognized as distinct entities within a country. Although both nations have federal constitutions, their origins and evolution are very different. The U.S. Constitution (1787), created out of a violent struggle with the powerful British nation, specifically limited centralized authority, giving all unspecified powers to the states or the people. In contrast, the Canadian Constitution (1867), which created a federal parliamentary state, was an act of the British Parliament. Originally, it reserved most powers to central authorities and maintained many political links between Canada and the British Crown. Ironically, the evolution of the United States as a federal republic produced an increasingly powerful central government, while Canada's geopolitical balance of power shifted toward more provincial autonomy and a relatively weak national government. For example, the federal government largely controls U.S. public lands, but in Canada provincial authorities retain power over public Crown lands.

Quebec's Challenge The political status of Quebec remains a major issue within Canada (see Figure 3.31). Economic disparities between the Anglo and French populations have reinforced cultural differences between the two groups, with the French Canadians often suffering when compared with their wealthier neighbors in Ontario. Beginning in the 1960s, a separatist political party in Quebec (the Parti Quebecois) increasingly voiced French Canadian concerns. When the party won provincial elections in 1976, it declared French the official language of Quebec. Formal provincial votes over the question of remaining within Canada were held in 1980 and 1995. Both measures failed. Since then, support for separation has ebbed in favor of a more modest strategy of increased "autonomy" within Canada. In elections held in 2007, for example, the Parti Quebecois ran a poor third in Quebec, suggesting its separatist rhetoric is no longer as popular within the province.

Native Peoples and National Politics

Another challenge to federal political power has come from North American Indian and Inuit populations, in both Canada and the United States. Within the United States, the renewed assertion of Native American political power began in the

Figure 3.31 Geopolitical Issues in North America Although Canada and the United States share a long and peaceful border, many political issues still divide the two countries. In addition, internal political conflicts cause tensions, particularly in multicultural Canada.

1960s and marked a decisive turn away from earlier policies of assimilation. Since passage of the Indian Self-Determination and Education Assistance Act of 1975, the trend has been toward increased Native American autonomy. The Indian Gaming Regulatory Act (1988) offered potential economic independence for many tribes. By 2006, Indian gaming operations nationally netted tribes almost $25 billion annually. In the western American interior, where Indians control roughly 20 percent of the land, tribes are solidifying their hold on resources, reacquiring former reservation acreage, and participating in political interest groups such as the Native American Fish and Wildlife Society and the Council of Energy Resource Tribes. In Alaska, native peoples acquired title to 44 million acres (18 million hectares) of land in 1971 under the Alaska Native Claims Settlement Act.

In Canada even more ambitious challenges to a weaker centralized government have yielded dramatic results. As natives pressed their claims for land and political power in the 1970s, Canada established the Native Claims Office (1975) and began negotiating settlements with various groups, particularly within the country's vast northern interior. Agreements with native peoples in Quebec, Yukon, and British Columbia turned over millions of acres of land to aboriginal control and increased native participation in managing remaining public lands. By far, the most ambitious agreement has been to create the territory of Nunavut out of the eastern portion of the Northwest Territories in 1999 (see Figure 3.31). Nunavut is home to 30,000 people (85 percent Inuit) and is the largest territorial/provincial unit within Canada (Figure 3.32). Its creation represents a new level of native self-government in North America, particularly significant in a part of the world witnessing rapid climate change. Recently, agreements between the federal Parliament and British Columbia tribes have initiated a similar move toward more native self-government in that western province. The 6,400 members of the Nisga'a tribe, for example, now control a 770-square-mile (1,992-square-kilometer) portion of the Nass River valley near the Alaska border (see Figure 3.25). Elsewhere, recent disputes over land claims in Ontario (near Toronto) and in oil-rich Alberta are reminders that Canadians, much like their neighbors to the south, are still struggling with defining the political status of their native populations.

The Politics of U.S. Immigration

Immigration policies are hotly contested within the United States. Four key issues remain at the center of the debate. First, there are ongoing disagreements concerning the overall numbers of legal immigrants that should be allowed into the country. Some groups, such as the Federation for American Immigration Reform (FAIR), suggest that sharply reduced numbers of immigrants would protect American jobs and allow for a more gradual assimilation of existing foreigners. But other groups such as the Cato Institute take the opposite position, proposing to loosen existing restrictions on immigrants in a move to spur economic growth and business expansion.

A second major issue, particularly along the border with Mexico, is how to tighten up on the daily flow of illegal immigrants. Since the terrorist attacks in 2001, many have argued that the country's wide-open southern border is a national security issue. Recent federal legislation has mandated an increased number of border patrol agents and more than 20,000 officers monitor the boundary. Almost 700 miles (1,125 kilometers) of fencing also have been built or improved (Figure 3.33). Others argue such measures are pointless and merely sour relations with Mexico. Meanwhile, hundreds of thousands of foreigners are apprehended along the border annually, although the economic recession of the late 2000s reduced the numbers from earlier in the decade.

Third, the growth of drug-related violence near the border has soured relations between the two countries. Mexico remains the leading source of methamphetamine, heroin, and marijuana for the United States and is a key transit nation for northward bound cocaine originating in South America. In addition, more than 10,000 deaths, mostly in northern Mexico, were tied to the drug business in 2008 and 2009. Some of that violence has spilled north into places such as El Paso and Phoenix and some American officials worry that the Mexican government has fundamentally lost effective political control of its northern border.

Finally, there is no political consensus on a fair policy to deal with millions of existing undocumented workers within the United States. Some policymakers have advocated stricter felony-level penalties for illegal immigrants, while others have proposed loosening requirements for citizenship or some form of amnesty to enable illegal immigrants to more easily enter the mainstream of American society. Suggested political compromises have included expanding the country's guest worker program or requiring illegal immigrants to return to their home countries ("touching base") before they can reapply for legal status within the United States.

Figure 3.32 Life in Nunavut This scene on Baffin Island is from Iqaluit, the largest urban center in the new Canadian province of Nunavut.

A Global Reach

The geopolitical reach of the United States, in particular, has taken its influence far beyond the bounds of North America. The Monroe Doctrine (1824) asserted that U.S. interests were hemispheric and transcended national boundaries, but it was not until after 1895 that the United States accelerated its global expansion. Two principal settings served as early laboratories for political imperatives in the United States. In the Pacific, the United States claimed the Philippines as a prize of the Spanish-American War (1898), and further annexations of Guam (1898) and the Hawaiian Islands (1898) began the country's 20th-century dominance of the region. In Central America and the Caribbean, the growing role of the American military between 1898 and 1916 shaped politics in Cuba, Puerto Rico, Panama, Nicaragua, Haiti, Mexico, and else-

Figure 3.33 International Border, Nogales, Arizona At Nogales, a crowded hillside on the Mexican side of the border contrasts with the lights, surveillance cameras, and patrol roads evident on the U.S. (Arizona) side of the international boundary. The southern border of the United States has still proven remarkably porous, annually allowing in hundreds of thousands of illegal immigrants into the country.

where. Further, the country's role in World War I raised its stakes in European affairs.

The 1920s and 1930s briefly returned the United States to isolationist policies, but World War II and its aftermath forever redefined the country's role in world affairs. Victorious in both the Atlantic and Pacific theaters, postwar America emerged from the conflict as the world's dominant political power. Quickly, however, a resurgent Soviet Union challenged the United States, and the Cold War began in the late 1940s. In response, the Truman Doctrine promised aid to struggling postwar economies and actively challenged communist expansion in Europe and elsewhere. The United States also fashioned multinational political and military agreements, such as those establishing the North Atlantic Treaty Organization (NATO) and the Organization of American States (OAS), which were designed to cast a broad umbrella of U.S. protection across much of the noncommunist world. Violent conflicts in Korea (1950–1953) and Vietnam (1959–1975) pitted U.S. political interests against communist attempts to extend their Asian dominance beyond the Soviet Union and China. Tensions also ran high in Europe as the Berlin Wall crisis (1961) and nuclear weapons deployments by NATO- and Soviet-backed forces brought the world closer to another global war. The Cuban missile crisis (1962) reminded Americans that traditional political boundaries provided little defense in a world uneasily brought closer together by technologies of potential mass destruction.

Even as the Cold War gradually receded during the late 1980s, the global political reach of the United States continued to expand. A few examples suggest the pattern. Interventionist policies in Central America favored regimes friendly to the United States. President Carter's successful Middle East Peace Treaty between Israel and Egypt (1979) guaranteed a continuing diplomatic and military presence in the eastern Mediterranean. When Iraq's Saddam Hussein threatened Persian Gulf oil supplies in 1990, the United States led a United Nations coalition to contain the aggression. In the late 1990s, Serbian aggression within Kosovo prompted an American- and NATO-led intervention, which included major air attacks on the Serbian capital of Belgrade (1999) and a peacekeeping presence (with the UN) in the disputed area of Kosovo. The ongoing conflicts in Afghanistan (2001-) and Iraq (2003-) suggest that the United States will continue playing a key, highly visible role in the world's political affairs.

The U.S. military has also changed its worldwide distribution substantially since 2000. Military planners argue that in the future less emphasis should be given to housing large numbers of troops in relatively friendly foreign "hub" settings such as Germany, South Korea, and Japan. A growing number of the U.S. military's 1.4 million active-duty personnel appear headed for more spartan assignments located near source regions for terrorists and in arenas of potential conflict. Iraq and Afghanistan continue to be primary areas of interest. From the drug- and rebel-filled jungles of Colombia to the terrorist training camps of the Philippines, the region of growing global deployment also includes much of Africa, the Middle East, Central Asia, and Southeast Asia.

ECONOMIC AND SOCIAL DEVELOPMENT: Geographies of Abundance and Affluence

Along with its global political clout, North America possesses the world's most powerful economy and its most affluent population. Its 340 million people consume huge quantities of global resources but also produce some of the world's most sought-after manufactured goods and services. North America's size, geographic diversity, and resource abundance have all contributed to the region's global dominance in economic affairs. More than that, however, the region's human capital—the skills and diversity of its population—has enabled North Americans to achieve high levels of economic development (Table 3.2).

An Abundant Resource Base

North America is blessed with a varied storehouse of natural resources. The region's climatic and biological diversity, its soils and terrain, and its abundant energy, metals, and forest resources have provided a variety of raw materials for development. Indeed, the direct extraction of natural resources

TABLE 3.2 Development Indicators

Country	GNI Per Capita PPP 2008	GDP Average Annual % Growth (2000–08)	Human Development Index (2007)[a]	Percent Population Living Below $2 a Day	Life Expectancy (2010)[b]	Under Age 5 Mortality Rate (1990)	Under Age 5 Mortality Rate (2008)	Gender Equity (2008)[c]
Canada	38,710	2.5	0.966		81	6	6	99
United States	48,430	2.4	0.956		78	11	8	100

[a]*United Nations, Human Development Report, 2009.*

[b]*Population Reference Bureau, World Population Data Sheet, 2010.*

[c]*Gender Equity—Ratio of female to male enrollments in primary and secondary school. Numbers below 100 have more males in primary/secondary school, numbers above 100 have more females in primary/secondary schools.*

Source: World Bank, World Development Indicators, 2010.

still makes up 3 percent of the U.S. economy and 6 percent of the Canadian economy. Some of these North American resources are then exported to global markets, while other raw materials are imported to the region.

Opportunities for Agriculture North Americans have created one of the most efficient food-producing systems in the world and agriculture remains a dominant land use across much of the region (Figure 3.34). Farmers practice highly commercialized, mechanized, and specialized agriculture. The system emphasizes the importance of efficient transportation, global markets, and large capital investments in farm machinery. Today, agriculture employs only a small percentage of the labor force in both the United States (1 percent) and Canada (2 percent). At the same time, changes in farm ownership have sharply reduced the number of operating units while average farm sizes have steadily risen.

The geography of North American farming represents the combined impacts of (1) diverse environments, (2) varied continental and global markets for food, (3) historical patterns of settlement and agricultural evolution, and (4) the growing role of **agribusiness**, or corporate farming, in which large-scale business enterprises control closely-integrated segments of food production, from farm to grocery store. In the Northeast, dairy operations and truck farms take advantage of their nearness to major cities in Megalopolis and southern Canada. Corn and soybeans dominate the Midwest and western Ontario, where a tradition of mixed farming combines the growing of feed grains with the production and fattening of livestock (see "Exploring Global Connections: Ginseng Heartland"). To the south, only remnants of the old Cotton Belt remain, largely replaced by varied subtropical specialty crops; poultry, catfish, and livestock production; and commercial logging (Figure 3.35). Farther west, extensive, highly mechanized commercial grain-growing operations stretch from Kansas to Saskatchewan and Alberta. Depending on surface and groundwater resources, irrigated agriculture across western North America also offers opportunities for farming. Indeed, California's agricultural output, nourished particularly by large agribusiness operations in the irrigated Central Valley, accounts for more than 10 percent of the nation's farm economy.

Industrial Raw Materials North Americans produce and consume huge quantities of other natural resources. While the region is well endowed with a variety of energy and metals resources, the scale and diversity of the North American economy have led to the need to import additional raw materials. Petroleum use in the United States exemplifies the pattern. The country produces about 12 percent of the world's oil but consumes about 25 percent. As a result, the United States imports about 60 percent of its oil, much of it coming from the Americas (Venezuela, Mexico, and Canada), the Middle East (Saudi Arabia), and Africa (Nigeria and Angola). Within the region, the major areas of oil and gas production are the Gulf coast of Texas and Louisiana; the Central Interior, including West Texas, Oklahoma, and eastern Kansas; Alaska's North Slope; and Central Canada (especially Alberta).

The most abundant fossil fuel in the United States is coal, but its relative importance in the overall energy economy declined in the 20th century as industrial technologies changed and environmental concerns grew. Nonetheless, the country's 400-year supply of coal reserves (27 percent of the world's total) will be an important energy resource, both for domestic consumption as well as for export. The nation's leading coal-producing region is Appalachia, but demand has been increasing for the environmentally cleaner low-sulfur coals of the western interior (Great Plains, Intermountain West, and Alberta).

North America also remains a major producer of metals resources, although global competition, rising extraction costs, and environmental concerns pose challenges for this sector of the economy. Still, the region is endowed with more than 20 percent of the world's copper, lead, and zinc reserves, and it accounts for more than 20 percent of global gold, silver, and nickel production.

Creating a Continental Economy

The timing of European settlement in North America was critical in its rapid economic transformation. The region's abundant resources came under the control of Europeans possessing new technologies that reshaped the landscape and reorganized its economy. By the 19th century, North Americans actively contributed to those technological

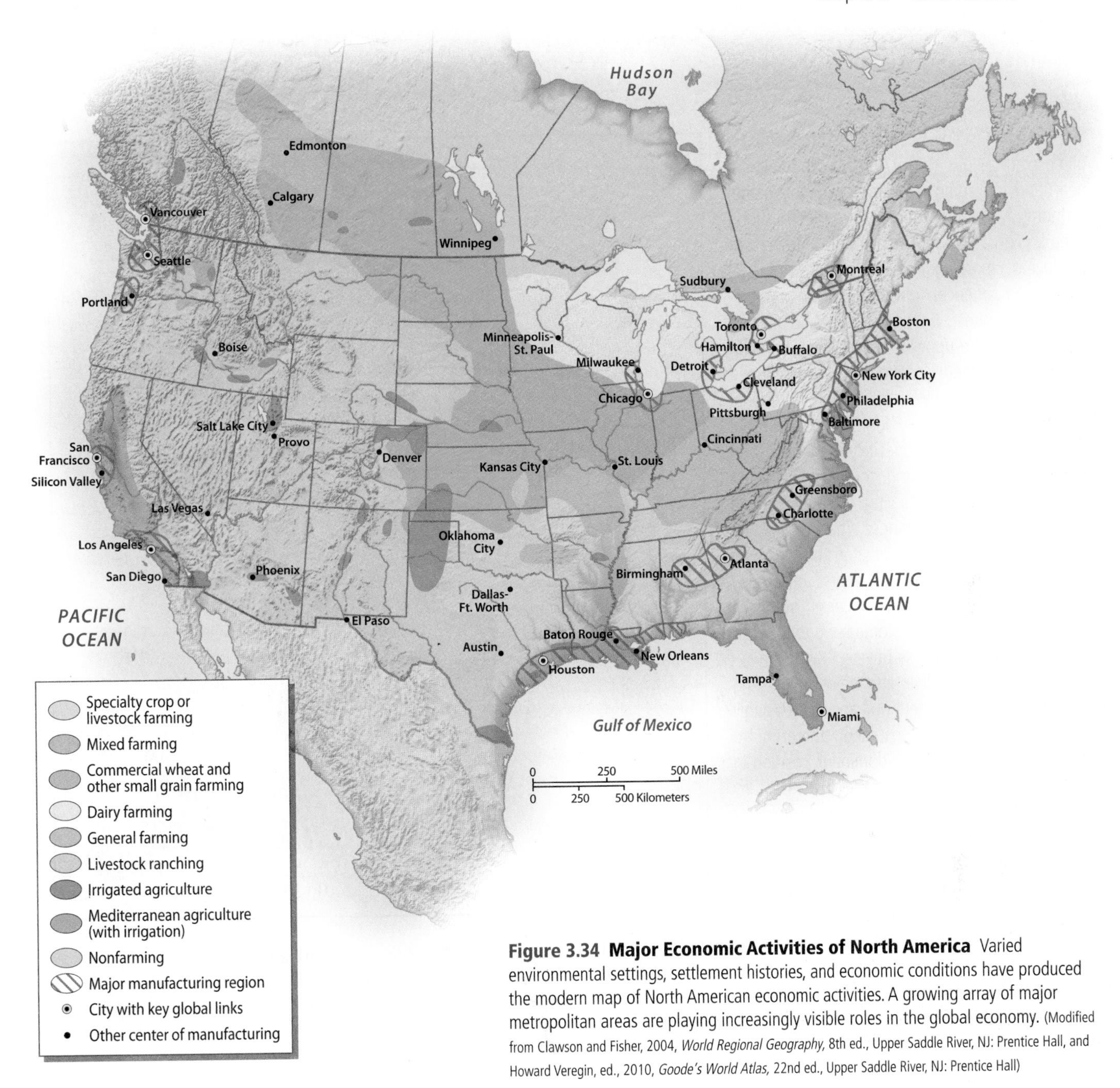

Figure 3.34 Major Economic Activities of North America Varied environmental settings, settlement histories, and economic conditions have produced the modern map of North American economic activities. A growing array of major metropolitan areas are playing increasingly visible roles in the global economy. (Modified from Clawson and Fisher, 2004, *World Regional Geography*, 8th ed., Upper Saddle River, NJ: Prentice Hall, and Howard Veregin, ed., 2010, *Goode's World Atlas*, 22nd ed., Upper Saddle River, NJ: Prentice Hall)

Figure 3.35 Specialty Agriculture in the American South Today, the agriculture of the American South focuses on specialty operations engaged in subtropical cropping, commercial forestry, and poultry, catfish, and livestock production. This Florida farm grows basil, a mint plant widely used as a cooking herb.

EXPLORING GLOBAL CONNECTIONS Ginseng Heartland

The quiet landscape of rural Wisconsin offers a glimpse into how North America's agricultural heartland is forging new links with the global economy. Marathon County, Wisconsin, has long been known as North America's ginseng capital, producing more than 75 percent of the continent's American ginseng (*Panax quinquefolius, L.*) (Figures 3.5.1 and 3.5.2). Wild ginseng grows across the upper Midwest and beginning in the 1860s, dried ginseng roots were shipped to China, where they were used in traditional Chinese medicine. Since the 1970s, commercial farmers in the region, specifically in Marathon County, began cultivating the root in larger quantities and solidifying their trading relationship with the Chinese. The Schumacher family, for example, has marketed their ginseng to Asia for more than 40 years and today promotes their family-run business over the Internet (with sales to customers worldwide) as well as directly to Chinese buyers.

But all is not well in the heart of ginseng country. Global competition and increasingly demanding Chinese consumers threaten to unseat Marathon County's dominant position in the business. Recently, Chinese buyers, many representing large pharmaceutical and consumer products companies in Asia, have complained that they are not getting the quality product they need to sell to their discriminating overseas customers. Today, American ginseng in China is sold in a wide variety of powder capsules, crystal extracts, toothpaste, tea, candy, and chewing gum. Wisconsin growers have assured their buyers that they can meet the challenge. But falling prices, competition from government-subsidized growers in Canada,

Figure 3.5.1 The Ginseng Heartland Wisconsin's Marathon County has emerged as a key global player in supplying Chinese consumers with high-quality ginseng used in a variety of medicinal and household products.

changes. In addition, new natural resources were developed in the interior and new immigrant populations arrived in large numbers. In the 20th century, although natural resources remained important, industrial innovations and more jobs in the service sector added to the economic base and extended the country's global reach.

Connectivity and Economic Growth Dramatic improvements in North America's transportation and communication systems laid the foundation for urbanization, industrialization, and the commercialization of agriculture. Indeed, the region's economic success was a function of its **connectivity**, or how well its different locations became linked with one another through improved transportation and communications networks. Those links greatly facilitated the potential for interaction between locations and dramatically reduced the cost of moving people, products, and information over long distances. Before 1830, North American connectivity gradually improved with the help of better roads, but it still took a week to travel from New York City to Ohio. For commercial traffic, canal construction also facilitated the movement of bulky goods. More than 1,000 miles (1,600 kilometers) of canals crisscrossed the eastern United States, including the Erie Canal (completed in 1825), which linked New York City with the North American interior.

Tremendous technological breakthroughs revolutionized North America's economic geography between 1830

Figure 3.5.2 Ginseng Farm, Marathon County, Wisconsin This ginseng farm in Marathon County, Wisconsin, illustrates the intensive nature of this agricultural operation. Increasingly affluent Chinese consumers are demanding high-quality American ginseng in a myriad of products.

and cheap low-quality ginseng grown in Asia are all pressuring the Wisconsin producers. In 1990, there were 1,500 ginseng farmers in the state, but that number now has dwindled to only 150–200 growers.

Meanwhile, the Chinese are becoming a more visible economic presence elsewhere in the state. In addition to jawboning the American ginseng producers to improve the quality of their crops, the Chinese are snatching up cheap pieces of real estate, including an empty mall northwest of Milwaukee and an inexpensive industrial park south of the city. Chinese buyers have also increased their purchases of Wisconsin soybeans and livestock products. The University of Wisconsin has also become a destination for hundreds of Chinese students, particularly those specializing in science and engineering.

This new global relationship between the Wisconsin heartland and the growing power of Chinese capital, Chinese consumers, and Chinese students suggests a reworking of North America's linkages with that key Asian nation. Hopefully, the Schumachers and other ginseng producers in Marathon County can profit in the process, perhaps spending their hard-earned dollars in the Chinese-owned mall down the road.

Source: Adapted from John Pomfret, "The Chinese are changing us," *The Washington Post,* November 14, 2009.

and 1920. By 1860 more than 30,000 miles (48,387 kilometers) of railroad track had been laid in the United States, and the network grew to more than 250,000 miles (403,226 kilometers) by 1910. Farmers in the Midwest and Plains found ready markets for their products in cities hundreds of miles away. Industrialists collected raw materials from faraway places, processed them, and shipped manufactured goods to their final destinations. The telegraph brought similar changes to information: Long-distance messages flowed across eastern North America by the late 1840s, and 20 years later undersea cables linked the region to Europe, another milestone in the process of globalization.

North America's transportation and communications systems were modernized further after 1920. Automobiles, mechanized farm equipment, paved highways, commercial air links, national radio broadcasts, and dependable transcontinental telephone service reduced the cost of distance across the region. After World War II, continental connectivity also benefited from the St. Lawrence Seaway between the Atlantic Ocean and Great Lakes (1959), vast improvements in jet airline connections, and the increasing prevalence of television. Perhaps most importantly, the region has taken the lead in the global information age, integrating computer, satellite, telecommunications, and Internet technologies in a web of connections that assists the flow of knowledge both within the region and beyond (Figure 3.36). Since 1990, North America's transformative role in information technologies has been one of its most profound, innovative contributions to globalization and global economic change.

The Sectoral Transformation Changes in employment structure signaled North America's economic modernization just as surely as its increasingly interconnected society. The **sectoral transformation** refers to the evolution of a nation's labor force from one dependent on the *primary* sector (natural resource extraction) to one with more employment in the *secondary* (manufacturing or industrial), *tertiary* (services), and *quaternary* (information processing) sectors. For example, with agricultural mechanization, lower demands for primary-sector workers are replaced by new opportunities in the growing industrial sector. In the 20th century, new services (trade, retailing) and information-based activities (education, data processing, research) created other employment opportunities. Today, the tertiary and quaternary sectors employ more than 70 percent of the labor force in both Canada and the United States. These recent trends also reveal the tangible imprint of globalization on the North American labor force. Much of North America's sustained growth in the tertiary and quaternary sectors is directly tied to the ability of those industries to export innovations in services, financial management, and information processing to a worldwide clientele.

Regional Economic Patterns North America's industries show important regional patterns. **Location factors** are the varied influences that explain *why* an economic activity is located where it is. Many influences, both within and beyond the region, shape patterns of economic activity. Patterns of industrial location illustrate the concept (see Figure 3.34). The historical manufacturing core includes Megalopolis (Boston, New York, Philadelphia, and Baltimore), southern Ontario (Toronto and Hamilton), and the industrial Midwest. The region's proximity to *natural resources* (farmland, coal, and iron ore); increasing *connectivity* (canals and railroad networks, highways, air traffic hubs, and telecommunications centers); a ready supply of *productive labor*; and a growing national, then global, *market demand* for its industrial goods encouraged continued *capital investment* within the core.

Traditionally, the core has dominated in the production of steel, automobiles, machine tools, and agricultural equipment and played a critical role in producer services such as banking and insurance. In the last half of the 20th century, industrial and service-sector growth shifted to the South and West. Cities of the South's Piedmont manufacturing belt (Greensboro to Birmingham) grew after 1960, partly because lower labor costs and Sun Belt amenities attracted new investment. By 2007, for example, the North Carolina "research triangle" area between Raleigh, Durham, and Chapel Hill emerged to become the nation's third largest biotech cluster behind California and Massachusetts. The Gulf Coast industrial region is strongly tied to nearby fossil fuels that provide raw materials for its many energy-refining and petrochemical industries (Figure 3.37).

The varied West Coast industrial region stretches from Vancouver, British Columbia, to San Diego, California, (and beyond into northern Mexico), and it demonstrates the increasing importance of Pacific Basin trade. Large aerospace operations in the West also suggest the role of *government spending* as a location factor. Silicon Valley is now one of North America's leading regions of manufacturing exports. Its proximity to Stanford, Berkeley, and other universities demonstrates the importance of *access to innovation and research* for many fast-changing high-technology industries. Silicon Valley's location also shows the advantages of *agglomeration economies,* in which many companies with similar and often integrated manufacturing operations locate near one another (Figure 3.38). Smaller places such as Provo, Utah, and Austin, Texas, also specialize in high-technology industries and demonstrate the growing role of *lifestyle amenities* in shaping industrial location decisions, both for entrepreneurs and for the skilled workers who need to be attracted to such opportunities.

Persisting Social Issues

Profound economic and social problems shape the human geography of North America. Even with its continental wealth, great differences persist and have increased between rich and poor. High per capita incomes in the United States and Canada fail to reveal the differences in wealth within the two countries (see Table 3.2). Broader measures of social well-being suggest disparities in health care and education. Race, particularly within the United States, continues to be an issue of overwhelming impor-

Figure 3.36 Cell Phone Nation These college graduates stay connected with friends and family through picture-phone technology. The telecommunications and information services industries promise to continue reshaping North American lifestyles in the decades to come.

Figure 3.37 Gulf Coast Petroleum Refining Petroleum-related manufacturing has transformed many Gulf coast settings. Much of Houston's 20th-century growth has been fueled by the dramatic expansion of oil-related industries. The port of Houston remains a major center of North America's refining and petrochemical operations.

tance. In addition, both nations face problems associated with gender inequity and aging populations. One consequence of globalization is that many of these economic and social challenges are increasingly defined beyond the region. Poverty in the rural American South may be related to low Asian wage rates, for example, and a viral outbreak in Hong Kong might be only a plane flight away from suburban Vancouver.

The global economic downturn of the late 2000s rippled through the economy of both the United States and Canada. Unemployment levels soared in 2008 and 2009 and hovered between 8.5 and 10 percent by the end of the decade. Young and poor African Americans and Hispanics typically fared the worst. Among young Hispanic men, U.S. unemployment rates in 2009 reached a staggering 19 percent. The unemployment rate for young African American men stood at 30 percent. Many people also lost health-care coverage and deferred medical procedures to save money. Real estate values and home ownership rates fell. Many people lost their homes to foreclosure. Poverty rates rose for the first time in years. Tax revenues in both nations fell. States such as California, Nevada, and Arizona, once home to real estate and construction booms, witnessed some of the most dramatic economic declines.

Wealth and Poverty The North American landscape vividly displays contrasting signatures of wealth and poverty. Elite suburbs, gated and guarded neighborhoods, upscale shopping malls, and posh resorts are all geographical expressions of the spatial exclusivity that characterizes many wealthier North American communities, particularly within the United States (Figure 3.39). On the other hand, signatures of poverty are displayed in a great range of local and regional settings. Substandard housing, abandoned property, aging infrastructure, and unemployed workers are visual reminders of the gap between rich and poor within the region. Specifically, within the United States, black household incomes remain only 67 percent of the national average, while Hispanic incomes fare slightly better at 80 percent of the national average.

The distribution of wealth and poverty varies widely across the United States and Canada. Many of America's wealthiest communities are suburbs on the edge of large metropolitan areas. Just outside New York City, Fairfield, Connecticut, is one of the nation's wealthiest counties; similar settings are found in suburban New Jersey, Maryland, and Virginia, as well as on the peripheries of dozens of southern and western cities. Resort and retirement communities are havens for the rich as well: Palm Beach, Florida, and Aspen, Colorado, have some of the nation's most desirable real estate and costliest housing. In terms of average income, the Northeast and West remain the richest regions in the United States. In Canada, Ontario, Alberta and British Columbia are the country's wealthiest provinces, with Vancouver's high housing prices vying with those of San Francisco.

Poverty levels have declined overall in both countries since 1980, but poor populations are still clustered in a variety of geographical settings. About 13 percent of the U.S. population and 16 percent of the Canadian population live in poverty. Poverty rates for children (18 percent of children lived in poverty in the U.S., 15 percent in Canada) vary considerably across North America and act as one regional

Figure 3.38 Silicon Valley The high-technology industrial landscape of California's Silicon Valley contrasts sharply with the look of traditional manufacturing centers. Here, similar industries form complex links, benefiting from their proximity to one another and to nearby universities such as Stanford and Berkeley.

measure of overall economic development (Figure 3.40). The problems of the rural poor remain major regional social issues in Appalachia, the Deep South, and the Southwest. In the United States, ethnicity is often linked with rural poverty, particularly in the case of southern blacks, Hispanics in the Borderlands, and Native Americans on reservation lands in the Southwest. Many poor people in the United States, however, live in central-city locations, and the links between ethnicity and poverty are even stronger in these communities. Nationally, 25 percent of the country's African Americans and 22 percent of its Hispanic population live below the poverty line.

Additional 21st-Century Challenges Measures of social well-being in North America compare favorably with those of most other world regions (see Table 3.2), although many economic and social challenges confront the region. North Americans compete globally, and the declining percentage of stable, high-paying jobs with long-term security and benefits suggests that companies and employees are scrambling to adjust to the uncertainties of the world economy.

Education is also a major public policy issue in Canada and the United States. Although political parties differ in their approach, most public officials agree that more investment in education can only improve North America's chances for competing successfully in the global marketplace. In the United States, dropout rates average about 13 percent for 18- to 24-year-olds, but rates are much higher in many poor urban neighborhoods and rural areas. In addition, race plays a key role: American whites are two or three times more likely than African Americans or Hispanics to hold a college degree. Another challenge, particularly in the United States, is debating an effective national education policy in the face of a very strong tradition of local control of education.

Gender remains a key social issue in both countries. Since World War II, both nations have seen great changes in the role that women play in society, but the **gender gap** is yet to be closed when it comes to differences between men and women in salary issues, working conditions, and political power. Women widely participate in the workforce in both countries and are as educated as men (see Table 3.2), but they still earn only about 78¢ for every $1 that men earn. Although the gap is shrinking, corporate America's "glass ceiling" still makes it extraordinarily difficult for women to advance to top managerial positions and pay scales. Even more daunting is the fact that women head the vast majority of poorer single-parent families in the United States and many of these women are unwed mothers. Indeed, in 2009, 39 percent of all births in the United States were to unwed mothers, up from only 12 percent in 1972.

Canadian women, particularly single mothers who work full-time, are also greatly disadvantaged, averaging only about two-thirds of the salaries of Canadian men. In fact, more than half the single women who head families in Canada are classified as low-income workers. In addition, political power remains largely in male hands, even though women make up the majority of the population and electorate. Although Canadian

Figure 3.39 Gated America An automatic gate protects the entrance of a gated community in Apollo Beach, near Tampa, Florida.

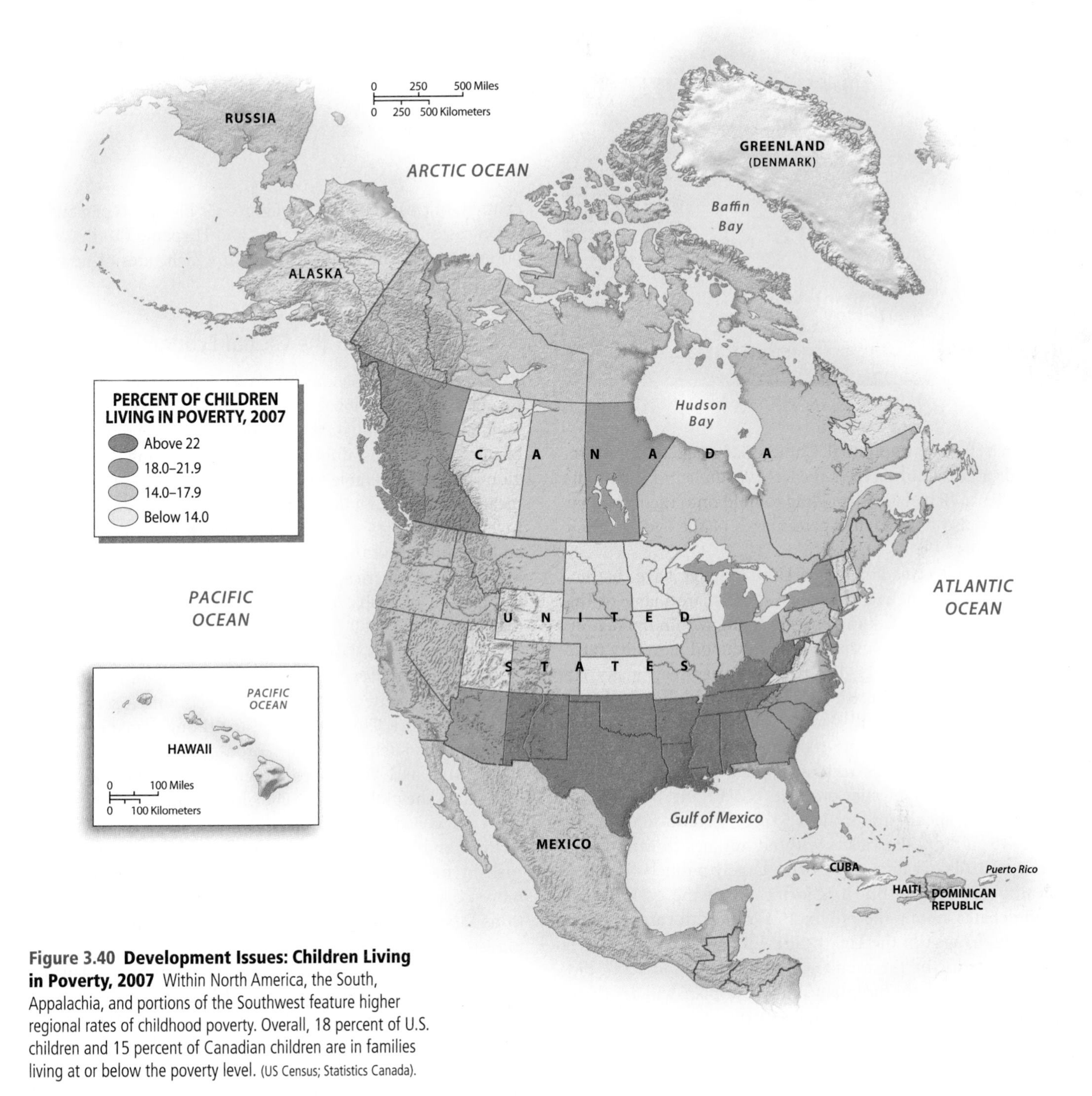

Figure 3.40 Development Issues: Children Living in Poverty, 2007 Within North America, the South, Appalachia, and portions of the Southwest feature higher regional rates of childhood poverty. Overall, 18 percent of U.S. children and 15 percent of Canadian children are in families living at or below the poverty level. (US Census; Statistics Canada).

women have voted since 1918 and U.S. women since 1920, females in the early 21st century remain in the minority in the Canadian Parliament and the U.S. Congress.

Health care and aging are also key concerns within a region of graying baby boomers. A recent report on aging in the United States predicted that 20 percent of the nation's population will be older than 65 by 2050. Today, the most elderly senior citizenry (age 85+) constitute more than 12 percent of all seniors and they are the fastest-growing part of the population, particularly among women. Poverty rates are also higher for seniors. With fewer young people to support their parents and grandparents, officials debate the merits of reforming social security programs. Whatever the outcome of such debates, the geographical consequences of aging are already abundantly clear. Whole sections of the United States—from Florida to southern Arizona—have become increasingly oriented around retirement (Figure 3.41). Communities cater to seniors with special assisted-living arrangements, health-care facilities, and recreational opportunities.

Aging baby boomers will no doubt experience a growing variety of specialized retirement communities. Today, one can spend one's leisure time in gay and lesbian retirement communities, a new facility (near Nashville) that caters to country music lovers, or a California community

Figure 3.41 Tomorrow's Baby Boom Landscape? Businesses at this Sun City minimall near Phoenix suggest the age-related services increasingly in demand by retiring baby boomers. As North Americans grow older, entire subregions may be devoted to age-dependent demands for housing, recreation, and health services.

oriented around traditional Chinese language and culture. Florida and California have the largest number of senior citizens; Farm Belt states struggling to hold onto their younger residents also have higher percentages of the aged in their populations.

Although Canada and the United States have distinctive health-care delivery systems, they share many of the health-related benefits and costs of living within the region. As long average lifespans and low childhood mortality rates suggest (see Table 3.2), the two countries reap many rewards from modern health-care systems that offer the latest in high technology. Still, North Americans voice concern over escalating health-care costs and uneven access to premium care. While U.S. residents spend more than 15 percent of gross domestic product (GDP) on health care (Canadians spend slightly less), there were in 2009 more than 45 million Americans without any health-care insurance. Canada has long possessed a largely publicly financed health-care system, but it is facing a growing cost crunch and debates among the provinces over health-care policies. Recent changes in the U.S. health care system are targeted towards reducing the number of uninsured Americans, but many critics of the program argue the plan will result in higher costs and more federal government intrusion into the health care sector.

The rising incidence of chronic diseases associated with aging (heart disease, cancer, and stroke are the three leading causes of death) will continue to pressure both health-care systems. In addition, hectic lives, often oriented around fast food and more meals eaten out (the average American consumed 603 more calories daily in 2000 than 20 years earlier), have contributed to rapidly growing rates of obesity. On a typical day, more than 30 percent of American children and adolescents report eating fast food. The long-term results are sobering: almost two-thirds of adult Americans are overweight and the trend is contributing to higher rates of heart disease and diabetes. Another cost has been the care and treatment of the region's 1.1 million HIV/AIDS victims. The price of the disease will be broadly borne in the 21st century, but particularly among poorer black (46 percent of AIDS cases in the United States) and Hispanic (18 percent) populations in the United States, in which rates of new infection are still on the rise.

North America and the Global Economy

Together with Europe and East Asia, North America plays a pivotal role in the global economy. In prosperous times, the region benefits from global economic growth, but in periods of international instability, globalization means that the region is more vulnerable to economic downturns. These links mean more to the region than abstract trade flow and foreign investment statistics. Increasingly, North American workers and localities find their futures directly tied to export markets in Latin America, the rise and fall of Asian imports, or the pattern of global investments in U.S. stock and bond markets.

The region is home to a growing number of truly "global cities" that serve as key connecting points and decision-making centers in the world economy (see Figure 3.34). New York City is the largest, although smaller metropolitan areas such as Miami, Toronto, Chicago, Los Angeles, Seattle, and Vancouver are emerging as pivotal urban players on the global stage. The global status of these urban centers has had profound local consequences. Life changes as thousands of new jobs are created; suburbs grow with a rush of new and ethnically diverse migrants; and a new, more cosmopolitan culture replaces older regional traditions (see "People on the Move: Skilled Immigrants Fuel Big City Growth").

The United States, with Canada's firm support, played a formative role in creating much of this new global economy and in shaping many of its key institutions. In 1944 allied nations met at Bretton Woods, New Hampshire, to discuss economic affairs. Under U.S. leadership, the group set up the International Monetary Fund (IMF) and the World Bank and gave these global organizations the responsibility for defending the world's monetary system and making key postwar investments in infrastructure. The United States also spurred the creation (1948) of the General Agreement on Tariffs and Trade (GATT). Renamed the **World Trade Organization (WTO)** in 1995, its 153 member states are dedicated to reducing global barriers to trade. In addition, the United States and Canada participate in the **Group of Eight (G-8)**—a collection of powerful countries (including Japan, Germany, Great Britain, France, Italy, and Russia)—that confers regularly on key global economic and political issues.

Patterns of Trade North America is prominent in both the sale and purchase of goods and services within the international economy. Both countries import diverse products from many global sources and the post-1990 growth of the Asian trade (particularly with China, South Korea, and southeast Asia) has fundamentally changed the North American economy.

Dominated overwhelmingly by the United States, Canada imports large quantities of manufactured parts, vehicles, computers, and foodstuffs. For the United States, imports continue to grow, creating a persistent global trade deficit for the country. Canada, Mexico, China, Japan, and world oil exporters supply the United States with a diversity of raw materials, low-cost consumer goods, and high-quality vehicles and electronics products (Figure 3.42).

Outgoing trade flows suggest what North Americans produce most cheaply and efficiently. Canada's exports include large quantities of raw materials (grain, energy, metals, and wood products), but manufactured goods are becoming increasingly important, particularly in its pivotal trade with the United States. Since 1994, trade initiatives with the Pacific Rim have offered Canadians new opportunities for export growth. The United States also enjoys many lucrative global economic ties, and its geography of exports reveals particularly strong links to other portions of the more developed world. Sales of automobiles, aircraft, computer and telecommunications equipment, entertainment, financial and tourism services, and food products contribute to the nation's flow of exports.

Patterns of Global Investment Patterns of capital investment and corporate power place North America at the center of global money flows and economic influence. Given its relative stability, the region attracts huge inflows of foreign capital, both as investments in North American stocks and bonds and as foreign direct investment (FDI) by international companies. For Canada, U.S. wealth and proximity have meant that 80 percent of foreign-owned corporations in the country are based in the United States. In the United States, sustained economic growth and supportive government policies have encouraged large foreign investments, particularly since the late 1970s. Today, the United States is the largest destination of foreign investment in the world.

The impact of U.S. investments in foreign stock markets also suggests how flows of out-bound capital are transforming the way business is done throughout the world. Aging U.S. baby boomers have poured billions of pension fund and investment dollars into Japanese, European, and "emerging" stock markets such as Brazil, Russia, and China. U.S. investments in foreign countries also flow through direct investments made by multinational corporations based in the United States.

But the geography of 21st-century multinational corporations is changing, illustrating three recent shifts in broader patterns of globalization. These shifts have important consequences for North Americans. First, traditional American-based multinational corporations are adopting a new, more globally integrated model. For example, IBM now has more than 50,000 employees in India. According to Samuel J. Palmisano, IBM's Chairman of the Board, the company is now committed to putting people and jobs anywhere in the world "based on the right cost, the right skills and the right business environment . . . work flows to the places where it will be done best."

Second, a growing array of multinational corporations based elsewhere in the world, especially in places such as China, India, Russia, and Latin America, are buying up companies and assets once controlled by North American or European capital. Brazilian investors, for example, spend billions annually in buying overseas assets (especially in North America). In Asia, China's Lenovo purchased IBM's huge personal computer business within that country. Similarly, Indian multinational companies are also buying hun-

Figure 3.42 Container Shipping, Port of Seattle Major North American ports such as Seattle are key links in facilitating global trade between the region and the rest of the world. Standard-sized container-shipping modules can easily be stored, stacked, and moved in such settings.

PEOPLE ON THE MOVE Skilled Immigrants Fuel Big City Growth

While broader questions surrounding immigration continue to foster debate in both Canada and the United States, one economic trend is beyond dispute: the region's growing inflow of skilled immigrants is providing a powerful economic stimulus for many of North America's largest and most dynamic cities. Part of the story is simply the numerical impact of immigrant populations: the New York City, Los Angeles, San Francisco, and Boston metropolitan areas, for example, would all have actually lost population between 2000 and 2008 without substantial numbers of new immigrants. The economic consequences of those movements are also huge. William Frey, a demographer with the Brookings Institution, notes that "a lot of cities rely on immigration to prop up their housing market and prop up their economies."

Many immigrants become urban entrepreneurs, starting new businesses that cater both to their own communities as well as to larger metropolitan populations (Figure 3.6.1). Whether it is the Chinese in Vancouver or the Cubans in Miami, immigrants in many of North America's largest, most global cities have made huge capital and human investments in their adopted communities. Almost 30 percent of the Korean-born and 20 percent of the Iranian-born populations in the United States are self-employed, a strong indicator of business ownership. A 2007 report issued by the Center for an Urban Future argues that "immigrants have been the entrepreneurial spark plugs of cities from New York to Los Angeles." Is it any surprise that first-generation immigrants created 22 of the 100 fastest-growing companies in Los Angeles or that 25 percent of all U.S. engineering and technology firms founded between 1995 and 2005 were headed by immigrant entrepreneurs?

Additional statistics gathered by the U.S. Department of Homeland Security also point to the unique contributions of highly skilled immigrants. So-called H-1B visas are granted to special "temporary skilled workers" to encourage computer programmers, doctors, and other professionals to work in the United States. More than 400,000 such visas were issued in 2008, enabling these individuals to work within the United States, adding immeasurably to that country's creative human capital. The geography of the top 20 contributing countries offers yet another snapshot of the economic impact of globalization and is a powerful reminder of another way in which the United States benefits in the process (Figure 3.6.2). While there are predictable ties to many European nations, the linkages with India (particularly through the computer,

Figure 3.6.1 Immigrant Entrepreneurs This Russian immigrant couple operates a deli in Los Angeles, California.

dreds of foreign assets based in the more developed world. One sign of the times can be measured in the home country of the world's largest multinational companies. Measured by their market value (2009), only 10 of the top 20 global corporations are based in North America (such as Exxon Mobil, Wal-Mart Stores, and Google). Four are European (such as Royal Dutch Shell and Nestlé), five are Chinese (such as PetroChina, Bank of China, and China Mobile), and one is Brazilian (Petrobas).

Third, many of these same multinational companies are making huge investments of their own in other portions of the less developed world, from Africa to Southeast Asia, bypassing North American control altogether. Simply put, the late-20th-century top-down model of multinational corporate control and investment, traditionally based in North America, Europe, and Japan, is being replaced by a more globally distributed model of corporate control. This new model has many origins, many destinations, and new patterns of labor, capital, production, and consumption.

North Americans have certainly experienced direct consequences from these shifts in global capitalism. There has been a growing reaction in the United States, for example, to corporate **outsourcing**, a business practice that transfers portions of a company's production and service activities to lower-cost settings, often located overseas. In addition, millions of jobs in manufacturing, textiles, semiconductors, and electronics have effectively migrated to settings such as China, India, and Mexico as those localities offer low-cost, less regulated settings for production, both for local and for foreign firms. The results are complex: North American consumers benefit from buying cheap imports, but they may find their own jobs are threatened in the corporate restructurings that make such bargains possible.

software, and electronics industries) are dramatic. Continental connections to nearby Mexico and Canada are also notable, along with growing links to Asian nations such as China, South Korea, Japan, the Philippines, and Taiwan. The patterns offer a powerful reminder that the economic evolution of both the United States and Canada remain intimately connected to skilled immigrant populations and that the region's large urban centers will continue to be a destination for these creative and talented individuals.

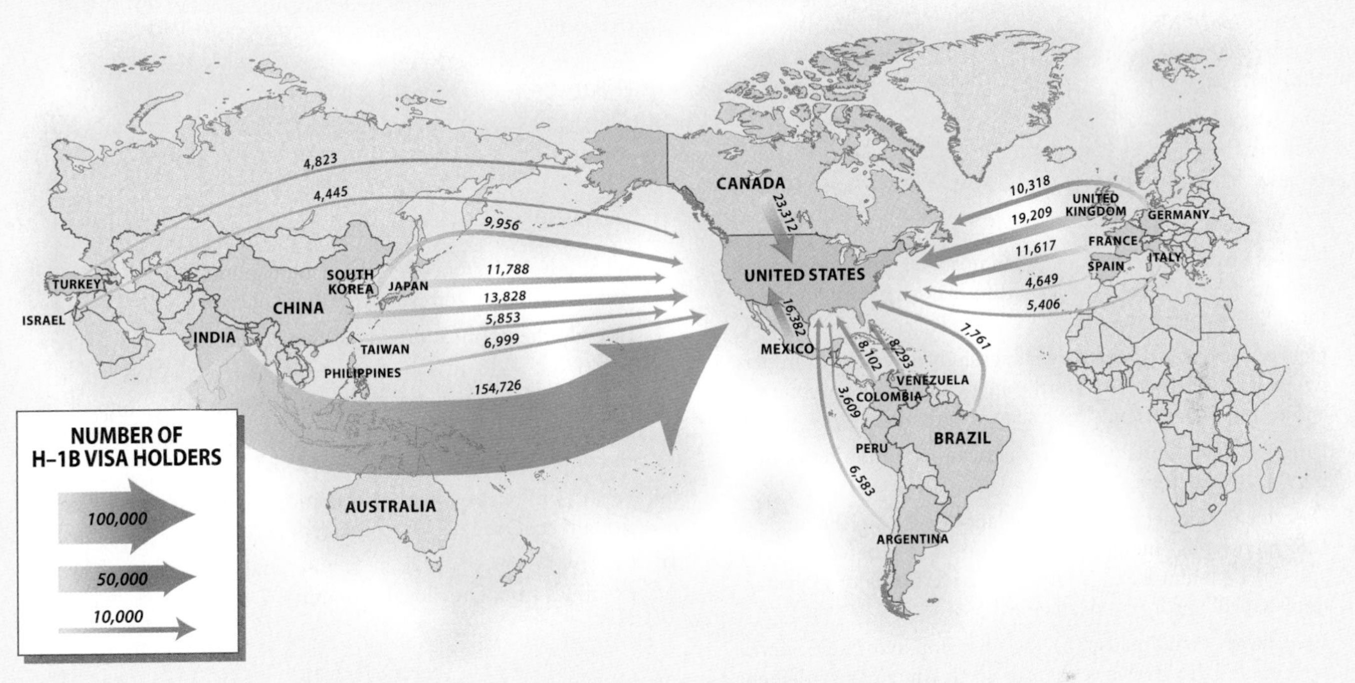

Figure 3.6.2 Origins of Temporary Skilled Workers (H-1B Visas) in the United States, 2008 (Top 20 Contributing Countries) India's contribution to the skilled human capital of the United States is particularly notable. Both Mexico and Canada contribute many skilled workers, as do other nations of East Asia and western Europe. (U.S. Department of Homeland Security, 2008)

Summary

- North Americans have reaped the natural abundance of their region, and in the process they have transformed the environment, created a highly affluent society, and extended their global economic, cultural, and political reach.
- North America's affluence has come with a considerable price tag, and today the region faces significant environmental challenges, including soil erosion, acid rain, and air and water pollution.
- In a remarkably short time period, a unique mix of varied cultural groups from around the world has contributed to the settlement of a huge and resource-rich continent that is now one of the world's most urbanized regions.
- North Americans produced two societies that are closely intertwined, yet face distinctive national political and cultural issues. In Canada, the nation's identity remains problematic as it works through the persisting challenges of its multicultural character and the costs and benefits of its proximity to its continental neighbor.
- For the United States, social problems linked to ethnic diversity, immigration issues, and enduring poverty and racial discrimination remain central concerns, particularly in many of the nation's largest cities.
- The recent global economic downturn profoundly impacted North America's economic geography, particularly in many regions that were hit hardest by the housing crisis and by rising rates of unemployment.

Key Terms

acid rain (*page 78*)
agribusiness (*page 110*)
boreal forest (*page 82*)
cap-and-trade energy policy (*page 84*)
concentric zone model (*page 90*)
connectivity (*page 112*)
cultural assimilation (*page 94*)
ethnicity (*page 94*)
federal states (*page 106*)
gender gap (*page 116*)
gentrification (*page 91*)
Group of Eight (G-8) (*page 118*)
location factors (*page 114*)
Megalopolis (*page 84*)
new urbanism (*page 91*)
nonmetropolitan growth (*page 89*)
North American Free Trade Agreement (NAFTA) (*page 106*)
outsourcing (*page 120*)
postindustrial economy (*page 74*)
prairie (*page 82*)
renewable energy sources (*page 80*)
sectoral transformation (*page 114*)
Spanglish (*page 103*)
sustainable agriculture (*page 80*)
tundra (*page 82*)
unitary states (*page 106*)
urban decentralization (*page 89*)
urban realms model (*page 90*)
World Trade Organization (WTO) (*page 118*)

Review Questions

1. Describe North America's major landform regions, and suggest ways in which their physical setting has shaped patterns of human settlement.
2. How are patterns of commercial agriculture related to underlying climatic patterns in North America? What key economic and technological trends have shaped agriculture?
3. Describe the dominant North American migration flows during the 20th century.
4. Describe the principal patterns of land use within the modern U.S. metropolis. Include a discussion of (a) the central city and (b) the suburbs/edge city. How have forces of globalization shaped North American cities?
5. How have (a) railroads, (b) the township-and-range survey system, and (c) freeways shaped North America's settlement geography?
6. What are the distinctive eras of immigration in U.S. history, and how do they compare with those of Canada?
7. How do the political origins of the United States and Canada differ?
8. Cite examples of globalization that illustrate the impact of North American cultural, political, and economic influence elsewhere in the world.
9. What is the sectoral transformation, and how does it help explain economic change in North America?
10. Cite five types of location factors, and illustrate each with examples from your local economy.

Thinking Geographically

1. Compare and contrast the African American experience in the Black Belt and in urban America. Assess the unique challenges faced by each group.
2. Explain how "natural hazards" can be "culturally" defined. In other words, what role do humans play in shaping the distribution of hazards?
3. Summarize and map the ethnic background and migration history of your own family. Discuss how these patterns parallel or depart from larger North American trends.
4. Describe the strengths and weaknesses of federalism, and cite examples from both Canada and the United States.
5. The environmental price for North American development has often been steep. Suggest why it may or may not be worth the price, and defend your answer.
6. Immigration remains a key issue for North America. Argue the pros and cons of sharply curtailing immigration in the future. What about opening up the region to even larger flows of immigrants?
7. Who will America's leading trade partner be in 2050? Explain the reasons for your choice.

Regional Novels and Films

Novels

Willa Cather, *My Antonia* (1918, Houghton Mifflin)

Ivan Doig, *This House of Sky* (1978, Harcourt Brace Jovanovich)

Frederick Philip Grove, *Settlers of the Marsh* (1925, Ryerson Press)

David Guterson, *Snow Falling on Cedars* (1995, Vintage Books)

William Kennedy, *Ironweed* (1983, Viking Press)

Annie Proulx, *The Shipping News* (1993, Scribner's)

Richard Russo, *Empire Falls* (2001, Knopf)

Wallace Stegner, *Angle of Repose* (1971, Doubleday)

John Steinbeck, *The Grapes of Wrath* (1939, Viking Press)

Nathaniel West, *The Day of the Locust* (1939, Random House)

Films

Avalon (1990, U.S.)

Atanarjuat: The Fast Runner (2001, Canada)

Black Robe (1991, Canada)

Chinatown (1974, U.S.)

Dances with Wolves (1990, U.S.)

The Deer Hunter (1978, U.S.)

Fargo (1996, U.S.)

Gran Torino (2008, U.S.)

Mystic River (2003, U.S.)

Thunderheart (1992, U.S.)

To Kill a Mockingbird (1962, U.S.)

Traffic (2000, U.S.)

Bibliography

Agnew, John A., and Smith, Jonathan M., eds. 2002. *American Space/American Place: Geographies of the Contemporary United States*. New York: Routledge.

Clay, Grady. 1994. *Real Places: An Unconventional Guide to America's Generic Landscape*. Chicago: University of Chicago Press.

Conzen, Michael P., ed. 2010. *The Making of the American Landscape*. New York: Routledge.

Ford, Larry R. 1994. *Cities and Buildings: Skyscrapers, Skid Rows, and Suburbs*. Baltimore: Johns Hopkins University Press.

Hardwick, Susan W., Shelley, Fred M., and Holtgrieve, Donald G. 2008. *The Geography of North America: Environment, Political Economy, and Culture*. Upper Saddle River, NJ: Prentice Hall.

Hudson, John C. 2002. *Across This Land: A Regional Geography of the United States and Canada*. Baltimore: Johns Hopkins University Press.

McIlwraith, Thomas F., and Muller, Edward K., eds. 2001. *North America: The Historical Geography of a Changing Continent*. Lanham, MD: Rowman and Littlefield.

Miller, Char, ed. 2003. *The Atlas of U.S. and Canadian Environmental History*. New York: Routledge.

Nostrand, Richard L., and Estaville, Lawrence E., eds. 2001. *Homelands: A Geography of Culture and Place Across America*. Baltimore: Johns Hopkins University Press.

Zelinsky, Wilbur. 1992. *The Cultural Geography of the United States: A Revised Edition*. Upper Saddle River, NJ: Prentice Hall.

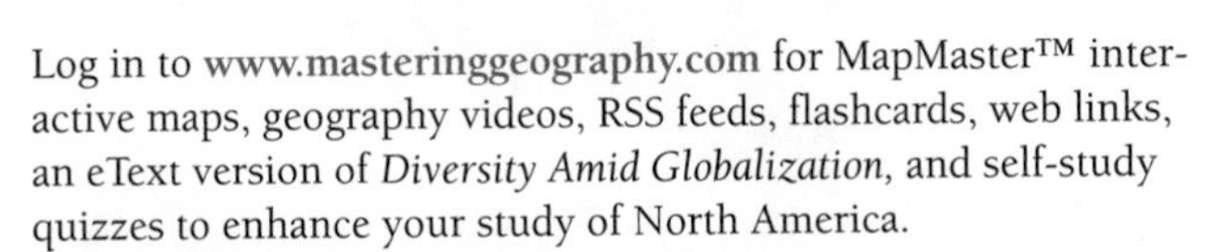

Log in to www.masteringgeography.com for MapMaster™ interactive maps, geography videos, RSS feeds, flashcards, web links, an eText version of *Diversity Amid Globalization*, and self-study quizzes to enhance your study of North America.

MapMaster™ presents 13 Place Name and 13 Layered Thematic interactive maps to help students practice and master their geographic literacy, spatial reasoning, and critical thinking skills.

4 Latin America

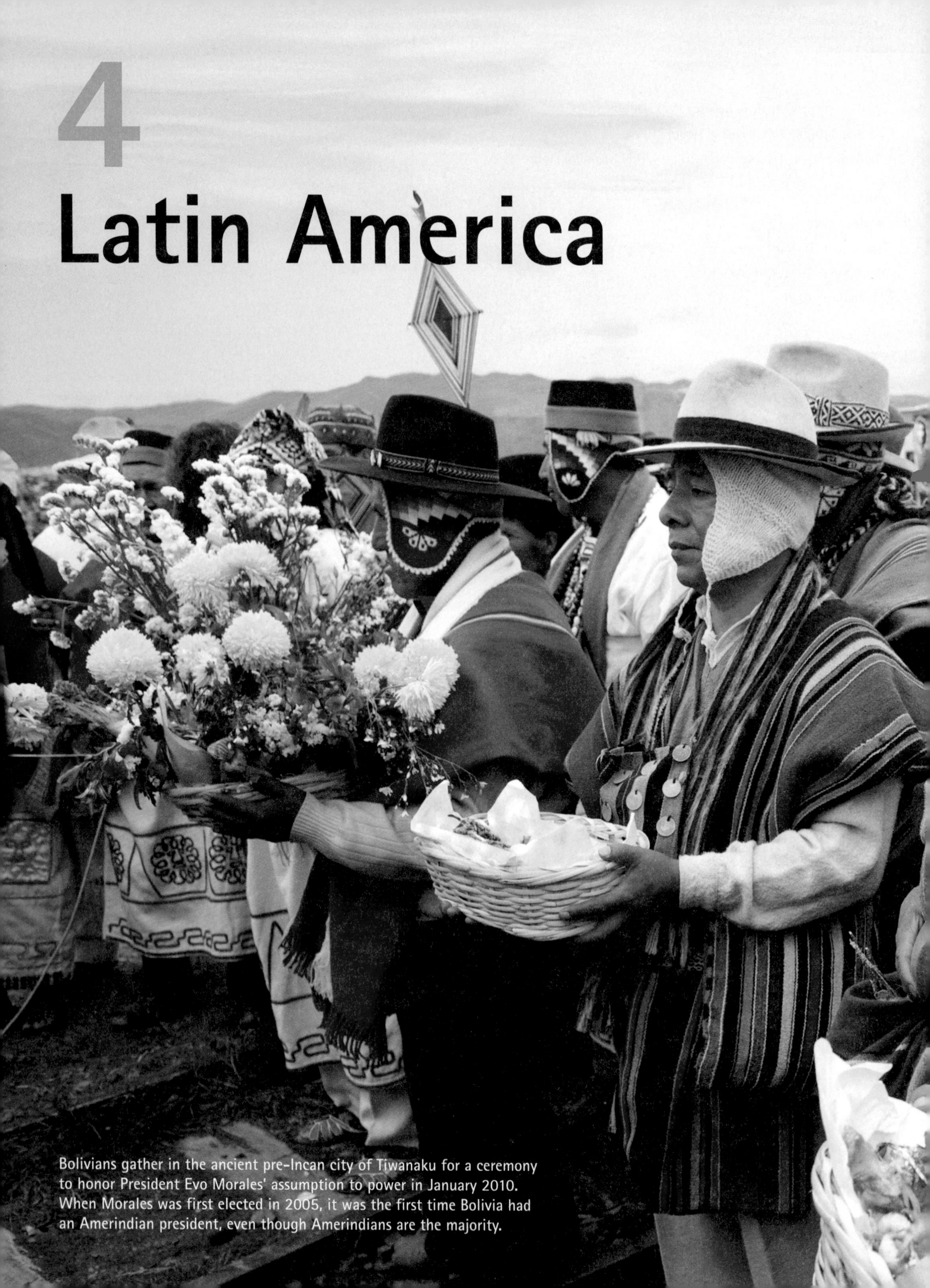

Bolivians gather in the ancient pre-Incan city of Tiwanaku for a ceremony to honor President Evo Morales' assumption to power in January 2010. When Morales was first elected in 2005, it was the first time Bolivia had an Amerindian president, even though Amerindians are the majority.

DIVERSITY AMID GLOBALIZATION

Neoliberal policies have profoundly changed the way Latin American economies and societies function. Foreign investment and international trade have intensified, forging new connections with other world regions.

ENVIRONMENTAL GEOGRAPHY

Tropical forests in Latin America, especially in the Amazon Basin, are one of the planet's greatest reserves of biological diversity. How this diversity will be managed is a critical question especially with increasing pressure to extract mineral wealth or convert forests into farms or pasture.

POPULATION AND SETTLEMENT

Latin America is the most urbanized region of the developing world with 77 percent of the population living in cities. Four megacities (10 million people or more) are found there.

CULTURAL COHERENCE AND DIVERSITY

Amerindian activism is on the rise in Latin America. Indigenous people from Central America to the Andes are finding their political voice by demanding territorial and cultural recognition.

GEOPOLITICAL FRAMEWORK

As Latin American states begin to celebrate two centuries of independence from Spain, most are fully democratic. Recent elections in the region have seen liberal democrats and populists gain power. These same politicians are skeptical of the ability of neoliberal reforms and economic globalization to reduce poverty in the region.

ECONOMIC AND SOCIAL DEVELOPMENT

From NAFTA to Mercosur, regional economic integration is changing the way Latin Americans trade with each other and the world. For many South American states, trade with Europe and East Asia has increased dramatically in recent years. Although exports are up, many of the poor have not benefited from this growth.

Beginning with Mexico and extending to the tip of South America, Latin America's regional coherence stems largely from its shared colonial history, rather than from the different levels of development seen today. More than 500 years ago, the Iberian countries of Spain and Portugal began their conquest of the Americas. Iberia's mark is still visible throughout Latin America: Officially, two-thirds of the inhabitants speak Spanish, and the rest speak Portuguese. Iberian architecture and town design add homogeneity to the colonial landscape. The vast majority of the population is Catholic. These European traits blended with those of different Amerindian peoples. The Indian presence remains especially strong in Bolivia, Peru, Ecuador, Guatemala, and southern Mexico, where large and diverse indigenous populations maintain their native languages, dress, and traditions. After the initial conquest, other cultural groups were added to this mix of indigenous and Iberian peoples. The legacy of slavery imparted a strong African influence, primarily on the coasts of Colombia and Venezuela and throughout Brazil. In the 19th and 20th centuries new waves of settlers came from Spain, Italy, Germany, Japan, and Lebanon. The result is one of the world's most racially mixed regions. The modern states of Latin America are multiethnic, with distinct indigenous and immigrant profiles and very different rates of social and economic development. Given the significant outflows of emigrants from this region today, the peoples of Latin America are found throughout the world but especially in North America, Europe, and Japan.

The concept of Latin America as a distinct region has been popularly accepted for nearly a century. The boundaries of this region are straightforward, beginning at the Rio Grande (called the *Rio Bravo* in Mexico) and ending at Tierra del Fuego. French geographers are credited with coining the term *Latin America* in the 19th century to distinguish the Spanish- and Portuguese-speaking republics of the Americas plus Haiti from the English-speaking territories. There is nothing particularly "Latin" about the area, other than the predominance of romance languages. The term stuck because it was vague enough to be inclusive of different colonial histories while also offering a clear cultural boundary from Anglo-America, the region referred to as North America in this text (Figure 4.1).

This chapter describes Latin America, consisting of the Spanish- and Portuguese-speaking countries of Central and South America, including Mexico. This division emphasizes the important Indian and Iberian influences affecting mainland Latin America, and separates it from the unique colonial and demographic history of the Caribbean, discussed in Chapter 5.

Through colonialism, immigration, and trade, the forces of globalization have been embedded in the Latin American landscape. The early Spanish Empire concentrated on extracting precious metals, sending galleons laden with silver and gold across the Atlantic. The Portuguese became prominent producers of dyewoods, sugar products, gold, and, later, coffee. In the late 19th and early 20th centuries, exports to North America and Europe fueled the region's economy. Most countries specialized in one or two products: bananas and coffee, meats and wool, wheat and corn, petroleum and copper. Such a primary export tradition, according to Latin American economists, led to an unhealthy economic dependence. They argued in the 1960s that Latin American economies were too specialized and faced unequal terms of trade that inhibited overall development.

Since then the countries of the region have industrialized and diversified their production, but they continue to be major producers of primary goods for North America, Europe, and East Asia. Today, neoliberal policies that encourage foreign investment, export production, and privatization have been adopted by many states. These policies exemplify the current impact of economic globalization on Latin America. The results are mixed, with some states experiencing impressive economic growth but increased disparity between rich and poor. Intraregional trade within Latin America stimulated by Mercosur (the Southern Common Market found in South America), as well as the impact of the North American Free Trade Agreement (NAFTA) (see Chapter 3) and the newly formed **Central American Free Trade Association (CAFTA)**, are indicators of heightened economic integration in the hemisphere.

Roughly equal in area to North America, Latin America has a much larger and faster-growing population of 542 million people. Its most populous state, Brazil, has 193 million people, making it the fifth largest country in the world. The next largest state, Mexico, has a population of 110 million and in 2010 one of its citizens, telecom mogul Carlos Slim, was identified as the richest man in the world. Collectively, many Latin American states fall into the middle-income category and support a significant middle class. But national debt, political scandal, currency devaluation, and triple-digit inflation triggered grave economic hardships throughout the region, especially during the 1980s and early 1990s. Despite many social and economic gains in the past decade, poverty remains a major concern for this region. It is estimated that 22 percent of the people in the region live on less than $2 per day. Yet unlike most areas of the developing world today, Latin America is decidedly urban. Prior to World War II, most people lived in rural settings and worked as farmers. Today three-quarters of Latin Americans are city dwellers. Even more startling is the number of **megacities**. São Paulo, Mexico City, Buenos Aires, and Rio de Janeiro all have more than 10 million inhabitants. In addition, more than 40 cities have at least 1 million residents. Residents of these cities aspire for global recognition. A dozen Brazilian cities will host the World Cup in 2014 and Rio de Janeiro will proudly host the Olympics in 2016.

Despite the region's growing industrial capacity, extractive industries will continue to prevail, in part because of the area's impressive natural resources. Latin America is home to Earth's largest rain forest, the greatest river by volume, and massive reserves of natural gas, oil, and copper. With its vast territory, its tropical location, and its relatively low population density (Latin America has half the population of India

Figure 4.1 Latin America Roughly equal in size to North America, Latin America supports a larger population and far greater ecological diversity. The 17 countries included in this region share a history of Iberian colonization. Three-quarters of the region's 542 million people live in cities, making it the most urbanized region of the developing world. It is noted for its production of primary exports and manufactured goods, although rates of economic development vary greatly among states.

Figure 4.2 Tropical Wilderness The biological diversity of Latin America—home to the world's largest rain forest—is increasingly seen as a genetic and economic asset. These forested mesas (called *tepuis*) in southern Venezuela are representative of the wild lands that many conservationists seek to protect.

in nearly seven times the area), the region is also recognized as one of the world's great reserves of biological diversity. How this diversity will be managed in the face of global demand for natural resources is an increasingly important question for the countries of this region (Figure 4.2).

ENVIRONMENTAL GEOGRAPHY: Neotropical Diversity and Urban Degradation

Much of the region is characterized by its tropicality. Travel posters of Latin America showcase verdant forests and brightly colored parrots. Naturalists eager to understand the unique flora and fauna have long been attracted to the diversity and uniqueness of the neotropics, which are the tropical ecosystems of the Western Hemisphere. It is no accident that Charles Darwin's insights on evolution were inspired by his two-year journey in tropical America. Even today, scientists throughout the region work to understand complex ecosystems, discern new species, conserve genetic resources, and interpret the impact of human settlement, especially in neotropical forests. Not all of the region is tropical. Important population centers extend below the tropic of Capricorn, most notably Buenos Aires and Santiago. Much of northern Mexico, including the city of Monterrey, is north of the tropic of Cancer. Yet it is Latin America's tropical climate and vegetation that prevail in popular images of the region. Given the territory's large size and relatively low population density, Latin America has not experienced the same levels of environmental degradation witnessed in Europe and East Asia. The region's biggest environmental concerns are related to deforestation, the loss of biodiversity, and the livability of urban areas (Figure 4.3).

Huge areas of Latin America still remain relatively untouched, supporting an incredible diversity of plant and animal life. Throughout the region, national parks offer some protection to unique communities of plants and animals. A growing environmental movement in countries such as Costa Rica and Brazil has yielded both popular and political support for "green" initiatives. In short, Latin Americans enter the 21st century with a real opportunity to avoid many of the environmental mistakes seen in other regions of the world. At the same time, global market forces are driving governments to exploit minerals, fossil fuels, forests, and soils. The region's biggest natural resource challenge is to balance the economic benefits of extraction with the ecological soundness of conservation (see "Global to Local: Can Eco-Friendly Fair Trade Coffee Promote Biodiversity and Rural Development?" on page 131). Another major challenge is to improve the environmental quality of Latin American cities.

Rural Environmental Issues: Declining Forests and Degraded Farmlands

Perhaps the environmental issue most commonly associated with Latin America is deforestation. In rural areas, other pressing environmental issues occur on farmlands, and result from modern agricultural practices and soil erosion.

Forest Destruction and Biodiversity Loss The Amazon Basin and portions of the eastern lowlands of Central America and Mexico still maintain unique and impressive stands of tropical forest. But other woodland areas, such as the Atlantic coastal forests of Brazil and the Pacific forests of Central America, have nearly disappeared as a result of agriculture,

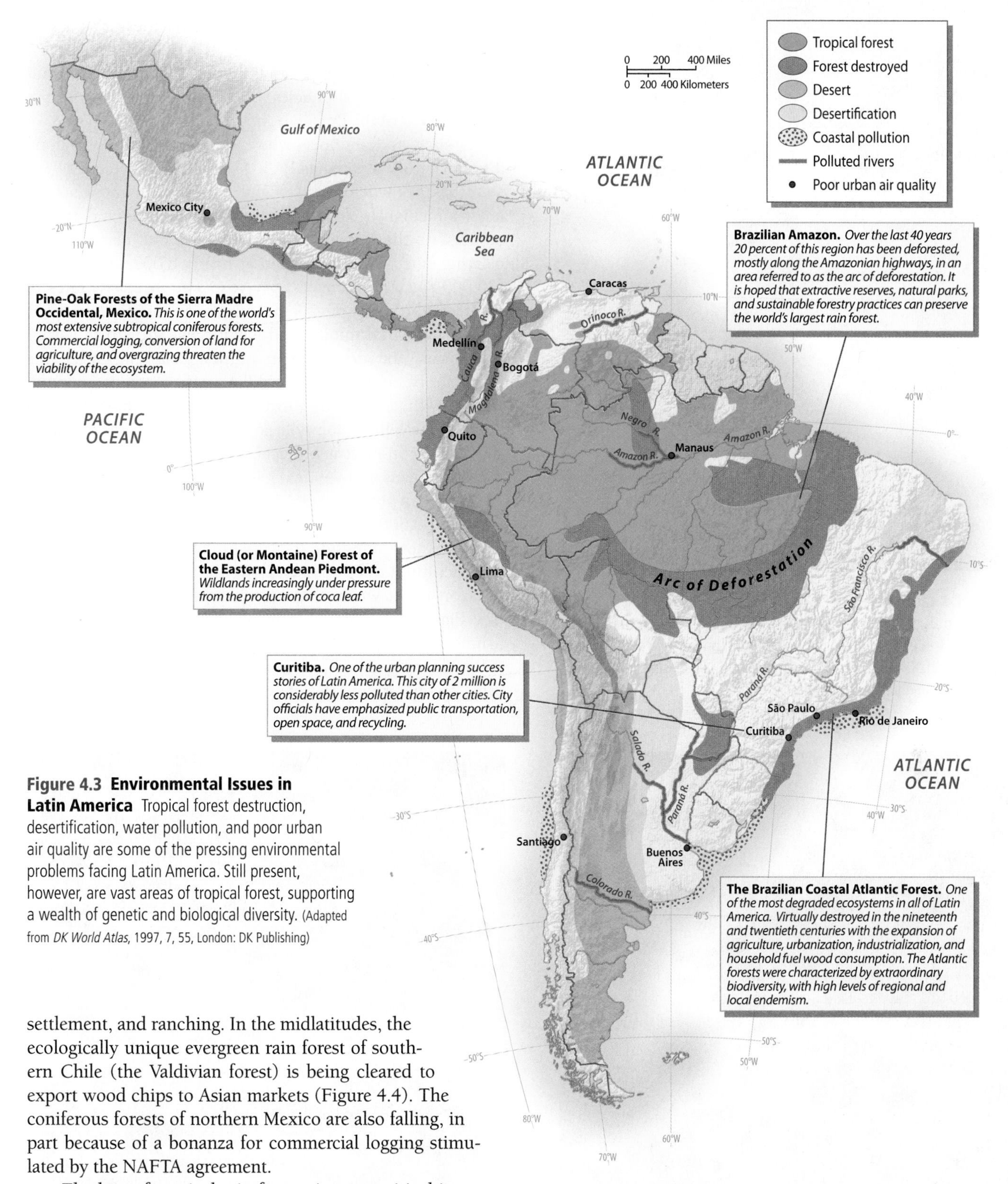

Figure 4.3 Environmental Issues in Latin America Tropical forest destruction, desertification, water pollution, and poor urban air quality are some of the pressing environmental problems facing Latin America. Still present, however, are vast areas of tropical forest, supporting a wealth of genetic and biological diversity. (Adapted from *DK World Atlas*, 1997, 7, 55, London: DK Publishing)

settlement, and ranching. In the midlatitudes, the ecologically unique evergreen rain forest of southern Chile (the Valdivian forest) is being cleared to export wood chips to Asian markets (Figure 4.4). The coniferous forests of northern Mexico are also falling, in part because of a bonanza for commercial logging stimulated by the NAFTA agreement.

The loss of tropical rain forests is most critical in terms of biological diversity. Tropical rain forests account for only 6 percent of Earth's landmass, but at least 50 percent of the world's species are found in this biome. Moreover, the Amazon Basin contains the largest undisturbed stretches of rain forest in the world. Unlike Southeast Asian forests, where hardwood extraction drives forest clearance, Latin American forests are usually seen as an agricultural frontier that state governments divide in an attempt to give land to the landless or reward political cronies. Thus, forests are cut and burned, with settlers and politicians carving them up to create permanent settlements, slash-and-burn plots, or large cattle ranches. In addition, some tropical forest cutting has been motivated by the search for gold (Brazil, Venezuela,

Figure 4.4 Chilean Wood Chips A mountain of wood chips awaits shipment to Japanese paper mills from the southern Chilean port of Punta Arenas. The exploitation of wood products from both native and plantation forests supports Chile's booming export economy. Increasingly these wood exports are bound for markets in East Asia.

and Costa Rica) and the production of coca leaf for cocaine (Peru, Bolivia, and Colombia).

Brazil has incurred more criticism than other countries for its Amazon forest policies. During the past 40 years close to 20 percent of the Brazilian Amazon has been deforested. In states such as Rondônia, where settlers streamed in along a popular road known as BR364, close to 60 percent of the state has been deforested (Figure 4.5). What most alarms environmentalists and forest dwellers (Indians and rubber tappers) is the dramatic increase in the rate of rain-forest clearing since 2000, estimated at nearly 8,000 square miles (20,000 square kilometers) per year. The increased rates of deforestation in the Brazilian Amazon are due to the expansion of industrial mining and logging, the growth in corporate farms, the development of new road networks, the incidence of human-ignited wildfires, and continued population growth. Under the *Advance Brazil* program started in 2000, some $40 billion will go to new highways, railroads, gas lines, hydroelectric projects, power lines, and river canalization projects that will reach into remote areas of the basin. The Brazilian government, under President Lula da Silva, has created 150,000 square kilometers of new conservation areas, many of them alongside the "arc of deforestation"—a swath of agricultural development along the southern edge of the Amazon Basin where the worst deforestation has occurred (see Figure 4.3). Government officials hope that the establishment of these protected areas, along with more aggressive enforcement by Brazil's environmental protection agency, will bring down current rates of deforestation.

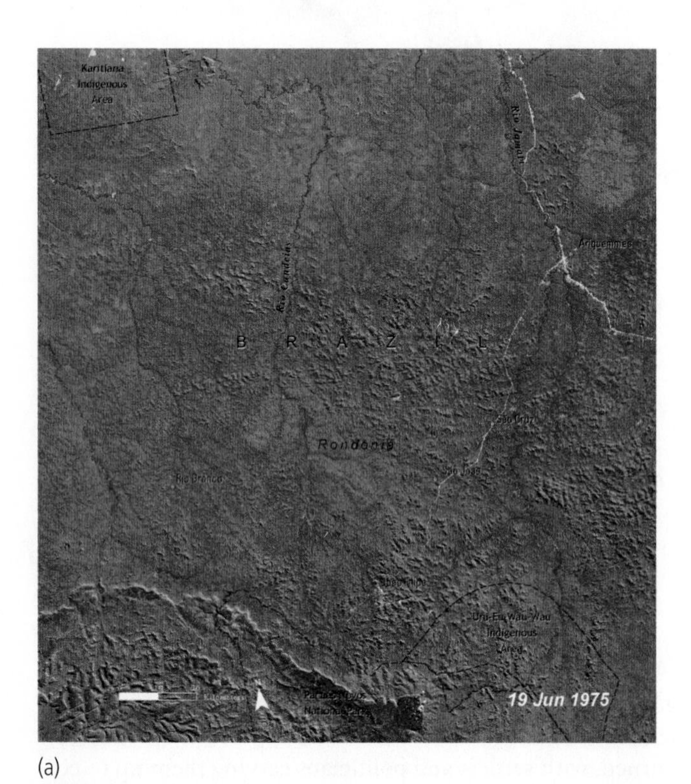

(a)

(b)

Figure 4.5 Tropical Forest Settlement in the Amazon These images of Rondônia, Brazil, illustrate the dramatic change in forest cover. In 1975 (a) there was one major road, and most of the region was tropical forest. By 2001 (b), roads, pastures, and farms had replaced the forest.

GLOBAL TO LOCAL Can Eco-Friendly Fair Trade Coffee Promote Biodiversity and Rural Development?

Most of the coffee consumed in North America and Europe comes from Latin America (especially Central America, Mexico, Colombia, and Brazil). Environmental groups believe that if a small percentage of North American and European consumers demanded Fair Trade and organic shade-grown coffee (Figure 4.1.1) a vast area of habitat could be saved in Northern Latin America and small farmers will earn more for their coffee harvest. But what is the link between organic shade-grown coffee and biodiversity?

As traditionally grown, under a canopy of shade trees and other vegetation, coffee is an environmentally friendly crop. The coffee bushes grow for years without the need for farmers to disturb the soil, and the surrounding vegetation offers valuable habitat to wildlife, especially birds. One study by the Smithsonian Migratory Bird Center found more than 150 species of birds on shade-coffee farms in Chiapas, Mexico. Many of these birds are migratory species such as the Baltimore oriole or the Tennessee warbler that summer in North America and winter in Latin America. In the last two decades, populations of these familiar migratory birds began a drastic decline, due in part to a loss of winter habitat in Latin America.

Figure 4.1.1 Shade-grown Coffee Shade-grown organic coffee in Latin America provides a biologically diverse habitat for many species and provides peasant farmers with an opportunity to earn more money from their harvest while being better stewards of their environment.

From the 1970s until the early 1990s, thousands of acres of traditional shade-coffee plants were replaced by high-yielding varieties grown in full sun and requiring extensive chemical inputs to be productive. This monocrop production system resulted in very high yields but created a green desert that supported few other species. Millions of dollars were spent turning shade-grown plantations into so-called technified ones. With the loss of tropical forests and shade-coffee farms, scientists noted annual declines of migratory songbirds. Over time, farmers also found their profits dwindling because of the high price of chemical fertilizers and pesticides and the declining price of coffee. Coffee is an extremely important export crop for rural Latin America, second only to oil as the most valuable commodity in terms of world trade.

Recently, environmentalists and nongovernmental organizations (NGOs) have promoted coffee certified as **Fair Trade** which means the growers (often smallholders) receive a higher price for their crop (at least 10 percent above market price). Coffee farmers from Central America, Mexico, and Haiti are finding that the use of organic methods lowers their costs and Fair Trade certification increases their earnings. An added benefit of shade-grown organic coffee is that it helps maintain an economically productive and biologically diverse habitat. Even Starbucks now markets shade-grown and organic coffee in some of its stores. Labels identifying coffee as Fair Trade, organic, or bird-friendly are making consumers aware of the connection between their daily cup of coffee and a more biologically diverse and economically sustainable environment.

Grassification The conversion of tropical forest into pasture, called **grassification**, is another practice that has contributed to deforestation. Particularly in southern Mexico, Central America, and the Brazilian Amazon, a hodgepodge of development policies from the 1960s through the 1980s encouraged deforestation to make room for cattle. The preference for ranching as a status-conferring occupation seems to be a carryover from Iberia. The image of the *vaquero* (cowboy) looms large in the region's history. Even poor farmers appreciate the value of having livestock. Like a savings account, cattle can be quickly sold for cash.

There are many natural grasslands suitable for grazing in Latin America, such as the Llanos in Colombia and Venezuela, and the Chaco and Pampas in Argentina. However, the rush to convert forest into pasture made ranching a scourge on the land. Even in cases in which domestic demand for beef increased, ranching in remote tropical frontiers is seldom economically self-sustaining (Figure 4.6).

Problems on Agricultural Lands The pressure to modernize agriculture has produced a series of environmental problems. As peasants were encouraged to adopt new hybrid varieties of corn, beans, and potatoes, an erosion of genetic diversity occurred. Efforts to preserve dozens of native domesticates are underway at agricultural research centers in the central Andes and Mexico. Nonetheless, many useful native plants may have been lost. Modern agriculture also depends on chemical fertilizers and pesticides that eventually run off into surface and groundwater supplies. Consequently, many rural areas suffer from contamination of local water supplies. Even more troublesome is the direct exposure of farmworkers to toxic agricultural chemicals. Mishandling of pesticides and fertilizers

Figure 4.6 Converting Forest into Pasture Cattle graze in northern Guatemala's Peten region. Clearing of this tropical forest lowland began in the 1960s and continues today. Ranching is a status-conferring occupation in Latin America with serious ecological costs. The beef produced from this region is for domestic and export markets.

can lead to exposure, resulting in rashes and burns. And in some areas, such as Sinaloa, Mexico, the widespread application of chemicals parallels a rise in serious birth defects.

Soil erosion and fertility decline occur in all agricultural areas. Certain soil types in Latin America are particularly vulnerable to erosion, most notably the volcanic soils and the reddish *oxisols* found in the humid tropical lowlands. The productivity of the Paraná basalt plateau in Brazil, for example, has declined over the decades due to the ease with which these volcanic soils erode and the failure to apply soil conservation methods. By contrast, the oxisols of the tropical lowlands can quickly degrade into a baked claypan surface when the natural cover is removed, making permanent agriculture nearly impossible. Ironically, the consolidation of the large-scale modern farms in the basins and valleys of the highlands tends to push peasant subsistence farmers into marginal areas. On these hillside farms, gullies and landslides reduce productivity. Lastly, the sprawl of Latin American cities consumes both arable land and water, eliminating some of the region's best farmland.

Urban Environmental Challenges

Some of the most pressing environmental issues in Latin America are found in the region's urban areas. Mexico City, in particular, suffers from a host of environmental problems and is a good example of the kinds of environmental challenges facing modern Latin American cities. The severe problems facing this urban ecosystem underscore how environment, population, technology, and politics are interwoven in the effort to keep Mexico City livable.

The Valley of Mexico's Environmental Problems At the southern end of the great Central Plateau of Mexico lies the Valley of Mexico, the cradle of Aztec civilization and the site of Mexico City, a metropolitan area of approximately 18 million people. This high-altitude basin with mild temperatures, fertile soils, and ample water surrounded by snow-capped volcanoes was an early center of Amerindian settlement and plant domestication, and the site of one of Spain's most important colonial cities. Given these features, it is no wonder that Mexico City has retained its primacy into the 21st century, even though the environmental setting that made it attractive centuries ago is now severely degraded. Some of the most pressing problems are air quality, adequate water, and subsidence (soil sinkage) caused by overdrawing the valley's aquifer (groundwater).

Air quality has been a major issue for Mexico City since the 1960s, driven in part by the city's phenomenal rate of growth, larger number of cars, and physical setting. It is difficult to imagine a better setting for creating pollution. The city sits in a bowl 7,400 feet (2,250 meters) above sea level, and a layer of warm air traps a layer of cold air near the surface (called a thermal inversion) that is filled with exhaust, industrial smoke, garbage, and fecal matter (Figure 4.7). During the worst pollution emergencies in the winter, when a layer of warm air traps pollutants, schoolchildren are required to stay indoors, and high-polluting vehicles are barred from the streets. The health costs of breathing such contaminated air are real, as elevated death rates due to heart disease, influenza, and pneumonia suggests. Steps were finally taken in the late 1980s to reduce emissions from factories and cars. Unleaded gas is now widely available for the 4 million cars in the metropolitan area, cars manufactured for the Mexican market must have catalytic converters, and some of the worst polluting factories in the Valley of Mexico have closed. In 2008 a program that restricts driving on certain days based on license plate numbers was expanded to include Saturdays. The mayor of Mexico City has expanded a low-emissions Metrobus systems, which eliminated thousands of tons of carbon monoxide. A suburban train system is to replace hundreds of thousands of vehicles. The payoff is real, Mexico City no longer ranks in the top polluted cities in the world and it appears to have cut most of its pollutants at least by half.

One of Mexico City's most relentless environmental problems is water. Ironically, it was the abundance of water that made this site attractive for settlement. Large shallow lakes once filled the valley, but over the centuries most were drained to expand agricultural land. As surface water became scarce, wells were dug to tap the basin's massive freshwater aquifer. Today approximately 70 percent of the water used in the metropolitan area is drawn from the valley's aquifer. There is troubling evidence that the aquifer is being overdrawn and at risk of contamination, especially in areas where unlined drainage canals can leak pollutants into the surrounding soil, which then leach into the aquifer. To reduce reliance on groundwater, the city now pumps water

Figure 4.7 Air Pollution in Mexico City Pollution blankets Mexico City against a backdrop of Ixtaccihuatl volcano. Mexico City is notorious for its smog. Its high elevation and immense size make air quality management difficult. Improvements in mass transportation, better regulations, and unleaded gasoline have steadily improved the city's air quality in the past decade.

nearly a mile uphill from more than 100 miles (160 kilometers) away.

A lesser-known but equally vexing problem is that Mexico City is sinking. As the metropolis grows and pumps more water from its aquifer, subsidence worsens. Mexico City sank 30 feet (9 meters) during the 20th century. By comparison, Venice, an Italian city knows for its subsidence problems, sank only 9 inches (23 centimeters) during the same period. Although the amount of subsidence varies across the metropolitan area, its impact is similar throughout the city. Building foundations are destroyed, and water and sewer lines rupture. Cherished landmarks, such as the national cathedral in the central plaza, have cracks, list to one side, and are supported by scaffolding. Because of subsidence, groundwater is no longer pumped from the city center, so that sinking has slowed to 1 inch (2.5 centimeters) per year. In the periphery, where the aquifer is actively tapped, some areas are sinking as much as 20 inches (50 centimeters) per year.

Ongoing Urban Challenges and Responses For most Latin Americans, air pollution, inadequate water, and garbage removal are the pressing environmental problems of everyday life. In this most urbanized region of the developing world, city dwellers do have better access to water, sewers, and electricity than their counterparts in Asia and Africa. Yet the inevitable environmental problems that come from dense urban settings ultimately require expensive infrastructural remedies. Since many urban dwellers tend to reside in unplanned squatter settlements, servicing these communities with utilities after they are built is difficult and costly.

Industrialization is a major cause of pollution. Factories, electricity generation, and transportation all contribute to urban pollution. A lax attitude about enforcing environmental laws tends to be the norm. In the worst cases, the consequences pose a serious threat to people and the environment. In the 1980s the Brazilian industrial center of Cubatão, not far from São Paulo, was derogatively referred to as the "Valley of Death" because of its severe industrial contamination and a notorious pipeline fire that destroyed a squatter settlement. The events in Cubatão, more than the destruction of rain forest, were credited with invigorating the environmental movement in Brazil. While residents in the city still suffer from higher rates of cancer, there is widespread agreement that pollution has improved since the 1990s with stronger enforcement of laws and cleaner technology.

Not far from Cubatão in Brazil is Curitiba, the capital of Paraná state. Curitiba is the celebrated "green city" of Brazil because of some relatively simple yet innovative planning decisions. More than 2 million people inhabit this industrial and commercial center, yet it is significantly less polluted than other similar-sized cities. Because the location was vulnerable to flooding, city planners built drainage canals and set aside the remaining natural drainage areas as parks in the 1960s, well before explosive growth would have made such a policy difficult. Next, public transportation became a top priority. An innovative and clean bus system made Curitiba a model for transportation in the developing world. And last, a low-tech but effective recycling program has greatly reduced solid waste. Cities such as Curitiba demonstrate that designing with nature makes sense both ecologically and economically.

Western Mountains and Eastern Shields

Latin America is a region of diverse landforms, including high mountains and extensive upland plateaus. The movement of tectonic plates explains much of the region's basic topography, including the formation of its geologically young and tectonically active western mountain ranges (see Figure 4.1). In contrast, the Atlantic side of South America is characterized by humid lowlands interspersed with large upland plateaus called **shields**. Across these lowlands meander some of the great rivers of the world, including the Amazon.

Historically the most important areas of settlement in tropical Latin America were not along the region's major rivers, but across its shields, plateaus, and fertile intermontane basins. In these localities the combination of arable land, mild climate, and sufficient rainfall produced the region's most productive agricultural areas and its densest settlement. The Mexican Plateau, for example, is a massive upland area ringed by the Sierra Madre Mountains. The Valley of Mexico is located at the southern end of the plateau. Similarly, the elevated and well-watered basins of Brazil's southern mountains provide an ideal setting for agriculture. These especially fertile areas are able to support high population densities, so it is not surprising that the two largest cities, Mexico City and São Paulo, emerged in these settings.

Figure 4.8 Altiplano and Lake Titicaca The Altiplano is an elevated plateau straddling the Peruvian and Bolivian Andes. One of its striking features is beautiful Lake Titicaca at an elevation of 12,500 feet (3,810 meters). Amerindians inhabit this stark and windswept land, which is one of the poorer areas of the Andes.

The Andes Beginning in northwestern Venezuela and ending at Tierra del Fuego, the Andes are relatively young mountains that extend nearly 5,000 miles (8,000 kilometers). Created by the collision of oceanic and continental plates, the mountains are a series of folded and faulted sedimentary rocks with intrusions of crystalline and volcanic rock. The Andes are still forming, so active volcanism and regular earthquakes are common in this zone. This was made clear in February 2010 when an 8.8 magnitude earthquake struck near the Chilean city of Concepción, killing some 400 people and unleashing tsunami warnings across the Pacific. The earthquake was so powerful that geologists estimate that the entire city of Concepción moved 10 ft. (3 m) to the west. Because the epicenter was not near a major population center and Chilean buildings are engineered to withstand earthquakes, the death toll was remarkably light given the intensity of the event. Even so, it is estimated that nearly 1.5 million Chileans were temporarily displaced by this natural disaster.

The subduction of Nasca plate under the South American plate has produced an impressive chain of some 30 peaks higher than 20,000 ft (6,000 meters). Due to the violent and complex origins of the Andes, many rich veins of precious metals and minerals are found here. The initial economic wealth of many Andean countries came from mining silver, gold, tin, copper, and iron.

Given the length of the Andes, the mountain chain is typically divided into northern, central, and southern components. In Colombia, the northern Andes actually split into three distinct mountain ranges before merging near the border with Ecuador. High-altitude plateaus and snow-covered peaks distinguish the central Andes of Ecuador, Peru, and Bolivia. The Andes reach their greatest width here. Of special interest is the treeless high plain of Peru and Bolivia, called the **Altiplano**. The floor of this elevated plateau ranges from 11,800 feet (3,600 meters) to 13,000 feet (4,000 meters) in altitude, and it has limited usefulness for grazing. Two high-altitude lakes, Titicaca on the Peruvian and Bolivian border and the smaller Poopó in Bolivia, are located in the Altiplano, as well as many mining sites (Figure 4.8). The southern Andes are shared by Chile and Argentina. The highest peaks of the Andes are found in the southern Andes, including the highest peak in the Western Hemisphere, Aconcagua, at almost 23,000 feet (6,958 meters). South of Santiago, Chile, the mountains are lower and the chain less compact. The impact of glaciation is most evident in the southernmost extension of the Andes.

The Uplands of Mexico and Central America The Mexican Plateau and the Volcanic Axis of Central America are the most important Latin American uplands in terms of long-term settlement. Most major cities of Mexico and Central America are located here. The Mexican Plateau is a large, tilted block that has its highest elevations in the south, about 8,000 feet (2,500 meters) around Mexico City, and its lowest, just 4,000 feet (1,200 meters), at Ciudad Juárez. The southern end of the plateau, the Mesa Central, contains a number of flat-bottomed basins interspersed with volcanic peaks. It also contains Mexico's megalopolis—a concentration of the largest population centers such as Mexico City, Guadalajara, and Puebla. (The Valley of Mexico, discussed earlier, is one of the basins of the Mesa Central.) Across the Mexican Plateau are rich seams of silver, copper, and zinc. The quest for silver drove much of the economic activity of colonial Mexico.

Along the Pacific coast of Central America lies a chain of volcanoes that stretches from Guatemala to Costa Rica. The Volcanic Axis of Central America is a handsome landscape of rolling green hills, elevated basins with sparkling lakes, and conical volcanic peaks. More than 40 volcanoes are found here, many of them still active. Their legacy is a rich volcanic soil that yields a wide variety of domestic and export crops. Most of Central America's population is also concentrated in this zone, in the capital cities or the surrounding rural villages. The bulk of the agricultural land is tied up in large holdings that yield export products including beef, cotton, and coffee. Yet most of the farms are small subsistence properties that produce corn, beans, squash, and assorted fruits. A major rift (a large crustal fracture) lies east of the Volcanic Axis in Nicaragua. In this low-lying valley are Lakes Managua and Nicaragua, the largest in Central America.

The Shields South America has three major shields—large upland areas of exposed crystalline rock that are similar to upland plateaus found in Africa and Australia. (The Guiana shield will be discussed in Chapter 5.) The Brazilian shield is the largest and more important in terms of natural resources and settlement. Far from a uniform land surface, the Brazilian shield covers much of Brazil from the Amazon Basin in the north to the Plata Basin in the south. It is studded with isolated low ranges and flat-topped plateaus in the north. In southeastern Brazil a series of mountains (Serra da Mantiqueira and Serra do Mar) reach elevations of 9,000 feet (2,700 meters). In between these ranges are elevated basins that offer a mild climate and fertile soils. In one of these basins is the city of São Paulo, the largest urban conglomeration in South America.

The Paraná basalt plateau, located on the southern end of the Brazilian shield, is celebrated for its fertile red soils (*terra roxa*), which yield coffee, oranges, and soybeans. The basalt plateau, much like the Deccan plateau in India, is an ancient lava flow that resulted from the breakup of Gondwanaland. So fertile is this area that the economic rise of São Paulo is attributed to the expansion into this area of commercial agriculture (Figure 4.9).

Figure 4.9 Brazilian Oranges Most estate-grown oranges in Brazil are processed into frozen concentrate and exported. São Paulo and Paraná have some of the finest soils in Brazil. In addition to oranges, coffee and soybeans are widely cultivated.

Figure 4.10 Patagonian Wildlife Guanacos thrive on the thin steppe vegetation found throughout Patagonia. Native to South America, the numbers of guanacos fell dramatically due to hunting and competition with introduced livestock.

The vast low-lying Patagonian shield lies in the southern tip of South America. Beginning south of Bahia Blanca and extending to Tierra del Fuego, the region to this day remains sparsely settled and hauntingly beautiful. It is treeless, covered by scrubby steppe vegetation, and home to wildlife such as the condor and guanaco (Figure 4.10). Sheep were introduced to Patagonia in the late 19th century, spurring a wool boom. More recently, offshore oil production has renewed the economic importance of this area.

River Basins and Lowlands

Three great river basins drain the Atlantic lowlands of South America: the Amazon, Plata, and Orinoco (Figure 4.11). Within these basins are vast interior lowlands, less than 600 feet (200 meters) in elevation, that lie over young sedimentary rock. From north to south they are the Llanos, the Amazon lowlands, the Pantanal, the Chaco, and the Pampas. With the exception of the Pampas, most of these lowlands are sparsely settled and offer limited agricultural potential except for grazing livestock. Yet the pressure to open new areas for settlement and to exploit natural resources has created pockets of intense economic activity in the lowlands. Areas such as the Amazon and the Chaco have witnessed marked increases in resource extraction, soy cultivation, and settlement since the 1970s.

Amazon Basin The Amazon drains an area of roughly 2.4 million square miles (6.6 million square kilometers), making it the largest river system in the world by volume and area and the second longest by length. Everywhere in the basin, annual rainfall is more than 60 inches (150 centimeters), and in many places more than 80 inches (200 centimeters). The basin's largest city, Belém, averages close to 100 inches (250 centimeters) a year. Although there is no real dry season,

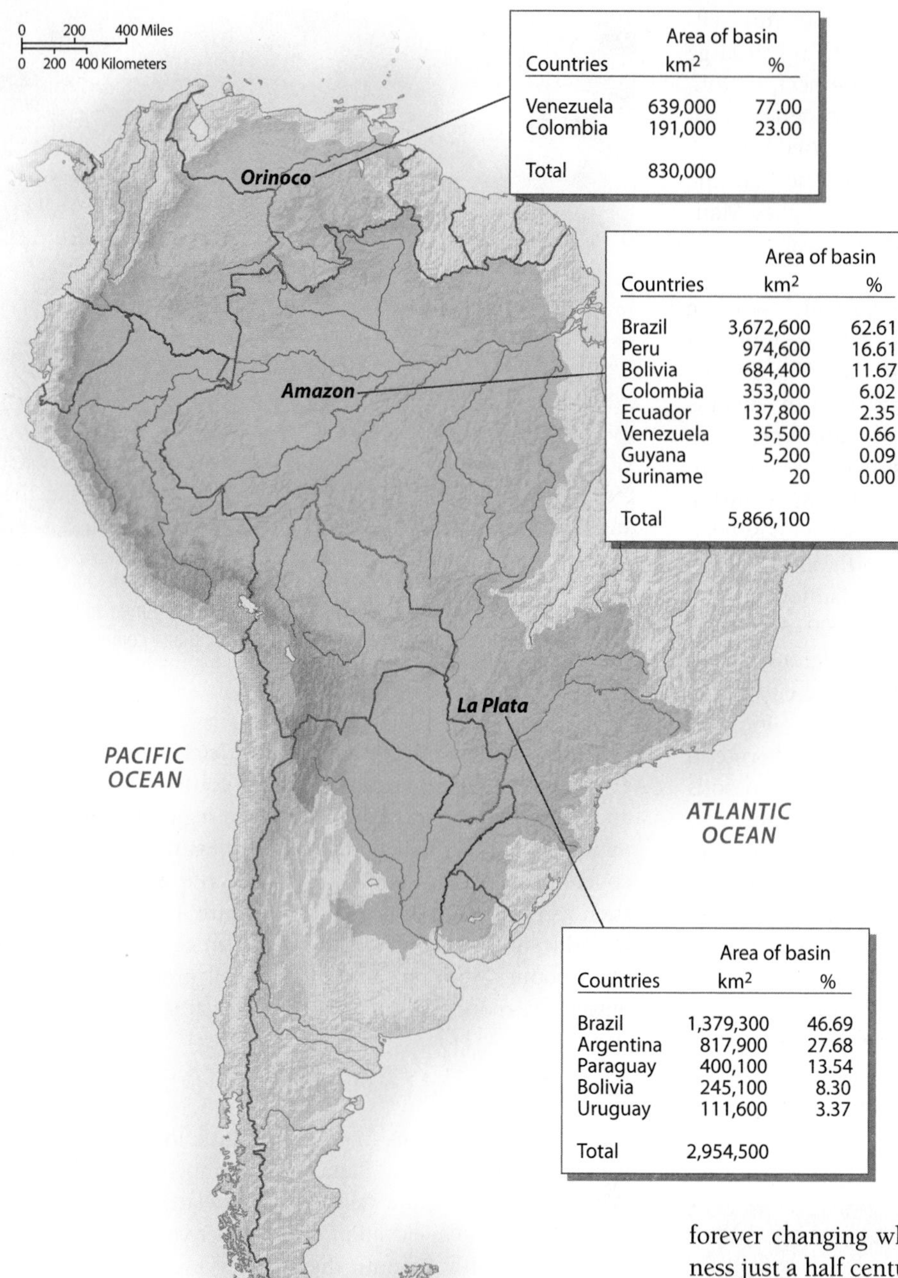

Countries	Area of basin km²	%
Venezuela	639,000	77.00
Colombia	191,000	23.00
Total	830,000	

Countries	Area of basin km²	%
Brazil	3,672,600	62.61
Peru	974,600	16.61
Bolivia	684,400	11.67
Colombia	353,000	6.02
Ecuador	137,800	2.35
Venezuela	35,500	0.66
Guyana	5,200	0.09
Suriname	20	0.00
Total	5,866,100	

Countries	Area of basin km²	%
Brazil	1,379,300	46.69
Argentina	817,900	27.68
Paraguay	400,100	13.54
Bolivia	245,100	8.30
Uruguay	111,600	3.37
Total	2,954,500	

Figure 4.11 South American River Basins The three great river basins of the region are the Amazon, Plata, and Orinoco. The Amazon basin covers 6 million square kilometers, including portions of eight countries, but the majority of the basin is within Brazil. The Amazon is the largest river system in the world in terms of volume of water and area. The Plata Basin drains nearly 3 million square kilometers across five countries and is intensely farmed and used for hydroelectricity. The Orinoco basin is shared by Venezuela and Colombia, covering nearly 1 million square kilometers.

there are definitely drier and wetter times of year, with August and September the driest months. The immensity of this watershed and its hydrologic cycle is underscored by the fact that 20 percent of all freshwater discharged into the oceans comes from the Amazon.

Since the Amazon Basin draws from eight countries, this watershed is an ideal network to integrate the northern half of South America. In fact, the river is navigable to Iquitos, Peru, some 2,100 miles (3,600 kilometers) upstream. Ironically, compared with the other great rivers of the world, settlement in the basin continues to be sparse, in large part due to the poor quality of the forest soils. The best soils are found on the floodplain, where natural levees reach heights of 20 feet (6 meters). Thousands of years of alluvium deposited on these levees have made them extremely fertile, as well as safe from normal flooding. Consequently, most of the older settlements, such as Manaus, are found on the levees. Active colonization of the Brazilian portion of the Amazon since the 1960s has boosted the population. Roughly 15 million people live in the Brazilian Amazon, which is about 8 percent of the country's total population. Still, the population in the Amazonian states is increasing at nearly 4 percent a year. The development of the basin, most notably through towns, roads, dams, farms, and mines, is forever changing what was viewed as a vast tropical wilderness just a half century ago.

Plata Basin The region's second largest watershed begins in the Tropics and discharges into the Atlantic in the midlatitudes. Three major rivers make up this system: the Paraná, the Paraguay, and the Uruguay. The Paraguay River and its tributaries drain the eastern Andes of Bolivia, the Brazilian shield, and the Chaco. The Paraná primarily drains the Brazilian uplands before the Paraguay River joins it in northern Argentina. The Paraná and the considerably smaller Uruguay empty into the Rio de la Plata estuary, which begins north of Buenos Aires.

Unlike the Amazon, much of the Plata Basin is now economically productive through large-scale mechanized agriculture, especially soybean production. Arid areas such as the Chaco and inundated lowlands such as the Pantanal support livestock. The Plata Basin contains several major dams, including the region's largest hydroelectric plant, the Itaipú on the Paraná, which generates electricity for all of Paraguay and much of southern Brazil (Figure 4.12). As agricultural output in this watershed grows, sections of the Paraná River are being canalized and dredged to enhance the river's capacity for barge and boat traffic.

Orinoco Basin The third largest river basin by area is the Orinoco in northern South America. The Orinoco River meanders through much of southern Venezuela and part of eastern Colombia, giving character to a tropical grassland called the *Llanos*. Although it is only one-sixth the size of the Amazon watershed, its discharge roughly equals that of the Mississippi River. Like the Amazon, this basin is home to very few individuals; 90 percent of Venezuela's population lives north of the basin. With the exception of the industrial developments between Ciudad Guayana and Ciudad Bolívar, cities are few. Much of the Orinoco drains the Llanos, which are inundated by several feet of water during the rainy season. Since the colonial era, these grasslands have supported large cattle ranches. Although cattle are still important, the Llanos have also become a dynamic area of petroleum production for both Colombia and Venezuela.

Climate Patterns

In tropical Latin America average monthly temperatures in localities such as Managua, Quito, or Manaus show little variation (Figure 4.13). Precipitation patterns do vary, however, and create distinct wet and dry seasons. In Managua, for example, January is typically a dry month, and June is a wet one. The tropical lowlands of Latin America, especially east of the Andes, are usually classified as tropical humid climates that support forest or savanna, depending on the amount of rainfall. The region's desert climates are found along the Pacific coasts of Peru and Chile, Patagonia, northern Mexico, and the Bahia of Brazil. Because of the extreme aridity of the Peruvian coast, a city such as Lima, Peru, which is clearly in the tropics, averages only 1.5 inches (4 centimeters) of annual rainfall.

Midlatitude climates, with hot summers and cold winters, prevail in Argentina, Uruguay, and parts of Paraguay and Chile. Of course, the midlatitude temperature shifts in the Southern Hemisphere are the inverse of those in the Northern Hemisphere, meaning cold Julys and warm Januarys. Chile's climate is a mirror image of the west coast of Mexico and the United States, with the Atacama Desert in the north (like Baja California), a Mediterranean dry summer around Santiago (similar to Los Angeles), and a marine west coast climate with no dry season south of Concepción (like the coasts of Oregon and Washington). In the mountain ranges, complex climate patterns result from changes in elevation. To appreciate how humans adapt to tropical mountain ecosystems, one must understand the concept of **altitudinal zonation**, the relationship between cooler temperatures at higher elevations and changes in vegetation.

Altitudinal Zonation First described in the scientific literature by Prussian naturalist Alexander von Humboldt in the early 1800s, altitudinal zonation has practical applications that are intimately understood by all the region's native inhabitants. Humboldt systematically recorded declines in temperature as he ascended to higher elevations, a phenomenon known as the **environmental lapse rate**. According to Humboldt's description of the environmental lapse rate, temperature declines approximately 3.5°F for every 1,000 feet in elevation or 6.5°C for every 1,000 meters. Humboldt also noted changes in vegetation by elevation, demonstrating

Figure 4.12 Itaipú Dam The largest dam in Latin America, the Itaipú blocks the flow of the Paraná River on the border between Paraguay and Brazil. The power station at Itaipú generates all of Paraguay's electricity needs and much of southern Brazil's power.

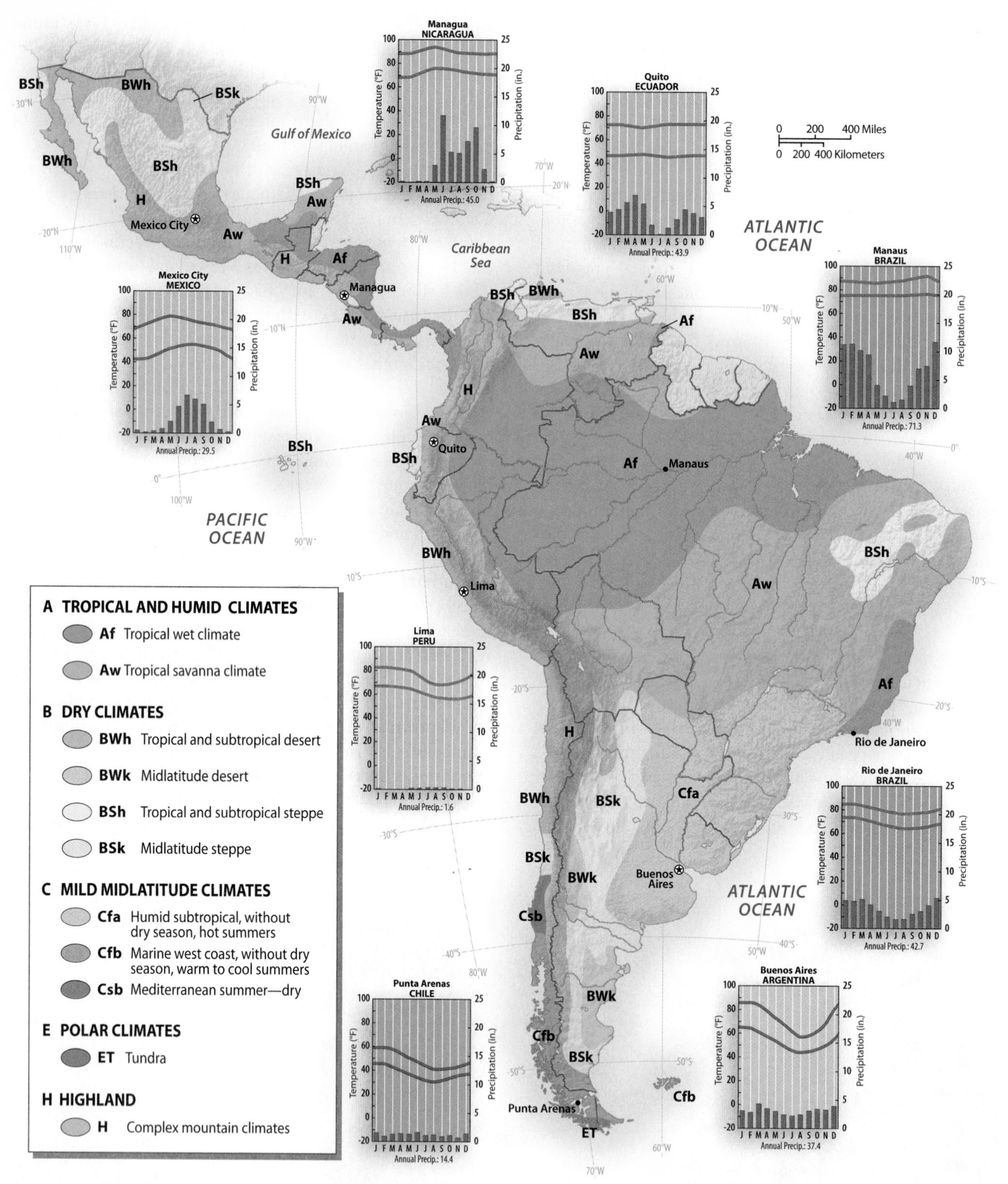

Figure 4.13 Climate Map of Latin America Latin America includes the world's largest rain forest (Af) and driest desert (BWh), as well as nearly every other climate classification. Latitude, elevation, and rainfall play important roles in determining the region's climates. Note the contrast in rainfall patterns between humid Quito and arid Lima. (Temperature and precipitation data from Pearce and Smith, 1984, *The World Weather Guide*, London: Hutchinson)

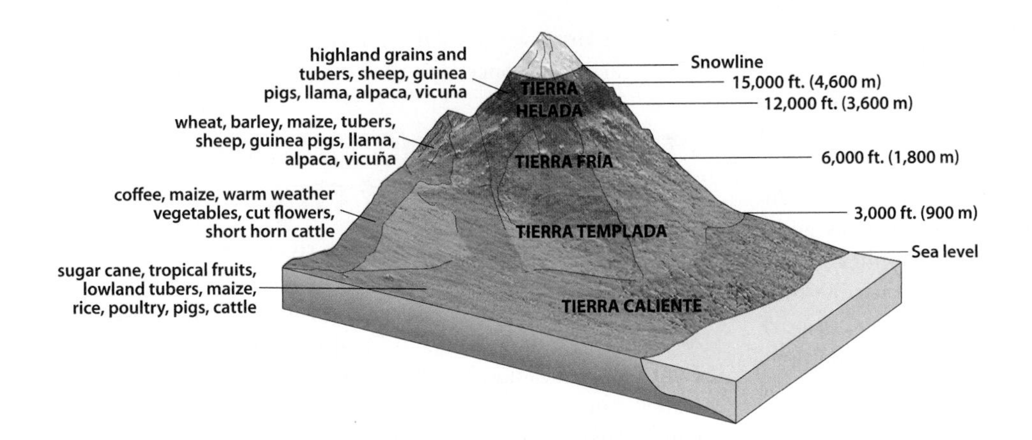

Figure 4.14 Altitudinal Zonation Tropical highland areas support a complex array of ecosystems. In the *tierra fría* zone (6,000 to 12,000 feet; 1,800 to 3,700 meters), for example, midlatitude crops such as wheat and barley can be grown. This diagram depicts the range of crops and animals found at different elevations in the Andes.

that plant communities common to the midlatitudes could thrive in the tropics at higher elevations. These different altitudinal zones are commonly referred to as the *tierra caliente* (hot land) from sea level to 3,000 feet (900 meters); the *tierra templada* (temperate land) at 3,000 to 6,000 feet (900 to 1,800 meters); the *tierra fría* (cold land) at 6,000 to 12,000 feet (1,800 to 3,600 meters); and the *tierra helada* (frozen land) above 12,000 feet (3,600 meters). Exploitation of these zones allows agriculturists, especially in the uplands, access to a great diversity of domesticated and wild plants (Figure 4.14).

The concept of altitudinal zonation is most relevant for the Andes, the highlands of Central America, and the Mexican Plateau. For example, traditional Andean farmers might use the high pastures of the Altiplano for grazing llamas and alpacas, the tierra fría for potato and quinoa production, and the lower temperate zone to produce corn. All the great precontact civilizations, especially the Incas and the Aztecs, systematically extracted resources from these zones, thus ensuring a diverse and abundant resource base. Yet these complex ecosystems are extremely fragile and have become important areas of research for the effects of climate change in the tropics.

El Niño Probably the most talked-about weather phenomenon in the world, **El Niño** (a reference to the Christ child) occurs when a warm Pacific current arrives along the normally cold coastal waters of Ecuador and Peru in December, around Christmastime. This change in ocean temperature happens every few years and produces torrential rains, signaling the arrival of an El Niño year. The 2009–10 El Niño was especially bad for Latin America; scores of people were killed by floods or storms attributed to El Niño–related disturbances. Devastating floods occurred in Peru and Brazil. In Peru, heavy rains and flooding damaged the railroad leading to the ancient Incan site of Machu Picchu, temporarily limiting access to this popular tourist destination until the railroad could be rebuilt.

El Niño's impacts vary, however, across Latin America. While the Pacific coast of South and North America experienced record rainfall in the 1997–98 El Niño, Colombia, Venezuela, northern Brazil, Central America, and Mexico battled drought. Crop and livestock losses were estimated to be in the billions of dollars, and hundreds of brush and forest fires left their mark. In addition, extreme weather events unrelated to El Niño also affect Central America and Mexico, such as hurricanes and their associated heavy rain, flooding, and landslides.

Latin America and Global Warming

Latin America represents about 8 percent of the world's population and produces about 6 percent of global greenhouse gas emissions. The rate of growth of greenhouse gas emissions in Latin America is dramatically lower than in all other regions of the developing world, except for Sub-Saharan Africa. The region's relatively low emissions can be explained by lower average energy consumption, higher reliance on renewable energy (especially hydropower and biofuels), and greater dependence upon public transportation. The burning of forest and brush, a common practice in the region, does produce spikes in carbon dioxide ($C0_2$) emissions, but the regrowth of vegetation also absorbs vast amounts of $C0_2$.

Global warming has both immediate and long-term implications for Latin America. Of greatest immediate concern is how climate change will influence agricultural productivity, water availability, changes in the composition and productivity of ecosystems, and incidence of vector-born diseases such as malaria and dengue fever. Changes attributable to global warming are already apparent in higher elevations, making these concerns more pressing. The long-term effects of global climate change on lowland tropical forest systems is less clear: for example, some areas may experience more rainfall, others less. Other long-term impacts, such as rising sea level, will not cause the same levels of displacement in Latin America as predicted for the Caribbean or Oceania.

Climate change research indicates that highland areas are particularly vulnerable to global warming. Tropical mountain systems are projected to experience increased temperatures of 1° to 3°C, as well as lower rainfall. This will raise the altitudinal limits of various ecosystems, impacting the range of crops and arable land available to farmers and pastoralists. Research over the past 50 years has documented the dramatic retreat of Andean glaciers—some no longer exist and others will cease to exist in the next 10 to 15 years

(Figure 4.15). While this is a visible indicator of global warming, it also has pressing human repercussions. Many Andean villages, as well as metropolitan areas such as La Paz, Bolivia, get much of their water from glacial runoff. A major Bolivian glacier, Chacaltaya, has lost 80 percent of its area in the last 20 years. Thus as average temperatures increase in the highlands, and glaciers recede, there is widespread concern about future drinking water supplies.

Another immediate concern brought on by warmer temperatures is the sudden rise in dengue fever, a mosquito-born virus. Once considered relatively uncommon in highland Latin America, the number of cases has risen sharply in the past few years. Ten of thousands now suffer from the fever, headache, nausea, joint pain, and, in rare cases, external and internal bleeding that can result in death. The World Health Organization has estimated that dengue is now widespread in more than 100 tropical and subtropical countries around the world, but it is in Latin America where the sudden rise in cases suggests that warmer highland temperatures have placed millions more at risk.

Scientists are not yet sure whether or how global warming will impact the frequency and strength of El Niño cycles. If El Niño cycles intensify and occur more often as a result of global warming, increased flooding in western South America and a declining fishery (which is supported by nutrient upwelling in the normally cold currents) off the coast of Peru and Chile will result. Moreover, some evidence indicates that hurricane intensity will increase as ocean temperatures warm with climate change, heightening the impacts of these natural disasters. The effects of global warming on El Niño and extreme storm events are the subject of ongoing research, but are of particular concern to the Latin American region.

Figure 4.15 Glacial Retreat. Research in the Cordillera Blanca of Peru over the past four decades indicate a dramatic decline in the size of Andean glaciers. An indicator of the pace of climate change in the tropics, highland glaciers are also sources of water for mountain villages and cities. Their demise has serious repercussions for the communities that depend upon them.

POPULATION AND SETTLEMENT: The Dominance of Cities

Latin America did not have great river basin civilizations like those in Asia. In fact, the great rivers of the region are surprisingly underutilized as areas of settlement or corridors for transportation. While the major population clusters of Central America and Mexico are in the interior plateaus and valleys, the interior lowlands of South America are relatively empty. Historically, the highlands supported most of the region's population during the pre-Hispanic and colonial eras. In the 20th century, population growth and migration to the Atlantic lowlands of Argentina and Brazil, along with continued growth of Andean coastal cities such as Guayaquil, Barranquilla, and Maracaibo, have reduced the demographic importance of the highlands. Major highland cities such as Mexico City, Guatemala City, Bogotá, and La Paz still dominate their national economies, but the majority of large cities are on or near the coasts (Figure 4.16).

Like the rest of the developing world, Latin America experienced dramatic population growth in the 1960s and 1970s. In 1950 its population totaled 150 million people, which equaled the population of the United States at that time. By 1995 the population had tripled to 450 million; in comparison, the United States reached 300 million in 2006. Latin America outpaced the United States because its birthrate remained consistently higher as infant mortality rates dropped and life expectancy soared. In 1950 Brazilian life expectancy was only 43 years; by the 1980s it was 63 and by 2010 it was 73. In fact, between 1950 and 1980 most countries in the region experienced a 15- to 20-year improvement in life expectancy, which pushed up growth rates. Four countries account for over 70 percent of the region's population: Brazil with 193 million, Mexico with 110 million,

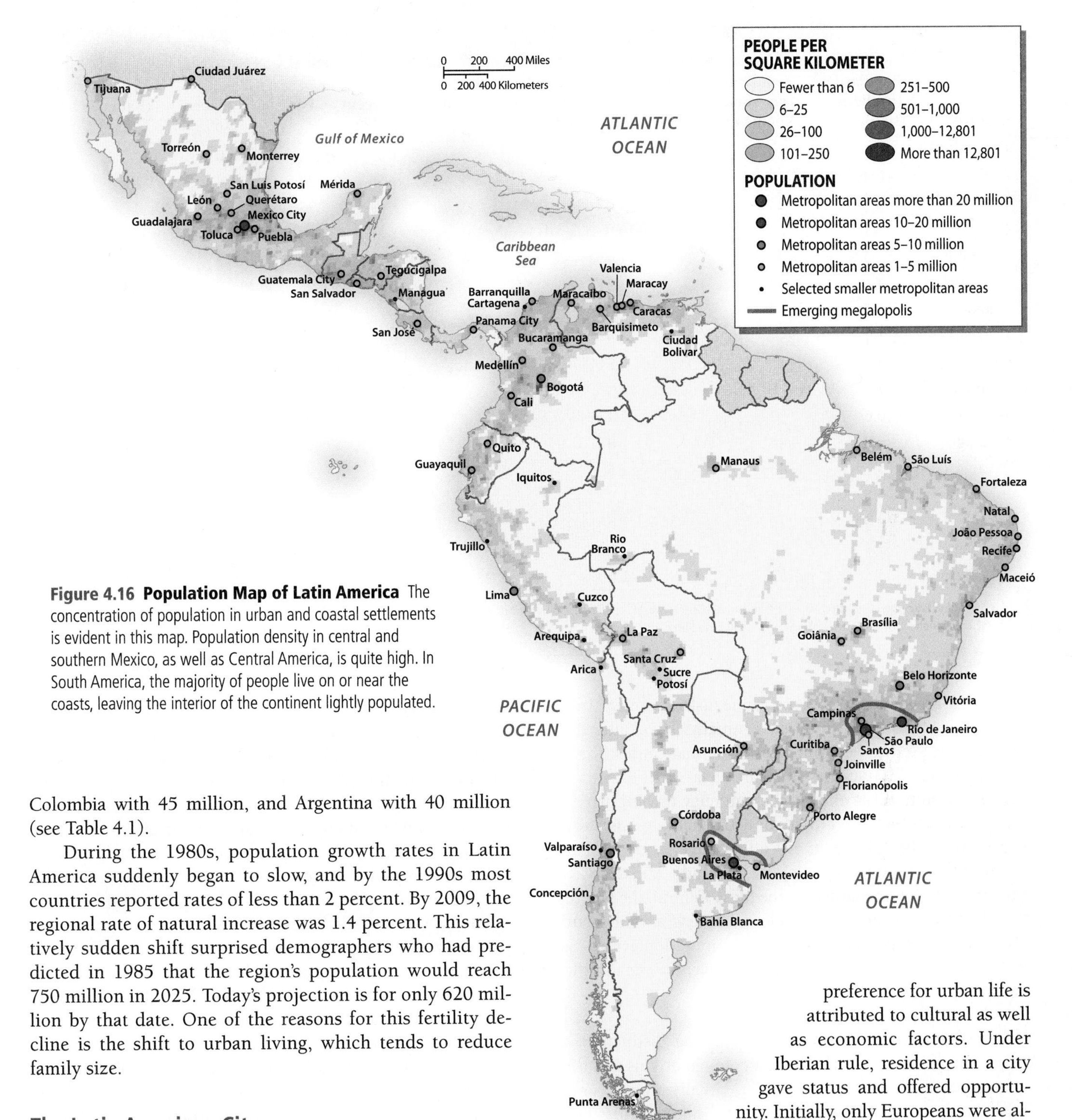

Figure 4.16 Population Map of Latin America The concentration of population in urban and coastal settlements is evident in this map. Population density in central and southern Mexico, as well as Central America, is quite high. In South America, the majority of people live on or near the coasts, leaving the interior of the continent lightly populated.

Colombia with 45 million, and Argentina with 40 million (see Table 4.1).

During the 1980s, population growth rates in Latin America suddenly began to slow, and by the 1990s most countries reported rates of less than 2 percent. By 2009, the regional rate of natural increase was 1.4 percent. This relatively sudden shift surprised demographers who had predicted in 1985 that the region's population would reach 750 million in 2025. Today's projection is for only 620 million by that date. One of the reasons for this fertility decline is the shift to urban living, which tends to reduce family size.

The Latin American City

A quick glance at the population map of Latin America shows a concentration of people in cities. One of the most significant demographic shifts has been the movement from rural areas to cities, which began in earnest in the 1950s. In 1950, only one-quarter of the region's population was urban; the rest lived in small villages and the countryside. Today the pattern is reversed, with three-quarters of the population living in cities. In the most urbanized countries, such as Argentina, Chile, Uruguay, and Venezuela, more than 90 percent of the population live in cities (see Table 4.1). This preference for urban life is attributed to cultural as well as economic factors. Under Iberian rule, residence in a city gave status and offered opportunity. Initially, only Europeans were allowed to live in the colonial cities, but this exclusivity was not strictly enforced. Over the centuries colonial cities became the hubs for transportation and communication, underscoring their primary role in structuring regional economies.

Latin American cities are noted for high levels of **urban primacy**, a condition in which a country has a primate city three to four times larger than any other city in the country. Examples of primate cities are Lima, Caracas, Guatemala City, Santiago, Buenos Aires, and Mexico City. Primacy is often viewed as a liability, as too many national

TABLE 4.1 Population Indicators

Country	Population (millions) 2010	Population Density (per square kilometer)	Rate of Natural Increase (RNI)	Total Fertility Rate	Percent Urban	Percent <15	Percent >65	Net Migration (Rate per 1,000) 2005–10[a]
Argentina	40.5	15	1.0	2.3	91	26	10	0.2
Bolivia	10.4	9	2.0	3.5	65	37	4	–2.1
Brazil	193.3	23	1.0	2.0	84	27	7	–0.2
Chile	17.1	23	0.9	1.9	87	24	8	0.4
Colombia	45.5	40	1.4	2.4	75	30	7	–0.5
Costa Rica	4.6	90	1.3	1.9	59	23	7	1.3
Ecuador	14.2	50	1.6	2.6	65	33	7	–5.2
El Salvador	6.2	294	1.4	2.4	63	31	6	–9.1
Guatemala	14.4	132	2.8	4.4	47	42	4	–3.0
Honduras	7.6	68	2.3	3.3	50	38	4	–2.8
Mexico	110.6	57	1.4	2.2	77	29	6	–4.5
Nicaragua	6.0	46	1.8	2.5	56	35	3	–7.1
Panama	3.5	46	1.6	2.5	64	30	6	0.7
Paraguay	6.5	16	1.9	3.1	58	34	5	–1.3
Peru	29.5	23	1.6	2.6	76	31	6	–4.4
Uruguay	3.4	19	0.5	2.0	94	23	14	–3.0
Venezuela	28.8	32	1.6	2.6	88	30	6	0.3

[a]Net Migration Rate from the United Nations, Population Division, World Population Prospects: The 2008 Revision Population Database.
Source: Population Reference Bureau, World Population Data Sheet, 2010.

resources are concentrated into one urban center. In an effort to decentralize, some governments have intentionally built new cities far from existing primate cities (for example, Ciudad Guayana in Venezuela and Brasília in Brazil). Despite these efforts, the tendency toward primacy remains. Moreover, the growth of urbanized regions has inspired the label of *megalopolis* for three areas in Latin America. Emerging megalopolises include Mexico City–Puebla–Toluca–Cuernavaca on the Mesa Central, the Niterói–Rio de Janeiro–Santos–São Paulo–Campinas axis in southern Brazil, and the Rosario–Buenos Aires–Montevideo–San Nicolás corridor in Argentina and Uruguay's lower Rio Plata Basin (see Figure 4.16).

Urban Form Latin American cities have a distinct urban morphology that reflects both their colonial origins and their contemporary growth (Figure 4.17). Usually a clear central business district (CBD) exists in the old colonial core. Radiating out from the central business district is older middle- and lower-class housing found in the zones of maturity and *in situ* accretion. In this model, residential quality declines as one moves from the center to the periphery. The exception is the elite spine, a newer commercial and business strip that extends from the colonial core to newer parts of the city. Along the spine one finds superior services, roads, and transportation. The city's best residential zones, as well as shopping malls, are usually on either side of the spine. Close to the elite residential sector, a limited area of middle-class housing is typically found. Most major urban centers also have a *periférico* (a ring road or beltway highway) that circumscribes the city. Industry is located in isolated areas of the inner city and in larger industrial parks outside the ring road.

Straddling the periférico is a zone of peripheral squatter settlements where many of the urban poor live in the worst housing. Services and infrastructure are extremely limited: roads are unpaved; water is often trucked in; and sewer systems are nonexistent. The dense ring of squatter settlements (variously called ranchos, favelas, barrios jovenes, or pueblos nuevos) that encircle Latin American cities reflect the speed and intensity with which these zones were created. The squatter settlements are also found in disamenity zones near the core of the city—such as steep hillsides or narrow gorges prone to flooding. In some cities more than one-third of the population live in these self-built homes of marginal or poor quality. These kinds of dwellings are recognizable throughout the developing world, yet the practice of building one's home on the "urban frontier" has a longer history in Latin America than in most Asian and African cities. The combination of a rapid inflow of migrants, the inability of governments to meet pressing housing needs, and the eventual official recognition of many of these neighborhoods with land titles and utilities meant that this housing strategy was rarely discouraged. Each successful colonization encouraged more.

Among the inhabitants of these neighborhoods, the **informal sector** is a fundamental force that houses, services, and

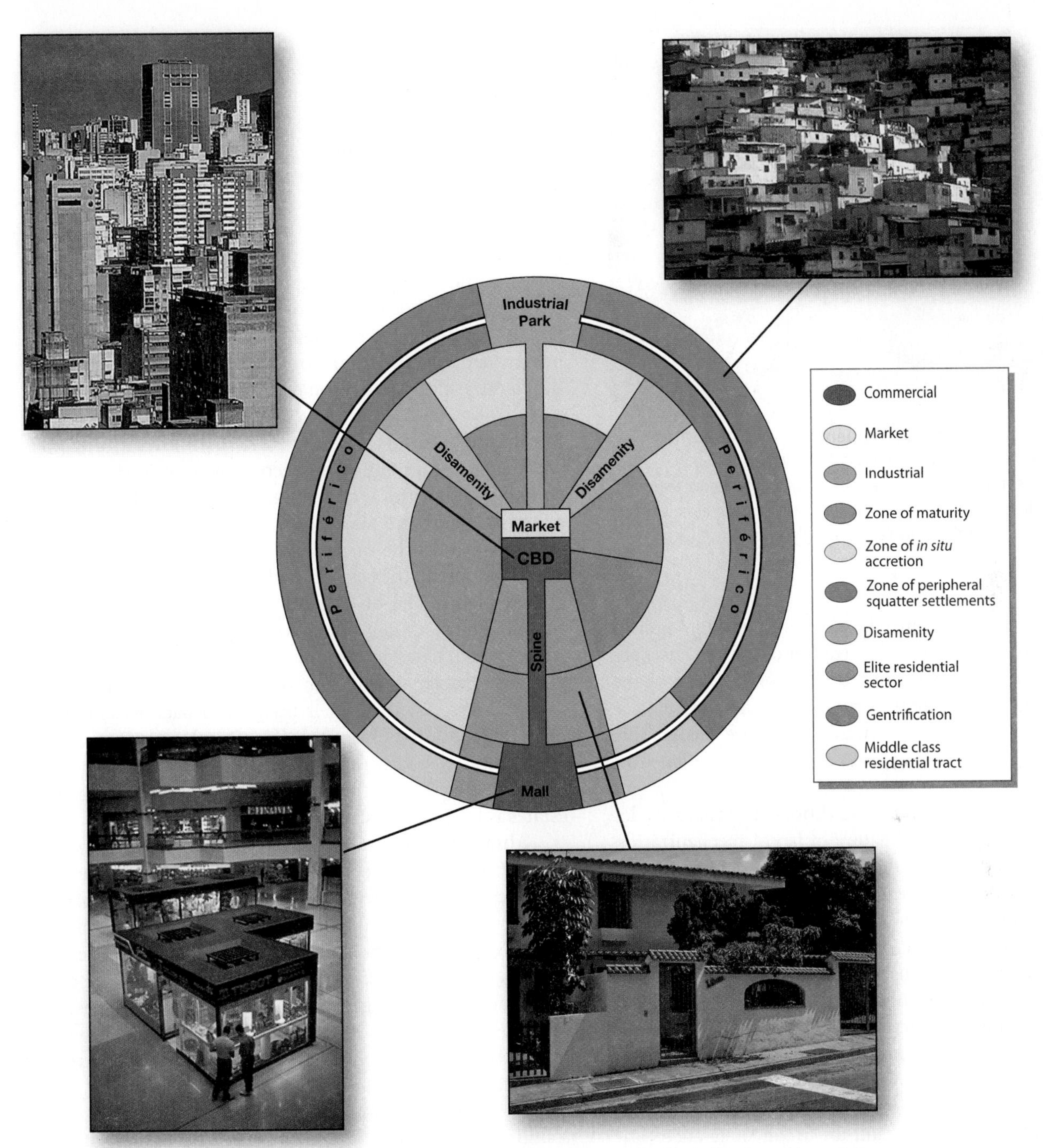

Figure 4.17 Latin American City Model This urban model highlights the growth of Latin American cities and the class divisions within them. While the central business district (CBD), elite spine, and residential sectors may have excellent services and utilities, life in the zone of peripheral squatter settlements is much more difficult. In many Latin American cities, one-third of the population resides in squatter settlements. (Model reprinted from Ford, 1996, "New and Improved Model of Latin American City Structure," *Geographical Review* 86[3], 437–40.)

employs them. Definitions of the informal sector are much debated. The term usually refers to the economic sector that relies on self-employed, low-wage jobs (such as street vending, shoe shining, and artisan manufacturing) that are virtually unregulated and untaxed. Some scholars include as part of the informal sector illegal activities such as drug smuggling, prostitution, and sale of contraband items such as illegally copied movie or music CDs and tapes. One of the most interesting expressions of informality is the housing in the squatter settlements. In Lima, Peru, an estimated 40 percent of the population live in self-built housing, often of very poor quality. Typically these settlements begin as illegal invasions of open spaces that are carefully planned and timed to avoid the risk of eviction by city authorities. If the hastily built

Figure 4.18 Lima Squatter Settlement In arid Lima, squatters initially build their homes using straw mats. As settlements become established, residents invest in adobe and cinder block to improve their homes. Life on the urban frontier is harsh. Water is trucked in; electricity is irregular; and travel toward the city center is costly and slow.

communities go unchallenged, squatters steadily improve their houses (Figure 4.18). The creation of these landscapes reflects a conscious and organized effort on the part of the urban poor, many of whom have rural origins, to make a place for themselves in Latin America's cities. (See Geographic Tools: Participatory Mapping in the Slums of Rio de Janeiro.)

Rural-to-Urban Migration As conditions in rural areas deteriorated due to the consolidation of lands, mechanization of agriculture, and increased population pressure, peasants began to pour into the cities of Latin America in a process referred to as **rural-to-urban migration**. The strategy of rural households sending family members to the cities for employment as domestics, construction workers, artisans, and vendors has been well established since the 1960s. Once in the cities, rural migrants generally found conditions better, especially access to education, health care, electricity, and clean water.

It was not poverty alone that drove people out of rural areas, but individual choice and an urban preference. Migrants believed in, and often realized, greater opportunities in cities, especially the capital cities. Those who came were usually young (in their twenties) and better educated than those who stayed behind. Women slightly outnumbered men in this migrant stream. The move itself was made easier by extended kin networks formed by earlier migrants who settled in discrete areas of the city and aided new arrivals. The migrants maintained their links to their rural communities by periodically sending remittances and making return visits.

Patterns of Rural Settlement

Although the majority of Latin Americans live in cities, some 125 million people do not. Throughout the region a distinct rural lifestyle exists, especially among peasant subsistence farmers. In Brazil alone, more than 35 million people live in rural areas. Interestingly, the absolute number of people living in rural areas today is roughly equal to the number in the 1960s. Yet rural life has changed dramatically. In addition to subsistence agriculture, in most rural areas highly mechanized capital-intensive farming occurs. The links between rural and urban areas are much improved, meaning that rural folks are less isolated. Also, as international migration increases, many rural communities are directly connected to cities in North America and Europe, with immigrants sending back remittances and supporting home town associations. Much like the region's cities, the rural landscape is divided by extremes of poverty and wealth. The root of social and economic tension in the countryside is the uneven distribution of arable land.

Rural Landholdings The control of land in Latin America was the basis for political and economic power. Historically, colonial authorities granted large tracts of land to the colonists, who were also promised the service of Indian laborers as part of the *encomienda* system. These large estates typically took up the best lands along the valley bottoms and coastal plains. The owners were often absentee landlords, spending most of their time in the city and relying on a mixture of hired, tributary, and slave labor to run their rural operations. Passed down from one generation to the next, many estates can trace their ownership back a century or two. The allocation of large blocks of land to one owner also denied peasants access to land, so they were forced to labor on the estates. This entrenched practice of maintaining large estates is referred to as **latifundia**.

Although the pattern of estate ownership is well documented, peasants have always farmed small plots for their subsistence. This practice of **minifundia** can lead to permanent or shifting cultivation. Small farmers typically plant a mixture of crops for subsistence as well as for trade. Peasant farmers in Colombia or Costa Rica, for example, grow corn, fruits, and various vegetables alongside coffee bushes that produce beans for export. Strains on the minifundia system occur when demographic pressures create land scarcity or political elites reallocate land for their needs.

Much of the turmoil in 20th-century Latin America surrounded the question of land, with peasants demanding its redistribution through the process of **agrarian reform**. Governments have addressed these concerns in different ways. The Mexican Revolution in 1910 yielded a system of communally held lands called *ejidos*. In the 1950s Bolivia crafted agrarian reform policies that led to the expropriation of estate lands and their reallocation to small farmers. As part of the Sandinista revolution in Nicaragua in 1979, lands were expropriated from the political elite and converted into collective farms. In 2000, President Hugh Chavez ushered in a new era of agrarian reform in Venezuela and the Bolivian President, Evo Morales, introduced an agrarian reform program in 2006 aimed at giving land title to indigenous communities in the eastern lowlands. These programs have met with resistance and, at times, have proven to be costly politically. Eventually the path chosen by most governments was

GEOGRAPHIC TOOLS Participatory Mapping in the Slums of Rio de Janeiro

Participatory mapping is a catchall label that refers to an array of community-based research approaches used to map places. These approaches rely upon the collaboration of local people with mapping technicians. Today with advances such as Google Earth, up loading via wiki platforms, and cell phones with GPS, more people can map than ever before. The value of participatory mapping is that places that were once barely known or "off the map" can be better known. Marginalized groups, such as indigenous peoples or urban slum dwellers, can create maps of their environments using these new technologies and in the process lay claim to their space.

It was the notion of mapping the unmapped that drew a Brazilian NGO called Rede Jovem (Youth Net) to the *favelas,* or slums, of Rio de Janeiro. The favelas are areas of informal housing wedged into narrow valleys or on steep hillsides throughout Rio. Many favelas have existed for more than 50 years with schools, churches, shops, and clubs. Yet due to their informal beginnings, these densely settled neighborhoods seldom have formal street names, limited formal services, and often are not shown on city maps. Moreover, these hillside communities that are seen by all, tend to be known by few people other than their residents.

Many *Cariocas* (residents of Rio) fear favelas as high crime localities where the poor reside and drug gangs rule. With major events like the World Cup and Olympics coming to Rio de Janeiro, programs are in place to wall in the favelas in an effort to contain them from spreading and keep them isolated from the formal city. Yet favelas are also iconic symbols of Rio, much like Sugarloaf Mountain itself, and people are curious to see them. There are tour operators who take tourists for favela sightseeing and to attend *Carioca Funk* or *Baile Funk* dance parties, whose unique sound has a global following.

Rede Jovem's mission was to make the favelas known as real places with schools, institutions, clubs, and restaurants and not just areas of marginalization and violence. They believe that by putting favelas on the map, they will be less ignored. Through funding received from a Brazilian telecommunications firm, they hired local residents to map five favelas. In the end they hired young women, between the ages of 17 and 25, because they could walk more freely in the favelas than young men who were more likely to be stopped by police. Each wiki-reporter is paid a monthly stipend and given a GPS-equipped cell phone. Rafaela Concalves de Silva, one of the wiki-reporters, says "I start up the GPS at the beginning of the street, then I walk without stopping until the end." The information is then uploaded to the wikimap site. To learn the names of streets, she talks to older residents and then taps the street name into her cell phone.

Photos are also taken of shops, restaurants and schools, and then uploaded to the emerging map. These wiki-reporters are from the favelas and want to show that their communities are places of life. Putting them on the map is one way of making this claim real.

Source: Adapted from AFP, 2009, "Rio youth use GPS phones to put favelas on map," Oct 17.

Figure 4.2.1 Favela Santa Marta Hillside favelas, such as Santa Marta, can be seen throughout the city of Rio de Janeiro. Santa Marta is one of neighborhoods being mapped through the Rede Jovem project.

Figure 4.2.2 Mapping Favelas Favela resident, Alini dos Santos Silva, shows the map she is developing for her neighborhood located near Rio de Janeiro's Ipanema beach.

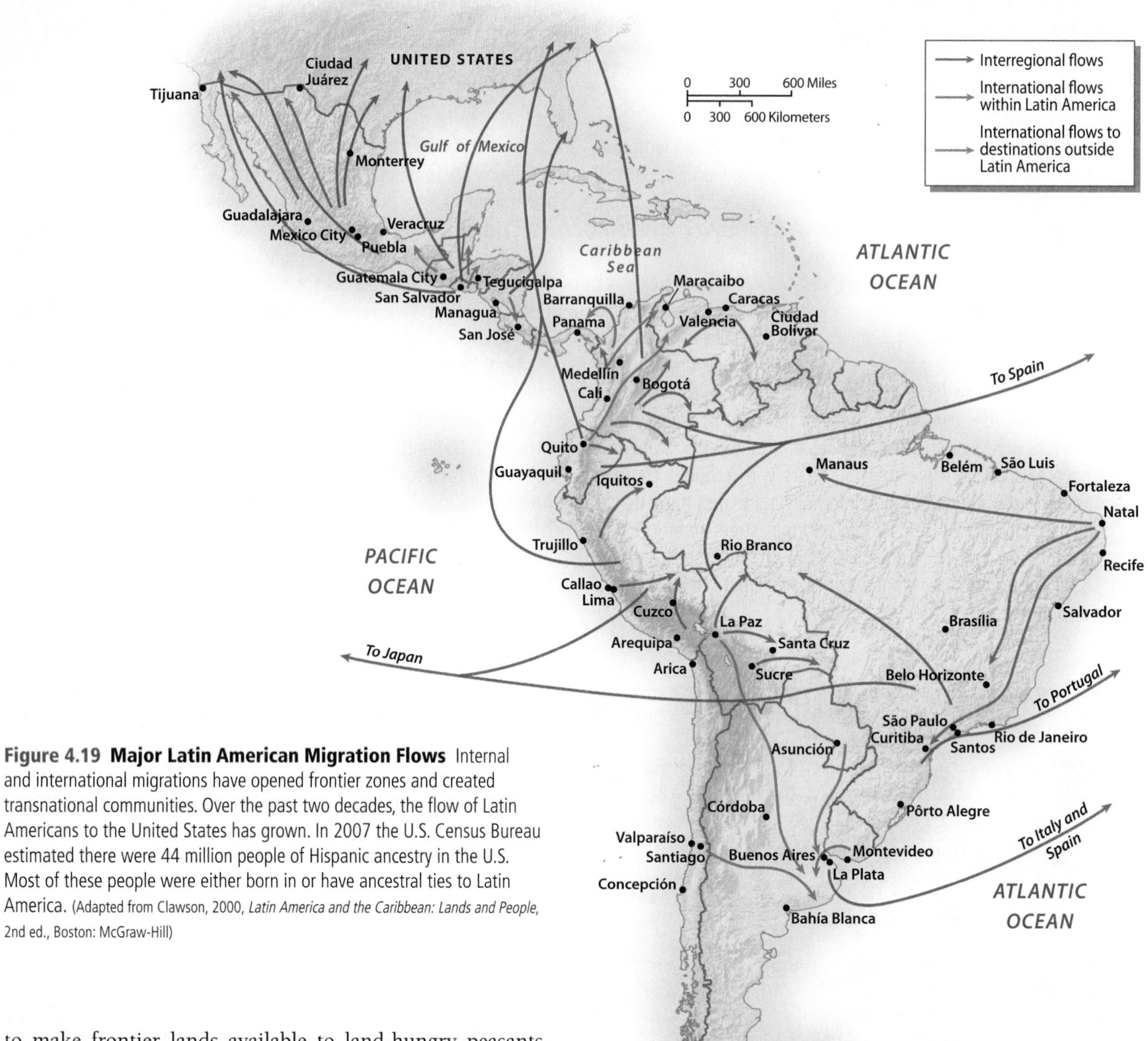

Figure 4.19 Major Latin American Migration Flows Internal and international migrations have opened frontier zones and created transnational communities. Over the past two decades, the flow of Latin Americans to the United States has grown. In 2007 the U.S. Census Bureau estimated there were 44 million people of Hispanic ancestry in the U.S. Most of these people were either born in or have ancestral ties to Latin America. (Adapted from Clawson, 2000, *Latin America and the Caribbean: Lands and People*, 2nd ed., Boston: McGraw-Hill)

to make frontier lands available to land-hungry peasants. The opening of tropical frontiers, especially in South America, was a widely practiced strategy that changed national settlement patterns and began waves of rural-to-rural and even urban-to-rural migration.

Agricultural Frontiers The creation of agricultural frontiers serves several purposes: providing peasants with land, tapping unused resources, and shoring up political boundaries. A number of frontier colonization efforts in South America are noteworthy. In addition to settlement along Brazil's Trans-Amazon Highway, Peru developed its Carretera Marginal (Marginal Highway) in an effort to lure colonists into the cloud and rain forests of eastern Peru. In Bolivia, Colombia, and Venezuela, agricultural frontier schemes in the lowland tropical plains attracted peasant farmers and large-scale investors. Mexico sent colonists, some displaced by dam construction, into the forests of Tehuantepec. Guatemala developed its northern Petén region. El Salvador had no frontier left, but many desperately poor Salvadorans poured into the neighboring states of Honduras and Belize in search of land. In short, although the dominant demographic trend has been a rural-to-urban movement, an important rural-to-rural flow has turned previously virgin areas into agricultural zones (Figure 4.19).

The opening of the Brazilian Amazon for settlement was the most ambitious frontier colonization scheme in the region. In the 1960s, Brazil began its frontier expansion by constructing several new Amazonian highways, a new capital (Brasília), and state-sponsored mining operations. It was the Brazilian military who directed the opening of the

Amazon to provide an outlet for landless peasants and to extract the region's many resources. Yet the generals' plans did not deliver as intended. Throughout the basin, thin forest soils were incapable of supporting permanent agricultural colonies and, in the worst cases, degraded into baked claylike surfaces devoid of vegetation. Government-promised land titles, agricultural subsidies, and credit were slow to reach small farmers, even if they were fortunate enough to be given land on *terra roxa*, a nutrient-rich purple clay soil. Instead, a disproportionate amount of money went to subsidizing large cattle ranches through tax breaks and improvement deals where "improved" land meant cleared land. Because of the many competing factions (miners, loggers, ranchers, peasants, and corporate farmers) and the uncertainty of land title and political authority, the Amazon can be a violent place. A victim of this violence was Sister Dorothy Stang, an American nun who had spent her life in Pará state working with poor peasant farmers towards a model of sustainable development. In 2005, she was killed by hired gunmen after trying to stop ranchers from clearing a remote area of forest. Her assailants were eventually caught and convicted, but most of the murders that occur due to resource conflicts remain unsolved.

Despite a concerted effort by Brazilian officials to settle the entire region, most commercial activities and residents are concentrated in a few sites and much resource extraction occurs illegally. Four times more people lived in the Amazon in the 1990s than in the 1960s. The population of 12 million people in the Brazilian Amazon in 2000 is projected to grow to 15 million by 2010; thus, increased human modification of the Brazilian Amazon is inevitable (Figure 4.20).

Population Growth and Movement

The high growth rates in Latin America throughout the 20th century are attributed to natural increase as well as immigration. The 1960s and 1970s were decades of tremendous growth resulting from high fertility rates and increasing life expectancy. In the 1960s, for example, a Latin American woman typically had six or seven children. Today the average number of children per woman is between 2 and 3. The 2010 TFR is 2.0 for Brazil, 2.4 for Colombia, and 2.2 for Mexico (see Table 4.1). A number of factors explain this trend: more urban families, which tend to be smaller than rural ones; increased participation of women in the workforce; higher education levels of women; state support of family planning; and better access to birth control. Even the more rural countries with a high percentage of Amerindians are experiencing smaller families—Bolivia's TFR is 3.5 and Ecuador's is 2.6.

Even with family sizes shrinking, and in the cases of Chile and Costa Rica falling below replacement rates, there is built-in potential for continued growth because of the relative demographic youth of these countries. Approximately 30 percent of the population are below the age of 15. In North America the similar cohort is 20 percent of the population, and in Europe it is just 15 percent. This means that a proportionally larger segment of the population has yet to enter into its childbearing years.

Waves of immigrants into Latin America and migrant streams within Latin America have influenced population size and patterns of settlement. Beginning in the late 19th century, new immigrants from Europe and Asia added to the region's size and ethnic complexity. Important population shifts within countries have also occurred in recent decades, as witnessed by the growth of Mexican border towns and the demographic expansion of the Bolivian plains. In an increasingly globalized economy, even more Latin Americans live and work outside the region, especially in the United States and Europe.

European Migration After gaining their independence from Iberia in the 19th century, Latin America's new leaders sought to develop their territories through immigration. Firmly believing in the dictum "to govern is to populate," many countries set up immigration offices in Europe to attract hardworking peasants to till the soils and

Figure 4.20 Amazonian Settlement The village of Boca da Valeria in Amazonas state is a small village on the Amazon River in Brazil. Like other riverine settlements in the Amazon, people rely on the river for transportation and fish. The surrounding forest and areas of dark earth are used for food production and resource extraction. Interestingly, as more tourists come to experience the Amazon on small cruise ship, scenic Boca da Valeria is a popular port of call.

"whiten" the **mestizo** population, those people of mixed European and Indian ancestry. The Southern Cone countries of Argentina, Chile, Uruguay, and southern Brazil were the most successful in attracting European immigrants from the 1870s until the depression of the 1930s. During this period, some 8 million Europeans arrived (more than came during the entire colonial period), with Italians, Portuguese, Spaniards, and Germans the most numerous. Some of this immigration was state-sponsored, such as the nearly 1 million laborers (including entire families) brought to the coffee estates surrounding São Paulo at the start of the 20th century. Other migrants came seasonally, especially the Italian peasants who left Europe in the winter for agricultural work in Argentina and were thus nicknamed "the swallows." Still others paid their own passage, intending to settle permanently and prosper in the growing commercial centers of Buenos Aires, São Paulo, Montevideo, and Santiago.

Asian Migration Less well known are the Asian immigrants who also arrived during the late 19th and 20th centuries. Although considerably fewer, over time they established an important presence in the large cities of Brazil, Peru, Argentina, and Paraguay. Beginning in the mid-19th century, the Chinese and Japanese who settled in Latin America were contracted to work on the coffee estates in southern Brazil and the sugar estates and guano mines of Peru. A son of Japanese immigrants, Alberto Fujimori was president of Peru from 1990–2000.

The Japanese in Brazil are the most studied Asian immigrant group. Between 1908 and 1978, a quarter-million Japanese immigrated to Brazil; today the country is home to more than 1.5 million people of Japanese descent. Initially, most Japanese were landless laborers, yet by the 1940s they had accumulated enough capital so that three-quarters of the migrants had their own land in the rural areas of São Paulo and Paraná states. As a group, the Japanese have been closely associated with the expansion of soybean and orange production. Today Brazil leads the world in exports of orange juice concentrate, with most of the oranges grown on Japanese-Brazilian farms. Increasingly, second- and third-generation Japanese have taken professional and commercial jobs in Brazilian cities. South America's economic turmoil in the 1990s encouraged many ethnic Japanese to emigrate to Japan in search of better opportunities. Nearly one-quarter of a million ethnic Japanese left South America in the 1990s (mostly from Brazil and Peru) and now work in Japan (Figure 4.21).

The latest Asian immigrants are from South Korea. Unlike their predecessors, most of the Korean immigrants came with enough capital to invest in small businesses, and they settled in cities, rather than in the countryside. According to official South Korean statistics, 120,000 Koreans emigrated to Paraguay between 1975 and 1990. While many have stayed in Paraguay, there seems to be a pattern of secondary immigration to Brazil and Argentina. Recent Korean immigrants in São Paulo have created more than 2,500 small businesses. Unofficial estimates of the number of Koreans living in Brazil range from 40,000 to 120,000. As a group they are decidedly commercial in orientation and urban in residence; their cities of choice are Asunción and Ciudad del Este in Paraguay, São Paulo in Brazil, and Buenos Aires in Argentina.

Latino Migration and Hemispheric Change Movement within Latin America and between Latin America and North America has had a significant impact on sending and receiving communities alike. Within Latin America, international migration is shaped by shifting economic and political realities. Thus, Venezuela's oil wealth during the 1960s and 1970s attracted between 1 and 2 million Colombian immigrants who tended to work as domestics or agricultural laborers. Argentina has long been a destination for Bolivian and Paraguayan laborers. And, of course, farmers in the United States have depended on Mexican laborers for over a century (see Figure 4.19).

Figure 4.21 Japanese Brazilians Brazilian youth of Japanese ancestry perform in Curitiba, Brazil, to mark the 100th anniversary of Japanese immigration to Brazil in 2008. In 1908 the first Japanese immigrants arrived as agricultural workers, choosing Brazil as a destination after countries such as the United States and Canada had banned Japanese immigration. Today there are over 1.5 million ethnic Japanese in Brazil, especially in the states of São Paulo and Parana, and they have distinguished themselves as major farmers and urban professionals.

Figure 4.22 Mexican–U.S. Border Pedestrians and cars stream across the Gateway to the Americas Bridge between Laredo, Texas and Nuevo Laredo, Mexico. Like many of the paired border towns, Nuevo Laredo is the larger settlement. Mexicans cross to Laredo for shopping and jobs, and Americans go to Nuevo Laredo for shopping and lower cost medical care.

Political turmoil also sparked waves of international migrants. Chilean intellectuals fled to neighboring countries in the 1970s when General Pinochet wrested power from the socialist government led by Salvador Allende. Nicaraguans likewise fled when the socialist Sandanistas came to power in 1979. The bloody civil wars in El Salvador and Guatemala sent waves of refugees into neighboring countries, such as Mexico and the United States. Violence and low-intensity conflict have internally displaced over 2.5 million Colombians in the past two decades, and official statistics suggest another 3 million Colombians live abroad. With democratization on the rise in the region, many of today's immigrants are classified as economic migrants, not political asylum seekers.

Presently, Mexico is by far the largest country of origin of legal immigrants to the United States (Figure 4.22). In 2007, 28 million people claimed Mexican ancestry, and of those, approximately 12 million were immigrants. Mexican labor migration to the United States dates back to the late 1800s when relatively unskilled labor was recruited to work in agriculture, mining, and railroads. This practice was formalized in the 1940s through the 1960s with the *Bracero* program, which granted temporary employment residence to 5 million Mexican laborers (much like the "guest workers" that West Germany recruited from southern Europe and Turkey). Today roughly two-thirds of the Hispanic population (both foreign-born and native-born) in the United States claims Mexican ancestry. Mexican immigrants are most concentrated in California and Texas, but increasingly they are found throughout the country. Although Mexicans continue to have the greatest presence among Latinos in the United States, the number of immigrants from El Salvador, Guatemala, Nicaragua, Colombia, Ecuador, and Brazil has steadily grown. The U.S. Census Bureau estimated that there were 44 million Hispanics in the United States (both foreign-and native-born) in 2007. By 2050, the U.S. Census Bureau estimates that nearly one-in-four Americans will be Hispanic. Most of this population has ancestral ties with peoples from Latin America and the Caribbean (see Chapter 5).

Today, Latin America is a region of emigration rather than one of immigration. The majority of states have negative rates of annual net migration, which means they are losing more people each year than they are gaining through immigration. Mexico has an annual rate of –4.5 per 1,000 and El Salvador's is –9.1. By comparison the average rate of net migration for the developing world as a whole is a much smaller figure of –0.5 (see Table 4.1). Both skilled and unskilled workers from Latin America are an important source of labor in North America, Europe, and Japan. Many of these immigrants send monthly **remittances** (monies sent back home) to sustain family members. In 2008, it was estimated that immigrants sent nearly $70 billion to Latin America. Most of this money came from workers in the United States, but Latino immigrants in Spain, Portugal, Japan, Canada, and Italy also sent money back to the region. The global financial crisis greatly impacted remittance flows, which after going up for years dropped 15 percent in 2009 as immigrants had less money to send to their families.

Through remittances and technological advances that make communication faster and cheaper, immigrants maintain close contact with their home countries in ways that earlier generations could not. Scholars have labeled this ability to straddle livelihoods between two countries as **transnationalism**. A cultural and an economic outcome of globalization, transnationalism highlights the social and economic links that form between home and host countries. Transnational migrants maintain a dual or hybrid identity, which is also seen as a cultural expression of globalization. Salvadorans working in Washington, DC, for example, maintain regular contract with their rural villages in El Salvador through sending goods and forming hometown associations while also developing vital immigrant social networks in their new home (Figure 4.23).

Figure 4.23 Salvadorans in the Suburbs Day laborers wait for employment in a Maryland suburb outside Washington, DC. War and economic hardship drove many Salvadorans from their country in the 1980s and 1990s. Today, Salvadorans are the largest immigrant group in the Washington, DC, metropolitan area.

PATTERNS OF CULTURAL COHERENCE AND DIVERSITY: Repopulating a Continent

The Iberian colonial experience (1492 to the 1800s) imposed a political and cultural coherence on Latin America that makes it distinguishable today as a world region. Yet this was not a simple transplanting of Iberia across the Atlantic. Often a syncretic process unfolded in which European and Indian traditions blended as indigenous groups were incorporated into either the Spanish or the Portuguese empires. In some areas such as southern Mexico, Guatemala, Bolivia, Ecuador, and Peru, Amerindian cultures have showed remarkable resilience, as evidenced by the survival of indigenous languages. Yet the prevailing pattern is one of forced assimilation in which European religion, languages, and political organization were imposed on the surviving fragments of native society. Later, other cultures, arriving as both forced and voluntary migrants, added to the region's cultural mix. Perhaps the single most important factor in the dominance of European culture in Latin America was the demographic collapse of native populations.

Demographic Collapse

Beginning in 1492 and lasting until the 19th century, the cultural change and human loss resulting from the cataclysmic encounter between European and Indian worlds was enormous. Throughout the region archaeological sites are poignant reminders of the complexity of precontact civilizations. Dozens of stone temples found throughout Mexico and Central America, where the Mayan and Aztec civilizations flourished, attest to the ability of these societies to thrive in the area's tropical forests and upland plateaus (Figure 4.24). The Mayan city of Tikal flourished in the lowland forests of the Petén, supporting tens of thousands, before its mysterious collapse centuries before the arrival of Europeans. In the Andes, stone terraces built by the Incas are still being used by Andean farmers; earthen platforms for village sites and raised fields for agriculture are still being discovered and mapped. Evidence of the complexity of precontact civilizations are ceremonial centers such as Cuzco—the core of the great Incan empire—and the Incan site of Machu Picchu—unknown to most of the world until American archaeologist Hiram Bingham investigated the site in the early 1900s. The Spanish, too, were impressed by the sophistication and wealth they saw around them, especially in the incomparable Tenochtitlán, site of present-day Mexico City. Tenochtitlán was the political and ceremonial center of the Aztecs, supporting a complex metropolitan area with some 300,000 residents. The largest city in Spain at the time was considerably smaller.

The most telling figures of the impact of European expansion are demographic. Experts believe that the precontact Americas had 54 million inhabitants; by comparison, western Europe in 1500 had approximately 42 million. Of the 54 million, about 47 million were in what is now Latin America, and the rest were in North America and the Caribbean. The region had two major population centers: one in Central Mexico with 14 million people and the other in the Central Andes (highland Peru and Bolivia) with nearly 12 million. By 1650, after a century and a half of colonization, the indigenous population was one-tenth its precontact size. The human tragedy of this population loss is difficult to comprehend. The relentless elimination of 90 percent of the indigenous population was largely caused by epidemics of influenza and smallpox, but warfare, forced labor, and starvation due to a collapse of food production systems also contributed to the death rate.

Figure 4.24 Tikal, Guatemala This ancient Mayan city located in the lowland forests of the Peten was part of a complex network of cities located in the Yucatan and northern Guatemala. At its height Tikal supported over 100,000 residents before its collapse in late 10th century. Today Tikal is a major tourist destination.

The population low point for Amerindians was in 1650, but the tragedy continued throughout the colonial period and to a much lesser extent continues today. After the indigenous population began its slow recovery in the Central Andes and Central Mexico, there were still tribal bands in southern Chile (the Mapuche) and Patagonia (Araucania) that experienced the ravages of disease three centuries after

Columbus landed. Even now, the isolation of some Amazonian tribes has made them vulnerable to disease. Conflicts with outsiders who invade their territories in search of land or gold still occur. In an all too familiar story, a common cold can prove deadly to forest dwellers who lack the needed immunities.

The Columbian Exchange Historian Alfred Crosby likens the contact period between the Old World (Europe, Africa, and Asia) and the New World (the Americas) as an immense biological swap, which he terms the **Columbian exchange**. According to Crosby, Europeans benefited greatly from this exchange, and Amerindian peoples suffered the most from it. On both sides of the Atlantic, however, the introduction of new diseases, peoples, plants, and animals forever changed the human ecology. Consider, for example, the introduction of Old World crops. The Spanish brought their staples of wheat, olives, and grapes to plant in the Americas. Wheat did surprisingly well in the highland tropics and became a widely consumed grain over time. Grapes and olive trees did not fare as well, but eventually grapes were produced commercially in the temperate zones of South America. The Spanish grew to appreciate the domestication skills of Indian agriculturalists who had developed valuable starch crops such as corn, potatoes, and bitter manioc, as well as condiments such as hot peppers, tomatoes, pineapple, cacao, and avocados. Corn never became a popular food for Europeans, but many African peoples adopted it as a vital staple. After initial reluctance, Europeans and Russians widely consumed the potato as a basic food. Domesticated in the highlands of Peru and Bolivia, the humble potato has an impressive ability to produce a tremendous volume of food in a very small area, especially in cool climates. This root crop is credited with driving Europe's rapid population increase in the 18th century when peasant farmers from Ireland to Russia became increasingly dependent on it as a basic food. This potato dependence also made them vulnerable to potato blight, a fungal disease that emerged in the 19th century and came close to unraveling Irish society.

Tropical crops transferred from Asia and Africa reconfigured the economic potential of the region. Sugarcane, an Asian transfer, became the dominant cash crop of the Caribbean and the Atlantic tropical lowlands of South America. With sugar production came the importation of millions of African slaves. Coffee, a later transfer from East Africa, emerged as one of the leading export crops throughout Central America, Colombia, Venezuela, and Brazil in the 19th century. Introduced African pasture grasses enhanced the forage available to livestock.

The movement of Old World animals across the Atlantic had a profound impact on the Americas. Initially, these animals hastened Indian decline by introducing animal-borne diseases and by producing feral offspring that consumed everything in their paths. The utility of swine, sheep, cows, and horses was eventually appreciated by native survivors. Draft animals were adopted, as was the plow, which facilitated the preparation of soil for planting. Wool became a very important fiber for indigenous communities in the uplands. And slowly, pork, chicken, and eggs added protein and diversity to the staple diets of corn, potatoes, and cassava. With the major exception of disease, many transfers of plants and animals ultimately benefited both worlds. Still, it is clear that the ecological and material basis for life in Latin America was completely reworked through the exchange process initiated by Columbus.

Indian Survival Presently, Mexico, Guatemala, Ecuador, Peru, and Bolivia have the largest indigenous populations. Not surprisingly, these areas had the densest native populations at contact. Indigenous survival also occurs in isolated settings where the workings of national and global economies are slow to penetrate. The isolated Miskito Coast of Honduras is home to Miskito, Pech, and Garífuna. In eastern Panama the Kuna and Emberá are present (Figure 4.25). The Brazilian state of Roraima in the northern

Figure 4.25 The Emberá of Panama An Emberá couple from Drúa, Panama, show one of their handmade baskets, which they sell to tourists. Their small village is in Chagres National Park in eastern Panama on the Chagres River and is accessible only by canoe. The Emberá people, like the Kuna, live in the lowlands of eastern Panama and western Colombia.

Amazon, some 15,000 indigenous Pemong speakers organized in 2004 to create the Raposa/Serra do Sol Indian reservation. In these relatively isolated areas, small groups of people have managed to maintain a distinct way of life despite the pressures to assimilate.

In many cases Indian survival comes down to one key resource—land. Indigenous peoples who are able to maintain a territorial home, formally through land title or informally through long-term occupancy, are more likely to preserve a distinct ethnic identity. Because of this close association between identity and territory, native peoples are increasingly insisting on a recognized space within their countries. Some of Panama's indigenous groups have organized indigenous territories called *comarcas* where they assert local authority and have limited autonomy. These efforts to define indigenous territory are seldom welcomed by the state but they are occurring throughout the region.

From Amazonia to the highlands of Chiapas, many native groups are demanding formal political and territorial recognition as a means to redress centuries of injustice. Whether these efforts will actually reshape political and cultural space for Latin America's Amerindians is still uncertain. Yet there are hopeful signs of increased political participation by Amerindian peoples. In 2001, Peruvians elected President Alejandro Toledo, an Amerindian who rose from acute poverty to obtain a doctorate in economics from Stanford University. In 2005, Bolivians elected a former union organizer, Evo Morales, as their first Amerindian president. More so than Toledo, Morales is seen as a powerful symbol of the rise of indigenous political power in the Andes. President Morales won reelection in 2009.

Patterns of Ethnicity and Culture

The Indian demographic collapse enabled Spain and Portugal to refashion Latin America into a European likeness. Yet instead of a neo-Europe rising in the Tropics, a complex ethnic blend evolved. Beginning with the first years of contact, unions between European sailors and Indian women began the process of racial mixing that over time became a defining feature of the region. The courts of Spain and Portugal officially discouraged racial mixing but were unable to prevent it. Spain, which had a far larger native population to oversee than did the Portuguese in Brazil, became obsessed with the matter of race and maintaining racial purity among its colonists. An elaborate classification system was constructed to distinguish emerging racial castes. Thus, in Mexico in the 18th century a Spaniard and an Indian union resulted in a *mestizo* child. A child of a mestizo and a Spanish woman was a *castizo*. However, the children from a castizo woman and a Spanish man were considered Spanish in Mexico but a quarter mestizo in Peru. Likewise, *mulattoes* were the progeny of European and African unions, and *zambos* were the offspring of Africans and Indians.

After generations of intermarriage, such a classification system collapsed under the weight of its complexity, and four broad categories resulted: *blanco* (European ancestry), *mestizo* (mixed ancestry), *indio* (Indian ancestry), and *negro* (African ancestry). The *blancos* (or Europeans) continue to be well represented among the elites, yet the vast majority of people are of mixed racial ancestry. In Venezuela, for example, the phrase "cafe con leche" (coffee with milk) is used to describe the racial makeup of the majority of the population who share European, African, and Indian characteristics. Dia de la Raza, the region's observance of Columbus Day, recognizes the emergence of a new mestizo race as the legacy of European conquest. Throughout Latin America, more than other regions of the world, miscegenation—or racial mixing—is the norm, which makes the process of mapping racial or ethnic groups especially difficult.

Enduring Amerindian Languages Roughly two-thirds of Latin Americans are Spanish speakers, and one-third speak Portuguese. These colonial languages were so prevalent by the 19th century that they were the unquestioned languages of government and instruction for the newly independent Latin American republics. In fact, until recently many countries actively discouraged, and even repressed, Indian tongues. It took a constitutional amendment in Bolivia in the 1990s to legalize native-language instruction in primary schools and to recognize the country's multiethnic heritage (more than half the population is Indian; and Quechua, Aymara, and Guaraní are widely spoken) (Figure 4.26).

Because Spanish and Portuguese dominate, there is a tendency to neglect the influence of indigenous languages in the region. Mapping the use of indigenous languages, however, reveals important pockets of Indian resistance and survival. In the Central Andes of Peru, Bolivia, and southern Ecuador, more than 10 million people still speak Quechua and Aymara, along with Spanish. In Paraguay and lowland Bolivia there are 4 million Guaraní speakers, and in southern Mexico and Guatemala at least 6 to 8 million speak Mayan languages. Small groups of native-language speakers are found scattered throughout the sparsely settled interior of South America and the more isolated forests of Central America, but many of these languages have fewer than 10,000 speakers.

Blended Religions Like language, the Roman Catholic faith appears to have been imposed upon the region without challenge. Most countries report 90 percent or more of their population as Catholic. Every major city has dozens of churches, and even the smallest hamlet maintains a graceful church on its central square (Figure 4.27). In some countries, such as El Salvador and Uruguay, a sizable portion of the population attend Protestant evangelical churches, but the Catholic core of this region is still intact.

Figure 4.26 Language Map of Latin America The dominant languages of Latin America are Spanish and Portuguese. Nevertheless, there are significant areas in which indigenous languages persist and, in some cases, are recognized as official languages. Smaller language groups exist in Central America, the Amazon Basin, and southern Chile. (Adapted from *The Atlas of the World's Languages*, 1994, New York: Routledge)

Figure 4.27 The Catholic Church Churches, such as the Dolores Church in Tegucigalpa, Honduras, are important religious and social centers. The vast majority of people in Latin America define themselves as Catholic. Many churches built in the colonial era are valued as architectural treasures and are beautifully preserved.

Exactly what native peoples absorbed of the Christian faith is unclear. Throughout Latin America **syncretic religions**, blends of different belief systems, enabled pre-Hispanic religious practices to be folded into Christian worship. These blends took hold and endured, in part because Christian saints were easy surrogates for pre-Christian gods and because the Catholic Church tolerated local variations in worship as long as the process of conversion was underway. The Mayan practice of paying tribute to spirits of the underworld seems to be replicated today in Mexico and Guatemala via the practice of building small cave shrines to favorite Catholic saints and leaving offerings of fresh flowers and fruits. One of the most celebrated religious icons in Mexico is the Virgin of Guadalupe, a dark-skinned virgin seen by an Indian shepherd boy in the 16th century who became the patron saint of Mexico.

Syncretic religious practices also evolved and endured among African slaves. By far the greatest concentration of slaves was in the Caribbean, where slaves were used to replace the indigenous population, which was wiped out by disease (see Chapter 5). Within Latin America, the Portuguese colony of Brazil received the most Africans—at least 4 million. In Brazil, where the volume and the duration of the slave trade were the greatest, the transfer of African-based religious and medical systems is most evident. West African-based religious systems such as Batuque, Umbanda, Candomblé, and Shango are often mixed with, or ancillary to, Catholicism, and are widely practiced in Brazil. So accurate were some of these religious transfers that it is common to have Nigerian priests journey to Brazil to learn forgotten traditions. In many parts of southern Brazil, Umbanda is as popular with people of European ancestry as with Afro-Brazilians. Typically a person becomes familiar with Umbanda after falling victim to a magician's spell by having some object of black magic buried outside his or her home. To regain control of his or her life, the victim needs the help of a priest or priestess.

The syncretic blend of Catholicism with African traditions is most obvious in the celebration of carnival, Brazil's most popular festival and one of the major components of Brazilian national identity. The three days of carnival, known as the Reign of Momo, combine Christian Lenten beliefs with pagan influences and feature African musical traditions epitomized by the rhythmic samba bands. Although the street festival was banned for part of the 19th century, Afro-Brazilians in Rio de Janeiro resurrected it in the 1880s with nightly parades, music, and dancing. Within 50 years the street festival had given rise to formalized samba schools and helped break down racial barriers. By the 1960s, carnival became an important symbol for Brazil's multiracial national identity. Today the festival—which is most associated with Rio—draws thousands of participants from all over the world (Figure 4.28).

Figure 4.28 Carnival in Rio de Janeiro A samba band marches in the streets of Rio de Janeiro. Samba, the quintessential music of carnival, draws inspiration from African rhythmic traditions.

The Global Reach of Latino Culture

Latin American culture, vivid and diverse as it is, is widely recognized throughout the world. Whether it is the sultry pulse of the tango or the fanaticism with which Latinos embrace soccer as an art form, aspects of Latin American culture have been absorbed into a globalizing world culture. A dramatic example of the reach of Latino culture can be seen every Saturday on the Univision television network, based in Miami. There, a charismatic Chilean, Don Francisco, has hosted *Sabado Gigante* since 1986. This 4-hour-long Spanish variety show is viewed in 28 countries and draws a weekly audience of 120 million viewers. In the arts, Latin American writers such as Gabriel García Marquez and Isabel Allende have obtained worldwide recognition. In terms of popular culture, new musical artists such as Colombia's Shakira and Brazil's hip-hop samba singer Max de Castro reach international audiences, while Latino superstars Jennifer Lopez and Ricky Martin are icons of world pop culture. Through music, literature, and even *telenovelas* (soap operas), Latino culture is being transmitted to an eager worldwide audience.

Telenovelas Popular nightly soap operas are a mainstay of Latin American television. These tightly plotted series are filled with intrigue and double deals. Unlike their counterparts in the United States, they end, usually after 100 episodes. Once standard fare for the working class, many telenovelas take hold and absorb an entire nation. During particularly popular episodes, the streets are noticeably calm as millions of people tune in to catch up on the lives of their favorite heroines. Brazil, Venezuela, and Mexico each produce scores of telenovelas, but the Mexican ones are international mega-hits.

Televisa, a Mexican production agency, has aggressively marketed its inventory of soap operas to an eager global public. Mexican telenovelas are avidly watched in countries as diverse as Croatia, Russia, China, South Korea, Iran, the United States, and France, as well as throughout Latin America. Predictably scripted as Mexican Cinderella stories, these

PEOPLE ON THE MOVE National Identity and Japanese Brazilians

Over the last two decades tens of thousands of Japanese Brazilians have moved to Japan for factory employment. But in 2009, in response to the global economic downturn, some began to return to Brazil. Many left because they were unemployed due to factory layoffs, others complained of discrimination, some took advantage of a controversial Japanese policy to pay for their return to Brazil, and in the process many discovered that they were more Brazilian than Japanese.

About 1.5 million Brazilians are of Japanese descent, making them the largest Japanese population outside of Japan. The flow of Japanese migrants began in 1908 and stopped during WWII. For nearly 50 years there was relatively little contact with Japan, and the immigrants distinguished themselves in Brazilian agriculture and commerce. Yet for many of these immigrants and their descendants there was a persistent feeling that in Brazil they were always seen as Japanese. In 2008 when Prince Naruhito of Japan visited Brazil to celebrate the centennial of Japanese emigration there, thousands came out to enthusiastically greet him.

Figure 4.3.1 **Japanese Brazilians Protest** Japanese-Brazilians living in Japan display a banner that reads "Don't Throw Away Japanese-Brazilian Workers" during a rally in 2009. As unemployment jumped due to the global downturn in 2009, the Japanese government started offering return flights to jobless migrants from Brazil.

It was in 1990 that Japan invited people of Japanese descent to "return" to Japan with special work visas in an attempt to address Japanese labor shortages brought on, in part, by their aging population. Some 300,000 Brazilians responded to the opportunities in a homeland most had never known. For ethnically homogenous Japan it seemed like a win-win situation, they would get the immigrant labor they needed but would bring in people who were ethnically Japanese. Yet despite their Japanese appearance, Brazilians in Japan are culturally foreign, usually speaking Japanese poorly, if at all. Facing discrimination for their poor language skills and "manners" (the curse of looking Japanese but not acting so), Japanese Brazilians began to embrace samba music, carnival, and even soccer with renewed passion. Portuguese language schools and newspapers were also created. And Brazilian influence began to grow; Tokyo has a major carnival parade and Bossa Nova music has become popular with Japanese youth.

Yet the real shock to the identity of Japanese Brazilians came in 2009 when the Japanese government encouraged them to leave. An emergency program was introduced whereby unemployed Brazilians were offered $3,000 toward air fare plus $2,000 for each dependent. The catch was that they would be stripped of their work visa and would find it very difficult to ever return to Japan. After decades of feeling they were Japanese, many Japanese descendants are rediscovering their Brazilianness.

sagas of poor underclass women (often domestics) falling in love with members of the elite, battling jealous rivals, and ultimately emerging triumphant seem to resonate with fans around the world. In addition to their broad appeal, telenovelas are big business, perhaps Mexico's largest cultural export. While Hollywood and Mumbai grind out movies, much of Mexico's entertainment industry is geared toward producing this popular art form.

Soccer Perhaps the quintessential global sport, soccer has a fanatical following throughout much of the world. Yet it is Latin America, and especially in South America, where "*futbol*" is considered a cultural necessity. Still largely a male game, although girls are beginning to play, young boys and men are constantly seen on fields, beaches, and blacktops playing soccer, especially in late afternoons and on the weekends. The great soccer stadiums of Buenos Aires (Bombonera) and Rio de Janeiro (Maracaña) are regarded as shrines to the game. Many individuals use the victories and losses of their national soccer teams as the important chronological markers of their lives.

It was Pelé, the Brazilian soccer phenomenon of the 1960s and 1970s, who introduced the free-flowing acrobatic style that became known as "the beautiful game." Today Latino soccer stars (especially Brazilians) play for corporate clubs all over Europe and earn millions. Latin Americans also fill up the few slots allotted to foreign players on the U.S. Major League Soccer teams. Yet the dream of many Latino soccer players is to be on the national team and bring home the World Cup. Visit any Latin American country when their team is playing a World Cup qualifying match and the streets are eerily quiet. During the 2006 World Cup, the business and financial centers of Argentina and Brazil virtually came to a standstill during the entire tournament. Of the 19 World Cups awarded between 1930 and 2010, South American teams have won 9. Somehow, as if by magic, it is believed that a World Cup victory will make things better, improve the economy, and even reduce crime. In short, soccer is regarded with near religious significance. And as Latin Americans emigrate (both as players and as laborers) they bring their enthusiasm for the sport with them.

National Identities Viewed from the outside, there is considerable homogeneity to this region; yet distinct national identities and cultures flourish in Latin America. Since the early days of the republics, countries celebrated particular elements from their pasts when creating their national histories. In the case of Brazil, the country's interracial characteristics were highlighted to proclaim a new society in which the color lines between Europeans and Africans ceased to matter. Mexico celebrated the architectural and cultural achievements of its Aztec predecessors while at the same time forging an assimilationist strategy that discouraged surviving indigenous culture and language. Yet national identities are fluid and context dependent. A compelling example of this is the case of the Japanese Brazilians (see People on the Move: The National Identify of Japanese Brazilians on page 155).

Musical and dance traditions evolved and became emblematic of these new societies: The tango in Argentina, caporales in Bolivia, the vallenato and cumbia in Colombia, the mariachi in Mexico, and the samba in Brazil are easily distinguished styles that are representative of distinct national cultures (Figure 4.29). Literature also reflects the distinct identities found in Latin America. Writers such as Isabel Allende, Gabriel García Marquez, Mario Vargas Llosa, Carlos Fuentes, and Jorge Amado situate their stories in their native countries and, in so doing, celebrate the unique characteristics of Chileans, Colombians, Peruvians, Mexicans, and Brazilians. Distinct political cultures also evolved, which at times led to expansionist policies that brought neighbors into conflict.

(a)

(b)

Figure 4.29 Musical and Dance Traditions (a) Caporales is a popular Bolivian folkloric dance. During carnival and other holidays, caporales dance groups take to the streets in friendly competition. (b) Tango is the signature dance and music of Buenos Aires, Argentina. Today tango clubs are popular with tourists who want to see the dance and even learn how to do it.

GEOPOLITICAL FRAMEWORK: Redrawing the Map

Latin America's colonial history, more than its present condition, unifies this region. For the first 300 years after Columbus's arrival, Latin America was a territorial prize sought by various European countries but effectively settled by Spain and Portugal. By the 19th century, the independent states of Latin America had formed, but they continued to experience foreign influence and, at times, overt political pressure, especially from the United States. At other times a more neutral Pan-American vision of American relations and hemispheric cooperation has held sway, represented by the formation of the **Organization of American States (OAS)**. The present organization was chartered in 1948, but its origins date back to 1889. Yet there is no doubt that U.S. policies toward trade, economic assistance, political development, and at times military intervention are often seen as compromising to the sovereignty of these states.

Within Latin America there have been cycles of intraregional cooperation and antagonism. Neighboring countries have fought over territory, closed borders, imposed high tariffs, and cut off diplomatic relations. Even today there are a dozen long-standing border disputes in Latin America that occasionally erupt into armed conflict. The 1990s witnessed a revival in the trade block concept with the formation of **Mercosur**, the Southern Common Market, which now includes Brazil, Uruguay, Argentina, Paraguay, and Venezuela as full members and Bolivia, Chile, Colombia, Ecuador and Peru as associate members; and NAFTA, which includes Mexico, the United States, and Canada. As economic ties strengthen, these trade blocks could form the basis for a new alignment of political and economic interests in the region.

Iberian Conquest and Territorial Division

Because it was Christopher Columbus who claimed the Americas for Spain, the Spanish were the first active colonial agents in the Western Hemisphere. In contrast, the Portuguese presence in the Americas was the result of the **Treaty of Tordesillas**, brokered by the pope in 1493–94. By that time Portuguese navigators had charted much of the coast of Africa in an attempt to find an ocean route to the Spice Islands (Moluccas) in Southeast Asia. With the help of Christopher Columbus, Spain sought a western route to the Far East. When Columbus discovered the Americas, Spain and Portugal asked the pope to settle how these new territories should be divided. Without consulting other European powers, the pope divided the Atlantic world in half—the eastern half containing the African continent was awarded to Portugal, the western half with most of the Americas was given to Spain. The line of division established by the treaty actually cut through the eastern edge of South America, placing it under Portugese control. This treaty was never recognized by the French, English, or Dutch, who also asserted territorial claims in the Americas, but it did provide the legal apparatus for the creation of a Portuguese territory in America—Brazil—which would later become the largest and most populous state in Latin America (Figure 4.30).

Six years after the treaty was signed, Portuguese navigator Alvares Cabral inadvertently reached the coast of Brazil on a voyage to southern Africa. The Portuguese soon realized that this territory was on their side of the Tordesillas line. Initially they were unimpressed by what Brazil had to offer; there were no spices or major indigenous settlements. Quickly, however, they came to appreciate the utility of the coast as a provisioning site as well as a source for brazilwood, used to produce a valuable dye. Portuguese interest in the territory intensified in the late 16th century with the development of sugar estates and the expansion of the slave trade, and in the 17th century with the discovery of gold in the Brazilian interior.

Spain, in contrast, aggressively pursued the conquest and settlement of its new American territories from the very start. After discovering little gold in the Caribbean, by the mid-16th century Spain's energy was directed toward developing the silver resources of Central Mexico and the Central Andes (most notably Potosí in Bolivia). Gradually the economy diversified to include some agricultural exports, such as cacao and sugar, as well as a variety of livestock. In terms of foodstuffs, the colonies were virtually self-sufficient. Some basic manufactured items, such as crude woolen cloth and agricultural tools, were also produced, but, in general, manufacturing was forbidden in the Spanish American colonies to keep them dependent on Spain.

Revolution and Independence It was not until the 1800s, with a rise of revolutionary movements between 1810 and 1826, that Spanish authority on the mainland was challenged. Ultimately, elites born in the Americas gained control, displacing the representatives of the crown. In Brazil, the evolution from Portuguese colony to independent republic was a slower and less violent process that spanned eight decades (1808–89). In the 19th century Brazil was declared a separate kingdom from Portugal with its own monarch, and later it became a republic.

The territorial division of Spanish and Portuguese America into administrative units provided the legal basis for the modern states of Latin America (see Figure 4.30). The Spanish colonies were first divided into two viceroyalties (New Spain and Peru) and within these were various subdivisions that later became the basis for the modern states. (In the 18th century the Viceroyalty of Peru, which included all of Spanish South America, was divided to form three viceroyalties: La Plata, Peru, and New Granada). Unlike Brazil, which evolved from a colony into a single republic, the former Spanish colonies experienced fragmentation in the 19th century. Prominent among revolutionary leaders was Venezuelan-born Simon Bolívar (Figure 4.31), who

Figure 4.30 Shifting Political Boundaries The evolution of political boundaries in Latin America began with the 1494 Treaty of Tordesillas, which gave much of the Americas to Spain and a slice of South America to Portugal. The larger Spanish territory was gradually divided into viceroyalties and audiencias, which formed the basis for many modern national boundaries. The 1830 borders of these newly independent states were far from fixed. Bolivia would lose its access to the coast; Peru would gain much of Ecuador's Amazon; and Mexico would be stripped of its northern territory by the United States. (From Lombardi, Cathryn L., and Lombardi, John V. *Latin American History: A Teaching Atlas*, © 1993. Reprinted by permission of the University of Wisconsin Press)

advocated his vision for a new and independent state of Gran Colombia. For a short time (1822–30) Bolívar's vision was realized, as Colombia, Venezuela, Ecuador, and Panama were combined into one political unit. Similarly, in 1823 the United Provinces of Central America was formed to avoid annexation by Mexico. By the 1830s, this union also broke apart into the states of Guatemala, Honduras, El Salvador, Nicaragua, and Costa Rica. Today the former Spanish mainland colonies include 16 states plus 3 Caribbean ones, with a total population of nearly 370 million. If the Spanish colonial territory had remained a unified political unit, it would now have the third largest population in the world, following China and India.

Persistent Border Conflicts As the colonial administrative units turned into states, it became clear that the territories were not clearly delimited, especially the borders that stretched into the sparsely populated interior of South America. This would later become a source of conflict as new states struggled to demarcate their territorial boundaries. Numerous border wars erupted in the 19th and 20th centuries, and the map of Latin America has been redrawn many times. Some of the more noted conflicts were the War of the Pacific (1879–82) in which Chile expanded to the north and Bolivia lost its access to the Pacific; warfare between Mexico and the United States in the 1840s, which resulted in the present border under the Treaty of Hidalgo (1848); and the War of the Triple Al-

Figure 4.31 Simon Bolívar Heroic likenesses of Simon Bolívar (the Liberator) are found throughout South America, especially in his native country of Venezuela. This statue stands in the central plaza of the Andean city of Mérida, Venezuela.

liance (1864–70), the bloodiest war of the postcolonial period, which occurred when Argentina, Brazil, and Uruguay allied themselves to defeat Paraguay in its claim to control the upper Paraná River Basin. It is estimated that this conflict resulted in the reduction of Paraguay's adult male population by nine-tenths. Sixty years later, the Chaco War (1932–35) resulted in a territorial loss for Bolivia in its eastern lowlands and a gain for Paraguay. In the 1980s Argentina lost a war with Great Britain over control of the Falkland, or Malvinas, Islands in the South Atlantic. And as recently as 1998, Peru and Ecuador skirmished over a disputed boundary in the Amazon Basin.

Outright war in the region is less common than ongoing and nagging disputes over international boundaries. Any of a dozen dormant claims erupt from time to time, given the political climate between neighbors. These include Venezuela and Guyana's border dispute, Ecuador's claims to the Peruvian eastern lowlands, and the maritime boundary dispute between Honduras and Nicaragua as well as one between Venezuela and Colombia. Many of these disputes are based on historical territorial claims dating back to the colonial period.

One of the more interesting geopolitical conflicts in the Southern Cone is based on territorial claims to Antarctica. Seven claimant states—Argentina, Chile, Norway, the United Kingdom, France, New Zealand, and Australia—plus five nonclaimant states (including the United States) seek resource rights to Antarctica. Even though an Antarctic Treaty has existed since 1959 stating that the landmass should be used for peaceful purposes, Chile and Argentina have incorporated their Antarctic territorial claims on national maps and even postage stamps. In the 1970s, Chilean President Augusto Pinochet spent a week touring the Chilean claim. Argentina held a national cabinet meeting on its portion of Antarctica in the 1970s and then sent a pregnant Argentine woman to give birth to the continent's first child. For the most part, these nationalist claims are symbolic. Recent treaties show an international inclination toward cooperation and the banning of commercial exploitation of Antarctica.

The Trend Toward Democracy Early in the 21st century, most of the 17 countries in this region will celebrate their bicentennials. Compared with most of the developing world, Latin Americans have been independent for a long time. Yet political stability is not a hallmark of the region. Among the countries in the region, some 250 constitutions have been written since independence, and military coups have been alarmingly frequent. Since the 1980s, however, the trend has been toward democratically elected governments, the opening of markets, and broader popular participation in the political process. Where dictators once outnumbered elected leaders, by the 1990s each country in the region had a democratically elected president. (Cuba, the one exception, will be discussed in Chapter 5.)

Democracy may not be enough for the millions frustrated by the slow pace of political and economic reform. In survey after survey, Latin Americans register their dissatisfaction with politicians and governments. Most of the newly elected democratic leaders are also free-market reformers who are quick to eliminate state-backed social safety nets, such as food subsidies, government jobs, and pensions. Many of the poor and middle class have grown skeptical about whether this brand of democracy could improve their lives. But the political left has not yet produced an alternative to privatization and market-driven policies, although there has definitely been a trend of leftist leaning politicians winning presidential elections in Brazil, Bolivia, Nicaragua, Ecuador, Peru and Venezuela. For now, the status quo continues, although popular frustration with falling incomes, rising violence, corruption, and chronic underemployment are a recipe for political instability throughout the region.

Regional Organizations

Even as democratically elected leaders struggle to address the pressing needs of their countries, political developments at the supranational and subnational levels pose new challenges to their authority. The most discussed **supranational organizations**—governing bodies that include several states—are the trade blocs. **Subnational organizations**, which are groups that represent areas or people within the state, often form along ethnic or ideological lines and can provoke serious internal divisions. Indigenous groups seeking territorial or political recognition and insurgent groups espousing Marxist ideas (such as the FARC in Colombia) have challenged the authority of the states. Finally, the financial and political force of drug cartels (mafia-like organization in charge of the drug trade) can undermines judicial systems and influence areas beyond state boundaries.

Trade Blocs Beginning in the 1960s, regional trade alliances were attempted in an effort to foster internal markets and reduce trade barriers. The Latin American Free Trade Association (LAFTA), the Central American Common Market, and the Andean Community have existed for decades, but their ability to influence economic trade and growth is limited at

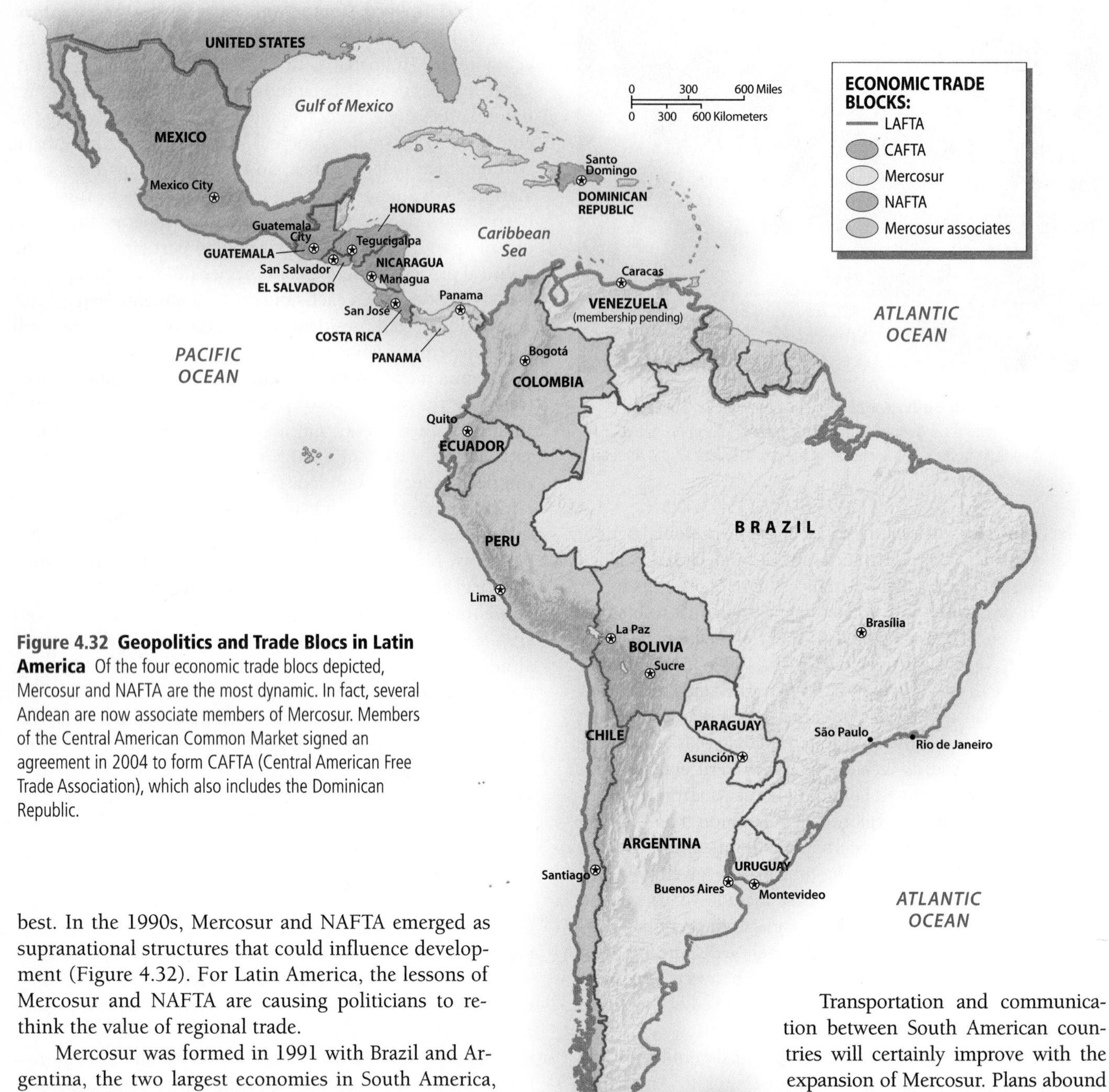

Figure 4.32 Geopolitics and Trade Blocs in Latin America Of the four economic trade blocs depicted, Mercosur and NAFTA are the most dynamic. In fact, several Andean are now associate members of Mercosur. Members of the Central American Common Market signed an agreement in 2004 to form CAFTA (Central American Free Trade Association), which also includes the Dominican Republic.

best. In the 1990s, Mercosur and NAFTA emerged as supranational structures that could influence development (Figure 4.32). For Latin America, the lessons of Mercosur and NAFTA are causing politicians to rethink the value of regional trade.

Mercosur was formed in 1991 with Brazil and Argentina, the two largest economies in South America, and the smaller states of Uruguay and Paraguay as members. Since its formation, trade among these countries grew tremendously, so much so that Chile, Bolivia, Peru, Ecuador and Colombia have joined the group as associate members and Venezuela is awaiting ratification as a full member. This is significant in two ways: It reflects the growth of these economies and the willingness to put aside old rivalries (especially long-standing antagonisms between Argentina and Brazil) for the economic benefits of regional cooperation. The size and productivity of this market have not gone unnoticed. Mercosur countries are the European Union's largest trading partner in Latin America, and there have been renewed efforts to form a free-trade agreement between Mercosur and the EU. (See Exploring Global Connections: Latin American Trade with Europe.)

Transportation and communication between South American countries will certainly improve with the expansion of Mercosur. Plans abound to fundamentally rework the flow of goods and communication in this area. As privatization of the telephone companies has benefited telecommunication, major joint engineering projects are being considered. Efforts are under way to improve navigation along the Paraná and Paraguay rivers, which are already important arteries for the transport of grains. Other schemes include building a bridge across the Plata River estuary to form a more direct link between the capital cities of Uruguay and Argentina, a tunnel through the Andes between Chile and Argentina, and a regional network of natural gas pipelines that would link Bolivia, Argentina, Peru, and southern Brazil. In short, the elaboration of Mercosur could change the way individuals in the mem-

EXPLORING GLOBAL CONNECTIONS The Growth in European and Asia Investment in Latin America

Throughout the 20th century the United States was Latin America's largest trading partner, and it still is. But trade with and investment from Europe and Asia is diversifying Latin America's trade relationships and lessening its dependence upon U.S. markets. These trends also suggest an erosion of Washington's influence in the region. This is especially true in South America where trade with the European Union is rapidly catching up, and in some countries surpassing, trade with the United States

Collectively, the European Union is Latin America's second largest trading partner after the United States. Between 1990 and 2006 trade between Latin America and the European Union doubled. In 2008 the trade volume between Latin America and the European Union amounted to €178 billion, representing nearly 15 percent of Latin America's trade. Latin American countries still tend to export agricultural and energy products to Europe. They import European machinery, transport equipment, and chemicals.

The European Union now has the position of being the leading investor in the region. The presence of European businesses in Latin America has been an important source of new direct foreign investment. Perhaps not surprisingly given its colonial ties to the region, Spain has been the single largest European investor, especially in the 1990s. Under neoliberal reforms as state-owned industries in Latin America were sold, Spain was poised to invest in telecommunications, banking, public utilities, oil, and natural gas (Figure 4.4.1). And over the last few years the European Union has established Free Trade Agreements with Chile and Mexico. Moreover, the Union has free trade agreement negotiations with Central America, the Andean Community, and Mercosur.

Coming up fast is China, whose investments in Latin America soared in the past decade. In terms of individual states, China is now Latin America's second largest trading partner after the United States. Chinese investors are focused on energy and mineral resources. During the economic crisis of 2009 the Chinese backed loans to the region, brokering deals with Venezuela, Ecuador, Argentina, and Brazil. For example, China's loan of $10 billion to Brazil's national oil company will go to offshore oil exploration. Similarly, China added $6 billion to a development fund in Venezuela. These monies give Venezuela access to hard currency, in exchange the volume of Venezuelan oil shipments to China nearly tripled.

Source: Adapted from "Deals Help China Expand Sway in Latin America," *New York Times,* April 16, 2009; Commission of the European Communities, 2009, "The European Union and Latin America: Global Players in Partnership."

Figure 4.4.1 Spanish Investment in Latin America Pedestrians stroll by a Banco Santander branch in Mexico City. Banco Santander is one of the largest banks in the world in terms of market capitalization. It originated in Santander, Spain. Today, Spain's largest bank has branches throughout Europe and is heavily invested in Latin America.

ber countries relate to one another and think about themselves. Much like the European Union, which has fostered a sense of European identity, Mercosur may over time shape a South American identity.

NAFTA took effect in 1994 as a free trade area that would gradually eliminate tariffs and ease the movement of goods among the member countries (Mexico, the United States, and Canada). As a trade block, it includes more than 450 million people that collectively produce more than $14 trillion in goods and services each year. NAFTA has increased intraregional trade; but considerable controversy rages about who are the beneficiaries and who are the losers in this trade block (see Chapter 3). NAFTA did stimulate trade and overall growth, but it also brought dislocation and change to the economic landscape of the countries involved. Although NAFTA did not deter immigration to the United States by creating jobs in Mexico, it did prove that a free trade area combining industrialized and developing states was possible. By some measures the success of NAFTA inspired other free trade agreements, the newest one being the Central American Free Trade Agreement (CAFTA) signed in 2004 to include the Dominican Republic along with Guatemala, El Salvador, Honduras, Nicaragua, and Costa Rica. The push for CAFTA emerged after it became clear that an overly ambitious Free Trade Area of the Americas (including all states in the hemisphere except Cuba) would fail, in part due to Mercosur's opposition to the plan. Like NAFTA, the CAFTA agreement requires opening markets for goods and services, dismantling protections of local industries, and enforcement of intellectual property rights. In return, the United States promises greater market access for Central American and Dominican products, including textiles and agricultural commodities. CAFTA is likely to increase direct foreign investment and jobs in the block, although detractors worry about the rise of low-wage sweatshop employment and not more skilled labor. Interestingly, it was resistance from Costa Rica—the most prosperous member of the group—that delayed its full implementation until 2009.

Insurgencies and Drug Trafficking Guerrilla groups such as the FARC (Revolutionary Armed Forces of Colombia) in Colombia have controlled large territories of their countries through the support of those loyal to their cause, along with theft, kidnapping, and violence. The FARC, along with the ELN (National Liberation Army), gained wealth and might through the drug trade in the 1980s and 1990s. The level of violence in Colombia escalated further with the rise of paramilitary groups—well-armed vigilante groups that seek to "cleanse" areas of insurgency sympathizers. The paramilitary groups are blamed for the rise in politically motivated murders and there are clear indications that they received help from individuals in the government. As many as 2.5 million Colombians have been internally displaced by violence since the late 1980s, most fleeing rural areas for towns and cities. A decade ago the level of violence in Colombia had become intolerable. In the major cities of Bogotá and Medellín, people did not go out at night for fear of crime and kidnapping, economic activity declined, and many people left the country. Colombian statistics suggest that over 3 million Colombians now live abroad. Fortunately, the situation in the last 7 years has improved considerably. Under President Uribe, the police presence increased throughout the state and negotiations with insurgencies stopped, ultimately reducing their power, and levels of violence in Colombia. Yet drug cartels and gangs in states as diverse as Mexico, El Salvador, and Brazil have been blamed for increases in violence and lawlessness. In 2009 the Mexican government was forced to bring in the army to quell the levels of violence associated with Mexican drug traffickers in Ciudad Juarez.

Initially, most Latin governments cared little about controlling the drug trade as it brought in much-needed hard currency. Within the region, drug consumption was scarcely a problem. Some drug lords even became popular folk heroes, lavishly spending money on their communities for housing, parks, and schools. The social costs of the drug trade to Latin America became evident by the 1980s, when the region was crippled by a badly damaged judicial system. By paying off police, the military, judges, and politicians, the drug syndicates wield incredible political power that threatens the civil fabric of the states in which they work. And eventually drug consumption and addiction became a problem, especially in Latin America's large cities.

Years of counternarcotics work have done little to reduce the overall flow of drugs to North America and Europe, but the programs have changed the areas of production. In the 1980s, the major production centers were in the cloud forests of Peru and Bolivia, and the producers were peasants working small plots. During the 1990s, the production of coca leaf (the main ingredient for cocaine) shifted to the rain forests of Colombia, where coca was grown on large, modern estates. Since 1990, the United States has spent around $4 billion to reduce Andean coca and cocaine production. Yet as Figure 4.33 shows, the overall area of coca production

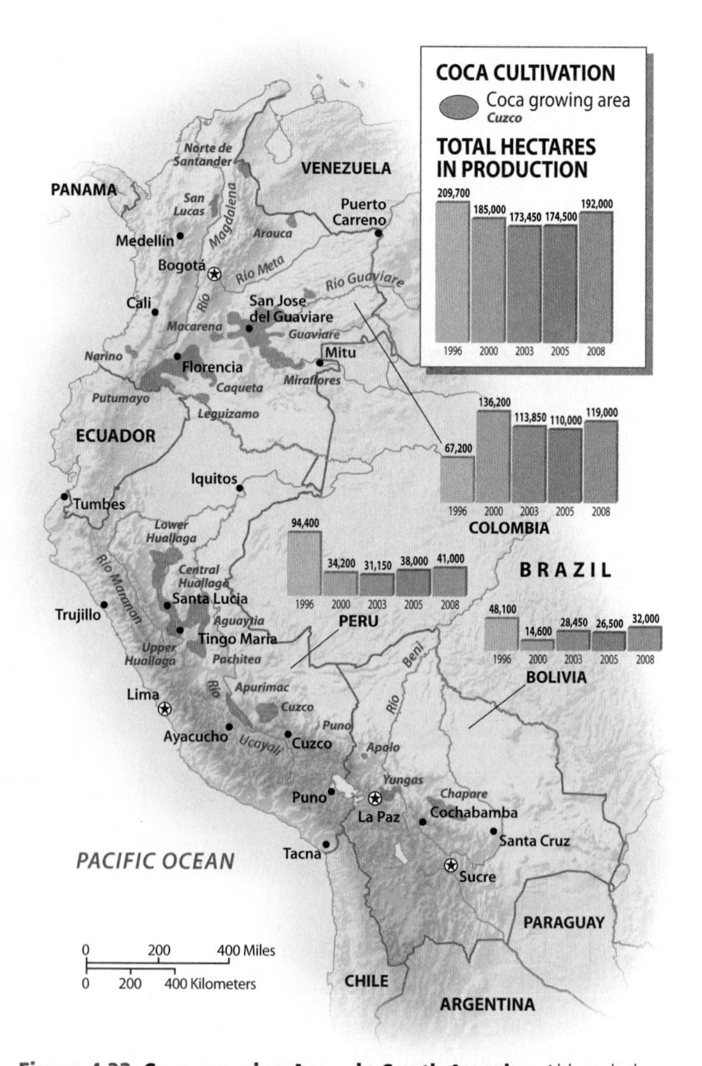

Figure 4.33 Coca-growing Areas in South America Although the oldest coca-growing regions are in Bolivia and Peru, Colombian traffickers turned the processing and distribution of cocaine into an international narcotics trade. By the late 1990s the bulk of coca production had shifted out of Bolivia and Peru and to Colombia. (From U.S. Government, 2002, 2003, 2005, 2008, "Latin American Narcotics Cultivation and Production Estimates," Washington, DC: CIA Crime and Narcotics Center)

has not fallen significantly between 1996 and 2008. The main difference is that more coca is now grown in Colombia than in Peru and Bolivia. Mexico's role in the hemispheric drug trade has also steadily increased. It is both a transshipment area for cocaine and a production region of marijuana, heroine, and methamphetamine, almost all of which is bound for consumers in the United States.

Less appreciated are the environmental consequences of coca production and coca eradication. Geographer Ken Young estimates that in Peru alone as many as 2.5 million acres (1 million hectares) of tropical forest were degraded by coca cultivation. Delicate tropical ecosystems and their waterways are being contaminated by the tons of chemicals used to produce cocaine. Unfortunately, serious ecological damage may also be occurring as a result of eradication efforts. The preferred method in southern Colombia is aerial spraying, which causes serious health problems, destruction of legal crops, increased social tension, and damage to both terrestrial and aquatic ecosystems (Figure 4.34).

It is profitability that drives the cocaine trade. As long as there is demand, the supply will be met. Although still confined to South America, coca can grow in other tropical areas, including Southeast Asia, the Pacific, and Africa. The poppy industry, the base for heroin, offers an instructive comparison. Once poppies were almost an exclusive crop of Southeast Asia, but today Central Asia as well as Colombia and Mexico are major producers of poppies for the heroin trade. Given the global nature of the drug trade, it is likely that a victory in Colombia against coca cultivation will result in another "coca war" elsewhere. This doesn't mean that nothing should be done; it only underscores the inherent problems associated with a supply-oriented drug policy.

Indigenous Groups At the grassroots level, scores of indigenous organizations have formed to protest the consequences of neoliberalism and globalization. The Zapatista rebellion in southern Mexico was in part a reaction to globalization. The rebellion began on January 1, 1994, in Chiapas, the day NAFTA took effect. Although Zapatista supporters—largely Amerindian peasants—are mostly interested in local issues, such as land and basic services, their movement reflects a general concern about how increased foreign trade and investment hurt rural peasants.

Amerindian groups in the Andes have flexed their political muscle as well. In January 2000, thousands of Indians, allied with dissident army officers, forced the resignation of Ecuadorian President Jamil Mahuad. Two years later, the Indian movement in Ecuador helped to elect Lucio Gutiérrez, one of the dissident army colonels. Bolivia witnessed the organized protests of Amerindians who inhabit the city of El Alto, on the Altiplano overlooking La Paz. Angry with their promining and protrade president, the Indians set up road blocks in 2004 that cut off La Paz from supplies. In the end, indigenous-led protests forced two presidents to resign in two years and led to the election of Bolivia's first Amerindian president, Evo Morales, in 2005. Such experiences show the greater involvement of organized Indian groups in national politics and, some might argue, a deepening of democracy. Still others express concern that ethnically driven politics could lead to national fragmentation.

Figure 4.34 Coca Eradication A member of an elite antidrug unit of Colombia's National Police watches as an aircraft dumps herbicide on an illicit coca crop in Colombia. Coca leaf, the raw material for cocaine, has become the leading cash crop in the area. With financial assistance from the United States, Colombians use aerial spraying in an effort to eradicate the crop.

ECONOMIC AND SOCIAL DEVELOPMENT: From Dependency to Neoliberalism

Most Latin American economies fit into the broad middle-income category set by the World Bank. Clearly part of the developing world, their people are much better off than those in Sub-Saharan Africa, South Asia, and China. Still the economic contrasts are sharp both between states and within them. Table 4.2 compares the purchasing power parity (PPP) per capita of the countries in the region. The highest figures are for Mexico and Argentina ($14,320 and $13,990 respectively). Next is Chile, with a PPP per capita figure of over $13,200, closely followed by Venezuela, Panama and Uruguay. The poorest countries in the region are in Central America. Nicaragua and Honduras have the region's lowest PPP per capita figure of under $4,000. In terms of annual growth, all the countries have positive annual growth rates for the period 2000 to 2008. In fact, several countries had average annual growth rates of 4 percent or higher. Given the grave financial crises in Ecuador, Argentina, and Uruguay in 2000 and 2001, the positive growth figures for these countries are especially impressive. What is not captured in this data is the sharp downtown experienced in 2009 due to the economic crisis; signs of economic recovery were evident in 2010. Even with its middle-income status, extreme poverty is evident throughout the region as more than one in five people live on less than $2 a day (Table 4.2).

The economic engines of Latin America are the two largest countries, Brazil and Mexico. According to the World

TABLE 4.2 Development Indicators

Country	GNI per capita, PPP 2008	GDP Average Annual % Growth 2000–08	Human Development Index (2006)[a]	Percent Population Living Below $2 a Day	Life Expectancy (2010)[b]	Under Age 5 Mortality Rate (1990)	Under Age 5 Mortality Rate (2008)	Gender Equity (2008)[c]
Argentina	13,990	5.3	0.866	7.3	75	29	16	104
Bolivia	4,140	4.1	0.729	21.9	66	122	54	99
Brazil	10,070	3.6	0.813	12.7	73	56	22	102
Chile	13,240	4.4	0.878	2.4	79	22	9	99
Colombia	8,430	4.9	0.807	27.9	74	35	20	104
Costa Rica	10,950	5.4	0.854	4.3	79	22	11	102
Ecuador	7,770	5.0	0.806	12.8	75	53	25	100
El Salvador	6,630	2.9	0.747	13.2	71	62	18	98
Guatemala	4,690	3.9	0.704	24.3	70	77	35	94
Honduras	3,830	5.3	0.732	29.7	72	55	31	107
Mexico	14,340	2.7	0.854	8.2	76	45	17	101
Nicaragua	2,620	3.5	0.699	31.8	71	68	27	102
Panama	12,620	6.6	0.840	17.8	76	31	23	101
Paraguay	4,660	3.7	0.761	14.2	72	42	28	99
Peru	7,940	6.0	0.806	17.8	73	81	24	101
Uruguay	12,540	3.8	0.865	4.3	76	24	14	98
Venezuela	12,840	5.2	0.844	10.2	74	32	18	102

[a] *United Nations, Human Development Index, 2009.*

[b] *Population Reference Bureau, World Population Data Sheet, 2010.*

[c] *Gender Equity—Ratio of female to male enrollments in primary and secondary school. Numbers below 100 have more males in primary/secondary school, numbers above 100 have more females in primary/secondary schools.*

Source: World Bank, World Development Indicators, 2010.

Bank in 2007, Brazil's economy was the 10th largest in the world, and Mexico's was the 13th largest based on gross national income. Although gains have been made throughout the region in terms of income averages over the past 20 years, income inequality for Latin America remains the highest in the world. For example in Brazil, the Gini index is 57 (0 represents absolute equality and 100 represents absolute inequality) whereas in Mexico the figure is 46. In 2009 the country with the worst income inequality was Bolivia; it is also one of the poorest in Latin America.

Development Strategies

The Latin American reality today was not the future envisioned for the region in the mid-1960s, when Brazil, Mexico, and Argentina all seemed poised to enter the ranks of the industrialized world. Multilateral agencies such as the World Bank and the Inter-American Development Bank loaned money for big development projects: continental highways, dams, mechanized agriculture, and power plants. All sectors of the economy were radically transformed. Agricultural productions increased with the application of "green revolution" technology and mechanization. State-run industries reduced the need for important goods, and the service sector ballooned as a result of new government and private sector jobs. Yet the rush to modernize produced victims. In rural areas, poverty, landlessness, and inadequate systems of credit hindered the productivity of small farmers. As rural people were pushed off the land, there were not enough jobs in industry to absorb them, so they created their own niche in the urban informal economy. In the end, most countries made the transition from predominantly rural and agrarian economies dependent on one or two commodities to more economically diversified and urbanized countries with mixed levels of industrialization.

In the 1990s, most Latin American governments radically changed their economic development strategies. National industries and tariffs were jettisoned for a set of free-market policy reforms that came to be known as the Washington Consensus. Through tough fiscal policy, increased trade, privatization, and reduced government spending, many countries in the 1990s saw their economies grow and poverty decline. Yet a series of economic downturns from 1999–2002 made these neoliberal policies highly unpopular with the masses, causing major political and economic turmoil. In particular, the value of increased trade and foreign investment was criticized as benefiting only a minority of the people in the region. In the past decade, political leaders such as Venezuela's Hugo Chavez, Brazil's Lula da Silva, and Bolivia's Morales have openly challenged the Washington Consensus as a path to development. However, an alternative Latin American approach, perhaps driven by more regional trade and greater state involvement, is still being articulated.

Industrialization Since the 1960s, most government development policies have emphasized manufacturing. Various strategies have been employed, from planned industrial centers and nationalized industries to import substitution. Since the 1990s, much of the growth in manufacturing in Mexico and Central America has been driven by investment by foreign companies, especially along the U.S.–Mexico border. The results have been mixed. Today at least 25 percent of the formally employed male labor force in Argentina, Brazil, Bolivia, Chile, Costa Rica, Ecuador, El Salvador, Mexico, Peru, Uruguay, and Venezuela is employed in industry (including mining, construction, and energy). Yet this is short of the hoped-for levels of industrial manufacturing, especially when the size of urban populations is taken into account. Moreover, the most industrialized areas tend to be around the capitals or planned industrial cities, such as Ciudad Guyana in Venezuela and the Mexican border cities of Ciudad Juárez and Tijuana.

In several noncapital cities, industry has thrived without direct state support. The cities of Monterrey, Mexico; Medellín, Colombia; and São Paulo, Brazil, all developed important industrial sectors initially from local investment. Long before Medellín (4 million people) was associated with cocaine, it was a major center of textile production, and was more industrialized than the larger capital of Bogotá. A popular image of Medellín's inhabitants—the South American Yankees—surfaced as early as the 1950s. Celebrated as hardworking and entrepreneurial to the core, residents developed a strong sense of regional pride so that being from the department of Antioquia (where Medellín is capital) meant more than being Colombian. Similar perceptions exist for metropolitan Monterrey (3.7 million people), a city that is not well known outside Mexico but is perceived within Mexico as innovative, resourceful, and solidly middle class.

The industrial giant of Latin America is metropolitan São Paulo in Brazil. Rio de Janeiro has greater name recognition and was the capital before Brasília was built, but it does not have the economic muscle of São Paulo. This city of 18 million, which vies with Mexico City for the title of Latin America's largest, began to industrialize in the early 1900s when the city's coffee merchants started to diversify their investments. Since then, a combination of private and state-owned industries have agglomerated around São Paulo. Within a 60-mile radius of the city center, automobiles, aircraft, chemicals, processed foods, and construction materials are produced. There are also heavy industry and industrial parks. With the port of Santos nearby and the city of Rio de Janeiro a few hours away, São Paulo is the uncontested financial center of Brazil. A stunning sight to most first-time visitors to São Paulo is the forest of high rises that greets them, a tropical version of Manhattan, only larger.

Maquiladoras and Foreign Investment The Mexican assembly plants that line the border with the United States, called **maquiladoras**, are characteristic of manufacturing systems in an increasingly globalized economy. Nearly 4,000 maquiladoras exist, employing 1.1 million people who assemble automobiles, consumer electronics, and apparel. Between 1994 and 2000, 3 of every 10 new jobs in Mexico were in the maquiladoras, which account for nearly half of Mexico's exports. Maquiladora employment peaked in 2001 with 1.3 million people employed. Since 2003, China has become a favorite destination for the labor-intensive assembly work that Mexico has specialized in for the last three decades. As Mexican wages have gone up, some companies have relocated factories to East Asia. Northern Mexico is still an attractive location but competition from China and even Central America may erode Mexico's various locational and structural advantages (Figure 4.35).

The policy of border industrialization began in the 1960s, long before offshore production was widely practiced. To accelerate industrialization, the Mexican government allowed the duty-free import of machinery, components, and supplies from the United States to be used for manufacturing goods for export back to the United States. Initially, all products had to be exported, but changes in the law in 1994 now allow up to half of the goods to be sold in Mexico. The program was slow to develop, but it took off in the 1980s as foreign companies realized tremendous profits from inexpensive Mexican labor.

Figure 4.35 Mexican Maquiladora Workers These women manufacture car radios at Delphi Delco Electronics in Matamoros, Mexico. Delphi, which makes parts for General Motors cars, has about 11,000 Mexican workers in seven factories near Matamoros. Many maquiladoras rely upon women for tedious and demanding assembly work. Matamoros is across the border from Brownsville, Texas, near the Gulf of Mexico.

An autoworker in a General Motors plant in Mexico earns in a day what the same worker in the United States earns in an hour and with far fewer benefits.

Considerable controversy surrounds this form of industrialization on both sides of the border. Organized labor in the United States complains that well-paying manufacturing jobs are being lost to low-cost competitors, while environmentalists decry serious industrial pollution resulting from lax government regulation. Mexicans worry that these plants are poorly integrated with the rest of the economy and that many of the workers are young unmarried women who are easily exploited. With NAFTA, foreign-owned manufacturing plants are no longer restricted to the border zone, which may result in slower rates of industrialization for the border towns. Foreign-owned plants are increasingly being constructed near the population centers of Monterrey, Puebla, and Veracruz. However, Mexican workers and foreign corporations continue to locate in the border zone because of the unique advantages of proximity to the U.S. border.

Mexico's competitive advantage is twofold: its location along the U.S. border and its membership in NAFTA. However, other Latin American states are attracting foreign companies through tax incentives and low labor costs. Assembly plants in Honduras, Guatemala, and El Salvador are drawing foreign investors, especially in the apparel industry. A recent report from El Salvador claims that not one of its apparel factories has a union. Making goods for American labels such as the Gap, Liz Claiborne, and Nike, many Salvadoran garment workers complain they do not make a living wage, work 80-hour weeks, and face mandatory pregnancy tests. With the signing of the CAFTA agreement, Central American states are hopeful that more foreign investment will flow into their countries and that wages and conditions will improve.

The situation in Costa Rica, which is now a major chip manufacturer for Intel, is quite different. With a well-educated population, low crime rate, and stable political scene, Costa Rica is now attracting other high-tech firms. Costa Rica is transitioning from a banana republic (bananas and coffee were the country's long-standing exports) to a high-tech manufacturing center—Intel processors made up 20 percent of Costa Rica's exports in 2008. Uruguay is another example of a small country with a well-educated population that has recently emerged as the leader in Latin American **outsourcing** operations. Outsourcing, most commonly associated with India, is the practice of moving service jobs such as tech support, data entry, and programming to cheaper locations. Partnered with the Indian multinational company Tata, in the last few years TCS Iberoamerica in Uruguay has crated the largest outsourcing operation in the region. Uruguay takes advantage of being in a similar time zone as the eastern United States. While India's top engineers sleep, Uruguayan engineers and programmers can serve their customers from Montevideo and no one is the wiser.

The Entrenched Informal Sector Even in prosperous Montevideo, Uruguay, a short drive to the urban periphery shows large neighborhoods of self-built housing filled with street traders and family-run workshops. Such activities make up the informal sector, the provision of goods and services without the benefit of government regulation, registration, or taxation. Most people in the informal economy are self-employed and receive no wages or benefits except the profits they clear. The most common informal activities are housing construction (in many cities as many as half of all residents live in self-built housing), manufacturing in small workshops, street vending, transportation services (messenger services, bicycle delivery, and collective taxis), garbage picking, street performing, and even line-waiting (Figure 4.36). These activities are legal. Illegal informal activities also exist: drug trafficking, prostitution, and money laundering, for example. The vast majority of people who rely on informal livelihoods produce legal goods and services.

No one is sure how big this economy is, in part because separating formal activities from informal ones is difficult. Visitors to Lima, Belém, Guatemala City, or Guayaquil could easily get the impression that the informal economy *is* the economy. From self-help housing that dominates the landscape to hundreds of street vendors that crowd the sidewalks, it is impossible to avoid. There are advantages in the informal sector—hours are flexible, children can work with their par-

Figure 4.36 Peruvian Street Vendors Street vending plays a critical role in the distribution of goods and the generation of income. Some aspects of street vending (such as access to space) are regulated by local governments and the street vendors themselves. These street vendors are selling produce in Huancayo, Peru.

ents, and there are no bosses. Peruvian economist Hernando de Soto even argues that this most dynamic sector of the economy should be encouraged and offered formal lines of credit. As important as this sector may be, however, widespread dependence on it signals Latin America's poverty, not its wealth. It reflects the inability of the formal economies of the region, especially in industry, to absorb labor. For millions of urban dwellers, formal employment that offers benefits, safety, and a living wage is still a dream. With nowhere else to go, the ranks of the informally employed continue to grow.

Primary Export Dependency

Historically, Latin America's abundant natural resources were its wealth. In the colonial period, silver, gold, and sugar generated tremendous riches for the colonists. With independence in the 19th century, a series of export booms introduced to an expanding world market commodities such as bananas, coffee, cacao, grains, tin, rubber, copper, wool, and petroleum. One of the legacies of this export-led development was a tendency to specialize in one or two major commodities, a pattern that continued into the 1950s. During that decade, Costa Rica earned 90 percent of its export earnings from bananas and coffee; Nicaragua earned 70 percent from coffee and cotton; 85 percent of Chilean export income came from copper; half of Uruguay's export income came from wood. Even Brazil generated 60 percent of its export earnings from coffee in 1955; by 2000, coffee accounted for less than 5 percent of the country's exports, yet Brazil remained the world leader in coffee production.

The economies of the region have industrialized and diversified. Since the 1990s, however, an increased demand from Asia for primary exports (from fossil fuels to metals and soy beans) have driven up commodity prices, creating another boom in primary export commodities. While many Latin America states are riding this wave, from oil-rich Venezuela to copper-laden Chile, there is concern that Latin America may once again become too reliant on its bountiful natural resources.

Agricultural Production The trend in Latin America has been to diversify and to mechanize agriculture. Nowhere is this more evident than in the Plata Basin, which includes southern Brazil, Uruguay, northern Argentina, Paraguay, and eastern Bolivia. Soybeans, used for oil and animal feed, transformed these lowlands in the 1980s and 1990s. Brazil is now the second largest producer of soy in the world (following the United States) and is already the world's largest exporter of soy. Argentina is the third largest producer, and production is still ascending. Between the late 1990s and 2009 soy production tripled in Argentina. The speed with which the shields are being converted into soy fields alarms many; it is eliminating forest and savanna, negatively impacting biodiversity and greenhouse gases. But with soy prices high, the rush to plant continues (Figure 4.37). In addition to soy, acres of rice, cotton, and orange trees, as well as the more traditional plantings of wheat and sugar continue in the Plata Basin.

Similar large-scale agricultural frontiers exist along the piedmont zone of the Venezuelan Llanos (mostly grains), the Pacific slope of Central America (cotton and some tropical fruits), and the Central Valley of Chile and the foothills of Argentina (wine and fruit production). In northern Mexico, water supplied from dams along the Sierra Madre Occidental has turned the valleys in Sinaloa into intensive producers of fruits and vegetables for consumers in the United States. The relatively mild winters in northern Mexico allow growers to produce strawberries and tomatoes during the winter months.

In each of these cases, the agricultural sector is capital-intensive and dynamic. By using machinery, high-yielding hybrids, chemical fertilizers, and pesticides, many corporate farms are extremely productive and profitable. What these operations fail to do is employ many rural people, which is especially problematic in countries where a third or more of the population depends on agriculture for its livelihood. As industrialized agriculture becomes the norm in Latin America, subsistence peasant producers are further marginalized. The overall trend is for agricultural production to increase while employing proportionally fewer people. In Peru, 45 percent of the male labor force worked in agriculture in 1980; by 2002 the figure had plummeted to just 11 percent. In absolute terms, however, the number of people living in rural areas is about the same as it was in 1960 (roughly 125 million). Thus the modernization of agriculture has left behind many subsistence producers who make up the ranks of Latin America's most impoverished people.

Mining and Forestry The exploitation of silver, zinc, copper, iron ore, bauxite, and gold is an economic mainstay for many countries in the region. Moreover, many commodities prices reached record levels in 2005 through 2008, boosting foreign exchange earnings. Chile is the world leader in copper production; in 2009 it produced five times the amount of copper of the next two largest producers, Peru and the United States. Peru, however, led in global silver production in 2009; and regional rivals Chile, Bolivia, and Mexico were each top-10 producers. Peru was also Latin America's top

Figure 4.37 Soy Production in Brazil Fartura Farm in the state of Mato Grosso, Brazil, embodies the large-scale industrial agriculture that has transformed much of South America into one of the world's largest producers and exporters of soy products.

gold producer. A metal gaining world interest in lithium, this soft silver-white metal is ideal for making lightweight batteries, like those in cell phones and laptops. It will also be a key metal for electric car batteries. Today the largest producer of lithium is Chile, but the world's largest reserves are in Bolivia under the Salar de Uyuni in the Altiplano. These reserves are so immense Bolivia has been dubbed the Saudi Arabia of lithium. But it remains to be seen how and under what terms this critical resource will be extracted.

Like agriculture, mining has become more mechanized and less labor intensive. Even Bolivia, a country long dependent on tin production, cut 70 percent of its miners from the payrolls in the 1990s. The measure was part of a broad-based austerity program, yet is suggests that the majority of the miners were not needed. Similarly, the vast copper mines of northern Chile are producing record amounts of copper with few miners. In contrast, gold mining continues to be labor intensive, offering employment for thousands of prospectors.

Logging is another important, and controversial, extractive activity. Ironically, many of the forest areas cleared for cattle were not systematically harvested. More often than not, all but the most valuable trees were burned. Logging concessions are commonly awarded to domestic and foreign timber companies, which export boards and wood pulp. These one-time arrangements are seen as a quick means for foreign exchange, particularly if prized hardwoods such as mahogany are found. Logging can mean a short-term infusion of cash into a local economy. Yet rarely do long-term conservation strategies exist, making this system of extraction unsustainable. There is growing interest in certification programs that designate wood products that have been produced sustainably. This is due to consumer demand for certified wood, mostly in Europe. Unfortunately, such programs are small and the lure of profit usually overwhelms the impulse to conserve for future generations.

Several countries rely on plantation forests of introduced species of pines, teak, and eucalyptus to supply domestic fuelwood, pulp, and board lumber. These plantation forests grow single species and fall far short of the complex ecosystems occurring in natural forests. Nonetheless, growing trees for paper or fuel reduces the pressure on other forested areas. Leaders in plantation forestry are Brazil, Venezuela, Chile, and Argentina. Considered Latin America's economic star in the 1990s, Chile relied on timber and wood chips to boost its export earnings. Thousands of hectares of nonnative trees (eucalyptus and pine) have been planted, systematically harvested, cut into boards, or chipped for wood pulp (see Figure 4.4). Japanese capital is heavily involved in this sector of the Chilean economy. The recent expansion of the wood chip business, however, has led to a dramatic increase in the logging of native forests.

The Energy Sector The oil-rich nations of Venezuela, Mexico, and Colombia are able to meet their own fuel needs and to earn vital state revenues from oil exports. In 2008 Mexico was the 7th largest producer of oil in the world and Venezuela was 10th (Figure 4.38). Venezuela is most dependent on revenues from oil, earning up to 90 percent of its foreign exchange from crude petroleum and petroleum products. Vast oil reserves exist in the llanos of Colombia, but a costly and vulnerable pipeline that connects the oil fields to the coast has been a regular target of guerrilla groups, thus reducing the country's production. Colombia, however, continues to be a major exporter of coal. The largest oil discovery in recent years has been off the coast of Brazil. Although not in production yet, the reserves have the potential to turn Brazil, an oil importer, into one of the world's leading oil exporters.

Venezuela and Bolivia have the largest reserves of natural gas in the region. A network of gas pipelines links Bolivia's lowlands to southern Brazil and Argentina. In 2006, Bolivian President Morales nationalized the country's oil and gas industry in an effort to retain more of the industry's profits. Brazilian and Spanish firms had built much of the network, and in the end were forced to accept new contracts that were less favorable but still profitable. A new network of pipelines from Venezuela's gas fields in the llanos through the Brazilian Amazon has also been proposed. Yet this idea has been deemed ecologically disruptive and too expensive. As demands for liquefied gas grow, especially from China, both Peru and Bolivia seek to fill this demand through pipelines and a dedicated port on the Pacific.

In the area of biofuels, Brazil offers a story of sweet success. In the 1970s when oil prices skyrocketed, oil-poor Brazil decided to convert its abundant sugarcane harvest into ethanol. It also induced its car industry to build cars that run on ethanol. Over the years, even when oil prices plummeted, Brazil continued to invest in ethanol, building mills and a distribution system that delivered ethanol to gas stations. One of Brazil's major technological successes was

Figure 4.38 Oil and Gas Production A gas compression platform at work on Lake Maracaibo, Venezuela. Oil and gas production has impacted the pace of development in countries such as Venezuela, Mexico, and Ecuador. These economies struggled in the 1990s when oil prices were down but benefited from higher prices over the past decade.

ENVIRONMENT AND POLITICS The Political Ecology of Extraction

The Western Amazon Basin is lightly settled and in many areas indigenous people are the majority. Yet the reality of neoliberalism and expanding resource extraction is transforming this region (that includes Ecuador, Peru, Colombia, Brazil and Bolivia) and exposing the competing geographies for resource control.

Figure 4.5.1 shows areas leased for oil and gas production. Leases are controlled by the state but can go to national and foreign companies. Many of these leased lands abut or overlap protected areas. For example, concessions in Bolivia overlap with significant areas of two national parks and one biosphere reserve. In Ecuador large sections of Yasuni' National Park are leased for oil extraction.

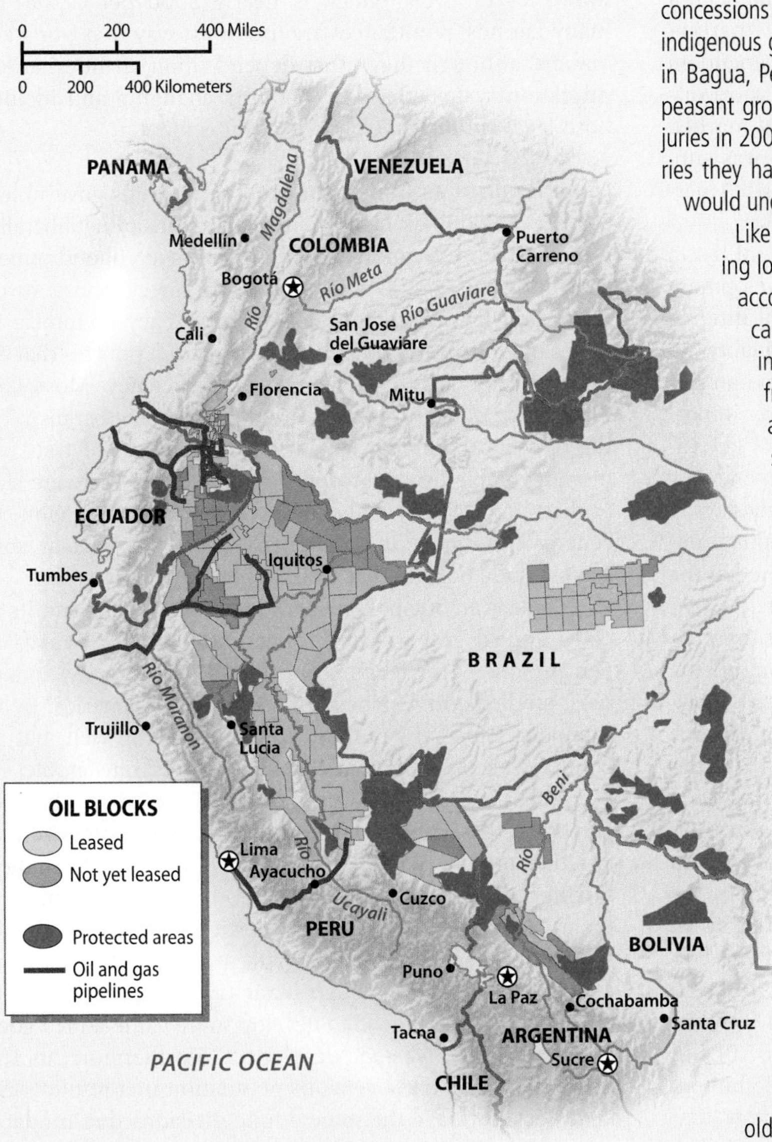

What the map does not show, however, is that many of indigenous and peasant communities have been using these lands for generations or have asserted claims to these lands. In Ecuador, concessions overlap with the ancestral or titled land of 10 different indigenous groups. The immediacy of local concerns was manifested in Bagua, Peru, when violence between police and indigenous and peasant groups ended with nearly three dozen deaths and 200 injuries in 2009. Community leaders were demanding access to territories they had long occupied, fearing that hydrocarbon concessions would undermine their resource base and livelihood.

Like hydrocarbons, access to mining rights is also undermining local control of resources. At the beginning of the 1990s, according to geographer Antony Bebbington, Latin Americans received about 12 percent of the global investments in mining; today the share is around one-third. Estimates from Peru and Ecuador suggest that up to half of peasant communities may be affected by mining concessions. Mined areas, often in the heads of watersheds, are degraded areas that impact local communities as well as more distant downstream communities.

Yet the current presidents of each of these countries, from Peru's Alan Garcia to Ecuador's Rafael Correa, are adamant that these resources belong to the nation and that will be developed in an effort to enhance the social and economic growth of the nation. Moreover, they contend that with national control, cleaner technologies, and better management, the positive benefits of resource extraction will outweigh the negative costs.

The competing geographies of scale are exquisitely evident in this region, with national interests trumping local concerns about water quality and territorial access for traditional farming or extraction. In a climate of resource nationalism, the rights of indigenous groups or peasant communities are easily suspended. Moreover, long-term efforts to establish and designate projected lands can be easily revoked. This is a familiar story, as old as mining itself. But for many indigenous and peasant communities who have fought for territorial recognition and control, the extraction boom has challenged their understanding of citizenship.

Figure 4.5.1 Leased Hydrocarbon Lands in Western Amazonia Lands in western Amazonia are being leased for oil and gas extraction at an accelerated pace as states seek to extract more revenue from this lightly populated but biologically diverse region. (Source: Finer M, Jenkins CN, Pimm SL, Keane B, Ross C, 2008, "Oil and Gas Projects in the Western Amazon: Threats to Wilderness, Biodiversity, and Indigenous Peoples." PLoS ONE 3(8): e2932. doi:10.1371/journal.pone.0002932 (Figure 2))

Source: Adapted from Bebbington, Tony, 2009, "The New Extraction: Rewriting the Political Ecology of the Andes?," *NACLA Report on the Americas*, September/October.

inventing flex-fuel cars that run on any combination of ethanol and gasoline. At the time, Brazil was motivated by its limited oil reserves but today, with interest in biofuels growing as a way to reduce CO_2 emissions, Brazil's support of its ethanol program looks visionary.

Latin America in the Global Economy

To conceptualize Latin America's place in the world economy, **dependency theory** in the 1960s by scholars from the region. The premise of the theory is that expansion of European capitalism created the region's underdevelopment. For the developed "cores" of the world to prosper, the "peripheries" became dependent and impoverished. Dependent economies, such as those in Latin America, were export oriented and vulnerable to fluctuations in the global market. Even when they experienced economic growth, it was subordinate to the economic demands of the core (North America and Europe).

Economists who accepted this interpretation of Latin America's history were convinced that economic development could occur only through self-sufficiency, growth of internal markets, agrarian reform, and greater income equality. In short, they argued for vigorous state intervention and an uncoupling from the economic cores. Policies such as import substitution, industrialization, and nationalization of key industries were partially influenced by this view. Dependency theory has its detractors. In its simplest form, it becomes a way to blame forces external to Latin America for the region's problems. Implicit in dependency theory is also the notion that the path to development taken by Europe and North America cannot be easily replicated. This was a radical idea for its time.

Latin America's century-long dependence upon the United States as its major trading partner is still evident. However, growing trade with, as well as investment from, Europe and East Asia, means that a more complex and less U.S.-dependent pattern of trade is emerging. Latin America is linked to the world economy in ways other than trade. Figure 4.39 shows the changes in foreign direct investment (FDI) from 1995 to 2007. For nearly every country in the region, the value of foreign investment went up in this time period, with the exceptions of Bolivia, Ecuador and Venezuela. The two largest economies in the region, Brazil and Mexico, received the most foreign investment. In 1995, Brazil's FDI was less than $5 billion and Mexico's was $9.5 billion. By 2007 FDI in Brazil was valued at $35 billion and Mexico's was $25 billion.

Remittances are another important financial flow that reflects the integration of Latin America migrants into labor markets around the world, especially North America. In 2009, the value of remittances to Latin America was estimated at $59 billion, down significantly from the figure of $69 billion in 2008 due to the financial crisis. Mexico is the regional leader, receiving over $21 billion in remittance income in 2009 (which is equal of over $200 per capita). Brazil was a distant second at $5 billion. The value of remittances began to soar in the late 1990s as many households saw their personal income decline and unemployment rise.

To cope with these changes, family members abroad began to send more financial resources home. Reliance on remittances is an indicator of globalization from below. Rather than corporate-based foreign investment, millions of individual migrants are transferring money (often $200 to $300 a month) to support families and communities left behind. Scholars debate whether this flow of capital can actually lead to sustained development or if it is simply a survival strategy of last resort. The economic impact of remittances shown on a per capita basis is real (see Figure 4.39). El Salvador, a county of 7 million people, received over $3.5 billion in remittances in 2009, which is nearly $500 per capita. For many Latinos, remittances are the surest way to poverty alleviation, although they must depend upon an international migration system that is constantly changing and includes both legal and illegal channels of movement.

Neoliberalism as Globalization By the 1990s governments and the World Bank had become champions of neoliberalism as a sure path to economic development. **Neo-liberal policies** stress privatization, export production, direct foreign investment, and few restrictions on imports. They epitomize the forces of globalization by turning away from policies that emphasize state intervention and self-sufficiency. Most Latin American political leaders are embracing neoliberalism and the benefits that come with it, such as increased trade and more favorable terms for debt repayment. Yet there are signs of discontent with neoliberalism throughout the region. Recent protests in Peru and Bolivia reflect the popular anger against trade policies that seem to benefit only the elite.

Chile is an outspoken champion of neoliberalism. Its average annual growth rate between 2000 and 2008 was 4.4 percent, one of the region's healthiest. Consequently, it is the most studied and watched country in Latin America. By the numbers, Chile's 17 million people are doing well, but the country's accomplishments are not readily transferable. For example, the radical move to privatize state-owned business and open the economy occurred under an oppressive military dictatorship that did not tolerate opposition. Such dictatorships are rare in Latin America today, but in a democracy the pace of reform must be slower. Much of Chile's export-led growth has been based on primary products: fruits, seafood, copper, and wood. Although many of these products are renewable, Chile will need to develop more value-added goods before it will be labeled "developed." Furthermore, its relatively small and homogeneous population in a resource-rich land does not have the same ethnic divisions that hinder so many states in Latin America. While neoliberalism has worked in the Chilean case, for many Latin American states the social and environmental disruptions associated with neo-liberal policies have led to political upheaval, and an active search for alternatives is underway.

Dollarization As financial crises spread through Latin America in the late 1990s, governments began to consider the economic benefits of **dollarization**, a process by which a country adopts—in whole or in part—the U.S. dollar as its official cur-

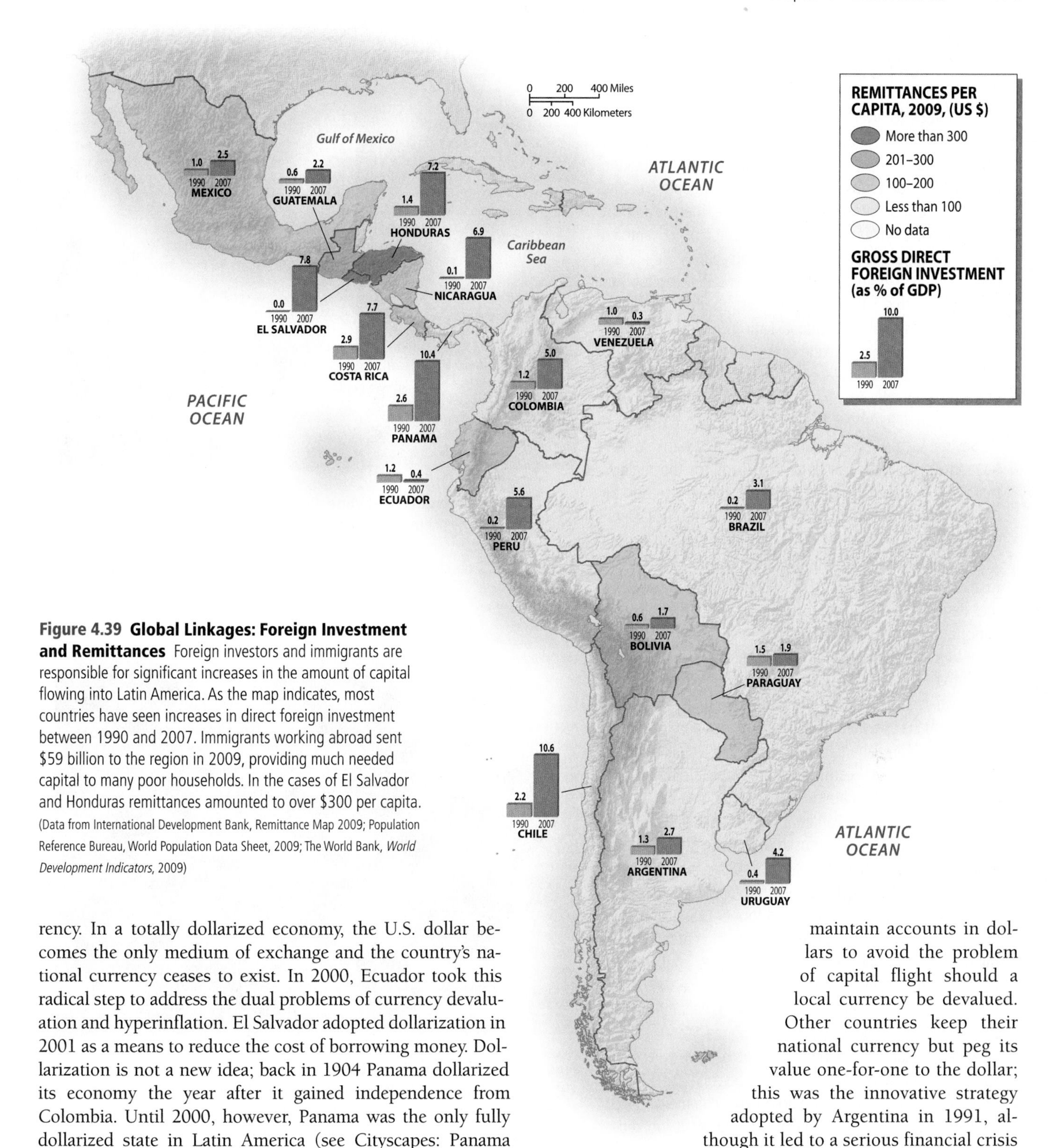

Figure 4.39 Global Linkages: Foreign Investment and Remittances Foreign investors and immigrants are responsible for significant increases in the amount of capital flowing into Latin America. As the map indicates, most countries have seen increases in direct foreign investment between 1990 and 2007. Immigrants working abroad sent $59 billion to the region in 2009, providing much needed capital to many poor households. In the cases of El Salvador and Honduras remittances amounted to over $300 per capita. (Data from International Development Bank, Remittance Map 2009; Population Reference Bureau, World Population Data Sheet, 2009; The World Bank, *World Development Indicators*, 2009)

rency. In a totally dollarized economy, the U.S. dollar becomes the only medium of exchange and the country's national currency ceases to exist. In 2000, Ecuador took this radical step to address the dual problems of currency devaluation and hyperinflation. El Salvador adopted dollarization in 2001 as a means to reduce the cost of borrowing money. Dollarization is not a new idea; back in 1904 Panama dollarized its economy the year after it gained independence from Colombia. Until 2000, however, Panama was the only fully dollarized state in Latin America (see Cityscapes: Panama City Latin America's Aspiring Crossroads).

A more common strategy in Latin America is limited dollarization, in which U.S. dollars circulate and are used alongside the country's national currency. Limited dollarization exists in many countries around the world, but most notably in Latin America. Since the economies of Latin America are prone to currency devaluation and hyperinflation, limited dollarization is a type of insurance. Many banks in Latin America, for example, allow customers to maintain accounts in dollars to avoid the problem of capital flight should a local currency be devalued. Other countries keep their national currency but peg its value one-for-one to the dollar; this was the innovative strategy adopted by Argentina in 1991, although it led to a serious financial crisis in 2001 and was eventually stopped. Dollarization, partial or full, tends to reduce inflation, eliminate fears of currency devaluation, and reduce the cost of trade by eliminating currency conversion costs.

Dollarization has its drawbacks. The obvious one is that a country no longer has control of its monetary policy, making it reliant on the decisions of the U.S. Federal Reserve. Foreign governments do not have to ask permission to dollarize their economies. At the same time, the United

CITYSCAPES Panama City—Latin America's Aspiring Crossroads

One might be forgiven for thinking they have arrived in the wrong place when they see Panama City for the first time. This is a small country of 3.5 million, and even metropolitan Panama City only has about 1.2 million people. Yet the city's skyline suggests a much bigger place. There is a forest of skyscrapers in various stages of completion jammed along the city's Pacific shore. Many are banks, others are condominiums, hotels, and offices. Foreign capital has poured into this isthmian nation, especially in the past three decades, making Panama Latin America's most important international banking center. Its biggest competitor is Miami. It's a chaotic place, where building is far ahead of planning. Even the financial crisis of 2009 barely slowed things down. (Figure 4.6.1).

Several factors explain Panama's role as a banking center and its stunning rise as a crossroads for the region's trade and finance. Obviously the canal, the very reason for the country's creation in 1904, is a unique resource. Operating since 1914, over 14,000 ships pass through it each year and the country is currently building a new set of locks to allow passage of larger ships. (Figure 4.6.2) It also helps that Panama uses the U.S. dollar as its only currency. More importantly, in the 1970s when Panama signed the agreement with Jimmy Carter for the return of the canal, it also instituted several key banking reforms to make the country attractive to offshore and international banks. The banks came, from Europe, East Asia, Latin America and the United States. Although Panamanians do not like to talk about it, their banks and real estate benefited from Colombian investment, both legitimate and laundered narco-dollars.

The aspirations of Panamanians to make their city a global destination were infused when the city itself was made whole with the return of the U.S.-controlled canal zone on December 31, 1999. Nearby places that were once off limits could be developed, and former U.S. government buildings could be repurposed as cultural centers such as the City of Knowledge. Panama also stepped up its promotion as a tourist destination and central place to conduct business in Latin America. Like Costa Rica, Panama courted foreign retirees, especially from the United States, to consider a tropical seaside retirement at bargain prices. The salsa star and actor, Ruben Blades, became the Minister of Tourism and proudly promoted the commercial, ecological, and cultural wonders of his country. It worked—between 1996 and 2004, tourism doubled. Like many cities aspiring to world city status some showpiece architecture was needed. The BioMuseo designed by the celebrated architect Frank Gehry, is located at the entrance of the canal and can be seen from the city. When it opens it will celebrate Panama as a transoceanic and transcontinental bridge of life.

Figure 4.6.1 Panama City Skyline Due to Panama City's important role as a center of trade and finance for Latin America, the city's real estate market has been booming. High rises seem to spring up over night. Many of the new buildings are banks. Even U.S. developer Donald Trump has invested in up-scale apartment complexes.

Figure 4.6.2 The Miraflores Locks on the Panama Canal First opened in 1914, the Panama Canal has been a vital link for global trade for nearly a century. Constructed as part of a U.S.-controlled Canal Zone, the Panama Canal has been in complete authority of the Panamanian government since 2000. New locks for larger ships are under construction. It is hoped that they will be operating in time for the centennial celebration of the canal in 2014.

Source: Adapted from Barney Warf, 2002, "Tailored for Panama: Offshore banking at the crossroads of the Americas," *Geografiska Annaler, Series B, Human Geography* 84(1): 33–47.

States insists that all its monetary policies be based exclusively on domestic considerations, regardless of the impact such decisions may have on foreign countries. The political impact of eliminating a national currency is serious. The case of Ecuador is instructive. In 1999, when President Jamil Mahuad announced his plan to dollarize the economy to head off hyperinflation, he was quickly forced out of office by a coalition of military and Indian activists. When Vice-president Gustavo Naboa became president and the economic situation worsened, the country's political leadership went ahead with dollarization. In short, dollarization may help in a time of economic duress, but is not a popular policy.

Social Development

Over the past three decades Latin America has experienced marked improvements in life expectancy, child survival, and educational equality. One telling indicator is the steady decline in mortality rate for children below the age of 5 between 1990 and 2008 (see Table 4.2). In 1990, Brazil had 58 deaths per 1,000 children under age 5; in 2008, the number was 22. Nicaragua dropped from 68 deaths per 1,000 in 1990 to 27 in 2008. This indicator is important because when children younger than 5 years of age are surviving, it suggests that basic nutritional and health-care needs are being met. One can also infer that resources are being used to sustain women and their children. By comparison, the United States has a mortality rate of 8 per 1,000 children under age 5, and Japan's rate is 4. Most other developing countries have much higher under-5 death rates than Latin America; India's is 69 per 1,000 and Kenya's is 128. Despite economic downturns, the region's social networks have been able to mitigate some of the negative effects on children.

A combination of government policies and grassroots and nongovernmental organizations (NGOs) play a fundamental role in contributing to social well-being. In the past few years the Brazilian poverty reduction program, *Bolsa Família,* has grown to benefit over 46 million people. Families that qualify receive a monthly check from the state but are required to keep their children in school and take them to clinics for health checkups. Such conditional cash transfer programs have the immediate impact of reducing extreme poverty and the long-term hope of improving educational attainment and health care. Mexico has adopted a similar program. For states with far fewer resources than Brazil and Mexico, international humanitarian organizations, church organizations, and community activists provide many services that state and local governments cannot. Catholic Relief Services and Caritas, for example, work with the rural poor throughout the region to improve their water supplies, health care, and education. Other groups lobby local governments to build schools or recognize squatters' claims. Grassroots organizations also develop cooperatives that market anything from sweaters to cheeses. Cooperative organizations are able to realize some economies of scale as well as improve access to credit.

Other important gauges for social development are life expectancy, gender educational equity, and access to improved water sources. In aggregate, 84 percent of the people in the region have access to an adequate amount of water from an improved source; slightly more girls receive education than boys; and life expectancy (men and women) is 73 years (see Table 4.2). Masked by this aggregate data are extreme variations between rural and urban areas, between regions, and along racial and gender lines.

Within Mexico and Brazil tremendous internal differences exist in socioeconomic indicators. The northeastern part of Brazil lags behind the rest of the country in every social indicator. The country has a literacy rate of over 85 percent, but in the northeast it is only 60 percent. Moreover, within the northeast, literacy for city residents is 70 percent, but for rural residents is only 40 percent. In Mexico, the levels of poverty are highest in the more Indian south. In contrast, Mexico City, the states of Nuevo Leon (Monterrey is the capital), and Quintana Roo (home to Mexico's largest resort, Cancun) and Campeche have the highest GDP per capita. Noting this north-south economic divide, President Calderón explained, "there is one Mexico more like North America and another Mexico more like Central America. It is a very clear challenge for me to make them more alike." All countries have spatial inequities regarding income and availability of services, but the contrasts tend to be sharper in the developing world. In the cases of Mexico and Brazil, it is hard to ignore ethnicity and race when explaining these patterns.

Race and Inequality There is much to admire about race relations in Latin America. The complex racial and ethnic mix that was created in Latin America fostered tolerance for diversity. However, Indians and blacks are disproportionately represented among the poor of the region. More than ever, racial discrimination is a major political issue in Brazil. Headlines report organized killings of street children, most of them Afro-Brazilian. For decades, Brazil put forward its vision of a color-blind racial democracy. True, residential segregation by race is rare in Brazil and interracial marriage is common, but certain patterns of social and economic inequity seem best explained by race.

Assessing racial inequities in Brazil is problematic. The Brazilian census asks few racial questions, and all are based on self-classification. In the 2000 census, less than 11 percent of the population called itself black. Some Brazilian sociologists, however, claim that more than half the population is of African ancestry, making Brazil the second largest "African state" after Nigeria. Racial classification is always highly subjective and relative, but certain patterns support the existence of racism. Evidence from northeastern Brazil, where Afro-Brazilians are the majority, shows death rates approaching those of some of the world's poorest countries. Throughout Brazil, blacks suffer higher rates of homeless-

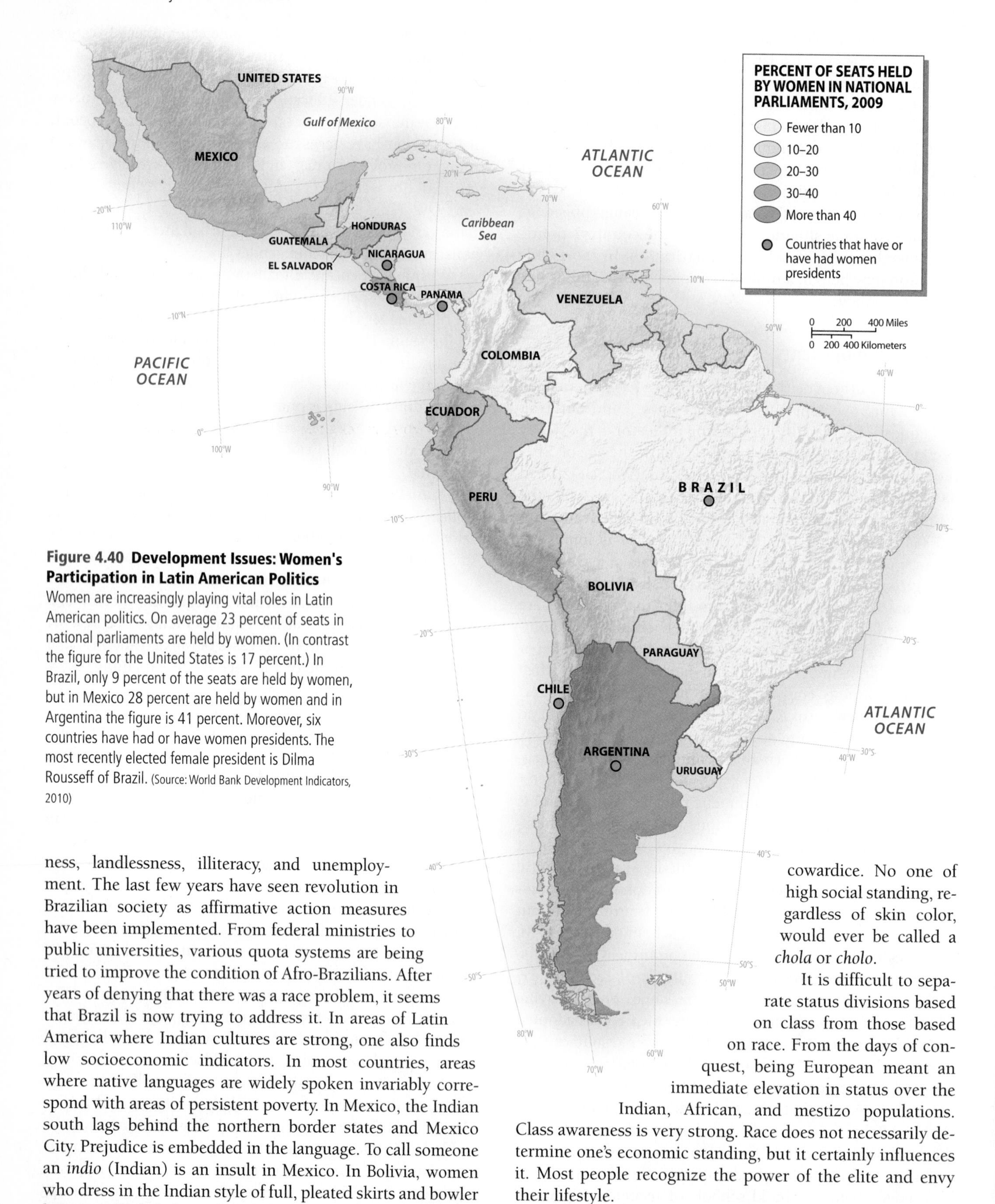

Figure 4.40 Development Issues: Women's Participation in Latin American Politics Women are increasingly playing vital roles in Latin American politics. On average 23 percent of seats in national parliaments are held by women. (In contrast the figure for the United States is 17 percent.) In Brazil, only 9 percent of the seats are held by women, but in Mexico 28 percent are held by women and in Argentina the figure is 41 percent. Moreover, six countries have had or have women presidents. The most recently elected female president is Dilma Rousseff of Brazil. (Source: World Bank Development Indicators, 2010)

ness, landlessness, illiteracy, and unemployment. The last few years have seen revolution in Brazilian society as affirmative action measures have been implemented. From federal ministries to public universities, various quota systems are being tried to improve the condition of Afro-Brazilians. After years of denying that there was a race problem, it seems that Brazil is now trying to address it. In areas of Latin America where Indian cultures are strong, one also finds low socioeconomic indicators. In most countries, areas where native languages are widely spoken invariably correspond with areas of persistent poverty. In Mexico, the Indian south lags behind the northern border states and Mexico City. Prejudice is embedded in the language. To call someone an *indio* (Indian) is an insult in Mexico. In Bolivia, women who dress in the Indian style of full, pleated skirts and bowler hats are called *cholas,* a descriptive term referring to the rural mestizo population that suggests backwardness and even cowardice. No one of high social standing, regardless of skin color, would ever be called a *chola* or *cholo*.

It is difficult to separate status divisions based on class from those based on race. From the days of conquest, being European meant an immediate elevation in status over the Indian, African, and mestizo populations. Class awareness is very strong. Race does not necessarily determine one's economic standing, but it certainly influences it. Most people recognize the power of the elite and envy their lifestyle.

An emerging middle class also exists that is formally employed, aspires to own a home and car, and strives to give its children a university education. The vast majority of people, however, are the working poor who struggle to meet basic food, shelter, clothing, and transport needs. These class differences are evident in the landscape. Go to any large Latin American city and find handsome suburbs, country clubs, and trendy shopping centers. High-rise luxury apartment buildings with beautiful terraces offer all the modern amenities, including maids' quarters. The elite and the middle class even show a preference for decentralized suburban living and dependence on automobiles, as do North Americans. Yet near these same residences are shantytowns where urban squatters build their own homes, create their own economy, and eke out a living.

The Status of Women Many contradictions exist with regard to the status of women in Latin America. Many Latina women work outside the home. In most countries the formal figures hover between 30 and 40 percent of the workforce, not far off from many European countries, but lower than in the United States. Legally speaking, women can vote, own property, and sign for loans, although they are less likely than men to do so, which reflects the patriarchal tendencies in the society. At the same time, women in the region are increasingly being elected to parliaments and even the presidencies of their countries at rates higher than the U.S. Even though Latin America is predominantly Catholic, divorce is legal and family planning is promoted. In most countries, however, abortion remains illegal (Figure 4.40).

Overall, access to education in Latin America is good compared to other developing regions, and illiteracy rates tend to be low. In terms of gender parity, there are slightly more girls than boys getting primary and secondary educations. Throughout higher education in Latin America today, male and female students are equally represented. Consequently, in the fields of education, medicine, and law, women are regularly employed.

The biggest changes for women are the trends toward smaller families, urban living, and educational parity with men. These factors have greatly improved the participation of women in the labor force. And in nearly every country, women's participation in the labor force has climbed by at least 10 percent since 1980. In the countryside, however, serious inequities remain. Rural women are less likely to be educated and tend to have larger families. In addition, they are often left to care for their families alone as husbands leave seasonally in search of employment. In most cases, the conditions and prejudices facing rural women have been slow to improve.

Women play an increasingly active role in Latin American politics. A Mayan woman from Guatemala, Rigoberta Menchú, won the Nobel Peace Prize in 1992 for denouncing human rights abuses in her native country. In 1990 Nicaragua elected the first woman president in Latin America, Violeta Chamorro, the owner of an opposition newspaper. And, in 2005 Chileans elected Dr. Michelle Bachelet. A pediatrician and single mother, one of President Bachelet's first acts in office was to extend the availability of state-subsidized child care. Two year later, Cristina Fernadez de Kirchner was elected president of Argentina, following the presidency of her husband Nestor Kirchner (Figure 4.41). And in 2010 Costa Ricans elected Laura Chinchilla. Women are also active organizers and participants in cooperatives, microenterprises, and unions. In a relatively short period, urban women have won a formal place in the economy and a political voice. Moreover, evidence suggests that this trend will continue, reflecting an improved status for women in the region.

Figure 4.41 Three Latin American Presidents Brazilian President Luiz Inacio Lula da Silva, Argentinian President Cristina Fernandez de Kirchner, and Bolivian President Evo Morales met in 2008 to discuss the development of natural gas in Bolivia. Collectively they are representative of greater diversity among political leaders in Latin America.

Summary

- Latin America and the Caribbean were the first world regions to be fully colonized by Europe. In the process, perhaps 90 percent of the native population died from disease, cruelty, and forced resettlement. The slow demographic recovery of native peoples and the continual arrival of Europeans and Africans resulted in an unprecedented level of racial and cultural mixing.
- Unlike other developing areas, most Latin Americans live in cities. This shift started early and reflects a cultural bias toward urban living with roots in the colonial past. The cities are large and combine aspects of the formal industrial economy with an informal one.
- Compared to Europe and Asia, this region is still rich in natural resources and relatively lightly populated. Yet as populations continue to grow and trade in natural resources increases, there is a growing concern for the state of the environment. Of particular concern is the relentless cutting of tropical forest.
- Uneven development and economic frustration have led many Latin Americans (both highly skilled and low skilled) to consider emigration as an economic strategy. Today, Latin America is a region of emigration, with migrants going to North America, Europe, and Japan to work. Collectively they send billions of dollars in remittances back to Latin America each year.
- Latin American governments were early adopters of neoliberal economic policies. While some states prospered, others faltered, sparking popular protests against the negative effects of globalization. Out of this frustration new political actors are emerging—from indigenous leaders to women—who are challenging old ways of doing things.

Key Terms

30 words

agrarian reform (*page 144*)
Altiplano (*page 134*)
altitudinal zonation (*page 137*)
Central American Free Trade Agreement (CAFTA) (*page 126*)
Columbian Exchange (*page 151*)
dependency theory (*page 170*)
dollarization (*page 170*)
El Niño (*page 139*)
environmental lapse rate (*page 137*)
fair trade (*page 131*)
grassification (*page 131*)
informal sector (*page 142*)
latifundia (*page 144*)
maquiladora (*page 165*)
megacity (*page 126*)
Mercosur (*page 157*)
mestizo (page 148)
minifundia (*page 144*)
neoliberal policies (*page 176*)
Organization of American States (OAS) (*page 157*)
outsourcing (*page 166*)
remittances (*page 149*)
rural-to-urban migration (*page 144*)
shields (*page 134*)
subnational organizations (*page 159*)
supranational organizations (*page 159*)
syncretic religions (*page 154*)
transnationalism (*page 149*)
Treaty of Tordesillas (*page 157*)
urban primacy (*page 141*)

Review Questions

1. Why is this region called *Latin America*?
2. Discuss the different environmental problems facing Mexico City. What special problems must megacities, such as Mexico City, have to contend with?
3. What is altitudinal zonation, and how is it relevant to agricultural practices and global climate change?
4. What was the Columbian Exchange? Discuss ways in which it still occurs today.
5. What are the origins of latifundia? Discuss the relationship between latifundia and minifundia.
6. What is the role of coca production in the region? Where is it produced and how has it impacted Latin American states?
7. Discuss the origins of the current political boundaries in Latin America. Where are boundary disputes occurring and why?
8. How are trade blocs reshaping international relations in Latin America?
9. What is dollarization, and how is it being used in Latin America?
10. What social and demographic shifts account for improvements in the lives of women in Latin America?

Thinking Geographically

1. Discuss the processes driving tropical deforestation in Latin America and how they compare with deforestation in other areas of the world.
2. Agrarian reform has been advocated throughout Latin America as a means to reduce social and economic inequalities. Has it worked in the region? How have land distribution programs fared? Where else in the world have agrarian reforms been attempted?
3. How is neoliberalism influencing the way Latin America interacts with the rest of the world? What are the social and environmental costs of neoliberalism? Is this model appropriate for understanding the impact of globalization in the developing world?
4. After examining the language map, what conclusions can you draw about the patterns of Indian survival in Latin America?
5. Given the dominance of cities in this region, describe the particular urban environmental problems facing cities in the developing world? How might Latin America's megacities use their size and density to reduce the environmental problems associated with urbanization?

6. Participatory mapping is being used by indigenous peoples and urban slum residents in Latin America. Who gains from participatory mapping? How might participatory mapping be used in other developing regions of the world? Could it be useful in rural settings as well as urban ones?
7. Discuss the social, environmental, and economic consequences behind the modernization of agriculture. How does Latin America's experience with modern agricultural systems compare with North America's?
8. In this multiracial region, what is the importance of race in assessing social and economic development? How have Amerindian groups faired in comparison to Afro-Latino communities?
9. Latin America is a region of emigration. Discuss the impact mass emigration has had on this region. Why is it higher here than in other regions of the developing world? Which countries are producing the most immigrants and where are they going?
10. Is the dependency theory useful in understanding Latin America's position in the world economy? Can the theory be applied to other regions in the developing world? Explain.

Regional Novels and Films

Novels

Isabel Allende, *House of the Spirits* (1982, Barcelona Plaza & Janes)

Jorge Amado, *Dona Flor and Her Two Husbands* (1967, Losada)

Jorge Amado, *Gabriela, Clove and Cinnamon* (1958, Biblioteca Ayacucha)

Mario Bencastro, *Odyssey to the North* (1998, Arte Público)

Jorge Luis Borges, *Labyrinths* (1962, New Directions)

Carlos Fuentes, *The Old Gringo* (1985, Fondo de Cultura Economica)

Gabriel García Marquez, *Love in the Time of Cholera* (1988, Penguin)

Gabriel García Marquez, *One Hundred Years of Solitude* (1967, Biblioteca Ayacucha)

Edmundo Paz Soldán, *The Matter of Desire* (2004, Mariner Books)

Mario Vargas Llosa, *The Real Life of Alejandro Mayta* (1986, Farrar, Straus & Giroux)

Films

Black Orpheus (1958, Brazil)

City of God (2002, Brazil)

The Devil's Miner (2005, U.S./Germany)

El Norte (1983, U.S.)

Amores Perros (2000, Mexico)

La Americana (2008, U.S.)

Like Water for Chocolate (1993, U.S.)

Maria Full of Grace (2004, Colombia)

Traffic (2000, U.S.)

The Motorcycle Diaries (2004, Brazil/Argentina)

Transnational Fiesta (1992, U.S.)

Bibliography

Crosby, Alfred. 1972. *The Columbian Exchange, Biological and Cultural Consequences of 1492*. Westport, CT: Greenwood Press.

Denevan, William M. 1992. *The Native Population of the Americas in 1492*. 2nd ed. Madison, WI: University of Wisconsin Press.

Ezcurra, Exequiel, et al. 1999. *The Basin of Mexico: Critical Environmental Issues and Sustainability*. Tokyo: United Nations Press.

Gade, Daniel W. 1999. *Nature and Culture in the Andes*. Madison: University of Wisconsin Press.

Gilbert, Alan. 1998. *The Latin American City*. 2nd ed. New York: Monthly Review Press.

Hecht, Susanna, and Cockburn, Alexander. 1989. *The Fate of the Forest: Developers, Destroyers and Defenders of the Amazon*. London: Verson Press.

Jackiewicz, Edward L., and Bosco, Fernando J. 2008. *Placing Latin America: Contemporary Themes in Human Geography. Landham,* MD: Rowman and Littlefield.

Mann, Charles. 2006. *1491: New Revelations of the Americas Before Columbus*. New York: Vintage Books.

Price, Marie, and Cooper, Catherine. 2007. "Competing Visions, Shifting Boundaries: The Construction of Latin America as a World Region." *Journal of Geography* 106:113–122.

Roberts, J. Timmons, and Thanos, Nikki Demetria. 2003. *Trouble in Paradise: Globalization and Environmental Crises in Latin America*. New York: Routledge.

Voeks, Robert. 1997. *Sacred Leaves of Candomblé: African Magic, Medicine and Religion in Brazil*. Austin: University of Texas Press.

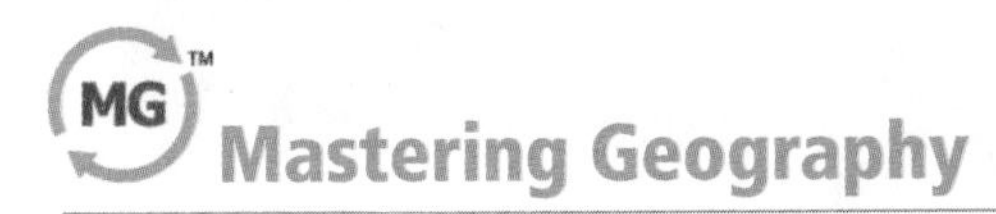

Log in to www.masteringgeography.com for MapMaster™ interactive maps, geography videos, RSS feeds, flashcards, web links, an eText version of *Diversity Amid Globalization*, and self-study quizzes to enhance your study of Latin America.

MapMaster™ presents 13 Place Name and 13 Layered Thematic interactive maps to help students practice and master their geographic literacy, spatial reasoning, and critical thinking skills.

5 The Caribbean

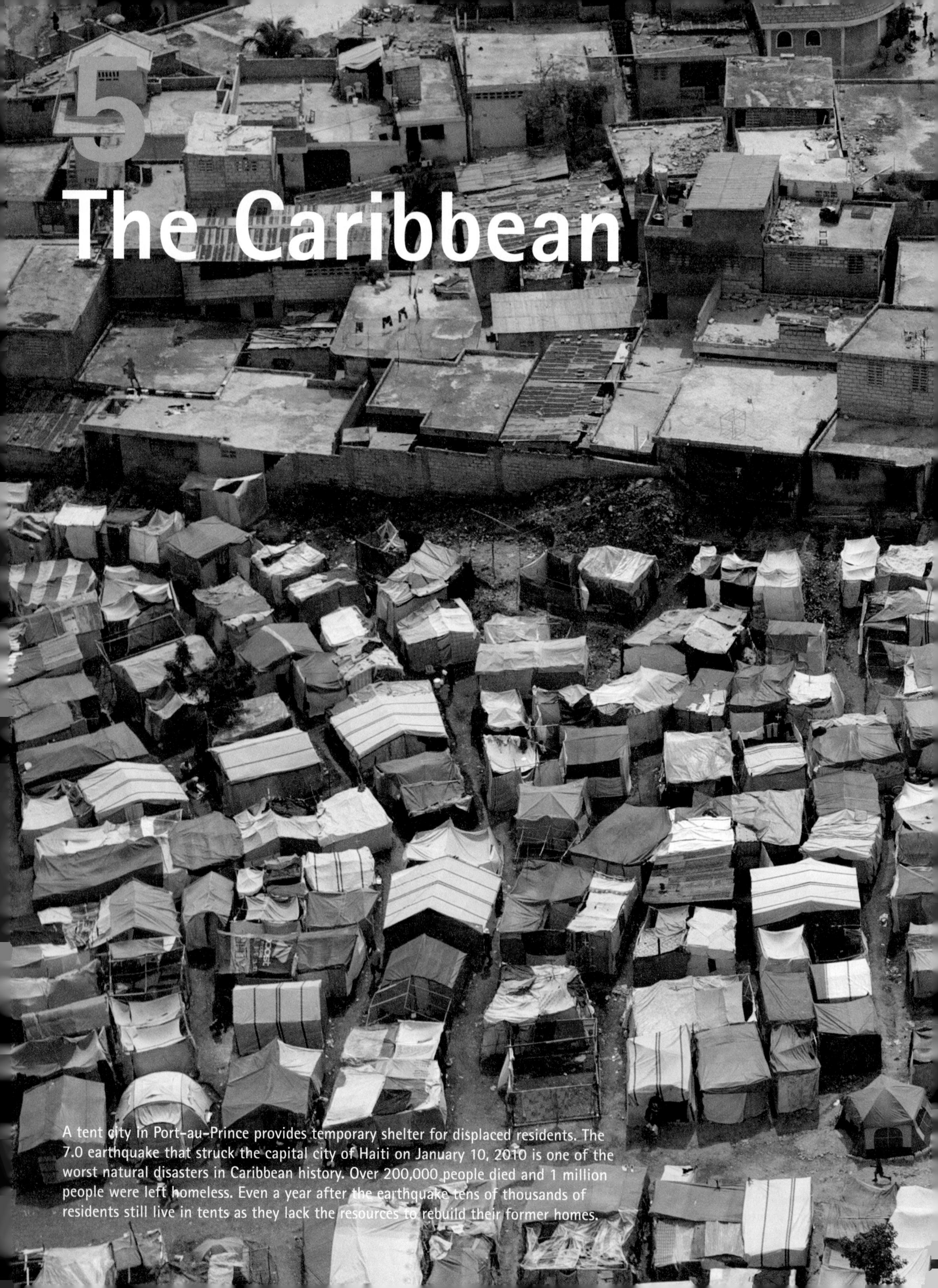

A tent city in Port-au-Prince provides temporary shelter for displaced residents. The 7.0 earthquake that struck the capital city of Haiti on January 10, 2010 is one of the worst natural disasters in Caribbean history. Over 200,000 people died and 1 million people were left homeless. Even a year after the earthquake tens of thousands of residents still live in tents as they lack the resources to rebuild their former homes.

DIVERSITY AMID GLOBALIZATION

Named for some of its former inhabitants—the Carib Indians—the modern Caribbean is a culturally complex and economically peripheral world region. Settled by various colonial powers, the residents of this region are mostly a mix of African, European, and South Asian peoples. Greatly impacted by the global economy but with relatively little influence on it, the livelihoods of Caribbean peoples are filled with uncertainty.

ENVIRONMENTAL GEOGRAPHY

Climate change threatens the Caribbean, with the potential for stronger and more frequent hurricanes, loss of territory due to sea level rising, and destruction of coral reefs.

POPULATION AND SETTLEMENT

The earthquake that leveled Port-au-Prince, Haiti, in 2010 showed the vulnerability of this large but very poor Caribbean city. Throughout the region, large numbers of people have emigrated in search of economic opportunity and are sending back millions of dollars in remittances.

CULTURAL COHERENCE AND DIVERSITY

Creolization, the blending of African, European, and even Amerindian elements, has resulted in many unique Caribbean expressions of culture, such as *rara, reggae,* and steel drum bands.

GEOPOLITICAL FRAMEWORK

The first area of the Americas to be extensively explored and colonized by Europeans, the region has seen many rival European claims and, since the early 20th century, has experienced strong U.S. influence.

ECONOMIC AND SOCIAL DEVELOPMENT

Environmental, locational, and economic factors make tourism a vital component of this region's economy, particularly in Puerto Rico, Cuba, the Dominican Republic, and the Bahamas.

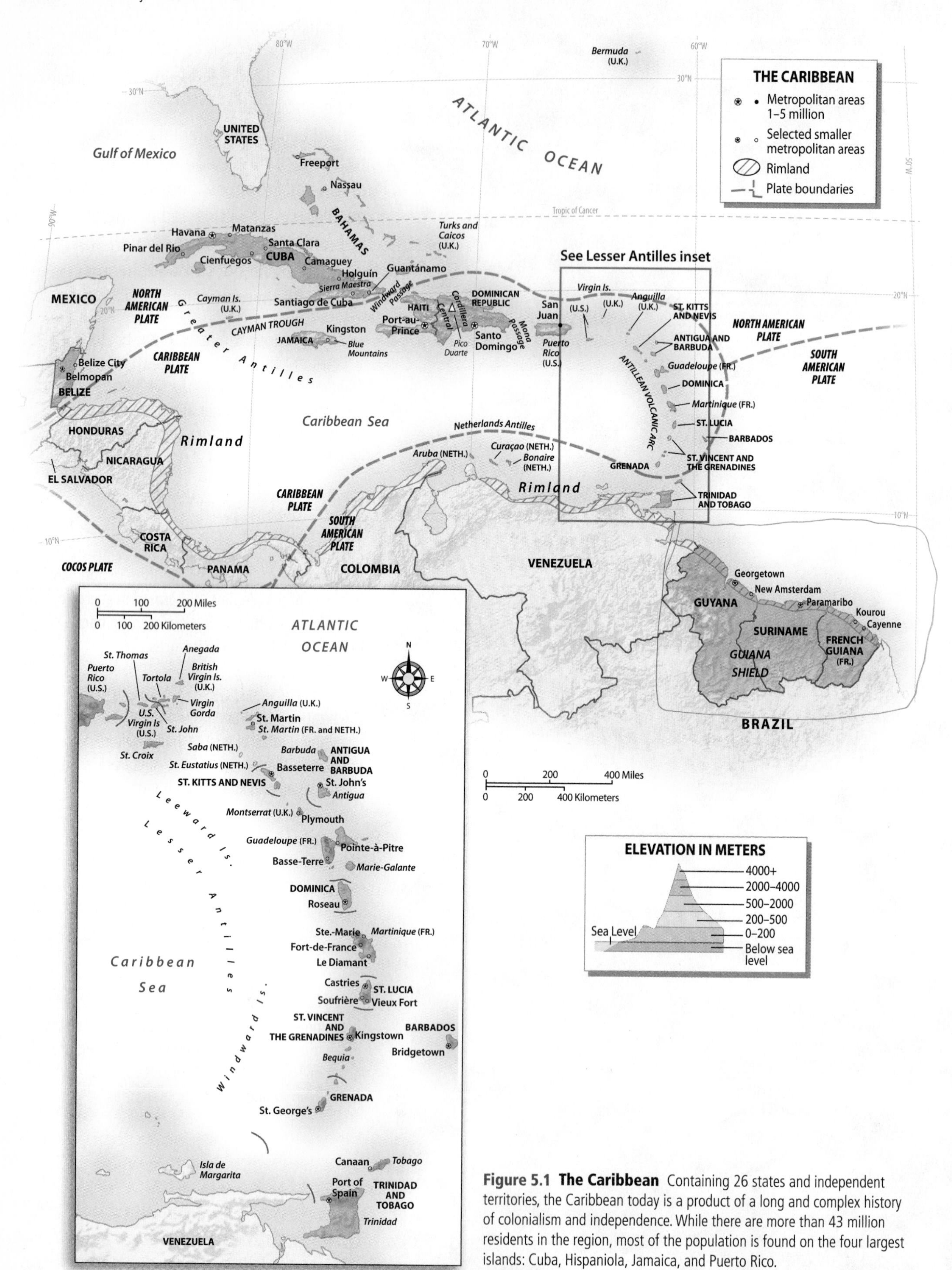

Figure 5.1 The Caribbean Containing 26 states and independent territories, the Caribbean today is a product of a long and complex history of colonialism and independence. While there are more than 43 million residents in the region, most of the population is found on the four largest islands: Cuba, Hispaniola, Jamaica, and Puerto Rico.

The Caribbean was the first region of the Americas to be extensively explored and colonized by Europeans. Yet its modern regional identity is unclear, often merged with Latin America but also viewed as apart from it. Today the region is home to 43 million inhabitants scattered across 26 countries and dependent territories. They range from the small British dependency of the Turks and Caicos, with 12,000 people, to the island of Hispaniola, with nearly 20 million. In addition to the Caribbean Islands, Belize of Central America and the three Guianas—Guyana, Suriname, and French Guiana—of South America are included as part of the Caribbean. For historical and cultural reasons, the peoples of these mainland states identify with the island nations and are thus included in this chapter (Figure 5.1).

Historically, the Caribbean was a battleground between rival European powers competing for territorial control of these tropical lands. In the early 1900s, the United States took over as the dominant geopolitical force in the region, regularly sending troops and maintaining what some have called a neocolonial presence. As in many developing areas, external control of the Caribbean produced highly dependent and inequitable economies. The plantation was the dominant production system and sugar the leading commodity. Over time other products began to have economic importance, as did the international tourist industry. Increasingly, governments have sought to diversify their economies by expanding nontraditional exports (such as flowers, nuts, and processed fruits), banking services, information processing services, and manufacturing to reduce the region's dependence on agriculture and tourism.

The basis for treating the Caribbean as a distinct area lies within its particular cultural and economic history. Culturally, this region can be distinguished from the largely Iberian-influenced mainland of Latin America because of its more diverse European colonial history and its strong African imprint due to the region's historical reliance upon slavery. In terms of economic production, the dominance of export-oriented plantation agriculture explains many of the social, economic, and environmental patterns in the region.

Generally, when one thinks of the Caribbean, images of white sandy beaches and turquoise tropical waters come to mind. Tucked between the tropic of Cancer and the equator, with year-round temperatures averaging in the high 70s, the hundreds of islands and picturesque waters of the Caribbean often have inspired comparisons to paradise. Columbus began the tradition by describing the islands of the New World as the most marvelous, beautiful, and fertile lands he had ever known, filled with flocks of parrots, exotic plants, and friendly natives. Writers today are still lured by the sea, sands, and swaying palms of the Caribbean. Since the 1960s, the Caribbean has earned much of its international reputation as a playground for northern vacationers. There is, of course, another Caribbean that is far poorer and economically more dependent than the one portrayed on travel posters. Haiti, by many measures the poorest country in the Western Hemisphere, has the third largest population in the region with 9.8 million people. The two largest countries—Cuba (11.2 million) and the Dominican Republic (9.9 million)—also suffer from serious economic problems.

The majority of Caribbean people are poor, living in the shadow of North America's vast wealth. The concept of **isolated proximity** has been used to explain the region's unusual and contradictory position in the world. The *isolation* of the Caribbean sustains the area's cultural diversity (Figure 5.2) but also explains its limited economic opportunities. Caribbean writers note that this isolation fosters a strong sense of place and an inward orientation by the people. Yet the relative *proximity* of the Caribbean to North America (and, to a lesser extent, Europe) ensures its transnational connections and economic dependence. For example, each year Dominican workers abroad send nearly $3 billion to family and friends in the Dominican Republic who rely on this money for their sustenance. Through the years, the Caribbean has evolved as a distinct but economically marginal world region. This status expresses itself today as workers flee the region in search of employment, while

Figure 5.2 Carnival Dancers Carnival participants take a break from the festivities in Port of Spain, Trinidad. Trinidad and Tobago celebrate carnival before the beginning of Lent with elaborate parades and costuming.

foreign companies are attracted to the Caribbean for its cheap labor. The economic well-being of most Caribbean countries is precarious. Despite such uncertainty, an enduring cultural richness and attachment to place is witnessed here that may explain a growing countercurrent of immigrants back to the region.

ENVIRONMENTAL GEOGRAPHY: Paradise Undone

The levels of environmental degradation and poverty found in Haiti are unmatched in the Western Hemisphere, and that was true well before a devastating earthquake leveled the capital city of Port-au-Prince in 2010. Haiti's story, although extreme, illustrates how social and economic inequities contribute to unsound human and natural resource choices, which in turn lead to more poverty. What was once considered France's richest colony—its tropical jewel—now has a per capita income (PPP) of $1,050. Haiti's life expectancy (61 years) is the lowest in the Americas, while its levels of child malnutrition and infant mortality are the highest. The reasons for Haiti's misfortune are complex, but political economy, demography, natural resources, and natural hazards are each critical factors.

In the late 18th century, Haiti's plantation economy yielded vast wealth for several thousand Europeans (mostly French) who exploited the labor of half a million African slaves. During the colonial period, the lowlands were cleared to make way for cane fields, but most of the mountains remained forested. The inequities of Haiti's plantation economy led to the world's first successful slave uprising, followed by Haitian independence in 1804. Even though Haiti was the second state in the Americas to achieve independence, social and economic inequities persisted. Most Haitians were subsistence farmers, planting food crops along with some coffee and cacao. In 1915, the United States sent in the marines to quell political unrest and safeguard U.S. interests in the region. U.S. troops and advisors remained for nearly 20 years, during which time they rewrote the country's laws so that foreign companies could own land. The U.S. occupation resulted in improvements in Haitian infrastructure, but it also caused increased land pressure as foreign sugar companies purchased the best lands, forcing the majority of rural peasants onto the marginal soils of the hillsides.

By the mid-20th century, a destructive cycle of environmental and economic impoverishment was established that still confronts Haiti. While Haiti was under the rule of corrupt dictators from 1957 to 1986, Haiti's elite benefited as the conditions for the country's poor worsened. Half of Haiti's people are peasants who work small hillside plots and seasonally labor on large estates. As the population grew, people sought more land. They cleared the remaining hillsides, subdivided their plots into smaller units, and abandoned the practice of fallowing land in an effort to eke out an annual subsistence. When the heavy tropical rains came, the exposed and easily eroded mountain soils were washed away. As sediment collected in downstream irrigation ditches and behind dams, agriculture suffered, electricity production declined, and water supplies were degraded throughout the country.

Deforestation was further aggravated by the dependence of the population on wood for fuel. Because of their poverty and limited electricity supplies, most Haitians use charcoal (made from trees) to cook meals and heat water. By the late 1990s, only an estimated 3 percent of Haiti remained forested. Ongoing political turmoil hampers reforestation programs, and the effects of forest removal have only worsened. In less than a lifetime, hills that were once covered in forest now support shrubs and grasses (Figure 5.3).

The tragic 7.0 earthquake that leveled Port-au-Prince on January 12, 2010, is one of the worst natural disasters

Figure 5.3 Deforestation in Haiti The denuded slopes outside of the city of Gonaives in northern Haiti illustrate the severe soil erosion problems that beset this country. Heavy rains, especially during hurricane season, have resulted in landslides and flooding that haved killed scores of people in Gonaives and destroyed thousands of homes.

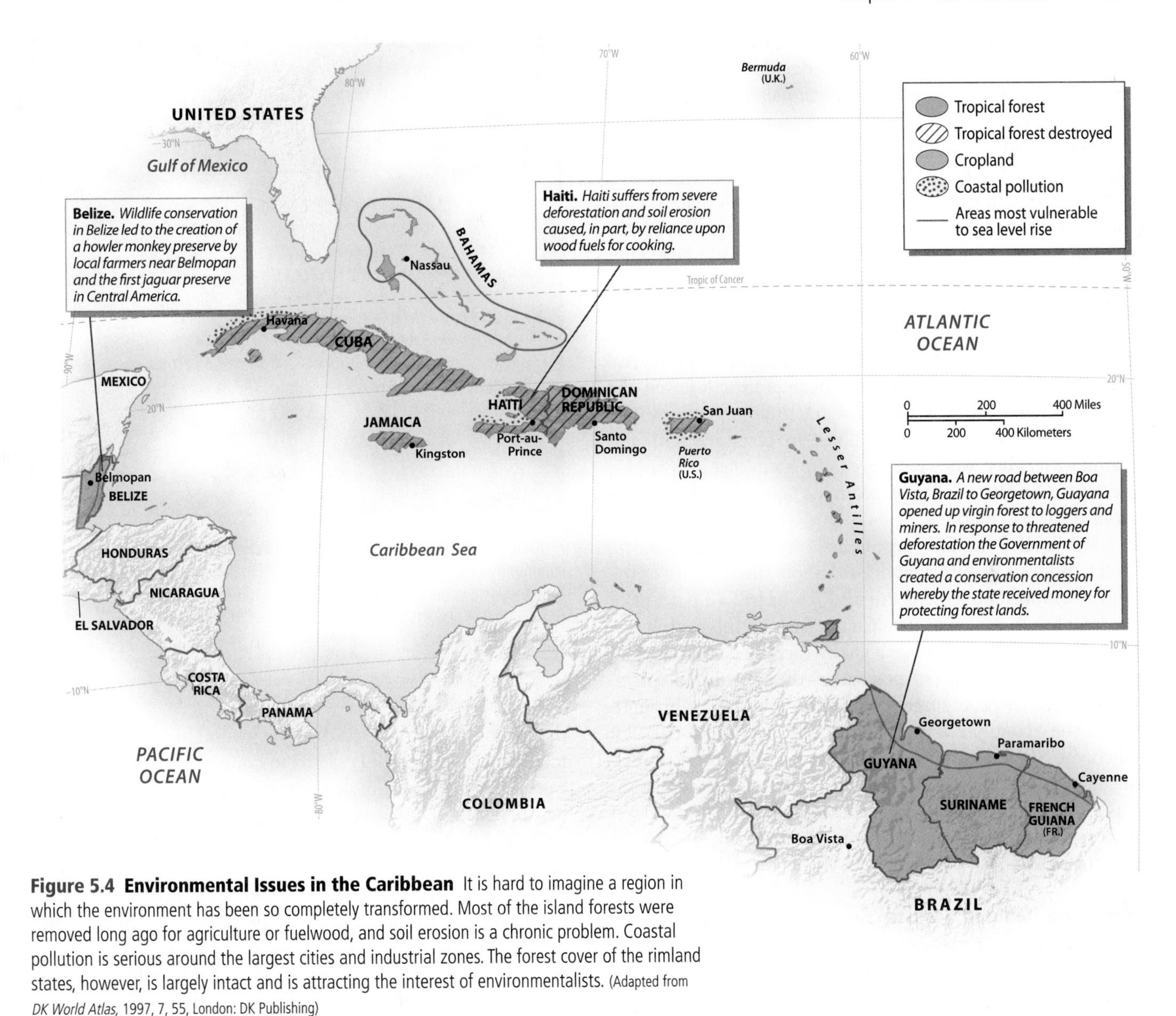

Figure 5.4 Environmental Issues in the Caribbean It is hard to imagine a region in which the environment has been so completely transformed. Most of the island forests were removed long ago for agriculture or fuelwood, and soil erosion is a chronic problem. Coastal pollution is serious around the largest cities and industrial zones. The forest cover of the rimland states, however, is largely intact and is attracting the interest of environmentalists. (Adapted from *DK World Atlas*, 1997, 7, 55, London: DK Publishing)

in the region's history. In the densely settled and poor capital streets, over 200,000 died, and anywhere from 2 to 3 million people were affected as homes were rendered unsafe and water and electricity supplies were disrupted. Many of the government buildings and institutions that would assist in the relief response were also ruined. The tragedy of the Haitian earthquake was compounded by the state's poverty and corruption, as most buildings were not built to standards that would withstand an earthquake of this magnitude. Under such circumstances, the needs of Haiti are especially acute and visible for the world to see. There is hope the renewed international interest and aid may lead to a rebuilding of the city and the nation, but such efforts will take decades and can easily be undermined by corruption.

Environmental Issues

Ecologically speaking, it is difficult to imagine an area so completely reworked through colonization and global trade as the Caribbean. For nearly five centuries, the destruction of forests and the unrelenting cultivation of soils resulted in the extinction of many endemic Caribbean plants and animals, including various shrubs and trees, songbirds, large mammals, and monkeys. The beauty of the Caribbean belies its current economic and social instability, the result of severe depletion of biological resources (Figure 5.4). Most of the environmental problems in the region are associated with agricultural practices, soil erosion, excessive reliance on wood and charcoal for fuel, and pollution associated with sprawling and impoverished cities.

Deforestation on the Islands Much of the Caribbean was covered in tropical rain forests and deciduous forests prior to the arrival of Europeans. The great clearing of its forests began in earnest on the smaller islands of the eastern Caribbean in the 17th century and spread westward. The island forests fell to make room for sugarcane planting as well as to provide lumber for housing, fences, and ships. In contrast with soils in the midlatitudes, the newly exposed tropical soils easily eroded and ceased to be productive after several harvests, a situation that led to two distinct land-use strategies. On the larger islands of Cuba and Hispaniola and on the mainland, new lands were constantly cleared and older ones abandoned or fallowed (left untilled for several seasons) in an effort to keep up sugar production. On the smaller islands, such as Barbados and Antigua, where land was limited, labor-intensive efforts to conserve soil and maintain fertility were employed. In both cases, however, the island forests were replaced by a landscape devoted to crops for world markets.

While Haiti has lost most of its forest cover, on Jamaica, the Dominican Republic, and Cuba about one-quarter of the land is in forest cover. On Puerto Rico nearly 40 percent of the land has forest cover. The Cuba case is an interesting one because the forest cover has actually increased from lows in the 1990s when only 20 percent of the island had forest. Cuba experienced a surge in charcoal production brought on by the economic and energy crises that began in 1990 when the Soviet Union collapsed and Cuba lost its subsidized imported fuel source. In the 2000s, Cuba addressed this issue with a reforestation campaign which has increased its forest cover to 25 percent with the goal of reaching 30 percent by 2015.

Figure 5.5 Protecting Habitat and Wildlife Tourists visit the community "Baboon" Sanctuary in Bermudian Landing, Belize. The sanctuary is a community-run project to preserve the habitat and increase the number of black howler monkeys (locally referred to as baboons). The sanctuary, established in 1985, attracts domestic and foreign visitors.

Managing Rimland Forests The Caribbean **rimland** is the coastal zone of the mainland, beginning with Belize and extending along the coast of Central America to northern South America. In general, the biological diversity and stability of the rimland states are less threatened than in the rest of the Caribbean. Thus, current conservation efforts could produce important results. Even though much of Belize was selectively logged for mahogany in the 19th and 20th centuries, healthy forest cover still supports a diversity of mammals, birds, reptiles, and plants. Public awareness of the negative consequences of deforestation is also greater now. Many protected areas have been established in Belize. In the mid-1980s, villagers in Bermudian Landing, Belize, established a community-run sanctuary for black howler monkeys (locally referred to as *baboons*). The villagers banded together to maintain habitat for the monkeys and commit to land management practices that accommodate this gregarious species. The success of the project has resulted in tourists visiting the area to see these indigenous primates up close (Figure 5.5).

In Guyana and Suriname, the relatively pristine interior forests are becoming a battleground among conservationists, indigenous peoples, and developers. For over a decade, the Guyanese and Surinamese governments have made the wood processing industry a priority by encouraging private investment and granting concessions to companies from Malaysia, Europe, and China for logging and sawmill operations. A new dry-season highway traverses the length of Guyana, connecting Boa Vista, Brazil, with Georgetown, Guyana. Not only does this road improve trade between Guyana and Brazil, giving Brazil access to the North Atlantic, it also provides access to the forest and mineral resources (chiefly gold) in southern Guyana. While governments in Guyana and Brazil are encouraged by the economic possibilities of this road, an alliance of conservationists (local and foreign) and indigenous peoples established the region's first conservation concession in the 2002, whereby the government of Guyana received money for not clearing the concession lands. Meanwhile, as the pace of logging and mining accelerates in Suriname, the government is meeting resistance from its maroon population, made up of escaped slaves who have lived in the forests for more than 200 years. They claim legal rights to the land based on old treaties with the Dutch.

Urban Environmental Problems The growth of Caribbean cities has led to environmental issues concerning water quality and proper waste disposal. The urban poor are the most vulnerable to health problems associated with overtaxed or nonexistent water and sewage services. According to 2009 World Bank estimates, 70 percent of Haiti's urban population had access to improved water sources, typically through shared neighborhood faucets. By comparison only half the residents in rural Haiti had access to improved water sources. The figures for Jamaica and the Dominican Republic have steadily improved so that nearly all urban residents have access to improved water supplies. The expense of improving basic urban infrastructure far exceeds the capabilities of most island economies. Several dams exist on the larger islands to supply water, while some of the smaller islands rely on expensive desalination plants. Still, existing freshwater supplies fall far short of domestic needs. As tourism, offshore manufacturing, and a growing urban population demand more water and produce more waste, Caribbean countries will be forced to make hard decisions about their infrastructure.

In addition to being a public health concern, water contamination from improper waste disposal poses serious economic problems for states dependent on tourism. Governments are caught between a desire to fix the problem before tourists notice and a tendency not to discuss it at all. Doing nothing, however, may not be an option for countries that rely heavily on tourism. In the more industrialized Puerto Rico, it is estimated that half the country's coastline is unfit for swimming, mostly due to contamination from sewage.

Island and Rimland Landscapes

It is the Caribbean Sea itself—that body of water enclosed between the Antillean islands (the arc of islands that begins with Cuba and ends with Trinidad) and the mainland of Central and South America—that links the states of the Caribbean region. Historically the sea connected people through its trade routes and sustained them with its marine resources of fish, green turtle, manatee, lobster, and crab. While the sea is noted for its clarity and biological diversity, the quantities of any one species are not great, so it has never supported large commercial fishing. The surface temperature of the sea ranges from 73° to 84°F (23° to 29°C), over which forms a warm tropical marine air mass that influences daily weather patterns. This warm water and tropical setting continue to be key resources for the region, attracting millions of tourists to the Caribbean each year (Figure 5.6).

The arc of islands that stretches across the sea is the region's most distinguishing feature. The Antillean islands are divided into two groups: the Greater and Lesser Antilles. The rimland (the Caribbean coastal zone of the mainland) includes Belize and the Guianas, as well as the Caribbean shoreline of Central and South America.

Greater Antilles The four large islands of Cuba, Jamaica, Hispaniola (shared by Haiti and the Dominican Republic), and Puerto Rico make up the **Greater Antilles**. On these islands are found the bulk of the region's population, arable lands, and large mountain ranges. Given the popular interest in the Caribbean coasts, it still surprises many people that Pico Duarte on the Dominican Republic is more than 10,000 feet tall (3,000 meters), Jamaica's Blue Mountains top 7,000 feet (2,100 meters), and Cuba's Sierra Maestra is more than 6,000 feet tall (1,800 meters). The mountains of the Greater Antilles were of little economic interest to plantation owners, who preferred the coastal plains and valleys. Yet the mountains were an important refuge for runaway slaves and subsistence farmers, and thus figure prominently in the cultural history of the region.

The best farmlands are found in the central and western valleys of Cuba, where a limestone base contributes to the formation of a fertile red clay soil (locally called *matanzas*) and a gray or black soil type called *rendzinas* (also found in Antigua, Barbados, and lowland Jamaica). The rendzinas

Figure 5.6 Caribbean Sea Noted for its calm, turquoise waters, steady breezes, and treacherous shallows, the Caribbean Sea has both sheltered and challenged sailors for centuries. This aerial photograph shows the southern Caribbean islands of Los Roques, off the Venezuelan coast.

soils, which consist of a gravelly loam with a high organic content ideal for sugar production, are actively exploited for agriculture wherever found. Surprisingly, given the area's agricultural orientation, many of the soils are nutrient-poor, heavily leached, and acidic. These poor, *ferralitic* soils are found in the wetter areas where crystalline base rock exists (as in parts of Hispaniola, the Guianas, and Belize). They are characterized by heavy accumulations of red and yellow clays and offer little potential for permanent intensive agriculture.

Lesser Antilles The **Lesser Antilles** form a double arc of small islands stretching from the Virgin Islands to Trinidad. Smaller in size and population than the Greater Antilles, they were important early footholds for rival European colonial powers. The islands from St. Kitts to Grenada form the inner arc of the Lesser Antilles. These mountainous islands, with peaks ranging from 4,000 to 5,000 feet (1,200 to 1,500 meters), have volcanic origins. In this subduction zone, the heavier North and South American plates dive underneath the Caribbean Plate, producing volcanic activity and earthquakes. Erosion of the island peaks and the accumulation of ash from eruptions have created small pockets of arable soils, although the steepness of the terrain limits agricultural development. The latest round of volcanic activity began in July 1995 on Montserrat. A series of volcanic eruptions of ash and rock took several lives and forced most of the island's 10,000 inhabitants to nearby islands and even to London. Plymouth, the capital, was abandoned in 1997, although interim government buildings now exist on the northwest corner of the island and some residents have returned.

Just east of this volcanic arc are the low-lying islands of Barbados, Antigua, Barbuda, and the eastern half of Guadeloupe. Covered in limestone that overlays volcanic rock, these lands were inviting for agriculture, especially sugar cane. Trinidad and Tobago are on the South American Plate and consist of sedimentary rather than volcanic rock. These islands include alluvial soils and, more important, sedimentary basins that contain oil reserves.

The Rimland Unlike the rest of the Caribbean, the rimland states of Belize and the Guianas still contain significant amounts of forest cover. As on the islands, agriculture in these states is closely tied to local geology and soils. Much of low-lying Belize is limestone. Sugarcane dominates in the drier north, while citrus is produced in the wetter central portion of the state. The Guianas, however, are characterized by the rolling hills of the Guiana Shield. The shield's crystalline rock explains the area's overall poor soil quality. Most agriculture in the Guianas occurs on the narrow coastal plain, where sugar and rice are produced. Timber continues to be an important export for these rimland states. Metal extraction (bauxite and gold) also is vital to the economies of Guyana and Suriname. French Guiana, which is an overseas territory of France, relies mostly on French subsidies but exports shrimp and timber. It also is home to the European Space Center at Kourou (Figure 5.7).

Climate and Vegetation

Coconut palms framed by a blue sky evoke the postcard image of the Caribbean. This tropical region is warm year round, and rainfall is abundant. Much of the Antillean islands and rimland receive more than 80 inches (200 centimeters) of rainfall annually and can support tropical forests. Amid the forests are pockets of naturally occurring grasslands in parts of Cuba, Hispaniola, and southern Guyana. Distinctly dry areas exist, such as the rain shadow basin in western Hispaniola. As explained earlier, much of the natural vegetation on the islands has been removed to accommodate agriculture and fuel needs, and only small forest fragments remain today.

As in many tropical lowlands, seasonality in the Caribbean is defined by changes in rainfall more than temperature. Although some rain falls throughout the year, the rainy season is from July to October, when the Atlantic high-pressure cell is farthest north and easterly winds generate moisture-laden and unstable atmospheric conditions that sometimes yield hurricanes. In Belize City and Havana, October is the wettest month. In Bridgetown and San Juan, the wettest month is November (Figure 5.8). During the slightly cooler months of December through March, rainfall declines. This time of year corresponds with the peak tourist season.

The Guianas have a different rainfall cycle. These territories, on average, receive more rain than the Antillean islands. In Cayenne, French Guiana, an average of 126 inches (320 centimeters) falls each year. Unlike the Antilles, the Guianas experience a brief dry period in late summer (September to October). Also, January tends to be a wet period

Figure 5.7 Kourou, French Guiana The European Space Agency regularly launches rockets from its center in Kourou, French Guiana. This French territory, near the equator and on the coast, makes an ideal launching site. In this photo, the *Ariane 5* rocket is being moved to the launch pad.

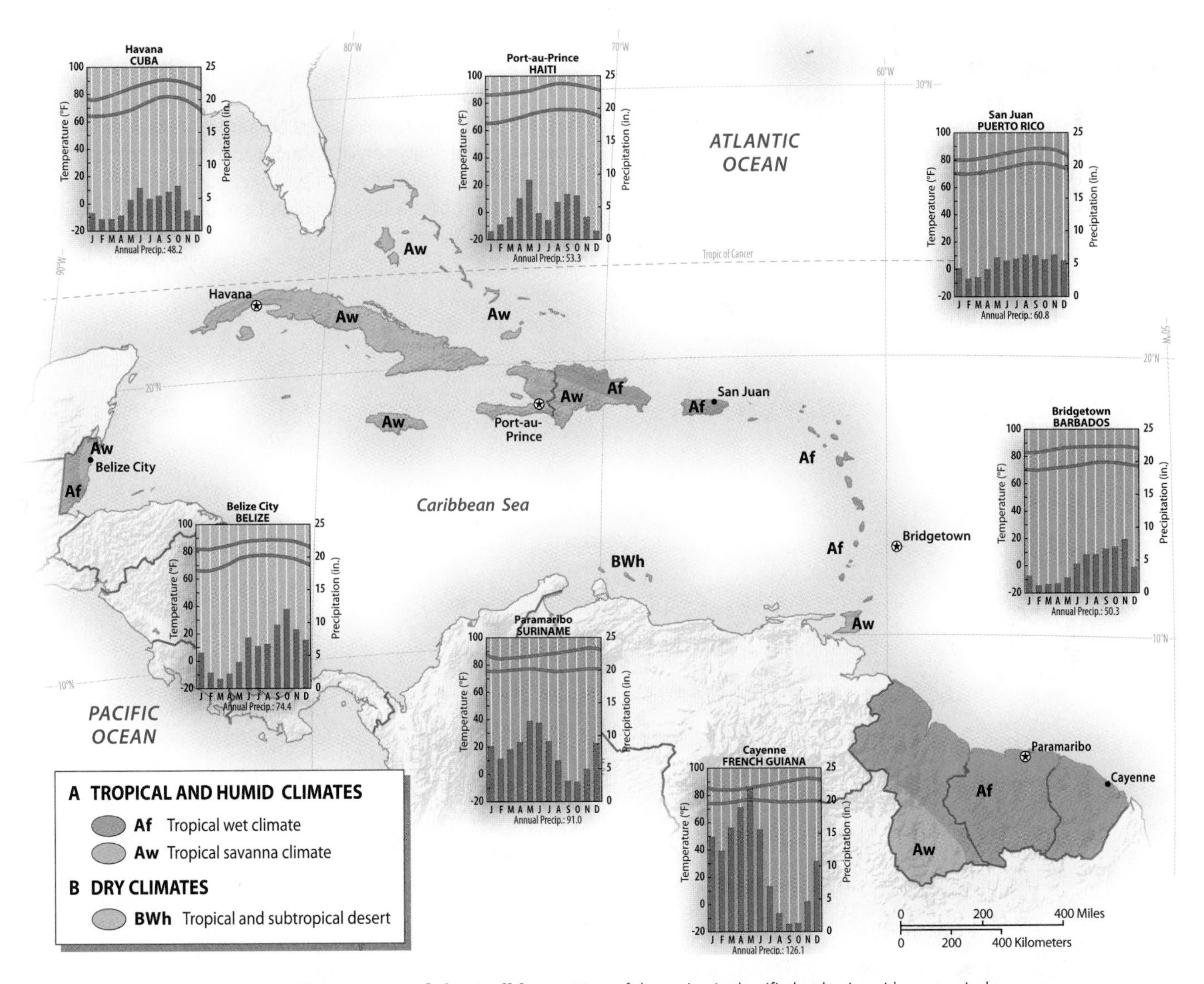

Figure 5.8 Climate Map of the Caribbean Most of the region is classified as having either a tropical wet (Af) or tropical savanna (Aw) climate. Temperature varies little across the region, with highs slightly above 80 degrees and lows around 70 degrees. Important differences in total rainfall and the timing of the dry season distinguish different places. (Temperature and precipitation data from Pearce and Smith, 1984, *The World Weather Guide*, London: Hutchinson)

for the mainland, while it is a dry time for the islands. Climatically, the Guianas also are distinguishable from the rest of the region because they are not affected by hurricanes.

Hurricanes Each year several **hurricanes** form, pounding the Caribbean as well as Central and North America with heavy rains and fierce winds. In July through October, westward-moving low-pressure disturbances form off the coast of West Africa and pick up moisture and speed as they move across the Atlantic. These tropical disturbances are usually no more than 100 miles across, but to achieve hurricane status they must reach velocities of more than 74 miles per hour. Hurricanes may take several paths through the region, but they typically enter through the Lesser Antilles. They then arc north or northwest and collide with the Greater Antilles, Central America, Mexico, or southern North America before moving to the northeast and dissipating in the Atlantic Ocean. The hurricane zone lies just north of the equator on both the Pacific and Atlantic sides of the Americas. Typically a half dozen to a dozen hurricanes form each season and move through the region; most of them never hit land or, if they do, inflict little damage (see "Geographic Tools: Tracking Hurricanes").

Most long-time residents of the Caribbean have felt the full force of at least one major tropical storm in their lifetimes. The destruction caused by these storms is not just

GEOGRAPHIC TOOLS Tracking Hurricanes

The National Hurricane Center in Miami tracks and predicts hurricanes in the Atlantic, the Caribbean, the Gulf of Mexico, and the eastern Pacific. During hurricane season (July–October), surface reports and satellite data are constantly monitored to detect and track tropical depressions that may achieve hurricane status. The National Hurricane Center uses a combination of sophisticated models, geostationary satellites, and actual measurements of the approaching hurricanes through specially equipped aircraft that fly into hurricanes before they approach land. Using supercomputers that integrate dynamic data from a particular tropical depression

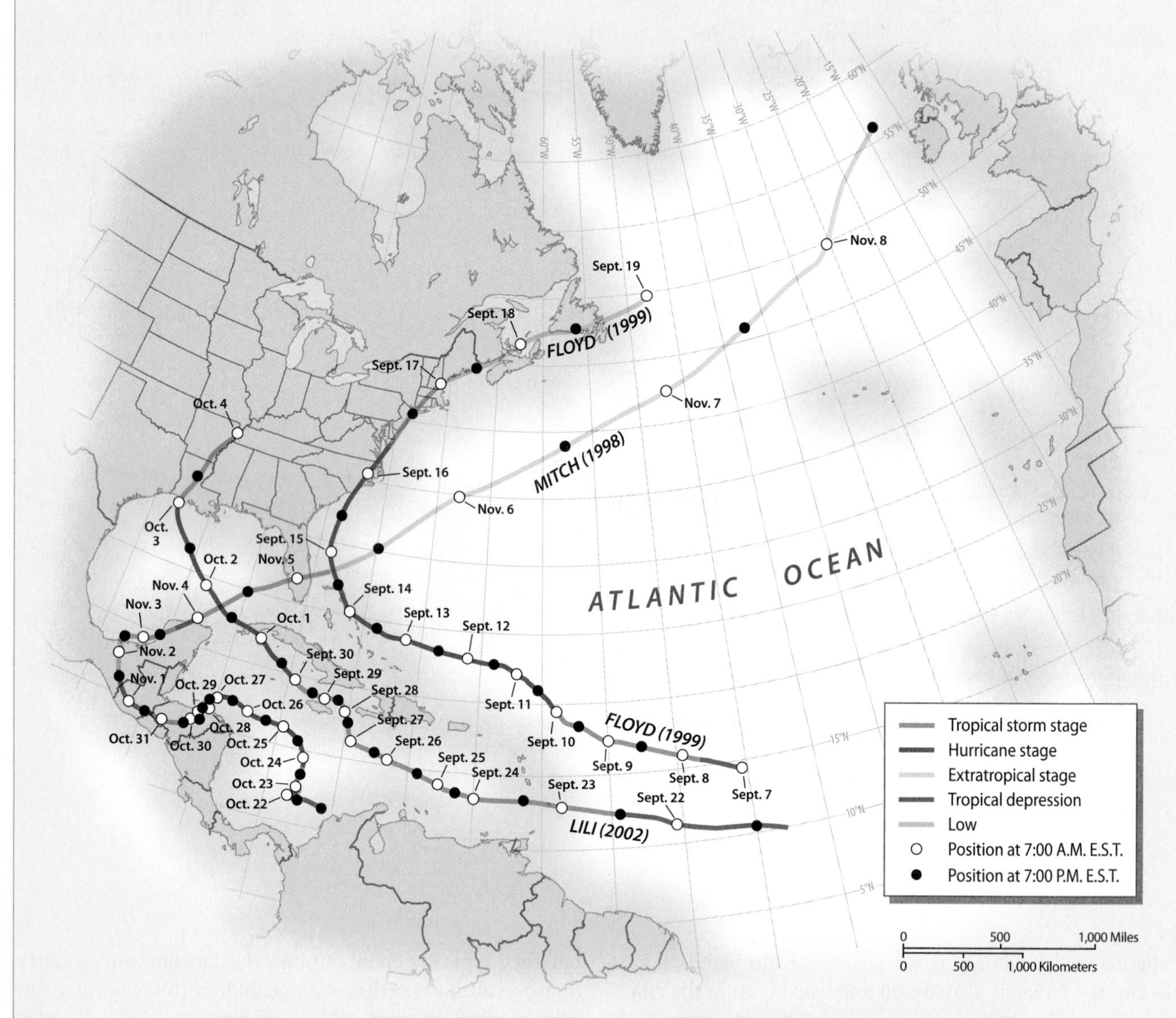

Figure 5.1.1 Tracking the Paths of Hurricanes Mitch, Floyd, and Lili (From Aguado and Burt, 2004, *Understanding Weather and Climate*, 3rd ed., Upper Saddle River, NJ: Prentice Hall)

from the high winds, but also from the heavy downpours that can cause severe flooding and deadly coastal tidal surges. In 2004 the island nation of Grenada was pummeled by Hurricane Ivan; although only a few people were killed, the island's infrastructure suffered 85 percent devastation (Figure 5.9).

Forests, Savannas, and Mangroves Tropical forests, once common throughout the region, are today found almost exclusively in the rimland, and even these areas are increasingly under pressure. As for the rest of the Caribbean, small pockets of forest remain broken by fields and savannas. On the island of Puerto Rico, for example, the small El Yunque

along with laws of physics and historical patterns, meteorologists project the storm position, direction, wind speed, and air pressure every few hours. Figure 5.1.1 shows the actual path of three hurricanes over the course of 2 weeks in 12-hour intervals. A particular storm can take an irregular jog or have periods of slower or faster movement through an area. One reason Hurricane Mitch was so deadly was that it slowed down and lingered over a small portion of Central America for nearly a week, dropping huge amounts of rain (Figure 5.1.2).

As good as the models are, scientists are unable to forecast the path of a hurricane more than 72 hours in advance. Within 24 hours of a hurricane's reaching land, forecasts of its path and where it might make landfall improve considerably, but the margin of error still is about 100 miles (160 kilometers). When one considers the difficulty and expense of evacuating a heavily settled coastal area, this degree of error is far from ideal. Within 6 hours of landfall, the immediate path of a hurricane is much more predictable, but by then the time for taking precautions is limited.

Atmospheric scientists and geographers believe that we have entered a more active hurricane period. Since 1995, a change in the multiyear cycle of sea surface temperatures in the Atlantic Ocean may be contributing to more intense tropical depressions. With settlement and population growth intensifying in coastal zones of North America and the Caribbean, major storms are likely to do more damage. Apart from deaths resulting from Hurricanes Mitch and Katrina, fatalities from hurricanes over the last 30 years have declined because of better forecasting of hurricane movement, resulting in initiation of evacuation procedures at least 24 hours ahead of a storm's landing. Forecasting cannot reduce the damage to crops, forests, homes, or infrastructure. An unfortunately timed storm can destroy a banana harvest or shut down a resort for a season or more. The sheer force of a storm can radically change the landscape as well, turning a palm-lined sandy beach into a barren rocky shore covered with debris.

Source: Adapted from Edward Aguado and James E. Burt, 2004, *Understanding Weather and Climate,* 3rd ed., Upper Saddle River, NJ: Prentice Hall, 367–74.

Figure 5.1.2 Hurricane Mitch Satellite imagery shows the immense scale of one of the most destructive hurricanes to hit the region. Mitch achieved wind speeds of over 200 mph and unleashed torrential rains. The storm and subsequent floods killed more than 10,000 people, left 2.5 million people homeless, and caused over $5 billion in damage. This image was taken on October 26, 1998, three days before Mitch made landfall. (Hal Pierce/Laboratory of Atmospheres/ NASA/ Goddard Space Flight Center)

rain forest offers a glimpse of the Caribbean's forested past. Protected by the Spanish in the 19th century, today El Yunque is the smallest national forest in the U.S. system, but it is also one of the most diverse.

The palm savannas, found mostly in the tropical savanna (Aw) zones, are important biomes in the Caribbean. The palm savannas of Hispaniola and Cuba are quite fertile and easily adapted to agriculture. In central Cuba, the natural grasslands studded with palms have the best soils in the entire country and are planted mostly in sugarcane and citrus. The savanna soils of southern Guyana, however, are acidic with little agricultural potential.

Figure 5.9 Granada after Hurricane Ivan A resident of St. Georges in Grenada sits by the remains of his home after Hurricane Ivan destroyed it in 2004. One of the most destructive tropical storms to hit the Antilles in years, Ivan damaged most of Grenada's infrastructure.

For centuries the coastal mangrove swamps of the region were largely left undisturbed. These wet environments are poorly suited to human settlement, although they are vital nurseries for young crustaceans and fish. Found throughout the Caribbean, especially on the calmer leeward shores, the tangled stands of woody mangrove are now often cleared to create open beaches. Removal of mangroves, besides eliminating vital marine habitat, exposes coasts to increased erosion. In addition, mangroves decline when disturbances such as increased soil erosion and runoff from areas upstream increase the silt load in the water. Moreover, mangrove swamps and other coastal wetlands are increasingly vulnerable to the long-term effects of global warming, especially rises in sea level.

The Caribbean and Global Warming

The effects of global warming on the Caribbean region include sea level rise, increased intensity of storms, variable rainfall leading to both floods and droughts, and loss of biodiversity (both in forests and coral reefs). The scientific consensus is that global warming could promote a sea level rise of 3 to 10 feet (1–3 m) in this century. In terms of land loss to inundation, the low-lying Bahamas would be the most impacted country—losing nearly 30 percent of its land with a 10 foot (3 m) sea level rise. Yet in terms of people affected by inundation, Suriname, French Guiana, Guyana, Belize, and the Bahamas would be the most severely impacted—just a 3 foot (1 m) sea level rise would be devastating because most of the population lives near the coast. Should sea level rise reach 10 feet (3 m), 30 percent of Suriname's population and 25 percent of Guyana's would be displaced (see Figure 5.4).

The Caribbean has not been a major contributor of greenhouse gases, but this maritime region is extremely vulnerable to the negative impacts of climate change. In addition to land loss and population displacement due to sea level rise, other concerns focus on changes in rainfall patterns, leading to declines in agricultural yields and freshwater supplies, and increases in storm intensity—especially hurricanes. All of these changes would negatively impact tourism and thus the gross domestic income of countries in the region. Some of the worst-case scenarios are catastrophic.

In terms of biodiversity, continued warming of ocean temperatures will further negatively impact the Caribbean's coral reefs, which are the most biologically diverse ecosystems of the marine world. These reefs, particularly those of the rimland, are already threatened by water pollution and subsistence fishing practices. Now there is mounting evidence of coral bleaching and die off due to higher sea temperatures. Coral reefs are diverse and productive ecosystems that function as nurseries for many marine species. Healthy reefs also serve as barriers to protect populated coastal zones as well as mangroves and wetlands. As the reefs become more ecologically vulnerable, so too do the human populations that depend upon the many services that the reefs provide.

A division of the UN Environment Program that focuses on the world's coral reefs has supported a regional conservation effort in the Caribbean. The longest barrier reef in the Americas lies off the coast of Belize, where international researchers monitor the impact of warmer waters, disease, and pollution on the health of the coral. Examples of local conservation efforts are seen in Bonaire, where the National Parks Foundation actively manages the Bonaire Marine Park, recognized as one of the most effectively managed marine reserves in the Caribbean.

Throughout the Caribbean, protecting the environment and preparing for the effects of global warming are increasingly being recognized not as a luxury but as a question of economic livelihood. In fact, the Caribbean regional organization CARICOM has monitored the threat of climate change for over a decade. To address the issue of greenhouse gases regionally, Guyana entered an innovative agreement with Norway in 2009. Norway provided Guyana with an initial payment of $30 million into its Reduction of Emissions from Deforestation and Forest Degradation (REDD) development fund. If this initial investment succeeds in reducing emissions and tackling poverty, Norway has agreed to investment of up to $250 million in this project.

POPULATION AND SETTLEMENT: Densely Settled Islands and Rimland Frontiers

In the Caribbean, the population density is generally quite high and, as in neighboring Latin America, increasingly urban. Eighty-seven percent of the region's population is concentrated on the four islands of the Greater Antilles (Figure 5.10). Add to this Trinidad and Tobago's 1.3 million and Guyana's 800,000, and most of the population of the Caribbean is accounted for by six countries and one U.S. territory (Puerto

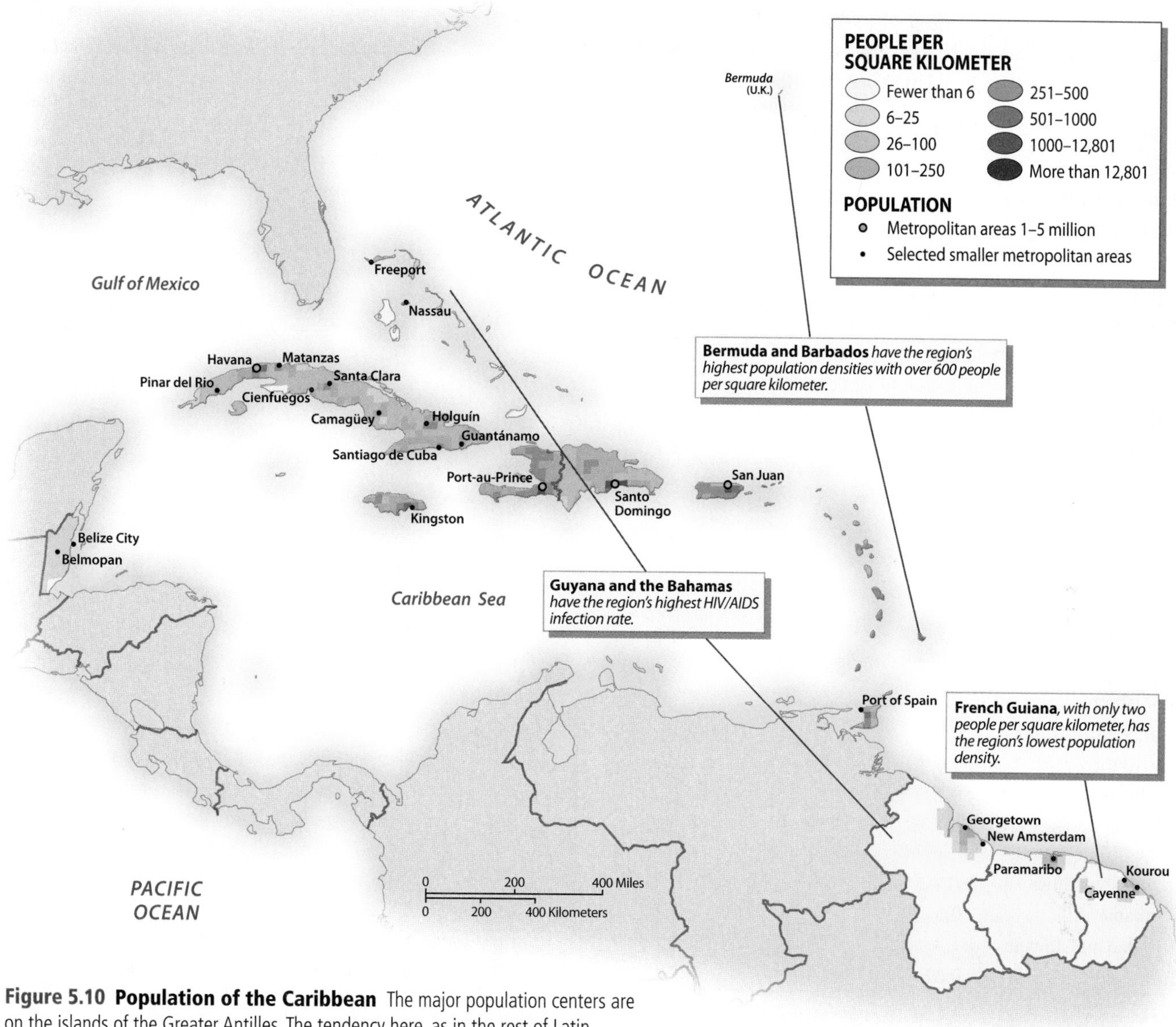

Figure 5.10 Population of the Caribbean The major population centers are on the islands of the Greater Antilles. The tendency here, as in the rest of Latin America, is toward greater urbanism. The largest city of the region is Santo Domingo, followed by Havana. In comparison, the rimland states are very lightly settled.

Rico). Of these, Puerto Rico has the greatest population density, with 448 people per square kilometer, followed by Haiti with 353 people per square kilometer (Table 5.1).

In absolute numbers, few people inhabit the Lesser Antilles; nevertheless, some of these microstates are densely settled. The small island of Barbados is an extreme example. With only 166 square miles (430 square kilometers) of territory, it has 1,643 people per square mile (637 people per square kilometer). Bermuda, which is one-third of the size of the District of Columbia, has over 1,260 people per square kilometer. Population densities on St. Vincent, Martinique, and Grenada, while not as high, are still more than 700 people per square mile (270 people per square kilometer). If one takes into consideration the scarcity of arable land on some of these islands, it is clear that access to land is a basic resource problem for many inhabitants of the Caribbean. The growth in the region's population coupled with its scarcity of land has forced many people into the cities or abroad. It also has forced many Caribbean states to be net importers of food.

In contrast to the islands, the mainland territories of Belize and the Guianas are lightly populated; Guyana averages 10 people per square mile (4 people per square kilometer), Suriname only 8 (3 per square kilometer), and Belize 36 (14 per square kilometer). These areas are sparsely settled in part because the relatively poor quality and accessibility of arable land made them less attractive to colonial enterprises.

Demographic Trends

During the years of slave-based sugar production, mortality rates were extremely high because of disease, inhumane treatment, and malnutrition. Consequently, the only way

TABLE 5.1 Population Indicators

Country	Population (millions) 2010	Population Density (per square kilometer)	Rate of Natural Increase (RNI)	Total Fertility Rate	Percent Urban	Percent <15	Percent >65	Net Migration (Rate per 1,000) 2005–10[a]
Anguilla	0.01	162		1.8	100	24	8	13.5
Antigua and Barbuda	0.1	205	0.9	1.9	31	28	7	
Bahamas	0.3	25	0.9	1.9	83	26	6	1.2
Barbados	0.3	637	0.5	1.7	38	19	9	
Belize	0.3	15	2.3	3.1	51	37	5	–0.7
Bermuda	0.07	1264		2.0	100	18	15	2.1
Cayman	0.05	189		1.9	100	19	10	16.0
Cuba	11.2	101	0.3	1.6	75	18	12	–3.5
Dominica	0.1	96	0.7	2.0	73	23	10	
Dominican Republic	9.9	203	1.7	2.7	67	32	6	–2.8
French Guiana	0.2	3	2.4	3.6	81	35	4	5.5
Grenada	0.1	320	0.8	2.2	31	31	10	–9.7
Guadeloupe	0.4	239	0.7	2.0	100	22	13	–1.5
Guyana	0.8	4	1.6	2.8	28	33	5	–10.5
Haiti	9.8	353	1.8	3.5	48	37	4	–2.9
Jamaica	2.7	246	1.2	2.4	52	28	8	–7.4
Martinique	0.4	368	0.6	1.9	89	20	14	–1.0
Montserrat	0.005	50		1.2		27	7	
Netherlands Antilles	0.2	255	0.7	2.1	92	22	10	8.7
Puerto Rico	4.0	448	0.4	1.6	94	20	14	–1.1
St. Kitts and Nevis	0.1	203	0.7	1.8	32	24	7	
St. Lucia	0.2	327	0.7	1.7	28	25	9	–1.2
St. Vincent and the Grenadines	0.1	276	0.9	2.1	40	28	7	–9.2
Suriname	0.5	3	1.3	2.4	67	30	7	–2.0
Trinidad and Tobago	1.3	257	0.6	1.6	12	25	7	–3.0
Turks and Caicos	0.02	25		2.9	92	30	4	8.6

[a]*Net Migration Rate from the United Nations, Population Division, World Population Prospects: The 2008 Revision Population Database.*
Source: Population Reference Bureau, World Population Data Sheet, 2010.
Additional data from CIA World Factbook, 2010.

population levels could be maintained was through the continual importation of African slaves. With the end of slavery in the mid-to-late 19th century and the gradual improvement of health and sanitary conditions on the islands, natural population increase began to occur. In the 1950s and 1960s, many states achieved peak growth rates of 3.0 or higher, causing population totals and densities to soar. Over the past 30 years, however, growth rates have steadily come down and stabilized. As noted earlier, the current population of the Caribbean is 43 million. However, the population is now growing at an annual rate of 1.2 percent, and projected population in 2025 is 48 million (see Table 5.1).

Fertility Decline The most significant demographic trend in the Caribbean is the decline in fertility (Figure 5.11). Cuba has the region's lowest rate of natural increase (0.3), followed by Puerto Rico (0.4). In socialist Cuba, the education of women combined with the availability of birth control and abortion has resulted in 1.6 children (compared to 2.1 in the United States). Yet in capitalist Puerto Rico, similarly low rates of natural increase (0.4) have also been achieved, along with a total fertility rate of 1.6. In general, educational improvements, urbanization, and a preference for smaller families have contributed to slower growth rates. Even states with relatively high total fertility rates, such as Haiti, have seen a decline in family size. Haiti's total fertility rate fell from 6.0 in 1980 to 3.5 in 2010.

The Rise of HIV/AIDS The rate of HIV/AIDS infection in the Caribbean has come down in the past few years, but it is still twice that of North America, making the disease an important regional issue. Although nowhere near the infection rates in Sub-Saharan Africa (see Chapter 6), slightly more than 1 percent of the Caribbean population between the ages

Figure 5.11 Smaller Caribbean Families A family in Bermuda enjoys a day at the beach. The average Caribbean family is far smaller now than 30 years ago. Higher levels of education, improved availability of contraception, an increase in urban living, and a larger percentage of women in the labor force contribute to slower population growth rates in the region.

of 15 and 49 had HIV/AIDS in 2008. In Haiti, one of the earliest locations where AIDS was detected, 2.2 percent of the population between the ages of 15 and 49 is infected with the virus. The infection rate in Guyana is 2.5 percent, Belize is 2.1 percent, and the Bahamas is 3 percent. AIDS is already the largest single cause of death among young men in the English-speaking Caribbean.

Many Caribbean countries have taken steps to educate their populations about the spread of HIV/AIDS, and as a result infection rates have come down over the past decade. Officials in the Bahamas have introduced drug treatments to help prevent mother-to-child transmission. And nearly every country has launched educational campaigns to bring infection rates down. Cuba, which witnessed a surge in both tourism and prostitution in the 1990s, has a very low infection rate of 0.1 percent among its 15- to 49-year-old population. Education programs and an effective screening and reporting system have kept Cuba's infection rates down.

Emigration Driven by the region's limited economic opportunities, a pattern of emigration to other Caribbean islands, North America, and Europe began in the 1950s. For more than 50 years, a Caribbean **diaspora**—the economic flight of Caribbean peoples across the globe—has defined existence and identity for much of the region (Figure 5.12). Barbadians generally choose England, most settling in the London suburb of Brixton with other Caribbean immigrants. In contrast, one out of every three Surinamese has moved to the Netherlands, with most residing in Amsterdam. As for Puerto Ricans, only slightly more live on the island than reside on the U.S. mainland. In the 1980s, roughly 10 percent of Jamaica's population legally emigrated to North America (some 200,000 to the United States and 35,000 to Canada). Cubans have made the city of Miami their destination of choice since the 1960s. Today they make up the majority of that city's population.

Intraregional movements also are important. Perhaps one-fifth of all Haitians do not live in their country of birth, with their most common destination being the neighboring Dominican Republic, followed by the United States, Canada, and French Guiana. Dominicans are also on the move; the vast majority come to the United States, settling in New York City, where they are the single largest immigrant group. Others, however, simply cross the Mona Passage and settle in Puerto Rico. As a region, the Caribbean has one of the highest annual rates of net migration in the world at −3.0 per thousand. That means for every 1,000 people in the region, 3 annually leave. Individual countries have much higher rates, such as Guyana and Grenada at −10 per 1,000 and Jamaica at −7 per 1,000. The economic implications of this labor-related migration are significant and will be discussed later.

Most migrants, with the exception of the Cubans, are part of a **circular migration** flow. In this type of migration, a man or woman typically leaves children behind with relatives in order to work hard, save money, and return home. Other times a **chain migration** begins, in which one family member at a time is brought over to the new country. In some cases, large numbers of residents from a Caribbean town or district send migrants to a particular locality in North America or Europe. Thus, chain migration can account for the formation of immigrant enclaves. Caribbean immigrants have increasingly practiced **transnational migration**—the straddling of livelihoods and households between two countries. Dominicans are probably the most transnational of all the Caribbean groups. They regularly move back and forth between two islands, Hispaniola and Manhattan. Dominican President Leonel Fernandez was first elected in 1996 for a 4-year term and was reelected in 2004 and in 2008. He grew up in New York City, still holds a green card, and has said he intends to return when his presidential term is over.

The Rural–Urban Continuum

Initially, plantation agriculture and subsistence farming shaped Caribbean settlement patterns. Low-lying arable lands were dedicated to export agriculture and controlled by wealthy colonial landowners. Only small amounts of land

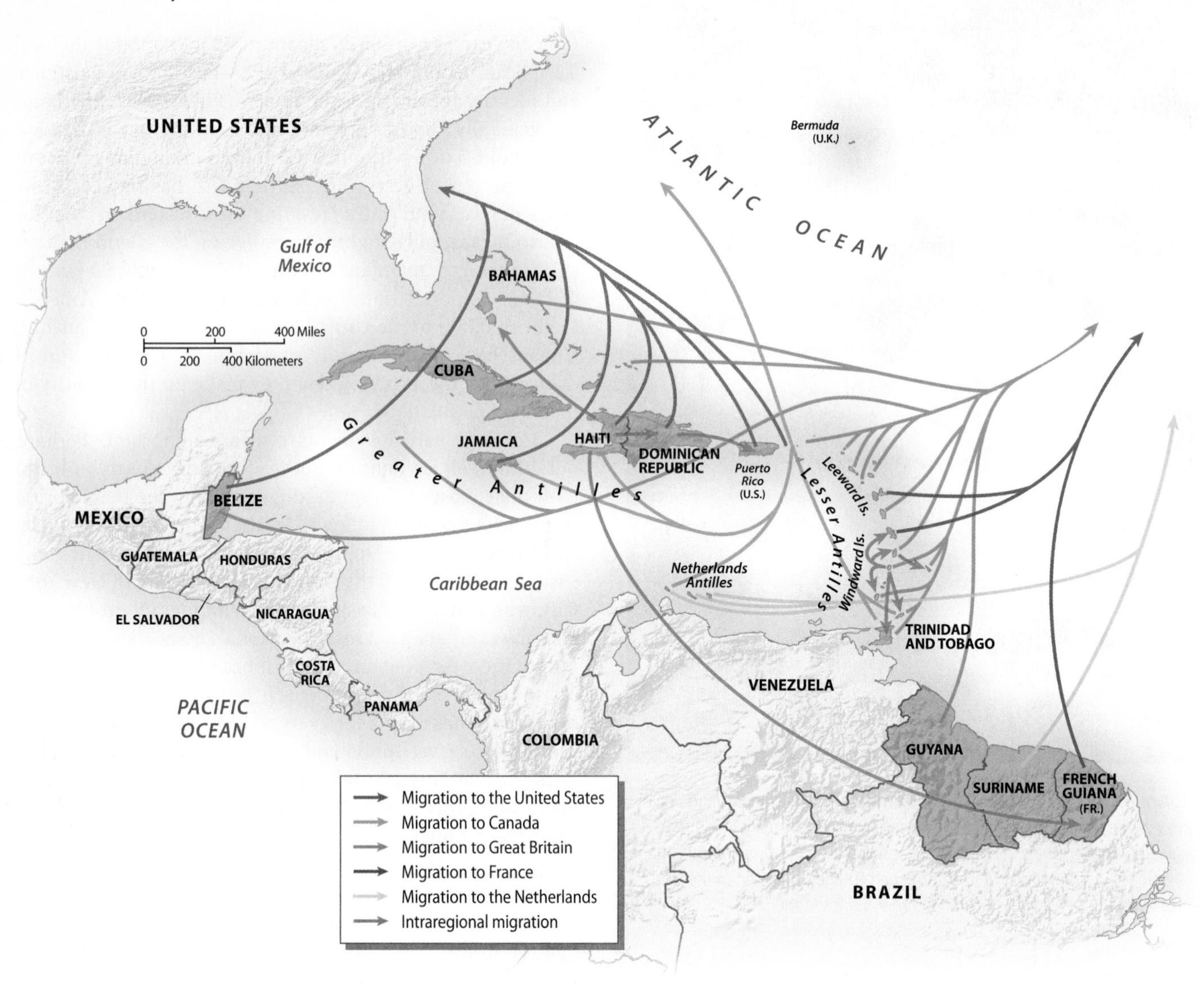

Figure 5.12 Caribbean Diaspora Emigration has long been a way of life for Caribbean peoples. With relatively high education levels but limited professional opportunities, migrants from the region head to North America, Great Britain, France, and the Netherlands. Intraregional migrations between Haiti and the Dominican Republic or the Dominican Republic and Puerto Rico also occur. (Data from Levin, 1987, *Caribbean Exodus*, Westport, CT: Praeger Publishers)

were set aside for subsistence production. Over time, villages of freed or runaway slaves were established, especially in remote areas of the interior. But the vast majority of people continued to live on estates as owners, managers, or slaves. Cities were formed to serve the administrative and social needs of the colonizers, but most were small, containing a small fraction of a colony's population. The colonists who linked the Caribbean to the world economy saw no need to develop major urban centers.

Even today the structure of Caribbean communities reflects the plantation legacy. Many of the region's subsistence farmers are ancestors of former slaves who continue to work their small plots and seek seasonal wage-labor on estates. The social and economic patterns generated by slavery still mark the landscape. Rural communities tend to be loosely organized; labor is transient; and small farms are scattered on available pockets of land. Because men tended to leave home for seasonal labor, matriarchal family structures and female-headed households are common.

Caribbean Cities Since the 1960s, the mechanization of agriculture, offshore industrialization, and rapid population growth have caused a surge in rural-to-urban migration. Cities have grown accordingly, and today 63 percent of the region is classified as urban. Of the large countries, Cuba is the most urban (76 percent) and Haiti the least (48 percent). Caribbean cities are not large by world standards, as only four have more than 1 million residents: Santo Domingo, Havana, Port-au-Prince, and San Juan. All but Port-au-Prince were laid out by the Spanish.

Like their counterparts in Latin America, the Spanish Caribbean cities were laid out on a grid with a central plaza.

Vulnerable to raids by rival European powers and pirates, these cities were usually walled and extensively fortified. The oldest continually occupied European city in the Americas is Santo Domingo in the Dominican Republic, settled in 1496, and today is a metropolitan area of 2.5 million (see Cityscapes: Santo Domingo). Havana emerged as the most important colonial city in the region, serving as a port for all incoming and outgoing Spanish galleons. Strategically situated on Cuba's north coast at a narrow opening to a natural deep-water harbor, Havana became an essential city for the Spanish empire. Consequently, Havana possesses a handsome collection of colonial architecture, especially from the 18th and 19th centuries (Figure 5.13). The largest city in the region is now San Juan, estimated at 2.8 million. It too has a restored colonial core that is dwarfed by the modern sprawling city that supports the island's largest port. San Juan is the financial, political, manufacturing, and tourism hub of Puerto Rico.

Other colonial powers left their mark on the region's cities. For example, Paramaribo, the capital of Suriname, has been described as a tropical, tulipless extension of Holland. In the British colonies, a preference for wooden white-washed cottages with shutters is evident. Yet the British and French colonial cities tended to be unplanned afterthoughts; these port cities were built to serve the rural estates rather than serve the needs of all residents. Most of them have grown dramatically over the last 40 years. No longer small ports for agricultural exports, increasingly these cities are oriented to welcoming cruise ships and sun-seeking tourists.

Figure 5.14 Belize City Cottage Residents of Belize City build their wooden cottages on stilts as protection against flooding. Shuttered cottages with metal roofs are typical throughout the British Caribbean.

Figure 5.13 Old Havana A Cuban family strolls down a narrow cobblestoned street near the Cathedral in Old Havana, Cuba. The best examples of 18th- and 19th-century colonial architecture in the Caribbean are found in Havana. In 1982, UNESCO declared Old Havana a World Heritage Site, and new funds became available for its restoration.

Caribbean cities and towns do have their charms and reflect a variety of cultural influences. Most people still get around by foot, bicycle, or public transportation; neighborhoods are filled with small shops and services that are within easy walking distance. Streets are narrow, and the pace of life is markedly slower than in North America and Europe. Even when space is tight in town, most settlements are close to the sea and its cooling breezes. An afternoon or evening stroll along the waterfront is a common activity. Flowering shrubs and swaying palms add to the tropical ambiance.

Housing Throughout the region, houses are often simple structures (made of wood, brick, or stucco), sometimes raised off the ground a few feet to avoid flooding, and painted in soft pastels (Figure 5.14). The sudden surge in urbanization in the Caribbean is best explained by an erosion of rural jobs, rather than a rise in urban opportunities. Thousands poured into the cities as economic migrants, erecting shantytowns and filling the ranks of the informal sector. Squatter settlements in Port-au-Prince and Santo Domingo are especially bad, with residents living in wretched housing without the benefit of sewers and running water. Electricity is usually pirated from nearby power lines.

The one place that dramatically breaks from this pattern is Cuba. Forged in a socialist mode, Cubans are housed in

CITYSCAPES The Caribbean Metropolis of Santo Domingo

As the oldest European city in the Western Hemisphere, Santo Domingo is an important symbol of the historical role of the Caribbean in the conquest of the Americas, the complex political rivalries that unfolded in the region, and its metropolitan vitality today. This sprawling metropolis of 2.5 million is home to 1 out of every 4 Dominicans. It is the political, economic and cultural capital of the Dominican Republic. And while its historical roots are deep, much of the city's growth has occurred in the past 50 years.

Santo Domingo was founded in 1496 by Bartholomew Columbus, the younger brother of Christopher, along the Ozama River on the southern coast of Hispaniola. Given the city's early prominence as a colonial administrative center for the Spanish empire, it has an impressive colonial core that was declared a UN World Heritage Site in 1990. Recently, a modern port for cruise ships was built near the colonial zone so that visitors can easily disembark and experience the charms of the old city (Figure 5.2.1).

From the beginning this was contested space, as evidenced by the city's standing fortress and former walls on the west bank of the Ozama. Control of the city was not assured, and in the late 16th century pirates captured it and controlled they port for nearly 50 years, forcing Spanish officials to abandon it. In the late 18th century Spain lost control again, first to France, later to Haitian slaves in 1801, and did not regain control until 1809 before losing it again in 1821. The city continued to be a prize as Dominicans fought off several attempts by Haiti in the 19th century to take control.

Figure 5.2.1 Colonial Santo Domingo A statue of Christopher Columbus is the focal element of this plaza in the city's colonial zone near the Ozama River. Behind the statue is the Santa Maria Cathedral, the oldest cathedral in the Americas. The city of Santo Domingo was founded in 1496 by Bartholomew Columbus, the younger brother of Christopher. Not far from this site is the Alcazar, a stone fortress which was home to Diego Columbus (son of Christopher), and is the oldest Viceregal residence in the Americas.

uniform government-built apartment blocks like those seen throughout Russia and eastern Europe (Figure 5.15). These unimaginative complexes contain hundreds of identical one- and two-bedroom apartments where basic modern amenities (plumbing, electricity, and sewage) are provided. Compared with other large cities of the developing world, the lack of squatter settlements makes Havana unusual.

Figure 5.15 Government Housing in Havana Cubans in a state-built apartment block on the outskirts of greater Havana. Under socialism, thousands of standardized apartment blocks were built to ensure that all Cubans had access to basic modern housing.

The modernization of Santo Domingo began in earnest in the 1960s, and merengue (a fast-paced highly danceable music that originated in the Dominican Republic) is the soundtrack that pulses through the metropolis day and night. As thousands of rural migrants poured into the city in search of employment and opportunity, the city steadily grew. Most of these newcomers settled in self-built homes in the periphery, often without running water and electricity pirated from overhead powerlines. Yet gradually a middle class also emerged, industrial parks were created, planned suburbs were built, tourism expanded, and millions of dollars in remittances flowed into the city from immigrants working in the United States. In 2009 the city's first high-speed Metro opened, and several more lines are planned to link the downtown with the suburbs and reduce the crushing traffic. The country has experienced solid growth in the 2000s but there is still inadequate housing, electricity, employment, and schooling for a large portion of Santo Domingo's residents. Some critics argue that an expensive underground Metro was ill advised given the city's other pressing needs (Figure 5.2.2).

In recognition of the 500th anniversary of Columbus's landing in the Americas, Santo Domingo sought to promote its historical significance by building the *Faro de Colon* (Columbus Lighthouse), an enormous lighthouse in the shape of a cross that can illuminate the heavens. The lighthouse and surrounding park were built near the downtown in a slum area that was cleared for the project—much to the objection of the displaced residents. Dominicans then placed the remains of Christopher Columbus (which were in the city's cathedral) in the structure. Spain, however, claims that Columbus is buried in Seville and disputes Dominican claims that he is buried in the Caribbean.

Figure 5.2.2 Slum Life in Santo Domingo A woman washes clothes in the riverside settlement of Capotillo, Santo Domingo. One of the city's poorer neighborhoods, houses here lack running water and sanitation.

CULTURAL COHERENCE AND DIVERSITY: A Neo-Africa in the Americas

Linguistic, religious, and ethnic differences abound in the Caribbean. A score of former European colonies, millions of descendents of ethnically distinct Africans and indentured workers from India and China, and isolated Amerindian communities on the mainland challenge any notion of cultural coherence.

Common historical and social processes hold the region together. European colonies, with their plantation-based economies, reproduced similar social structures throughout the region. The imprint of more than 7 million African slaves, creating a neo-Africa in the Americas, reflects the important linkages between the Caribbean and the wider Atlantic world. Last, in a process called **creolization,** African and European cultures were blended in the Caribbean. Through this mixing, European languages were transformed into vibrant local dialects, and at times entirely new languages were created (Papiamento and French Creole). This melding also produced the rich and diverse musical traditions now heard throughout the world: reggae, salsa, merengue, rara, and calypso. Contemporary Caribbean identity is also shaped by sports, especially baseball, where a large percentage of major league players (from Alex Rodriquez to Sammy Sosa and David Ortiz) trace their roots to this region.

The Cultural Imprint of Colonialism

European colonization of the Caribbean destroyed indigenous societies and imposed completely different social systems and cultures. The arrival of Columbus in 1492 triggered a devastating chain of events that depopulated the region within 50 years. A combination of Spanish brutality, enslavement, warfare, and disease reduced the densely settled islands, which supported up to 3 million Caribs and Arawaks, into an uninhabited territory ready for the colonizer's hand. The demographic collapse of Amerindian populations occurred throughout the Americas (see Chapter 4), but the death rates were highest in the Caribbean. Only fragments of Amerindian communities survive, mostly on the rimland.

By the mid-16th century, as rival European states vied for Caribbean territory, the lands they fought for were virtually uninhabited. In many ways, this simplified their task, as they did not have to acknowledge indigenous land claims or work amid Amerindian societies. Instead, the Caribbean territories were reorganized to serve a plantation-based production system. The critical missing element was labor. Once slave labor from Africa, and later indentured labor from Asia, was secured, the small Caribbean colonies became surprisingly profitable. Much of Caribbean culture and society today can be traced to the same processes that created plantation America.

Figure 5.16 Sugar Plantation A historical illustration (1823) depicts slaves harvesting sugarcane on a plantation in Antigua. Sugar production was profitable but arduous work. Several million Africans were enslaved and forcibly relocated into the region.

Plantation America The term **plantation America** was coined by anthropologist Charles Wagley to designate a cultural region that extends from midway up the coast of Brazil through the Guianas and the Caribbean into the southeastern United States. Ruled by a European elite dependent on an African labor force, this society was primarily coastal and produced agricultural exports. Other characteristics included a reliance on **monocrop production** (a single commodity, such as sugar) under a plantation system that concentrated land in the hands of elite families. Such a system created rigid class lines, as well as the formation of a multiracial society in which people with lighter skin were privileged. The term *plantation America* is not meant to describe a race-based division of the Americas, but rather a production system that relied upon export commodities, coerced labor, and limited access to land (Figure 5.16).

Asian Immigration Before detailing the pervasive influence of Africans on the Caribbean, the lesser-known Asian presence deserves discussion. By the mid-19th century, most colonial governments in the Caribbean had begun to free their slaves. Fearful of labor shortages, they sought **indentured labor** (workers contracted to labor on estates for a set period of time, often several years) from South and Southeast Asia.

The legacy of these indentured arrangements is clearest in Suriname, Guyana, and Trinidad and Tobago. In Suriname, a former Dutch colony, more than one-third of the population is of South Asian descent, and 16 percent are Javanese (from Indonesia). Guyana and Trinidad were British colonies, and most of their contract labor came from India. Today nearly half of Guyana's population and 40 percent of Trinidad and Tobago's claim South Asian ancestry. Hindu temples are found in the cities and villages, and many families speak Hindi in the home (Figure 5.17). Such racial diversity has led to tension. For the past decade Guyana has had an Afro-Guyanese Prime Minister (Sam Hinds) and an Indo-Guyanese President (Bharrat Jadgeo) governing the country, which seems to have reduced racial tensions. Most

Figure 5.17 South Asian Indians in Trinidad Trinidadians of Indian ancestry celebrate a Hindu festival by sending burning boats down the Marianne River in Trinidad. About 40 percent of Trinidad and Tobago's population is ethnically Indian and adheres to the Hindu faith.

of the former English colonies have Chinese populations of not more than 2 percent. Once these East Asian immigrants fulfilled their agricultural contracts, they often became merchants and small-business owners, positions they still hold in Caribbean society.

Creating a Neo-Africa

African slaves were first introduced to the Americas in the 16th century, partly in response to the demographic collapse of the Amerindians. The flow of slaves continued into the 19th century. This forced migration of mostly West Africans to the Americas was only part of a much more complex African diaspora—the forced removal of Africans from their native area. The slave trade also crossed the Sahara to include North Africa and linked East Africa with a slave trade in the Middle East (see Chapter 6). The best documented slave route is the transatlantic one; at least 10 million Africans landed in the Americas, and it is estimated that another 2 million died en route. More than half of these slaves were sent to the Caribbean (Figure 5.18).

This influx of slaves, combined with the extermination of nearly all the native inhabitants, recast the Caribbean as the area with the greatest concentration of African transfers in the Americas. The African source areas extended from Senegal to Angola, and slave purchasers intentionally mixed tribal groups in order to dilute ethnic identities. Consequently, intact transfer of African religions and languages into the Caribbean did not occur; instead languages, customs, and beliefs were blended.

Maroon Societies Communities of runaway slaves—termed **maroons**—offer the most compelling examples of African cultural diffusion across the Atlantic. Hidden settlements of escaped slaves existed wherever slavery was practiced. Called *maroons* in English, *palenques* in Spanish, and *quilombos* in Portuguese, many of these settlements were short-lived, but others have endured and allowed for the

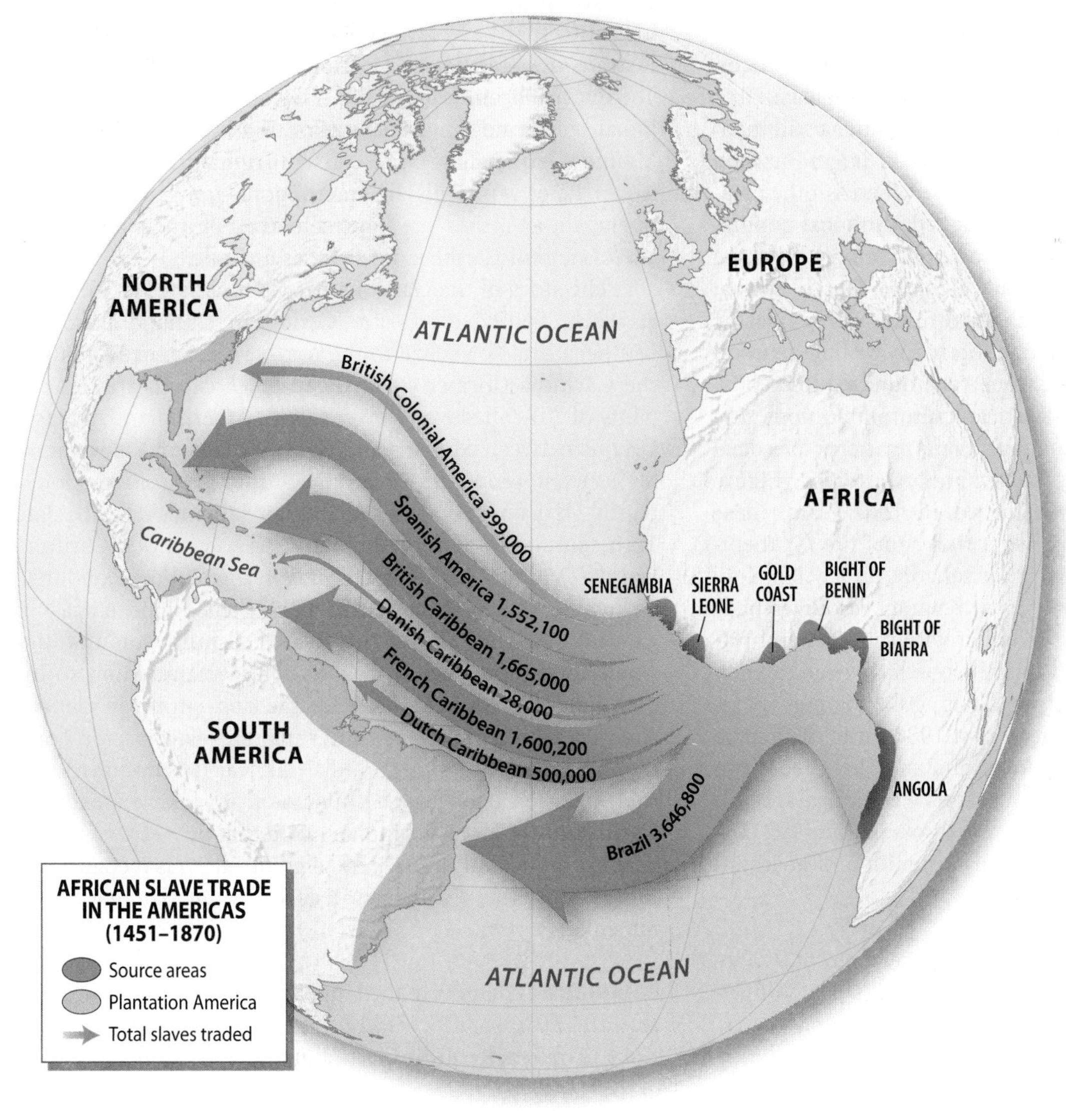

Figure 5.18 Transatlantic Slave Trade At least 10 million Africans landed in the Americas during the four centuries in which the Atlantic slave trade operated. Most of the slaves came from West Africa, especially the Gold Coast (now Ghana) and the Bight of Biafra (now Nigeria). Angola, in southern Africa, was also an important source area. (Data based on Curtin, 1969, *The Atlantic Slave Trade, A Census,* Madison: University of Wisconsin Press, 268)

Figure 5.19 Maroon Village in Suriname A maroon woman carries a pail of water to her home in the village of Stonuku, Suriname. Maroon communities existed throughout the Caribbean as slaves ran away from plantations and formed villages in remote locations. The maroon communities in Suriname still retain many African traditions.

survival of African traditions, especially farming practices, house designs, community organization, and language.

The maroons of Suriname and French Guiana still manifest clear links to West Africa. They form the largest maroon population in the Western Hemisphere. Whereas other maroon societies gradually were assimilated into local populations, to this day the maroons of Suriname maintain a distinct identity. These runaways fled the Dutch coastal plantations in the 17th and 18th centuries, forming riverine settlements amid the interior rain forest. Six distinct maroon tribes formed, ranging in size from a few hundred to 20,000. Clear manifestations of West African cultural traditions persist, including religious practices, crafts, patterns of social organization, agricultural systems, and even dress (Figure 5.19). Living relatively undisturbed for 200 years, these rain-forest inhabitants fashioned a rich ritual life for themselves involving oracles, spirit possession, and witch doctors. Yet pressures to modernize and extract resources have placed the maroons in direct conflict with the state and private investors. The maroons in Suriname have been directly affected by the construction of dams, gold mining operations, and logging concessions. From 1986 until 1992 there was a civil war between the maroons and the Creole-run military, in which hundreds of maroons were killed and villages destroyed. Although peace was brokered in 1992, the maroons continue to fight for legal recognition of ancestral claims to land and resources.

African Religions Linked to maroon societies, but more widely diffused, is the transfer of African religious and magical systems to the Caribbean. These patterns, another reflection of neo-Africa in the Americas, are most closely associated with northeastern Brazil and the Caribbean. In Chapter 4 we discussed how millions of Brazilians practice the African-based religions of Umbanda, Macuba, and Candomblé, along with Catholicism. Likewise, Afro-religious traditions in the Caribbean have evolved into unique forms that have clear ties to West Africa. The most widely practiced are Voodoo (also Vodoun) in Haiti, Santería in Cuba, and Obeah in Jamaica. These religions have their own priesthood and unique patterns of worship. Their impact is considerable; the father and son dictators of Haiti, the Duvaliers, were known to hire Voodoo priests to scare off government opposition during their rule from 1957 to 1986. Moreover, as Figure 5.20 shows, many of these religions have diffused from their areas of origin. Santería is practiced in Florida and New York by some Cuban immigrants. Likewise, belief in Obeah diffused when Jamaicans migrated to Panama and Los Angeles.

Creolization and Caribbean Identity

Creolization refers to the blending of African, European, and even some Amerindian cultural elements into the unique sociocultural systems found in the Caribbean. The Creole identities that have formed over time are complex; they illustrate the cultural and national identities of the region. Today Caribbean writers (V. S. Naipaul, Derek Walcott, and Jamaica Kincaid), musicians (Bob Marley, Celia Cruz, and Manno Charlemagne), and artists (Trinidadian costume designer Peter Minshall) are internationally regarded. Collectively, these artists are representative of their individual islands and of Caribbean culture as a whole.

The story of the Garifuna people illustrates creolization at work. Settled along the Caribbean rimland from the southern coast of Belize to the northern coast of Honduras, the Garifuna (formerly called the *Black Carib*) are descendants of African slaves who speak an Amerindian language. Unions between Africans and Carib Indians on the island of St. Vincent produced an ethnic group that was predominantly African but spoke an Indian language. In the late 18th century, Britain forcibly resettled some 5,000 Garifuna from St. Vincent to the Bay Islands in the Gulf of Honduras. Over time, the Garifuna settled along the Caribbean coast of Central America, living in isolated fishing communities from Honduras to Belize. In addition to maintaining an Indian language, the Garifuna are the only group in Central America who regularly eat bitter manioc—a root crop common in lowland tropical South America. It is assumed that they acquired their taste for manioc from their exposure to Carib culture. The Afro-Indian blend that the Garifuna manifest is unique, but the process of creolization is recognizable throughout the Caribbean, especially in language and music.

Language The dominant languages in the region are European: Spanish (25 million speakers), French (10 million), English (6 million), and about half a million Dutch

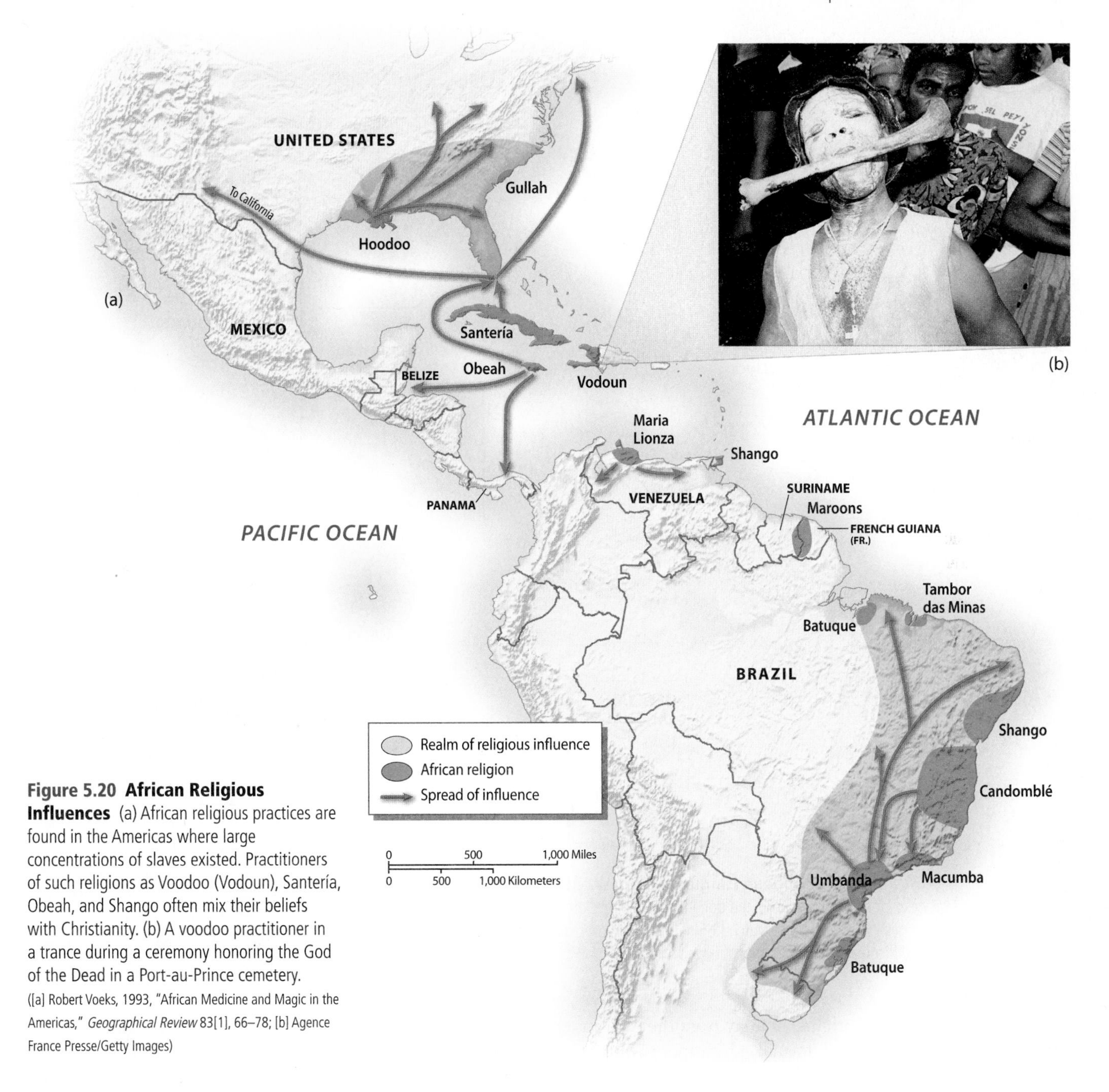

Figure 5.20 African Religious Influences (a) African religious practices are found in the Americas where large concentrations of slaves existed. Practitioners of such religions as Voodoo (Vodoun), Santería, Obeah, and Shango often mix their beliefs with Christianity. (b) A voodoo practitioner in a trance during a ceremony honoring the God of the Dead in a Port-au-Prince cemetery.
([a] Robert Voeks, 1993, "African Medicine and Magic in the Americas," *Geographical Review* 83[1], 66–78; [b] Agence France Presse/Getty Images)

(Figure 5.21). Yet these figures tell only part of the story. In Cuba, the Dominican Republic, and Puerto Rico, Spanish is the official language, and it is universally spoken. As for the other countries, colloquial variants of the official language exist, especially in spoken form, that can be difficult for a nonnative speaker to understand. In some cases, completely new languages emerge; in the ABC islands, Papiamento (a trading language that blends Dutch, Spanish, Portuguese, English, and African languages) is the *lingua franca*, with usage of Dutch declining. In Suriname the vernacular language is Sranan Tongo (an amalgam of Dutch and English with many African words). Similarly, French Creole or *patois* in Haiti has constitutional status as a distinct language. In practice, French is used in higher education, government, and the courts, but patois (with clear African influences) is the language of the street, the home, and oral tradition. Most Haitians speak patois, but only the formally educated know French.

With independence in the 1960s, Creole languages became politically and culturally charged with national meaning. During the colonial years, Creole was negatively viewed as a corruption of standard European forms. Those who spoke Creole were considered uneducated or backward. As linguists began to study these languages, they found that while the vocabulary came from Europe, the syntax or se-

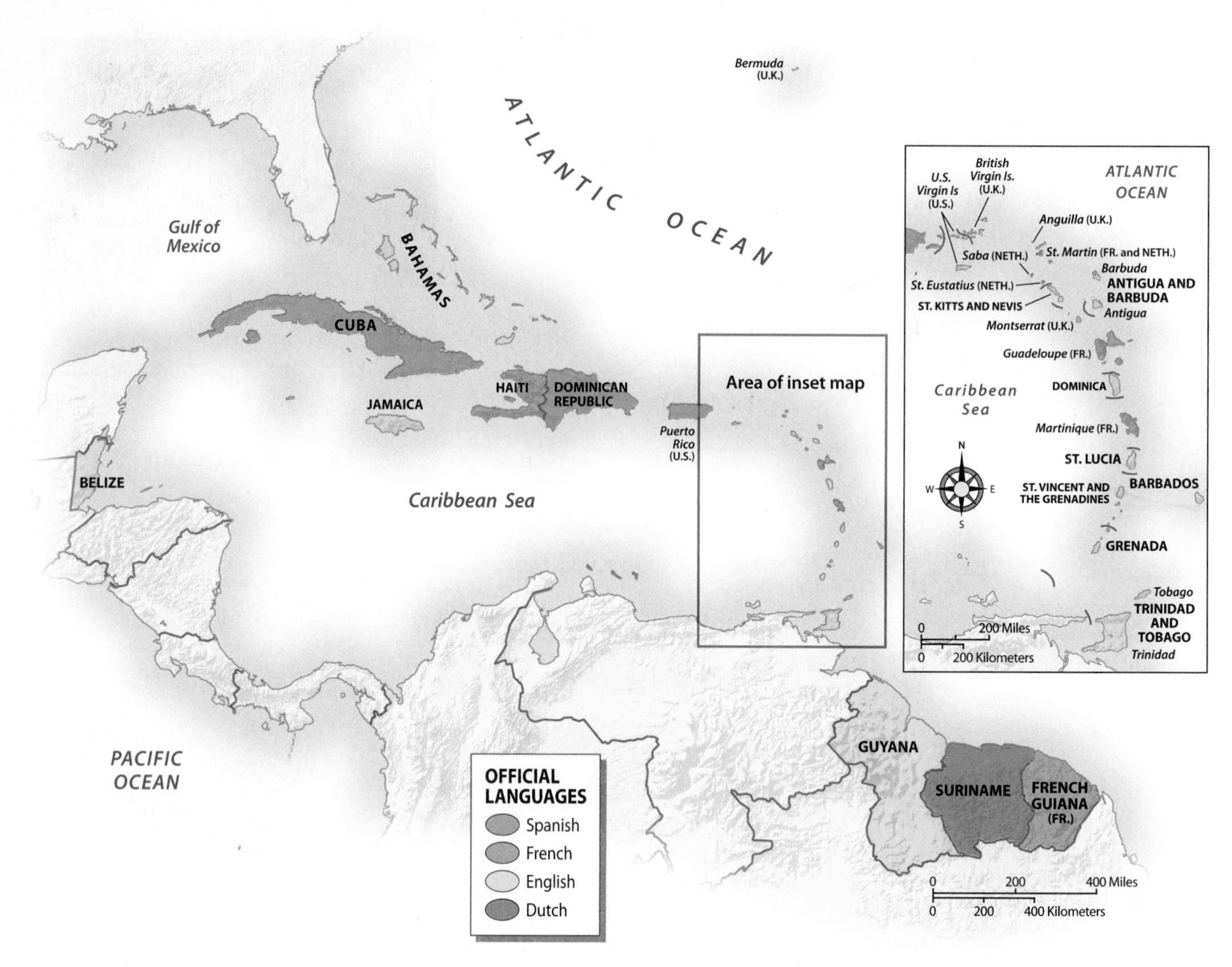

Figure 5.21 Caribbean Language Map Since this region has no significant Amerindian population (except on the mainland), the dominant languages are European: Spanish (25 million), French (10 million), English (6 million), and Dutch (0.5 million). However, many of these languages have been creolized, making it difficult for outsiders to understand them.

mantic structure had other origins, notably from the African language families. While most formal education is taught using standard language forms, the richness of vernacular expression and its ability to instill a sense of identity are appreciated. Locals rely on their ability to switch from standard to vernacular forms of speech. Thus, a Jamaican can converse with a tourist in standard English and then switch to a Creole variant when a friend walks by, effectively excluding the outsider from the conversation. While this ability to switch is evident in many cultures, it is widely used in the Caribbean.

Music The rhythmic beats of the Caribbean might be the region's best-known product. This small area is the hearth of reggae, calypso, merengue, rumba, zouk, and scores of other musical forms. The roots of modern Caribbean music reflect a combination of African rhythms with European forms of melody and verse. These diverse influences coupled with a long period of relative isolation sparked distinct local sounds. As circulation among Caribbean inhabitants increased, especially during the 20th century, musical traditions were grafted onto each other, but characteristic sounds remained (see "Global to Local: Caribbean Carnival Around the World").

The famed steel pan drums of Trinidad were created from oil drums discarded from a U.S. military base there in the 1940s. The bottoms of the cans are pounded with a sledge hammer to create a concave surface that produces different tones. During carnival, racks of steel pans are pushed through the streets by dancers while the panmen sound off. So skilled are these musicians that they even perform classical music, and government agencies encourage troubled teens to learn steel pan.

The eclectic sound and the ingenious rhythms make Caribbean music very popular. It is much more than good dancing music; the music is closely tied to Afro-Caribbean

GLOBAL TO LOCAL Caribbean Carnival Around the World

The origins of modern carnival combine Christian pre-Lenten festivities and African musical traditions. The term is Latin in origin, a reference to giving up meat eating (*carnivale* means to put meat away) in observance of Lent (the 40-day period of spiritual preparation prior to Easter). Throughout the Christian world, festive gatherings with ample food and drink are observed prior to Lent. In the context of the Caribbean and Brazil, former slaves imbued carnival with special meaning because it became their opportunity to break the monotony of their daily lives as well as provide a chance to mock the white elite while appearing to observe Christian traditions. In the Americas, carnival was always associated with African rhythms, dancing, and costuming. Afro-Trinidadians, for example, might dress as European royalty and mawkishly put on white face while dancing to rhythmic music. In the 19th century, local governments made several attempts to ban carnival, but the music and the celebration were hard to resist; over time carnival celebrations became part of the national identity for Brazil and Trinidad, and to a lesser extent for the Dominican Republic and Cuba. In each of these cases, the musical styles and method of celebration were quite different. Brazilians preferred samba, Trinidadians liked steel pan, and Dominicans moved to merengue (Figure 5.3.1).

Today, carnival is celebrated on nearly every island of the Caribbean as a national street party that can go on for weeks and attract thousands of tourists. Many official carnivals are no longer tied to Lent. In Barbados, carnival runs from late May to early August; in the Cayman Islands, it takes place in May; and in Cuba, it is in July. Stripped of its religious significance, carnival is now a celebration of life in the Caribbean, in which people express themselves through music, costumes, dance, beauty contests, and parades. Over time, the celebrations have become more and more alike, with popular musical styles and elaborate, revealing costumes.

Carnival celebrations have followed the Caribbean diaspora to new settings in North America and Europe. More than a dozen U.S. cities such as Atlanta, Baltimore, and Boston have carnivals that last a day or two in the summer. One of the biggest carnivals in North America is in Toronto, where a large Caribbean immigrant population maintains the tradition every July. In London, Birmingham, and Leicester, Caribbean carnivals are celebrated annually. The same is true for the Netherlands, where a large number of Surinamese have introduced carnival to Amsterdam and Rotterdam. Carnival has even caught on in places where there are relatively few Caribbean migrants. Helsinki, Finland, now has a summer carnival with samba bands, steel drums, and women parading around in sequined outfits (Figure 5.3.2). As a cultural measure of globalization, the diffusion of carnival shows that a great party easily translates into many cultural settings.

Figure 5.3.1 Carnival Drummer A steel pan drummer performs while his drum cart is pushed through the streets during carnival in Port of Spain, Trinidad.

Figure 5.3.2 Finnish Samba Band A sign of globalization, Finns celebrate carnival in Helsinki with their own samba band and carnival queen.

religions and is a popular form of political protest. In Haiti, rara music mixes percussion instruments, saxophones, and bamboo trumpets, while weaving in funk and reggae bass lines (Figure 5.22). The songs always are performed in French Creole and typically celebrate Haiti's African ancestry and the use of Voodoo. The lyrics address difficult issues, such as political oppression or poverty. Consequently, rara groups and other musicians have been banned from performing and even forced into exile—most notably, folk singer Manno Charlemagne, who later returned to Haiti and was elected mayor of Port-au-Prince in the 1990s.

The music of the late Bob Marley also takes a political stand. Jamaican-born Marley sang of his life in the Kingston ghetto of Trenchtown. He was a devout Rastafarian who be-

Figure 5.22 Haiti's Rara Music Performed in procession, rara music is sung in patois. It is considered the music of the poor and used to express risky social commentary. This rara band performs at a folk festival in Washington, DC.

lieved that Jah was the living force, that New World Africans should look to Africa for a prince to emerge (determined to be Haile Selassie of Ethiopia), and that *ganja* (marijuana) should be consumed regularly. It was Marley's political voice as peacemaker, however, that touched so many lives. His first hit, "Simmer Down," was written to quell street violence that had erupted in Kingston in 1964. Other songs, such as "Get Up, Stand Up" and "No Woman No Cry," had a message of social unity and freedom from oppression that resonated in the 1970s. Commercial success never dulled Marley's political edge. He was wildly popular in Africa, and before his death in 1981, one of his last concerts was in Zimbabwe to mark its independence.

From Baseball to Béisbol Latin Americans are known for their love of soccer, but baseball is the dominant sport for much of the Caribbean. A by-product of early U.S. influence in the region, baseball is the sport of choice in Cuba, Puerto Rico, and the Dominican Republic. Even in socialist Cuba, baseball is embraced with a fervor that would humble many U.S. fans. Yet it is the Dominican Republic that sends more players to the major leagues than any other country outside of the United States. In 2009 Dominican-born players made up 17 percent of the major league and minor league rosters.

The Dominican Republic became a talent pipeline for major league baseball due to complex mix of talent, economic inequality, and greed. This small country has produced many baseball legends and over the decades franchises have invested millions into training camps there. In the past two decades, however, Dominican pride in its baseball prowess has been tinged by the realities of a merciless feeder system that depends on impoverished kids, performance enhancing drugs, fake documents, and scouts who skim a percentage of the signing bonuses. Yet the reality is that more and more young boys, who can sign contracts at age 16, see their future in baseball rather than schooling. Even a modest signing bonus of $10,000 to $20,000 can build a nice home for a teen's family (Figure 5.23).

San Pedro de Macoris, not far from Santo Domingo, epitomizes this field of dreams. This humble sugarcane town has produced many baseball legends. This is a place of cane fields, kids on bicycles with bats and gloves, sugarcane factories, dusty baseball diamonds, and large homes of former players such as George Bell, Pedro Guerrero, and Sammy Sosa. These houses are silent testaments to what is possible with baseball. In an effort to clean up baseball's image, Major League Baseball has officials in the Dominican Republic investigating drug use and fraudulent papers. But as long as there are families pushing their teenage boys and a talent pool that delivers, this transnational system is self perpetuating.

Figure 5.23 Caribbean Baseball A young Cuban batter takes aim during a pick-up game in rural Cuba. Several Caribbean states have adopted baseball as their national sport, most notably the Dominican Republic and Cuba.

GEOPOLITICAL FRAMEWORK: Colonialism, Neocolonialism, and Independence

Caribbean colonial history is a patchwork of rival powers dueling over profitable tropical territories. By the 17th century, the Caribbean had become an important proving ground for European colonial ambitions. Spain's grip on the region was tentative, and rivals felt confident that they could win territory by gradually moving from the eastern edge of the sea to the west. Many territories, especially islands in the Lesser Antilles, changed hands several times (Figure 5.24). In a few instances, contested colonial holdings have produced contemporary border disputes. Only recently did the Guatemalan government give up its claim to Belize, arguing that the British illegally acquired it. Also, there are several long-standing border disputes among the Guianas. Yet there are indications of newfound regional cooperation, especially with the expansion of the Caribbean Community membership beyond the English-speaking countries.

Europeans viewed the Caribbean as a strategic and profitable region in which to produce sugar, rum, and spices in the 17th and 18th centuries. Geopolitically, rival European powers also felt that their presence in the Caribbean limited Spanish authority there. Yet Europe's geopolitical dominance in the Caribbean began to wane by the mid-19th century just as the U.S. presence increased. Inspired by the **Monroe Doctrine**, which claimed that the United States would not tolerate European military involvement in the Western Hemisphere, the U.S. government made it clear that it considered the Caribbean to be within its sphere of influence. Even though several English, Dutch, and French colonies persisted after this date, the United States indirectly (and sometimes directly) asserted its control over the region, ushering in a period of **neocolonialism**. In an increasingly global age, however, even neocolonial interest can be short-lived or sporadic. The Caribbean has not attracted the level of private foreign investment seen by other regions. Moreover, as the Caribbean's strategic importance in a post–Cold War era fades, the leaders of the region openly worry about their areas becoming more marginal.

Life in the "American Backyard"

To this day, the United States maintains a controlling attitude toward the Caribbean, which was commonly referred to as the "American backyard" in the early 20th century. The initial foreign policy objectives were to free the region from European authority and foster democratic governance. Yet time and again, American political and economic ambitions undermined those goals. President Theodore Roosevelt made his priorities clear with imperialistic policies that extended the influence of the United States beyond its borders. Policies and projects such as the construction of the Panama Canal and the maintenance of open sea-lanes benefited the United States but did not necessarily support social, economic, or political gains for the Caribbean people. The United States later offered benign-sounding development packages such as the Good Neighbor Policy (1930s), the Alliance for Progress (1960s), and the Caribbean Basin Initiative (1980s). The Caribbean view of such initiatives has been wary at best. Rather than feeling liberated, many residents believe that one kind of political dependence was being traded for another—colonialism for neocolonialism.

In the early 1900s, the role of the United States in the Caribbean was overtly military and political. The Spanish-American War (1898) secured Cuba's freedom from Spain and also resulted in Spain's ceding the Philippines, Puerto Rico, and Guam to the United States; the latter two are still U.S. territories. The U.S. government also purchased the Danish Virgin Islands in 1917, renaming them the U.S. Virgin Islands and developing the harbor of St. Thomas. French, English, and Dutch colonies were tolerated as long as these allies recognized the supremacy of the United States in the region. Avowedly against colonialism, the United States had become much like an imperial force.

One of the requirements of an empire is the ability to impose one's will, by force if necessary. When a Caribbean state refused to abide by U.S. trade rules, U.S. Navy vessels would block its ports. Marines landed and U.S.-backed governments were installed throughout the Caribbean basin. These were not short-term engagements; U.S. troops occupied the Dominican Republic from 1916 to 1924, Haiti from 1913 to 1934, and Cuba from 1906 to 1909 and 1917 to 1922 (Figure 5.25). Even today the United States maintains several important military bases in the region, including Guantánamo in eastern Cuba. There is greater reluctance to commit troops in the area now, but as recently as 1994 and 2004 U.S. troops were sent to Haiti to suppress political violence and prevent a mass exodus of Florida-bound refugees. And after the Haitian earthquake in 2010, U.S. naval vessels and troops were deployed to assist in the relief effort.

Many critics of U.S. policy in the Caribbean complain that business interests overwhelm democratic principles when foreign policy is determined. U.S. banana companies settled the coastal plain of the Caribbean rimland and operated as if they were independent states. Sugar and rum manufacturers from the United States bought the best lands in Cuba, Haiti, and Puerto Rico. Meanwhile, truly democratic institutions remained weak, and there was little improvement in social development. True, exports increased, railroads were built, and port facilities improved; but levels of income, education, and health remained abysmally low throughout the first half of the 20th century.

The Commonwealth of Puerto Rico Puerto Rico is both within the Caribbean and apart from it because of its status as a commonwealth of the United States. Throughout the 20th century, various Puerto Rican independence movements sought to uncouple the island from the United States. Even today, residents of the island are divided about their island's political future. At the same time, Puerto Rico depends on U.S. investment and welfare programs; U.S. food stamps are a major source of income for many Puerto Rican families. Commonwealth status also means that Puerto Ricans can freely move between the island and the U.S. mainland, a right they actively assert. In other ways, Puerto Ricans symbolically manifest their independence; for example, they support their own "national" sports teams and send a Miss Puerto Rico to international beauty pageants. The dispute over Vieques Island also has showcased a current of Puerto Rican independence (see Environment and Politics: Los Vieques: From Naval Testing Ground to Upscale Resort and Superfund Site).

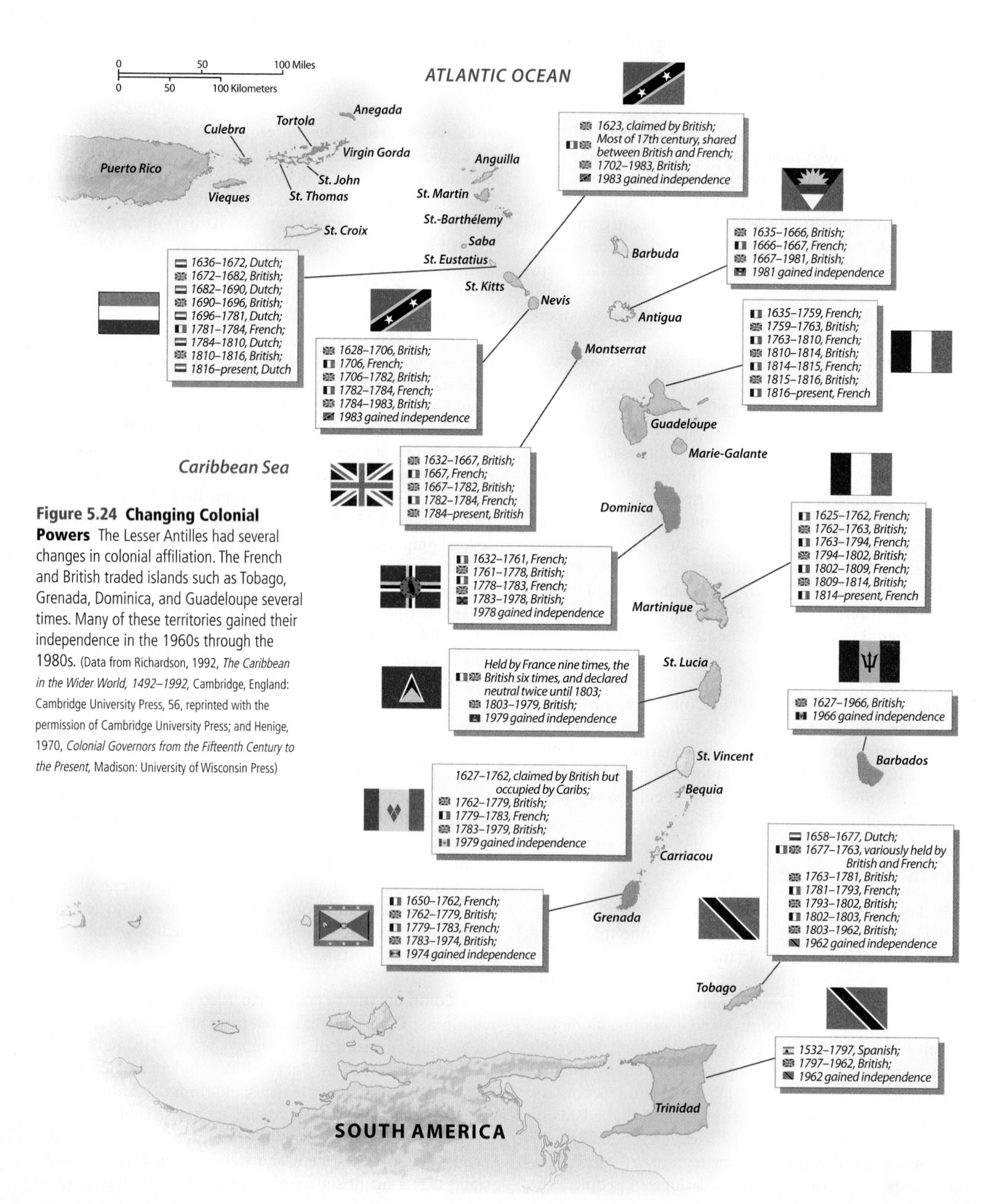

Figure 5.24 Changing Colonial Powers The Lesser Antilles had several changes in colonial affiliation. The French and British traded islands such as Tobago, Grenada, Dominica, and Guadeloupe several times. Many of these territories gained their independence in the 1960s through the 1980s. (Data from Richardson, 1992, *The Caribbean in the Wider World, 1492–1992,* Cambridge, England: Cambridge University Press, 56, reprinted with the permission of Cambridge University Press; and Henige, 1970, *Colonial Governors from the Fifteenth Century to the Present,* Madison: University of Wisconsin Press)

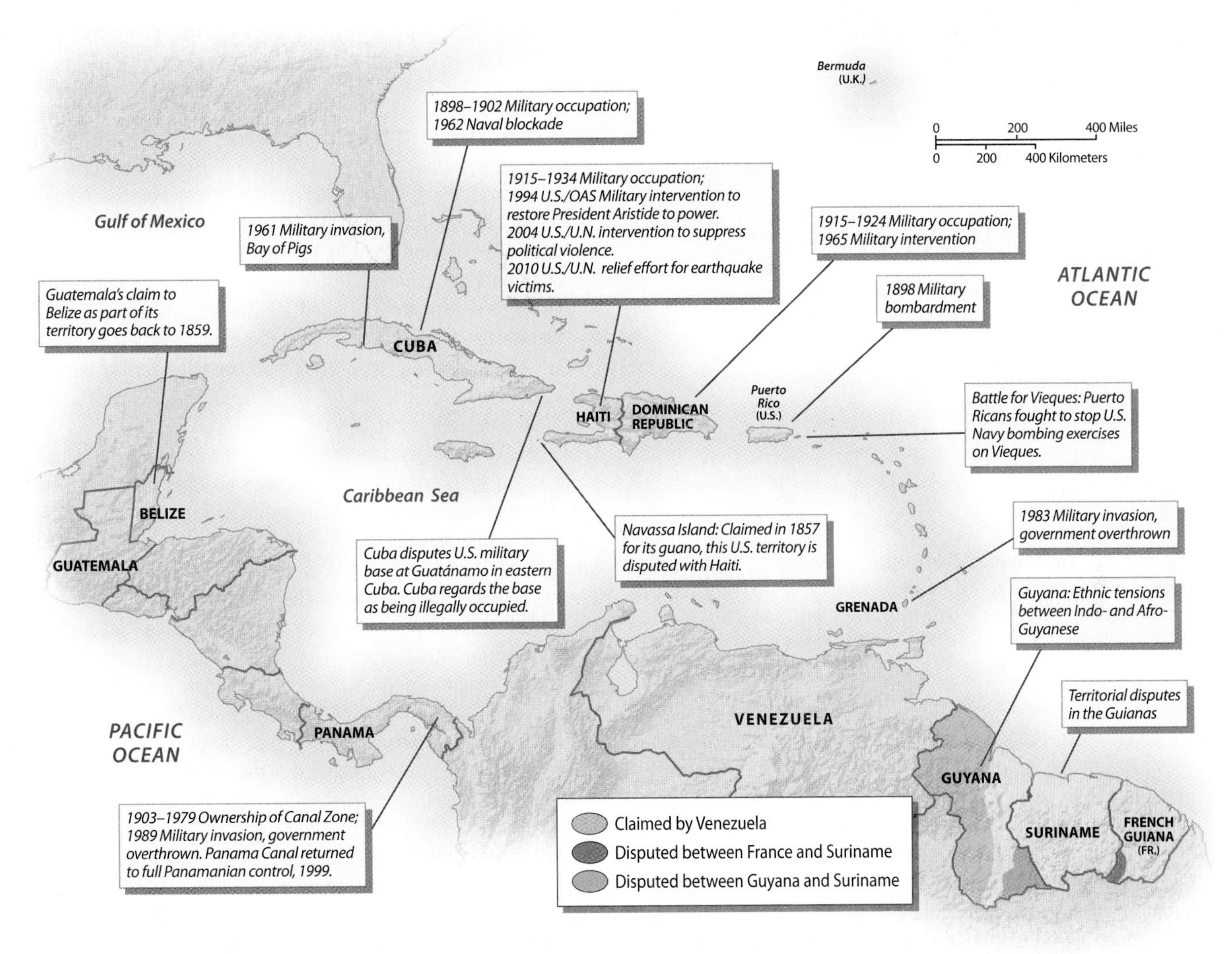

Figure 5.25 Caribbean Geopolitics: U.S. Military Involvement and Regional Disputes The Caribbean was labeled the geopolitical backyard of the United States, and U.S. military occupation was a common occurrence in the first half of the 20th century. Border and ethnic conflicts also exist, most notably in the Guianas. (Data from Tenenbaum, ed., 1996, *1996 Encyclopedia of Latin American History and Culture*, vol. 5, 296, with permission of Charles Scribner's Sons; and Allcock, 1992, *Border and Territorial Disputes*, 3rd ed., Harlow, Essex, UK: Longman Group)

Puerto Rico led the Caribbean in the transition from an agrarian economy to an industrial one beginning in the 1950s. For some U.S. officials Puerto Rico became the model for the rest of the region. Puerto Rican President Muñoz Marín championed an industrialization program called "Operation Bootstrap." Through tax incentives and cheap labor, hundreds of U.S. textile and apparel firms relocated to Puerto Rico. Over the next two decades, 140,000 industrial jobs were added, resulting in a marked increase in per capita GNI. In the 1970s, when Puerto Rico faced stiff competition from Asian apparel manufacturers, the government encouraged petrochemical and pharmaceutical plants to relocate to the island (Figure 5.26). By the 1990s, Puerto Rico was one of the most industrialized places in the region, with a significantly higher per capita income than its neighbors. Yet it still shows many signs of underdevelopment, including rampant out-migration, low rates of educational attainment, and widespread poverty and crime.

Cuba and Regional Politics The most profound challenge to U.S. authority in the region came from Cuba and its superpower ally, the former Soviet Union. In the 1950s, a revolutionary effort led by Fidel Castro began in Cuba against the pro-American Batista government. Cuba's economic productivity had soared but its people were still poor, uneducated, and increasingly angry. The contrast between the lives of average cane workers and the foreign elite was stark. Castro tapped a deep vein of Cuban resentment against six decades of American neocolonialism. In 1959, Castro took power.

ENVIRONMENT AND POLITICS Los Vieques: From Naval Testing Ground to Upscale Resort and Superfund Site

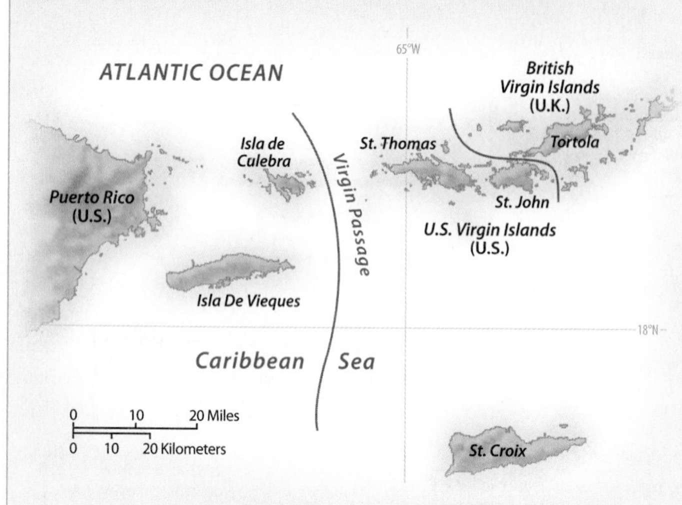

Figure 5.4.1 Los Vieques This relatively small island is about 20 miles long and 5 miles wide. For nearly six decades the U.S. Navy used the territory to conduct bombing exercises. Now small hotels are popping up as Los Vieques reinvents itself as a Carribean resort.

The small tropical island of Los Vieques, off Puerto Rico's eastern coast, was the center of a political and environmental controversy in the late 20th century. An island of marginally productive sugarcane estates, the U.S. Navy took over three-quarters of the island in the 1940s to conduct live bombing exercises. The local residents, mostly fishermen, had insisted that the U.S. Navy should go, but during World War II and the Cold War, the issue did not gain traction (Figure 5.4.1). Puerto Rican nationalists deeply resented hosting the largest training area for the United State Atlantic Fleet Forces on their territory and sought supporters at home and abroad to denounce it.

A turning point in the battle for Vieques came in 1999 when a resident, working as a security guard at the Atlantic Fleet Weapons Training Facility, was killed by a bomb dropped during target practice. This led then Puerto Rican gubernatorial candidate, Sila Calderon, to include an end to Navy-led live-fire exercises on Vieques in her 2000 platform. Her victory, added to years of protest that included Puerto Rican activists' placing themselves in the line of fire, led President George W. Bush to end all naval exercises in 2003 and close the facility. And so began the transformation of Los Vieques from Naval testing ground to upscale resort.

Instead of hearing ordnance explode in the distance, one is likely to hear the sound of construction crews building upscale resorts and restaurants. Since the military controlled most of the island, there was little development. The island is blessed with scores of white sandy beaches. Sea turtles nest there in the winter, wild horses roam the interior, and manatee swim in the shallow lagoons. An added attraction is a bay with bioluminescence (caused by micro-organisms) that glows neon-blue when the water is disturbed by a swimmer or a kayak in the evening. For nightlife, one of the former military bunkers at Green Beach has been transformed into a 10,000 sq ft. megadisco. In short, islanders and newcomers alike are poised to market Vieques as a new destination for tourists seeking a posh new setting to enjoy pristine Caribbean beaches and wildlife (Figure 5.4.2).

What does not make the tourist brochures is that the island was designated a superfund site in 2005 and the Navy is still looking for and exploding discarded ordnance in an effort to clean up the island. It turns out that turning a former naval testing ground into a tropical paradise is not so easy. Under the dense tropical vegetation lie unexploded cluster bombs—which first have to be found, then detonated. A typical cleanup at former military training ranges is on-site detonations, as long as limited amounts are exploded at a time. The islanders complain that detonation produces smoke and contaminants, leading to respiratory problems.

In 2009 the federal agency that assesses health hazards at superfund sites reversed its 2003 conclusion that contamination at Vieques posed no health risks to residents. The EPA is now investigating claims against the U.S. government for contamination and illness, and insists that more monitoring is required to determine health risks. For many of the 9,000 residents of Vieques, the long resistance to the Navy's operations is over but the fight to clean up their island continues.

Figure 5.4.2 The harbor of Isabel, Los Vieques The small settlement of Isabel, on the north coast of Los Vieques, serves the growing tourist population. Formally a fishing community, tourists come to enjoy the island's relaxed atomsphere.

Source: Adapted from Mireya Navarro, "New Battle on Vieques, Over Navy's Cleanup of Munitions," *New York Times*, August 7, 2009.

After Castro's government nationalized American industries and took ownership of all foreign-owned properties, the United States responded by refusing to buy Cuban sugar and ultimately ending diplomatic relations with the state. Various U.S. trade embargoes against Cuba have existed for nearly five decades. What sealed Cuba's fate as a geopolitical enemy was its establishment of diplomatic relations with the USSR in 1960 during the height of the Cold War. With the Soviet Union financially and militarily backing Castro, a direct U.S. invasion of Cuba was too risky. The fall of 1962 produced one of the most dangerous episodes of the Cold War when Soviet missiles were discovered on Cuban soil. Ultimately, the Soviet Union removed its weapons; in return, the United States promised not to invade Cuba.

Castro wanted to extend his socialist vision abroad, supporting at different times revolutionaries in Colombia, Venezuela, Guyana, Suriname, Bolivia, and across the Atlantic in Angola. Other Caribbean states watched the Cuba socialist experiment with interest, especially its remarkable strides in literacy and public health. By the mid-1970s, Cuba had the lowest infant mortality rate in all of the Caribbean and Latin America. Cuban-style socialism did not readily transfer to other countries, in part because the United States was determined to prevent it, and Soviet support of Cuba proved too expensive. Even with the end of the Cold War, when Cuba lost its financial support from the Soviet Union, it managed to reinvent itself by growing its tourism sector and courting foreign investment, especially from Spain. More recently, Castro and Venezuelan President Hugo Chavez have become close political allies, signing an important trade agreement in 2004 to exchange medical resources (from Cuba) and petroleum (from Venezuela). Still today, the United States maintains its tough trade sanctions against Cuba and forbids U.S. tourists from visiting the island.

A new political era for Cuba appears imminent. In 2008 Fidel Castro, 82 and in poor health, left office and his younger brother, Raúl Castro, assumed the duties of president. Although many consider Raúl more willing to encourage private enterprise in Cuba, as of 2010 no major political and economic changes have occurred (see Exploring Global Connections: Cuba's Medical Diplomacy).

Independence and Integration

Given the repressive colonial history of the Caribbean, it is no wonder that the struggle for political independence began more than 200 years ago. Haiti was the second colony in the Americas to gain independence, in 1804; the United States was the first, in 1776. However, the political independence of many states in the region has not guaranteed economic independence. Many Caribbean states struggle to meet the basic needs of their people. Surprisingly, today some Caribbean territories maintain their colonial status as an economic asset. For example, the French territories of Martinique, Guadeloupe, and French Guiana are overseas departments of France; residents have full French citizenship and social welfare benefits.

Figure 5.26 Female Workers in a Puerto Rican Pharmaceutical Plant Since the 1980s, U.S. pharmaceutical companies have located plants in Puerto Rico due to lower costs and tax advantages offered by the U.S. government.

Independence Movements Haiti's revolutionary war began in 1791 and ended in 1804. Spanish, French, and British forces were involved, as well as factions within Haiti that had formed along racial lines. During this conflict the island's population was cut in half by casualties and emigration; ultimately, the former slaves became the rulers. Independence, however, did not allow this crown of the French Caribbean to prosper. "Plantation America" watched in horror as Haitian slaves used guerrilla tactics to gain their freedom. Fearing that other colonies might follow Haiti's lead, plantation owners in other countries were on guard for the slightest hint of revolt. For its part, Haiti did not become a leader in liberation. Slowed by economic and political problems, it was shunned by the European powers and never embraced by the states of the Spanish mainland when they became independent in the 1820s.

Several revolutionary periods followed in the 19th century. In the Greater Antilles, the Dominican Republic finally gained independence in 1844 after wresting control of the territory from Spain and Haiti. Cuba and Puerto Rico were freed from Spanish colonialism in 1898, but their independence was compromised by greater U.S. involvement. The British colonies also faced revolts, especially in the 1930s, yet it was not until the 1960s that independent states emerged from the English Caribbean. First the larger colonies of Jamaica, Trinidad and Tobago, Guyana, and Barbados gained their independence. Other British colonies followed throughout the 1970s and early 1980s. Suriname, the only Dutch colony on the rimland, became an autonomous territory in 1954 but remained part of the Kingdom of the Netherlands until 1975, when it declared itself an independent republic.

Present-day Colonies Britain still maintains several crown colonies in the region: the Cayman Islands, the Turks and Caicos, Anguilla, Montserrat, and Bermuda. The combined population of these islands is about 120,000 people, yet their standard of living is high, due in part to their specialization in the recently developed industry of offshore financial services. French Guiana, Martinique, and Guadeloupe are each departments of France and thus, technically speaking, not colonies. Together they total 1 million people. The Dutch islands in the Caribbean are considered autonomous countries that are part of the Kingdom of the Netherlands. Curaçao, Bonaire, St. Martin, Saba, and St. Eustatius make up the federation of the Netherlands Antilles. Aruba left the federation in 1986 and governs without its influence. Together, the population of the Dutch islands is a quarter million people.

Regional Integration Perhaps the most difficult task facing the Caribbean is to increase economic integration. Scattered islands, a divided rimland, different languages, and limited economic resources hinder the formation of a meaningful regional trade bloc. It is more common to see economic cooperation between groups of islands with a shared colonial background than between, for example, former French and English colonies.

During the 1960s, the Caribbean began to experiment with regional trade associations as a means to improve its economic competitiveness. The goal of regional cooperation was to improve employment rates, increase intraregional trade, and ultimately reduce economic dependence. The countries of the English Caribbean took the lead in this development strategy. In 1963, Guyana proposed an economic integration plan with Barbados and Antigua. In 1972, the integration process intensified with the formation of the **Caribbean Community and Common Market (CARICOM)**. Representing the former English colonies, CARICOM proposed an ambitious regional industrialization plan and the creation of the Caribbean Development Bank to assist the poorer states. CARICOM also oversees the University of the West Indies, with campuses in Trinidad, Jamaica, and Barbados. As important as this trade group is as an institutional symbol of collective identity, it has produced limited improvements in intraregional trade.

Today there are 15 full-member states in CARICOM—all of the English Caribbean, French-speaking Haiti, and Dutch-speaking Suriname. Other dependencies, such as Anguilla, Bermuda, Turks and Caicos, Cayman Islands, and the British Virgin Islands, are associate members. As CARICOM's membership grows, so does its services. In 2005 CARICOM passports began to be issued to facilitate travel between member nations. And, during the 2007 Cricket World Cup, a CARICOM Visa was issued to facilitate travel between the nine host countries for this global sporting event.

The dream of regional integration as a way to produce a more stable and self-sufficient Caribbean has yet to be realized. One scholar of the region argues that a limiting factor to this regional integration is a "small-islandist ideology." For example, islanders tend to keep their backs to the sea, oblivious to the needs of neighbors. At times, such isolationism results in suspicion, distrust, and even hostility toward nearby states. Yet economic necessity dictates engagement with partners outside the region. And so this peculiar status of isolated proximity unfolds in the Caribbean, expressing itself in uneven social and economic development trends.

ECONOMIC AND SOCIAL DEVELOPMENT: From Cane Fields to Cruise Ships

Collectively, the population of the Caribbean, although poor by U.S. standards, is economically better off than most of Sub-Saharan Africa, South Asia, and China. Most Caribbean countries fall in the lower middle-income range set by the World Bank. And with the exception of Haiti where the majority of the population is living on less than $2 per day, extreme poverty is relatively rare. Despite periods of economic stagnation in the Caribbean, social gains in education, health, and life expectancy are significant (Table 5.2). Historically, the Caribbean's links to the world economy were

EXPLORING GLOBAL CONNECTIONS Cuba's Medical Diplomacy

Cuba is widely recognized as a poor country with good health care. Yet is also has a plan for expanding health-care access to the developing world. Today, 30,000 Cuban doctors and health-care professionals work overseas in some 40 countries among the rural and urban poor. There are hundreds of Cuban doctors in South Africa, Gambia, and Ghana. There are thousands of doctors in Venezuela, El Salvador, Guatemala, Nicaragua, Bolivia, and Haiti. After Hurricane Mitch devastated Central America in 1998, Cuba sent medical personnel to Honduras and El Salvador. In 2005, over 2,000 Cuban doctors traveled to northern Pakistan to attend earthquake victims. When Hurricane Katrina battered New Orleans that same year, some 1,500 medical personnel from Cuba offered to help, but the U.S. declined Cuba's support (Figure 5.5.1).

Cuba's health-care successes have been the pride of the revolution. In the past decade, Cuba has expanded its medical diplomacy as a means to barter for oil with Venezuela but also as an attempt to showcase something it does well. "Operation Miracle" is an outreach program that relies upon Cuba's human capital and Venezuela's petrodollars to provide free eye surgery. This joint venture began in 2004 and has performed over 750,000 surgeries on people who could not afford it in their own countries. In Venezuela, President Chavez also began the "Barrio Aldentro" program to have Cuban health-care professionals live and work in Venezuela's shantytowns. Venezuelan doctors have protested this invasion of Cuban doctors but at the same time the Cubans work in places that are typically not served.

Perhaps the most lasting impact of Cuba's medical diplomacy is the creation of ELAM—the Latin American School of Medicine. ELAM students come from scores of countries and thousands have been trained. The program selects ethnically diverse students from underserved areas and lower income groups. The students do not have to pay but they are asked to go back to their countries and work in needy areas. As an indication of the school's reach, there are scholarships available for students from East Timor and Pakistan. The majority of students are from the Americas, including students from the United State and Puerto Rico. After completing 6 years of demanding training, in Spanish and English, in rural and urban areas, with limited resources, the doctors are sent back out to the world. And, in an ironic twist, the U.S. government recently recognized medical degrees from Cuba.

Source: Adapted from Robert Huish and John M. Kirk, 2007, "Cuban Medical Internationalism and the Development of the Latin American School of Medicine," *Latin American Perspectives* 34(6): 77–92, and Sarah Blue, 2010, "Cuban Medical Internationalism: Domestic and International Impacts," *Journal of Latin American Geography* 9 (1): 31–49.

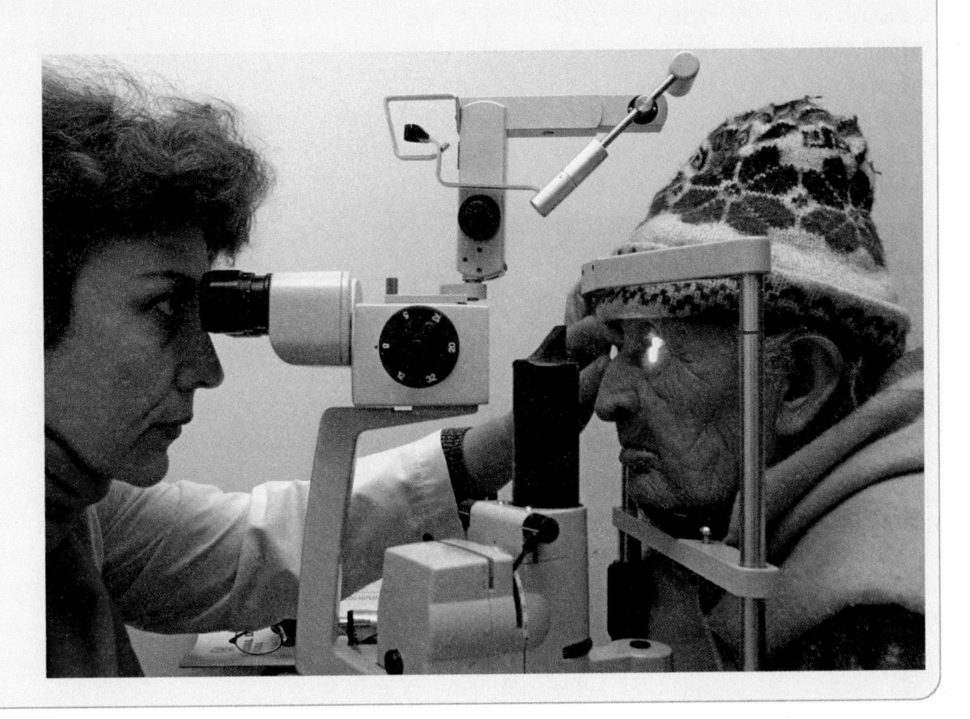

Figure 5.5.1 Cuban Doctor Working in Bolivia A Cuban doctor examines the eyes of a Bolivian woman in a Bolivian public hospital as part of the Operation Miracle outreach program. Many Bolivian doctors, however, protested this form of Cuban assistance, claiming that it hurt their business. Meanwhile, Bolivia's urban poor eagerly took advantage of this free medical service.

tropical agricultural exports, yet several specialized industries, such as tourism, offshore financial services, and assembly plants, have challenged the dominance of agriculture. These industries grew because of the region's proximity to North America and Europe, the availability of cheap labor, and the presence of policies that created a nearly tax-free environment for foreign-owned companies. Unfortunately, growth in these sectors does not employ all the region's displaced rural workers, so the lure of jobs in North America and Europe is still strong.

From Fields to Factories and Resorts

Agriculture used to dominate the economic life of the Caribbean. Decades of turbulent commodity prices and decline in preferential trade agreements with former colonial

TABLE 5.2 Development Indicators

Country	GNI per capita, PPP 2008	GDP Average Annual % Growth 2000–08	Human Development Index (2007)[a]	Percent Population Living Below $2 a Day	Life Expectancy (2010)[b]	Under Age 5 Mortality Rate (1990)	Under Age 5 Mortality Rate (2008)	Gender Equity (2008)[c]
Anguilla	12,200				81			
Antigua and Barbuda	19,650		.868		75			95
Bahamas	31,100		.856		74	29	13	101
Barbados	19,200		.903		74	17	12	101
Belize	5,940		.772		73	43	25	102
Bermuda	69,900				81			
Cayman	43,800				81			90
Cuba			.863		78	14	6	99
Dominica	8,290		.814		75	18	11	101
Dominican Republic	7,800	5.4	.777	12.3	72	62	33	103
French Guiana					78			
Grenada	8,430		.813		70	37	19	99
Guadeloupe					80			
Guyana	3,020		.729	16.8	66	88	60	96
Haiti	1,300	0.5	.532	72.1	61	151	72	
Jamaica	7,360	1.8	.766	5.8	72	33	31	100
Martinique					80			
Montserrat	3,400				73			
Netherlands Antilles	16,000				76			
Puerto Rico	17,900				79			
St. Kitts and Nevis	15,480		.838		74	36	18	97
St. Lucia	9,020		.821	41	73	21	18	103
St. Vincent and the Grenadines	8,560		.772		72	22	19	101
Suriname	6,680		.769	27	69	51	29	114
Trinidad and Tobago	24,230	8.4	.837	13.5	69	34	35	101
Turks and Caicos	11,500				76			

[a] *United Nations, Human Development Report, 2009.*

[b] *Population Reference Bureau, World Population Data Sheet, 2010.*

[c] *Gender Equity—Ratio of female to male enrollments in primary and secondary school. Numbers below 100 have more males in primary/secondary school, numbers above 100 have more females in primary/secondary schools.*

Additional data from the CIA World Factbook, 2010.

Source: World Bank, World Development Indicators, 2010.

states have produced more hardship than prosperity. Ecologically, the soils are overworked, and there are no frontiers in which to expand production, except for areas of the rimland. Moreover, agricultural prices have not kept pace with rising production costs, so wages and profits remain low. With the exception of a few mineral-rich territories, such as Trinidad, Guyana, Suriname, and Jamaica, most countries have systematically tried to diversify their economies, relying less on their soils and more on manufacturing and services.

Comparing export figures over time demonstrates the shift away from monocrop dependence. In 1955, Haiti earned more than 70 percent of its foreign exchange through the export of coffee; by 1990, coffee accounted for only 11 percent of its export earnings. Similarly, in 1955 the Dominican Republic earned close to 60 percent of its foreign exchange through sugar, but 35 years later sugar earned less than 20 percent of the country's foreign exchange, and pig iron exports surpassed that of sugar.

Sugar and Coffee The economic history of the Caribbean cannot be separated from the production of sugarcane. Even relatively small territories such as Antigua and Barbados yielded fabulous profits because there was no limit to the demand for sugar in the 18th century. Once considered a luxury crop, it became a popular necessity for European and North American laborers by the 1750s. It sweetened tea and coffee and made jams a popular spread for stale bread. In short, it made the meager and bland diets of ordinary people tolerable, and it also boosted caloric intake. Distilled into rum, sugar produced a popular intoxicant. Though it is hard to imagine today, individual consumption of a pint of rum a day was not uncommon in the 1800s.

Sugarcane is still grown throughout the region for domestic consumption and export. Its economic importance has declined, however, mostly because of increased competition from corn and sugar beets grown in the midlatitudes. The Caribbean and Brazil are the world's major sugar exporters. Until 1990, Cuba alone accounted for more than 60 percent of the value of world sugar exports. The country earned 80 percent of its foreign exchange through sugar production. Cuba's dominance in sugar exports had more to do with its subsidized and guaranteed markets in eastern Europe and the Soviet Union than with exceptional productivity. Since 1990, the value of the Cuban sugar harvest has plummeted.

Coffee is planted in the mountains of the Greater Antilles. Haiti has been the most dependent on coffee, relying on peasant sharecroppers to tend the plants and harvest the beans. For other countries, coffee is a valued specialty commodity. Beans harvested in the Blue Mountains of Jamaica, for example, fetch two to three times the going price for similar highland coffee grown in Colombia. Puerto Rico and Cuba also are trying to develop a niche in the gourmet coffee market. An important production distinction with coffee, in contrast to sugar, is that it is mostly grown on small farms and sold to buyers or delivered to cooperatives. Typically, farmers plant other crops between the coffee bushes so that they can meet their subsistence needs as well as produce a cash crop. Even with this self-provisioning system, peasants often seasonally abandon their farms for work elsewhere as laborers. The instability of coffee prices, which were especially low in 2000 through 2004, makes the economics of growing coffee unpredictable.

The Banana Wars The major banana exporters are in Latin America, not the Caribbean. In fact, the success of banana plantations is mixed in this region, as banana plants are especially vulnerable to hurricanes. Still, several small states in the Lesser Antilles, most notably Dominica, St. Vincent, and St. Lucia, have become dependent on bananas, earning as much as 60 percent of their export earnings from the yellow fruit. Bananas have not made people rich, yet their production for export has fostered greater economic and social development. In the eastern Caribbean, where most bananas are grown on small farms of five acres, the landowners are the laborers and thus earn two to four times more than banana plantation workers in Ecuador and Central America. Moreover, for these small states, banana exports are their link with the global economy. Yet with legal pressure on the European Union (EU) to stop the guaranteed prices paid to banana growers from the former colonies, the economic viability of this crop in the Lesser Antilles is in doubt (Figure 5.27). Where once a market and a minimum price were guaranteed, under the new global order neither is certain.

The case of the eastern Caribbean underscores what happens to the losers in globalization. In 1996, the United States, Ecuador (the world's leading banana exporter), Mexico, Guatemala, and Honduras challenged the EU's banana

Figure 5.27 Caribbean Bananas A laborer in St. Lucia harvests bananas on a small farm. Farmers in the Eastern Caribbean have long depended on preferential markets in Europe to sell their bananas. A recent trade decision by the World Trade Organization court opens up Caribbean banana farmers to new competition, jeopardizing their economic viability.

trade agreement in the World Trade Organization (WTO) court. The agreement was denounced as unfair, and the EU was told to eliminate it by 1998. To make matters worse, consumer tastes had changed, with buyers preferring the uniformly large and unblemished yellow banana typical of the Latin American plantations rather than those grown in places like St. Lucia. To survive in the newly competitive market, the growers of the eastern Caribbean will have to produce a more standardized fruit and increase their yield per acre. While island governments are trying to aid in this transition, local farmers have experimented with new crops, such as okra, tomatoes, avocados, and marijuana. A few well-tended marijuana plants will earn 30 times more per pound than bananas. Just what will happen to family-run banana farms is hard to tell, but many Caribbean growers fear that the days of the banana economy are numbered. The banana story also reflects another major shift in the region: the decline in the economic importance of agriculture as well as the number of people employed by it.

Assembly-plant Industrialization One regional strategy to deal with the unemployed was to invite foreign investors to set up assembly plants and thus create jobs. This strategy first succeeded in Puerto Rico in the 1950s and was copied throughout the region. In Puerto Rico's "Operation Bootstrap," island leaders encouraged U.S. investment by offering cheap labor, local tax breaks, and, most importantly, federal tax exemptions (something only Puerto Rico can do because of its special status as a commonwealth of the United States). Initially the program was a tremendous success, and by 1970 nearly 40 percent of the island's gross domestic product (GDP) came from manufacturing. Today, 25 percent of male labor force and 11 percent of the female labor force in Puerto Rico are employed in industry, and this sector accounts for nearly half of the island's GDP. Yet competition from other states with even lower wages and the U.S. Congress's decision in 1996 to phase out many of the tax exemptions may threaten Puerto Rico's ability to maintain its specialized industrial base.

Through the creation of **free trade zones (FTZs)**, duty free and tax-exempt industrial parks for foreign corporations, the Caribbean is an increasingly attractive location to assemble goods for North American consumers. The Dominican Republic developed its Free Trade Zones by taking advantage of tax incentives and guaranteed access to the U.S. market first offered through the Caribbean Basin Initiative. The number of operational free trade zones in the Dominican Republic in 2001 was 16, including an FTZ on the island's north shore, near the Haitian border, optimistically named "Hong Kong of the Caribbean." More have been added since then. Firms from the United States and Canada are the most frequent investors in these zones, followed by Dominican, South Korean, and Taiwanese firms. In the Dominican Republic FTZs produce garments and textiles for export (Figure 5.28). In 2006 output from these factories accounted for 70 percent of all exports. Recently, these manufacturing centers have been hit hard by competition from Central America and China, as well as the global recession, forcing layoffs in 2009.

The growth in manufacturing depended on national and international policies that support export-led development through foreign investment. Certainly, new jobs are being created and national economies are diversifying in the process, but critics believe that foreign investors gain more than the host countries. Because most goods are assembled from imported materials, there is little development of national suppliers. Often higher than local averages, wages are still miserable compared with those in the developed world—sometimes just $3 to $5 dollars a day. Moreover, as other developing countries compete with the Caribbean for the establishment of FTZs, this strategy may become less significant over time.

Figure 5.28 Free Trade Zones in the Dominican Republic Another sign of globalization is the proliferation of duty-free and tax exempt industrial parks in the Caribbean. Currently 16 FTZs are operational in the Dominican Republic, with foreign investors from the United States, Canada, South Korea, and Taiwan. (Modified from Warf, 1995, "Information Services in the Dominican Republic," *Yearbook, Conference of Latin American Geographers*, 21, 15)

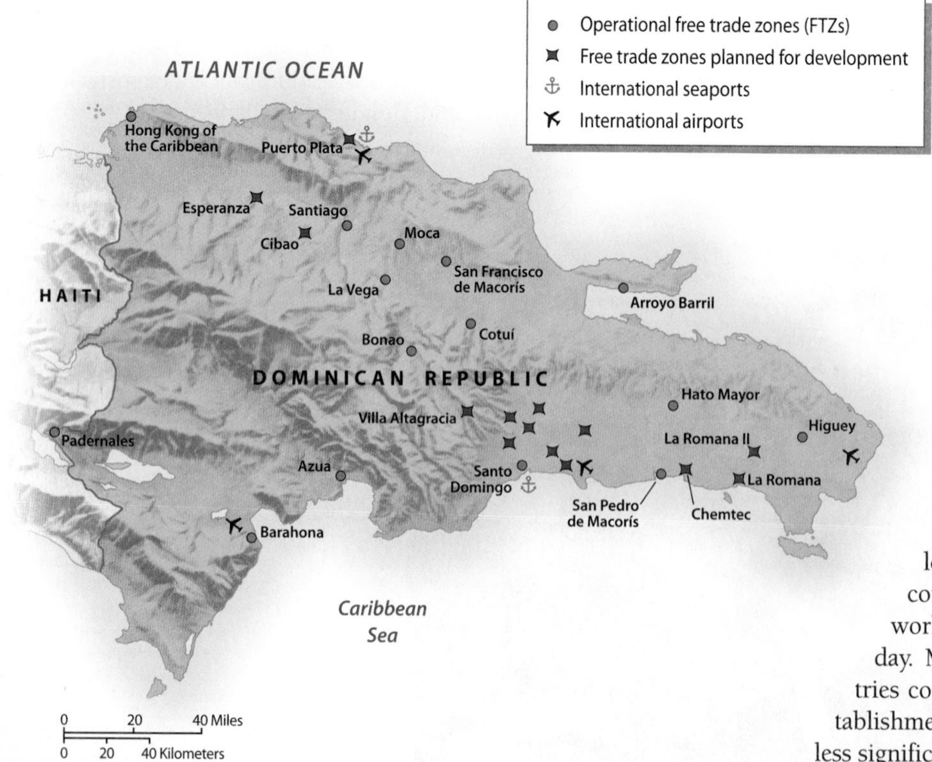

Offshore Banking and Online Gambling The rise of offshore banking in the Caribbean is most closely associated with the Bahamas, which began this industry back in the 1920s. **Offshore banking** centers appeal to foreign banks and corporations by offering specialized services that are confidential and tax exempt. Places that provide offshore banking make money through registration fees, not taxes. The Bahamas were so successful in developing this sector that by 1976 the country was the third largest banking center in the world. Its dominance began to decline because of competitors from the Caribbean, Hong Kong, and Singapore and because there was no longer a tax advantage in booking large international loans offshore. Concerns about corruption and laundering of drug money also hurt the islands' financial status in the 1980s, and major reforms were introduced to reduce the presence of funds gained from illegal activities. By 1998, the Bahamas' ranking among global financial centers had dropped to 15th. Still, offshore banking remains an important part of the Bahamian economy. In the 1990s, the Cayman Islands emerged as the region's leader in financial services. With a population of 45,000, this crown colony of Britain has some 50,000 registered companies and a per capita PPP of $43,800. In 2007, it was estimated that the Caymans were the fifth largest banking center in the world after New York, London, Hong Kong, and Tokyo.

Each of the offshore banking centers in the Caribbean tries to develop special financial services to attract clients, such as banking, functional operations, insurance, or trusts. Bermuda, for example, is the global leader in the reinsurance business that makes money from underwriting part of the risk of other insurance companies (Figure 5.29). The Caribbean is an attractive location for such services because of its closeness to the United States (home of many of their registered firms), client demand for these services in different countries, and the steady improvement in telecommunications that make this industry possible. The resource-poor islands of the region see providing financial services as a way to bring foreign capital to state treasuries and to find a way to be part of economic globalization.

As offshore financial services were expanding in the 1990s in the Caribbean, Southeast Asia, the Pacific, and Europe (Liechtenstein and Jersey), international efforts sought to curb money laundering by threatening the suspension of privacy provisions whenever criminal activities were suspected. After the terrorist attacks on the United States in September 2001, interest in promoting know-your-customer laws as a means of tracking offshore assets of terrorist organizations was renewed. The upshot has been a more stringent regulatory environment that is making offshore banking less attractive to newcomers. Grenada recently announced that it is no longer a site for offshore banking. The major banking centers, such as the Cayman Islands and the Bahamas, are likely to endure while smaller operations such as those on St. Kitts or Barbados will have a more difficult time surviving.

Online gambling is the newest industry for the microstates of the Caribbean. Antigua and St. Kitts were the leaders of the region, beginning legal online gambling services

Figure 5.29 **Financial Services in Bermuda** Front Street in Hamilton, Bermuda, is a reflection of the territory's ties to the United Kingdom and its prosperity. Tourism and financial services in the reinsurance business explain Bermuda's high standard of living.

in 1999. Other states soon followed; as of 2003 Dominica, Grenada, Belize, and the Cayman Islands had gambling domain sites. Although the online gambling business is illegal in the United States, nothing can stop Americans from betting in cyberspace. In only a decade, online gambling services have taken root throughout the Caribbean. The countries of the region have offered casino-based resort gambling for decades.

In 2007, the WTO deemed restrictions imposed on overseas Internet gambling sites by the United States as illegal. The tiny nation of Antigua is currently seeking $3 billion in compensation from the United States in lost revenue due to illegal restrictions placed on Antigua's business. Meanwhile, sensing a lucrative business opportunity, lobby efforts are underway to legalize Internet gambling in the United States, which would have repercussions on the Internet gambling business in the Caribbean.

Tourism Environmental, locational, and economic factors converge to support tourism in the Caribbean. The earliest visitors to this tropical sea admired its clear and sparkling turquoise waters. By the 19th century, wealthy North Americans were fleeing winter to enjoy the healing warmth of the Caribbean during its dry season. Developers later realized that the simultaneous occurrence of the Caribbean dry season with the Northern Hemisphere winter was ideal for beach resorts. By the 20th century, tourism was well established, with both destination resorts and cruise lines. By the 1950s, the leader in tourism was Cuba, and the Bahamas was a distant second. Castro's rise to power, however, eliminated this sector of the island's economy for nearly three decades and opened the door for other islands to develop tourism economies.

Six countries or territories hosted two-thirds of the 21 million international tourists who came to the Caribbean in 2007: Puerto Rico, the Dominican Republic, Cuba, the Bahamas, Jamaica and the British Virgin Islands (Figure 5.30). Puerto Rico saw its tourist sector begin to grow with commonwealth status in 1952. San Juan is now the largest home port for cruise lines and the second largest cruise-ship port in the world in terms of total visitors. The cruise business, in combination with stay-over visitors, accounted for almost 3.7 million arrivals in 2007. The Bahamas attributes most of its economic development and high per capita income to tourism. With more than 1.5 million stay-over visitors in 2007, and almost as many cruise-ship passengers, the Bahamas is another major hub for tourism in the region. Some 30 percent of the Bahamian population is employed in tourism, and tourism represents nearly half the country's GDP (Figure 5.31).

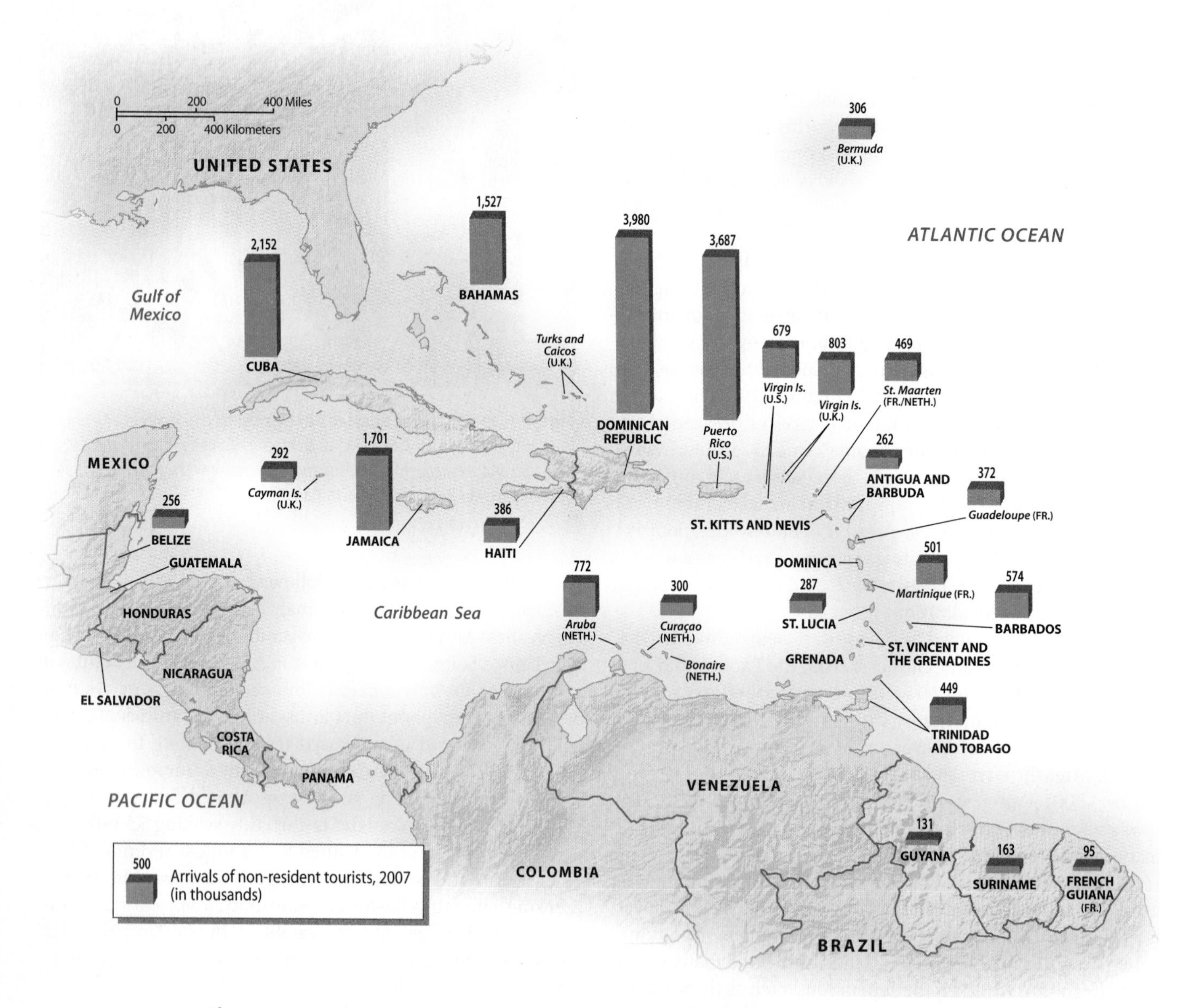

Figure 5.30 Global Linkages: International Tourism in the Caribbean Tourism directly links the Caribbean to the global economy. Each year more than 20 million tourists come to the islands, mostly from North America, Latin America, and Europe. The most popular destinations are the Dominican Republic, Puerto Rico, Cuba, the Bahamas, and Jamaica. (Source: *Yearbook of Tourism Statistics*, 2009 edition)

Figure 5.31 Caribbean Cruise Ship A ship from the Carnival Cruise Lines, a British-American owned company, anchors at Grand Turk Island in the Turks and Caicos. Tourism is vital to many Caribbean states, but most cruise ships are owned by companies outside the region and offer relatively little direct employment for Caribbean workers.

The Dominican Republic is the region's largest tourist destination, receiving 4 million visitors in 2007, many of them Dominican nationals who live overseas. Since 1980, tourist receipts have increased 20-fold, making tourism the leading foreign-exchange earner at more than $3.0 billion. Jamaica has become similarly dependent on tourism for hard currency, earning more than $1.5 billion in tourist receipts in 2005. The vast majority of tourists to Jamaica are from the United States and the United Kingdom. While drawing 1.7 million stayovers in 2007, Jamaica was a port-of-call for another half-million cruise ship passengers (see Figure 5.30).

After years of neglect, tourism is being revived in Cuba in an attempt to earn badly needed hard currency. Tourism represented less than 1 percent of the national economy in the early 1980s. By 2007, over 2 million tourists (mostly Canadians and Europeans) yielded gross receipts approaching $2 billion. Conspicuous in their absence are travelers from the United States, forbidden to travel to Cuba because of the U.S.-imposed sanctions. With U.S. investors out of the picture, Spanish and other European investors are busy building up Cuba's tourist capacity, anticipating that some day the U.S. ban will be lifted.

As important as tourism is for the larger islands, it is often the principal source of income for smaller ones. The Virgin Islands, Barbados, Turks and Caicos, and, recently, Belize all greatly depend on international tourists. To show how quickly this sector can grow, consider this example: Belize began promoting tourism in the early 1980s, when just 30,000 arrivals came each year. Close to North America and English-speaking, it specialized in a more rustic ecotourism that showcased its interior tropical forests and coastal barrier reef. In the mid-1990s, the number of land-based tourists peaked at 300,000, and tourism was credited with employing one-fifth of the workforce. Yet in 2000, arrival figures had plummeted to just 100,000. By 2007 the tourist figures were up to 250,000. Belize City became a port of call for day visitors from cruise ships in 2000, making it the fastest-growing tourist port in the Caribbean. Yet the influx of day visitors has done little to improve the city's infrastructure or high unemployment (Figure 5.32).

For more than four decades, tourism has been one of the Caribbean's mainstays. Yet this regional industry has grown more slowly in recent years compared with the industries in the Middle East, southern Europe, and even Central America. It seems that Americans are favoring domestic destinations, such as Hawaii, Florida, and Las Vegas, or are going to more "exotic" localities, such as Costa Rica. European tourists also seem to be staying closer to home or venturing to new locations, such as Dubai on the Persian Gulf or Goa in India. A destination's reputation can suddenly deteriorate as a result of social or natural forces. Increasingly, foreign tourists are opting to experience the Caribbean from the decks of cruise ships rather than land-based resorts. This trend undermines the local benefits of tourism, directing capital to large cruise lines rather than island economies.

Tourism-led growth has detractors for other reasons. It is subject to the overall health of the world economy and current political affairs. Thus, when North America experiences a recession or international tourism declines because of heightened fears of terrorism, the flow of tourist dollars to the Caribbean dries up. Where tourism is on the rise, local resentment may build as residents confront the disparity between their own lives and those of the tourists. There is also a serious problem of **capital leakage**, which is the huge gap between gross receipts and the total tourist dollars that remain in the Caribbean. Since many guests stay in hotel chains with corporate headquarters outside the region, leakage of profits is inevitable. On the plus side, tourism tends to promote stronger environmental laws and regulation. Coun-

Figure 5.32 Resort Life Young tourists enjoy the laid-back scene on the island of Caye Caulker, Belize. Once a village of fishermen, the main business on this small Belizean cay is serving the growing number of tourists coming to enjoy the barrier reef and stay in locally owned hotels.

tries quickly learn that their physical environment is the foundation for success. And while tourism does have its costs (higher energy and water consumption and demand for more imports), it is environmentally less destructive than traditional export agriculture and at present more profitable.

Social Development

While the record for economic growth in the region is inconsistent, measures of social development are generally strong. For example, most Caribbean peoples have an average life expectancy of more than 70 years (see Table 5.2). Literacy levels are high, and in most countries there is near parity in terms of school enrollment by gender. Indeed, high levels of educational attainment and out-migration have contributed to a marked decline in natural increase rates over the last 30 years, so that the average Caribbean woman has two or three children, making the regional Total Fertility Rate 2.5 children per woman.

These demographic and social indicators explain why Caribbean nations fare well in the Human Development Index (Figure 5.33). All ranked states (territories are not ranked) are in the high and medium human development categories. The island nation of Barbados, ranked 37th in the world in terms of human development, is in the very high human development category along with states such as Norway, South Korea, Australia, and the United States. In the Caribbean, Haiti has the lowest ranking at 149th in the world, but that still places it in the medium human development level. Figure 5.33 also shows that many of these well-ranked states have significant annual per capita flows of remittances entering the economy. In Jamaica, nearly $800 per capita arrives each year via remittances. It has been argued that remittances have become extremely important in boosting the overall level of social and economic development in the region. Despite real social gains, many inhabitants are chronically underemployed, poorly housed, and perhaps overly dependent on foreign remittances. For rich and poor alike, the temptation to leave the region in search of better opportunities remains.

Status of Women The matriarchal basis of Caribbean households is often singled out as a distinguishing characteristic of the region. The rural custom of men leaving home for seasonal employment tends to nurture strong and self-sufficient female networks. Women typically run the local street markets. With men absent for long periods of time, women tend to make household and community decisions. While giving women local power, this position does not always confer status. In rural areas, female status is often undermined by the relative exclusion of women from the cash economy; men earn wages while women provide subsistence.

As Caribbean society urbanizes, more women are being employed in assembly plants (the garment industry, in particular, prefers to hire women), in data-entry firms, and in tourism. With new employment opportunities, female laborforce participation has surged; in countries such as the Bahamas, Barbados, Jamaica, and Puerto Rico more than 40 percent of the workforce is female. Increasingly, women are the principal earners of cash, and they are more likely to complete secondary education than men. There are also signs of greater political involvement by women. In recent years Jamaica, Dominica, and Guyana have all had women prime ministers.

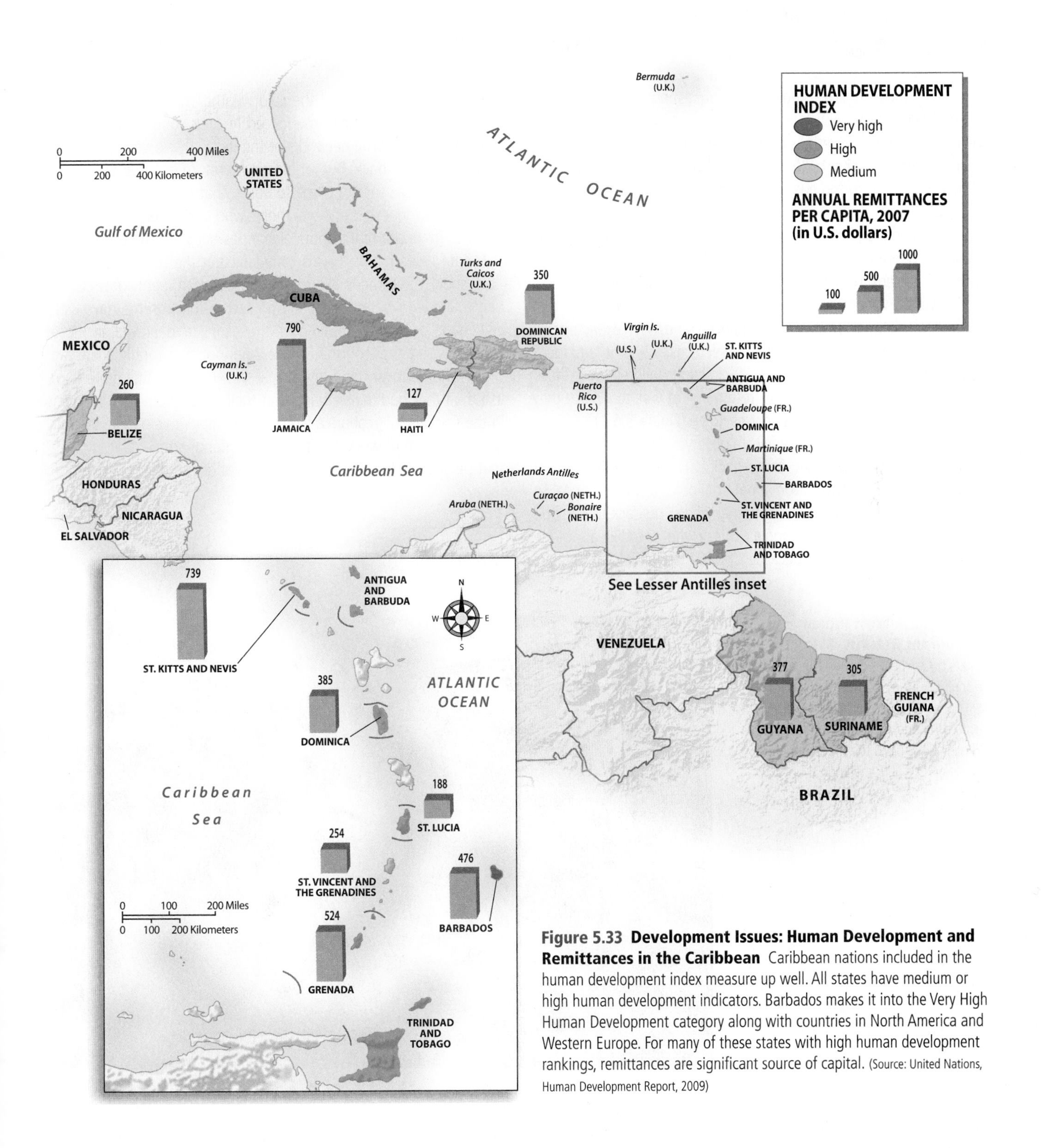

Figure 5.33 Development Issues: Human Development and Remittances in the Caribbean Caribbean nations included in the human development index measure up well. All states have medium or high human development indicators. Barbados makes it into the Very High Human Development category along with countries in North America and Western Europe. For many of these states with high human development rankings, remittances are significant source of capital. (Source: United Nations, Human Development Report, 2009)

Education Many Caribbean states have excelled in educating their citizens. Literacy is the norm, and the expectation is for most people to receive at least a high school degree. In many respects, Cuba's educational accomplishments are the most impressive, given the size of the country and its high illiteracy rates in the 1960s. Today nearly all adults are literate. Hispaniola is the obvious contrast to Cuba's success. While the Dominican Republic has made strides in improving adult literacy (87 percent of adults are literate), half of all Haitian adults are illiterate. Political stability and economic growth have helped the Dominican Republic better its social conditions over the past decade. In fact, many Haitians crossed the border into the Dominican Republic because conditions, although far from ideal, are much better than in their homeland.

PEOPLE ON THE MOVE The Diaspora Drifts Back

Throughout the English Caribbean an interesting process of return is unfolding. Across the islands, crates are arriving, new houses are being built, and associations of returnees are being formed. Just as their departure in the 1960s and 1970s produced changes in the United Kingdom, so too has their return to the Caribbean 30 or 40 years later.

A common strategy for many immigrants is to work in the United Kingdom and retire in the Caribbean with a pension and savings that ensure a comfortable old age. Many returnees are finally living a long-deferred dream of return, bringing with them cash, new skills, distinct life experiences, and new expectations. They left after years of work to fulfill their sense of home and identity. And, interestingly, sometimes their adult children, who have never lived in the Caribbean, are "returning" as well to a place they have never lived in but were told was home (Figure 5.6.1).

As Kenneth Bruney from Dominica observes, "England was good to me but I always yearned to be back... From a family point of view, England is home but from a national point of view, it's not. So when I go to Dominica I'm going home; when I go back to England I'm going back to England. People can say what they like about me there, but they can't here because I'm 101% Dominican." Others return because they tire of the cold weather and want to retire in the tropical climate they experienced in their youth. For many returnees it is the social networks or the distinctly slower pace of life that draws them back. But for most who return it is that emotional connection to a sense of place, to home, that motivates them.

Tensions do exit between those who never left and the returnees. There are real income disparities between the two groups that spur resentments. Locals complain that the returnees act like "big shots." Returnees express frustration with inept local governance and legal systems. Those who successfully resettle have usually made numerous visits over the years, and stayed in close contact with people on the islands. And, not surprisingly, there are those who discover upon return that they have become more English than they had realized and that their "real" home is no longer in the Caribbean. Economic globalization may drive people to emigrate in search of work. Yet it is the emotional attachment to place that persists and explains, in part, the Caribbean diaspora's return.

Source: Adapted from "Homeward Bound," *Guardian Weekend,* November 28, 2009, 16–29.

Figure 5.6.1 West Indian Immigrant Returns A returnee tends his home garden on the island of Dominica. Many West Indian immigrants who spent decades working in the United Kingdom are returning to the English Caribbean to enjoy retirement in their tropical homeland.

Education is expensive for these nations, but it is considered essential for development. Ironically, many states express frustration about training professionals for the benefit of developed countries in a phenomenon called **brain drain**. In the early 1980s, the prime minister of Jamaica complained that 60 percent of his country's newly university-trained workers left for the United States, Canada, and Britain, representing a subsidy to these economies far greater than the foreign aid Jamaica received from them. A 2007 study by the World Bank of skilled migrants revealed that 40 percent of Caribbean immigrants living abroad were college educated. No other region in the world has this many educated people leaving. To be fair, some of these migrants moved when they were young and received their education in North America and Europe. Still, other immigrants, especially health professionals, received their educations in the Caribbean and were recruited abroad because of higher wages and better opportunities. Brain drain occurs throughout the developing world, especially between former colonies and the mother countries. Given the small population of many Caribbean territories, each professional person lost to emigration can negatively impact local health care, education, and enterprise. While the outflow of professionals continues to be high, many countries are experiencing a return migration of Caribbean peoples from North America and Europe as a **brain gain**. Brain gain refers to the potential of returnees to contribute to the social and economic development of a home country with the experiences they have gained abroad (see "People on the Move: The Diaspora Drifts Back").

Labor-related Migration Given the region's high educational rates and limited employment opportunities, Caribbean countries have seen their people emigrate for decades. Besides influencing regional economies, this strategy impacts community and household structures as well. Historically, labor circulated within the Caribbean in pursuit of the sugar harvest—Haitians went to the Dominican Republic; residents of the Lesser Antilles journeyed to Trinidad. During the construction of the Panama Canal in the early 1900s, thousands of Jamaicans and Barbadians journeyed to the Canal Zone, and many remained.

After World War II, better transportation and political developments in the Caribbean produced a surge of migrants to North America. This trend began with Puerto Ricans going to New York in the early 1950s and intensified in the 1960s with the arrival of nearly half a million Cubans. Since then, large numbers of Dominicans, Haitians, Jamaicans, Trinidadians, and Guyanese also have migrated to North America, typically settling in Miami, New York, Los Angeles, and Toronto.

Crucial in this exchange of labor from north to south is the flow of cash **remittances** (monies sent back home). Immigrants are expected to send something back, especially when immediate family members are left behind. Collectively, nearly $3 billion was sent to the Dominican Republic by immigrants in 2009, making remittance income the country's second leading industry. Jamaicans and Haitians remit nearly $2 billion annually to their countries. Governments and individuals alike depend on these transnational family networks. Families carefully select the household member most likely to succeed abroad in the hope that money will flow back and a base for future immigrants will be established. A Caribbean nation in which no one left would soon face crisis. Economic opportunities are limited, and the populations have expanded far faster than the economies. Many of the people who leave would be jobless at home, or would force someone else to be jobless if they stayed.

Such labor-related migration is a standard practice for tens of thousands of households in the region, as well as a clear expression of the globalization of labor flows. Still, it is unlikely that migrants and their remittances can challenge the larger forces that make migration attractive in the first place. The impact of emigration and remittances is experienced primarily at the household level and is still too fragmented to represent a national development force.

Summary

- The Caribbean is more integrated into the global economy than most areas in the developing world, even though it remains on the economic periphery. Although small in area, the region offers some of the clearest examples of the long-term effects of globalization from plantation agriculture to offshore banking.
- This tropical region has been exploited to produce export commodities such as sugar, coffee, and bananas. The region's warm waters and mild climate attract millions of tourists. Yet serious problems with deforestation, soil erosion, and water contamination have degraded urban and rural environments. Global warming poses a serious threat to the region.
- Population growth in the Caribbean has slowed over the past two decades. The average woman now has two or three children. Many people leave the region for employment opportunities in North America and Europe. Life expectancy is quite high, as are literacy rates, and most people live in cities. In 2010 an earthquake devastated Port-au-Prince, one of the largest and certainly poorest cities in the Caribbean. The rebuilding of this capital city with foreign assistance will take time; the long-term effects of the rebuilding of Haiti are yet to be determined.
- The Caribbean was forged through European colonialism and the labor of millions of Africans. The blending of African and European elements, referred to as creolization, has resulted in many unique cultural expressions in music, language, and religion.
- Today the region contains 20 independent countries and several dependent territories. With the end of the Cold War, many microstates in the region fear that their lack of strategic significance may result in neglect by the United States and Europe, which may limit their ability to participate in world trade.
- In terms of economic development, employment in free trade zones, tourism, and offshore banking have steadily replaced agricultural jobs in the Caribbean. The region's positive strides in social development, namely in education, health, and the status of women, distinguish it from other developing areas.

Key Terms

brain drain (*page 220*)
brain gain (*page 221*)
capital leakage (*page 217*)
Caribbean Community and Common Market (CARICOM) (*page 210*)
chain migration (*page 193*)
circular migration (*page 193*)
creolization (*page 197*)
diaspora (*page 193*)
free trade zone (FTZ) (*page 214*)
Greater Antilles (*page 185*)
hurricanes (*page 187*)
indentured labor (*page 198*)
isolated proximity (*page 181*)
Lesser Antilles (*page 186*)
maroons (*page 199*)
monocrop production (*page 198*)
Monroe Doctrine (*page 205*)
neocolonialism (*page 205*)
offshore banking (*page 215*)
plantation America (*page 198*)
remittances (*page 221*)
rimland (*page 184*)
transnational migration (*page 193*)

Review Questions

1. What characteristics define the Caribbean as a world region?
2. What environmental, economic, and locational factors contributed to the growth of tourism in the Caribbean?
3. What is the typical path of a hurricane in the Caribbean, how are they tracked, and when are these meteorological disturbances likely to occur?
4. How will global warming impact the Caribbean over the next century?
5. What factors contributed to the deforestation of the Caribbean islands? Why are the rimland forests, in comparison, still intact?
6. What happened to the Amerindian population in the Caribbean?
7. What cultural traditions transferred from Africa to the Caribbean via the slave trade?
8. What is the history of carnival in the Caribbean? How is it important today? Why are carnival traditions spreading around the world?
9. Which territories in the region are still colonies? Why have they not sought independence?
10. What are the banana wars, and how do they illustrate the tensions brought on by globalization?

Thinking Geographically

1. After looking at the map of tourism in the Caribbean, what conclusions can you draw about where this occurs?
2. Contrast the historical African diaspora to the contemporary Caribbean one. What patterns are formed by these two distinct population movements? What social and economic forces are behind them? Are they comparable?
3. What are the advantages and disadvantages of offshore banking as a development strategy?
4. What advantages might Caribbean free trade zones retain over their competitors in Southeast and East Asia? What disadvantages might they face?
5. Why have U.S. actions in the region been considered neocolonial? Are there other regions in the world where the United States exerts neocolonial tendencies?
6. Compare and contrast the industrial development of Cuba with Puerto Rico. How were these two islands integrated into the world economy during the 20th century?
7. Why does chain migration occur between the Caribbean and North America or Europe? How does a migrant benefit from being part of a chain migration? How does chain migration affect sending communities?
8. In the 21st century, will agricultural exports continue to be important for Caribbean economies?
9. How has the January 2010 earthquake in Port-au-Prince impacted Haiti and Haitians? What is the best way for the international community to support Haitian reconstruction?
10. Are remittances a sign of the Caribbean's isolation or integration with the global economy? How might remittances be used for development?

Regional Novels and Films

Novels

Reinaldo Arenas, *Farewell to the Sea: A Novel of Cuba* (1994, Penguin)

Aimé Césaire, *Notebook on a Return to the Native Land* (2001, Wesleyan University Press)

Patrick Chamoiseau, *Texaco* (1988, Vintage International)

Edwidge Danticat, ed., *The Butterfly's Way: Voices from the Haitian Dyaspora in the United States* (2001, Soho)

Zee Edgell, *Beka Lamb* (1982, Heinemann)

Cristina Garcia, *Dreaming in Cuba* (1992, Random House)

Jamaica Kincaid, *A Small Place* (1988, Penguin)

William Luis, ed., *Dance Between Two Cultures: Latino Caribbean Literature Written in the United States* (1997, Vanderbilt)

V. S. Naipaul, *The Middle Passage* (1963, Macmillan)

Derek Walcott, *Omeros* (1990, Farrar, Straus & Giroux)

Films

Bitter Cane (1986, Haiti)

Buena Vista Social Club (1999, Cuba)

Casi Casi (2006, Puerto Rico)

Circle of Voodoo (2002, Haiti)

The Emperor's Birthday: The Rastafarians Celebrate (1997, Jamaica)

The Harder They Come (1972, Jamaica)

Inside Castro's Cuba (1995, Cuba)

The King Does Not Lie: The Initiation of a Shango Priest (1993, Trinidad)

No Seed (2002, Antigua and Barbuda)

Sugar (2008, U.S. and Dominican Republic)

Bibliography

Duval, David Timothy, ed. 2004. *Tourism in the Caribbean: Trends, Development, Prospects.* London: Routledge.

Grossman, Lawrence. 1998. *The Political Ecology of Bananas: Contract Farming, Peasants, and Agrarian Change in the Eastern Caribbean.* Chapel Hill, NC: University of North Carolina.

Hillstrom, Kevin, and Hillstrom, Laurie Collier. 2004. *Latin America and the Caribbean: A Continental Overview of Environmental Issues.* Santa Barbara, CA: ABC Clio.

Klak, Thomas, ed. 1998. *Globalization and Neoliberalism: The Caribbean Context.* Lanham, MD: Rowman and Littlefield.

McNeill, J.R. 2010. *Mosquito Empire: Ecology and War in the Greater Caribbean,* 1620–1914. Cambridge, England: Cambridge University Press.

Mintz, Sidney. 1985. *Sweetness and Power: The Place of Sugar in Modern History.* New York: Viking Penguin.

Richardson, Bonham C. 1992. *The Caribbean in the Wider World, 1492–1992.* Cambridge, England: Cambridge University Press.

Roberts, Susan. 1995. "Small Place, Big Money: The Cayman Islands and the International Financial System." *Economic Geography* 71(3), 237–56.

Scarpaci, Joseph; Segre, Roberto; and Coyula, Mario. 2002. *Havana, Two Faces of the Antillean Metropolis.* Chapel Hill, NC: University of North Carolina Press.

Voeks, Robert. 1993. "African Medicine and Magic in the Americas." *Geographical Review* 83(1), 66–78.

Wilson, Mark. 2003. "Chips, Bits and the Law: An Economic Geography of Internet Gambling." *Environment and Planning A* 35(7), 1245–60.

Log in to www.masteringgeography.com for MapMaster™ interactive maps, geography videos, RSS feeds, flashcards, web links, an eText version of *Diversity Amid Globalization,* and self-study quizzes to enhance your study of The Carribean.

MapMaster™ presents 13 Place Name and 13 Layered Thematic interactive maps to help students practice and master their geographic literacy, spatial reasoning, and critical thinking skills.

6
Sub-Saharan Africa

Young boys play soccer on a field overlooking Mazeppa Bay in Eastern Cape Province, South Africa. Sub-Saharan Africa is a youthful region, with 43 percent of the population under 15 years old.

DIVERSITY AMID GLOBALIZATION

Struggling with extreme poverty, Sub-Saharan Africans have experienced few of the benefits of globalization. Yet over the past decade, foreign assistance to the region grew substantially, access to mobile phone technology enhanced communication, and many in the region expressed pride in the fact that Africa hosted its first World Cup in the summer of 2010.

ENVIRONMENTAL GEOGRAPHY

Wood is a main source of energy for this region. Kenyan Professor Wangari Maathai won the Noble Peace Prize for her Greenbelt Movement, which led to the planting of millions of trees by rural women. In areas such as the Sahel, policies that provided ownership or incentives for the protection of trees have resulted in an increase in tree cover.

POPULATION AND SETTLEMENT

Sub-Saharan Africa has the highest infection rates of HIV/AIDS. As a result of this disease, many countries have seen life expectancy rates plummet into the 40s. Yet the population continues to grow at a 2.5 percent rate of natural increase, which is the highest rate for any world region.

CULTURAL COHERENCE AND DIVERSITY

This is a region with large and growing numbers of Muslims, Christians, and animists. With a few exceptions, religious diversity and tolerance has been a distinctive feature of the region.

GEOPOLITICAL FRAMEWORK

Most countries gained their independence in the 1960s. Since then, many ethnic conflicts have resulted as governments struggle for a sense of national unity within the boundaries drawn by European colonialists.

ECONOMIC AND SOCIAL DEVELOPMENT

The Millennium Development Goals created by the United Nations to reduce extreme poverty by 2015 will not be met by most of the states of this region. Yet as global demand for natural resources grows, investors from China, Europe, and North America are eager to extract the region's metals and fossil fuels.

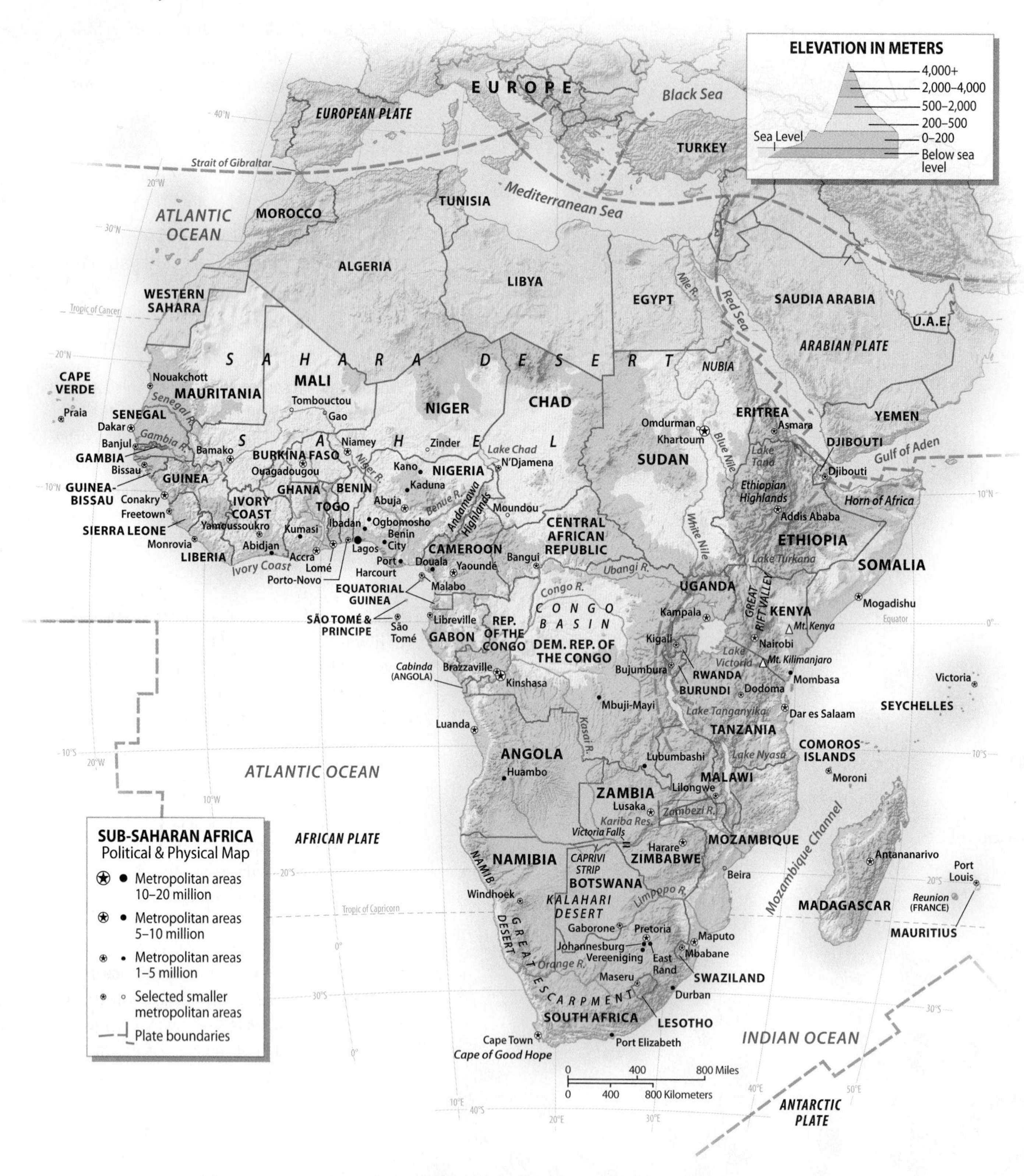

Figure 6.1 Sub-Saharan Africa Africa south of the Sahara includes 48 states and one territory. This vast region of rain forest, tropical savanna, and desert is home to 865 million people. Much of the region consists of broad plateaus ranging from 1,600 to 6,500 feet (500 to 2,000 meters) in elevation. Although the population is growing rapidly, the overall population density of Sub-Saharan Africa is low. Considered one of the least developed regions of the world, it remains an area rich in natural resources.

Compared with Latin America and the Caribbean, Africa south of the Sahara is poorer and more rural, and its population is very young. Nearly 865 million people reside in this region, which includes 48 states and one territory. Demographically, this is the world's fastest-growing region (2.5 percent rate of natural increase); in most countries, nearly half the population (43 percent) is younger than 15 years old. Income levels are extremely low: 74 percent of the population lives on less than $2 per day. Life expectancy is only 52 years. This part of the world is known for poverty, disease, violence, and refugees. Overlooked in the all-too-frequent negative headlines are initiatives by private, local, and state groups to improve the region's quality of life. Many states have reduced infant mortality, expanded basic education, and increased food production in the past two decades, yet many economic and political crises still hinder the region's development.

Sub-Saharan Africa—that portion of the African continent lying south of the Sahara Desert—is a commonly accepted world region (Figure 6.1). The unity of this region has to do with similar livelihood systems and a shared colonial experience. No common religion, language, philosophy, or political system ever united the area. Instead, loose cultural bonds developed from a variety of lifestyles and idea systems that evolved here. The impact of outsiders also helped to determine the region's identity. Slave traders from Europe, North Africa, and Southwest Asia treated Africans as chattel; up until the mid-1800s millions of Africans were taken from the region and sold into slavery. In the late 1800s, the entire African continent was divided by European colonial powers, imposing political boundaries that remain to this day. In the postcolonial period, which began in the 1960s, Sub-Saharan African countries faced many of the same economic and political challenges.

When setting this particular regional division, the major question is how to treat North Africa. Some scholars argue for treating the African continent as one world region because the Sahara has never formed a complete barrier between the Mediterranean north and the rest of the African landmass. Regional organizations such as the African Union are modern expressions of this continental unity. Yet North Africa is generally considered more closely linked, both culturally and physically, to Southwest Asia. Arabic is the dominant language and Islam the dominant religion of North Africa. Consequently, North Africans feel more closely connected to the Arab hearth in Southwest Asia than to the Sub-Saharan world.

In this chapter we focus on the states south of the Sahara. We preserve political boundaries when delimiting the region so that the Mediterranean states of North Africa are discussed with Southwest Asia. Sudan—Africa's largest state in terms of area—will be discussed in both Chapters 6 and 7 because it shares characteristics common to both regions. In the more populous and powerful north, Muslim leaders have crafted an Islamic state that is culturally and politically oriented toward North Africa and Southwest Asia. Southern Sudan, however, has more in common with the animist and Christian groups in Sub-Saharan Africa.

The example of Sudan underscores the cultural complexity of the states in this region. In one large state it is not usual for 20 or more languages to be spoken. Consequently, most Africans understand and speak several languages. Ethnic identities do not conform to the political divisions of Africa, sometimes resulting in bloody ethnic warfare, as witnessed in Rwanda in the mid-1990s and in Sudan today. Nevertheless, throughout the region peaceful coexistence among distinct ethnic groups is the norm. The cultural significance of European colonizers cannot be ignored: European languages, religions, educational systems, and political ideas, were adopted and modified. Yet the daily rhythms of African life are often far removed from the industrial or post-industrial world. Most Africans engage in subsistence and cash-crop agriculture. Women in particular are charged with tending crops and procuring household necessities (Figure 6.2). Male roles revolve around tending livestock and the public life of the village and market.

The influence of African peoples outside the region is great especially when one considers that human origins are traceable to this part of the world. In historic times, the legacy of the slave trade resulted in the transfer of African peoples, religious systems and even musical traditions throughout the Western Hemisphere. Even today, African-based religious systems are widely practiced in the Caribbean and Latin America, especially in Brazil. American jazz, Brazilian samba, and Cuban rumba would not exist but for the influence of Africans. The popularity of "world music" has created a large international audience for performers such as Senegal's Youssou N'Dour or South Africa's Ladysmith Black Mambazo. Through music and the arts,

Figure 6.2 Women Farmers in Senegal Senegalese women grow manioc, a drought-resistant staple introduced from the Americas. Throughout Sub-Saharan Africa, subsistence agriculture is still widely practiced, and women are often responsible for tending these crops.

Sub-Saharan Africa has exerted a significant influence on global culture.

The African economy, however, is marginal when compared with the rest of the world. According to the World Bank, Sub-Saharan Africa's economic output in 2008 amounted to less than 2 percent of global output, even though the region contains 12 percent of the world's population. Moreover, the gross national product of just one country, South Africa, accounts for one-third of the region's total economic output. It is the region's grinding poverty and the unnecessary vulnerability of its people that draws the concern of the global community.

Many scholars feel that Sub-Saharan Africa has benefited little from its integration (both forced and voluntary) into the global economy. Slavery, colonialism, and export-oriented mining and agriculture served the needs of consumers outside the region but undermined domestic production for those within. Foreign assistance in the post-independence years initially improved agricultural and industrial output, but also led to mounting foreign debt and corruption, which over time undercut the region's economic gains. Ironically, many of these same scholars and politicians worry that negative global attitudes about the region have produced a pattern of neglect. Private capital investment in Sub-Saharan Africa lags far behind other developing regions. The idea of debt forgiveness for Africa's poorest states is steadily being applied as a strategy to reduce human suffering in the region.

The past few years have witnessed a surge in philanthropic outreach to the region. Rock star Bono of U2 and the Bill and Melinda Gates Foundation have led the One Campaign, directing millions of dollars to support health care, disease prevention, education, and poverty reduction in Sub-Saharan Africa and other developing areas. Through a combination of internal reforms, better governance, foreign assistance, and foreign investment in infrastructure and technology, many believe that social and economic gains are possible for the region.

ENVIRONMENTAL GEOGRAPHY: The Plateau Continent

The largest landmass straddling the equator, Sub-Saharan Africa is vast in scale and its physical environment is remarkably beautiful. Called the *plateau continent,* the African interior is dominated by extensive uplifted areas that resulted from the breakup of **Gondwanaland**, an ancient mega-continent that included Africa, South America, Antarctica, Australia, Madagascar, and Saudi Arabia. Some 250 million years ago, it began to split apart through the forces of continental drift. As this process unfolded, the African landmass experienced a series of continental uplifts that left much of the area with vast elevated plateaus. The highest areas are found on the eastern edge of the continent, where the Great Rift Valley forms a complex upland area of lakes, volcanoes, and deep valleys. In contrast, lowlands prevail in West Africa (see Figure 6.1).

The landscape of Sub-Saharan Africa offers a palette of intense colors: deep red soils studded with bright green food crops; the blue tropical sky; golden savannas that ripple with the movement of animal herds; dark rivers meandering through towering rain forests; and sun-drenched deserts. Amid this beauty, however, one finds relatively poor soils, persistent tropical diseases, and frequent droughts. Large areas exist with greater potential for agricultural development, especially in southern Africa; and throughout the continent significant water resources, biodiversity, and mineral wealth abound.

Plateaus and Basins

A series of plateaus and elevated basins dominate the African interior and explain much of the region's unique physical geography. Generally, elevations increase toward the south and east of the continent. Most of southern and eastern Africa lies well above 2,000 feet (600 meters), and sizable areas sit above 5,000 feet (1,500 meters). These areas are typically referred to as High Africa; Low Africa includes West Africa and much of Central Africa. The high plateaus in countries such as Kenya, Zimbabwe, and Angola are noted for their cooler climates and relatively abundant moisture. Steep escarpments form where plateaus abruptly end, as illustrated by the majestic Victoria Falls on the Zambezi River (Figure 6.3). Much of southern Africa is rimmed by a landform called the **Great Escarpment** (a steep cliff separating the coastal lowlands from the plateau uplands), which begins in southwestern Angola and ends in northeastern South Africa, creating a natural barrier to coastal settlement. South Africa's Drakensberg Mountains (with elevations

Figure 6.3 Victoria Falls The Zambezi River descends over Victoria Falls. A fault zone in the African plateau explains the existence of a 360-foot (110-meter) drop. The Zambezi has never been important for navigation, but it is a vital supply of hydroelectricity for Zimbabwe, Zambia, and Mozambique.

Figure 6.4 The Rift Valley Tucked into a fertile area of the Rift Valley in Kenya, this Kikuyu village of small landholders produces subsistence and market crops. Parts of the Rift Valley have East Africa's highest population densities.

reaching 10,000 feet, or 3,100 meters) rise up from the Great Escarpment. Because of this landform, coastal plains tend to be narrow, with few natural harbors, and river navigation is impeded by a series of falls.

Though Sub-Saharan Africa is an elevated landmass, it has few significant mountain ranges. The one extensive area of mountainous topography is in Ethiopia, which lies in the northern portion of the Rift Valley zone. Yet even there the dominant features are high plateaus intercut with deep valleys rather than actual mountain ranges. Receiving heavy rains in the wet season, the Ethiopian Plateau is densely settled and forms the headwaters of several important rivers, most notably the Blue Nile, which joins the White Nile at Khartoum, Sudan.

A discontinuous series of volcanic mountains, some of them quite tall, are associated with the southern half of the Rift Valley. Kilimanjaro at 19,000 feet (5,900 meters) is the continent's tallest mountain, and nearby Mount Kenya (17,000 feet, or 5,200 meters) is the second tallest. The Rift Valley itself reveals the slow but inexorable progress of geological forces. Eastern Africa is slowly being torn away from the rest of the continent, and within some tens of millions of years it will form a separate landmass. Such motion has already produced a great gash across the uplands of eastern Africa, much of which is occupied by elongated and extremely deep lakes (most notably Nyasa, Malawi, and Tanganyika). In central East Africa this rift zone splits into two separate valleys, each of which is flanked by volcanic uplands. Between the eastern and western rifts lies a bowl-shaped depression, the center of which is filled by Lake Victoria—Africa's largest body of water. Not surprisingly, some of the densest areas of settlement are found amid the fertile and well-watered soils that border the Rift Valley (Figure 6.4).

Watersheds Africa south of the Sahara conspicuously lacks the broad, alluvial lowlands that influence patterns of settlement throughout other regions. The four major river systems are the Congo, Nile, Niger, and Zambezi. Smaller rivers—such as the Orange in South Africa; the Senegal, which divides Mauritania and Senegal; and the Limpopo in Mozambique—are locally important but drain much smaller areas. Ironically, most people think of Africa south of the Sahara as suffering from water scarcity and tend to discount the size and importance of the watersheds (or catchment areas) that these river systems drain.

The Congo River (or Zaire) is the largest watershed in the region in terms of drainage and volume of flow. It is second only to South America's Amazon River in terms of annual flow. The Congo flows across a relatively flat basin that lies more than 1,000 feet (300 meters) above sea level, meandering through Africa's largest tropical forest, the Ituri (Figure 6.5). Entry from the Atlantic into the Congo Basin is prevented by a series of rapids and falls, making the Congo River only partially navigable. Despite these limitations, the Congo River has been the major corridor for travel within the Republic of the Congo and the Democratic Republic of the Congo (formerly Zaire); the capitals of both countries, Brazzaville and Kinshasa, rest on opposite sides of the river.

The Nile River, the world's longest, is the lifeblood of Egypt and Sudan. Yet this river originates in the highlands of the Rift Valley zone, and is an important link between North and Sub-Saharan Africa. The Nile begins in the lakes of the rift zone (Victoria and Edward) before descending into a vast wetland in southern Sudan known as the Sudd. Agricultural development projects in the 1970s greatly increased the agricultural potential of the Sudd, especially its peanut crop. Unfortunately, the past three decades of civil war in Sudan ravaged this area, turning farmers and herders into refugees and undermining the productive capacity of this important ecosystem. A more extensive discussion of the Nile River is in Chapter 7.

Like the Nile, the Niger River is the critical source of water for two otherwise arid countries: Mali and Niger. Originating in the humid Guinea highlands, the Niger flows first to the northeast and then spreads out to form a huge inland delta in Mali before making a great bend southward at the margins of the Sahara near Gao. On the banks of the Niger River are the capitals of Mali (Bamako) and Niger (Niamey), as well as the historic city of Tombouctou (Timbuktu). After flowing through the desert, the Niger River returns to the humid lowlands of Nigeria, where the Kainji Reservoir temporarily blocks its flow to produce electricity for Africa's most populous state.

The considerably smaller Zambezi River originates in Angola and flows east, spilling over an escarpment at Victoria Falls, and finally reaching Mozambique and the Indian Ocean. More than other rivers in the region, the Zambezi is a major supplier of commercial energy. Sub-Saharan Africa's two largest hydroelectric installations, the Kariba on the border of Zambia and Zimbabwe and the Cabora Bassa in Mozambique, are on this river.

Figure 6.5 Congo River A small fishing village on the banks of the Congo River. The Congo River is an important transportation corridor and source of fish. Behind the village lies the immense Ituri rain forest.

Soils With a few major exceptions, Sub-Saharan Africa's soils are relatively infertile. Generally speaking, fertile soils are young soils, those deposited in recent geological time by rivers, volcanoes, glaciers, or windstorms. In older soils—especially those located in moist tropical environments—natural processes tend to wash out most plant nutrients over time. Over most of Sub-Saharan Africa, the agents of soil renewal have largely been absent; the region has few alluvial lowlands where rivers periodically deposit fertile silt, and it did not experience significant glaciation in the last ice age, as did North America.

Portions of Sub-Saharan Africa are, however, noted for their natural soil fertility, and not surprisingly these areas support denser settlement. Some of the most fertile soils are in the Rift Valley, enhanced by the volcanic activity associated with the area. The population densities of rural Rwanda and Burundi, for example, are partially explained by the highly productive volcanic soils. The same can be said for highland Ethiopia, which supports the region's second largest population, with more than 80 million people. The Lake Victoria lowlands and central highlands of Kenya also are noted for their sizable populations and productive agricultural bases.

In the drier grasslands and semi-desert areas one finds a soil type called *alfisols*. High in aluminum and iron, these red soils have greater fertility than comparable soils found in wetter zones. This helps to explain the tendency of farmers to plant in drier areas, such as the Sahel, even though they risk exposure to drought. With irrigation, many agronomists suggest that the southern African countries of Zambia and Zimbabwe could greatly increase commercial grain production on these soils.

Climate and Vegetation

Sub-Saharan Africa lies in the tropical latitudes. Beginning just north of the Tropic of Cancer, crossing the equator, and extending past the Tropic of Capricorn in the south, it is the largest tropical landmass on the planet. Only the far south of the continent extends into the subtropical and temperate belts. Much of the region averages high temperatures from 70°F to 80°F (22° to 28°C) year-round. The seasonality and amount of rainfall, more than temperature, determine the different vegetation belts that characterize the region. As Figure 6.6 shows, Addis Ababa, Ethiopia, and Walvis Bay, Namibia, have similar average temperatures, but the former is in the moist highlands and receives nearly 50 inches (127 centimeters) of rainfall annually, while Walvis Bay rests on the Namibian Desert and receives less than 1 inch (2.5 centimeters).

To understand the relationship between climate and vegetation in Sub-Saharan Africa (see Figure 6.6), imagine a series of concentric vegetation belts that begin in the western equatorial zone as forest (Am and Af), followed by woodlands and grasslands (Aw), semidesert (BSh), and finally desert (BWh). The montane zones of East Africa, Cameroon, Guinea, and South Africa (especially along the Drakensberg Range) exhibit the forces of altitudinal zonation discussed in Chapter 4. Capturing more rainfall than the surrounding lowlands, these mountains often support unique flora with large numbers of native species. The only midlatitude climates of the region are found in South Africa. The southwestern corner of South Africa contains a small zone of Mediterranean climate that is comparable to that of California and noted for its production of fine wine (Csb). The eastern coast of South Africa has a moist subtropical climate, not unlike that of Florida (Cfa).

Tropical Forests The core of Sub-Saharan Africa is remarkably moist. The world's second-largest expanse of humid equatorial rain forest, the Ituri, lies in the Congo Basin, extending from the Atlantic Coast of Gabon two-thirds of the way across the continent, including the Republic of the Congo and northern portions of the Democratic Republic of the Congo (Zaire). The conditions here are constantly warm to hot, and precipitation falls year-round (see the graph for Kisangani in Figure 6.6).

Commercial logging and agricultural clearing have degraded the western and southern fringes of this vast forest, but much of the northeastern section of the forest is still intact. Considering the high rates of tropical deforestation in Southeast Asia and Latin America, the Central African case is a pleasant exception. Major national parks such as Okapi and Virunga have been created in the Democratic Republic of the Congo. Virunga National Park—one of the oldest in Africa—is on the eastern fringe of the Ituri forest. It is home to the endangered mountain gorillas. Poor infrastructure and political chaos in the Democratic Republic of the Congo over the past 15 years has made large-scale logging impossible but it has

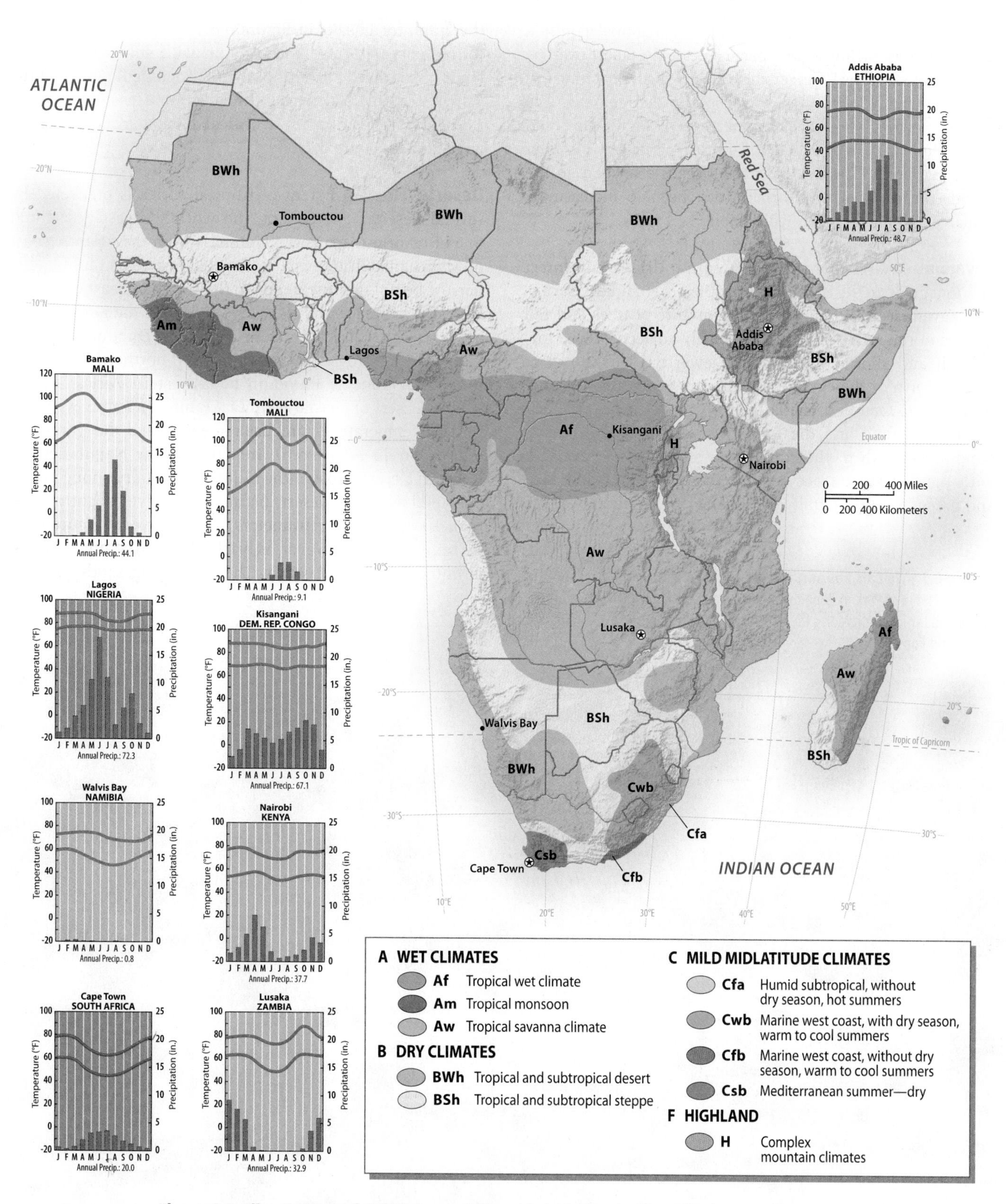

Figure 6.6 Climate map of Sub-Saharan Africa Much of the region lies within the tropical humid and tropical dry climatic zones; thus, the seasonal temperature changes are not great. Precipitation, however, varies significantly. Compare the distinct rainy seasons in Lusaka and Lagos: Lagos is wettest in June and Lusaka receives most of its rain in January. Although there are important tropical forests in West and Central Africa, much the territory is tropical savanna. (Temperature and precipitation data from *The World Weather Guide* [1984] by E. A. Pearce and C. G. Smith. London: Hutchinson)

also made conservation difficult. Due to regional conflict, parks such as Virunga have been repeatedly taken over by rebel groups and park rangers have been killed; poaching has become a means of survival for people in the region. In the future, it seems likely that Central Africa's rain forests could suffer the same kind of degradation experienced in other equatorial areas. As deforestation proceeds elsewhere in the world, the trees of equatorial Africa become increasingly valuable and, hence, more vulnerable.

Savannas Wrapped around the Central African rain forest belt in a great arc lie Africa's vast tropical wet and dry savannas. Savannas are dominated by a mixture of trees and tall grasses in the wetter zones immediately adjacent to the forest belt and shorter grasses with fewer trees in the drier zones. North of the equatorial belt, rain generally falls only from May to October. The farther north one travels, the less the total rainfall and the longer the dry season. Climatic conditions south of the equator are similar, only reversed, with the wet season occurring between October and May and precipitation generally decreasing toward the south (see the graph for Lusaka in Figure 6.6). A larger area of wet savanna exists south of the equator, with substantial woodlands in southern portions of the Democratic Republic of the Congo, Zambia, northern Zimbabwe and eastern Angola. These savannas also are a critical habitat for the region's large fauna (Figure 6.7).

Deserts The vast extent of tropical Africa is bracketed by several deserts. The Sahara, the world's largest desert and one of its driest, spans the landmass from the Atlantic coast of Mauritania all the way to the Red Sea coast of Sudan. A narrow belt of desert extends to the south and east of the Sahara, wrapping around the **Horn of Africa** (the northeastern corner that includes Somalia, Ethiopia, Djibouti, and Eritrea) and pushing as far as eastern and northern Kenya. An even drier zone is found in southwestern Africa. In the Namib Desert of coastal Namibia, rainfall is a rare event, although temperatures are usually mild (see the graph for Walvis Bay in Figure 6.6). Inland from the Namib lies the Kalahari Desert. Most of the Kalahari is not dry enough to be classified as a true desert, because it receives slightly more than 10 inches (25 centimeters) of rain a year (Figure 6.8). Its rainy season, however, is brief. Most of the precipitation is immediately absorbed by the underlying sands. Surface

Figure 6.7 African Savannas Buffalo gather at a watering hole in Zambezi National Park in Zimbabwe. The savannas of southern Africa are a noted habitat for the region's larger mammals such as buffalo, elephant, zebra, and lions.

Figure 6.8 Kalahari Desert A San man teaches his four-year-old son to hunt with a bow and arrow. The Kalahari, though not a true desert, does not get enough rain to support agriculture. Yet this ecosystem is home to abundant wildlife and hunter-gatherers, such as the San (formerly known as the Bushmen).

water is thus scarce, giving the Kalahari a desertlike aspect for most of the year.

Africa's Environmental Issues

The prevailing perception of Africa south of the Sahara is one of environmental scarcity and degradation, no doubt fostered by televised images of drought-ravaged regions and starving children. Single explanations such as rapid population growth or colonial exploitation cannot fully capture the complexity of Africa's environmental issues or the ingenious ways that people have adapted to living in stressed ecosystems. Because much of Sub-Saharan Africa's population is rural and poor, earning its livelihood directly from the land, sudden environmental changes are keenly felt and can cause mass migrations, famine, and even death. As Figure 6.9 illustrates, **desertification**, the expansion of desertlike conditions as a result of human-induced degradation, and deforestation are commonplace. Sub-Saharan Africa is also vulnerable to drought, most notably in the Horn of Africa, parts of southern Africa, and the Sahel. Many scientists fear that droughts will come more often and be prolonged with global warming.

As Sub-Saharan cities grow in size and importance, urban environments increasingly face problems of air and water pollution as well as sewage and waste disposal. At the same time, wildlife tourism is an increasingly important source of foreign exchange for many African states. Throughout the region, national parks have been created in an effort to strike a balance between humans' and animals' competing demands for land.

The Sahel and Desertification In the 1970s, the **Sahel** became an emblem for the dangers of unchecked population growth and human-induced environmental degradation when a relatively wet period came to an abrupt end and 6 years of drought (1968–74) ravaged the land. The Sahel is a zone of ecological transition between the Sahara to the north and wetter savannas and forest in the south. Life depends on a delicate balance of limited rain, drought-resistant plants, and a pattern of animal **transhumance**, the movement of animals between wet-season and dry-season pasture. Parts of the Sahel were important areas of settlement long before European colonization. What appears to be desert wasteland in April or May is transformed into a lush garden of millet, sorghum, and peanuts after the drenching rains of June. Relatively free of the tropical diseases found in the wetter zones to the south, Sahelian soils also are quite fertile, which helps to explain why people continue to live there despite the unreliable rainfall patterns (Figure 6.10). During the drought of the early 1970s, rivers in the area diminished and desert-like conditions began to move south. Unfortunately, tens of millions of people lived in the areas, and farmers and pastoralists whose livelihoods had come to depend on the more abundant precipitation of the relatively wet period were temporarily forced out.

Considerable disagreement continues over the basic causes of desertification and drought in the Sahel. Was it a matter of too many humans degrading their environment, unsound settlement schemes encouraged by European colonizers, or a failure to understand global atmospheric cycles? Agricultural practices certainly contributed to desertification. Throughout most of the Sahel, French colonial authorities forced villagers to grow peanuts as an export crop, a policy continued by the newly independent states of the region. However, peanuts tend to deplete several key soil nutrients, which means that peanut farms are often abandoned after a few years as cultivators move on to fresh sites. Since peanuts grow underground, moreover, the soil must be overturned at harvest time. This typically occurs at the onset of the dry season, when dry winds from the Sahara can carry away the fine dirt of the newly harvested field. With the loss of topsoil, the ground no longer absorbs the precipitation that does fall—often in drenching torrents—during the brief rainy season. The end result of this grim process is the spread of desertification regardless of actual changes in precipitation.

Overgrazing also has been implicated in Sahelian desertification, as is true in other regions of the world. Livestock is a traditional product of the region, and animal production dramatically expanded after World War II as the population

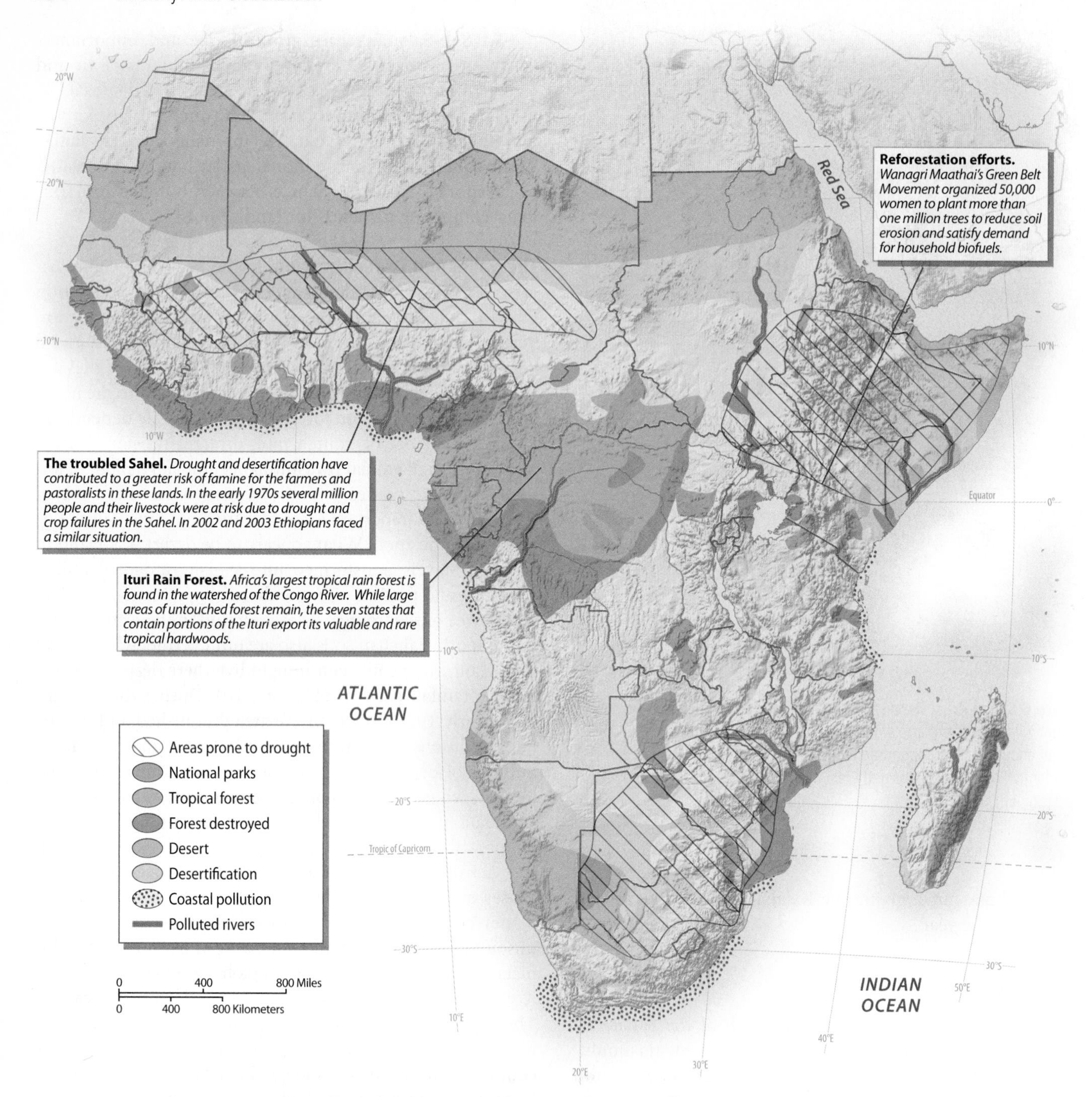

Figure 6.9 Environmental Issues in Sub-Saharan Africa Given the immense size of Sub-Saharan Africa, it is difficult to generalize about environmental problems. Dependence on trees for fuel places strains on forests and wooded savannas throughout the region. In semiarid regions, such as the Sahel and Horn of Africa, population pressures, climate change, and land-use practices seem to have exacerbated desertification. Yet Sub-Saharan Africa also supports the most impressive array of wildlife, especially large mammals, on Earth. (Adapted from *DK World Atlas*, 1997, 75. London: DK Publishing)

swelled and the need for export earnings grew. In many areas, natural pasture was greatly reduced by intensified grazing pressure, leading to increased wind erosion. National and international developmental agencies, hoping to increase production further, began to dig deep wells in areas that previously had been unused by herders through most of the year. The new supplies of water, in turn, allowed year-round grazing in places that, over time, could not withstand it. Large barren circles around each new well began to appear even on satellite images.

Expansion of agriculture and overgrazing are often cited as major factors leading to the loss of natural vegetation and declines in soil fertility in the Sahel. Yet in the case of the Sahelian portion of Niger, local agronomists have documented

Figure 6.10 Sahel in Bloom A woman prepares millet grains grown near the city of Maradi, Niger. The soils of the Sahel are fertile and peasant farmers can produce a surplus when adequate rain falls. Yet in times of drought, crop failures can lead to famine in this region.

an unanticipated increase in tree cover the past 30 years. More interesting still, increases in tree cover have occurred in some of the most densely populated rural areas. The relative greening of this area is attributed to simple actions taken by farmers, a change in government policy and better rainfall (see "Geographic Tools: Monitoring Land Cover and Conservation Changes in Niger"). Regardless of the causes given or action taken, many of the areas experiencing desertification are the same ones most vulnerable to drought and famine. These conditions may become more frequent with global climate change.

Deforestation Although Sub-Saharan Africa still contains extensive forests in some areas, much of the region has relatively little tree cover. Unlike tropical America, forest clearance in the wet and dry savanna is of greater local concern than the limited commercial logging of the rain forest. North of the equator, for example, only a few wooded areas remain in a landscape dominated by grasslands, savanna (grassland with scattered trees and shrubs), and cropland. Highland Ethiopia once was covered with lush forests, but these have long since been reduced to a few remnant patches. Loss of woody vegetation has resulted in extensive hardship, especially for women and children who must spend many hours a day scrounging for wood for home consumption or to sell in local markets (Figure 6.11). Deforestation, especially in the woodlands of the savannas, aggravates problems of increased runoff, soil erosion, and shortages of **biofuels** (wood and charcoal used for household energy needs, especially cooking).

In the southern tropical savanna, where human population has historically been lighter, extensive tracts of dry woodland remain intact. The trees of the dry forest have little commercial value compared to some of the species found in the rain forest, but they provide vital subsistence resources for local people as well as wildlife habitat. Even in these more abundant areas, biofuel scarcity is common around the larger towns and villages.

In some countries, village women have organized into community-based nongovernmental organizations (NGOs) to plant trees and create greenbelts to meet future fuel needs. One of the most successful efforts is in Kenya under the leadership of Wangari Maathai. Maathai's Green Belt Movement has more than 50,000 members, mostly women, organized into 2,000 local community groups. Since the movement's inception in 1977, these local groups have successfully planted millions of trees. In those areas, village women now spend less time collecting fuel, and local environments have im-

Figure 6.11 Wood Fuel in Ethiopia A women walks to Addis Ababa with a large bundle of firewood to sell in the city market. Collecting firewood is the work of women, both for household subsistence needs and for sale in local markets.

GEOGRAPHIC TOOLS Monitoring Land Cover and Conservation Changes in Niger

Much to everyone's surprise, Niger is becoming greener. Through satellite imagery and field inventories at the village level, local agronomists have documented that tree-covered areas have increased by 7 million acres (Figure 6.1.1). These basic geographic tools were used to reveal unexpected improvements in tree cover. It appears that better local conservation practices and improved rainfall have led to a dramatic increase in vegetation over the past 30 years. More interesting still, increases in tree cover have occurred in some of the most densely populated rural areas. Given that the Sahel is an area noted for drought, famine, and desertification, documenting land cover change and explaining how such changes happen offers valuable lessons for other drought-prone areas.

The greening of Niger was not the result of a large-scale government effort or a foreign assistance program. The impetus for change came from farmers themselves who realized after the drought of 1984 that maintaining tree cover was important for the long-term sustainability of farming practices. Farmers began actively protecting saplings instead of clearing them from their fields, including the nitrogen-fixing goa tree which had disappeared from many villages. During the rainy season, the goa tree loses its leaves, so it does not compete with crops for water or sun. The leaves themselves are added fertilizer to the soil. In general, trees and shrubs reduce wind erosion, and their presence also reduces runoff and erosion when the rains do come. Sahelian farmers also use branches, pods, and leaves from trees for fuel and animal fodder. Any excess can be sold at market.

Given the utility of native trees, why were farmers removing them? Much vegetation disappeared in the 1970s when the region suffered prolonged drought. Until the 1990s, all trees were considered property of the state of Niger, thus giving farmers little incentive to protect them. Since then, the government has recognized the value of allowing individuals to own trees. Not only can farmers sell branches, pods, fruit, and bark, they also conserve them to insure their agricultural livelihood.

For forest regeneration to work it requires the effort of the whole community. If farmers do not take action themselves and the community does not support them, forest regeneration will not be successful. Consequently, some villages in Niger are much greener than others. In the village of Moussa Bara, where regeneration has been a success, not one child died of malnutrition in the famine of 2005, largely because farmers were able to earn income from selling firewood when their crops failed.

The Sahel is a poor region that has always been vulnerable to drought. It is uncertain how global climate changes will impact this region; some years may be wetter than average but there is the possibility of more prolonged droughts. For now, at least, Niger offers a provocative example of how small and relatively simple conservation practices can have a positive impact.

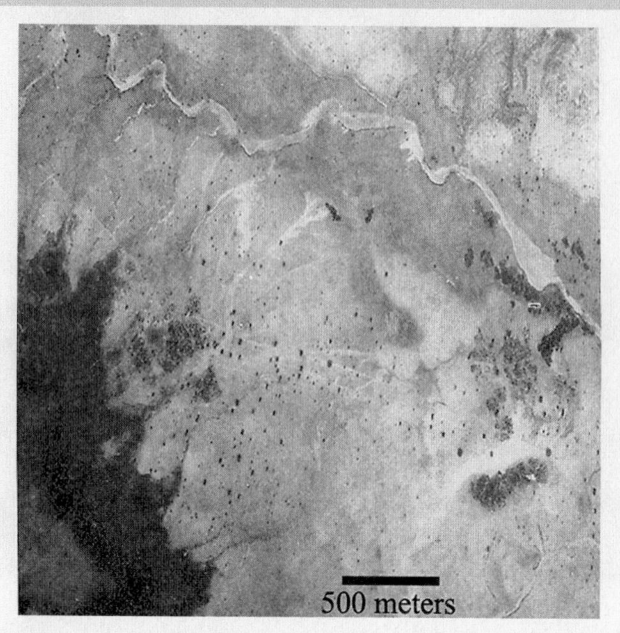

(a)

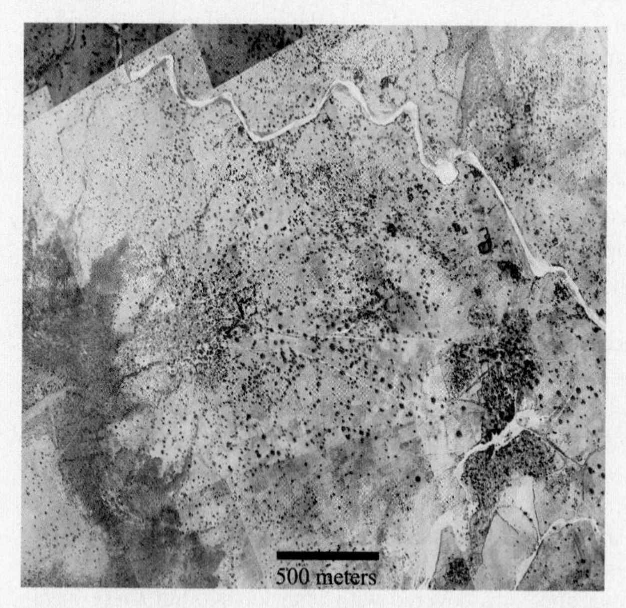

(b)

Figure 6.1.1 More Trees in Southern Niger These satellite images demonstrate an increase in tree cover in Niger. In 1975 (a), there were relatively few trees in this part of the Sahel. By 2003 (b), tree cover had increased substantially.

Source: Adapted from Lydia Polgreen, "In Niger, Trees and Crops Turn Back the Desert," *New York Times,* Feb, 11, 2007.

Figure 6.12 Deforestation in Madagascar A man fills a bag with charcoal made from burning tropical forest. Much of the island has been deforested, due to a reliance on slash-and-burn agriculture and biofuels. Loss of forest cover has led to serious problems with soil erosion and species extinction for the island nation. About 3 percent of the national territory is in protected areas.

proved. As a result of these efforts, the total forest area of Kenya remained about the same between 1990 and 2007 even though the population greatly increased. Maathai's success has drawn interest from other African countries, spurring a Pan-African Green Belt Movement largely organized through nongovernmental organizations interested in biofuel generation, protection of the environment, and the empowerment of women. In 2004 Professor Maathai was awarded a Nobel Peace Prize for her contribution to sustainable development, democracy, and peace.

Destruction of rain forest for logging is most evident in the southern fringes of Central Africa's Ituri forest (see Figure 6.9). Given the vastness of this forest and the relatively small number of people living there, however, it is less threatened than other forest areas. Two smaller rain forests, one along the Atlantic coast from Sierra Leone to western Ghana and the other along the eastern coast of the island of Madagascar, have nearly disappeared. These rain forests have been severely degraded by commercial logging and agricultural clearance. At present rates, several countries in West Africa will be completely deforested early in this century. For example, according to World Bank estimates, only 5 percent of the West African nation of Togo is forested. Off the coast of southern Africa, Madagascar's forests have suffered serious degradation in the past four decades (Figure 6.12). Deforestation in Madagascar is especially worrisome because the island forms a unique environment with a large number of native species. Many species of fauna (including birds, reptiles, and lemurs) now are endangered, and several face imminent extinction. Thankfully, over the past decade deforestation rates have slowed, which is a good sign, but less than a quarter of the island remains forested.

Wildlife Conservation Sub-Saharan Africa is justly famous for its wildlife. In no other region of the world can one find such abundance and diversity, especially of large mammals. The survival of wildlife here reflects, to some extent, the historically low human population density and the fact that sleeping sickness (transferred by the tsetse fly) and other diseases have kept people and their livestock out of many areas. In addition, many African peoples have developed various means of successfully coexisting with wildlife and about 12 percent of the region is under nationally protected lands.

But as is true elsewhere in the world, wildlife is quickly declining in much of Sub-Saharan Africa. West Africa now has few areas of prime habitat, and what remains is rapidly shrinking. The most noted wildlife reserves are in East Africa; in Kenya and Tanzania these reserves are economically important major tourist attractions. Even there, however, population pressure, political instability, and poverty make the maintenance of large wildlife reserves difficult. Poaching is a major problem, particularly for rhinoceroses and elephants; the price of a single horn or tusk in distant markets represents several years' wages for most Africans. Ivory is avidly sought in East Asia, especially in Japan. In China, powdered rhino horn is used as a traditional medicine, whereas in Yemen rhino horn is prized for dagger handles. During the 1980s, the region's elephant population fell by more than half, to 600,000, in part due to the ivory trade.

The most secure wildlife reserves now are located in southern Africa. Zoologists and park guides have employed handheld global positioning systems, called *CyberTrackers*, to record wildlife observations and map their findings. In Botswana, the San (a tribal group) have been given these devices to record the animal movements that only these expert hunters usually observe. In fact, elephant populations are considered to be too large in countries such as Botswana and South Africa. Some wildlife experts here contend that herds should be culled to prevent overgrazing and that the ivory should be legally sold in the international market in order to generate revenue for further conservation. Only such a market-oriented approach, they argue, will give African countries a long-term incentive to preserve habitat. Many environmentalists, not surprisingly, disagree strongly.

In 1989, a worldwide ban on the legal ivory trade was imposed as part of the Convention on International Trade in Endangered Species (CITES). While several African states, such as Kenya, lobbied hard for the ban, others, such as Zimbabwe, Namibia, and Botswana, complained that their herds were growing and the sale of ivory had helped to pay for conservation efforts. Conservationists feared that lifting the ban would bring on a new wave of poaching and illegal trade. In the late 1990s, the ban was lifted so that some southern African states could sell down their substantial inventories of ivory and continue with limited sales. The long-term repercussions for elephant survival are not yet known, although political leaders from Kenya and India have voiced strong opposition to all ivory sales. In 2000 elephant population estimates ranged from 400,000 to 600,000, with the majority of these animals found in protected areas. Rhino populations are far smaller at 10,000 to 15,000. The black rhino is a critically endangered species and the white rhino is more numerous but endangered as well (Figure 6.13).

Both are protected under CITES and rhino horn trade is illegal. The ivory controversy, however, shows how global markets and international conservation policies impact elephant survival in Sub-Saharan Africa.

Global Warming and Vulnerability in Sub-Saharan Africa

Global climate change poses extreme risks for this region due to its widespread poverty, recurrent droughts, and overdependence upon rain-fed agriculture. Sub-Saharan Africa is the lowest emitter of greenhouse gases in the world, but it is likely to experience greater human vulnerability to global warming because of the region's limited resources to both respond and adapt to environmental change. The areas most vulnerable are arid and semiarid regions such as the Sahel and Horn of Africa, some grassland areas, and the coastal lowlands of West Africa and Angola.

Climate change models suggest that parts of highland East Africa and equatorial Central Africa may receive more rainfall. Thus some lands that are currently marginal for farming might be made more productive. These effects are likely to be negated by the decline in agricultural productivity in the Sahelian belt as well as the grasslands of southern Africa, especially in Zambia and Zimbabwe. Drier grassland areas could deplete wildlife populations, which are a major source behind the growing tourist economy. As in Latin America, higher temperatures in the tropics may also result in the expansion of vector-born diseases such as malaria and dengue fever into the highlands where they were once relatively rare. Given the relatively high elevations in the region, the negative consequences of sea-level rising would be mostly felt on the West African coast (Senegal, Gambia, Sierra Leone, Nigeria, Cameroon, and Gabon).

The World Food Program estimates that millions of Africans struggle against starvation each year. Famine occurs when there is a surge in hunger-related mortality. Most global warming scenarios show that prolonged drought will become more common in parts of the region over time. As critical resources such as water and food become scarce, resource-driven conflicts and human displacement become more likely. Some scholars argue that the ethnic conflict in Sudan's Darfur region is driven, in part, by resource competition between pastoralists and agriculturalists in an ecosystem that is experiencing water scarcity.

Even without the threat of global warming, famine stalks many areas of Africa. The Famine Early Warning Systems (FEWS) network is in place to monitor food insecurity throughout the developing world, but especially in Sub-Saharan Africa. By tracking rainfall, vegetation cover, food production, food prices, and conflict, the network maps food security along a continuum from food secure to famine. In 2005, crops in Niger failed due to drought and an estimated 2.5 million people were vulnerable to famine but early warning systems were taken seriously and relief efforts resulted in relatively low mortality rates. Figure 6.14 shows the status

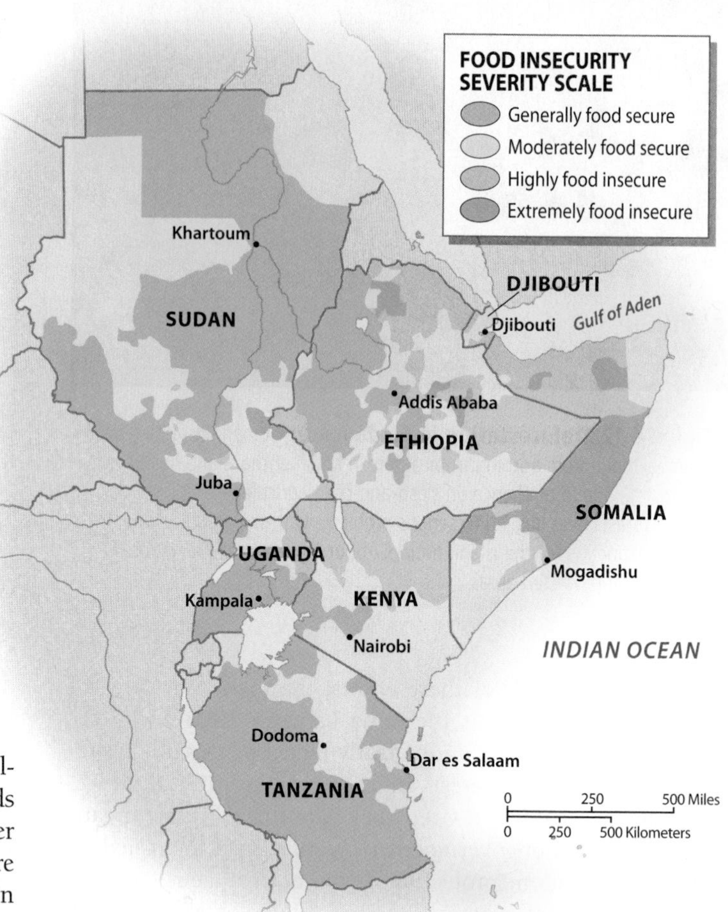

Figure 6.14 Map of Food Insecurity in Eastern Africa Anticipating areas of food insecurity based upon the timing and amount of rainfall and changes in vegetation/crop cover is the mission of FEWS network. FEWS has existed since the late 1980s to map areas of potential famine, especially in Sub Saharan Africa. (Source: Famine Early Warning System (FEWS) Network, April to September 2010 Report, U.S. Agency for International Development)

Figure 6.13 South African Wildlife A white rhinoceros grazes the savannas of Kruger National Park in South Africa, one of the region's oldest wildlife parks. A protected and endangered species, white rhinos are found in South Africa, Zimbabwe, Botswana, Angola and Kenya.

of food security in eastern Africa during a relatively wet year in 2010 when food insecurity conditions were expected to remain stable or improve slightly and there were no areas under famine status. Still, even in a good year, the FEWS report estimated that 19 million people faced food insecurity in eastern Africa, especially in Somalia and eastern Ethiopia.

POPULATION AND SETTLEMENT: Young and Restless

Sub-Saharan Africa's population is growing quickly. While the global population is projected to increase by nearly 40 percent by 2050, the projected population growth for Africa south of the Sahara is over 100 percent over the same time period. It is also a very young population with 43 percent of the people younger than age 15, compared to just 17 percent for more-developed countries. Only 3 percent of the region's population is over 65 years of age, whereas in Europe 16 percent is over 65. Families tend to be large, with a woman having an average of five or six children (Table 6.1). Yet child and maternal mortality rates are high, reflecting disturbingly low access to basic health services. The most troubling indicator for the region is its stubbornly low life expectancy of 52 years. Life expectancy in other developing nations is much better; India's is 64 years and China's 74 years. The growth of cities also is a major trend. In 1980 it was estimated that 23 percent of the population lived in cities; now the figure is 35 percent.

Behind these demographic facts lie complex differences in settlement patterns, livelihoods, belief systems, and access to health care. Although the region is experiencing rapid population growth, Sub-Saharan Africa is not densely populated. The entire region holds 865 million persons—roughly half the population that is crowded into the much smaller land area of South Asia. In fact, the overall population density of the region (36 people per square kilometer) is similar to that of the United States (32 people per square kilometer). Just six states account for over half of the region's population: Nigeria, Ethiopia, the Democratic Republic of Congo, South Africa, Tanzania, and Sudan (see Table 6.1). Some states have very high population densities (such as Rwanda or Mauritius) while others are sparsely settled. Namibia and Mauritania have just 3 people per square kilometer.

Crude population density is an imperfect indicator of whether a country is overpopulated. Geographers are often more interested in **physiological density**, the number of people per unit of arable land. The physiological density in Chad, where only 3 percent of the land is arable, is much higher than its crude population density of 9 people per square kilometer. Perhaps a more telling indicator of population pressure and potential food shortages is **agricultural density**, the number of farmers per unit of arable land. Because the majority of people in Sub-Saharan Africa earn their livings from agriculture, agricultural density indicates the number of people who directly depend on each arable square kilometer. The agricultural density of many Sub-Saharan countries is 10 times greater than their crude population density.

Population Trends and Demographic Debates

The growth in particular areas of Sub-Saharan Africa with high agricultural densities and rates of natural increase has combined with reversals in some economic and social indicators to make demographers concerned about the region's overall well-being. The Horn of Africa, for example, is not crowded by European or Asian standards, but it may already contain too many people for the land to support given the unpredictability of its limited rainfall.

This assertion remains controversial, however—as does the entire field of African demography. Some believe that the region could support many more people than it presently does and, as a whole, it is still underpopulated. Others argue that the region is a demographic time bomb and that unless fertility is quickly reduced, Sub-Saharan Africa will face continued food insecurity and suffering. The majority of African states officially support lowering rates of natural increase and are steadily promoting modern contraception practices both to reduce family size and to protect people from sexually transmitted diseases. One thing is certain: the region is, demographically, the fastest growing world region.

Family Size A preference for large families is the basis for the region's demographic growth where most women have 5 to 6 live births in a lifetime. In the 1960s many areas in the developing world had total fertility rates (TFR) of 5.0 or higher. Today Sub-Saharan Africa is the only region with such a high TFR at 5.2. A combination of cultural practices, rural lifestyles, high child mortality, and economic realities encourages large families (Figure 6.15).

Throughout the region, large families guarantee a family's lineage and status. Even now, most women marry young, typically when they are teenagers, which increases

Figure 6.15 Large Families This family in Senegal includes five children. Large families are still common in Sub-Saharan Africa. The average total fertility rate for the region is 5.2 children per woman.

TABLE 6.1 Population Indicators

Country	Population (millions) 2010	Population Density (per square kilometer)	Rate of Natural Increase (RNI)	Total Fertility Rate	Percent Urban	Percent <15	Percent >65	Net Migration (Rate per 1000) 2005–10[a]
Angola	19.0	15	2.5	5.8	57	45	2	0.9
Benin	9.8	87	3.0	5.6	41	45	3	1.2
Botswana	1.8	3	1.9	3.2	60	33	5	1.6
Burkina Faso	16.2	59	3.4	6.0	23	46	3	–0.9
Burundi	8.5	306	2.1	5.4	10	41	3	8.1
Cameroon	20.0	42	2.3	4.7	53	41	4	–0.2
Cape Verde	0.5	128	2.0	2.9	61	35	5	–5.1
Central African Republic	4.8	8	2.2	4.8	38	41	4	0.2
Chad	11.5	9	2.9	6.2	27	46	3	–1.4
Comoros	0.7	309	2.6	4.1	28	38	3	–3.1
Congo	3.9	12	2.5	5.0	60	42	4	–2.8
Dem. Rep. of Congo	67.8	29	2.9	6.4	33	48	3	–0.3
Djibouti	0.9	38	1.8	4.0	76	37	3	0.0
Equatorial Guinea	0.7	25	2.3	5.5	39	42	3	3.1
Eritrea	5.2	44	2.9	4.7	21	42	2	2.3
Ethiopia	85.0	77	2.7	5.4	16	44	3	-0.8
Gabon	1.5	6	1.9	3.6	84	39	4	0.7
Gambia	1.8	155	2.7	5.3	54	43	3	1.8
Ghana	24.0	101	2.2	4.0	48	39	4	–0.4
Guinea	10.8	44	3.0	5.7	28	43	3	–6.1
Guinea-Bissau	1.6	46	2.4	5.8	30	43	3	–1.6
Ivory Coast	22.0	68	2.4	4.9	61	40	2	–1.4
Kenya	40.0	69	2.7	4.6	18	42	3	–1.0
Lesotho	1.9	63	0.9	3.2	23	34	6	–3.5
Liberia	4.1	37	3.3	5.9	58	44	3	13.3
Madagascar	20.1	34	2.7	4.8	31	43	3	–0.1
Malawi	15.4	130	2.9	6.0	14	46	3	–0.3
Mali	15.2	12	3.1	6.6	33	48	3	–3.2
Mauritania	3.4	3	2.3	4.5	40	40	3	0.6
Mauritius	1.3	628	0.5	1.5	42	22	7	0.0
Mozambique	23.4	29	2.3	5.1	31	44	3	–0.2
Namibia	2.2	3	1.9	3.4	35	38	4	–0.1
Niger	15.9	13	3.5	7.4	20	49	2	–0.4
Nigeria	158.3	171	2.4	5.7	47	43	3	–0.4
Reunion	0.8	333	1.3	2.4	92	26	8	0.0
Rwanda	10.4	395	2.9	5.4	17	42	2	0.3
São Tomé and Principe	0.2	170	2.9	4.9	58	44	4	–8.8
Senegal	12.5	64	2.8	4.9	41	44	2	–1.7
Seychelles	0.1	193	1.0	2.3	53	22	10	
Sierra Leone	5.8	81	2.4	5.1	36	43	2	2.2
Somalia	9.4	15	3.0	6.5	34	45	3	–5.6
South Africa	49.9	41	0.9	2.4	52	31	5	2.8
Sudan	43.2	17	2.2	4.5	38	41	3	0.7
Swaziland	1.2	69	1.5	3.7	22	40	3	–1.0
Tanzania	45.0	48	3.0	5.6	25	45	3	–1.4
Togo	6.8	119	2.5	4.8	40	41	3	–0.2
Uganda	33.8	140	3.4	6.5	13	49	3	–0.9
Zambia	13.3	18	2.5	6.2	37	46	3	–1.4
Zimbabwe	12.6	32	1.3	3.7	37	42	4	–11.1

[a]Net Migration Rate from the United Nations, Population Division, World Population Prospects: The 2008 Revision Population Database.

Source: Population Reference Bureau, World Population Data Sheet, 2010

their opportunities to have more children. Demographers often point to the limited formal education available to women as another factor contributing to high fertility. Stubbornly high child mortality rates may encourage women to have more children. In most states one in ten children die before their fifth birthday; in the poorest and war-torn countries, such as Chad and Somalia, one in five children perish. Interestingly, religious affiliation seems to have little bearing on the region's fertility rates; Muslim, Christian, and animist communities all have similarly high birthrates.

The everyday realities of rural life make large families an asset. Children are an important source of labor; from tending crops and livestock to gathering fuel, they add more to the household economy than they take. Also, for the poorest places in the developing world, such as Sub-Saharan Africa, children are seen as social security. When parents' health falters, they expect their grown children to care for them.

Government policies toward family size have shifted dramatically in the past four decades. During the 1970s, population growth was not perceived as a problem by many African governments; in fact, many equated limiting population size with a neocolonial attempt to slow regional development. By the 1980s, a shift in national policies occurred. Following the United Nations International Conference on Population and Development in Cairo, Egypt, African governments announced their intent to bring natural increase rates down to 2.0 and to increase the rate of contraceptive use to 40 percent by 2020. They may reach their goal: As of 2010, the regional rate of natural increase was 2.5 percent, and nearly one-quarter of married women used some method of contraception.

Migration patterns within the region and out of it are also influencing growth rates. Africans are mobile people and movement across borders within the region has long been a part of livelihood systems and survival strategies during times of conflict. As Table 6.1 shows, some African states such as Burundi and Liberia have very high rates of net migration either due to receiving refugees or former refugees returning home. Similarly, the states with some of the highest negative net migration rates are places experiencing conflict such as Zimbabwe, or are states extremely dependent upon foreign labor migration such as Cape Verde. In general, many of the international flows in Africa are from one poor country to another and most are not recorded. Sub-Saharan Africans are also leaving the region in small but substantial numbers, especially to Europe and North America.

Other factors are converging to slow the growth rate. As African states slowly become more urban, there is a corresponding decline in family size—a pattern seen throughout the world. Tragically, declines in natural increase also are occurring as a result of AIDS.

The Impact of AIDS on Africa Now entering its fourth decade, HIV/AIDS may become one of the deadliest epidemics in modern human history. As of 2008, two-thirds of the HIV/AIDS cases in the world were found in Sub-Saharan Africa, where 22.4 million adults and children are living with HIV/AIDS. In 2008 there were 1.9 million new HIV infections, down from 2.3 million in 2001. This is the world region most heavily affected by HIV. Yet there are some hopeful signs as the number of newly infected people and the number of deaths due to AIDS have started to decline slowly.

The virus is thought to have originated in the forests of the Congo, possibly crossing over from chimpanzees to humans sometime in the 1950s. Yet it was not until the 1980s that the impact of the disease was widely felt. In Sub-Saharan Africa, as in much of the developing world, the disease is transmitted by unprotected heterosexual transmission. Mother-to-child transmission through the birthing process or breast feeding also contributes to the number of new cases, although strides have been made reduce these numbers. Women bear a disproportionate burden of the HIV/AIDS epidemic. They account for approximately 60 percent of the HIV infections and they are usually the caregivers for those who are infected. Until the late 1990s, many African governments were unwilling to acknowledge publicly the severity of the situation or to discuss frankly the measures necessary for prevention. The consequences of this inaction were deadly.

Southern Africa is ground zero for the AIDS epidemic that is ravaging the region (Figure 6.16). The nine countries with the highest HIV prevalence are all located there. In South Africa, the most populous state in southern Africa, 5.7 million people (nearly one in five people aged 15–49) are infected with HIV/AIDS. The rate of infection in neighboring Botswana is 24 percent (down from a staggering 36 percent in 2001) for the same age group. Although the rate of infection is highest in southern Africa, East Africa and West Africa are seeing increases. In Ivory Coast, it is estimated that 4 percent of the 15–49 age group has HIV or AIDS. For Kenya the figure is 6 percent. By comparison, only 0.6 percent of the same age group in North America is infected.

Sadly, the social and economic implications of this epidemic are profound. Life expectancy rates have tumbled in the last decade, in a few places dropping to the early 40s. AIDS typically hits the portion of the population that is most active economically. Time lost to care for sick family members and the outlay of workers' compensation benefits has reduced economic productivity and overwhelmed public services in hard-hit areas. The disease makes no class distinctions, so that countries are losing peasant farmers and educated professionals (doctors, engineers, and teachers). Infection rates among newborns are high but several countries have been successful at using antiviral medicines to block the transmission of HIV from mothers to children at birth. Many countries are struggling to care for nearly 12 million children orphaned by AIDS.

There was a scaling-up of access to antiretroviral therapies and AIDS prevention education in the past decade. For some countries, such as Botswana, Kenya, and Uganda, this has resulted in significantly lower death rates due to the disease (Figure 6.17). Prevention education has resulted in a significant decline in HIV incidence in Tanzania. HIV testing, counseling, and prevention services in prenatal settings

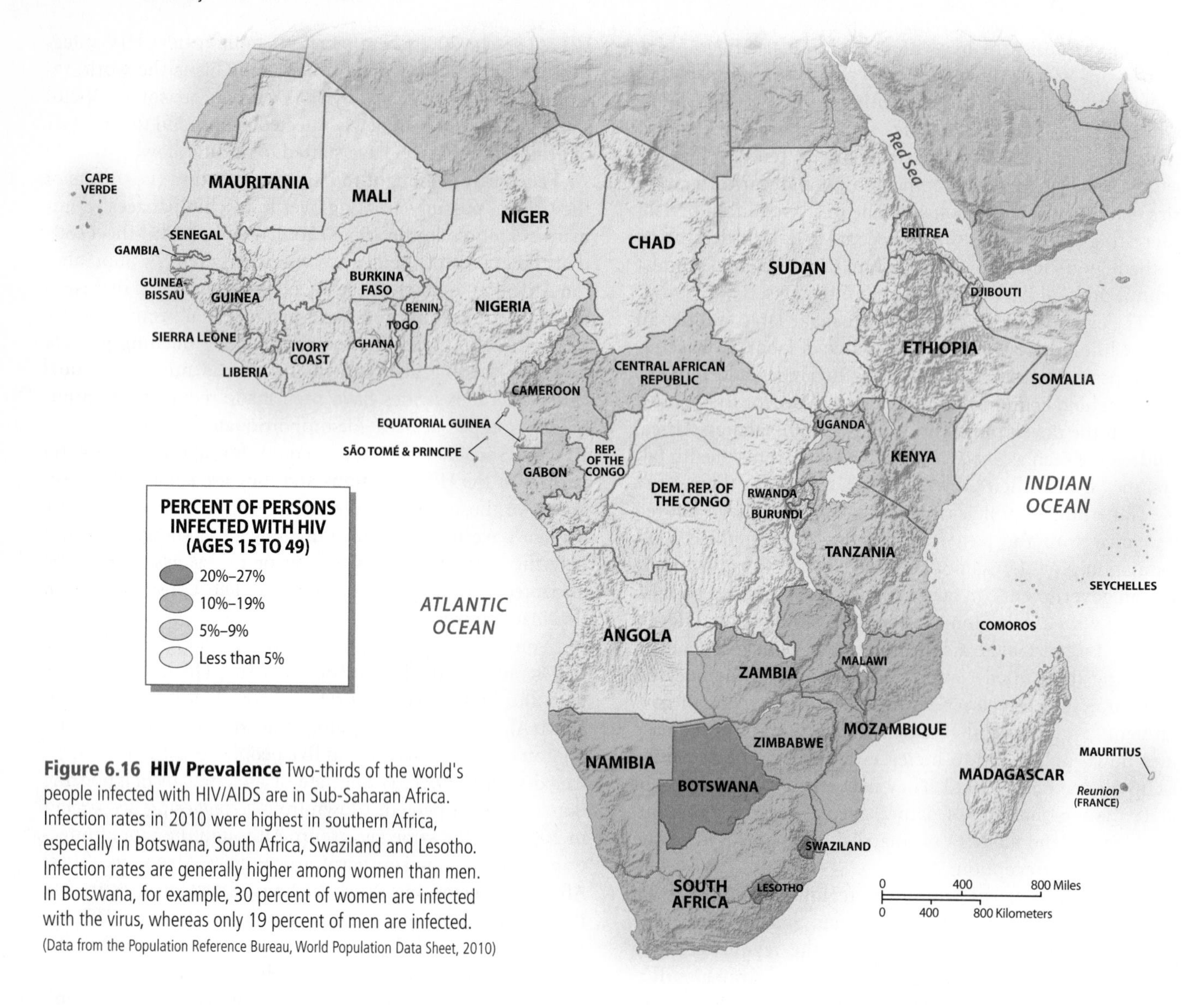

Figure 6.16 HIV Prevalence Two-thirds of the world's people infected with HIV/AIDS are in Sub-Saharan Africa. Infection rates in 2010 were highest in southern Africa, especially in Botswana, South Africa, Swaziland and Lesotho. Infection rates are generally higher among women than men. In Botswana, for example, 30 percent of women are infected with the virus, whereas only 19 percent of men are infected. (Data from the Population Reference Bureau, World Population Data Sheet, 2010)

have proven to be important in preventing newborns from being infected and increasing the life expectancy of HIV-infected women in the region.

For now, the surest way to stem the epidemic is through prevention, mostly through educating people about how the virus is spread and persuading them to change their sexual behavior. In Uganda, state agencies, along with NGOs, began a national no-nonsense campaign for AIDS awareness in the 1980s that focused on the schools. President Museveni of Uganda pioneered the catchy phrase "Abstain-Be faithful-use a Condom" (ABC). As a result of explicit materials, role-playing games, and frank discussion, the prevalence of HIV among women in prenatal clinics had declined by the late 1990s.

Patterns of Settlement and Land Use

Because of the dominance of rural settlements in Sub-Saharan Africa, people are widely scattered throughout the region (Figure 6.18). Population concentrations are the highest in West Africa, highland East Africa, and the eastern half of South Africa. The first two areas have some of the region's best soils, and indigenous systems of permanent agriculture developed there. In South Africa, the more densely settled east results from an urbanized economy based on mining, as well as the forced concentration of black South Africans into eastern homelands.

West Africa is more heavily populated than most of Sub-Saharan Africa, although the actual distribution pattern is patchy. Density in the far west, from Senegal to Liberia, is moderate; this area is characterized by broad lowlands with decent soils, and in many areas the cultivation of wet rice has enhanced agricultural productivity. Greater concentrations of people are found along the Gulf of Guinea, from southern Ghana through southern Nigeria, and again in northern Nigeria along the southern fringe of the Sahel. Nigeria is moderately to densely settled through most of its extensive territory; with nearly 160 million inhabitants, it stands as the demographic core of Sub-Saharan Africa. The next largest country, Ethiopia, has a population of 85 million.

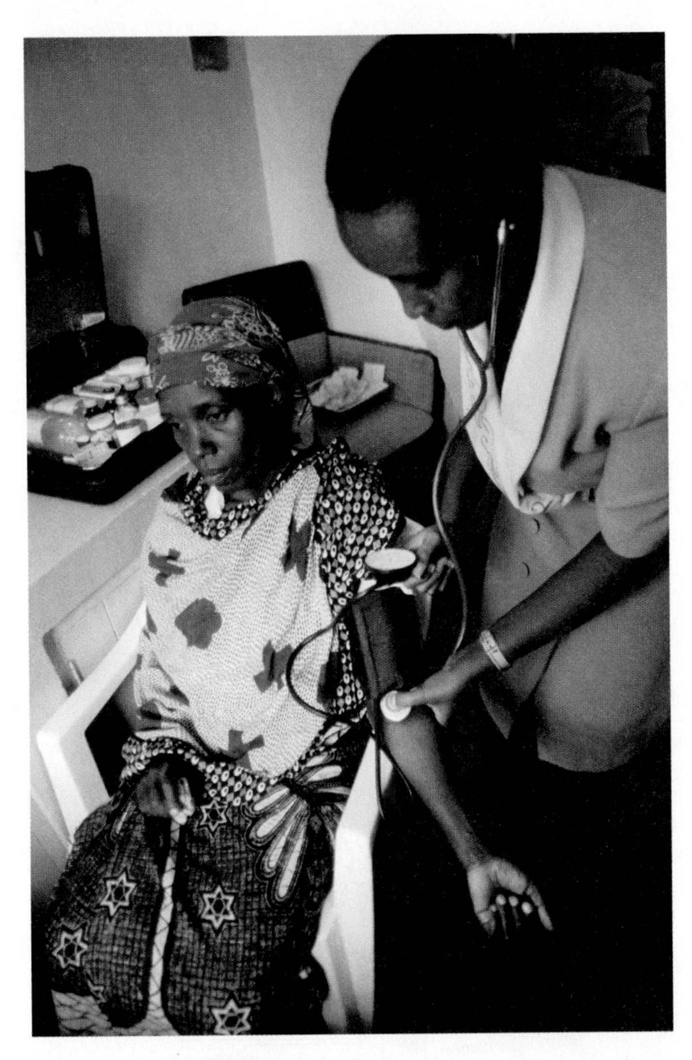

Figure 6.17 HIV/AIDS Clinic in Tanzania A woman with AIDS receives a check-up in a Tanzanian clinic. Greater availability of antiretroviral medication in the region has improved the life expectancy of people living with HIV/AIDS. Still the infection rates are high, especially among women.

As more Africans move to cities, patterns of settlement are evolving into clusters of higher concentration. Towns that were once small administrative centers for colonial elites grew into major cities. The region even has its own megacity, Lagos, which recently topped 10 million residents. Throughout the continent, African cities are growing faster than rural areas. But before examining the Sub-Saharan urban scene, a more detailed discussion of rural subsistence is needed.

Agricultural Subsistence The staple crops over most of Sub-Saharan Africa are millet, sorghum, rice, and corn (maize), as well as a variety of tubers and root crops such as yams. Irrigated rice is widely grown in West Africa and Madagascar. Geographer Judith Carney in her book *Black Rice* documents how African slaves introduced rice cultivation to the Americas. Corn, in contrast, was introduced to Africa from the Americas by the slave trade and quickly grew to become a basic food. In higher elevations of Ethiopia and South Africa, wheat and barley are grown. Intermixed with subsistence foods are a variety of export crops—coffee, tea, rubber, bananas, cocoa, cotton, and peanuts—that are grown in distinct ecological zones and often in some of the best soils.

In areas that support annual cropping, population densities are greater. In parts of humid West Africa, for example, the yam became the king of subsistence crops. The Ibos' mastery of yam production allowed them to procure more food and live in denser permanent settlements in the southeastern corner of present-day Nigeria. Much of traditional Ibo culture is tied to the arduous tasks of clearing the fields, tending the delicate plants, and celebrating the harvest.

Over much of the continent, African agriculture remains relatively unproductive, and rural population densities tend to be low. Amid the poorer tropical soils, cropping usually entails shifting cultivation, or **swidden**. This process involves burning the natural vegetation to release fertilizing ash and planting crops such as maize, beans, sweet potatoes, banana, papaya, manioc, yams, melon, and squash. Each plot is temporarily abandoned once its source of nutrients has been exhausted. Swidden cultivation often is a very finely tuned adaptation to local environmental conditions, but it is unable to support high population densities.

Women are the subsistence farmers of the region, producing for their household needs as well as for local and foreign markets. In the densely settled country of Rwanda, most farms are small but the mountainous volcanic soil is ideal for growing coffee. In an effort to rebuild the country after the ethnic genocide in the 1990s, the government targeted improving the quality of the country's coffee through the formation of cooperatives that would bring ethnic groups together and improve farmers' incomes. The cooperatives have taken off and farmers, many of them war widows, have seen their incomes improve. For decades the country's coffee production languished, and farmers earned very little for their low-quality beans. Yet across Rwanda's hillsides were older varieties of coffee plants with high value in today's premium coffee market. Farmers just needed a better way to prepare the beans and market them. Through community-run cooperatives the quality of their washed and sorted beans improved. So too did their ability to bargain with major buyers such as Starbucks or Green Mountain, leaving out the middleman. For many small farmers in Rwanda, their premium coffee is now a source of pride (Figure 6.19).

Plantation Agriculture Designed to produce crops on a large scale for export, plantation agriculture is also critical to the economies of many states. If African countries are to import the modern goods they require, they must sell their own products on the world market. Because the region has few competitive industries, the bulk of its exports are primary products derived from farming, mining, and forestry.

A number of African countries rely heavily on one or two export crops. Coffee, for example, is vital for Ethiopia,

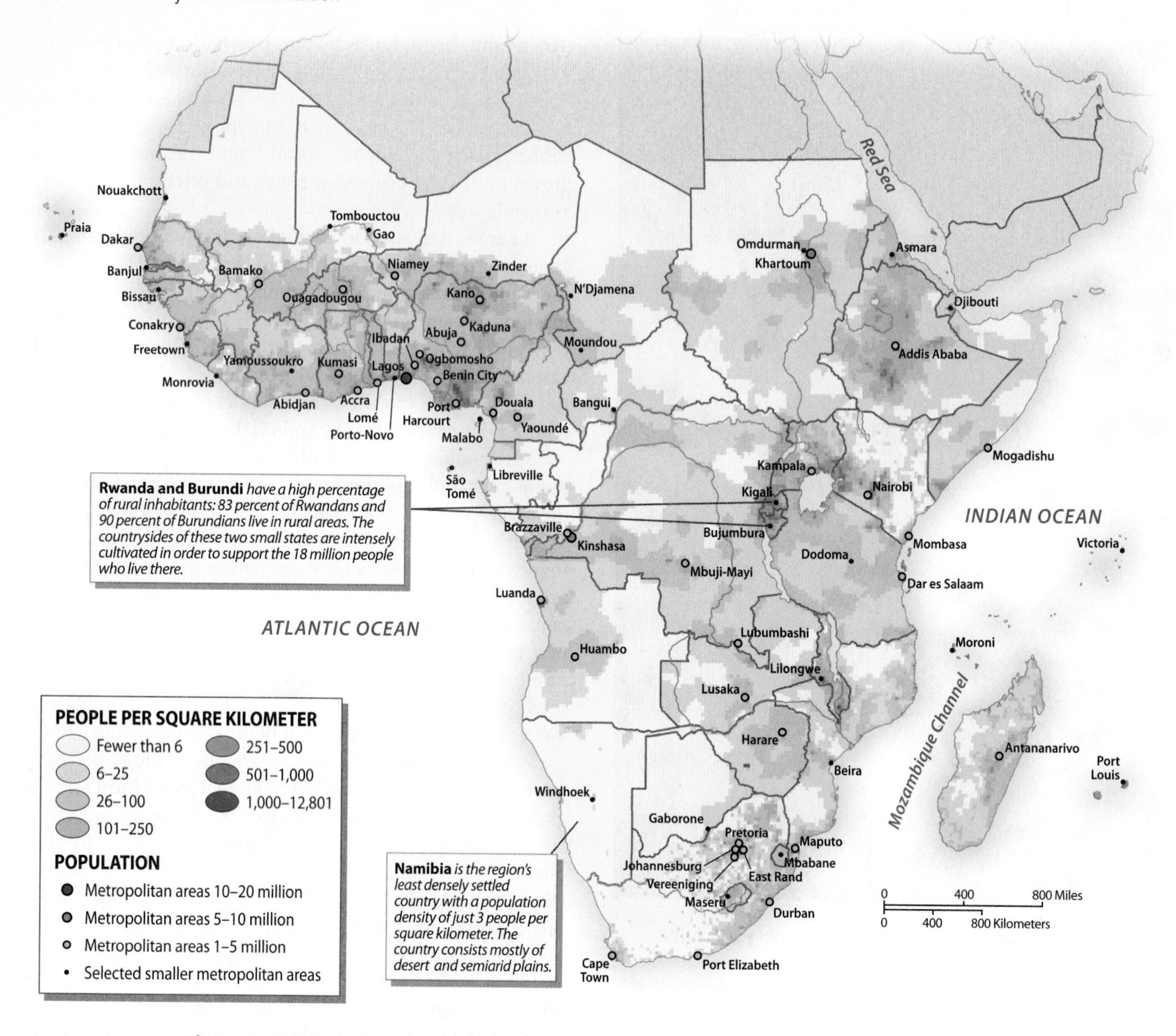

Figure 6.18 Population Distribution The majority of people in Sub-Saharan Africa live in rural areas. Some of these rural zones, however, are densely settled, such as West Africa and the East African highlands. Major urban centers, especially in South Africa and Nigeria, support millions. There is only one megacity in the region with more than 10 million (Lagos, Nigeria), but over two dozen cities have more than 1 million residents.

Kenya, Ivory Coast, and Rwanda. Tea is significant in the East African countries of Kenya, Uganda, and Tanzania. Peanuts have long been the primary foreign exchange earner in the Sahel, while cotton is tremendously important for Mali, Benin, Sudan, Zimbabwe, Ivory Coast, and Cameroon. Ghana and the Ivory Coast have long been the world's main suppliers of cacao (the source of chocolate); Liberia produces plantation rubber; and many farmers in Nigeria specialize in palm oil. The export of such products can bring good money when commodity prices are high as they are now, but when prices periodically collapse, economic devastation may follow.

Nontraditional agricultural exports have emerged in the last two decades that depend upon significant capital inputs and refrigerated air transportation. One industry is floriculture for the plant and cut flower industry. Here the highland tropical climate of Kenya, Ethiopia, and South Africa is advantageous. Each of these states has experienced growth in floriculture. Similarly, the European market for fresh vegetables and fruits in the winter is being met by some African producers in both West and East Africa.

Pastoralism Animal husbandry (the care of livestock) is extremely important in Sub-Saharan Africa, particularly in the semiarid zones. Camels and goats are the principal animals in the Sahel and the Horn of Africa, but farther south cattle are king (Figure 6.20). Many African peoples have traditionally specialized in cattle raising and are often tied into

Figure 6.19 Rwandan Coffee Farmer A farmer picks coffee berries at her farm outside of Kigali, Rwanda. Coffee was first domesticated in Ethiopia. Today, several African states, such as Rwanda, depend upon coffee as an important export crop.

mutually beneficial relationships with neighboring farmers. Such **pastoralists** typically graze their stock on the stubble of harvested fields during the dry season and then move them to drier uncultivated areas during the wet season when the pastures turn green. Farmers thus have their fields fertilized by the manure of the pastoralists' stock, while the pastoralists find good dry-season grazing. At the same time, the nomads can trade their animal products for grain and other goods of the sedentary world. Several pastoral peoples of East Africa, however, are noted for their extreme reliance on cattle and general (but never complete) independence from agriculture. The Masai of the Tanzanian–Kenyan borderlands traditionally derive a large percentage of their nutrition from drinking a mixture of milk and blood. The blood is obtained by periodically tapping the animal's jugular vein, a procedure that evidently causes little harm.

Large expanses of Sub-Saharan Africa have been off-limits to cattle because of infestations of **tsetse flies**, which spread sleeping sickness to cattle, humans, and some wildlife. Where wild animals, which harbor the disease but are immune to it, were present in large numbers, especially in environments containing brush or woodland (which are necessary for tsetse fly survival), cattle simply could not be raised. Some evidence suggests that tsetse fly infestations dramatically increased in the late 1800s, greatly harming African societies dependent on livestock but benefiting wildlife populations. At present, tsetse fly eradication programs are reducing the threat, and cattle are spreading into areas that were previously unsuitable. When people and their livestock move into new areas in sizable numbers, wildlife almost inevitably declines.

Sub-Saharan Africa thus presents a difficult environment for raising livestock because of the virulence of its animal diseases. In the tropical rain forests of Central Africa, cattle have never survived well and the only domestic animal that thrives is the goat. Raising horses, moreover, has historically been feasible only in the Sahel and in South Africa. As we shall see later, the disease environment of tropical Africa has presented a variety of problems for humans as well.

Figure 6.20 Pastoralists in the Horn of Africa Two Ethiopian girls walk along a herd of camels. Camels are well suited for the arid conditions of eastern Ethiopia and Somalia. Pastoralists rely on their camels for milk, transport and trade.

Urban Life

Although Sub-Saharan Africa is considered the least urbanized region in the developing world, most Sub-Saharan cities are growing at twice the national growth rates. If present trends continue, half of the region's population may well be living in cities by 2030. One of the consequences of this surge in city living is urban sprawl. Rural-to-urban migration, industrialization, and refugee flows are forcing the cities of the region to absorb more people and use more resources. As in Latin America, the tendency is toward urban primacy, the condition in which one major city is dominant and at least three times larger than the next largest city. Luanda, the capital of Angola, was built for half a million people but now has at least 4 million residents (many of them were displaced during the country's long years of war). Most of the city's inhabitants live in the growing slums that surround the colonial core (Figure 6.21). In such rapidly growing places, city officials struggle to build enough roads and provide electricity, water, trash collection, and employment for all these people.

European colonialism greatly influenced urban form and development in the region. Africans, however, had an urban tradition prior to the colonial era, although a very small percentage of the population lived in cities. Ancient cities, such as Axum in Ethiopia, thrived 2,000 years ago. Similarly, in the Sahel, prominent trans-Saharan trade centers, such as Timbuktu (Tombouctou) and Gao, have existed for more than a millennium. In East Africa, an urban mercantile culture emerged that was rooted in Islam and the Swahili language. The prominent cities of Zanzibar, Tanzania, and Mombasa, Kenya, flourished by supporting a trade network that linked the East African highlands with the Persian Gulf. The stone ruins of Great Zimbabwe in southern Africa are testimony to the achievements of stone working, metallurgy, and religion achieved by Bantu groups in the 14th century. West Africa, however, had the most developed precolonial urban network, reflecting both indigenous and Islamic traditions. Today it supports some of the region's largest cities.

West African Urban Traditions The West African coastline is dotted with cities, from Dakar, Senegal, in the far north to Lagos, Nigeria, in the east. Half of Nigerians live in cities, and seven of Nigeria's metropolitan areas have populations of more than 1 million. Historically, the Yoruba cities in southwestern Nigeria are the best documented. Developed in the 12th century, cities such as Ibadan were walled and gated, with a palace encircled by large rectangular courtyards at the city center. An important center of trade for an extensive hinterland, Ibadan also was a religious and political center. Lagos was another Yoruba settlement. Founded on a coastal island on the Bight of Benin, most of the modern city has spread onto the nearby mainland. Its coastal setting and natural harbor made this relatively small, indigenous city attractive to colonial powers. When the British took control in the mid-19th century, the city's size and importance grew.

Whereas in 1960, Lagos was a city of only 1 million, today it is well over 10 million residents, making it the largest city in the region. Unable to keep up with the surge of rural migrants, Lagos's streets are clogged with traffic; for those living on the city's periphery, three- and four-hour commutes (one way) are common. For many, the informal sector (unregulated services and trade) provides employment. Crime is another major problem. The chances of being attacked on the streets or robbed in one's home are quite high, even though most windows are barred and houses are fenced. To deal with

Figure 6.21 Urban slums of Luanda, Angola The Roche Santiero market on the outskirts of Luanda serves the city's growing urban poor. Like many Sub-Saharan African cities, Luanda is growing much faster than rural areas in the country.

Figure 6.22 Downtown Lagos, Nigeria A sea of shoppers, vendors and buses fill the streets of the Oshodi Market, the commercial hub of Lagos. Lagos is the largest city in Sub-Saharan Africa with over 10 million inhabitants.

this problem, Lagos has the highest density of police in Nigeria, but the weak social bonds between urban migrants, widespread poverty, and the gap between rich and poor seem to encourage lawlessness (Figure 6.22).

Most West African cities are hybrids, combining Islamic, European, and national elements such as mosques, Victorian architecture, and streets named after independence leaders. Accra is the capital city of Ghana and home to nearly 2 million people. Originally settled by the Ga people in the 16th century, it became a British colonial administrative center by the late 1800s. Here again, the modern city is being transformed through neoliberal policies introduced in the 1980s that attracted international corporations. A growth in foreign investment in financial and producer services led to the creation of a "Global CBD" on the east side of the city away from the "National CBD" (Figure 6.23). Here foreign companies clustered in areas with secure land title, new modern roads, parking, and access to the airport. Upper-income gated communities have also formed near the Global CBD. Accra, like other cities in the region, is rapidly changing due to an influx of foreign capital. The result is highly segregated urban spaces that reflect a global phase in urban development in which world market forces, rather than colonial or national ones, are driving the change.

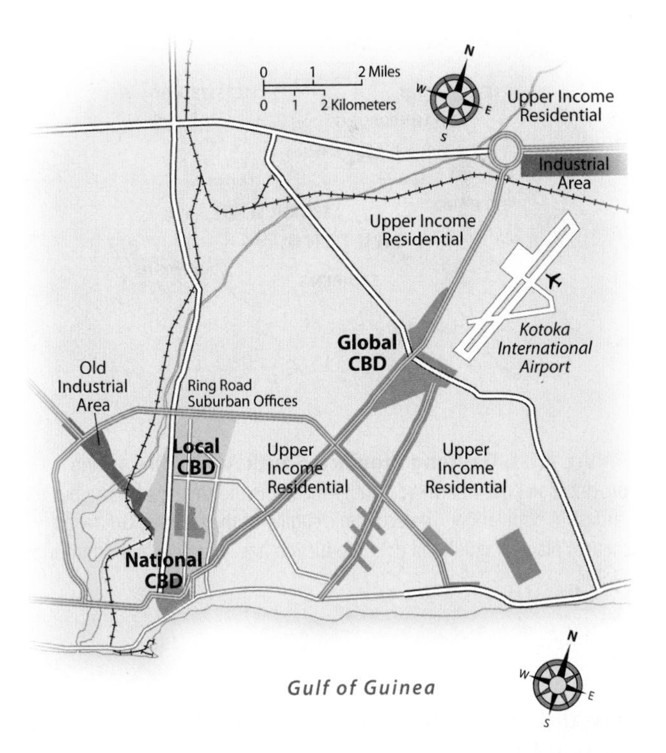

Figure 6.23 Contemporay Accra, Ghana This model represents the global phase of Accra's development with three distinct CBDs. The national CBD developed from the former European Town. The local CBD is nearby and formed from the Native Town. The Global CBD, fostered by international investment, is removed from the other two and close to the international airport. (From Grant and Nijman, 2002, "Globalization and the Corporate Geography of Cities in the Less-Developed World," *Annals of the Association of American Geographers* 92(2): 320–40.)

Urban Industrial South Africa The major cities of southern Africa, unlike those of West Africa, are colonial in origin. Cities such as Lusaka, Zambia, and Harare, Zimbabwe, grew as administrative or mining centers. South Africa is one of the most urbanized states in the entire region, and it is cer-

CITYSCAPES Africa's Golden and Global City

Johannesburg is a city of contradictions, where business executives and peasants, artists and refugees, blacks and whites, converge on a sprawling chaotic plateau 5700 ft (1750 m) above sea level in South Africa's *highveld.* Part of Gauteng Province, South Africa's smallest and most urbanized province, over 10 million people reside here in three major metropolitan areas. At the core is the City of Johannesburg (3.6 million), then there is Ekurhuleni (a recently consolidated metropolitan area of former townships to the east with 3.2 million), and Pretoria in the north (1.3 million and one of South Africa's three capitals). Gauteng is a Sesotho word meaning "place of gold," which is how the world came to know Johannesburg when immense gold deposits were discovered there in 1886 (Figure 6.2.1).

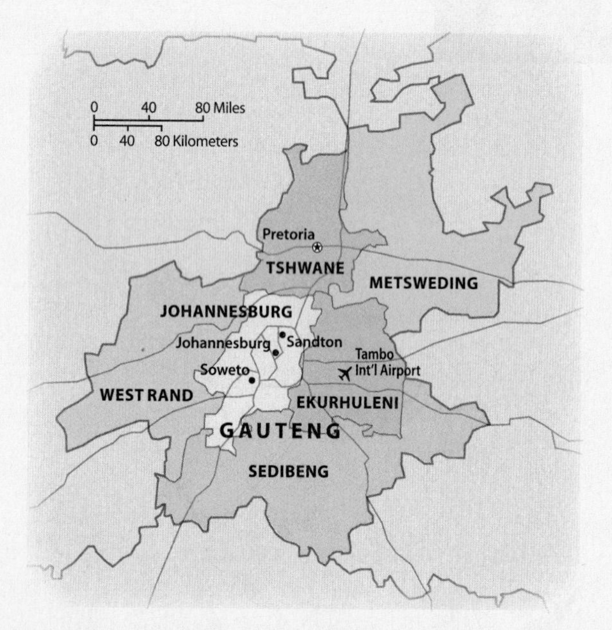

Figure 6.2.1 Gauteng Province, South Africa One of nine provinces in post-Apartheid South Africa, it is the smallest by size but the largest in population. The economic engine of the country, Guateng means 'place of gold' and gold production has long been important in this region.

During apartheid, Johannesburg was reworked and segregation intensified. Blacks were forcibly removed from the city center and into all-black townships (segregated neighborhoods on the outskirts of cities) to the east and southwest of the downtown. To the north, wealthier white suburbs formed. In between the commercial center and the townships were the gold mines, where black laborers worked. In fact, Soccer City, the premier stadium site during South Africa's World Cup, is located on a former mine, near the township of Soweto.

Like other townships, Soweto was linked to the mines and the city proper by trains—today's metrorail. A rail for the black masses, this old and over-crowded system was created so that townships would be far from white areas but black and coloured laborers could get to their places of employment. Through in-migration Soweto became one of the biggest townships in the country, and a center for political resistance. Both Nelson Mandela and Bishop Desmond Tutu, leaders against apartheid, had their homes in Soweto. In the post-apartheid era, Soweto is a mixture of lean-tos and middle class housing that attracts mostly black migrants from the former homelands and immigrants from neighboring countries such as Zimbabwe and Mozambique. Soweto was formally incorporated into the City of Johannesburg in the 1990s. And times are changing for Soweto; neat townhouses are replacing squatter homes and Maponya Mall, one of the largest in South Africa, opened in 2007 (Figure 6.2.2).

As apartheid ended, Johannesburg became infamous for its high crime rate. Many white-owned businesses fled the central business district for the northern suburb of Sandton. By the late 1990s, Sandton emerged as the new financial and business center of the City of Johannesburg. When Gauteng province invested $3 billion dollars in a new high-speed commuter rail called Gautrain, it was no accident that the first functioning line linked Sandton with Tambo International Airport. Eventually these luxurious trains will link other parts of Johannesburg as well as the city of Pretoria (Figure 6.2.3).

Whatever your image of Johannesburg may be—be it city of gold, sprawling slums, violent gangsters or cosmopolitan playground—this is the African metropolitan area that aspires to global city status. By successfully hosting the World Cup in 2010—the most watched international sporting event—South Africa, and especially Johannesburg, showed the world the economic vibrancy of this golden city.

tainly the most industrialized. The foundations of South Africa's urban economy rest largely on its incredibly rich mineral resources (diamonds, gold, chromium, platinum, tin, uranium, coal, iron ore, and manganese). Seven of its metropolitan areas have more than 1 million people; the largest of them are Johannesburg, Durban, and Cape Town.

The form of South African cities continues to be imprinted by the legacy of **apartheid**, an official policy of racial segregation that shaped social relations in South Africa for nearly 50 years. Even though apartheid was abolished in 1994, it is still evident in the landscape. Under apartheid rules, all cities were divided into residential areas according to racial categories: white, **coloured** (a South African term describing people of mixed African and European ancestry), Indian (South Asian), and African (black). In Cape Town, whites occupied the largest and most desirable portions of the city, especially the scenic areas below Table Mountain (Figure 6.24). Blacks were crowded into

Figure 6.2.2 Maponya Mall, Soweto Soweto residents leaving Maponya Mall after shopping. This shopping center, which opened in 2007, is emblematic of the new Soweto where first-world comforts meet the gritty reality of former township life. Soweto is home to some of South Africa's most important landmarks in the struggle over white-minority rule. There is still tremendous poverty here but increasingly shopping centers, middle class housing and landscaped parks are also part of the landscape.

Figure 6.2.3 Sandton, Johannesburg. Just north of Johannesburg's Central Business District, Sandton has emerged as the financial and business center of the New South Africa. Many businesses relocated here, fleeing high crime rates in the old CBD. With world-class shopping, hotels, a convention center, and even Nelson Mandela Square, Sandton has become the part of Johannesburg that most tourists and business people experience.

the least desired areas, forming squatter settlements called **townships** in places such as Gugulethu. Today blacks, coloureds, and Indians are legally allowed to live anywhere they want. Yet the economic differences between racial groups, as well as deep-rooted animosity, hinder residential integration. Post-apartheid settlements, such as Cape Town's Delpht South, promote residential integration of black and coloured households, something that was forbidden during the apartheid years.

Post-Apartheid South African cities, such as Cape Town and Johannesburg, are consciously being developed to reflect a multi-racial and globalized reality (see Cityscapes: Africa's Golden and Global City). As evidenced by the many construction projects completed to host the 2010 World Cup, the country sought to project a modern and progressive image. The relative economic strength of South Africa also means that thousands of immigrants from other African states have settled there in the last decade in search of eco-

Figure 6.24 Elite Cape Town Landscape Europeans settled in this prize farmland to the east and south of Table Mountain. Today this area is home to South Africa's best vineyards.

nomic opportunities not available in their countries of origin. This has produced tensions, as many black South Africans fear that African immigrants are competing with them for jobs and scare resources. Yet the economic clout and ethnic diversity of Johannesburg—along with its sister metropolitan area Ekurhuleni—make this urban space extremely significant in shaping cultural and urban trends for the rest of Sub-Saharan Africa.

CULTURAL COHERENCE AND DIVERSITY: Unity through Adversity

No world region is culturally homogeneous, but most have been partially unified in the past by widespread systems of belief and communication. Traditional African religions, however, were largely limited to local areas, and the religions that did become widespread, Islam and Christianity, are primarily associated with other world regions. A handful of African trade languages have long been understood over vast territories (Swahili in East Africa, Mandingo and Hausa in West Africa), but none spans the entire Sub-Saharan region. Sub-Saharan Africa also lacks a history of widespread political union or even of an indigenous system of political relations. The powerful African kingdoms and empires of past centuries all were limited to distinct subregions of the landmass.

The lack of traditional cultural and political coherence across Sub-Saharan Africa is not surprising if one considers the region's huge size. Sub-Saharan Africa is more than four times larger than Europe or South Asia. Had foreign imperialism not impinged on the region, it is quite possible that West Africa and southern Africa would have developed into their own distinct world regions.

An African identity south of the Sahara was forged through a common history of slavery and colonialism as well as struggles for independence and development. More telling, the people of the region often define themselves as African, especially to the outside world. That Sub-Saharan Africa is poor by most economic measures, no one will argue. And yet the cultural expressions of its people—its music, dance, and art—are joyous. Africans share an extraordinary resilience and optimism that visitors to the region often comment on. The cultural diversity of the region is obvious, yet there is unity among the people drawn from surviving many adversities.

Language Patterns

In most Sub-Saharan countries, as in other former colonies, multiple languages are used that reflect tribal, ethnic, colonial, and national affiliations. Indigenous languages, many from the Bantu subfamily, often are localized to relatively small rural areas. More widely spoken African trade languages, such as Swahili or Hausa, serve as a lingua franca over broader areas. Overlaying native languages are Indo-European (French and English) and Afro-Asiatic ones (Arabic). Figure 6.25 illustrates the complex pattern of language families and major languages found in Africa today. Contrast the larger map with the inset that shows current "official" languages. A comparison of the two shows that most African countries are multilingual, which can be a source of tension within states. In Nigeria, for example, the official language is English, yet there are millions of Hausa, Yoruba, Igbo (or Ibo), Ful (or Fulani), and Efik speakers, as well as speakers of dozens of other languages.

African Language Groups Three of the six language groups mapped in Figure 6.25 are unique to the region (Niger-Congo, Nilo-Saharan, and Khoisan), while the other three (Afro-Asiatic, Austronesian, and Indo-European) are more closely associated with other parts of the world. Afro-Asiatic languages, especially Arabic, dominate North Africa and are understood in Islamic areas of Sub-Saharan Africa as well. Amharic in Ethiopia and Somali of Somalia are also Afro-Asiatic languages. The Malayo-Polynesian language family is limited to the island of Madagascar, which many believe was first settled by seafarers from Indonesia some 1,500 years ago. Indo-European languages, especially French, English, Portuguese, and Afrikaans, are a legacy of colonialism and widely used today.

Of the three language groups found exclusively in the region, the Niger-Congo language group is by far the most influential. This linguistic group originated in West Africa and includes Mandingo, Yoruba, Ful (ani), and Igbo, among others. Around 3,000 years ago a people of the Niger-Congo stock began to expand out of western Africa into the equatorial zone. This group, called the Bantu, commenced one of

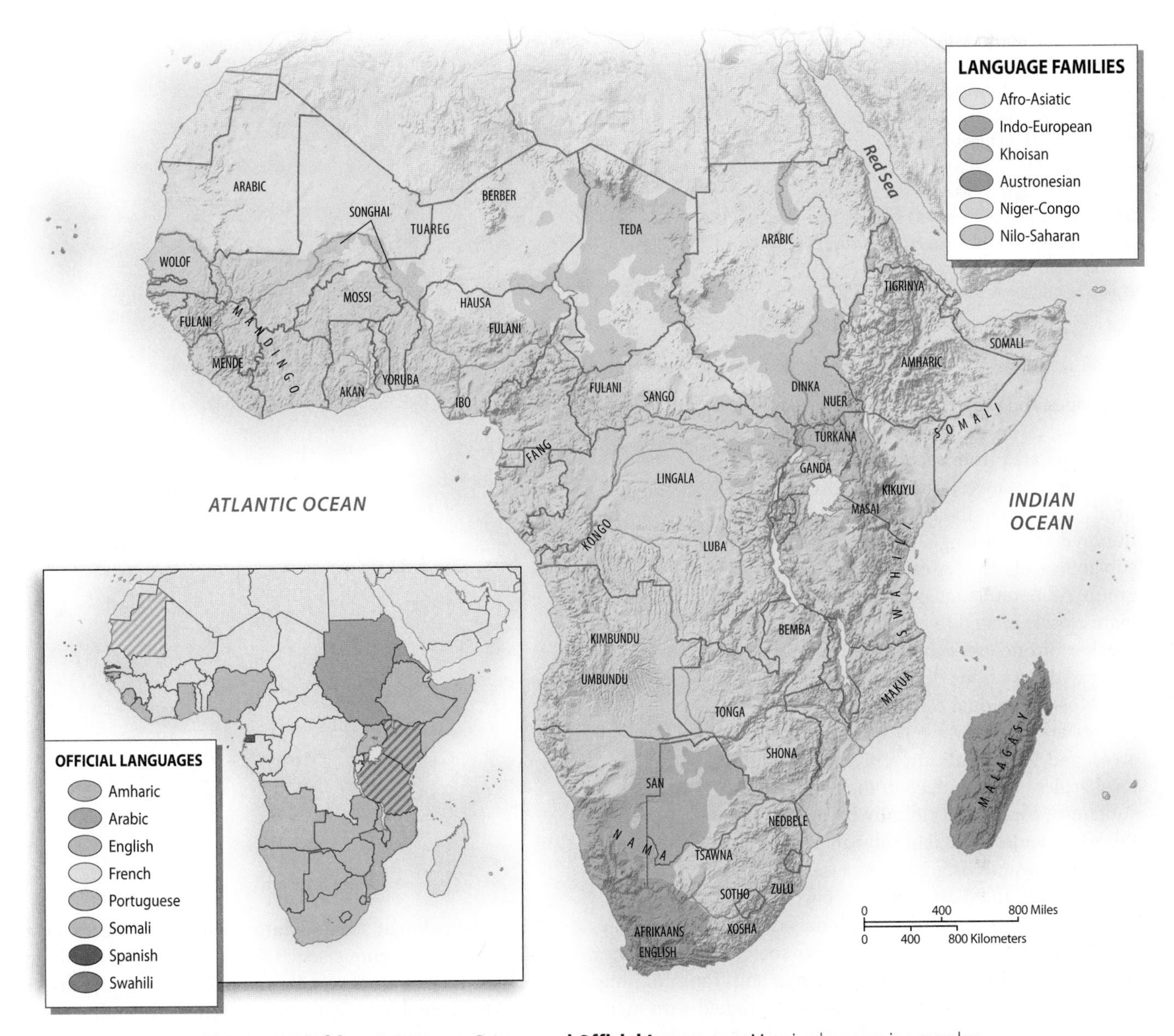

Figure 6.25 African Language Groups and Official Languages Mapping language is a complex task for Sub-Saharan Africa. There are languages with millions of speakers, such as Swahili, and there are languages spoken by a few hundred people living in isolated areas. Six language families are represented in the region. Among these families are scores of individual languages (see the labels on the map). Because most modern states have many indigenous languages, the colonial language often became the "official" language because it was less controversial than picking from among several indigenous languages. English and French are the most common official languages in the region (see inset).

the most far-ranging migrations in human history, which introduced agriculture into large areas of central and southern Africa. One Bantu group migrated east across the fringes of the rain forest to settle in the Lake Victoria basin in East Africa, where they formed an eastern Bantu core that later pushed south all the way to South Africa (Figure 6.26). Another group moved south, into the rain forest proper. The equatorial rain-forest belt immediately adjacent to the original Bantu homeland had been very sparsely settled by the ancestors of the modern pygmies (a distinct people noted for their short stature and hunting skills). Pygmy groups, having entered into close trading relations with the Bantu newcomers, eventually came to speak Bantu languages as well. While several pygmy populations have persisted to the present, their original languages disappeared long ago.

Once the Bantu migrants had advanced beyond the rain forest into the savannas and woodlands, their agricultural techniques proved highly successful and their influence expanded (see Figure 6.26). Around 650 CE, Bantu-speaking peoples reached South Africa. Over the centuries, the various

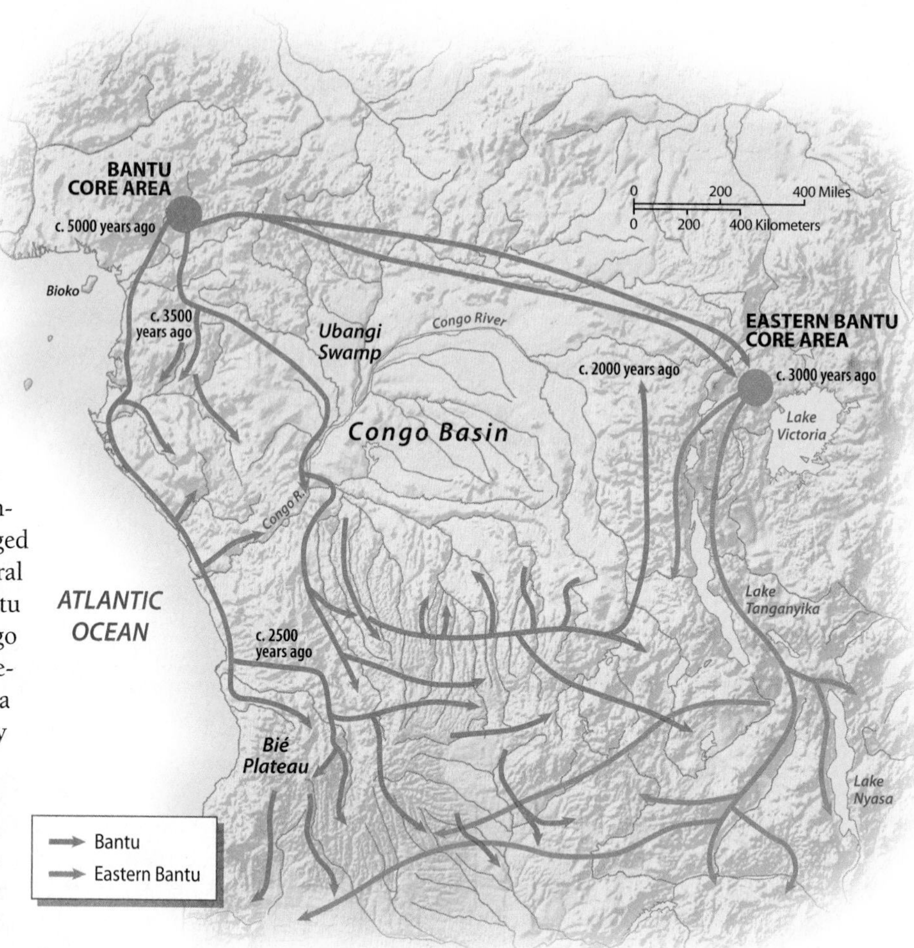

Figure 6.26 Bantu Migrations Bantu languages, a subfamily of the Niger-Congo language family, are widely spoken throughout Sub-Saharan Africa. The out-migration of Bantu tribes from an original core in West Africa and a secondary core in East Africa helps to explain the diffusion of Bantu languages, which include Zulu, Swahili, Bemba, Shona, Lingala, and Kikuyu, among others. (From Newman, 1995, *The Peopling of Africa*, New Haven: Yale University Press, 141)

languages and dialects of the many Bantu-speaking groups, which often were separated from each other by considerable distances, gradually diverged from each other. Today there are several hundred distinct languages in the Bantu subfamily of the great Niger-Congo group. All Bantu languages, however, remain closely related to each other, and a speaker of one can generally learn any other without undue difficulty.

Most individual Sub-Saharan languages are limited to relatively small areas and are significant only at the local scale. One language in the Bantu subfamily, Swahili, eventually became the most widely spoken Sub-Saharan language. Swahili originated as a trade language on the East African coast, where a number of merchant colonies from Arabia were established around 1100 CE. A hybrid society grew up in a narrow coastal band of modern Kenya and Tanzania, one speaking a language of Bantu structure enriched with many Arabic words. While Swahili became the primary language only in the narrow coastal belt, it spread far into the interior as the language of trade. After independence, both Kenya and Tanzania adopted Swahili as an official language. Swahili, with an estimated 60–90 million speakers, is the lingua franca of East Africa. It has generated a fairly extensive literature and is often studied in other regions of the world.

Language and Identity Ethnic identity as well as linguistic affiliation have historically been highly unstable over much of Sub-Saharan Africa. The tendency was for new groups to form when people threatened by war fled to less-settled areas, where they often mixed with peoples from other places. In such circumstances, new languages arise quickly, and divisions between groups are blurred. Nevertheless, distinct **tribes** formed that consisted of a group of families or clans with a common kinship, language, and definable territory. The impetus to formalize tribal boundaries came from European colonial administrators who were eager to establish a fixed indigenous social order to better control native peoples. In this process, a flawed cultural map of Sub-Saharan Africa evolved. Some tribes were artificially divided, meaningless names were applied, and territorial boundaries were often misinterpreted.

Social boundaries between different ethnic and linguistic groups have become more stable in recent years, and a number of individual languages have become particularly important for communication on a national scale. Wolof in Senegal; Mandingo in Mali; Mossi in Burkina Faso; Yoruba, Hausa, and Igbo in Nigeria; Kikuyu in central Kenya; and Zulu, Xhosa, and Sesotho in South Africa all are nationally significant languages spoken by millions of people (Figure 6.27). None, however, has the status of being the official language of any country. With the end of apartheid in South Africa, the country officially recognized 11 languages, although English is still the lingua franca of business and government. Indeed, a single language has a clear majority status in only a handful of Sub-Saharan countries. The more linguistically homogeneous states include Somalia (where virtually everyone speaks Somali) and the very small states of Rwanda, Burundi, Swaziland, and Lesotho.

Figure 6.27 Multilingual South Africa A South African sign warns pedestrians not to cross a highway in three languages: English, Afrikaans, and Zulu. Throughout Sub-Saharan Africa, many people are multilingual because of the diversity of languages that exist in each state.

European Languages In the colonial period, European countries used their own languages for administrative purposes in their African empires. Education in the colonial period also stressed literacy in the language of the imperial power. In the postindependence period, most Sub-Saharan African countries have continued to use the languages of their former colonizers for government and higher education. Few of these new states had a clear majority language that they could employ, and picking any minority tongue would have aroused the opposition of other peoples. The one exception is Ethiopia, which maintained its independence during the colonial era. The official language is Amharic, although other indigenous languages are also spoken.

Two vast blocks of European languages exist in Africa today: Francophone Africa, encompassing the former colonies of France and Belgium, where French serves as the main language of administration; and Anglophone Africa, where the use of English prevails (see inset, Figure 6.25). Early Dutch settlement in South Africa resulted in the use of Afrikaans (a Dutch-based language) by several million South Africans. Portuguese is spoken in Angola and Mozambique, former colonies of Portugal. In Mauritania and Sudan, Arabic serves as the main language.

Religion

Indigenous African religions generally are classified as *animist,* a somewhat misleading catchall term used to classify all local faiths that do not fit into one of the handful of "world religions." Most animist religions are centered on the worship of nature and ancestral spirits, but the internal diversity within the animist tradition is vast. Classifying a religion as animist says more about what it is not than what it actually is.

Both Christianity and Islam actually entered the region early in their histories, but they advanced slowly for many centuries. There are even ancient Jewish populations in the region–especially in Ethiopia–but many Ethiopian Jews resettled in Israel in the 1980s. Since the beginning of the 20th century, both Christianity and Islam have spread rapidly—more rapidly, in fact, than in any other part of the world. But tens of millions of Africans still hold animist beliefs, and many others combine animist practices and ideas with their observances of Christianity and Islam.

The Introduction and Spread of Christianity Christianity came first to northeast Africa. Kingdoms in both Ethiopia and central Sudan were converted by 300 CE—the earliest conversions outside of the Roman Empire. The peoples of Ethiopia adopted the Coptic form of Christianity and have thus historically looked to Egypt's Christian minority for their religious leadership (Figure 6.28). Today, roughly half of the population of both Ethiopia and Eritrea follow Coptic Christianity; most of the rest are Muslim, but there are still some animist communities.

European settlers and missionaries introduced Christianity to other parts of Sub-Saharan Africa beginning in the 1600s. The Dutch, who began to colonize South Africa

Figure 6.28 Eritrean Christians at Prayer Coptic Christians gather in front of St. Mary's Church in Asmara, Eritrea, on Good Friday, a holy day for Christians. Half of the populations in Eritrea and Ethiopia belong to the Coptic Church, which has ties to the Christian minority in Egypt.

at this time, brought their Calvinist Protestant faith. Later European immigrants to South Africa brought Anglicanism and other Protestant creeds, as well as Catholicism. A substantial Jewish community also emerged, concentrated in the Johannesburg area. Most black South Africans eventually converted to one or another form of Christianity as well. In fact, churches in South Africa were instrumental in the long fight against white racial supremacy. Religious leaders such as Bishop Desmond Tutu were outspoken critics of the injustices of apartheid and worked to bring down the system.

Elsewhere in Africa, Christianity came with European missionaries, most of whom arrived after the mid-1800s. As was true in the rest of the world, missionaries had little success where Islam had preceded them, but they eventually made numerous conversions in animist areas. As a general rule, Protestant Christianity prevails in areas of former British colonization, while Catholicism is important where France, Belgium, and Portugal had staked their empires. In the postcolonial era, African Christianity has diversified, at times taking on a life of its own independent from foreign missionary efforts. Increasingly active in the region are various Pentecostal, Evangelical, and Mormon missionary groups, mostly from the United States. Yet many areas also have seen the emergence of syncretic faiths, in which Christianity is complexly intertwined with traditional belief systems. It is difficult to map the distribution of Christianity in Africa, however, because it has spread irregularly across the entire non-Islamic portion of the region.

The Introduction and Spread of Islam Islam began to advance into Sub-Saharan Africa 1,000 years ago (Figure 6.29). Berber traders from North Africa and the Sahara introduced the religion to the Sahel, and by 1050 the Kingdom of Tokolor in modern Senegal emerged as the first Sub-Saharan Muslim state. Somewhat later, the ruling class of the powerful Mande-speaking mercantile empires of Ghana and Mali converted as well. In the 14th century, the emperor of Mali astounded the Muslim world when he and his huge entourage made the pilgrimage to Mecca, bringing with them so much gold that they set off an inflationary spiral throughout Southwest Asia.

Mande-speaking traders, whose networks spanned the area from the Sahel to the Gulf of Guinea, gradually introduced the religion to other areas of West Africa. There, however, many peoples remained committed to animism, and

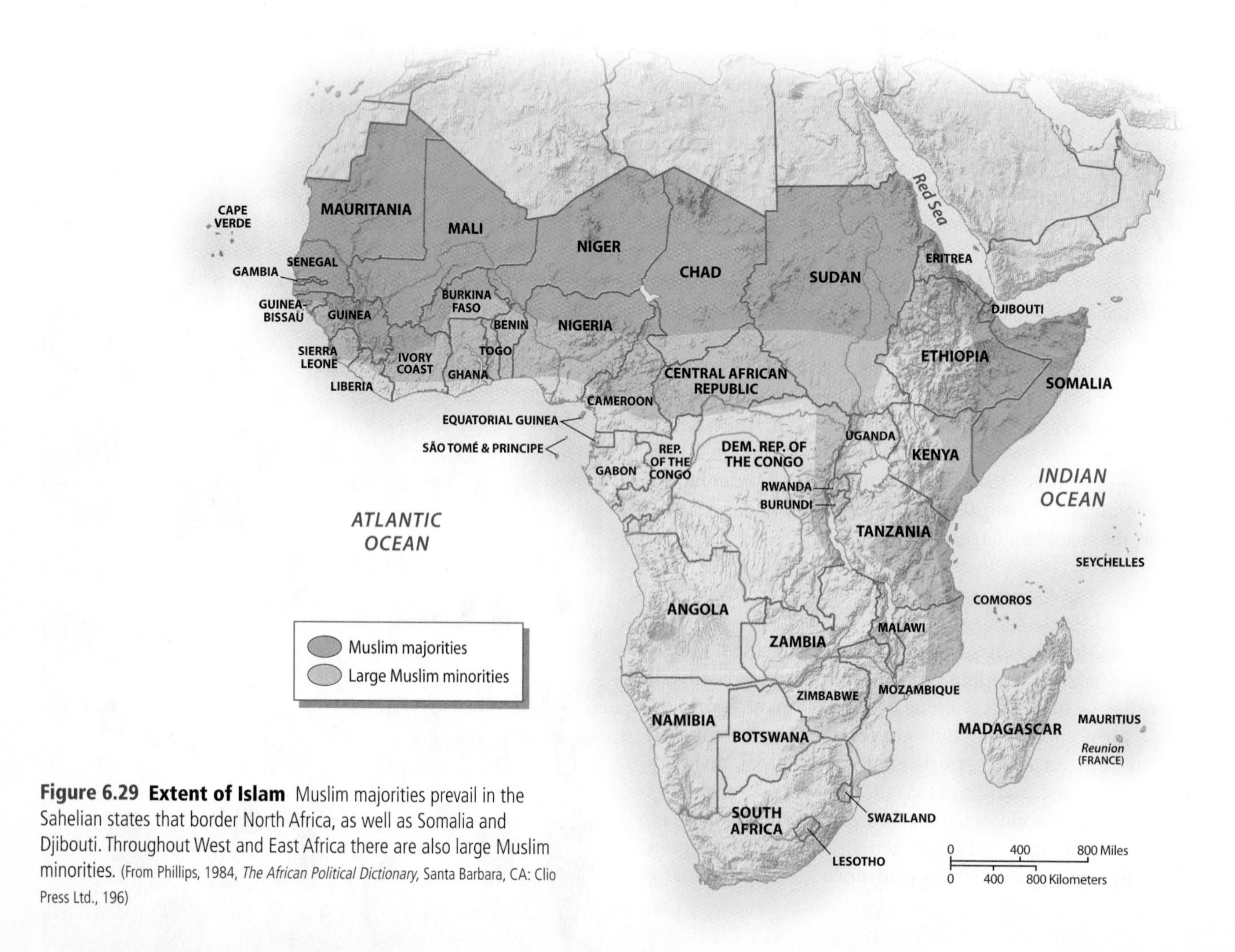

Figure 6.29 Extent of Islam Muslim majorities prevail in the Sahelian states that border North Africa, as well as Somalia and Djibouti. Throughout West and East Africa there are also large Muslim minorities. (From Phillips, 1984, *The African Political Dictionary*, Santa Barbara, CA: Clio Press Ltd., 196)

Figure 6.30 West African Muslims Villagers leave the Larabanga mosque after Friday prayers. The Muslim village of Larabanga in northern Ghana is home to one of the oldest mosques in West Africa and was recently declared a UN World Heritage Site.

Islam made slow and fitful progress. Even in the Sahel, syncretic forms of Islam prevailed through the 1700s. In the early 1800s, however, the pastoral Fulani people launched a series of successful holy wars designed to shear away animist practices and to establish pure Islam. Today orthodox Islam prevails through most of the Sahel. Farther south Muslims are mixed with Christians and animists, but their numbers continue to grow and their practices tend to be orthodox as well (Figure 6.30).

Interaction Between Religious Traditions The southward spread of Islam from the Sahel, coupled with the northward dissemination of Christianity from the port cities, has generated a complex religious frontier across much of West Africa. In Nigeria, the Hausa are firmly Muslim, while the southeastern Igbo are largely Christians. The Yoruba of the southwest are divided between Christians and Muslims. In the more remote parts of Nigeria, moreover, animist traditions remain vital. But despite this religious diversity, religious conflict in Nigeria has been relatively rare until recently. In 2000, seven Nigerian states imposed Muslim sharia laws, which triggered intermittent violence ever since, especially in the northern cities of Kano and Kaduna. More recently in 2010 near the central city of Jos, hundreds of Christian and Muslim villagers were slaughtered in violent clashes and retaliatory killings. Many of the victims were women and children. Still, most of West Africa's regional conflicts continue to be framed more along ethnic terms, rather than religious ones.

Religious conflict historically has been far more acute in northeastern Africa, where Muslims and Christians have struggled against each other for centuries. Islam came early to the coastal areas of the Horn, and soon it had virtually isolated the Ethiopian highlands from the rest of the Christian world. In recent years, religious conflict in the Horn has remained muted and has been largely replaced by ethnic and ideological struggles. The new country of Eritrea, which is roughly half Christian and half Muslim, has in the past few years emerged as something of a model of peaceful coexistence between members of different faiths.

Sudan, on the other hand, is currently the scene of an intense conflict that is both religious and ethnic in origin. Islam was introduced to Sudan in the 1300s by an invasion of Arabic-speaking pastoralists who destroyed the indigenous Coptic Christian kingdoms of the areas. Within a few hundred years, central and northern Sudan became completely Islamic. The southern equatorial provinces of Sudan, however, where tropical diseases and extensive wetlands stalled Arab advances, remained animist or converted to Christianity under British colonial rule. In the 1970s, the Arabic-speaking Muslims of north and central Sudan emerged as the country's dominant group and began to build an Islamic state. Experiencing both religious discrimination and economic exploitation, the Christian and animist peoples of the south subsequently launched a massive rebellion. Fighting began in the 1980s and was intense until 2003 when an agreement was reached between Sudan People's Liberation Army (SPLA) and the government of Khartoum. During the conflict the government generally controlled the main towns and roads and the rebels maintained power in the countryside. Southern Sudan has excellent soils and oil reserves, which have been only partially exploited because of the years of conflict. As part of the peace agreement southern Sudan was also promised an opportunity to vote on secession from the north in 2011 (although many doubt that the north will allow the vote to happen). Just as peace was finally brokered in southern Sudan, a brutal ethnic conflict broke out in the western province of Darfur, which will be discussed in detail later in the chapter.

With the exception of Sudan, a pattern of peaceful coexistence among different faiths is a distinguishing feature of Sub-Saharan Africa that deserves further study. Sub-Saharan Africa is a land of religious vitality. Both Christianity and Islam are spreading rapidly, but animism continues to hold widespread appeal. Many new syncretic (blended) forms of religious expression are also emerging. With such diversity of faiths, it is fortunate that religion is not typically the cause of overt conflict.

Globalization and African Culture

The slave triangle that linked Africa to the Americas and Europe set in process patterns of cultural diffusion that transferred African peoples and practices across the Atlantic. Tragically, slavery damaged the demographic and political strength of African societies, especially in West Africa, from where the most slaves were taken. An estimated 12 million Africans were shipped to the Americas as slaves from the 1500s until 1870 (Figure 6.31). Slavery impacted the entire

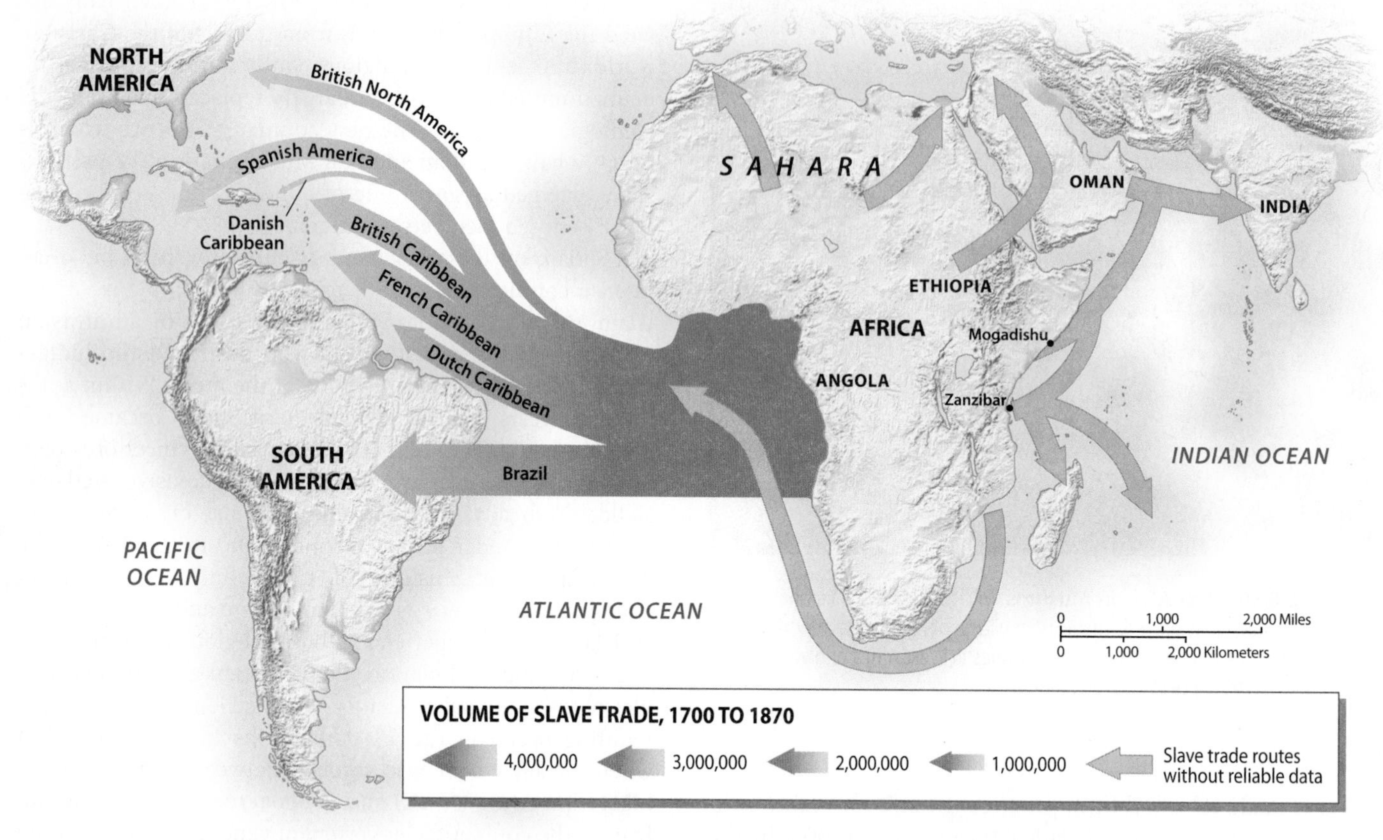

Figure 6.31 African Slave Trade The slave trade had a devastating impact on Sub-Saharan societies. From ship logs, it is estimated that 12 million Africans were shipped to the Americas to work as slaves on sugar, cotton, and rice plantations; the majority went to Brazil and the Caribbean. Yet other slave routes existed, although the data are less reliable. Africans from south of the Sahara were used as slaves in North Africa. Others were traded across the Indian Ocean into Southwest Asia and South Asia.

region, sending Africans not just to the Americas, but to Europe, North Africa, and Southwest Asia. The vast majority, however, worked on plantations across the Americas.

Out of this tragic diaspora came a blending of African cultures with Amerindian and European ones. African rhythms are at the core of various musical styles from rumba to jazz, the blues, and rock and roll. Brazil, the largest county in Latin America, is claimed to be the second largest "African state" (after Nigeria) because of the huge Afro-Brazilian population. Thus the forced migration of Africans as slaves had a huge cultural influence on many areas of the world.

Cultural exchanges are never one-way. Foreign language, religion, and dress were absorbed by Africans who remained in the region. With independence in the 1960s and 1970s, several states sought to rediscover their ancestral roots by openly rejecting European cultural elements. Ironically, the search for traditional religions found African scholars traveling to the Caribbean and Brazil to consult with Afro-religious practitioners about ceremonial elements lost to West Africa.

Africans are also emigrating from the region in record numbers, and in the process they are influencing cultures of many world regions. Whether it is the selection of African soccer stars for prestige European clubs, to the election in 2007 of Nigerian-native, Rotimi Adebari, as mayor of Porlaoise, Ireland, to the settlement of Somali refugees in Minneapolis, Minnesota, contemporary African migration is creating new transnational connections and reshaping the ethnic composition of localities around the world. Perhaps one of the most celebrated persons of African decent is President Barack Obama, the son of a Kenyan man and a woman from Kansas. Obama's heritage and upbringing embody the forces of globalization. In Kenya, he is hailed as part of the modern African diaspora—young professionals and their offspring who leave the continent and make their mark somewhere else. As an indicator of our global age, Kenyan President Mwai Kibaki declared a Kenyan national holiday after Obama won the U.S. election in November 2008 (People on the Move: The New African Americans).

Popular culture in Africa, like everywhere else in the world, is a dynamic mixture of global and local influences. Kwaito, the latest musical craze in South Africa, sounds a lot like rap from the United States. A closer listen, however, reveals an incorporation of local rhythms, lyrics in Zulu and Xhosa, and themes about life in the post-apartheid town-

PEOPLE ON THE MOVE The New African Americans

Since the 1990s tens of thousands of African immigrants have settled in the US, making Africa one of the fastest growing source regions for new immigrants. Like other immigrants, they come for economic opportunity, family reunification and as resettled refugees. There are about 1.5 million African-born immigrants in the U.S., the majority (75 percent) settled here since 1990. Nigeria and Ethiopia are the biggest sending countries, which is not terribly surprising since they are also the biggest countries in terms of population.

The latest arrivals from Africa are a mix of highly educated college graduates and refugees seeking asylum. Over 40 percent of African immigrants have college degrees, a figure that is much higher than the U.S. population as a whole. In particular, many women come as health care professionals, especially nurses. While Africans are found throughout the country, one-third of them are concentrated in three metropolitan areas: New York, Washington DC, and Atlanta. For example, Metropolitan Washington hosts the largest concentration of Ethiopians in the United States. They have established scores of restaurants and businesses in the Shaw Neighborhood, a historic African American section of the District of Columbia that is now often referred to as "Little Ethiopia" (Figure 6.3.1). Similarly in New York and Atlanta, African foods, music, and cultural events have been incorporated into the urban-cultural landscape.

Refugees are also an important part of the immigrant flow. Through resettlement programs, concentrations of Africans have emerged in Minneapolis–St. Paul, Boston, and Columbus. In Minneapolis–St. Paul, where many Somali refugees were resettled, Africans account for 20 percent of all immigrants in the Twin Cities. In general, refugees are a more needy population, arriving with less social capital and economic resources. Unlike economic migrants, refugees are assisted in their settlement and job placement as part of the federally funded Refugee Act of 1980. The major African refugee populations in the U.S. include Ethiopians, Eritreans, Somalis, Liberians, and Sudanese.

People of African ancestry have lived in the United States for nearly as long as people of European ancestry, and account for 12 percent of the total U.S. population. The percentage of African Americans in the US population was significantly higher in 1800. The percentage declined, in part, due to the abolition of slavery. Yet in the late 19th century restrictive measures were also put into place that made it virtually impossible for Africans to emigrate to America until 1965, when the Immigration and Nationality Act opened up immigration to peoples from Africa and Asia.

In the United States, African immigrants often find the term African American confusing. As newcomers from Africa, some have proposed the term *Neo* African Americans because their cultural and social history is so different from native-born African Americans. African immigrants have also maintained important ties with their countries of origin through remittances and transnational businesses. In fact, several African governments actively court this modern diaspora in the hope that some will return and invest in their home countries.

Figure 6.3.1 Ethiopian businessman in Washington, DC An Ethiopian entrepreneur stands in front of his shop in the Shaw Neigborhood of Washington DC. Metropolitan Washington attracts many African immigrants, and is home to the largest concentration of Ethiopians in the U.S.

ships. Lagos, Nigeria, is Africa's film capital. Referred to as Nollywood because over 500 films a year are made there, most are shot in a few days and with budgets of $10,000. These pulp movies are often criticized for being exploitive imitations of Hollywood B-movies. Yet the African movies touch on themes that resonate with Africans such as religion, ethnicity, corruption, witchcraft, and ritual killings. They can be seen all over English-speaking Africa.

Music in West Africa Nigeria is the musical center of West Africa, with a well-developed and cosmopolitan recording industry. Modern Nigerian styles such as juju, highlife, and Afro-beat are influenced by jazz, rock, reggae, and gospel, but they are driven by an easily recognizable African sound. Further up the Niger River, lies the country of Mali. Bamako, the capital, is also a music center that has produced scores of recording artists. Many Malian musicians descend from a traditional caste of musical storytellers performing on either the traditional kora (a cross between a harp and a lute) or the guitar. The musical style is strikingly similar to that of Blues from the Mississippi Delta. So much so that Ali Farka Touré, from northern Mali, was referred to as the Bluesman of Africa because of his distinctive, yet familiar, guitar work. Each January, not far from Timbuktu, music fans from West Africa and Europe gather for the Festival in the Desert. In this remote Saharan locale, a celebration of Malian music and Touareg nomadic culture draws together western tourists, African musicians, and nomads (Figure 6.32).

Contemporary African music can be both commercially and politically important. Nigerian singer Fela Kuti became a voice of political conscience for Nigerians struggling for true democracy. From an elite family and educated in England, Kuti borrowed from jazz, traditional, and popular music to produce the Afro-beat sound in the 1970s. Yet it was his searing and angry lyrics that attracted the most attention. Acutely critical of the military government, he sang of police harassment, the inequities of the international economic order, and even Lagos's infamous traffic. Singing in English and Yoruba, his message was transmitted to a larger audience, and he became a target of state harassment. At times self-exiled in Ghana, in the 1980s Kuti was jailed briefly by the Nigerian government, but widespread protests eventually led to his early release. Though Fela Kuti's protest music is unpopular with the state, it has been copied by other groups, making music an important form of political expression in Nigeria as well as in other Sub-Saharan states. Sadly, Kuti died in Lagos in 1997 from complications of AIDS; he was 58 years old. His son, however, is now an important recording artist in Nigeria.

Figure 6.32 Festival in the Desert Every January, thousands of Touareg nomads, Malian musicians, and Western tourists gather in the oasis town of Essakane, Mali, to attend the Festival in the Desert. In this image a Western tourist sits among the Touareg at the festival.

For many African emigrants, music can also be deeply nostalgic. No singer expresses this longing for home better than Cesária Evora from Cape Verde. A Grammy-winning singer, Evora is known for a style of singing called *mornas*, ballads that express sadness and yearning. Her need to sing is rooted in the emotion of *sodade*—a longing to find a better life elsewhere combined with the hope of returning to live among one's people. This need to leave one's home in order to support one's family is an experience that has shaped Cape Verdeans since the late 1800s when mass emigration began. As more Africans emigrate from their countries of origin in response to dire economic or political conditions, their music is reflective of this experience.

The Pride of Runners Ethiopia and Kenya have produced many of the world's greatest distance runners. Abebe Dikila won Ethiopia's and Africa's first Olympic gold medal running barefoot at the Rome games in 1960. Since then nearly every Olympic games has yielded medals for Ethiopia and Kenya. At the Beijing Olympics in 2008, Kenyan runners won 14 medals and Ethiopians 7 (4 of them gold). These states were the top medal winners for Africa in Beijing (Figure 6.33).

Running is a national pastime in Kenya and Ethiopia, where elevation—Addis Ababa sits at 7,300 feet (2,200 meters) and Nairobi at 5,300 feet (1,600 meters)—increases oxygen carrying capacity. Past medalists Haile Gebrselassie and Derartu Tulu are national celebrities in Ethiopia who are idealized by the country's youth. Tulu, the first black African woman to win a gold medal in distance running, is a forceful voice for women's rights in a country where women are discouraged from putting on running shorts. There is talk of a political career for Gebrselassie after he hangs up his running shoes.

Sub-Saharan Africa has shared many of its cultural practices with the world, and the creative and athletic vitality of its people continues to be recognized. In contrast to the fairly peaceful coexistence of diverse religious traditions in Sub-Saharan Africa, ethnic tensions have often boiled over into violent disputes. To understand the roots of ethnic conflict in the region, it is necessary first to examine the region's political history, paying particular attention to the legacy of European colonization.

Figure 6.33 Ethiopian or Kenyan Distance Runners Kenenisa Bekele of Ethiopia, right, leads Eliud Kipchoge and Edwin Soi of Kenya on the last lap of the 5000 meters race at the Beijing Olympic Games. Kenyans and Ethiopians take pride in their elite distance runners.

GEOPOLITICAL FRAMEWORK: Legacies of Colonialism and Conflict

The duration of human settlement in Sub-Saharan Africa is unmatched by any other region. Evidence shows that humankind originated there, evidently evolving from a rather apelike *Australopithicus* all the way to modern *Homo sapiens*. Over the millennia, many diverse ethnic groups formed in the region. Although conflicts among these groups have existed, cooperation and coexistence among different peoples has also occurred.

With the arrival of Europeans, patterns of human relatedness and ethnic relations were changed forever. As Europeans rushed to carve up the continent to serve their imperial ambitions, they instituted various policies that heightened ethnic tensions and promoted hostility. Many of the region's modern conflicts can trace their roots back to the colonial era, especially the arbitrary drawing of political boundaries. Others are attributed to struggles over national identity and political control over limited resources among different ethnic groups.

Indigenous Kingdoms and European Encounters

The first significant state to emerge in Sub-Saharan Africa was Nubia, which controlled a large territory in central and northern Sudan some 3,000 years ago (Figure 6.34); 1,000 years later the Kingdom of Axum arose in northern Ethiopia and Eritrea. Both of these states were strongly influenced by political models derived from Egypt and Arabia. The first wholly indigenous African states were founded in the Sahel around 700 CE. Kingdoms such as Ghana, Mali, Songhai, and Kanem-Bornu grew rich by exporting gold to the Mediterranean and importing salt from the Sahara, and they maintained power over lands to the south by monopolizing horse breeding and mastering cavalry warfare.

Over the next several centuries, a variety of other states emerged in West Africa. Some were large but diffuse empires organized through elaborate hierarchies of local kings and chiefs; others were centralized states focused on small centers of power. The Yoruba of southwestern Nigeria, for example, developed a city-state form of government, and their homeland is still one of the most urbanized and densely populated parts of Africa. The most powerful Sub-Saharan states continued to be located in the Sahel until the 1600s, when European coastal trade undercut the lucrative trans-Saharan networks. Subsequently, the focus of power moved to the Gulf of Guinea, where well-organized African states (such as Dahomey and Ashanti) as well as Europeans took advantage of the lucrative opportunities presented by the slave trade and in the process increased their military and economic power.

Early European Encounters Unlike the relatively rapid colonization of the Americas, Europeans needed centuries to gain effective control of Sub-Saharan Africa. Portuguese traders arrived along the coast of West Africa in the 1400s, and by the 1500s they were well established in East Africa as well. Initially, the Portuguese made large profits, converted a few local rulers to Christianity, established several fortified trading posts, and acquired dominion over the Swahili trading cities of the east. They stretched themselves too thin, however, and many of their settlements failed or were lost to other colonizers. Only where a sizable population of mixed African and Portuguese descent emerged, as along the coasts of modern Angola and Mozambique as well as the islands of Cape Verde, could Portugal maintain power. Along the Swahili, or eastern, coast they were eventually expelled by Arabs from Oman, who subsequently established their own mercantile empire in the area.

The Disease Factor One of the main reasons for the Portuguese failure was the disease environment of Africa. With no resistance to malaria and other tropical diseases, roughly half of all Europeans who remained on the African mainland died within a year. Protected both by their formidable armies and by the diseases of their native lands, African states were able to maintain an upper hand over European traders and adventurers well into the 1800s. Unlike the Americas, where European conquest was facilitated by the introduction of Old World diseases that devastated native populations (see Chapters 4 and 5), Sub-Saharan Africa's native diseases limited European settlement until the mid-19th century.

The hazards of malaria and other tropical diseases such as sleeping sickness were compensated by the lure of profit, and other European traders soon followed the Portuguese.

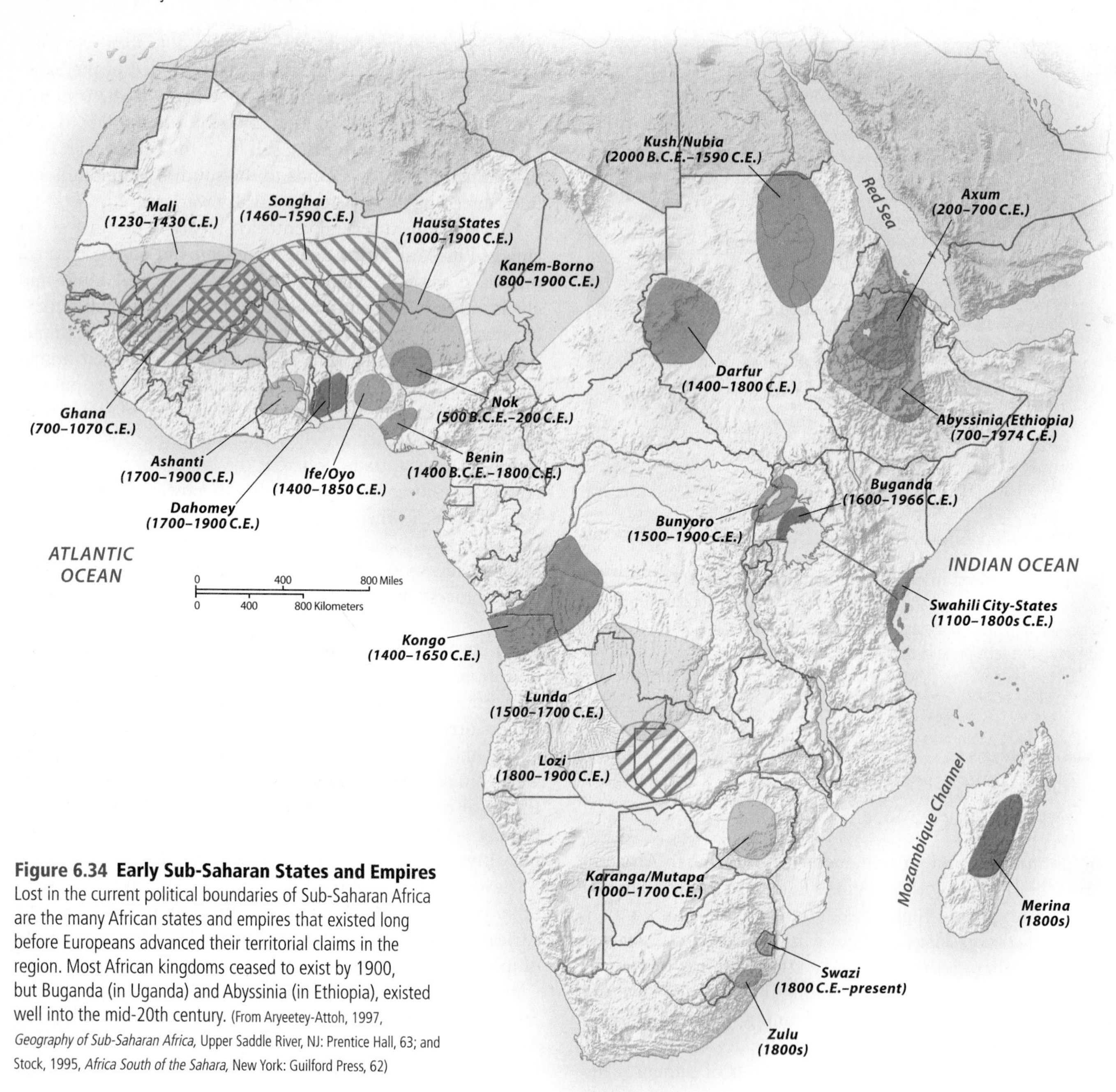

Figure 6.34 Early Sub-Saharan States and Empires Lost in the current political boundaries of Sub-Saharan Africa are the many African states and empires that existed long before Europeans advanced their territorial claims in the region. Most African kingdoms ceased to exist by 1900, but Buganda (in Uganda) and Abyssinia (in Ethiopia), existed well into the mid-20th century. (From Aryeetey-Attoh, 1997, *Geography of Sub-Saharan Africa,* Upper Saddle River, NJ: Prentice Hall, 63; and Stock, 1995, *Africa South of the Sahara,* New York: Guilford Press, 62)

By the 1600s, Dutch, British, and French firms dominated the lucrative export of slaves, gold, and ivory from the Gulf of Guinea. The Dutch also established a settler colony in South Africa, safely outside of the tropical disease zone, to supply their ships bound for Indonesia. For the next 200 years, European traders came and went, occasionally building fortified coastal posts, but they almost never ventured inland and seldom had any real influence on African rulers. By exporting millions of slaves, however, they had a profoundly negative impact on African society.

In the 1850s, European doctors discovered that a daily dose of quinine would offer protection against malaria, radically changing the balance of power in Africa. Explorers immediately began to penetrate the interior of the continent, while merchants and expeditionary forces began to move inland from the coast. The first imperial claims soon followed. The French quickly grabbed power along the easily navigated Senegal River, while the British established protectorates over the indigenous states of the Gold Coast (modern-day Ghana), in which the British guaranteed protection for these territories in exchange for various trade preferences.

European Colonization

In the 1880s, European colonization of the region quickly accelerated, leading to the so-called scramble for Africa. By this time, after the invention of the machine gun, no African state could long resist European force. The exact reasons for the abrupt division of Africa among the colonial powers remain controversial, but several developments seem to have been crucial. One was the British seizure in 1882 of Egypt, a territory that the French had long coveted. In return, the infuriated French began to seize additional lands in West Africa, equatorial Africa, and Madagascar.

Another contributing factor was the desire of several new European countries to join the game of empire-building. Because Asia was either occupied by established European powers or controlled by still-formidable indigenous empires, Africa emerged as the main arena of rivalry and expansion. Even though Belgium had been a country only since 1830, its king quickly began to carve out a personal empire along the Congo River, using particularly brutal techniques. The German government—which itself dated back only to 1871—began to claim territories wherever German missionaries were active, and it soon staked out the colonies of Togo, Cameroon, Namibia, and Tanganyika (modern Tanzania minus the island of Zanzibar). The Italians eyed the Horn of Africa, while Spain acquired a small coastal foothold in equatorial West Africa. Alarmed by such activity, the Portuguese began to push inland from their coastal possessions in Angola and Mozambique.

Also in the early 1800s, two small territories were established in West Africa so that freed and runaway slaves would have a place to return to in Africa. The territory that was to become Liberia was set up by the American Colonization Society in 1822 to settle former African American slaves. By 1847, it was the independent and free state of Liberia. Sierra Leone served a similar function for ex-slaves from the British Caribbean, but it remained a protectorate of Britain until the 1960s. Despite the good intentions behind the creation of these territories, they too were colonies. Liberia, in particular, was imposed on existing indigenous groups who viewed their new "African" leaders with contempt.

The Berlin Conference As the scramble intensified, tensions among the participating countries mounted. Rather than risk war, 13 countries convened in Berlin at the invitation of the German chancellor Bismarck in 1884 in a gathering known as the **Berlin Conference**. During the conference, which no African leaders attended, rules were established as to what constituted "effective control" of a territory, and Sub-Saharan Africa was carved up and traded like properties in a game of Monopoly® (Figure 6.35). Exact boundaries in the interior, which was still poorly known, were not determined, and a decade of "orderly" competition remained as imperial armies marched inland. Although European arms were by the 1880s far superior to anything found in Africa, several indigenous states did mount effective resistance campaigns. For example, in central Sudan, an Islamic-inspired force held out against the British until 1900, and as late as 1914, the Darfur region of western Sudan maintained tenuous independence.

Eventually European forces prevailed everywhere, with one major exception: Ethiopia. The Italians had conquered the Red Sea coast and the northern highlands (modern Eritrea) by 1890, and they quickly set their sights on the large Ethiopian kingdom, called Abyssinia, which had itself been vigorously expanding for several decades. In 1896, however, Abyssinia defeated the invading Italian army, earning the respect of and recognition by the European powers. In the 1930s, fascist Italy launched a major invasion of the country, now renamed Ethiopia, to redeem its earlier defeat, and with the help of poison gas and aerial bombardment, it quickly prevailed. By 1942, Ethiopia had regained its freedom.

Although Germany was a principal instigator of the scramble for Africa, it lost its own colonies after suffering defeat in World War I. Britain and France then partitioned most of Germany's African empire between them. Figure 6.35 shows the colonial status of the region in 1913, prior to Germany's territorial loss. The French held most of West Africa, but the British controlled populous Nigeria and several other coastal territories. The French also colonized Gabon in western equatorial Africa, Madagascar, and the small but strategic enclave of Djibouti at the southern end of the Red Sea. The British holdings were larger still, covering a continuous swath of territory in the east from Sudan to South Africa. Belgium and Portugal were the other main colonial powers. The government of Belgium had taken direct control over the personal domain of King Leopold II and had extended it to the south, eventually reaching the mineral-rich copper belt. Portugal, the weakest European power, controlled huge territories in southwestern and southeastern Africa (Angola and Mozambique). Britain's drive to the north from South Africa, organized by imperial dreamer and diamond magnate Cecil Rhodes, thwarted the Portuguese effort to bridge the continent.

Establishment of South Africa While the Europeans were cementing their rule over Africa after World War I, South Africa was inching toward political freedom, at least for its white population. South Africa was one of the oldest colonies in Sub-Saharan Africa, and it became the first to obtain its political independence from Europe in 1910. Its economy is the most productive and influential of the region, yet the legacy of apartheid made it an international outcast, especially within Africa.

The original body of Dutch settlers in South Africa grew slowly, expanding to the north and east of its original nucleus around Cape Town. As a farming and pastoral people largely isolated from the European world, these Afrikaners, or Boers, developed an extremely conservative cultural outlook marked by an intensifying belief in their racial superiority over the native population. In 1806 the British, then the world's unchallenged maritime power, seized the Cape district from the Dutch. Relations between the British rulers and their new Afrikaner subjects quickly soured, in part because the British attempted to restrict the enslavement of Africans as part of a larger abolitionist movement. In the 1830s, the bulk of the Afrikaner community opted to leave British territory and strike out for new lands. After being rebuffed by the powerful Zulu state in the Natal area, they set-

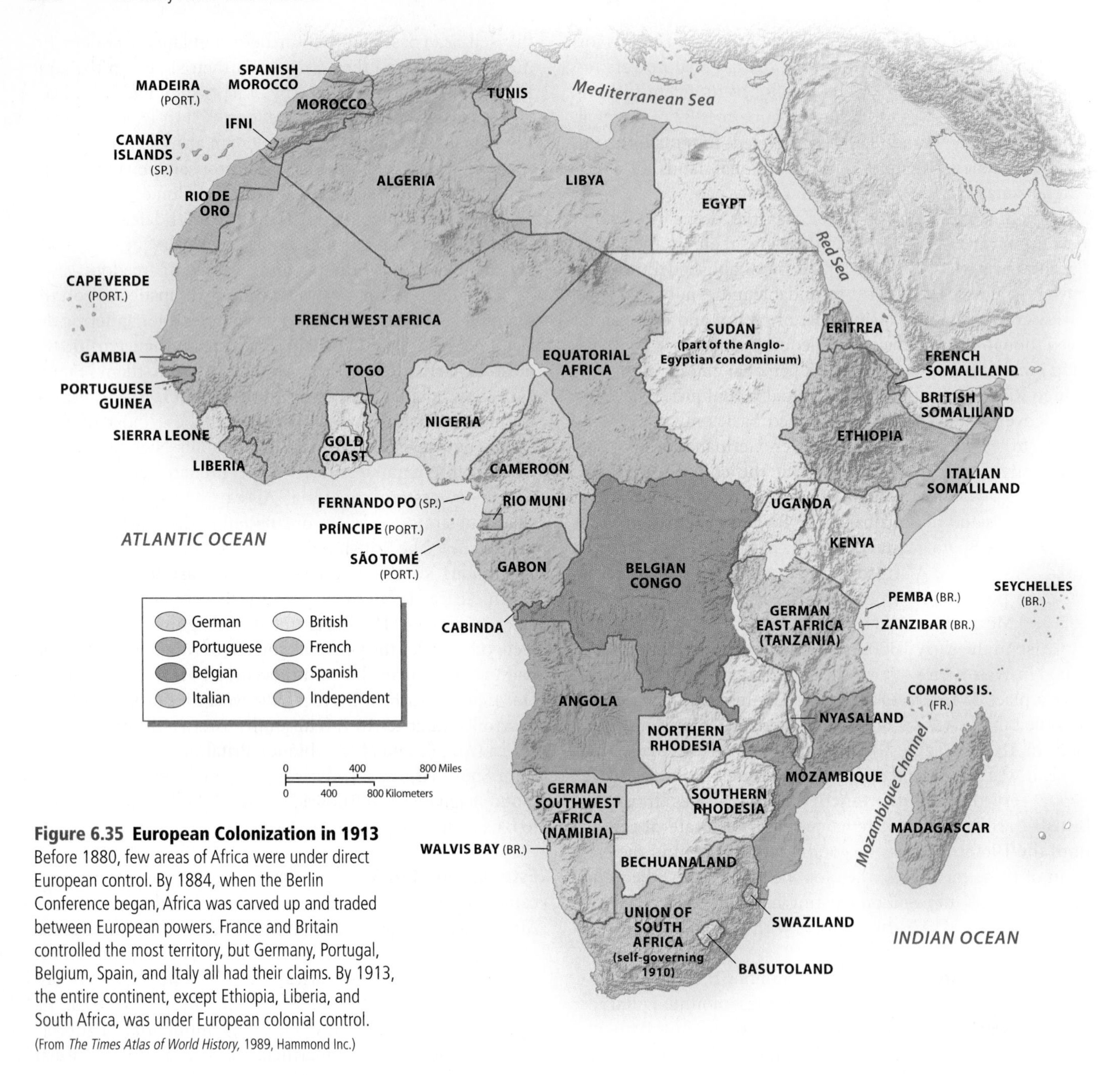

Figure 6.35 European Colonization in 1913 Before 1880, few areas of Africa were under direct European control. By 1884, when the Berlin Conference began, Africa was carved up and traded between European powers. France and Britain controlled the most territory, but Germany, Portugal, Belgium, Spain, and Italy all had their claims. By 1913, the entire continent, except Ethiopia, Liberia, and South Africa, was under European colonial control. (From *The Times Atlas of World History*, 1989, Hammond Inc.)

tled on the northeastern plateau, known as the highveld. By the 1850s, they had established two "republics" in the area: the South African Republic (commonly called the *Transvaal*) and the Orange Free State (Figure 6.36).

Once relocated on the highveld, the Afrikaners found themselves threatened by Zulu armies. The Zulus had been a relatively minor group until a Zulu king introduced centralized rule and new military techniques in the 1820s. From then on Zulu armies were virtually invincible. After a series of inconclusive wars, the British intervened in 1878 on behalf of the Dutch settlers. The first British army sent into Zululand was annihilated; the second, equipped with machine guns, prevailed. By 1900, the British had incorporated the Zulu into South Africa's Natal province.

The British had more difficulty subduing the Afrikaners. By the late 1800s, it had become clear that the two Boer republics were sitting above one of the world's greatest troves of mineral wealth. The British were increasingly drawn to the area, and they grew resentful of Afrikaner power. The result was the Boer War, which turned into a protracted and brutal guerrilla struggle between the British Army and mobile Afrikaner bands. By 1905, the Boers relented, after which the British joined the two former republics to Cape Province and Natal to form the Union of South Africa. Five years later South Africa was given its independence from Britain. Britons and other Europeans continued to settle there, but the Afrikaners remained the majority white group. Black and coloured South Africans, however, greatly outnumbered whites.

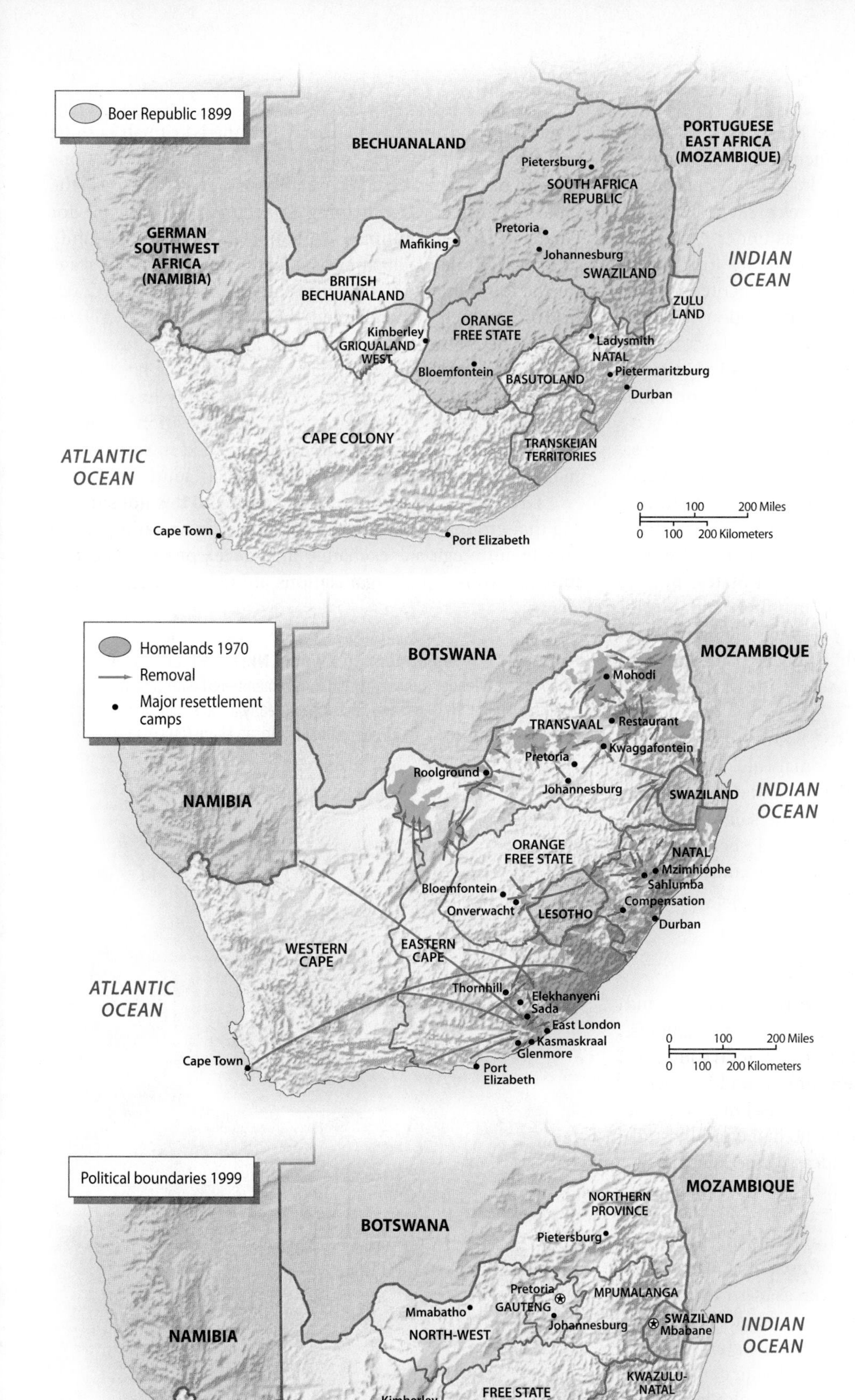

Figure 6.36 The Evolution of South African Political Boundaries South Africa's internal political boundaries have been redrawn several times in the past century. To appease the Afrikaans (Dutch) settlers, a Boer republic was established in the late nineteenth century with British approval. It was later reabsorbed into the Union of South Africa when gold and diamonds were discovered. During the apartheid years, homelands were created with the intent of having all black South Africans live in these ethnic-based rural settings. In order to accomplish this, 3 million black South Africans were forcibly relocated. Lastly, in the new South Africa, homelands were eliminated and new province boundaries were drawn. (From Christopher, 1994, *The Atlas of Apartheid,* London: Routledge, 16, 83)

It was not until 1948, when the Afrikaner's National Party gained control of the government, that they introduced the policy of "separateness" known as *apartheid*. British South Africans had enacted a series of laws that were prejudicial to nonwhite groups, but it was under Afrikaner leadership that racial separation become more formalized and systematic. Operating on three scales—petite, meso-, and grand—apartheid managed social interaction by controlling space. Petite apartheid, like Jim Crow laws in the United States, created separate service entrances for government buildings, bus stops, and restrooms based on skin color. Meso-apartheid divided the city into residential sectors by race, giving whites the best and largest urban zones. Finally, grand apartheid was the construction of black **homelands** by ethnic group. Technically, blacks were to become citizens of the nominally independent homelands (see the 1970 map in Figure 6.36). Likened to the reservations created for Native Americans in the United States, homelands were rural, overcrowded, and on marginal land. Moreover, to ensure the notion that every black had a homeland, some 3 million blacks were forcibly relocated into homelands during apartheid, and residence outside of the homelands was strictly regulated.

Granted its independence in 1910, South Africa was the first state in the region freed from colonial rule. Yet because of its formalized system of discrimination and racism, it was hardly a symbol of liberty. Ironically, at the same time that the Afrikaners tightened their political and social control over the nonwhite population, the rest of the continent was preparing for political independence from Europe.

Decolonization and Independence

Decolonization of the region happened rather quickly and peacefully beginning in 1957. Independence movements, however, had sprung up throughout the continent, some dating back to the early 1900s. Workers' unions and independent newspapers became voices for African discontent and the hope for freedom. Black intellectuals, who had typically studied abroad, were influenced by the ideas of the **Pan-African Movement** led by W. E. B. Du Bois and Marcus Garvey in the United States. Founded in 1900, the movement's slogan of "Africa for Africans" encouraged a trans-Atlantic liberation effort. Nevertheless, Europe's hold on Africa remained secure through the 1940s and early 1950s, even as other colonies in South and Southeast Asia gained their independence.

By the late 1950s, Britain, France, and Belgium decided that they could no longer maintain their African empires and began to withdraw. (Italy had already lost its colonies during World War II, and Britain gained Somalia and Eritrea.) Once started, the decolonization process moved rapidly. By the mid-1960s, virtually the entire region had achieved independence. In most cases, the transition was relatively peaceful and smooth with the exception of southern Africa.

Dynamic African leaders put their mark on the region during the early independence period. Men such as Kenya's Jomo Kenyatta, Ivory Coast's Felix Houphuët-Boigney, Julius Nyerere of Tanzania, and Ghana's Kwame Nkrumah became powerful father figures who molded their new nations (Figure 6.37). President Nkrumah's vision for Africa was the most expansive. After helping to secure independence for Ghana in 1957, his ultimate aspiration was the political unity of Africa. While his dream was never realized, it set the stage for the founding of the Organization of African Unity (OAU) in 1963, which was renamed the **African Union** (AU) in 2002. The AU is a continent-wide organization headquartered in Addis Ababa, Ethiopia, whose main role has been to mediate disputes between neighbors. Certainly in the 1970s and 1980s it was a constant voice of opposition to South Africa's minority rule, and the AU intervened in some of the more violent independence movements in southern Africa.

Given the scale of the African continent, it is not surprising that groups of states formed regional organizations to facilitate intraregional exchange and development. The two most active regional organizations are the Southern African

Figure 6.37 A Monument to Kwame Nkrumah Charismatic independence leader Kwame Nkrumah is remembered with this monument in Accra, Ghana. Nkrumah led Ghana to an early independence in 1957; he was also a founder of the Organization of African Unity (OAU), now called the African Union.

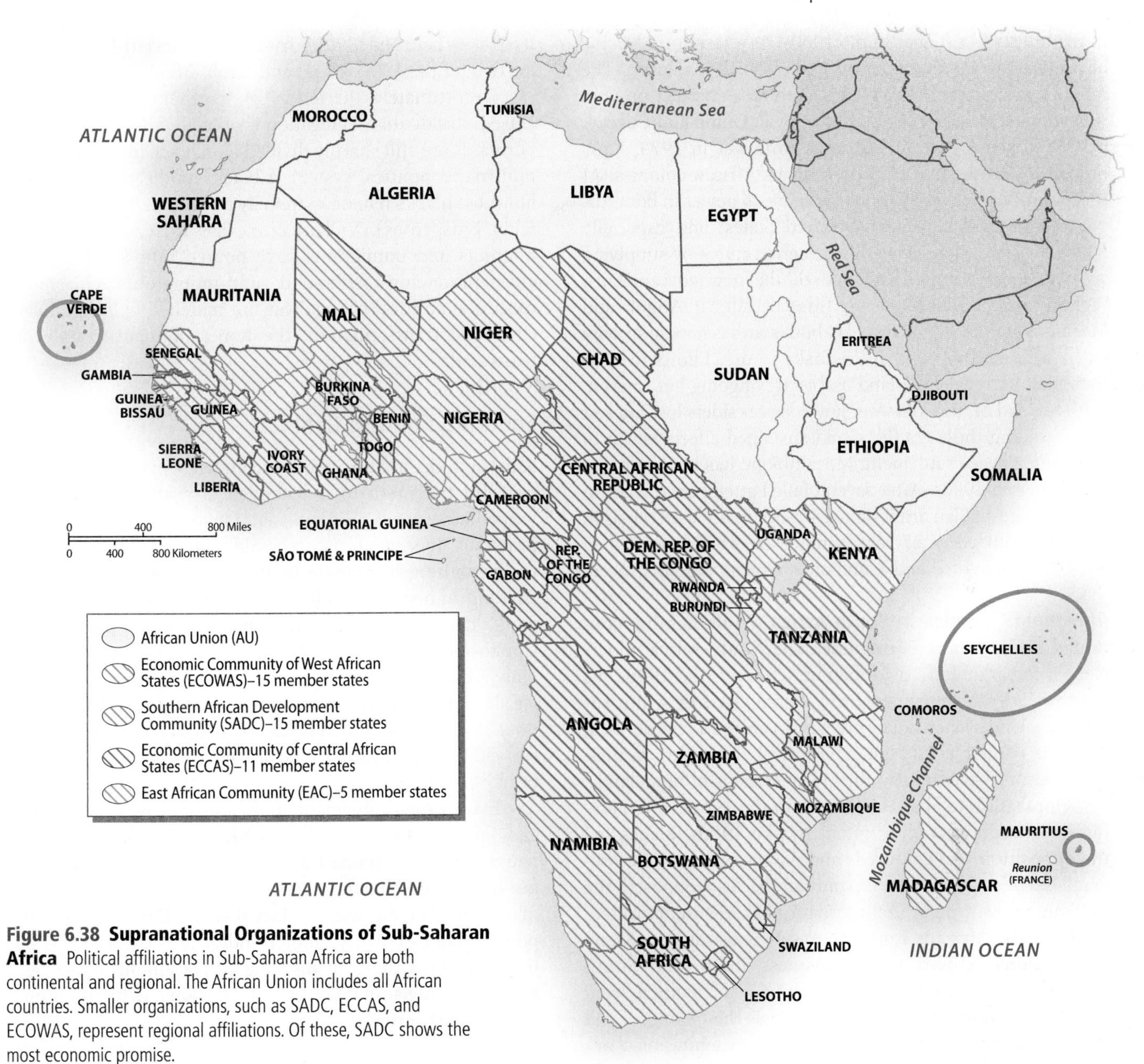

Figure 6.38 Supranational Organizations of Sub-Saharan Africa Political affiliations in Sub-Saharan Africa are both continental and regional. The African Union includes all African countries. Smaller organizations, such as SADC, ECCAS, and ECOWAS, represent regional affiliations. Of these, SADC shows the most economic promise.

Development Community (SADC) and the Economic Community of West African States (ECOWAS). Both were founded in the 1970s but became more prominent in the 1990s (Figure 6.38). SADC and ECOWAS are anchored by the region's two largest economies: South Africa and Nigeria. The Economic Community of Central African States (ECCAS) was founded in the mid-1980s and is headquartered in Libreville, Gabon. Its effectiveness has been hampered by the political instability of the area. Several ECCAS members also are members of SADC.

Southern Africa's Independence Battles

Independence did not come easily to southern Africa. In Southern Rhodesia (modern-day Zimbabwe), the problem was the presence of some 250,000 white residents, most of whom owned large farms. Unwilling to see power pass to the country's black majority, then some 6 million strong, these settlers declared themselves the rulers of an independent, white-supremacist state in 1965. The black population continued to resist, however, and in 1978 the Rhodesian government was forced to give up power. The renamed country of Zimbabwe was, henceforth, ruled by the black majority, although the remaining whites still formed an economically privileged community. Since the mid-1990s, disputes over government land reform (splitting up the large commercial farms mostly owned by whites and giving the land to black farmers) and President Robert Mugabe's strongman politics have resulted in serious racial and political tensions as well as the collapse of the country's economy.

In the former Portuguese colonies, independence came violently. Unlike the other imperial powers, Portugal refused to relinquish its colonies in the 1960s. As a result, the people of Angola and Mozambique turned to armed resistance. The most powerful rebel movements adopted a socialist orientation and received support from the Soviet Union and Cuba. A new Portuguese government came to power in 1974, however, and it withdrew abruptly from its African colonies. At this point, Marxist regimes quickly came to power in both Angola and Mozambique. The United States, and especially South Africa, responded to this perceived threat by supplying arms to rebel groups that opposed the new governments. Fighting dragged on for nearly three decades in Angola and Mozambique. The countryside in both states is now so heavily laden with land mines that it is risky to use. Efforts to clear the mines and make the land usable are ongoing but uneven. With the end of the Cold War, however, outsiders lost interest in continuing these conflicts, and sustained efforts began to negotiate a peace settlement. Mozambique has been at peace since the mid-1990s. After several failed attempts at peace in Angola, the Angolan army signed a peace treaty with rebels in 2002 that ended a 27-year conflict in which more than 300,000 people died and 3 million Angolans were displaced.

Apartheid's Demise in South Africa While fighting continued in the former Portuguese zone, South Africa underwent a remarkable transformation. Through the 1980s, its government had remained firmly committed to white supremacy. Under apartheid only whites enjoyed real political freedom, while blacks were denied even citizenship in their own country—technically, they were citizens of homelands.

Opposition to apartheid began in the 1960s, intensifying and becoming more violent by the 1980s. Blacks led the opposition, but coloureds and Asians (who suffered severe, but less extreme, discrimination) also opposed the Afrikaner government. As international pressure mounted, white South Africans found themselves ostracized. Many corporations refused to do business there, and South African athletes (regardless of color) were banned from most international competitions, such as the Olympics and World Cup Soccer. Increasing numbers of whites also opposed the apartheid system, and many businesspeople began to believe that apartheid threatened to undermine their economic endeavors.

The first major change came in 1990, when South Africa withdrew from Namibia, which it had controlled as a protectorate since the end of World War I. South Africa now stood alone as the single white-dominated state in Africa. A few years later, the leaders of the Afrikaner-dominated political party decided they could no longer resist the pressure for change. In 1994, free elections were held in which Nelson Mandela, a black leader who had been imprisoned for 27 years by the old regime, emerged as the new president. Black and white leaders pledged to put the past behind them and work together to build a new, multiracial South Africa. The homelands themselves were the first to be eliminated from the political map of the new South Africa (see the 1999 map in Figure 6.36). Since Mandela's presidency, orderly elections have been held and South Africans have elected Thabo Mbeki for two terms (1999–2009) and Jacob Zuma in 2009.

Unfortunately, the legacy of apartheid is not so easily erased. Residential segregation is officially illegal, but neighborhoods are still sharply divided along race lines. Under the multiracial political system, a black middle class emerged, but most blacks remain extremely poor (and most whites remain prosperous). Violent crime has increased, and rural migrants and immigrants have poured into South African cities, producing a xenophobic anti-immigrant backlash. Because the political change was not matched by a significant economic transformation, the hopes of many people are frustrated.

Enduring Political Conflict

Although most Sub-Saharan countries made a relatively peaceful transition to independence, virtually all of them immediately faced a difficult set of institutional and political problems. In several cases, the old authorities had done virtually nothing to prepare their colonies for independence. Lacking an institutional framework for independent government, countries such as the Democratic Republic of the Congo confronted a chaotic situation from the beginning. Only a handful of Congolese had received higher education, let alone been trained for administrative posts. The indigenous African political framework had been essentially destroyed by colonization, and in most cases very little had been built in its place.

Even more problematic in the long run was the political geography of the newly independent states. Civil servants could always be trained and administrative systems built, but little could be done to rework the region's basic political map. The problem was the fact that the European colonial powers had essentially ignored indigenous cultural and political patterns, both in dividing Africa among themselves and in creating administrative subdivisions within their own imperial territories.

The Tyranny of the Map All over Africa, different ethnic groups found themselves forced into the same state with peoples of different linguistic and religious backgrounds, many of whom had recently been their enemies. At the same time, a number of the larger ethnic groups of the region found their territories split between two or more countries. The Hausa people of West Africa, for example, were divided between Niger (formerly French) and Nigeria (formerly British), each of which they had to share with several former enemy groups.

Given the imposed political boundaries, it is no wonder that many African countries struggled to generate a common sense of national identity or establish stable political institutions. **Tribalism**, or loyalty to the ethnic group rather than to the state, has emerged as the bane of African political life. Especially in rural areas, tribal identities usually supersede national ones. Because virtually all of Africa's countries

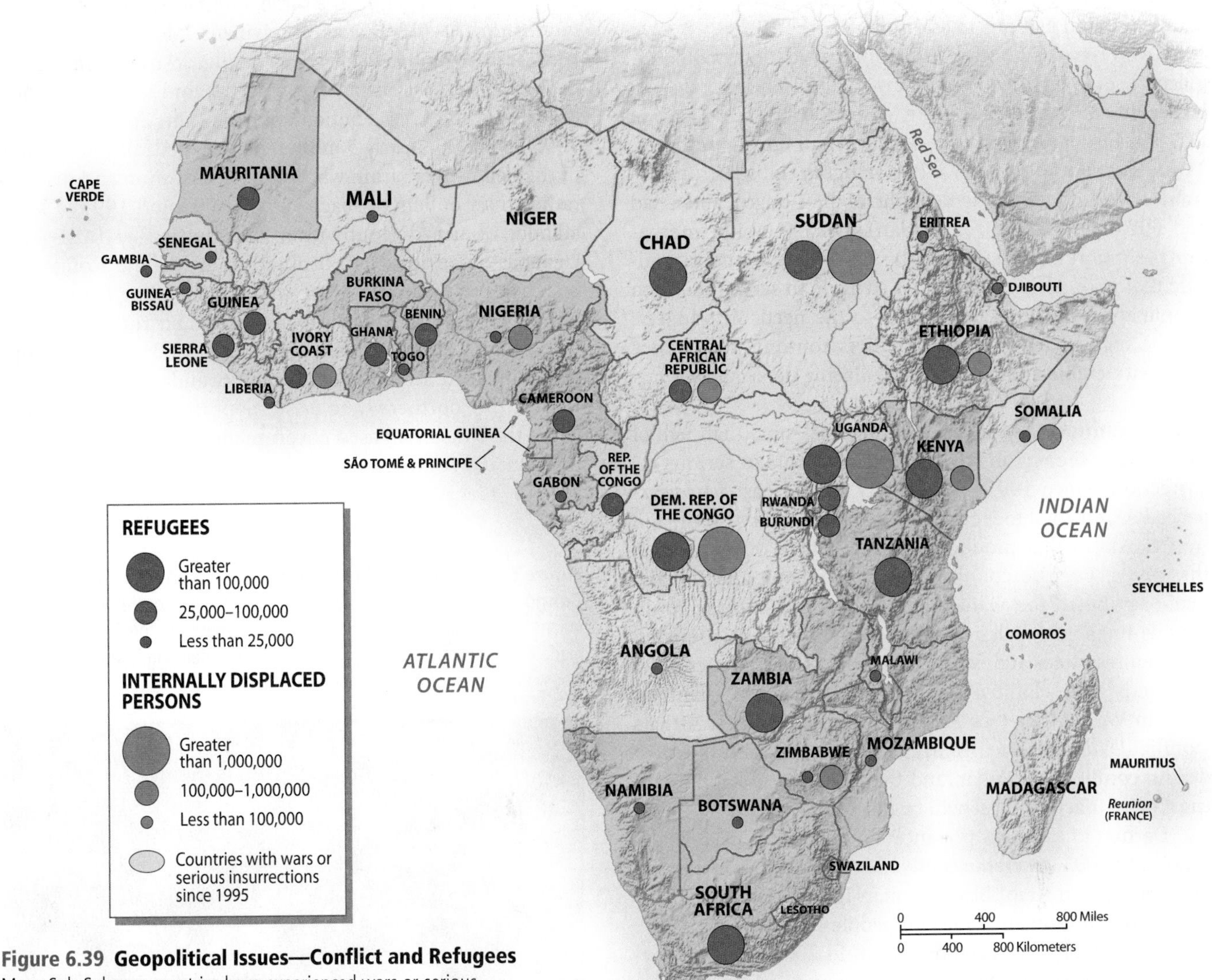

Figure 6.39 Geopolitical Issues—Conflict and Refugees Many Sub-Saharan countries have experienced wars or serious insurrections since 1995. These same states are also likely to produce refugees (red circles) and internally displaced persons (yellow circles). As of 2009, nearly 3 million Africans were refugees and 13 million were internally displaced. (Data from U.S. Committee for Refugees, 2009, *World Refugee Survey*)

inherited an inappropriate set of colonial borders, one might assume that they would have been better redrawing a new political map based on indigenous identities. Such a strategy was impossible, as all the leaders of the newly independent states realized. Any new territorial divisions would have created winners and losers, and thus would have resulted in more conflict. Moreover, because ethnicity in Sub-Saharan Africa was traditionally fluid, and because many groups were territorially intermixed, it would have been difficult to generate a clear-cut system of division. Finally, most African ethnic groups were considered too small to form viable countries. With these complications in mind, the new African leaders meeting in 1963 to form the Organization of African Unity agreed that colonial boundaries should remain. The violation of this principle, they argued, would lead to pointless wars between states and endless civil struggles within them.

Despite the determination of Africa's leaders to build their new nations within existing boundaries, challenges to the states began soon after independence. Figure 6.39 maps the ethnic and political conflicts that have disabled parts of Africa since 1995. The human cost of this turmoil is several million refugees and internally displaced persons. **Refugees** are people who flee their state because of a well-founded fear of persecution based on race, ethnicity, religion, or political orientation. Nearly 3 million Africans were considered refugees in 2009. Added to this figure are another 13 million internally displaced persons. **Internally displaced persons** have fled from conflict but still reside in their country of origin. Sudan had the largest number of internally displaced people (5 to 6 million), followed by the Democratic Republic of the Congo and Uganda (over 1 million each). These populations are not technically considered refugees, making it difficult for humanitarian nongovernmental organizations and the United Nations to assist them.

Over the past decade, the number of internally displaced people has risen, while the total number of refugees has declined. This decline is due, in part, to the reluctance of neighboring states to take on the burden of hosting refugees. Tanzania had more than 300,000 refugees (mostly from Burundi and the Democratic Republic of the Congo), and Chad had 300,000 (mostly from Sudan) in 2009. With international support of refugees on the decline, it is understandable that poor African states that struggle to serve their own people are disinclined to respond to the needs of refugees. Yet the region's many ethnic conflicts continue to produce many vulnerable and needy people fleeing persecution.

Ethnic Conflicts As Figure 6.39 suggests, more than half of the states in the region have experienced wars or serious insurrections since 1995. Fortunately, in the past few years peace has returned to Sierra Leone, Liberia, Ivory Coast, and Angola, states that produced large numbers of refugees in the 1990s. For the countries of Sierra Leone and Liberia, many attributed the availability of diamonds in these countries as financing cycles of violence. While the relationship between resources and conflict is complex, the term conflict diamonds was employed when discussing the diamond trade in West Africa (see Environment and Politics: Africa's Conflict Diamonds). Currently, the two states with the most deadly conflicts are Sudan and the Democratic Republic of the Congo. Each of these will be discussed in turn.

Darfur, a Sahelian province in Sudan, is the size of France. In 2003, an ethnic conflict between Arab nomads and herders and non-Arab "black" farmers ignited. Since the fighting began, an estimated 300,000 people have died and more than 2.5 million have been displaced. In 2004, the U.S. government accused the Sudanese government of **genocide**—the deliberate and systematic killing of a racial, political, or cultural group. The African Union and the United Nations have sent peacekeeping troops to assist the displaced and reduce the violence. In 2009, the International Criminal Court issued a warrant for the arrest of Sudanese President Bashir for war crimes, but as of 2010 Bashir remains in office and denies the charges.

The conflict began when a non-Arab rebel group (later called the Sudan Liberation Movement) attacked a Sudanese military base. The SLM complained of economic marginalization and wanted a power-sharing arrangement with the Sudanese government. The Sudanese government responded to the rebels by encouraging Arab militiamen (called *janjaweed*—an Arabic term meaning "horse and gun") to attack non-Arab civilians in Darfur. The janjaweed storm villages on horseback and camelback, destroying grain supplies and livestock, killing villagers, and setting buildings on fire. Human rights groups are especially alarmed by the systematic use of rape by the janjaweed as a weapon of ethnic cleansing and terror. The government of Khartoum claims it has been unable to stop the janjaweed atrocities, yet international observers continue to produce evidence that the government is supporting the Arab militiamen through aerial bombardment and attacks on civilians. More than 500 villages were completely or substantially destroyed in Darfur.

The deadliest ethnic and political conflict in the region has been in the Democratic Republic of the Congo. Between 1998 and 2008, nearly 5 million people died, although many of the deaths were from war-induced starvation and diseases rather than bullets or machetes. In 1996 and 1997, a loose alliance of armed groups from Rwanda (led by Tutsis) and Uganda joined forces with other groups in the Congo and marched their way across the country, installing Laurent Kabila as president. Under Kabila's rocky and ruthless leadership, which ended in assassination in 2001, rebel groups again invaded from Uganda and Rwanda and soon controlled the northern and eastern portion of the country while the Kinshasa-based government loosely controlled the western and southern portions.

Figure 6.40 Chukudus in the Democratic Republic of the Congo A Congolese boy with his chukudu (wooden scooter) delivers onions to a Goma market in the Democratic Republic of the Congo. Locally made chukudus are rugged symbols of improvisation and invention in this war-weary region of Central Africa.

With Kabila's death in 2001, his son Joseph took power and signed a peace accord with the rebels in 2002. In 2003 rebel leaders were part of a transitional government and an unsteady peace was in place with help from the UN, the African Union, and support by western donors. Remarkably, elections were held in 2006, Joseph Kabila was elected president. The country has no real experience with democracy, a civil service that barely functions, and virtually no roads or working infrastructure for the 70 million people who live there. In 2008, foreign aid accounted for 15 percent of the country's GNI, down from 20 percent in 2005. For now, continuing peace and the ability of the Congo's people to improvise and cope under such adversity may gradually bring about social and economic improvements. A symbol of this improvisation is the rugged wooden scooters (called *chukudus*) that move cargo and people around in the eastern Congo (Figure 6.40).

Secessionist Movements Problematic African political boundaries have occasionally led to attempts by territories to secede and form new states. The Shaba (or Katanga) province in what was then the state of Zaire tried to leave its national territory soon after independence. The rebellion was crushed a couple of years after it started with the help of France and Belgium. Zaire was unwilling to give up its copper-rich territory, and the former colonialists had economic interests in the territory worth defending. Similarly, the Igbo in oil-rich southeastern Nigeria declared an independent state of Biafra in 1967. After a short but brutal war during which Biafra was essentially starved into submission, Nigeria was reunited.

In 1991, the government of Somalia disintegrated. The territory has since been ruled by clan-based warlords and their militias, who have informally divided the country into clan-based units. **Clans** are social units that are branches of a tribe or ethnic group larger than a family. Early in the conflict, the northern portion of the country declared its independence as a new country—Somaliland (Figure 6.41). Somaliland has a constitution, a functioning parliament, government ministries, a police force, a judiciary, and a president. The territory produces its own currency and passports. Yet no country has recognized this territory, in part because no government exists in Somalia to negotiate the secession. In 1998, neighboring Puntland also declared itself an autonomous state. Although it does not seek outright independence like Somaliland, Puntland is creating its own administration. Meanwhile, Islamic courts with their well-armed militias control the south, where Al Qaeda operatives are believed to reside, along with pirates who seize ships for ransom in the Indian Ocean and Gulf of Aden.

Only one territory in the region has successfully seceded. In 1993, Eritrea gained independence from Ethiopia after two decades of civil conflict. This territorial secession is striking because Ethiopia forfeited its access to the Red Sea, making it landlocked. Yet the creation of Eritrea still did not bring about peace. After years of fighting, the transition

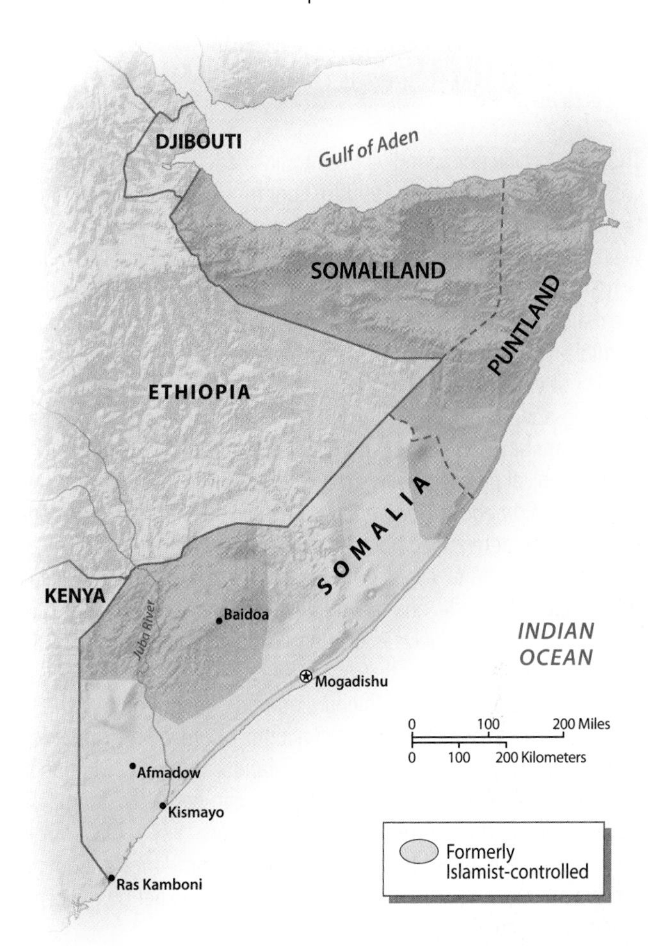

Figure 6.41 Somalia Divided A state in name only, Somalia has had no central government since 1991. Two areas in the north—Somaliland and Puntland—seek political autonomy or independence. Meanwhile, Islamic courts have controlled areas in the south.

to Eritrean independence began remarkably well. Unfortunately, border disputes between the two countries erupted in 1998, resulting in some 100,000 troop deaths. In 2000, a peace accord was reached and the fighting stopped. As Ethiopia comes to accept Eritrea's independence, it might be possible for other areas torn by ethnic warfare to find peace as well. Still, major transformations of Africa's political map should not be expected.

ECONOMIC AND SOCIAL DEVELOPMENT: The Struggle to Rebuild

By almost any measure, Sub-Saharan Africa is the poorest and least-developed world region. While other regions have experienced significant improvements in life expectancy and per capita income since 1990, many Sub-Saharan African states have experienced declines in life expectancy, and income levels have remained stubbornly low. The average GNI per capita figure using purchasing power parity is about $1,950. Some states such as Botswana, Equatorial

ENVIRONMENT AND POLITICS Africa's Conflict Diamonds

The term 'conflict diamond' is used to link African diamond exports with armed conflict. In the 1990s in countries such as Liberia, Sierra Leone, Angola and the Democratic Republic of the Congo, there was clear evidence that diamonds were used to purchase arms and perpetuate conflict. Yet, there are also several African states that are major diamond producers, such as South Africa, Botswana, and Namibia, where a link between diamond extraction and armed conflict is not clear.

Geographer Philippe Le Billon has researched the linkages between natural resources such as diamonds and armed conflict from a geographical perspective in a recent study in the *Annals of the Association of American Geographers.* He concludes that the relationship is not predetermined but has much to do with the materiality of production, locational aspects of the resource, the commodity chains that support the trade, and the political economic context. That said there are particular aspects in diamond extraction and trade that are well suited for funding conflict.

In the media the idea of 'blood diamonds' caught on in the late 1990s and was used to explain the actions of greedy warlords manipulating easily looted resources for nefarious purposes. Diamonds are small in size and highly valuable. They can be easily smuggled, traded, and used as currency. Moreover, where they come from is not easily determined. Most rough diamonds from Africa are bound for two places, Antwerp and London, so controls higher up the commodity chain could be effective if applied.

Yet subtleties of diamond extraction were lost in most media portrayals. One important distinction is the extraction of diamonds from primary deposits where capital-intensive industrial mining techniques are used, versus secondary alluvial deposits were artisanal mining occurs that is spatially more diffuse and politically harder to regulate. These artisanal practices also can have deleterious environmental impacts as watersheds are disturbed and contaminated. It is usually artisanal mining that is most associated with conflict diamonds (Figure 6.4.1)

Figure 6.4.1 Diamond Miners in Sierra Leone Men sieve gravel looking for diamonds in the Kono District of Sierra Leone. Alluvial diamond deposits exist throughout Sierra Leone and are often extracted using simple artisanal methods.

The relationship between diamonds and conflict is most acute in the West African countries of Sierra Leone and Liberia. Sierra Leone has significant alluvial deposits of diamonds and a long tradition of artisanal mining that goes back to the 1930s. In the 1990s, the Revolutionary United Front (RUF) supported by former Liberian president and warlord Charles Taylor, gained control of the eastern and northern sections of Sierra Leone where there are significant diamond deposits. During the war in Sierra Leone that lasted from 1991 to 2000,

Guinea, Mauritius, the Seychelles, and South Africa have much higher per capita GNI-PPP (Table 6.2). Purchasing power parity is a relative income figure; when absolute income is measured, over 70 percent of the population have average incomes of less than $2 a day. Nearly all the states in the region are ranked at the bottom of most lists comparing per capita GNI or the Human Development Index.

In the 1990s, most countries in Sub-Saharan Africa saw their economic productivity decline, with low or negative growth rates between 1990 and 1999. The economic and debt crisis of the 1980s and 1990s prompted the introduction of **structural adjustment programs**. Promoted by the International Monetary Fund (IMF) and the World Bank, structural adjustment programs typically reduce government spending, cut food subsidies, and encourage private-sector initiatives. Yet these same policies have caused immediate hardships for the poor, especially women and children, and have led to social protest, most notably in the cities. The idea of debt forgiveness for Africa's poorest states, argued for by the unlikely duo of economist Jeffrey Sachs and U2 rock star Bono, has gradually gained acceptance as a strategy to reduce human suffering by redirecting monies that would have gone to debt repayment to the provision of services and the building of infrastructure. Sachs argues that in order for the region to get out of the poverty trap, it will need substantial sums of new foreign aid and investment.

On the positive side, there are signs of economic growth since 2000. For most states, the average annual growth rates from 2000 to 2008 are positive. Several countries have seen average annual growth rates of 5 percent or more (see Table 6.2). While good news overall, such figures can be deceptive. Some are due to soaring oil prices during this time period (notably so for Angola, Chad, Nigeria, and Sudan), whereas others are due to countries beginning from a very low base after years of conflict (Mozambique, Sierra Leone, Rwanda). Still, for the first time in many years, Sub-Saharan African economies are growing at a faster rate than their populations. It appears that the global financial crisis of 2008–09 did not greatly impact these growth trends.

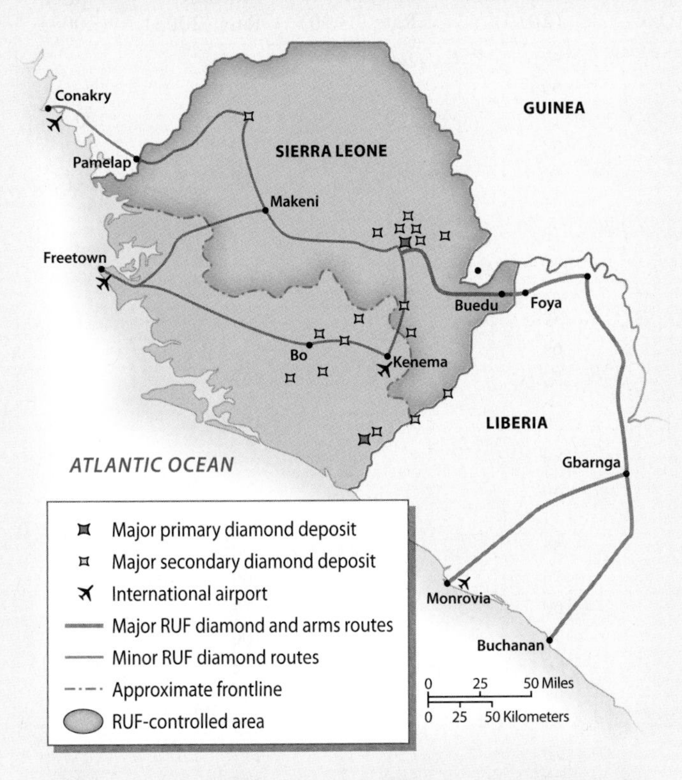

Figure 6.4.2 Armed Conflict and Diamonds in Sierra Leone During the 1990s the Revolutionary United Front (RUF) lead by Charles Taylor from Liberia, controlled much of northern Sierra Leone and its diamond trade. Funds from illicit diamond sales funded armed conflict in Sierra Leone and Liberia. (Source: Philippe Le Billion, 2008, "Diamond Wars? Conflict Diamonds and Geographies of Resource Wars" *Annals of the Association of American Geographers* 98(2): p 362)

it is estimated that diamond sales funneled $25 to 75 million per year to rebel forces. Liberia was also engulfed in conflict, with arms bought by RUF from diamond sales. War-related deaths in these two states approached 300,000 from 1989 to 2000. Diamonds were a significant resource that provided capital and arms to fuel the conflicts (Figure 6.4.2). As of 2010 Charles Taylor, who was president of Liberia from 1997 until 2003, is detained and under investigation through United Nations courts for his role in the civil wars.

One result of public concern about conflict diamonds was the establishment of the Kimberly Process Certification Scheme, adopted in 2002, to keep conflict diamonds out of the market. Corporate leaders, such as DeBeers, were initially resistant to the certification process but later came to support it as a means protect this valuable luxury product from the possibility of wider boycotts due to its tainted image. The process is voluntary and requires participating nations to agree not to trade in diamonds from non-participating countries.

Overall the certification process has been seen as a positive step for diamond-dependent countries in Africa. As a result of the Kimberly Process, diamond exports flow from Sierra Leone and are viewed as critical for the country's post-conflict development. Ironically, industrial approaches to mining (with tighter controls) are preferred over artisanal methods when it comes to certification. This has the unintended consequence of making it difficult for small-scale producers to have their diamonds certified even when there is no evidence of conflict.

Adapted from Philippe Le Billon, 2008, "Diamond Wars? Conflict Diamonds and Geographies of Resource Wars" *Annals of the Association of American Geographers* 98(2): 345-372.

Roots of African Poverty

In the past, outside observers often attributed Africa's poverty to its environment. Favored explanations included the infertility of its soils, the erratic patterns of its rainfall, the lack of navigable rivers, and the virulence of its tropical diseases. Most contemporary scholars, however, argue that such handicaps are not prevalent throughout the region and that even where they do exist they can be—and have often been—overcome by human labor and ingenuity. The favored explanations for African poverty today look much more to historical and institutional factors than to environmental circumstances.

Numerous scholars have singled out the slave trade for its debilitating effect on Sub-Saharan African economic life. Large areas of the region were effectively depopulated, and many people were forced to flee into poor, inaccessible refuges. Colonization was another blow to Africa's economy. European powers invested little in infrastructure, education, and public health and were instead interested mainly in extracting mineral and agricultural resources for their own benefit. Several plantation and mining zones did achieve a degree of prosperity under colonial regimes, but strong national economies failed to develop. In almost all cases, the rudimentary transport and communications systems were designed to link administration centers and zones of extraction directly to the colonial powers rather than to their own surrounding areas. As a result, upon achieving independence, Sub-Saharan African countries thus faced economic and infrastructural challenges that were as daunting as their political problems (Figure 6.42). Today, the average amount of paved road per person in countries such as Ghana, Ethiopia, Kenya, Tanzania, and Uganda is just 33 feet (10 meters) per person. The average for the non-African developing world is 2.8 miles (4.5 kilometers) per person. South Africa is the only African state with a developed national road network.

Failed Development Policies The first decade or so of independence was a time of relative prosperity and optimism for many African countries. Most of them relied heavily on

TABLE 6.2 Development Indicators

Country	GNI per capita, PPP 2008	GDP Average Annual % Growth 2000–08	Human Development Index (2007)[a]	Percent Population Living Below $2 a Day	Life Expectancy (2010)[b]	Under Age 5 Mortality Rate (1990)	Under Age 5 Mortality Rate (2008)	Gender Equity (2008)[c]
Angola	4,820	13.5	0.564		47	260	220	
Benin	1,470	3.9	0.492		59	184	121	
Botswana	13,300	4.5	0.699	49.4	55	50	31	100
Burkina Faso	1,160	5.6	0.389	81.2	53	201	169	85
Burundi	380	2.9	0.394	93.4	50	189	168	91
Cameroon	2,170	3.5	0.523	57.7	51	149	131	84
Cape Verde	3,080		0.708		73			
Central African Republic	730	0.5	0.369	81.9	49	178	173	
Chad	1,070	11.9	0.392		49	201	209	64
Comoros	1,170		0.576		64			
Congo	2,800	3.9	0.601		53	104	127	
Dem. Rep. of Congo	280	5.2	0.389		48	199	199	76
Djibouti	2,320		0.520	41.2	55			
Equatorial Guinea	21,700		0.719		49			
Eritrea	640	1.3	0.472		59	150	58	77
Ethiopia	870	8.2	0.414	77.5	55	210	109	85
Gabon	12,390	2.2	0.755		60	92	77	
Gambia	1,280	5.1	0.456	56.7	55	153	106	102
Ghana	1,320	5.6	0.526	53.6	60	118	76	96
Guinea	970	3.2	0.435	87.2	57	231	146	77
Guinea-Bissau	520	0.6	0.396	77.9	46	240	195	
Ivory Coast	1,580	0.5	0.484		73	150	114	
Kenya	1,550	4.5	0.541	39.9	57	105	128	96
Lesotho	1,970	3.9	0.514	62.2	41	101	79	105
Liberia	310	−1.1	0.442		56	219	145	86
Madagascar	1,050	3.8	0.543	89.6	60	167	106	97
Malawi	810	4.2	0.493	90.4	49	225	100	99
Mali	1,090	5.2	0.371	77.1	51	250	194	78
Mauritania	1,990	5.1	0.520	44.1	57	129	118	104
Mauritius	12,570	3.7	0.804		73	24	17	100
Mozambique	770	8.0	0.402	90.0	48	249	130	87
Namibia	6,240	5.6	0.686		61	72	42	104
Niger	680	4.4	0.340	85.6	48	305	167	74
Nigeria	1,980	6.6	0.511	83.9	47	230	186	85
Reunion					78			
Rwanda	1,110	6.7	0.460	90.3	51	174	112	100
São Tomé and Principe	1,790		0.651		66			
Senegal	1,780	4.5	0.464	60.3	55	149	108	96
Seychelles	19,630		0.845	<2	73			
Sierra Leone	770	10.3	0.365	76.1	47	278	194	84
Somalia					49	200	200	
South Africa	9,780	4.3	0.683	42.9	55	56	67	100
Sudan	1,920	7.4	0.531		58	124	109	89
Swaziland	5,000	2.6	0.572	81.0	46	84	83	92
Tanzania	1,260	6.8	0.530	96.6	55	157	104	
Togo	830	2.4	0.499		61	150	98	75
Uganda	1,140	7.5	0.514	75.6	52	186	135	99
Zambia	1,230	5.3	0.481	81.5	42	172	148	95
Zimbabwe		−5.7			43	79	96	97

[a] *United Nations, Human Development Report, 2009.*

[b] *Population Reference Bureau, World Population Data Sheet, 2010.*

[c] *Gender Equity—Ratio of female to male enrollments in primary and secondary school. Numbers below 100 have more males in primary/secondary school, numbers above 100 have more females in primary/secondary schools.*

Source: World Bank, World Development Indicators, 2010.

Figure 6.42 Lack of Infrastructure Aid workers in the Republic of the Congo push their vehicles out of the mud along the main road from Kinkala to the capital of Brazzaville. Basic infrastructure, such as paved roads, is limited in Central Africa. During the rainy season, many dirt roads are not passable.

the export of mineral and agricultural products, and through the 1970s commodity prices generally remained high. Some foreign capital was attracted to the region, and in many cases the European economic presence actually increased after decolonization. (This is one reason critics refer to the period as one of "neocolonialism.")

In the 1980s, as most commodity prices began to decline, foreign debt began to weigh down many Sub-Saharan countries. By the end of the 1980s, most of the region was in serious economic decline. By the early 1990s, the region's foreign debt was around $200 billion. Although low compared to that of other developing regions (such as Latin America), Sub-Saharan Africa's debt was the highest in the world as a percentage of its economic output.

Many economists contend that Sub-Saharan African governments enacted counterproductive economic policies and thus brought some of their misery on themselves. Keen to build their own economies and reduce their dependency on the former colonial powers, most African countries followed a course of economic nationalism. More specifically, they set about building steel mills and other forms of heavy industry in which they were simply not competitive (Figure 6.43). Local currencies also were often maintained at artificially elevated levels, which benefited the elite who consume imported products but undercut exports. Some former French colonies kept their currencies pegged to the franc, which made successful export strategies almost impossible.

The largest blunders made by Sub-Saharan leaders were in agricultural and food policies in the post-independence period. The main objective was to retain a cheap supply of staple foods in the urban areas. A modern industrial economy could emerge, or so it was argued, only if manufacturing wages remained lower, which in turn was feasible only if food could remain inexpensive. The main problem with this policy was that the majority of Africans were farmers, who could not make money from their crops because prices were kept artificially low. Thus they opted to grow food mainly for subsistence, rather than to sell—often at a loss—to national marketing boards. At the same time, they were encouraged to shift into export crops, such as coffee, cocoa, peanuts, and cotton. The end result was the failure to meet staple food needs at a time when the population was growing explosively. In the 1980s famine occurred in 22 African states, partly as a result of failed agricultural policies com-

Figure 6.43 Industrialization Heavy industry, such as this chemical plant in Kafue, Zambia, has failed to deliver Sub-Saharan Africa from poverty. In the worst cases, these industrial enterprises were unable to produce competitive products for world and domestic markets.

bined with drought. As a result of these events, most countries began to reform their agricultural policies in the 1980s, but they met with resistance from the urban poor who had grown dependent upon low prices for staple goods.

Millennium Development Goals By the 1990s, after two decades of stymied economic development, a different approach was sought. The implementation of the **Millennium Development Goals** is one example of global effort to foster development in Sub-Saharan Africa, as well as other poor areas of the developing world. The Millennium Development Goals (MDG) are part of a United Nations effort to reduce extreme poverty by 2015. The goals are to:

- Halve the proportion of people living on incomes of less than $1 a day between 1990 and 2015;
- Ensure universal primary schooling and promote gender equity in school enrollment by 2015;
- Reduce child mortality by two-thirds from 1990 to 2015;
- Reduce by three-quarters the maternal mortality rate between 1990 and 2015;
- Halt the growing prevalence of HIV/AIDS and malaria, and begin to reduce their spread by 2015;
- Halve the proportion of the population without sustainable access to safe drinking water and sanitation by 2015.

As of 2010, several regions of the world are on track to meet many of these goals, although it appears that Sub-Saharan Africa will not meet many of its 2015 goals, especially in terms of reducing extreme poverty and achieving universal primary education. The region has made gains in fighting disease, especially HIV/AIDS. That said, individual countries in Sub-Saharan Africa have made impressive gains towards achieving some of these goals.

International interest in Africa's overall development seems to be growing: Between 2000 and 2008 net development assistance to the region tripled from $13 billion to $40 billion. Some argue that aid increased as a result of having stated MDG benchmarks and agreed-upon measures for assessment. Most of the new aid is directed towards fighting disease, improving health care, and providing basic education. There are still pressing needs for infrastructure development. Now, as in the past, some donated funds do end up in the hands of corrupt officials.

Corruption Although prevalent throughout the world, corruption also seems to have been particularly rampant in several African countries. Civil servants typically are not paid a living wage and are thus virtually forced to solicit bribes. Teachers solicit informal fees from students in order to get paid. According to a recent poll of international businesspeople, Nigeria ranks as the world's most corrupt country. (Skeptical observers, however, point out that several Asian nations with highly successful economies, such as China, also are noted for high levels of corruption, so that corruption alone may not be an explanation for Africa's economic problems.)

With millions of dollars in loans and aid pouring into the region, officials at various levels were tempted to skim from the top. Some African states, such as the Democratic Republic of the Congo (DRC), were dubbed kleptocracies. A **kleptocracy** is a state in which corruption is so institutionalized that politicians and government bureaucrats siphon off a huge percentage of the country's wealth. President Mobutu of the DRC was a legendary kleptocrat. While his country was saddled with an enormous foreign debt, he reportedly skimmed several billion dollars and deposited them in Belgian banks during his presidency from 1965 to 1997.

Links to the World Economy

Sub-Saharan Africa's trade connection with the world is limited, accounting for less than 2 percent of global trade. The level of overall trade is low both within the region and outside it. Traditionally, most exports went to the European Union, especially the former colonial powers. The United States is the second most common destination. These trade patterns are changing, especially as trade with China and India grows. As of 2007, exports to Asia nearly equaled exports to the United States. The pattern of imports is also changing. The majority of African countries still turn to Europe for imports but Asia (particularly China) now accounts for one-third of all imports.

By most measures of connectivity, Sub-Saharan Africa lags behind other developing regions. Telephone lines are scarce; Nigeria has only one landline per 100 people. The regional average is slightly better, at two lines per 100 people. Communication via cellular phones, however, is soaring. According to World Bank estimates for every 100 Nigerians, there were 42 mobile phone subscriptions in 2008, and for the region as a whole there were 36 mobile phone subscriptions for every 100 people. By comparison, in 2000 the regional average was just 2 subscriptions per 100 people. No longer reliant on expensive fixed telephone lines for communication, multinational providers now are competing for mobile-phone customers. Internet use is also on the rise; recent World Bank estimates show that 6.5 percent of the population uses the Internet. The growth of mobile telephones and Internet access are a powerful indicator of Africans' desire to acquire modern telecommunication systems long denied to them due to limited landlines (Figure 6.44).

Africa lacks the infrastructure to facilitate more trade. Only southern Africa has a telecommunications and road network of any note. Of the few roads that exist in Central and West Africa, the majority are not paved. This, too, is a legacy of colonialism, and one that keeps Africans from nurturing economic linkages with each other and with the outside world. Trade groups, such as the Southern African Development Community, have the potential to build up regional economies and infrastructure but have yet to achieve the influence of similar groups such as Mercosur in South America.

Figure 6.44 Mobile Phones for Africa A young Maasai man in Eremit, Kenya uses his cell phone. Since 2000 mobile phone access and use has surged in the region, greatly improving telecommunications.

Aid Versus Investment In many ways, Sub-Saharan Africa is linked to the global economy more through the flow of financial aid and loans than through the flow of goods. As Figure 6.45 reveals, aid for several states equals 20 percent or more of the GNI. In the extreme case of Liberia, official development assistance in 2008 was estimated to be 185 percent of the GNI; in Burundi it was 44 percent in the same time period. Most of this aid comes from a handful of developed countries (see the inset in Figure 6.45). The United States and the European Commission provided the most aid to the region in 2008, followed by France, the United Kingdom, and Germany.

While aid is extremely important for many African states, foreign direct investment in the region substantially increased from only $4.5 billion in 1995 to $34 billion in 2008. The largest recipients of this foreign investment were the region's major mineral and oil producers. South Africa received nearly 30 percent of the region's foreign investment, but Nigeria, the Republic of the Congo, Sudan, and Ghana also experienced foreign investment growth. The poorer states—without minerals or oil to export—were largely bypassed by foreign investors. Yet the overall level of foreign investment remains low. Consider that all of the foreign direct investment in Sub-Saharan Africa in 2008 was just three-quarters of the foreign investment in Brazil alone.

The reasons foreign investors avoid Africa are fairly obvious. The region is generally perceived to be too poor and unstable to merit much attention, and most foreign investors eager to take advantage of low wages put their money in Asia or Latin America. Sub-Saharan Africa, some economists contend, is therefore starved for capital. Other scholars, however, see foreign investment as more of a trap—and one that the region would be wise to avoid. Recently in countries such as the Republic of the Congo, Chad, Equatorial Guinea, and Angola, investment in oil exploration and production has attracted considerable overseas interest. It is not just the oil producing states attracting foreign capital, copper-rich Zambia saw its foreign investment increase 10-fold between 1995 and 2008. Optimists interpret this as a sign of Africa's economic renewal, but others counter that in the past, mineral extraction has failed to lead to broad-based and lasting economic gains (see "Exploring Global Connections: China's Investment in Africa").

Debt Relief Another development strategy making a difference in the region is debt relief. The World Bank and the IMF proposed in 1996 to reduce debt levels for heavily indebted poor countries, many of which are in Sub-Saharan Africa. Most Sub-Saharan states are indebted to official creditors such as the World Bank, not to commercial banks (as is the case in Latin America and Southeast Asia). Under this World Bank/IMF program, substantial debt reduction is given to Sub-Saharan countries that are determined to have "unsustainable" debt burdens. Mauritania, for example, spends six times more money on repaying its debts than it does on health care.

States qualify for different levels of debt relief provided they present a poverty reduction strategy. Uganda was the first state to qualify for the program, using the money it saved on debt repayment to expand primary schooling. Ghana qualified for debt relief in 2004 and received a $3.5 billion relief package. Other countries that have benefited from debt reduction are Tanzania, Mozambique, Ethiopia, Mauritania, Mali, Niger, Nigeria, Senegal, Burkina Faso, and Benin. More African countries may qualify in the future to redirect debt payments toward building infrastructure and improving basic health and education.

Economic Differentiation Within Africa

As in most regions, considerable differences in levels of economic and social development persist in Sub-Saharan Africa. In many respects, the small island nations of Mauritius and the Seychelles have little in common with the mainland. With a high per capita GNI, life expectancies averaging in the low 70s, and economies built on tourism, they could more easily fit into the Caribbean were it not for their Indian Ocean location. Only a few states, such as Botswana, South Africa, Gabon, Namibia, and Equatorial Guinea, have a per

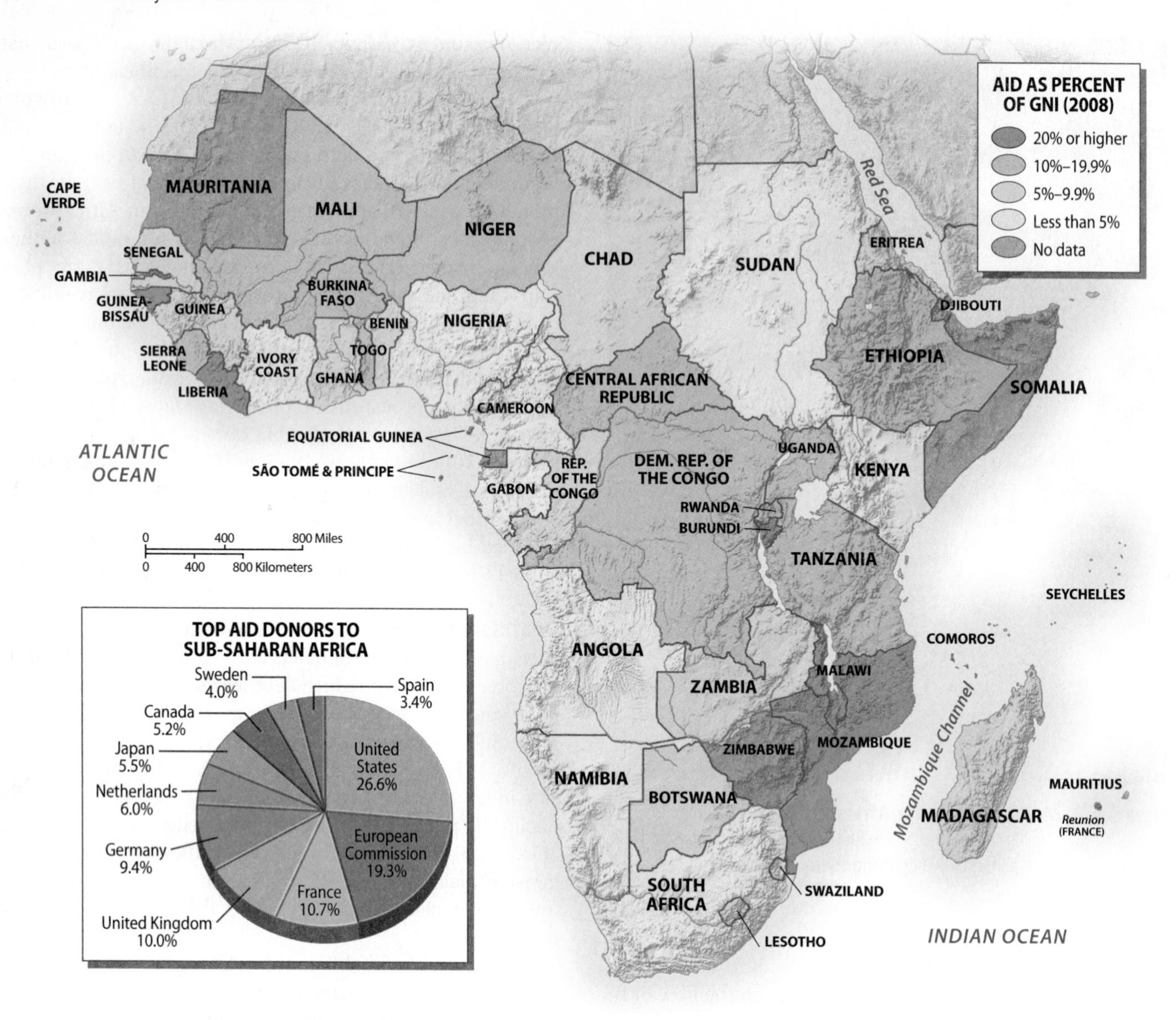

Figure 6.45 Global Linkages: Aid Dependency Many states in Sub-Saharan Africa are dependent on foreign aid as their primary link to the global economy. This figure maps aid as a percentage of GNI, which ranges from less than 1 percent to 186 percent in Liberia. (From *World Development Indicators,* 2010, Washington, DC: World Bank)

capita GNI-PPP of over $5,000. In the developed world, where a latte at Starbucks costs $3, it is hard to imagine existence on $2 a day, but such is the case for nearly three-quarters of all people living South of the Sahara.

There are millions of Africans living in extreme poverty in every state in the region. In rural areas many of the very poor still practice subsistence farming and are barely a part of the formal economy. In the rapidly growing urban areas, massive slums are homes to millions with inadequate shelter and no regular access to clean water. Most make their living through the informal economy. These are the people that the Millennium Development Goals hope to reach, although securing the resources and finding the right programs to ease this poverty is both a difficult and long-term proposition.

There are places and people experiencing real social and economic gains in Sub-Saharan Africa. South Africa is the unchallenged economic powerhouse of the region. The per capita GNI-PPP of Nigeria, the next largest economy, is just one-fifth of South Africa's. Only South Africa has a well-developed and well-balanced industrial economy. It also boasts a healthy agricultural sector and, more importantly, it stands as one of the world's mining superpowers. South Africa remains unchallenged in gold production, and is a leader in many other minerals and precious gems, including diamonds. In the summer of 2010 South Africa hosted the World Cup, the first African country to do so, symbolizing its arrival as a developed and modern nation.

While South Africa is undeniably a wealthy country by African standards, and while its white minority is prosperous by any standard, it is also a country beset by severe and widespread poverty. In the former townships lying on the outskirts of the major cities and in the rural districts of the former homelands, employment opportunities remain limited and living standards marginal. Despite the end of apartheid in 1994, South Africa still has one of the most unequal income distributions in the world measured by the Gini index.

Oil and Mineral Producers Another group of relatively prosperous Sub-Saharan countries benefits from substantial oil and mineral reserves and small populations. The prime example is Gabon, a country of noted oil wealth that is inhabited by nearly 2 million people. Its neighbors, the Republic of the Congo and Equatorial Guinea, have also experienced steady growth in income and foreign investment due to oil finds in the 1980s and 1990s respectively. Equatorial Guinea, a former Spanish colony, began producing significant amounts of oil in 1998. Today it has the distinction of having the region's highest per capita GNI at $21,700. Yet after a decade of oil production, these newfound revenues have not been invested in the country's citizens but, as is often the case, seem to have fallen into the hands of few elites (Figure 6.46). Farther south, Namibia and Botswana also have the advantage of small populations and abundant mineral resources, especially diamonds. Over the past few years, both countries have enjoyed sound government and experienced significant economic and social development reflected in their relatively high human development index figures.

The Leaders of ECOWAS The most populous country in Africa, Nigeria is the core member of the Economic Community of West African States (ECOWAS). Nigeria has the

EXPLORING GLOBAL CONNECTIONS China's Investment in Africa

In 2006, China hosted nearly every African head of state at the China–Africa Forum. The event signaled China's growing economic investment in the region at a time when U.S. and European interests were (and still are) focused on proffering aid or fighting terrorism. China wants to secure the oil and ore it needs for its massive industrial economy. In exchange, it offers African nations money for roads, railways, and schools, with relatively few strings attached. Some African leaders see China as a new kind of global partner, one that wants straight commercial relations without an ideological or political agenda. China's trade with Africa is growing faster than with any other region except Southwest Asia, increasing 10-fold in the past decade. Angola, a country where China has invested heavily, is now China's largest foreign supplier of oil—beating out Saudi Arabia. Sudan, with its largely untapped oil reserves, has also received substantial Chinese investment, despite international pressure not to negotiate with the Sudanese government due to the genocide in Darfur (Figure 6.5.1).

Figure 6.5.1 Chinese investment in Sudan Liu Guijin, the Chinese special representative for African Affairs, shakes hands with a man in a market in al-Fashir, Sudan. China has invested heavily in Sudan arguing that economic development is the best way to end the bloodshed there. Meanwhile, western states disagree with this approach.

China's engagement with Africa is not new. In the 1960s and 1970s, China supported the socialist governments of postcolonial Africa with development assistance. In the 1970s, Beijing funded and built the Tan-Zam railroad that gave landlocked Zambia an outlet to the sea via Tanzania, bypassing the need for Zambia to use apartheid South Africa's ports. China's investment today is strictly commercial and substantially larger. In addition to buying up oil, China purchases copper from Zambia and wood from the Democratic Republic of the Congo. It also sells cheap manufactured goods and clothing in record quantities.

The biggest complaint about Chinese investment in Africa comes from the West. There is frustration that China readily invests in states that have wretched human rights records (Sudan and Zimbabwe). China mostly ignores the lending standards that have been used in the region to curb corruption. Some Africans complain that China is only interested in exporting raw materials and importing manufactured goods back to Africa. Yet through its investments, China has also gained diplomatic friends, as well as access to raw materials. Interestingly, Chinese officials see themselves as the developing world's biggest beneficiary of globalization, and by partnering with the region most ignored by globalization, they will foster more development in Sub-Saharan Africa.

Figure 6.46 Oil Production in Equatorial Guinea A tanker approaches the liquefied natural gas (LNG) loading jetty in Equatorial Guinea. Gas and oil exports from Equatorial Guinea have soared in the past decade, contributing to the country's high per capita GNI.

largest oil reserves in the region and it is an OPEC member. Yet despite its natural resources, its per capita GNI-PPP is a relatively low $1,980. It has been argued that oil money has helped to make Nigeria notoriously inefficient. A small minority of its population has grown fantastically wealthy more by manipulating the system than by engaging in productive activities. Sixty percent of Nigerians, however, remain trapped in poverty earning less than $2 per day. Oil money led to the explosive growth of the former capital city of Lagos, which by the 1980s had become one of the most expensive—and least livable—cities in the region. As a result, the Nigerian government opted to build a new capital city in Abuja, located near the country's center, a move that has proved tremendously expensive. In 1991 Abuja became the national capital and it now has 1.5 million residents. The southwestern corner of the country, in the Niger delta where most of the oil is produced, has seen few of the benefits of oil production but bears the burden of the environmental costs and related social unrest.

The second and third most populous states in ECOWAS, Ivory Coast and Ghana, are also important West African commercial centers. These states rely upon a mix of agricultural and mineral exports. In the mid-1990s, the Ivorian economy began to take off. Boosters within the country called it an emerging "African elephant" (comparing it to the successful "economic tigers" of eastern Asia). Yet a destructive civil war began in 2002 in which rebel forces controlled the northern half of the county and over half a million Ivorians were displaced. A peace agreement was signed in 2007, but the economic growth of the 1990s has yet to return. Ghana, a former British colony, began its economic recovery in the 1990s. In 2001, it negotiated with the IMF and World Bank for debt relief and in 2004 it qualified for $3.5 billion in relief. The country's average annual growth from 2000 to 2008 was an impressive 5.6 percent.

East Africa Long the commercial and communications center of East Africa, Kenya experienced economic decline and political tension throughout the 1990s. In recent years the economy has grown at 4 percent per year and per capita GNI-PPP is at $1,550. Kenya boasts good infrastructure by African standards, and about 1 million foreign tourists come each year to marvel at its wildlife and natural beauty. Traditional agricultural exports of coffee and tea, as well as nontraditional exports such as cut flowers, dominate the economy (see "Global to Local: Roses for Europe, Thanks to Kenya"). As for social indicators, Kenyan women are having fewer children now than in the late 1970s (down from an average of 8 to 5 children per woman), and those children are better educated and less likely to be in extreme poverty compared to other countries in the region. If Kenya can avoid political implosion due to ethnic rivalries, it could lead East Africa into better economic integration with the southern and western parts of the continent.

The political and economic indicators for Kenya's neighbors, Uganda and Tanzania, are also improving with average annual growth rates in the 2000s of 7.5 percent and 6.8 percent respectively (see Table 6.2). When compared with Kenya, both countries have a much higher proportion of their populations living on less than $2 a day. Both coun-

GLOBAL TO LOCAL Roses to Europe, Thanks to Kenya

When Europeans give roses on Valentine's Day, they often have Kenyans to thank. Kenya produces about one-third of the cut flowers sold in Europe, especially roses. In the middle of the Northern Hemisphere winter, fresh roses usually come from tropical Africa or America. Cut flowers are now Kenya's second largest source of foreign exchange in agriculture after traditional tea exports. The industry employs some 50,000 workers, the majority of them women.

This was not always the case. Beginning the 1970s the highland tropical flower business began to bloom. Kenya's floriculture trade took off in the 1990s. Today the two largest producers of cut flowers in the world are Colombia and Kenya. Near the equator, with growing areas above 5,000 ft, year-round spring-like conditions and 12-hour days perfect for growing roses, these environmental and locational attributes contribute to their prominence in the global flower trade. Environmental conditions are not enough; to succeed in floriculture a country needs near-by international airports, refrigerated warehouses and transport, capital intensive greenhouse technology, and knowledge of the cultural demands of different national markets (for example Mother's Days are celebrated on different dates throughout the world).

Due to the seasonality of the cut flower trade (with peaks at Christmas, Valentine's Day and various Mother's Days) hours can be irregular and peak-season days long. Women dominate this industry, both in the greenhouses and in the assembly areas where flowers are sorted and arranged (Figure 6.6.1). There have been many reports of abuses in this industry. The workers (often young women) are poorly paid, work long-hours with limited benefits, and are exposed to dangerous agricultural chemicals. After numerous national and international campaigns, it appears that some of the more abusive aspects of this industry are being addressed.

Figure 6.6.1 Kenyan Greenhouse Workers Workers cut roses for export at a flower farm in Naivasha. Floriculture is a vital contributor to Kenya's agricultural exports and has grown dramatically in the past decade.

Figure 6.6.2 Lake Naivasha Landscape Irrigated vegetables and flowers grow on the shore of Lake Naivasha, Kenya. A combination of elevation, good soils, irrigation, and access to Nairobi International Airport, makes Naivasha the hub of Kenya's floriculture industry.

Around the shores of Lake Naivasha, in Kenya's Rift Valley and about a one-hour drive from Nairobi International Airport, lies the country's major production area (Figure 6.6.2). Greenhouses now dominate the lakeshore, and water pollution has become a serious issue. Consequently, growers are now being forced to recycle wastewater and purify it for reuse in their operations, but environmentalists are still concerned about the decline in water levels and quality in Naivasha. Also, as European demand for socially and economically sustainable *fair trade* products has increased, some Kenya producers have switched over to fair trade programs in which workers receive more benefits (housing, childcare, and medical treatment) in exchange for premium prices in the market.

In April 2010 when a volcanic eruption in Iceland closed airspace over Europe for several days, Kenyan flower producers lost tens of thousands of dollars. Unable to ship their crop, spoiled flowers were returned to growers for composting. Fortunately, air traffic returned to normal before various Mother's Day celebrations in Europe. Under normal conditions, harvested, packed and chilled flowers are trucked to Nairobi airport, air freighted to Europe, distributed (usually through the Netherlands), and in shops for sale within 72 hours of being cut in Kenya. Such commodity chains were unthinkable several decades ago but are now a part of everyday consumption.

For further study see Lone Riisgaard (2009) "Global Value Chains, Labor Organization and Private Social Standards: Lessons from East Africa Cut Flower Industries" *World Development* 37(2): 326-340.)

tries rely heavily upon agricultural exports and mining (especially gold). Uganda and Tanzania both benefited from debt reduction agreements that redirected funds from debt repayment to education and health care.

Measuring Social Development

By world standards, measures of social development in Africa are extremely low. Yet there are some positive trends, especially with regard to child survival, education, and gender equity. Also, many governments in the region have reached out to the modern African diaspora (economic migrants who have left the region for employment in Europe and North America). As in other parts of the world, African immigrant organizations have worked to improve schools and health care, and immigrants themselves have returned to invest in businesses and real estate. The economic impact of remittances on this region is relatively small when compared to Latin America, South Asia, or the Caribbean, but it is growing.

Reductions in child mortality are often a surrogate measure for improved social development because if most children make it to their fifth birthday there is usually adequate primary care and nutrition. In a country where the child mortality rate is above 200, it means that one out of five children die before their fifth birthday. As Table 6.2 shows, most of the states in the region saw modest to significant improvements in child survival between 1990 and 2008. Angola, Ethiopia, and Malawi actually experienced dramatic gains in child survival rates. Yet in countries undergoing prolonged conflict (Sierra Leone) or with high HIV/AIDS infection rates (Swaziland), child mortality increased or barely improved over the time period. With a regional child mortality rate of 144 per 1,000 in 2008, the region is a long way from cutting child mortality by two-thirds, as stated in the MDG goals.

Life Expectancy and Health Issues Live expectancy for Sub-Saharan Africa is only 52 years, whereas life expectancy for Latin America is 73. Countries hit hard by HIV/AIDS or conflict have seen life expectancies tumble to 40 years. Despite these statistics, there are indications that access to basic health care is improving and, eventually, so will life expectancies. Keep in mind that while high infant- and childhood-mortality figures depress the overall life expectancy figures; average life expectancies for people who make it to adulthood are much better. States such as Botswana, Kenya, and Zimbabwe achieved relatively good life expectancy figures in the 1980s but the AIDS epidemic undercut those achievements.

The causes of short life expectancy generally are related to extreme poverty, environmental hazards (such as drought), and various environmental and infectious diseases (cholera, measles, malaria, schistosomiasis, and AIDS). Often these factors work in combination. The effects of disease are exacerbated by poverty, with undernourished children being the most vulnerable to the effects of high fevers. For example, cholera outbreaks occur in crowded slums and villages where food or water is contaminated by the feces of infected persons and basic infrastructure is lacking. Tragically, diseases that are preventable, such as measles, occur when people have no access to or cannot afford vaccines. Funding from the Gates Foundation in particular has targeted fighting such diseases as a critical first step in overall development, and some diseases have seen dramatic declines in incidence due to the availability of relatively simple treatments.

The scarcity of doctors and health facilities, especially in rural areas, also helps to explain Sub-Saharan Africa's short life expectancies and high child mortality levels. The more successful countries in the region have discovered that inexpensive rural clinics dispensing basic treatments, such as oral rehydration therapy for infants with severe diarrhea, can substantially improve survival rates. Another problem is the severity of the African disease environment itself. Malaria, which kills a half-million African children each year, is reappearing in many areas as the disease-causing organisms develop resistance to common drugs and the mosquito carriers develop resistance to insecticides. While most Africans have partial immunity to malaria, it remains a major killer, and many who survive infections remain debilitated. Collectively, national and international health agencies, along with local NGOs, have improved the availability of basic health care in Sub-Saharan Africa, but much work remains to be done.

Meeting Educational Needs Basic education is another obstacle confronting the region. The goal of universal access to primary education is a daunting one for a region in which 43 percent of the population is less than 15 years old. World Bank estimates show that 62 percent of African children have access to primary education (up from just 51 percent in 1991). The region is home to one-sixth of the world's children under 15 but half of the world's uneducated children. Traditionally girls were less likely than boys to attend school. But many West African countries, such as Niger, have improved their gender equity ratios (Niger's was 53 percent in 1999 and rose to 74 percent in 2008) (see Table 6.2). For the region as a whole, the ratio of girls to boys in school is 88 to 100, which is an improvement in terms of greater equity of access for those with schools to attend.

In part, the Millennium Development Goals contributed to a greater emphasis on universal education since 2000. More resources have been directed to schools from the state (through restructured foreign debt) and private donors and organizations. One example of the transformative influence of better resourced schools comes from the village of Bumwalukani, Uganda. This remote village in eastern Uganda has government schools that accommodate hundreds of children but classes are crowded, teachers are poorly trained and paid, and educational materials are scarce. Malaria and hunger keep overall attendance low. A new school in Bumawalukani, the Arlington Academy of

Figure 6.47 Ugandan School Children Cheering at a school event, these children in rural Bumwalukani, Uganda, are receiving a good education as a result of efforts by a U.S.-based NGO founded by a Ugandan emigrant from this village. Educating the region's many children is one of the most pressing challenges.

Hope, was built in 2003 and funded by a U.S. NGO started by a Ugandan migrant who grew up in this rural mountain village. Students pay a small fee to attend the school and are given hot lunches, uniforms, and incomparable educational resources compared to the government schools (Figure 6.47). Teachers are recruited from the best teaching schools in Kampala and are well paid by Ugandan standards. All classes are in English. In Uganda, students who excel receive scholarships to attend residential high schools in the capital. Many of the students from Arlington Academy are able to continue their education because of the foundation received in their model village school.

Women and Development

Development gains cannot be made in Africa unless the economic contributions of African women are recognized. Officially, women are the invisible contributors to local and national economies. In agriculture, women account for 75 percent of the labor that produces more than half the food consumed in the region. Tending subsistence plots, taking in extra laundry, and selling surplus produce in local markets all contribute to household income. Yet because many of these activities are considered informal economic activities, they are not counted. For many of Africa's poorest people, however, the informal sector is the economy, and within this sector women dominate.

Status of Women The social position of women is difficult to gauge for Sub-Saharan Africa. Women traders in West Africa, for example, have considerable political and economic power. By such measures as female labor force participation, many Sub-Saharan African countries show relative gender equality. And women in most Sub-Saharan societies do not suffer the kinds of traditional social liabilities encountered in much of South Asia, Southwest Asia, and North Africa. In 2006, Mrs. Johnson-Sirleaf, was sworn in as Liberia's president, making her Africa's first elected female leader. A former World Bank economist and grandmother, she beat out a Liberian soccer star in a runoff election. In fact, throughout the region women occupied 18 percent of all seats in national parliaments in 2009. In the country of Rwanda, over half the parliamentary seats are filled by women (Figure 6.48).

By other measures, however, such as the prevalence of polygamy, the practice of the "bride-price," and the denial of property inheritance, African women do suffer discrimination. Perhaps the most controversial issue regarding women's status is the practice of female circumcision, or genital mutilation. In Sudan, Ethiopia, Somalia, and Eritrea, as well as parts of West Africa, almost 80 percent of girls are subjected to this practice, which is extremely painful and can have serious health consequences. Yet because the practice is considered traditional, most African states are unwilling to ban it.

Regardless of their social position, most African women still live in remote villages where educational and wage-earning opportunities remain limited and bearing numerous children remains a major economic contribution to the family. As educational levels increase and urban society expands—and as reduced infant mortality provides greater security—one can expect fertility in the region to gradually decrease. Governments can speed up the process by providing birth control information and cheap contraceptives—and by investing more money in health and educational efforts aimed at women. As the economic importance of

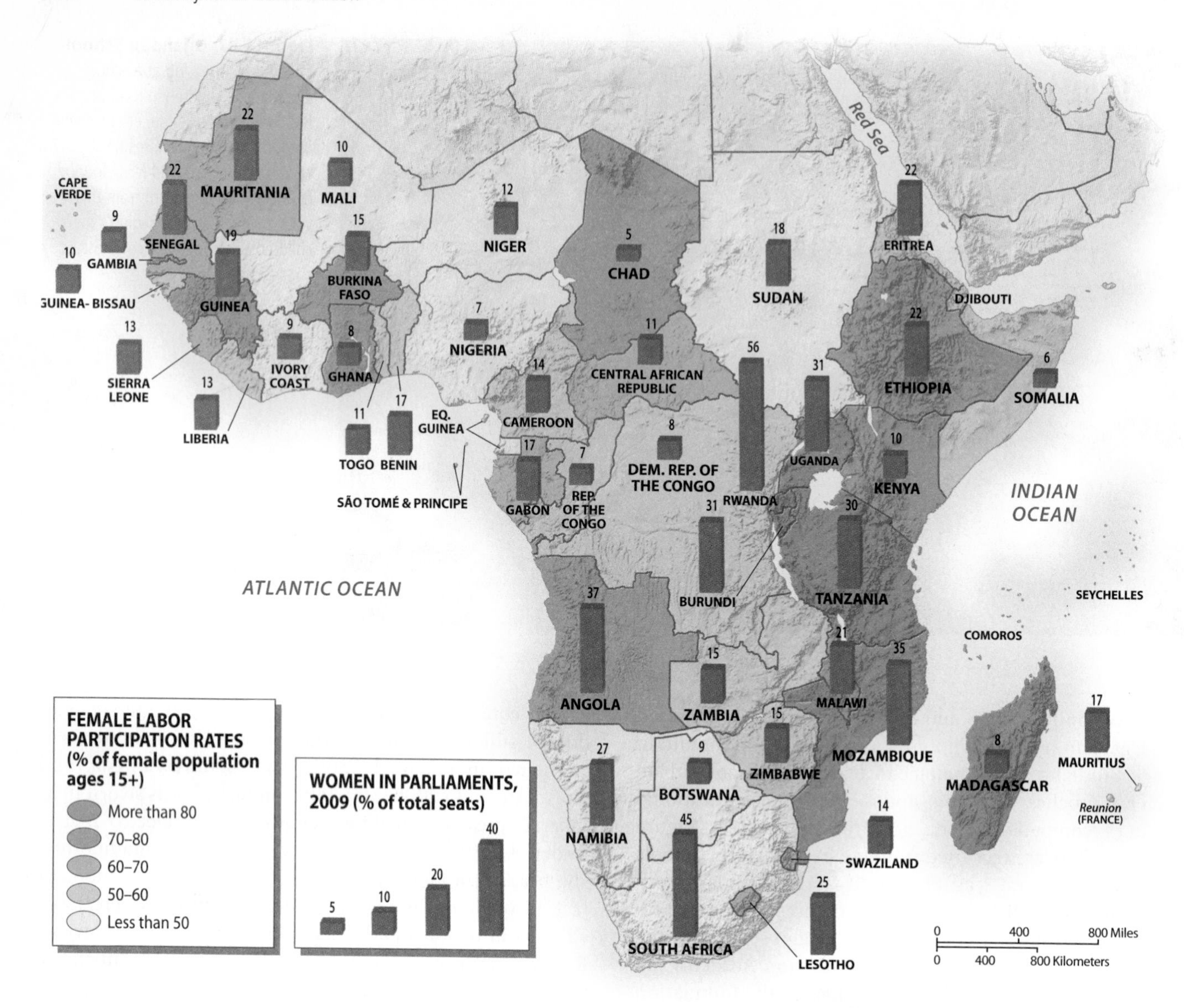

Figure 6.48 Development Issues: African Women in the Work Force and Politics Female participation in the work force is comparable to that of developed countries, as 70 percent of women over the age of 15 are in the labor force. Another significant change for the region is the increase of women holding seats in national parliaments. While the national average is 18 percent, in South Africa 45 percent of parliamentary seats are held by women, and 56 percent of the seats in Rwanda are held by women. In Africa's largest country, Nigeria, only 7 percent of the parliamentry seats are held by women.

women receives greater attention from national and international organizations, more programs are being directed exclusively toward them.

Building from Within Major shifts in the way development agencies view women and women view themselves have the potential to transform the region. All across the continent, support groups and networks have formed, raising women's consciousness, offering women micro-credit loans for small businesses, and harnessing their economic power. From farm-labor groups to women's market associations, investment in the organization of women has paid off. In Kenya, for example, hundreds of women's groups organize tree plantings to prevent soil erosion and ensure future biofuel supplies.

Whether inspired by feminism, African socialism, or the free market, community organizations have made a difference in meeting basic needs in sustainable ways. No doubt the majority of the groups fall short of all of their objectives. Yet for many people, especially women, the message of creating local networks to solve community problems is an empowering one.

Summary

- The largest landmass straddling the equator, Africa is called the plateau continent because it is dominated by extensive uplifted plains. Key environmental issues facing this tropical region are desertification, deforestation, and drought. Global warming is likely to make these problems worse. At the same time, the region supports a tremendous diversity of wildlife, especially large mammals.
- With 865 million people, Sub-Saharan Africa is the fastest growing region in terms of population. Yet it is also the poorest region, with three-quarters of the population living on less than $2 a day. In addition, it has the lowest average life expectancy at 52 years. Disease, especially the scourge of HIV/AIDS, has driven down life expectancy in the region.
- Culturally, Sub-Saharan Africa is an extremely diverse region, where multi-ethnic and multi-religious societies are the norm. With a few exceptions, religious diversity and tolerance has been a distinctive feature of the region. Most states have been independent for 50 years, and in that time, pluralistic but distinct national identities have been forged. Many African cultural expressions such as music, dance, religion, and film have had an influence well beyond the region.
- Since 1995 there have been numerous bloody ethnic and political conflicts in the region. Fortunately, peace now exists in many former conflict-ridden states such as Angola, Liberia, Sierra Leone, and the Democratic Republic of the Congo. However, ongoing ethnic and territorial disputes in Darfur, Sudan, northern Uganda, and Somalia have produced millions of internally displaced persons and refugees.
- In terms of contemporary economic globalization, Sub-Saharan Africa's connections to the global economy are weak. With 12 percent of the world's population, the region only accounts for less than 2 percent of the world's economic activity. Most of the region's economic ties are through international aid and loans rather than trade. Foreign direct investment, especially from China, is increasing in the region. Economic growth also improved in the 2000s, from the low to negative rates in the 1990s.
- Poverty is the region's most pressing issue. Since 2000, economic growth has occurred, led in part by debt forgiveness policies, higher commodity prices, and the end of some of the longest running conflicts in the region. Foreign aid to the region tripled from 2000 to 2008, in part due to the targets set by the Millennium Development Goals. It remains to be seen if the international community is truly committed to helping millions of Africans out of extreme poverty and into longer and healthier lives.

Key Terms

African Union (AU) *(page 264)*
agricultural density *(page 239)*
apartheid *(page 248)*
Berlin Conference *(page 261)*
biofuels *(page 235)*
clan *(page 269)*
coloured *(page 248)*
desertification *(page 233)*
genocide *(page 268)*
Gondwanaland *(page 228)*
Great Escarpment *(page 228)*
homelands *(page 264)*
Horn of Africa *(page 232)*
internally displaced persons *(page 267)*
kleptocracy *(page 274)*
Millennium Development Goals *(page 274)*
Pan-African Movement *(page 264)*
pastoralists *(page 245)*
physiological density *(page 239)*
refugee *(page 267)*
Sahel *(page 233)*
structural adjustment programs *(page 270)*
swidden *(page 243)*
township *(page 249)*
transhumance *(page 233)*
tribalism *(page 266)*
tribe *(page 252)*
tsetse fly *(page 245)*

Review Questions

1. Describe the relationship between soil types and patterns of settlement in Sub-Saharan Africa.
2. Is desertification a natural or a human-induced process? Where does it occur? How might global warming impact desertification trends?
3. What are the demographic, social, and economic consequences of AIDS in Sub-Saharan Africa?
4. Explain the factors that contribute to high population growth rates in the region.
5. What was the political significance of the 1884 Berlin Conference for Africa?
6. Why have there been relatively few boundary changes since African nations won their independence?
7. How are private and grassroots organizations shaping African development?
8. Explain the history and policies that make South Africa distinct from other states in the region.
9. How might the food policies of African governments have undermined the region's ability to feed itself?
10. What are the region's key economic resources that may facilitate its integration into the global economy?

Thinking Geographically

1. What factors might explain why European conquest and settlement occurred much earlier in tropical America than in tropical Africa?
2. What are some of the cultural and political ties that unite Africa with the Americas? With Southwest Asia? With South Asia?
3. Discuss how increasing urbanization in the 21st century might affect the overall structure of the population of Sub-Saharan Africa.
4. More than any other region, Sub-Saharan Africa is noted for its wildlife, especially large mammals. What environmental and historical processes explain the existence of so much fauna? Why are there relatively fewer large mammals in other world regions?
5. Compare and contrast the role of tribalism in Sub-Saharan Africa with that of nationalism in Europe.
6. Historically, how was Sub-Saharan Africa integrated into the global economy? Was its role similar to or different from that of other developing regions?
7. The northern boundary of the region of Sub-Saharan Africa is particularly problematic because of its relationship to North Africa. Should the Sahara be considered a cultural divide between these two regions? Why or why not?
8. Compare the contrasting development model put forward by the United States and Europe with that of China. Will Chinese influence in the region alter the course of development for the region?
9. Why is the political map of Africa blamed for the region's many ethnic conflicts? Is it the map at fault or are other factors more significant?
10. How useful are the Millennium Development Goals in measuring development? What are the problems in tracking Sub-Saharan Africa's relative progress given the lack of infrastructure and data?

Regional Novels and Films

Novels

Chinua Achebe, *Things Fall Apart* (1959, Fawcett Crest)

Ishmael Beah, *A Long Way Gone* (2007, Sarah Crichton Books)

Chris Cleave, *Little Bee: A Novel* (2010, Simon & Schuster)

Nuruddin Farah, *Knots* (2007, Riverhead Books)

Nadine Gordimer, *Burger's Daughter* (1979, Viking)

Barbara Kingsolver, *The Poisonwood Bible* (1998, Harper)

Doris Lessing, *African Laughter: Four Visits to Zimbabwe* (1992, Harper)

Alexander McCall Smith, *The No. 1 Ladies' Detective Agency* (2002, Anchor Books)

Alan Paton, *Cry, the Beloved Country* (1948, Scribner's)

Ngugi Wa Thiong'o, A Grain of Wheat (1967, Heinemann)

Ngugi Wa Thiong'o, *Petals of Blood* (1977, Heinemann)

Films

Daresalam (2000, Chad)

The Gods Must Be Crazy (1981, Botswana/South Africa)

Hotel Rwanda (2004, U.S.)

The Last King of Scotland (2006, U.S./U.K)

Live and Become (2005, France/Israel)

Lumumba (2000, France)

Masai: The Rain Warriors (2005, Kenya/France)

Thunderbolt (2000, Nigeria)

Tsotsi (2005, South Africa)

Youssou N'Dour: Return to Gorée (2006, Senegal/Switzerland)

Bibliography

Aryeetey-Attoh, Samuel, ed. 2009. *Geography of Sub-Saharan Africa, 3rd ed.* Upper Saddle River, NJ: Prentice Hall.

Christopher, A. J. 1984. *Colonial Africa: A Historical Geography*. Totowa, NJ: Barnes & Noble.

Carney, Judith. 2002. *Black Rice: The African Origins of Rice Cultivation in the Americas*. Cambridge, MA: Harvard University Press.

Curtin, Philip D. 1969. *The Atlantic Slave Trade*. Madison: University of Wisconsin Press.

Grant, Richard. 2009. *Globalizing City: The Urban and Economic Transformation of Accra, Ghana.* Syracuse: Syracuse University Press.

Le Billion, Philippe. 2008. "Diamond Wars? Conflict Diamonds and Geographies of Resource Wars." *Annals of the Association of American Geographers* 98(2): 345–372.

Ndegwa, Stephen N. 1996. *The Two Faces of Civil Society: NGOs and Politics in Africa.* West Hartford, CT: Kumarian Press.

Newman, James L. 1995. *The Peopling of Africa: A Geographical Interpretation.* New Haven: Yale University Press.

Rain, David. 1999. *Eaters of the Dry Season: Circular Labor Migration in the West African Sahel.* Boulder, CO: Westview Press.

United Nations and World Health Organization. 2009. *'09 AIDS Epidemic Update.* Geneva, Switzerland: United Nations.

Stock, Robert. 2004. *Africa South of the Sahara: A Geographical Interpretation,* 2nd ed. New York: Guilford Press.

Log in to www.masteringgeography.com for MapMaster™ interactive maps, geography videos, RSS feeds, flashcards, web links, an eText version of *Diversity Amid Globalization,* and self-study quizzes to enhance your study of Sub-Saharan Africa.

MapMaster™ presents 13 Place Name and 13 Layered Thematic interactive maps to help students practice and master their geographic literacy, spatial reasoning, and critical thinking skills.

7
Southwest Asia and North Africa

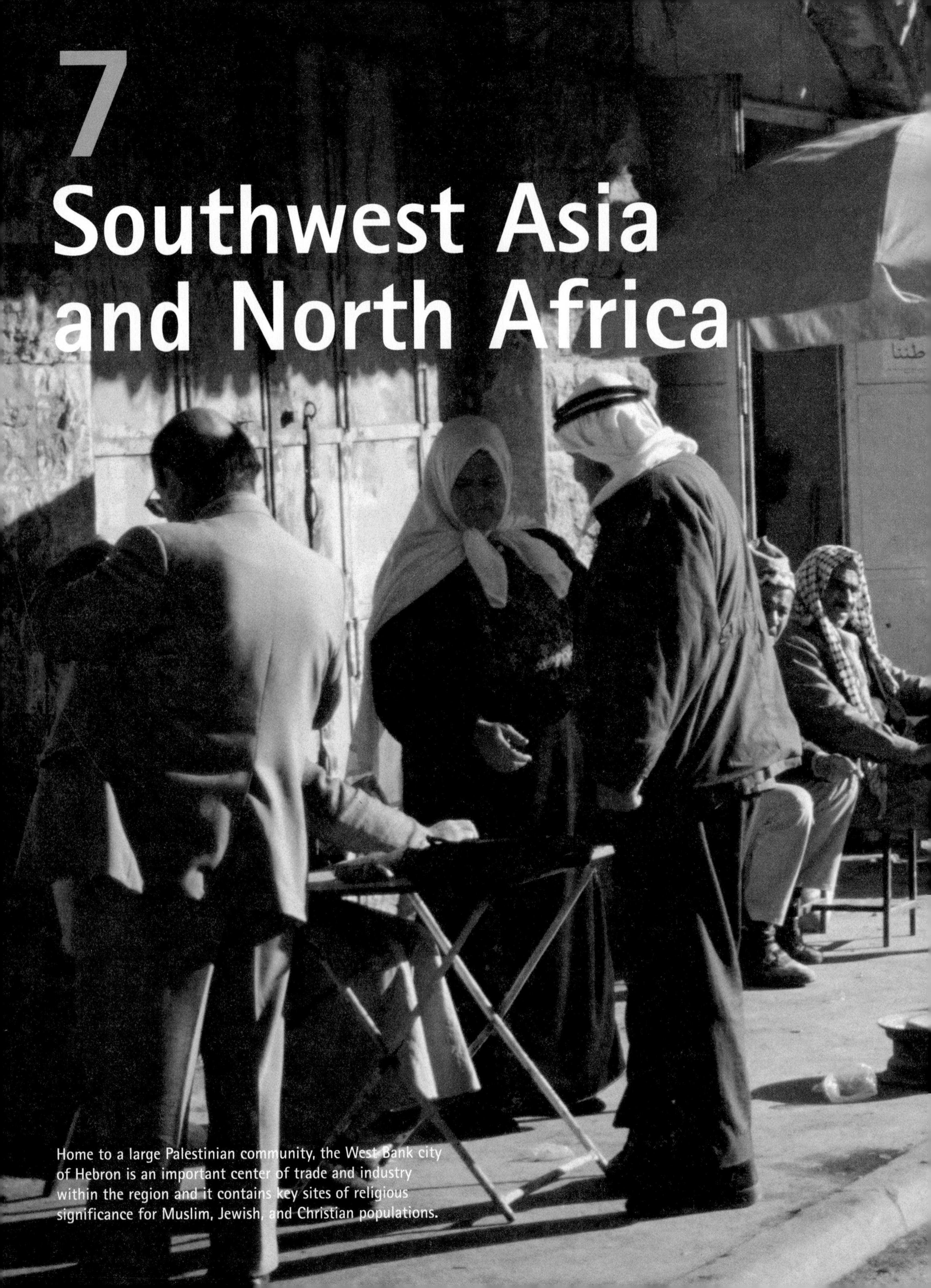

Home to a large Palestinian community, the West Bank city of Hebron is an important center of trade and industry within the region and it contains key sites of religious significance for Muslim, Jewish, and Christian populations.

DIVERSITY AMID GLOBALIZATION

Globalization has played out across Southwest Asia and North Africa in complex ways. Energy-rich nations have reaped the greatest economic rewards in satisfying global demands for oil and natural gas, but varied cultural and political elements of the outside world have not been embraced by many residents of the region.

ENVIRONMENTAL GEOGRAPHY

Water shortages are likely to increase across this arid region in the early 21st century as growing populations, rapid urbanization, and increasing demands for agricultural land consume limited water supplies.

POPULATION AND SETTLEMENT

Rapid population growth in North African cities such as Algiers and Cairo is far outpacing the ability of these urban places to supply adequate housing and services.

CULTURAL COHERENCE AND DIVERSITY

The heart of the Islamic world, this region finds itself at the center of the global rise of Islamist movements that often come into conflict with Western values and traditions.

GEOPOLITICAL FRAMEWORK

Ongoing political instability is a fact of life across much of the region. Religious and ethnic differences, shifting political allegiances, and persistent economic problems all contribute to the region's geopolitical challenges.

ECONOMIC AND SOCIAL DEVELOPMENT

Changing world oil prices have a tremendous economic impact on this region, which holds almost two-thirds of all petroleum reserves on the planet. The economic crisis of 2008–2010 produced widespread unemployment in the energy, tourism, and manufacturing sectors of the regional economy and selectively contributed to more political instability across the region.

Climate, culture, and oil all help define the Southwest Asia and North Africa world region. Straddling the historic meeting ground between Europe, Asia, and Africa, the region sprawls across thousands of miles of parched deserts, rugged plateaus, and oasis-like river valleys. It extends 4,000 miles (6,400 kilometers) between Morocco's Atlantic coastline and Iran's eastern boundary with Pakistan. More than two dozen nations are included within its borders, with the largest populations found in Egypt, Turkey, and Iran (Figure 7.1). Generally, its climates are arid, although the region's diverse physical geography causes precipitation to vary considerably.

Regional boundaries and terminology defy easy definition. Often the same broad area is simply called the *Middle East,* but most experts would exclude the western parts of North Africa as well as Turkey and Iran from such a region. In addition, the *Middle East* carries with it a peculiarly European vantage point—Lebanon is in the "middle of the east" only from the perspective of the western Europeans who colonized the region. Instead, *Southwest Asia and North Africa* offers a useful and straightforward way to describe the general limits of the region.

There also are problems with simply defining the geographical limits of the region. To the northwest, a small piece of Turkey actually sits west of the Bosporus Strait, generally considered to be the dividing line between Europe and Asia (see Figure 7.1). The addition of Cyprus (off the coast of Turkey) to the European Union in 2004 effectively removed it from the region, and that country is treated in the chapter on Europe (see Chapter 8). To the northeast, the largely Islamic peoples of Central Asia share many religious and linguistic ties with Turkey and Iran, but we have chosen to treat these groups in a separate chapter on Central Asia (see Chapter 10).

African borders also are problematic. The conventional regional division of "North Africa" from "Sub-Saharan Africa" often cuts directly through the middle of modern Mauritania, Mali, Niger, Chad, and Sudan. We place all of these transitional countries in Chapter 6, Sub-Saharan Africa, but also include some discussion of Sudan within our Southwest Asia and North Africa material because of its status as an "Islamic republic," a political and cultural designation that ties it strongly to the Muslim world.

Culturally, diverse languages, religions, and ethnic identities have molded land and life within the region for centuries. One traditional zone of conflict is in the eastern Mediterranean, where Jewish, Christian, and Islamic peoples have yet to resolve long-standing cultural tensions and political differences, particularly as they relate to the state of Israel and various Palestinian groups within the region. Iraq also has been a setting for recent violence as an American-led invasion toppled the regime of Saddam Hussein and various religious, ethnic, and political factions have divided the

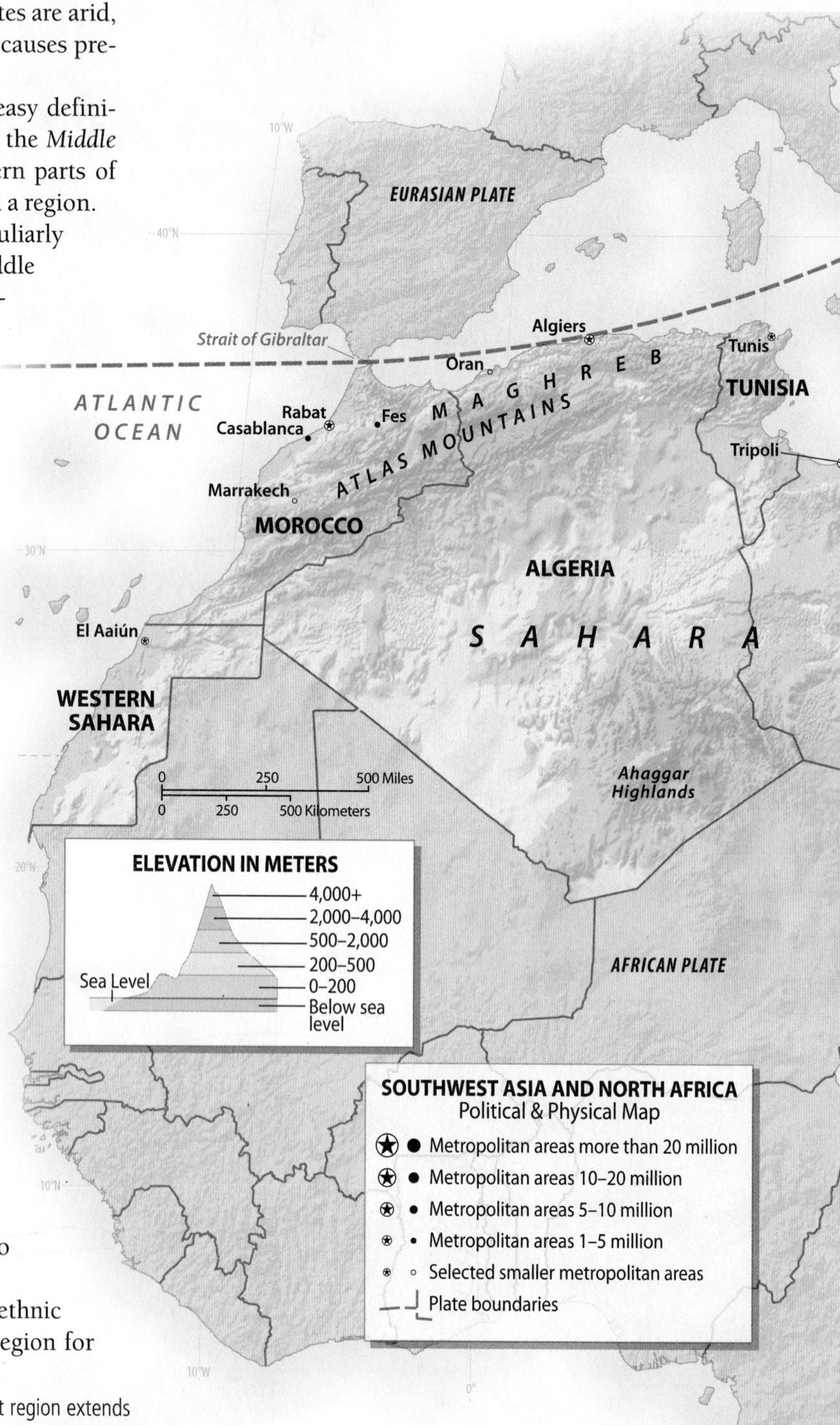

Figure 7.1 Southwest Asia and North Africa This vast region extends from the shores of the Atlantic Ocean to the Caspian Sea. Within its boundaries, major cultural differences and globally important petroleum reserves have contributed to recent political tensions.

country. Nearby Iran and its ongoing development of a nuclear capability have also increased tensions. In addition, Southwest Asia and North Africa's extraordinary petroleum resources place the area in the global economic spotlight. The strategic value of oil has combined with ongoing ethnic and religious conflicts to produce one of the world's least stable political settings, one prone to geopolitical conflict both within and between the countries of the region.

No world region has better exemplified the theme of globalization throughout history than Southwest Asia and North Africa. A key global **culture hearth**, the region witnessed many cultural innovations that subsequently diffused widely to other portions of the world. As an early center for agriculture, several great civilizations, and three major world religions, the region has been a key human crossroads for thousands of years. Important trade routes have connected North Africa with the Mediterranean and Sub-Saharan Africa. Southwest Asia also has had historical ties to Europe, the Indian subcontinent, and Central Asia. As a result, innovations within the region

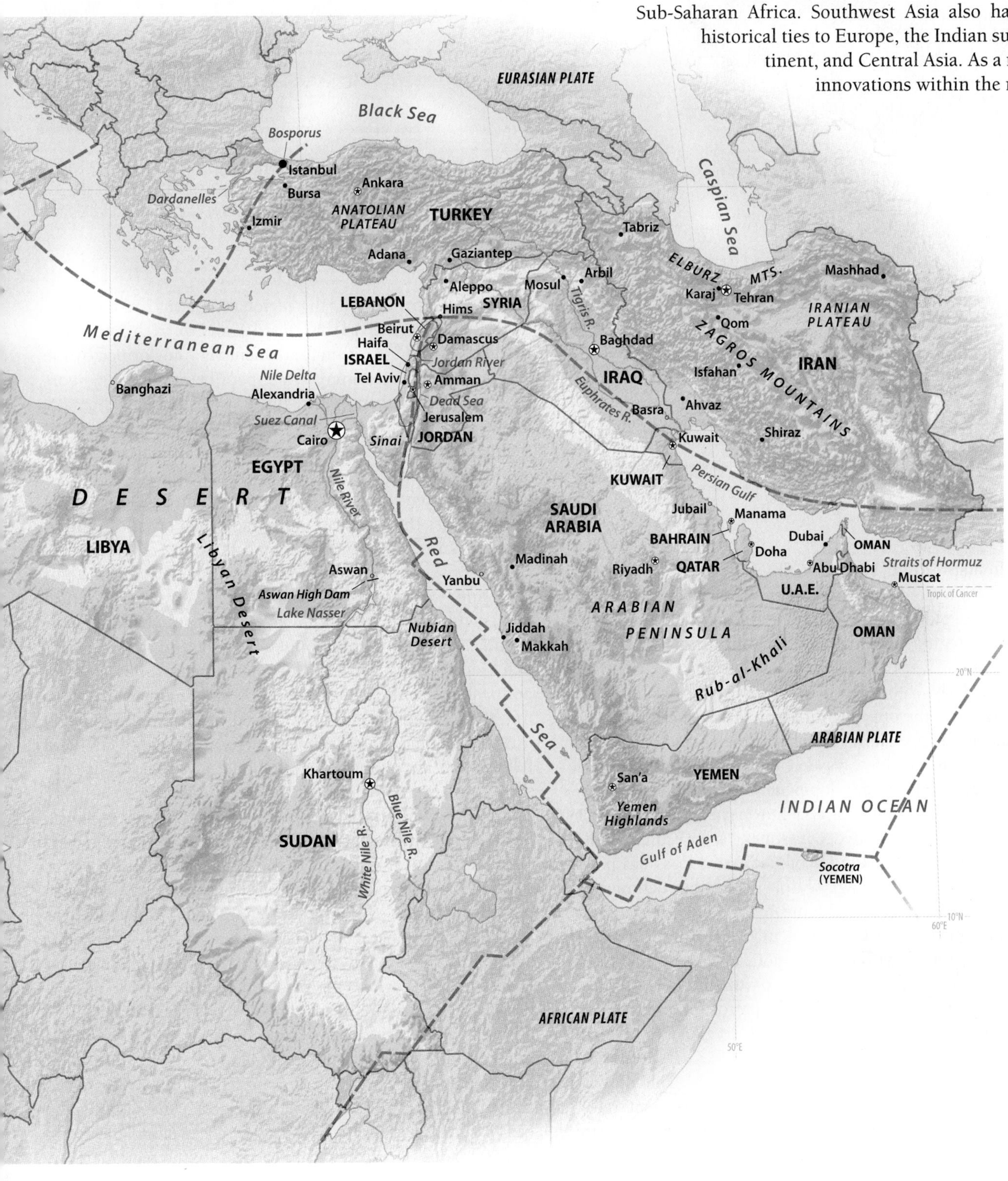

Figure 7.2 Man at Livestock Market, Al Ayn, United Arab Emirates Longtime residents of Southwest Asia have witnessed incredible change in their lifetimes. In many localities, traditional ways have been radically transformed by urbanization, foreign investment, and external cultural influences.

have often spread far beyond its bounds. For example, the domestication of wheat and cattle in Southwest Asia had far-reaching global impacts. In addition, religions born within the region (Judaism, Christianity, and Islam) have shaped many other parts of the world.

Particularly within the past century, globalization also has operated in the opposite direction: The region's strategic importance has made it increasingly vulnerable to outside influences. Traditional lifeways have been transformed (Figure 7.2). The 20th-century development of the petroleum industry, largely initiated by U.S. and European investment, has had enormous, but selective, consequences for economic development. Global demand for oil and natural gas has powered rapid industrial change within the region, defining its pivotal role in world trade (Figure 7.3). Many key members of the **Organization of the Petroleum Exporting Countries** (OPEC) are found within the region, and these countries strongly influence global prices and production levels for petroleum. Regional dependence on the petroleum industry has also made the area especially vulnerable to global recessions, and the recent downturn of 2008–2010 was no exception. Once-booming cities such as Dubai in the United Arab Emirates (U.A.E.) witnessed soaring unemployment rates, stalled construction projects, and real financial crises. Future growth in such settings continues to be wed to the health of

Figure 7.3 Saudi Arabian Oil Refinery Eastern Saudi Arabia's Ras Tanuna Oil Refinery links the oil-rich country to the world beyond. Huge foreign and domestic investments since 1960 have dramatically transformed many other settings in the region.

global oil demand. Still, the regional patterns are complex. Oil-rich nations such as the United Arab Emirates, Saudi Arabia, and Kuwait have been fundamentally transformed by the global fossil fuel economy, while petroleum-poor neighbors such as Jordan, Lebanon, and Tunisia have seen less dramatic impacts. Politically, petroleum resources also have elevated the region's strategic importance, adding to many traditional cultural tensions.

More recently, **Islamic fundamentalism** has advocated a return across the region to more traditional practices within the Muslim religion. Fundamentalists in any religion advocate a conservative loyalty to enduring beliefs within their faith and they strongly resist change. A related political movement within Islam known as **Islamism** challenges the encroachment of global popular culture and blames colonial, imperial, and Western elements for many of the region's political, economic, and social problems. Islamists resent the role they claim the West has played in creating poverty in their world, and many Islamists advocate merging civil and religious authority and rejecting modern, Western-style consumer culture. While often demonized in the Western press, particularly when associated with terrorist acts of violence, Islamic fundamentalism and Islamism can be seen as cultural and political reactions to the disruptive role that forces of globalization have played across the region.

The region's environment provides additional challenges for the populations of Southwest Asia and North Africa. The availability of water in this largely dry portion of the world has shaped the region's physical and human geographies. Biologically, the region's plants and animals must adapt to the aridity of long dry seasons and short, often unpredictable rainy periods. Human settlement is linked to water in similar ways. Whether it comes from precipitation, underground aquifers, or rivers, water has shaped settlement patterns and has placed severe limits on agricultural development across huge portions of the region. In the future, the region's growing population of more than 400 million people will stress these available resources even further, and water issues will no doubt increase economic and political instability.

ENVIRONMENTAL GEOGRAPHY: Life in a Fragile World

In the popular imagination, much of Southwest Asia and North Africa is a land of shifting sand dunes, searing heat, and scattered oases. Although examples of those stereotypes certainly can be found across the region, the actual physical setting, in both landforms and climate, is considerably more complex (see Figure 7.1). In reality, the regional terrain varies greatly, with rocky plateaus and mountain ranges more common than sandy deserts. Even the climate, although dominated by aridity, varies remarkably from the dry heart of North Africa's Sahara Desert to the well-watered highlands of northern Morocco, coastal Turkey, and western Iran. One theme is pervasive; however, a lengthy legacy of human settlement has left its mark on a fragile environment, and the entire region will be faced with increasingly daunting ecological problems in the decades ahead.

Legacies of a Vulnerable Landscape

The island of Socotra illustrates the region's fragile and vulnerable environment and suggests how processes of globalization may threaten the area's long-term ecological health. Socotra's stony slopes rise out of the shimmering waters of the Indian Ocean about 230 miles (368 kilometers) southeast of Yemen. The island's unique natural and cultural history has recently caught the world's attention. Separated for millions of years from the mainland of the Arabian Peninsula, Socotra's environment evolved in isolation. More than 30 percent of the island's 850 plants are found nowhere else on Earth. Exotic dragon's blood trees dot many of the island's dry and rocky hillsides, and dozens of other species have only recently been catalogued by botanists (Figures 7.4 and 7.5). Offshore, equally rare coral reefs and unusual fish populations have evolved as a part of the island's unique environmental setting in the wet–dry tropics of the northwestern Indian Ocean.

Figure 7.4 Socotra's Dragon's Blood Tree Unique to Socotra, the rare dragon's blood tree reflects the island's environmental isolation. Evolving as a part of the island's ecosystem, the tree survives in the region's dry tropical climate.

Socotra's unique environmental heritage now hangs in the balance. Global participants in the 1992 environmental summit at Rio de Janeiro established a $5 million international biodiversity project on Socotra, and Edinburgh's Royal Botanic Garden has conducted extensive studies on the island. In 2008, the island was recognized as a United Nations World Natural Heritage Site, and the European Union has officially supported the idea of preserving the area's unique biogeography.

At the same time, global pharmaceutical companies have already been allowed to harvest some of the island's unique plant species and other plants and animals have been illegally taken by visitors. In addition, the Yemeni government has invited international petroleum companies to explore the island's offshore oil and gas potential, and it has considered a grand scheme for developing a series of luxury tourist hotels that would forever change the island's character. In 2009, an environmental expert noted that even well-meaning ecotourists on the island are already damaging coral reefs and their growing numbers have led to increased firewood gathering and road construction. Only time will tell how Socotra, which is often cited as the Galápagos of the Indian Ocean, will fare in the 21st century. A recently developed mixed-use zoning plan and a newly paved airstrip suggest development is inevitable.

The larger environmental history of Southwest Asia and North Africa contains many examples of the clever but shortsighted legacies of its human occupants. Lengthy human settlement in a marginal land has resulted in deforestation, soil salinization and erosion, and depleted water resources (see Figure 7.5). Indeed, the pace of population growth and technological change during the past century suggests that the region's vulnerable environment is destined to face even greater challenges in the 21st century.

Deforestation and Overgrazing Deforestation is an ancient problem in Southwest Asia and North Africa. Although much of the region is too dry for trees, the more humid and elevated lands that ring the Mediterranean once supported heavy forests. Included in these woodlands are the cedars of Lebanon, cut during ancient times and now reduced to a few scattered groves in a largely denuded landscape. In many settings, growing demands for agricultural land caused upland forests to be removed and replaced with grain fields, orchards, and pastures.

Human activities have conspired with natural conditions to reduce most of the region's forests to grass and scrub. Mediterranean forests grow slowly, are highly vulnerable to fire, and usually fare poorly if subjected to heavy grazing. Browsing by goats in particular often has been blamed for much of the region's forest loss, but other live-

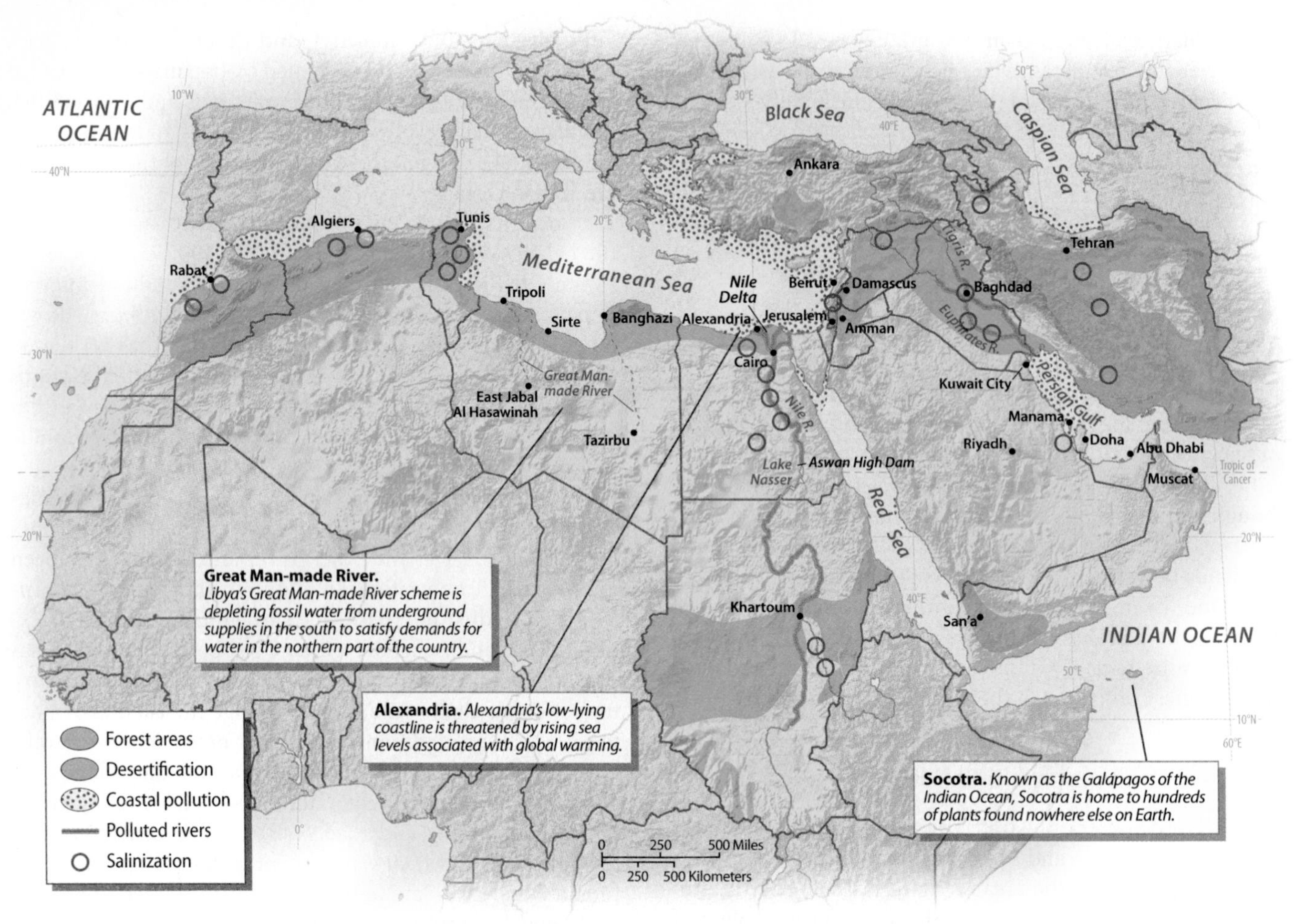

Figure 7.5 Environmental Issues in Southwest Asia and North Africa Growing populations, pressures for economic development, and pervasive aridity combine to create environmental hazards across the region. Long human occupance has contributed to deforestation, irrigation-induced salinization, and expanding desertification. Saudi Arabia's deep-water wells, Egypt's Aswan High Dam, and Libya's Great Man-made River are all recent technological attempts to expand settlement, but they may carry a high long-term environmental price tag.

stock have also impacted the vegetative cover, especially in steeply sloping semiarid settings that are slow to recover from grazing. Deforestation also has resulted in a millennia-long deterioration of the region's water supplies and in accelerated soil erosion.

Some forests survive in the mountains of northern and southern Turkey, and northern Iran also retains considerable tree cover. Scattered forests can be found in western Iran, the eastern Mediterranean, and the Atlas Mountains. Moreover, several governments have launched reforestation drives and forest preservation efforts in recent years. For example, both Israel and Syria have expanded their coverage of wooded lands since the 1980s. In nearby Lebanon, more than 5 percent of the country's total area was included in the Shouf Cedar Reserve in 1996 to protect old-growth cedars, junipers, and oak forests. An even larger area (including the Shouf Cedar Reserve) was declared the Shouf Biosphere Reserve by the United Nations in 2005 and is designed to protect both the rare trees as well as endangered mammals such as the wolf and Lebanese jungle cat.

Salinization Salinization, or the buildup of toxic salts in the soil, is another ancient environmental issue in a region where irrigation has been practiced for centuries (see Figure 7.5). The accumulation of salt in the topsoil is a common problem wherever desert lands are subjected to extensive irrigation. All freshwater contains a small amount of dissolved salt, and when water is diverted from streams into fields, salt remains in the soil after the moisture is absorbed by the plants and evaporated by the sun. In humid climates, accumulated salts are washed away by saturating rains, but in arid climates this rarely occurs. Where irrigation is practiced, salt concentrations build up over time, leading to lower crop yields and eventually to land abandonment.

Hundreds of thousands of acres of once-fertile farmland within the region have been destroyed or degraded by salinization. The problem has been particularly acute in Iraq, where centuries of canal irrigation along the Tigris and Euphrates rivers has seriously degraded land quality. Similar conditions plague central Iran, Egypt, and irrigated portions of the Maghreb.

Managing Water Residents of the region are continually challenged by many other problems related to managing water in one of the driest portions of Earth. As technological change has accelerated, the scale and impact of water management schemes have had a growing effect on the region's environment. Sometimes the costs are justified if large new areas are brought into productive and sustainable agricultural use. In other cases, the long-term environmental price will far outweigh any immediate and perhaps short-lived gains in agricultural production.

Regional populations have been modifying drainage systems and water flows for thousands of years. The Iranian **qanat system** of tapping into groundwater through a series of gently sloping tunnels was widely replicated on the Arabian Peninsula and in North Africa. With simple technology, farmers directed the downslope flow of underground water to fields and villages where it could be efficiently utilized.

In the past half century, however, the scope of environmental change has been greatly magnified. Water management schemes have become huge engineering projects, major budget expenditures, and even political issues, both within and between countries of the region. One remarkable example is Egypt's Aswan High Dam, completed in 1970 on the Nile River south of Cairo (see Figure 7.5). Many benefits came with the dam's completion. Most important, greatly increased storage capacity in the upstream reservoir promoted more year-round cropping and an expansion of cultivated lands along the Nile, critical changes in a country that continues to experience rapid population growth. The dam also generates large amounts of clean electricity for the region. But the environmental costs have been high. The dam has changed methods of irrigation along the Nile, greatly increasing salinization. While fresh sediments and soil nutrients once annually washed across the valley floor with high river flows, more controlled irrigation has meant greatly increased inputs of costly fertilizers. Sediments that used to move downstream now accumulate behind the dam, infilling Lake Nasser. Other problems with the dam include an increased incidence of schistosomiasis (a debilitating parasitic disease spread by waterborne snails in irrigation canals) and the collapse of the Mediterranean fishing industry near the Nile Delta, an area previously nourished by the river's silt.

Israel's "Peace Corridor" project is another potential massive engineering venture within the region. Designed to stimulate the regional economy and bring Red Sea water north into the Dead Sea, the effort received increased government backing in 2007. The ambitious plan calls for a 120-mile (200-kilometer) conduit to be built from near Aqaba to the southern end of the Dead Sea (Figure 7.6). The water could be desalinated, generate hydroelectricity, and help save the Dead Sea and its major tourist resorts from literally drying up.

Elsewhere, **fossil water**, or water supplies stored underground during earlier and wetter climatic periods, has also been put to use by modern technology. Libya's "Great Man-made River" scheme taps underground water in the southern part of the country, transports it 600 miles (965 kilometers) to northern coastal zones, and uses the precious resource to expand agricultural production in one of North Africa's driest countries (see Figure 7.5). Similarly, Saudi Arabia has invested huge sums to develop deep-water wells,

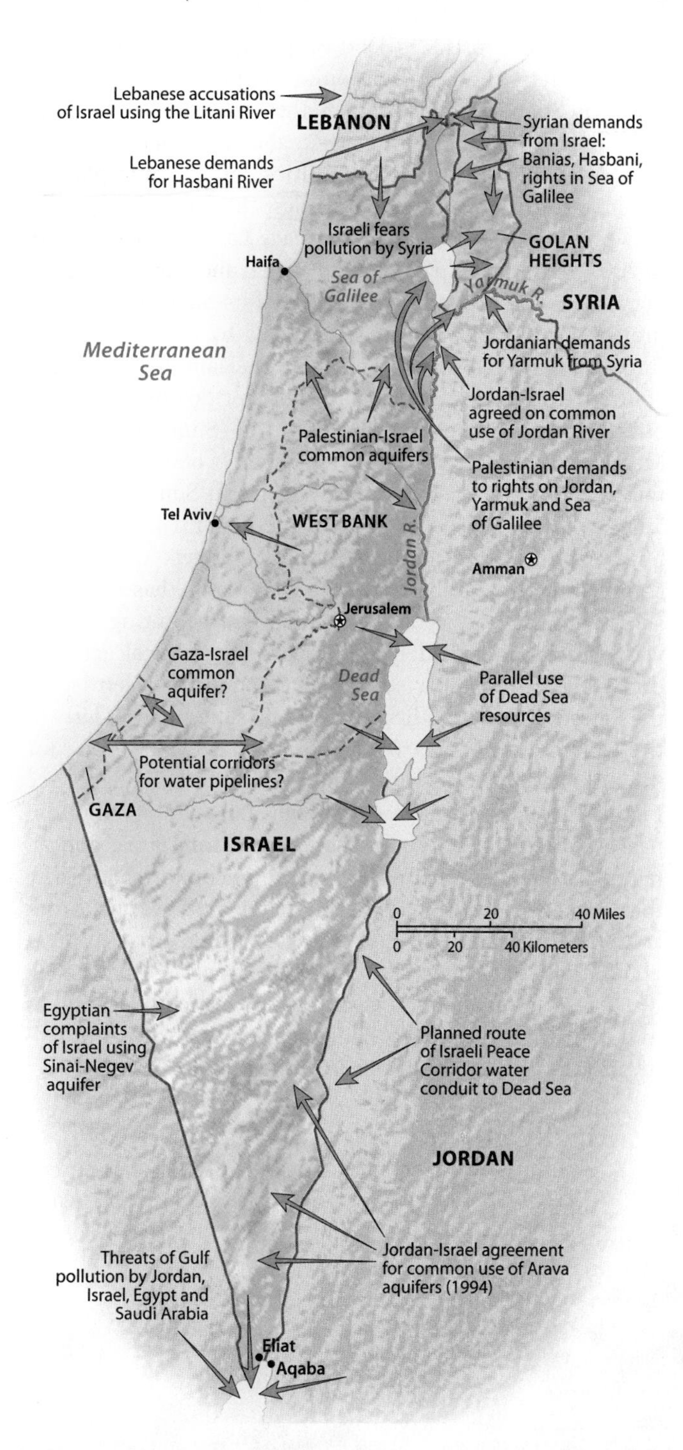

Figure 7.6 Water Issues in the Middle East Many water-related issues complicate the geopolitical setting in the Middle East. The Jordan River system has been a particular focus of conflict. (Modified from *Soffer, Arnon, Rivers of Fire: The Conflict over Water in the Middle East,* 1999, p. 180. Reprinted by permission of Rowman and Littlefield Publishers, Inc.)

allowing it to expand its food output greatly (Figure 7.7). Unfortunately, these underground supplies are being depleted much more rapidly than they are being recharged, thus limiting the long-term sustainability of such ventures.

Most dramatically, **hydropolitics**, or the interplay of water resource issues and politics, has raised tensions between countries that share drainage basins. For example, in 2007, with the help of Chinese capital and engineering expertise, Ethiopia began construction of its huge Tekeze Dam project on the Nile. Already referred to by its promoters as the equivalent of "China's Three Gorges Dam in Africa," the controversial project threatens to disrupt downstream fisheries and irrigation in North Africa. Similarly Sudan's Merowe Dam project (on another stretch of the Nile) has raised major concerns in nearby Egypt. In Southwest Asia, Turkey's growing development of the upper Tigris and Euphrates rivers (the Southeast Anatolia Project or "GAP"), complete with 22 dams and 19 power plants, has raised issues with Iraq and Syria, who argue that capturing "their" water might be considered a provocative political act. Hydropolitics also has played into negotiations between Israel, the Palestinians, and other neighboring states, particularly in the Golan Heights area (an important headwaters zone for multiple nations) and within the Jordan River drainage (see Figure 7.6). Israelis fear Palestinian and Syrian pollution; nearby Jordanians argue for more water from Syria; and all regional residents must deal with the uncomfortable reality that, regardless of their political differences, they must drink from the same limited supplies of freshwater. In addition, a 2009 report by Amnesty International suggested that Israeli water restrictions harshly discriminated against Palestinians in both Gaza and the West Bank (see "Environment and Politics: Water and the Palestinians in the Middle East").

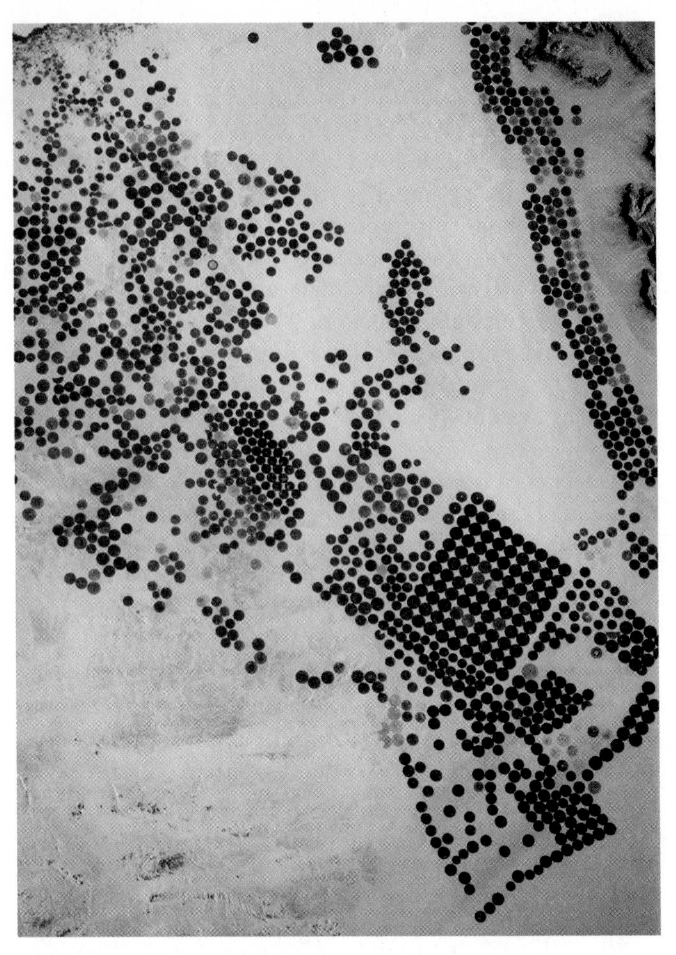

Figure 7.7 Saudi Arabian Irrigation These irrigated fields in the Saudi Desert draw from wells over 4,000 feet deep. While significantly expanding the country's food production, such efforts are rapidly depleting underground supplies of fossil water.

Figure 7.8 Atlas Mountains The rugged Atlas Mountains dominate a broad area of interior Morocco. Steep slopes and narrow canyons can make travel slow and hazardous across the region.

ENVIRONMENT AND POLITICS Water and the Palestinians in the Middle East

Inevitably, water and politics are intertwined in the Middle East. Not surprisingly, fundamental disagreements over water have trickled into the long-standing conflict between the Israelis and the Palestinians, both in Gaza and in the West Bank. Palestinian authorities argue that the Israelis have used water as a weapon of control and that Palestinian settlements are often left high and dry because of Israel's discriminatory policies. On the other hand, Israelis argue that the Palestinians often waste water or fail to utilize it fully and that they have no right to complain about Israel's own brand of "hydropolitics."

Take the small, overcrowded, poverty-stricken lands of the Gaza Strip (see Figure 7.6 and 7.1.1). In a 2009 report, Amnesty International observed that most Gaza residents survived on only 20–70 liters of water per day, compared to the Israeli average of 300 liters. Much of Gaza's water storage and delivery infrastructure has been damaged or destroyed in recent military operations, and Israeli blockades along the Gaza border have severely limited efforts to rebuild the system. The report noted that both the area's water and sewage systems had reached a "crisis point." But it also criticized local Palestinian leaders, suggesting their own management of the existing system was in "total chaos."

In the nearby West Bank area (see Figure 7.6), water is also at the center of many political tensions between Palestinian and Jewish settlers. Palestinians point to lush gardens and swimming pools in Israeli-controlled areas that stand in sharp contrast to thirsty Palestinian towns that often end up severely restricted in their access to underground aquifers or to water flowing in the nearby Jordan River (which is largely appropriated for Israeli uses). Palestinians also complain that well-drilling permits are difficult to obtain from Israeli authorities and that cisterns used to store rainwater are often used for target practice by the Israeli army. Israeli authorities argue that many of the criticisms leveled against them are nonsense. They point to the fact that many well permits have not even been used by Palestinians and that international aid that Palestinians have received for investments in water infrastructure has been spent elsewhere.

Regardless of whose claims can be believed, water remains a politically charged resource in the Middle East. There will never be enough of it, and its resulting scarcity creates inevitable conflicts between those who control it and those who need it.

Figure 7.1.1 Gaza Strip Basic water storage and delivery systems are often heavily damaged or destroyed in settings such as this Palestinian refugee camp north of Gaza City in the war-torn Gaza Strip.

Regional Landforms

A quick tour of the region reveals a surprising diversity of environmental settings and landforms (see Figure 7.1). In North Africa, the **Maghreb** region (meaning "western island") includes the nations of Morocco, Algeria, and Tunisia and is dominated near the Mediterranean coastline by the Atlas Mountains. The rugged flanks of the Atlas rise like a series of islands above the narrow coastal plains to the north and the vast stretches of the lower Saharan deserts to the south (Figure 7.8). South and east of the Atlas Mountains, interior North Africa varies between rocky plateaus and extensive lowlands. In northeast Africa, the Nile River dominates the scene as it flows north through Sudan and Egypt.

Southwest Asia is more mountainous than North Africa. In the **Levant**, or eastern Mediterranean region, mountains rise within 20 miles (32 kilometers) of the sea, and the highlands of Lebanon reach heights of more than 10,000 feet (3,048 meters). Farther south, the Arabian Peninsula forms a massive tilted plateau, with western highlands higher than 5,000 feet (1,524 meters) gradually sloping eastward to extensive lowlands in the Persian Gulf area. North and east of the Arabian Peninsula lie the two great upland areas of Southwest Asia: the Iranian and Anatolian plateaus (*Anatolia* refers to the large peninsula of Turkey, sometimes called Asia Minor) (see Figures 7.1 and 7.9). Both of these plateaus, averaging between 3,000 and 5,000 feet (915 to 1,524 meters) in elevation, are geologically active and prone to earthquakes. One dramatic quake in western Turkey (1999) measured 7.8 on the Richter scale, killed more than 17,000 people, and left 350,000 residents homeless. Another quake near the Iranian city of Bam (2003) claimed more than 30,000 lives.

Figure 7.9 Satellite View of Turkey This satellite image of Turkey suggests the varied, quake-prone terrain encountered across the Anatolian Plateau. The Black Sea coastline is visible near the top of the image, and the island-studded Aegean Sea borders Turkey on the west.

Smaller lowlands characterize other portions of Southwest Asia. Narrow coastal strips are common in the Levant, along both the southern (Mediterranean) and northern (Black Sea) Turkish coastlines, and north of the Iranian Elburz Mountains near the Caspian Sea. Iraq contains the most extensive alluvial lowlands in Southwest Asia, dominated by the Tigris and Euphrates rivers, which flow southeast to empty into the Persian Gulf. Although much smaller, the distinctive Jordan River Valley is also a notable lowland that straddles the strategic borderlands of Israel, Jordan, and Syria and drains southward to the Dead Sea (Figure 7.10).

Patterns of Climate

Although often termed the *dry world,* a closer look at Southwest Asia and North Africa reveals a more complex climatic pattern (Figure 7.11). Both latitude and altitude come into play. Aridity dominates large portions of the region. A nearly continuous belt of desert lands stretches eastward from the Atlantic coast of southern Morocco across the continent of Africa, through the Arabian Peninsula, and into central and eastern Iran. Throughout this vast dry zone, plant and animal life have adapted to extreme conditions. Deep or extensive root systems and rapid life cycles allow desert plants to benefit from the limited moisture they receive. Similarly, animals adjust by efficiently storing water, hunting nocturnally, or migrating seasonally to avoid the worst of the dry cycle.

Some of the driest conditions are found across North Africa. Away from the Atlas Mountains, the Sahara Desert dominates much of the region. Much of the central Sahara receives less than 1 inch (2.5 centimeters) of rain a year and can thus support only the most meager forms of vegetation. Saharan summers feature extremely hot days and warm evenings (the world's record high, 136°F [58°C] was recorded in Libya), but winters are generally pleasant, even cool at night. Only in the tropical latitudes of southern

Figure 7.10 Jordan Valley This view of the Jordan Valley shows a fertile mix of irrigated vineyards and date palm plantations.

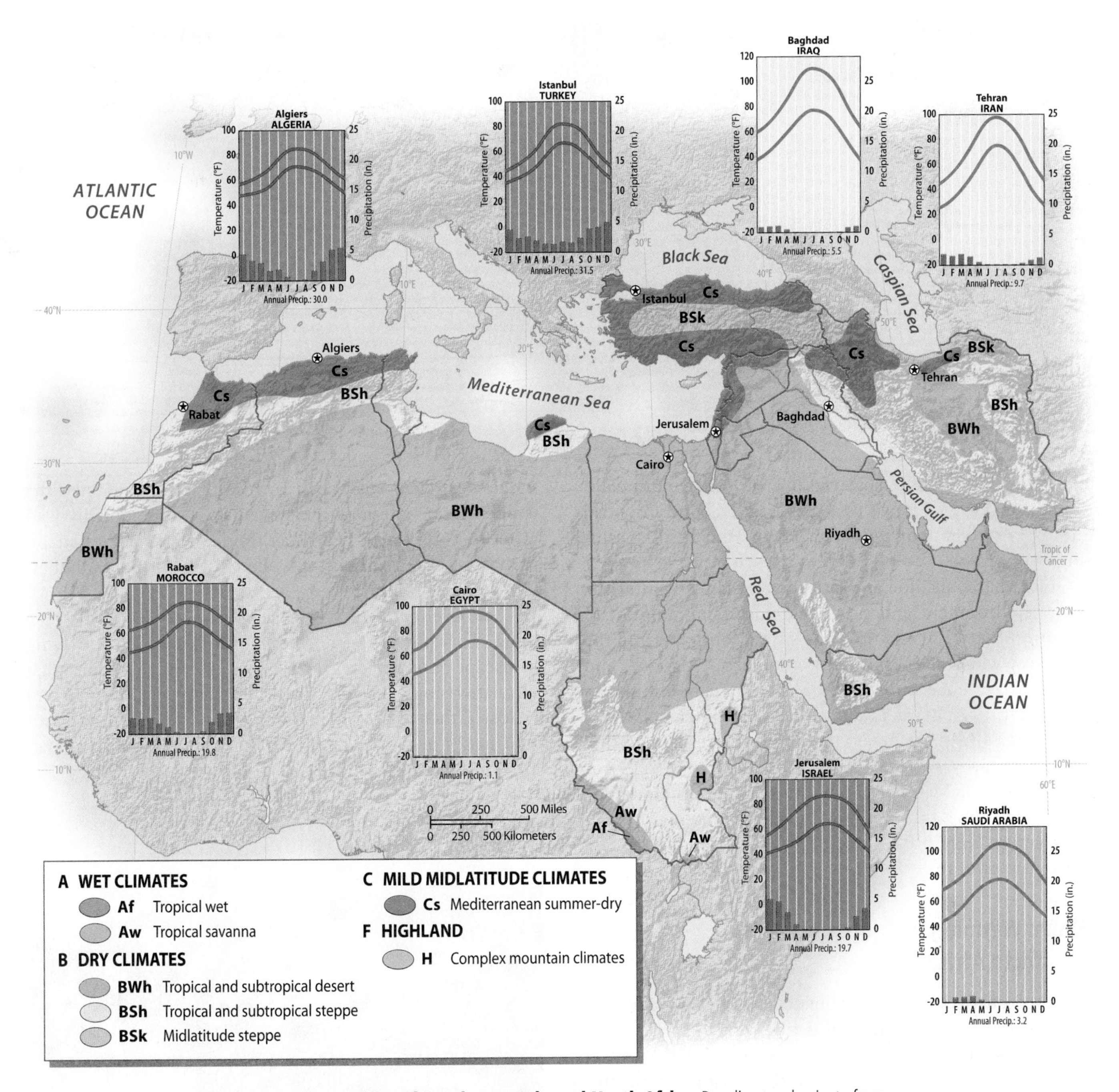

Figure 7.11 Climate Map of Southwest Asia and North Africa Dry climates dominate from western Morocco to eastern Iran. Within these zones, persistent subtropical high-pressure systems offer only limited opportunities for precipitation. Elsewhere, mild midlatitude climates with wet winters are found near the Mediterranean Basin and Black Sea. To the south, tropical savanna climates provide summer moisture to southern Sudan.

Sudan do precipitation levels rise significantly. Here, summer rains produce between 20 and 50 inches (51 to 127 centimeters) of precipitation, and tropical savannas and woodlands replace the desert vegetation to the north.

Across the Red Sea, deserts also dominate Southwest Asia. Most of the Arabian Desert is not quite as dry as the Sahara proper, although the Rub-al-Khali along Saudi Arabia's southern border is one of the world's most desolate areas. On the fringe of the summer-monsoon belt, the Yemen Highlands along the southwestern edge of the Arabian Peninsula receive much more rain than the rest of the region and thus form a more favorable site for human habitation. To the northwest, precipitation also increases slightly in the central Tigris and Euphrates river valleys, with Baghdad, Iraq, aver-

Figure 7.12 Arid Iran Only sparse vegetation dots this arid scene from central Iran, a landscape characterized by isolated mountain ranges and dry interior plateaus.

aging just over 5 inches (13 centimeters) of rain annually. Another major arid zone lies across Iran, a country divided into a series of minor mountain ranges and desert basins (Figure 7.12).

Elsewhere, altitude and latitude dramatically alter the desert environment and produce a surprising amount of climatic variety. For example, the Atlas Mountains and the nearby lowlands of northern Morocco, Algeria, and Tunisia experience a distinctly Mediterranean climate in which hot, dry summers alternate with cooler, relatively wet winters. In these areas, the landscape resembles that found in nearby southern Spain or Italy (Figure 7.13). A second zone of Mediterranean climate extends along the Levant coastline into the nearby mountains and northward across large portions of northern Syria, Turkey, and northwestern Iran.

Global Warming in Southwest Asia and North Africa

Projected changes in global climate will tend to aggravate already-existing environmental issues within North Africa and Southwest Asia. Temperature changes are predicted to have a greater impact on the region than changes in precipitation. The already arid and semiarid region will likely remain relatively dry, but warmer average temperatures are likely to have several major consequences. First, higher overall evaporation rates and lower overall soil moisture across the region will more likely stress crops, grasslands, and other vegetation. Semiarid lands are particularly vulnerable, especially dryland cropping systems that cannot depend on irrigation. Even in irrigated zones, higher temperatures are likely to reduce yields for crops such as wheat and maize. Second, warmer temperatures will likely reduce net runoff into the region's already stressed streams and rivers, potentially reducing hydroelectric potential and water that is available for the region's increasingly urban population. Third, there will be a higher likelihood of extreme weather events, such as record-setting summertime temperatures. These extreme events will undoubtedly lead to more heat-related deaths as residents struggle to adapt, particularly in urban settings where many people cannot afford air conditioning.

Sea level changes will pose special threats to the Nile Delta. This portion of northern Egypt is a vast, low-lying landscape of settlements, farms, and marshland. Studies that simulate rising sea levels reveal that much of the region will be lost, either to inundation, erosion, or salinization. Farmland losses of more than 250,000 acres (100,000 hectares) are quite possible with even modest sea level changes. For

Figure 7.13 Algerian Orange Harvest The Mediterranean moisture in northern Algeria produces an agricultural landscape similar to that of southern Spain or Italy. Winter rains create a scene that contrasts sharply with deserts found elsewhere in the region.

Figure 7.14 Alexandria, Egypt This beachside view along northern Egypt's low-lying coastline at Alexandria could change significantly if global sea levels were to rise.

example, estimates are that a 1-meter (3.3 feet) sea-level rise could affect 15 percent of Egypt's habitable land and displace 8 million Egyptians in coastal and delta settings. In the Egyptian city of Alexandria, $30 billion in losses have been projected because sea level changes will devastate the city's huge resort industry as well as nearby residential and commercial areas (Figure 7.14).

Some experts have also attempted to estimate the broader political and economic costs associated with potential climate changes within the region. For example, given the political instability of the Middle East, even relatively small changes in water supplies, particularly where they might involve several nations, could significantly add to the potential for conflict within the region. Projected economic implications of climate change may depend on the relative importance of agriculture within different nations. In addition, wealthier nations such as Israel and Saudi Arabia may have more available resources to plan, adjust, and adapt to climate shifts and extreme events versus poorer, less-developed countries such as Yemen, Syria, or Sudan.

POPULATION AND SETTLEMENT: Changing Rural and Urban Worlds

The geography of human population across Southwest Asia and North Africa demonstrates the intimate tie between water and life in this part of the world. The pattern is complex: Large areas of the population map remain almost devoid of permanent settlement, while more moisture-favored lands suffer increasingly from problems of crowding and overpopulation (Figure 7.15). Almost everywhere across the region, humans have uniquely adapted themselves to living within arid or semiarid settings.

The Geography of Population

Today, about 450 million people live in Southwest Asia and North Africa (Table 7.1). The distribution of that population is strikingly varied. In countries such as Egypt, large zones of almost empty desert land stand in sharp contrast to crowded, well-watered locations, such as those along the Nile River. While the overall population density in such countries appears modest, the **physiological density**, which is the number of people per unit of arable land, is among the highest on Earth. Patterns of urban geography also are highly uneven: Although less than two-thirds of the overall population is urban, many nations are overwhelmingly dominated by huge and sprawling cities that produce the same problems of urban crowding found elsewhere in the developing world. Rates of recent urban growth have been phenomenal: Cairo, a modest-sized city of 3.5 million people in the 1960s, has quadrupled in population in the past 40 years.

Across North Africa, two dominant clusters of settlement, both shaped by the availability of water, account for most of the region's population (see Figure 7.15). In the Maghreb, the moister slopes of the Atlas Mountains and nearby better-watered coastal districts have accommodated denser populations for centuries. Today, concentrations of both rural and urban settlement extend from south of Casablanca in Morocco to Algiers and Tunis on the shores of the southern Mediterranean. Indeed, most of the populations of Morocco, Algeria, and Tunisia crowd into this crescent, a stark contrast to the almost empty lands south and east of the Atlas Mountains. Casablanca and Algiers are the largest cities

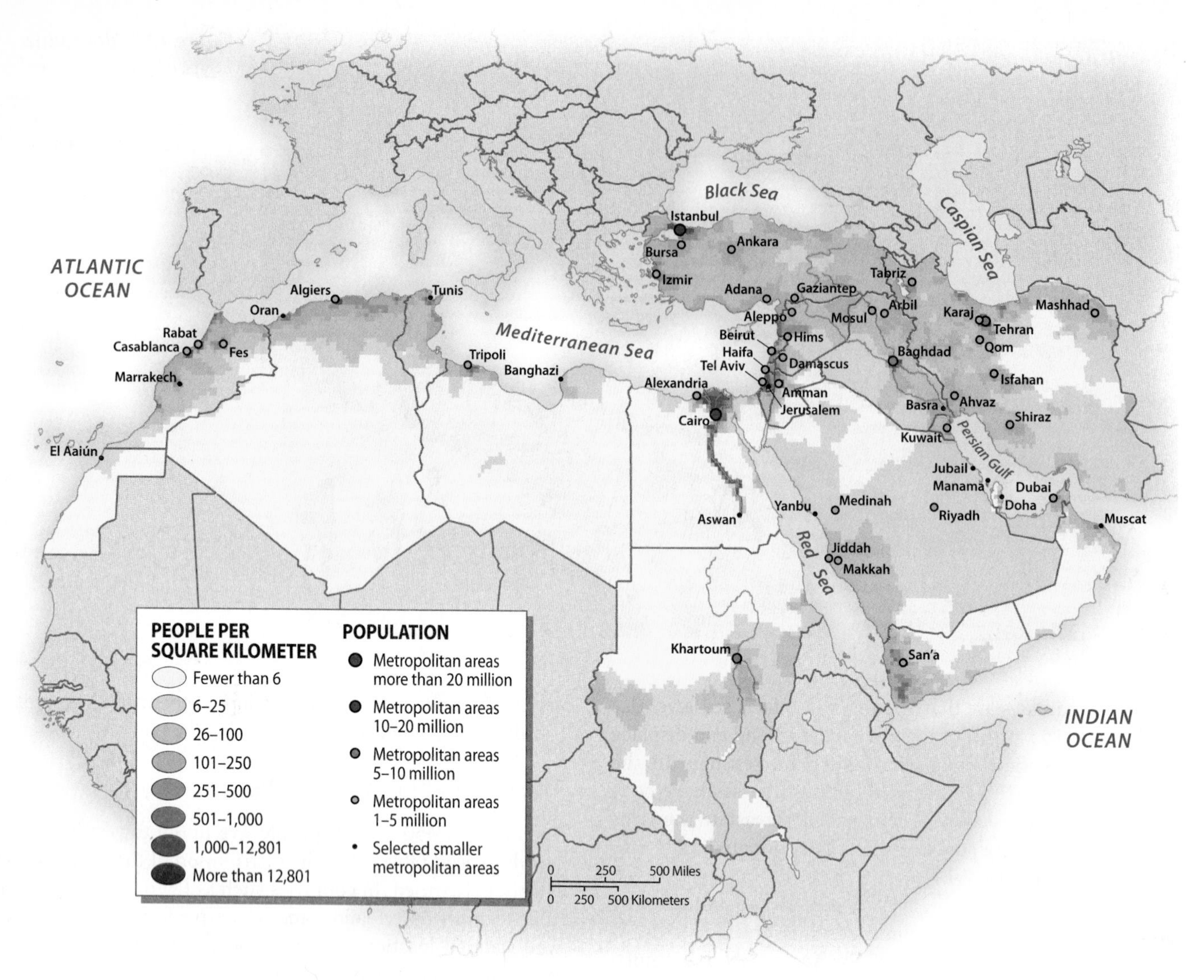

Figure 7.15 Population Map of Southwest Asia and North Africa The striking contrasts between large, sparsely occupied desert zones and much more densely settled regions where water is available are clearly evident. The Nile Valley and the Maghreb region contain most of North Africa's people, while Southwest Asian populations cluster in the highlands and along the better-watered shores of the Mediterranean.

in the Maghreb. They have rapidly growing metropolitan populations of 3 to 4 million residents each. Farther east, much of Libya and western Egypt is very thinly settled. Egypt's Nile Valley, however, is home to the other great North African population cluster. The vast majority of Egypt's 80 million people live within 10 miles of the river. Optimistic politicians and planners are creating a second corridor of denser settlement in Egypt's "New Valley" west of the Nile by diverting water from Lake Nasser into the Western Desert.

Most Southwest Asian residents are clustered in favored coastal zones, moister highland settings, and desert localities where water is available from nearby rivers or subsurface aquifers. High population densities are found in better-watered portions of the eastern Mediterranean (Israel, Lebanon, and Syria) and Turkey. Nearby Iran is home to more than 70 million residents, but population densities vary considerably from thinly occupied deserts in the east to more concentrated settlements near the Caspian Sea and across the more humid highlands of the northwest. Turkey's Istanbul, formerly Constantinople, (12.8 million) and Iran's Tehran (8.4 million) are Southwest Asia's largest urban areas, and both have grown in recent years as rural populations gravitate toward the economic opportunities of these cities (Figure 7.16). Elsewhere, sizable populations are scattered through the Tigris and Euphrates valley, the Yemen Highlands, and near oases where groundwater can be tapped to support agricultural or industrial activities.

Water and Life: Rural Settlement Patterns

Water and life are closely linked across the rural settlement landscapes of Southwest Asia and North Africa. Indeed, the diverse environments of Southwest Asia, in particular, are home to one of the world's earliest hearths of **domestication**,

TABLE 7.1 Population Indicators

Country	Population (Millions) 2010	Population Density (per Square Kilometer)	Rate of Natural Inrease (RNI)	Total Fertility Rate	Percent Urban	Percent <15	Percent >65	Net Migration (per 1,000, 2005–10)[a]
Algeria	36.0	15	1.8	2.3	63	28	5	–0.8
Bahrain	1.3	1,807	1.3	1.9	100	20	2	5.2
Egypt	80.4	80	2.1	3.0	43	33	4	–0.8
Gaza and West Bank	4.0	672	2.8	4.6	83	44	3	–0.5
Iran	75.1	46	1.3	1.8	69	28	5	–1.4
Iraq	31.5	72	2.6	4.1	67	41	3	–3.9
Israel	7.6	342	1.6	3.0	92	28	10	2.4
Jordan	6.5	73	2.6	3.8	83	37	3	8.3
Kuwait	3.1	175	2.0	2.2	98	23	2	8.3
Lebanon	4.3	409	1.5	2.3	87	25	10	–0.6
Libya	6.5	4	1.9	2.7	77	30	4	0.6
Morocco	31.9	71	1.5	2.4	57	29	5	–2.7
Oman	3.1	10	1.8	2.6	72	29	2	1.7
Qatar	1.7	152	0.8	1.8	100	15	1	93.9
Saudi Arabia	29.2	14	2.6	3.8	81	38	2	1.2
Sudan	43.2	17	2.2	4.5	38	41	3	0.7
Syria	22.5	122	2.5	3.3	54	36	3	7.7
Tunisia	10.5	64	1.2	2.1	66	24	7	–0.4
Turkey	73.6	94	1.2	2.1	76	26	7	–0.1
United Arab Emirates	5.4	64	1.4	2.0	83	19	1	15.6
Western Sahara	0.5	2	2.5	4.5	81`	40	3	19.6
Yemen	23.6	45	3.0	5.5	29	45	3	–1.2

[a]*Net Migration Rate from the United Nations, Population Division, World Population Prospects: The 2008 Revision Population Database.*
Source: Population Reference Bureau, World Population Data Sheet, 2010.

where plants and animals were purposefully selected and bred for their desirable characteristics. Beginning around 10,000 years ago, increased experimentation with wild varieties of wheat and barley led to agricultural settlements that later included domesticated animals, such as cattle, sheep, and goats. Much of the early agricultural activity focused on the **Fertile Crescent**, an ecologically diverse zone that stretches from the Levant inland through the fertile hill country of northern Syria into Iraq. Between 5,000 and 6,000 years ago, better knowledge of irrigation techniques and increasingly centralized political states promoted the spread of agriculture into nearby valleys such as the Tigris and Euphrates (Mesopotamia) and North Africa's Nile Valley. Since then, different peoples of the region have adapted to its environmental diversity and limitations in distinctive ways. In the process, they have practiced forms of agriculture appropriate to their settings and have left their own unique imprints upon the landscape (Figure 7.17).

Pastoral Nomadism Most common in the drier portions of the region, **pastoral nomadism** is a traditional form of subsistence agriculture in which practitioners depend on the seasonal movement of livestock for a large part of their livelihood. Arabian Bedouins, North African Berbers, and Iranian Bakhtiaris provide surviving examples of nomadism within the region. Today, however, with fewer than 10 million nomads remaining, the lifestyle is in decline, the victim of constricting political borders, a reduced demand for traditional beasts of burden such as camels, competing land uses, and selective overgrazing. In addition, government resettlement programs in Saudi Arabia, Syria, Egypt, and elsewhere are actively promoting a more settled lifestyle for many nomadic groups (see "People on the Move: Jordanian Bedouins and the Badia Project").

The settlement landscape of pastoral nomads reflects their need for mobility and flexibility as they seasonally move camels, sheep, and goats from place to place. Near highland zones such as the Atlas Mountains or the Anatolian Plateau, nomads practice **transhumance** (the seasonal movement of animals between wet-season and dry-season pastures) by herding their livestock to cooler, greener high country pastures in the summer and then returning them to valley and lowland settings for fall and winter grazing. Elsewhere, seasonal movements often involve huge territories of desert to support small groups of a few dozen families. In addition, nomads trade with sedentary agricultural populations, a mutu-

Figure 7.16 Tehran, Iran With a population of more than 8 million people, Iran's capital city of Tehran sprawls in all directions and is facing many of the same urban challenges as cities in North America.

ally beneficial relationship in which they exchange meat, milk, hides, and wool for cereal and orchard crops available at desert oases.

Oasis Life Permanent oasis settlements dot the arid landscape where high groundwater levels or modern deep-water wells provide reliable moisture in otherwise arid locales (Figure 7.18). Tightly clustered, often walled villages, their sun-baked mud houses blending into the surrounding scene, sit adjacent to small but intensely utilized fields where underground water is carefully applied to tree and cereal crops. In more recently created oases, concrete blocks and prefabricated housing add a contemporary look. Surrounded by large zones of desert, these green islands of rural

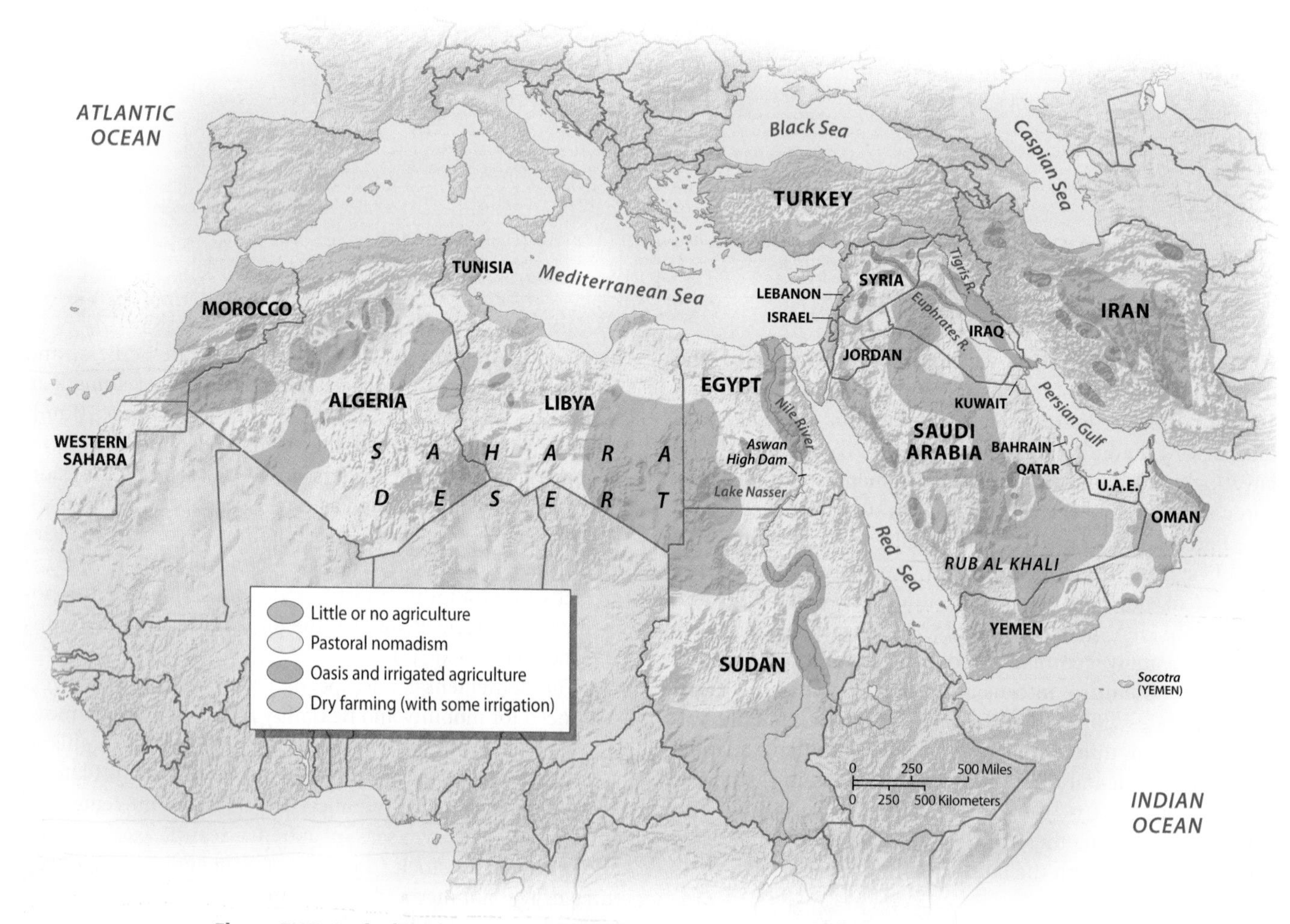

Figure 7.17 Agricultural Regions of Southwest Asia and North Africa Important agricultural zones include oases and irrigated farming where water is available. Elsewhere, dry farming supplemented with irrigation is practiced in midlatitude settings. (Modified from Clawson and Fisher, 2004, *World Regional Geography,* 8th ed., Upper Saddle River, NJ: Prentice Hall, and Bergman and Renwick, 1999, *Introduction to Geography,* Upper Saddle River, NJ: Prentice Hall)

Figure 7.18 Oasis Agriculture The green fields and trees of Morocco's Tinghir oasis contrast dramatically with the surrounding desert landscape.

activity stand out in sharp contrast to the sand- and rock-strewn landscape.

Traditional oasis settlements are composed of close-knit families who work their own irrigated plots or, more commonly, work for absentee landowners. Although oases are usually small, eastern Saudi Arabia's Hofuf oasis covers more than 30,000 acres (12,150 hectares). While some crops are raised for local consumption, commercial trade always has played a role in such settings. In the past century, the expanding world demand for products such as figs and dates has included even these remote oases in the global economy, and many products end up on the tables of hungry Europeans or North Americans. New drilling and pumping technologies, particularly in Saudi Arabia, have added to the size and number of oasis settlements. But oasis life across the region faces major challenges. Population growth, groundwater depletion, and the pressures of global cultural change threaten the economic and social integrity of these settlements.

Settlement Along Exotic Rivers For centuries, the region's densest rural settlement has been tied to its great river valleys and their seasonal floods of water and enriching nutrients. In such settings, **exotic rivers** transport precious water and nutrients from distant, more humid lands into drier regions, where the resources are utilized for irrigated farming (Figure 7.19). The Nile and the Tigris and Euphrates rivers are the largest regional examples of such activity, and both systems have large, densely settled deltas. Other linear irrigated settlements can be found near the Jordan River in Israel and Jordan, along short streams originating in North Africa's Atlas Mountains, and on the more arid peripheries of the Anatolian and Iranian plateaus. These settings, although capable of supporting sizable rural populations, also are among the most vulnerable to overuse, particularly if irrigation results in salinization.

Farming in such localities supports much higher population densities than is the case with pastoral nomadism or traditional desert oases. Fields are small, intensely utilized, and connected with closely managed irrigation systems designed to store and move water efficiently through the settlement. In Egypt, farmers of the Nile Valley live in densely settled, clustered villages near their fields, work much of their land with the same tools and technologies their ancestors used, and grow a mix of cotton, rice, wheat, and forage crops. The scale and persistence of their agricultural imprint is dramatically visible from space (Figure 7.19). But rural

Figure 7.19 Nile Valley This satellite image of the Nile Valley dramatically reveals the impact of water on the North African desert. Cairo lies at the southern end of the delta, where it begins to widen toward the Mediterranean Sea. The coastal city of Alexandria sits on the northwest edge of the low-lying Nile Delta.

PEOPLE ON THE MOVE Jordanian Bedouins and the Badia Project

Jordan's Prince Hassan Bin Talal (uncle of the country's ruling King Abdullah) has recently created the Badia Project, an initiative designed to keep the country's mobile Bedouin population viable and free. The "Badia" refers to the nation's large desert and semidesert zones which sprawl south and east of the capital city of Amman. For centuries, the Badia has been home to Bedouin populations who depend on a nomadic lifestyle and seasonal pastures to survive with their herds of camels, sheep, and goats (Figure 7.2.1).

But pressures upon the Bedouin lifestyle have steadily mounted in recent years. Droughts have reduced the amount of pasture land available for grazing. The attraction of urban amenities and the promise of better jobs have also drawn younger Bedouins away to the bright lights of Amman. Finally, political tensions have conspired to limit Bedouin mobility, further restricting their access to potential grazing pastures. For example, the eastern border zone with Iraq has often been closed and subject to political instability. The Saudi government has also gotten more restrictive and no longer permits unregulated Bedouin movements across the border to the south and southeast. The result is a growing crisis in the Bedouin community.

Prince Hassan has called attention to the problem and has vowed to support a traditional lifestyle he believes is worth preserving throughout the region. He has worked with religious leaders to cultivate Islamic environmentalism, a movement that uses traditional passages and interpretations from the Quran to encourage cooperation, better pasture management, water conservation, and the revegetation of overgrazed areas. Indeed, the Muslim concept of "haema" or protected area has been applied to encourage better use of the land. Local imams detail practices designed to help the Bedouins survive, and they use traditional texts to encourage cooperation among the clans. In addition the Prince is working with Jordan's financial institutions to increase Bedouin access to capital to build up their herds and improve the efficiency of their operations, thus encouraging them to remain on the land. Optimistically, the Prince even believes the initiative can lead towards more pan-Arab cooperation. He is lobbying nearby Arab nations to allow for more trans-boundary movement between countries, an initiative the Prince believes might help Bedouin populations throughout the region and might contribute to greater cooperation among the region's Arab countries.

Longer term, the survival of the Bedouin's mobile lifestyle remains in doubt, but Prince Hassan's initiative suggests some of the ways in which ancient cultural practices might be preserved and enhanced through a creative blend of tradition and innovation that combines local and global knowledge.

Figure 7.2.1 Jordan's Bedouins This Bedouin weaver is helping to preserve important cultural traditions in her Jordanian community.

life is also changing in such settings. New dam- and canal-building schemes in Egypt, Israel, Syria, Turkey, and elsewhere are increasing the storage capacity of river systems, which allows for more year-round agricultural activity. Higher-yielding rice and wheat varieties and more mechanized agricultural methods also have raised food output, particularly in places such as Egypt and Israel. Some of the most efficient farms in the region are associated with Israeli **kibbutzim**, collectively worked settlements that produce grain, vegetable, and orchard crops irrigated by waters from the Jordan River and from the country's elaborate feeder canals. For example, Israel's National Water Carrier system takes water from the Sea of Galilee and moves it south and west, where intensively worked agricultural operations produce key food crops for nearby Tel Aviv and Jerusalem.

The Challenge of Dryland Agriculture Mediterranean climates in portions of the region permit varied forms of dryland agriculture that depend largely on seasonal moisture to support farming. These zones include the better-watered valleys and coastal lowlands of the northern Maghreb, lands along the shore of the eastern Mediterranean, and favored uplands across the Anatolian and Iranian plateaus. A regional variant of dry farming is practiced across Yemen's terraced highlands in the moister corners of the southern Arabian Peninsula (Figure 7.20).

Often, a variety of crops and livestock surrounds the Mediterranean villages of the region. Drought-resistant tree crops are common, with olive groves, almond trees, and citrus orchards producing output for both local consumption and commercial sale. Elsewhere, favored locations support grape vineyards, while more marginal settings are used to grow wheat and barley or to raise forage crops to feed cattle, sheep, and goats. Vulnerability to drought and the availability of more sophisticated water management strategies are leading some Mediterranean farmers to utilize more irrigated cropping, thus improving production of cotton, wheat, citrus fruits, and tobacco across portions of the region. More mechanization, crop specialization, and fertilizer use are also transforming such agricultural settings, following a pattern set earlier in nearby areas of southern Europe. One commercial adaptation of growing regional and global importance is Morocco's flourishing hashish crop. More than 200,000 acres (80,000 hectares) of cannabis are cultivated in the hill country near Ketama in northern Morocco, generating more than $2 billion annually in illegal exports (mostly to Europe).

Many-Layered Landscapes: The Urban Imprint

Cities have played a pivotal role in the region's human geography. Indeed, some of the world's oldest urban places are located in the region. Today, enduring political, religious, and economic ties link the city and countryside.

A Long Urban Legacy Cities in the region traditionally have played important functional roles as centers of political and religious authority, as well as key focal points of local and long-distance trade. Urbanization in Mesopotamia (modern Iraq) began by 3500 BCE, and cities such as Eridu and Ur reached populations of 25,000 to 35,000 residents. Similar centers appeared in Egypt by 3000 BCE, with Memphis and Thebes assuming major importance amid the dense populations of the middle Nile Valley. These ancient cities were key centers of political and religious control. Temples, palaces, tombs, and public buildings dominated the urban landscapes of such settlements, and surrounding walls (particularly in Mesopotamia) offered protection from outside invasion. By 2000 BCE, however, a different kind of city was emerging, particularly along the shores of the eastern Mediterranean and at the junction points of important caravan routes. Centers such as Beirut, Tyre, and Sidon, all in modern Lebanon, as well as Damascus in nearby Syria, exemplified the growing role of trade in creating urban landscapes. Expanding port facilities, warehouse districts, and commercial thoroughfares suggested how trade and commerce shaped these urban settlements, and many of these early Middle Eastern trading towns have survived to the present.

Islam also left a lasting mark because cities traditionally served as centers of Islamic religious power and education. By the 8th century, Baghdad had emerged as a religious center, followed soon thereafter by the appearance of Cairo as a seat of religious authority and expansion. Urban settlements from North Africa to Turkey felt the influences of Islam. Indeed, the Islamic Moors carried its characteristic Muslim signature to Spain, where it shaped urban centers such as Córdoba and Málaga. Islam's impact upon the settlement landscape merged with older urban traditions across the region and established a characteristic Islamic cityscape that

Figure 7.20 Rural Yemen This isolated hilltop village in the rugged mountains of northern Yemen is surrounded by small, terraced fields and livestock pastures.

exists to this day. Its traditional attributes include a walled urban core, or **medina**, dominated by the central mosque and its associated religious, educational, and administrative functions (Figure 7.21). A nearby bazaar, or *suq*, functions as a marketplace where products from city and countryside are traded (Figure 7.22). Housing districts feature an intricate maze of narrow, twisting streets that maximize shade and accentuate the privacy of residents, particularly women. Houses have small windows, frequently are situated on dead-end streets, and typically open inward to private courtyards often shared by extended families with similar ethnic or occupational backgrounds.

More recently, European colonialism added another layer of urban landscape features in selected cities. Particularly in North Africa, coastal and administrative centers during the late 19th century added dozens of architectural features from Britain and France. Victorian building blocks, French mansard roofs, suburban housing districts, and wide European-style commercial boulevards complicated the settlement landscapes of dozens of cities, both old and new. Centers such as Algiers (French), Fes (French), and Cairo (British) vividly displayed the effects of colonial control, and many of these urban signatures remain today (Figure 7.23).

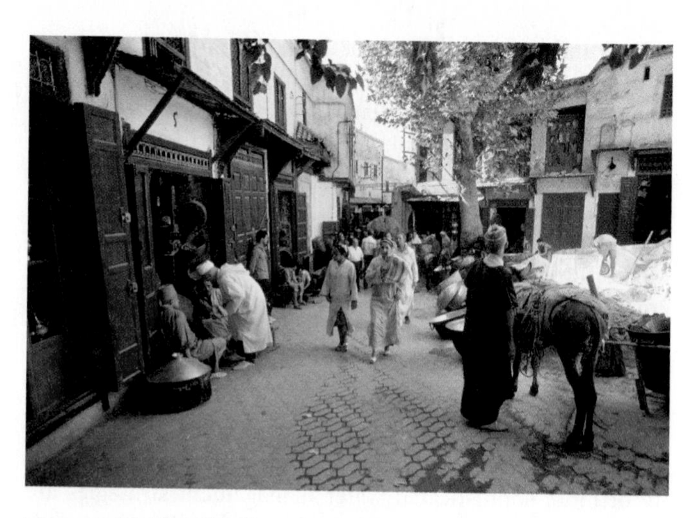

Figure 7.22 The Old City of Fes, Morocco These narrow streets in the old city of Fes are lined with shops and commercial stalls. The winding labyrinth of alleyways and cul-de-sacs forms an almost impenetrable maze to visitors, and it exemplifies the complex urban landscapes of the traditional Islamic city.

Figure 7.21 The Islamic Landscape Iranian mullahs discuss the religious questions of the day beneath the minarets of the Moussavi Mosque in Qom. Islam has left a widespread mark on the region's cultural landscapes.

Signatures of Globalization Since 1950, dramatic new forces have transformed the urban landscape. Cities have become key gateways to the global economy. As the region has been opened to new investment, industrialization, and tourism, the urban landscape reflects the fundamental changes taking place. Expanded airports, commercial and financial districts, industrial parks, and luxury tourist facilities all mark the impress of the global economy.

Further, as urban centers become focal points of economic growth, surrounding rural populations are drawn to the new employment opportunities, thus fueling rapid population increases. The results are both impressive and problematic. Many traditional urban centers, such as Algiers and Istanbul, have more than doubled in size in recent years. Booming demand for homes has produced ugly, cramped high-rise apartment houses in some government-planned neighborhoods, while elsewhere sprawling squatter settlements provide little in the way of quality housing or municipal services.

Crowded Cairo, now with 15 million people, exemplifies the pattern of urban expansion. The city is now much more densely settled, as well as larger in size. Its legendary "City of the Dead" neighborhood, home to almost 1 million residents, is an urban cemetery intermingled with homes and apartment houses, a vivid acknowledgment of the premium put on living space. Many central-city neighborhoods have become so crowded and congested that wealthier urbanites have left, moving to outlying suburbs with large villas and lower population densities. Other suburbs are sites for major industrial expansion, often financed through foreign investment. In some cases, entirely new industrial centers are being established, part of Egypt's ambitious New-Town Program.

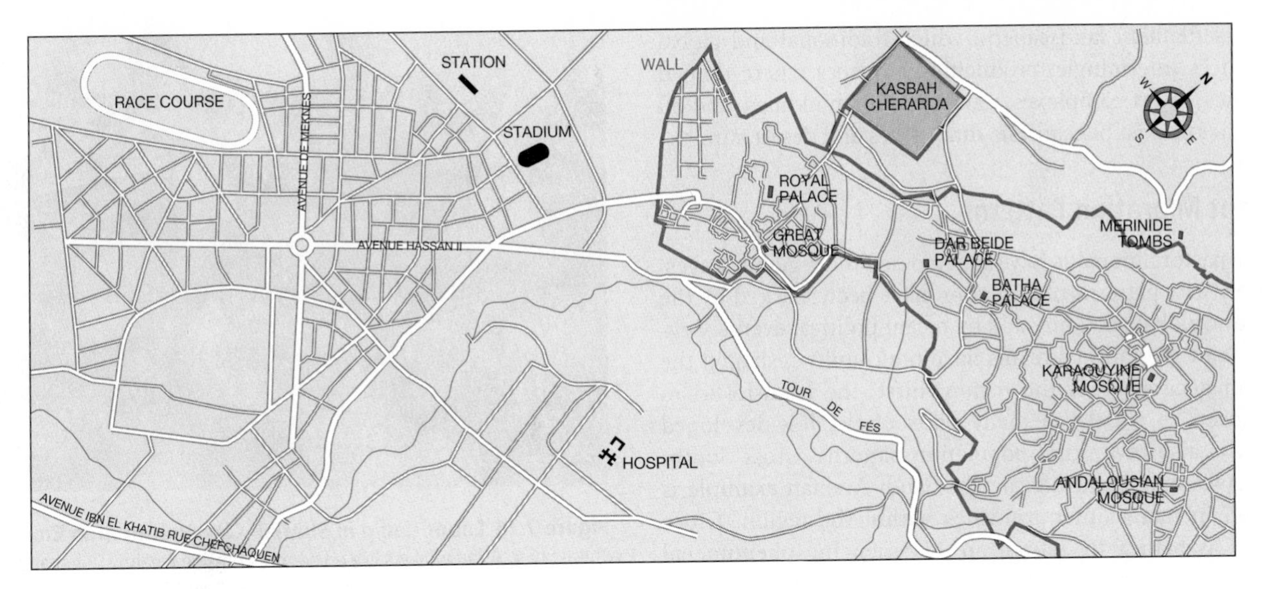

Figure 7.23 Map of Fes, Morocco The tiny neighborhoods and twisting lanes of the old walled city reveal features of the traditional Islamic urban center. To the west, however, the rectangular street patterns, open spaces, and broad avenues suggest colonial European influences.

Certainly, the oil-rich states of the Persian Gulf display the most extraordinary changes in the urban landscape (Figure 7.24). Before the 20th century, urban traditions were relatively weak in the area, and even as late as 1950 only 18 percent of Saudi Arabia's population lived in cities. All that changed, however, as the global economy's demand for petroleum mushroomed. Today, the Saudi Arabian population is more urban than those of many industrialized nations, including the United States, and the capital city of Riyadh has grown to about 5 million people. Particularly after 1970, other cities, such as Abu Dhabi (United Arab Emirates), Doha (Qatar), and Kuwait City (Kuwait), grew in size and took on modern Western characteristics, including futuristic architecture and new transportation infrastructure. In addition, investments in petrochemical industries have fueled the creation of new urban centers, such as Jubail along Saudi Arabia's Persian Gulf coastline. The result is an

Figure 7.24 Modern Doha, Qatar The rapidly changing urban landscapes of the oil-rich Persian Gulf region are illustrated in the modern architecture seen in Doha's West Bay district.

urban settlement landscape in which traditional and global influences intermingle, producing cityscapes where domed mosques, resort complexes, mirrored bank buildings, and oil refineries coexist beneath the dusty skies and desert sun.

Recent Migration Patterns

While pastoral nomads have crisscrossed the region for ages, entirely new patterns of migration have been sparked by the dynamic global economy and by recent political events. Several major migration streams have profoundly reshaped the human geography of the region. First, the rural-to-urban shift seen so widely in many parts of the less-developed world is also reworking population patterns across Southwest Asia and North Africa. The Saudi Arabian example is echoed in many other countries within the region. Cities from Casablanca to Tehran are experiencing phenomenal growth rates, spurred by in-migration from rural areas.

Second, large numbers of workers are migrating within the region to areas with growing job opportunities. Labor-poor countries such as Saudi Arabia, Kuwait, and the United Arab Emirates, for example, have attracted thousands of Jordanians, Palestinians, Yemenis, and Iranians to construction and service jobs within their borders. Indeed, the demand for workers has been so strong that large numbers of South Asians, Southeast Asians, and East Africans, particularly from Muslim nations, have flooded into the region. Today, for instance, more than 85 percent of Saudi Arabia's labor force is foreign and they annually export more than $20 billion in wages. In Dubai (United Arab Emirates), Pakistani cab drivers, Filipino nannies, and Indian shop clerks typify a foreign workforce that currently makes up a large majority of the city's 1.4 million inhabitants (Figure 7.25). Job-related movements, however, can be fickle. In the economic downturn of 2008–2010, many workers left Dubai and returned to their home countries, leaving thousands of abandoned cars in the airport parking lot.

A third major pattern of movement involves residents migrating to job opportunities elsewhere in the world. Because of its strong economy and close location, Europe has been a particularly powerful draw. More than 2 million Turkish guest workers live in Germany. Both Algeria and Morocco also have seen large out-migrations to western Europe, particularly France. Only 8 miles separates Africa from Europe, and illegal northbound boat traffic has reached unprecedented levels. Many Lebanese, often skilled workers and business owners, have also emigrated. Thousands of migrants have journeyed to North America, particularly more educated professionals who seek high-paying jobs.

Political forces have also encouraged migrations. Thousands of wealthier residents, for example, left Lebanon and Iran during the political turmoil of the 1980s and are today living in cities such as Toronto, Los Angeles, and Paris. Elsewhere, more recent political instability has provoked other refugee movements. Although Iraq's political situation has stabilized some since 2009, hundreds of thousands of Iraqi refugees are still in nearby Syria and Jordan and several million remain displaced from their homes within Iraq. More than 35,000 Iraqis have resettled in the United States since 2003.

Figure 7.25 Labor Camp in Shariah City, United Arab Emirates These South Asian workers live a few miles outside wealthy Dubai and enjoy a card game during their time off from work. Over 80 percent of the country's population is foreign-born and many immigrants work in the construction and service sectors of the economy.

Elsewhere, thousands of displaced Afghans remain in eastern Iran. In the Middle East, the movement of more than 500,000 Russian Jewish immigrants to Israel after the 1991 breakup of the Soviet Union has also reshaped the cultural and political geography of that country. To the south, huge numbers of people in western Sudan's unsettled Darfur region have been forced from their land and forced to move to dozens of ill-prepared refugee camps in nearby Chad.

Shifting Demographic Patterns

While high population growth remains a critical issue throughout the region, the demographic picture is shifting. Uniformly high rates of population growth in the 1950s and 1960s have been replaced by more varied regional patterns, and many nations are seeing birthrates fall fairly rapidly. For example, women in Tunisia and Turkey are now averaging fewer than three births, representing a large decline in the total fertility rate (see Table 7.1). Indeed, some experts now predict that fertility rates on the European and North African sides of the Mediterranean will almost converge (at just over two births per woman) by 2030. Many factors help explain the changes. More urban, consumer-oriented populations have opted for fewer children. Many Arab women now are delaying marriage into the middle 20s and even early 30s. Even in more traditional Bahrain, the average marriage age of women has risen from 15 to 23 years of age since the early 1970s.

Intriguingly, fundamentalist Iran has witnessed the fastest decline in fertility in the past two decades, producing a pronounced "youth bulge" in its population pyramid (Figure 7.26). Fertility has declined by more than two-thirds since the mid-1970s (from an average of 6.6 births per woman to 2.0 births per woman). While Iran's family planning program was initially dismantled after the fundamentalist revolution in 1979 (it was seen as a Western idea), more recent Iranian leaders have recognized the wisdom in containing the country's large population. As a result, Iran has one of the most successful family planning programs in the world, according to a 2009 report by the Population Reference Bureau.

Still, demographic challenges loom. Areas such as the West Bank, Gaza, and Yemen experience some of the highest rates of natural increase on the planet. In some localities, persisting patterns of poverty and traditional ways of rural life contribute to the large rates of population increase, and even in more urban and industrialized Saudi Arabia, annual growth rates remain very high. Such growth rates result from the combination of high birthrates and low death rates.

Elsewhere, even as new births are declining, already existing populations pose problems. For most countries in the region, more than 35 percent of the present population is younger than 15 years old, promising another large generation of people in need of food, jobs, and housing. In Egypt, for example, even though birthrates are likely to decline, the labor market will need to absorb more than 500,000 new workers annually over the next 10 to 15 years just to keep up with the country's large, youthful population. As that population ages in the mid-21st century, it will continue to demand jobs, housing, and social services. Given recent trends in job growth and migration, the region's large cities probably will bear the brunt of future population increases. Growing populations will also impose new demands on the region's already limited water resources.

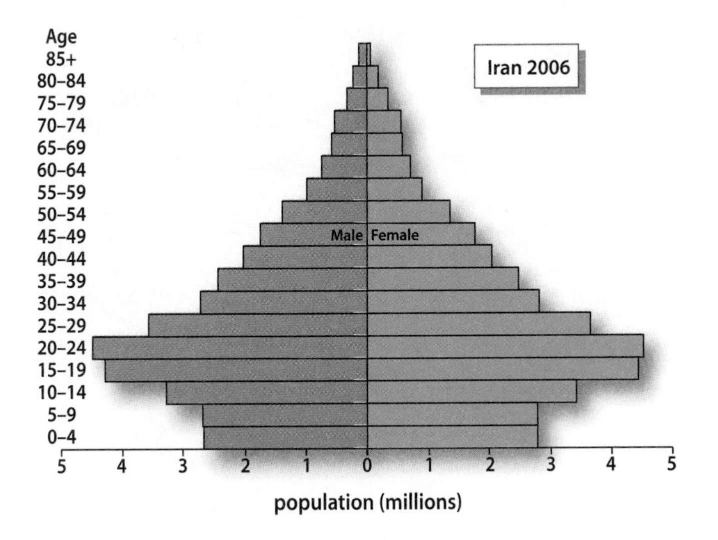

Figure 7.26 Iran's Population Pyramid, 2006 Iran's annual population growth rates have slowed considerably, thanks to an effective national family planning program. A pronounced youth bulge (ages 15–29) has been followed more recently by much lower birthrates. (Population Reference Bureau, 2009)

CULTURAL COHERENCE AND DIVERSITY: Signatures of Complexity

While Southwest Asia and North Africa clearly define the heart of the Islamic and Arab worlds, a surprising degree of cultural diversity characterizes the region. Muslims practice their religion in varied ways, often disagreeing profoundly on basic religious views as well as on how much of the modern world and its mass consumer culture should be incorporated into their daily lives. In addition, diverse religious minorities complicate the region's contemporary cultural geography. Linguistically, Arabic languages form an important cultural core historically centered on the region. However, different Arab dialects can mean that a Syrian may struggle to comprehend an Algerian. In addition, many non-Arab peoples, including Persians, Kurds, and Turks, populate important homelands. Understanding these varied patterns of cultural geography is essential to comprehending many of the region's political tensions, as well as appreciating why many of its residents resist processes of globalization and celebrate the lasting cultural identity of their home neighborhoods and communities.

Patterns of Religion

Religion permeates the lives of most people within the region. Its centrality stands in sharp contrast to largely secular cultures in many other parts of the world. Whether it is the quiet ritual of morning prayers or profound discussions about contemporary political and social issues, religion is part of the daily routine of most regional residents from Casablanca to Tehran. The geographies of religion—their points of origin, paths of diffusion, and patterns of modern regional distribution—are essential elements in understanding cultural and political conflicts within the region.

Hearth of the Judeo-Christian Tradition Both Jews and Christians trace their religious roots to the eastern Mediterranean, and while neither group is numerically dominant across the area, each plays a key cultural role. The roots of Judaism lie deep in the past: Abraham, an early patriarch in the Jewish tradition, lived some 4,000 years ago and led his people from Mesopotamia to Canaan (modern-day Israel), near the shores of the Mediterranean. From Jewish history, recounted in the Old Testament of the Holy Bible, springs a rich religious heritage focused on a belief in one God (or **monotheism**), a strong code of ethical conduct, and a powerful ethnic identity that continues to the present. Around 70 CE, during the time of the Roman Empire, most Jews were forced to leave the eastern Mediterranean after they challenged Roman authority. The resulting forced migration, or diaspora, of the Jews took them to the far corners of Europe and North Africa. Only in the past century have many of the world's far-flung Jewish populations returned to the religion's hearth area, a process that accelerated greatly with the formation of the Jewish state of Israel in 1948.

Christianity also emerged in the vicinity of modern-day Israel and has left a lasting legacy across the region. An outgrowth of Judaism, Christianity was based on the teachings of Jesus and his disciples, who lived and traveled in the eastern Mediterranean about 2,000 years ago. While many Christian traditions became associated with European history, some forms of early Christianity remained strong near the religion's original hearth. To the south, one stream of Christian influences linked with the Coptic Church diffused into northern Africa, shaping the culture of places such as Egypt and Ethiopia. In the Levant, another group of early Christians, known as the Maronites, retained a separate cultural identity that survives today. The region remains the "Holy Land" in the Christian tradition and attracts millions of the faithful annually, as they visit the most sacred sites in the Christian world.

The Emergence of Islam Islam originated in Southwest Asia in 622 CE, forming yet another cultural hearth of global significance. Muslims can be found today from North America to the southern Philippines; however, the Islamic world is still centered on its Southwest Asian origins. Most Southwest Asian and North African peoples still follow its religious teachings and moral doctrines. Muhammad, the founder of Islam, was born in Makkah (Mecca) in 570 CE and taught in nearby Medinah (Medina) (Figure 7.27). In many respects, his creed parallels the Judeo-Christian tradition. Muslims believe that both Moses and Jesus were true prophets and that both the Hebrew Bible (or Old Testament) and the Christian New Testament, while incomplete, are basically accurate. Ultimately, however, Muslims hold that the **Quran** (or Koran), a book of revelations received by Muhammad from Allah (God), represents God's highest religious and moral revelations to humankind.

The basic teachings of Islam offer an elaborate blueprint for leading an ethical and religious life. Islam literally means "submission to the will of God," and the creed rests on five essential pillars: (1) repeating the basic creed ("There is no God but God, and Muhammad is his prophet"); (2) praying facing Makkah five times daily; (3) giving charitable contributions; (4) fasting between sunup and sundown during the month of Ramadan; and (5) making at least one religious pilgrimage, or **Hajj**, to Muhammad's birthplace of Makkah (Figure 7.28). Islam is a more austere religion than most forms of Christianity, and its modes of worship and forms of organization are generally less ornate. Muslims avoid the use

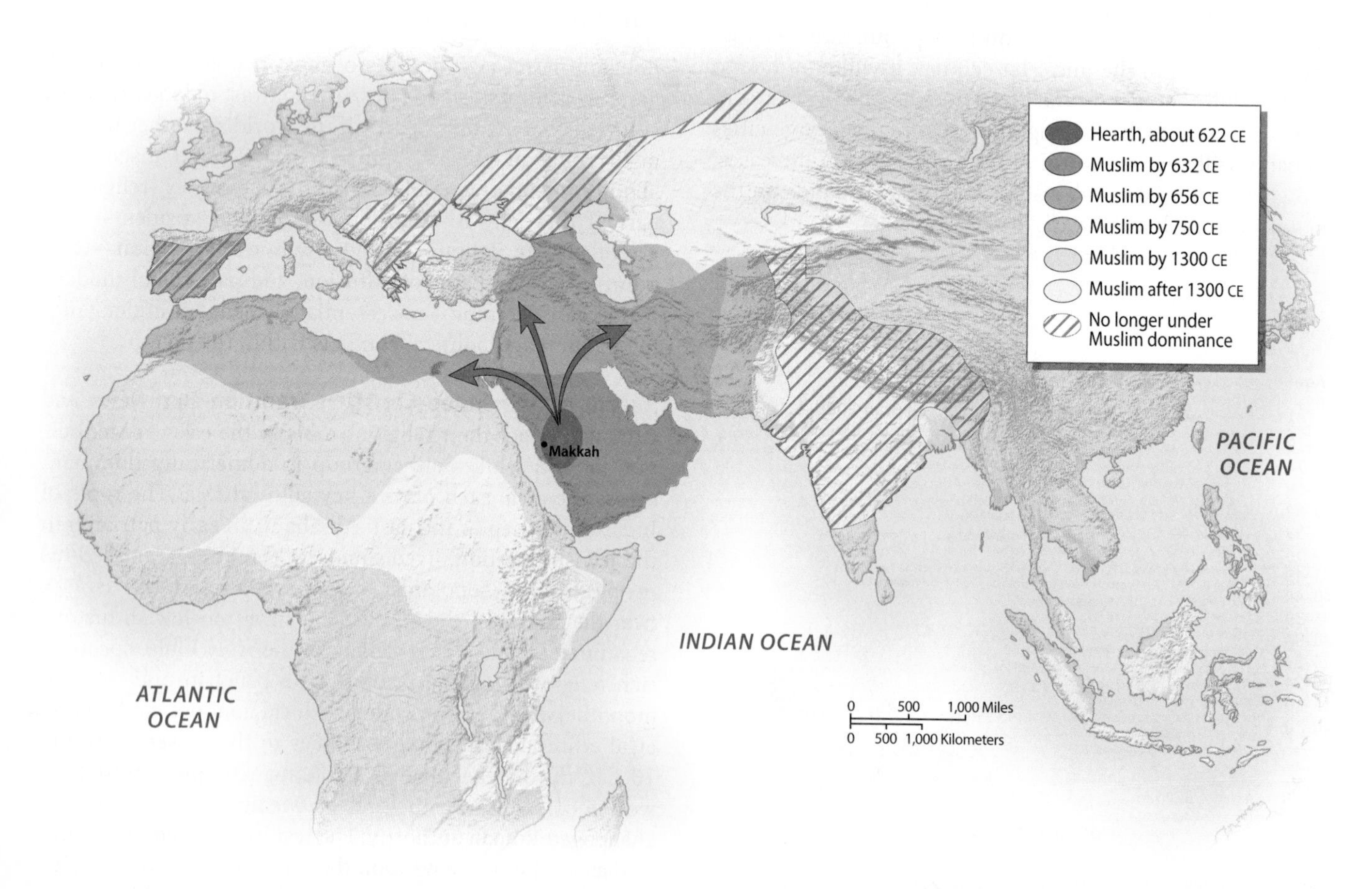

Figure 7.27 Diffusion of Islam The rapid expansion of Islam that followed its birth is shown here. From Spain to Southeast Asia, Islam's legacy remains strongest nearest its Southwest Asian hearth. In some settings, its influence has ebbed or has come into conflict with other religions, such as Christianity, Judaism, and Hinduism. (Modified from Rubenstein, 2005, *An Introduction to Human Geography,* 8th ed., Upper Saddle River, NJ: Prentice Hall)

Figure 7.28 Makkah Thousands of faithful Muslims gather at the Grand Mosque in central Makkah, part of the pilgrimage to this sacred place that draws several million visitors annually. Portions of the city's commercial district are in the distance.

of religious images, and the strictest interpretations even forbid the depiction of the human form. Followers of Islam are prohibited from drinking alcohol and are instructed to lead moderate lives, avoiding excess. Many Islamic fundamentalists still argue for a theocratic state, such as modern-day Iran, in which religious leaders (ayatollahs) guide policy.

A major religious schism divided Islam early on, and the differences endure to the present. The breakup occurred almost immediately after the death of Muhammad in 632 CE. Key questions surrounded the succession of religious power. One group, now called the **Shiites**, favored passing on power within Muhammad's own family, specifically to Ali, his son-in-law. Most Muslims, later known as **Sunnis**, advocated passing power down through the established clergy. This group emerged victorious. Ali was killed, and his Shiite supporters went underground. Ever since, Sunni Islam has formed the mainstream branch of the religion, to which Shiite Islam has presented a recurring, and sometimes powerful, challenge. The Shiites argue that a successor to Ali will someday return to reestablish the pure, original form of Islam. The Shiites also are more hierarchically organized than the Sunnis. Iran's ayatollahs, for example, have overriding religious and political power.

In a very short period of time, Islam diffused widely from its Arabian hearth, often following camel caravan routes and Arab military campaigns as it expanded its geographical range and converted thousands to its beliefs. By the time of Muhammad's death in 632 CE, the peoples of the Arabian Peninsula were united under its banner. Shortly thereafter, the Persian Empire fell to Muslim forces and the Eastern Roman (or Byzantine) Empire lost most of its territory to Islamic influences. By 750 CE, Arab armies had swept across North Africa, conquered most of Spain and Portugal, and established footholds in Central and South Asia. At first, only the Arab conquerors followed Islam, but the diverse inhabitants of Southwest Asia and North Africa gradually were absorbed into the religion, although with many distinct local variants. By the 13th century, most people in the region were Muslims, while older religions such as Christianity and Judaism became minority faiths or disappeared altogether.

Between 1200 and 1500, Islamic influences expanded in some areas and contracted in others. The Iberian Peninsula (Spain and Portugal) returned to Christianity by 1492, although many Moorish (Islamic) cultural and architectural features remained behind and still shape the region today. At the same time, Muslims expanded their influence southward and eastward into Africa. In addition, Muslim Turks largely replaced Christian Greek influences in Southwest Asia after 1100. One group of Turks moved into the Anatolian Plateau and finally conquered the last vestiges of the Byzantine Empire in 1453. These Turks soon created the vast **Ottoman Empire** (named after one of its leaders, Osman), which included southeastern Europe (including modern-day Albania, Bosnia, and Serbia) and most of Southwest Asia and North Africa. The legacy of the Ottoman Empire was considerable. It offered a new, distinctly Turkish interpretation of Islam, and it provided a focus of Muslim political power within the region until the empire's disintegration in the late 19th and early 20th centuries.

Modern Religious Diversity Today, Muslims form the majority population in all of the countries of Southwest Asia and North Africa except Israel, where Judaism is the dominant religion (Figure 7.29). Still, divisions within Islam have created key cultural differences within the region. While most (73 percent) of the region is dominated by Sunni Muslims, the Shiites (23 percent) remain an important element in the contemporary cultural mix. In Iraq, for example, the majority Shiite population in the southern portion of the country (around Najaf, Karbala, and Basra) set its own cultural and political course following the fall of Saddam Hussein. Many Shiites saw the departure of Hussein (who gathered most of his support from Sunnis in the central and western parts of Iraq) as an opportunity for increasing their cultural and political influence within the country. Shiites claim a majority in nearby Iran, as well, where their religious fervor has fundamentally shaped government policies since the late 1970s. In Yemen, important Shiite minorities claim the government has neglected their needs and that has led to increased political instability in that impoverished nation. In addition, they form a major religious minority in Lebanon, Bahrain, Algeria, and Egypt.

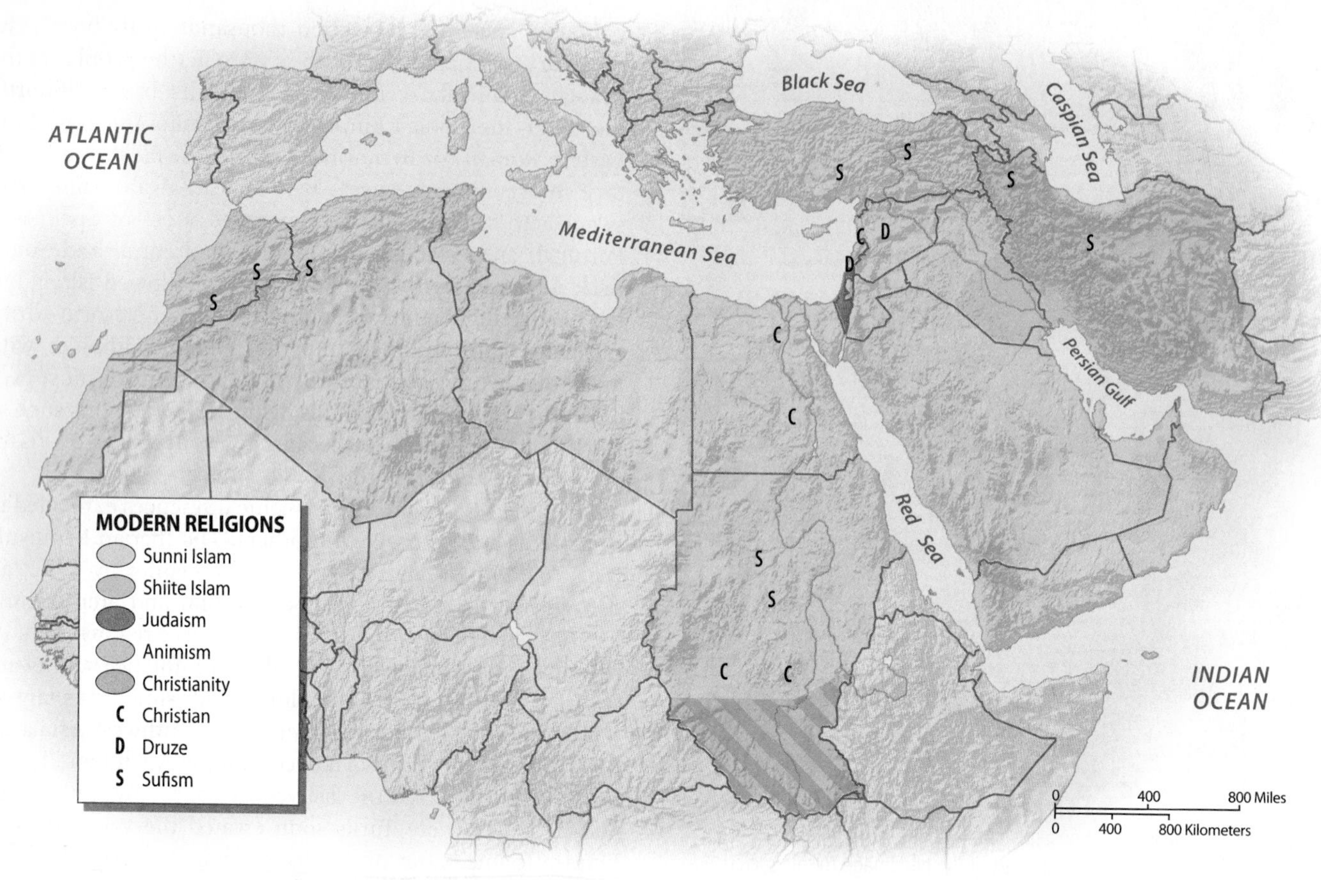

Figure 7.29 Modern Religions Islam remains the dominant religion across the region. Most Muslims are tied to the Sunni branch, while Shiites are found in places such as Iran and southern Iraq. In some locales, however, Christianity and Judaism remain important. African animism is found in southern portions of Sudan. (Modified from Rubenstein, 2011, *An Introduction to Human Geography*, 10th ed., Upper Saddle River, NJ: Prentice Hall, and National Geographic Society, 2003, *Atlas of the Middle East*, Washington, DC)

Strongly associated with the recent flowering of Islamist movements, the Shiites also have benefited from rapid growth rates because their brand of Islam is particularly appealing to the poorer, powerless, and more rural populations of the region. While some Sunnis have been attracted to Islamist tendencies as well, many reject its more radical cultural and political teachings and argue for a more modern Islam that accommodates some Western values and traditions.

Whereas the Sunni–Shiite split is the great divide within the Muslim world, other variations of Islam also can be found in the region. One division, for example, separates the mystically inclined form of Islam—known as *Sufism*—from the mainstream tradition (see Figure 7.29). Sufism is especially prominent in the peripheries of the Islamic world, including the Atlas Mountains of Morocco and Algeria. Other well-established examples of Sufism are found across northwestern Iran and portions of Turkey. The Druze of Lebanon practice another variant of Islam, and they form a cohesive religious minority in the Shouf Mountains east of the Lebanese capital of Beirut.

Southwest Asia also is home to many non-Islamic communities. Israel has a Muslim minority (16 percent) that is dominated by that nation's Jewish population (77 percent), while Christians make up another 2 percent of the total. Even within Israel's Jewish community, increasing cultural differences divide Jewish fundamentalists from more reform-minded Jews. Indeed, this cultural diversity within Judaism has shaped the way Israel has dealt with political and social issues involving Muslims within and beyond its borders. In neighboring Lebanon, there was a slight Christian (Maronite and Orthodox) majority as recently as 1950. Christian outmigration and differential birthrates, however, have created a nation that today is 60 percent Muslim. Christians also form approximately 10 percent of Syria's population; Iraqi Christians, concentrated mostly in the rugged northern uplands, make up about 3 percent of its population.

Within the Middle East, the city of Jerusalem (now the Israeli capital) holds special religious significance for several groups and also stands at the core of the region's political problems (Figure 7.30). Indeed, the sacred space of this ancient Middle Eastern city remains deeply scarred and divided as different groups argue for more control of contested neighborhoods and nearby suburbs. Historically, Jews particularly revere the city's old Western Wall (the site of a Roman-era temple); Christians honor the Church of the Holy Sepulchre (the burial site of Jesus); and Muslims hold sacred religious sites in the city's eastern quarter (including the place from which the prophet Mohammad reputedly ascended to heaven).

Important Jewish and Christian communities also have left a long legacy across North Africa. Roman Catholicism was once dominant in much of the Maghreb, but it disappeared several hundred years after the Muslim conquest. The Maghreb's Jewish population, on the other hand, remained prominent until the post–World War II period, when most of the region's Jews migrated to the new state of Israel. In Egypt, Coptic Christianity has maintained a stable presence through the centuries and today includes approximately 9 percent of the country's population. In earlier years the Coptic community had a secure place in Egyptian society, and numerous Copts held high-level posts in government and business. Today, however, Egypt's Christians are being increasingly marginalized, and some of their communities have been put under pressure, even subjected to physical attack, by extremist Islamic elements. Other Christian communities are located in southern Sudan, but unlike those of Egypt, these are mostly recent converts from traditional African religions.

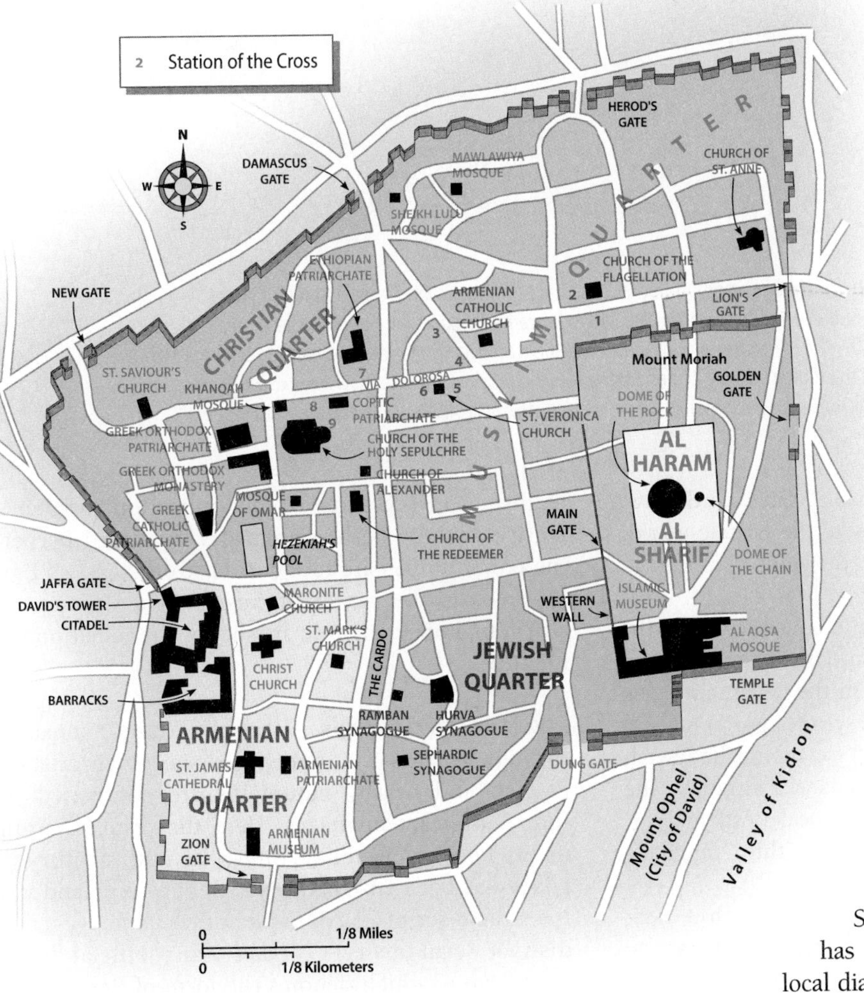

Figure 7.30 Old Jerusalem Jerusalem's historic center reflects its varied religious legacy. Sacred sites for Jews, Christians, and Muslims all cluster within the Old City. The Western Wall, a remnant of the ancient Jewish temple, stands at the base of the Dome of the Rock and Islam's Al Aqsa Mosque. (Reprinted from Rubenstein, 2011, *An Introduction to Human Geography*, 10th ed., Upper Saddle River, NJ: Prentice Hall)

Geographies of Language

Although the region is often referred to as the "Arab World," linguistic complexity creates many important cultural divisions across Southwest Asia and North Africa (Figure 7.31). The geography of language offers insights into regional patterns of ethnic identity and potential cultural conflicts that exist at linguistic borders. While the language map is useful in identifying the major families found across the region, it is important to remember that there are many local variations in language not seen on the map that represent distinctive dialects and well-defined islands of cultural and ethnic identity. For example, more than 70 separate languages are recognized in Iran, 36 in Turkey, and 23 in Iraq.

Semites and Berbers Afro-Asiatic languages dominate much of the region. Within that family, Arabic-speaking Semitic peoples can be found from Morocco to Saudi Arabia. Before the expansion of Islam, Arabic was limited to the Arabian Peninsula. Today, however, Arabic is spoken from the Persian Gulf to the Atlantic and reaches southward into Sudan, where it borders the Nilo-Saharan speaking peoples of Sub-Saharan Africa. As the language has diffused, it has slowly diverged into local dialects. As a result, the everyday Arabic spoken on the streets of Fes, Morocco, is quite distinct from Arabic spoken in the United Arab Emirates. The Arabic language also has a special religious significance for all Muslims because it was the sacred language in which God delivered his message to Muhammad. While most of the world's Muslims do not speak Arabic, the faithful often memorize

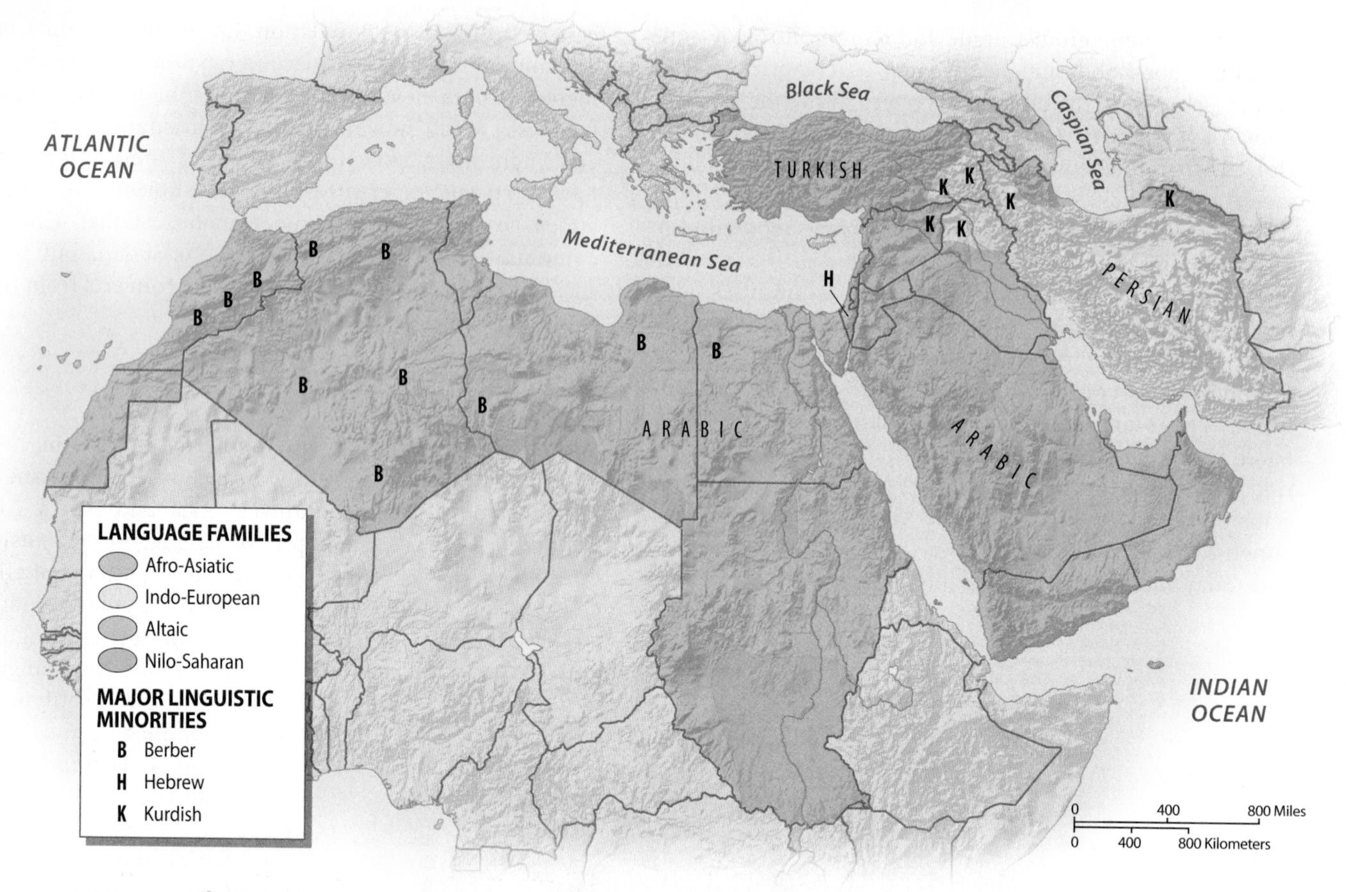

Figure 7.31 Modern Languages Arabic is a Semitic Afro-Asiatic language, and it dominates the region's cultural geography. Turkish, Persian, and Kurdish, however, remain important exceptions, and such differences within the region often have had lasting political consequences. Israel's more recent reintroduction of Hebrew further complicates the region's linguistic geography. (Modified from Rubenstein, 2011, *An Introduction to Human Geography,* 10th ed., Upper Saddle River, NJ: Prentice Hall and National Geographic Society, 2003, *Atlas of the Middle East,* Washington, DC)

certain prayers in the language, and many Arabic words have entered the other important languages of the Islamic world. Advanced Islamic learning, moreover, demands competence in Arabic.

Hebrew, another traditional Semitic language of Southwest Asia, was recently reintroduced into the region with the creation of Israel. Hebrew originated in the Levant and was used by the ancient Israelites 3,000 years ago. Today its modern version survives as the sacred tongue of the Jewish people and is the official language of Israel, although the country's non-Jewish population largely speaks Arabic. English is also widely used as a second or third language throughout the country.

While Arabic eventually spread across North Africa, several older languages survive in more remote areas. Older Afro-Asiatic tongues endure in the Atlas Mountains and in certain parts of the Sahara. Collectively known as Berber, these languages are related to each other but are not mutually intelligible. Most Berber languages have never been written, and none has generated a significant literature. Indeed, a Berber-language version of the Quran was not completed until 1999. The decline in pastoral nomadism and pressures of modernization threaten the integrity of these Berber languages. Scattered Berber-speaking communities are found as far to the east as Egypt, but Morocco is the center of this language group, where it plays an important role in shaping that nation's cultural identity.

Persians and Kurds Although Arabic spread readily through portions of Southwest Asia, much of the Iranian Plateau and nearby mountains are dominated by older Indo-European languages. Here the principal tongue remains Persian, although, since the 10th century, the language has been enriched with Arabic words and written in the Arabic script. Persian, like other languages, developed distinct local dialects. Today Iran's official language is called *Farsi,* which denotes the form of Persian spoken in Fars, the area around the city of Shiraz. Thus, although both Iran and neighboring Iraq are Islamic nations, their ethnic identities spring from quite different linguistic and cultural traditions.

The Kurdish speakers of northern Iraq, northern Iran, and eastern Turkey add further complexity to the regional pattern of languages. Kurdish, also an Indo-European language, is spoken by 10 to 15 million people in the region. Kurdish has not historically been a written language, but the Kurds do have a strong sense of shared cultural identity. Indeed, "Kurdistan" has sometimes been called the world's largest nation without its own political state because the group remains a minority in several countries of the region. In the 2003 war in Iraq, the Kurds emerged as a cohesive group in the northern part of the country that was opposed to Saddam Hussein. Iraqi Kurds have gained more political autonomy in a post-Saddam Iraq state and are also intent on maintaining control of important oil resources found in their portion of the country. Their leading city of Kirkuk has been called the "Kurdish Jerusalem" because its history and settlement so richly capture the group's cultural identity. Nearby Kurds in eastern Turkey, however, still complain that their ethnic identity is frequently challenged by the majority Turks, leaving some to wonder if they will attempt to join forces with their Iraqi neighbors.

The Turkish Imprint Turkish languages provide more variety across much of modern Turkey and in portions of far northern Iran. The Turkish languages are a part of the larger Altaic language family that originated in Central Asia. Turkey remains the largest nation in Southwest Asia dominated by that family. Tens of millions of people in other countries of Southwest and Central Asia speak related Altaic languages, such as Azeri, Uzbek, and Uighur. During the era of the Ottoman Empire, Turkish speakers ruled much of Southwest Asia and North Africa, but Iran is the only other large country in the region today where Turkic languages persist, particularly in the northwest part of the country.

Regional Cultures in Global Context

Many cultural connections tie the region with the world beyond. Islam links the region with a Muslim population that now lives in many different settings around the world. In addition, European and American cultural influences have multiplied greatly across the region since the mid-19th century. Colonialism, the boom in petroleum investment, and the growing presence of Western-style popular culture have had enduring impacts from the Atlas Mountains to the Indian Ocean. These global cultural connections are nuanced and complex, linking this region with the rest of the world in fascinating, often unanticipated ways (see "Global to Local: How Morocco Went Hollywood in Ouarzazate").

Islamic Internationalism While Islam is geographically and theologically divided, all Muslims recognize the fundamental unity of their religion. This religious unity extends far beyond Southwest Asia and North Africa. Islamic communities are well established in such distant places as central China, European Russia, central Africa, and the southern Philippines. Today, Muslim congregations also are expanding rapidly in the major urban areas of western Europe and North America, largely through migration but also through local conversions. Islam is thus emerging as a truly global religion. Even with its global reach, however, Islam remains centered on Southwest Asia and North Africa, the site of its origins and its most holy places. As Islam expands in number of followers and geographical scope, the religion's tradition of pilgrimage ensures that Makkah will become a city of increasing global significance in the 21st century. The global growth of Islamist fundamentalism and Islamism also focuses attention on the region, where these contemporary movements recently burst upon the scene. In addition, the oil wealth accumulated by many Islamic nations is used to sustain and promote the religion. Countries such as Saudi Arabia and Libya invest in Islamic banks and economic ventures and make donations to Islamic cultural causes, colleges, and hospitals worldwide. Indeed, geographers have studied new flows of global capital in an "Islamic financial services sector" that promotes economic development through the application of Islamic religious principles (see "Geographic Tools: Measuring the Urban Impact of Islamic Banking").

Globalization and Cultural Change The region also is struggling with how its growing role in the global economy is changing traditional cultural values. European colonialism left its own cultural legacy not only in the architectural landscapes still found in the old colonial centers, but also in the widespread use of English and French among the Western-educated elite across the region. In oil-rich countries, huge capital investments have had important cultural implications as the number of foreign workers has grown and as more affluent young people have embraced elements of Western-style music, literature, and clothing. The expansion of Islamic fundamentalism and Islamism are in many ways reactions to the threat posed by external cultural influences, particularly the supposed evils of European colonialism, American and Israeli power, and local governments and cultural institutions that are seen as selling out to the West.

Technology has also contributed to cultural change. Educated and more affluent residents in the region have been embracing the Internet and its power to access a global wealth of information and entertainment. Cell phone use also has increased rapidly. In Morocco, for example, while less than 6 percent of the country's population is served by fixed phone lines, more than 85 percent is within the reach of cellular phones. Widespread television-viewing has transformed the region, with viewers offered everything from Islamist religious programming to American-style reality TV and talent shows. For some, these technological transformations bring problems, particularly in conservative Islamic nations. Sudan's government has attempted to limit the influx of global cultural media and television programming.

GLOBAL TO LOCAL How Morocco Went Hollywood in Ouarzazate

The global movie-making industry has carved out a unique niche in the North African desert. Filmmakers have been drawn to Morocco for over a century, beginning during its French colonial period and accelerating in the 1980s. Orson Welles filmed *Othello* there in the 1940s, and David Lean shot the epic *Lawrence of Arabia* in Morocco in the 1960s. More recently, Martin Scorsese filmed *Kundun* and *The Last Temptation of Christ,* while Ridley Scott filmed *Gladiator, Black Hawk Down,* and *Kingdom of Heaven* in this North African country.

In the early days of cinematography it was Morocco's fabulous light and dry desert climate that attracted filmmakers dependent upon outdoor locations and natural lighting. The country's rugged coastline, high Atlas Mountains, desert dunes, arid valleys, and historic architecture also became a lure, as its varied physical and cultural landscapes inspired location scouts to transform exotic Moroccan settings into other places in the world such as Saudi Arabia, Somalia, the biblical Holy Land, or even Tibet (Figure 7.3.1).

Economics and regional politics also contribute to Morocco's attractiveness as a destination. Close to Europe and easily accessible from North America, the cost of crews and local "extras" is considerably less in Morocco. For directors who want to set their films in desert environments, other countries in North Africa and Southwest Asia (such as Saudi Arabia, Iran, or Libya) are off limits to Western production companies. In contrast, the Moroccan government has courted foreign filmmakers and created government agencies that facilitate logistics. Over the years, Morocco has financially supported its own burgeoning film industry and film schools. Indeed, the country has reaped the economic benefits of the policy. In 2009 global movie-making productions injected over $100 million into the Moroccan economy, employing numerous local workers.

In the center of the country, just south of the high Atlas Mountains on a barren plateau, is the city of Ouarzazate. This modern Berber city of only 70,000 residents is now the center of the country's movie industry and nearby landscapes such as the traditional mud-brick Kasbah of Ait Benhaddou have ended up as part of the backdrop in both *Lawrence of Arabia* and *Gladiator* (Figure 7.3.2). Isolated from the distractions of major cities, with lots of open arid space, the location enables production companies to build the massive sets they need as well as exploit nearby historic structures. A gateway city to the Sahara desert with modern amenities, Ouarzazate caters to filmmakers and desert-seeking tourists. Ouarzazate is the Hollywood of Morocco, but it is relatively unknown to those outside of the country. Yet it is likely that moviegoers have seen this city's environs, only they might have thought they were looking at Tibet, Somalia, Jerusalem, Egypt, or the Arabian Peninsula.

Figure 7.3.1 Movie Production in the Desert, Ouarzazate, Morocco Visitors tour a movie set at Atlas Film Corporation Studios in Ouarzazate, Morocco.

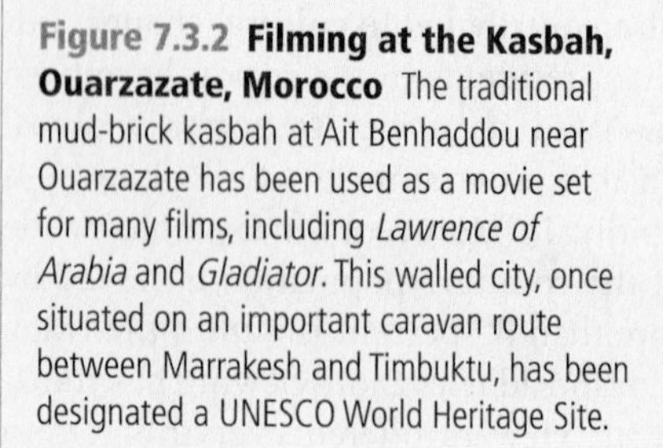

Figure 7.3.2 Filming at the Kasbah, Ouarzazate, Morocco The traditional mud-brick kasbah at Ait Benhaddou near Ouarzazate has been used as a movie set for many films, including *Lawrence of Arabia* and *Gladiator*. This walled city, once situated on an important caravan route between Marrakesh and Timbuktu, has been designated a UNESCO World Heritage Site.

Figure 7.32 Desert Heat Rappers Salim and Abdullah Dahman from the United Arab Emirates created the group Desert Heat, and they celebrate many aspects of Arab culture in their music.

Similarly, Iran has restricted the use of satellite dishes and has prevented access to many Internet Websites.

Hybrid forms of popular culture also express the shaping power of globalization. Recently, for example, both within the region and beyond, Arab hip-hop music has offered a form of artistic and political expression that represents the fusion of several cultural traditions. Tune in MTV Arabia and you are liable to hear the newest cut from Desert Heat (United Arab Emirates), Malikah (Lebanon), or MWR (Palestine). These young artists represent a generation of rappers that spits out lyrics challenging many of the cultural and political stereotypes within the region. The global vibe is hard to miss: The African-American roots of rap remain a fixture in the strong beats and bops that set the tone amid the Arabic ouds (lutes), clangs of Asian pop, and soulful lyrics that might emerge in Arabic, English, or French.

The social roots of hip-hop help explain its resonance in the region. Just as urban African Americans used rap to vent frustration and offer commentary on their lives, beginning in the 1970s, this newer generation of Arab artists tells a similar social and cultural story. For Salim and Abdullah Dahman (aka Desert Heat), their pioneering venture into hip-hop within the United Arab Emirates was designed to cultivate Arab unity and to offer to the world a more positive view of the Arab experience after the painful events of September 11, 2001 (Figure 7.32). They were a recent star attraction at the Red Bull Air Race in Abu Dhabi and their first album, "When the Desert Speaks," was released in 2008 along with their debut video, "Keep It Desert." Their aim is clear. As Salim noted in 2009 to a *Gulf News* (published in the United Arab Emirates) reporter: "We just want to fight two stereotypes. Firstly that Arabs and Muslims are not all terrorists or ignorant people, and secondly that hip-hop is not all negative. Hip-hop was born out of positivity; out of struggle and freedom of speech." These new connections forged in music between younger generations of Arabs, Jews, Europeans, North Americans, and others may be powerful stimulants to redefining traditional cultural and political narratives. In this way, the global leanings and nontraditional messages of Arab hip-hop may be much more than entertainment.

But cultural changes among younger residents of the region are not going unchallenged. In the culturally conservative Persian Gulf states of Qatar and the United Arab Emirates, a new penchant for cross-dressing among young women has produced a considerable backlash amongst more orthodox Muslims. Many young women in those countries (particularly on university campuses) have donned western-style baggy trousers and wear short-cropped hair. Termed *boyats*, or "tomboys," the women have been accused of adopting a "foreign trend" in dress brought about by globalization. Their "deviant behavior" has been seen as a menace to society. The same clothing style and look that would go unnoticed on North American college campuses has provoked calls among conservatives for the death penalty, "medical treatment," or reeducation workshops in femininity.

Elsewhere, conservative cultural influences are gaining strength and visibility. For example, in Tunisia, long known for its more Western-leaning, modernized brand of Islam, the growing use of headscarves and veils among conservative women has caused a stir. A ban on such traditional practices was enacted in 1981 by a Tunisian government that described it as a "sectarian form of dress." More recently, such clothing has been seen by some as promoting the political agenda of extremist Islamist elements within the country. Interestingly, European human rights advocates have come to the aid of the women as they are increasingly harassed by police to remove the scarves and sign pledges that they will not go back to wearing them. In Turkey, a growing

GEOGRAPHIC TOOLS Measuring the Urban Impact of Islamic Banking

Urban and economic geographers often utilize various forms of "network analysis" to measure how effectively one locality is connected to another. A study published in 2010, completed by a group of geographers in Belgium, illustrates the technique and suggests how the unique cultural character of the Islamic financial sector is linked in special ways with the rest of the world's cities and how it may be contributing to urban and economic growth within the Islamic world.

Figure 7.4.1 Persian Gulf City of Manama, Bahrain The 2010 view of Manama's Financial Harbor was taken from Muharraq Island. Towering construction cranes in the distance suggest the city's role as one of the centers of the Islamic financial world is likely to continue.

Unlike dominant banks and financial institutions based in Europe or North America, the Islamic financial services (IFS) sector operates on religious principles that are consistent with the Quran and the Muslim religion. For example, based on teachings within Sunni Islam, conventional interest-bearing loans are prohibited because the simple idea that money should make money is considered immoral. Instead, the loan recipient may agree to pay back the principal in the form of profits. In addition, these institutions ban anything that could be considered as gambling or speculation since these activities are also prohibited in Islam. Indeed, the IFS sector would not advocate participating in any activity associated with prohibitions within traditional Islam.

The Belgian geographers wanted to know how these special characteristics influenced how the IFS interacted at the global level and within the Islamic world. They also wanted to explore a different way to examine the world-city system which often is based on the singular dominance of localities such as New York City, London, or Tokyo. The geographers identified dozens of leading IFS sector firms and mapped their headquarters as well as their linkages to various global branch offices. They analyzed the resulting networks to assess where these firms fit into more traditional interpretations of world city rankings and where their activities seemed most directly linked with economic development. Drawing upon spatial interaction models to measure linkages between places, the geographers examined the top 100 IFS firms and their urban-based location strategies.

The results of the study show how the IFS sector is strongly centered in Southwest Asia. The Gulf State of Bahrain and its capital city of Manama emerged as the undisputed global leader of the IFS (Figure 7.4.1). Fully one-quarter of all the major IFS firms were headquartered there, and many other firms had branch offices in the city. Indeed, Manama has been a well-established center of regional banking and relative economic stability for decades, making it an attractive financial center for the Saudis and many other regional Islamic interests. Dubai and Abu Dhabi, both within the nearby United Arab Emirates, were also major participants in the global IFS network. Both Tehran and Istanbul were key participants, although the Iranian city's greater dominance was somewhat surprising and probably indicates the impact of its explicitly Islamist regime. Beyond the region, London and Paris are important European participants in the system as is Kuala Lumpur (in Muslim-friendly Malaysia) in Southeast Asia. New York, however, barely made it into the top 20 cities on the list.

The innovative use of network analysis offered these geographers a fresh way to look at how the world city-system works and it provided a reminder how a cultural lens can often inform traditional economic analysis and complicate our understanding of how future development in the Islamic world will inevitably be tied to both Western flows of capital as well as to the blossoming IFS sector that remains firmly based within the region as well as its religion.

Source: Adapted from David Bassens, Ben Derudder, and Frank Witlox, "Searching for the Mecca of finance: Islamic financial services and the world city network," *Area* 42, 1 (2010): 35–46.

traditional Islamist presence in the government and in the civil service bureaucracy has allowed for more religious study and prayers in that country's public school system. Science textbooks are also being changed, deemphasizing Western interpretations in favor of more traditional religiously based explanations.

GEOPOLITICAL FRAMEWORK: Never-Ending Tensions

Geopolitical tensions remain high in Southwest Asia and North Africa (Figure 7.33). Some of the tensions surround the struggle of different ethnic, religious, and linguistic

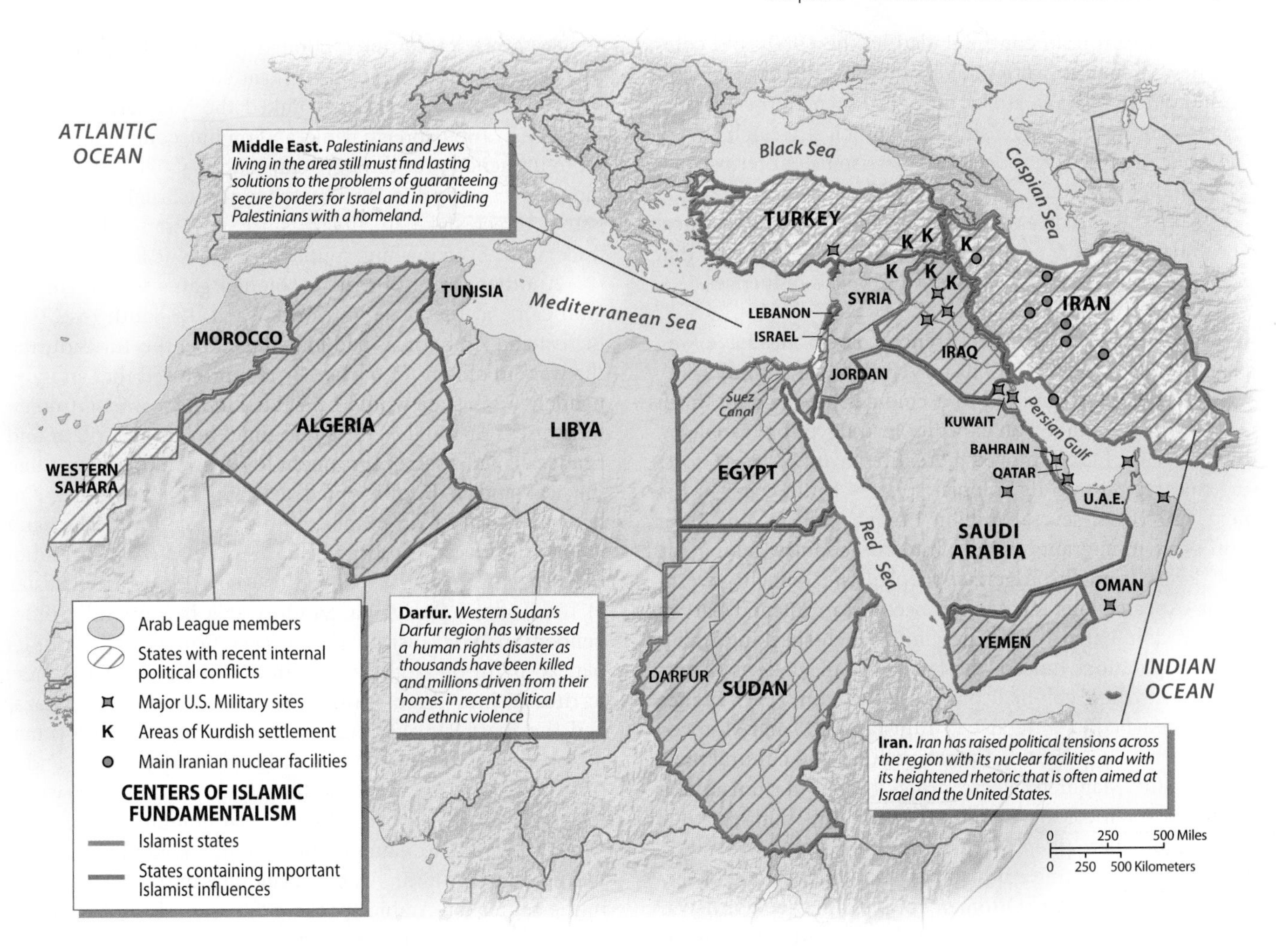

Figure 7.33 Geopolitical Issues in Southwest Asia and North Africa Political tensions continue across much of the region, including those surrounding events in Iraq and Iran. Although regional conflict remains oriented around Israel, its neighboring states, and the rights of resident Palestinians, other regional trouble spots periodically erupt into violence. Islamist political movements have challenged political stability in settings from Algeria to Saudi Arabia. Several large American military bases are found in the region.

groups to live with one another in a rapidly changing world of political relationships. The region's brief but complex ties to the era of European colonialism also contribute to present difficulties, because the boundaries of many countries were imposed by colonial powers. American political power has been repeatedly exercised throughout the region, evident in its multiple military bases and in its recent involvement in Iraq. Geographies of wealth and poverty also enter the geopolitical mix: Some residents profit from petroleum resources and industrial expansion, while others struggle just to feed their families. Islamist elements in many countries such as Iran, Egypt, Algeria, Turkey, and Saudi Arabia have changed the political atmosphere dramatically in the past 20 years. The result is a political climate charged with tension, a region in which the sounds of bomb blasts and gunfire remain for many a common characteristic of everyday life.

The Colonial Legacy

European colonialism arrived relatively late in Southwest Asia and North Africa, but the era left an important impact upon the modern political landscape. The Turks were one reason for Europe's late participation in imposing colonial rule. Between 1550 and 1850, much of the region was dominated by the Turkish Ottoman Empire, which expanded from its Anatolian hearth to engulf much of North Africa as well as nearby areas of the Levant, the western Arabian Peninsula, and modern-day Iraq. Ottoman rule imposed an outside political order and economic framework that had lasting consequences. The tide began to turn, however, in the early 19th century as the European presence increased and Ottoman power ebbed. Still, it took a century for Ottoman influences to be replaced with largely European colonial dominance after World War I (1918). While much of that direct European control ended by the 1950s, old colonial ties persist in a variety of economic, political, and cultural contexts. It is still common to encounter British English on the streets of Cairo, and French still can be heard in Algiers and Beirut. Indeed, the persistence of terms such as "Middle East" and "Levant" (a French term meaning "east" or "the Orient") within our 21st-century geographical vocabulary is a reminder of the long-standing linkages that tied this "eastern" land to western European influences.

Imposing European Power French colonial ties have long been a part of the region's history. Beginning around 1800, France became committed to a colonial presence in North Africa which included an expedition to Egypt by Napoleon. By the 1830s, France moved more directly into Algeria, forcing that region into its colonial sphere of influence. In the next 120 years, several million French, Italian, and other European immigrants poured into the country, taking the best lands from the Algerian people. The French government expected this territory to become an integral part of France, dominated by the growing French-speaking immigrant population. Indeed, the landscape of modern Algiers still reflects these colonial connections (Figure 7.34). France also established its influence in Tunisia (1881) and Morocco (1912), ensuring a lasting French political and cultural presence in the Maghreb. These **protectorates** in Tunisia and Morocco retained some political autonomy, but remained under a broader sphere of French influence and protection from other competing colonial powers. Finally, France's victory over the German–Ottoman Turk alliance in World War I produced additional territorial gains in the Levant, as France gained control of the northern zone, encompassing the modern nations of Syria and Lebanon.

Great Britain's colonial fortunes also grew within the region before 1900. To control its sea-lanes to India, Britain established a series of 19th-century protectorates over the small coastal states of the southern Arabian Peninsula and the Persian Gulf. In this manner, such places as Kuwait, Bahrain, Qatar, the United Arab Emirates, and Aden (in southern Yemen) were loosely incorporated into the British Empire. Nearby Egypt also caught Britain's attention. Once the European-engineered **Suez Canal** linked the Mediterranean and Red seas in 1869, foreign banks and trading companies gained more influence over the Egyptian economy. The British took more direct control in 1883. In the process, Britain also inherited a direct stake in Sudan, since Egyptian soldiers and traders had been pushing south along the Nile for decades.

Another series of British colonial gains within the region came at the close of World War I. In Southwest Asia, British and Arab forces had joined to expel the Turks during the war. To obtain Arab trust, Britain promised that an independent Arab state would be created in the former Ottoman territories. At roughly the same time, however, Britain and France signed a secret agreement to partition the area. When the war ended, Britain opted to slight its Arab allies and honor its treaty with France, with one exception: The Saud family convinced the British that a smaller country (Saudi Arabia) should be established, focused on the desert wastes of the Arabian Peninsula. Saudi Arabia became fully independent in 1932. Elsewhere, however, Britain carried out its plan to partition lands with France. Britain divided its new territories into three entities: Palestine (now Israel, Gaza, and the West Bank) along the Mediterranean coast; Transjordan to the east of the Jordan River (now Jordan); and a third zone that later became Iraq. Iraq, in particular, was a territory contrived by the British that combined three dissimilar former Ottoman provinces. It included the centers of Basra in the south (an Arabic-speaking Shiite area), Baghdad in the center (an Arabic-speaking Sunni area), and Mosul in the north (a Kurd-dominated zone).

Other settings within the region felt more marginal colonial impacts. Libya, for example, was long regarded by Europeans as a desert wasteland. Italy, never a dominant colonial power, expelled Turkish forces from the coastal districts by 1911 and its colonial influence continued until 1947. Spain also carved out its own territorial stake within the region, gaining control over southern Morocco (now Western Sahara) in 1912. To the east, Persia and Turkey never were directly oc-

Figure 7.34 Algiers European colonial influences still abound in the old French capital of Algiers in northern Algeria. The city's modern streets have witnessed periodic political and religious tensions since 1992.

cupied by European powers. In Persia, the British and Russians agreed to establish mutual spheres of economic influence (the British in the south, the Russians in the north), while respecting Persian independence. In 1935, Persia's modernizing ruler, Reza Shah, changed the country's name to Iran.

In nearby Turkey, the old core of the Ottoman Empire was almost partitioned by European powers following World War I. After several years of fighting, however, the Turks expelled the French from southern Turkey and the Greeks from western Turkey. The key to the successful Turkish resistance was the spread of a new modern, nationalist ideology under the leadership of Kemal Ataturk. Ataturk decided to emulate the European countries and establish a culturally unified and resolutely secular state. He was successful, and Turkey was quickly able to stand up to European power.

Decolonization and Independence European colonial powers began their withdrawal from several Southwest Asian and North African colonies before World War II. By the 1950s, most of the countries in the region were independent, although many maintained political and economic ties with their former colonial rulers. In North Africa, Britain finally withdrew its troops from Sudan and Egypt in 1956. Libya (1951), Tunisia (1956), and Morocco (1956) achieved independence peacefully during the same era, but the French colony of Algeria became a major problem. Since several million French citizens resided there, France had no intention of simply withdrawing. A bloody war for independence began in 1954, and France agreed to an independent Algeria in 1962, but the two nations continued to share a close—if not always harmonious—relationship thereafter.

Southwest Asia also lost its colonial status between 1930 and 1971, although many of the imposed colonial-era boundaries continue to shape the regional geopolitical setting. While Iraq became independent from Britain in 1932, its later instability in part resulted from its artificial borders, which never recognized much of its cultural diversity. Similarly, the French division of its Levant territories into the two independent states of Syria and Lebanon (1946) greatly angered local Arab populations and set the stage for future political instability in the region. As a favor to its small Maronite Christian majority, France carved out a separate Lebanese state from largely Arab Syria, even guaranteeing the Maronites constitutional control of the national government. The action created a culturally divided Lebanon as well as a Syrian state that repeatedly has asserted its influence over its Lebanese neighbors.

Modern Geopolitical Issues

The geopolitical instability in Southwest Asia and North Africa continues in the 21st century. It remains difficult to predict political boundaries that seem certain to change as a result of negotiated settlements or political conflict. A quick regional transect from the shores of the Atlantic to the borders of Central Asia suggests how these forces are playing out in different settings early in the century.

Across North Africa Varied North African settings threaten the region's political stability. Libya's leader, Colonel Muammar al-Qaddafi, has intermittently fomented regional tensions since he took power in 1969. Libya has financed violent political movements directed against Israel, western Europe, and the United States. Since 2004, however, Qaddafi has vowed to disarm, winning praise and growing diplomatic recognition from both the European Union and the United States.

Elsewhere in North Africa, Islamist political movements have reshaped the political landscape in several states. Most notably, Algeria was plunged into an escalating cycle of Islamist-led violence and protests for much of the 1990s, and more than 150,000 Algerians and foreign visitors were killed in the process. Since 1999, an amnesty has encouraged some rebel groups to lay down their arms, but the government still faces active opposition from Islamist extremists who are linked with Al Qaeda. Nearby Egypt has found itself ensnared in Islamist-initiated instability, and groups such as the Muslim Brotherhood have pushed for more radical political change.

Sudan faces some of the most daunting political issues in North Africa. A Sunni Islamist state since a military coup in 1989, Sudan imposed Islamic law across the country and in the process antagonized both moderate Sunni Muslims as well as the nation's large non-Muslim (mostly Christian and animist) population in the south. A long civil war between the Muslim north and the Christian and animist south produced more than 2 million casualties (mostly in the south) in the past 20 years. The carnage in the south has lessened since 2001 and a tentative peace agreement was signed in 2005. The agreement opened the way for a 2011 referendum on independence in southern Sudan, but intertribal squabbles and government corruption in the south have clouded hopes for a new, stable, independent state of South Sudan. Meanwhile, a newer conflict erupted in the Darfur region in the western portion of the country (see Figure 7.33). Ethnicity, race, and control of territory seem to be at the center of the struggle in the largely Muslim region as a well-armed Arab-led militia group (with many ties to the central government in Khartoum) has attacked hundreds of black-populated villages, driving more than 2.5 million people from their homes and leading to the deaths (through violence, starvation, and disease) of more than 300,000 Sudanese.

The Arab–Israeli Conflict The 1948 creation of the Jewish state of Israel produced another enduring zone of cultural and political tensions within the eastern Mediterranean. Jewish migration to Palestine increased after the defeat of the Ottoman Empire. In 1917 Britain issued the Balfour Declaration, a pledge to encourage the creation of a Jewish homeland in the region. After World War II, the United Nations divided the region into two states, one to be predominantly Jewish, the other primarily Muslim (Figure 7.35). Indigenous Arab Palestinians rejected the partition, and war erupted. Jewish forces proved victorious, and by 1949 Israel had actually grown in size. The remainder of Palestine, in-

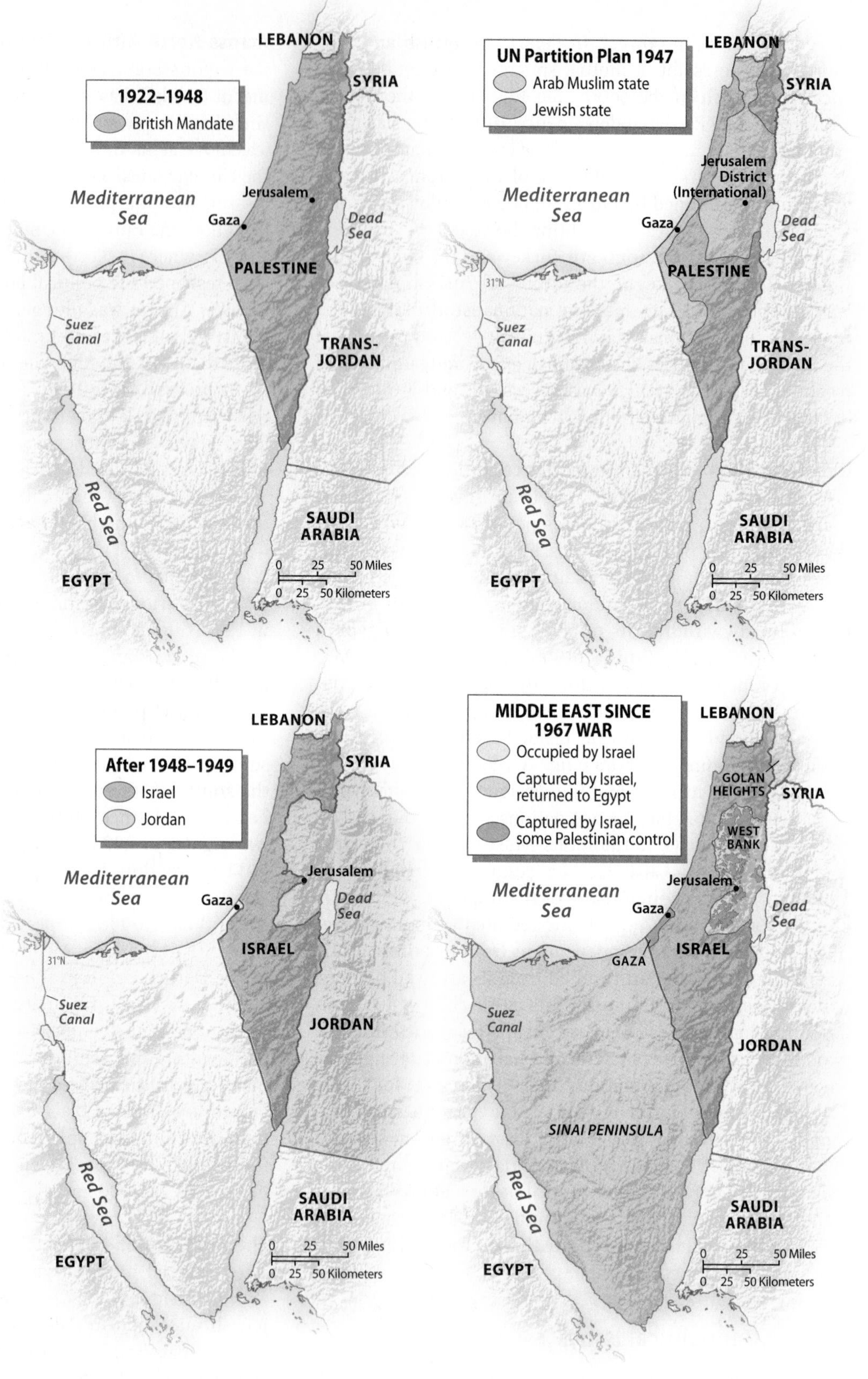

Figure 7.35 Evolution of Israel Modern Israel's complex evolution began with an earlier British colonial presence and a UN partition plan in the late 1940s. Thereafter, multiple wars with nearby Arab states produced Israeli territorial victories in settings such as Gaza, the West Bank, and the Golan Heights. Each of these regions continues to figure importantly in the country's recent relations with nearby states and with resident Palestinian populations. (Modified from Rubenstein, 2011, *An Introduction to Human Geography*, 10th ed., Upper Saddle River, NJ: Prentice Hall)

cluding the West Bank and the Gaza Strip, passed to Jordan and Egypt, respectively. Hundreds of thousands of Palestinian refugees fled from Israel to neighboring countries, where many of them remained in makeshift camps. Under these conditions, Palestinians nurtured the idea of creating their own state on land that had become Israel.

Israel's relations with neighboring countries remained poor. Supporters of Arab unity and Muslim solidarity sympathized with the Palestinians, while their antipathy toward Israel grew. Israel fought additional wars in 1956, 1967, and 1973. In territorial terms, the Six-Day War of 1967 was the most important conflict (see Figure 7.35). In this struggle

Figure 7.36 West Bank Portions of the West Bank were returned to Palestinian control in the 1990s, but Israel has partially reasserted its authority in some of these areas since 2000, citing the increased violence in the region. New Israeli settlements also are scattered through the West Bank in areas still under their nominal control. (Modified from Rubenstein, 2011, *An Introduction to Human Geography,* 10th ed., Upper Saddle River, NJ: Prentice Hall)

against Egypt, Syria, and Jordan, Israel occupied substantial new territories in the Sinai Peninsula, the Gaza Strip, the West Bank, and the Golan Heights. Israel annexed the eastern part of the formerly divided city of Jerusalem, arousing particular bitterness among the Palestinians, since Jerusalem is a sacred city in the Muslim tradition (it contains the Dome of the Rock and the holy Al-Aqsa Mosque) (see Figure 7.30). Jerusalem is sacred in Judaism as well (it contains the Temple Mount and the holy Western Wall of the ancient Jewish temple), and Israel remains adamant in its claims to the entire city. A peace treaty with Egypt resulted in the return of the Sinai Peninsula in 1982, but tensions focused on other occupied territories that remained under Israeli control. To strengthen its geopolitical claims, Israel also built additional Jewish settlements in the West Bank and in the Golan Heights, further angering Palestinian residents.

Palestinians and Israelis began to negotiate a settlement in the 1990s. Preliminary agreements called for a quasi-independent Palestinian state in the Gaza Strip and across much of the West Bank. A tentative agreement late in 1998 strengthened the potential control of the ruling **Palestinian Authority (PA)** in the Gaza Strip and portions of the West Bank (Figure 7.36). But a new cycle of heightened violence erupted late in 2000 as Palestinian attacks against Jews increased and as the Israelis continued with the construction of new settlements in occupied lands (especially in the West Bank) (Figure 7.37). In 2010, with about 500,000 Israeli settlers already living in these West Bank settlements, the Israeli government pledged to continue its support for new construction, including developments in and near Jerusalem. Palestinian authorities continue to strongly protest these new West Bank Jewish settlements.

Even more daunting has been construction of the Israeli security barrier, a partially completed series of concrete walls, electronic fences, trenches, and watchtowers designed to effectively separate the Israelis from Palestinians across

Figure 7.37 Jewish Settlement, West Bank The new houses and well-planned neighborhoods of the West Bank Jewish settlement of Eli (foreground) contrast with the older Palestinian settlement in the distance. Many Palestinians resent the construction of these controversial Israeli settlements in the West Bank region.

much of the West Bank region (Figure 7.38). Israeli supporters of the barrier (to be more than 400 miles long when completed) see it as the only way they can protect their citizens from suicide bombings and more terrorist attacks. Palestinians see it as a land grab, an "apartheid wall" designed to socially and economically isolate many of their settlements along the Israeli border.

Nearby, Israel withdrew troops and settlers from the Gaza Strip in 2005, but political instability thereafter resulted in repeated and violent Israeli incursions and reoccupations of the small, war-torn region. Today, Gaza remains a poor and isolated portion of the Middle East. Some of its most effective links to the outside world are a series of tunnels that connect it with settlements in nearby Egypt. But its hostile Palestinian leadership combined with Israeli fears of continued terrorist attacks suggest tensions in the area will remain very high.

The political fragmentation of the Palestinians has added further uncertainty. In 2006, control of the Palestinian government was split between the Fatah and Hamas political parties. Hamas has long been seen by many Israelis as an extremist and violent Palestinian political party, whereas the Fatah party has shown more of a willingness to work peacefully with the state of Israel. In 2007, the split between these rival Palestinian factions became violent, with Hamas gaining effective control of the PA within Gaza and with Fatah maintaining its greatest influence across the West Bank region. Further complicating this political geography is the emergent Islamist movement among some Palestinians. Often, Islamists have challenged the more secular orientation of both Hamas and Fatah. These growing complexities within the Palestinian movement make an eventual peace treaty with Israel seem even more difficult to achieve.

Instability also continues along Israel's northern border with Lebanon. In 2006, a Shiite militia group known as Hezbollah increased its rocket attacks into northern Israel, prompting an armed response by the Israelis. Along with Hamas, Hezbollah represents a radicalized Arab political element within the region that shows little inclination to negotiate with Israel (or vice versa).

One thing is certain: Geographical issues will remain at the center of the regional conflict. Palestinians hope for a land they can call their own, and Israelis continue their search for more secure borders that guarantee their political integrity in a region where they are surrounded by potentially hostile neighbors. Ultimately, the sacred geography of Jerusalem, a mere 220 acres of land within the Old City, stands at the center of the conflict (see Figure 7.30). Imaginative compromises in defining that political space will need to recognize its special value to both its Israeli and Palestinian residents.

Devastated Iraq Iraq is another multinational state born during the colonial era that has yet to escape the consequences of its geopolitical origins. When the country was carved out of the British Empire in 1932, it contained the cultural seeds of its later troubles decades before Saddam Hus-

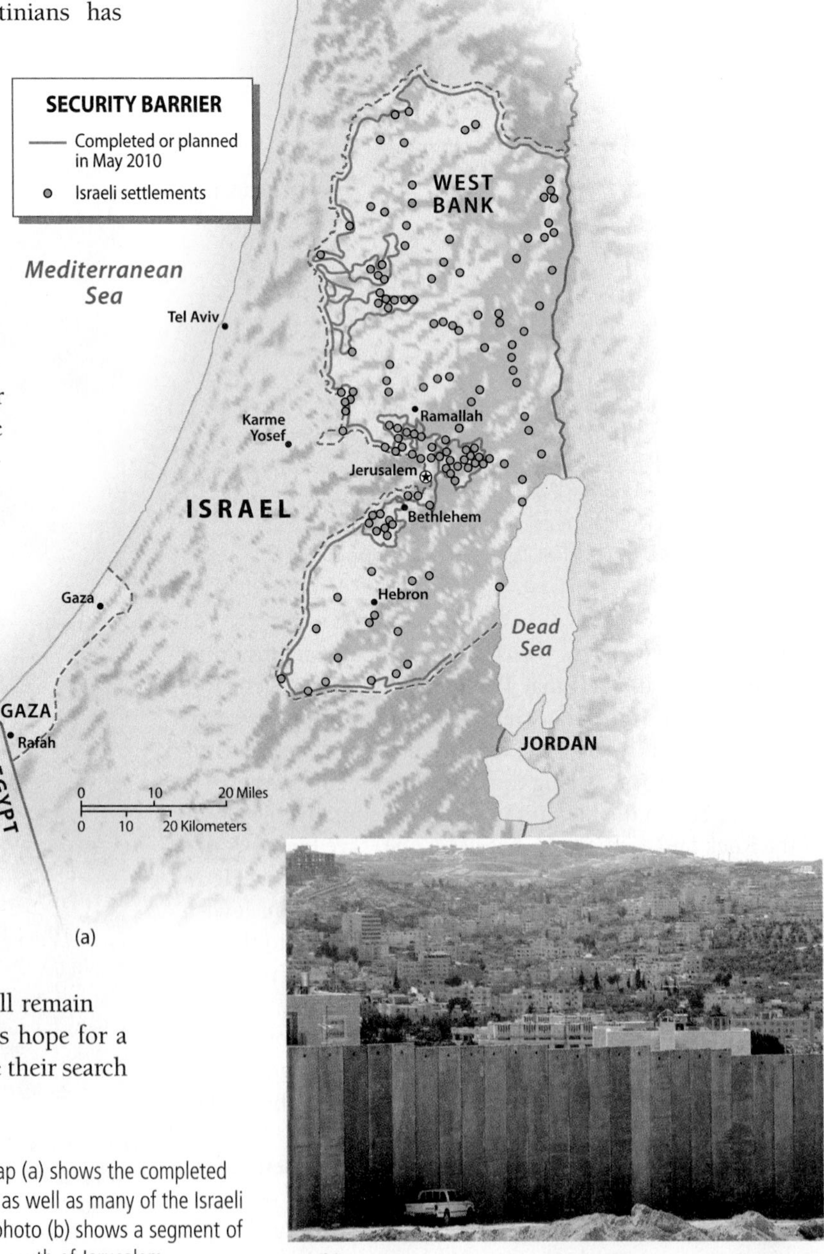

Figure 7.38 Israeli Security Barrier The map (a) shows the completed and planned portions of the Israeli security barrier as well as many of the Israeli settlements located in the West Bank region. The photo (b) shows a segment of the Israeli security barrier passing near Bethlehem, south of Jerusalem.

sein came to power. Iraq remains culturally complex today (Figure 7.39). Most of the country's Shiite population lives in the lower Tigris and Euphrates valleys near and south of Baghdad. Indeed, the region focused around the city of Basra contains some of the holiest Shiite shrines in the world. In northern Iraq, the culturally distinctive Kurds have their own ethnic identity and political aspirations. Many Kurds want complete independence from Baghdad, and they have managed to establish a federal region that already enjoys a great deal of autonomy from the central Iraqi government. A third major subregion is dominated by the Sunnis and includes part of the Baghdad area as well as a triangle of territory to the north and west that includes Sunni strongholds such as Falujah and Tikrit. After enjoying a great deal of power under Saddam Hussein, many Sunnis feel they may get the short end of any future power-sharing agreement within the country.

Led by Saddam Hussein, Iraq created a major source of instability in the region before he was removed in 2003. In 1980, Hussein invaded oil-rich but politically weakened Iran to gain a better foothold in the Persian Gulf. Eight years of bloody fighting resulted in a stalemate, and the conflict left Iraq's finances in disarray. Iraq then invaded and overran Kuwait in 1990, claiming it as an Iraqi province. A U.S.-led UN coalition, receiving substantial support from Saudi Arabia, expelled Iraq from Kuwait in early 1991. Twelve years later, the 2003 American-led invasion of the country replaced one set of uncertainties with another.

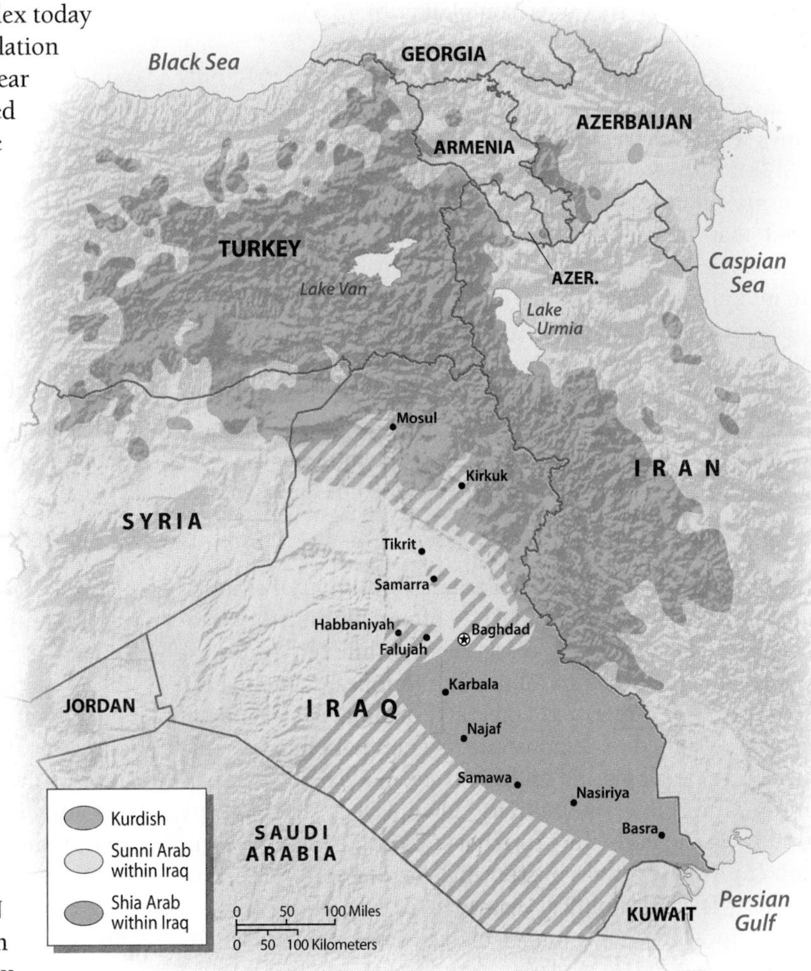

Figure 7.39 Multicultural Iraq Iraq's complex colonial origins produced a state with varying ethnic characteristics. Shiites dominate south of Baghdad, Sunnis hold sway in the western triangle zone, and Kurds are most numerous in the north near oil-rich Kirkuk and Mosul.

When Iraqi leaders assumed control of their new state on June 28, 2004, more than 135,000 American troops remained in the country. Thereafter, growing religious and ethnic violence between different Iraqi factions threw portions of the nation into civil war. Rival Sunni and Shiite groups forced many Iraqis from their communities. American troops within the country in 2007 and 2008 worked with Iraqi officials to reduce the level of violence. New parliamentary elections in 2010 confirmed the ongoing political uncertainties within Iraq. Both the leading Shiite and Sunni candidates proved about equally popular, largely splitting the vote along regional lines (Shiites winning in the south and narrowly in Baghdad; Sunnis winning in central and western provinces). Complicating matters further, the Kurdish Alliance remained dominant in many of the country's Kurdish areas. An enduring geopolitical solution for a stable Iraqi state has yet to take shape, but it will need to recognize these cultural differences that continue to divide the country, as well as allow for an ongoing revival of the country's oil industry. One challenge to the notion of shared political power and economic wealth is the uneven distribution of oil resources within the country. By far, the country's dominant oil fields are located in Shiite- and Kurdish-controlled portions of Iraq, an uneasy fact of resource geography that has long troubled the nation's sizable Sunni population.

Instability in the Arabian Peninsula Saudi Arabia, the region's greatest oil power, is led by a conservative monarchy (the Saud family) that has been unwilling to promote many democratic reforms. On the surface, it has supported U.S. efforts in the region to provide stable flows of petroleum, but beneath the surface certain elements of the regime may have financed radically anti-American groups such as Al Qaeda. The Saudi people themselves, largely Sunni Arabs, are torn between an allegiance to their royal family (and the economic stability it brings), the lure of a more democratic, open Saudi society, and an enduring distrust of foreigners, particularly Westerners. Furthermore, the Sunni majority includes many Wahhabi sect members, whose radical Islamist philosophy has fostered anti-American sentiment. Add to this a large number of foreign laborers and a persisting American military and economic presence within the country (one of the chief complaints of Al Qaeda leader Osama bin Laden) and you have a setting ripe for political instability. Economic declines between 2008 and 2010 only made matters worse as many Saudis saw fewer job opportunities within their energy-dependent economy.

In nearby Yemen, a growing rebel movement along the country's northwest border with Saudi Arabia has destabilized that rugged, isolated portion of the peninsula (see Figure 7.33). Since 2004, the area's Houthis, a group of Shiite militants, have protested the policies and dominance of the Yemeni government. Both Yemen and Saudi Arabia claim the Houthis have received extensive support from Iran and that the disputed area plays host to Iranian-sponsored terrorist training camps. The Saudi government has stepped up its own attacks on the Houthis across the ill-defined border zone and American experts have monitored the area with increased concern.

Tensions in Turkey Turkey continues to find itself strategically positioned between diverse, often contradictory geopolitical forces. Many pro-Westerners within Turkey, for example, are committed to joining the European Union (EU). To do so, the country has embarked on an active agenda of reforms designed to demonstrate its commitment to democracy. Press freedoms and multiparty elections have been permitted, the role of the military downplayed, and minority groups (particularly the Kurds) have gained greater recognition from the central government. At the same time, anti-Western Islamist political elements have been on the rise, linked to a growing number of terrorist bombings in the country since 2003. Moderate Turks are hoping that its road to the EU will not be interrupted with the violence, but Islamists and ultra-nationalists are suspicious of the EU pledge to move the country in a radically different direction. Turkey also has continuing tensions with nearby Greece and Cyprus, two countries that are now both in the EU. Final approval of Turkey's admission still remains clouded by all of these concerns.

Iranian Geopolitics Nearby Iran also garners a great deal of international attention. Islamic fundamentalism dramatically appeared on the political scene in 1978 as Shiite Muslim clerics overthrew the shah, an authoritarian, pro-Western ruler friendly to U.S. political and economic interests. The new leaders proclaimed an Islamic republic in which religious officials ruled both clerical and political affairs.

Today, Iran supports Shiite Islamist elements throughout the region (such as Hezbollah) and has repeatedly threatened the state of Israel. The country has also been seen as a growing threat by many moderate Arab states. In 2009, the nearby United Arab Emirates, for example, invited the French to open their first new military base in decades on foreign soil and French leaders pledged to defend against Iranian expansion within the region. Adding uncertainty has been Iran's ongoing nuclear development program, an initiative its government claims is solely related to the peaceful construction of power plants (see Figure 7.33). Many in the West, however, remain unconvinced of the government's motives and demand that its program be stopped or opened to more outside inspections before allowing for greater international cooperation with the Iranian regime. Israel, in particular, has suggested it may destroy these Iranian nuclear sites before Iranian leaders have an opportunity to attack Israel with a nuclear-tipped missile. In addition, hotly contested and controversial presidential elections in 2009 and further street demonstrations in 2010 reveal growing frustration with the status quo within the country, suggesting more internal political instability is on the horizon. Many of Iran's middle-class intellectuals, well-educated youth, and human-rights advocates are harsh critics of a government they feel is increasingly intolerant and guilty of isolating the country on the world stage.

An Uncertain Political Future

Few areas of the world pose more geopolitical questions than Southwest Asia and North Africa. Twenty years from now, the region's political map could look quite different from the one today. The region's strategic global importance increased greatly after World War II, propelled into the international spotlight by the creation of Israel, the tremendous growth in the world's petroleum economy, Cold War tensions between the United States and the Soviet Union, and, more recently, the rise of Islamic fundamentalism and Islamist political movements. Now that the Cold War between global superpowers has ended, the region is experiencing a geopolitical reorientation.

One key question that remains is what geopolitical role should the United States play in Southwest Asia? That simple question continues to confound policymakers in the United States as well as residents of that ever-troubled part of the world. The 2003 Iraqi invasion put large numbers of American troops in the region. Even as many of those troops gradually return home, the American role remains central to the region's geopolitical future. How does the United States identify and deal with hostile elements? Should the United States go it alone in these efforts ("unilateralism"), or depend on sharing the responsibilities with other countries or international organizations ("multilateralism")? Regional resistance movements directed against the United States also have grown more powerful in many countries within the region, drawing on growing popular suspicion of American motives and on the heated religious rhetoric of Islamist elements.

Four interrelated settings appear destined to hold continuing U.S. attention within the region. First, the U.S. involvement in Iraq remains important even as the country struggles to find its own, more independent political identity. Iraq's fragile political structure and its globally significant oil economy, all cast amid the country's complex ethnic and religious geography, seem likely to remain key concerns for American policymakers. A second question mark is oil-rich Saudi Arabia. Of particular concern are domestic challenges to the all-powerful al-Saud royal family. U.S. observers fret that a sudden change in Saudi Arabian politics would have global ramifications, particularly if oil exports are disrupted. Third, the growing geopolitical presence of Iran challenges American, Israeli, and many moderate Arab (particularly Saudi Arabian) interests within the region. Iran's regional aspirations and its always murky nuclear installations remain key question marks for this portion of the world. Finally, recent events in Israel, Gaza, and the West Bank continue to make global headlines. Palestinian infighting, continuing attacks on Israel, and Israeli reprisals (including the construction of its security barrier) continue to fracture and enflame the region. What is an appropriate role for the United States in tactically improving short-term relationships between Israel and its Palestinian populations and in crafting a longer-term strategic policy aimed at achieving a permanent peace in the region? How the United States responds to Southwest Asia's shifting geopolitics seems destined to shape the region for decades to come.

ECONOMIC AND SOCIAL DEVELOPMENT: Lands of Wealth and Poverty

Southwest Asia and North Africa is a region of both incredible wealth and disheartening poverty. While a few of its countries enjoy great prosperity, due mainly to rich reserves of petroleum and natural gas, other nations within the region are among the least developed in the world (Table 7.2). Overall, recent economic growth rates have lagged behind those of many other parts of the developing world. The global recession of 2008–2010 also hit many economies within the region hard. Persistent political instability also has contributed greatly to the regional economic malaise. These economic stumbling blocks have had profound social consequences: Investments in education, health care, and new employment opportunities have slowed considerably in many countries from the heady

TABLE 7.2 Development Indicators

Country	GNI per Capita, PPP (2008)	GDP Average Annual % Growth (2000–08)	Human Development Index (2007)[a]	Percent Population Living Below $2 a Day	Life Expectancy (2007)[b]	Under Age 5 Mortality Rate (1990)	Under Age 5 Mortality Rate (2005)	Gender Equity[c]
Algeria	7,880	4.3	0.754	23.6	72	64	41	102
Bahrain	33,400		0.895		75			
Egypt	5,470	4.7	0.703	18.4	72	90	23	
Gaza and West Bank		−0.9	0.737		72	40	23	
Iran	10,840	5.9	0.782	8.0	71	73	32	116
Iraq		−11.4			67	53	44	
Israel	27,450	3.5	0.935		81	11	5	101
Jordan	5,710	7.2	0.775	3.5	73	38	20	103
Kuwait	53,430	8.4	0.916		78	15	11	100
Lebanon	11,740	4.0	0.803		72	40	13	103
Libya	16,260	5.6	0.847		74	38	17	105
Morocco	4,180	5.0	0.654	14.0	71	88	36	88
Oman	22,150	4.0	0.846		72	31	12	99
Qatar		9.0	0.910		76	20	10	120
Saudi Arabia	24,490	4.1	0.843		76	43	21	91
Sudan	1,920	7.4	0.531		58	124	109	89
Syria	4,490	4.4	0.742		74	37	16	97
Tunisia	7,450	4.9	0.769	12.8	74	50	21	103
Turkey	13,420	5.7	0.806	8.2	72	84	22	89
United Arab Emirates		7.8	0.903		77	17	8	101
Western Sahara					60			
Yemen	2,220	3.9	0.575	46.6	63	127	69	

[a] *United Nations, Human Development Report, 2009.*

[b] *Population Reference Bureau, World Population Data Sheet, 2010.*

[c] *Gender Equity—Ratio of female to male enrollments in primary and secondary school. Numbers below 100 have more males in primary/secondary. school, numbers above 100 have more females in primary/secondary schools.*

Sources: World Bank, World Development Indicators, 2010.

gains of the late 1970s and 1980s. Petroleum will no doubt figure prominently into the region's future economic relationships with the rest of the world, but many countries in the area also have focused on increasing agricultural output, investing in new industries, and promoting tourism as important ways to broaden the regional economic base. In addition, development within the region will be shaped by its access to information and how well it is connected with the rest of the world. Cell phone and Internet use, for example, vary widely across North Africa and Southwest Asia (Figure 7.40). Lower penetration of cell phone technology in some settings (Egypt, Sudan, and Iran) suggests basic investments in infrastructure remain lacking, while areas of very high cell phone usage (Israel, many Persian Gulf states) are correlated with greater affluence as well as highly mobile, foreign-born populations. Internet use is also highly variable, ranging from 1 percent in Yemen to more than half the population of the United Arab Emirates. In some settings such as Iran (18 percent), official government resistance to the Internet plays a role, as well.

The Geography of Fossil Fuels

The striking global geographies of oil and natural gas reveal the region's persistent importance in the world oil economy, as well as the extremely uneven distribution of these resources within the region (Figure 7.41). Higher oil prices early in the new century have once again highlighted the economic clout of this region of the world. Saudi Arabia remains one of the major producers of petroleum in the world, and Iran, the United Arab Emirates, Libya, and Algeria also contribute significantly. The region plays an important though less dominant role in natural gas production. The distribution of fossil fuel reserves suggests that regional supplies will not be exhausted anytime soon. Overall, with only 7 percent of the world's population, the region holds a staggering 60 percent of the world's proven oil reserves. Saudi Arabia's pivotal position, both regionally and globally, is clear: Its 30 million residents live atop 20 percent of the planet's known oil supplies.

Several major geological zones supply much of the region's output of fossil fuels. The world's largest concentration

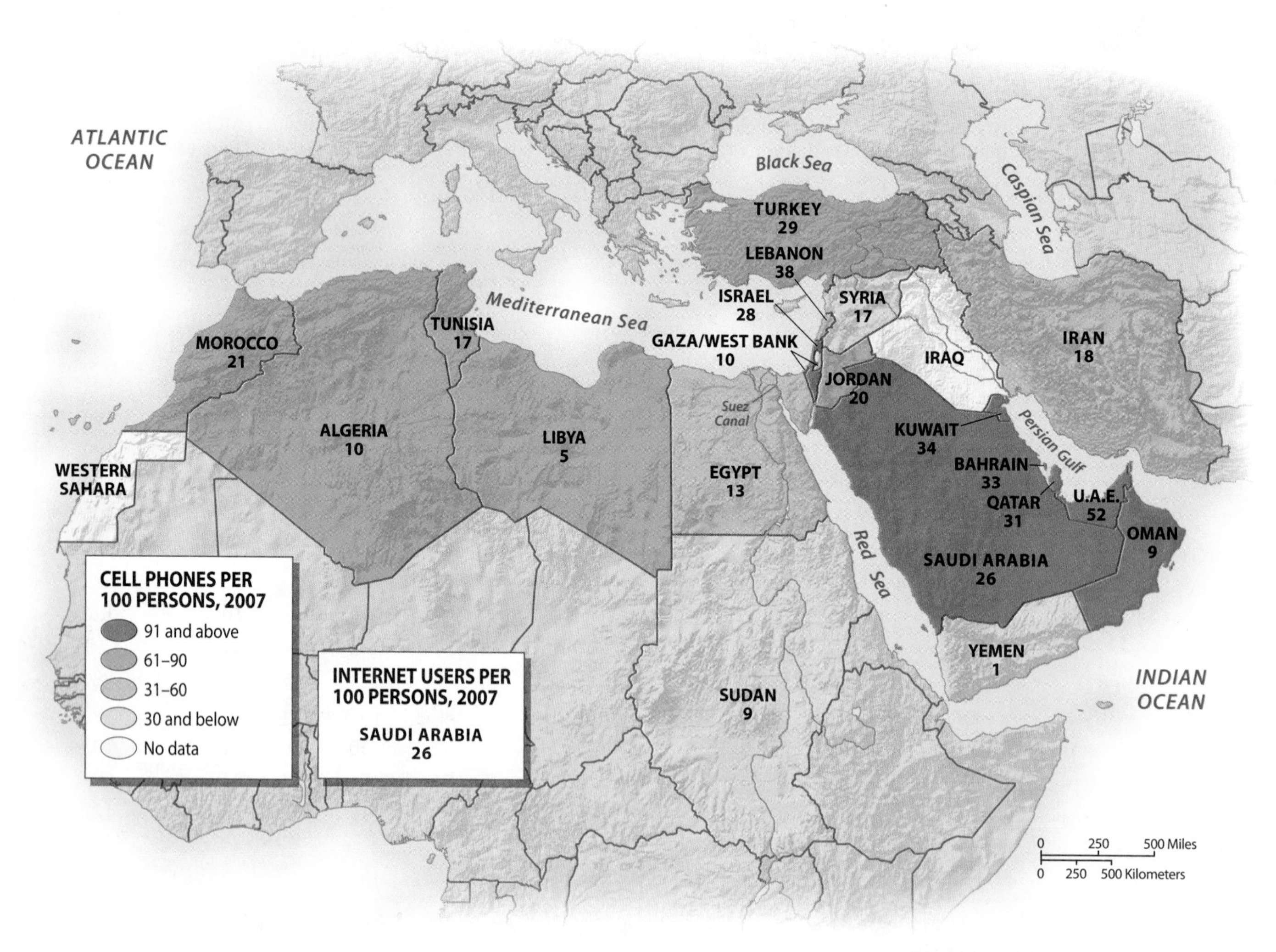

Figure 7.40 Development Issues: Cell Phone and Internet Use (per 100 in the population), 2007 The map illustrates how access to information technology can be used as a measure of development within the region. The oil-rich United Arab Emirates emerge as far more connected to global information flows than are Egypt, Yemen, or Sudan. (World Bank Development Indicators, 2007)

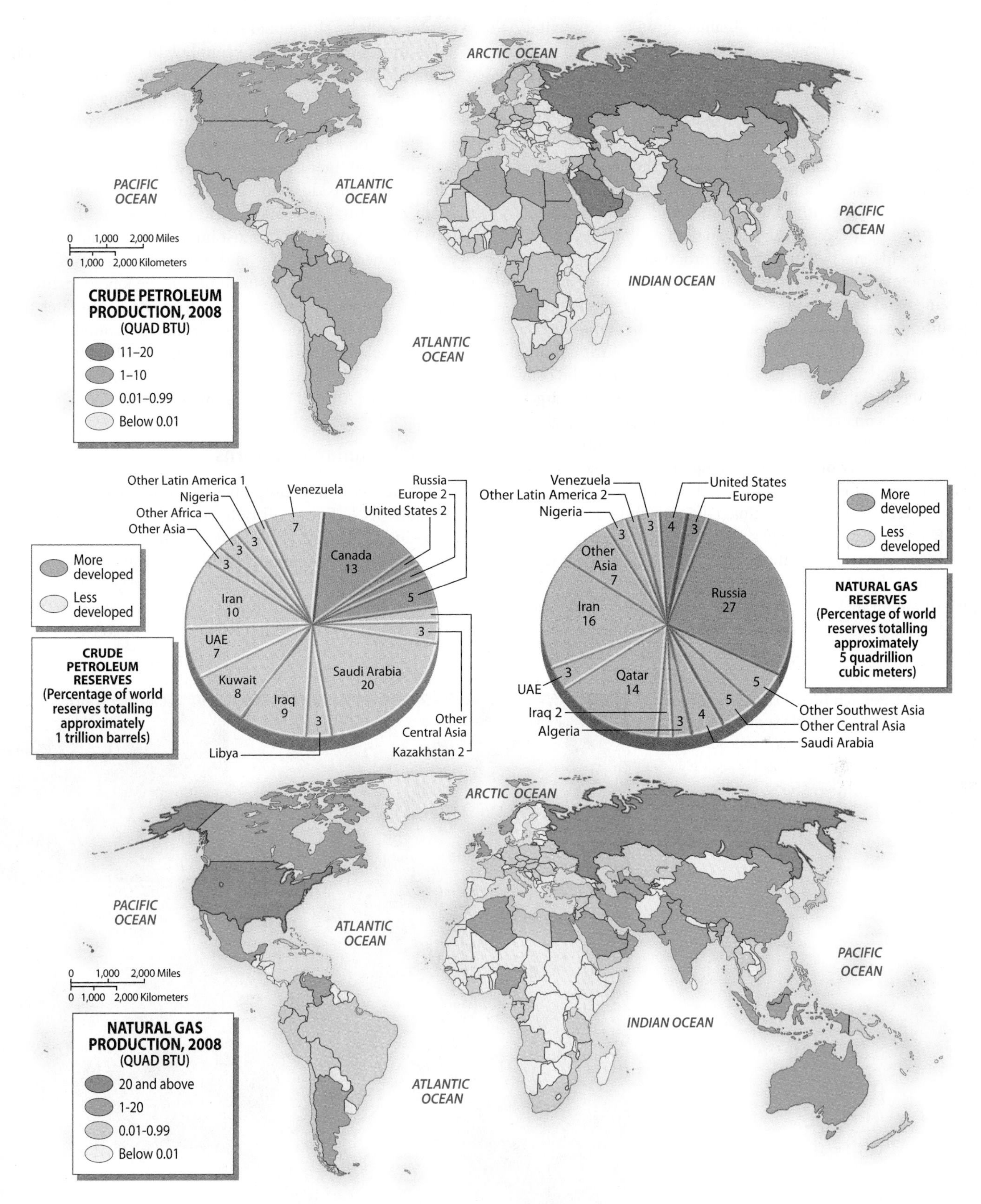

Figure 7.41 Crude Petroleum and Natural Gas Production and Reserves The region plays a central role in the global geography of fossil fuels. Abundant regional reserves suggest that the pattern will continue. (Modified from Rubenstein, 2011, *An Introduction to Human Geography,* 10th ed., Upper Saddle River, NJ: Prentice Hall)

of petroleum lies within the Arabian–Iranian sedimentary basin, a geological formation that extends from northern Iraq and western Iran to Oman and the lower Persian Gulf (Figure 7.42). All of the states bordering the Persian Gulf reap the benefits of oil and gas deposits within this geological basin, and it is not surprising that the world's densest concentration of OPEC members is found in the area. A second important zone of oil and gas deposits includes eastern Algeria, northern and central Libya, and scattered developments in northern Egypt. As in the Persian Gulf region, these North African fields are tied to regional processing points and to global petroleum markets by a complex series of oil and gas pipelines and by networks of technologically sophisticated oil shipping facilities. Third, a zone of growing importance is being developed in the African nation of Sudan, including areas near and south of Khartoum. Near Khartoum, the commercial center of Alsunut has become one of Africa's largest commercial construction sites. Home to regional oil giants Petrodar and the Greater Nile Petroleum Operating Company, the new city of Alsunut is being built on Sudan's petroleum riches and is financed by a global set of investors from the Gulf region, Malaysia, China, Pakistan, and elsewhere.

Even with all these riches, the geography of fossil fuels is strikingly uneven across the region. Some states—even those with tiny populations (Bahrain, Qatar, and Kuwait, for example)—contain incredible fossil fuel reserves, especially when considered on a per capita basis. Many other countries, however, and millions of regional inhabitants, reap relatively few benefits from the oil and gas economy. In North Africa, for example, Morocco possesses few developed petroleum reserves, and even the fruits of Sudan's blossoming wealth will likely remain very concentrated within a small segment of a large, poor population. In oil-rich Southwest Asia, the distribution of fossil fuels is also uneven: Israel, Jordan, and Lebanon all lie outside favored geological zones for either petroleum or natural gas. While Turkey has some developed fields in the far southeast, it must import substantial supplies to meet the needs of its large and industrializing population.

The region's abundant sunshine may be key to its future energy production in the centuries to come. Long after the last drop of oil has been pumped from the ground, the region's vast solar potential will be critical to the world's energy budget (see "Exploring Global Connections: Sunny Side Up: North Africa's Desertec Initiative").

Regional Economic Patterns

Remarkable economic differences characterize the region (see Table 7.2). Some oil-rich countries have prospered greatly since the early 1970s, but in many cases fluctuating oil prices, political disruptions, and rapidly growing populations have reduced prospects for economic growth. Other nations, although poor in oil and gas reserves, have seen brighter prospects through moves toward greater economic

Figure 7.42 Persian Gulf This satellite view of the Persian Gulf reveals one of the world's richest sources of petroleum. Sedimentary rocks, both on land and offshore, contain additional reserves that can sustain production for decades.

diversification. Finally, some countries in the region are subject to persistent poverty, where rapid population growth and the basic challenges of economic development combine with political instability to produce very low standards of living.

Higher-Income Oil Exporters The richest countries of Southwest Asia and North Africa owe their wealth to massive oil reserves. Nations such as Saudi Arabia, Kuwait, Qatar, Bahrain, and the United Arab Emirates benefit from fossil fuel production, as well as from their relatively small populations. Large investments in transportation networks, urban centers, and in other petroleum-related industries have reshaped the cultural landscape. Billions of dollars have also poured into new schools, medical facilities, low-cost housing, and modernized agriculture, significantly raising the standard of living in the past 40 years.

The Saudi petroleum-processing and shipping centers of Jubail (on the Persian Gulf) and Yanbu (on the Red Sea) are examples of this commitment to expand the region's economic base beyond the simple extraction of crude oil. Other oil-rich cities have cultivated an even more cosmopolitan image, with globalized Dubai in the United Arab Emirates appealing both to international tourists and business interests (see "Cityscapes: Dubai's Rocky Road to Urbanization").

Still, problems remain. Dependence on oil and gas revenues produces economic pain in times of falling prices, such as those seen in the recent economic downturn (2008–2010). Such fluctuations in world oil markets will inevitably continue in the future, disrupting construction projects, producing large layoffs of immigrant populations, and slowing investment in the region's economic and social infrastructure. In addition, countries such as Bahrain and Oman are faced with the problem of rapidly depleting their reserves over the next 20 to 30 years.

Lower-Income Oil Exporters Other states in the region are important secondary players in the oil trade, but different political and economic variables have hampered sustained economic growth. In North Africa for example, Algerian oil and natural gas overwhelmingly dominate its exports, but the past 15 years have also brought political instability and increasing shortages of consumer goods. While the country contains some excellent agricultural lands in the north, the overall amount of arable land has increased little over the past 25 years, even as the country's population has grown by more than 50 percent.

In Southwest Asia, Iraq faces huge economic and political challenges. The war crippled much of Iraq's already deteriorated infrastructure, and continuing political instability has made the task of rebuilding its economy more difficult. Iraq suffers from extremely high unemployment, more than 20 percent of the population remains malnourished, and only 25 percent of the country is served by dependable electricity. The country's economic leaders do have an ambitious plan for increasing oil output, and in 2009 the Iraqi cabinet endorsed a plan to reorganize its domestic oil industry under a single national oil company.

Figure 7.43 Israeli High-Tech Industry These Israeli workers are employees of the Telrad factory, a high-tech division of the Koor conglomerate.

The situation in Iran also is challenging. The country is large and populous, and has a relatively diverse economy. Iran's oil and gas reserves are huge and have seen active commercial development since 1912. The country also has a sizable industrial base, much of it built in the 20 years prior to the fundamentalist revolution of 1979. But today Iran is relatively poor, burdened with a stagnating if not declining standard of living. Since 1980, the country's fundamentalist leaders have downplayed the role of international trade in consumer goods and services, fearing they would import unwanted cultural influences from abroad. There are economic bright spots, as well: The country benefits from energy developments in Central Asia. In addition, Iran's literacy rate (particularly for women) has risen, reflecting a new emphasis on rural education in the country. Political differences with potential Western trading partners are vast, however, and Iran's recent nuclear development program isolated the country from many nations, including the United States.

Prospering Without Oil Some countries, while lacking petroleum resources, have nevertheless found paths to increasing economic prosperity. Israel, for example, supports one of the highest standards of living in the region, even with its political challenges (see Table 7.2). The Israelis and many foreigners have invested large amounts of capital to create a highly productive agricultural and industrial base. The country also is emerging as a global center for high-tech computer and telecommunications products. The country is known for its fast-paced and highly entrepreneurial business culture, which resembles California's Silicon Valley (Figure 7.43). Israel also has daunting economic problems. Its persistent struggles with the Palestinians and with neighboring states have sapped much of its potential vitality. Defense spending absorbs a large share of total gross national

EXPLORING GLOBAL CONNECTIONS Sunny Side Up: North Africa's Desertec Initiative

One of the planet's most ambitious renewable energy schemes, coordinated by the Desertec Foundation, is taking shape across North Africa and the Middle East. If successful, it will supply nearby portions of western and central Europe with at least 15 percent of its energy needs by 2050. While the initiative also includes wind, biomass, geothermal, and hydroelectric components, the key source in the green energy network will be thousands of concentrating solar thermal power (CSP) collectors that will be built across vast portions of North Africa and southwest Asia (Figure 7.5.1). The parabolic mirrors concentrate the sun's energy to heat a liquid that can be used to power steam turbines to generate safe and clean electricity. Once the electricity is produced, a sprawling network of long-distance transmission cables will take the energy to local sources of consumption within the region as well as to distant European users.

One plan calls for a major axis of collecting panels to extend through Morocco with other portions of the network gathering energy from Algeria, Tunisia, Libya, and Egypt (Figure 7.5.2). Eventually, the network might be extended further into the Sahara Desert as well as across the Arabian Peninsula and other portions of the Middle East. The transmission network will include huge underwater connections across the bottom of the Mediterranean Sea. Once in Europe, the electricity will be targeted to consumers in settings such as Germany, Spain, Italy, France, and Great Britain.

Backers of the proposal formed the Desertec Industrial Initiative in 2009 and include help from the German government as well as a growing collection of European and North American renewable energy companies. Both public and private utility interests in settings such as Morocco, Tunisia, and Israel are also keen to see their por-

Figure 7.5.1 Solar Energy Collectors If the Desertec project is successful, North Africa's sun-drenched deserts will be populated by thousands of cylindrical parabolic solar collectors, such as these in southern Spain.

income, necessitating high tax rates. Poverty among the Palestinians is also widespread, and the gap between rich and poor within the country has widened considerably.

Turkey also has a diversified economy even though its per capita income is modest by regional standards. Lacking petroleum, Turkey produces varied agricultural and industrial goods for export. About 35 percent of the population remains employed in agriculture, and the country's principal commercial products include cotton, tobacco, wheat, and fruit. The industrial economy has grown since 1980, including exports of textiles, food, and chemicals. Turkey remains an important tourist destination in the region, as well, attracting more than 6 million visitors annually in recent years. Some Turkish leaders hope to bolster the economy by eventually joining the EU.

Regional Patterns of Poverty Poorer countries of the region share the problems of much of the less-developed world. For example, Sudan, Egypt, and Yemen each face unique economic challenges. For Sudan, continuing political problems have stood in the way of progress. Civil war has resulted in major food shortages. The country's transportation and communications systems have seen little new investment and secondary school enrollments remain very low. On the other hand, Sudan's fertile soils could support more farming, and its new oil production suggests petroleum's expanding role in the economy. However, the country's sustained economic development appears delayed by continuing political instability.

Egypt's economic prospects are also unclear. On the one hand, the country experienced real economic growth during

tion of the world connected to Europe in this new way. Germany's Deutsche Bank is helping to coordinate the $400 billion investment that will be necessary to fund the first phase of the project. Pilot projects are already underway and larger flows of solar-generated electricity are planned by 2015. Renewable energy enthusiasts argue the technology will soon be available to make every part of the plan feasible, even the challenging task of economically transmitting electricity across long distances. Ironically, if they are right, the region's greatest contribution to the planet's long-term energy needs may have nothing to do with fossil fuels.

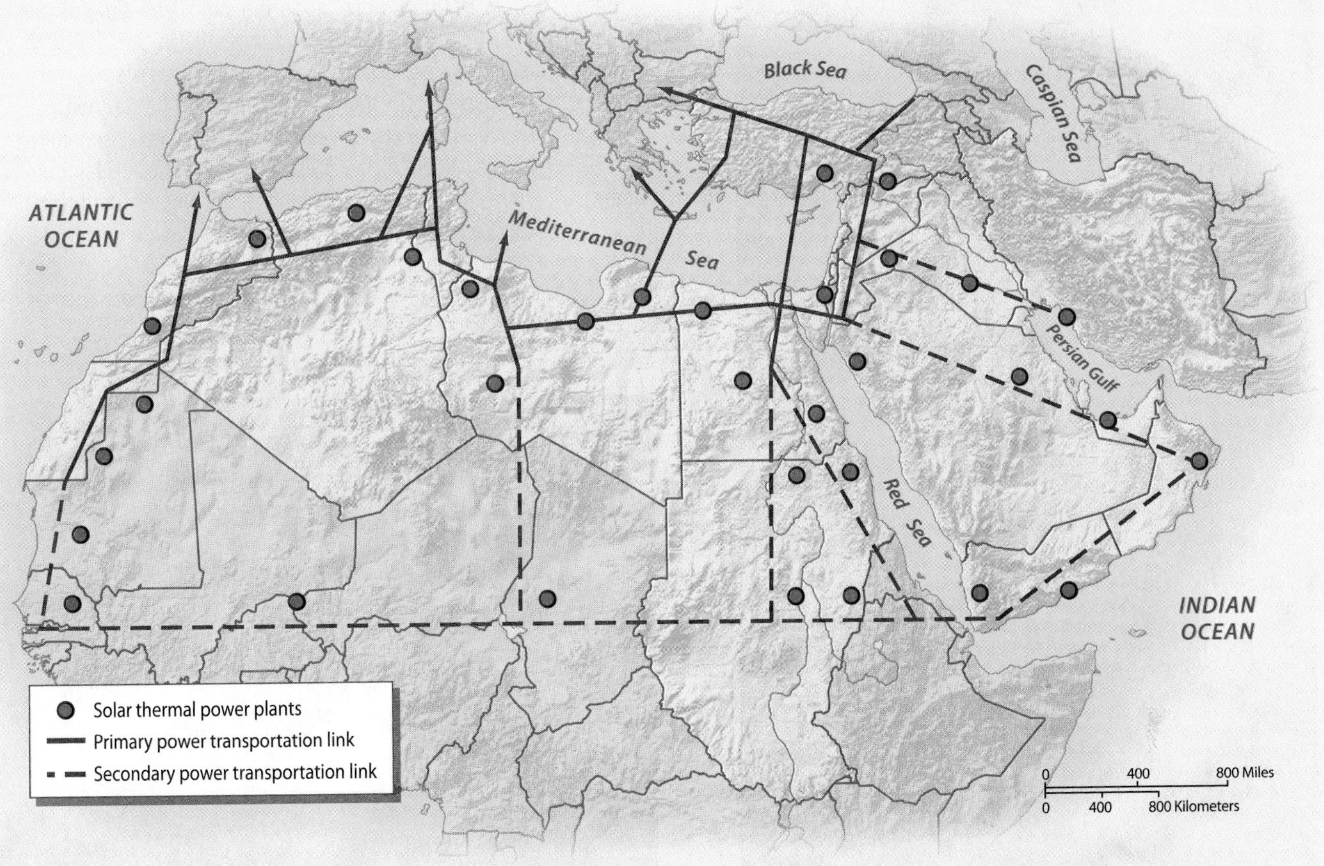

Figure 7.5.2 Desertec Solar Energy Project This map of North Africa and nearby portions of Europe suggests one planned scenario by which desert solar energy can contribute to Europe's future energy budget.

Figure 7.44 Cairo Slums One distinctive group of Cairo's population is the Zabbaleen, a community of Coptic Christians who collect the city's trash and live along the narrow streets of Manshiet Nasser, one of the city's most poverty-stricken neighborhoods.

the 1990s as President Hosni Mubarak pushed for more foreign investment, smaller government deficits, and a multibillion-dollar privatization program to put government-controlled assets under more efficient management. Even so, many Egyptians live in poverty, and the gap between rich and poor continues to widen (Figure 7.44). Illiteracy is widespread, and the country suffers from the **brain drain** phenomenon as some of its brightest young people leave for better jobs in western Europe or the United States. Egypt's 80 million people already make it the region's most heavily

CITYSCAPES Dubai's Rocky Road to Urbanization

It was a somewhat tarnished celebration. Early in 2010, the Burj Dubai was officially completed in the United Arab Emirates. The tallest free-standing structure in the world at over 2,625 feet (800m), the 160-story building dominates the urban skyline of a city that has struggled mightily in the global economic downturn between 2008 and 2010 (Figure 7.6.1). After booming for years, the city witnessed a spectacular crash in real estate values. Construction projects folded, workers exited the city as fast as they had arrived, and the entire regional economy teetered, awash in a debt crisis that threatened the very stability of regional banks and governments.

Before the crash, Dubai's rise along the sandy shoreline of the Persian Gulf had been nothing short of spectacular. Studded with eye-popping skyscrapers and famous for its fabricated islands in the shapes of palm trees and the world map, Dubai got its start with oil dollars (Figure 7.6.2). The city became a major financial and commercial hub of the Middle East. Its leaders, the ruling al-Maktoum family, still aspire to make it a global city that competes with London, Tokyo, New York, and Hong Kong. In less than a lifetime, Dubai grew from a small fishing village of 40,000 in 1960, to a metropolis of 1.4 million residents today. Clustered developments such as

Figure 7.6.1 Burj Dubai, Dubai, United Arab Emirates Dubai's soaring Burj Khalifa, completed in 2010, became the world's tallest free-standing structure, but it towered over a city still struggling with profound economic challenges.

Figure 7.6.2 Dubai's Offshore Real Estate Dubai's island-studded shoreline includes the ambitious "World" real estate venture, but many of the archipelago's pricey properties (which resemble an outline of the world's continents from above) remain vacant, a testimony to the deep global economic downturn that hit this region hard between 2008 and 2010.

populated state, and recent efforts to expand the nation's farmland have met with numerous environmental, economic, and political problems.

In Southwest Asia, Yemen remains the poorest country on the Arabian Peninsula. Positioned far from most of the region's principal oil fields, Yemen's low per capita GNI puts it on par with many nations in impoverished Sub-Saharan Africa. The largely rural country relies mostly on marginally productive subsistence agriculture, and much of its mountain and desert interior lacks effective links to the outside world. The present state emerged in 1990 with the political union of North and South Yemen, but political instability also remains a fact of life for many within the country. Coffee, cotton, and fruits are commercial agricultural products and modest oil exports bring in needed foreign currency. Overall, however, high unemployment and marginal subsistence farming remain widespread across the country.

Unique problems afflict the Palestinian populations of Gaza and the West Bank. Continued declines have devastated the economy as political disruptions discourage investment and conflicts with the Israelis destroy infrastructure. In many

Figure 7.6.3 Landscapes of Luxury, Dubai, United Arab Emirates The canals of the elite Madinat Jumeirah Arabian Resort are visible in the foreground. In the background, the telltale sail of the luxurious Burj Al Arab Hotel is one of the most dramatic visual symbols of affluence in the region. The hotel bills itself as the world's first seven-star hotel.

Internet City, Media City, Healthcare City (partnered with Harvard University), and the Dubai International Financial Center blossomed to serve 21st-century needs of the Middle East.

Millions of tourists from the Middle East, Russia, and Europe still pour into the city. The Burj al Arab, a seven-star hotel in the shape of a sail, serves a jet-set elite from Moscow to Riyadh (Figure 7.6.3). International tourists, glitzy sporting events, and major corporation headquarters (including Haliburton) find this Arab boomtown irresistible. In 2005, Dubai added indoor skiing to the Mall of the Emirates (the largest mall in the Middle East, naturally). City billboards proclaim, in both English and Arabic, "If you can dream it, it can happen."

But Dubai's rise and subsequent stumble are emblematic of the integrating yet uneven forces of globalization. Oil and ambition created the city and it also benefited indirectly from post–September 11 geopolitics. Arab investors flush with cash began to invest more in the Middle East—Dubai was ready and waiting. In addition, Dubai promoters attracted global investors who were keen to get in on the appeal of ever-rising real estate prices. When the tide turned in 2008, Dubai investors and government-controlled companies had run up a debt of over $80 billion. The resulting downturn has meant unfinished golf courses, half-empty office buildings, and an exit of thousands of foreign workers, many of them from South Asia.

Dubai's promoters argue persuasively that the city is down, but not out. Recovering oil prices have improved the region's economic prospects. Investment capital is once again flowing into the city. Immigrant workers are returning. Longer term, the city's exclusive offshore island real estate market may bloom again. In the process, Dubai is creating one of the world's most globalized cities, a place where four out of five residents are foreign-born, where wealthy people from around the world come to shop and play, and where the price of oil continues to shape the fortunes of a landscape that reflects both the best and worst of an unpredictable global economy.

cases, when Israeli forces suspect political dissidents and terrorist elements in a given Palestinian town or urban neighborhood, they destroy the settlement, leveling houses and shops in the process. Poverty grips two-thirds of the Palestinian population, unemployment hovers above 40 percent, and the ongoing construction of the Israeli security barrier promises to further disrupt the Palestinian economy in the years to come. The isolated Gaza Strip has been particularly hard hit, although local Hamas leaders have proven effective in helping to organize the informal economy and providing basic social services.

A Woman's Changing World

The role of women in the largely Islamic region also remains a major social issue. Female labor participation rates in the workforce are among the lowest in the world, and large gaps typically exist between male and female literacy (see Table 7.2). In the most conservative parts of the region, few women are allowed to work outside the home. Even in parts of Turkey, where Western influences are widespread, it is rare to see rural women selling goods in the marketplace or driving cars in the street. More orthodox Islamic states impose legal restrictions on the activities of women. In Saudi Arabia, for example, women are not allowed to drive, although a growing flurry of recent protests (including highly publicized convoys of renegade women drivers) may lead to lifting that limitation. In Iran, full veiling remains mandatory in more conservative parts of the country and women are restricted from serving as judges and as political leaders. Generally, Islamic women lead more private lives than men: Much of their domestic space is shielded from the world by walls and shuttered windows, and their public appearances are filtered through the use of the face veil or chador (full-body veil).

Yet in some places women's roles are changing, even within the norms of more conservative, Islamic societies. Many young Algerian women demonstrate the pattern (Figure 7.45). Most studies suggest the younger generation are more religious than their parents and are more likely to cover their heads and drape their bodies with traditional religious clothing. At the same time, they are more likely to be educated and employed than ever before. Today, 70 percent of Algeria's lawyers and 60 percent of its judges are women. A majority of university students are women and women dominate the health-care field. These new social and economic roles also help explain why the birthrates in such settings are declining and are likely to continue doing so. The Algerian case also demonstrates how complex social processes actually unfold, rarely following simple models for how "traditional" societies might evolve.

Women's roles are shifting elsewhere in the region. In Sudan and Saudi Arabia, a growing number of women pursue high-level careers. Education may be segregated, but it is available. Libya's Qaddafi has singled out the modernization of women as a high priority, and today more women than men graduate from the country's university system. In the Western Sahara, Saharawi women play a leading role in the country's political fight for independence from Morocco and their broader educational backgrounds and social freedoms (including the right to divorce their husbands) further separate them from Moroccan women. Women also have a more visible social position in Israel, except in fundamentalist Jewish communities where conservative social customs require more traditional domestic roles.

Global Economic Relationships

Southwest Asia and North Africa share close economic ties with the world. While oil and gas remain critical commodities that dominate international economic linkages, the growth of manufacturing and tourism are also redefining the region's role in the world.

OPEC's Changing Fortunes While OPEC no longer controls oil and gas prices globally, it still influences the cost and availability of these pivotal products within the developed and less-developed worlds. Western Europe, the United States, Japan, China, and many less industrialized countries depend on the region's fossil fuels. In the case of Saudi Arabia, for example, petroleum and its related products make up more than 90 percent of its exports. One recent trend evident in oil-producing countries (such as Saudi Arabia) is their willingness to form partnerships with foreign corporations, accelerating the economic integration of the region with the rest of the world. Even so, the region's major energy producers will be particularly vulnerable to global-scale recessions, such as the downturn which hit the region hard between 2008 and 2010.

Beyond key OPEC producers, other economic activities have also contributed to global economic integration. Turkey, for example, ships textiles, food products, and manufactured goods to its principal trading partners: Germany, the United

Figure 7.45 Algerian Women Even within the norms of a more conservative Islamic society, Algerian women are increasingly visible and highly productive contributors to that nation's workforce.

States, Italy, France, and Russia. Tunisia sends more than 60 percent of its exports (mostly clothing, food products, and petroleum) to nearby France and Italy. Israeli exports emphasize the country's highly skilled workforce: Products such as cut diamonds, electronics, and machinery parts are exported to the United States, western Europe, and Japan.

Regional and International Linkages Future interconnections with the global economy may depend increasingly on cooperative economic initiatives far beyond OPEC. Relations with the EU are critical. Since 1996, Turkey has enjoyed closer economic ties with the EU, but recent attempts at full membership in the organization have failed. Other so-called Euro-Med agreements also have been signed between the EU and countries across North Africa and Southwest Asia that border the Mediterranean Sea.

Most Arab countries, however, are wary of too much European dominance. They formed a regional political organization known as the Arab League in 1945. Seventeen league members established the Greater Arab Free-Trade Area (GAFTA) in 2005, designed to eliminate all intraregional trade barriers and spur economic cooperation. In addition, Saudi Arabia plays a pivotal role in regional economic development through organizations such as the Islamic Development Bank and the Arab Fund for Economic and Social Development.

The Geography of Tourism Tourists are another link to the global economy. Traditional magnets such as ancient historical sites and globally significant religious areas (for a multitude of faiths) draw millions of visitors annually. As the developed world becomes wealthier, there is also a growing global demand for recreational spots that can offer beaches, sunshine, and novel entertainment. Indeed, many miles of the Mediterranean, Black, and Red Sea coastlines are now lined with the upscale, but often ticky-tacky, landscapes of resort hotels and condominiums dedicated to serving travelers' needs. More adventurous travelers seek ecotourist activities such as snorkeling in Naama Bay on Egypt's Sinai Coast or four-wheeling among the Berbers in the Moroccan backcountry. Endangered wildlife also beckons photographers and poachers hoping to catch a glimpse of a South Arabian grey wolf, Nubian ibex, or a darting Persian squirrel.

All of this activity means big business to many countries in the region, and the economic impacts of tourism seem likely to grow during the 21st century. Tourism already is a huge part of the regional economy in settings such as Turkey, Israel, and Egypt. Elsewhere, smaller inflows of tourist dollars still have an extraordinary impact on the economies of, for example, northern Morocco and Tunisia, Lebanon (near Beirut), and the holy Saudi Arabian cities of Makkah and Medinah. In some cases, tourism remains much less developed, but the small existing inflows in such places as Libya and Jordan may be the beginning of a much larger economic trend in future years, depending on the region's political stability.

The growth of tourism, while further connecting the region to the global economy, has come at a considerable price. In addition to the visual blight of high-rise hotels, real environmental damage is increasingly the product of the growing human presence in many fragile regional settings. Various archeological and sacred sites have been negatively impacted across the region. More broadly, the economic realities of the tourist economy have produced a local underclass of poorly paid service workers, a growing sense of social distance between the haves and have-nots in tourist settings, and a consumption-oriented lifestyle that contrasts sharply with many traditional ways of life. Still, these changes appear inevitable for a region that, even in uncertain political times, seems to irresistibly attract tourists from every corner of the globe.

Summary

- Positioned at the meeting ground of Earth's largest land masses, the Southwest Asia and North Africa region historically has played a central role in world history and in processes of globalization. In ancient times, the region's inhabitants were early contributors to the agricultural revolution, a long process of plant and animal domestication destined to reshape world cultures. The region also served as a birthplace for some of the world's earliest urban civilizations, offering in the process a fundamentally new way of human settlement that continues to reshape the distribution of global populations today. Three of the world's great religions—Judaism, Christianity, and Islam—emerged beneath its desert skies.

- Many nations within the region suffer from significant environmental challenges. Twentieth-century population growth across the region was dramatic. At the same time, it has been difficult and costly to expand the region's limited supplies of agricultural land and water resources. The results, apparent from the eroded soils of the Atlas Mountains to overworked garden plots along the Nile, are a classic illustration of the environmental price paid when population growth outstrips the ability of the land to support it.

- Culturally, the region remains the hearth of Christianity, the spatial and spiritual core of Islam, and the political and territorial salvation of modern Judaism. Muslims worldwide are influenced by its cultural and political evolution, and the Arab–Israeli conflict will continue to be a critical element in global geopolitics.

- Political conflicts have disrupted economic development across the region. Civil wars, conflicts between states, and regional tensions have worked against initiatives for greater cooperation and trade. Perhaps most important, the region must deal with the conflict between modernity and more fundamentalist interpretations of Islam. One thing is certain: Future cultural change will be guided by a complex response to Western influences.

- Abundant reserves of oil and natural gas, coupled with the global economy's reliance on fossil fuels, dictate that the region will remain prominent in world petroleum markets. Also likely are moves toward economic diversification and integration, which will gradually draw the region closer to Europe and other participants in the global economy.

Key Terms

brain drain (*page 333*)
culture hearth (*page 289*)
domestication (*page 300*)
exotic rivers (*page 303*)
Fertile Crescent (*page 301*)
fossil water (*page 293*)
Hajj (*page 310*)
hydropolitics (*page 294*)
Islamic fundamentalism (*page 290*)
Islamism (*page 290*)
kibbutzim (*page 304*)
Levant (*page 295*)
Maghreb (*page 295*)
medina (*page 306*)
monotheism (*page 309*)
Organization of the Petroleum Exporting Countries (OPEC) (*page 290*)
Ottoman Empire (*page 311*)
Palestinian Authority (PA) (*page 323*)
pastoral nomadism (*page 314*)
physiological density (*page 299*)
protectorates (*page 320*)
qanat system (*page 293*)
Quran (*page 310*)
Shiites (*page 311*)
Suez Canal (*page 320*)
Sunnis (*page 299*)
theocratic state (*page 311*)
transhumance (*page 301*)

Review Questions

1. Why do Southwest Asia and North Africa form a useful world region? What are some of the problems associated with defining the region?
2. Describe the climatic changes you might experience as you traveled from the eastern Mediterranean coast to the highlands of Yemen. What are some of the key climatic variables that explain these variations?
3. Discuss five important human modifications of the Southwest Asian and North African environment, and assess whether these changes have benefited the region.
4. Discuss how pastoral nomadism, oasis agriculture, and dryland wheat farming represent distinctive adaptations to the regional environments of Southwest Asia and North Africa. How do these rural lifeways create distinctive patterns of settlement?
5. Compare the modern maps of religion and language for the region, and identify three major examples where Islam dominates non-Arabic-speaking areas. Explain why that is the case.
6. Describe the role played by the French and British in shaping the modern political map of Southwest Asia and North Africa. Provide specific examples of their lasting legacy.
7. Define Islamic fundamentalism and Islamism. Outline three regional examples where Islamic fundamentalism or Islamism redefined the domestic geopolitical setting since the late 1970s.
8. Explain how ethnic differences have shaped Iraq's political conflicts in the past 50 years.
9. Describe the basic geography of oil reserves across the region, and compare the pattern with the geography of natural gas reserves.
10. What strategies for economic development have recently been employed by nations such as Turkey, Israel, Egypt, and Morocco? How successful have they been, and how do they relate to the theme of globalization?

Thinking Geographically

1. How might a major project for transferring water from Turkey to the Arabian Peninsula affect the development of Saudi Arabia? What would be some of the potential political and ecological ramifications of such a project?
2. Why are birthrates declining in selected countries within the region? Despite the cultural differences with North America, what common processes seem to be at work in both regions that have contributed to this demographic transition?
3. What economic changes could occur if Israel and the Palestinians were to reach a lasting peace? What kinds of general connections might be found between political conflict and economic conditions throughout the region?
4. Are relations between the region's Muslim societies and the United States likely to improve or worsen? What factors would you cite to defend your answer?
5. What might be some of the reasons for Southwest Asia and North Africa's general failure to match the rates of economic and industrial growth found in China or India?
6. Why has the idea of Arab nationalism failed to achieve any lasting geopolitical changes in the region?
7. Imagine you are a ruler of a conservative, Islamist Arab state. What might be the advantages and disadvantages of opening up your country to the Internet?

Regional Novels and Films

Novels

Shmuel Agnon, *Shira* (1989, Schocken)

Hanan Al-Shaykh, *Beirut Blues: A Novel* (1995, Doubleday)

Murid Barghuthi, *I Saw Ramallah* (2003, Doubleday)

Simin Daneshvar, *Savushun: A Novel About Modern Iran* (1990, Mage)

Kahlil Gibran, *Broken Wings: A Novel* (1957, Citadel)

Barbara Hodgson, *The Tattooed Map* (1995, Chronicle Books)

Yasar Kemal, *Salman the Solitary* (1998, Harvill)

Naguib Mahfouz, *Palace of Desire* (1991, Doubleday)

Naguib Mahfouz, *Palace Walk* (1991, Doubleday)

Abdelrahman Munif, *Cities of Salt: A Novel* (1987, Random House)

Films

Casablanca (1942, U.S.)

The Children of Heaven (1999, Iran)

The English Patient (1996, U.S.)

The House on Chelouche Street (1973, Israel)

Kadosh (1999, Israel)

Lawrence of Arabia (1962, U.S.)

Lion of the Desert (1981, U.K.)

The Lizard (Marmoulak) (2004, Iran)

The Ten Commandments (1956, U.S.)

Time of Favor (2000, Israel)

Bibliography

Cramer, Richard Ben. 2004. *How Israel Lost: The Four Questions*. New York: Simon and Schuster.

Esposito, John L., ed. 1999. *The Oxford History of Islam*. Oxford, UK: Oxford University Press.

Ibrahim, Fouad, and Ibrahim, Barbara. 2003. *Egypt: An Economic Geography*. New York: Macmillan.

Keyder, Caglar, ed. 1999. *Istanbul: Between the Global and the Local*. Lanham, MD: Rowman and Littlefield.

Lewis, Bernard. 2004. *The Crisis of Islam: Holy War and Unholy Terror*. New York: Random House.

Rodenbeck, Max. 1999. *Cairo: The City Victorious*. New York: Knopf.

Rogan, Eugene. 2009. *The Arabs: A History*. New York: Basic Books.

Smith, Dan. 2006. *The State of the Middle East: An Atlas of Conflict and Resolution*. Berkeley: University of California Press.

Soffer, Arnon. 1999. *Rivers of Fire: The Conflict over Water in the Middle East*. Lanham, MD: Rowman and Littlefield.

Swearingen, Will D., and Bencherifa, Abdellatif, eds. 1996. *The North African Environment at Risk*. Boulder, CO: Westview Press.

MG™ Mastering Geography

Log in to www.masteringgeography.com for MapMaster™ interactive maps, geography videos, RSS feeds, flashcards, web links, an eText version of *Diversity Amid Globalization*, and self-study quizzes to enhance your study of Southwest Asia and North Africa.

MapMaster™ presents 13 Place Name and 13 Layered Thematic interactive maps to help students practice and master their geographic literacy, spatial reasoning, and critical thinking skills.

8 Europe

Restoration of buildings on the town square, or *rynek*, of Wroclaw, Poland, attest to the resurgence of economic and cultural vitality of this historic trade center on the Oder River. After recent history as the German city of Breslau, Wroclaw was returned to Poland in 1945 with the realignment of national borders following WW II.

DIVERSITY AMID GLOBALIZATION

Historically, Europe was one of the world's earliest promoters of globalization as it spread its languages, laws, religion, politics, and economic systems throughout the world during the colonial period. Today, however, Europe struggles to adjust to the social and ethnic diversity sown by these earlier seeds as migrants from former colonies seek a share of Europe's high standard of living.

ENVIRONMENTAL GEOGRAPHY

Europe is one of the "greenest" world regions, with strong environmental laws and regulations about recycling, energy efficiency, and pollution. Europe is also a leader in promoting solutions to the problems associated with global warming.

POPULATION AND SETTLEMENT

With low (or no) natural population growth, immigration into Europe from other world regions is both a solution to Europe's labor needs and also a troublesome social issue.

CULTURAL COHERENCE AND DIVERSITY

Europe has a long history of internal cultural tensions linked to cultural differences in language and religion. Although these differences have been largely resolved through the integrating force of the European Union (EU), other cultural tensions have surfaced associated with immigration from Africa and South Asia.

GEOPOLITICAL FRAMEWORK

After half a century of the Cold War, which divided Europe into two parts, east and west, the region is now experiencing political integration of former adversaries at a national level. However, outbreaks of micro-nationalism, with agendas of autonomy, are still problematic in the Balkans, Spain, and the British Isles.

ECONOMIC AND SOCIAL DEVELOPMENT

Through the past half century, the EU has achieved remarkable success in bringing peace and prosperity to this diverse region. Nonetheless, serious economic and social challenges still remain, particularly in southern and eastern Europe.

Figure 8.1 Europe Stretching from Iceland in the Atlantic to the Black Sea, Europe includes 41 countries, ranging in size from large states, such as France and Germany, to the microstates Liechtenstein, Andorra, San Marino, and Monaco. Currently, the population of the region is about 531 million.

Europe is one of the most diverse regions in the world, encompassing a wide assortment of people and places in an area considerably smaller than North America. More than half a billion people reside in this region, living in 41 countries that range in size from giant Germany to microstates such as Andorra and Monaco (Figure 8.1).

The region's remarkable cultural diversity produces a geographic mosaic of languages, religions, and landscapes. Commonly, a day's journey finds travelers speaking two or three languages, crossing several political borders, and possibly changing money as they travel through different countries.

While a traveler may revel in Europe's cultural and environmental diversity, these distinct regional differences are also responsible for Europe's troubled past. In the 20th century alone, Europe was the principal battleground for two world wars, followed by the 45-year **Cold War** (1945–1990) that divided the continent and the world into two hostile, highly armed camps—Europe and the United States against the former Soviet Union (now Russia).

Figure 8.2 Anti-EU Feeling Not everyone is happy with the European Union, for some groups feel its standardized regulations trample on the historic uniqueness of different regions. Here, a Frenchman lobbies for a "no" vote on a recent EU agricultural issue.

Today, however, a spirit of cooperation prevails as Europe sets aside nationalistic pride and works toward regional economic, political, and cultural integration through the **European Union (EU)**. This supranational organization is made up of 27 countries, anchored by the western European states of Germany, France, Italy, and the United Kingdom, but also including many eastern European countries that were former Soviet satellites. Furthermore, the geographic reach and economic policies of the EU will continue to integrate the region during the next decades as it expands its membership. While controversial at times, the EU has unquestionably emerged as an unparalled regional organization (Figure 8.2).

ENVIRONMENTAL GEOGRAPHY: Human Transformation of a Diverse Landscape

Despite its small size, Europe's environmental diversity is extraordinary. Within its borders are found a startling array of landscapes, from the Arctic tundra of northern Scandinavia to the barren hillsides of the Mediterranean islands, and from the explosive volcanoes of southern Italy to the glaciated seacoasts of western Norway and Iceland.

Four factors explain this environmental diversity:

- The complex geology of this western extension of the Eurasian land mass has both the the newest, as well as the oldest, landscapes in the world.
- Europe's latitudinal extent from the Arctic to the Mediterranean subtropics affects climate, vegetation, and many human activities (Figure 8.3).
- These latitudinal controls are modified by the moderating influences of the Atlantic Ocean and the Black, Baltic, and Mediterranean seas.
- Last, the long history of human settlement has transformed and modified Europe's natural landscapes in fundamental ways over thousands of years (see "Geographic Tools: Landscape Restoration in England").

Landform and Landscape Regions

European landscapes can be organized into four general topographic regions: the European Lowland, forming an arc from southern France to the northeast plains of Poland, but also including southeastern England; the Alpine mountain system, extending from the Pyrenees in the west to the Balkan Mountains of southeastern Europe; the Central Uplands, positioned between the Alps and the European Lowland; and the Western Highlands, which include mountains in Spain, portions of the British Isles, and the highlands of Scandinavia.

The European Lowland This lowland, also known as the North European Plain, is the unquestioned economic focus of western Europe with its high population density, intensive agriculture, large cities, and major industrial regions. Though

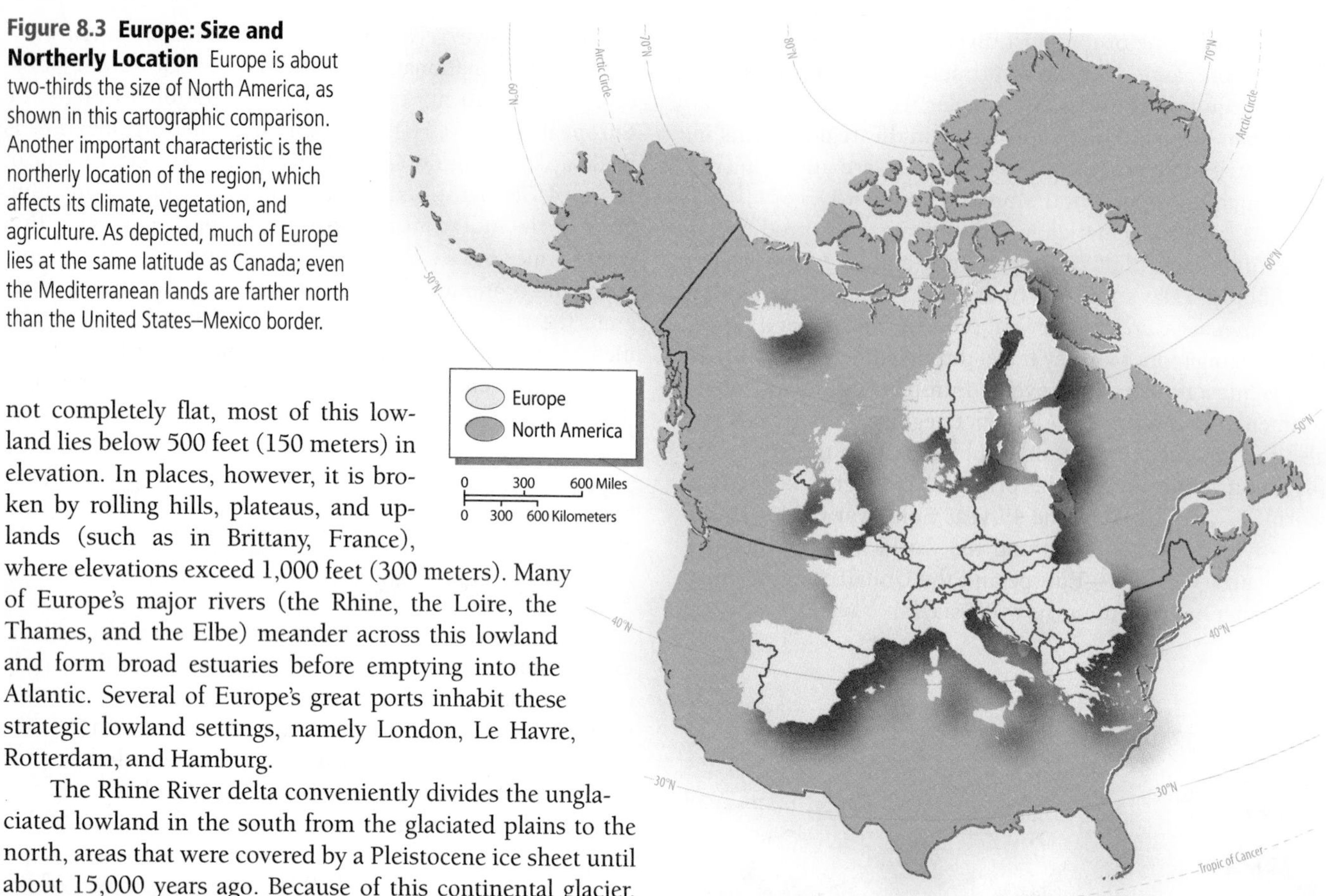

Figure 8.3 Europe: Size and Northerly Location Europe is about two-thirds the size of North America, as shown in this cartographic comparison. Another important characteristic is the northerly location of the region, which affects its climate, vegetation, and agriculture. As depicted, much of Europe lies at the same latitude as Canada; even the Mediterranean lands are farther north than the United States–Mexico border.

not completely flat, most of this lowland lies below 500 feet (150 meters) in elevation. In places, however, it is broken by rolling hills, plateaus, and uplands (such as in Brittany, France), where elevations exceed 1,000 feet (300 meters). Many of Europe's major rivers (the Rhine, the Loire, the Thames, and the Elbe) meander across this lowland and form broad estuaries before emptying into the Atlantic. Several of Europe's great ports inhabit these strategic lowland settings, namely London, Le Havre, Rotterdam, and Hamburg.

The Rhine River delta conveniently divides the unglaciated lowland in the south from the glaciated plains to the north, areas that were covered by a Pleistocene ice sheet until about 15,000 years ago. Because of this continental glacier, the North European Lowland, including the Netherlands, Germany, Denmark, and Poland, is far less fertile for agriculture than the unglaciated portion in Belgium and France (Figure 8.4). Rocky clay materials in Scandinavia were eroded and transported south by glaciers. As the glaciers retreated with a warming climate, piles of glacial debris known as **moraines** were left on the plains of Germany and Poland. Elsewhere in the north, glacial meltwater created infertile outwash plains that have limited agricultural potential.

The Alpine Mountain System The alpine mountain system forms the topographic spine of Europe. It consists of a series of mountains running east to west from the Atlantic to the Black Sea and the southeastern Mediterranean. Though these mountain ranges carry distinct regional names, such as the Pyrenees, Alps, Apennines, Carpathians, Dinaric Alps, and Balkan Ranges, they share geologic traits. All were created more recently (about 20 million years ago) than other upland areas of Europe, and all are constructed from a complex arrangement of rock types.

The *Pyrenees* form the political border between Spain and France and include the microstate of Andorra. This rugged range extends almost 300 miles (480 kilometers) from the Atlantic to the Mediterranean. Within the mountain range, glaciated peaks reaching to 11,000 feet (3,350 meters) alternate with broad glacier-carved valleys. The Pyrenees are home to the Basque people in its western reaches, as well as to the distinctive Catalan-speaking minorities in the east. Both groups have strong separatist traditions.

Figure 8.4 The European Lowland Also known as the North European Plain, this large lowland extends from southwestern France to the plains of northern Germany and into Poland. Although this landform region has some rolling hills, most of it is less than 500 feet (150 meters) in elevation.

GEOGRAPHIC TOOLS Landscape Restoration in England

Because thousands of years of European settlement have dramatically altered the landscape, there is great interest today in restoring certain treasured landscapes to their earlier, more natural state. In England, landscape restoration efforts are focused on two environments, ancestral woodlands and the heath lands. To accomplish this, a number of geographic tools are used in these projects.

A first step in landscape restoration is to recreate the ecology of the lost environment. This is often done by using a combination of science and social science methods, namely palynology (pollen analysis) and sediment dating, but also using archival sources with historical descriptions to reconstruct the earlier landscape.

Once a baseline is established based upon these methods, land restoration efforts begin and other tools are then used to monitor progress. Botanical and wildlife surveys are done regularly to tabulate plant and animal species. Often these data are located and mapped with a Global Positioning Systems (GPS) and then fed into geographic information system (GIS) files. Using these computerized maps, geographers and ecologists can document how landscape changes over time. In addition, both land and aerial photography is used to track environmental change in the restored landscapes.

One of the restoration efforts in England focuses on the original or ancestral woodlands of England, which were deciduous forests of oak, beech, elm, and similar broadleaf trees. Drawing upon historical land-use maps and local tax records, a restoration group called the Woodland Trust calculated that these forests once covered more than a million acres (335,000 hectares). Today, however, the broadleaf woodland is about half that amount, primarily because it has been replaced by planted nonnative conifer trees. Another restoration effort in England focuses on the fast disappearing lowland heath land, a tundra-like environment composed primarily of grasses, heather, and gorse. Unlike England's ancestral forest that is primarily natural, heath lands were created over thousands of years by human activity. More specifically, they resulted from a long history of grazing sheep and cattle on these environments. Because these animals tend to eat young saplings, tree growth was inhibited and the heath lands expanded as treeless meadows.

However, in the last 200 years these heath lands have disappeared as livestock grazing has become more specialized and localized on more productive lands. As a result, trees have taken over many former heath lands (Figure 8.1.1). Using historical maps from 1800 to recreate England's landscape of that period, restoration groups calculate that fully 82 percent of the heath lands have been lost to forests. As a result, restoration efforts focus on removing trees to allow native heath to regrow.

Since similar projects also quite common in North America, geographers who learn the tools and methods of ecological restoration find employment opportunities throughout the United States and Canada working with local and government agencies.

Figure 8.1.1 Oxshott Heath in England This heath landscape west of London embodies both the beauty and the problems of heath restoration. Note the willow trees (to the right) encroaching on the heath. In the background are ancestral woodlands fringing the heath.

The centerpiece of Europe's geologic system is the prototypical mountain range, the Alps, reaching more than 500 miles (800 kilometers) from France to eastern Austria. These impressive mountains are highest in the west, reaching to more than 15,000 feet (4,575 meters) in Mt. Blanc on the French–Italian border; in Austria, to the east, the Alps are much more subdued where few peaks exceed 10,000 feet (3,050 meters). Though easily crossed today by car or train through a system of long tunnels and valley-spanning bridges, these mountains have historically formed an important cultural divide between the Mediterranean lands to the south and central and western Europe in the north.

The Apennine Mountains are located south of the Alps; the two ranges, however, are physically connected by the hilly coastline of the French and Italian Riviera. Forming the mountainous spine of Italy, the Apennines are lower and lack the spectacular glaciated peaks and valleys of the true Alps. Farther to the south, the Apennines take on their own distinct character with the impressive and explosive volcanoes of Mt. Vesuvius (just over 4,000 feet, or 1,200 meters) outside Naples and the much higher (almost 11,000 feet, or 3,350 meters) Mt. Etna off Italy's toe on the island of Sicily. These two active volcanoes are products of the meeting between the African and Eurasian tectonic plates.

To the east, the Carpathian Mountains define the limits of the alpine system in eastern Europe. They are a plow-shaped upland area that extends from eastern Austria to the Iron Gate gorge, a narrow passage for Danube river traffic where the borders of Romania and Serbia intersect. While about the same length as the main alpine chain, the

Figure 8.5 Fjord in Norway During the Pleistocene continental ice sheets and glaciers carved deep U-shaped valleys along what is now Norway's coastline. As the ice sheets melted and sea level rose, these valleys were flooded by Atlantic waters, creating spectacular fjords. Many fjord settlements are accessible only by boat, linked to the outside world by Norway's extensive ferry system.

Carpathians are not nearly as high. The highest summits in Slovakia and southern Poland are less than 9,000 feet (2,780 meters).

Central Uplands In western Europe, a much older highland region occupies an arc between the Alps and the European Lowland in France and Germany. These mountains are much lower in elevation than the alpine system, with their highest peaks at 6,000 feet (1,830 meters). Dated to about 100 million years ago, much of this upland region is characterized by rolling landscapes of about 3,000 feet (less than 1,000 meters).

Their importance to western Europe is great because they contain the raw materials for Europe's industrial areas. In both Germany and France, for example, these uplands have provided the iron and coal necessary for each country's steel industry; in the eastern reaches, mineral resources from the Bohemian highlands have also fueled major industrial areas in Germany, Poland, and the Czech Republic.

Western Highlands The Western Highlands define the western edge of the European subcontinent, extending from Portugal in the south, through the portions of the British Isles in the northwest, to the highland backbone of Norway, Sweden, and Finland in the far north. These are Europe's oldest mountains, formed about 300 million years ago.

As with other upland areas that traverse many separate countries, specific place-names for these mountains differ from country to country. A portion of the Western Highlands forms the highland spine of England, Wales, and Scotland, where picturesque glaciated landscapes are found at modest elevations of 4,000 feet (1,220 meters) or less. These U-shaped glaciated valleys also appear in Norway's uplands where they produce a spectacular coastline of **fjords**, or flooded valley inlets similar to the coastlines of Alaska and New Zealand (Figure 8.5).

Though less elevated, the Fenno-Scandian Shield of Sweden and northern Finland is noteworthy because it is made up of some of the oldest rock formations in the world, dated conservatively at 600 million years. This **shield landscape** was eroded to bedrock by Pleistocene glaciers and, because of the cold climate and sparse vegetation, still has extremely thin soils that severely limit agricultural activity. As with other heavily glaciated areas like the Canadian Shield of North America, numerous small lakes dot the countryside, giving clues to the impressive erosional power of ice sheets.

Europe's Climates

Three principal climates characterize Europe (Figure 8.6). Along the Atlantic coast, a moderate and moist **maritime climate** dominates, modified by oceanic influences. Farther inland, continental climates prevail, with hotter summers and colder winters. Finally, dry-summer Mediterranean climates are found in southern Europe, from Spain to Greece.

One of the most important climate controls is that of the Atlantic Ocean. Though most of Europe is at a relatively high latitude (London, England, for example, is slightly farther north than Vancouver, British Columbia), the mild North Atlantic current, which is a continuation of the warm Atlantic Gulf Stream, moderates coastal temperatures from Iceland and Norway to Portugal, and even inland to the western reaches of Germany. As a result, this maritime influence gives Europe a climate 5 to 10°F (2.8 to 5.7°C) warmer than comparable latitudes lacking this ocean current's effect. In the **marine west-coast climate** region, no winter months average below freezing, though cold rain, sleet, and an occasional

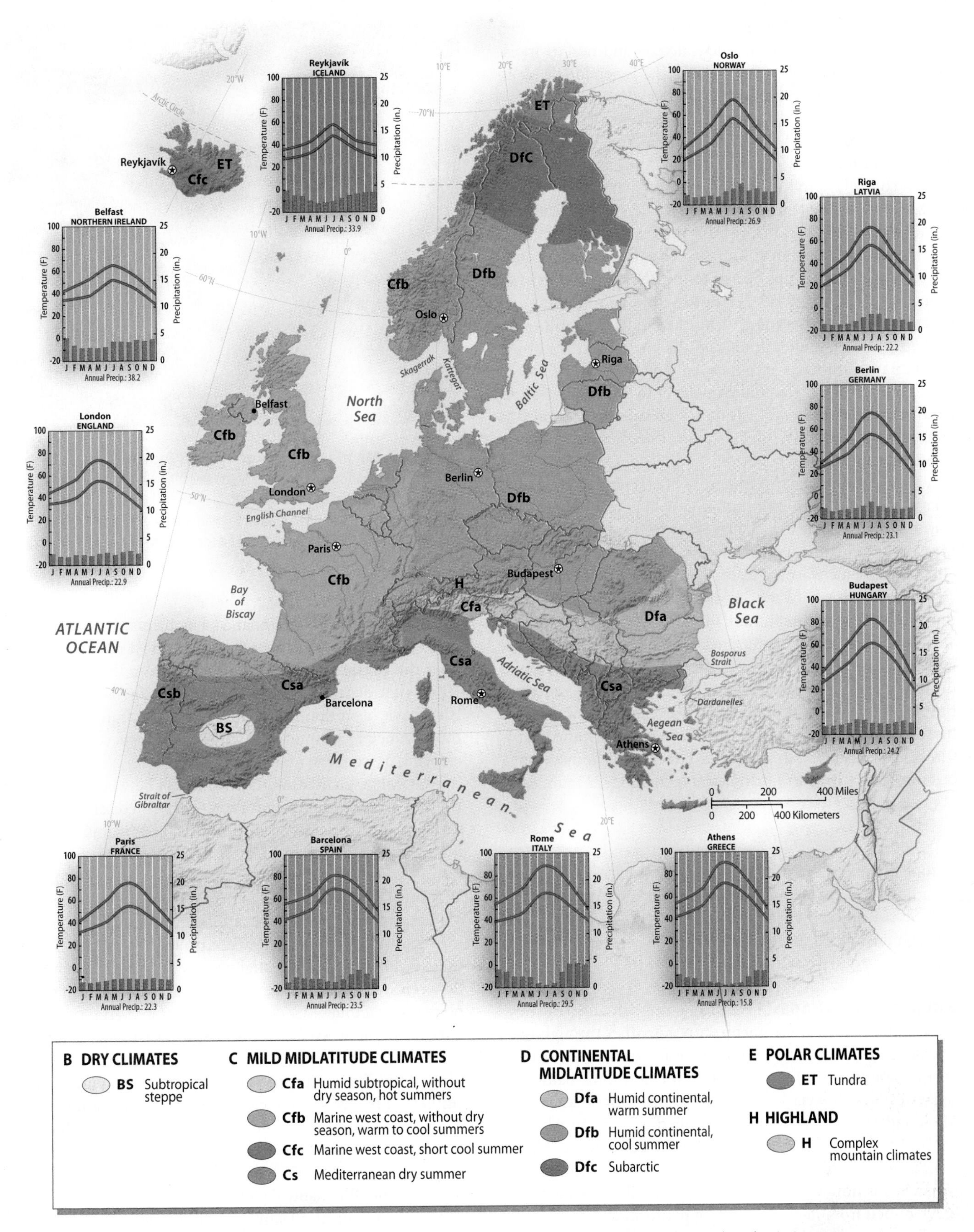

Figure 8.6 Climate Map of Europe Three major climate zones dominate Europe. Close to the Atlantic Ocean, the marine west-coast climate has cool seasons and steady rainfall throughout the year. Farther inland, continental climates have at least one month averaging below freezing, as well as hot summers, with a precipitation maximum occurring during the warm season. Southern Europe has a dry-summer Mediterranean climate.

blizzard are common winter visitors. Summers are often cloudy and overcast with frequent drizzle and rain as moisture flows in from the ocean. Ireland, the Emerald Isle, is an appropriate icon of this maritime climate.

With increasing distance from the ocean (or where a mountain chain limits the maritime influence, as in Scandinavia), land-mass heating and cooling becomes a strong climatic control, producing hotter summers and colder winters. Indeed, all **continental climates** average at least one month below freezing during the winter.

In Europe, the transition between maritime and continental climates takes place close to the Rhine River border of France and Germany. Farther north, although Sweden and other nearby countries are close to the moderating influence of the Baltic Sea, high latitude and the blocking effect of the Norwegian mountains produce cold winter temperatures characteristic of continental climates. Most precipitation in continental climates comes as rain from summer frontal systems and local thundershowers, with lesser amounts falling as winter time snow.

The **Mediterranean climate** is characterized by a distinct dry season during the summer, which results from the warm-season expansion of the Atlantic (or Azores) high-pressure area. This high pressure is produced by the global circulation of air warmed in the equatorial tropics that subsides (descends) between latitudes of 30 and 40 degrees, thus inhibiting summer rainfall. This same phenomenon also produces the Mediterranean climates of California, western Australia, parts of South Africa, and Chile. While these rainless summers may attract tourists from northern Europe, the seasonal drought can be problematic for agriculture. It is no coincidence that traditional Mediterranean cultures, such as the Arab, Moorish, Greek, and Roman, have been major innovators of irrigation technology (Figure 8.7).

Figure 8.7 Mediterranean Agriculture Because of water scarcity during the hot, dry summers in the Mediterranean climate region, local farmers have evolved agricultural strategies for making the best use of soil and water resources. In this photo from Portugal, the terraces used to prevent soil erosion on steep slopes and the mixture of tree and ground crops are evident.

Seas, Rivers, and Ports

Europe remains a maritime region with strong ties to its surrounding seas; even landlocked countries like Austria, Hungary, Serbia, and the Czech Republic have access to the ocean through an interconnected network of navigable rivers and canals.

Europe's Ring of Seas Five major seas encircle Europe; these water bodies are connected to each other through narrow straits with strategic importance for controlling waterborne trade and naval movement. In the north, the Baltic Sea separates Scandinavia from north-central Europe and Denmark and Sweden have long controlled the narrow Skagerrak and Kattegat straits that connect the Baltic to the North Sea. Besides its historic role as a major fishing ground, the North Sea is now well known for its rich oil and natural gas fields mined from deep-sea drilling platforms.

The English Channel (in French, *La Manche*) separates the British Isles from continental Europe, separating the two at its narrowest point, the Dover Straits, only 20 miles (32 kilometers) wide. Although England has thought of the channel as a protective moat, it has primarily been a symbolic barrier, for it deterred neither the French Normans from the continent nor Viking raiders from the north. Since 1993, after decades of resistance by the English, the British Isles have been connected to France through the 31-mile (50-kilometer) Eurotunnel, with its high-speed rail system carrying passengers, autos, and freight.

Gibraltar guards the narrow straits between Africa and Europe at the western entrance to the Mediterranean Sea, and Britain's stewardship of this passage remains an enduring symbol of its once great sea-based empire. Finally, on Europe's southeastern flanks are the Straits of Bosporus and the Dardanelles, the narrows connecting the eastern Mediterranean with the Black Sea. Disputed for centuries, these pivotal waters are now controlled by Turkey. Though these straits are often thought of as the physical boundary between Europe and Asia, they are easily bridged in several places to facilitate truck and train transportation within Turkey and between Europe and Southwest Asia.

Rivers and Ports Europe is also a region of navigable rivers connected by a system of canals and locks that allow inland barge travel from the Baltic and North seas to the Mediterranean, and between western Europe and the Black Sea. Many rivers on the European Lowland, namely the Loire, Seine, Rhine, Elbe, and Vistula, flow into Atlantic and Baltic waters. However, the Danube, Europe's longest river, flows east and south, rising in the Black Forest of Germany only a few miles from the Rhine River and running southeastward to the Black Sea, offering a connecting artery between central and eastern Europe. Similarly, the Rhône headwaters rise

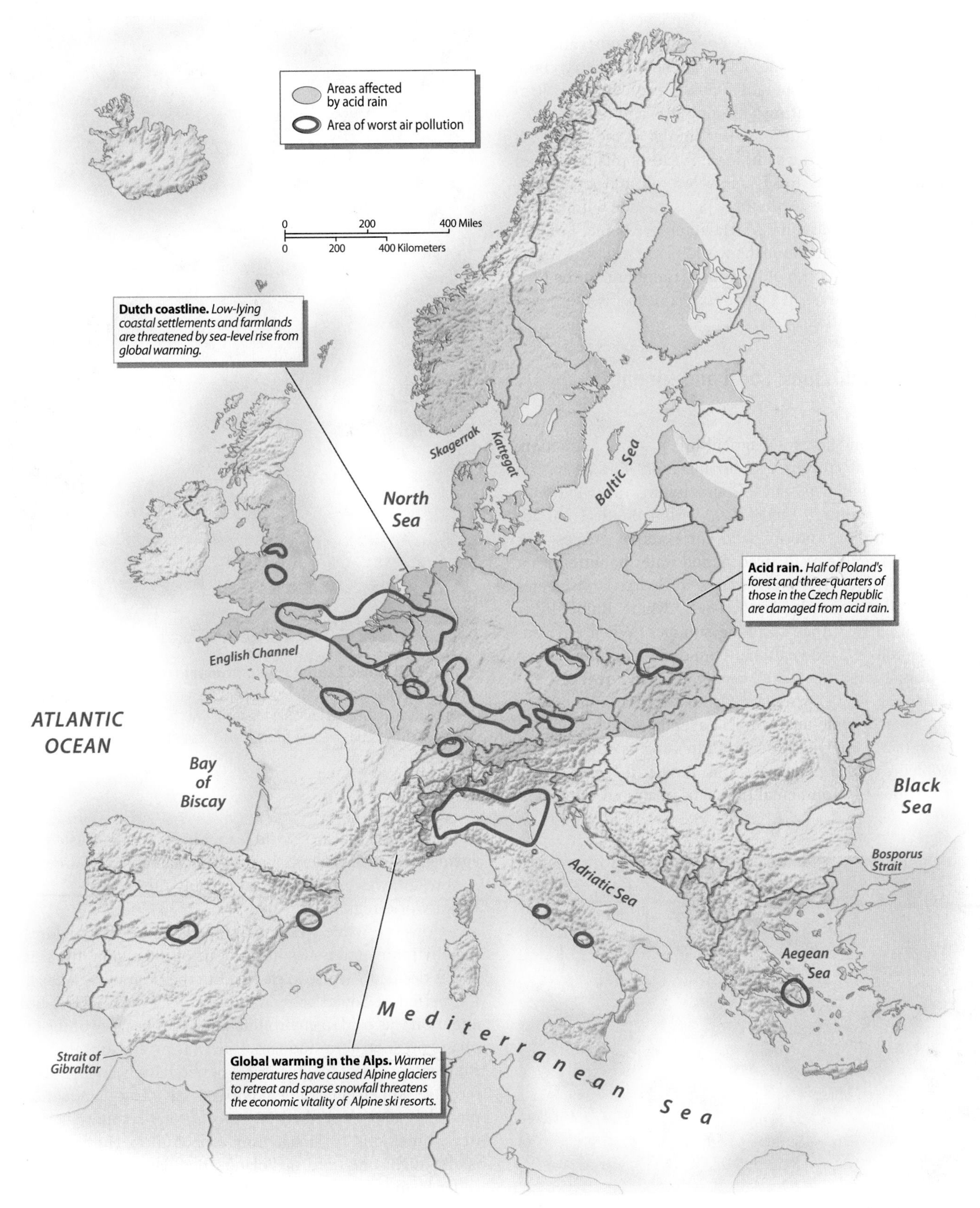

Figure 8.8 Environmental Issues in Europe While western Europe has worked energetically over the past 50 years to solve environmental problems such as air and water pollution, eastern Europe lags behind because environmental protection was not a high priority during the postwar communist period. Current efforts, though, show great promise.

close to those of the Rhine in Switzerland, yet it flows southward into the Mediterranean. Both the Danube and the Rhône are connected by locks and canals with the rivers of the European Lowland, making it possible for barge traffic to travel between all of Europe's fringing seas and oceans.

Major ports are found at the mouths of most western European rivers, serving as transshipment points for inland waterways as well as focal points for rail and truck networks. From south to north, these ports include Bordeaux at the mouth of the Garonne, Le Havre on the Seine, London on the Thames, Rotterdam (the world's largest port in terms of tonnage) at the mouth of the Rhine, Hamburg on the Elbe River, and, to the east in Poland, Szcezin on the Oder and Gdansk on the Vistula.

Environmental Issues: Local and Global, East and West

Because of its long history of agriculture, resource extraction, industrial manufacturing, and urbanization, Europe has its share of environmental issues. Compounding the situation is the fact that pollution rarely stays within political boundaries. Air pollution from England, for example, creates serious acid rain problems in Sweden, and water pollution from the upper Rhine River by factories in Switzerland creates major problems for the Netherlands, where Rhine River water is used for urban drinking supplies. When environmental problems cross national boundaries, solutions must come from intergovernmental cooperation (Figure 8.8).

Since the 1970s, when the EU added environmental issues to its economic and political agenda, western Europe has been increasingly effective in addressing its environmental problems with regional solutions. Besides focusing on the more obvious environmental problems of air and water pollution, the EU is also a world leader in recycling, waste management, reduced energy use, and sustainable resource use. As a result, western Europe is probably the "greenest" of the major world regions.

Figure 8.10 Toxic Landscape in Romania A troublesome legacy of Soviet communism in eastern Europe is the numerous toxic dump sites and polluted landscapes found in the region, such as this site in Romania. Money from the EU is helping to clean up these sites now that Romania is a member.

Figure 8.9 Acid Rain and Forest Death Acid precipitation has taken a devastating toll on eastern European forests, such as those shown here in Bohemia, Czech Republic. In this country, three-quarters of the forests have been damaged by acid precipitation, which is caused by industrial and auto emissions both in and outside the country.

However, while western Europe is successfully addressing environmental issues in that part of the region, the situation is decidedly worse in eastern Europe (Figure 8.9). During the period of Soviet economic planning (1945–1990), little attention was paid to environmental issues because of the emphasis on short-term industrial output. As a result, the environmental costs of that historical period have been high. Until recently, 90 percent of Poland's rivers have had no aquatic or plant life, and more than 50 percent of the country's forest trees show signs of damage from air pollution. Human illness is also high; one-third of Poland's population is reported to suffer from environmentally induced diseases such as cancer or respiratory illness.

The future, however, appears more positive with the expansion of EU environmental policies for new member states in eastern Europe (Figure 8.10). Even before EU membership was granted, countries such as Poland, Romania, and Slovakia had to enact environmental legislation comparable to that of western European states, and the same is true of the current array of former Yugoslavian countries (Croatia, Serbia, Macedonia) aspiring to EU membership.

Global Warming in Europe: Problems and Prospects

The fingerprints of global warming are everywhere in Europe, from dwindling sea ice, melting glaciers, and sparse snow cover in arctic Scandinavia to increasingly frequent droughts in the water-starved Mediterranean area (Figure 8.11). Furthermore, the projections for future climate change are ominous: World-class ski resorts in the Alps are vulnerable to warmer winters and reduced snow pack, while in the lowlands higher summertimes could result in more frequent heat waves that devastate farmers and urban dwellers alike. As well, rising sea levels from melting polar ice sheets will threaten the Low Countries, where much of the population lives in diked lands below sea level. Because of these threats, Europe has become a world leader in addressing global warming, and has creating numerous policies and programs to reduce greenhouse gas (GHG) emissions and adapt to changing environmental conditions.

Figure 8.11 Glacial Retreat in the Austrian Alps The blue sign in the foreground marks where the snout of the Pasterze glacier was in 1985, and illustrates how far the glacier has retreated in the last 25 years, shrinkage that is probably a result of global warming. This glacier is the largest in Austria and has its origin on Austria's highest mountain, the Grossglockner.

Europe and the Kyoto Protocol The EU entered the 1997 Kyoto negotiations with an innovative scheme underscoring its philosophy that regional action to solving environmental problems was superior to that of efforts by individual countries. More specifically, the EU proposed setting a collective goal of an 8 percent reduction below 1990 GHG emission levels as an umbrella for its individual member states. Under this umbrella, or EU collective as it's called, some European states were asked to make larger reductions while others were actually allowed an increase in GHG emissions. The point of the umbrella was to allow growth and industrial development in poorer countries, namely Spain, Greece, and Portugal, which were allowed GHG increases; at the same time more developed states such as Germany, Denmark, England, and France, were required to make emission reductions. As a result of this collective approach, ideally, Europe as a whole would produce an 8 percent reduction in GHG emissions. Noteworthy is that this umbrella approach has remained in place as the EU has grown from the 15 members at Kyoto in 1997 to its current 27 member states.

The EU's Emission Trading Scheme In 2005 the EU inaugurated the world's largest carbon trading scheme as a main pillar of its global warming policy. Under this plan specific yearly emission caps are set for the EU's largest GHG emitters; if these emitters exceed those limits they must compensate for their excesses either by purchasing carbon equivalences from a factory below its own cap or by buying credits from the EU carbon market. The point of this cap and trade system is to make business more expensive for those who pollute while rewarding those industries that stay under their carbon quota.

Currently, the plan sets caps for industries responsible for about 50 percent of the EU's GHG emissions, installations made up primarily of power plants, steel factories, and cement manufacturers. In 2012 aviation will be added to the cap and trade program since that sector of the economy is through to be responsible for at least 5 percent of the region's emissions.

Results to Date Despite its good intentions, recent data show that the EU is not on target to meet its 2012 goal of an 8 percent reduction of GHG. Instead, the EU umbrella has currently reduced emissions less than 1 percent below its

1990 baseline. The main reason for this failure, EU experts say, is the unanticipated growth in truck transport over the past decade, coupled with excessive emissions from industrial development in Spain and the new EU countries of eastern Europe. However, it should be also noted that the major industrial powers, England and Germany, have made dramatic reductions in their emissions of over 10 percent, primarily by reducing their coal usage. Further GHG reductions have come from increased energy generation from wind and solar power.

To compensate for this slow start, the EU recently agreed to set a new target of a 20 percent reduction below the 1990 baseline for the umbrella, a goal the EU hopes to achieve by 2020. To achieve that reduction Europe is promoting an aggressive program of alternative energy that will draw heavily on both wind and solar power (Figure 8.12). Currently there are over 25,000 wind farms in Europe; this amount is expected to double by 2015, with Germany and Spain leading the way.

POPULATION AND SETTLEMENT: Slow Growth and Rapid Migration

The map of Europe's population distribution shows that, in general, population densities are higher in the historical industrial core areas of western Europe (England, the Netherlands, northern France, northern Italy, and western Germany) than in the periphery to the east and north (Figure 8.13). While this generalization overlooks important urban clusters in Mediterranean Europe, it does express a sense of a densely settled European core set apart from a more rural, agricultural periphery.

Much of this distinctive population pattern is linked to areas of early industrialization, yet there are also many modern consequences of this core–periphery distribution. For example, economic subsidies from the wealthy, highly urbanized core to the less affluent, agricultural periphery have been an important part of the EU's development policies for several decades. Further, while the urban-industrial core is characterized by extremely low natural growth rates, it is also the target area for migrants—both legal and illegal—from Europe's peripheral countries, as well as from outside Europe (see "People on the Move: Smuggling Illegal Immigrants into Europe").

Natural Growth: Beyond the Demographic Transition

Probably the most striking characteristic of Europe's population is its slow natural growth (Table 8.1). More to the point, in many European countries, the death rate exceeds the birthrate, meaning that there is simply no natural growth at all. Instead, many countries are experiencing negative growth rates, and were it not for in-migration from other countries and other world regions, these countries would record declines in population over the next few decades.

There seem to be several reasons for the lack of natural population growth in Europe. First of all, recall from Chapter 1 that the concept of the demographic transition was based on the historical change in European growth rates as the population moved from rural settings to more urban and industrial locations. What we see today is an extension of that model: the continued expression of the low-fertility/low-mortality fourth stage of the demographic transformation. Some demographers suggest adding a fifth stage to the model—a "post-industrial" phase, in which population falls below replacement levels, which results when parents have fewer than two children. Evidence for this fifth stage comes from

Figure 8.12 Wind Power in Northern Europe Not only is Europe trying to reduce its carbon dioxide emissions, but it is also a world leader in generating renewable energy from wind, sun, and biofuels. This large wind farm is in Denmark.

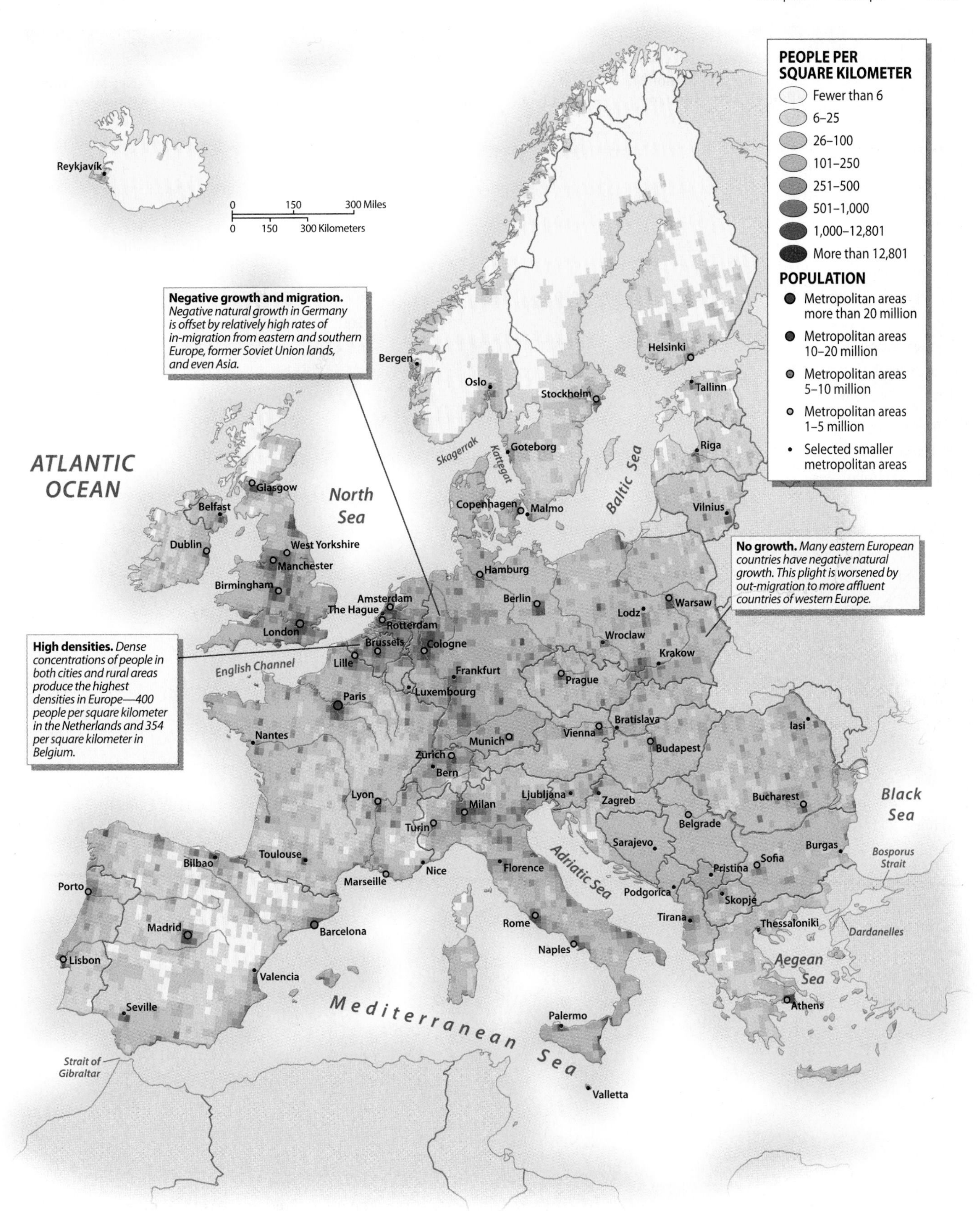

Figure 8.13 Population Map of Europe The European region includes more than 531 million people, many of them clustered in large cities in both western and eastern Europe. As can be seen on this map, the most densely populated areas are in England, the Netherlands, Belgium, western Germany, northern France, and south across the Alps to northern Italy.

TABLE 8.1 Population Indicators

Country	Population (millions) 2010	Population Density (per square kilometer)	Rate of Natural Increase (RNI)	Total Fertility Rate	Percent Urban	Percent <15	Percent >65	Net Migration (Rate per 1,000) 2005–2010[a]
Western Europe								
Austria	8.4	100	0.0	1.4	67	15	17	3.9
Belgium	10.8	354	0.2	1.7	99	17	17	3.8
France	63.0	114	0.4	2.0	77	18	17	1.6
Germany	81.6	229	–0.2	1.3	73	14	20	1.3
Liechtenstein	0.04	225	0.4	1.4	15	16	13	
Luxembourg	0.5	196	0.4	1.6	83	18	14	8.4
Monaco	0.04	35,835	0.0		100	13	24	
Netherlands	16.6	400	0.3	1.7	66	18	15	1.2
Switzerland	7.8	190	0.2	1.5	73	15	17	2.7
United Kingdom	62.2	256	0.4	1.9	80	18	16	3.1
Eastern Europe								
Bulgaria	7.5	68	–0.4	1.6	71	14	18	–1.3
Czech Republic	10.5	133	0.1	1.5	74	14	15	4.4
Hungary	10.0	108	–0.3	1.3	67	15	16	1.5
Poland	38.2	122	0.1	1.4	61	15	13	–0.6
Romania	21.5	90	–0.2	1.3	55	15	15	–1.9
Slovakia	5.4	111	0.2	1.4	55	15	12	0.7
Southern Europe								
Albania	3.2	112	0.5	1.6	49	25	9	–4.8
Bosnia and Herzegovina	3.8	75	0.0	1.2	46	16	14	–0.5
Croatia	4.4	78	–0.2	1.5	56	15	17	0.5
Cyprus	1.1	118	0.6	1.5	62	18	10	5.8
Greece	11.3	86	0.1	1.5	73	14	19	2.7
Italy	60.5	201	0.0	1.4	68	14	20	5.6
Macedonia	2.1	80	0.2	1.5	65	19	11	–1.0
Malta	0.4	1,326	0.2	1.4	94	16	14	2.5
Montenegro	0.6	46	0.4	1.8	64	20	13	–1.6
Portugal	10.7	116	–0.1	1.3	55	15	18	3.8
San Marino	0.03	522	0.3	1.2	84	15	16	
Serbia	7.3	94	–0.5	1.4	58	15	17	0.0
Slovenia	2.1	101	0.2	1.5	50	14	16	2.2
Spain	47.1	93	0.3	1.4	77	15	17	7.9
Northern Europe								
Denmark	5.5	129	0.1	1.8	72	19	17	1.1
Estonia	1.3	30	0.0	1.6	69	15	17	0.0
Finland	5.4	16	0.2	1.9	65	17	17	2.1
Iceland	0.3	3	0.9	2.1	93	21	12	12.8
Ireland	4.5	64	1.0	2.1	60	21	11	9.1
Latvia	2.2	35	–0.4	1.3	68	14	17	–0.9
Lithuania	3.3	51	–0.1	1.5	67	15	16	–6.0
Norway	4.9	13	0.4	2.0	80	19	15	5.7
Sweden	9.4	21	0.2	1.9	84	17	18	3.3

[a] *Net Migration Rate from the United Nations Population Division, World Population Prospects: The 2008 Revision Population Database.*
Source: Population Reference Bureau, World Population Data Sheet, 2010.

PEOPLE ON THE MOVE Smuggling Illegal Immigrants into Europe

Europe is the destination of choice for many thousands of immigrants each year, both legal and illegal. As is the case in so many other parts of the world, the plight of illegal immigrants can be a tragic one. Often these people are taken advantage of by smugglers who promise much and deliver little, charging exorbitant rates for uncertain outcomes.

Worldwide, people smuggling is now big business and, as a result, organized crime has become increasingly involved. More specifically, European immigration officials estimate that fully 90 percent of those attempting to enter EU countries were helped by international crime networks.

Commonly immigrants pay around $15,000 to be smuggled into Europe, a life savings for most, yet organized crime appears to be more interested in accumulating riches than in the immigrants' success. Many immigrants end up broke, sick, and hopelessly stranded in a jail or detention center in Africa rather than Europe. Or worse, they drown in the ocean or die in refugee camps.

A common smuggling route takes immigrants from all parts of the world to the West African coast where they are loaded on decrepit boats hoping to make landfall on the Canary Islands, an autonomous community of Spain (Figure 8.2.1). If the boats make it that far, immigrants are usually told to swim ashore and ask for political asylum. If successful, the immigrants will be taken to a refugee center in mainland Spain where, after a period of questioning, they could be released and ordered out of the country. Instead, though, most immigrants would begin an underground and illegal life in Europe, moving from country to country, across the soft borders of the EU.

In the last few years the Canary Islands have been overwhelmed by some 7,000 illegal immigrants who made it to shore. Uncounted others were not so lucky, either ending up back in Africa or dead from drowning or exposure in ill-equipped boats. Those reaching the Canary Islands are not just from African countries but also from Sri Lanka, Afghanistan, Myanmar, even India and Pakistan.

The fact that many of these immigrants come from great distances and from a mix of countries is taken as another indication that organized crime systems now control the people-smuggling business in Europe. Further evidence comes from the fact that the ships smuggling these immigrants fly a variety of foreign flags and have international crews. North Korean ships with Georgian or Russian crews seem particularly popular with the smugglers.

Figure 8.2.1 Illegal Migration to Europe This boatload of would-be immigrants to Europe was intercepted by Spanish authorities as it approached the Canary Islands.

the highly urban and industrial populations of Germany, France, and England, all of which are below zero population growth when immigration is excluded.

These low-birthrate countries, however, are attempting to increase their rates of natural growth. In Germany, for example, couples are given an outright monetary gift if they produce a child. In Austria, the government gives a modest cash award to couples when they marry, and still more follows when they give birth. France provides parents with monthly payments that last until their children reach school age. In addition, most EU countries have liberal maternity benefits that guarantee both parents time off from their jobs without penalty.

Yet despite these incentives most European countries still have natural growth rates below replacement levels, and several, most notably in the former communist countries of eastern Europe, show negative growth. In these countries the political and economic turmoil associated with the relatively recent transition from communism to capitalism may explain these very low birthrates. An example is Bulgaria, with a Rate of Natural Increase (RNI) of −0.4

Migration To and Within Europe

Migration is one of the most challenging population issues facing Europe today because the region is caught in a web of conflicting policies and values (Figure 8.14). Currently, there is widespread resistance to unlimited migration into Europe because of high unemployment in the European industrial countries. Many Europeans argue that scarce jobs should go first to European citizens, not to immigrants. This topic has become so controversial that the EU is working toward a common immigration policy for its member countries.

During the 1960s postwar recovery, when western Europe's economies were booming, many countries looked to migrant workers to ease labor shortages. The former West Germany, for example, depended on workers from Europe's periphery—Italy, the former Yugoslavia, Greece, and Turkey—for industrial and service jobs. These *gastarbeiter*, or **guest workers**, arrived by the thousands. As a result, large ethnic enclaves of foreign workers became a common part of the German urban landscape. Today, there are more than 2 million Turks in Germany, most of whom are there because of the open-door worker policies of past decades.

Figure 8.14 Migration into Europe Historically, Europe opened its doors to migrants to help solve post–World War II labor shortages, but now immigration is a contentious and controversial issue, as culturally homogeneous countries such as France, England, and Germany confront tensions resulting from large numbers of immigrants, both legal and illegal.

Figure 8.15 **Mosque in Germany** Girls from the Turkish Muslim community stand in front of their new mosque in Moers, Germany. Because of West Germany's need for immigrant labor as the economy expanded in the 1960s, large numbers of "guest workers" settled in the country, beginning a migration pattern still evident today.

These foreign workers, however, are now the target of considerable ill will as Europe suffers from economic recession and stagnation. As a result, unemployment rates approaching 25 percent are common for young people. In addition, the region has witnessed a massive flood of migrants from former European colonies in Asia, Africa, and the Caribbean as these people seek a better life. The former colonial powers of England, France, and the Netherlands have been particular magnets for immigration. England, for example, has inherited large numbers of former colonial citizens from India, Pakistan, Jamaica, and Hong Kong. Indonesians (from the former Dutch East Indies) are common in the Netherlands, while migrants in France are often from former colonies in both northern and Sub-Saharan Africa.

More recently, political and economic troubles in eastern Europe and the former Soviet Union have generated another wave of migrants to Europe. With the total collapse of Soviet border controls in 1990, emigrants from Poland, Bulgaria, Romania, Ukraine, and other former Soviet satellite countries poured into western Europe. This flight from the post-1989 economic and political chaos of eastern Europe has also included refugees from war-torn regions of former Yugoslavia, particularly from Bosnia and Kosovo.

As a result of these different migration streams, Germany has become a reluctant land of migrants that now receives 400,000 newcomers each year. Currently, about 7.5 million foreigners live in Germany, making up about 9 percent of the population (Figure 8.15). This is about the same percentage as the foreign-born segment of the U.S. population. However, while the United States celebrates its history of immigration, Germany, like most other European countries, is ambivalent about this ethnic mixture and struggles with its new cultural diversity.

Fortress Europe An important aggravating ingredient in Europe's issue with foreign migration is the political agreement between the heartland countries of Europe to facilitate free movement for its citizens across national borders without passport inspections or checks. This agreement has reshaped the political geography of Europe by creating divisions between insiders and outsiders. To those on the inside, it creates a long-dreamed-of Europe without borders, while to those on the outside, it presents a "Fortress Europe," a defensive perimeter perceived as hostile to migrants from Asia, Africa, and the Russian domain. At the so-called "hard" borders on the EU's perimeter, especially in Spain, Italy, Slovakia, and Poland, foreigners and foreign goods are subject to lengthy passport checks, visa requirements, and searches, whereas crossing a "soft" border within the European heartland involves little more than a visual check and a wave from the border police, if there even is a border station. Often these internal borders are marked only with signs.

Taking its name from the city of Schengen, Luxembourg, where the original declaration of intent was signed in 1985, the EU's **Schengen Agreement** has as one of its goals the gradual reduction of border formalities for travelers moving between EU countries (Figure 8.16). Though reduced border formalities within western Europe seemed like reasonable and desirable goals in 1985, today the situation is much more

Figure 8.16 **Czech Republic Border Station** This border station between Germany and the Czech Republic was once a highly fortified Iron Curtain checkpoint. But today, with the Czech Republic's membership in the EU, most people pass through with a cursory wave after showing their passport.

CITYSCAPES Berlin Reinvents Itself

Berlin is in many ways a microcosm of Europe, showing considerable scar tissue from the wars and political divisions of the 20th century, yet now healing by reinventing itself for the 21st century with its promise of political and economic integration. As a result, the city's landscape is a fascinating mosaic of the past and present, along with hopes for the future.

Berlin has been the historical capital of the north German region since the fifteenth century. In 1871, with the formation of the German state, it became the capital of Germany until the end of World War II in 1945. At that time, Berlin, along with all of Germany, was occupied by the four Allied powers—the United States, France, Britain, and the Soviet Union. Berlin was in the Soviet zone, and, as Cold War tensions increased, the city became isolated from western Germany.

Dividing the Soviet part of the city from the American, French, and British sectors was the infamous Berlin Wall, built in August 1961 to prevent the East Berlin population from fleeing East Germany, and torn down in November 1989 when the East Berliners revolted against their government.

Today, the remnants of The Wall are major attractions that remind inhabitants and tourists alike of the city's divided past (Figure 8.3.1). The entire course of The Wall is marked in the streets by a continuous trace of inlaid bricks. In some places, actual remnants of

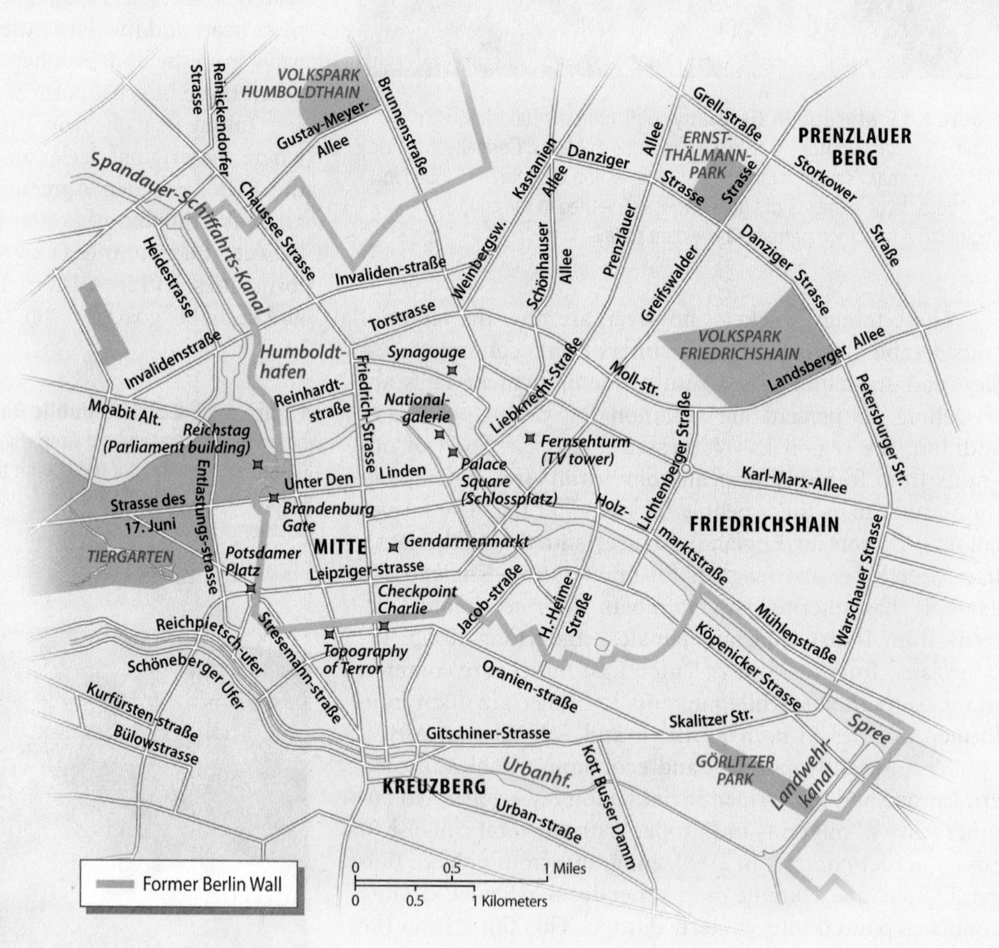

Figure 8.3.1 Map of Berlin This map shows the former extent of the infamous Berlin Wall that divided the city into two distinct political units until 1989. Note that the traditional city center, Mitte, was in former East Berlin.

complicated because of the large number of illegal migrants from the former Soviet lands, Africa, and South Asia.

The Landscapes of Urban Europe

One of the major characteristics of Europe's population is its high level of urbanization. All European countries except several Balkan countries have more than half their population in cities, while several western European countries, namely the United Kingdom and Belgium, are more than 90 percent urbanized. Once again, the gradient from the highly urbanized European heartland to the less-urbanized periphery reinforces the distinctions between the affluent, industrial core area and the more rural, less-well-developed periphery.

The Past in the Present North American visitors often find European cities far more interesting than our own because of the mosaic of historical and modern landscapes, of me-

The Wall remain, such as along Muehlenstrasse, where the concrete slabs have been painted by artists from throughout the world and form the extensive open-air "East Side Gallery." In other areas, vacant lots and boarded-up buildings provide mute testimony to The Wall's ominous presence (Figure 8.3.2).

The communist past also is found in the huge (1,207 feet, or 368 meters) TV tower on Alexanderplatz that dominates the city. It was built in the 1970s to block out Western television programs and broadcast state propaganda. The large square itself is ringed by huge, unattractive buildings that were formerly state department stores, mixed in with the gigantic multistory apartment houses that were the communist solution to postwar housing problems.

The historical city center (Mitte) of Berlin was in the former east zone and languished under the communists; now, with the Wall's demise and the city unified, this area is once more the vibrant heart of Berlin. Today, restored historic landmarks include the parliament, the Brandenburg Gate, the Berlin cathedral, and several opera houses, along with numerous museums. These buildings now form an attractive backdrop to the dynamic street life of this lively area. Rather surprisingly, within two years after communism fell, world-class department stores and hotels returned to the Mitte, proving wrong the many skeptics who thought the former East Berlin would never recover from its communist past.

Berlin's future is seen downtown around Potsdamer Square, where skyscrapers designed by world-famous architects blossom. These skyscrapers house corporate headquarters for companies that wager Berlin will become the new economic capital of the expanded EU.

Figure 8.3.2 The Berlin Wall Zone Still today, the former Wall zone is marked by fences, vacant lots, weeds, and boarded-up buildings.

dieval quarters and churches, interspersed with high-rise buildings and other contemporary structures. Three historical eras dominate most European city landscapes. The medieval (900–1500), Renaissance–Baroque (1500–1800), and industrial (1800–present) periods have each left characteristic marks on the European urban scene. Learning to recognize these stages of historical growth provides visitors to Europe's cities with fascinating insights into both past and present landscapes (Figure 8.17).

The **medieval landscape** is one of narrow, winding streets, crowded with three- or four-story masonry buildings with little setback from the street. This is a dense landscape with few open spaces, except around churches and public buildings. Here and there, public squares or parks are clues to medieval open-air marketplaces where commerce was transacted.

As picturesque as we find medieval-era districts today, they nevertheless present challenges to contemporary inhabitants because of their narrow, congested streets and old

Figure 8.17 Urban Landscapes This aerial view of Grosseto, Italy shows how the historic medieval city was encircled by the Renaissance-Baroque fortifications built to protect the settlement. Today, parks and public buildings are found in place of the former walls and moat.

housing. Often, modern plumbing and heating are lacking, and rooms and hallways are small and cramped compared to present-day standards. Because these medieval districts usually lack modern facilities, the majority of people living in them have low or fixed incomes. In many areas, this majority is made up of the elderly, university students, and ethnic migrants.

Many cities in Europe, though, are enacting legislation to upgrade and protect their historic medieval landscapes. This movement began in the late 1960s and has become increasingly popular as cultures have worked to preserve the uniqueness of special urban centers as their suburbs increasingly surrender to sprawl and shopping centers (Figure 8.18).

In contrast to the cramped and dense medieval landscape, those areas of the city built during the **Renaissance–Baroque** period produce a landscape that is much more open and spacious, with expansive ceremonial buildings and squares, monuments, ornamental gardens, and wide boulevards lined with palatial residences. During this period (1500–1800), a new artistic sense of urban planning arose in Europe that resulted in the restructuring of many European cities, particularly the large capitals. These changes were primarily for the benefit of the new urban elite—royalty and successful merchants. City dwellers of lesser means remained in the medieval quarters, which became increasingly crowded and cramped as more city space was devoted to the ruling classes.

During this period, city fortifications limited the outward spread of these growing cities, thus increasing density and crowding within. With the advent of assault artillery, European cities were forced to build an extensive system of defensive walls. Once encircled by these walls, the cities could not expand outward. Instead, as the demand for space increased within the cities, a common solution was to add several new stories to the medieval houses.

Industrialization dramatically altered the landscape of European cities. Historically, factories clustered together in cities beginning in the early 19th century, drawn by their large markets and labor force and supplied by raw materials shipped by barge and railroad. Industrial districts of factories and worker tenements grew up around these transportation lines. In continental Europe, where many cities retained their defensive walls until the late 19th century, the new industrial districts were often located outside the former city walls, removed from the historic central city. In Paris, for example, when the railroad was constructed in the 1850s, it was not allowed to enter the city walls. As a result, terminals and train stations for the network of tracks were located beyond the original fortifications. Although the walls of Paris are long gone, this pattern of train stations lying beyond the city's historical city walls exists still today.

Not to be overlooked are the post-WW2 changes to European cities as they rebuilt from the war's destruction, and later, have adapted to the political and economic demands of more recent decades (see "Cityscapes: Berlin Reinvents Itself"). Like North American cities, suburban sprawl has become an issue as people seek lower density housing in nearby rural environments. Unlike most North American cities, though, European urban areas generally have well-developed public transportation systems that offer alternatives to commuting by auto.

Figure 8.18 Historic Preservation of Cities Most European countries have laws and regulations designed to preserve and protect their cultural heritage, as embodied in historical cities. One of the earliest historic preservation projects in Germany was the inner city of Regensburg, on the Danube River.

CULTURAL COHERENCE AND DIVERSITY: A Mosaic of Differences

The rich cultural geography of Europe demands our attention for several reasons. First, the highly varied and fascinating mosaic of languages, customs, religions, ways of life, and landscapes that characterizes Europe has also strongly shaped regional identities that have all too often stoked the fires of conflict.

Second, European cultures have played leading roles in processes of globalization as European colonialism has brought about changes in languages, religion, economies, and values in every corner of the globe. If you question this, consider cricket games in Pakistan, high tea in India, Dutch architecture in South Africa, and the millions of French-speaking inhabitants of equatorial Africa. Before modern media technologies, European culture spread across the world, changing the speech, religion, belief systems, dress, and habits of millions of people on every continent.

Today, though, new waves of global culture are spreading back into Europe (Figure 8.19). While some European cultures embrace (or simply condone) these changes, other cultures actively resist. France, for example, struggles against both U.S.-dominated popular culture and the multicultural influences of its large migrant population.

Figure 8.19 Global Culture in Europe U.S. popular culture is received with great ambivalence in Europe. While many people embrace everything from fast food to Hollywood movies, at the same time they protest the loss of Europe's traditional regional cultures. This photo was taken in Prague, Czech Republic.

Geographies of Language

Language has always been an important component of nationalism and group identity in Europe (Figure 8.20). Today, while some small ethnic groups such as the Irish or the Bretons work hard to preserve their local language in order to reinforce their cultural identity, millions of Europeans are also busy learning multiple languages so they can communicate across cultural and national boundaries. The EU itself, while primarily an integrating force in Europe, honors this linguistic mosaic by recognizing more than 20 official languages.

As their first language, 90 percent of Europe's population speaks Germanic, Romance, or Slavic languages, all of which are linguistic groups of the Indo-European family. Germanic and Romance speakers each number almost 200 million in the European region. There are far fewer Slavic speakers (about 80 million) when Europe's boundaries are drawn to exclude Russia, Belarus, and Ukraine.

Germanic Languages Germanic languages dominate Europe north of the Alps. Today, about 90 million people speak German as their first language. This is the dominant language of Germany, Austria, Liechtenstein, Luxembourg, eastern Switzerland, and several small areas in Alpine Italy. Until recently, there were also large German-speaking minorities in Romania, Hungary, and Poland, but many of these people left eastern Europe when the Iron Curtain was lifted in 1990 and resettled in Germany.

English is the second-largest Germanic language, with about 60 million speakers using it as their first language. In addition, a large number of Europeans learn English as a second language, particularly in the Netherlands and Scandinavia, where many are as fluent as native speakers. Linguistically, English is closest to the Low German spoken along the coastline of the North Sea, which reinforces the linguistic theory that an early form of English evolved in the British Isles through contact with the coastal peoples of northern Europe. However, one of the distinctive traits of English that sets it apart from German is that almost one-third of the English vocabulary is made up of Romance words brought to England during the Norman French conquest of the 11th century.

Elsewhere in this region, Dutch (the Netherlands) and Flemish (northern Belgium) account for another 20 million people, and roughly the same number of Scandinavians speak the closely related languages of Danish, Norwegian, and Swedish. Icelandic, though, is a distinct language because of its geographic separation from its Scandinavian roots.

Romance Languages Romance languages, such as French, Spanish, and Italian, evolved from the vulgar (or everyday) Latin used within the Roman Empire. Today, Italian is the largest of these regional dialects, with about 60 million Europeans speaking it as their first language. In addition, Italian is

Figure 8.20 Language Map of Europe Ninety percent of Europeans speak an Indo-European language. These languages can be grouped into the major categories of Germanic, Romance, and Slavic languages. Ninety million Europeans speak German as a first language, which places it ahead of the 60 million who list English as their native language. However, given the large number of Europeans who speak fluent English as a second language, one could make the case that English is the dominant language of modern Europe.

an official language of Switzerland and is also spoken on the French island of Corsica.

French is spoken in France, western Switzerland, and southern Belgium, where it is known as *Walloon*. Today, there are about 55 million native French speakers in Europe. As with other languages, French also has very strong regional dialects. Linguists differentiate between two forms of French in France itself: that spoken in the north (the official form because of the dominance of Paris) and the language of the south, or *langue d'oc*. This linguistic divide expresses long-standing tensions between Paris and southern France. In the past decade, this strong regional awareness of the southwest (centered on Toulouse and the Pyrenees) has led to a rebirth of its own distinct language, *Occitanian*.

Spanish also has very strong regional variations. About 25 million people speak Castilian Spanish, the country's official language, which dominates the interior and northern areas of that large country. However, the *Catalan* form that some argue is a completely separate language is found along the eastern coastal fringe, centered on Barcelona, Spain's major city in terms of population and economy. This distinct language reinforces a strong sense of cultural separateness that has led to the state of Catalonia being given autonomous status within Spain.

Portuguese is spoken by another 12 million speakers in that country and in the northwestern corner of Spain, although considerably more people speak the language in Brazil, a former Portuguese colony in Latin America. Finally, Romanian represents the most eastern extent of the Romance language family; it is spoken by 24 million people in Romania. Though unquestionably a Romance language, Romanian also contains many Slavic words.

Figure 8.21 The Alphabet of Ethnic Tension With the departure of many of Kosovo's Serbs during recent ethnic unrest, signs using the Cyrillic alphabet were taken off many stores, shops, and restaurants as Kosovars of Albanian ethnicity restated their claims to the region. Here the Albanian owner of a restaurant in the capital city of Pristina scrapes off Cyrillic letters.

The Slavic Language Family Slavic is the largest European subfamily of the Indo-European languages. Traditionally, Slavic speakers are separated into northern and southern groups, divided by the non-Slavic speakers of Hungary and Romania.

To the north, Polish has 35 million speakers, and Czech and Slovakian about 14 million each. These numbers pale in comparison, however, with the number of northern Slav speakers in nearby Ukraine, Belarus, and Russia, where one can easily count more than 150 million. Southern Slav languages include three groups: 14 million Serbo-Croatian speakers (now considered separate languages because of the political differences between Serbs and Croats), 11 million Bulgarian-Macedonian, and 2 million Slovenian.

The use of two alphabets further complicates the geography of Slavic languages (Figure 8.21). In countries with a strong Roman Catholic heritage like Poland and the Czech Republic, the Latin alphabet is used in writing. In contrast, countries with close ties to the Orthodox church—Bulgaria, Montenegro, Macedonia, parts of Bosnia-Herzegovina, and Serbia—use the Greek-derived **Cyrillic alphabet**.

Geographies of Religion, Past and Present

Religion is an important component of the geography of cultural coherence and diversity in Europe because many of today's ethnic tensions result from historical religious events. To illustrate, strong cultural borders in the Balkans and eastern Europe are based upon the 11th-century split of Christianity into eastern and western churches, or between Christianity and Islam; in Northern Ireland, blood is still shed over the tensions between the 17th-century division of Christianity into Catholicism and Protestantism; and much of the ethnic cleansing terrorism in former Yugoslavia resulted from the historical struggle between Christianity and Islam in that part of Europe. In addition, there is considerable tension regarding the large Muslim migrant populations in England, France, and Germany. Understanding these important contemporary issues involves taking a brief look at the historical geography of Europe's religions (Figure 8.22).

The Schism Between Western and Eastern Christianity In southeastern Europe, early Greek missionaries spread Christianity through the Balkans and into the lower reaches of the Danube. Progress was slower than in western Europe, perhaps because of continued invasions by peoples from the Asian steppes. As well, Greek missionaries refused to accept the control of Roman bishops from western Europe.

This tension with western Christianity led to an official split of the eastern church from Rome in 1054. This eastern church subsequently splintered into Orthodox sects closely linked to specific nations and states. Today, for example, we find Greek Orthodox, Bulgarian Orthodox, and Russian Orthodox churches, all with different rites and rituals.

Another factor that distinguished eastern Christianity from western was the Orthodox use of the Cyrillic alphabet instead of the Latin. Because Greek missionaries were primarily responsible for the spread of early Christianity in southeastern Europe, it is not surprising that they used an alphabet based on Greek characters. More precisely, this alphabet is attributed to the missionary work of St. Cyril in the 9th century. As a result, the division between western and eastern churches, and between the two alphabets, remains one of the most prominent and problematic cultural boundaries in Europe.

Figure 8.22 Religions of Europe This map shows the divide in western Europe between the Protestant north and the Roman Catholic south. Historically, this distinction was much more important than it is today. Note the location of the former Jewish Pale, which was devastated by the Nazis during World War II. Today, ethnic tensions with religious overtones are found primarily in the Balkans, where adherents of Roman Catholicism, Eastern Orthodoxy, and Islam are found in close proximity to one another.

The Protestant Revolt Besides the division between western and eastern churches, the other great split within Christianity occurred between Catholicism and Protestantism. This division arose in Europe during the 16th century and has divided the region ever since, although with the exception of the Troubles in Northern Ireland, tensions today between these two major groups are far less problematic than in the past.

Conflicts with Islam Both eastern and western Christian churches struggled with challenges from Islamic empires to Europe's south and east. Even though historical Islam was reasonably tolerant of Christianity in its conquered lands, Christian Europe was far less accepting of Muslim imperialism. The first crusade to reclaim Jerusalem from the Turks took place in 1095. After the Ottoman Turks conquered Constantinople in 1453 and gained control over the Bosporus strait and the Black Sea, they moved rapidly to spread their Muslim empire throughout the Balkans and arrived at the gates of Vienna in the middle of the 16th century. There, Christian Europe stood firm and stopped Islam from expanding into western Europe.

Ottoman control of southeastern Europe, however, lasted until the empire's end in the early 20th century. This historical presence of Islam explains the current mosaic of religions in the Balkans, with intermixed areas of Muslims, Orthodox, and Roman Catholics.

Today, there are still tensions with Muslims as Europe struggles to address the concerns of its fast-growing immigrant population. Although many of these tensions transcend pure religious issues, there is nevertheless a strong undercurrent of concern about the varying attitudes of Muslims toward cultural assimilation in Europe. In addition, the EU must address the issue of Turkey's application to join the EU. Although Turkey is an avowed secular state, its population is predominantly Muslim, thus its membership in the EU is a contentious issue for Europe.

A Geography of Judaism Europe has long been a difficult homeland for Jews forced to leave Palestine during the Roman Empire. At that time, small Jewish settlements were located in cities throughout the Mediterranean. Later, by 900 CE, about 20 percent of the Jewish population was clustered in the Muslim lands of the Iberian Peninsula where Islam showed greater tolerance for Judaism than had Christianity. Furthermore, Jews played an important role in trade activities both within and outside the Islamic lands. After the Christian reconquest of Iberia, however, Jews once more faced severe persecution and fled from Spain to more tolerant countries in western and central Europe.

One focus for migration was the area in eastern Europe that became known as the Jewish Pale. In the late Middle Ages, at the invitation of the Kingdom of Poland, Jews settled in cities and small villages in what is now eastern Poland, Belarus, western Ukraine, and northern Romania (see Figure 8.22). Jews collected in this region for several centuries, in the hope of establishing a true European homeland, despite the poor natural resources of this marshy, marginal agricultural landscape.

Until emigration to North America began in the 1890s, 90 percent of the world's Jewish population lived in Europe, and most were clustered in the Pale. Even though many emigrants to the United States and Canada came from this area, the Pale remained the largest grouping of Jews in Europe until World War II. Tragically, Nazi Germany used this ethnic clustering to its advantage by focusing its extermination activities on the Pale.

In 1939, on the eve of World War II, there were 9.5 million Jews in Europe, or about 60 percent of the world's Jewish population. During the war, German Nazis murdered some 6 million Jews in the horror of the Holocaust. Today, fewer than 2 million Jews live in Europe. Since 1990 and the lifting of quotas on Jewish emigration from Russia, Belarus, and Ukraine, more than 100,000 Jews have emigrated to Germany, giving it the fastest-growing Jewish population outside Israel (Figure 8.23).

The Patterns of Contemporary Religion In Europe today, there are about 250 million Roman Catholics and fewer than 100 million Protestants. Generally, Catholics are found in the southern half of the region, except for significant numbers in Ireland and Poland, while Protestants dominate in the north. In addition, since World War II, there has been a noticeable loss of interest in organized religion, mainly in western Europe, which has led to declining church attendance in many areas. This trend is so marked that the term **secularization** is used, referring to the widespread movement away from the historically important organized religions of Europe.

Not to be overlooked are the 13 million Muslims in Europe. Most are migrants from Africa and southwestern Asia, although some are recent converts to Islam. More than half of

Figure 8.23 Jewish Synagogue in Berlin Before World War II, Berlin had a large and thriving Jewish population, as attested to by this synagogue built in 1866. However, during the Nazi period, the Jewish population was forcibly removed and largely exterminated. This synagogue was then used by the Nazis to store military clothing until it was heavily damaged by Allied bombing. It was restored and opened again in 1995 as a synagogue and museum.

Figure 8.24 Religious Tensions in Northern Ireland Despite considerable and continual efforts to forge peace between Protestants and Catholics in Northern Ireland, outbreaks of violence are still common. Here in Belfast, sectarian fighting broke out only hours after groups from both religions met to discuss ways to end the fighting.

the Muslim population (7.5 million) is found in France, with the second largest number (around 4 million) in Germany.

Catholicism dominates the religious geography and cultural landscapes of Italy, Spain, France, Austria, Ireland, and southern Germany. In these areas, large cathedrals, monasteries, monuments to Christian saints, and religious place-names draw heavily on pre-Reformation Christian culture. Because visible beauty is part of the Catholic tradition, elaborate religious structures and monuments are more common here than in Protestant lands.

Protestantism is most widespread in northern Germany, the Scandinavian countries, and England, and it is intermixed with Catholicism in the Netherlands, Belgium, and Switzerland. Because of its reaction against ornate cathedrals and statues of the Catholic Church, the landscape of Protestantism is much more sedate and subdued. Large cathedrals and religious monuments in Protestant countries are associated primarily with the Church of England, which has strong historical ties to Catholicism; St. Paul's Cathedral and Westminster Abbey in London are examples.

Tragically, another sort of religious landscape has emerged in Northern Ireland, where barbed-wire fences and concrete barriers separate Protestant and Catholic neighborhoods in an attempt to reduce violence between warring populations (Figure 8.24). Religious affiliation is a major social force that influences where people live, work, attend school, shop, and so on. Of the 1.6 million inhabitants of Northern Ireland, 54 percent are Protestant and 42 percent Roman Catholic. The Catholics feel that they have been discriminated against by the Protestant majority and have been treated as second-class citizens. As a result, they seek a closer relationship with the Republic of Ireland across the border to the south. The Protestant reaction is to forge even stronger ties with the United Kingdom. Unfortunately, these differences have been expressed in prolonged violence that has led to about 4,000 deaths since the 1960s, largely resulting from paramilitary groups from both factions who promote their political and social agendas through terrorist activities.

European Culture in a Global Context

Europe, like all other world regions, is currently caught up in a period of profound cultural change; in fact, many would argue that the pace of cultural change in Europe has been accelerated because of the complicated interactions between globalization and Europe's internal agenda of political and economic integration. While pundits celebrate the "New Europe" of integration and unification, other critics refer to a more tension-filled New Europe of foreign migrants and guest workers troubled by ethnic discrimination and racism.

Migrants and Culture Migration patterns are influencing the cultural mix in Europe. Historically, Europe spread its cultures worldwide through aggressive colonialization. Today, however, the region is experiencing a reverse flow, as millions of migrants move into Europe, bringing their own distinct cultures from the far-flung countries of Africa, Asia, and Latin America. Unfortunately, in some areas of Europe, the products of this cultural exchange are highly problematic (see "Global to Local: European Hip Hop and Hyperlocalism").

Ethnic clustering leading to the formation of ghettos is now common in the cities and towns of western Europe (Figure 8.25). The high-density apartment buildings of suburban Paris, for example, are home to large numbers of French-speaking Africans and Arab Muslims caught in the crossfire of high unemployment, poverty, and racial discrim-

Figure 8.25 Turkish Store in Germany This Turkish store in the Kreuzberg district of Berlin, sometimes referred to as "Little Istanbul," serves not only the large Turkish population, but also German shoppers who have found this store an appealing aspect of the country's broadening ethnicity. Unfortunately, to some, Europe's migrant cultures are more problematic.

GLOBAL TO LOCAL European Hip Hop and Hyperlocality

From its New York City roots in the 1970s, where hip hop blended the rapping narrative of traditional western African music with Jamaican-American influences, this dynamic musical form became a true cultural hybrid during its global travels as it was transformed and adapted to local languages and cultures (Figure 8.4.1). Despite these changes, the core of hip hop retains its original voice of protest by articulating the concerns and issues of subcultures and ethnic groups marginalized by global society.

North American hip hop first came to Europe in the early 1980s as a cultural transplant, dominated by English-language rap artists, their CDs, and music videos; however, within a decade hip hop had been completely transformed by young Europeans as it became the preferred voice for building cultural identities in specific places—-what sociologists refer to as "hyperlocality." By rapping in local languages and dialects, hip hop artists and their audiences created local communities of shared concerns, issues, outrages, alienation, values, and, in some parts of Europe, micro-nationalism. While uniquely European in many ways, this hyperlocal hip hop resonated with its American roots as the oppositional voice of racial and cultural discrimination.

No surprise, then, that in Europe hip hop first became popular with immigrants—South Asians and Africans in England, Italy, and France; Turkish youth in Germany; Surinamese in the Netherlands—as they reshaped it into their own music by rapping in their native languages. By the 1990s, youth from Europe's regional cultures folded in their micro-national concerns, resulting in a Basque hip hop in Spain and southwestern France, Occitanian rap in southern France, Sami in northern Scandinavia, Neapolitan in southern Italy, Roma in Hungary, and on and on. In the Balkans, with the breakup of the former Yugoslavia, hyperlocal rap became central to forging new cultural identities for the youth of Serbia, Croatia, Albania, and other newly independent countries.

In many cases European hip hop was not only rapped in local languages and dialects but also combined with traditional local musical expressions, leading to Barcelona's hip hop, for example, folding in a flamenco flavor, and Roma rap drawing upon traditional gypsy violin riffs.

Although hyperlocal hip hop is still a major part of the current European music scene as in North America, rap has moved far beyond its roots as ethnic protest music. Today, for example, one of the more popular Finnish groups from the Helsinki suburbs raps about middle-class (and middle-aged) *angst*. Although one might argue that hip hop has lost much of its earlier cultural power now that it's mainstreamed, hip hop continues to evolve and renew itself by rapping on topics drawn from current events. Examples would be the hip hop that arose during recent ethnic violence in French cities, or—more extreme—the use of hip hop as a recruiting tool by Islamic jihadists.

Figure 8.4.1 French Rapper, MC Solaar Claude M'Barail, aka MC Solaar, one of the French's biggest rap stars, was born in Senegal and was six months old when his parents emigrated to France where they lived in a Paris suburb populated by other migrants. MC Solaar released his first CD in 1990 when he was a philosophy graduate student, rapping about African migrant issues. In 2008 he won France's best urban music record of the year award.

ination. As a result, cultural battles have emerged in many European countries. For example, French leaders, unsettled by the country's large Muslim migrant population, attempted to speed assimilation of female high school students into French mainstream culture by banning a key symbol of conservative Muslim life: the head scarf. This rule triggered riots, demonstrations, and counterdemonstrations. As a result of these kinds of conflicts, the political landscape of many European countries now contains far-right, nationalistic parties with thinly disguised platforms of excluding migrants from their countries (Figure 8.26).

Figure 8.26 Neo-Nazis in Germany Purporting to embrace "pure and true" Aryan values, these neo-Nazi Germans provoke police at a rally against foreigners. A rise in extreme forms of nationalism has become increasingly problematic in many European countries, with discrimination against immigrants a major focus of their activities.

Figure 8.27 Geopolitical Issues in Europe While the major geopolitical issue of the early 21st century remains the integration of eastern and western Europe into the European Union, numerous issues of micro- and ethnic nationalism also engender geopolitical fragmentation. In other parts of Europe, such as Spain, France, and Great Britain, questions of local ethnic autonomy within the nation-state structure challenge central governments.

GEOPOLITICAL FRAMEWORK: A Dynamic Map

One of Europe's unique characteristics is its dense fabric of 41 independent states within a relatively small area; no other world region demonstrates such a mosaic of geopolitical division. Conventional wisdom posits that Europe invented the nation-state. Later, these political ideas founded in Europe fueled the flames of political independence and democracy worldwide, replacing Europe's colonial rule in Asia, Africa, and the Americas.

But Europe's geopolitical landscape has been as much problem as promise. Twice in the past century, Europe shed blood to redraw its political borders, and within the past several decades nine new states have appeared in Europe, more than half through war. Further, today's map of geopolitical troubles suggests that still more political fragmentation is possible in the near future (Figure 8.27).

Redrawing the Map of Europe Through War

Two world wars radically reshaped the geopolitical maps of 20th-century Europe (Figure 8.28). By the early 20th century, Europe was divided into two opposing and highly armed camps that tested each other for a decade before the outbreak of World War I in 1914. France, Britain, and Russia were allied against Germany and the Austro–Hungarian and Ottoman Empires, both of which controlled a complex assort of ethnic groups in central Europe and the Balkans. While World War I was referred to as the "war to end all wars," it fell far short of solving Europe's geopolitical problems. Instead, according to many experts, it made another European war unavoidable.

When Germany and Austria-Hungary surrendered in 1918, the Treaty of Versailles peace process set about redrawing the map of Europe with two goals in mind: first, to punish the losers through loss of territory and severe financial reparations and, second, to recognize the nationalistic aspirations of unrepresented peoples by creating new nation-states. As a result, the new states of Czechoslovakia and Yugoslavia were born. In addition, Poland was reestablished, as were the Baltic states of Finland, Estonia, Latvia, and Lithuania.

Though the goals of the treaty were admirable, few European states were satisfied with the resulting map. New states were resentful when their ethnic citizens were left outside the new borders and became minorities in other newly shaped states. This created an epidemic of **irredentism**, or state policies for reclaiming lost territory and peoples. Examples are the large German population in the western portion of the newly created state of Czechoslovakia and the Hungarians stranded in western Romania by border changes.

This imperfect geopolitical solution was greatly aggravated by the global economic depression of the 1930s that brought high unemployment, food shortages, and political unrest to Europe. Three competing ideologies promoted their own solutions to Europe's pressing problems: Western democracy (and capitalism), communism from the Soviet revolution to the east, and a fascist totalitarianism promoted by Mussolini in Italy and Hitler in Germany. With industrial unemployment commonly approaching 25 percent in western Europe, public opinion fluctuated wildly between the extremist solutions of fascism and communism. In 1936, Italy and Germany joined forces through the Rome–Berlin "axis" agreement. As in World War I, this alignment was countered with mutual protection treaties between France, Britain, and the Soviet Union. When an imperialist Japan signed a pact with Germany, the scene was set for a second global war.

Nazi Germany tested Western resolve in 1938 by first annexing Austria, the country of Hitler's birth, and then Czechoslovakia, under the pretense of providing protection for ethnic Germans located there. After signing a nonaggression pact with the Soviet Union, Hitler invaded Poland on September 1, 1939. Two days later, France and Britain declared war on Germany. Within a month, the Soviet Union moved into eastern Poland, the Baltic states, and Finland to reclaim territories lost through the peace treaties of World War I. Nazi Germany then moved westward and occupied Denmark, the Netherlands, Belgium, and France, and then began preparations to invade England.

In 1941 the war took several startling new turns. In June, Hitler broke the nonaggression pact with the Soviet Union and, catching the Red Army by surprise, took the Baltic states and drove deep into Soviet territory. When Japan attacked Pearl Harbor, Hawaii, in December, the United States entered the war in both the Pacific and Europe.

By early 1944, the Soviet army had recovered most of its territorial losses and moved against the Germans in eastern Europe, beginning the long communist domination in that region. By agreement with the Western powers, the Red Army stopped when it reached Berlin in April 1945. At that time, Allied forces crossed the Rhine River and began their occupation of Germany. With Hitler's suicide, Germany signed an unconditional surrender on May 8, 1945, ending the war in Europe. But with Soviet forces firmly entrenched in the Baltic states, Poland, Czechoslovakia, Bulgaria, Romania, Hungary, Austria, and eastern Germany, the military battles of World War II were quickly replaced by the ideological Cold War between communism and democracy that lasted 45 years, until 1990.

A Divided Europe, East and West

From 1945 until 1990, Europe was divided into two geopolitical and economic blocs, east and west, separated by the infamous **Iron Curtain** that descended shortly after the peace agreement of World War II. East of the Iron Curtain border, the Soviet Union imposed the heavy imprint of communism on all activities—political, economic, military, and cultural. To the west, as Europe rebuilt from the destruction of the war, new alliances and institutions were created to counter the Soviet presence in Europe.

Cold War Geography The seeds of the Cold War are commonly thought to have been planted at the Yalta Conference of February 1945, when Britain, the Soviet Union, and the

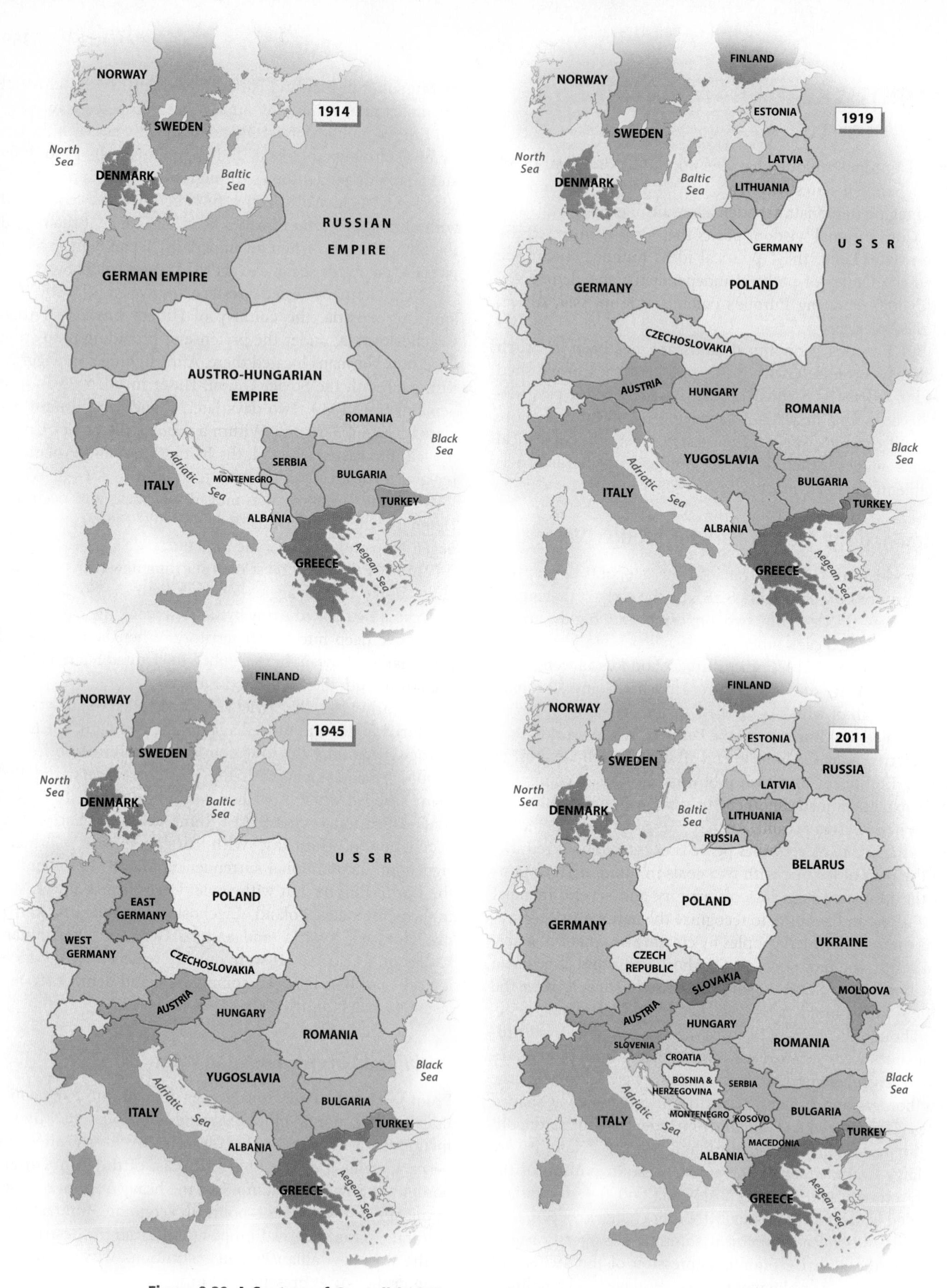

Figure 8.28 A Century of Geopolitical Change At the outset of the 20th century, central Europe was dominated by the German, Austro-Hungarian (or Hapsburg), and Russian empires. Following World War I, these empires were largely replaced by a mosaic of nation-states. More border changes followed World War II, largely as a result of the Soviet Union's turning that area into a buffer zone between itself and western Europe. With the demise of Soviet hegemony in 1989, further political change took place.

United States met to plan the shape of postwar Europe. Because the Red Army was already in eastern Europe and moving quickly on Berlin, Britain and the United States agreed that the Soviet Union would occupy eastern Europe and the Western allies would occupy parts of Germany.

The larger geopolitical issue, though, was the Soviet desire for a **buffer zone** between its own territory and western Europe. This buffer zone consisted of an extensive bloc of satellite countries, dominated politically and economically by the Soviet Union, that could cushion the Soviet heartland against possible attack from western Europe. In the east, the Soviet Union took control of the Baltic states, Poland, Czechoslovakia, Hungary, Bulgaria, Romania, Albania, and Yugoslavia. Austria and Germany were divided into occupied sectors by the four (former) allied powers. In both cases, the Soviet Union dominated the eastern portion of each country, areas that contained the capital cities of Berlin and Vienna. Both capital cities, in turn, were divided into French, British, U.S., and Soviet sectors.

In 1955, with the creation of an independent and neutral Austria, the Soviets withdrew from their sector, effectively moving the Iron Curtain eastward to the Hungary–Austria border. However, Germany quickly evolved into two separate states, West Germany and East Germany, that remained separate until 1990.

Along the border between east and west, two hostile military forces faced each other for almost half a century. Both sides prepared for and expected an invasion by the other across the barbed wire of a divided Europe (Figure 8.29). In small military units from satellite countries, both NATO (North Atlantic Treaty Organization) and the Warsaw Pact countries were armed with nuclear weapons, making Europe a tinderbox for a devastating third world war.

Berlin was the flashpoint that brought these forces close to a fighting war on two occasions. In winter 1948, the Soviets imposed a blockade on the city by denying Western powers access to Berlin across its East German military sector. This attempt to starve the city into submission by blocking food shipments from western Europe was thwarted by a nonstop airlift of food and coal by NATO. Then, in August 1961, the Soviets built the Berlin Wall to curb the flow of East Germans seeking political refuge in the west. The Wall became the concrete-and-mortar symbol of a firmly divided postwar Europe. For several days while the Wall was being built and the West agonized over destroying it, NATO and Warsaw Pact tanks and soldiers faced each other with loaded weapons at point-blank range. Though war was avoided, the Wall stood for 28 years, until November 1989.

Figure 8.29 The Iron Curtain During the Cold War, when Germany was divided into east and west, villages were often split in half by the Iron Curtain border with communist East Germany. Here, the village is truncated by the communist "death strip" in the foreground, where those who attempted to flee East Germany were shot on sight.

The Cold War Thaw The symbolic end of the Cold War in Europe came on November 9, 1989, when East and West Berliners joined forces to rip apart the Berlin Wall with jackhammers and hand tools (Figure 8.30) (see "Politics and the Environment: The Iron Curtain Green Zone").

By October 1990, East and West Germany were officially reunified into a single nation-state. During this period, all other Soviet satellite states, from the Baltic Sea to the Black Sea, also underwent major geopolitical changes that have resulted in a mixed bag of benefits and problems.

The Cold War's end came as much from a combination of problems within the Soviet Union (discussed in Chapter 9) as from rebellion in eastern Europe. By the mid-1980s, the Soviet leadership was advocating an internal economic restructuring and also recognizing the need for a more open dialogue with the West. Financial problems from supporting a huge military establishment, along with heavy losses from an unsuccessful war in Afghanistan, lessened the Soviet appetite for occupying other countries.

In August 1989, Poland elected the first noncommunist government to lead an eastern European state since World War II. Following this, with just one exception, peaceful revolutions with free elections spread throughout eastern

(a)

(b)

Figure 8.30 The Berlin Wall In August 1961, the East German and Soviet armies built a concrete and barbed-wire structure, known simply as "the Wall," to stem the flow of East Germans leaving the Soviet zone. It was the most visible symbol of the Cold War until November 1989, when Berliners physically dismantled The Wall after the Soviet Union renounced its control over eastern Europe.

Europe as communist governments renamed themselves and broke with doctrines of the past. In Romania, though, street fighting between citizens and military resulted in the violent overthrow and execution of communist dictator Nicolae Ceausescu.

As a result of the Cold War thaw, the map of Europe began changing once again. Germany reunified in 1990. Elsewhere, political separatism and ethnic nationalism, long suppressed by the Soviets, were unleashed in southeastern Europe as the former Yugoslavia has fractured into a number of independent states. Similarly, on January 1, 1993, Czechoslovakia was replaced by two separate states, the Czech Republic and Slovakia.

The Balkans: Waking from a Geopolitical Nightmare

The Balkans have been a troublesome area for several centuries with their complex mixture of languages, religions, and ethnic allegiances, which, throughout history, has led to a political and often changing mosaic of small countries (Figure 8.31). Indeed, the term **balkanization** is used to describe the geopolitical processes of small-scale independence movements based upon ethnic fault lines. Following the fall of two competing empires in the early 20th century, much of the region was unified under the political umbrella of the former Yugoslavian state. In the 1990s, however, Yugoslavia broke apart as ethnic factionalism and nationalism produced a decade of violence, turmoil and wars of independence, creating a geopolitical nightmare for Europe, the EU, NATO, and the world. Today, though, despite lingering tensions in several areas, there are signs that the Balkan countries are moving toward a new era of peace and stability.

Geo-Historical Background As World War I ended in 1918, so did the Austro-Hungarian and Ottoman Turk empires' control of the Balkans. In its place the independent kingdom of Serbia became the core for creating the new state of Yugoslavia in 1929. Attached to Serbia were a collection of former Hapsburg lands, including what are now the states of Slovenia, Croatia, Bosnia and Herzegovina, Montenegro, Macedonia, and Kosovo. The result was a new political entity made up of three major religions (Islam, Catholicism, and Eastern Orthodoxy), a handful of languages, at least eight different ethnic groups, and two distinct alphabets. From the beginning, a Serbian king ruled over this new country, and Serbs dominated political and economic activity. Because Belgrade and the Serbian core amassed wealth at the expense of the ethnic periphery, Yugoslavia was marked by internal tension and hostility from its earliest years.

In 1941 Yugoslavia was invaded by Hitler and his Nazi army. While German troops were welcomed as liberators in Croatia, a region that was particularly restless under Serbian rule, resistance to the Nazi occupation came from two fronts: Serbian royalists who supported the king and opposed Hitler, and Josip Tito's band of resistance fighters who were supported by the Allies in their battle against the Nazis.

Following World War II, Britain, the United States, and the Soviet Union rewarded Tito for his efforts by backing him as the leader of a new socialist Yugoslavia. Conventional wis-

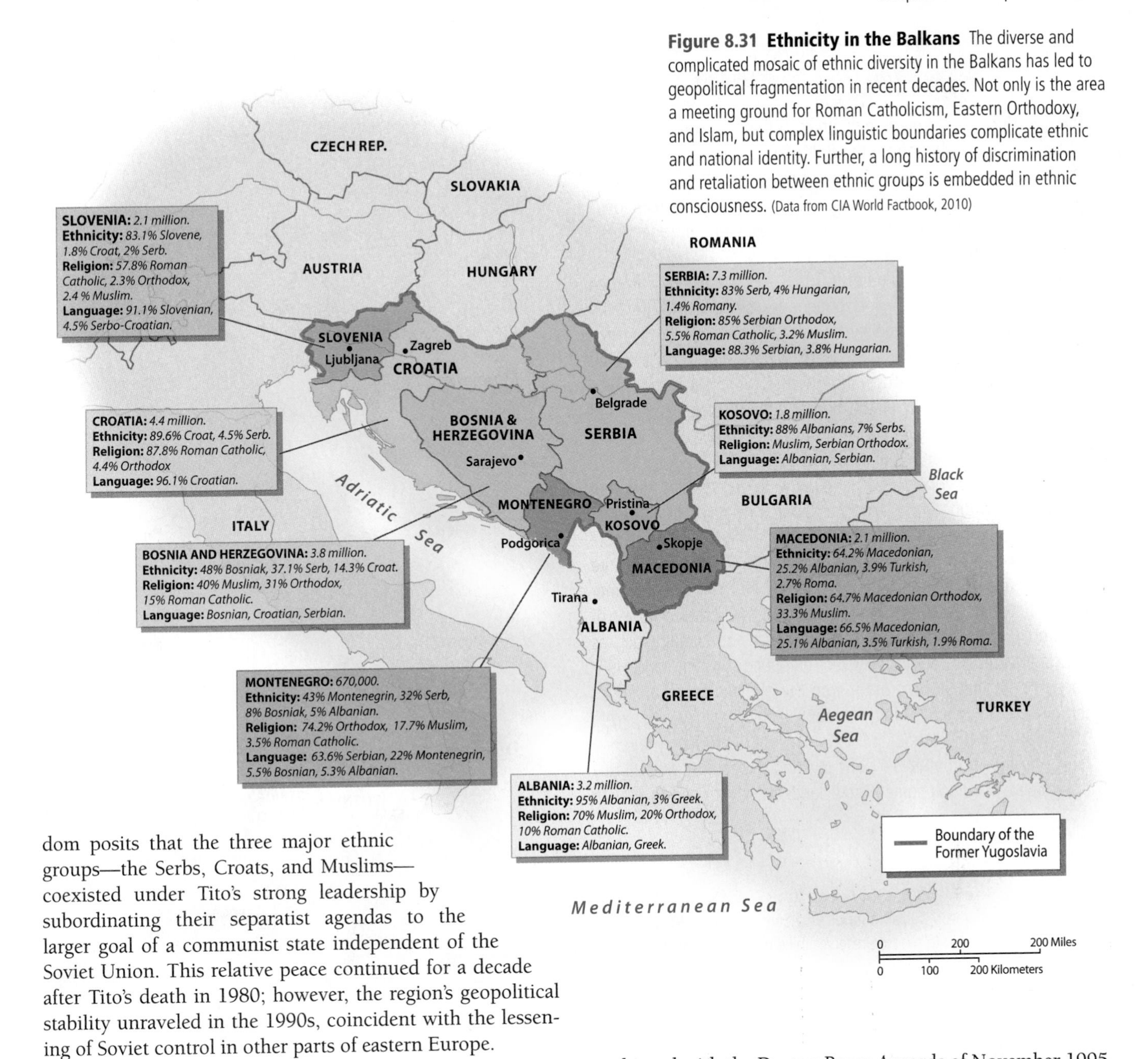

Figure 8.31 Ethnicity in the Balkans The diverse and complicated mosaic of ethnic diversity in the Balkans has led to geopolitical fragmentation in recent decades. Not only is the area a meeting ground for Roman Catholicism, Eastern Orthodoxy, and Islam, but complex linguistic boundaries complicate ethnic and national identity. Further, a long history of discrimination and retaliation between ethnic groups is embedded in ethnic consciousness. (Data from CIA World Factbook, 2010)

dom posits that the three major ethnic groups—the Serbs, Croats, and Muslims—coexisted under Tito's strong leadership by subordinating their separatist agendas to the larger goal of a communist state independent of the Soviet Union. This relative peace continued for a decade after Tito's death in 1980; however, the region's geopolitical stability unraveled in the 1990s, coincident with the lessening of Soviet control in other parts of eastern Europe.

Balkan Wars of Independence During 1990 elections were held in Yugoslavia's different republics over the issue of secession from the mother state. While secessionist parties gained control in Slovenia and Croatia, Serbian voters opted for continued Yugoslav unification in what observers considered to be a government-controlled election..

Nevertheless, Slovenia and Croatia declared independence in 1991, Macedonia in January 1992, and Bosnia and Herzegovina in April 1992. When the Yugoslavian army attacked Slovenia, the Balkan situation got Europe's full attention and a negotiated settlement resulted in that country's independence. But in Bosnia Serb paramilitary units waged a ferocious war of ethnic cleansing against Muslims and Croats, a war that lasted until 1995. In Croatia, fighting between Serbs and Croats continued until a fragile peace was achieved with the Dayton Peace Accords of November 1995. In this treaty Serbia agreed to respect a new set of boundaries between itself and its newly independent neighbors. Sixty thousand NATO troops, including 20,000 Americans, were brought in to enforce the peace treaty. In Bosnia and Herzegovina two complex political entities were created, a Serb republic and a Muslim–Croat federation, both ruled by the same legislature and president.

Kosovo, to the south of Serbia (as the remnant Yugoslavia was now called), was another trouble spot with longstanding tensions between Serbs and Muslims Because of this complicated history, Kosovo had enjoyed differing degrees of autonomy in the former Yugoslavia. In 1990, however, this autonomy was removed by Belgrade to protect the Serb minority population, and, not surprisingly, the Muslim Kosovar rebels responded by proclaiming Kosovo's

ENVIRONMENT AND POLITICS The Iron Curtain Green Zone

The Iron Curtain—formerly a death strip that separated countries, villages, and neighbors from the Baltic to the Black Seas—is now becoming a 5,000 mile (8,500 km) long greenbelt of nature preserves, hiking trails, bike paths, and picnic spots, a living testimonial to replacing past politics and animosities with an agenda of east–west cooperation (Figure 8.5.1).

During the Cold War the Iron Curtain was not merely a barbed wire fence marking East–West boundaries but a heavily fortified border zone several hundred yards wide (200 meters), consisting of military patrol trails and roads, guard towers, and, on the Soviet side, minefields. With the opening of eastern Europe in the 1990s and the demilitarization of the border zone, ecology and nature preservation groups lobbied the EU to convert the border zone into a pan-European greenbelt and eco-zone.

While the greenbelt agenda is moving forward at different speeds and in different ways throughout the former Iron Curtain border zone, from Scandinavia to the Balkans, considerable progress has been made already in Germany, where 150 nature preserves are found along the 862 miles (1393 km) of the former border between East and West Germany. Unlike other portions of the Iron Curtain that followed international boundaries (such as that between Austria and the Czech Republic, Slovakia, and Hungary), the German portion of the Iron Curtain was drawn along internal state boundaries when Germany was formally partitioned in 1949. Upon reunification in 1990 what had previously been a highly fortified potential war zone dividing the country was quickly dismantled. Although some border lands were returned to private

Figure 8.5.1 The Iron Curtain Green Zone This map depicts the extent of the former Iron Curtain that divided Europe until 1990, and is now being converted into nature preserves, bike paths, and parks.

independence in 1991. This act was resisted vigorously by Serbia, which responded with a violent ethnic cleansing program designed to oust the Muslims and make Kosovo a pure Serbian province (Figure 8.32). Diplomatic efforts failed to resolve this violence, and as warfare escalated in Kosovo, NATO (which included the United States) began bombing Belgrade in 1999 to force Serbia to accept a negotiated settlement. From 1999 until 2008, Kosovo was administered by the UN as a protectorate, enforced by some 50,000 peacekeepers from 30 countries.

In 2008 the UN-sanctioned Kosovo parliament once again declared the country's independence. Although Serbia refuses to recognize Kosovo as an independent state, over 60 countries (including the United States) have. However, Russia, a staunch ally of Serbia, has not, adding to the tensions between Russia and the United States. At Serbia's request the UN World Court reviewed the legality of Kosovo's independence and ruled in July 2010 that it was indeed legal. Serbia, however, does not accept this ruling.

Moving Toward Stability Although Serbia remains steadfast about retaining Kosovo, in most other matters a more moderate and less nationalistic government has led to Serbia's being reinstated in the UN and the Council of Europe. Except for the matter of Kosovo, Serbian diplomatic relations with the United States and other countries have been normalized, ending a long period of isolation and offering hope that the country might peacefully resolve its long history of internal ethnic tensions. Of the former Yugoslav republics only Slovenia has achieved membership in the European Union; all others—including Serbia—have now initiated formal talks requesting EU membership.

landowners in 1996, 90 percent of the boundary land is currently publicly owned, an important fact facilitating the creation of nature preserves and parks. A hike and bike trail (the Iron Curtain Trail, or ICT) runs the full length of the past border (Figure 8.5.2).

Not that all traces of the Cold War were completely erased for at several places guard towers and border markers have been left as museum pieces and reminders of the Cold War. At Point Alpha in the Fulda Gap of southern Germany, the former flashpoint where western military strategists expected a Soviet invasion, an open air military museum, complete with an intact section of the Iron Curtain, testifies to the tense decades of the Cold War when armies faced each other at pointblank distance.

Farther south, in the Balkan region, where the landscape shows scars of more recent wars, the Europe greenbelt follows the boundaries of the former Yugoslavia, encircles Albania (which, although a communist country until 1990, was wary of the Soviet Union and fortified its boundaries against invasion with bunkers and tank traps), and then follows the Bulgaria–Greek and Turkey boundary to the Black Sea. While less formalized than in Germany, the greenbelt seems to be acting as a wildlife corridor for bears, lynx, and wolves as they expand their territory. Ecologists report, however, that deer, unlike other species, are still wary of the Iron Curtain, and use their traditional pathways that parallel—but do not cross—the former boundary.

Figure 8.5.2 The Iron Curtain Trail This photo shows a part of the former Iron Curtain in Germany that is now a hike and bike trail. On the right side of the photo are younger trees that have grown over the last 20 years on the former East German death strip.

Figure 8.32 Kosovar Refugees Thousands of Muslim Kosovars were forced to flee the Yugoslavian province of Kosovo during a period of ethnic cleansing by Serbian forces in the late 1990s. This violence led to aerial bombardment of former Yugoslavia by NATO planes to force Serbian leaders to accept UN peacekeepers in the province. After a long period as a NATO Protectorate, Kosovo declared its independence in 2008, an act Serbia has challenged in the World Court.

ECONOMIC AND SOCIAL DEVELOPMENT: Integration and Transition

As the acknowledged birthplace of the Industrial Revolution, Europe in many ways invented the modern economic system of industrial capitalism. Though Europe was the world's industrial leader in the early 20th century, it was later eclipsed by both Japan and the United States as the region struggled to cope with the effects of two world wars, a decade of global depression, and the more recent Cold War.

In the past 50 years, however, economic integration guided by the EU has been increasingly successful. In fact, western Europe's success at blending national economies has given the world a new model for regional cooperation, an approach that in the near future may be imitated in Latin America and Asia. Eastern Europe, however, has not fared as well. The results of four decades of Soviet economic planning were, at best, mixed. The total collapse of that system in 1990 cast eastern Europe into a period of chaotic economic, political, and social transition that has resulted in a highly differentiated pattern of rich and poor regions. While some countries, such as the Czech Republic and Slovenia, prosper, the future prospects for Albania, Hungary, and Romania are uncertain (Table 8.2).

Accompanying western Europe's economic boom has been an unprecedented level of social development as measured by worker benefits, health services, education, literacy, and gender equality. Though the improved social services set an admirable standard for the world, cost-cutting politicians and businesspeople now argue that these services increase the cost of business so that European goods cannot compete in the global marketplace. As a result, and aggravated by the recent global economic recession, many of these traditional benefits, namely job security and long vacation periods, have been eroded, causing considerable social tension in many European countries.

Europe's Industrial Revolution

Europe is the cradle of modern industrialism. Two fundamental changes were associated with this industrial revolution: first, machines replaced human labor in many manufacturing processes, and, second, inanimate energy sources, such as water, steam, electricity, and petroleum, powered the new machines. Though we commonly apply the term *Industrial Revolution* to this transformation, implying rapid change, in reality, it took more than a century for the interdependent pieces of the industrial system to come together. This new system emerged first in England between 1730 and 1850. Later in the 19th century, this new industrialism had spread to throughout Europe, and, within decades, to the rest of the world.

Centers of Change England's textile industry, located on the flanks of the Pennine Mountains, was the center of early industrial innovation, which took place in small towns and villages away from the rigid control of the urban guilds. The county of Yorkshire, on the eastern side of the Pennines, had been a center of woolen textile making since medieval times, drawing raw materials from the extensive sheep herds of that region, and using the clean mountain waters to wash the wool before it was spun. By the 1730s, water wheels were used to power mechanized looms at the rapids and waterfalls of the Pennine streams (Figure 8.33). By the 1790s, though, the steam engine had become the preferred source of energy to drive the new looms. However, steam engines needed fuel, and only with the building of railroads after 1820 could coal be moved long distances at a reasonable cost.

Development of Industrial Regions in Continental Europe

By the 1820s, the first industrial districts had begun appearing in continental Europe. These areas were, and largely still are, near coalfields (Figure 8.34). The first area outside Britain was the Sambre-Meuse region, named for the two river valleys straddling the French–Belgian border. Like the English Midlands, it also had a long history of cottage-based wool textile manufacturing that quickly converted to the new technology of steam-powered mechanized looms. In addition, a metalworking tradition drew on charcoal-based iron foundries in the nearby forests of the Ardenne Mountains. Coal was also found in these mountains, and in 1823, the first blast furnace outside Britain started operation in Liege, Belgium.

Figure 8.33 Water Power and Textiles Europe's industrial revolution began on the flanks of England's Pennine Mountains, where swift-running streams and rivers were used to power large cotton and wool looms. Later, many of these textile plants switched to coal power.

TABLE 8.2 Development Indicators

Country	GNI per capita, PPP 2008	GDP Average Annual % Growth 2000–08	Human Development Index (2007)[a]	Percent Population Living Below $2 a Day	Life Expectancy (2010)[b]	Under Age 5 Mortality Rate (1990)	Under Age 5 Mortality Rate (2008)	Gender Equity (2008)[c]
Western Europe								
Austria	36,360	2.2	0.955		80	9	4	97
Belgium	35,380	2.0	0.953		80	10	5	98
France	33,280	1.8	0.961		81	9	4	100
Germany	35,950	1.2	0.947		80	9	4	98
Liechtenstein			0.951		80			
Luxembourg	52,770		0.960		80			
Monaco								
Netherlands	40,620	1.9	0.964		80	8	5	98
Switzerland	39,210	1.9	0.960		82	8	5	97
United Kingdom	36,240	2.5	0.947		80	9	6	102
Eastern Europe								
Bulgaria	11,370	5.8	0.840	<2	73	18	11	97
Czech Republic	22,890	4.6	0.903	<2	77	12	4	101
Hungary	18,210	3.6	0.879	<2	74	17	7	99
Poland	16,710	4.4	0.880	<2	76	17	7	99
Romania	13,380	6.4	0.837	4.1	73	32	14	99
Slovakia	21,460	6.3	0.880	<2	75	15	8	100
Southern Europe								
Albania	7,520	5.4	0.818	7.8	75	46	14	
Bosnia and Herzegovina	8,360	5.4	0.812	<2	75	23	15	100
Croatia	17,050	4.5	0.871	<2	76	13	6	102
Cyprus	24,980		0.914		79			
Greece	28,300	4.2	0.942		80	11	4	97
Italy	30,800	1.0	0.951		82	10	4	99
Macedonia	9,250	3.2	0.817	5.3	74	36	11	98
Malta	20,580		0.902		80			
Montenegro	13,420		0.834	<2	82			
Portugal	22,330	0.9	0.909		74	15	4	101
San Marino					79			
Serbia	10,380	5.4	0.826	<2	74	29	7	102
Slovenia	27,160	4.4	0.929	<2	79	10	4	99
Spain	30,830	3.3	0.955		81	9	4	103
Northern Europe								
Denmark	37,530	1.6	0.955		79	9	4	102
Estonia	19,320	7.4	0.883	<2	74	18	6	101
Finland	35,940	3.0	0.959		80	7	3	102
Iceland	25,300		0.969		81			
Ireland	35,710	5.0	0.965		79	9	4	103
Latvia	16,010	8.2	0.866	<2	73	17	9	100
Lithuania	17,170	7.7	0.870	<2	72	16	7	100
Norway	59,250	2.4	0.971		81	9	4	99
Sweden	37,780	2.8	0.963		81	7	3	99

[a] *United Nations, Human Development Report, 2009.*

[b] *Population Referene Bureau, World Population Data Sheet, 2010.*

[c] *Gender Equity—Ratio of female to male enrollments in primary and secondary scchool. Numbers below 100 have more males in primary/secondary school, numbers above 100 have more females in primary/secondary school.*

Source: World Bank, World Development Indicators, 2010.

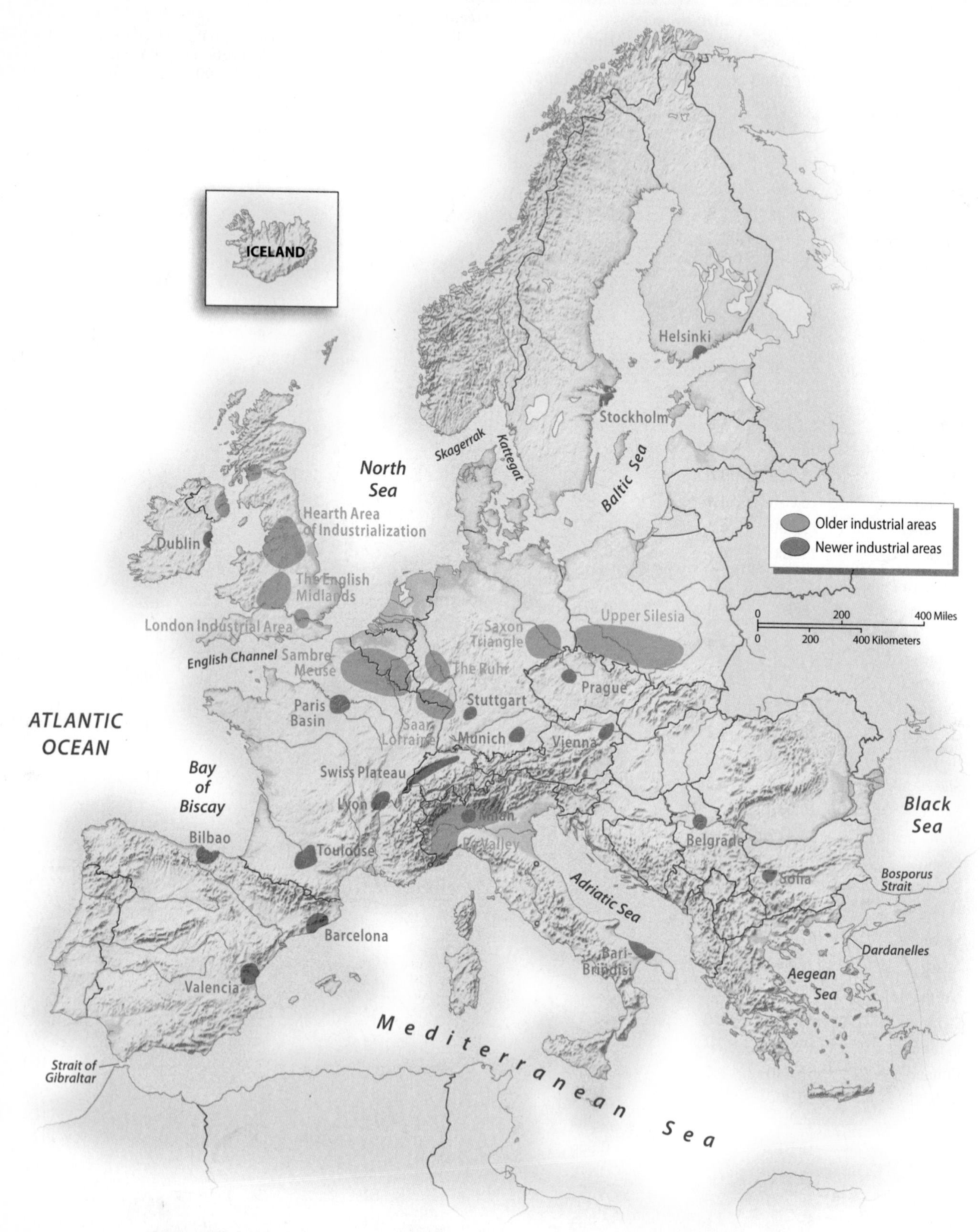

Figure 8.34 Industrial Regions of Europe From England, the Industrial Revolution spread to continental Europe, starting with the Sambre-Meuse region on the French–Belgian border and then diffusing to the Ruhr area in Germany. Readily accessible surface coal deposits powered these new industrial areas. Early on, iron ore for steel manufacture came from local deposits, but later it was imported from Sweden and other areas in the shield country of Scandinavia. Most of the newer industrial areas are closely linked to urban areas.

Figure 8.35 The Ruhr Industrial Landscape Long the dominant industrial region in Europe, the Ruhr region was bombed heavily during World War II because of its central role in providing heavy arms to the German military. After the war the area was rebuilt and modernized, yet more recently has faced difficult times with the reorientation of the global steel industry.

By the second half of the 19th century, the dominant industrial area in all Europe (including England) was the Ruhr district in northwestern Germany, near the Rhine River. Rich coal deposits close to the surface fueled the Ruhr's transformation from a small textile region to a region oriented around heavy industry, particularly iron and steel manufacturing. By the early 1900s, the Ruhr had used up its modest iron ore deposits and was importing ore from Sweden, Spain, and France. Several decades later, the Ruhr industrial region became synonymous with the industrial strength behind Nazi Germany's war machine and thus was bombed heavily in World War II. Although rebuilt during the 1950s, the Ruhr district has more recently lost much of its industrial vitality due to a shortage of coal resources, the reduced demand for German steel, and vexatious pollution issues (Figure 8.35).

Rebuilding Postwar Europe: Economic Integration in the West

As noted, Europe was unquestionably the leader of the industrial world in the early 20th century when it produced 90 percent of the world's manufactured output. However, four decades of political and economic chaos left Europe divided and in shambles. By the mid-20th century, its cities were in ruins; industrial areas were destroyed; vast populations were dispirited, hungry, and homeless; and millions of refugees moved about Europe, looking for safety and stability.

ECSC and EEC In 1950, western Europe began discussing a new form of economic integration that would avoid both the historical pattern of nationalistic independence through tariff protection and the economic inefficiencies resulting from this duplication of industrial effort. Robert Schuman, France's foreign minister, proposed that German and French coal and steel production be coordinated by a new supranational organization. In May 1952, France, Germany, Italy, the Netherlands, Belgium, and Luxembourg ratified a treaty that joined them in the European Coal and Steel Community (ECSC). Because of the immediate success of the ECSC, these six states soon agreed to work toward further integration by creating a larger European common market that would encourage the free movement of goods, labor, and capital. In March 1957, the Treaty of Rome was signed, establishing the European Economic Community (EEC).

From the EEC to the EU The EEC reinvented itself in 1965 with the Brussels Treaty that laid the groundwork for adding a political union to the already successful economic community. In this "second Treaty of Rome," aspirations for more than economic integration were clearly stated, with the creation of an EEC council, court, parliament, and political commission. The EEC also changed its name to the European Community (EC).

In 1991, the EC became the European Union (EU) as it expanded its goals yet again, this time with the Treaty of Maastricht (named after the town in the Netherlands in which delegates met). While economic integration remains an underlying theme in the new constitution, particularly with its commitment to a single currency through the European Monetary Union (discussed below), the EU has moved further into supranational affairs with discussion of common foreign policies and mutual security agreements.

More recently, in May 2004, the EU added 10 new states, including a core group of former Soviet-controlled communist countries from eastern Europe. These new members—Latvia, Estonia, Lithuania, Poland, Slovakia, the Czech Republic, Hungary, Slovenia, Malta, and Cyprus—brought the total to 25; Bulgaria and Romania were then admitted in January 2007, resulting in a total of 27 EU countries (Figure 8.36). As of this writing (late 2010), Iceland, several Balkan countries, and Turkey have formally applied for EU membership.

Figure 8.36 The European Union The driving force behind Europe's economic and political integration has been the European Union (EU), which was formed in the 1950s as an organization with six members focused solely on rebuilding the region's coal and steel industries. As of 2011, the EU has 27 members. Besides the official applicants of Turkey, Macedonia, Croatia, and Iceland, several Balkan countries are in earlier stages of applications. Note that Norway is not a member of the EU, primarily because membership would restrict that country's fishing industry. Those same fishing restrictions may also become problematic for Iceland's membership.

Euroland: The European Monetary Union As any world traveler knows, each state usually has its own monetary system. As a result, crossing a political border often means changing money and becoming familiar with a new system of bills and coins. In the recent past, travelers bought pounds sterling in England, marks in Germany, francs in France, lira in Italy, and so on. Today, however, much of Europe has moved from individual state monetary systems to a common currency. On January 1, 1999, in a major advance toward a united Europe, 11 of the then 15 EU member states joined to form the European Monetary Union (EMU). As of that day, cross-border business and trade transactions began taking place in the new monetary unit, the *euro*. Then, on January 1, 2002, new euro coins and bills became available for everyday use. This currency completely replaced the different national currencies of **Euroland**, those countries belonging to the EMU, in July 2002.

By adopting a common currency, Euroland members increased the efficiency and competitiveness of both domestic and international business. Formerly, when products were traded across borders, there were transaction costs associated with payments made in different currencies. These have now been eliminated within Euroland. Germany, for example, exports two-thirds of its products to other EU members. With a common currency, this business now becomes essentially domestic trade, protected from the fluctuations of different currencies and not subject to transaction costs.

However, some EU member countries, most notably the United Kingdom, have reservations about joining Euroland; as a result, membership remains a controversial political and economic topic for all European countries. During the economic meltdown of 2008–2009, the experiences of using a common currency were mixed. While some states in Euroland benefited from the use of a common currency, others suffered greatly because they could not adjust their national currencies to local economic conditions. Greece and Spain, for example, suffered more heavily from the global recession than did France and Germany, yet because of their Euroland linkages, the weak economies in those two Mediterranean countries caused fiscal and political crises not just in all other EMU countries but also rippled through the global economy.

Economic Integration, Disintegration, and Transition in Eastern Europe

Historically, eastern Europe has been less developed economically than its western counterpart (see Table 8.2). This is partially explained by the fact that eastern Europe is not as rich in natural resources as is western Europe, having only modest amounts of coal, less iron ore, and little oil and natural gas. In addition, the few resources found in the region have been historically exploited by outside interests, including the Ottomans, Hapsburgs, Germans, and, most recently, the Soviet Union.

The Soviet-dominated economic planning of the postwar period (1945–1990) was an attempt to develop eastern Europe's economy by coordinating resource usage in a way that also served Soviet interests. When the Soviets took control of eastern Europe, their goal was complete economic, political, and social integration through a **command economy**, one that was centrally planned and controlled. However, the collapse of that centralized system in 1991 threw many eastern European countries into deep economic, political, and social chaos that has handicapped the region (Figure 8.37).

Figure 8.37 Post-1989 Hardship in Eastern Europe With the fall of communism in eastern Europe after 1989, economic subsidies and support from the Soviet Union ended. Accompanying this transition has been a high unemployment rate, resulting from the closure of many industries, such as this plant in Bulgaria.

EXPLORING GLOBAL CONNECTIONS A New Cold War over Heating Europe

The politics of heating Europe with natural gas involves complicated interactions with countries in at least three other world regions—the Russian domain, Central Asia, and Southwest Asia. Currently, one-quarter of Europe's natural gas supply comes from Russia, a fact that makes many Europeans nervous because Russia has used this resource as a political weapon by shutting off the gas and freezing out countries as political temperatures soar.

At present, two major gas pipelines transect Ukraine on their route from Russia to Europe. A loyal member of the former Soviet Union, Ukraine became a West-leaning independent country with aspirations to join both the European Union and NATO, actions that vexed Russia and, as a result, caused heating problems for Europe. Basically, Russia would like to reel in its former satellite countries through membership in a Russia-dominated customs union; however, until a February 2010 change of government in Ukraine, one that initially appears more friendly to Russia, the country refused to join. To counter that opposition, Gazprom, the powerful Russian company that controls this resource, would periodically shut down the gas flow to Ukraine, accusing Ukraine of stealing gas from the pipelines illegally, or saying that the country hasn't paid its bill, or, simply, that there are "technical problems" with the pipeline.

But it's not just Ukraine that suffered from these shutdowns. Europe—particularly Germany—does, too. For example, in January 2009, when political tempers flared between Russia and Ukraine and the gas was turned off, European householders froze, forcing the German government to ration its scarce supply.

Russia's response to the Ukraine problem has been to build a new pipeline that avoids the country altogether, routing a new pipeline northward through Finland and then under the Baltic Sea to Germany. While highly controversial from an environmental perspective, this project is currently under construction, with an expected completion date of 2012 (Figure 8.6.1). In addition, Russia is also promoting a southern route that also avoids Ukraine by going under the Black Sea to Bulgaria, and then on to central Europe. If this line is built, gas will flow into Europe by 2015.

However, many EU politicians are promoting a plan that doesn't rely on Russia and, instead, draws natural gas from fields in Azerbaijan and Turkmenistan in Central Asia. From there, the pipeline route would go through Georgia, southwest across Turkey, entering Europe in Bulgaria. While still in the planning stages, if built this southern pipeline could deliver Central Asian natural gas as early as 2014. Not surprisingly, Gazprom is outraged at this proposal and accuses the United States of promoting this plan as a way of undermining Russia. Clearly, a new Cold War between Russia, Europe, and the U.S. has begun. Fortunately, it's more about heating Europe than destroying each other through nuclear weapons.

Figure 8.6.1 Russian Natural Gas Pipeline into Europe The 4.5 ft (1.4 meter) pipeline that will transport Russian natural gas into Germany beginning in 2012 is shown at a Baltic Sea construction site. This northern route into Europe circumvents Ukraine in an attempt to reduce delivery disruptions caused by political and economic tensions between Russia and its former Soviet state.

The Results of Soviet Economic Planning After 40 years of communist economic planning, the results were mixed within eastern Europe. In Poland and former Yugoslavia, for example, many farmers strongly resisted the national ownership of agriculture; thus, most productive land remained in private hands. However, in Romania, Bulgaria, Hungary, and the former Czechoslovakia, 80 to 90 percent of the agricultural sector was converted to state-owned communal farms. But across eastern Europe, despite these changes in agricultural structure, food production did not increase dramatically, and, in fact, food shortages were commonplace.

Perhaps most notable during the Soviet period were dramatic changes to the industrial landscape with many new factories built in both rural and urban areas that were fueled by cheap energy and raw materials imported from the Soviet Union. As a means to economic development in eastern Europe, Soviet planners chose heavy industry, such as steel plants and truck manufacturing, over consumer goods, and, as in the Soviet Union, the bare shelves of retail outlets in eastern Europe became both a sign of communist shortcomings and, more importantly, a source of considerable public tension. As western Europeans enjoyed an increasingly high standard of living, with an abundance of consumer goods, eastern Europeans struggled to make ends meet as the utopian vision promised by Soviet communists became increasingly elusive.

Transition and Change Since 1991 As Soviet domination over eastern Europe collapsed in the 1990s, so did the forced economic integration of the region. In place of Soviet coordination and subsidy came a painful period of economic transition that was outright chaos in many eastern European countries. The causes of this economic pain were complex. As the Soviet Union turned its attention to its own economic and political turmoil, it stopped exporting cheap natural gas and petroleum to eastern Europe. Instead, Russia sold these fuels on the open global market to gain hard currency (see "Exploring Global Connections: A New Cold War Over Heating Europe"). Without cheap energy, many eastern European industries were unable to operate and shut down operations, laying off thousands of workers. In the first two years of the transition (1990–1992), industrial production fell 35 percent in Poland and 45 percent in Bulgaria. In addition, markets guaranteed under a command economy, many of them in the Soviet Union, evaporated. Consequently, many factories and services closed because they lacked markets for their goods.

Given these problems, eastern European countries began redirecting their economies away from Russia and toward western Europe with the goal of joining the EU. This meant moving from a socialist-based economy of state ownership and control to a capitalist economy predicated on private ownership and free markets. To achieve this, eastern European countries such as the Czech Republic, Hungary, and Poland have gone through a period of **privatization**, or the transfer to private ownership of firms and industries previously owned and operated by state governments (Figure 8.38).

A good deal of hardship came with this transition. Price supports, tariff protection, and subsidies were removed

Figure 8.38 New Supermarket in Bulgaria Consumer items, including food, were often in short supply in Bulgaria during the post-war Soviet period. Then, with the collapse of Soviet Russia's support in 1989, and after a decade of hardship, Bulgaria shifted its orientation toward the European Union, becoming a member in 2007. As a result, European commerce, such as this German supermarket chain, has moved into the country. Note that while Bulgaria uses the Cyrillic alphabet many of the supermarket's signs are in Roman script, illustrating yet another influence of western Europe.

from consumer goods as countries moved to a free market. For the first time, goods from western Europe and other parts of the world became plentiful in eastern European retail stores. The irony, though, is that this prolonged period of economic transition took a toll on many eastern European consumers because not many people could afford these long-dreamed-of products. Unemployment was commonly in the double digits, and underemployment was widespread. Financial security was elusive, with irregular paychecks for those with jobs and uncertain welfare benefits for those without.

Despite this economic and social chaos it is now quite clear that the preferred economic and political trajectory of eastern Europe is toward increased integration with western Europe and the EU, and not, as was once thought, with Russia.

Regional Disparities in Eastern Europe

As mentioned, most eastern European countries have either already joined the EU or are in the process of gaining membership. While the details of EU application and membership are extraordinarily complex, the economic, political, legal, environmental, and educational changes necessary for EU membership should not be underestimated because the transformation of eastern Europe over the last 20 years has been significant. Further, even though the first eastern European countries joined the EU in 2004, it is wise to think of the whole region as not as an end state but, instead, still in transition as they work out the numerous issues and problems associated with EU membership.

Clearly, some eastern European countries are doing better than others, so much so that EU officials express concern about a pattern of dual economies emerging, one characterized by haves and have-nots, of winners and losers, so to speak (Figure 8.39). While recent economic data suggest Slovenia and the Czech Republic are doing very well, the situation is less clear for the new EU members of Hungary, Bulgaria, and Romania. Similarly, the future prospects of EU applicants from the former Yugoslavia, and Albania are of additional concern.

Social development data adds credence to these assessments. Slovenia and the Czech Republic have the highest rates of in-migration, for example, while Lithuania and Albania have the highest rate of out-migration. Another data source, the *CIA World Fact Book,* estimates the portion of each country's population that lives in poverty. Kosovo and Macedonia top the list with 37 and 30 percent respectively, followed by Albania, Bosnia and Herzegovina, and Romania with a quarter of their population living in poverty. (For comparison's sake the same source reports these figures for the impoverished population of other countries: United States, 12 percent; United Kingdom, 14 percent; and France, 6 percent.)

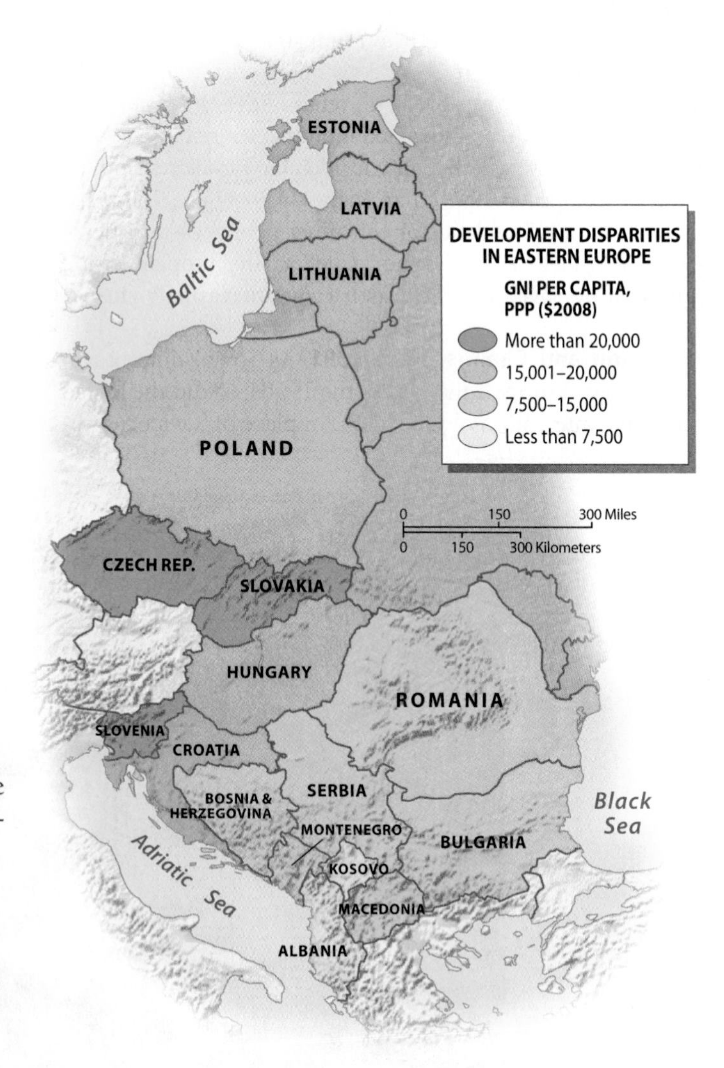

Figure 8.39 Development Issues: Development Disparities in Eastern Europe While some countries in eastern Europe are doing fairly well economically, others are lagging behind as depicted by one measure of economic vitality: Per capita gross national income. More data on these economic and social disparities are found in Table 8.2.

Summary

- In terms of environmental protection, Europe has made great progress over the last several decades. Not only have individual countries enacted strong environmental legislation, but the EU has played an important role in solving transboundary problems such as air and water pollution. In addition, the EU has extended its environmental standards to new members in eastern Europe.
- Europe continues to face challenges related to population and migration. The phenomena of slow and no growth have profound implications for the labor force and fiscal vitality of many countries. Another pressing problem is dealing with immigration from Asia, Africa, Latin America, the former Soviet lands, and even its own underdeveloped areas within Europe itself.
- Political and cultural tensions are high regarding migration. As the bulk of Europeans meld into a common culture, the differences increase between Europeans and migrants from outside the region. While European countries pay lip service to building multicultural societies, the pathways to this goal are still unclear because of racism and discrimination.
- Major changes have taken place in Europe's geopolitical structure during the last 20 years with the end of the Cold War. In eastern Europe countries formerly allied with the Soviet Union have shifted their alliances to the west through the EU, while in the Balkans a handful of newly independent countries have replaced the former Yugoslavia.
- Although the EU celebrated its 50th birthday in 2007 in a self-congratulatory mood, that spirit of optimism was severely challenged by the economic downturn and stagnation of 2008–2010 that brought many countries to the brink of bankruptcy. Iceland, Ireland, Hungary, Spain, and Greece became unwitting poster children for that crisis as this financial crisis underscored both the strengths and weaknesses of EU monetary policies.

Key Terms

balkanization *(page 372)*
buffer zone *(page 371)*
Cold War *(page 343)*
command economy *(page 381)*
continental climate *(page 346)*
Cyrillic alphabet *(page 363)*
Euroland *(page 381)*
European Union (EU) *(page 343)*
fjord *(page 346)*
guest worker *(page 355)*
Iron Curtain *(page 361)*
irredentism *(page 369)*
marine west-coast climate *(page 346)*
maritime climate *(page 346)*
medieval landscape *(page 359)*
Mediterranean climate *(page 348)*
moraine *(page 344)*
privatization *(page 383)*
Renaissance–Baroque landscape *(page 360)*
Schengen Agreement *(page 357)*
secularization *(page 365)*
shield landscape *(page 346)*

Summary and Review

1. How does the European Lowland differ north and south of the Rhine River? Why is this?
2. Describe the different upland and mountain regions of Europe.
3. What are the major factors influencing Europe's weather and climate?
4. List those European countries that are below replacement in natural population increase, as well as those that will grow in the next 20 years. Explain the factors behind these differences.
5. Explain what is meant by "Schengenland" and "Fortress Europe." What are the geographic advantages and disadvantages of the new arrangement behind these two terms?
6. What are the major tongues of the Germanic language family? Which have the most speakers?
7. Map the interface boundaries, both historical and modern, between Christianity and Islam in different parts of Europe. Where is this interface still a problem? Why?
8. List the new nation-states that appeared on Europe's map in 1919, 1945, and 1992.
9. Trace the evolution of the EU since 1952, noting how its goals and membership have changed. Be prepared to make a map of its membership at different points in time.
10. Discuss the kinds of economic, social, and political changes that have taken place in eastern European countries since 1990.

Thinking Geographically

1. To compare the scale of Europe to North America, draw a circle 500 miles (800 kilometers) in diameter, which is about day's journey by car, around Frankfurt, Germany. Then do the same for Chicago. What does this exercise tell you about the differences in scale between the two regions?
2. Investigate the implications of sea level rise from global warming for different parts of Europe, including the Dutch coastline. How might this influence the EU's policy on control of atmospheric emissions?
3. How and why has western Europe become more energy efficient than North America?
4. Map the different rates of natural increase in Europe, noting which countries will grow and which will decline in the next 20 years. Then, link this map to a discussion of migration in Europe. Given these two factors, how might the population map change in 20 years?
5. Find good maps of several mid-sized European cities, and deduce the historical development of the cities based on their street patterns, arrangement of open spaces, boulevards, parks, and so on. Draw on the differences between the medieval and Renaissance–Baroque periods for your interpretation.
6. Look at the ads in a French or German news magazine and list the "globalized" English words used. Discuss your findings in terms of what sectors of society seem to use these foreign terms.
7. Critically examine the geographic aspects of the Dayton Peace Accords, and explain whether you think the boundaries were drawn wisely.
8. Research the fiscal, political, and social changes that had to take place in an eastern Europe country such as Poland, Hungary, Bulgaria or Romania before they became EU members.

Regional Novels and Films

Novels

John Berger, *Once in Europa* (1983, Random House)

Emilie Carles, *A Life of Her Own: A Countrywoman in Twentieth-Century France* (1991, Rutgers University Press)

Aleksandar Hemon, ed., *Best European Fiction, 2010* (2010, Archive Press)

James Joyce, *The Dubliners* (1967, Viking Press)

Milan Kundera, *Ignorance* (2002, Harper & Row)

Sándor Márai, *Embers* (2001, Random House)

V. S. Naipal, *The Enigma of Arrival* (1987, Knopf)

Charles Power, *In the Memory of the Forest* (1997, Scribner's)

Peter Schneider, *The Wall Jumper: A Berlin Story* (1998, University of Chicago Press)

Dimitur Talev, *The Iron Candlestick* (1964, Foreign Language Press)

Films

Ashes and Diamonds (1958, Poland)

Au Revoir, Les Enfants (1987, France)

Before the Rain (1995, Macedonia)

Bend It like Beckham (2002, U.S., UK)

Cabaret Balkan (1998, Serbia)

Gomorra (2008, Italy)

Goodbye, Lenin (2003, Germany)

Head On (2004, Turkey, Germany)

Kontroll (2003, Hungary)

L'Auberge Espanole (2003, France)

One Day You'll Understand (2008, France)

The Russian Dolls (Sequel to *L'Auberge Espanole* 2005 France)

Storm (2009; World Court and Balkan wars)

The Lives of Others (2006, Germany)

The Pianist (2002, U.S., Poland)

The Postman (1994, Italy)

The Remains of the Day (1994, UK)

The White Ribbon (2009, Germany)

Underground (1995, Yugoslavia [Bosnia])

Volver (2006, Spain)

4 months, 3 weeks, and 2 days (2007, Romania)

Bibliography

Allcock, John. 2000. *Explaining Yugoslavia.* New York: Columbia University Press.

Barnes, Ian, and Hudson, Robert. 1998. *The Historical Atlas of Europe: From Tribal Societies to a New European Unity.* New York: Macmillan.

Caldwell, Christopher. 2009. *Reflections on the Revolution in Europe: Immigration, Islam, and the West.* New York: Doubleday.

Denitch, Bogdan. 1996. *Ethnic Nationalism: The Tragic Death of Yugoslavia.* Minneapolis: University of Minnesota Press.

Doherty, Paul, and Poole, Michael. 1997. "Ethnic Residential Segregation in Belfast, Northern Ireland, 1971–1991." *Geographical Review* 87(4), 520–36.

Martin, Philip L. 1998. *Germany: Reluctant Land of Immigration.* Washington, DC: American Institute for Contemporary German Studies.

Pells, Richard. 1997. *Not Like Us: How Europeans Have Loved, Hated, and Transformed American Culture Since World War II.* New York: Basic Books.

Reid, T.R. 2004. *The United States of Europe: The New Superpower and the End of American Supremacy.* New York: Penguin Press.

Sarotte, Mary Elise. 2009. *1989: The Struggle to Create Post–Cold War Europe.* Princeton, N.J. Princeton University Press.

Log in to www.masteringgeography.com for MapMaster™ interactive maps, geography videos, RSS feeds, flashcards, web links, an eText version of *Diversity Amid Globalization*, and self-study quizzes to enhance your study of Europe.

MapMaster™ presents 13 Place Name and 13 Layered Thematic interactive maps to help students practice and master their geographic literacy, spatial reasoning, and critical thinking skills.

9

The Russian Domain

This view of Moscow's growing International Business Center suggests the new connections being forged between the Russian domain and the global economy.

DIVERSITY AMID GLOBALIZATION

Russia's reemergence on the global stage has been obvious since 2000. Its growing role is evident in the global energy economy, in its participation in global economic meetings such as the G-8, and in its increasing geopolitical involvement in settings such as North Korea, Venezuela, and Iran. Internal economic challenges, cultural differences, and political instability, however, continue to shape prospects for the region in the early 21st century.

ENVIRONMENTAL GEOGRAPHY

Many areas within the Russian domain suffered severe environmental damage during the Soviet era (1917–91), and today air, water, toxic chemical, and nuclear pollution plague large portions of the region.

POPULATION AND SETTLEMENT

Urban landscapes within the Russian domain reflect a fascinating mix of imperial, socialist, and post-communist influences. Recently, many larger urban areas within the region increasingly reflect parallel trends towards sprawl and decentralization seen elsewhere in North America and western Europe.

CULTURAL COHERENCE AND DIVERSITY

Although Slavic cultural influences dominate the region, many non-Slavic minorities shape the cultural and political geography of the domain, including varied indigenous peoples in Siberia and a complex collection of ethnic groups in the Caucasus Mountains.

GEOPOLITICAL FRAMEWORK

The centralization of Russian political power has increased its political dominance within the region, but democratic freedoms have suffered amid crackdowns on the press and personal liberties.

ECONOMIC AND SOCIAL DEVELOPMENT

Russia's large supplies of oil and natural gas have made it a major player in the global economy, but future prosperity may increasingly hinge on unpredictable world prices for fossil fuels. The recent global economic downturn resulted in large increases in unemployment across the entire region.

The Russian domain sprawls across the vast northern half of Eurasia and includes not only Russia itself, but also the nations of Ukraine, Belarus, Moldova, Georgia, and Armenia (Figure 9.1). The boundaries of the Russian domain have shifted over time. For decades the regional definitions were relatively easy because the highly centralized Soviet Union (or Union of Soviet Socialist Republics) dominated the region's political geography. The country, born in a communist revolution in 1917, dwarfed all other states in the world and was powerfully united by the Soviet government and largely controlled by ethnic Russians. Most geographers agreed that the Soviet Union could thus be viewed as a single world region. Although the country contained many cultural minorities, the Soviet Union wielded remarkable political and economic power from Leningrad (now St. Petersburg) on the Baltic Sea to Vladivostok on the Pacific Ocean. After World War II, some geographers even included much of Soviet-dominated eastern Europe within the region in response to the country's expanded military role in nations such as East Germany, Poland, and Hungary.

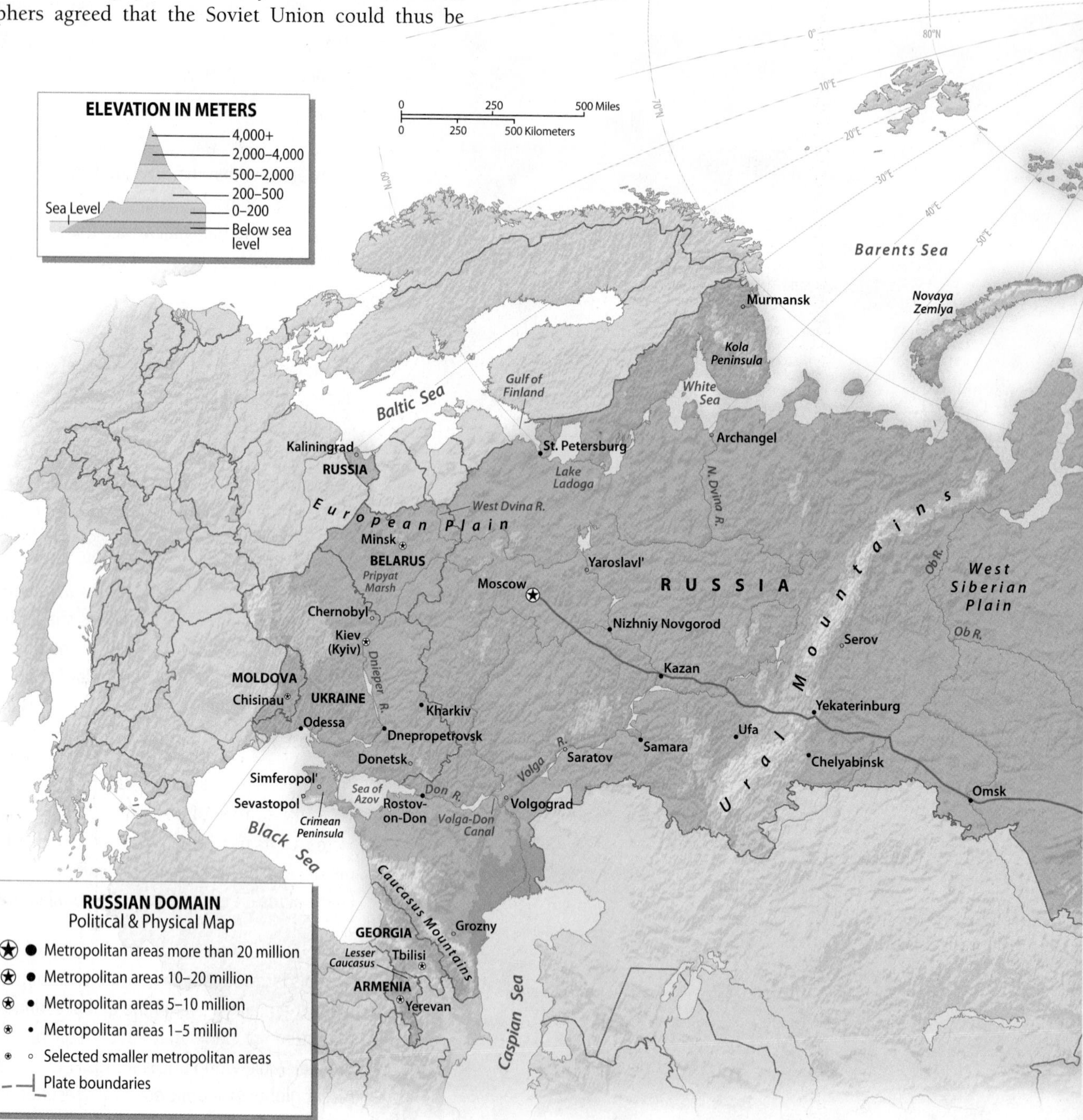

Figure 9.1 The Russian Domain Russia and its neighboring states of Belarus, Ukraine, Moldova, Georgia, and Armenia make up a dynamic and unpredictable world region. Sprawling from the Baltic Sea to the Pacific, the region includes huge industrial centers, vast farmlands, and almost-empty stretches of tundra.

The maps were suddenly redrawn late in 1991. The once-powerful Soviet state was officially dissolved, and in its place stood 15 former "republics" that had once been united under the Soviet Union. Now independent, each of these republics has tried to make its own way in a post-Soviet world. While some geographers initially treated the region as the "Former Soviet Union," it quickly became clear that diverse cultural forces, economic trends, and political orientations were taking the republics in different directions. Even so, the Russian Republic remained dominant in size and area and thus came to form the nucleus of a new Russian domain.

The new regional definition reflects the changing political and cultural map since the breakup of the Soviet Union. The

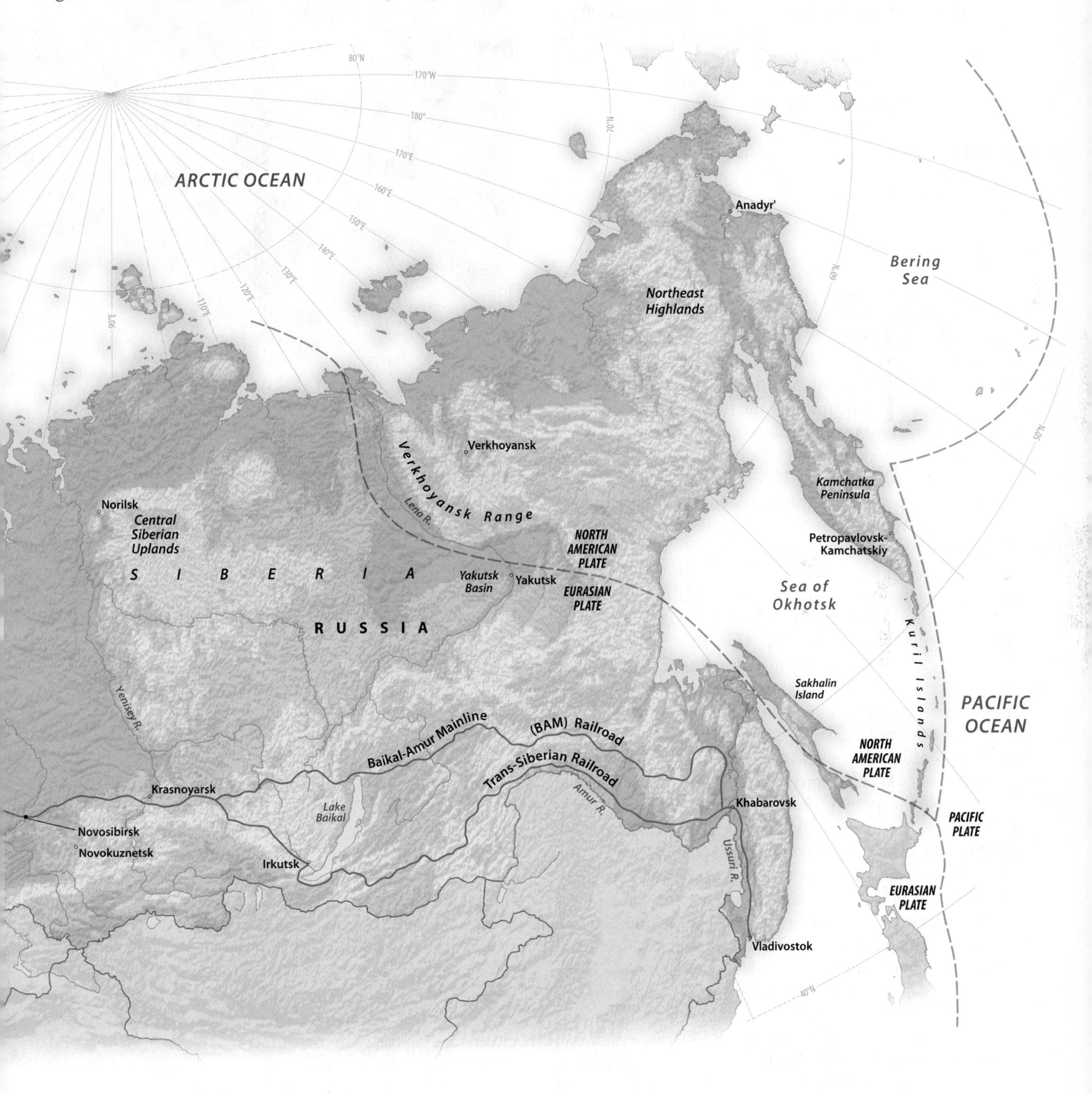

term *domain* suggests persisting Russian influence within the five other nations included in the region. Russia, Ukraine, and Belarus make up the core of the region. Enduring cultural and economic ties closely connect these three countries. Nearby Moldova and Armenia also broadly remain within Russia's geopolitical orbit, although forces within both countries are advocating closer ties with the European Union. Relations between Russia and Georgia remain very strained and recent political conflicts in Georgia brought Russian troops into that small country. Two significant areas that were once a part of the Soviet Union have been eliminated from the domain. The mostly Muslim republics of Central Asia and the Caucasus (Kazakhstan, Uzbekistan, Kyrgyzstan, Turkmenistan, Tajikistan, and Azerbaijan) have become aligned with a Central Asia world region (Chapter 10), while the Baltic republics (Estonia, Latvia, and Lithuania), all now in the EU and NATO, are best grouped with Europe (Chapter 8).

The Russian domain is rich with superlatives: endless Siberian spaces, unlimited natural resources, legends of ruthless Cossack warriors, and tales of epic wars and revolutions are all part of the region's geographical and historical mythology (Figure 9.2). Indeed, the rise of Russian civilization remarkably parallels the story of the United States. Both cultures grew from small beginnings to become imperial powers that benefited from the fur trade, gold rushes, and transcontinental railroads during the 19th century. In addition, both countries experienced dramatic change resulting from industrialization in the 20th century.

Recently, however, the Russian domain has witnessed particularly breathtaking change. With the fall of the Soviet Union in 1991, new political and economic institutions reshaped everyday life. Economic collapse in the late 1990s produced steep declines in living standards throughout the region. Political instability also grew between neighboring states as well as within countries. After 2000, strong and increasingly centralized leadership within Russia set the region on a different course. With the help of higher energy prices (Russia is a major exporter of oil and natural gas), real economic improvements benefited most of the region, helping stabilize its economy and its larger political role in the world.

That remarkable record of recovery, however, proved vulnerable in the recent global economic decline. Both Russia and Ukraine, for example, witnessed sharp economic contractions during 2009 as regional and global demand declined for many of the raw materials and industrial goods they produce. In fact, the region experienced some of the steepest economic declines in the developed world between 2008 and 2010. In addition, many of the democratic reforms welcomed with the fall of communism in 1991 are now being eroded by a powerful Russian government committed to more direct state control. This new assertiveness on the part of the Russian government is increasing tensions within that country as well as between Russia and its neighboring states.

Globalization is also shaping the Russian domain in complex ways. The region's relationship with the rest of the world shifted dramatically during the last 10 years of the 20th century. Until the end of 1991, all six countries belonged to the Soviet Union, the world's most powerful communist state. Under Soviet control, the region's 20th-century economy saw large increases in industrial output that made the nation a major global producer of steel, weaponry, and petroleum products. Its communist system offered a powerful set of ideas that promised economic prosperity and hope to residents within the region and around the world. Indeed, the political and military reach of the Soviet Union spanned the globe, making it a superpower on par with the United States. The Soviet presence dominated many eastern European countries, and nations from Cuba to Vietnam enjoyed close strategic ties with the country.

Figure 9.2 Cossack Natives of the Ukrainian and Russian steppe, the highly mobile Cossacks played a pivotal role in aiding Russian expansion into Siberia during the 16th century. Many modern descendants retain their skills of horsemanship and are proud of their distinctive ethnic heritage.

Suddenly, as the old communist order evaporated early in the 1990s, the now-independent republics of Russia, Ukraine, Belarus, Moldova, Georgia, and Armenia had to carve out new regional and global relationships. With the breakdown of Soviet control, the region also felt the growing presence of western European and American influences. Westernized popular culture combined with radical economic changes to transform the region. These new global relationships have not been easy. Social tensions have risen within the region as people grapple with fundamental economic changes. Political relationships with neighboring regions in Europe and Asia remain uncertain. The Russian domain has also been more fully exposed to both the opportunities and the competitive pressures of the global economy. The result is a world region that has seen its global linkages redefined in the recent past. Today, fluctuating world oil markets, shifting patterns of foreign investment, new patterns of migration, and the shadowy flows of illegal drugs and Russian mafia money all demonstrate the unpredictable nature of the Russian domain's global connections.

Slavic Russia (population 140 million) dominates the region. Although only about three-quarters the size of the former Soviet Union, Russia's dimensions still make it the largest state on Earth. West of Moscow, the country's European front borders Finland and Poland, while far to the east Mongolia and China share a sparsely populated boundary with sprawling Russian Siberia. Its area of 6.6 million square miles (17 million square kilometers) dwarfs even Canada, and its 9 time zones are a reminder that dawn in Vladivostok on the Pacific Ocean is still only evening in Moscow.

With the demise of the Soviet Union, the Russians ended almost 75 years of Marxist rule. After a decade of political and economic instability (1991–2000), Russia has made impressive progress in the new century. Russian president Vladimir Putin built a reputation for strong leadership (2000–08) and that tradition has continued since 2008 as Putin has assumed the newly powerful role of Prime Minister. Much of the economic growth has come to Russian cities, where expanding middle and professional classes are enjoying better living standards. Many rural areas, however, remain deeply mired in poverty. Another concern is that Putin's desire for wealth and power has been matched by his need for more centralized political control. In addition, a 2010 agreement to form a new customs union and ultimately a common economic trade bloc between Russia, Belarus, and Kazakhstan suggests that the Russian desire for more centralized authority did not end with the demise of the Soviet Union. Such an organization may also act as a counterbalance on the east to an expanding European Union (EU).

The bordering states of Ukraine, Belarus, Moldova, Georgia, and Armenia will inevitably be linked to the evolution of their giant neighbor, even as they attempt to make their own way as independent nations. Emerging from the shadows of Soviet dominance has been difficult. Ukraine, in particular, has the size, population, and resource base to become a major European nation, but it has struggled to create real political and economic change since independence. With 45 million people and a rich storehouse of resources, Ukraine's size of 233,000 square miles (604,000 square kilometers) is similar to that of France. Nearby Belarus is smaller (80,000 square miles or 208,000 square kilometers), and its population of 10 million is likely to remain more closely tied economically and politically to Russia. Presently, its strikingly authoritarian and anti-foreign leadership reflects many aspects of the old Soviet empire.

Moldova, with 4 million people, shares many cultural links with Romania, but its economic and political connections have kept it more closely tied to the Russian domain than to nearby portions of central Europe (Figure 9.3). South of Russia and beyond the bordering Caucasus Mountains, the Transcaucasian countries of Armenia and Georgia are similar in size to Moldova. Their populations differ culturally from their Slavic neighbor to the north. In addition, these two nations face significant political challenges: Armenia shares a hostile border with Azerbaijan (see Chapter 10), and Georgia's ethnic diversity and contentious relations with neighboring Russia threaten its political stability.

Figure 9.3 Chisinau, Moldova With a population of almost one million people, the Moldovan capital of Chisinau (often called by its Russian name, *Kishinev*, during the Soviet period) is in the most economically affluent part of this small eastern European nation. Its landscape still reflects the influence of Soviet-era planning and urban design.

ENVIRONMENTAL GEOGRAPHY: A Vast and Challenging Land

The Kamchatka Peninsula dangles dramatically into the waters of the North Pacific Ocean, far from the hectic streets of St. Petersburg or Moscow (see Figure 9.1). Unlike much of the Russian domain, which has been poisoned by almost a century of reckless development and environmental exploitation, the Kamchatka region remains relatively untouched by the outside world (Figure 9.4). But that may be changing soon and Russia's ability to preserve the region will say a great deal about that country's future commitment to maintaining and restoring its environmental health.

Six native species of Pacific salmon thrive in the region, spawning in the millions in the area's free-flowing rivers. A move to preserve the salmon habitat is growing among both Russian and global environmental organizations. The salmon are at the center of a complex environmental web: They provide food for the brown bears, seals, and Stellar's sea eagles that are abundant here; they also remain part of the subsistence diet of the Koryak and other native peoples of these coastal zones; and they offer ecotourism and sports fishing opportunities found nowhere else in the world.

But outside threats are rapidly appearing. Canadian and South Korean oil companies have acquired drilling rights in the region and the huge Russian oil consortium, Rosneft, has similar leases in the offshore waters. Illegal poaching of salmon and salmon eggs is also on the rise. A concerted global effort is now underway to prevent development in seven sensitive spawning areas of Kamchatka wilderness. One Russian-supported plan calls for protecting an area nearly triple the size of Yellowstone National Park and a recent consolidation of several national parks on the peninsula in 2010 should make management of the region's natural resources more efficient and less costly.

A growing number of North American universities, conservation groups, and sports-fishing organizations are also adding their research support and assistance. If such initiatives on the Kamchatka Peninsula are successful, it will mark a new era of environmental consciousness in a part of the world that has produced some of the most toxic landscapes on the planet. It will also mark an important marriage between a fragile local setting and a larger collection of national and international environmental organizations dedicated to seeing the wild salmon continue to run.

A Devastated Environment

The Russian domain has no shortage of fragile and endangered environments. The breakup of the Soviet Union and subsequent opening of the region to international public scrutiny revealed some of the world's most severe environmental degradation (Figure 9.5). Even official studies commissioned by the Russian government estimate that almost two-thirds of the Russian population live in an environment harmful to their health and that the country's environmental problems continue to worsen. The studies also suggest that nearly 65 million Russians live in areas of chronically poor air quality and that the drinking water is unsafe in half of the country.

The frenetic pace of seven decades of Soviet industrialization took its toll across the region. Even in some of the most remote reaches of Russia, careless mining and oil drilling, the spread of nuclear contamination, and rampant forest cutting have resulted in frightening environmental damage. New Russian environmental and antinuclear movements have protested these ecological disasters, but to date these movements remain a minor political voice in a region dominated by the desire for economic growth. Indeed, the magnitude of many of these environmental challenges is so great that they have global implications and may affect world climate patterns, water quality, and nuclear safety. For example, since the 1980s, the global environmental costs of Siberian forests lost to lumbering and pollution may have exceeded the more widely publicized destruction of the Brazilian rain forest.

Figure 9.4 Kamchatka Peninsula The Kamchatka Peninsula offers many spectacular natural settings as well as a habitat for a variety of Pacific salmon.

Air and Water Pollution Poor air quality plagues hundreds of cities and industrial complexes throughout the region (see Figure 9.5). The traditional Soviet practice of building large clusters of industrial processing and manufacturing plants in concentrated areas, often with minimal environmental controls, has produced an ongoing legacy of fouled air that stretches from Belarus to Russian Siberia. A traditional reliance on abundant but low-quality coal also contributes to pollution problems. The air quality in dozens of cities within the region typically fails to meet health standards, particularly in the winter when cold-air inversions trap the polluted atmosphere for days on end. Large numbers of urban residents across the region suffer from chronic respiratory problems. Siberia's northern mining and smelting city of Norilsk is one of Russia's most polluted urban areas. In addition, a large swath of larch-dominated forest has died in a huge zone of contamination that stretches more than 75 miles (120 kilometers) east of the city. Elsewhere, growing rates of private car ownership have greatly increased automobile-related pollution. Today, 90 percent of Moscow's air pollution has been linked to the city's growing automobile traffic.

Degraded water is another hazard that residents of the region must cope with daily (see Figure 9.5). Municipal water supplies are constantly vulnerable to industrial pollution, flows of raw sewage, and demands that increasingly exceed capacity. For example, the Baltic Sea near the city of St. Petersburg has reached a critical level of pollution that has killed fish and threatens to permanently damage the region's ecosystem. The biggest problem is that 30 percent of all the residential and industrial waste that enters the sea via the nearby Neva River is unfiltered raw sewage, a toxic mix of heavy metals and human waste that is rapidly killing nearby portions of the Baltic. Elsewhere, the extensive industrialization and dam-building along Russia's Volga Valley has produced a corridor of degraded water that stretches

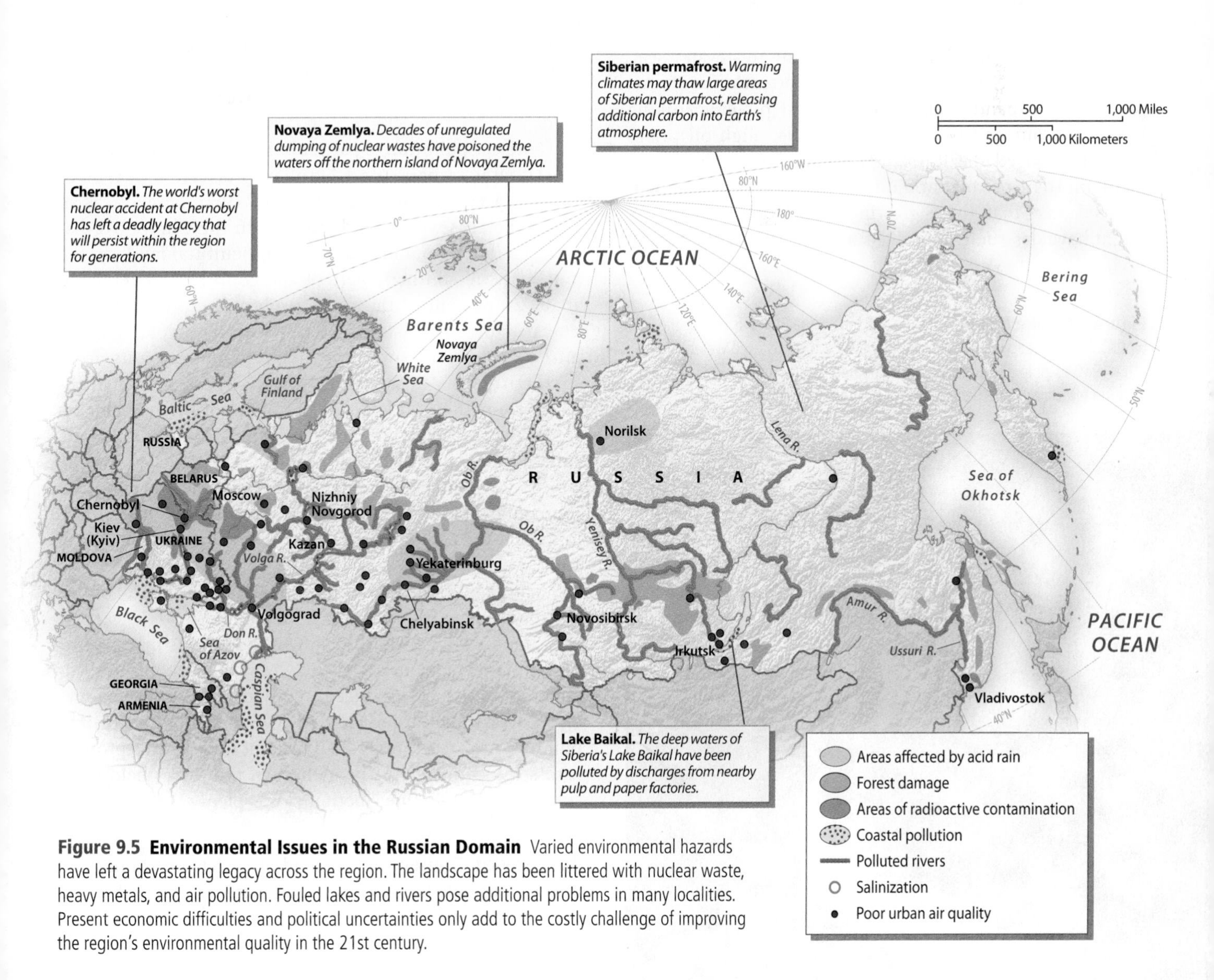

Figure 9.5 Environmental Issues in the Russian Domain Varied environmental hazards have left a devastating legacy across the region. The landscape has been littered with nuclear waste, heavy metals, and air pollution. Fouled lakes and rivers pose additional problems in many localities. Present economic difficulties and political uncertainties only add to the costly challenge of improving the region's environmental quality in the 21st century.

for hundreds of miles. Water pollution has also affected much of the northern Black Sea, large portions of the Caspian Sea shoreline, and even Arctic waters off Russia's northern coast.

The fate of Siberia's Lake Baikal appears brighter (Figure 9.6). In the 1950s and 1960s, the Soviet government built pulp and paper factories along its shores, attracted by both the lake's clean water (useful for producing high-quality wood fibers) and the abundant forests in the surrounding uplands. With factory discharges, the lake's purity rapidly declined. By the 1970s, Russian ecologists warned that Baikal's entire ecosystem was threatened. Owing in part to the resulting international pressure and to the growth of a large lakeshore tourism economy, pollution from the paper mills was reduced and the lake's water quality has improved.

Today, Lake Baikal, the world's largest reserve of freshwater, is a remarkable natural feature. Not only is the lake almost 400 miles (644 kilometers) long, but it is also 5,300 feet (1,615 meters) deep, occupying a structural rift in the continental crust. It remains home to a large array of unique species, including the world's only freshwater seal. Recently, the lake became the center of attention as Russia planned to expand a major Siberian oil pipeline linking Russian resources to East Asian markets. High oil prices have encouraged the Russians to make large new investments in their petroleum industry, but many environmentalists feared that these growing global demands for oil might have destructive local consequences for Lake Baikal.

Figure 9.6 Lake Baikal Southern Siberia's Lake Baikal is one of the world's largest deep-water lakes. Industrialization devastated water quality after 1950 as pulp and paper factories poured wastes into the lake. Recent cleanup efforts have helped, but many environmental threats remain.

In 2006, major protests and petition-signing drives opposed the planned pipeline's close proximity to the north shore of the lake. The initiative caught the attention of Russian President Putin who dramatically ordered that the pipeline be directed further away from the lake's fragile ecosystem.

The Nuclear Threat The nuclear era brought its own particularly deadly dangers to the region. The Soviet Union's aggressive nuclear weapons and nuclear energy programs expanded greatly after 1950, and issues of environmental safety were often ignored. Northeast Siberia's Sakha region, for example, suffered regular nuclear fallout in the era of above-ground nuclear testing. In other areas, nuclear explosions were widely utilized for seismic experiments, oil exploration, and dam-building projects. The once-pristine Russian Arctic has also been poisoned. During the Soviet era, the area around the northern island of Novaya Zemlya served as a huge and unregulated dumping ground for nuclear wastes. Nearby, dozens of atomic submarines have been abandoned to rust away among the fjords of the Kola Peninsula. Aging nuclear reactors also dot the region's landscape, often contaminating nearby rivers with plutonium leaks.

Nuclear pollution is particularly pronounced in northern Ukraine, where the Chernobyl nuclear power plant suffered a catastrophic meltdown in 1986. Large areas of nearby Belarus were also devastated in the Chernobyl disaster. Fallout contaminated fertile agricultural lands across much of the southern part of the country and rendered about 20 percent of the nation unhealthy to even live in. The reactor burned for 16 days, pouring smoke 2 miles into the sky and spreading nuclear contaminants from southern Russia to northern Norway. Thousands died directly from the accident, and millions of other residents across the region and in nearby affected portions of Europe suffered longer-term health problems. It proved to be the world's worst nuclear accident and one of the greatest environmental disasters of the modern age, and it will continue to impact the ecological health of the region for decades to come.

The region's involvement with the nuclear age continues to take new forms. In 2001 Russia's Rostov Nuclear Energy Station went on-line, making it the region's first nuclear power plant since the Soviet era. Growing blackouts and energy shortages have produced a new government drive to revive Russia's nuclear industry. In 2010, an ambitious plan to build 26 new reactors by 2020 produced a rash of protests across the country. At one gathering, protestors held signs simply declaring "One Chernobyl was plenty."

Post-Soviet Challenges The end of Soviet control has had mixed consequences for the region's environment. The demise of the Soviet Union initially brought about environmental improvement in some areas. Many factories shut down because they were no longer economically viable, which itself reduced pollution. The huge steel mills in the Russian city of Magnitogorsk, for example, produce less than half the raw steel they did twenty years ago, and global competition threatens to reduce demand further. Ironically, this decreased production has resulted in cleaner air. Although costly, advanced pollution control equipment is also beginning to be imported from western Europe. Elsewhere in Russia, nuclear warhead storage facilities have been consolidated, and government authorities are responsible for maintaining control over the nation's nuclear weapons. An environmental consciousness also is growing among young, educated Russians and Ukrainians, although public opinion polls across the region demonstrate that for most residents environmental concerns remain very low priorities compared to issues such as economic opportunity and access to decent education and health care.

Russia has also depended heavily on its natural resources to finance its return to economic health, but this response to growing global demand has produced new environmental problems for the region. State-run petroleum and natural-gas companies, for example, have tremendous power and freedom to produce profits as quickly as possible, regardless of the consequences. A similar mentality shapes the timber industry. With huge demand for lumber from nearby Japan and China, Siberian forests are disappearing at an alarming rate (Figure 9.7). At present rates of cutting, much of this natural resource will vanish in the next 10 to 20 years. Thus, growing East Asian appetites for new homes and furniture are directly contributing to the destruction of Siberia's forests. The forests of northwestern Russia also continue to disappear at an alarming rate, responding to similar market-driven pressures from Europe (see "Geographic Tools: Mapping Russian Forests").

Figure 9.7 Siberian Lumber A growing supply of Siberian lumber is flowing toward markets in both Japan and China, threatening the environmental quality and sustainability of Russia's forest resources.

Russia also faces an impending crisis of wildlife extinction. The old Soviet regime had some success in protecting both endangered species and sizable areas of natural habitat noted for their biological diversity. Unregulated hunting and trapping now are on the rise, however, and in some areas are virtually uncontrolled. It is questionable whether such animals as the Siberian tiger (only 350–450 still survive in the wild) and the Amur leopard (25–40 wild animals remain), both inhabiting the endangered forests of Russia's Far East, will survive the next few decades.

A Diverse Physical Setting

The region's northern latitudinal position is critical to understanding basic geographies of climate, vegetation, and agriculture (see Figure 9.1). Indeed, the Russian domain provides the world's largest example of a high-latitude continental climate, where seasonal temperature extremes and short growing seasons greatly limit opportunities for human settlement. In terms of latitude, Moscow is positioned as far north as Ketchikan, Alaska, and even the Ukrainian capital of Kiev (Kyiv) would sit north of the Great Lakes in Canada. Thus, apart from a subtropical zone near the Black Sea, much of the region experiences a classic continental climate with hard, cold winters and marginal agricultural potential (Figure 9.8).

The European West An airplane flight over the western portions of the Russian domain would reveal a vast, barely changing landscape below. European Russia, Belarus, and Ukraine cover the eastern portions of the vast European Plain, which runs from southwest France to the Ural Mountains. One major geographical advantage of European Russia is that different river systems, all now linked by canals, flow into four separate drainages. The result is that trade goods can easily flow in many directions within and beyond the region. The Dnieper and Don rivers flow into the Black Sea; the West and North Dvina rivers drain into the Baltic and White seas, respectively; and the Volga River runs to the Caspian Sea (see Figure 9.1).

Most of European Russia experiences cold winters and cool summers by North American standards. Moscow, for example, is about as cold as Minneapolis in January, yet not nearly as warm in July. In Ukraine, Kiev is milder, however, and Simferopol', near the Black Sea, offers wintertime temperatures that average more than 20°F (11°C) warmer than those of Moscow.

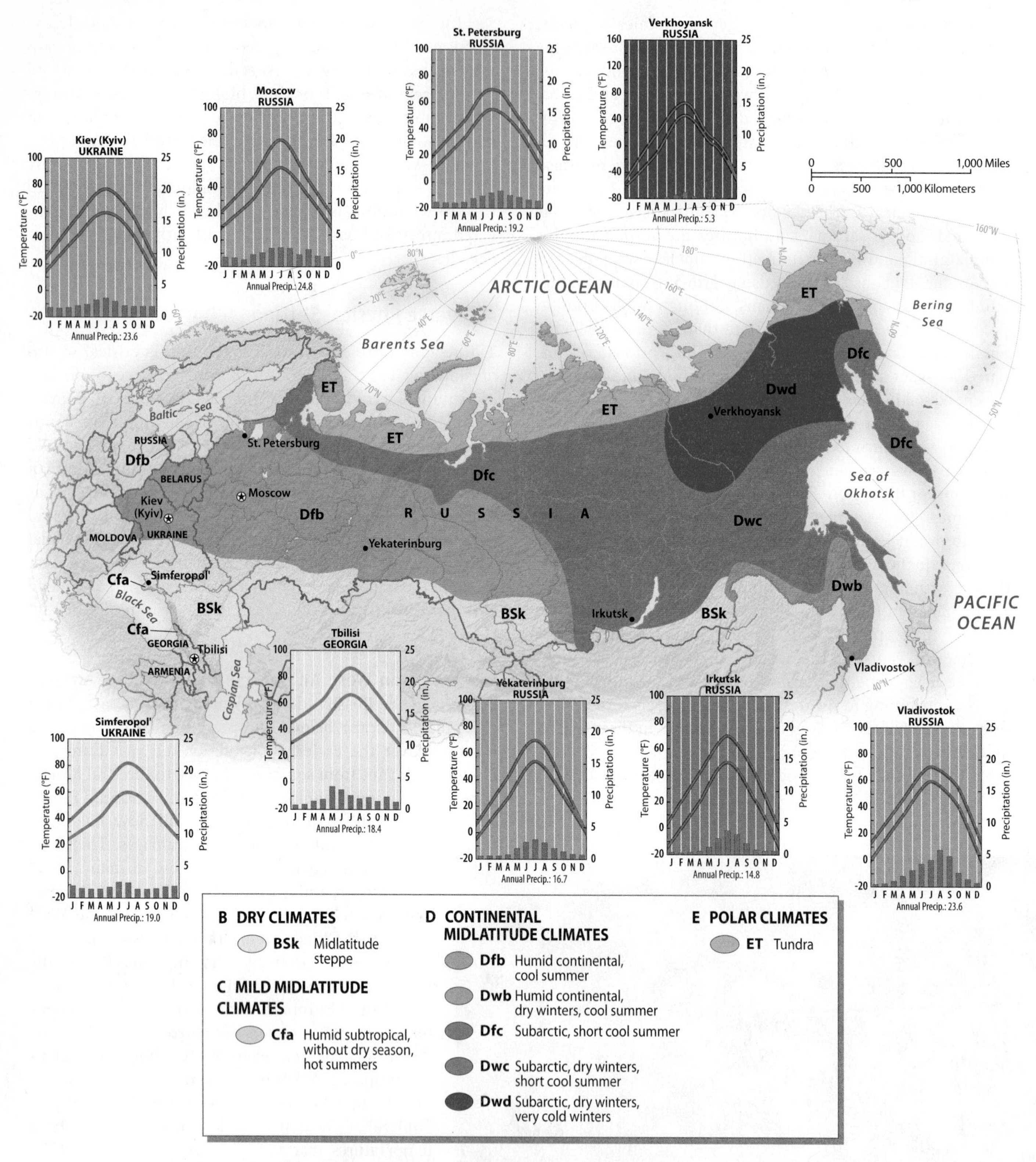

Figure 9.8 Climate Map of the Russian Domain The region's northern latitude and large landmass suggest that continental climates dominate. Indeed, farming is greatly limited by short growing seasons across much of the region. Aridity imposes limits elsewhere. Only a few small zones of mild midlatitude climates are found on the warming shores of the Black Sea in the far southwestern corner of the region, producing subtropical conditions in western Georgia.

GEOGRAPHIC TOOLS Mapping Russian Forests

Greenpeace Russia and the World Resources Institute recently sponsored one of the most ambitious satellite mapping efforts ever. Their goal was to accurately locate and assess the extent of remaining intact boreal forest lands in northwestern Russia (Figure 9.1.1). It was the largest project of its kind ever in Russia and it demonstrated how geographers and remote sensing experts can use satellite imagery to assess forest health and disturbance in high-latitude environments where existing maps are sketchy and where fieldwork is expensive and time-consuming.

The results of the study were sobering. Only the most remote portions (14 percent of the total) of the vast 563-million-acre (228-million-hectare) region, stretching from Finland to the Urals, retained their status as "intact forest landscapes." Such intact areas, critical for forest health and wildlife preservation, were defined as roadless natural areas of at least 123,000 acres (50,000 hectares) that were basically undisturbed by development. Most of the region, however, has felt the impact of a commercial timber industry that is busily supplying the never-ending demands for lumber in nearby western Europe.

Measuring and mapping the remaining intact forests involved five different sources. Geographers on the research team accumulated general topographic maps of the region to get basic information on human infrastructure, and they consulted Soviet-era historical maps to get perspective on changes in regional forest cover. Next, they collected middle resolution (with a moderate amount of detail) satellite imagery from an MSU-SK scanner on an orbiting Russian satellite. These images were used to identify larger clear-cuts, agricultural fields, and areas of obvious second-growth forest. Then, researchers used more costly, but higher-resolution (more detailed) Landsat images to make final decisions on the extent of smaller-scale human disturbances. Finally, several dozen research teams were sent to make field observations across the zone and verify the accuracy of the satellite data. Final maps of the intact and disturbed forests were then completed using an ArcView GIS software package that matched the images with base maps and other layers showing drainage and human activity.

The study demonstrates how geographical expertise can combine many different methodologies to explore real-world problems. The study also suggests that popular images of unending Russian forests are more myth than reality. Only a fragment of the great boreal forests of the Russian north remains. Authors of the study hope that their findings will prompt Russian officials to preserve much of the surviving intact forest, particularly because many of these tracts contain only marginally valuable timber.

Source: Alexey Yaroshenko, Peter Potapov, and Svetlana Turubanova, 2001, *The Last Intact Forest Landscapes of Northern European Russia*, Moscow: Greenpeace Russia.

Figure 9.1.1 Boreal Forest in Northwestern Russia The once endless boreal forests of northern Russia are quickly being harvested for commercial lumber markets in nearby western Europe. Only a small percentage of roadless forest remains today.

Three distinctive environments shape agricultural potential in the European West (Figure 9.9). North of Moscow and St. Petersburg, poor soils and cold temperatures severely limit farming. Belarus and central portions of European Russia possess longer growing seasons, but acidic **podzol soils**, typical of northern forest environments, limit output and the ability of the region to support a productive agricultural economy. Still, diversified agriculture includes grain (rye, oats, and wheat) and potato cultivation, swine and meat production, and dairying. South of 50° latitude, agricultural conditions improve across much of southern Russia and Ukraine. Forests gradually give way to steppe environments dominated by grasslands and by fertile "black earth" **chernozem soils**, which have proven valuable for

Figure 9.9 Agricultural Regions Harsh climate and poor soils combine to limit agriculture across much of the Russian domain. Better farmlands are found in Ukraine and in European Russia south of Moscow. Portions of southern Siberia support wheat production but yield marginal results. In the Russian Far East, warmer climates and better soils translate into higher agricultural productivity. *(Modified from Clawson and Fisher, 2004,* World Regional Geography, *8th ed., Upper Saddle River, NJ: Prentice Hall)*

commercial wheat, corn, and sugar beet cultivation and for commercial meat production (Figure 9.10).

The Ural Mountains and Siberia The Ural Mountains (see Figure 9.1) mark European Russia's eastern edge, separating it from Siberia. Despite their geographical significance as the traditional division between continents, the Urals are not a particularly impressive range; several of their southern passes are less than 1,000 feet (305 meters) high. Still, its ancient rocks contain many valuable mineral resources, and it traditionally marked Russia's eastern cultural boundary.

Figure 9.10 Commercial Wheat Production This wheat field in southern Ukraine typifies some of the region's most productive farmland. Longer growing seasons and good soils are important assets in this portion of the Russian domain.

East of the Urals, Siberia unfolds across the landscape for thousands of miles. The great Arctic-bound Ob, Yenisey, and Lena rivers (see Figure 9.1) drain millions of square miles of northern country that includes the flat West Siberian Plain, the hills and plateaus of the Central Siberian Uplands, and the rugged and isolated Northeast Highlands. Along the Pacific, the Kamchatka Peninsula offers spectacular volcanic landscapes. Wintertime climatic conditions, however, are severe across the entire region.

Siberian vegetation and agriculture reflect the climatic setting. The northern portion of the region is too cold for tree growth and instead supports tundra vegetation, which is characterized by mosses, lichens, and a few ground-hugging flowering plants. Much of the tundra region is associated with **permafrost**, a cold-climate condition of unstable, seasonally frozen ground that limits the growth of vegetation and causes problems for railroad construction. South of the tundra, the Russian **taiga**, or coniferous forest zone, dominates a large portion of the Russian interior.

The Russian Far East The Russian Far East is a distinctive subregion characterized by proximity to the Pacific Ocean, a more southerly latitude, and a pair of fertile river valleys. Located at about the same latitude as North America's New England, the region features longer growing seasons and milder climates than those found to the west or north. Here, the continental climates of the Siberian interior meet the seasonal monsoon rains of East Asia. It is a fascinating zone of ecological mixing: Conifers of the taiga mingle with Asian hardwoods; and reindeer, Siberian tigers, and leopards also find common ground.

The Caucasus and Transcaucasia In European Russia's extreme south, flat terrain gives way first to hills and then to the Caucasus Mountains, a large range stretching between the Black and Caspian seas (Figure 9.11). Many delicate and relatively pristine mountain environments within the region have been recognized as extraordinarily fragile. Indeed, environmental concerns over the upcoming 2014 Winter Olympics at Sochi (in the Russian Caucasus) have already prompted Russian leaders to claim this will be "greenest" Olympics ever held (see "Environment and Politics: Sochi 2014: Russia's Struggle for a Green Winter Olympics").

Farther south lies Transcaucasia and the distinctive natural setting of Georgia and Armenia. Patterns of both climate and terrain in the Caucasus and Transcaucasia are very complex. Rainfall is generally higher in the western zone, while the area's eastern valleys are semiarid. In areas of adequate rainfall or where irrigation is possible, agriculture can be quite productive. Georgia in particular has long been an important producer of subtropical fruits, vegetables, flowers, and wine.

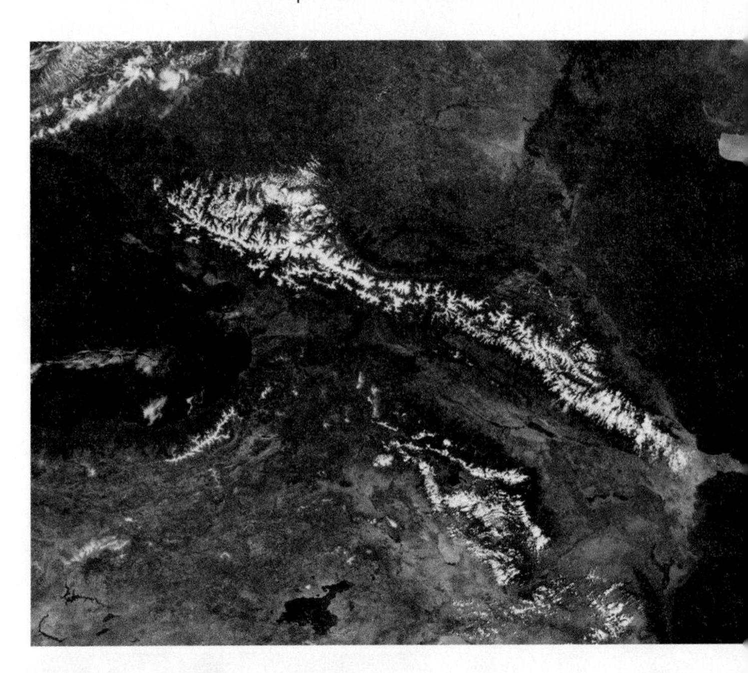

Figure 9.11 Satellite Image of the Caucasus Mountains Aligned between the Black Sea (left) and the Caspian Sea (right), the rugged snow-capped Caucasus Mountains prevent easy movement across the culturally diverse and politically contested borderlands between southern Russia and Georgia.

Global Warming in the Russian Domain

Given its latitude and continental climates, the Russian domain is often cited as a world region that would benefit from a warmer global climate. But such an interpretation oversimplifies the complex natural and human responses to global climate change, some of which are already occurring across the region.

Potential Benefits Optimists point to some of the economic benefits that may result from warmer Eurasian climates. Some models predict, for example, that the northern limit of spring cereal cultivation in northwestern Russia will shift 60–90 miles (100–150 kilometers) poleward for every 1°C (1.8°F) of warming. Elsewhere, warmer springs in settings such as Belarus might allow for higher yields of maize and sunflowers. Areas once in tundra vegetation may be more suitable for boreal forest expansion. Less severe winters may make energy and mineral development in subarctic settings less costly. In nearby portions of the Arctic Ocean and Barents Sea, less sea ice will translate to easier navigation, more high-latitude commerce, and easier drilling for oil and gas. Less ice on northward-flowing Siberian rivers such as the Ob and Yenisey may also make these waterways more valuable as corridors of commerce. Indeed, many Russian scientists and political leaders have publically dragged their feet at climate-change conferences such as the one held in Copenhagen, Denmark, in late 2009. After all, they argue, warming could benefit the region. They also point out that the region, unlike China or India, is not witnessing a rapid increase in its greenhouse gas emissions. Indeed, Russian output of carbon emissions actually fell (because of declines in industrial production) more than 30 percent between 1990 and 2010.

Potential Hazards Even with such rosy scenarios within the Russian domain, might the long-term regional and global costs outweigh the benefits? Climate experts point to several

ENVIRONMENT AND POLITICS Sochi 2014: Russia's Struggle for a "Green" Winter Olympics

What does it take to host an Olympic Games? Recent venues including Beijing and Vancouver were a reminder of the magnitude of the construction effort, even in and near large metropolitan areas. The Russians, scheduled to play host to the 2014 Winter Games, are now finding out how complicated the process is, particularly in an environmentally sensitive region lacking in many of the basic improvements required to host thousands of athletes and visitors. The mountainous area near Sochi, along southern Russia's Black Sea coastline, is the site of the 2014 Winter Olympics (Figure 9.1; Figure 9.2.1). While the Russians have already pledged to put on the "greenest" Olympics in history, environmental groups have unleashed a torrent of protest over the ecologically destructive nature of the project. Why the fuss? The Sochi venues include some of the most pristine mountain settings in Eurasia as well as potentially vulnerable coastal environments near the city itself.

Protesters point to a multitude of environmental issues. In the vicinity of Sochi, huge garbage dumps have swelled with untreated waste that has sometimes slumped into nearby neighborhoods. Local rivers run thick with brown sediment as construction projects accelerate. Critics also point to delicate marshlands near the Black Sea that may fall victim to plans for port expansion. Sochi will house the main Olympic Village, a stadium designed to seat 40,000 spectators, ice skating and speed skating arenas, and it will function as a regional transport hub amid a new array of roads, rail links, and expanded airport infrastructure.

The nearby Krasnaya Polyana region, within the rugged and spectacular Caucasus Mountains, is the venue for the alpine skiing events

Figure 9.2.1 Sochi The southern Russian city of Sochi plays host to the Winter Olympic Games in 2014.

areas of particular concern. First, hotter summers may increase the risk of wildfire. In what could be a sign of things to come, hundreds of wildfires broke out in Russia in the summer of 2010, mostly south and east of Moscow. The fires scorched more than 484,000 acres (196,000 hectares) of land, burning wheat fields and hundreds of structures and filling the skies of Moscow with smoke as visitors and residents experienced record heat (Figure 9.12).

Figure 9.12 Moscow's Red Square People in Russia's capital city of Moscow suffered with record summer heat and the smoke from nearby wildfires in the summer of 2010.

and where another Olympic village is being built (Figure 9.2.2). Critics, citing the vulnerable ecological settings within the nearby Caucasus Biosphere Reserve and Sochi National Park, have already forced the relocation of the bobsled and luge runs. But one spokesperson for the World Wildlife Fund noted that "massive damage had already been caused and more is expected." Logging of endangered species such as the Caucasian wingnut tree, threatens to cause irreparable environmental damage. The organization complained that environmental planning is sorely lacking as the Russians rush to complete construction, suggesting that one impact survey "for the combined railway and highway . . . is based on a couple of weeks of zoological and botanical research done by fewer than 10 people."

The Russians are sensitive to the environmental issues being raised. Prime Minister Putin (who owns a summer home in the area) has pledged that "in setting our priorities and choosing between money and the environment, we're choosing the environment." The Russians have also enlisted the support of the United Nations Environment Program (UNEP) in their effort to "green up" the planned event. In 2009, Olympic organizers signed a "Memorandum of Understanding" with UNEP, promising to make the games the most environmentally friendly ever. Supporters have adopted four slogans: Sochi will exemplify the "Games in harmony with nature," the "Climate Neutral Games," the "Zero Waste Games," and the "Enlightenment Games." Whether the actual Olympics affirm the green rhetoric or not remains an open question, but the controversy is a reminder of the central role that politics can play, both in the planning of international sports events and in the management of complex and fragile environments.

Figure 9.2.2 Olympic Construction in the Krasnaya Polyana Region Near Sochi Environmentalists fear that major investments in Olympic infrastructure will cause permanent damange in one of Eurasia's most pristine mountain settings.

Second, changes in ecologically sensitive arctic and subarctic ecosystems are already leading to major disruptions in wildlife and indigenous human populations in those settings of northern Russia. Take the example of the polar bear. The shrinking volume of arctic sea ice habitat for the bears has meant that they are forced to widen their search for food. This has brought them into closer contact with arctic villages and also disrupted traditional hunting practices. Poachers have also profited, increasing their illegal harvests. Paradoxically, Russian officials and a number of wildlife management experts have argued that increasing legal and controlled hunting (versus illegal poaching) of the bears may be the best way to more effectively maintain their populations in this period of environmental flux.

Finally, the largest potential change within the region, with truly global implications, relates to the thawing of the Siberian permafrost. Substantial areas of northern Russia are covered with permafrost that is already close to thawing. This same region of the world has witnessed some of the largest, most persistent global warming since 1950 (Figure 9.13). Thus, even minor increases in temperature could have large and irreversible consequences for the region. Major changes in topography (mud flows, slumping, and erosion), drainage (lake coverage, rivers), and vegetation will be the result. Existing fish and wildlife populations will need to adjust in order to survive. Human infrastructure such as buildings, roads, and pipelines will also require substantial modification.

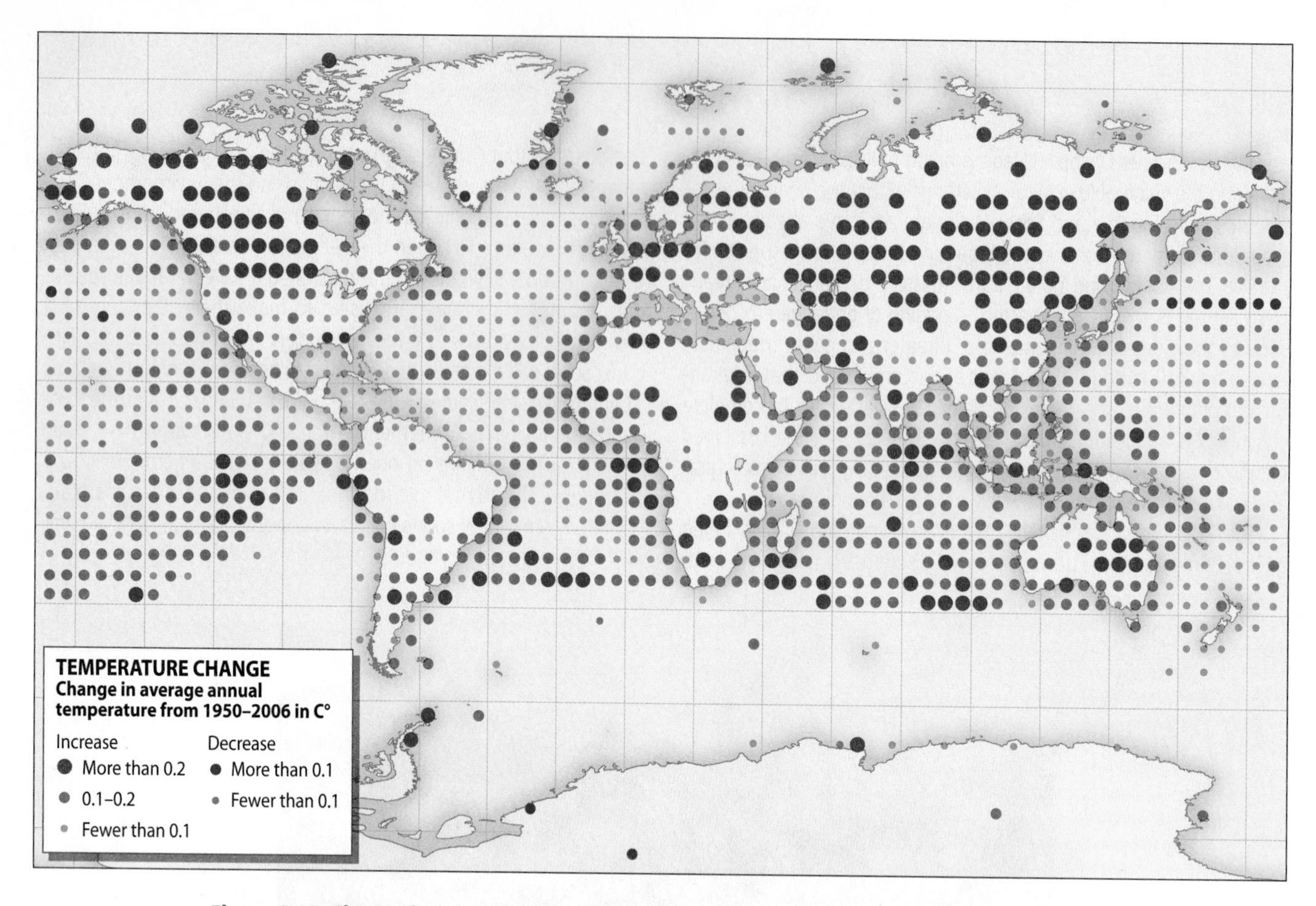

Figure 9.13 Change in Average Annual Temperature (C°), per Decade, 1950–2006 This map shows global temperature trends for different portions of the Earth's surface between 1950 and 2006. Much of the Russian interior, including Siberia, exhibits some of the greatest examples of surface warming on the planet, portending major changes for the region's permafrost, vegetation patterns, and human population. *(Modified from* Goode's World Atlas, *2010, 22nd ed., Upper Saddle River, NJ: Pearson Prentice Hall)*

But the greatest potential global impact may come with the huge release of carbon that is currently stored in existing permafrost environments. The soils frozen in permafrost contain large amounts of organic material that decomposes quickly when thawed. Most of the planet's permafrost could release its carbon reservoir within the next century, the equivalent of 80 years of burning fossil fuels. Such a contribution to the world's carbon budget, which would likely further warm Earth, is only beginning to be incorporated into models of global climate change. Thus, the survival of the Siberian permafrost may hold one of the keys to slowing or quickening further global warming.

POPULATION AND SETTLEMENT: An Urban Domain

The six states of the Russian domain are home to about 200 million residents (Table 9.1). While they are widely dispersed across a vast Eurasian land mass, most live in cities. The region's distinctive distributions of natural resources and changing migration patterns continue to shape its population geography. Although government policies have encouraged migration into the eastern portions of the domain, the population remains strongly concentrated in the traditional centers of the European West. Relatively low birthrates and higher death rates also remain a critical concern as these trends threaten to shrink future regional populations.

Population Distribution

The favorable agricultural setting of the European West has offered a home to more people than do the inhospitable conditions found across central and northern Siberia. Although Russian efforts over the past century have encouraged a wider dispersal of the population, it remains heavily concentrated in the west (Figure 9.14). European Russia is home to more than 100 million persons, while Siberia, although far larger, holds only some 35 million. When one adds the 60 million inhabitants of Belarus, Moldova, and Ukraine, the imbalance between east and west becomes even more striking.

TABLE 9.1 Population Indicators

Country	Population (Millions) 2010	Population Density (per square kilometer)	Rate of Natural Increase (RNI)	Total Fertility Rate	Percent Urban	Percent <15	Percent >65	Net Migration (Rate per 1,000), 2005–10[a]
Armenia	3.1	104	0.6	1.7	64	20	10	-4.9
Belarus	9.5	46	-0.3	1.4	74	15	14	0.0
Georgia	4.6	67	0.3	1.7	53	17	14	-11.5
Moldova	4.1	122	0.0	1.3	41	17	10	
Russia	141.9	8	-0.2	1.5	73	15	13	0.4
Ukraine	45.9	76	-0.4	1.5	69	14	16	-0.3

[a]*Net Migration Rate from the United Nations, Population Division, World Population Prospects: The 2008 Revision Population Database.*
Source: Population Reference Bureau, World Population Data Sheet, 2010.

The European Core The region's largest cities, biggest industrial complexes, and most productive farms are located in the European Core, a subregion that includes Belarus, much of Ukraine, and Russia west of the Urals (see Table 9.1). The sprawling city of Moscow and its nearby urbanized region clearly dominate the settlement landscape with a metropolitan area of more than 10 million people (Figure 9.15).

On the shores of the Baltic Sea, St. Petersburg (Leningrad in the Soviet period, 4.7 million people) has traditionally had a great deal of contact with western Europe. Between 1712 and

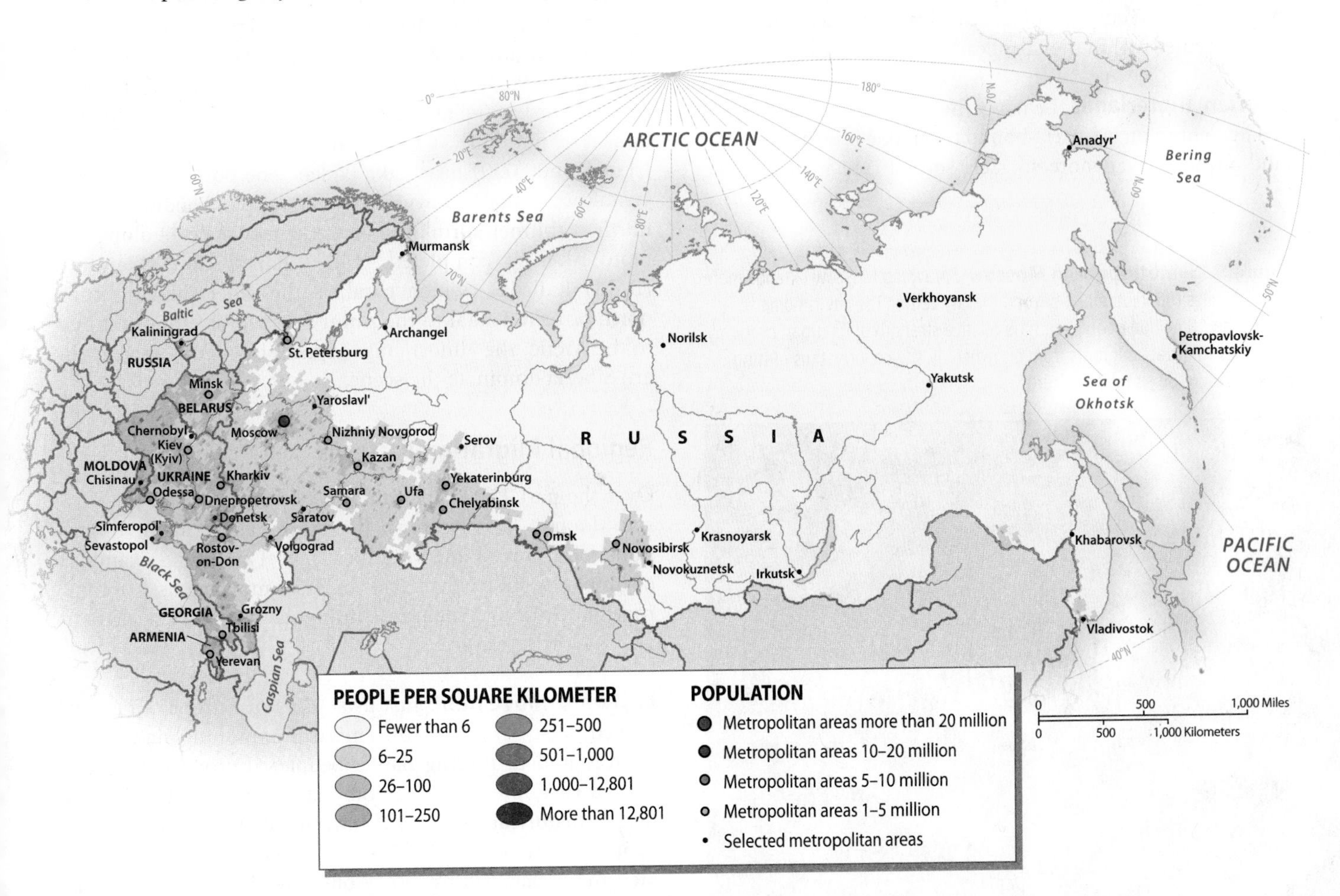

Figure 9.14 Population Map of the Russian Domain Population within the region is strongly clustered west of the Ural Mountains. Dense agricultural settlements, extensive industrialization, and large urban centers are found in Ukraine, much of Belarus, and across western Russia south of St. Petersburg and Moscow. A narrower chain of settlements follows the better lands and transportation corridors of southern Siberia, but most of Russia east of the Urals remains sparsely settled.

1917 it served as the capital of the Russian Empire. Its rich assortment of handsome buildings, bridges, and canals gives the city an urban landscape many have compared to the great cities of western Europe (see "Cityscapes: St. Petersburg").

Other urban clusters are oriented along the lower and middle stretches of the Volga River and include the cities of Kazan, Samara, and Volgograd. Industrialization within the region accelerated during World War II, as the region lay somewhat removed from German advances in the west. Today, the highly commercialized river corridor, also containing important petroleum reserves, supports a diverse industrial base strategically located to serve the large populations of the European Core. Nearby, the resource-rich Ural Mountains include the gritty industrial landscapes of Serov (1.0 million) and Chelyabinsk (1.1 million).

Beyond Russia, major population clusters within the European Core are also found in Belarus and Ukraine (see Table 9.1). The Belorussian capital of Minsk (1.7 million) is the dominant urban center in that country, and its landscape recalls the drab Soviet-style architecture of an earlier era. In nearby Ukraine, the capital of Kiev (Kyiv, 2.8 million) straddles the Dnieper River, and the city's old and beautiful buildings are a visual reminder of its historical role as a cultural and trading center within the European interior (Figure 9.16).

Figure 9.16 Kiev Kiev's historic Podil district features some of the city's most distinctive architecture, including the ornate green domes of St. Andrew's Orthodox Church (upper right).

Siberian Hinterlands Leaving the southern Urals city of Yekaterinburg on a Siberia-bound train, one is aware that the land ahead is ever more sparsely settled (see Figure 9.14). The distance between cities grows, and the intervening countryside reveals a landscape shifting gradually from farms to forest. The Siberian hinterland is divided into two characteristic zones of settlement, each of which can be linked to railroad lines. To the south, a collection of isolated, but sizable, urban centers follows the **Trans-Siberian Railroad**, a key railroad passage to the Pacific completed in 1904 (see Figure 9.1). The eastbound traveler encounters Omsk (1.1 million) as the rail line crosses the Irtysh River, Novosibirsk (1.4 million) at its junction with the Ob River, and Irkutsk (600,000) near Lake Baikal. The port city of Vladivostok (640,000) provides access to the Pacific. To the north, a thinner sprinkling of settlement appears along the more recently completed (1984) **Baikal-Amur Mainline (BAM) Railroad**, which parallels the older line but runs north of Lake Baikal to the Amur River. From the BAM line to the Arctic, the almost empty spaces of central and northern Siberia dominate the scene.

Figure 9.15 Metropolitan Moscow Sprawling Moscow extends more than 50 miles (80 kilometers) beyond the city center. The city is home to more than 10 million people, and the relative strength of its urban economy continues to attract migrants from elsewhere in the country, thus putting more pressure on its infrastructure.

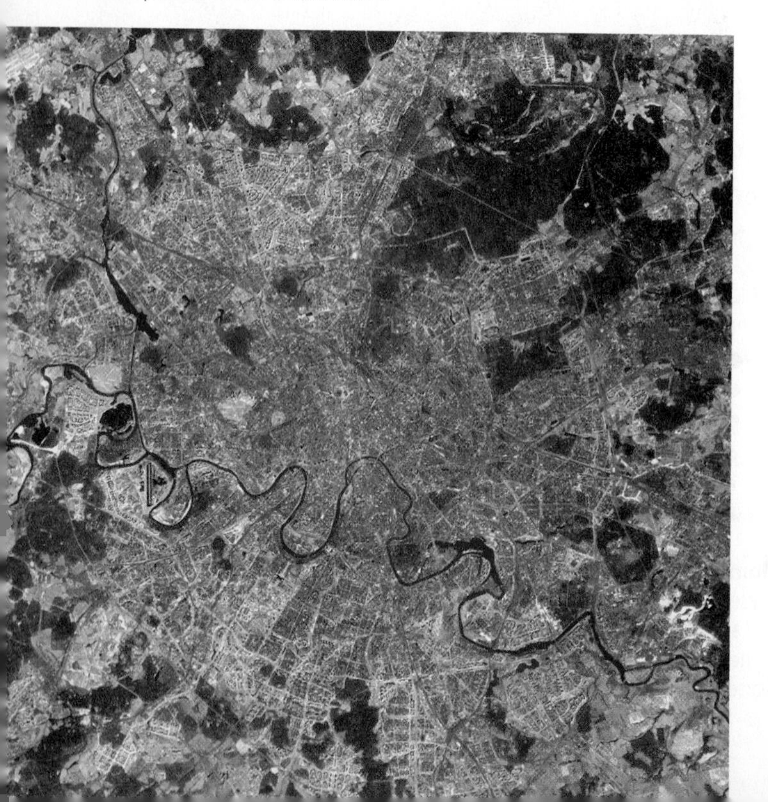

Regional Migration Patterns

Over the past 150 years, millions of people within the Russian domain have been on the move. These major migrations, both forced and voluntary, reveal sweeping examples of human mobility that rival the great movements from Europe and Africa or the transcontinental spread of settlement across North America.

Eastward Movement Just as settlers of European descent moved west across North America, exploiting natural resources and displacing native peoples, European Russians moved east across the vast Siberian frontier. Although the deeper historical roots of the movement extend back several centuries, the pace and volume of the eastward drift accelerated in the late 19th century once the Trans-Siberian Railroad was completed. Peasants were attracted to the region by its agricultural opportunities (in the south) and by greater political freedoms than they traditionally enjoyed under the **tsars** (or czars; Russian for "Caesar"), the authoritarian leaders who dominated politics during the pre-1917 Russian

CITYSCAPES St. Petersburg

It was an improbable place for one of Europe's great cities. Literally rising from swamplands in 1703, St. Petersburg grew from the creative imagination of its founder, Tsar Peter the Great. Seven feet tall, Peter towered above the swamps he envisioned to become Russia's great window to the West. While it took time, St. Petersburg gradually assumed that role in later years. It became the westward-looking tsarist capital, anchored on a protected arm of the Baltic Sea, until the Soviet takeover in 1917 (when the capital returned to Moscow). Peter's imagination helped create a Renaissance-style city built around grand avenues, palaces, and a network of canals that still draws comparisons with Venice and Amsterdam (Figure 9.3.1). Water remains a central element in the cityscape today. Miles of canals, crossed by a latticework of scenic bridges, connect the Neva, Fontanka, and Moyka rivers.

Near the city center, elaborate baroque and neoclassical mansions (many now converted to other uses) and commercial buildings recall earlier days when Russians such as Tchaikovsky and Dostoyevsky walked the streets. Cutting through the heart of the city and bordered by the bright lights of shops, cafes, and theaters, Nevsky Prospekt ends near the Neva River, just two blocks from the world-famed Hermitage Museum, home to 2.8 million pieces of art. No wonder native son Vladimir Putin chose St. Petersburg as the site of the 2003 G-8 summit meeting, which took place only three days after the city's grand 300-year birthday celebration.

But there is another side to St. Petersburg, a cityscape beyond the ken of most casual visitors. During most of the Soviet period, the city, known then as Leningrad (the name St. Petersburg was restored after 1991), also came to reflect some of the modern drabness of Soviet life. Block after block of dreary concrete apartment houses appeared in many of the city's suburban districts. Many of the city's graceful old buildings fell into disrepair, victims of a Soviet mindset that valued modern progress over tsarist architecture. Worsening economic times in the 1980s also did little for the city's appearance as its 4.7 million residents struggled with the chores of daily life. Recently, however, investments in urban preservation and an inflow of foreign investment have perked up the city's economy and freshened its urban landscape. Downtown cafes are bustling, theaters are getting a facelift, and it is rumored that some central government ministries may relocate to the city from Moscow. Gradually, a new day seems to be dawning on an urban landscape that still manages to capture the full breadth and imagination of a people who were willing to build their fortunes in a swamp.

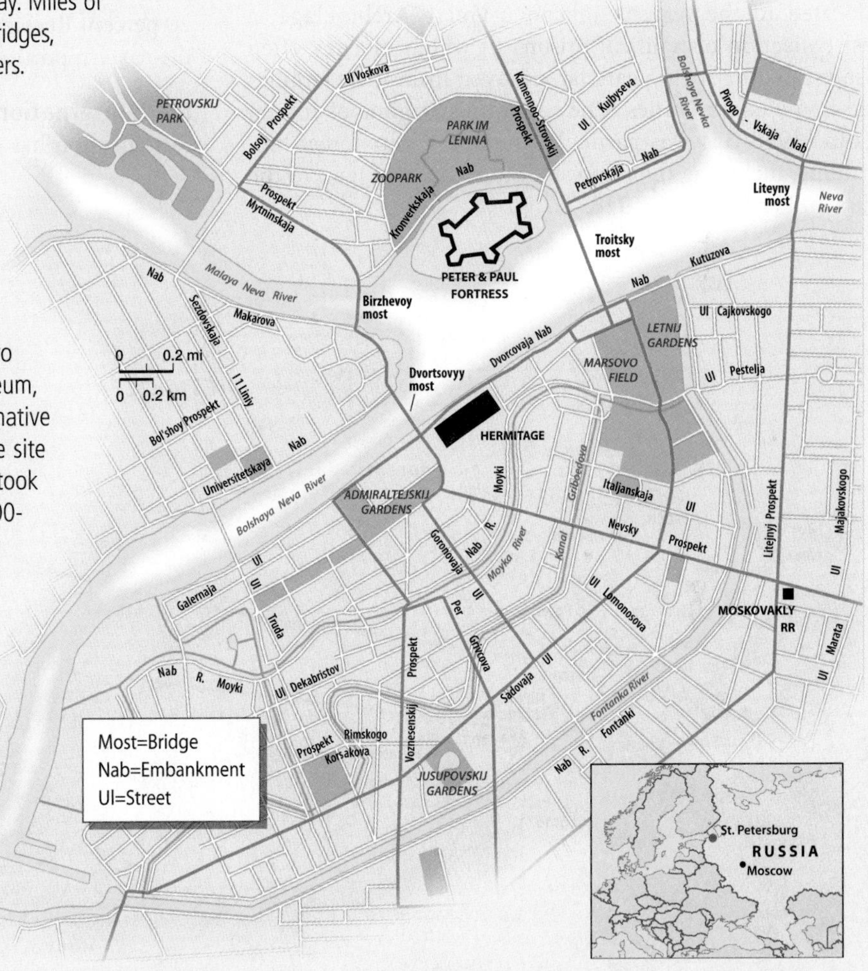

Figure 9.3.1 Map of St. Petersburg The grand avenues and commercial thoroughfares of downtown St. Petersburg are laid out amid nearby rivers and canals, creating an urban framework that many have compared to Venice and Amsterdam.

Empire. Almost 1 million Russian settlers moved into the Siberian hinterland between 1860 and 1914.

The eastward migration continued during the Soviet period, once communist leaders consolidated power during the late 1920s and saw the economic advantages of developing the region's rich resource base. The German invasion of European Russia during World War II demonstrated that there were also strategic reasons for settling the eastern frontier, and this propelled further migrations during and following the war. Indeed, by the end of the communist era, 95 percent of Siberia's population was classified as Russian (including Ukrainians and other western immigrants). With the completion of the

BAM Railroad in the 1980s, yet another corridor of settlement opened in Siberia, prompting new migrations into a region once remote from the outside world.

Political Motives Political motives also have shaped migration patterns. Particularly in the case of Russia, leaders from both the imperial and Soviet eras saw advantages in moving selective populations to new locations. The infilling of the southern Siberian hinterland had a political as well as an economic rationale. Both the tsars and the Soviet leaders saw their political power grow as Russians moved into the resource-rich Eurasian interior. For some, however, the move to Siberia was not voluntary as the region became a repository for political dissidents and troublemakers. Especially in the Soviet period, uncounted millions were forcibly relocated to the region's infamous **Gulag Archipelago**, a vast collection of political prisons in which inmates often disappeared or spent years far removed from their families and home communities. The communist regime of Joseph Stalin (1928–53) was particularly noted for its forced migrations, including the removal of thousands of Jews to the Russian Far East after 1928.

Russification, the Soviet policy of resettling Russians into non-Russian portions of the Soviet Union, also changed the region's human geography. Millions of Russians were given economic and political incentives to move elsewhere in the Soviet Union in order to increase Russian dominance in many of the outlying portions of the country. The migrations were geographically selective in that most of the Russians moved either to administrative centers where they took government positions or to industrial complexes focused on natural resource extraction. As a result, by the end of the Soviet period, Russians made up significant minorities within former Soviet republics (now independent nations) such as Kazakhstan (30 percent Russian), Latvia (30 percent), and Estonia (26 percent). Among its Slavic neighbors, Belarus remains 11 percent Russian and Ukraine more than 20 percent Russian, with concentrations particularly high in the eastern portions of those two countries.

New International Movements In the post-Soviet era, Russification has often been reversed (Figure 9.17). Several

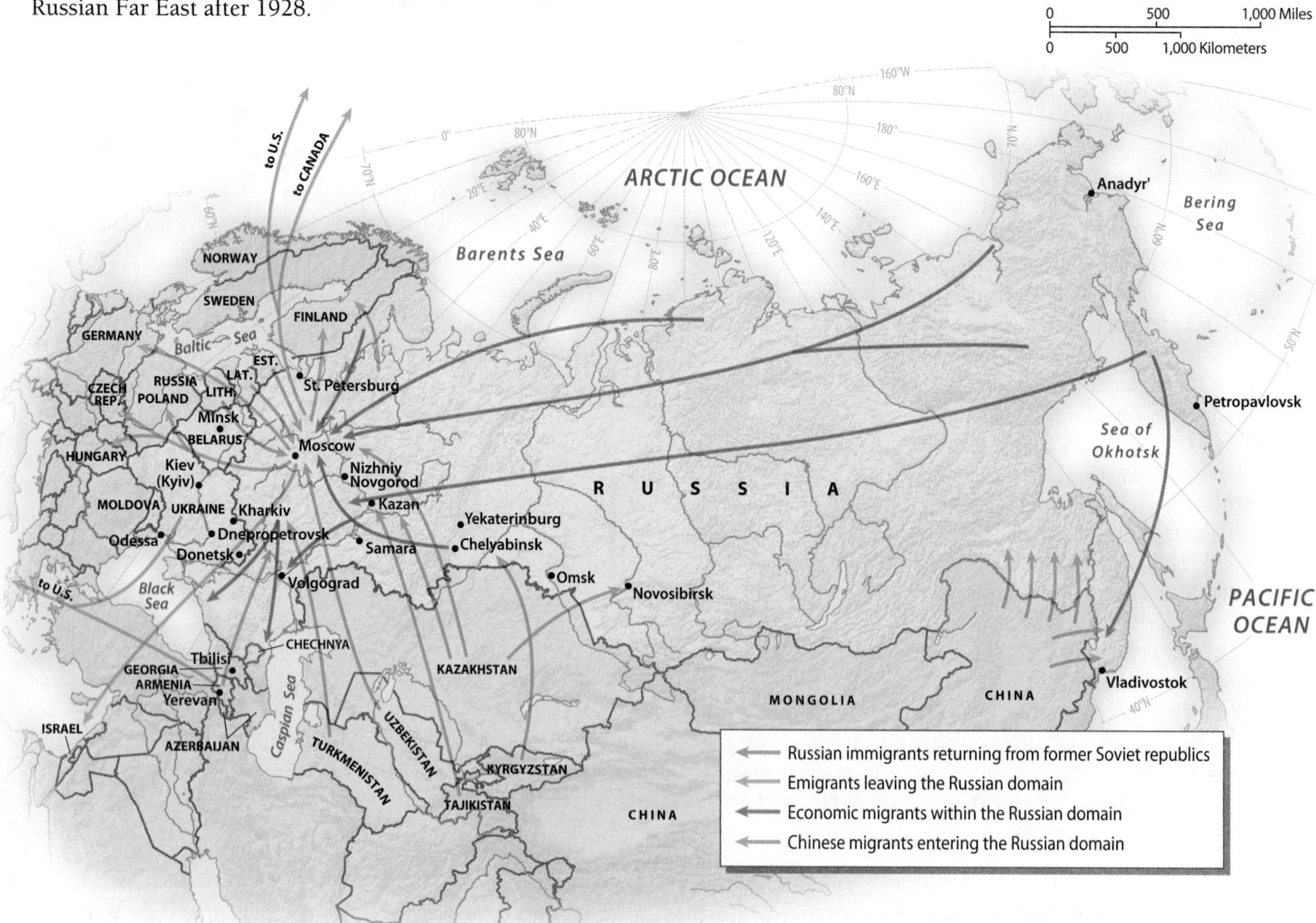

Figure 9.17 Recent Migration Flows in the Russian Domain Recent events are encouraging the return of ethnic Russians from former Soviet republics, while other Russians are emigrating from the domain for economic, cultural, and political reasons. Within Russia, economic forces also are at work encouraging people to be on the move.

of the newly independent non-Russian countries imposed rigid language and citizenship requirements, which encouraged many Russian residents to leave. In other settings, ethnic Russians simply experienced varied forms of social and economic discrimination. In addition, since 2005, the Russian government has vigorously promoted a repatriation program for ethnic Russians worldwide, offering incentives for Russian-speaking migrants to return or move to their cultural homeland. As a result, the Central Asia and Baltic regions, once a part of the Soviet Union, have seen their Russian populations decline significantly since 1991, often by 20 to 35 percent.

Russia has also experienced a growing immigrant population, many of them illegal migrants drawn to the region for work (Figure 9.18). The story has a familiar ring to it. More than 10 million illegal immigrants are suspected in the country. Most are young, upwardly mobile people, predominantly male, who come for better-paying jobs. Recently, the government has implemented tighter controls to restrict border crossings and enacted tougher penalties against businesses that hire illegal immigrants. There is a growing national debate concerning how many legal foreign workers should be allowed in the country and whether or not illegal immigrants should be granted amnesty. Indeed, as a percentage of total population, there are more illegal immigrants in Russia than in the United States.

The actual flows of illegal immigrants are complex. While 80 percent of Russia's immigrants—legal and illegal—come from portions of the former Soviet Union, a growing number of workers are from ethnically non-Slavic regions such as Central Asia. Street cleaners from Kyrgyzstan and shopkeepers from Uzbekistan are increasingly common sights in Moscow, where almost one-third of the country's illegal immigrants may live. One recent estimate suggests that about 20 percent of Tajikistan's economy is now based on funds sent to the country from émigrés living and working in Russia. But there is growing Russian resistance to these non-Slavic arrivals. For example, a new law in Moscow limits the number of non-Russians who can operate market stalls on the streets of the city. Growing violence against these non-Russian immigrants has also erupted in Moscow and elsewhere. In addition, just as in North America, migrant flows are strongly linked to economic opportunities. As a result, recessions in both Russia and the United States tend to produce similar declines in the incoming flows of migrants, both legal and illegal. Finally, population movements in Russia's Far East, principally the growing numbers of arriving Chinese immigrants, are reshaping the economic and cultural geographies of that portion of the region (see Figure 9.17) (see "People on the Move: Chinese Immigrants in Russia's Far East").

The domain's more open borders also have made it easier for other residents to leave the region (see Figure 9.17). Poor economic conditions and the region's unpredictable politics have encouraged many to emigrate. The "brain drain" of young, well-educated, upwardly mobile Russians has been considerable. Sometimes, ethnic links play a part in migration patterns. For example, many Russian-born ethnic Finns have moved to nearby Finland, much to the consternation of the Finnish government. Russia's Jewish population also continues to fall, a pattern begun late in the Soviet period. These emigrants have flocked mostly to Israel or the United States, where they locate in familiar Jewish neighborhoods such as Miami, Los Angeles, and New York City's Brighton Beach district.

Figure 9.18 Non-Russian Immigrants These immigrants from Tajikistan are standing in line in Moscow, applying for employment visas to work in the Russian capital city.

PEOPLE ON THE MOVE Chinese Immigrants in Russia's Far East

Chinese immigrants are fundamentally refashioning the economic and cultural geographies of the Russian Far East. Somewhere between five and seven million immigrants (many of them illegal) may now live in the region. Much of the transformation has occurred since the early 1990s and is in response to a large, job-hungry Chinese population to the south, a shrinking supply of Russian workers to the north, and a regional economy that is increasingly being shaped by powerful forces of globalization. Walk through the Russian cities of Vladivostok or Khabarovsk (Figure 9.4.1). Chinese immigrants are everywhere. Street signs feature both Russian and Chinese lettering. Entire neighborhoods are dominated by immigrant populations. Chinese children are learning Russian in school and Russians find themselves working for Chinese entrepreneurs.

Many Chinese arrived initially with tourist visas, while others have slipped casually over the border. Higher wages and better business opportunities have drawn the Chinese north. Many have entered the construction and forestry industries, others are in retailing or in market gardening, and a growing number have come with investment capital to open manufacturing plants. At the same time, Russian populations are aging and often leaving the region, adding demographic significance to the Chinese inflow.

What are the long-term consequences of this redefinition of the Russian Far East? Many experts see the Chinese as essential ingredients in fueling continued economic growth in a Russia in which native populations are steadily shrinking. Some Chinese officials smile and remind the Russians that the area was once part of imperial China and perhaps will be once again after several more decades of Chinese in-migration. While many Russians see the economic advantages in having hard-working Chinese living in Russia, some fear the growing "foreign" population in their country. A significant nationalist backlash has occurred: Chinese have been attacked by gangs in cities, Chinese shopkeepers complain of being rousted by Russian police, and recent legislation has made it harder for Chinese to operate businesses in the region. On the other hand, many younger Russians have welcomed their Chinese counterparts and there are a growing number of joint Russian-Chinese companies in the region as well as more intermarrying between the two groups. Thirty years from now, two things are clear: the "Russian" Far East is likely to be far more culturally diverse than it was during most of the 20th century. In addition, the region's successful integration with the global economy will in no small part be shaped by the new human geography taking shape there today.

Figure 9.4.1 Chinese Immigrants in the Russian Far East Many Chinese immigrants, such as these on Sakhalin Island in 2005, have become small-scale entrepreneurs in the region's commercial economy.

The Urban Attraction Regional residents also have been bound for the cities. The Marxist philosophy embraced by Soviet planners encouraged urbanization. In 1917, the Russian Empire was still overwhelmingly rural and agrarian; 50 years later, the Soviet Union was primarily urban. Planners saw great economic and political advantages in efficiently clustering the population, and Soviet policies dedicated to large-scale industrialization obviously favored an urban orientation. Today, Russian, Ukrainian, and Belorussian rates of urbanization are comparable to those of the industrialized capitalist countries (see Table 9.1).

Soviet cities grew according to strict governmental plans. Planners selected different cities for different purposes. Some were designed for specific industries, while others had primarily administrative roles. A system of internal passports prohibited people from moving freely from city to city. Instead, people generally went where the government assigned them jobs. Moscow, the country's leading administrative city, thrived under the Soviet regime. It formed the undisputed core of Soviet bureaucratic power as well as the center of education, research, and the media. Specialized industrial cities also grew at a rapid pace. In the mining and metallurgical zone of the southern Urals, centers such as Yekaterinburg and Chelyabinsk mushroomed into major urban centers. Another cluster of specialized industrial cities, including

Kharkiv and Donetsk, emerged in the coal districts of eastern Ukraine.

With the end of the Soviet Union, however, people gained basic freedoms of mobility. In addition, with a more open economy, especially in Russia, shifting urban employment opportunities increasingly reflect how effectively local economies could compete in the global market. This new economic reality has led to the depopulating of many older industrial areas, as they simply cannot produce raw materials or finished industrial goods at competitive global prices. Since 1989, for example, the Russian Northeast has seen its population decline as many workers have left harsh, unemployment-prone industrial centers for better opportunities elsewhere. In many cases, people are freely gravitating toward growth opportunities in locations of new foreign investment, principally in larger urban areas in western and southern Russia. Some also are receiving relocation assistance from a Russian government that admits many of these outlying settlements simply cannot be justified in a more market-oriented economy.

Inside the Russian City

Today, most people in the region live in a city, the product of a century of urban migration and growth (see Table 9.1). Large Russian cities possess a core area, or center, that features superior transportation connections; the best-stocked, upscale department stores and shops; the most desirable housing; and the most important offices (both governmental and private). In the largest urban centers, cities such as Moscow and St. Petersburg also feature extensive public spaces and examples of monumental architecture at the city center. Within the city, there is usually a distinctive pattern of circular land-use zones, each of which was built at a later date moving outward from the center. Such a ringlike urban morphology is not unique. However, as a result of the extensive power of government planners during the Soviet period, this urban form is probably more highly developed here than in most parts of the world.

At their very center, the cores of many older cities predate the Soviet Union. Pre-1900 stone buildings often dominate older city centers. Some of these are former private mansions that were turned into government offices or subdivided into apartments during the communist period but are now being privatized again. Many of these older buildings, however, are being leveled in rapidly growing urban settings such as downtown Moscow (Figure 9.19). Since 1992 urban preservation experts estimate that several thousand historic buildings have been destroyed, including many structures that had supposedly received official protection. Retailing malls replace many of these older structures in the city. Nearby, nightclubs and bars are filled with pleasure seekers as the city's professional elite mingles with foreign visitors and tourists. The Moscow International Business Center is also expanding into older industrial areas, offering new office space for Russian and international firms.

Figure 9.19 Downtown Moscow The city's downtown landscape has been rapidly transformed in the recent economic boom. It remains a complex and fascinating mix of imperial, Soviet, and post-Soviet influences.

Figure 9.20 Moscow Housing These drab apartment buildings in the Moscow suburbs date from the late Soviet period (1980s) and still offer housing to many middle-class residents of the city.

Farther out from the city centers are **mikrorayons**, large, Soviet-era housing projects of the 1970s and 1980s (Figure 9.20). Mikrorayons are typically composed of massed blocks of standardized apartment buildings, ranging from 9 to 24 stories in height. The largest of these supercomplexes contain up to 100,000 residents. While planners hoped that mikrorayons would foster a sense of community, most now serve as anonymous bedroom communities for larger metropolitan areas.

Some of Russia's most rapid urban growth has occurred on the metropolitan periphery, paralleling the North American experience. Moscow, for example, has seen its urban reach expand far beyond the city center. The surrounding administrative district (the Moscow Oblast) contains about 7 million residents and is home to more than 700 international companies. Land prices and tax rates are lower than in the central city; the bureaucracy is less onerous; and the transportation and telecommunications infrastructure is relatively new. New suburban shopping malls, paralleling the North American model, are also popping up on the urban fringe, allowing residents to shop and be entertained without having to visit the city center.

Elsewhere on Moscow's urban fringe, elite **dacha**, or cottage communities, appeal to many more well-to-do residents, particularly during the summer months. The tradition of rural retreats, dating back to the Russian Empire, also thrived during the Soviet era as Communist Party officials sought an escape from the dreary bureaucratic chores of the city. Today, there are about 300 cottage settlements on the Moscow periphery, including many new elite privatized developments northwest and southeast of the city that cater to the country's business class. The older dacha belt to the west of the city along the Moskva River also offers summer homes, spa retreats, and holiday hotels in a zone dubbed the "Sub-Moscow Switzerland."

The Demographic Crisis

Since 2005, the Russian government has identified population loss as a key issue of national importance. Some government and United Nations estimates have predicted that Russia's population could fall by a startling 45 million by 2050. Similar conditions are affecting the other countries within the region (see Table 9.1). Beginning during World War II, large numbers of deaths combined with low birthrates to produce sizable population losses. While population increases accelerated in the 1950s, growth slowed by 1970, and death rates began exceeding birthrates in the early 1990s. Two population pyramids tell the troubling tale (Figure 9.21). The first shows that, whereas adult-age populations are prominently represented (with the exception of

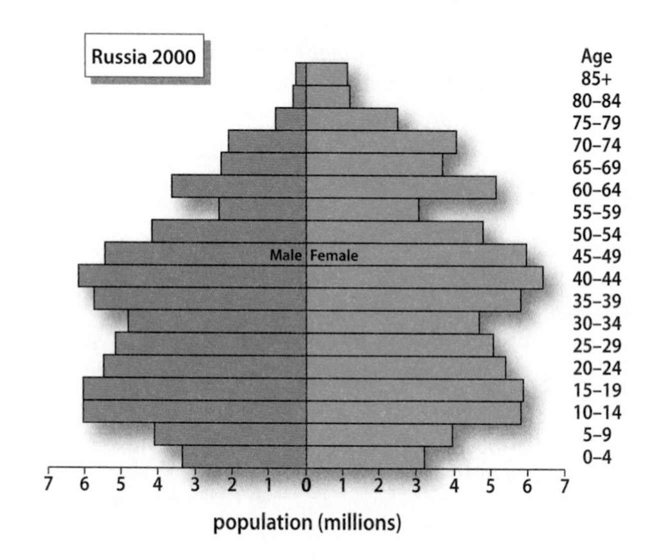

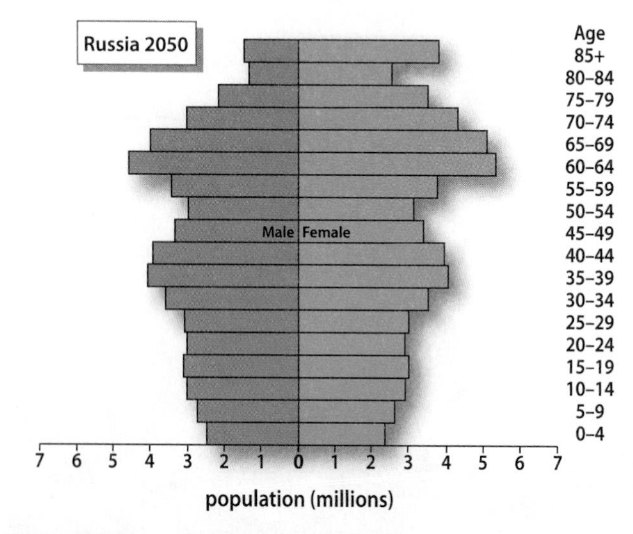

Figure 9.21 Russia's Changing Population These two population pyramids provide a recent glimpse (2000) as well as predicted patterns (2050) of Russia's population structure. Present trends suggest that Russia's population will continue to age, with relatively fewer young people supporting a relatively large elderly population. Also note the impact of earlier wars and higher death rates on older adult Russian males. *(Source: U.S. Census Bureau, International Data Base)*

older males and the birth crash of World War II), relatively few children are being added. Looking ahead to 2050, today's shrinking numbers take an even bigger toll as the pattern of small families is likely to continue.

In 2006, President Putin declared that demographic decline was Russia's "most acute problem." He and other Russian leaders have pushed for a long-term program aimed at raising birthrates. Under the plan, mothers of multiple children receive cash payments, extended maternity leave, and extensive day-care subsidies. Some Russian cities have even sponsored competitions, encouraging couples to have babies in the hopes of winning prizes such as automobiles. Recently, there have been higher birthrates within Russia, perhaps a function of changing government policies. In particular, rates among ethnic non-Russians in the country (for example, in the Caucasus region and portions of Siberia) have been significantly above those of ethnic Russians. In 2009, the country reported its first year of natural population growth in more than 15 years. Indeed, both Russian and Ukrainian birthrates are now significantly higher than those of Germany, Italy, and Japan and there are long waiting lists for urban kindergarten spots in Moscow (Figure 9.22). Employers are also offering more benefits to both mothers and fathers, making parenthood more appealing for some couples.

Still, many uncertainties remain concerning future demographic growth. Another severe and prolonged economic decline such as that experienced in the 1990s would likely depress birth rates. In addition, residents across the entire region still face significantly higher death rates than much of the developed world and they also have daunting health-care challenges. The region's demographic structure also forecasts fewer women of child-bearing age in the near future, making the current improvements difficult to maintain. One other wild card remains immigration: some Russian population experts urge leaders to encourage more migration to the region, but this continues to unsettle others who see such inflows as a threat to Russian cultural dominance.

Figure 9.22 Young Russian Parents A slight rise in Russian birthrates may help slow that country's population decline. Still, high death rates are a major issue, particularly for males throughout the region, and they are the product of many environmental and lifestyle variables.

CULTURAL COHERENCE AND DIVERSITY: The Legacy of Slavic Dominance

The Russian domain remains at the heart of the Slavic world. For hundreds of years, Slavic peoples speaking the Russian language expanded their influence from an early homeland in central European Russia. Eventually, this Slavic cultural imprint spread north to the Arctic Sea, south to the Black Sea and Caucasus, west to the shores of the Baltic, and east to the Pacific Ocean. In this process of diffusion, Russian cultural patterns and social institutions spread widely, influencing many non-Russian ethnic groups that continued to live under the rule of the Russian Empire. The legacy of that Slavic expansion continues today. It offers Russians a rich historical identity and sense of nationhood. It also provides a meaningful context in which to understand the way present-day Russians are dealing with forces of globalization and how non-Russian cultures have evolved within the region.

The Heritage of the Russian Empire

The expansion of the Russian Empire paralleled similar events in western Europe. As Spain, Portugal, France, and Britain carved out empires in the Americas, Africa, and Asia, the Russians expanded eastward and southward across Eurasia. Unlike other European empires, however, the Russians formed one single territory, uninterrupted by oceans or seas. Only with the fall of the Soviet Union in 1991 did this transformed empire finally begin to dissolve.

Origins of the Russian State The origin of the Russian state lies in the early history of the **Slavic peoples**, defined linguistically as a distinctive northern branch of the Indo-European language family. The Slavs originated in or near the Pripyat marshes of modern Belarus. Some 2,000 years ago they began to migrate to the east, reaching as far as modern Moscow by 200 CE. Slavic political power grew by 900 CE as Slavs intermarried with southward-moving warriors from Sweden known as *Varangians*, or *Rus*. Within a century, the state of Rus extended from Kiev (the capital) in modern Ukraine, to Lake Ladoga near the Baltic Sea. The new Kiev-Rus state interacted with the rich and powerful Byzantine Empire of the Greeks, and this influence brought Christianity to the Russian realm by 1000 CE. Along with the new religion came many other aspects of Greek culture, including the Cyrillic alphabet. Even as groups such as the Russians and Serbs converted to **Eastern Orthodox Christianity**, a form of Christianity historically linked to eastern Europe and church leaders in Constantinople (modern Istanbul), their Slavic neighbors to the west (the Poles, Czechs, Slovaks, Slovenians, and Croatians) accepted Catholicism. The resulting religious division split the

Slavic-speaking world into two groups, one oriented to the west, the other to the east and south. This early Russian state soon faltered and split into several principalities that were then ruled by invading Mongols and Tatars (a group of Turkish-speaking peoples).

Growth of the Russian Empire By the 14th century, however, northern Slavic peoples overthrew Tatar rule and established a new and expanding Slavic state (Figure 9.23). The core of the new Russian Empire lay near the eastern fringe of the old state of Rus. The former center around Kiev was now a war-torn borderland (or "Ukraine" in Russian) contested by the Orthodox Russians, the Catholic Poles, and the Muslim Turks. Gradually this area's language diverged from that spoken in the new Russian core, and *Ukrainians* and Russians developed into two separate peoples. A similar development took place among the northwestern Russians, who experienced several centuries of Polish rule and over time were transformed into a distinctive group known as the *Belorussians*.

The Russian Empire expanded remarkably in the 16th and 17th centuries (see Figure 9.23). Former Tatar territories in the Volga Valley (near Kazan) were incorporated into the Russian state in the mid-1500s. The Russians also allied with the seminomadic **Cossacks**, Slavic-speaking Christians who had earlier migrated to the region to seek freedom in the ungoverned steppes (see Figure 9.2). The Russian Empire granted them considerable privileges in exchange for their military service, an alliance that facilitated Russian expansion into Siberia during the 17th century. Premium furs were the chief lure of this immense northern territory. By the 1630s, Russian power was entrenched in central Siberia, and, by the end of the century, it had reached the Pacific Ocean. Chinese resistance, however, delayed Russian occupation of the Far East region until 1858, and the imperial designs of the Japanese halted further ex-

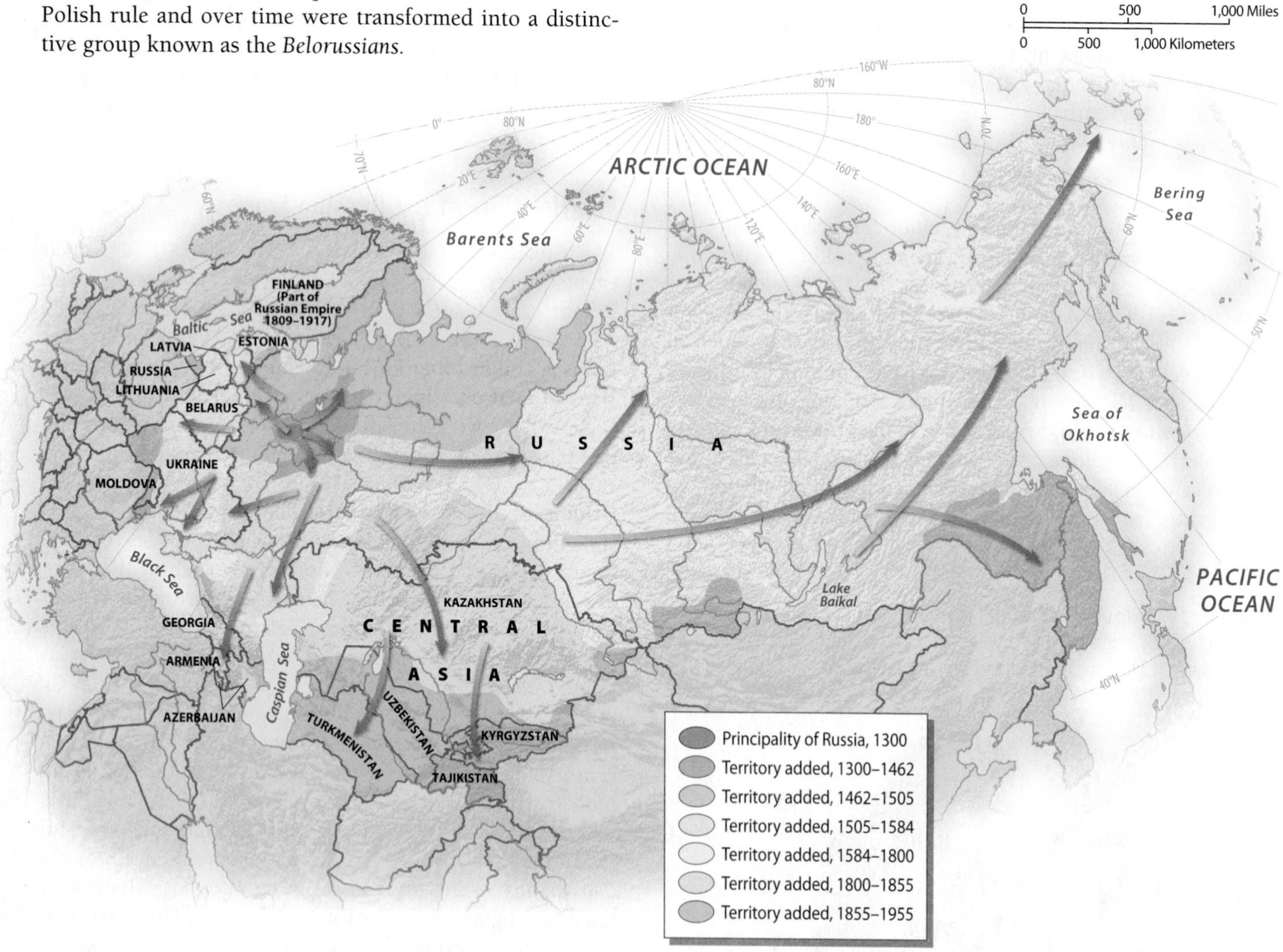

Figure 9.23 Growth of the Russian Empire Beginning as a small principality in the vicinity of modern Moscow, the Russian Empire took shape between the 14th and 16th centuries. After 1600, Russian influence stretched from eastern Europe to the Pacific Ocean. Later, portions of the empire were added in the Far East, Central Asia, and near the Baltic and Black seas. *(Modified from Bergman and Renwick, 1999,* Introduction to Geography, *Upper Saddle River, NJ: Prentice Hall)*

pansion to the southeast when the Russians lost the Russo-Japanese War in 1905.

While the Russian Empire expanded to the east with great rapidity, its westward expansion was slow and halting. In the 1600s, Russia still faced formidable enemies in Sweden, Poland, and the Ottoman (Turkish) Empire. By the 1700s, however, all three of these states had weakened, allowing the Russian Empire to gain substantial territories. After defeating Sweden in the early 1700s, Tsar Peter the Great (1682–1725) obtained a foothold on the Baltic, where he built the new capital city of St. Petersburg. Later in the 18th century, Russia defeated both the Poles and the Turks and gained all of modern-day Belarus and Ukraine. Tsarina Catherine the Great (1762–96) was particularly pivotal in colonizing Ukraine and bringing the Russian Empire to the warm-water shores of the Black Sea (Figure 9.24).

The 19th century witnessed the Russian Empire's final expansion. Large gains were made in Central Asia, where a group of once-powerful Muslim states was no longer able to resist the Russian army. The mountainous Caucasus region proved a greater challenge, as the peoples of this area had the advantage of rugged terrain in defending their lands.

Figure 9.24 Catherine the Great Tsarina Catherine the Great ruled the Russian Empire between 1762 and 1796 and expanded imperial influence over portions of southern Russia, Ukraine, and the warm-water coastline of the Black Sea.

South of the Caucasus, however, the Christian Armenians and Georgians accepted Russian power with little struggle because they found it preferable to rule by the Persian or Ottoman empires.

The Legacy of Empire The expansion of the Russian people was one of the greatest human movements Earth has ever witnessed. By 1900, a traveler going from St. Petersburg on the Baltic to Vladivostok on the Sea of Japan would everywhere encounter Russian peoples speaking the same language, following the same religion, and living under the rule of the same government. Nowhere else in the world did such a tightly integrated cultural region cover such a vast space.

The history of the Russian Empire also reveals points of ongoing tension with the world beyond. One of these tensions centers on Russia's ambivalent relationship with western Europe. Russia shares with the West the historical legacy of Greek culture and Christianity. Since the time of Peter the Great, Russia has undergone several waves of intentional Westernization. At the same time, however, Russia has long been suspicious of—even hostile to—European culture and social institutions. Although elements of this debate were transformed during the Soviet period, the central tension remains, and it influences Russia to this day. As Europe further unifies through the expansion of the European Union (EU), this gap between East and West may take new forms, but it is unlikely to disappear anytime soon.

Geographies of Language

Slavic languages dominate the region (Figure 9.25). The distribution of Russian-speaking populations is complicated. Russian, Belorussian, and Ukrainian are closely related languages. Some linguists argue that they ought to be considered separate dialects of a single Russian language, as they are all mutually intelligible. Most Ukrainians, however, insist that Ukrainian is a distinct language in its own right, and there is a well-developed sense of national distinction between Russians and Ukrainians. Belorussians, on the other hand, are more inclined to stress their close kinship with the Russians.

Patterns in Belarus, Ukraine, and Moldova The geographic pattern of the Belorussian people is relatively simple. The vast majority of Belorussians reside in Belarus, and most people in Belarus are Belorussians (see Figure 9.25). The country does, however, contain scattered Polish and Russian minorities. For the Russians, this presents few problems, since Russians and Belorussians can relatively easily assume each other's ethnic identity.

The situation in Ukraine, however, is more complex. Only about 67 percent of the population is Ukrainian. Russian speakers make up almost 25 percent of the population, but are strongly concentrated in eastern Ukraine, while they make up a much smaller portion of western

Figure 9.25 Languages of the Russian Domain Slavic Russians dominate the region, although many linguistic minorities are present. Siberia's diverse native peoples add cultural variety in that area. To the southwest, the Caucasus Mountains and the lands beyond contain the region's most complex linguistic geography. Ukrainians and Belorussians, while sharing a Slavic heritage with their Russian neighbors, add further variety in the west.

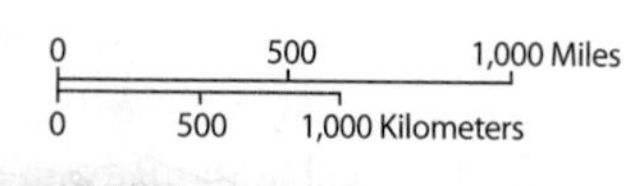

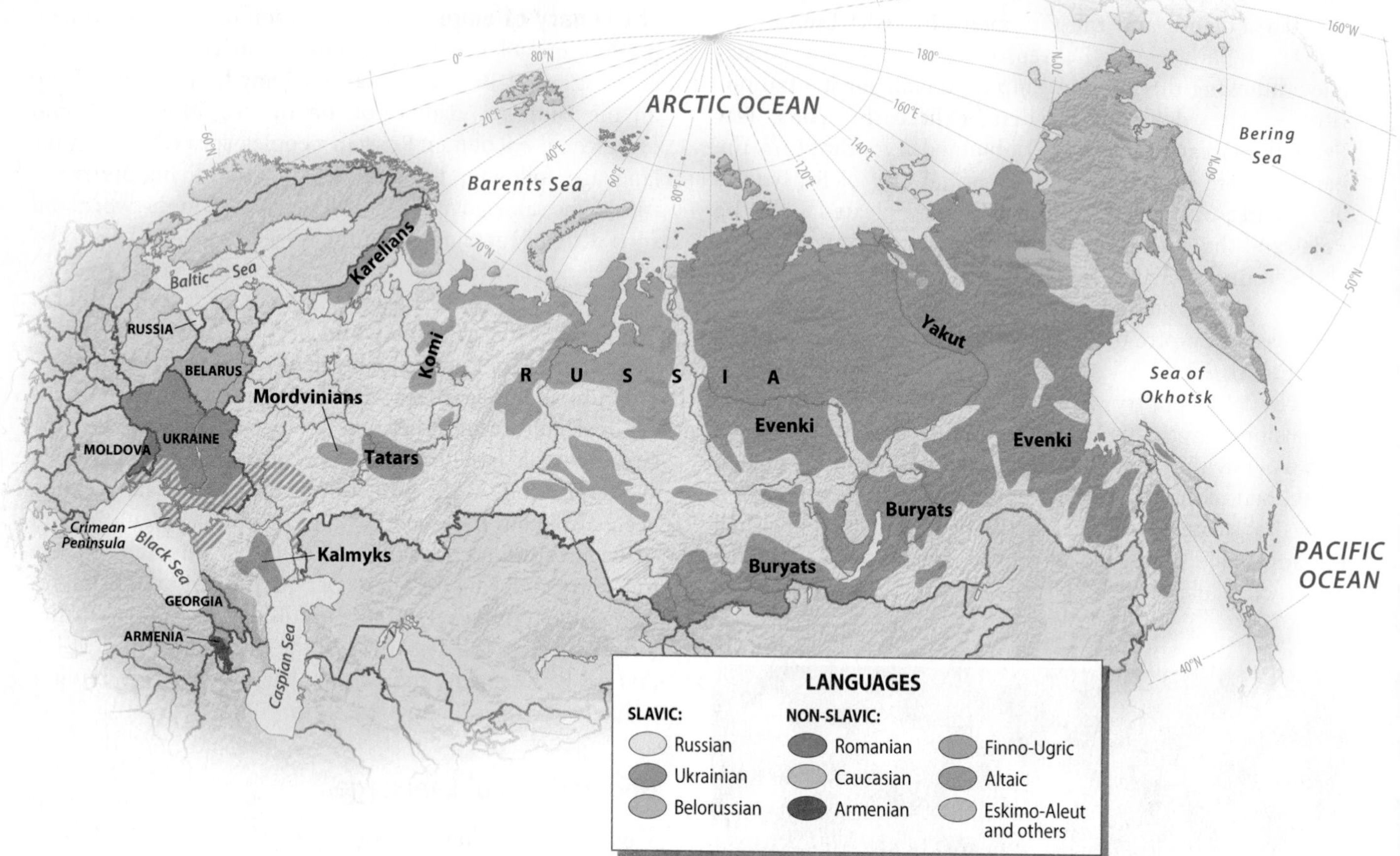

Ukraine's population (Figure 9.26). Similarly, the Crimean Peninsula, now a part of Ukraine, has long ethnic and political connections to Russia. As a result, there are many Ukrainian citizens in the eastern and southern parts of the country who rarely speak Ukrainian. Conversely, many Ukrainians born in the west never learn Russian. Kiev (Kyiv), the national capital, has been described as bilingual, "a Russian-speaking city whose people know how to speak Ukrainian." Since 2004, Ukrainian speakers have increased their efforts to enforce Ukrainian cultural and linguistic traditions, a campaign that has provoked popular and political resistance in the eastern portion of the country.

Figure 9.26 The Russian Language in Ukraine Both eastern Ukraine and the Crimean Peninsula retain large numbers of Russian speakers, a function of long cultural and political ties that continue to complicate Ukraine's contemporary human geography.

In nearby Moldova, Romanian (a Romance language) speakers are dominant, although ethnic Russians and Ukrainians each make up about 13 percent of the country's population. Use of the official Romanian language has been criticized by many Slavic speakers, particularly in the region east of the Dniester River (Transdniester) that borders Ukraine. During the Soviet period, such geographical complexities had little consequence, because the distinction between Russians, Ukrainians, and Moldavians was not viewed as important in official circles. Now that Russia, Ukraine, and Moldova are separate countries with a heightened sense of national distinction, this issue has emerged as a significant source of tension across the region.

Patterns Within Russia Approximately 80 percent of Russia's population claims a Russian linguistic identity. Russian speakers inhabit most of European Russia, but there are large enclaves of other peoples. The Russian linguistic zone extends across southern Siberia to the Sea of Japan. In sparsely settled lands of central and northern Siberia, Russians are numerically dominant in many areas, but they share territory with varied indigenous peoples.

Finno-Ugric peoples, though small in number, dominate sizable portions of the non-Russian north. The Finno-Ugric (part of the Uralic language family) speakers make up an entirely different language family from the Indo-European Russians. While many have been culturally Russified, distinct ethnic groups such as the Karelians, Komi, and Mordvinians remain a part of Russia's modern cultural geography. Karelians, in particular, identify strongly with their Finnish neighbors to the west.

Altaic speakers also complicate the country's linguistic geography. This language family includes the Volga Tatars, whose territory is centered on the city of Kazan in the middle Volga Valley. While retaining their ethnic identity, the Turkish-speaking Tatars have extensively intermarried with and borrowed from their Russian neighbors. Bilingual education is now common in the schools, with ethnic Tatars free to use their traditional language. Yakut peoples of northeast Siberia also represent Turkish speakers within the Altaic family. Other Altaic speakers in Russia belong to the Mongol group. In the west, examples include the Kalmyk-speaking peoples of the lower Volga Valley who migrated to the region in the 1600s. Far to the east, over 400,000 Buryats live in the vicinity of Lake Baikal and represent an indigenous Siberian group closely tied to the cultures and history of Central Asia.

The plight of many native peoples in central and northern Siberia parallels the situation in the United States, Canada, and Australia. Rural indigenous peoples in each of these settings remain distinct from dominant European cultures. Such groups also are internally diverse and often are divided into a number of unrelated linguistic clusters. One entire linguistic grouping of Eskimo-Aleut speakers is limited to approximately 25,000 people, who are widely dispersed through northeast Siberia. Another indigenous Siberian group is the Altaic-speaking Evenki, whose traditional territory covers a large portion of central and eastern Siberia (Figure 9.27). Many of these Siberian peoples have seen their traditional ways challenged by the pressures of Russification, just as indigenous peoples elsewhere in the world have been subjected to similar pressures of cultural and political assimilation. Unfortunately, other common traits seen within such settings are low levels of education, high rates of alcoholism, and widespread poverty.

Transcaucasian Languages Although small in size, Transcaucasia offers a bewildering variety of languages (Figure 9.28). From Russia, along the north slopes of the Caucasus, and to Georgia and Armenia east of the Black Sea, a complex history and a fractured physical setting have combined to produce some of the most complicated language patterns in the world. No fewer than three language families (Caucasian, Altaic, and Indo-European) are spoken within a region smaller than Ohio, and many individual languages are represented by small, isolated cultural groups. Not surprisingly, language remains a pivotal cultural and political issue within the fragmented Transcaucasian subregion.

Figure 9.27 Minority Evenki Russia's indigenous Evenki population speaks an Altaic language and they share many of the economic and social problems of native peoples in North America and Australia.

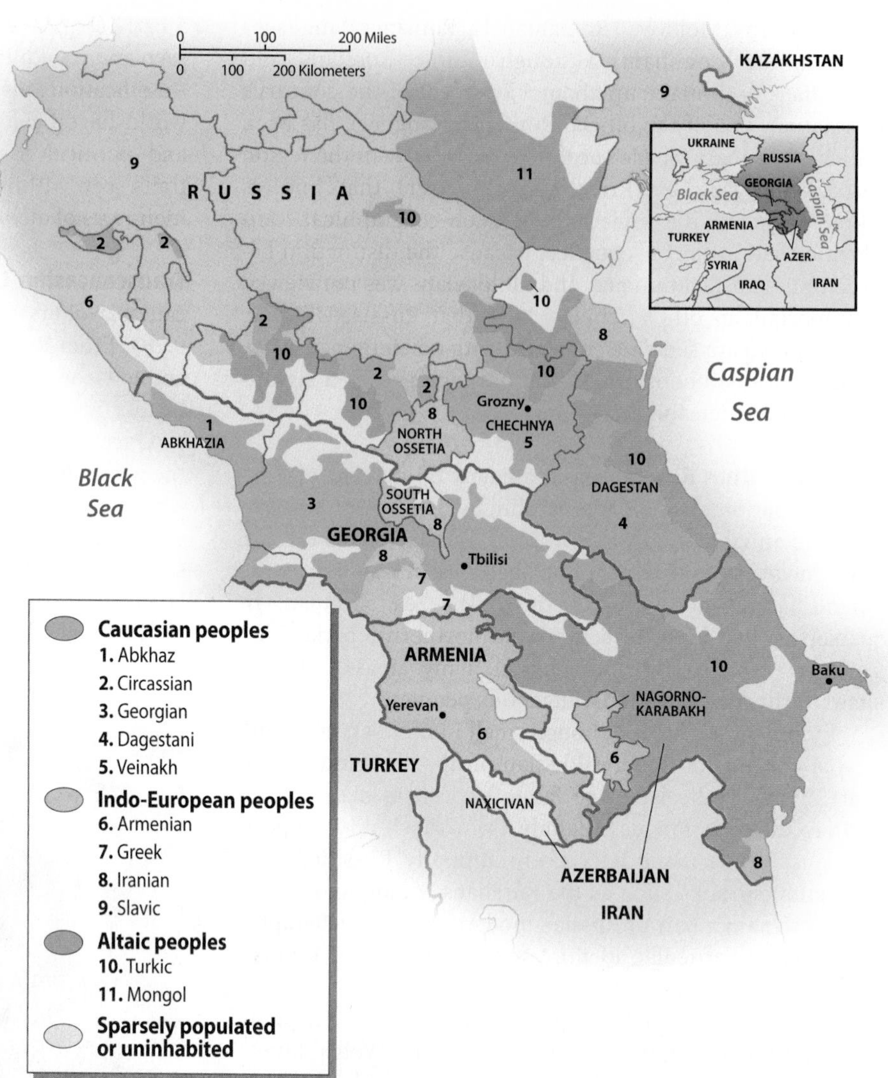

Figure 9.28 Languages of the Caucasus Region A complicated mosaic of Caucasian, Indo-European, and Altaic languages characterizes the Caucasus region of southern Russia and nearby Georgia and Armenia. Persisting political problems have erupted in the region as local populations struggle for more autonomy. Recent examples include independence movements in Chechnya and in nearby Dagestan. *(Source: U.S. Central Intelligence Agency)*

Geographies of Religion

Most Russians, Belorussians, and Ukrainians share a religious heritage of Eastern Orthodox Christianity. For hundreds of years, Eastern Orthodoxy served as a central cultural presence within the Russian Empire (Figure 9.29). Indeed, church and state were tightly fused, until the demise of the empire in 1917. Under the Soviet Union, however, religion in all forms was severely discouraged and actively persecuted. Most monasteries and many churches were converted into museums or other kinds of public buildings, and schools disseminated the doctrine of atheism.

Contemporary Christianity With the downfall of the Soviet Union, however, a religious revival has swept much of the Russian domain. In the past 20 years, more than 12,000 Orthodox churches have been returned to religious uses. Now, an estimated 75 million Russians are members of the Orthodox Church, including almost 500 monastic orders dispersed across the country. Within Russia, the Orthodox Church appears headed toward a return to its former role as state church. The government increasingly is using church officials to sanction various state activities, which is particularly ironic because many of Russia's current leaders played roles in the earlier Soviet period when religious observances were banned. For example, when he became president in 2000, former Soviet-era KGB agent Vladimir Putin professed his Orthodox beliefs and claimed that his mother baptized him during the height of Soviet repression. In turn, the Church has supported recent Russian political campaigns, both domestically and internationally.

Figure 9.29 Russian Orthodox Church A revival of interest in the Russian Orthodox Church followed the collapse of the Soviet Union in the early 1990s. This newly built church, completed in 2007, is located on energy-rich Sakhalin Island in the Russian Far East.

Other forms of Western Christianity also are present in the region. For example, the people of western Ukraine, who experienced several hundred years of Polish rule, eventually joined the Catholic Church. Eastern Ukraine, on the other hand, remained fully within the Orthodox framework. This religious split reinforces the cultural differences between eastern and western Ukrainians. Western Ukrainians have generally been far more nationalistic, hence, more firmly opposed to Russian influence than eastern Ukrainians.

Elsewhere, Christianity came early to the Caucasus, but modern Armenian forms—their roots dating to the 4th century CE—differ slightly from both Eastern Orthodox and Catholic traditions. Georgian Christianity, however, is more closely tied to the Orthodox faith. Evangelical Protestantism has also been on the rise since the demise of the Soviet Union.

Non-Christian Traditions Non-Christian religions also appear and, along with language, shape ethnic identities and tensions within the region. Islam is the largest non-Christian religion. In Russia, there are some 7,000 mosques and approximately 20 million adherents (Figure 9.30). Most are Sunni Muslims, and they include peoples in the North Caucasus, the Volga Tatars, and Central Asian peoples near the Kazakhstan border. Growth rates among Russia's Muslim population are three times that of the non-Muslim population. In addition, Moscow's job opportunities have attracted many Muslim immigrants. Recent estimates, including illegal immigrants, suggest that 20 percent of Moscow's population is now Islamic. Islamic fundamentalism also has been on the rise, particularly among Muslim populations in the Caucasus region, who increasingly resist what they see as strong-arm tactics and repressive actions on the part of the Russian government.

Figure 9.30 Moscow Mosque Moscow's growing immigrant population has included many Muslims from portions of the former Soviet Union, particularly from Central Asia. Today, their presence in the capital is an increasingly visible element in the city's cultural landscape.

Russia, Belarus, and Ukraine are also home to more than 1 million Jews, who are especially numerous in the larger cities of the European West. Jews suffered severe persecution under both the tsars and the communists. Recent out-migrations, prompted by new political freedoms, have further reduced their numbers in Russia, Belarus, and Ukraine. Buddhists also are represented in the region, associated with the Kalmyk and Buryat peoples of the Russian interior. Indeed, Buddhism has witnessed a recent renaissance and now claims approximately 1 million practitioners, mostly in Asiatic Russia.

Russian Culture in Global Context

Russian culture has interacted in varied ways with the world beyond. Russian cultural norms have for centuries embodied both an inward orientation toward traditional forms of expression and an outward orientation directed primarily to western Europe. By the 19th century, even as Russian peasants interacted rarely with the outside world, Russian high culture had become thoroughly Westernized, and Russian composers, novelists, and playwrights gained considerable fame in Europe and the United States. By the turn of the 20th century, Russian avant-garde artists were among the international leaders in devising the experimental styles of modernism.

Soviet Days During the Soviet period, a new mixture of cultural relationships unfolded within the socialist state. Initially, European-style modern art flourished in the Soviet Union, encouraged by the radical ideas of the new rulers. By the late 1920s, however, Soviet leaders turned against modernism, which they viewed as the decadent expression of a declining capitalist world. Many Soviet artists fled to the West, and others were exiled to Siberian labor camps. Increasingly, state-sponsored Soviet artistic productions centered on **socialist realism**, a style devoted to the realistic depiction of workers harnessing the forces of nature or struggling against capitalism. Still, traditional high arts, such as classical music and ballet, continued to receive lavish state subsidies, and to this day Russian artists regularly achieve worldwide fame.

Turn to the West By the 1980s, it was clear that the attempt to fashion a new Soviet culture based on socialist realism and working-class solidarity had failed. The younger generation instead adopted a rebellious stance, turning for inspiration to fashion and rock music from the West. The mass-consumer culture of the United States proved immensely popular, symbolized above all by brand-name chewing gum, jeans, and cigarettes. The Soviet government attempted to ward off this perceived cultural onslaught, but with little success. Soviet officials could more easily censor books and other forms of written expression than Western countries, but even here they were increasingly frustrated. By the end of the Soviet period, the persistent secrecy that separated the Soviet Union from the West had broken down, thus enabling more people and influence to flow into the region.

After the fall of the Soviet Union in 1991, basic freedoms brought an inrush of global cultural influences, particularly to the region's larger urban areas such as Moscow. Shops were quickly flooded with Western books and magazines; people pondered the financial mysteries of home mortgages and condominium purchases; and they reveled in the new-found pleasures of fake Chanel handbags and McDonald's hamburgers. English-language classes became even more popular in cities such as Moscow, where Russians hurried to embrace the world their former leaders had warned them about for generations. Cultural influences streaming into the country were not all Western in inspiration. Films from Hong Kong and Mumbai (Bombay), as well as the televised romance novels *(telenovelas)* of Latin America, for example, proved far more popular in the Russian domain than in the United States. This onrush of global cultural influences, however, has not equally spread throughout the region. While urban residents have access and money to explore such options, rural life across the Russian domain remains far more wedded to traditional cultural institutions and values.

The Music Scene Younger residents of the region also have embraced the world of popular music, and their enthusiasm for American and European performers, as well as their support of a budding home-grown music industry, symbolizes the changing values of an increasingly post-Soviet generation. MTV Russia went on the air at midnight on September 26, 1998. A year later, the network sponsored a huge open-air concert on Red Square. Headlined by the Red Hot Chili Peppers, the event was a far cry from Soviet-era tank parades in front of Communist Party officials. By 2005, Russian MTV reached more than 60 million viewers, particularly in large urban areas. Sony Music Entertainment also established its Russian operations in December 1999. It has been followed by several of the world's other major recording companies. Sony, BMG, and other labels have signed multiple Russian artists for domestic markets, helping boost a home-grown pop-music culture. Universal also has opened the way for Russian performers to go global. For example, it sponsored the English-language debut of Russian pop singer Alsou in 2001. Three years later, Ukrainian singer Ruslana won the coveted Eurovision 2004 award for her song "Wild Dance". In a superb example of cultural globalization, the high-tech performance of the song at the awards show in Istanbul, Turkey, was seen by more than 100 million television viewers worldwide, and it featured a mix of Ukrainian folk music traditions, leather-clad female dancers, and on-stage acrobatics. More recently, Moscow hosted the 2009 Eurovision awards, which featured a strong performance from Ukrainian singer, composer, and television personality Svetlana Loboda (Figure 9.31). She used her performance of "Be My Valentine (Anti-Crisis Girl)" and her popularity in both Russia and Ukraine as a platform to speak out against domestic violence against women, a widespread problem in her home country.

A Revival of Russian Nationalism Even as the Russian domain finds itself increasingly swept up in external cultural influences, countertrends that emphasize Russian nationalism persist. Extremely conservative nationalists resist foreign influences such as Western-style music, and they often have been supported by more traditional elements of the Russian Orthodox Church. Even Prime Minister Putin and the Russian government have carefully resurrected numerous symbols from the Russian past in order to cultivate a renewed sense of Russian identity within the region. Recently, Russian leaders proposed a return to the tsarist-era coat of arms (the double-headed eagle) to symbolize the ongoing connections between Russian society today and the earlier culture of Dostoyevsky and Pushkin. The glory of Soviet communist days (1917–91) also has been revived, with a strong Russian tone. The Soviet-era national anthem was reintroduced (with new lyrics) in 2001, and a number of Soviet holidays, celebrations, and national awards have been reinstated. Recent Russian textbooks have even resurrected the virtues of Joseph Stalin. Once discredited as a brutal, cruel leader, Stalin is now seen as an "effective manager" of Soviet affairs in an earlier era.

Figure 9.31 Svetlana Loboda Ukrainian singing star Svetlana Loboda was one of the top performers at Eurovision 2009, held in Moscow. In addition to a successful career in music and television, Loboda has championed the plight of battered women throughout the region.

GEOPOLITICAL FRAMEWORK: Resurgent Global Superpower?

The geopolitical legacy of the former Soviet Union still weighs profoundly upon the Russian domain. After all, the bold lettering of the "Union of Soviet Socialist Republics" dominated the Eurasian map for much of the 20th century, and the country's global political reach left no corner of the world untouched. Former Soviet republics still struggle to define new geopolitical identities for themselves. Neighboring states continue to view the region with nervousness left over from the days of Soviet political and military power. Present demands for more local and regional political control within countries such as Russia, Ukraine, Moldova and Georgia can still be understood in the context of the Soviet era, when a highly centralized government gave little voice to regional dissent. At the same time, Russia's new global visibility and its desire to recentralize authority in Moscow have raised concerns elsewhere within the domain, in Europe, and in the United States.

Geopolitical Structure of the Former Soviet Union

The Soviet Union rose from the ashes of the Russian Empire, which collapsed abruptly in 1917. The ultraconservative policies of the Russian tsar generated opposition among businesspeople and workers, while the peasants, who formed the majority of the population, had always resisted the powerful land-owning aristocracy. After the fall of the tsar and the aristocracy, a broad-based coalition government assumed authority. Several months later, however, the **Bolsheviks**, a faction of Russian communists representing the interests of the industrial workers, seized power within the country. The leader of these Russian communists was Vladimir Ilyich Ulyanov, usually known by his self-selected name, Lenin. Lenin was a key architect of the Soviet Union, which fully emerged between 1917 and 1922. It became the Russian Empire's successor state.

The new socialist state reconfigured Eurasian political geography. Although it resembled the territorial contours of the Russian Empire and centralized authority continued to be concentrated in European Russia, the political and economic structure of the country was radically transformed. Lenin and the other communist leaders were aware that they faced a major challenge in organizing the new state.

The Soviet Republics and Autonomous Areas Soviet leaders designed a geopolitical solution that maintained their country's territorial boundaries and acknowledged, at least theoretically, the rights of its non-Russian citizens. Each major nationality was to receive its own "union republic," provided it was situated on one of the nation's external borders (Figure 9.32). Eventually 15 such republics were established, thus creating the Soviet Union. A number of these republics were quite small, while the massive Russian Republic sprawled over roughly three-quarters of the Soviet terrain. Each republic was to have considerable political autonomy, vested with the right to withdraw from the union if it so desired. In practice, however, the Soviet Union remained a centralized state, with important decisions made in the capital of Moscow.

The Soviets came up with different geopolitical solutions to acknowledge smaller ethnic groups and nationalities that were not situated on the country's external borders. Indeed, dozens of significant minority groups pressed for recognition. One solution to this problem was the creation of **autonomous areas** of varying sizes that recognized special ethnic homelands but did so within the structure of existing republics. Thus, within the Russian Republic, the larger nationalities, such as the Yakut and the Volga Tatars, were granted their own "autonomous republics" (not to be confused with the 15 higher-level union republics, such as Russia or Ukraine). The system's main deficiency was that the autonomy it was designed to provide proved to be more of a charade than a reality.

Centralization and Expansion of the Soviet State In the early Soviet era, it appeared that the framework of separate republics and autonomous areas might allow non-Russian peoples to protect their own cultures and establish their own social and economic policies (provided, of course, that such policies embodied Marxist principles). From the beginning, however, it was clear that such self-determination would be a temporary measure. According to the official ideology of Marxist beliefs, the gradual development of a communist society would see the withering away of all significant ethnic differences and the disappearance of religion. In the future, a new classless Soviet society was supposed to emerge.

By the 1930s, it was clear that national autonomy would not have any real significance within an increasingly centralized Soviet state. The chief architect of this political consolidation was Soviet leader Joseph Stalin, who did everything he could to centralize power in Moscow and to assert Russian authority. Stalin launched a ruthless plan of state-controlled agricultural production and industrialization. Although many people initially resisted his policies, Stalin never hesitated to use force to bring about his vision of a purer socialist revolution within the Soviet Union. Stalin also utilized more subtle methods of control in his quest to produce a new communist society. The Soviet landscape itself came to reflect communist ideas. Soviet-style architecture, paintings, cinema, and many other visual elements in the everyday world came to symbolize the new world view.

The Stalin period also saw the geographical enlargement of the Soviet Union. As a victorious power in World War II, the country acquired southern Sakhalin and the Kuril Islands from Japan. It regained the Baltic republics (Lithuania, Latvia, and Estonia—independent between 1917 and 1940), as well as substantial territories formerly belonging to Poland, Romania, and Czechoslovakia.

After World War II, the Soviet Union also gained significant authority, although not actual sovereignty, over a broad swath of eastern Europe. As they pushed the German army

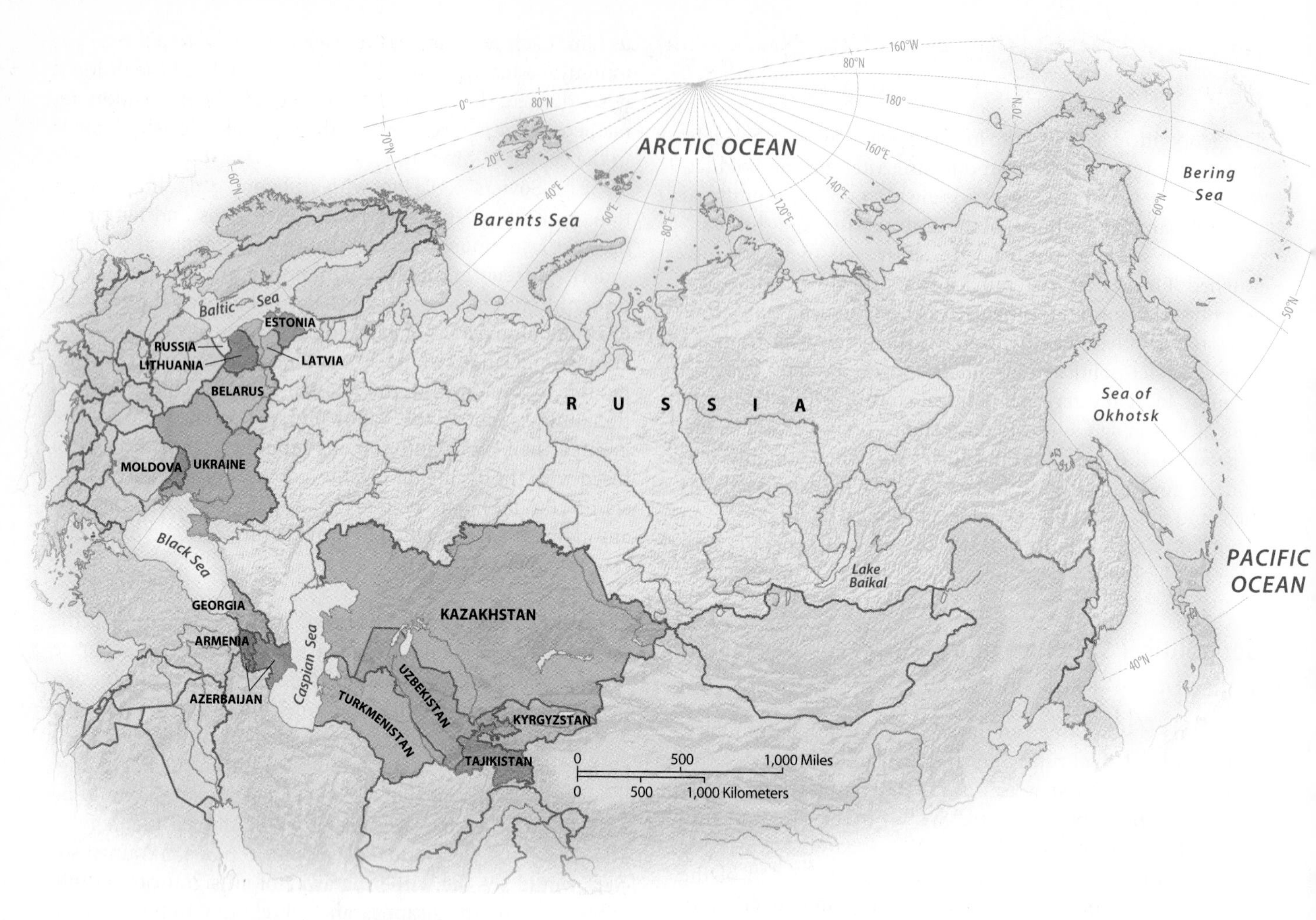

Figure 9.32 Soviet Geopolitical System During the Soviet period, the boundaries of the country's 15 internal republics often reflected major ethnic divisions. Ultimately, however, many of the ethnically non-Russian republics pressured the Soviet government for more political power. As the Soviet Empire disintegrated, the former republics became politically independent states and now form an uneasy ring of satellite nations around Russia. *(Modified from Rubenstein, 1999,* Introduction to Human Geography, *Upper Saddle River, NJ: Prentice Hall)*

west toward the end of the war, Soviet troops advanced across much of the region, actively working to establish communist regimes thereafter. In the words of British leader Winston Churchill, the Soviets extended an "**Iron Curtain**" between their eastern European allies and the more democratic nations of western Europe by restricting the free flow of people and information.

As eastern Europe disappeared behind the Iron Curtain, the Soviet Union and the United States, while allies during World War II, became antagonists in a global **Cold War** of escalating military competition that lasted from 1948 to 1991. In its post–World War II heyday, the Soviet Union became a global superpower, one of only two countries equipped with enough nuclear weapons to ensure global destruction. At the height of its power in the late 1970s, it enjoyed close military and economic alliances not only with eastern Europe, but also with certain countries in Asia (Mongolia, North Korea, Vietnam, Laos, and Cambodia), the Caribbean region (Cuba and Nicaragua), and Africa (Angola, Somalia, and several others).

End of the Soviet System Ironically, Lenin's system of culturally defined republics helped sow the seeds of the Soviet Union's downfall. Even though the nationally based republics and autonomous areas of the Soviet Union were never allowed real freedom, they did provide a lasting political framework for the maintenance of distinct cultural identities. Indeed, contrary to the expectations of Soviet leaders, ethnic nationalism intensified in the post–World War II era as the Soviet system grew less repressive. When Soviet president Mikhail Gorbachev initiated his policy of **glasnost**, or greater openness, during the 1980s, several republics—most notably the Baltic states of Lithuania, Latvia, and Estonia—demanded outright independence.

Other forces also worked toward the political end of the Soviet regime. A failed war in Afghanistan in the early 1980s frustrated both the Soviet leaders and population. In eastern Europe, multiple protests over Soviet dominance emerged after World War II from Hungary, Czechoslovakia, and Poland. Worsening domestic economic conditions, increasing food shortages, and a declining quality of life led to fundamental questions concerning the value of centralized planning within the country. In response, President Gorbachev introduced **perestroika**, or planned economic restructuring, aimed at making production more efficient and more responsive to the needs of Soviet citizens. In 1991, however, Gorbachev saw his authority slip away amid rising pressures for political decentralization and more dramatic economic reforms. During the summer, Gorbachev's regime was further imperiled by the popular election of reform-minded Boris Yeltsin as the head of the Russian Republic and by a failed military coup by communist hard-liners. By late December, all of the country's 15 constituent republics had become independent states, and the Soviet Union ceased to exist.

Current Geopolitical Setting

Post-Soviet Russia and the nearby independent republics have radically rearranged their political relationships since the collapse of the Soviet Union in 1991. All of the former republics still struggle to establish stable political relations with their neighbors, and tensions within Russia continue as recent leaders have pushed for more centralized control and more limits on media independence and personal liberties. For a time it seemed that a looser political union of most of the former republics, called the **Commonwealth of Independent States (CIS)**, would emerge from the ruins of the Soviet Union. All the former republics, with the exception of the three Baltic states, joined the CIS soon after the dismemberment of the old union (Figure 9.33). By the early 21st

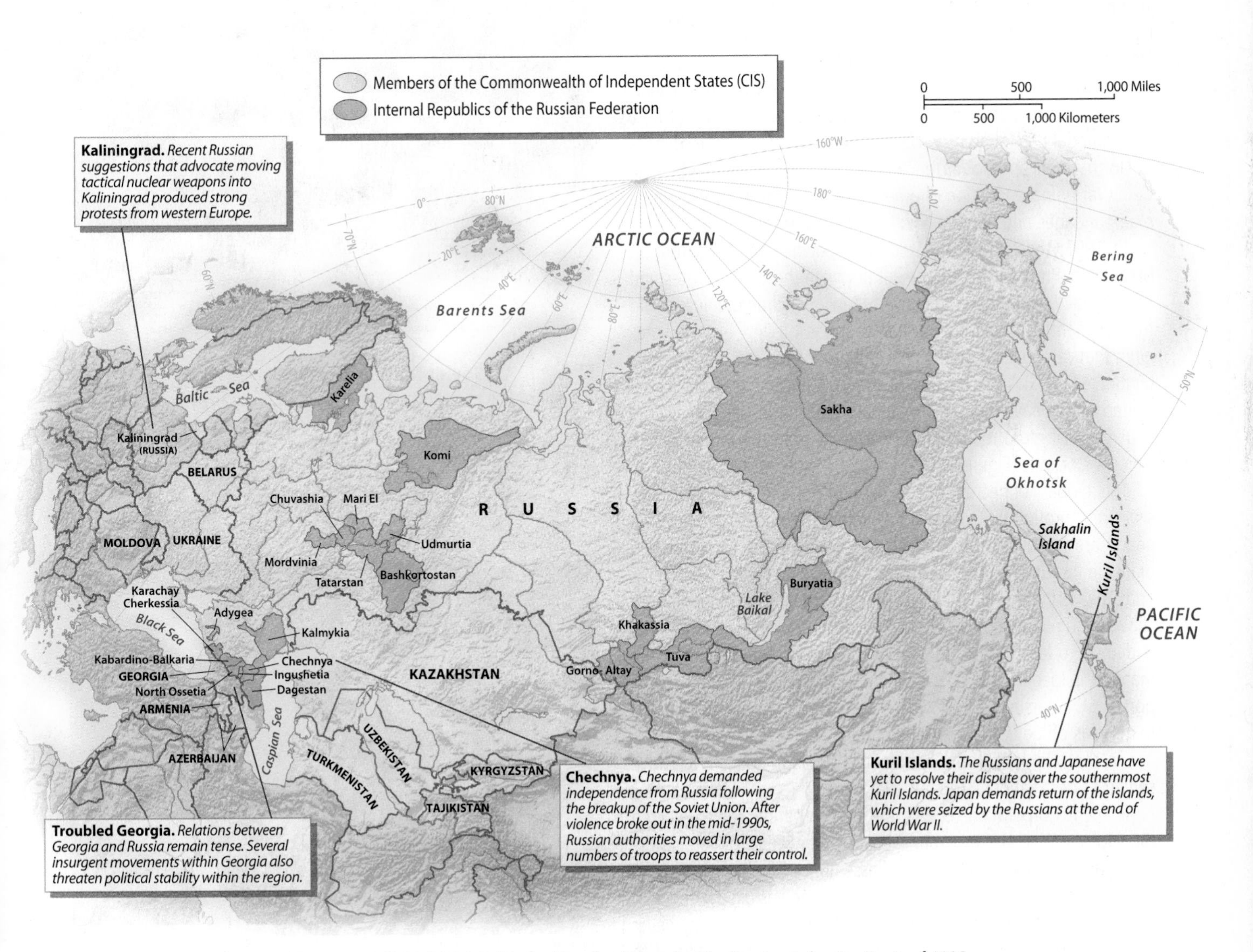

Figure 9.33 Geopolitical Issues in the Russian Domain The Russian Federation Treaty of 1992 created a new internal political framework that acknowledged many of the country's ethnic minorities. Recently, however, Russian authorities have moved to centralize power and limit regional dissent. Russia's relations with several nearby states remain strained. *(Modified from Rubenstein, 2011,* Introduction to Human Geography, *Upper Saddle River, NJ: Prentice Hall)*

century, however, the CIS had developed into little more than a forum for discussion, without real economic or political power. In addition, Georgia withdrew from participation in 2009 amid its sharpening disagreements with its Russian neighbor.

Russia continues to maintain a military presence in many of the former Soviet republics. **Denuclearization**, the return of nuclear weapons from outlying republics to Russian control and their partial dismantling, was initiated during the 1990s. Soviet-era nuclear arsenals of Kazakhstan, Ukraine, and Belarus were removed in the process (see "Exploring Global Connections: From Russia with Love: The Megatons to Megawatts Program"). Elsewhere, Tajikistan invited the Russian army in during the early 1990s to quell its own internal ethnic struggles, and Armenia has maintained close military ties with Moscow. Russia also maintains naval bases in Ukraine's Crimea region (Figure 9.34). The Soviet-era Black Sea fleet was traditionally based in the Crimean port of

EXPLORING GLOBAL CONNECTIONS

From Russia with Love: The Megatons to Megawatts Program

It is a supreme irony and a superb example of globalization that the very bombs once aimed at the heartland of North America are now lighting millions of its homes and fueling its 21st century economy. Experts estimate that fully ten percent of U.S. electricity is produced from reworked nuclear material that once sat on the tip of warheads in bomb silos across Russia and Ukraine. The story of how that material has found its way from the Russian domain to dozens of North American nuclear reactors is a tale both of the end of the Cold War and the creation of a new global energy economy where old boundaries and barriers now form new opportunities.

In 1993, the end of the Cold War spawned the Megatons to Megawatts Program, an agreement between the United States and Russia that was designed to convert highly-enriched uranium taken from dismantled Soviet-era nuclear weapons into low-enriched uranium that can be used for the peaceful production of nuclear power (Figure 9.5.1). By 2010, many of the 103 nuclear reactors in the United States have used this material, reducing in the process the equivalent of more than 15,000 warheads from the former Soviet Union's stockpile (Figure 9.5.2).

This global connection begins deep in nuclear storage facilities scattered throughout Russia and Ukraine. The nuclear-tipped warheads are removed from the missiles and shipped to processing plants in the Urals and western Siberia, including facilities in Ozersk and Yekaterinburg (see Figure 9.1). Warhead components are shaved and processed and the highly-concentrated nuclear material is diluted to less than 5 percent of its original concentration. The material is loaded into large cylinders and shipped by train to the port of St. Petersburg on the Baltic Sea. After an ocean voyage, the material is offloaded in the United States and sent to testing facilities in the Midwest (both Portsmouth, OH and Paducah, KY have served as destinations). The material is tested (by the United States Enrichment Corporation (USEC), the U.S. corporate agent selected for implementing the program) and then converted by fabricators into uranium oxide pellets and fuel assemblies. Finally, the material is shipped throughout the United States, where it is used in reactors to supply power and to light up the nighttime sky in a way never imagined during the Cold War. While the program is scheduled to wind down after 2013, future agreements may allow more opportunities for this virtuous global trade to continue in later years.

Figure 9.5.1 Soviet Nuclear-tipped Warheads Highly-enriched uranium within thousands of these Soviet-era nuclear warheads has been converted to less-enriched uranium that supplies the American nuclear power industry.

Figure 9.5.2 U.S. Nuclear Reactor Many nuclear power plants in the United States, such as the Crystal River unit near St. Petersburg, Florida, depend on nuclear material from dismantled Soviet-era warheads.

Figure 9.34 Crimean Peninsula Long an ethnic and political borderland, the subtropical Crimean Peninsula juts far into the Black Sea. Although it is a part of Ukraine, Crimea has a largely Russian population along with an important Tatar minority.

Sevastopol, now in Ukraine. After lengthy negotiations, Russia and Ukraine agreed in 1997 to share the naval base. Russia retains some 80 vessels and 15,000 service members in Ukraine. But the longer-term political sovereignty of the entire Crimean Peninsula remains in doubt. The peninsula has been contested real estate within the Russian domain for centuries. Some Ukrainians would like nothing better than to see the Russians leave. Many ethnic Russians living in the area, however, see the continuing Russian military presence as protecting their long-term interests.

Geopolitics in the South and West

Russia shares many cultural and political ties with Belarus and Ukraine and these connections have grown stronger since 2008. In Belarus, leaders have been slow to open to political and economic opportunities in western and central Europe. The country remains firmly within Russia's political orbit. In 2010, the two countries pledged to move towards a "union state" and expanded their common military maneuvers. Specifically, Russia and Belarus have protested the growth of NATO influence in central Europe and the two countries see that expansion as a potential military threat. As part of recent agreements, Russia also affirms its right to use its military within Belarus to defend against these regional threats.

Russia's political relationships with Ukraine have been less predictable since the breakup of the Soviet Union. While Russian investments in Ukraine have grown sharply since 2000, political tensions between the two countries increased. Ukraine remains highly dependent upon Russian energy supplies. In addition, major Russian pipelines to central and western Europe must pass through Ukraine. The result is that the two nations have often wrangled over energy prices and the availability of oil and natural gas supplies. Early in 2009, for example, Russia briefly shut off all natural gas supplies to Ukraine amid pricing and payment disputes between Ukraine and Gazprom, the Russian energy company.

Ukraine's internal politics have also riled Russia. In 2004, Viktor Yushchenko, a reformist leader and Ukrainian nationalist, came to power after a controversial election that even included an attempt to poison him. Yushchenko often criticized Russian leaders and excessive Russian interference with Ukraine's affairs. He has also explored Ukraine's admission into both the EU and NATO. On the other hand, Viktor Yanukovich, one of Ukraine's key opposition leaders, argued for closer relations with Moscow. Steep declines in Ukraine's economy in 2009 prompted widespread public protests in Kiev and calls for early elections, challenging Yushchenko's grip on power. Indeed, elections early in 2010 brought Yanukovich into power. Election returns virtually mirrored the map of Russian-speakers in the country (see Figure 9.26), but Yanukovich also succeeded in attracting other disgruntled Ukrainians weary of the nation's battered economy. The results appeared to signal closer relations with nearby Russia. However the political winds blow in Ukraine, it seems likely that neighboring Russia will continue to watch its neighbor and attempt to influence Ukrainian politics in any way it can to serve its own political and economic interests.

Tiny Moldova also has witnessed political tensions in the post-Soviet era. Conflict has repeatedly flared in the Transdniester region in the eastern part of the country where Russian troops remain and where Slavic separatists have pushed for independence from a central government dominated by Romanian-speaking Moldavians. Russia has encouraged the case for "independence," seeing it as a way to increase their influence in the region. On the other hand, a growing number of Romanians have suggested that Moldavians join with Romania (which is now in NATO and the EU) and dump troublesome Transdniester in the process.

Transcaucasia also remains unstable. Since 2003, the Georgian government has moved toward closer ties with the United States, even suggesting that it might join NATO. At the same time, Georgia's own internal politics recently provoked an invasion from neighboring Russia. In 2008, Georgia attempted to reassert its control over Abkhazia and South Ossetia, two breakaway regions that border Russia and that have wide support in that nearby country. The Russians responded with tanks, for a time occupying large sections of Georgia's territory. Almost 1,000 people died in the conflict and more than 30,000 people were displaced from their homes (Figure 9.35). The situation remains tense today with Georgia still claiming control over Abkhazia and South Ossetia and with Russia formally recognizing the "independence" of these microstates.

In nearby Armenia, the territories of the Christian Armenians and the Muslim Azeris interpenetrate one other in a complex fashion. The far southwestern portion of Azerbaijan (Naxicivan) is actually separated from the rest of the country by Armenia, while the important Armenian-speaking district of Nagorno-Karabakh is officially an autonomous portion of Azerbaijan. After Armenia successfully occupied much of Nagorno-Karabakh in 1994, fighting between the countries diminished. No final peace treaty has been signed, however, and Azerbaijan demands the return of the territory. Meanwhile, Armenia's traditionally close connections with Russia are increasingly counterbalanced with the country's interest in building ties with the United States and the European Union.

Geopolitics Within Russia Within Russia, further pressures for devolution, or more localized political control, produced the March 1992 signing of the Russian Federation Treaty. The treaty granted Russia's internal autonomous republics and its lesser administrative units greater political, economic, and cultural freedoms, including more control of their natural resources and foreign trade. Conversely, it weakened Moscow's centralized authority to collect taxes and to shape policies within its varied hinterlands. Defined essentially along ethnic lines, 21 regions possess status as republics within the federation and now have constitutions that often run counter to national mandates. Scattered from central Siberia to the north-facing slopes of the Caucasus, these republics reflect much of the linguistic and religious diversity of the nation (see Figure 9.33).

Figure 9.35 War in Georgia, 2008 When Georgia attempted in the summer of 2008 to assert tighter control over the breakaway regions of Abkhazia and South Ossetia, Russian troops and tanks briefly invaded the country. Tensions remain high between the two countries and Russia has recognized the independence of both the Abkhazia and South Ossetia regions.

But since 2002, Russian leaders, especially Vladimir Putin, have pushed for more centralized control within the country and region (Figure 9.36). Even after Putin left the presidency in 2008 (only to become Prime Minister thereafter), Russian President Dmitry Medvedev (a Putin ally) has maintained this strong trend towards a more centralized Russian state.

But cracks have appeared within Russia's more centralized political power structure. The Caucasus region and several of its internal republics (specifically Ingushetia, Chechnya, and Dagestan) remain very unstable (see Figure 9.33). While Russia officially claimed an end to hostilities within Chechnya in 2009 (where rebels have pushed for independence since 1994), the situation on the ground remains very unpredictable, particularly beyond major urban areas. In adjacent Ingushetia, observers describe a widespread lack of civil order and rebel groups have been increasingly bold in their attacks against government officials. Instability in the Caucasus also has wider implications: rebel groups from the region claimed responsibility for a deadly explosion on an express train between Moscow and St. Petersburg early in 2010 and more generally the region's instability challenges the overall authority of the central government in Moscow.

Russian Challenge to Civil Liberties A broader set of internal protests across Russia has also challenged Putin's hold on power since 2009 and may signal more trouble ahead for the Russian leadership. Human rights groups have organized several public protests in Russia and rallies advocating more free speech and a freer press have occurred in various cities, including Vladivostok, Kaliningrad, St. Petersburg, and Moscow. Russian police and military authorities have succeeded in breaking up the protests, but Putin's popularity has been challenged in the process. Elsewhere, economic uncertainties have also resulted in public demonstrations against the central government. While Putin and his allies remain popular nationally, the Russian leadership may have to confront more popular resistance in coming years.

These protests against the central government are in part a response to Russia's crackdown on civil liberties. Immediately after the fall of the Soviet Union, Russia enjoyed a

Figure 9.36 Vladimir Putin Since 2000, Vladimir Putin's influence across the Russian domain has been immense. Within Russia, Putin (first as President, then as Prime Minister) has managed an impressive economic recovery while limiting civil liberties. Beyond Russia, Putin has reestablished the region's geopolitical presence on the global stage.

genuine flowering of democratic freedoms during the initial post-Soviet years. A multiparty political system, independent media, and a growing array of locally and regionally elected political officials all signaled real change from the authoritarian legacy of the Soviet period.

Since 2002, however, many of these hard-won civil liberties have slipped away, victims of President (now Prime Minister) Putin's campaign to consolidate political power, increase the authority of the central government, and silence critics who disagreed with his policies. Many of these losses came amidst Putin's own declared "war on terror" within Russia and the increasingly antagonistic rhetoric he has leveled against the international community, including the United States. Indeed, many Russians argue that such adjustments are a tolerable price to be paid for a better standard of living and the country's newly elevated prestige on the global geopolitical stage.

Several events illustrate the changes afoot. First, a bill signed by President Putin in 2004 eliminated the popular election of dozens of Russian governorships and mayoral positions. Now, the Russian president nominates these officials, almost always receiving approval from weakened regional legislatures and municipal governments. Second, many of the country's media outlets—both television and print—have returned to state control or are now controlled by state-run companies such as Gazprom. This has silenced many critics of the government. Third, some of the country's most outspoken critics have simply been taken prisoner or mysteriously murdered. In 2003, the head of one of Russia's largest oil companies, Mikhail Khodorkovsky, was imprisoned for alleged tax violations. The assets of his company, Yukos Oil, were later auctioned off with most of the energy reserves ending up in the hands of state-controlled corporations. Three years later, one of Russia's leading journalists, Anna Politkovskaya, was murdered outside her apartment. Politkovskaya was a highly visible critic of President Putin, the central government, and Russia's harsh treatment of Chechnya. She is 1 of 13 journalists who were murdered or died under suspicious circumstances between 1999 and 2008. More print and television journalists have quit their jobs or been fired from key media positions. Recently, Russian officials have also increased their surveillance and regulation of the Internet, the latest chapter in a move to silence opposition.

The Shifting Global Setting

Since the fall of the Soviet Union, regional political tensions continue to challenge the Russians, in both the east and west. In East Asia, the boundary between Russia and China was imposed by the Russian Empire in 1858 and has never been fully accepted by Beijing. The two countries battled over the area in 1969, but since then relations have improved and an agreement signed in 2004 has clarified territorial claims along the Amur and Ussuri rivers. But the potential for renewed conflict remains. The Chinese want Russian oil, gas, and lumber and Chinese workers are increasingly slipping across the border to work in the Russian Far East.

Territorial disagreements also complicate Russia's relationship with Japan. Japan has long demanded the return of the four southernmost islands of the Kuril Archipelago seized by the Soviet Union in 1945. Although Russia could gain major financial benefits in any agreement, it has refused to return any territory to Japan. Growing economic ties between the two countries, however, are contributing to better political relationships.

To the west, Russia worries about the expansion of the North Atlantic Treaty Organization (NATO). Although most Russian leaders accepted the inevitable inclusion of Poland, Hungary, and the Czech Republic into NATO, they strongly opposed the recent addition of the Baltic republics (Estonia, Latvia, and Lithuania) to the increasingly powerful organization. In fact, the move provoked some Russian military leaders to recommend new troop and weapon deployments on the country's western borders. The plight of

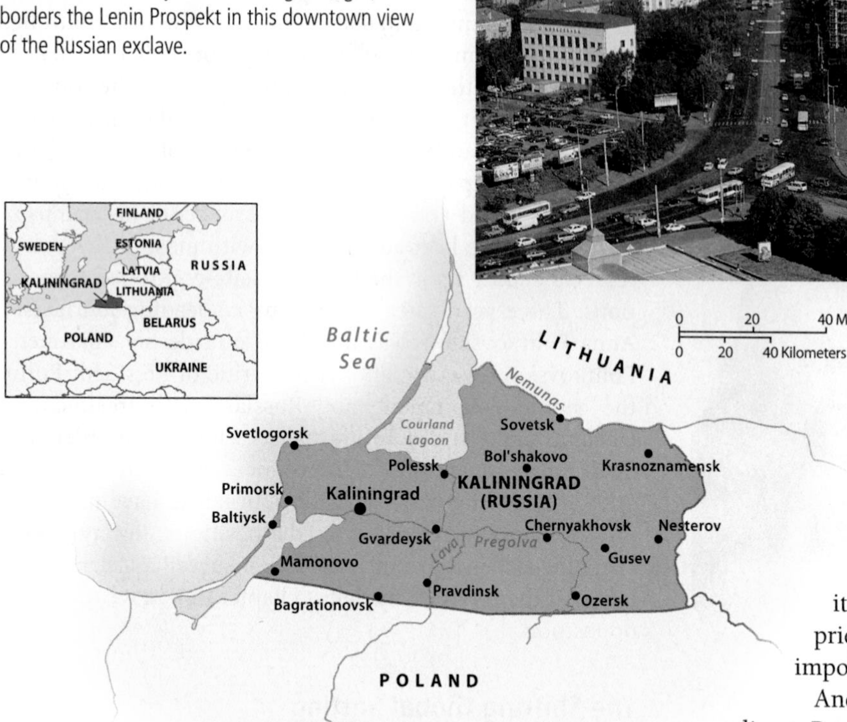

Figure 9.37 Kaliningrad Kaliningrad, a Russian exclave on the Baltic Sea, is now surrounded by Poland and Lithuania, both EU- and NATO-member states. The multistory Hotel Kaliningrad (right) borders the Lenin Prospekt in this downtown view of the Russian exclave.

Russian Kaliningrad is a related sore point (Figure 9.37). After World War II, the former Soviet Union added a small but strategic territory on the Baltic Sea. It was the northern portion of East Prussia (now the port of Kaliningrad), previously part of Germany. It still forms a small but highly strategic Russian **exclave**, which is defined as a portion of a country's territory that lies outside its contiguous land area. Kaliningrad (population 950,000) remains a part of the Russian Federation; but since 2004, neighboring Lithuania and Poland have both joined NATO and the European Union (EU). Russian officials fear that being in the midst of NATO and EU nations may encourage some residents of Kaliningrad to think about independence. In response, the Russians have beefed up their military presence in the exclave, making it clear that they have no intention of giving up one of their valuable port cities on the Baltic Sea.

Today, Russian leaders appear determined to reassert the nation's global political status. Russian officials have expressed their displeasure at the notion of seeing the United States as the world's sole superpower. Russian leaders have increasingly asserted their own interests in dealing with unpredictable nations such as North Korea, Iran, and Venezuela. They have also bristled at NATO's expansion into what they consider their own "post-Soviet sphere of influence," even accusing the alliance of sparking a new arms race in Europe. The country's nuclear arsenal, while reduced in size, remains a powerful counterpoint to American, European, and Chinese interests. Russia also retains a permanent seat on the United Nations Security Council, arguably the world's most important geopolitical body. Its inclusion in the G-8 economic meetings signifies its enduring international clout. Higher energy prices have recently increased Russia's economic importance on the world stage.

And despite their nominal independence from direct Russian control, the other nations within the Russian domain clearly feel the presence of their mammoth neighbor. Indeed, the long geopolitical history of the domain suggests that Russia's recent reemergence as a more powerful, centralized state upon the global stage is a sign of things to come.

ECONOMIC AND SOCIAL DEVELOPMENT: An Era of Ongoing Adjustment

The economic future of the Russian domain remains difficult to predict. Economic declines devastated the region for much of the 1990s. But between 2002 and 2008, particularly in the case of Russia, higher oil and gas prices brought significant but selective economic improvement (Table 9.2). Recently, however, the entire region was hard-hit in the global economic crisis and significant contractions in gross domestic product, exports, and employment were widely reported. For example, in 2009, Ukraine's economy shrunk by 14 percent and Russia's by about 8 percent. Although some economic stability returned to portions of the region in 2010, numerous challenges seem poised to limit future economic growth.

The true economic potential of the Russian domain always has been difficult to gauge. Optimists point to the vast size, abundant natural resources, and well-educated, urbanized populations of the region as significant assets. Indeed, in its heyday, the Soviet Union rose to become one of the great industrial powers in the world, and it did so in a remarkably short period of time.

TABLE 9.2 Development Indicators

Country	GNI per Capita, PPP 2008	GDP Average Annual % Growth (2000–08)	Human Development Index (2007)[a]	Percent of Population Living Below $2 a Day	Life Expectancy (2010)[b]	Under Age 5 Mortality Rate (1990)	Under Age 5 Mortality Rate (2008)	Gender Equity (2008)[c]
Armenia	5,060	12.4	0.798	21.0	72	56	23	104
Belarus	7,890	7.5	0.826	<2	70	24	13	101
Georgia	3,270	7.4	0.778	30.4	74	47	30	96
Moldova	2,150	7.1	0.720	11.5	70	37	17	102
Russia	10,640	6.2	0.817	<2	68	27	13	98
Ukraine	6,720	8.0	0.796	<2	68	21	16	99

[a]*United Nations, Human Development Report, 2009.*

[b]*Population Reference Bureau, World Population Data Sheet, 2010.*

[c]*Gender Equity—Ratio of female to male enrollments in primary and secondary school. Numbers below 100 have more males in primary/secondary school, numbers above 100 have more females in primary/secondary schools.*

Source: World Bank, World Development Indicators, 2010.

Skeptics note that size also brings its disadvantages, particularly by raising transportation costs within the region's economy. They also point out that Russia's northern location always has made food production problematic. Most notably, since the breakup of the Soviet Union, most economies within the region have struggled to evolve in a stable and predictable fashion toward greater productivity and output. Thus, the region's future path to prosperity remains clouded as it spends the early years of the 21st century reconfiguring an economy that was created largely during the Soviet period.

The Legacy of the Soviet Economy

The birth of the Soviet Union in 1917 initiated a radical change within the region's economy. Under the Russian Empire, most people were peasant farmers. Following the revolution, however, the Soviet Union quickly emerged to rival, and even surpass, many of the most powerful economies on Earth. During that era of unmatched growth, much of the region's present economic infrastructure was established, including new urban centers and industrial developments, as well as a modern network of transportation and communication linkages.

As communist leaders such as Stalin consolidated power in the 1920s and 1930s, they nationalized Russian industries and agriculture, creating a system of **centralized economic planning** in which the state controlled production targets and industrial output. The Soviets emphasized heavy, basic industries (steel, machinery, chemicals, and electricity generation), postponing demand for consumer goods to the future. By the late 1920s, Stalin also shifted agricultural land into large-scale collectives and state-controlled farms.

Much of the Russian domain's basic infrastructure—its roads, rail lines, canals, dams, and communications networks—originated during the Soviet period (Figure 9.38). Dam and canal construction, for example, turned the main rivers of European Russia into a virtual network of interconnected reservoirs. Invaluable links such as the Volga-Don Canal (completed in 1952), which connected those two key river systems, have greatly eased the movement of industrial raw materials and manufactured goods within the country (Figure 9.39). The Soviets also added thousands of miles of new railroad tracks in the European west, and the Trans-Siberian Line was modernized and complemented by the addition of the BAM link across central Siberia. Farther north, the Siberian Gas Pipeline was built to link the energy-rich fields of the Soviet arctic with growing demand in Europe. Overall, the postwar period produced real economic and social improvements for the Soviet people.

Despite these successes, problems increased during the 1970s and 1980s. Soviet agriculture remained inefficient and grain imports grew. Manufacturing efficiency and quality failed to match the standards of the West, particularly in regard to consumer goods. Equally troubling was the fact that the Soviet Union was failing to participate fully in the technological revolutions that were transforming the United States, Europe, and Japan. Disparities also visibly grew between the Soviet elite and an everyday population that still enjoyed few personal freedoms. By the late 1980s, the Soviet Union had reached both an economic and a political impasse.

The Post-Soviet Economy

Fundamental economic changes have shaped the Russian domain since the demise of the Soviet Union. Particularly within Russia itself, much of the highly centralized, state-controlled economy has been replaced by a mixed economy of state-run operations and private enterprise. The collapse of the communist state also meant that economic relationships between the former Soviet republics were no

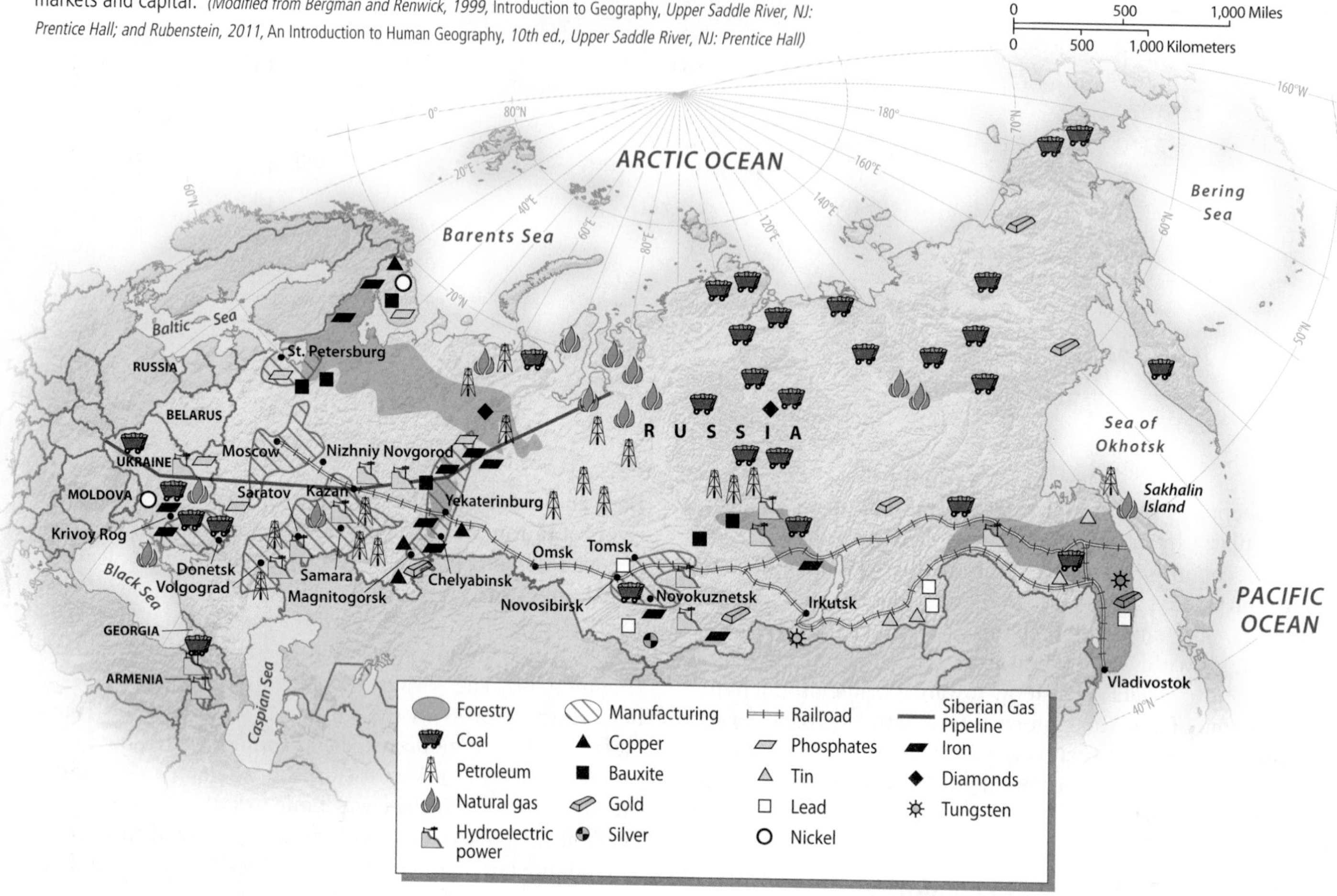

Figure 9.38 Major Natural Resources and Industrial Zones The region's varied natural resources and chief industrial zones are widely distributed. Fossil fuels are in abundance, although their distance from markets often imposes special costs. In southern Siberia, rail corridors offer access to many mineral resources. In the mineral-rich Urals and eastern Ukraine, proximity to natural resources sparked industrial expansion, while Moscow's industrial power is related to its proximity to markets and capital. *(Modified from Bergman and Renwick, 1999,* Introduction to Geography, *Upper Saddle River, NJ: Prentice Hall; and Rubenstein, 2011,* An Introduction to Human Geography, *10th ed., Upper Saddle River, NJ: Prentice Hall)*

longer controlled by a single, centralized government. Fundamental problems of unstable currencies, corruption, and changing government policies plagued the system for much of the 1990s. Higher oil and natural gas prices led to an impressive recovery in the Russian economy between 2002 and 2008 (see Table 9.2), only to be followed more recently by a regionwide economic downturn.

Redefining Regional Economic Ties Economic ties still link the six nations of the Russian domain, but the region has seen many of those relationships redefined in the post-Soviet era. The collapse of the communist state transformed planned internal exchanges between Soviet republics into

Figure 9.39 Volga-Don Canal This view near Volgograd suggests the enduring economic importance of the Volga-Don Canal. Built during the Soviet era, the canal remains a key commercial link that facilitates the economic integration of southern Russia.

less-predictable flows of foreign trade. Under the old regime, for example, government officials made certain that Ukraine received adequate supplies of oil from elsewhere in the country, just as Russia obtained cotton for its textile industry from Uzbekistan. Under the new economy, Ukraine must buy oil, or barter other products for it, from Russia, just as Russia must purchase more cotton from abroad.

Even with these disruptions, Russia retains an overwhelming economic presence within the region. Russian dominance is likely to continue, as suggested by that country's leading role in the creation of a new customs union with neighboring Belarus and Kazakhstan. Under the 2010 agreement, the three nations have pledged to carve out a single economic space, move towards the integration of the three countries, and apply as a unified bloc to the World Trade Organization (WTO). It still is unclear how effective this Russian-led response will be to the growth of the EU in coming years.

Overall, regional economic ties remain strong. Belarus, for example, is still dependent upon Russian economic assistance, particularly in the form of cheap energy exports. Similarly, Russia is by far Ukraine's largest trading partner. The country imports a great deal of its raw materials (particularly oil) from Russia, and its exports (principally metals, food, and machinery) are led by return flows to the north. Recently, cash-starved Ukrainian companies also have looked to Russia for help, and the result has been a Russian-led buying binge of cheap Ukrainian assets. Ukrainian aluminum smelters, television stations, and oil refineries now are controlled by large Russian corporations, certainly a new post-Soviet twist on a long-established Russian presence within the country.

Georgia has followed a somewhat different path as it has sought more economic connections with the United States and western Europe. Recent hostilities with Russia have only accelerated Georgia's move away from its giant neighbor. Turkey, Germany, and Canada remain more important trading partners than Russia. Georgian leaders have actively explored joining both NATO and the EU. Batumi, the nation's largest Black Sea port, has become a major focus of private foreign investment, and is home to new hotels and palm-lined boulevards.

Privatization and State Control For much of the region, the post-Soviet era has brought a great deal of economic uncertainty. A dramatic move came in October 1993, when the government initiated a massive program to privatize the Russian economy. Millions of Russians were given the option to buy into newly privatized agricultural lands and industrial companies. These initiatives greatly opened the economy to more private initiative and investment. Unfortunately, few legal and financial safeguards invited many abuses and often resulted in mismanagement and corruption within the new system. Elsewhere in the region, privatization proceeded more slowly. Much of the Belorussian economy, for example, remains mired in inflexible, corrupt state-controlled companies.

The agricultural sector continues to struggle. No matter its economic system, much of the region always will be challenged by short growing seasons, poor soils, and moisture deficiencies. Russia's best farmlands remain limited to a small slice of southern territory wedged between the Black and Caspian seas, leaving most of the rest of the country on the agricultural margins. In addition, the recent drastic economic changes have not been easy for the nation's farmers. Almost 90 percent of the country's farmland was privatized by 2003, with many farmers forming voluntary cooperatives or joint-stock associations to work the same acreage they did under the Soviet system. While crop prices have risen, costs have gone up even faster, and many farmers are not skilled in dealing with the uncertainties of a market-driven agricultural economy. Elsewhere, the government has held some agricultural land in the form of state farms, employing workers less willing to take risks in the fickle farm economy. Overall, agriculture employs about 10 percent of Russia's workforce, and the basic distribution of crops remains little changed from the Soviet era.

Russia has encouraged the rapid privatization of the service sector. Thousands of privatized retailing establishments have appeared, and they now dominate that portion of the economy. In addition, the long-established "informal economy" continues to flourish. Even during the Soviet era, millions of citizens earned extra money by informally selling Western consumer goods, manufacturing food and vodka, and providing skilled services such as computer and automobile repairs. Today, these barter transactions and informal cash deals form a huge part of the regional economy that is never reported to government authorities.

The natural resource and heavy industrial sectors of the economy were initially privatized in Russia, but in recent years, under President (and now Prime Minister) Putin's control, state-run enterprises have been taking more control of the nation's energy assets and infrastructure. Gazprom, the huge Russian natural gas company, exemplifies the process. Privatized in the spring of 1994, about one-third of the company was auctioned off to Russian citizens; 15 percent went to managers and workers; 10 percent remained in the company's treasury; and most of the rest was kept in government hands. Since 2005, Gazprom's activities have increasingly been controlled by the state and it is seen as a critical part of a newly emerging "state industrial policy" in which the central government is playing a more direct role in the economy. Nicknamed "Russia, Inc.," it remains one of the world's largest companies, employing more than 300,000 people and controlling a vast amount of Earth's known natural gas reserves.

Particularly in Russia, the uneven but spectacular successes of the new economy are increasingly visible on the landscape, especially in the country's cities. Luxury malls, office buildings, and more fashionable housing subdivisions are now part of the urban scene as the middle class grows in settings such as Moscow. Stroll the area near that city's Red Square and you will encounter Planet Sushi; the Moscow Bentley, Ferrari, and Maserati dealership; as well as trendy night clubs. On the other hand, the gap has grown between increasing urban affluence and grinding rural poverty. In many villages across southern Siberia, there are no telephones, jobs, or money. Vodka is easier to find than running water. Closed shops and a continued lack of services remain a part of everyday life in such rural settings.

The Challenge of Corruption Throughout the Russian domain, corruption also remains widespread. Doing business often means lining the pockets of government officials, company insiders, or trade union representatives. Much of the country's real wealth has been funneled into private Swiss bank accounts. Cumbersome national and regional tax policies also prevent efficient revenue collections from most private companies and individuals involved in the informal economy, and thus the government sees little benefit from such operations.

Organized crime remains pervasive in Russia and controls many aspects of the economy. The Russian mafia still controls significant portions of both the private economy and state-run enterprises. While organized crime certainly existed in the Soviet era, the more liberal economic and political environment that followed has allowed it to flourish even more widely and openly. The mafia provides critical capital and jobs. Today, high legal borrowing costs and a credit crunch for small businesses continue to make illegal sources of capital very attractive. Many close ties also remain between organized crime and Russian intelligence-collecting agencies. Various local and regional crime organizations have divided up much of the economy. One group might control the construction business in a Moscow suburb, whereas another syndicate oversees drug dealing and prostitution, and still another helps funnel illegal DVDs to eager consumers. The Russian mafia has also gone global. It has been implicated in huge money-laundering schemes that involve Russian, British, and U.S. banks as well as the flow of International Monetary Fund investments into the region. Banks in Switzerland, Liechtenstein, and Cyprus have been involved in similar schemes.

Ongoing Social Problems

Many residents of the Russian domain still struggle to make ends meet. Street crime remains high in many neighborhoods. Recently, higher rates of unemployment and declining social welfare expenditures have hit many families hard. Often, throughout the region, both husband and wife work multiple low-paying jobs with few benefits and long hours.

In many settings, women pay a particularly heavy price. Violence against women has been widely reported in the post-Soviet era. Beatings and rapes are common. A survey in Moscow suggested that one-third of divorced women had experienced domestic violence, while a women's rights group in Ukraine reported that rape was an all-too-common crime in many villages. In addition, Armenia and Moldova have been identified as major sources of young women who become involved with prostitution in Europe and the Middle East. Within Russia, many young women from rural areas are lured or forcibly brought to the country's large metropolitan areas in similar illegal operations. The region is also a major global source for Internet brides, a practice that invites additional exploitation and violence against women.

Health care is another major social problem within the Russian domain. Total annual expenditures on health care still remain only a fraction of what they were during the Soviet period (Figure 9.40). To put the problem in global perspective, Ukrainians spend only about 25 percent of what Japanese spend annually on health care ($542 vs. $2514) and Russians typically survive on barely 10 percent of what most Americans spend on health care ($638 vs. $6714). Mortality rates for Russian men are especially grim. One in three Russian men dies before they retire. Cardiovascular disease is a key contributor to these elevated death rates. One recent study suggests that half of all Russian men and one-third of Russian women are also plagued by long-term drinking problems.

Alcohol use in Russia (10.3 liters of pure alcohol per person annually) remains far above rates even in nearby Armenia or Georgia (see Figure 9.40). Some estimates that include illegal black market production put the Russian figure annually at more than 18 liters per person. For comparison, the average alcohol consumption in Japan (7.6 liters per year) is typical of many other developed-world settings, while India (.3 liters per year) and Egypt (.2 liters per year) exemplify the pattern in settings where religious and cultural traditions discourage alcohol use. In fact, Russian leaders initiated their own major anti-drinking campaign in early 2010, calling their country's plight "a national disaster." Prices for vodka were raised, although bootleg liquor is cheaper and widely available. An ambitious goal was set to reduce alcohol consumption by 50 percent within the next decade. In addition, the region is also plagued by a significant AIDS problem and by rapidly growing rates of tuberculosis. Infant mortality rates in rural parts of the region also are higher than in many less-developed parts of the world. Finally, toxic environmental conditions have extracted a huge price from the region, although precise estimates of their impacts are difficult to make.

Growing Economic Globalization

The relationship between the Russian domain and the world beyond has shifted greatly since the end of communism. During much of the Soviet era, the region was quite isolated from the world economic system. By the 1970s, however, the Soviet Union had begun to export large quantities of fossil fuels to the West while importing more food products. While connections with the global economy have grown with the downfall of the Soviet Union, the region remains one of the least globalized parts of the more developed world. Still, the Russian domain intersects with the rest of the world in a variety of ways that have refashioned life within the region and that have made the region's exports, particularly its natural resources, essential elements in the global economy.

A More Globalized Consumer Most visibly, a barrage of consumer imports now reaches many residents of the Russian domain, particularly those living in its larger cities. All

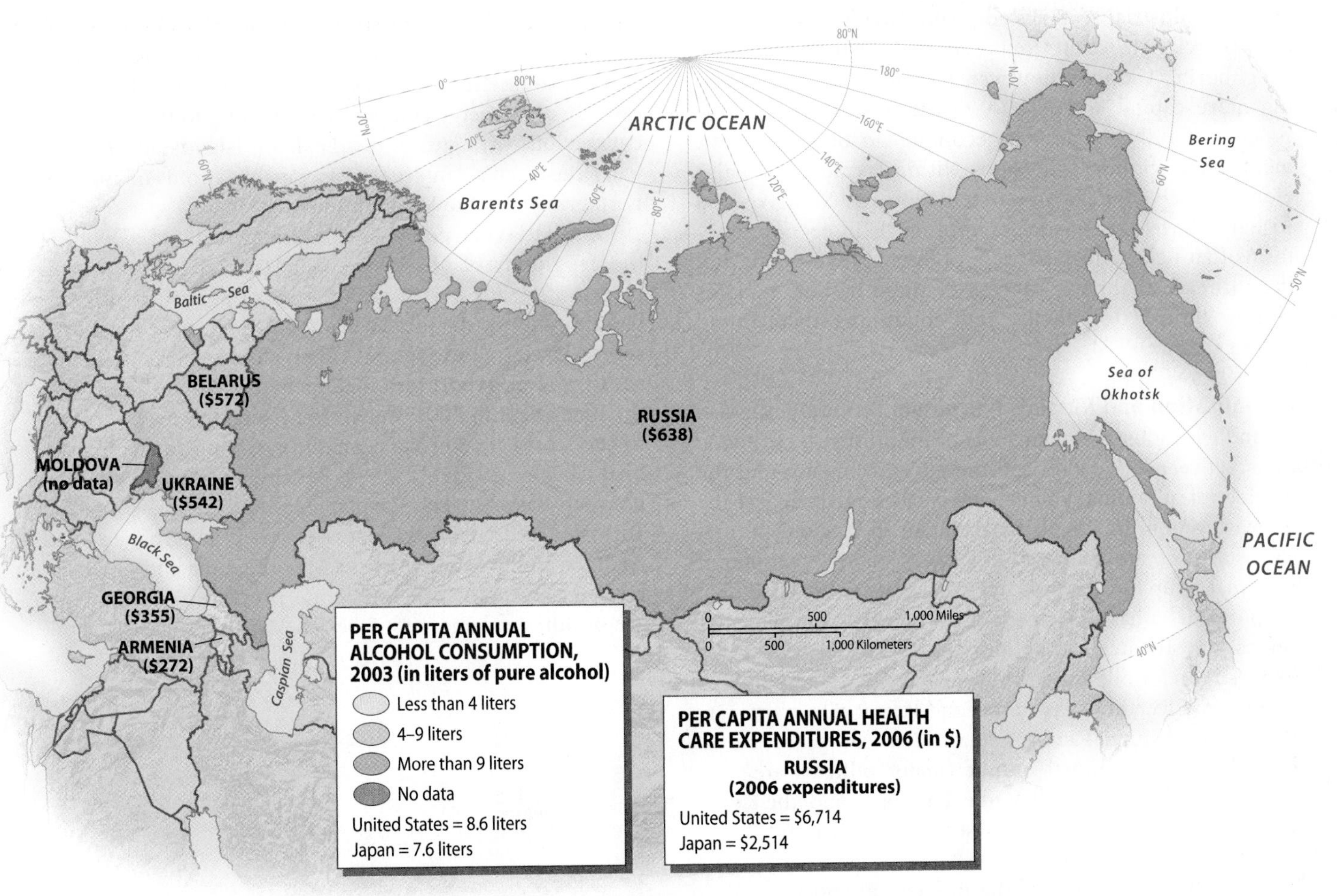

Figure 9.40 Development Issues: Health Care and Alcohol in the Russian Domain People in the Russian domain survive on less than $700 per year for health care compared with much higher expenditures in much of the rest of the developed world. Alcohol consumption remains highest in Russia, where it has been targeted as a key social problem by politicians and health-care experts. *(United Nations, WHO statistics, 2010)*

of the symbols of global capitalism are visible in the heart of Moscow and, increasingly, in many other settings throughout the region. In 2010, Russia was home to 170 KFC restaurants and more than 240 McDonald's. Luxury goods from the West have also found a small but enthusiastic, highly visible market among the Russian elite, a group noted for its devotion to BMW automobiles, Rolex watches, and other status emblems. Most Russians find such luxuries beyond their limited budgets, but they are interested in purchasing basic foods, cheap technology, and popular clothing and media from their Western neighbors or from eastern and southern Asia. Russia's largest sources of imports are Germany, China, and Japan, but many American consumer products are also in high demand.

Attracting Foreign Investment Despite all of the political and economic uncertainties, most countries of the Russian domain are attracting some foreign investment. While the global economic downturn dampened activity in 2008 and 2009, most countries within the region, including Russia, have courted the favor of global investors. As measured by total foreign investment, the strongest global ties by far have been with the United States, Japan and western Europe, particularly Germany and Great Britain. The success of Russian equity markets and the relative stability of its financial sector have been encouraging. Still, many potential investors are put off by continuing uncertainties with Russian legal frameworks, lingering problems with slow bureaucracies and red tape, and growing concerns about the ultimate political aims of Russian leaders. In a 2009 World Bank survey measuring the business climate for foreign investors, Russia ranked in 120th place (out of 181 nations surveyed), behind Nigeria. Some of the largest outside investments have been made in Russia's oil and gas economy, but recently some of those opportunities have cooled amid the country's desire to limit foreign ownership of its energy resources and infrastructure.

Elsewhere in the region, connections with the global economy vary. In Belarus, a lack of economic reforms continues to slow new investment. About 80 percent of the nation's industrial sector is controlled by the state and the political climate seems opposed to foreign investment. Neighboring Ukraine, however, has succeeded in attracting more capital since 2004, but it is still early in the process of fully opening its economy to outside investment. Sharp political divisions within Ukraine have also clouded the country's investment climate. Moldova and Armenia have both made some progress in economic reforms that help reduce barriers to foreign investment, but the ultimate economic success of these nations remains closely linked to policies and conditions that unfold within Russia.

Globalization and Russia's Petroleum Economy Russia's oil and gas industry remains one of the strongest economic links between the region and the global economy, and the diverse international connections it has forged suggest the increasing importance of the sector to the region's future. The statistics are impressive: Russia's energy production now makes up more than 25 percent of its entire economic output. Russia has 27 percent of the world's natural gas reserves (mostly in Siberia) and is the world's largest gas exporter. As for oil, Russia is by far the world's largest non-OPEC producer, and it is the second largest oil exporter in the world (behind Saudi Arabia). It far outpaces the United States in annual output (major oilfields are in Siberia, the Volga Valley, the Far East, and the Caspian Sea region), and it possesses more than 60 billion barrels of proven reserves.

The dynamic nature of the Russian oil and gas business exemplifies the forces of globalization and the region's changing economy. Prior to the breakup of the Soviet Union, about half of Russia's oil and gas exports went to other Soviet republics, such as Ukraine and Belarus. While these two nations still depend on Russian supplies, the primary destination for Russian petroleum products has overwhelmingly shifted to western Europe. Russia now supplies that region with more than 25 percent of its natural gas and 16 percent of its crude oil, and those linkages are likely to grow even stronger. An agreement between Russia and the EU in 2000 aimed at the rapid expansion of these East–West linkages. The Siberian Gas Pipeline already weds distant Asian fields with western Europe via Ukraine (see Figure 9.38). Those connections are being supplemented by new lines through Belarus (the Yamal-Europe Pipeline) and Turkey (the Blue Stream Pipeline). Planned underwater pipelines beneath the Baltic (Nord Stream) and Black (South Stream) seas will likely deliver even greater volumes of gas to northern and southern Europe respectively after 2012. Far to the east, Russian energy companies are adding to their pipeline connections with energy-rich Sakhalin Island and they also will construct a large new gas pipeline from Yakutia (in northern Siberia) to Vladivostok.

Similar expansions are also dramatically refashioning the geography of oil exports. The Druzhba Pipeline (replaced after 2011 with tanker traffic) as well as the expanding oil port facility at Primorsk (near St. Petersburg) serves western Europe and other global markets. To the south, a large export terminal opened at Novorossiysk (on the Black Sea) in 2001, delivering Caspian Sea oil supplies to the world market via a pipeline passing through troubled Chechnya (Figure 9.41). Nearby, another oil pipeline between Baku (on the Caspian Sea) and Ceyphan (on the Turkish Mediterranean) crosses both Azerbaijan and Georgia. In the Russian Far East, both China and Japan are lobbying hard for other pipeline projects. The Russians are building a large new Siberian Pacific Pipeline to link

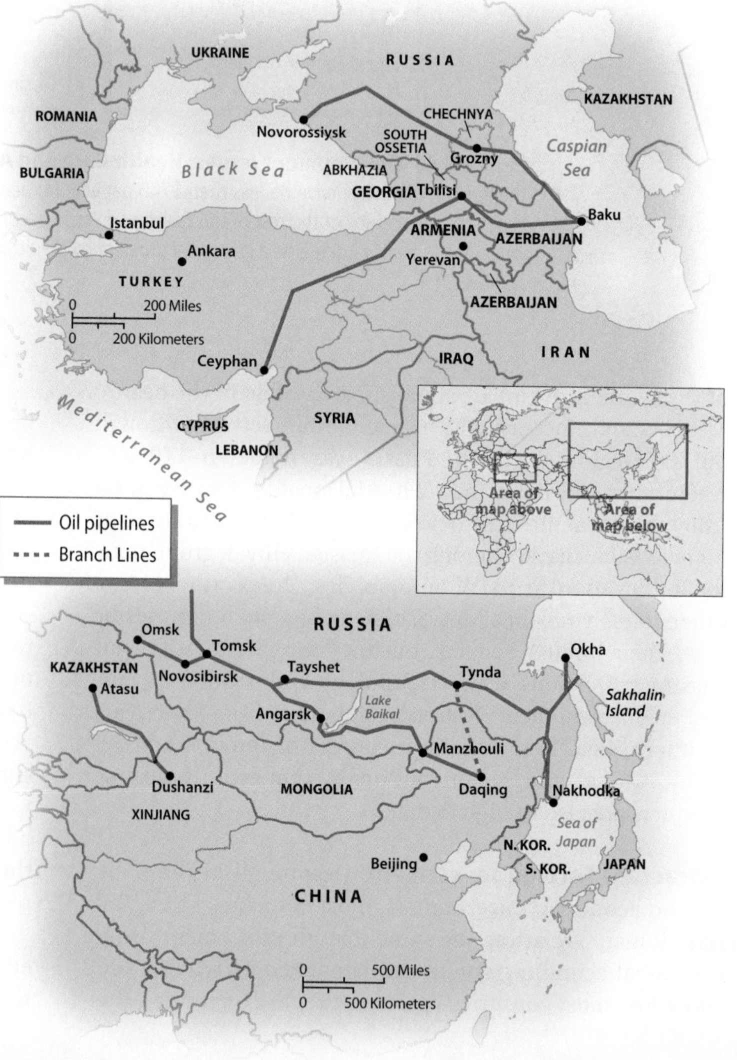

Figure 9.41 Russia's Expanding Pipelines These two maps show new and planned oil pipelines that are designed to expand Russia's presence in the global petroleum economy. Different pipeline projects in the Far East would benefit nearby China or Japan while projects near the Caspian Sea take pipelines through politically unstable portions of the region.

Figure 9.42 Sakhalin Island Large-scale investments by both foreign and Russian interests have concentrated on energy-rich Sakhalin Island. The area promises to be a producer of both oil and natural gas in the years to come.

its Siberian fields to Asian markets. The Chinese want Russian oil to flow to Daqing, where it could be refined for national and regional markets. Japan prefers a large new facility at the Pacific port of Nakhodka, well-positioned to supply Japan and offering Russia easy access to global markets via the Pacific Ocean. Other links connect the system with developments on Sakhalin Island, where several major energy projects are currently being completed (Figure 9.42).

While many global companies have participated in the development of Russia's energy infrastructure, recent events suggest that state-controlled Russian companies will play a larger role in the future. Some private companies that were based in Russia, but that also utilized foreign capital, have simply been nationalized. Other North American, European, and Japanese energy companies have seen their role in Russian projects limited or eliminated. On Sakhalin Island, for example, Gazprom, the state-controlled Russian company, has been allowed to take control of several ventures originally dominated by companies such as Royal Dutch Shell. Ironically, Russian authorities have stripped control of these projects from foreign companies by suggesting they failed to follow environmental regulations. Once in Russian hands, environmental concerns have faded away.

Local Impacts of Globalization As the geography of the Russian petroleum industry suggests, the local impacts of globalization are highly selective. Obviously, portions of the region that are close to oil and gas wells, pipeline and refinery infrastructure, and key petroleum shipping points are greatly affected (both economically and environmentally) by the changing global oil economy. The same is true more broadly: globalization has affected different locations within the Russian domain in very distinctive ways. Within Russia, capitalism has brought its most dramatic, though selective, benefits to Moscow. Luxury goods, fancy cars, and upscale restaurants cater to the city's newly rich. Indeed, much of the foreign investment in the country has gone to the Moscow area.

Beyond Moscow, St. Petersburg and the Siberian city of Omsk have seen growing global investment, while new oil and gas prospects in Siberia, the Far East (Sakhalin Island), and near the Caspian Sea have attracted other investments. Cities such as Nizhniy Novgorod and Samara also have succeeded in attracting global capital by building a pro-business economic environment. In addition, port cities such as Vladivostok are well positioned to take advantage of their accessibility to nearby markets, even though political red tape and corruption have hampered much growth in Pacific Basin trade (Figure 9.43).

Elsewhere, globalization clearly has imposed penalties (see "Global to Local: The Tale of Two Siberias"). Older, less globally competitive localities have been hit hard. Aging steel plants, for example, no longer have guaranteed markets for their high-cost, low-quality products as they did in the days of the planned Soviet economy. Instead, they must compete on the global market, a market that is increasingly prone to lower prices and weakening demand for many of the industrial goods the region produces. Many of the region's extractive centers are similarly vulnerable in a global economy of rapidly changing commodity prices.

Figure 9.43 Vladivostok The busy harbor of Vladivostok remains Russia's leading trade center in the Far East. With easy access to markets in Japan, China, and even the United States, this Pacific port is poised to grow as Russia's economy recovers.

GLOBAL TO LOCAL The Tale of Two Siberias

As the global economy has evolved over the past 50 years, many of Russia's mining and manufacturing towns have had either to change or die. Today, in the post-Soviet era, they are especially exposed to both the challenges and the opportunities of the global economy. The result is a dramatically varied geography of impacts between the winners and the losers. In some parts of Siberia, areas oriented around noncompetitive extractive activities and outdated heavy industries are seeing a mass exodus of jobs and population. In fact, Russians living in several northern industrial settlements recently qualified for a World Bank program designed to assist governments with "urban liquidation." The gold-mining center of Magadan, for example, will receive liquidation assistance from both the World Bank and the Russian government. Eleven settlements and 6,500 people will be affected, leaving only four towns occupied within the region. Many residents need little incentive to leave. Indeed, since 1991, more than 40 percent of Magadan's population emigrated from the region. The landscape is bleak, winters are long and brutal, and the program will offer a modest beginning to those lucky enough to receive assistance (Figure 9.6.1). Overall, more than 500,000 people may be relocated from such outdated settlements.

On the other hand, Siberian cities that have proven their competitive edge in the global economy have shown new life. Novosibirsk, a southern Siberian city of 1.4 million people, has witnessed real economic growth and sharply increased international trade since 1999 (Figure 9.6.2). The city serves important regional political functions and is home to the Siberian Branch of the Russian Academy of Sciences. In addition, it has attracted a lengthening list of foreign corporate offices, including representatives of PepsiCo, Pfizer, Honeywell, Volvo, and Hoffman-La Roche. The results are impressive: foreign exports (chemicals, machinery, processed food) more than doubled in the past seven years, and imports (machinery and chemicals) have risen by more than 70 percent. Ukraine and Kazakhstan remain the city's largest trading partners, but Germany, the United States, and even China are busily establishing new links with the city. Such localities, in Siberia and elsewhere, will reap persisting rewards as they find their profitable niche on the global stage, while other places lose people and close up shop, victims of a world economy that has passed them by.

Figure 9.6.1 Northeastern Siberia Bleak settlements in northeast Siberia have often lost population since the 1990s as migrants seek out better job opportunities elsewhere.

Figure 9.6.2 Novosibirsk The Russian city of Novosibirsk has become a major focus of investment and urban growth within Siberia since the disintegration of the Soviet Union.

Summary

- Huge environmental challenges remain for the Russian domain. The legacy of the Soviet era includes polluted rivers and coastlines, poor urban air quality, and a frightening array of toxic waste and nuclear hazards.
- Declining and aging populations are also part of the sobering reality for much of the region. While some localities see modest population growth related to in-migration (mostly toward expanding urban areas), many rural areas and less competitive industrial zones are likely to see continued outflows of people and very low birthrates.
- Much of the region's underlying cultural geography was formed centuries ago, the complex product of Slavic languages, Orthodox Christianity, and numerous ethnic minorities that continue to complicate the scene today. Further changing the country are

new global influences—a set of products, technologies, and attitudes that often clash with traditional cultural values.

- Much of the region's political legacy is rooted in the Russian Empire, a land-based system of colonial expansion that greatly enlarged Russian influence after 1600 and then reappeared as the Soviet Union expanded its own influence. Only large remnants of that empire survive on the modern map, yet it has stamped the geopolitical character of the region in lasting ways. Beyond the region, Russia's growing visibility on the international stage signals its reemergence as a truly global political power.
- The peoples of the Russian domain have endured immense challenges since 1991. Today, growing centralized power in Moscow increasingly limits democratic reforms. In addition, the entire region has suffered in the recent global economic downturn. The region's future economic geography, particularly in Russia, remains tied to the fortunes of the unpredictable global energy economy.

Key Terms

autonomous areas (*page 421*)
Baikal-Amur Mainline (BAM) Railroad (*page 406*)
Bolsheviks (*page 421*)
centralized economic planning (*page 429*)
chernozem soils (*page 399*)
Cold War (*page 422*)
Commonwealth of Independent States (CIS) (*page 423*)
Cossacks (*page 414*)
dacha (*page 412*)
denuclearization (*page 424*)
Eastern Orthodox Christianity (*page 413*)
exclave (*page 428*)
glasnost (*page 422*)
Gulag Archipelago (*page 408*)
Iron Curtain (*page 422*)
mikrorayons (*page 412*)
perestroika (*page 423*)
permafrost (*page 401*)
podzol soils (*page 399*)
Russification (*page 408*)
Slavic peoples (*page 413*)
socialist realism (*page 419*)
taiga (*page 401*)
Trans-Siberian Railroad (*page 406*)
tsars (*page 406*)

Review Questions

1. Compare the climate, vegetation, and agricultural conditions of Russia's European west with those of Siberia and the Russian Far East.
2. Describe some of the high environmental costs of industrialization within the Russian domain.
3. Discuss how major river and rail corridors have shaped the geography of population and economic development in the region. Provide specific examples.
4. Contrast Soviet and post-Soviet migration patterns within the Russian domain, and discuss the changing forces at work.
5. Describe some of the major land-use zones in the modern Russian city, and suggest why it is important to understand the impact of Soviet-era planning within such settings.
6. What were the key phases of colonial expansion during the rise of the Russian Empire, and how did each enlarge the reach of the Russian state?
7. What are some of the key ethnic minority groups within Russia and the neighboring states, and how have they been recognized in the region's geopolitical structure?
8. Describe how centralized planning created a new economic geography across the former Soviet Union. What is its lasting impact?
9. Briefly summarize the key strengths and weaknesses of the post-Soviet Russian economy and suggest how globalization has shaped its evolution.

Thinking Geographically

1. How might it be argued that Russia's natural environment is one of its greatest assets as well as one of its greatest liabilities?
2. In the future, how might the forces of capitalism and free markets reshape the landscapes and land uses of cities within the Russian domain?
3. What options do peoples such as the Volga Tatars or the Siberian Buryats have for preserving their cultural autonomy? What might be some of the advantages and disadvantages of such peoples' pressing for greater political independence?
4. On a base map of the Russian domain, suggest possible political boundaries 20 years from now. What forces will work for a larger Russian state? A smaller Russian state?
5. What were some of the greatest strengths and weaknesses of centralized Soviet-style planning between 1917 and 1991? How were Ukraine and Belarus impacted? Why did the system ultimately fail?
6. Why is organized crime such a critical problem in this part of the world?
7. How have the growing forces of globalization affected Russian culture?
8. From the perspective of a 22-year-old Russian college student and resident of Moscow, write a short essay that suggests how the economic and political changes over the past 15 years have changed your life.

Regional Novels and Films

Novels

Fyodor Dostoyevsky, *The Brothers Karamazov* (orig. pub. 1879; 1995, Bantam Books)

Fyodor Dostoyevsky, *Crime and Punishment* (orig. pub. 1866; 1984, Bantam Classics)

Eugenia Ginzburg, *Journey into the Whirlwind* (orig. pub. 1967; 1975, Harcourt Brace)

Nikolai Gogol, *Dead Souls* (orig. pub. 1842; 1961, Viking Press)

Mikhail Lermontov, *A Hero of Our Time* (orig. pub. 1840; 1966, Penguin)

Andrei Makine, *Dreams of My Russian Summers* (1998, Scribner's)

Andrei Makine, *Once Upon the River Love* (1999, Penguin)

Aleksandr Solzhenitsyn, *One Day in the Life of Ivan Denisovich* (orig. pub. 1962; 1998, Signet Classics)

Aleksandr Solzhenitsyn, *The Gulag Archipelago* (1973)

Leo Tolstoy, *War and Peace* (orig. pub. 1869; 1982, Viking Press)

Irene Zabytko, *The Sky Unwashed: A Novel* (2000, Algonquin Books)

Films

Brother (1997, Russia)

Burnt by the Sun (1994, France/Russia)

Dr. Zhivago (1965, U.S.)

Gorky Park (1983, U.S.)

Leo Tolstoy's Anna Karenina (1997, U.S.)

Onegin (1999, U.K., U.S.)

Prisoner of the Mountains (1996, Russia)

Reds (1981, U.S.)

Russia House (1990, U.S.)

Russian Ark (2002, Russia)

The Inner Circle (1991, Italy, USSR, U.S.)

The Return (2003, Russia)

The Thief (1997, Russia)

Bibliography

Billington, James. 2004. *Russia in Search of Itself*. Baltimore: Johns Hopkins University Press.

Cohen, Stephen F. 2009. *Soviet Fates and Lost Alternatives: From Stalinism to the New Cold War*. New York: Columbia University Press.

Crews, Robert D. 2006. *For Prophet and Tsar: Islam and Empire in Russia and Central Asia*. Cambridge, MA: Harvard University Press.

Dobrenko, Evgeny, and Naiman, Eric, eds. 2003. *The Landscape of Stalinism: The Art and Ideology of Soviet Space*. Seattle: University of Washington Press.

Griffin, Nicholas. 2001. *Caucasus: A Journey to the Land Between Christianity and Islam*. Chicago: University of Chicago Press.

Karny, Yo'av. 2000. *Highlanders: A Journey to the Caucasus in Quest of Memory*. New York: Farrar, Straus & Giroux.

Longworth, Philip. 2006. *Russia: The Once and Future Empire from Pre-History to Putin*. New York: St. Martin's Press.

Shaw, Denis, 1999. *Russia in the Modern World: A New Geography*. Oxford, UK: Blackwell.

Thornton, Judith, and Ziegler, Charles, eds. 2002. *Russia's Far East: A Region at Risk*. Seattle: University of Washington Press.

Wilson, Andrew. 2000. *The Ukrainians: Unexpected Nation*. New Haven, CT: Yale University Press.

Log in to www.masteringgeography.com for MapMaster™ interactive maps, geography videos, RSS feeds, flashcards, web links, an eText version of *Diversity Amid Globalization*, and self-study quizzes to enhance your study of The Russian Domain.

MapMaster™ presents 13 Place Name and 13 Layered Thematic interactive maps to help students practice and master their geographic literacy, spatial reasoning, and critical thinking skills.

10 Central Asia

Azerbaijan's capital of Baku is noted for its arresting contrasts. Here a forest of oil derricks can be seen behind the Bibi Heybat Mosque. The current mosque, built in the 1990s, was designed to replicate a structure from the 1200s that had been destroyed by the communist government of the Soviet Union in 1936.

DIVERSITY AMID GLOBALIZATION

Although it was once a power center as well as one of the main conduits of global trade, the landlocked region of Central Asia has been relatively isolated for most of the past 200 years, geopolitically dominated by countries in other world regions. Since the downfall of the Soviet Union in 1991, however, it has emerged as a key player in the global oil and natural gas trade and a focus of international geopolitical rivalry.

ENVIRONMENTAL GEOGRAPHY

Intensive agriculture along the rivers that flow into the deserts of Central Asia has resulted in water shortages, leading to the desiccation of many of the region's lakes and wetlands.

POPULATION AND SETTLEMENT

Pastoral nomadism, the traditional way of life across much of Central Asia, is gradually disappearing as people settle in towns and cities.

CULTURAL COHERENCE AND DIVERSITY

In much of eastern Central Asia, the growing Han Chinese population threatens the indigenous culture of the Tibetan and Uyghur people; in the west, the role of Islam in social and political life remains a major issue.

GEOPOLITICAL FRAMEWORK

Due to its rich resource base and strategic position, Central Asia has become a zone of geopolitical rivalry in which the United States, Russia, China, India, Pakistan, and Iran vie for influence.

ECONOMIC AND SOCIAL DEVELOPMENT

Despite its abundant resources, Central Asia remains a relatively poor region, although much of it does enjoy relatively high levels of social development.

Central Asia does not appear in most books on world regional geography (Figure 10.1). Although it covers a larger expanse than the United States, it is a remote and lightly populated area dominated by high mountains, barren deserts, and semiarid steppes (grasslands). Geopolitically, Central Asia was long ruled by external powers. Until 1991, most of the region was controlled by the Soviet Union. When the Soviet Union collapsed, Central Asia began to reappear in discussions of global geography. Suddenly a number of new countries appeared on the international scene, prompting scholars to reexamine the position of Central Asia in human affairs

The term *Central Asia* is defined in various ways by different writers. Most authorities agree that it includes five newly independent former-Soviet republics: Kazakhstan, Kyrgyzstan, Uzbekistan, Tajikistan, and Turkmenistan. This chapter also includes other territories—Mongolia, Afghanistan, Azerbaijan, and the autonomous regions of western China (Tibet, Xinjiang, and Inner Mongolia [Nei Mongol]). The inclusion of these additional territories within Central Asia is controversial. Azerbaijan is often classified with its neighbors in the Caucasus region (Georgia and Armenia); western China is obviously part of East Asia by political criteria; and Mongolia is also often placed within East Asia because of both its location and its historical connections with China. Afghanistan can alternatively be located within either South Asia or Southwest Asia.

But considering Central Asia's historical unity, its common environmental circumstances, and its reentry onto the stage of global geopolitics, we think that it deserves consideration in its own right. It also makes sense to define its limits rather broadly. For example, Azerbaijan is linked both culturally (language and religion) and economically (oil) more to Central Asia than to Armenia and Georgia.

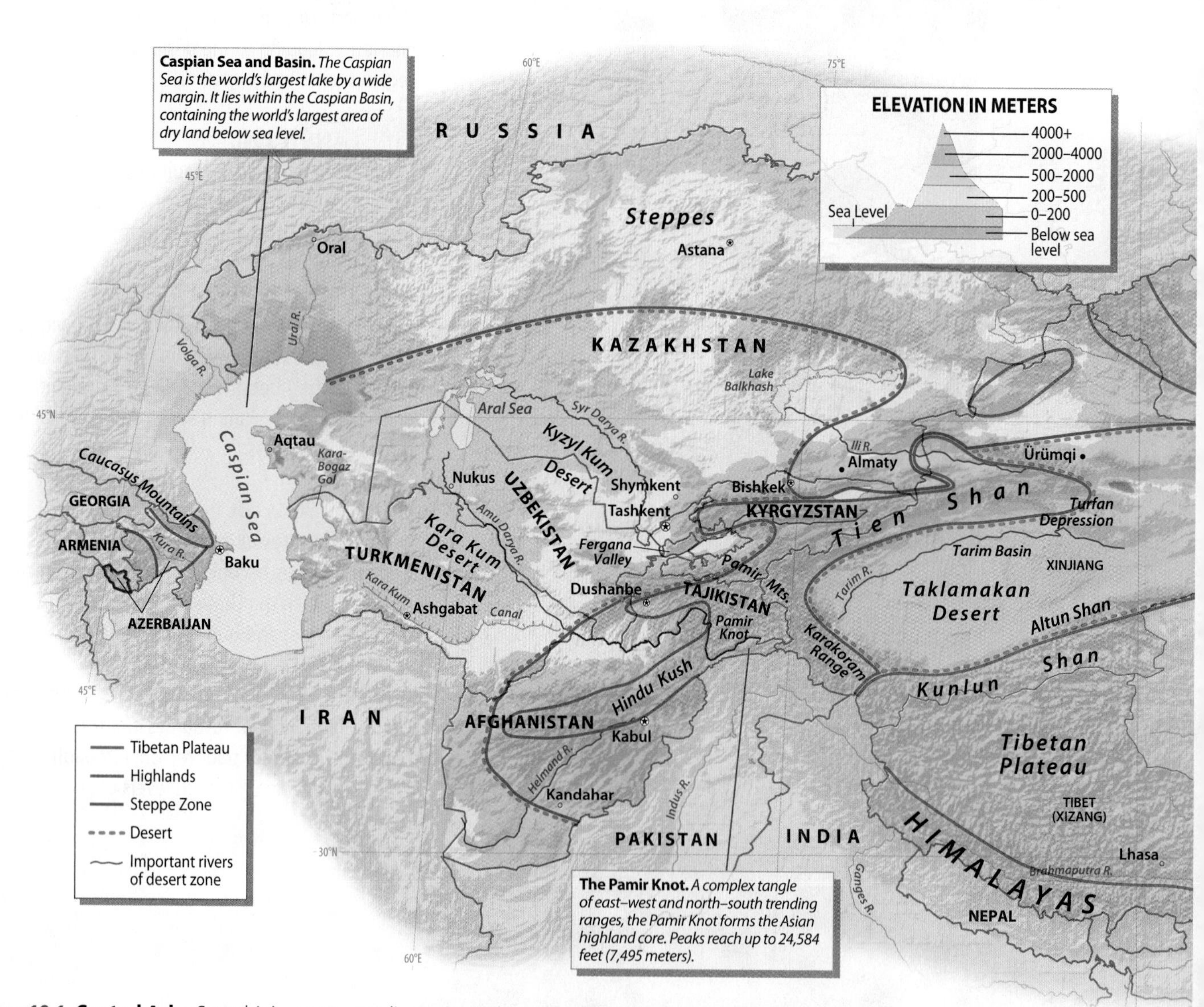

Figure 10.1 Central Asia Central Asia, a vast, sprawling region in the center of the Eurasian continent, is dominated by arid plains and basins and lofty mountain ranges and plateaus. Eight independent countries—Kazakhstan, Turkmenistan, Uzbekistan, Kyrgyzstan, Tajikistan, Azerbaijan, Afghanistan, and Mongolia—form Central Asia's core. China's lightly populated far west and north are often placed within Central Asia as well, due to patterns of cultural and physical geography.

At the same time, any unity that Central Asia possesses is far from stable. Continuing Chinese political control over southeastern Central Asia and Han Chinese (meaning people of Chinese cultural heritage) migration into that region threatens any claims for regional coherence. Moreover, Central Asia itself remains deeply divided along cultural lines. Most of the region is Muslim in religious orientation and Turkic in language, but both the northeastern and southeastern sections (Mongolia and Tibet) are traditionally Buddhist. Only time will tell whether Central Asia will continue to merit recognition as a distinct world region in its own right.

One reason for Central Asia's former lack of prominence is that it was poorly integrated into international trade networks. This situation began to change in the 1990s as large oil and gas reserves were found, especially in Kazakhstan, Turkmenistan, and Azerbaijan. Moreover, a number of external countries are seeking to exert geopolitical influence over Central Asia, including Iran, Pakistan, India, the United States, and Russia. Controversies surrounding China's strict control of its Central Asian lands also highlight the significance of the region.

Central Asia forms a large, compact region in the center of the Eurasian landmass. Alone among the world regions, it lacks ocean access. Owing to its continental position in the center of the world's largest landmass, Central Asia is noted for its rigorous climate. High mountains, deep basins, and extensive plateaus magnify its climatic extremes. The aridity of the region, as we shall see, has also contributed to some of the most severe environmental problems in the world.

ENVIRONMENTAL GEOGRAPHY: Steppes, Deserts, and Threatened Lakes of the Eurasian Heartland

Much of Central Asia has a relatively clean environment, owing mainly to low population density. Some parts of Central Asia, such as northwestern Tibet, remain almost pristine, with little human impact of any kind. Industrial pollution, however, is a serious problem in the larger cities, such as Uzbekistan's Tashkent and Azerbaijan's Baku. Elsewhere, the typical environmental dilemmas of arid environments plague the region: desertification (the spread of deserts), salinization (the accumulation of salt in the soil), and desiccation (the drying up of lakes and wetlands). The destruction of the Aral Sea, located on the boundary of Kazakhstan and Uzbekistan in western Central Asia, has been particularly tragic.

The gorge country of eastern Tibet. *Several extremely steep canyons alternate with lofty ridges, making eastern Tibet one of the most topographically forbidding places in the world.*

The Shrinking Aral Sea

In April 2010, U.N. Secretary General Ban-Ki Moon examined the remains of the Aral Sea by helicopter. He later told reporters that he had been shocked by what he saw. "It is clearly one of the worst environmental disasters of the world," he said. "It really left with me a profound impression, one of sadness that such a mighty sea has disappeared."

Until the late 20th century the Aral Sea was the world's fourth largest lake. Its only significant sources of water are the Amu Darya and Syr Darya rivers, which flow out of the Pamir Mountains, some 600 miles (960 kilometers) to the southeast. For thousands of years, water diversions for irrigation have lowered flows in both of these rivers, but after 1950 the scale of diversion vastly expanded. The valleys of the two rivers formed the southernmost farming districts of the Soviet Union, and thus the rivers are vital suppliers of water for warm-season crops, especially cotton. The largest of these projects was the Kara Kum Canal, which carries water from the Amu Darya across the deserts of southern Turkmenistan.

Unfortunately, the more river water was diverted for crop production, the less freshwater was available for the Aral Sea. The Aral proved to be particularly sensitive because it is relatively shallow. By the 1970s the shoreline was retreating at an unprecedented rate; eventually a number of "seaside" villages found themselves stranded more than 40 miles (64 kilometers) inland. As salinity levels increased, most fish species disappeared. New islands began to emerge, and by the 1990s the Aral Sea consisted of two separate lakes (see "Geographical Tools: Remote Sensing, GIS, and the Tragedy of the Aral Sea").

The destruction of the Aral Sea resulted in economic and cultural damage, as well as ecological devastation. Fisheries were wiped out and even local agriculture suffered. The retreating lake left large salt flats on its exposed beds; wind-

GEOGRAPHIC TOOLS Remote Sensing, GIS, and the Tragedy of the Aral Sea

Remote sensing technology is directly relevant to the study of the Aral Sea; satellite images clearly and immediately reveal the extensive and rapid reduction of the lake (Figure 10.1.1). But remote sensing and other technologically sophisticated geographical tools, such as geographic information system (or GIS, which allows spatially organized databases to generate numerous maps and map overlays), have also allowed geographers and other environmental scientists to more fully understand both the processes that have caused the tragedy and the broader social and environmental effects. Such tools are also invaluable for designing plans to reduce the damage.

Geographers Elena Lioubimtseva (Grand Valley State University) and Grigoriy Kapustin (Geography Institute, Russian Academy of Sciences) have recently used several sequences of medium-resolution satellite images to determine the spatial patterns and the underlying processes behind desertification in the Amu Darya Delta along the southern edge of the lake—one of the most devastated environments of the world. Their work shows the significance of overgrazing, wind erosion, and the salinization of soils, revealing as well local changes in microclimate caused by the shrinkage of the lake.

An even more comprehensive project, "GIS of the Aral Sea," is led by Rainer Ressl (German Aerospace Center), Andrei Ptichnikov (Institute of Geography, Russian Academy of Sciences), and Philip Micklin (USAID advisor on water management policy for the Uzbek government). As its Website (http://igras.geonet.ru/igras_e/project/aral/index.htm) explains, this project, under development since 1993, uses "dozens of georeferenced cartographic layers and a huge data base. It covers topics of physical geography and topography, social, demographic, health, water management and irrigation, ecological problems, desertification and biodiversity of the region around the Aral Sea, and the Sea itself. Satellite imagery of different resolution . . . is widely used to update the map layers. Models of optimal water and land use of irrigated lands, of ecosystem transformation and desertification are under construction."

If the problems of the Aral Sea are to be genuinely addressed at any level, the concerted efforts of a number of both national and international agencies will be required. Despite the overwhelming severity of those problems, it is heartening to learn that geographers have responded by devising such internationally focused projects, using the best and most up-to-date technologies at their disposal.

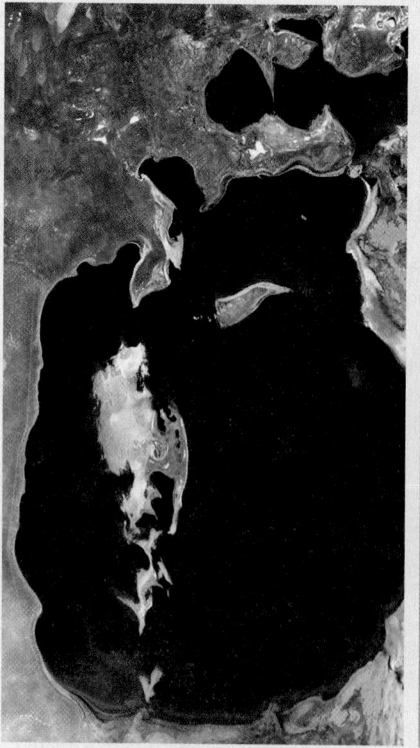

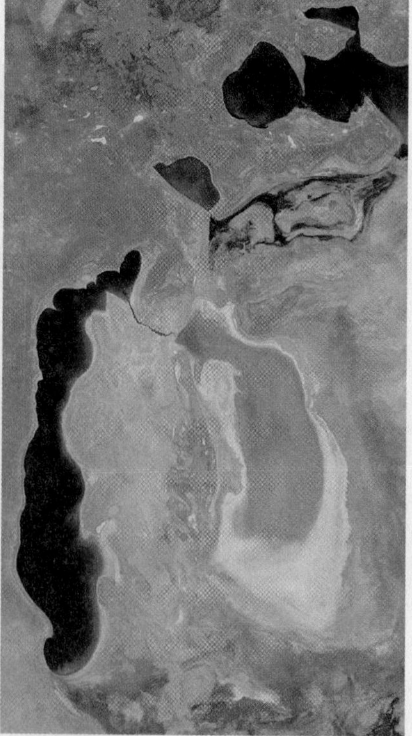

Figure 10.1.1 The Shrinking of the Aral Sea These satellite images show the dramatic shrinkage of the Aral Sea. A 90-percent reduction of the lake's size has occurred since the 1960s, resulting in severe economic damage and environmental degradation.

storms pick up the salt, along with the agricultural chemicals that have accumulated in the lake's shallows, and deposit it in nearby fields. Crop yields have thus declined; desertification has accelerated; and public health has been undermined. Particularly hard hit by this destruction are the Karakalpak people, members of a relatively powerless ethnic group who inhabit the formerly rich delta of the Amu Darya River.

In 2001, the oil-rich government of Kazakhstan decided to save what was left of its portion of the Aral Sea. By reconstructing canals, sluices, and other waterworks along the Syr Darya River, it doubled the flow of water reaching the northern Aral, which correspondingly began to rise (Figure 10.2 on page xxx). A large dam built across the lakebed prevented the extra water from flowing south into the salty waters and onto the extensive salt flats of the southern Aral Sea. By 2010, the area of the northern lake had increased by 50 percent, and its salt content had dropped low enough to allow the return of indigenous fish species. Currently, the World Bank and the government of Kazakhstan are discussing an extension of the project, hoping to extend the lake as far as the former port city of Aralsk.

The southern Aral Sea, meanwhile, continues to shrink, and its water quality continues to deteriorate. It too has been divided into two water bodies, and its eastern basin now dries up completely during the summer. Thus far, the oil-poor, agriculture-dependent government of Uzbekistan has done little to salvage what remains of its portion of the Aral Sea. It is worried, however, that Tajikistan will further reduce the flow of local rivers by its own water development projects, resulting in significant tension between the two countries (see "Environment and Politics: Tajikistan and Uzbekistan's Environmental Cold War").

ENVIRONMENT AND POLITICS Tajikistan and Uzbekistan's Environmental Cold War

In August 2008, the *New York Times* described Tajikistan and Uzbekistan as being in a "cold war." More recently, the conflict has warmed up; in early 2010, Uzbekistan imposed a partial blockade on Tajikistan. Uzbek authorities held freight cars at the border, stifling Tajikistan's economy. Uzbekistan's government claimed that technical and logistical difficulties merely slowed traffic over the border. Tajik authorities, however, grew so concerned that they threatened to take Uzbekistan to international courts, and they began talking about building alternative transportation routes. As anger against Uzbekistan mounted in Tajikistan, anger at Tajikistan mounted in Uzbekistan. In March 2010, around 1,000 university students and professors marched in the Uzbek city of Termez to protest Tajikistan's plans to expand a highly polluting aluminum smelter near the Uzbek border.

Figure 10.2.1 Blasting on the Vakhsh River in Preparation for the Construction of the Rogun Dam As Tajikistan moves ahead with ambitious water development projects, Uzbekistan fears that its own water resources, already in limited supply, will be further reduced.

A root cause of the Tajikistan-Uzbekistan feud is water. Former Soviet Central Asia is a dry region that depends on the snow-fed rivers flowing out of the Pamir and Tien Shan mountains. Mountainous Tajikistan and Kyrgyzstan contain its major headwaters, thus potentially controlling the water supply. Downstream Uzbekistan is water-short yet dependent on irrigated agriculture. It is thus hardly surprising that its government fears potential upstream water development diversions. Energy-poor Tajikistan wants to build dams primarily to generate hydroelectricity. In 2009, Uzbekistan partially cut-off natural gas supplies to debt-strapped Tajikistan, almost shutting down its aluminum smelters—one of the country's few profitable industries. Dam-building would not just protect the industry but would allow its expansion.

The dam in question is Rogun on the Vakhsh River, an important tributary of the Amu Darya (Figure 10.2.1). If and when it is ever completed, Rogun will be the world's highest dam. Work began in 1976, but the unfinished project sank with the Soviet Union in 1991. In 2007, Russia partnered with Tajikistan to complete the dam, but the two parties soon fell out. The Tajik government announced in early 2010 that it would try to raise by itself the $1.4 billion needed to finish the dam. Uzbekistan demanded an independent study to prove that its own water supplies would not be adversely affected, and when Tajikistan refused, the traffic came to a crawl. To complete Rogun Dam, Tajikistan needs to import vast quantities of construction supplies and heavy equipment through Uzbekistan.

Central Asia obviously has major water problems. The downstream countries, particularly agriculture-dependant Uzbekistan, have proposed the creation of a regional water authority that could coordinate development plans among all of the countries of the region. Water-rich Tajikistan and Kyrgyzstan, however, have been reluctant to enter in to such agreements. Considering as well the problems posed by global warming, Central Asia's hydropolitical problems are likely to become more extreme in the near future.

Figure 10.2 The Northern Aral Sea Due to water diversions from the Amu Darya and Syr Darya rivers, the Aral Sea has lost most of its volume and now forms two separate lakes. The government of Kazakhstan has recently built dikes to protect the water of the smaller Northern Aral, which is fed by the flow of the Syr Darya River.

Other Environmental Issues

Although the desiccation of the Aral Sea represents Central Asia's worst environmental crisis, other lakes in the region are also endangered. In addition, large areas of Central Asia suffer from extensive desertification and deforestation.

Fluctuating Lakes Western Central Asia supports large lakes because it forms a low-lying basin, without drainage to the ocean, that is virtually surrounded by mountains and other more humid areas. The world's largest lake, by a huge margin, is the Caspian Sea, located along the region's western boundary; and the 15th largest is Lake Balkhash in eastern Kazakhstan. Several other large lakes, such as Kyrgyzstan's spectacular Ysyk-Kol, are also located in the region.

Neither the Aral nor the Caspian are true seas because they are not connected with the ocean. The Caspian is less salty than the ocean (particularly in the north), while until the 1970s the Aral was only slightly brackish (salty). Lake Balkhash is almost fresh in the west but its long eastern extension is quite salty. Because none of these lakes are drained by rivers, all naturally fluctuate in size, depending on how much precipitation falls in their drainage basin in a given year. Like the Aral Sea, a number of Central Asia's lakes have suffered from reduced water flow and hence increasing salinity; Balkhash, for example, has recently shrunk by some 770 square miles (2,000 square kilometers).

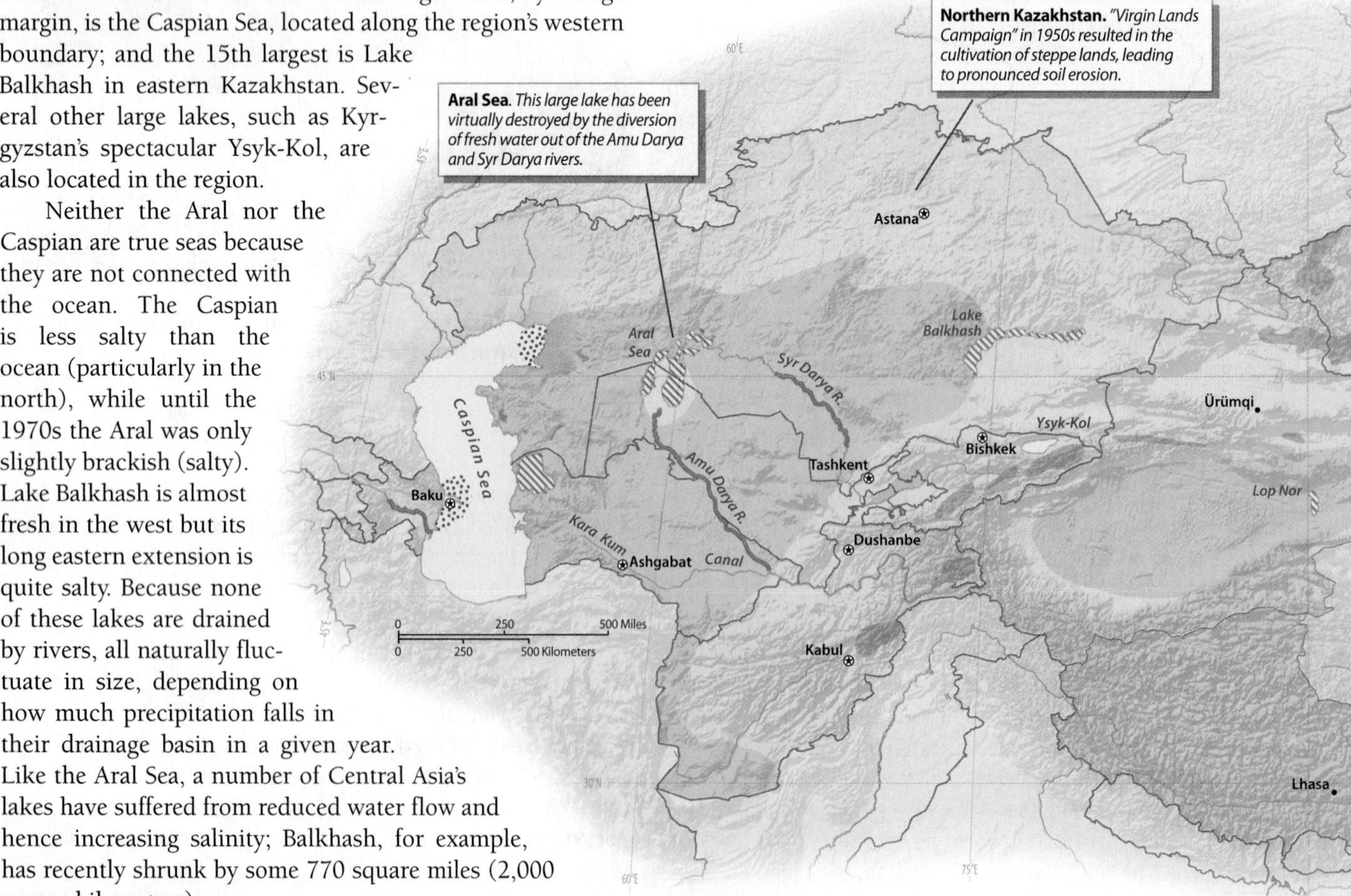

Figure 10.3 Environmental Issues in Central Asia Central Asia has experienced some of the world's most severe desertification problems. Soil erosion and overgrazing have led to the advance of desertlike conditions in much of western China and Kazakhstan. In western Central Asia, the most serious environmental problems are associated with the diversion of rivers for irrigation and the corresponding desiccation of lakes. Oil pollution is a particularly serious issue in the Caspian Sea area.

The story of the Caspian is more complicated. The Caspian Sea receives most of its water from the large rivers of the north, the Ural and the Volga, which drain much of European Russia. Owing to the development of extensive irrigation facilities in the lower Volga basin, the volume of freshwater reaching the Caspian began to decline in the second half of the 20th century. With a reduced influx of water, the level of the great lake dropped, exposing as much as 15,000 square miles (39,000 square kilometers) of former lake bed. A reduced volume of water resulted in increased salinity levels, undermining fisheries. The Russian caviar industry, centered in the northern Caspian, suffered extensive damage.

After reaching a low point in the late 1970s, the Caspian began to rise, presumably because of higher than normal precipitation in its drainage basin. By the mid 1990s it had risen some 8.2 feet (2.5 meters). This enlargement, too, has caused problems, inundating, for example, some of the newly reclaimed farmlands in the Volga Delta. Since 1995, the level of the lake has generally stabilized, experiencing only minor fluctuations. The most serious current environmental threat to the Caspian is pollution from the oil industry rather than fluctuation in size.

Water development projects have also begun to impact the Tibetan Plateau. Concerned about water shortages in the North China Plain, China is now diverting part of the flow of the headwaters of the Yangtze River into headwaters of the Huang He (or Yellow River). This and other transfer schemes now threaten several of the Tibetan Plateau's large wetlands, havens for many rare species of migrating waterfowl.

Desertification and Deforestation Desertification is another major concern in Central Asia. In the eastern part of the region, the Gobi and Taklamakan deserts have gradually spread southward, encroaching on settled lands in northeastern China. Farmers sometimes have to relocate their houses three or four times over the course of their lives in order to avoid the sands. The Chinese government has tried to stabilize dune fields with massive tree- and grass-planting campaigns, but such efforts have been only partially successful. As a result, sand and dust storms have increased in frequency, and now often affect much of northern China. In April 2010, sandstorms in the area were so severe that they were clearly visible on satellite images.

Former Soviet Central Asia has also seen extensive desertification. Northern Kazakhstan was one of the main sites of the ambitious Soviet "Virgin Lands Campaign" of the 1950s, in which semiarid grasslands were plowed and planted with wheat. Many of these lands have since returned to native grasses, but not before erosion stripped away much of their productivity. Some reports claim that up to 50 percent of Kazakhstan's farmland has been abandoned since 1990, due in part to desertification. In the irrigated lands of Uzbekistan, salinization has been a major problem, forcing the abandonment of some agricultural areas to the desert. Water quality is also poor through much of the region, due to dissolved agricultural chemicals as well as salts and other minerals.

Deforestation has also harmed the region, resulting in wood shortages and reduced dry-season water supplies. Although most of Central Asia is too dry to support forests, many of its mountains were once well wooded. Today extensive forests can be found only in the wild gorge country of the eastern Tibetan Plateau; in some of the more remote west- and north-facing slopes of the Tien Shan, Altai, and Pamir mountains; and in the mountains of northern Mongolia. Some of Central Asia's woodlands, particularly the walnut forests of Kyrgyzstan, are noted for their high biodiversity, and hence form specific target areas for conservation groups.

Central Asia's Physical Regions

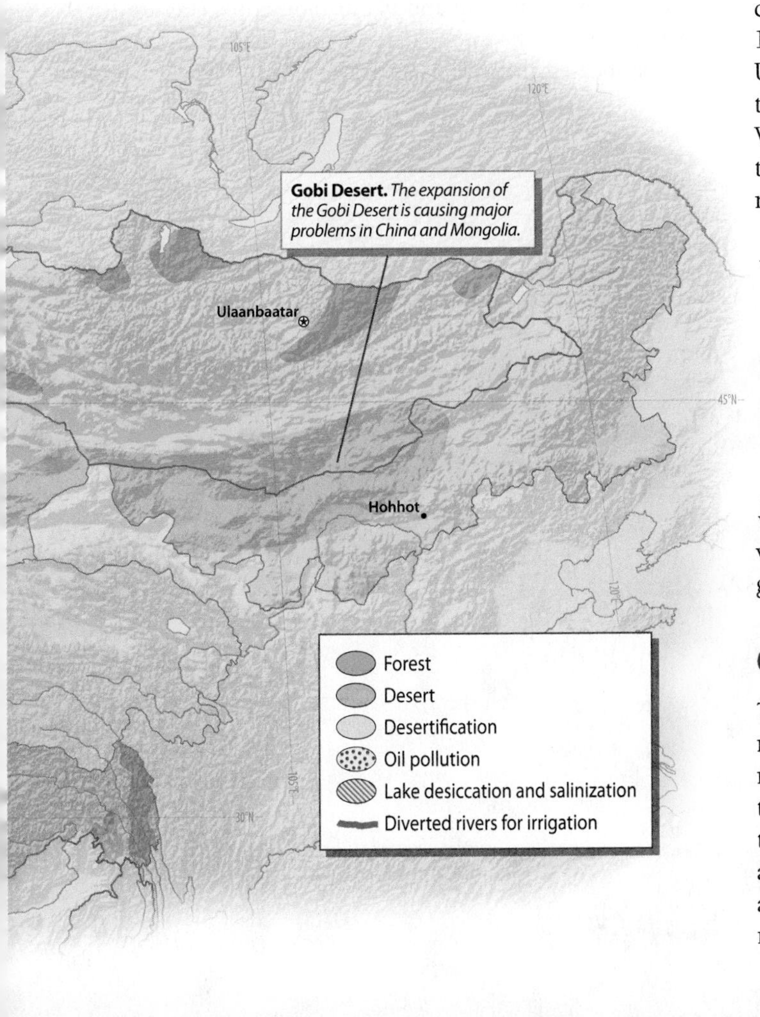

To understand why Central Asia suffers from such environmental problems as lake desiccation, an examination of the region's physical geography is necessary. In general, Central Asia is dominated by grassland plains (or steppes) in the north, desert basins in the southwestern and central areas, and high plateaus and mountains in the south-center and southeastern areas (Figure 10.3). Several mountain ranges also extend into the heart of the region, dividing the desert zone into a series of separate basins and giving rise to the rivers that flow into the deserts and hence into the imperiled lakes.

The Central Asian Highlands The highlands of Central Asia originated in one of the great tectonic events of Earth's history: the collision of the Indian subcontinent into the Asian mainland. This ongoing impact has created the highest mountain range in the world, the Himalayas, located along the boundary of South Asia and Central Asia.

Yet the Himalayas are merely one portion of a much larger network of high mountains and plateaus. To the northwest they merge with the Karakoram Range and then the Pamir Mountains. From the so-called Pamir Knot, a complex tangle of mountains situated where Pakistan, Afghanistan, China, and Tajikistan converge, other towering ranges radiate outward in several directions. The Hindu Kush sweeps to the southwest through central Afghanistan; the Kunlun Shan extends to the east (along the northern border of the Tibetan Plateau); and the Tien Shan swings out to the northeast into China's Xinjiang province. All of these ranges have peaks higher than 20,000 feet (6,000 meters) in elevation. Much lower but still significant ranges are found along Turkmenistan's boundary with Iran and Azerbaijan's boundaries with Russia, Armenia, and Iran.

More extensive than these mountain ranges, however, is the Tibetan Plateau (Figure 10.4). This massive upland extends some 1,250 miles (2,000 kilometers) from east to west and 750 miles (1,200 kilometers) from north to south. More remarkable than its size is its elevation; virtually the entire area is higher than 12,000 feet (3,700 meters) above sea level, and its average height is about 15,000 feet (4,600 meters). Most of the large rivers of South, Southeast, and East Asia originate on the Tibetan Plateau and adjoining mountains, including the Indus, Ganges, Brahmaputra, Mekong, Yangtze, and Huang He.

The greater part of the Tibetan Plateau, at about 15,000 feet (4,600 meters) of elevation, lies near the maximum height at which human life can exist. Rather than forming a flat, tablelike surface, the plateau is punctuated with east-west-running ranges alternating with undrained basins. Although the southeastern sections of the plateau receive ample precipitation, most of Tibet is arid. Cut off from any source of moisture by high ranges, large areas of the plateau receive only a few inches of rain a year. Winters on the Tibetan Plateau are cold, and while summer afternoons can be warm, summer nights remain chilly.

The Plains and Basins Although the mountains of Central Asia are higher and more extensive than those found anywhere else in the world, most of the region is characterized by plains and basins of low and intermediate elevation. This lower-lying zone can be divided into two main areas: a central belt of deserts punctuated by lush river valleys and a northern swath of semiarid steppe.

Central Asia's desert belt is itself divided into two segments by the Tien Shan and Pamir mountains (Figure 10.5). To the west lie the arid plains of the Caspian and Aral sea basins, located primarily in Turkmenistan, Uzbekistan, and

Figure 10.4 Tibetan Plateau The Tibetan Plateau is dominated by alpine grasslands and tundra interspersed with rugged mountains and saline lakes. In summer the sparse vegetation offers forage for the herds of nomadic Tibetan pastoralists. Much of the northern part of Tibet, however, is too high to sustain even low-intensity land use.

southern Kazakhstan (Figure 10.6). Most of this area is relatively flat and very low, with the surface of the Caspian Sea lying 92 feet (28 meters) below sea level. The climate of this region is continental; summers are dry and hot, whereas winter temperatures average well below freezing. Central Asia's eastern desert region extends for almost 2,000 miles (3,200 kilometers) from the extreme west of China to the southeastern edge of Inner Mongolia. It is conventionally divided into two deserts: the Taklamakan, found in the Tarim Basin of Xinjiang, and the Gobi, which runs along the border between Mongolia proper and the Chinese region of Inner Mongolia.

The environment of western Central Asia is distinguished from that of eastern Central Asia in part by its larger rivers. More snow falls on the western than the eastern slopes of the Pamir Mountains, giving rise to more abundant runoff. The largest of these rivers, as we have seen, flow to the Aral Sea. Others, such as the Helmand of Afghanistan, terminate in shallow lakes or extensive marshes and salt flats. In Xinjiang, one substantial river, the Tarim, flows out of the highlands onto the basin floor, where it frequently shifts course across the sandy lowlands. Before the completion of irrigation projects in the 1960s, it terminated in a salty lake called Lop Nor. Subsequently, the newly dried-out Lop Nor salt flat was used periodically by China for testing nuclear weapons.

North of the desert zone, rainfall gradually increases and desert eventually gives way to the great grasslands, or steppe, of northern Central Asia. Near the region's northern boundary, trees begin to appear in favored locales, outliers of the great Siberian taiga (coniferous forest) of the north. A nearly continuous swath of grasslands extends some 4,000 miles (6,400 kilometers) east to west across the entire region, only partially broken by Mongolia's Altai Mountains. Summers on the northern steppe are usually pleasant, but winters can be brutally cold.

Global Warming in Central Asia

Most climate change experts predict that Central Asia will be hard hit by global warming, largely because the region depends so heavily on snow-fed rivers. The Tibetan Plateau has already seen marked increases in temperature, resulting in the reduction of permafrost and the rapid drawback of mountain glaciers. Eighty percent of Tibet's glaciers are presently retreating, some at rates of up to 7 percent a year, leading to controversial predictions that the region could lose up to half of its ice fields within the next 50 years. It has also been predicted that Tajikistan could lose 20 percent of its ice cover by 2050. Tree-ring data from Mongolia indicate that the last century has been the warmest in more than a thousand years. Overall, Central Asia has experienced a 3.6°F (2°C) temperature increase since the early 1900s.

The retreat of Central Asia's glaciers is especially worrisome because of their role in providing dependable flows of water for local rivers. In some areas, the melting of ice has resulted in the temporary flooding of lowland basins, but the long-term result will be to further reduce freshwater resources in a region already burdened by aridity. Climate change from global warming could also reduce precipitation in the arid lowlands of western Central Asia, compounding the problem. Prolonged and devastating droughts have recently struck Afghanistan and several of its neighbors, indicating a possible shift to a drier climate. As a result, the U.N. Intergovernmental Panel on Climate Change (IPCC) has predicted a 30-percent crop decline for Central Asia as a whole by mid-century.

But as is true elsewhere in the world, global warming will not affect all parts of Central Asia in a uniform manner. Some areas, including the Gobi Desert and the Tibetan Plateau, could possibly see a long-term increase in precipi-

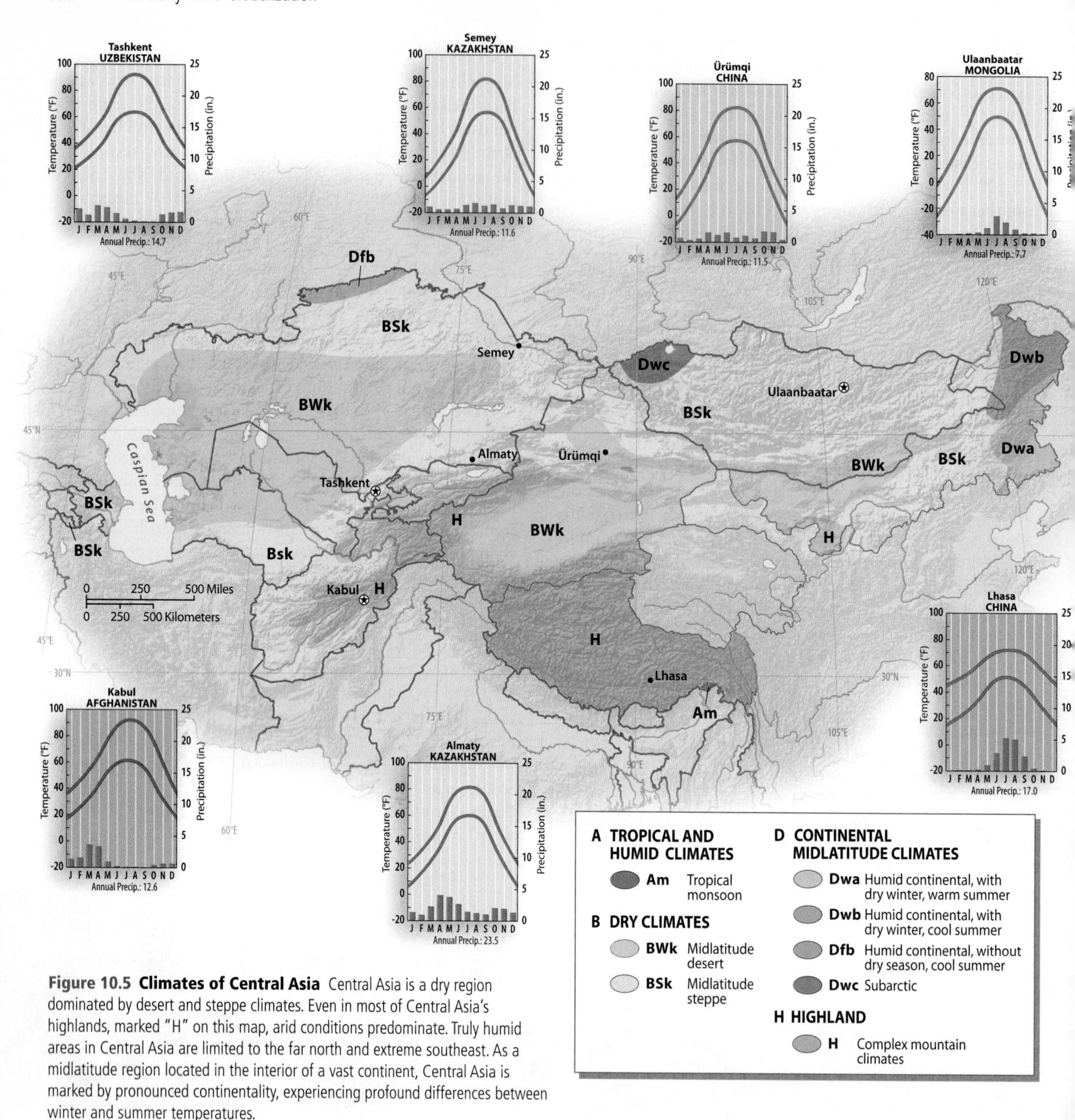

Figure 10.5 Climates of Central Asia Central Asia is a dry region dominated by desert and steppe climates. Even in most of Central Asia's highlands, marked "H" on this map, arid conditions predominate. Truly humid areas in Central Asia are limited to the far north and extreme southeast. As a midlatitude region located in the interior of a vast continent, Central Asia is marked by pronounced continentality, experiencing profound differences between winter and summer temperatures.

tation. Such forecasts are scant consolation, however, in a region that is experiencing rampant desertification. Some global climate models also indicate increased precipitation in the boreal forest zone of Siberia that extends into the mountains of northern Mongolia. But as the rivers coming out of these mountains generally flow north, few benefits would be reaped by the rest of Central Asia.

Overall, Central Asia is a moderate producer of greenhouse gasses. The oil-rich economies of Kazakhstan and Azerbaijan are noted for their inefficient energy use, as is the mechanized agriculture of Uzbekistan and Turkmenistan. The traditional nomadic pastoralism and peasant agriculture generate little excess carbon dioxide, but such modes of life are steadily declining across the region. Greenhouse gas emis-

Figure 10.6 Central Asian Desert Much of Central Asia is dominated by deserts and other arid lands. The Karakum Desert of Turkmenistan, seen in this photograph, is especially dry, supporting little vegetation.

sions from the region are rapidly increasing, due mainly to the expansion of the fossil fuel industry and the growth of cities and modern transportation systems, as we shall see in the following section.

POPULATION AND SETTLEMENT: Densely Settled Oases Amid Vacant Lands

Most of Central Asia is sparsely populated (Figure 10.7). Large areas are essentially uninhabited, either too arid or too high to support human life, while vast expanses are populated by widely scattered groups of nomadic **pastoralists** (people who raise livestock for subsistence purposes). Mongolia, which is more than twice the size of Texas, has only 2.6 million inhabitants. But as is common in arid environments, those few lowland areas with fertile soil and dependable water supplies can be thickly settled. Despite its overall aridity, Central Asia is well endowed with perennial rivers and productive oases. While the nomadic pastoralists of the steppe and desert zones have dominated the history of Central Asia, the sedentary peoples of the river valleys have always been more numerous.

Highland Population and Subsistence Patterns

The environment of the Tibetan Plateau is particularly harsh. Not only is the climate cold and water often scarce or brackish, but ultraviolet radiation in this high-elevation area is always pronounced. Only sparse grasses and herbaceous plants—so-called mountain tundra—can survive, and human subsistence is obviously difficult under such conditions. The only feasible way of life over much of the Tibetan Plateau is nomadic pastoralism based on the yak, an altitude-adapted relative of the cow. Several hundred thousand people manage to make a living in such a manner, roaming with their herds over vast distances.

Farming in Tibet is possible only in a few favorable locations, generally those that are *relatively* low in elevation and that have good soils and either adequate rain or a dependable irrigation system. The main zone of agricultural settlement lies in the far south, where protected valleys offer favorable conditions. The population of Tibet proper (the Chinese autonomous region of Xizang) is only 2.8 million, a small number indeed considering the vast size of the area.

Population densities are also low in the other highland areas of Central Asia, although densely settled agricultural communities can be found in protected valleys. The complex topography of the Pamir Range in particular offers a large array of small and nearly isolated valleys that are suitable for agriculture and intensive human settlement. Not surprisingly, this area is marked by great cultural and linguistic diversity. Many villages here are noted for their agricultural terraces and well-tended orchards.

Central Asia's mountains are vitally important for people living in the adjacent lowlands, whether they are settled farmers or migratory pastoralists. Many herders use the highlands for summer pasture; when the lowlands are parched, the high meadows provide rich grazing. The Kyrgyz (of Kyrgyzstan) are noted for their traditional economy based on **transhumance**, moving their flocks from lowland pastures in the winter to highland meadows in the summer. The farmers of Central Asia rely on the highlands for their wood supplies and, more important, for their water. Settled agricultural life in most of Central Asia is possible only because of the rivers and streams flowing out of the region's mountains.

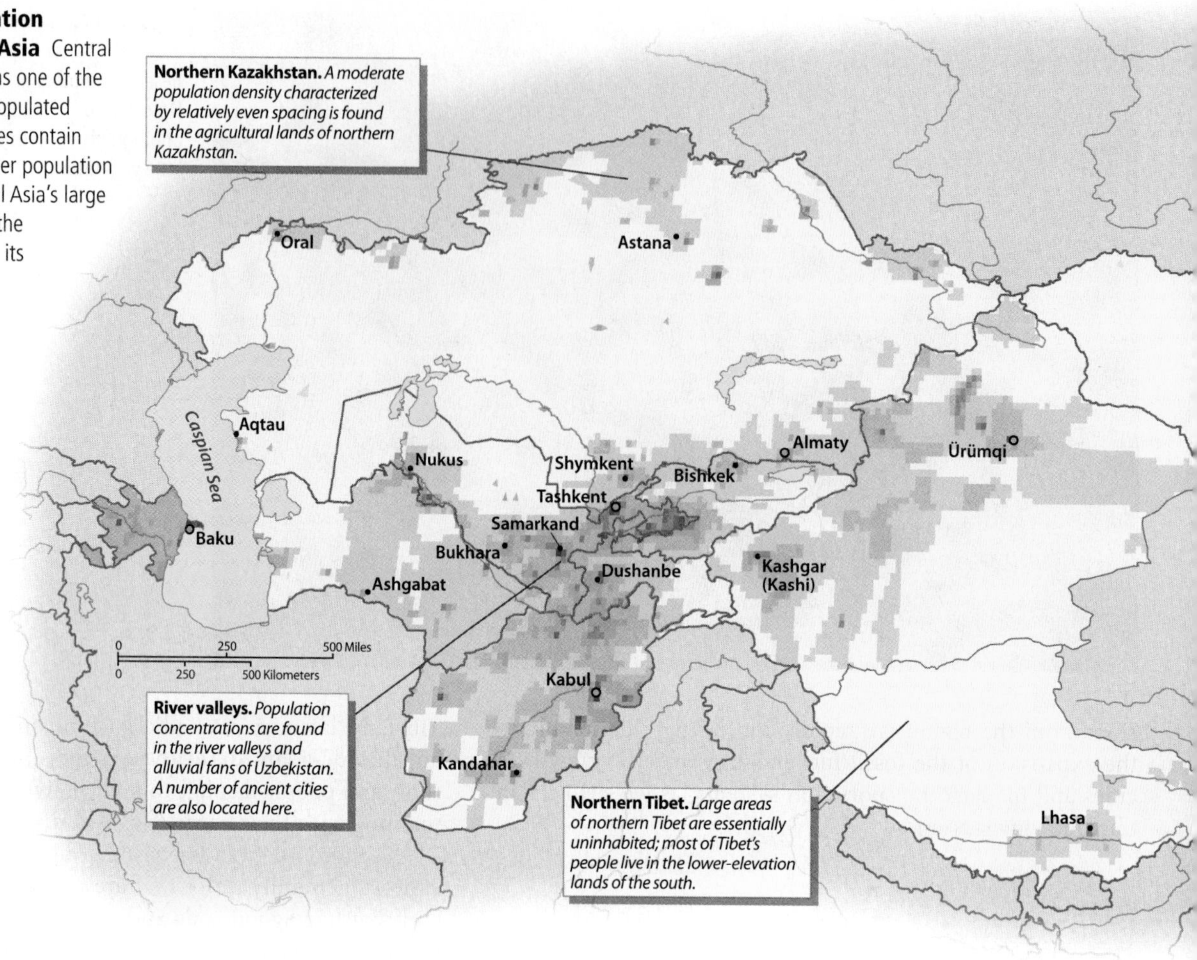

Figure 10.7 Population Density in Central Asia Central Asia as a whole remains one of the world's least densely populated regions, although it does contain distinct clusters of higher population density. Most of Central Asia's large cities are located near the region's periphery or in its major river valleys.

Lowland Population and Subsistence Patterns

Most of the inhabitants of Central Asia's deserts live in the narrow belt where the mountains meet the basins and plains. Here water supplies are adequate and soils are not contaminated with salt or alkali, as is often the case in the basin interiors. As a result, the population distribution pattern of China's Tarim Basin forms a ringlike structure (Figure 10.8). Streams flowing out of the mountains are diverted to irrigate fields and orchards in the narrow fertile band situated between the steep slopes of the mountains and the nearly empty deserts of the central basin.

The population west of the Pamir Range, in former Soviet Central Asia, is also concentrated in the transitional zone nestled between the highlands and the plains. A series of **alluvial fans**, fan-shaped deposits of sediments dropped by streams flowing out of the mountains, have long been devoted to intensive cultivation. Fertile **loess**, a silty soil deposited by the wind, is widespread, and in a few favored areas winter precipitation is high enough to allow rain-fed agriculture. Several large valleys in this area, such as the Fergana Valley of the upper Syr Darya River, offer fertile and easily irrigated farmland (Figure 10.9). In the far west of the region, Azerbaijan's Kura River Basin also supports intensive agriculture and concentrated settlement.

Unlike the other deserts of the region, the Gobi has few sources of permanent water. Rivers draining Mongolia's highlands flow to the north or terminate in interior basins, while only a few of the larger streams from the Tibetan Plateau reach the Gobi proper. (The Huang He, however, does swing north to reach the desert edge before turning south to flow through the Loess Plateau.) Owing to this scarcity of **exotic rivers** (those originating in more humid areas) and to its own aridity, the Gobi remains one of Asia's least-populated areas.

The steppes of northern Central Asia are the classical land of nomadic pastoralism. Until the 1900s, virtually none of this area had ever been plowed and farmed. To this day, pastoralism remains a common way of life across the grasslands. In northwestern China and the former Soviet republics, however, many pastoral peoples have been forced to adopt sedentary lifestyles. National governments in the region, like those in most other parts of the world, find migratory people hard to control and difficult to provide with social services. Migratory herders are also highly vulnerable to natural disasters. Severe cold in the winter of 2010, for example, killed millions of domesticated animals in Mongolia, resulting in a subsistence crisis over much of the country. Although roughly one-third of Mongolians are still nomadic pastoralists (Figure 10.10), the country's prime minister recently announced that he expected this way of life to essentially disappear over the next few decades.

In northern Kazakhstan, the Soviet regime converted the most productive pastures into farmland in the mid-1900s in order to increase the country's supply of grain. Some of these lands have since reverted to steppe, but large areas remain under the plow; Kazakhstan is a major producer of spring wheat. Consequently, northern Kazakhstan has the highest population density of the steppe belt.

Population Issues

Some portions of Central Asia are growing at a moderately rapid pace. The demographic data on the region show overall fertility rates near the global middle, while exhibiting pro-

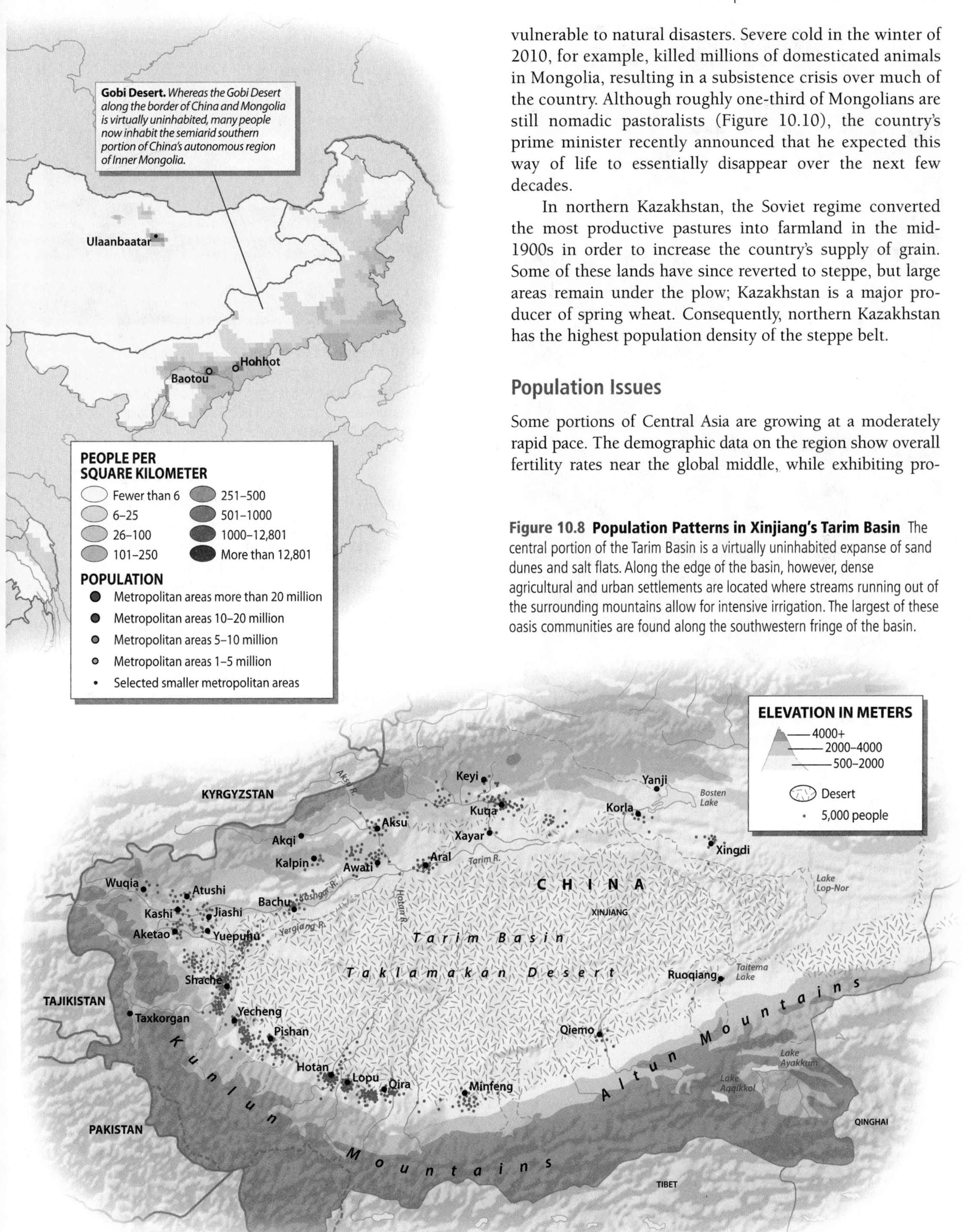

Figure 10.8 Population Patterns in Xinjiang's Tarim Basin The central portion of the Tarim Basin is a virtually uninhabited expanse of sand dunes and salt flats. Along the edge of the basin, however, dense agricultural and urban settlements are located where streams running out of the surrounding mountains allow for intensive irrigation. The largest of these oasis communities are found along the southwestern fringe of the basin.

Figure 10.9 Farmland in Uzbekistan The fertile river valleys of Uzbekistan have been intensely cultivated for many centuries, producing large harvests of fruits and vegetables in addition to cotton and grain.

Figure 10.10 Steppe Pastoralism The steppes of northern and central Mongolia offer lush pastures during the summer. Mongolians, some of the world's most skilled horse riders, have traditionally followed their herds of sheep, goats, and cattle, living in collapsible, felt-covered yurts. Many Mongolians still follow this way of life.

nounced variation between different areas (Table 10.1). Afghanistan, the least-developed and most male-dominated country of the region, has the highest birthrate by a substantial margin. Tajikistan, also noted for its low levels of economic development, also has an elevated birthrate.

Low fertility, however, can be counteracted by immigration, as is the case in much of western China. Here substantial population growth over the past 30 years has stemmed from the migration of Han Chinese into the area. While many of the indigenous inhabitants resent this influx, the Chinese government claims that it is necessary for economic development.

Migration has also complicated demographic patterns in the former Soviet zone (Figure 10.11). After the breakup of the

TABLE 10.1 Population Indicators

Country	Population (Millions, 2010)	Population Density (per Square Kilometer)	Rate of Natural Increase (RNI)	Total Fertility Rate	Percent Urban	Percent <15	Percent >65	Net Migration Rate (per 1,000, 2005–10)[a]
Afghanistan	29.1	45	2.1	5.7	22	44	2	7.5
Azerbaijan	9.0	104	1.1	2.2	54	23	7	−1.2
Kazakhstan	16.3	6	1.4	2.7	54	24	8	−1.3
Kyrgyzstan	5.3	27	1.6	2.8	35	29	5	−2.8
Mongolia	2.8	2	1.9	2.7	61	33	4	−0.8
Tajikistan	7.6	53	2.4	3.4	26	38	4	−5.9
Turkmenistan	5.2	11	1.4	2.5	47	31	4	−1.0
Uzbekistan	28.1	63	1.8	2.8	36	33	5	−3.0

[a]*Net Migration Rate from the United Nations, Population Division, World Population Prospects: The 2008 Revision Population Database.*
Source: Population Reference Bureau, World Population Data Sheet, 2010.

Soviet Union, millions of Russians and Ukrainians left Central Asia to return to their homelands (see "People on the Move: Migration In and Out of Kazakhstan"). More recently, Russia's economic boom has led many Azerbaijanis, Tajiks, Uzbeks, and other Central Asians to seek employment in Moscow and St. Petersburg, even though they face substantial discrimination and occasional violence. Russian neo-Nazi gangs with names such as "The White Wolves" have specifically targeted Central Asians for attacks. In early 2010, however, nine members of the White Wolvers were sentenced to long prison terms, leading some observers to hope that the Russian government was finally responding seriously to the problem.

Urbanization in Central Asia

Although the steppes of northern Central Asia had no real cities before the modern age, the river valleys and oases have been partially urbanized for millennia. Cities such as Samarkand and Bukhara in Uzbekistan were famous even in medieval Europe for their riches and their lavish architecture (see "Cityscapes: The Storied City of Samarkand"). This early urban development was built upon the region's economic and political position. The Amu Darya and Syr Darya valleys lay near the midpoint of the trans-Eurasian silk route, and they formed the core of a number of empires based on the cavalry forces of the steppe. The modern era of steamships and oceanic trade undermined this system, bringing hardship to the cities of Central Asia.

The conquest of Central Asia by the Russian and Chinese empires ushered in a new wave of urban formation. Cities slowly began to appear on the Kazakh steppe, where none had previously existed. The Manchu and Chinese conquerors of Inner Mongolia and Xinjiang built new administrative and garrison cities, often placing them only a few miles from indigenous urban sites. The old indigenous cities of the region are characterized by complex and almost mazelike networks of streets and alleyways, while the neighboring Chinese cities were constructed according to a strict geometrical order. The continuing growth of urban populations along with the establishment of new urban policies is now obscuring this old dualism. In 2009, Beijing announced that the old city of Kashgar (Kashi) in China's Far West would be largely demolished, due mainly to concerns about earth-

Figure 10.11 Recent Migration and Refugee Flows Since the early 1990s, large numbers of people have been moving out of the former Soviet Central Asia and into the Chinese portions of Central Asia. More recently, workers from Uzbekistan, Kyrgyzstan, and Tajikistan began moving to Kazakhstan. Afghanistan has one of the largest refugee problems in the world.

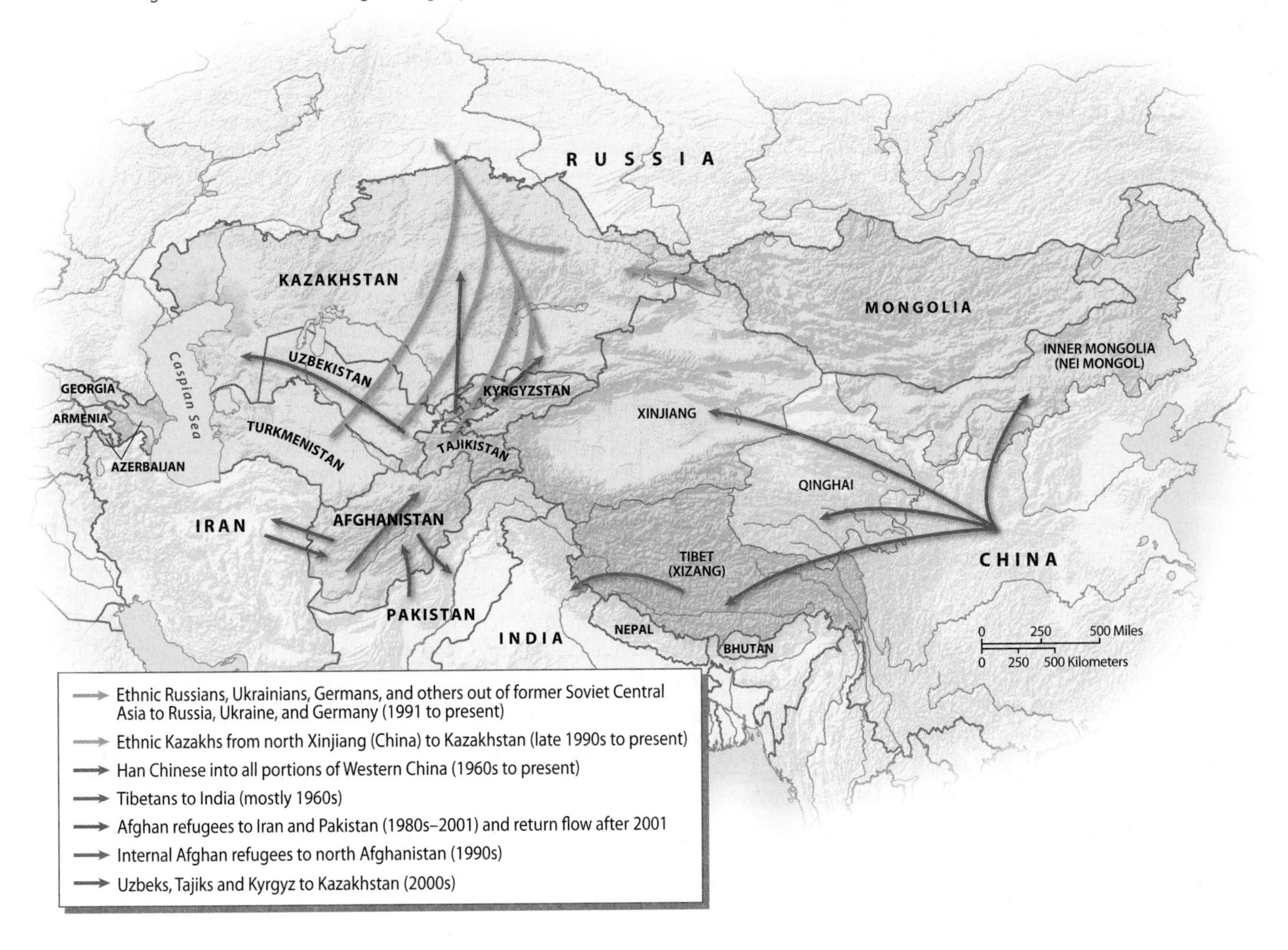

PEOPLE ON THE MOVE Migration In and Out of Kazakhstan

Kazakhstan has experienced several major migration waves over the past century. From the 1920s through the 1960s, large numbers of Russians, Ukrainians, and other people of European background migrated to the Kazakh Soviet Republic, settling in the cities of the south and the farming belt of the north. The 1930s also saw substantial out-migration of Kazakhs, fleeing from Soviet repression into China and Mongolia. At the same time, Soviet authorities exiled members of suspect ethnic groups to the region, including hundreds of thousands of Koreans and Crimean Tatars. Overall, many more people moved in than moved out, with Kazakhstan's population growing from 6 million in 1939 to almost 17 million in 1991.

Figure 10.3.1 Uzbek Migrant Laborers in Kazakhstan. Many poor workers from Uzbekistan, Kyrgyzstan, and Tajikistan have recently sought work in booming Kazakhstan. Here an Uzbek grandmother is seen holding her grandchild in a camp in Kazakhstan.

The independence of Kazakhstan in 1991 ushered in a period of substantial population loss as Russians and other peoples of European ancestry increasingly left the country. In 1989, Kazakhstan recorded 957,518 residents of German background; 10 years later, that number had dropped to 353,441. Russians accounted for 42 percent of the population in 1970 but only 24 percent in 2009. As a result of such emigration, Kazakhstan's total population dropped to under 15 million by 2002.

In the 1990s, the new government of Kazakhstan's grew increasingly concerned about depopulation. The country's birthrate was below the replacement level, yet large numbers of people continued to leave. As a result, it began to encourage the immigration of ethnic Kazakhs living in Mongolia and China, people called Oralmans in Kazakhstan. By 2001, around 180,000 Oralmans had moved into the country. The government continues to subsidize the migration of Oralmans, calling for the resettlement of another 20,000 such families in 2010. Overall, however, the immigration of ethnic Kazakhs is not adequate to replace the ethnic Russians, Ukrainians, and Germans who have emigrated from Kazakhstan in recent years.

Meanwhile, members of other ethnic groups have begun to move to Kazakhstan on their own accord, attracted by its booming economy. More than 500,000 people from other former Soviet countries have entered the Kazakhstan since the turn of the century. More than half are from Uzbekistan, and most the rest come from Kyrgyzstan and Tajikistan (Figure 10.3.1). Work opportunities in the Kazakh oilfields are plentiful, and even farm laborers in Kazakhstan are paid roughly five times more than they would receive in Uzbekistan. Illegal immigrants, however, do not always receive the going wage, and allegations of exploitation are numerous. Roughly three-quarters of the Uzbek migrants into the country are male, many of whom plan to stay only on a temporary basis. Increasingly, however, women and children are migrating as well.

The government of Kazakhstan generally has a positive attitude toward immigration, as it recognizes its need for additional labor. Most of the migrants from the south enter the country illegally, but permanent residency is relatively easy for them to gain. In 2006, the Kazakh Ministry of the Interior announced that it would legalize 100,000 immigrants, but ended up giving out 164,000 residency permits. Many Uzbek, Tajik, and Kyrgyz immigrants are also able to eventually obtain Kazakh citizenship.

The pathway to permanent residence is not so clear for Han Chinese migrants. According to one report, in 2006 alone an estimated 100,000 Chinese entered Kazakhstan for work, with only 5,000 doing so legally. Many Kazakhs are worried about Chinese migration, with only 7 percent of respondents to one poll saying that its effects were positive. Concerns mounted in December 2009, when Kazakh President Nursultan Nazarbaev announced that he was negotiating with Chinese investors who wanted to lease 1 million hectares of Kazakh farmland. Murat Auezov, Kazakhstan's former ambassador to China, was quick to urge caution, asking pointedly, "Who will grow the crops?" He went on to contend that, "If this question is decided in favor of the proposal from China, then it will be the very colonization of Kazakhstan."*

*Bruce Pannier, "Kazakhstan Mulls China Land Deal," *Asia Time Online,* December 19, 2009, http://www.atimes.com/atimes/Central_Asia/KL19Ag01.html)

quake safety. Although China promised to save many of Kashgar's most historic buildings and to build new residences for its inhabitants, some local people view the plan an assault on their cultural identity.

Although it is possible to distinguish Russian/Soviet cities from indigenous cities in the former Soviet zone, this dichotomy is not clear-cut. In Uzbekistan, for example, Tashkent is largely a Soviet creation, while many parts of Bukhara and Samarkand still reflect the older urban patterns. Several major cities, such as Kazakhstan's former capital of Almaty, did not exist before Russian colonization. In Azerbaijan, Baku emerged as a major city in the early 20th century as the first Caspian oil fields began to be intensively exploited. Almost everywhere, one can see the effects of centralized Soviet urban planning and design.

Urbanization is gradually but unevenly spreading across Central Asia. Northcentral Kazakhstan recently witnessed the rise of a major city, Astana, the country's new, centrally located capital. With fewer than 300,000 residents in 1999, Astana had grown to almost 700,000 people by 2010. Even Mongolia, long a land virtually without permanent settlements, now has more people living in cities than in the countryside. In some parts of the region, however, cities remain relatively few and far between. Only 26 percent of the people of Tajikistan, for example, are urban residents. Tibet similarly remains a predominantly rural society, but the influx of Han Chinese into the region is creating a larger urban system. To understand this movement and its broader ramifications, one must examine Central Asia's patterns of cultural geography.

CULTURAL COHERENCE AND DIVERSITY: A Meeting Ground of Different Traditions

Although Central Asia has a certain environmental unity, its cultural coherence is more questionable. The western half of the region is largely Muslim and is often classified as part of Southwest Asia. Northeastern and southeastern Central Asia—Mongolia and Tibet—are characterized by a distinctive form of Tibetan Buddhism. Tibet is culturally linked to both South and East Asia, and Mongolia is intimately associated with China, but neither fits easily within any world region. Such complexity can best be understood by examining the region's historical geography.

Historical Overview: Steppe Nomads and Silk Road Traders

The river valleys and oases of Central Asia were early sites of agricultural communities. Archaeologists have discovered abundant evidence of farming villages dating back to the Neolithic period (beginning circa 8000 BCE) in the Amu Darya and Syr Darya valleys and along the rim of the Tarim Basin. After the domestication of the horse around 4000 BCE, nomadic pastoralism emerged in the steppe belt as a new human adaptation. Eventually pastoral peoples gained power over the entire region, transforming not only the history of Central Asia but also that of virtually all of Eurasia. In the premodern period, highly mobile pastoral nomads enjoyed major military advantages over sedentary societies. Not until the age of gunpowder were the benefits of pastoralism offset by the demographic and economic advantages held by the more populous agriculturally based states.

The linguistic geography of Central Asia has undergone major changes over the ages. Into the first millennium CE, the inhabitants of the region spoke Indo-European languages closely related to Persian. Indo-European languages began to be replaced on the steppe by languages in the Altaic family, which include Mongolian and Turkic, more than a thousand years ago. By the second century BCE, a powerful nomadic empire of Turkic-speaking peoples arose in what today is Mongolia. As Turkic power spread through most of Central Asia, Turkic languages gradually began to replace Indo-European tongues in the agricultural communities. This process was never completed, however, and southwestern Central Asia remains a meeting ground of the Persian and Turkic languages. Uzbek, for example, is classified in the Turkic language family, but much of its basic vocabulary is of Persian origin.

The Turks were eventually replaced on the eastern steppes by another group of Altaic speakers, the Mongols. In the late 1100s the Mongols, under the leadership of Genghis Khan, united the pastoral peoples of Central Asia and used the resulting force to conquer nearby sedentary societies. By the late 1200s the Mongol Empire had grown into the largest contiguous empire the world had ever seen, stretching from Korea and southern China in the east to the Carpathian Mountains and the Euphrates River in the west (Figure 10.12). In today's Mongolia, Genghis Khan has reemerged as the country's national hero. China is also trying to claim the same legacy, having recently built a Genghis Khan theme park in Inner Mongolia.

Protected by mountain barriers and by the rigorous conditions of the plateau, Tibet has taken a different course from the rest of Central Asia. Tibet emerged as a strong, unified kingdom around 700 CE. Tibetan unity and power did not persist, however, and the region reverted to its former state of semi-isolation. Tibet was incorporated for a short period in the 1200s into the Mongol Empire, and in later centuries other Mongol states occasionally enjoyed limited powers over the Tibetans. These interactions resulted in the establishment of Mongolian communities in the northeastern portion of the plateau and in the eventual conversion of the Mongolian people to Tibetan Buddhism.

Contemporary Linguistic and Ethnic Geography

Today most of Central Asia is inhabited by peoples speaking Altaic languages, but patterns of linguistic geography remain complex. A few indigenous Indo-European languages are confined to the southwest, while Tibetan remains the main language of the plateau. Russian is also widely spoken

CITYSCAPES The Storied City of Samarkand

Like all cities of the former Soviet Union, Uzbekistan's Samarkand has its share of drab concrete buildings designed for public housing and government offices. And like all of the older cities of Central Asia, it has its share of crowded traditional neighborhoods with narrow streets (Figure 10.4.1) and houses facing inward toward private courtyards. But in other ways, Samarkand is unique; it contains more sites of historical and religious significance than any other city in the region. Life in the streets of central Samarkand, a city that poets once dubbed the "Rome of the East," forces one to confront and appreciate the past with an unusual intensity. In recognition of its significance, Samarkand was declared a United Nations Educational, Scientific and Cultural Organization (UNESCO) World Heritage Site in 2001.

Samarkand's historical riches derive in part from its physical geography. Located in a fertile valley at roughly the midway point of the Silk Road that once connected China to the Mediterranean, it was an important city as far back as the time of Alexander the Great. Many centuries later, it became the capital city of the famous and feared Turkic-Mongol conqueror, Tamerlane (Timur-i-leng; 1336–1405). Tamerlane and his descendants endeavored to beautify the city, building mosques, tombs, madrassahs (Islamic colleges), and a host of other public buildings. Many of these architectural marvels remain. Muslims from throughout Central Asia and beyond often make pilgrimages to visit the sacred sites of the city, some viewing the trip as second in importance only to the hajj to Mecca.

At the core of historical Samarkand sits Registan Square (Figure 10.4.2), described in the 19th century by Lord George Curzon (Britain's Viceroy of India) as simply "the best central city square in the world." Majestic buildings, including the famous Ulugbek Madrassah, line three sides of the square. Nearby, sitting in the midst of a crowded and colorful market, is the Bibi-Khanum Mosque, which was the largest in the world when built in the 15th century. But it is Tamerlane's final resting place, the Gur-Emir Mausoleum, that is considered by many as the most spectacular example of traditional Central Asian architecture.

As is true of many other Central Asian cities, Samarkand's historical legacy remains vulnerable. In 2008, UNESCO's World Heritage Committee requested "reinforced monitoring" of Samarkand, arguing that its traditional urban fabric was threatened by the poorly regulated construction of new roads and buildings.

Figure 10.4.1 The Narrow Streets of Samarkand The older districts of Samarkand are noted for their tightly packed buildings and narrow streets and alleys. Many of the old city's roads are too narrow for wheeled vehicles.

Figure 10.4.2 Traditional Architecture in Samarkand Samarkand is famous for its lavish Islamic architecture, some of it dating to the 1400s. This photograph shows some of the buildings arrayed around Registan Square, the city's traditional core.

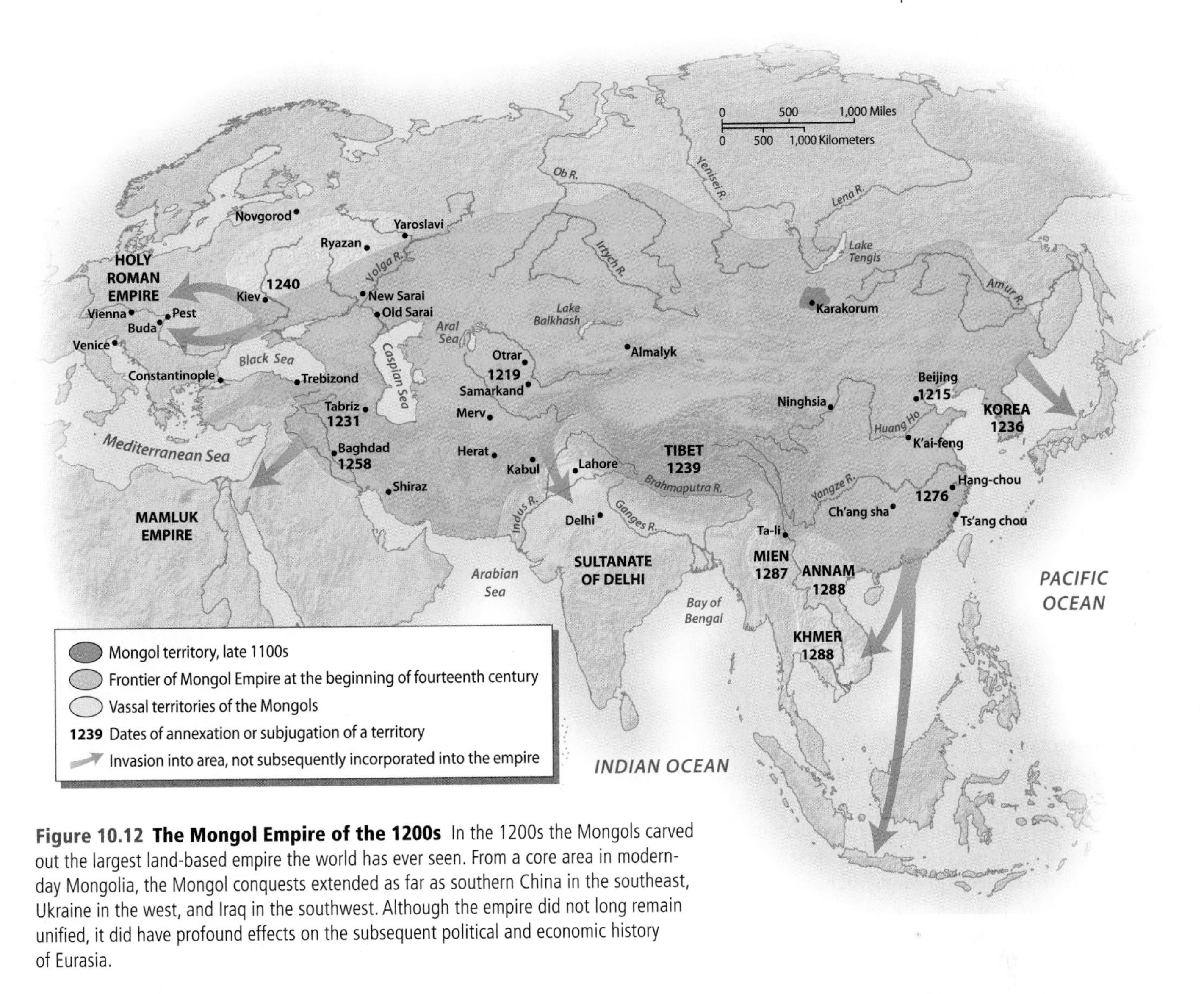

Figure 10.12 The Mongol Empire of the 1200s In the 1200s the Mongols carved out the largest land-based empire the world has ever seen. From a core area in modern-day Mongolia, the Mongol conquests extended as far as southern China in the southeast, Ukraine in the west, and Iraq in the southwest. Although the empire did not long remain unified, it did have profound effects on the subsequent political and economic history of Eurasia.

in the west, and in most former Soviet countries Russian retains an official status, whereas Mandarin Chinese is increasingly important in the east. In both Tibet and Xinjiang, Mandarin Chinese is now the basic language of higher education. In Tibet, most urban merchants are now Chinese-speaking immigrants, causing much concern among the indigenous people (Figure 10.13).

Tibetan Tibetan is divided into a number of distinct dialects that are spoken over almost the entire inhabited portion of the Tibetan Plateau. Approximately 6 million people speak Tibetan; of these, roughly 2.5 million live in Tibet proper, while most of the rest reside in China's provinces of Qinghai and Sichuan. Tibetan is usually placed in the Sino-Tibetan family, implying a shared linguistic ancestry between the Chinese and the Tibetan peoples, but some scholars argue that no definite relationship between the two has been established. Tibetan has an extensive literature written in its own script, most of which is devoted to religious topics.

Mongolian Mongolian is comprised of a cluster of closely related dialects spoken by approximately 5 million people. The standard Mongolian of both the independent country of Mongolia and China's Inner Mongolia is called Khalkha; other Mongolian dialects include Buryat (found in southern Siberia) and Kalmyk (found to the northwest of the Caspian Sea). Mongolian has its own distinctive script, which dates back some 800 years, but Mongolia itself adopted the Cyrillic alphabet of Russia in 1941. In China's region of Inner Mongolia, however, the old script is still widely used.

Although Mongolian speakers form about 90 percent of the population of Mongolia, in China's Inner Mongolian Autonomous Province they have been partly submerged by a wave of Han Chinese migrants over the past 50 years. Today only about 4 million out of the 25 million residents of Inner Mongolia identify themselves as Mongols. There are still more Mongols in China than in Mongolia, as Mongolia's total population is less that 3 million.

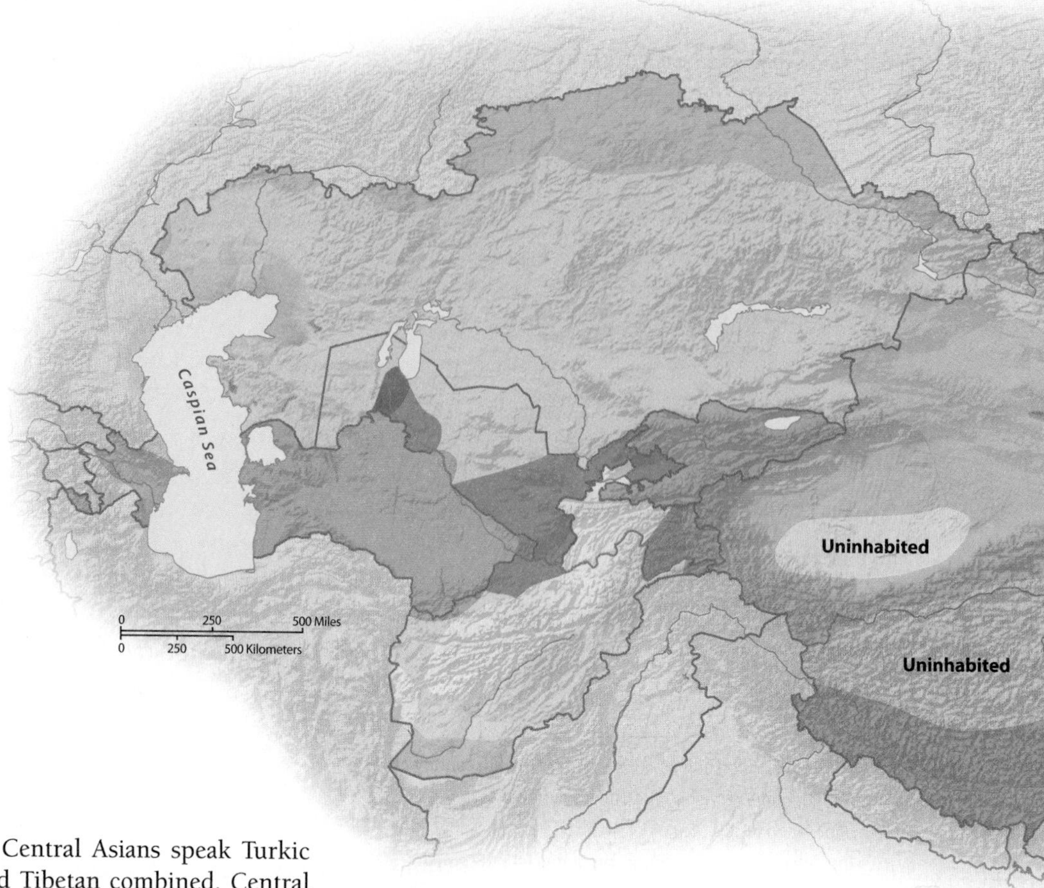

Figure 10.13 Linguistic Geography of Central Asia Most of Central Asia is dominated by languages in the Altaic family, which includes both the Turkic languages (found through most of the center and the west of the region) and Mongolian (found in Central Asia's northeast). Several Indo-European languages, however, are located in both the far northwestern and southcentral regions, while the Tibeto-Burman language of Tibetan covers most of the Tibetan Plateau in southeastern Central Asia.

Turkic Languages Far more Central Asians speak Turkic languages than Mongolian and Tibetan combined. Central Asia's Turkic linguistic sphere extends from Azerbaijan in the west through China's Xinjiang province in the east. The various Turkic languages are not as closely related to each other as are the dialects of Mongolian, but they are still obviously kindred tongues. Six main Turkic languages are found in Central Asia; five are associated with newly independent (former Soviet) republics of the west, while the sixth, Uyghur, is the main indigenous language of northwestern China.

Uyghur is an old and important language. The Uyghur number about 8 million, almost all of whom live in Xinjiang. As recently as 1949, the Uyghur formed about 90 percent of the population of Xinjiang; now, because of Han Chinese immigration, that figure has dropped to 45 percent. Eastern and northern Xinjiang are now largely Chinese-speaking. There are also about 1 million Kazakh speakers in far northern Xinjiang.

Of the six countries of the former Soviet Central Asia, five—Kazakhstan, Uzbekistan, Turkmenistan, Kyrgyzstan, and Azerbaijan—are named after the Turkic languages of their dominant populations. In three of these countries the indigenous people form a substantial majority. Some 82 percent of the people of Azerbaijan are classified as Azeri (there are more Azeris in northern Iran, however, than there are in Azerbaijan); some 80 percent of the people of Uzbekistan are classified as Uzbek; and some 77 percent of the people in Turkmenistan are classified as Turkmen. Most of these people speak their indigenous languages at home, although many Kazakhs are more comfortable speaking Russian than Kazakh.

With more than 23 million speakers, Uzbek is the most widely spoken Central Asian language. In the Amu Darya Delta in the far north of Uzbekistan, however, many people speak a Turkic language more closely related to Kazakh called Karakalpak. Kazakh speakers are widely scattered across Uzbekistan's sparsely populated deserts. Uzbekistan's older cities, such as Samarkand, are still partly Tajik (or Persian) speaking. Tajik activists, moreover, claim that the number of Tajik speakers in Uzbekistan is much greater than what the official statistics indicate, and that many local Tajiks have been pressured into assuming an Uzbek identity.

In the two other Turkic republics, the nationality for which the country is named forms less than two-thirds of the total population. Some 65 percent of the inhabitants of Kyrgyzstan are classified as Kyrgyz, while roughly 63 percent of the people of Kazakhstan are classified as Kazakh. The other residents of these countries identify themselves as Russians, Uzbeks, Ukrainians, and members of a wide assortment of other groups. In general, the Kazakhs and other Turkic peoples live in the center and south of Kazakhstan, while the people of European descent live in the agricultural districts of the north and in the cities of the southeast.

Linguistic Complexity in Tajikistan The sixth republic of the former Soviet Central Asia, Tajikistan, is dominated by people who speak an Indo-European rather than a Turkic language. Tajik is so closely related to Persian that it is considered to be a Persian (or Farsi) dialect. Roughly 5.8 million people in Tajikistan, about 80 percent of the total population, identify themselves as Tajiks. The remote mountains of eastern Tajikistan are populated by peoples speaking a variety of distinctive Indo-European languages, sometimes collectively referred to as "Mountain Tajik."

Tajikistan, like much of the rest of former Soviet portion of Central Asia, is noted for its complex mixture of language and ethnic groups. About 15 percent of its people, for example, are Uzbeks. Today ethnic suspicion pervades much of the region, in part because the governments of Tajikistan and Uzbekistan have poor relations. Before the Soviet period, however, mode of life was as important as language in determining identity; settled farmers and city-dwellers were usually classified as "Sarts" regardless of whether they spoke Tajik or a local Turkic language. Soviet planners essentially created the modern nationalities of Tajik and Uzbek, and they tried to force people into one category or the other. Because the two groups were generally mixed together, the Soviet Union also had to draw extremely convoluted boundaries around the republics of Uzbekistan and Tajikistan. The resulting political geography causes problems today for both countries.

ALTAIC
Turkic
Kazakh
Karakalpak
Uzbek
Uyghur
Kyrgyz
Turkmen
Azeri
Mongolian
Mongolian
INDO-EUROPEAN
Russian
Tajik and Dari (Persian)
Mountain Tajik (8 separate languages)
Pashtun
Baluchi
SINO-TIBETAN
Tibetan
Mandarin Chinese
OTHER
Other and uninhabited

Language and Ethnicity in Afghanistan The linguistic geography of Afghanistan is even more complex than that of Tajikistan (Figure 10.14). Afghanistan was never colonized by outside powers, and it is one of the few countries of the world to have inherited the

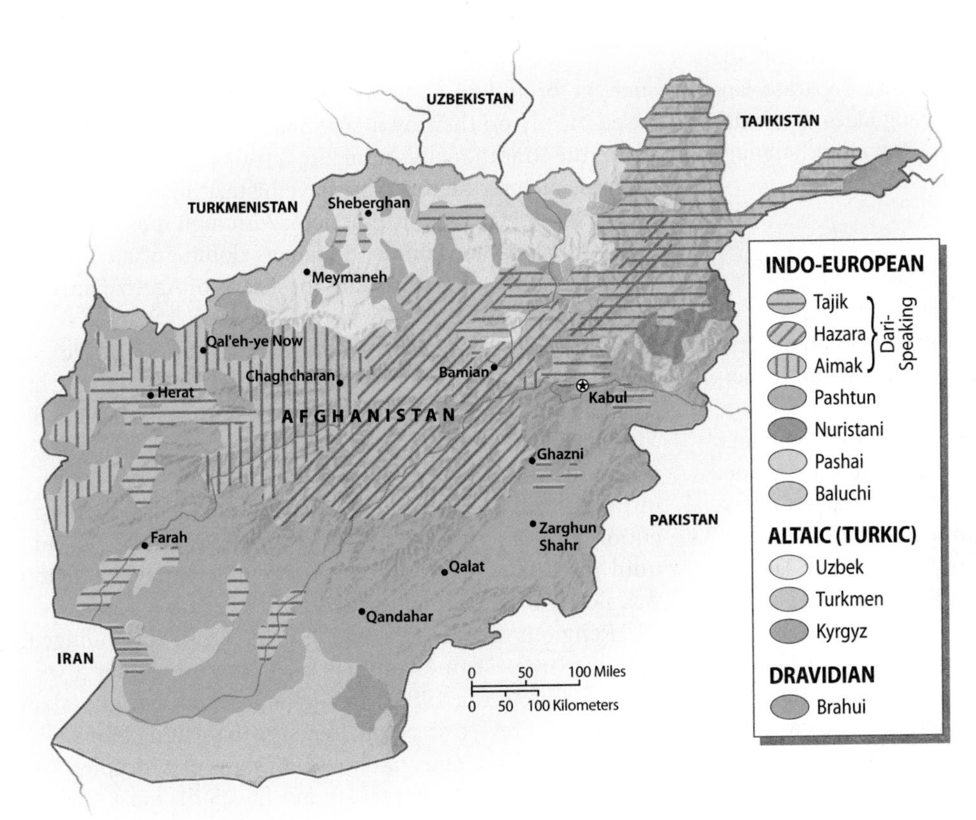

Figure 10.14 Afghanistan's Ethnolinguistic Patchwork
Afghanistan is one of the world's more ethnically complex countries. Its largest ethnic group is that of the Pashtuns, a people who inhabit most of the southern portion of the country as well as the adjoining borderlands of Pakistan. Northern Afghanistan is inhabited mostly by Uzbeks, Tajiks, and Turkmens—whose main population centers are located in Uzbekistan, Tajikistan, and Turkmenistan, respectively. The Hazaras of Afghanistan's central mountains, like the Tajiks, speak a form of Persian but are considered to be a separate ethnic group in part because they, unlike other Afghans, follow Shiite rather than Sunni Islam. Smaller groups are found elsewhere in the country.

boundaries of a premodern, indigenous kingdom. This kingdom emerged in the 1700s on traditional dynastic lines that did not reflect ethnic or linguistic divisions ("dynastic" linkages are those based on the family of the monarch). The modern nation-state ideal—that each country should be identified with a particular national group—was never very important in Afghanistan.

The 18th-century creators of Afghanistan were mostly members of the Pashtun ethnic group, but they did not attempt to build a nation-state around Pashtun identity. Indeed, more Pashtuns live in Pakistan than in Afghanistan. In Afghanistan, estimates of the proportion of the populace speaking the Pashto language of the Pashtuns vary from 40 to 60 percent. Most of them live to the south of the Hindu Kush Mountains.

Approximately half of the people of Afghanistan speak Dari, Afghanistan's variant of Persian, as their first language. Dari speakers are concentrated in the north. Three separate ethnicities are ascribed to the Dari-speaking people: the settled farmers and townspeople in the west and north are Tajiks; the traditionally seminomadic people of the west-center are Aimaks; and the East Asian–appearing villagers of the central mountains are Hazaras, reputed to be descendants of medieval Mongol conquerors. Although traditionally forming something of an underclass, Afghanistan's Hazaras are now showing more dedication to modern education than the country's other ethnic groups and as a result are beginning to prosper.

Although Pashto and Dari are Afghanistan's official languages, a variety of other languages are also spoken in the country, creating an especially complex ethnic patchwork. Approximately 11 percent of the people of Afghanistan speak Turkic languages, including both Uzbek and Turkmen. Other ethnolinguistic groups in Afghanistan include the Baluchis, the Nuristanis (who speak five separate languages), and the Pashai, all of whose languages are Indo-European, and the Brahui, who speak a Dravidian language related to the tongues of southern India.

Figure 10.15 Entrance to the Poi-Kalyan Mosque, Bukhara, Uzbekistan With the exceptions of Tibet and Mongolia, most of Central Asia is religiously dominated by Islam. The mosque architecture of Central Asia's older cities, such as Bukhara in Uzbekistan, are noted for their striking architecture.

Geography of Religion

An assortment of religions characterized ancient and medieval Central Asia. The major overland trading routes of premodern Eurasia crossed the region, giving easy access to both merchants and missionaries. Several varieties of Buddhism, Islam, Christianity, and Judaism, as well as several minor religions, have all thrived at various times and places within the region. Eventually, however, religious lines hardened, and Central Asia was divided into two opposed spiritual camps: Islam triumphed in the west and center (Figure 10.15), while Tibetan Buddhism prevailed in Tibet and Mongolia.

Islam in Central Asia As is true elsewhere in the Muslim world, different Central Asian peoples are known for their different interpretations of Islamic orthodoxy. The Pashtuns of Afghanistan are noted for their strict Islamic ideals—although critics contend that Pashtun religious practices, such as forbidding women's faces to be seen in public, are based mostly on their own customs (Figure 10.16). By contrast, the traditionally nomadic groups of the northern steppes, such as the Kazakhs, have historically been considered lax in their religious practices. Although most of the region's Muslims are Sunnis, Shiism is dominant among the Hazaras of central Afghanistan, the Azeris of Azerbaijan, and the mountain dwellers of eastern Tajikistan.

Under the communist rule of China, the Soviet Union, and Mongolia, all forms of religion were discouraged. Chinese authorities and student radicals suppressed Islam in Xinjiang during the Cultural Revolution of the late 1960s and early 1970s. Mosques were destroyed or converted to museums and religious schools were closed. Periodic persecution of Islam also occurred in Soviet Central Asia, and until the 1970s many observers thought that that religion was slowly diminishing throughout the region.

Religious expression was not so easily discouraged; however, interest in Islam began to grow in former Soviet Central Asia in the 1980s. In the post-Soviet period, Islam continues to revive as people reassert their indigenous heritage and identity. Thus far, however, signs of widespread Islamic political fundamentalism are few. Still, most Central

Figure 10.16 Afghan Women in Public Especially in the Pashtun areas of Afghanistan, women have traditionally been forced to cover their entire bodies when in public areas. In areas that were controlled by the Taliban, such dress codes were strictly enforced. Extremely modest dress still prevails in Afghanistan.

Asian leaders remain wary of any signs of radical Islamic practices. In 2005, for example, Tajikistan banned the wearing of Islamic head-scarves in public schools, and in 2009 it prohibited teachers under the age of 50 from growing beards. In response to widespread religious criticism, Tajikistan's government also banned revealing Western clothing styles—as well as graduation parties and the use of mobile phones and private cars at the country's secondary schools.

In China, Muslims now enjoy basic freedom of worship, but the state still closely monitors religious expression out of fear that it might become entangled with political separatism. Although some evidence suggests that Islam has served as a focal point of an anti-Chinese movement among the Uyghur people, most Uyghur leaders insist that their program is not radically fundamentalist. Still, ethnic relations in northwestern China remain very tense. In 2009, ethnic rioting pitting Uyghurs against Han Chinese took as many as 200 lives in Urumqi, the capital city of Xinjiang province.

In Afghanistan, however, Islamism (or politically radical Islamic fundamentalism) did emerge as a powerful political movement. From the mid-1990s until 2001, most of the country was controlled by the **Taliban**, an extremist organization that insisted that all aspects of society conform to its own harsh version of Islamic orthodoxy. This commitment was demonstrated in early 2001 when Afghanistan's religious authorities oversaw the destruction of Buddhist statues in the country—including some of the world's largest and most magnificent works of art—as symbols of a non-Islamic past. Although the Taliban regime fell in late 2001, Islamism remains a powerful force in Afghanistan, particularly among the Pashtun people of the south.

Islam is not the only religion represented in western Central Asia. Many Russians belong to the Russian Orthodox Church, and Uzbekistan has a small Jewish population. Kazakhstan also counts some 100,000 Catholics, mostly people of Polish or German descent.

Tibetan Buddhism Mongolia and Tibet stand apart from the rest of Central Asia in the adherence of their people to Tibetan Buddhism. Buddhism entered Tibet from India many centuries ago, where it merged with the indigenous religion of the area, called Bon. The resulting faith is more oriented toward mysticism than are other forms of Buddhism, and it is more hierarchically organized. Standing at the apex of Lamaist society is the Dalai Lama, considered to be a reincarnation of the **Bodhisattva** of Compassion (a Bodhisattva is a spiritual being who helps others attain enlightenment). Ranking below him is the Panchen Lama, followed by other religious officials. Until the Chinese conquest, Tibet was essentially a **theocracy**, or religious state, with the Dalai Lama enjoying political as well as religious authority. A substantial proportion of Tibet's male population has traditionally become monks, and monasteries once wielded both economic and political power (Figure 10.17).

Tibetan Buddhism suffered severe persecution in the 1960s. The Dalai Lama fled to India with many of his followers after China invaded Tibet in 1959, and has since been a powerful advocate for the Tibetan cause in international circles. During the 1960s and 1970s, an estimated 6,000 Tibetan Buddhist monasteries were destroyed and thousands of monks were killed. The number of active monks today is only about 5 percent of what it was before the Chinese occupation. Still, the Buddhist faith continues to form the basis of Tibetan identity, and in so doing helps keep alive the dream of independence or at least of real autonomy within China. In 2008, Tibetan monks led a series of protests against the Chinese government that resulted in ethnic riots between Tibetans and Han Chinese breaking out across the Tibetan Plateau. The Chinese government reacted harshly and later blamed the Dalai Lama for inciting the unrest.

Figure 10.17 Lamaist Buddhist Monastery Tibet is well known for its large Buddhist monasteries, buildings that in earlier years served as seats of political as well as religious authority. Lhasa's Jokhang Temple and Monastery, seen in this photograph, dates back to the 7th century.

In Mongolia the downfall of communism has allowed the Tibetan Buddhist faith to experience a renaissance. Several monasteries have been refurbished, and many people are returning to their national religion. The intensity of Buddhist belief, however, is not as strong in Mongolia as it is in Tibet. It is estimated that about half of the people of Mongolia actively follow the Buddhist religion today.

Central Asian Culture in International Context

During the Soviet period, the Russian language spread widely through western Central Asia. Russian served both as a common language and as a means of instruction in higher education. One had to be fluent in Russian in order to reach any position of responsibility. The Cyrillic (or Russian) script, moreover, replaced the modified Arabic script that was previously used for the indigenous languages. After the fall of the Soviet Union, however, most Russians migrated back to Russia, especially from the region's poorer countries. As a result, the use of Russian in education, government, business, and the media has declined in favor of local languages.

But despite its recent decline, Russian remains the common language of former Soviet Central Asia. Even some of the leaders of these countries speak their own languages imperfectly, and Russian is often the only way to communicate across Central Asia's national boundaries. The continuing prominence of Russia, however, irritates many people. Kazakh nationalists, for example, are upset that Russian remains the country's "official language of communication between different ethnic groups." In early 2010, hundreds of Kazakh students protested their country's language laws, demanding that governmental meetings, sessions of parliament, and presidential speeches be conducted only in Kazakh. In 2007, Tajikistan ordered its citizens to change their names if they sounded too Russian, and in 2009 it announced plans to ban the use of Russian in advertising, business communication, and governmental documents.

Figure 10.18 **U.S.-Style Restaurant in Baku, Azerbaijan.** Baku is a vibrant and increasingly cosmopolitan city closely linked to global economic and cultural networks. The coffee shop and wine bar seen in this photograph caters to both local and international tourists.

Although Central Asia is remote and poorly integrated into global cultural circuits, it is hardly immune to the forces of globalization (Figure 10.18). Even the tensions existing throughout the region between religious and secular orientations, and between ethnic nationalism and multiethnic inclusion, are aspects of the current global condition. So, too, the increased usage of English throughout Central Asia shows that this part of the world is not cut off from global culture. Such influences are especially marked in the oil cities of the Caspian Basin, such as Azerbaijan's Baku.

GEOPOLITICAL FRAMEWORK: Political Reawakening in a Power Void

Central Asia has played a minor role in global political affairs for the past 300 years. Before 1991 the entire region, except Mongolia and Afghanistan, lay under direct Soviet or Chinese control. Mongolia, moreover, was a Soviet satellite, and even Afghanistan came under Soviet domination in the late 1970s. The breakup of the Soviet Union, however, saw the emergence of six new countries, helping to reestablish Central Asia as a world region (Figure 10.19).

Partitioning of the Steppes

Before the 1700s, Central Asia had been a power center whose mobile armies threatened the far more populous, sedentary peoples of the Eurasian rim. Military developments, however, changed the balance of power, allowing the wealthier agricultural states to defeat the nomadic pastoralists and take over their lands by the late 1700s. The winners in this struggle were the two largest states bordering the steppes: Russia and China.

The Manchu conquest of China in 1644 undercut the autonomy of the peoples of the eastern steppe. The Manchus came from Manchuria, the eastern borderlands of Central Asia, and were themselves skilled in the arts of cavalry warfare. By the late 1700s China stood at its greatest territorial extent. Within its grasp lay not only Mongolia and Xinjiang but also Tibet and a slice of modern Kazakhstan. Although Manchu-ruled China declined rapidly after 1800, it was still able to retain most of its Central Asian territories. But by the early 1900s, Chinese authority had begun to diminish in Central Asia as well. When the Manchu (also called Ch'ing or Qing) dynasty fell in 1912, Mongolia became independ-

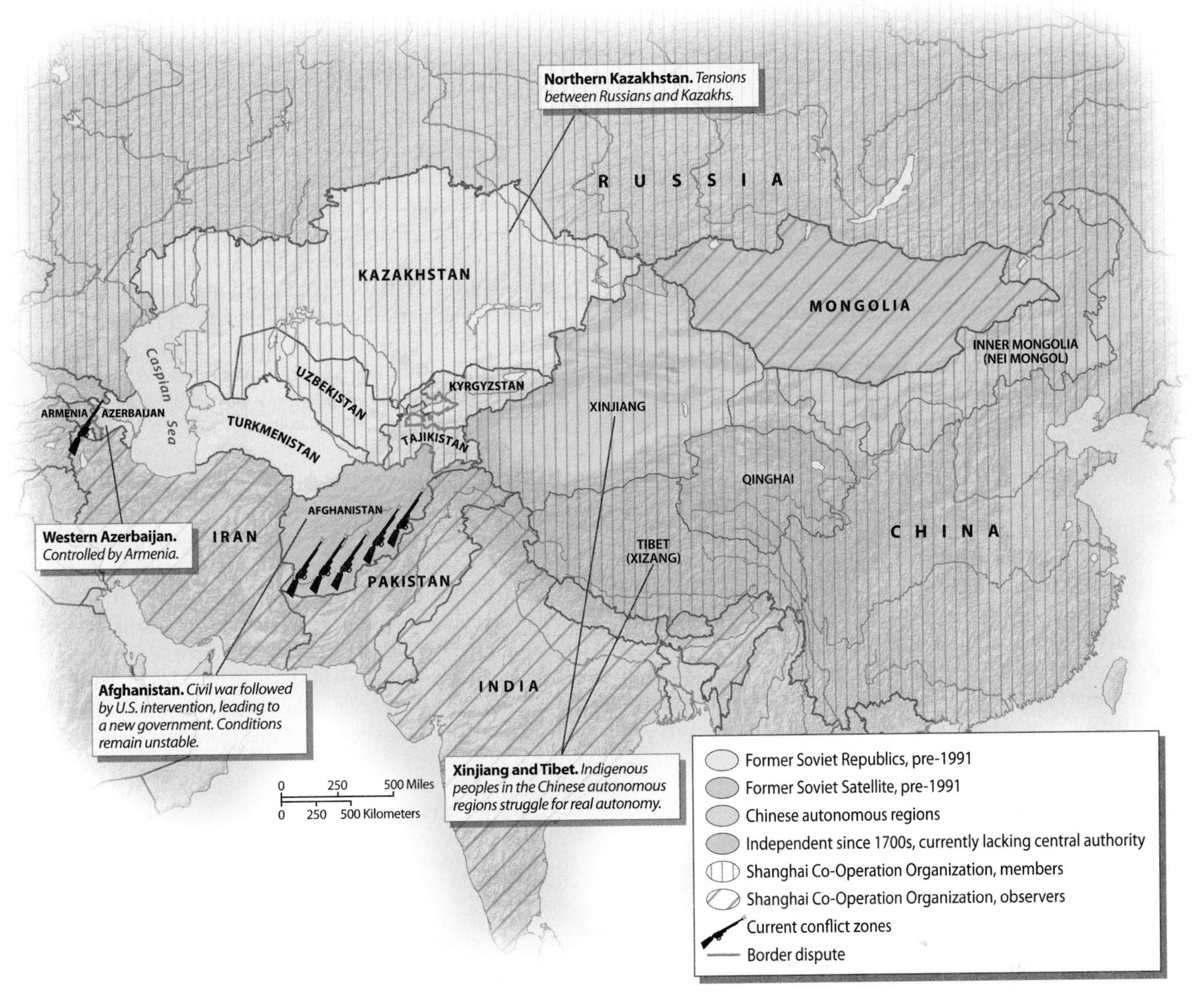

Figure 10.19 Central Asian Geopolitics Six of the eight independent states of Central Asia came into existence in 1991 with the dissolution of the Soviet Union. In eastern Central Asia, the most serious difficulties stem from China's maintenance of control over areas in which the indigenous peoples are not Chinese. Afghanistan, scene of a prolonged and brutal civil war, has experienced the most extreme forms of geopolitical tension in the region.

ent, although China did keep the extensive borderlands of Inner Mongolia (Nei Mongol). Tibet had earlier gained de facto independence, and even Xinjiang lay beyond the reach of effective Chinese authority during the 1920s.

Russia began to advance into Central Asia at roughly the same time as China. In the 1700s the Russian Empire undertook the systematic conquest of the Kazakh steppe. Expansion farther to the south, however, was blocked by the agricultural states of Uzbekistan. Only in the late 1800s, when European military techniques and materials raced ahead of those of Asia, was Russia able to conquer the Amu Darya and Syr Darya valleys. Its subjugation of this area was not completed until the early 1900s, just before the Soviet Union replaced the Russian Empire. The Russians advanced into Central Asia in part because of their concern over possible British influence in the area. Britain attempted to conquer Afghanistan, but was rebuffed by Afghan forces—and by the country's forbidding terrain. Subsequently, Afghanistan's position as an independent "buffer state" between the Russian Empire (later the Soviet Union) and the British Empire in South Asia remained secure.

Central Asia Under Communist Rule

Western Central Asia came under communist rule soon after the foundation of the Soviet Union in 1917, with Mongolia following in 1924. After the Chinese revolution of 1949, the communist system was imposed on Xinjiang, Tibet, and

Inner Mongolia. In all of these areas, major changes in the geopolitical order soon followed.

Soviet Central Asia Although the Soviet Union inherited the Russian imperial domain in Central Asia virtually intact, it enacted new policies. The Soviet regime sought to create a socialist economy and to build a new society that would eventually knit together all of the massive territories of the country. Central Asia's leaders were replaced by Communist Party officials loyal to the new state; Russian immigration was encouraged; and local languages could no longer be written in Arabic script.

Although Soviet authorities foresaw the emergence of a single Soviet nationality, they realized that local ethnic diversity would not disappear overnight. Early Soviet leaders such as Vladimir Lenin also hoped to insulate non-Russian peoples from Russian domination. The Soviet Union therefore divided the Soviet Union into a series of nationally defined "union republics" in which a certain degree of cultural autonomy would be allowed. They were uncertain, however, what the relevant units in Central Asia should be. Was there a single Turkic-speaking nationality, or were the Turkic peoples divided into a number of separate nationalities, with the Tajiks forming yet another? For several years boundaries shifted as new "republics" appeared on the map. Finally, in the 1920s the modern republics of Kazakhstan, Kyrgyzstan, Tajikistan, Uzbekistan, Turkmenistan, and Azerbaijan assumed their present configurations. In certain areas, such as the fertile Fergana Valley, the political boundaries so drawn remained extremely complex (Figure 10.20).

Some scholars argue that the Soviet nationalities policy backfired severely. Rather than forming a transitional step on the way to a Soviet identity, the constituent republics of the Soviet Union nurtured local nationalisms that ultimately undermined the Soviet system. Identities such as Turkmen, Uzbek, and Tajik that had been vague in the pre-Soviet period were now given real political significance. Another problem undercutting Soviet unity was the fact that the cultural and economic gaps separating Central Asians from Russians did not diminish as much as planned.

The Chinese Geopolitical Order After China reemerged as a united country in 1949, it too was able to reclaim most of its old Central Asia territories. China's communist leaders promised the non-Chinese peoples a significant degree of political self-determination and cultural autonomy, and thus found much local support in Xinjiang. Tibet, isolated behind its mountain walls and virtually independent for many years, presented a greater obstacle. China occupied Tibet in 1950, but the Tibetans launched a rebellion in 1959. When the rebellion was crushed, the Dalai Lama and some 100,000

Figure 10.20 Political Boundaries Around the Fergana Valley Some of the world's most convoluted political boundaries can be found in the vicinity of the Fergana Valley. The central portion of the valley belongs to Uzbekistan, which is otherwise separated from it by high mountains. The lower valley, on the other hand, is part of Tajikistan, the core area of which is likewise separated from the valley by highlands. The Fergana's upper periphery belongs to Kyrgyzstan. Note also the small exclaves of Uzbekistan within Kyrgyzstan.

followers found refuge in India, where they maintain the "Tibetan Government in Exile."

Loosely following the Soviet nationalities model, China established autonomous regions in areas occupied primarily by non-Han Chinese peoples, including Xinjiang, Tibet proper (called Xizang in Chinese), and Inner Mongolia. Such autonomy, however, often turned out to be more symbolic than real, and it did not prevent the massive immigration of Han Chinese into these areas. Nor were all parts of Chinese Central Asia granted autonomous status. The large and historically Tibetan and Mongolian province of Qinghai, for example, remained an ordinary Chinese province.

Current Geopolitical Tension

Although the former Soviet portion of Central Asia weathered the post-1991 transition to independence relatively smoothly (Figure 10.21), the region suffers from a number of ethnic conflicts as well as from the struggle between radical Islam and secular governments. Much of China's Central Asian territory is troubled, but China retains a firm grip. Such troubles are insignificant, however, compared to what is faced by Afghanistan, where brutal warfare has been going on for decades.

Figure 10.21 The End of Soviet Rule in Central Asia An old Tajik man looks over a toppled and decapitated statute of Lenin, founder of the Soviet Union, in Dushanbe in 1991. When the Soviet Union collapsed in that year, many emblems of the old regime were destroyed immediately.

Independence in Former Soviet Lands After 1991, the six newly independent countries of the former Soviet Union did not find it easy to chart their own political courses. They still had to cooperate with Russia on security issues, and all initially opted to join the Commonwealth of Independent States, the rather hollow successor of the Soviet Union. In most cases authoritarian rulers, rooted in the old order, retained power and sought to undermine opposition groups. All told, democracy made less progress in Central Asia than in other parts of the former Soviet Union. Kyrgyzstan seemed to form an exception to this pattern, especially in 2005 when citizens protesting corruption and authoritarianism forced the country's president out of office. Kyrgyzstan's new government, however, quickly followed in the footsteps of the old, resulting in renewed protest and unrest. In April 2010, demonstrators again brought down the government, forcing Kyrgyzstan's president to flee. Three months later, massive ethnic riots broke out in southern Kyrgyzstan as supporters of the country's ousted president attacked ethnic Uzbeks. As many as 2,000 people were killed in the rioting, and up to 300,000 were displaced.

Kazakhstan, Tajikistan, and Azerbaijan faced particular difficulties in the post-Soviet transition. Many Kazakhs wanted to create a national state centered on Kazakh identity, and they resented the presence and power of Russians and other peoples of European background. These Europeans feared what they considered alien Central Asian cultural standards, and for a time there was some concern that northern Kazakhstan, the country's breadbasket, might attempt to join Russia. In the late 1990s, Kazakhstan moved its capital from Almaty in the south to Astana in the Russian-dominated north, partly to forestall any plans for secession. The continued out-migration of Russians and other Europeans from Kazakhstan, however, has gradually made this issue irrelevant.

In Tajikistan war broke out almost immediately after independence in 1991. Many members of the smaller ethnic groups of the mountainous east resented the authority of the Tajiks and thus rebelled. They were joined by Islamist groups seeking to overthrow the secular state. The civil war was officially ended in 1997 when the Islamists agreed to work through normal political channels. However, continuing instability led Tajikistan's government to increase its pressure on the Islamic Renaissance Party, the only legal Islamist party in former Soviet Central Asia. The Islamic Renaissance Party responded by modernizing its image to appeal to educated people, youth, and women, but in the 2010 elections it received only 8 percent of the vote.

Azerbaijan also experienced strife following the breakup of the Soviet Union. Armenia invaded and occupied a portion of far-western Azerbaijan, allowing the Armenian-speaking highlands to form the "breakaway republic" of Nagorno-Karabakh. Hostile relations with Armenia make it difficult for Azerbaijan to govern its **exclave** of Nakhchivan, a piece of Azerbaijani territory separated from the rest of the country by Armenia and Iran. The situation remains tense; as recently as 2008, military skirmishes were taking place along the border between Azerbaijan and Nagorno-Karabakh.

Although Uzbekistan and Turkmenistan had relatively easy transitions out of the Soviet Union, both countries are ruled by repressive governments that allow little room for personal freedom or opposition movements. Uzbekistan has recently increased its censorship of the news and cracked down on human rights. Until his death in late 2006, Turkmenistan's authoritarian president, Saparmurat Niyazov—who called himself simply "Turkmenbashi" ("father of the Turkmens")—maintained a lavish personality cult, forcing all students in his country to read and honor his lengthy book on Turkmen history, philosophy, and mythology. Since Niyazov's death, Turkmenistan has gradually reduced its restrictions on daily life while tentatively opening up to the outside world. By 2010, for example, it had allowed the reopening of cinemas, theaters, and circuses, all of which had previously been banned as representing "alien culture."

Strife in Western China Local opposition to Chinese rule is pronounced in many of the region's indigenous communities. Protests by Tibetans have not been successful, but they have brought the attention of the world to their struggle. China maintains several hundred thousand troops in a region that has only 2.8 million civilian inhabitants. Such an overwhelming military presence is considered necessary because of both Tibetan resistance and the strategic importance of the region. Massive Tibetan protests in 2008 were handled harshly by the Chinese government; in 2009, four Tibetan activists were executed for inciting the unrest, and an estimated 670 were imprisoned.

China's control of Xinjiang is more secure than its hold on Tibet. Although fragmented by deserts and mountains, Xinjiang's terrain is less forbidding than that of Tibet and it is now the home of millions of Han Chinese immigrants. Xinjiang is also an economically vital part of China. It contains a variety of mineral deposits (including oil) essential for Chinese industry, and it has been the site of nuclear weapons tests. Many Uyghurs, not surprisingly, oppose such uses of their homeland, and they resent the periodic suppression of their religion by the communist regime. Ethnic strife between Han Chinese and Uyghurs resulted in massive unrest in 2009, leading to a number of deaths and large-scale destruction of property (Figure 10.22).

Figure 10.22 Ethnic Tension in Xinjiang In 2009, ethnic rioting pitting the indigenous Uyghurs against Han Chinese immigrants broke out in a number of cities in Xinjiang. Chinese security forces reacted harshly to Uyghur protesters.

China's position is that all of its Central Asian lands are integral portions of its national territory. Those who advocate independence are viewed as traitors and are sometimes considered to be in league with Western political forces that have sought to keep China weak and divided for the past 200 years.

War in Afghanistan None of the conflicts elsewhere in Central Asia compares in intensity to the situation in Afghanistan. Afghanistan's troubles began in 1978 when a Soviet-supported military "revolutionary council" seized power. The new Marxist-oriented government suppressed religion, which resulted in rebellion. When the government was about to collapse, the Soviet Union launched a massive invasion. Despite its power, the Soviet military was never able to gain control of the more rugged parts of the country. Moreover, Pakistan, Saudi Arabia, and the United States ensured that the anti-Soviet forces remained well armed and financed.

The exhausted Soviets finally withdrew their troops in 1989. The puppet government that they installed remained to face the insurgents alone. It managed to hold the country's core around the city of Kabul for a few years, largely because the opposition forces were themselves divided. Local warlords grabbed power in the countryside, destroying any semblance of central authority. Warlords and their forces committed many atrocities, including the shelling of civilian areas, the abduction of noncombatants, and the mass raping of women from enemy communities.

In 1995–96, the Taliban arrived on the Afghan scene. The Taliban was founded by young Muslim religious students disgusted with the anarchy that was consuming their country. They were convinced that only the firm imposition of Islamic law could end corruption and quell the disputes among the country's different ethnic groups. With large numbers of soldiers flocking to their standard and with military support from Pakistan, the Taliban won control of most of Afghanistan by the late 1990s. As their own power grew more secure, they offered refuge in Afghanistan to the radical Islamist organization Al Qaeda, led by Osama bin Laden.

The Taliban acquired power not only from its religious nature, but also from the ethnic divisions of Afghanistan. It gained most of its strength from the Pashtuns of the southern half of the country. The Pashtuns have a reputation for militarism, as well as good connections for military supplies among their relatives across the Pakistan border. The main opposition to the Taliban came from the country's other ethnic groups, especially the Uzbeks and Tajiks of the north and the Shiite Hazaras of the central mountains.

By the end of the century, most of the people of Afghanistan lost faith in the Taliban, largely because of the severe restrictions imposed on daily life. Such constraints were most pronounced for women, but even men were compelled to obey the Taliban's numerous decrees. Most forms of recreation were outlawed, including television, films, music, and even kite-flying. Little escaped the watchful eyes of the Taliban's religious police.

The aftermath of September 11, 2001, completely changed the balance of power in the region. The United States and Britain, working closely with anti-Taliban groups concentrated in northern Afghanistan, launched a war against Al Qaeda and the Taliban government. Within a few months, formal Taliban power collapsed, allowing the emphasis to shift to rebuilding Afghanistan's institutions and establishing an effective central government. Such projects were not very successful, however, as governmental corruption and ethnic animosity have prevented Afghanistan from becoming a stable country. By 2004, the Taliban had regrouped, operating from safe havens in Pakistan tribal areas (see Chapter 12).

Afghanistan's new government has had to rely on the military power of the North Atlantic Treaty Organization (NATO)-led International Security Assistance Force (ISAF). But despite its more than 58,000 troops, the ISAF has been unable to stop the Taliban resurgence (Figure 10.23). By 2008, many observers feared that Afghanistan was on the verge of becoming a failed state, and in 2009, the new administration of the United States responded by sending an additional 17,000 troops to the country. Early 2010 saw major offensives against Taliban strongholds in the Kandahar area. U.S. forces have also targeted high-level Taliban leaders, often by bombing their compounds. This strategy has weakened the Taliban command structure but has also resulted in large numbers of civilian casualties, reducing local support for the war effort. Although many areas of northern Afghanistan are relatively secure, the south remains an extremely dangerous and battle-scarred war zone.

Global Dimensions of Central Asian Geopolitics

As the previous discussion indicates, Central Asia has recently emerged as a key arena of geopolitical tension. Vying for power and influence in the region are a number of important countries, including China, Russia, Pakistan, India, Iran, Turkey, and the United States. In addition, the revival of Islam has generated international geopolitical repercussions.

A Continuing U.S. Role Afghanistan is not the only Central Asian country to have experienced a substantial U.S. armed presence. After 9/11, the United States established military operations in Uzbekistan, Tajikistan, and Kyrgyzstan (see "Exploring Global Connections: Foreign Military Bases in Central Asia"). These countries were eager to host the U.S. military because they feared that Islamic fundamentalist movements could overthrow their governments. Already by the late 1990s, the Islamic Movement in Uzbekistan (IMU), a group dedicated initially to the creation of an Islamic state in the Fergana Valley, was engaging the armies of Uzbekistan, Tajikistan, and Kyrgyzstan. As a result, military budgets increased throughout the region, straining governmental finances but also allowing Central Asian governments to get an upper hand on the IMU. The IMU suffered a major blow in early 2007 when a contingent of its fighters operating out of northern Pakistan had a falling-out with local Pashtun tribes, resulting in battles that left over 300 Uzbek militants dead. By 2010, experts had concluded that the IMU was no longer operating in former Soviet Central Asia

Since 2005, the influence of the United States has declined across much of the former Soviet portion of Central Asia. Mongolia, however, has sought closer ties with the United States (as well as with Japan), due largely to its concerns about falling under the influence of China or Russia.

Relations with China and Russia After the breakup of the Soviet Union, relations between the newly independent countries of Central Asia and China remained tense for a number of years. China objected to the boundary lines separating it from Tajikistan and Kyrgyzstan and asked for adjustments. These boundary disputes were peacefully settled in the early years of the new millennium, with Kyrgyzstan ceding 222,400 acres (90,000 hectares) of disputed land to China. China subsequently initiated a series of diplomatic maneuvers designed to quell tensions, fight Islamic separatism, and gain greater access to the natural resources of Central Asia.

After the fall of the Soviet Union, Russia continued to regard all of the former Soviet territories of Central Asia as lying within its sphere of influence, and as such has resented U.S. military initiatives. Russia has, moreover, retained or built military bases of its own in both Tajikistan and Kyr-

Figure 10.23 War in Afghanistan Warfare continues in Afghanistan as troops from the United States and other countries attempt to root out remnants of the Taliban and maintain order. Tense encounters with local residents often result when foreign soldiers search for weapons and insurgent fighters.

EXPLORING GLOBAL CONNECTIONS Foreign Military Bases in Central Asia

After the downfall of the Soviet Union, Russia sought to keep its strategic influence in Central Asia's newly independent countries. The Russian military has thus maintained installations in Azerbaijan and Kazakhstan, while in Tajikistan it retains a division of ground forces as well as a large airbase. After the United States built the Manas Air Base in Kyrgyzstan in 2001 to support its war in Afghanistan (Figure 10.5.1), Russia responded by constructing a new air force station only a few miles away from the American facility.

Outside Afghanistan, the United States has been less successful than Russia in maintaining its military influence. In 2001, the U.S. military established a sizable air base in eastern Uzbekistan, but in 2005, Uzbekistan ordered it to be shut down after the U.S. government harshly criticized that country for its violent response to an anti-government protest. In February 2009, the parliament of Kyrgyzstan voted overwhelmingly to expel U.S. forces from Manas Air Base. This move came after Russia agreed to provide Kyrgyzstan with $2 billion in new loans and $150 million in direct financial aid. Four months later, however, the government of Kyrgyzstan agreed that the United States could continue to use the facility in exchange for tripling its rent payments and referring to it as a "transit center" rather than an "air base." After the collapse of Kyrgyzstan's government in April 2010, some of the country's new leaders expressed anger at the United States, arguing that the U.S. had tolerated the corruption of the previous regime so long as it allowed the base to operate. After some negotiations, however, Kyrgyzstan agreed that the Manas facility could continue to operate for the time being.

Pakistan's military has also long been interested in extending its power into Central Asia, leading it to support the Taliban's efforts to rule Afghanistan before September 11, 2001. Pakistan's maneuvering, in turn, has prompted India to seek influence in the region. In 2004, the governments of India and Tajikistan ratified an agreement allowing India to construct an air base 80 miles (130 kilometers) south of Tajikistan's capital city, and only 1.2 miles (2 kilometers) from the Afghan border. Fully operational since 2007, the Farkhor Air Base, India's only foreign military installation, is thought to hold roughly dozen MiG-29 fighter-bombers. Pakistan, not surprisingly, sees this base as part of an Indian effort to surround Pakistan with hostile forces.

The expulsion of U.S. military bases from Uzbekistan and especially Kyrgyzstan has complicated the war effort in Afghanistan, as have continuing Indian–Pakistani competition and maneuvering in the region. Historians describe competing British and Russian efforts to gain influence in Central Asia in the late 1800s as the "Great Game"; some observers now argue that a "New Great Game" in the region will result in complex geopolitical tensions for some time to come.

Figure 10.5.1 Manas Transit Center A soldier is seen near the entrance of the Manas Transit Center, formerly called the Manas Air Base, in Kyrgyzstan.

gyzstan. Economics and infrastructure also tie Russia to the former Soviet Central Asian sphere. The main transportation line linking European Russia to central Siberia, for example, cuts across northern Kazakhstan, while former Soviet Central Asia's rail and pipeline links to the outside world are still largely oriented toward Russia. Kazakhstan also contains the huge Baikonur Cosmodrome, Russia's main site for rocket launches.

In the early years of the new millennium, most Central Asian leaders concluded that the potential economic and political advantages to be gained by cooperation with Russia and China outweighed the disadvantages. The major consequence of this attitude was the formation of the **Shanghai Cooperation Organization** (or SCO, commonly referred to as the "Shanghai Six"), composed of China, Russia, Kazakhstan, Kyrgyzstan, Tajikistan, and Uzbek-

Figure 10.24 Meeting of the Shanghai Cooperation Organization The leaders of the six countries that make up the Shanghai Cooperation Organization have gathered together to discuss their common concerns. Pictured here are the leaders (or former leaders) of Russia, China, Kazakhstan, Kyrgyzstan, Tajikistan, and Uzbekistan.

istan (Figure 10.24). The SCO seeks cooperation on such security issues as terrorism and separatism, aims to enhance trade, and serves as a counterbalance against the United States. Although not a member, Mongolia does have "observer" status in the SCO (as do Iran, Pakistan, and India); Turkmenistan, on the other hand, has rejected membership. By 2010, some observers were arguing that economic competition between Russia and China would weaken the Shanghai Cooperation Organization, reducing Russian influence in the region. They noted that none of the Central Asian countries supported Russia when it invaded Georgia in 2008, just as none has recognized the independence of the breakaway republics of South Ossetia and Abkhazia.

The Roles of Iran, Pakistan, and Turkey Iran is also interested in the new republics of Central Asia. It is a major trading partner, and it offers a good potential route to the ocean. Since the completion of a rail link between Iran and Turkmenistan in the late 1990s, some of Central Asia's global trade has been reoriented toward Iran's ports. Iran's cultural links with the region are old and deep, particularly in Tajikistan and northern Afghanistan. The Iranian government has gained some influence in Afghanistan, especially in the western region of Herat and among the Shiite Hazaras. Iran's relations with Azerbaijan, on the other hand, remain tense, in part because many Azerbaijanis refer to northwestern Iran—an Azeri-speaking region—as "Southern Azerbaijan."

Pakistan has also strived to gain influence in Central Asia, hoping that pipelines will eventually carry Central Asian oil and natural gas to its new deep-water port at Gwadar. Any such pipelines, however, would have to pass through the rugged topography and perilous political environment of Afghanistan. During the Taliban period, Pakistan enjoyed close relations with Afghanistan's government. The aftermath of 9/11 thus put Pakistan in a serious bind. In response to both U.S. pressure and promises, Pakistan joined the anti-Taliban coalition and supplied the U.S. military with valuable intelligence. The subsequent retreat of the Taliban into northwestern Pakistan, however, has resulted in extremely tense relations between the two countries.

Turkey's connections with Central Asia are potentially close. Most Central Asians speak Turkic languages, while Turkey also offers itself as the model—contrary to those of Iran and Pakistan—of the modern, secular state of Muslim heritage. It is unclear, however, whether the Turkish system can be exported easily, even to other Turkic-speaking states. But if Central Asia were to experience substantial Western-oriented economic growth, it seems likely that its relations with Turkey would strengthen.

ECONOMIC AND SOCIAL DEVELOPMENT: Abundant Resources, Struggling Economies

By most measures, Central Asia is one of the poorer regions of the world. Afghanistan in particular stands near the bottom of almost every list of economic and social indicators. Other Central Asian countries, however, have enjoyed relatively high levels of health and education, a legacy of the social programs enacted by their former communist regimes. These same governments, however, built inefficient economic systems, and after the fall of the Soviet Union, western Central Asia experienced a spectacular economic decline. Although much of the region is now economically expanding, growth has been largely based on the extraction of natural resources, particularly oil and natural gas.

Economic Development in Central Asia

Soviet economic planners sought to spread the benefits of economic development widely across their country. This required building large factories even in remote areas such as Tajikistan regardless of the costs involved. Such Central Asian industries relied heavily on subsidies from the Soviet central government. When those subsidies ended, the industrial base of the region began to collapse, leading to plummeting living standards. But as is true elsewhere in the former Soviet Union, certain well-connected individuals have grown very wealthy since the fall of communism. The growing gap between the rich and the poor remains a major concern across most of Central Asia.

The Postcommunist Economies As Table 10.2 shows, no Central Asian country or region could be considered prosperous by global standards. Kazakhstan generally stands as the most developed, and it probably has the best prospects. Kazakhstan has two of the world's largest underutilized deposits of oil and natural gas, the vast Tengiz and Kashagan fields in the northeastern Caspian Basin (Figure 10.25), as well

TABLE 10.2 Development Indicators

Country	GNI per Capita, PPP 2008	GDP Average Annual % Growth 2000–08	Human Development Index (2007)[1]	Percent Population Living Below $2 a Day	Life Expectancy (2010)[2]	Under Age 5 Mortality Rate (1990)	Under Age 5 Mortality Rate (2008)	Gender Equity[3] (2008)
Afghanistan	1,110	11.8	0.352		44	260	257	58
Azerbaijan	7,770	18.1	0.787	<2	72	98	36	98
Kazakhstan	9,710	9.5	0.804	<2	69	60	30	98
Kyrgyzstan	2,150	4.4	0.710	27.5	68	75	38	100
Mongolia	3,470	7.8	0.727	13.6	67	98	41	104
Tajikistan	1,860	8.6	0.688	50.8	67	117	64	91
Turkmenistan	6,120	14.5	0.739	49.6	65	99	48	
Uzbekistan	2,660	6.6	0.710		68	79	38	98

Source: World Bank, World Development Indicators, 2010 World Data Sheet, 2007; gender equity data from Millennium Development Goals.

[1] *United Nations, Human Development Report, 2009.*

[2] *Population Referene Bureau, World Population Data Sheet, 2010.*

[3] *Gender Equity—Ratio of female to male enrollments in primary and secondary school. Numbers below 100 have more males in primary/secondary school, numbers above 100 have more females in primary/secondary school.*

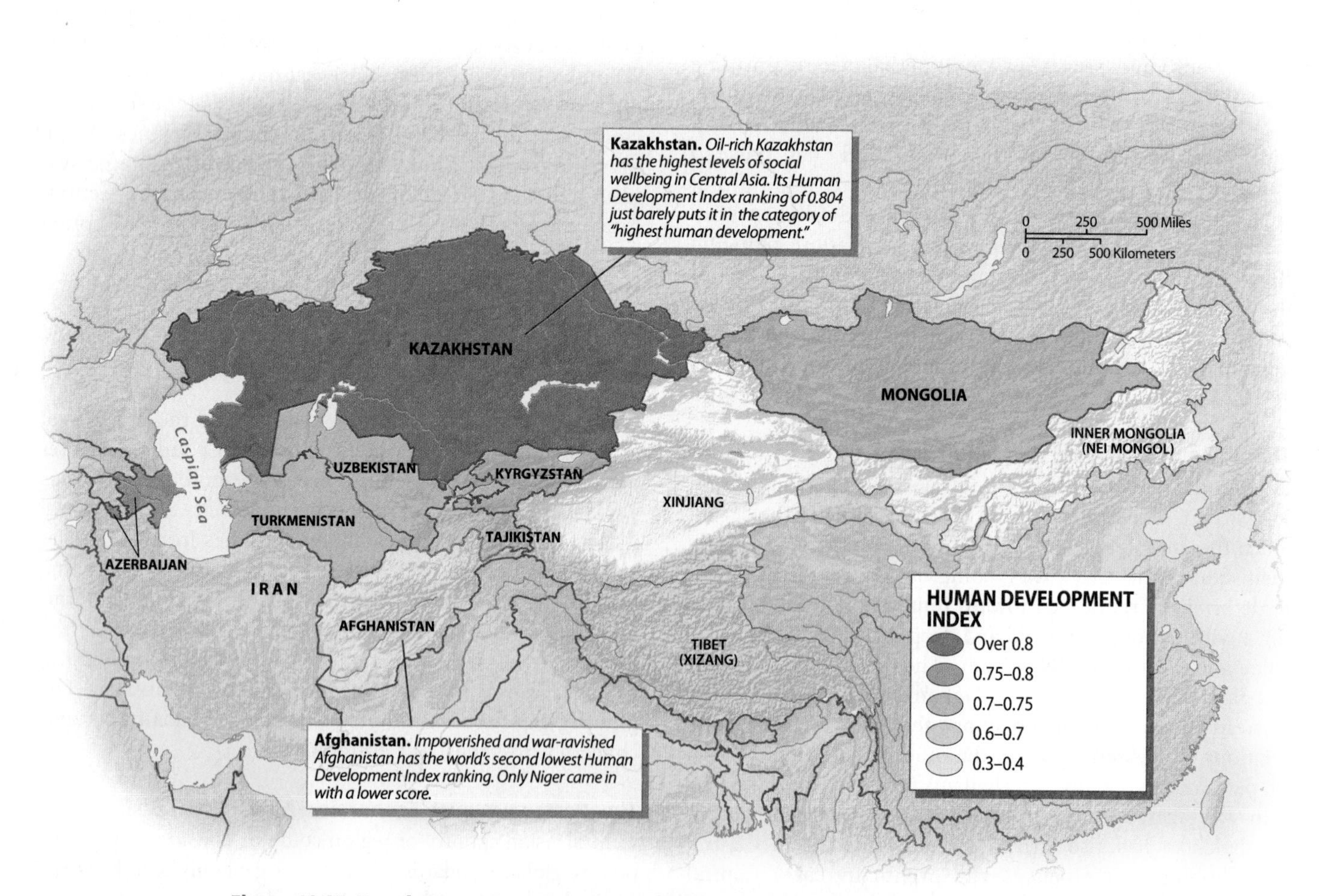

Figure 10.25 Development Issues: Human Development in Central Asia Levels of social development vary significantly across Central Asia. As this Human Development Index map shows, Afghanistan ranks as one of the least developed countries in the world, whereas both Azerbaijan and Kazakhstan rank relatively high.

Figure 10.26 Oil Development in Azerbaijan Although oil has brought a certain amount of wealth to Azerbaijan, it has also resulted in extensive pollution and visual blight. Because most of the petroleum is located either near or under the Caspian Sea, this sea—actually the world's largest lake—is now ecologically endangered.

as sizable deposits of other minerals. Kazakhstan has signed agreements with Western oil companies to exploit its oil reserves, and has received major investments from China. Between 2001 and 2008, Kazakhstan's oil and gas wealth led to one of the world's fastest growing economies, with annual growth averaging more than 8 percent. Like much of the rest of the world, Kazakhstan experienced an economic decline in 2009, but growth resumed in 2010.

Owing to its sizable population, Uzbekistan has the second largest economy in the region. Uzbekistan did not decline as sharply as its neighbors after the fall of the Soviet Union, largely because it retained many aspects of the old command economy (that is, an economy run by governmental planners rather than by private firms responding to the market). Uzbekistan remains the world's second largest exporter of cotton, and it has significant gold and natural gas deposits. However, environmental degradation threatens cotton production, and both inefficient state management and widespread corruption continue to hamper production.

Kyrgyzstan, in contrast, moved aggressively after the fall of the Soviet Union to privatize former state-run industries. Along with Mongolia, it is Central Asia's only member of the World Trade Organization (WTO). But Kyrgyzstan's economy is heavily agricultural, few of its industries are competitive, and political strife has discouraged foreign investment. Kyrgyzstan does, however, enjoy the largest supply of freshwater in Central Asia, which will likely become increasingly valuable in years to come, and its mineral reserves are substantial. Gold exports are particularly important; when gold exports slump, as they did in 2009, Kyrgyzstan experiences economic hardship. Political instability in early 2010 also hurt the country's economy.

Turkmenistan also has a substantial agricultural base, due mainly to Soviet irrigation projects, and it remains a major cotton exporter. It retains a state-run economy and has resisted pressure for economic liberalization. Government planners hope that the development of new oil and gas fields will bring prosperity. China has invested heavily, completing the Central Asia-China Gas Pipeline, which runs from Turkmenistan through Uzbekistan and Kazakhstan, in December 2009. In the same year, China lent Turkmenistan $3 billion to develop new natural gas fields.

Central Asia's oldest fossil fuel industry is located in Azerbaijan (Figure 10.26). Azerbaijan has attracted a great deal of international interest and investment, promising to revitalize its oil industry. Through the 1990s, however, its economy responded slowly, and Azerbaijan remains a poor country. In 2006, however, its oil-fueled economy reportedly grew by 34 percent, the fastest rate of expansion in the world. Even in the recession year of 2009, Azerbaijan's economy managed to expand by 9.3 percent. Critics, however, say that the country has not enacted adequate market reforms and remains far too dependent on its oil industry.

The most economically troubled of the former Soviet republics is Tajikistan. With a per capita GNI of only some $1,860, Tajikistan is poor indeed. It is burdened by its remote location, rugged topography, a lack of natural resources, and political strife. Almost half of Tajikistan's labor force works abroad, mostly in Russia and Kazakhstan, supporting the local economy through their remittances. Tajikistan's government hopes to revive its fortunes by opening new transportation routes to China and by further developing its generation of hydroelectricity. A large new dam was finished in 2009, and others are under construction. Enhanced electricity supplies are allowing

GLOBAL TO LOCAL The Mongolian Cashmere Industry Readjusts

For centuries, Mongolia has produced high-quality cashmere textiles derived from the soft undercoat of a particular variety of goat. Areas with particularly cold winters—such as Mongolia—produce especially soft and luxurious fibers. Herders in southern Mongolia have relied for hundreds of years on cashmere for their cash needs, raising small herds of goats in a sustainable manner and laboriously combing out the underfur of their animals each spring (Figure 10.6.1).

From the 1920s until 1990, most of Mongolia's goat herds were collectively owned, the cashmere fibers being transformed into textiles in state-owned factories. Everything changed, however, when communism collapsed. As unemployment skyrocketed, tens of thousands of urban Mongolians returned to the countryside to herd goats, figuring that they could always earn cash by selling cashmere. Lacking experience, however, many of these new herders stocked too many goats, leading to serious overgrazing hence the degradation of pasturelands. When a series of dry summers followed by brutally cold winters hit in 2000 and 2001, millions of goats died, forcing many of the new cashmere producers back into the urban economy.

Figure 10.6.1 Mongolia's Cashmere Industry A girl holding a young goat in southern Mongolia.

Economic as well as environmental disorder has plagued Mongolia's cashmere industry after the fall of communism. In the early 1990s, Mongolia's government sought to earn more foreign exchange by developing a larger textile industry, and was distressed that many herders were sending their cashmere combings to Chinese mills for processing. In 1994, Mongolia banned the export of raw cashmere, resulting in the rapid opening of 50 new Mongolian textile factories. In the following year, however, the Asian Development Bank informed Mongolia that essential new loans would be held back until the ban—regarded by the bank as anticompetitive—was dropped. Mongolia soon complied, putting its newly revived cashmere textile industry in deep trouble.

To revive its cashmere trade, Mongolia needed a new strategy. One problem was that its own textile styles were out-of-date, tending toward neon colors and stodgy designs from the 1950s. In response, a number of local firms joined together under the banner of the Mongolian Fibermark Society and sought assistance from international development organizations. When the United States Agency for International Development (USAID) began an economic reform initiative in Mongolia in 2003, the cashmere producers eagerly sought its support. By 2005, USAID was helping connect Mongolian producers with buyers in wealthy countries and providing information about the kinds of cashmere textiles that would satisfy demanding consumers in North America, Europe, and East Asia.

Within a year, these efforts appeared to be bearing fruit. In early 2006, the Mongolian Fibermark Society's exhibition at a Las Vegas textile trade fair generated considerable interest, as did an exhibit in New York in early 2007. It is still difficult, however, for Mongolia's cashmere textile manufacturers to compete with those of China, just as it is still hard for Mongolia's goat herders to earn a consistent income in an area noted for its periodic droughts and harsh winters. In April 2010, another cause for concern emerged when China announced that it had cloned goats genetically engineered to produce woollier coats.

Tajikistan to expand its aluminum smelting, one of its few competitive industries.

Mongolia, although never part of the Soviet Union, was a close Soviet ally run by a communist party. It too suffered an economic collapse in the 1990s when Soviet subsidies came to an end. (See "Global to Local: The Mongolian Cashmere Industry Readjusts") Mongolia thus emerged in the postcommunist period as a poor country. During the early 2000s, however, major mining investments by both Chinese and Western firms resulted in rapid economic expansion. In early 2010, international companies reached an agreement with the Mongolian government to develop the massive Oyu Tolgoi copper-gold project; with production scheduled to begin in 2013, it will be one of the world's largest mines. The mining boom, however, has generated problems of its own. Many Mongolians fear that their Chinese mining companies have too much power. Small-scale, unlicensed "ninja miners," moreover, have caused a significant amount of local environmental degradation.

The Economy of Western China The Chinese portions of Central Asia did not suffer the same economic crash that visited other parts of the region with the fall of the Soviet Union. China has one of the world's fastest-growing economies, although its centers of dynamism are located in the distant coastal zone. Still, even the more remote parts of the country have experienced a significant degree of development in recent years. From 2003 to 2008, Inner Mongolia's economy expanded at a scorching annual rate of around 20 percent, owing in large part to its extensive deposits of rare earth minerals and other valuable mineral resources.

Tibet in particular remains burdened by poverty. Most of the plateau is relatively cut off from the Chinese economy and is even more isolated from the global economy. Hoping to reduce Tibetan separatism, China has been investing large amounts of money in infrastructural projects. In 2006, it inaugurated the monumentally expensive Qinghai-Tibet Railway linking Tibet's capital city of Lhasa to the rest of the country. Reaching heights of over 16,400 feet (5,000 meters), the trains on this railway have to be equipped with supplemental oxygen. As a result of this and other transportation projects, Tibet's tourism economy is booming. Many Tibetans, however, argue that most of the benefits are flowing to Han Chinese immigrants rather than to the indigenous peoples of the region (Figure 10.27). Ethnic unrest in 2008 resulted in a marked slowdown in economic expansion.

Most observers agree that Xinjiang has tremendous economic potential. It boasts significant mineral wealth, including most of China's oil and natural gas reserves, and it has benefited from infrastructural development projects over the past several decades. The recent completion of oil and natural gas pipelines from Kazakhstan and Turkmenistan have also bolstered the region's petrochemical industries. But as is the case in Tibet, many of the indigenous peoples of Xinjiang believe that the wealth of their region is being monopolized by the Chinese state and the Han Chinese immigrants. Some point to the Xinjiang Production and Construction Corps, a quasi-military firm that runs hundreds of factories, farms, schools, and hospitals. This huge company was initially formed by the army that rejoined Xinjiang to China in 1949, and it is still run almost entirely by Han Chinese.

Figure 10.27 Chinese Business in Lhasa, Tibet An elderly Tibetan woman peers into a hair salon in Lhasa, Tibet, in 2006. As Han Chinese immigrants stream into Tibet's cities, rapid cultural change and economic modernization leave many Tibetans behind.

Economic Misery in Afghanistan Afghanistan is Central Asia's poorest country. Since the late 1970s, it has suffered nearly continuous war, undermining virtually all economic endeavors. Even before the war it had experienced little industrial or commercial development. Its only significant legitimate exports are animal products, handwoven carpets, and a few fruits, nuts, and semiprecious gemstones.

Through all of these years of chaos, Afghanistan remained economically competitive in one area: the production of illicit drugs for the global market. According to the CIA, Afghanistan had emerged by 1999 as the world's largest producer of opium. The country's drug economy has recently grown more sophisticated, as Afghan opium is increasingly processed into heroin within the country. From 2007 to 2009, however, opium production did decline, but pervasive corruption will make any further reductions very difficult.

Since 2003, the economy of Afghanistan has experienced some growth, although much of the gains have come through foreign assistance. From 2001 to 2009, international donors delivered roughly $15 billion to the country. Afghanistan does have substantial mineral deposits, however, and in 2008 China agreed to invest $2.8 billion in a massive copper mining scheme. Genuine economic development in Afghanistan seems unlikely, however, unless the country becomes politically secure.

Central Asian Economies in Global Context Afghanistan—despite its poverty and relative isolation—is thoroughly embedded in the global economy, albeit through illicit products. The overland flow of Afghan opium and heroin to Europe implicates several other Central Asian countries, especially Tajikistan, in the international drug economy. But with the exceptions of the drug trade and the oil and natural gas industries, the overall extent of globalization in Central Asia remains low (Figure 10.28).

In the former Soviet area, the most important international connections remain with Russia, although economic ties with China are rapidly expanding. In the early post-Soviet period, both Russia and the newly independent Central Asian countries were ambivalent about their continuing economic relationships. Disentangling Central Asian economies from that of Russia, however, proved difficult, largely because of infrastructural linkages inherited from the Soviet period. In 2007, two-thirds of Turkmenistan's natural gas exports flowed through Russia's state-owned Gazprom pipelines.

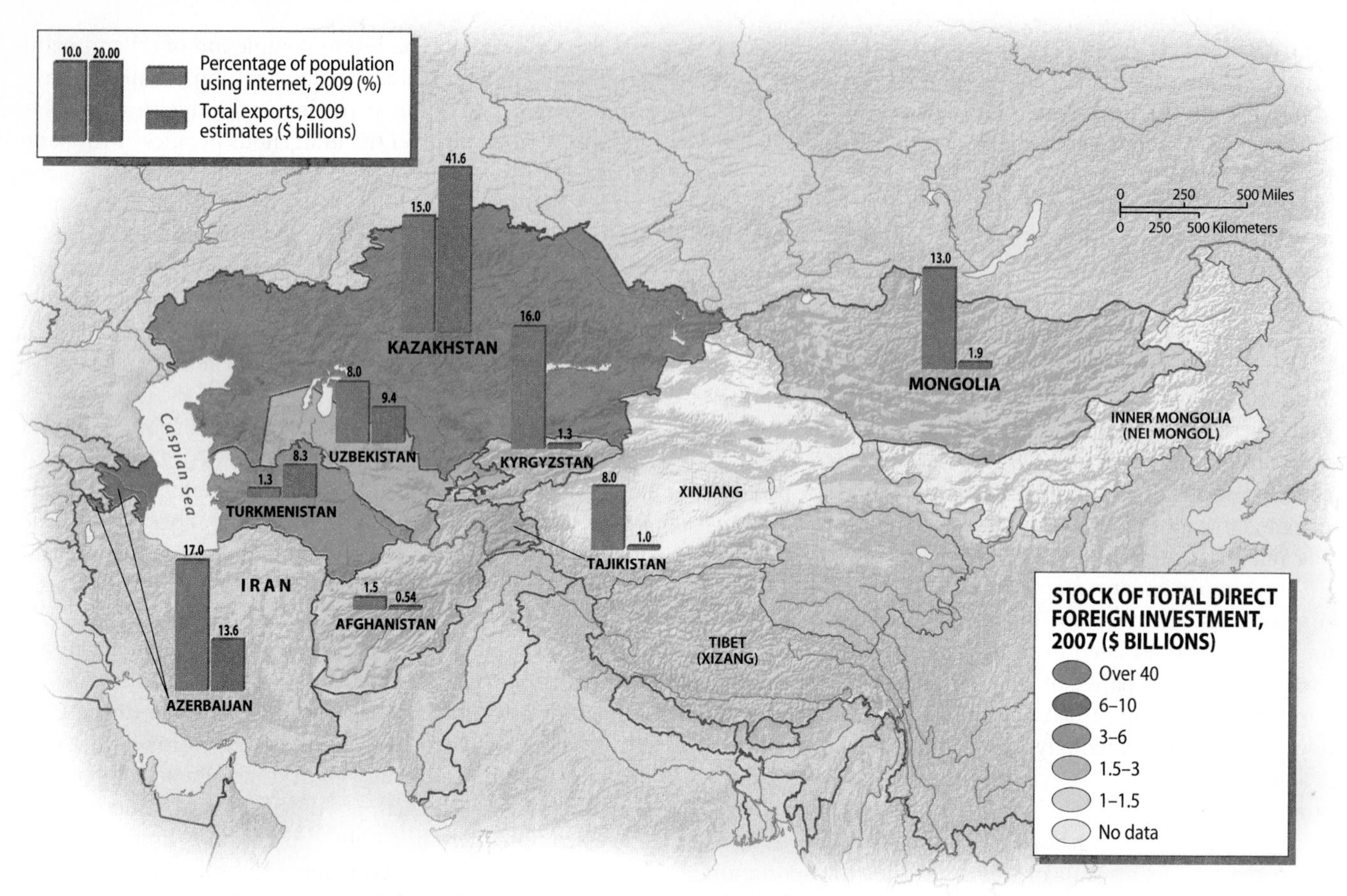

Figure 10.28 Global Linkages Although Azerbaijan and Kazakhstan are relatively well connected to the rest of the world through their exports of oil and natural gas, the other countries of Central Asia remain somewhat isolated from the global economy.

Since the creation of the SCO, moreover, Uzbekistan, Kazakhstan, Kyrgyzstan, and Tajikistan have eagerly sought closer linkages with both Russia and China. Connections to China were enhanced in 2006 with the opening of the Kazakhstan–China oil pipeline and again in 2009, with the completion of the Central Asia-China Gas Pipeline.

The United States and other Western countries, as well as India, are also drawn to the area by its oil and natural gas deposits. Many large oil companies operate in Azerbaijan and Kazakhstan. The delivery of Central Asian oil to the west was enhanced in 2006 with the opening of the extraordinarily expensive Baku-Tbilsi-Ceyhan Pipeline—the second longest in the world—which runs from the Caspian Sea to the Turkish Mediterranean at the port of Ceyhan. If the current Nabucco Pipeline plans come to fruition, Azerbaijan will be exporting large quantities of natural gas to Europe by 2015.

Social Development in Central Asia

Social conditions in Central Asia vary more than economic conditions. In the former Soviet territories, levels of health and education remain fairly high but are generally declining. Not surprisingly, Afghanistan stands at the bottom of the scale by every measure.

Social Conditions and the Status of Women in Afghanistan Social conditions in Afghanistan are no better than economic circumstances, although reliable information is difficult to obtain. As can be seen in Table 10.2, the average life expectancy in the country is a mere 44 years, one of the lowest figures in the world. Infant and childhood mortality levels remain extremely high. Not only does Afghanistan suffer from constant warfare, but its rugged topography hinders the provision of basic social and medical services. Illiteracy is commonplace and is notably gender-biased. Afghanistan's 21 percent adult female literacy figure is one of the lowest in the world.

Women in most parts of Afghanistan—and especially in Pashtun areas—lead highly constrained lives. In many areas they must completely conceal their bodies, including their faces, when venturing into public. Such restrictions intensified in the 1990s. Dress controls were strictly enforced where the Taliban gained authority. In most areas, Taliban forces prevented women from working, attending school, and often even from obtaining medical care. The assault on women by the Taliban did not go uncontested in Afghanistan. Many women risked their lives to document the horrific abuses that Afghan females were suffering. While Taliban leaders argued that they were upholding Islamic orthodoxy, their opponents contended that they were actually enforcing an extreme version of Pashtun customary law.

The fall of the Taliban brought temporary joy to most of the women of Afghanistan. Many promptly uncovered their faces and began working—or seeking work. The new Afghan constitution proclaimed that henceforth men and women would enjoy equal rights. But in most parts of the country, little actually changed with the coming of the new regime. Social customs continued to force most women into arranged marriages followed by domestic seclusion. Indeed, in some areas of the country the upsurge in both crime and political violence have made the position of women increasingly insecure. In recent years, Taliban forces have physically attacked several girls' schools, demonstrating their determination to thwart any movement toward gender equity. According to some Afghan women's groups, over 70 percent of women in the country face forced marriages, and roughly one third suffer physical, psychological, or sexual violence.

Social Conditions in the Former Soviet Republics and Mongolia Compared to women in Afghanistan, those in the former Soviet portion of Central Asia enjoy a high social position. Traditionally, women had much more autonomy among the northern pastoral peoples than among the farmers and city dwellers of the south, but under Soviet rule the position of women everywhere improved. In the former Soviet republics today, adult women's educational rates are not far behind those of men, and women are well represented in the workplace. Some evidence suggests, however, that the position of women has recently declined in many areas. In Kazakhstan, leaders have blamed feminism for the country's low birthrate, and in much of the region newly rich men are increasingly practicing polygamy. As education becomes more expensive, moreover, girls are increasingly staying at home. Women are also trafficked from the poorer countries of the region, particularly Uzbekistan and Tajikistan, into prostitution in Russia, the United Arab Emirates, and elsewhere.

The figures in Table 10.2 reveal generally favorable levels of social welfare overall for the former Soviet zone. This is especially notable when one considers the dismal state of the region's economy—or contrasts conditions with those of neighboring Afghanistan. Tajikistan, with a per capita GNI of only $1,860 has a female life expectancy of 69 years as well as almost universal adult literacy. The social successes of Tajikistan and its northern and western neighbors reflect investments made during the Soviet period. Health and educational facilities have declined in the face of economic and political turmoil. Certainly the region's relatively high levels of infant and childhood mortality do not bode well. Whereas Tajikistan might hope that its educated workforce will attract foreign investment, its remote location and continued instability make such a scenario unlikely.

Somewhat similar conditions prevail in Mongolia, although the preponderance of its women in professional positions is striking. This is a legacy of both the former socialist system and of traditional Mongol culture, which is noted for its female domestic authority. In 2000, one survey found that women made up 60 percent of Mongolia's lawyers, 77 percent of its doctors, and 84 percent of its university graduates (Figure 10.29). But despite such educational and professional advances, Mongolia's women have gained little political power.

Social Conditions in Western China Although China as a whole has made significant progress in health and education, many reports suggest that the indigenous peoples of Tibet and Xinjiang have been left behind. Some sources claim that more than half of the non-Han people of Xinjiang and Tibet remain illiterate. Chinese sources, on the other hand, maintain that illiteracy has been declining rapidly in both areas. In 2008, China's official media outlets reported that over 95 percent of the young people of Tibet had learned to read and write.

China's minority peoples have, however, been granted certain exemptions from the country's strict population control measures. But there are many reports from Xinjiang and elsewhere of forced sterilization and abortions, adding to the local political tensions.

Figure 10.29 Mongolian Women A female technician examines slides under a microscope at the Agricultural Institute in Ulaanbaatar, Mongolia.

Summary

- Central Asia, long hidden by Russian domination, has recently reappeared on the map of the world. Environmental problems, among others, have brought the region to global attention. The destruction of the Aral Sea is one of the worst environmental disasters the world has experienced, and many other lakes in the region have experienced similar problems. Desertification has devastated many areas, and the booming oil and gas industries have created their own environmental disasters.
- Large movements of people in Central Asia have attracted global attention. The migration of Han Chinese into Tibet and Xinjiang is steadily turning the indigenous peoples of these areas into minority groups. The migration of Russian speakers out of the former Soviet areas of Central Asia has also resulted in major transformations. People are currently moving from the poorer countries of the former Soviet zone to oil-rich Kazakhstan. In Afghanistan, war and continuing chaos have generated large refugee populations.
- Religious tension has recently emerged as a major cultural issue through much of western Central Asia. Radical Islamic fundamentalism remains a potent force in southern and western Afghanistan, and to a somewhat lesser extent in the Fergana Valley of Uzbekistan, Tajikistan, and Kyrgyzstan. Although moderate forms of Islam are prevalent through most of the region, Central Asian leaders have used the threat of religiously inspired violence to maintain repressive and antidemocratic policies.
- While China maintains a firm grip on Tibet and Xinjiang, the rest of Central Asia has emerged as a key area of geopolitical competition. Russia, the United States, China, Iran, Turkey, Pakistan, and India all contend for influence. Central Asian countries have attempted to play these powers against each other in order to maintain or increase their world status. Overall, political structures throughout the region remain largely authoritarian, although popular uprisings have brought down two governments in Kyrgyzstan.
- The economies of Central Asia are gradually opening up to global connections, largely because of their substantial fossil fuel reserves. However, Central Asia is likely to face serious economic difficulties for some time, especially since the region is not a significant participant in global trade and has attracted little foreign investment outside the fossil-fuel sector. The growing of opium and the manufacture of heroin remain a serious problem in Afghanistan. Drug trafficking, moreover, has weakened legitimate economic activities in other countries of the region.

Key Terms

alluvial fan *(page 452)*
Bodhisattva *(page 463)*
exclave *(page 467)*
exotic river *(page 452)*
loess *(page 452)*
pastoralist *(page 453)*
Shanghai Cooperation Organization (SCO); "Shanghai Six" *(page 470)*
Taliban *(page 471)*
theocracy *(page 463)*
transhumance *(page 451)*

Review Questions

1. Describe the three major regions of Central Asia as defined by physical geography.
2. Why is lake level a particularly serious concern in Central Asia?
3. Why is the distribution of population in Central Asia so uneven? Why are certain areas virtually uninhabited?
4. Describe the different agricultural patterns that are found on (a) the plains of northern Kazakhstan, (b) the major river valleys of Uzbekistan, and (c) the Tibetan Plateau.
5. Why is much of Central Asia often called "Turkestan"? Why does this label not apply to the entire region?
6. How does religion act as a political force in Central Asia? How does this situation vary in different parts of the region?
7. Why is the United States concerned about the fossil fuel resources of Central Asia? How do transportation routes play into this concern?
8. What role does Russia currently play in Central Asian geopolitics? How has Russian power historically influenced the region?
9. How do social conditions (health, longevity, literacy, etc.) vary across the border between Afghanistan and Uzbekistan? Why are the disparities so pronounced?
10. If western Central Asia is so rich in natural resources, why did the region experience such pronounced economic decline in the 1990s?

Thinking Geographically

1. To what extent will Central Asia's continental location be a disadvantage in years to come? Is coastal access truly significant in determining a country's competitive position in the global economy? Explain.
2. Will the countries of former Soviet Central Asia be able to maintain their high levels of education now that they are independent? If so, will they be able to use education to their own economic advantage? Explain.

3. Is Russia's influence in Central Asia bound to decline now that Russia has no direct political authority in the area? Explain.
4. Which external connection will prove most important for Central Asia in years to come—those based on religion, language, economic ties, or geopolitical connections?
5. Is China's political control over eastern Central Asia justified? Should Han Chinese migration to the area be a concern to the United States? Explain.
6. Is Chinese control over Tibet a legitimate concern of U.S. foreign policy? Should the United States use trade sanctions to attempt to influence Chinese policy in the area? Explain.
7. How might the desiccation of Central Asia's great lakes best be addressed?

Regional Novels and Films

Novels

Chinghiz Aitmatov, *The Day Lasts More Than a Hundred Years* (1983, Indiana University Press)

Rinjing Dorje, *Tales of Uncle Tompa, The Legendary Rascal of Tibet* (1997, Barrytown Ltd.)

Mark Frutkin, *Invading Tibet* (1993, Soho Press)

James Hilton, *Lost Horizon* (1996, Morrow)

Khaled Hosseini, *The Kite Runner* (2003, Riverhead Books)

Amin Maalouf, *Samarkand: A Novel* (1998, Interlink)

Hilary Roe Metternich, ed., *Mongolian Folktales* (1996, Avery Press)

Films

Kandahar (2001, Iran)

Kundun (1997, U.S.)

Mongol (2007, Germany, Kazakhstan, Mongolia, and Russia)

Osama (2003, Afghanistan)

Pure Coolness (2007, Kyrgyzstan)

The Road under the Heavens (2006, Uzbekistan)

The Story of the Weeping Camel (2003, Mongolia)

Tulpan (2008, Kazakhstan)

Urga: Close to Eden (1991, Mongolia)

Bibliography

Adshead, S. A. M. 1993. *Central Asia in World History*. New York: St. Martin's Press.

Beckwith, Christopher. 2009. *The Empires of the Silk Road: A History of Central Eurasia from the Bronze Age to the Present*. Princeton: Princeton University Press.

Bissell, Tom. 2003. *Chasing the Sea: Lost Among the Ghosts of Empire in Central Asia*. New York: Pantheon.

Drompp, Michael. 1989. "Centrifugal Forces in the Inner Asian 'Heartland': History Versus Geography." *Journal of Asian History* 23, 135–55.

Ewans, Martin. 2002. *Afghanistan: A Short History of Its People and Politics*. New York: Perennial.

Jones, Seth. 2010. *In the Graveyard of Empires: America's War in Afghanistan*. New York: W.W. Norton.

Kobori, Iwao, and Glantz, Michael. 1998. *Central Eurasian Water Crisis: Caspian, Aral, and Dead Sea*. Tokyo: United Nations University Press.

Mahnovski, Sergei. 2007. *Economic Dimension of Security in Central Asia*. Santa Monica, CA: RAND Corporation.

Rashid, Ahmed. 2001. *Taliban: Militant Islam, Oil, and Fundamentalism in Central Asia*. New Haven, CT: Yale University Press.

Sinor, Denis, ed. 1990. *The Cambridge History of Early Inner Asia*. Cambridge, UK: Cambridge University Press.

MG™ Mastering Geography

Log in to www.masteringgeography.com for MapMaster™ interactive maps, geography videos, RSS feeds, flashcards, web links, an eText version of *Diversity Amid Globalization*, and self-study quizzes to enhance your study of Central Asia.

MapMaster™

MapMaster™ presents 13 Place Name and 13 Layered Thematic interactive maps to help students practice and master their geographic literacy, spatial reasoning, and critical thinking skills.

11

East Asia

The Balinghe Bridge in China's Guizhou province was opened to traffic in December 2009. Located in the poorest province of China, it is one of the world's highest bridges.

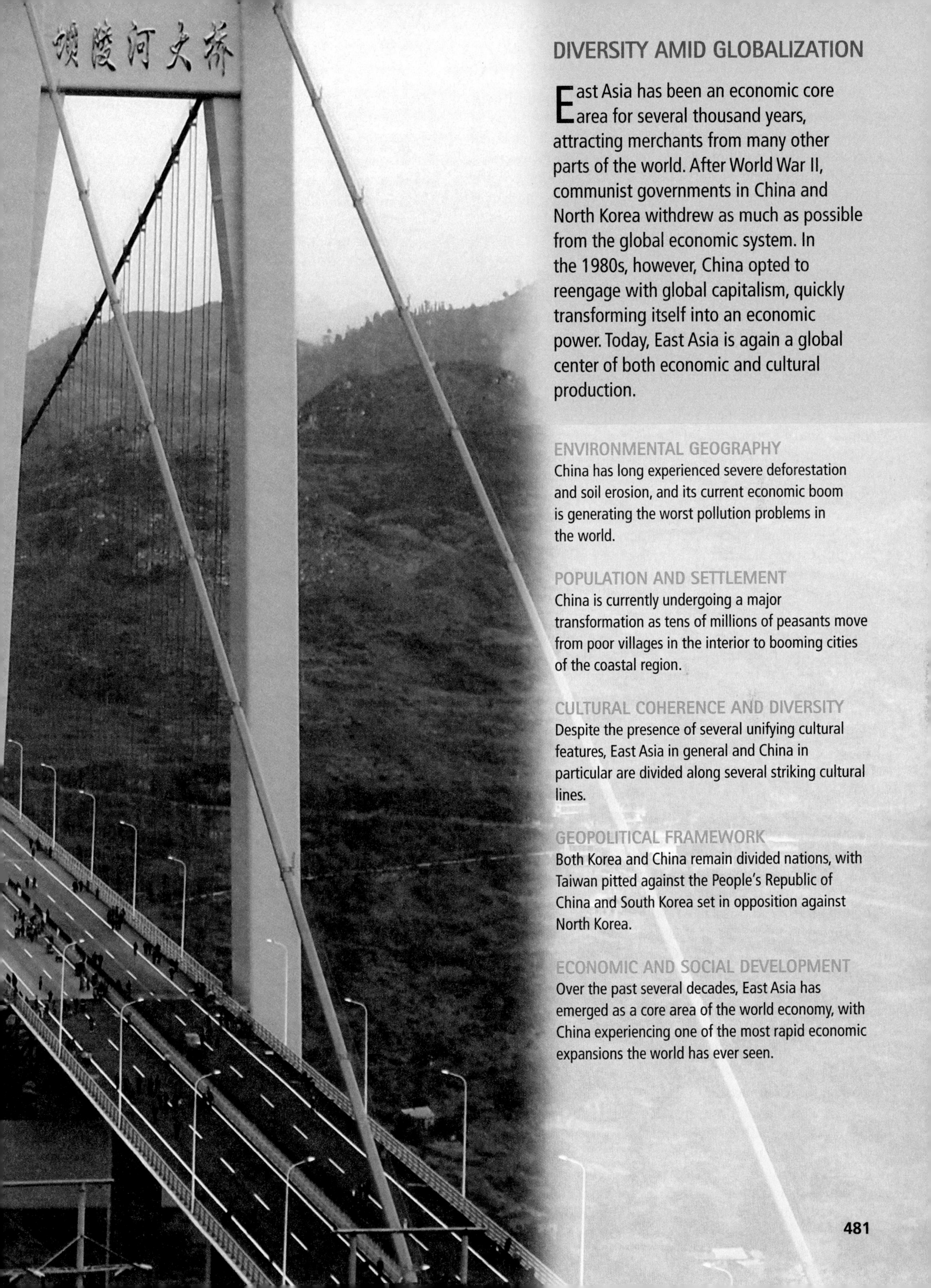

DIVERSITY AMID GLOBALIZATION

East Asia has been an economic core area for several thousand years, attracting merchants from many other parts of the world. After World War II, communist governments in China and North Korea withdrew as much as possible from the global economic system. In the 1980s, however, China opted to reengage with global capitalism, quickly transforming itself into an economic power. Today, East Asia is again a global center of both economic and cultural production.

ENVIRONMENTAL GEOGRAPHY

China has long experienced severe deforestation and soil erosion, and its current economic boom is generating the worst pollution problems in the world.

POPULATION AND SETTLEMENT

China is currently undergoing a major transformation as tens of millions of peasants move from poor villages in the interior to booming cities of the coastal region.

CULTURAL COHERENCE AND DIVERSITY

Despite the presence of several unifying cultural features, East Asia in general and China in particular are divided along several striking cultural lines.

GEOPOLITICAL FRAMEWORK

Both Korea and China remain divided nations, with Taiwan pitted against the People's Republic of China and South Korea set in opposition against North Korea.

ECONOMIC AND SOCIAL DEVELOPMENT

Over the past several decades, East Asia has emerged as a core area of the world economy, with China experiencing one of the most rapid economic expansions the world has ever seen.

East Asia, composed of China, Japan, South Korea, North Korea, and Taiwan (Figure 11.1), is the most populous region of the world. China alone is inhabited by more than 1.3 billion people, more than live in any other region of the world except South Asia. Although East Asia is historically unified by cultural features, in the second half of the 20th century it was divided ideologically and politically, with the capitalist economies of Japan, South Korea, Taiwan, and Hong Kong separated from the communist bloc of China and North Korea. Differences in levels of economic development also remained pronounced. As Japan reached the pinnacle of the global economy, much of China remained mired in extreme poverty.

Since the 1990s, however, divisions with East Asia have been reduced. While China is still governed by the Communist Party, it has embarked on a path of capitalist development. Business ties between its booming coastal zone and Japan, South Korea, and Taiwan have quickly strengthened. Animosity still exists between North Korea and South Korea and between China and Taiwan, but East Asia as a whole has witnessed a gradual reduction in political tension.

East Asia is now a core area of the world economy, and it is becoming a center of geopolitical power as well. Japan, South Korea, and Taiwan are among the world's key trading states. Tokyo, Japan's largest city, stands alongside New York and London as one of the financial centers of the globe, and is noted as both a consumer and producer of global culture. China has more recently emerged as a global trading power. Its coastal zone is now tightly integrated within global networks of information, commerce, and entertainment. China is also developing into a regional military power. This development has caused much concern among its neighbors—and among distant countries, such as the United States.

East Asia is easily marked off on the world map by the territorial extent of its constituent countries. Japan is an island country composed of four main islands and a chain of much smaller islands (the Ryukyus) that extends almost to Taiwan. Taiwan is basically a single-island country, but its political status is ambiguous since China claims it as part of its own territory. China itself is a vast continental state—the third most extensive in the world—that reaches well into Central Asia. It also includes the large island of Hainan in the South China Sea. Korea constitutes a compact peninsula between northern China and Japan, but it is divided into two sovereign states: North Korea and South Korea.

In political terms, such a straightforward definition of East Asia is appropriate. If one turns to cultural considerations, however, the issue becomes more complicated. The main cultural issue concerns the western half of China. This is a huge but lightly populated area; only about 5 percent of the residents of China live in the west. By certain criteria, only the populous eastern half of China (often called "China proper") fits into the East Asian world region. The indigenous inhabitants of western China are not Chinese by culture and language, and they have never accepted the religious and philosophical beliefs that give historical unity to East Asian civilization. In the northwestern quarter of China, called *Xinjiang* in Chinese, most indigenous inhabitants speak Turkic languages and are Muslim in religion. Tibet, in southwestern China, has a highly distinctive culture, and the Tibetans in general resent Chinese authority. Xinjiang and Tibet can thus be classified within both East Asia and Central Asia (covered in Chapter 10). In this chapter we will examine western China largely to the extent that it is politically part of China.

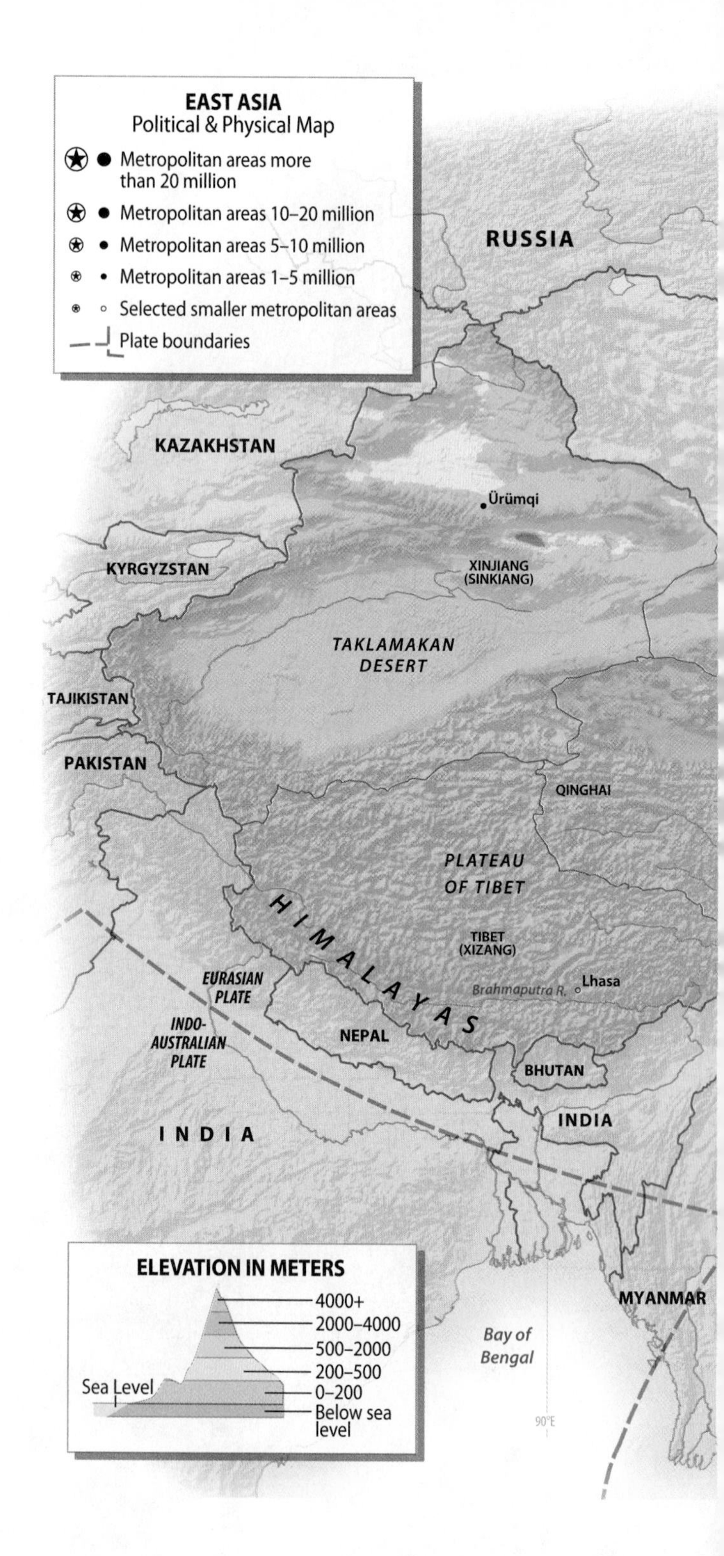

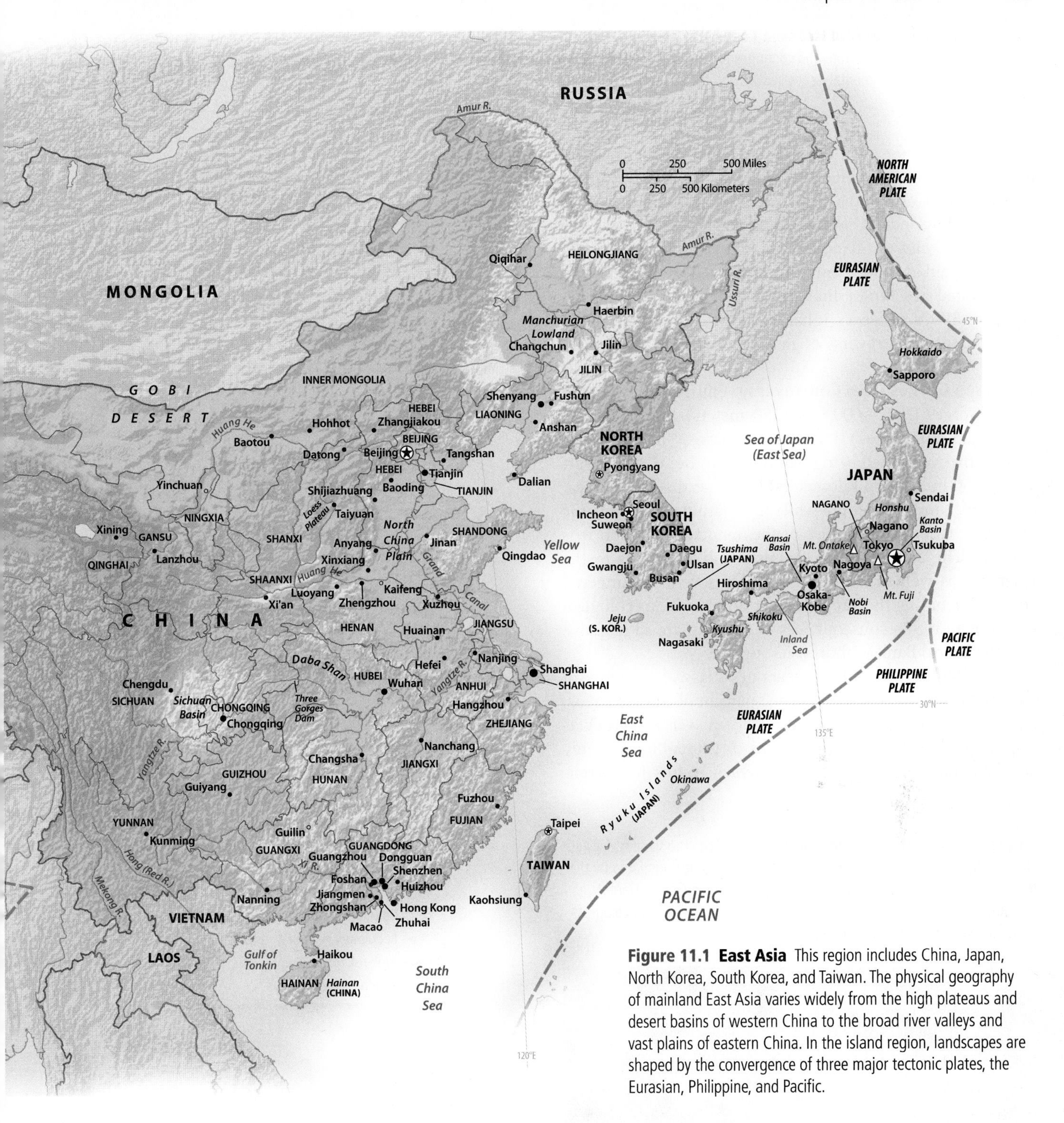

Figure 11.1 East Asia This region includes China, Japan, North Korea, South Korea, and Taiwan. The physical geography of mainland East Asia varies widely from the high plateaus and desert basins of western China to the broad river valleys and vast plains of eastern China. In the island region, landscapes are shaped by the convergence of three major tectonic plates, the Eurasian, Philippine, and Pacific.

ENVIRONMENTAL GEOGRAPHY: Resource Pressures in a Crowded Land

Environmental problems in East Asia are particularly severe owing to a combination of the region's large population, its massive industrial development, and its physical geography (Figure 11.2). The wealthier countries of the region, particularly Japan, have been able to invest heavily in environmental protection. China, in contrast, not only has less money available for conservation but also has more deeply rooted environmental problems. One of China's most controversial environmental issues has been the construction of the Three Gorges Dam on the Yangtze River.

Dams, Flooding, and Soil Erosion in China

The Yangtze River (also called the Chang Jiang) is one of the most important physical features of East Asia. This river, the third largest (by volume) in the world, emerges from the

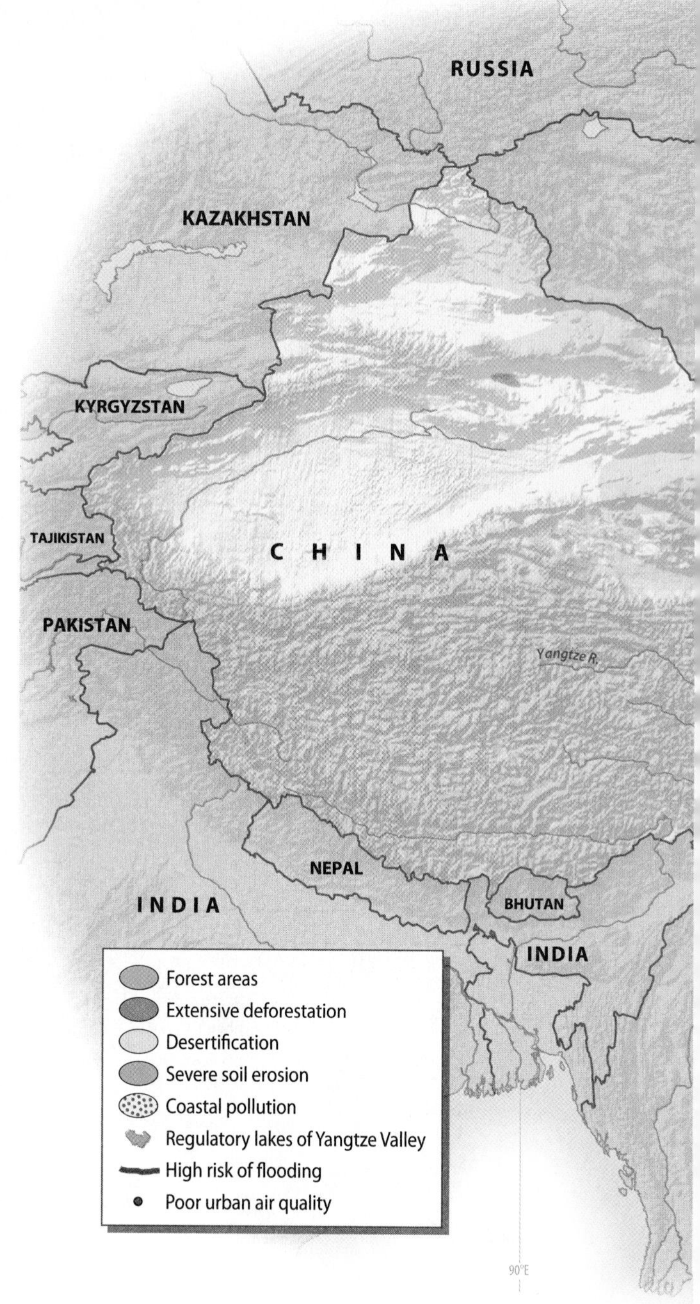

Figure 11.2 Environmental Issues in East Asia This vast world region has been almost completely transformed from its natural state and continues to have serious environmental problems. In China, some of the more pressing environmental issues involve deforestation, flooding, water control, and soil erosion.

Tibetan highlands onto the rolling lands of the Sichuan Basin, passes through what used to be a magnificent canyon in the Three Gorges area (Figure 11.3), and then meanders across the lowlands of central China before entering into the sea near the city of Shanghai. The Yangtze has historically been the main avenue of entry into the interior of China, and is celebrated in Chinese literature for its beauty and power. Since the 1990s, however, it has become the focal point of an environmental controversy of global proportions.

The Three Gorges Dam Controversy Finished in 2006, the Three Gorges Dam—600 feet (180 meters) high and 1.45 miles (2.3 kilometers) long—is the largest hydroelectric dam in the world, requiring $26 billion to build. The reservoir is 350 miles (563 kilometers) long, has displaced an estimated 1.2 million people while inundating a major scenic attraction (the Three Gorges of the Yangtze for which it is named). The dam will trap sediment and pollutants in the reservoir, as well as disrupt habitat for several endangered species, including the Yangtze River dolphin. The ecological and human rights consequences of the dam are so negative that the World Bank withdrew its support.

The Three Gorges Dam generates massive amounts of electricity, meeting approximately 3 percent of China's needs. As China industrializes, its demand for power is skyrocketing. At present, nearly three-fourths of China's total energy supply comes from burning coal, which results in horrendous air pollution and acid rain. But while dam-building may reduce air pollution, it will not be nearly adequate to supply China's vast energy needs. Most environmentalists therefore argue that the damage caused by the dams is greater than the benefits they confer.

Figure 11.3 The Three Gorges of the Yangtze The spectacular Three Gorges landscape of the Yangtze River is now gone, inundated by the reservoir that has formed behind the controversial Three Gorges Dam. Not only are the human costs high, but there may also be significant ecological costs to endangered aquatic species.

Chinese planners also praise the water-control benefits provided by dams. The lower and middle stretches of the Yangtze periodically suffer from devastating flooding, and planners hope that the Three Gorges Dam will store enough water to help protect 15 million people and over 3 million acres (1.2 million hectares) of farmland on the lower

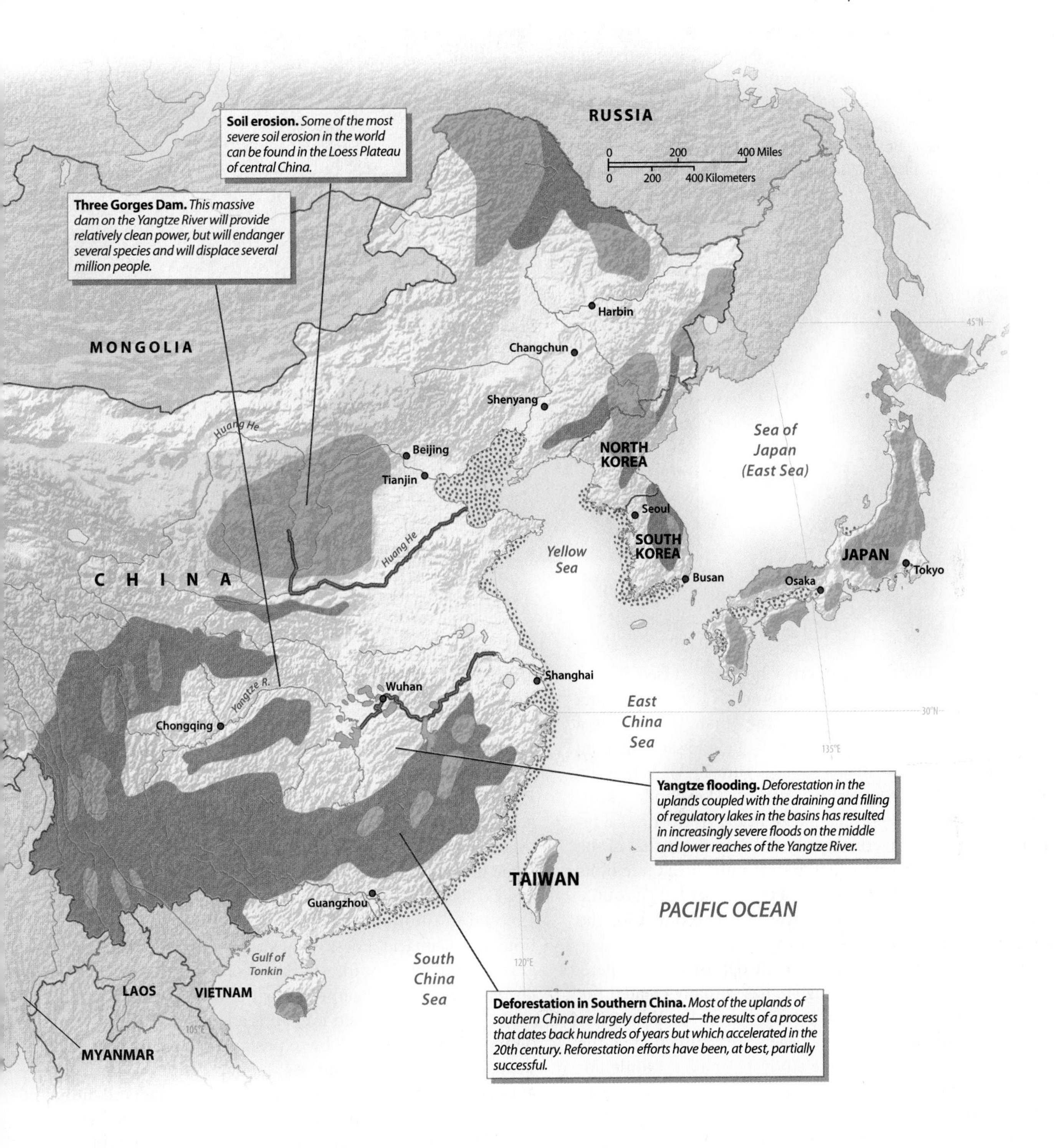

Yangtze floodplains. However, flooding has been exacerbated over the past several decades by the declining area of lakes on the Yangtze floodplain, which function as overflow areas during floods. East of Sichuan, the Yangtze passes through several basins (in Hunan, Hubei, and Jiangxi provinces), each containing a group of lakes. During periods of high water, river flows are diverted into these **regulatory lakes**, thus reducing the flow downstream. After the flood season is over, water drains from the lakes and therefore helps maintain the water level in the river downstream. In the 1950s and 1960s, however, the Chinese government drained many of these lakes in order to convert the lake beds to farmland. As a result, flooding downstream intensified.

Flooding in Northern China The North China Plain, which has been deforested for thousands of years, is plagued by both drought and flood. This area is dry most of the year, yet often experiences heavy downpours in summer. Since ancient times, large-scale hydraulic engineering projects have both controlled floods and allowed irrigation. But no matter how much effort has been put into water control, disastrous flooding has never been completely prevented.

Figure 11.4 The Huang He, the World's Muddiest River Tourists are shown visiting the Hukou waterfall, China's second largest cascade. Major flooding, sometimes inundating large sections of the North China Plain, has been a historical problem with the Huang He River. Dam-building is generally unsuccessful here because the reservoirs quickly fill up with silt.

The worst floods in northern China are caused by the Huang He, or Yellow River, which cuts across the North China Plain. Owing to upstream erosion, the Huang He carries a huge **sediment load**, or suspended clay, silt, and sand, making it the world's muddiest major river (Figure 11.4). When the river enters the low-lying plain, its velocity slows and its sediments begin to settle and accumulate in the riverbed. As a result, the level of the riverbed gradually rises above that of the surrounding lands. Eventually the river must break free of its course to find a new route to the sea over lower-lying ground. Twenty-six such course changes have been recorded for the Huang He throughout Chinese history.

Through the process of sediment deposition and periodic course changes, the Huang He has actually created the vast North China Plain. In prehistoric times, the Yellow Sea extended far inland. Even today the sea is retreating as the Huang He's delta expands; one study revealed a 6-mile advance of the land in a 3-year period. Such a process occurs in other alluvial plains, but nowhere else in the world is it so pronounced. Nowhere else, moreover, is it so destructive. The world's two deadliest recorded natural disasters, each of which killed between one and two million people, both involved flooding of the Yellow River (in 1887 and 1931). The North China Plain has been densely populated for millennia and is now home to more than 300 million people. Since ancient times, the Chinese have attempted to keep the river within its banks by building progressively larger dikes. Eventually, however, the riverbed rises so high that the flow can no longer be contained and catastrophic flooding results. While the river has not changed its course since the 1930s, most geographers think that another course correction is inevitable.

Erosion on the Loess Plateau The Huang He's sediment burden is derived from the eroding soils of the Loess Plateau, located to the west of the North China Plain. **Loess** consists of fine, windblown sediment that was deposited on this upland area during the last Ice Age, accumulating in some places to depths of several hundred feet.

Loess forms fertile soil, but it washes away easily when exposed to running water. At the dawn of Chinese civilization, the semiarid Loess Plateau was covered with tough grasses and scrubby forests that retained the soil. Chinese farmers, however, began to clear the land for the abundant crops that it yields when rainfall is adequate. Cultivation required plowing, which, by exposing the soil to water and wind, exacerbated erosion. As the population of the region gradually increased, the remaining areas of woodland diminished, leading to faster rates of soil loss. As the erosion process continued, great gullies cut across the plateau, steadily reducing the extent of arable land.

Today, the Loess Plateau is one of the poorest parts of China. Population is only moderately dense by Chinese standards, but good farmland is limited and drought is common. The Chinese government has long encouraged the construction of terraces to conserve the soil, but such efforts have not been effective everywhere. More recently, the World Bank and the government of China have pushed through the "Loess Plateau Watershed Rehabilitation Project," which has achieved considerable success in some

Figure 11.5 Denuded Hillslopes in China Because of the need to clear forests for wood products and agricultural lands, China's mountain slopes have long been deforested. Without forest cover, soil erosion is a serious issue.

areas. An estimated 133 square kilometers (51 square miles) of degraded land have been fully reforested.

Other East Asian Environmental Problems

The problems associated with the Yangtze and Huang He rivers are by no means the only environmental issues in East Asia. Deforestation, urban pollution, and the loss of wildlife are increasing problems in much of the region.

Deforestation and Desertification Many of the upland regions of China and North Korea support only grass, meager scrub, and stunted trees (Figure 11.5). China lacks the historical tradition of forest conservation that characterizes Japan. During earlier periods of Chinese history, hillsides were often cleared for fuelwood, and in some instances entire forests were burned for ash that could be used as fertilizer. In much of southern China, sweet potatoes, maize, and other crops have been grown on steep and easily eroded hillsides for hundreds of years. After centuries of exploitation, many hillslopes are now so degraded that they cannot easily regenerate forests.

Over the past 50 years, China, unlike North Korea, has initiated a number of reforestation efforts. Millions of people have been mobilized by the government to plant trees. In many areas such projects have failed, but the government claims that overall the area covered by forests has increased from 9 percent in 1949 to 13 percent in 2008. After severe flooding along the Yangtze River in 1998, reforestation efforts in central China were significantly enhanced. Still, it will be many years before such new forests can be logged. At present, substantial timber reserves are found only in China's far northeast, where a cold climate limits tree growth, and along the eastern slopes of the Tibetan Plateau, where rugged terrain restricts commercial forestry. As a result, China suffers a severe shortage of forest resources. China makes up for this shortfall by massively importing lumber and other wood products from other parts of the world.

Desertification is also a major problem for China. In many areas, sand dunes have pushed south from the Gobi Desert, smothering farmlands. Some reports claim that more desertification occurs in China than any other country. The Chinese government has responded with massive dune-stabilization schemes, which usually involve planting grasses to stop sand from blowing away (Figure 11.6). The most ambitious of such projects involved planting more than 300

Figure 11.6 Dune Stabilization in China Farmers are shown planting grass on a dune in an effort to keep the sand from blowing toward cultivated fields. China has one of the worst desertification problems in the world, but it has made some progress in slowing the advance of deserts in recent years.

million trees along a 3,000-mile (4,800-kilometer)-long swath of northern China. This "great green wall" was designed both to stop the southward expansion of the desert and reduce the dust storms that often hit northern China. Overall, the rate of desertification in China is reported to have declined from 3,400 square kilometers (1,300 square miles) per year in the 1990s to 1,300 square kilometers (502 square miles) in the early 2000s.

Mounting Pollution As China's industrial base expands, other environmental problems, such as water pollution and toxic-waste dumping, are growing more acute, particularly

ENVIRONMENT AND POLITICS China Embraces Green Energy

As China has undergone breakneck industrialization in recent years, it has emerged as the world's most polluted country. China's pollution extends well beyond its borders. Emissions of particulates, sulphur dioxide, and nitric oxide reach Japan and Korea; even mercury found in California rainwater has been traced to China. China is also the world's largest emitter of carbon dioxide. As a result, China has developed a poor reputation in international environmental circles. Recently, however, China's environmental standing has experienced something of a turnaround. Its government has embraced renewable energy with fervor, planning to become the world leader in green technology. By 2020, China hopes to meet 20 percent of its own energy needs through renewable sources.

China's push for renewable energy stems from political and economic considerations, as well as those of environmental protection. China is dependent on energy from Central and Southwest Asia, areas vulnerable to political instability. Governmental planners also see massive export potential in renewable energy. China is now the world leader in hydroelectric turbines, and produces over 30 percent of the world's solar photovoltaic panels. Chinese renewable energy firms respond to market signals, but the industry as a whole is guided by governmental subsidies and mandates. By law, utility companies in China must purchase renewable energy at prices above the market level.

China's accomplishments in wind power have been especially impressive. China is sometimes said to have the largest wind-power potential in the world, owing both to the southwest monsoon and to stiff breezes that blow out of Central Asia during much of the year. At the end of 2009, the United States produced more wind-generated electricity, but most experts expected that China would soon grab the top position. Between 2004 and 2009, China doubled its installed wind power every year (Figure 11.1.1). A 2009 study showed that China could economically meet all of its additional energy needs through 2030 by wind power alone.

Not surprisingly, China is emerging as a global leader in the manufacture of wind turbines. Chinese firms began by buying technology licenses from European companies but quickly began to develop their own expertise. The Beijing government has pushed forward the development of local technologies, mandating at first that 40 percent of the components used in wind farms be made in China, and then increasing that figure to 70 percent in 2009. China treats wind energy as strategic industry, and thus demands much secrecy. It is difficult for foreigners to obtain permission to visit not just wind turbine factories, but even wind farms themselves. The Chinese government sees the world's energy future as dominated by renewable sources, a market it hopes to dominate.

Figure 11.1.1 Chinese Wind Farm China is currently pushing hard for renewable energy. Wind power development is concentrated in the deserts and grasslands of Inner Mongolia.

in the booming coastal areas. The burning of high-sulfur coal has resulted in severe air pollution, a problem aggravated by the increasing number of automobiles driven by the Chinese. One recent study found that 16 of the world's 20 most polluted cities are located in China. According to a World Health Organization (WHO) report, air pollution kills 656,000 Chinese citizens each year, while another 95,000 die from polluted drinking water. Although Chinese citizens have organized thousands of protests against pollution and other forms of environmental destruction, the government has generally insisted that economic growth remain the top priority. Beginning in 2006, however, China's government has been supporting the development of clean, renewable sources of energy.

Considering its large population and intensive industrialization, Japan's environment is relatively clean. The very density of its population gives certain environmental advantages, allowing, for example, a highly efficient public transportation system. In the 1950s and 1960s, Japan did suffer from some of the world's worst pollution. Soon afterward the Japanese government passed stringent environmental laws covering both air and water pollution.

Japan's cleanup was aided by its insular location, because winds usually carry smog-forming chemicals out to sea. Equally important has been the phenomenon of pollution exporting. Because of Japan's high cost of production and its strict environmental laws, many Japanese companies have moved their dirtier factories overseas. In effect, Japan's pollution has been partially displaced to poorer countries. Of course, the same argument can also be made about the United States and western Europe, but to a somewhat lesser extent.

Serious air and water pollution problems also emerged early in other East Asian countries. But Taiwan and South Korea, which have large chemical, steel, and other heavy industries, responded by imposing more stringent environmental controls as they grew wealthier. These two countries have also followed Japan's footsteps by setting up new factories in poorer countries—including China—that have less strict environmental standards.

Figure 11.7 Coal-Fired Power Plant in China A coal-fired power plant emits large quantities of pollution in northwestern China. As China industrializes, it is building many such plants, damaging local air quality and contributing to global warming.

Endangered Species The growing number of endangered species in East Asia has been linked to the region's economic rise. Many forms of traditional Chinese medicine are based on products derived from rare and exotic animals. Deer antlers, bear gallbladders, snake blood, tiger penises, and rhinoceros horns are believed to have medical effectiveness, and certain individuals will pay fantastic sums of money to purchase them. As wealth has accumulated, trade in such substances has expanded. China itself has relatively little remaining wildlife, but other areas of the world help supply its demand for wildlife products.

China has, however, moved strongly to protect some of its remaining areas of habitat. Among the most important of these are the high-altitude forests and bamboo thickets of western Sichuan province, home of the panda bear. Efforts are also being made to preserve wildlands in northern Manchuria, where evidence of a surviving population of Siberian tigers was discovered in 1997, and in the canyon lands of northwestern Yunnan, which were declared off-limits to loggers in 1999.

Wildlife is also scarce in Korea. Ironically, however, the so-called demilitarized zone that separates North from South Korea functions as an unintentional wildlife sanctuary, supporting populations of several endangered species, including the Asiatic black bear.

Global Warming in East Asia

East Asia has come to occupy a central position in global warming debates, largely because of China's rapid increases in carbon emissions. From levels only about half those of the United States in 2000, China's total production of greenhouse gases surpassed those of the United States in 2007. This staggering rise has been caused both by China's explosive economic growth and by the fact that it relies on burning coal to meet most of its energy needs (Figure 11.7).

The potential effects of global warming in China have serious implications for human populations. According to the 2007 Chinese National Climate Change Assessment, the country's production of wheat, corn, and rice could fall by as much as 37 percent if average temperatures increase by 3.5 to 5.5°F (2 or 3°C) over the next 50 to 80 years. Increased evaporation rates, coupled with the melting of glaciers on the Tibetan Plateau, could greatly intensify the water shortages that already plague much of northern China. Officials believe that by the middle of the 21st century, even Shanghai, located in the humid Yangtze Delta, will either have to use desalinized water or build an expensive network of canals, dams, and pumps to import water from south-central China for municipal use. In the wet zones of southeastern China, global warming concerns center on the possibility of more intense storms in this already flood-prone region.

Despite these concerns, the Chinese government has insisted that economic growth take priority over reducing greenhouse gas emissions. Officials argue that China's per capita emissions are still far below those of the United States and most other wealthy countries, and that Chinese industries have been learning how to use energy more efficiently. Government officials also maintain that wealthier countries should take the lead on reducing emissions and figuring out other ways to combat global warming.

In June 2007, China released its first national plan on climate change, which called for a 20-percent gain in energy efficiency by 2010. China's new strategy also includes a major expansion of nuclear power, increased use of renewable energy sources, and the continuation of ambitious reforestation efforts. Current plans call for a target of producing 15 percent of the country's electricity through renewable sources by 2020 (see "Environment and Politics: China Embraces Green Energy"). While environmentalists were pleased to see China addressing these issues, most feel that the country has not gone nearly far enough, and they have been disappointed that it has not embraced the goal of halving carbon emissions by 2050.

China's contribution to the causes of global warming has overshadowed that of other East Asian countries. Japan, South Korea, and Taiwan are all major emitters of greenhouse gases, but all also have energy-efficient economies, releasing far less carbon than the United States on a per capita basis. Overall, Japan has been a strong proponent of international treaties, such as the Kyoto Protocol, designed to force reductions in greenhouse gas emissions. But while its total emissions have been relatively flat in recent years, Japan has not achieved the reductions that it promised in Kyoto. Increasingly, however, Japanese high-tech companies, like those of South Korea and Taiwan, realize that it could be very good business to develop more efficient forms of energy production. Japan is widely regarded as the world leader in a broad range of climate-friendly technologies.

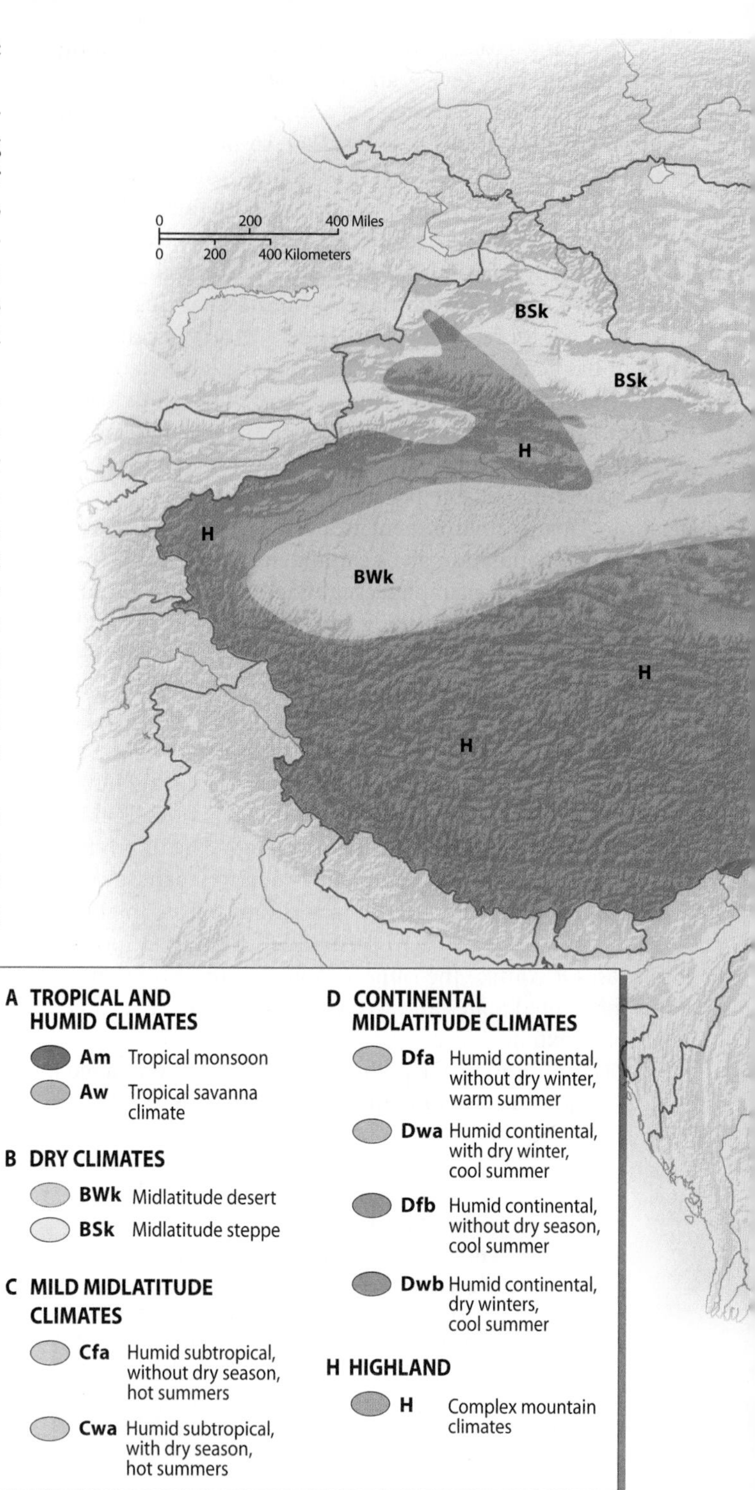

East Asia's Physical Geography

East Asia is situated in the same general latitudinal range as the United States, although it extends considerably farther north and south. The northernmost tip of China lies as far north as central Quebec, while China's southernmost point is at the same latitude as Mexico City. The climate of southern China is thus roughly comparable to southern Florida and the Caribbean, whereas that of northern China is more similar to south-central Canada (Figure 11.8).

The island belt of East Asia, extending from northern Japan through Taiwan, is situated at the intersection of three tectonic plates (the basic building blocks of Earth's crust): the Eurasian, the Pacific, and the Philippine. This area, particularly Japan, is therefore geologically active, experiencing numerous earthquakes and dotted with volcanoes (Figure 11.9).

Japan's Physical Environment Although slightly smaller than California, Japan extends farther north and south. As a result, Japan's extreme south, in southern Kyushu and the Ryukyu Archipelago, is subtropical, while northern Hokkaido is almost subarctic. Most of the country, however, is temperate. The climate of Tokyo is not unlike that of Washington, DC—although Tokyo is distinctly rainier.

Japan's climate varies not only from north to south, but also from southeast to northwest across the main axis of the

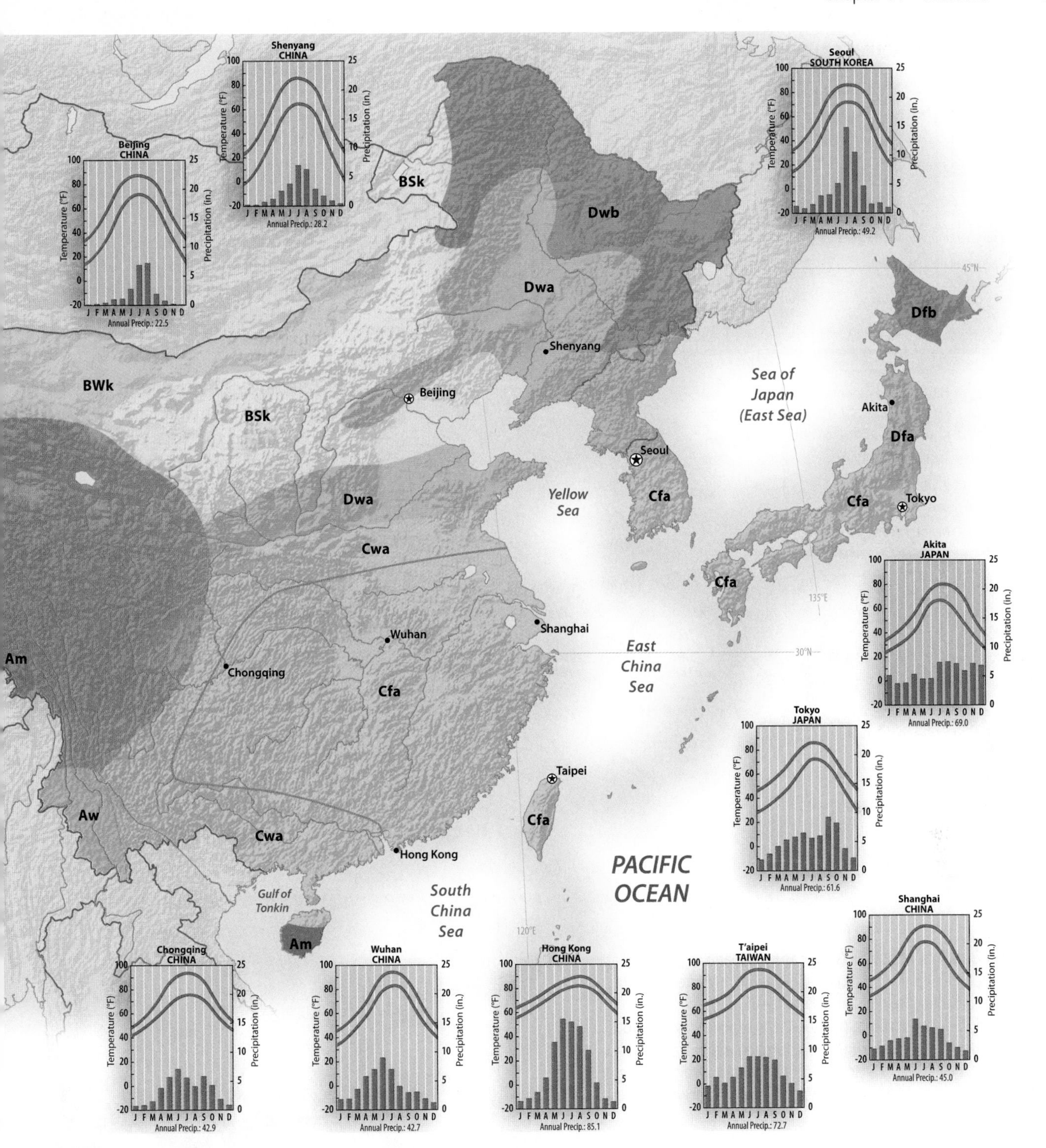

Figure 11.8 Climate Map of East Asia East Asia is located in roughly the same latitudinal zone as North America, and climatic parallels exist between the two world regions. The northernmost tip of China lies at about the same latitude as Quebec and shares a similar climate, whereas southern China approximates the climate of Florida. In Japan, maritime influences produce a milder climate.

archipelago. In the winter, the area facing the Sea of Japan receives much more snow than the Pacific Ocean coastline. During this time of the year, cold winds from the Asian mainland blow across the relatively warm waters of the Sea of Japan. The air picks up moisture over the sea and deposits it, usually as snow, when it hits land (Figure 11.10). The Pacific coast of Japan, on the other hand, is far more vulnerable to typhoons (hurricanes), which frequently strike the country.

The Pacific coast of Japan is separated from the Sea of Japan coast by a series of mountain ranges. Japan is one of the world's most rugged countries, with mountainous terrain covering some 85 percent of its territory. Most of these

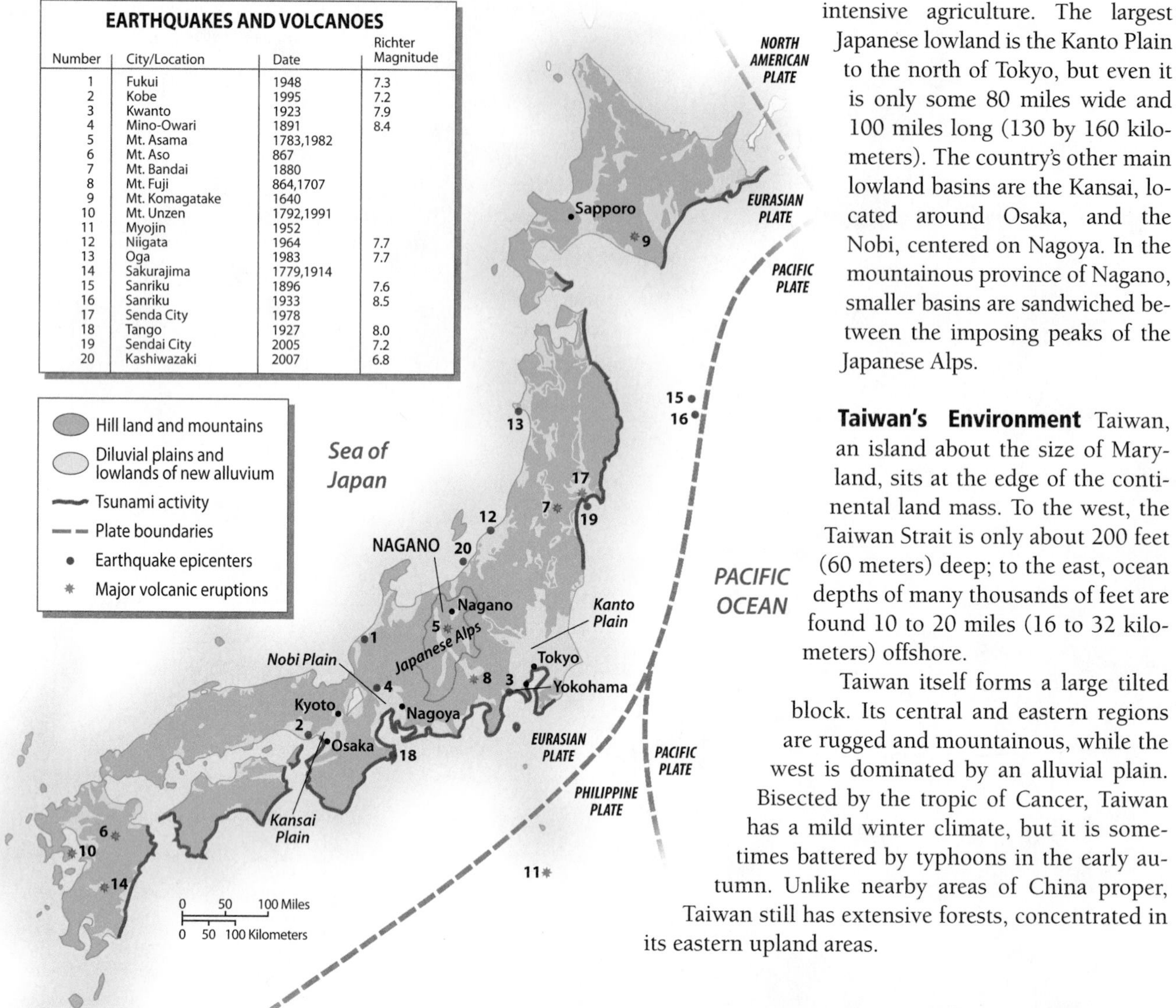

EARTHQUAKES AND VOLCANOES

Number	City/Location	Date	Richter Magnitude
1	Fukui	1948	7.3
2	Kobe	1995	7.2
3	Kwanto	1923	7.9
4	Mino-Owari	1891	8.4
5	Mt. Asama	1783,1982	
6	Mt. Aso	867	
7	Mt. Bandai	1880	
8	Mt. Fuji	864,1707	
9	Mt. Komagatake	1640	
10	Mt. Unzen	1792,1991	
11	Myojin	1952	
12	Niigata	1964	7.7
13	Oga	1983	7.7
14	Sakurajima	1779,1914	
15	Sanriku	1896	7.6
16	Sanriku	1933	8.5
17	Senda City	1978	
18	Tango	1927	8.0
19	Sendai City	2005	7.2
20	Kashiwazaki	2007	6.8

Figure 11.9 Japan's Physical Geography Japan has several sizable lowland plains, primarily along the coastline, but they are interspersed among rugged mountains and uplands. Because of its location at the convergence of three major tectonic plates, Japan experiences numerous earthquakes. Volcanic eruptions can also be hazardous and are linked directly to Japan's location on tectonic plate boundaries. Additionally, much of Japan's coast is vulnerable to devastating tsunamis (tidal waves) caused by earthquakes in the Pacific Basin.

uplands are thickly wooded, making Japan also one of the world's most heavily forested countries (Figure 11.11). Japan owes its lush forests both to its mild, rainy climate and to its long history of forest conservation. For hundreds of years, both the Japanese state and its village communities have enforced strict conservation rules, ensuring that timber and firewood extraction would be balanced by tree growth.

Along Japan's coastline and interspersed among its mountains are limited areas of alluvial plains (see Figure 11.9). Although these lowlands were once covered by forests and wetlands, they have long since been cleared and drained for intensive agriculture. The largest Japanese lowland is the Kanto Plain to the north of Tokyo, but even it is only some 80 miles wide and 100 miles long (130 by 160 kilometers). The country's other main lowland basins are the Kansai, located around Osaka, and the Nobi, centered on Nagoya. In the mountainous province of Nagano, smaller basins are sandwiched between the imposing peaks of the Japanese Alps.

Taiwan's Environment Taiwan, an island about the size of Maryland, sits at the edge of the continental land mass. To the west, the Taiwan Strait is only about 200 feet (60 meters) deep; to the east, ocean depths of many thousands of feet are found 10 to 20 miles (16 to 32 kilometers) offshore.

Taiwan itself forms a large tilted block. Its central and eastern regions are rugged and mountainous, while the west is dominated by an alluvial plain. Bisected by the tropic of Cancer, Taiwan has a mild winter climate, but it is sometimes battered by typhoons in the early autumn. Unlike nearby areas of China proper, Taiwan still has extensive forests, concentrated in its eastern upland areas.

Figure 11.10 Heavy Snow in Japan's Mountains Cold, moist air moving off the Sea of Japan produces heavy snows in northwestern Japan and along the country's mountainous spine. Numerous major ski areas dot the Japanese Alps, several of which have hosted world-class sports competitions.

Figure 11.11 Forested Landscapes of Japan Although much of Japan is heavily forested and supports a viable wood products industry, the yield is not large enough to satisfy demand. As a result, Japan imports timber extensively from North America, Southeast Asia, and Latin America.

Chinese Environments Even if one excludes its Central Asian provinces of Tibet and Xinjiang, China is a vast country with diverse environmental regions. For the sake of convenience, China proper can be divided into two main areas, one lying to the north of the Yangtze River valley, the other including the Yangtze and all areas to the south. As Figure 11.12 shows, each of these divisions can be subdivided into a number of distinctive regions.

Southern China is a land of rugged mountains and hills interspersed with lowland basins. The lowlands of southern China are far larger than those of Japan. One of the most distinctive is the former lake bed of central Sichuan (in the central Upper Yangtze region of Figure 11.12). Protected by imposing mountains, Sichuan is noted for its mild winter climate. Eastward from Sichuan, the Yangtze River passes through several other broad basins (in the Middle Yangtze region), partially separated from each other by hills and low mountains, before flowing into a large delta near Shanghai.

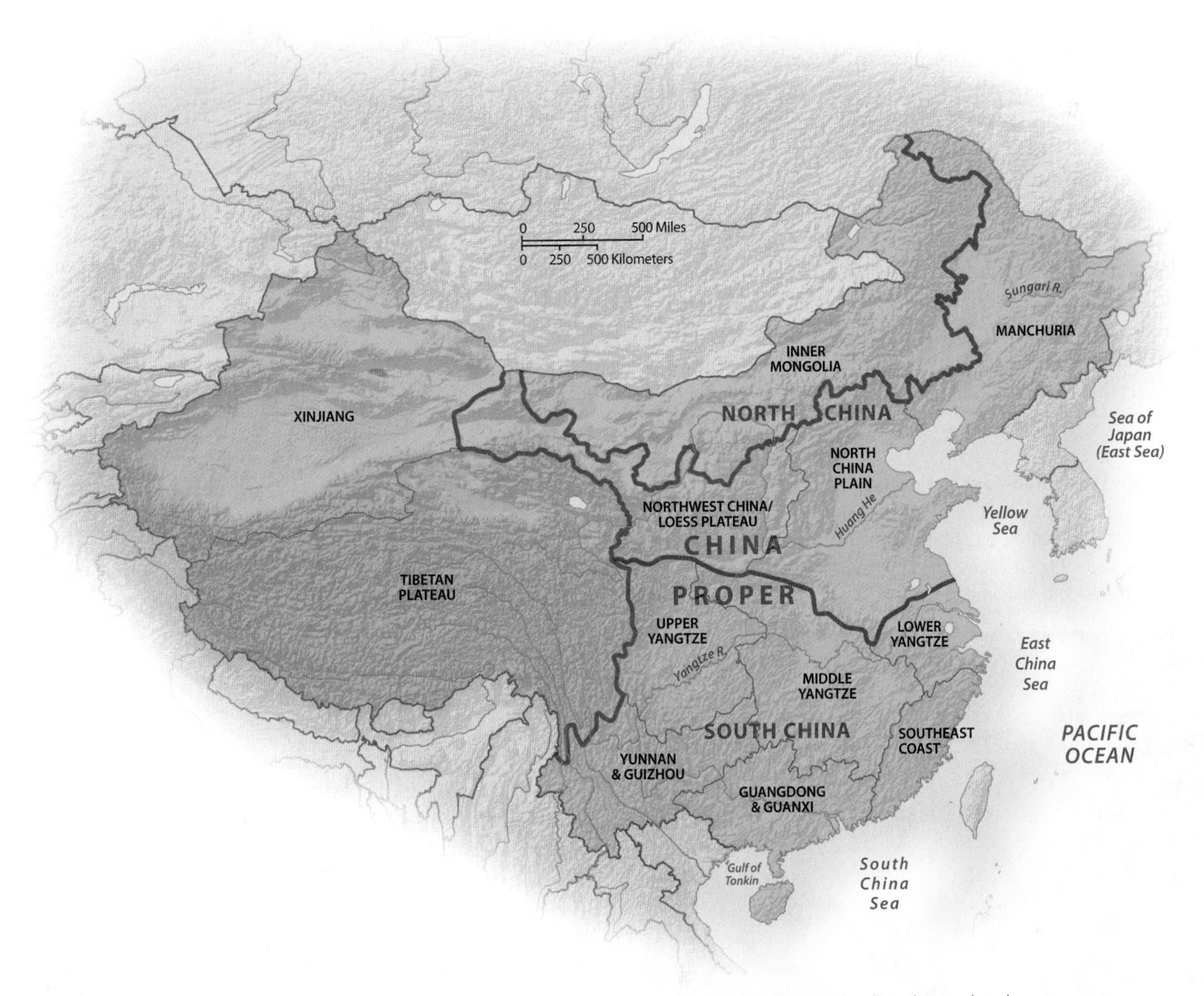

Figure 11.12 Landscape Regions of China The term *China proper* denotes the densely populated, culturally Han Chinese areas to the east of the blue line. The Yangtze Valley divides China proper into two general areas. Immediately to the north is the large fertile area of the North China Plain, bisected by the Huang He (or Yellow) River. To the west is the Loess Plateau, an upland area of soil derived from wind-deposited silt after the prehistoric glacial period, about 15,000 years ago.

South of the Yangtze Valley, the mountains are higher (up to 7,000 feet, or 2,150 meters, in elevation), but they are still interspersed with alluvial lowlands. Sizable valleys are found in the far south, such as the Xi Basin in Guangdong province. Here the climate is truly tropical, free of frost. West of Guangdong lie the moderate-elevation plateaus of Yunnan and Guizhou, the former noted for its perennial springlike weather. Finally, to the northeast of Guangdong lies the rugged coastal province of Fujian (Figure 11.13). With its narrow coastal plain and deeply indented coastline, Fujian's landscape does not support a productive agricultural economy; as a result, the Fujianese people have often sought a maritime way of life, many of them working in the fishing and shipping industries.

North of the Yangtze Valley, the climate is both colder and drier than it is to the south. Summer rainfall is generally abundant except along the edge of the Gobi Desert, but the other seasons are dry. Desertification is a major threat in parts of the North China Plain, which have experienced prolonged droughts in recent years. With the exception of a few low mountains in Shandong province, the entire area is a virtually flat plain. Seasonal water shortages in this area are growing increasingly severe as withdrawals for irrigation and industry increase, leading to concerns about an impending crisis.

West of the North China Plain sits the Loess Plateau. This is a fairly rough upland of moderate elevation and uncertain precipitation. It does, however, have fertile soil—as well as huge coal deposits. Farther west one finds the semiarid plains and uplands of Gansu province, situated at the foot of the great Tibetan Plateau.

China's far northeastern region is called *Dongbei* in Chinese and Manchuria in English. Manchuria is dominated by a broad, fertile lowland basin sandwiched between mountains stretching along China's borders with North Korea, Russia, and Mongolia. Although winters here can be brutal, summers are usually warm and moist. Manchuria's peripheral uplands have some of China's best-preserved forests and wildlife refuges.

Figure 11.14 The Mountains and Lowlands of Korea Like Japan, Korea has extensive uplands interspersed with lowland plains. The country's highest mountains are in the north, whereas its most extensive alluvial plains are in the south. South Korea's provinces, shown here, are culturally as well as physically distinctive.

Figure 11.13 The Fujian Coast of China In southeastern China lies the rugged coastal province of Fujian. Here the coastal plain is narrow and the shoreline deeply indented, producing a picturesque landscape. Because of limited agricultural opportunities along this rugged coastline, many Fujianese people work in maritime activities.

Korean Landscapes Korea forms a well-demarcated peninsula, partially cut off from Manchuria by rugged mountains and sizable rivers (Figure 11.14). Like Japan, its latitudinal range is pronounced. The far north, which just touches Russia's Far East, has a climate not unlike that of Maine, whereas the southern tip is more reminiscent of the Carolinas. Korea is a mountainous country with scattered alluvial basins, a landscape that has, as in many other parts of East Asia, deeply influenced its demographic and agricultural development. The lowlands of the southern portion of the peninsula are more extensive than those of the north, giving South Korea a distinct agricultural advantage over North Korea. However, North Korea has more abundant mineral deposits, forests, and hydroelectric resources.

POPULATION AND SETTLEMENT: A Realm of Crowded Lowland Basins

East Asia, along with South Asia, is the most densely populated region of the world (Table 11.1). The lowlands of Japan, Korea, and China are among the most intensely used portions of Earth, containing East Asia's major cities and most of its agricultural lands (Figure 11.15). Although the population density of East Asia is extremely high, the region's demographic growth rate has plummeted since the 1970s. In Japan especially, the current concern is impending population loss, hence an aging population that will need to be supported by a shrinking number of younger workers.

Agriculture and Settlement in Japan

Japan is a highly urbanized country, supporting two of the largest urban agglomerations in the world. Yet it is also one of the world's most mountainous countries, with lightly inhabited uplands. Agriculture must therefore share the limited lowlands with cities and suburbs, resulting in extremely intensive farming practices.

Japan's Agriculture Lands Japanese agriculture is largely limited to the country's coastal plains and interior basins. Japanese rice farming has long been one of the most productive forms of agriculture in the world, helping support a population of 127 million people on a relatively small and rugged land. Although rice is grown in almost all Japanese lowlands, the country's premier rice-growing districts lie along the Sea of Japan coast in Honshu. Vegetables and even rice crops are cultivated intensively on tiny patches within suburban and urban neighborhoods (Figure 11.16). The valleys of central and northern Honshu are famous for their temperate-climate fruit, while citrus comes from the milder southwestern reaches of the country. Crops that thrive in a cooler climate, such as potatoes, are produced mainly in Hokkaido and northern Honshu.

Settlement Patterns All Japanese cities—and the vast majority of the Japanese people—are located in the same lowlands that support the country's agriculture. Not surprisingly, the three largest metropolitan areas—Tokyo, Osaka, and Nagoya—sit near the centers of the three largest plains.

The fact that Japan's settlements are largely restricted to roughly 15 percent of its land area means that the country's effective population density—the actual crowding it experiences—is one of the highest in the world. This is especially notable in the main industrial belt, which extends from Tokyo through Nagoya and Osaka, hence along the Inland Sea (the maritime region sandwiched between Shikoku, western Honshu, and Kyushu) to the northern coast of Kyushu. This area is perhaps the most intensively used part of the world for human habitation, industry, and agriculture.

Japan's Urban–Agricultural Dilemma Due to such space limitations, all Japanese cities are characterized by dense settlement patterns. In the major urban areas, the amount of available living space is highly restricted for all but the most affluent families. Many observers—especially American ones—argue that Japan should allow its cities and suburbs to expand into nearby rural areas. However, because most uplands are too steep for residential use, such expansion would have to come at the expense of farmland.

As it now stands, the wholesale conversion of farmland to neighborhoods would be difficult. Although most farms are extremely small (the average is only several acres) and only marginally viable, farmers are politically powerful and croplands are often protected by the tax code. Moreover,

TABLE 11.1 Population Indicators

Country	Population (millions) 2010	Population Density (per square kilometer)	Rate of Natural Increase (RNI)	Total Fertility Rate	Percent Urban	Percent <15	Percent >65	Net Migration (Rate per 1,000) 2005–10[a]
China	1,338.1	140	0.5	1.5	47	18	8	−0.3
Hong Kong	7.0	6,410	0.6	1.0	100	12	13	3.3
Japan	127.4	337	−0.0	1.4	86	13	23	0.2
North Korea	22.8	189	0.5	2.0	60	22	9	0.0
South Korea	48.9	491	0.4	1.2	82	17	11	−0.1
Taiwan	23.2	644	0.2	1.0	78	16	11	

[a]*Net Migration Rate from the United Nations, Population Division, World Population Prospects: The 2008 Revision Population Database.*
Source: Population Reference Bureau, World Population Data Sheet, 2010.

Figure 11.15 Population Map of East Asia Parts of East Asia are extraordinarily densely settled, particularly in the coastal lowlands of China and Japan. This contrasts with the sparsely settled lands of western China, North Korea, and northern Japan. Although the total population of this world region is high, as is the overall density, the rate of natural population increase has slowed rather dramatically in the last several decades.

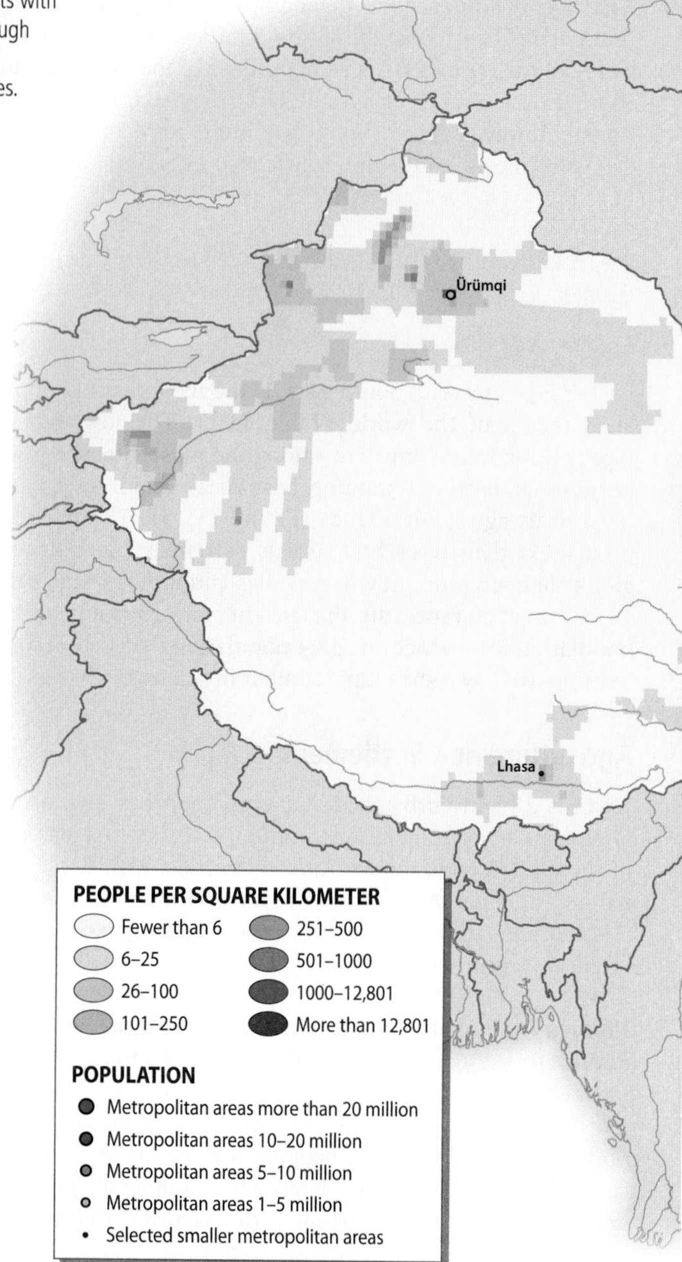

most Japanese citizens believe that it is vitally important for their country to remain self-sufficient at least in rice. Rice imports are therefore highly restricted.

U.S. economic interests, along with certain Japanese consumer advocates, stress the fact that the present system forces Japanese consumers to pay several times the world market price for rice, their staple food. Critics also point out that Japan relies on imports for more than half of its overall food needs as well as for the energy that is required to grow rice. But whatever arguments are made, change will only come gradually, if at all. For the moment, Japan must live with the tensions resulting from the fact that its affluent population is forced to live in tight proximity and to pay high prices for basic staples.

Agriculture and Settlement in China, Korea, and Taiwan

Like Japan, Taiwan and Korea are urban societies. Although China remains predominantly rural, with an urbanization rate of only some 47 percent, its cities are growing rapidly. Chinese cities are rather evenly distributed across the plains and valleys of China proper. As a result, the overall pattern of population distribution in China closely follows the geography of agricultural productivity.

China's Agricultural Regions A line drawn just to the north of the Yangtze Valley divides China into two main agricultural regions. To the south, rice is the dominant crop. To the north, wheat, millet, and sorghum are the most common.

In southern and central China, population is highly concentrated in the broad lowlands, which are famous for their fertile soil and intensive agriculture. More than 100 million people live in the Sichuan Basin, while more than 70 million reside in deltaic Jiangsu, a province smaller than Ohio. Crop-

Figure 11.16 Japanese Urban Farm Japanese landscapes often combine dense urban settlement with small patches of intensively farmed agriculture. Here a tiny farm coexists with an urban neighborhood.

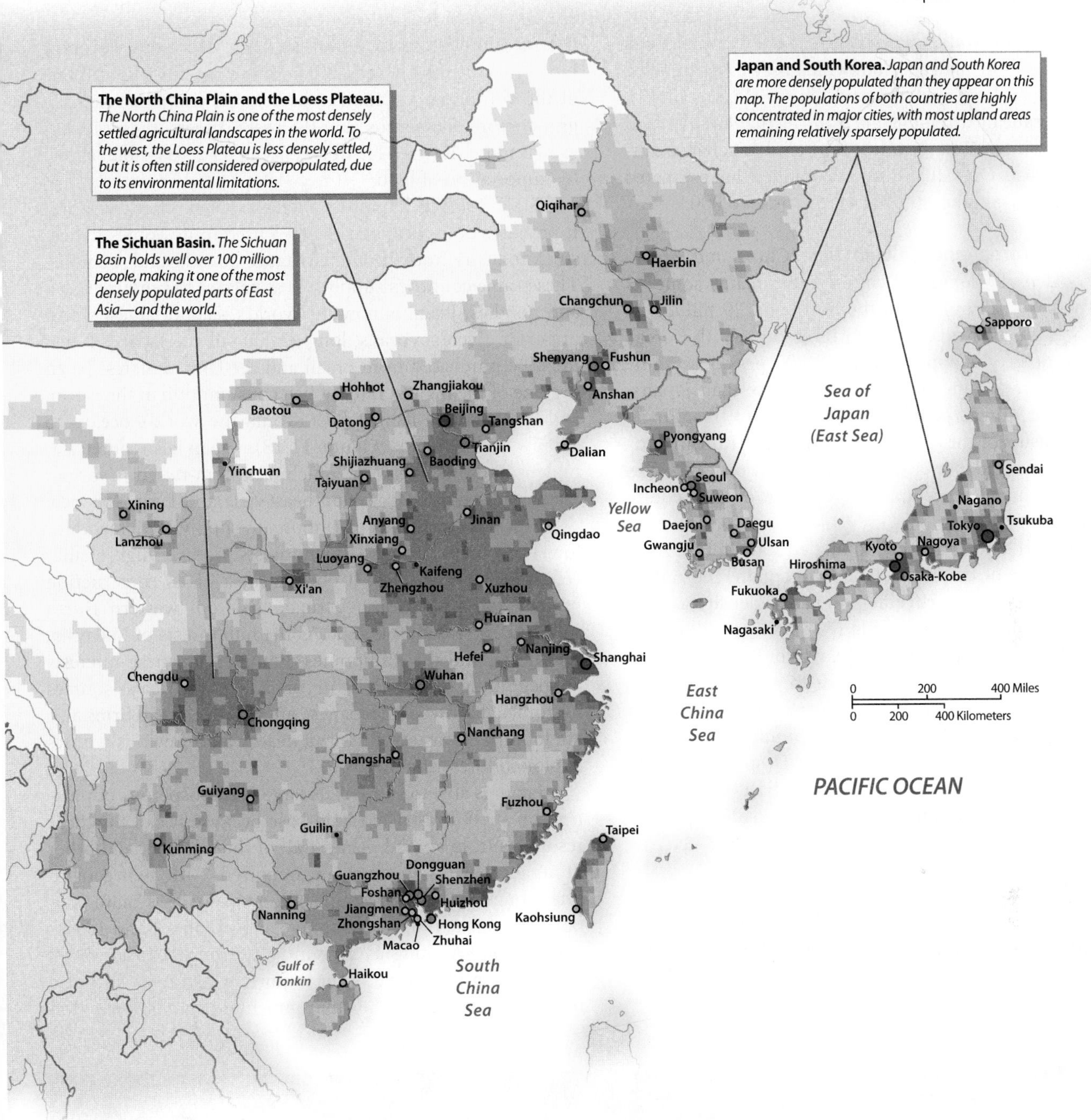

ping occurs year-round in most of southern and central China; summer rice alternates with winter barley or vegetables in the north, whereas two rice crops can be harvested in the far south. Southern China also produces a wide variety of tropical and subtropical crops, and moderate slopes throughout the area supply sweet potatoes, corn, and other upland crops.

The North China Plain is one of the world's most thoroughly **anthropogenic landscapes**; that is, it has been heavily transformed by human activities. Virtually its entire extent is either cultivated or occupied by houses, factories, and other structures of human society. Too dry in most areas for rice, the North China Plain supports the main crops of wheat, millet, and sorghum.

Manchuria was a lightly populated frontier zone as recently as the mid-1800s. Today, with a population of more than 100 million, its central plain is thoroughly settled. Still, Manchuria remains less crowded than many other parts of China, and it is one of the few parts of China to produce a consistent food surplus.

The Loess Plateau is more thinly settled yet, supporting only some 70 million inhabitants. But considering the area's aridity and widespread soil erosion, this is a high figure. Like other portions of northern China, the Loess Plateau produces wheat and millet. One of its unique settlement features is subterranean housing. Although loess erodes quickly under the impact of running water, it holds together well in other circumstances. For millennia, villagers have excavated pits on the surface of the plateau, from which they have tunneled into the earth to form underground houses (Figure 11.17). These subterranean dwellings are cool in the summer and warm in the winter. Unfortunately, they tend to collapse during earthquakes. One quake in 1920 killed an estimated 100,000 persons; another in 1932 killed some 70,000.

Patterns in Korea and Taiwan Like China and Japan, Korea is a densely populated country. It contains some 72 million persons (23 million in the north and 49 million in the south) in an area smaller than Minnesota. South Korea's population density is approximately 1,200 per square mile (491 per square kilometer), a higher density than Japan's. Most of South Korea's people are crowded into the alluvial plains and basins of the west and south. South Korean agriculture is dominated by rice. North Korea, in contrast, relies heavily on corn and other upland crops that do not require irrigation.

Taiwan is the most densely populated state in East Asia. Roughly the size of the Netherlands, it contains more than 22 million inhabitants. Its overall population density is more than 1,500 people per square mile (644 per square kilometer), one of the highest figures in the world. Because mountains cover most of central and eastern Taiwan, virtually the entire population is concentrated in the narrow lowland belt in the north and west. Large cities and numerous factories are scattered here amid lush farmlands. Despite its highly productive agriculture, Taiwan, like other East Asian countries, is forced to obtain much of its food from abroad.

Agriculture and Resources in Global Context

Japan, South Korea, and Taiwan are major food importers, and China has recently moved in the same direction. Other resources also are being drawn in from all quarters of the world by the powerful economies of East Asia.

Global Dimensions of Japanese Agriculture and Forestry Japan may be mostly self-sufficient in rice, but it is still one of the world's largest food importers. As the Japanese have grown more prosperous over the past 60 years, their diet has grown more diverse. That diversity is made possible largely by importation of food.

Japan imports food from a wide array of other countries. It procures both meat and the feed used in its domestic livestock industry from the United States, Canada, and Australia. These same countries supply wheat needed to produce bread and noodles. Japan is now the world's second largest wheat importer. Even soybeans, long a staple of the Japanese diet, must be purchased from Brazil and the United States. Japan has one of the highest rates of fish consumption in the world, and the Japanese fishing fleet scours the world's oceans to supply the demand (Figure 11.18). Japan also purchases prawns and other seafood, much of it farm-raised in former mangrove swamps, from Southeast Asia and Latin America.

Japan also depends on imports to supply its demand for forest resources. While its own forests produce high-quality cedar and cypress logs, it obtains most of its construction lumber and pulp (for papermaking) from western North America and Southeast Asia. As the rain forests of Malaysia, Indonesia, and the Philippines diminish, Japanese interests are beginning to turn to Latin America and Africa as sources of tropical hardwoods. Japanese and South Korean firms also are looking to eastern Russia, a nearby and previously little-exploited forest zone.

Japan is able to support its large and prosperous population on such a restricted land base because it can purchase resources from abroad. Almost all of the oil, coal, and other minerals that it consumes are imported. Although geographers once looked at resource endowments as a main support of each country's economy, such a view is no longer supportable. As long as a country can export products of value, it can obtain whatever imports it requires. This also means, however, that the environmental degradation gener-

Figure 11.17 Loess Settlement A typical subterranean dwelling carved out the soft loess sediment in central China. Approximately 70 million people live similarly in the Loess Plateau region. Unfortunately, this region is prone to major earthquakes that take a high toll on the local population because of dwelling collapse.

Figure 11.18 Japanese Fishing Boat A Japanese whaling ship leaves the port of Shimonoseki in southwestern Japan. While Japanese whaling is particularly controversial, the larger Japanese fishing fleet covers much of the world's ocean surface in search of high-quality seafood.

ated by a successful economy, such as that of Japan, becomes globalized as well.

Global Dimensions of Chinese Agriculture Until the late 1990s, China was self-sufficient in food, despite its huge population and crowded lands. But rapid economic growth has resulted in an increased consumption of meat, which requires large amounts of feed grain. Growth also has brought about the loss of agricultural lands to residential and industrial development. China is now the world's largest importer of soybeans and vegetable oil, and it imports significant quantities of wheat and other grains in most years. China's government aims to be 95 percent self-sufficient in food, but planners have recently discussed lowering that figure to 90 percent.

Although China is a net food importer when it comes to total calories, the situation is not so clear in regard to total value. China exports many high-value specialty crops and processed foods, such as garlic, apples, farm-raised fish, and many vegetables. Farmers in both wealthy countries such as the United States and poor countries such as the Philippines often complain that China sells its fruit and vegetables abroad at below-market rates, seeking to build export dominance. Consumers in foreign markets are concerned about the safety of Chinese exports. In 2007, the United States prohibited the import of a number of Chinese seafood products until they could be proved safe. China responded by banning chicken and pork products from Tyson Foods, a U.S. company. At the same time, however, the Chinese government shut down more than 100 food-processing plants, and went so far as to execute the former head of its own food and drug agency after he was convicted of corruption.

Korean Agriculture in a Global Context South Korea has also made the transition, like Japan and Taiwan, to a global food and resource procurement pattern. It is now one of the world's main importers of wheat, corn, and soybeans. South Korea also imports large quantities of beef, particularly from Australia. In 2003, it banned beef imports from the United States due to fears about mad cow disease. U.S. beef was again allowed into the country in 2008, but the decision prompted massive protests by South Korean farmers and others concerned about food safety.

North Korea, on the other hand, has pursued a goal of self-sufficiency. While relatively successful for a number of years, in the mid-1990s this policy resulted in widespread famine after a series of floods, followed by drought, destroyed most of the country's rice and corn crops. Since then, North Korea has relied heavily on international aid—much of it coming from South Korea—to feed its people, but malnutrition and even starvation remain widespread. North Korea also purchases food from China, buying more than 13,000 tons of grain in January 2010 alone.

Urbanization in East Asia

China has one of the world's oldest urban foundations, dating back more than 3,500 years. In medieval and early modern times, East Asia supported a well-developed system of cities. In the early 1700s, Tokyo, then called Edo, probably overshadowed all other cities, with a population of more than 1 million.

But despite this early start, East Asia was largely rural at the end of World War II. Some 90 percent of China's people then lived in the countryside, and even Japan was only about 50 percent urbanized. But as the region's economy began to grow after the war, so did its cities. Japan, Taiwan, and South Korea are now between 78 and 86 percent urban, which is typical for advanced industrial countries. Most of China's people still live in rural areas, but within a few years the country will have an urban majority.

Chinese Cities Traditional Chinese cities were clearly separated from the countryside by defensive walls. Most were planned in accordance with strict geometrical principles that were thought to reflect the cosmic order. The old-style Chinese city was dominated by low buildings and characterized by straight streets. Houses were typically built around courtyards, and narrow alleyways served both commercial and residential functions.

China's urban fabric began to change during the colonial period. A group of port cities was taken over by European interests, which proceeded to build Western-style buildings and modern business districts. By far the most important of these semicolonial cities was Shanghai, built near the mouth of the Yangtze River, the main gateway to interior China.

When the communists came to power in 1949, Shanghai, with a population of more than 10 million, was the second-largest city in the world. The new authorities, however, viewed it as a decadent, foreign creation. They therefore milked it for taxes, which they invested elsewhere. Most of the wealthy people fled to Hong Kong, and relatively few migrants were allowed in. As a result, much of the city began to decay.

Since the late 1980s, Shanghai has experienced a major revival and is again in many respects China's premier city. Migrants are now pouring into the city, even though the state still tries to restrict the flow, and building cranes crowd the skyline. Official statistics now put the population of the municipality at more than 19 million, but the actual number is greater, as Shanghai probably holds around 3 million undocumented migrants from other parts of China (see "People on the Move: Rural Migrants Flock to China's Cities). The new Shanghai is a city of massive high-rise apartments and concentrated industrial developments (Figure 11.19). In 2008, Shanghai had the second busiest port in the world (after Singapore) and boasted the world's fastest train. But despite Shanghai's revived economic fortunes, the city remains politically secondary to Beijing, China's capital.

Beijing was China's capital during the Manchu period (1644–1912), a status it regained in 1949. Under communist rule, Beijing was radically transformed; old buildings were razed and broad avenues were plowed through neighborhoods (Figure 11.20). Crowded residential districts gave way to large blocks of apartment buildings and massive government offices. Some historically significant structures were saved; the buildings of the Forbidden City, for example, where the Manchu rulers once lived, survived as a complex of museums. The area immediately in front of the Palace Museum, however, was cleared. The resulting plaza, Tiananmen Square, is reputed to be the largest open square in any city of the world. The Chinese government invested large amounts of money into Beijing's urban infrastructure in preparation for the city's 2008 Summer Olympics.

The Chinese Urban System China's urban system as a whole is fairly well balanced, with sizable cities relatively evenly spaced across the landscape and with no city overshadowing all others. This balance stems from China's heritage of urbanism, its vast size and distinctive physical–geographical regions, and its legacy of socialist planning. In the early 1960s, noted China scholar G. William Skinner argued that the distribution of Chinese cities is best explained through the use of **central place theory**. Central place theory holds that an evenly distributed rural population will give rise to a regular hierarchy of urban places, with uniformly spaced larger cities surrounded by constellations of smaller cities, each of which, in turn, will be surrounded by smaller towns.

In the 1990s, Beijing and Shanghai vied for the first position among Chinese cities, with Tianjin, serving as Beijing's port, coming in third. All three of these cities, along with Chongqing in the Sichuan Basin, have been removed from the regular provincial structure of the country and granted their own metropolitan governments. In 1997, another major city, Hong Kong, passed from British to Chinese control. Rather than becoming an ordinary part of China, Hong Kong was granted a unique status as an autonomous "special administrative region," and allowed to largely manage its own affairs. Although not as populous as Beijing or Shanghai, Hong Kong is far wealthier. The greater metropolitan area of the Xi Delta, composed of Hong Kong, Shenzhen, and Guangzhou (called Canton in the

Figure 11.19 Contemporary Shanghai This vibrant city of approximately 20 million embodies the new China with its massive high-rise apartments, industrial developments, and office towers.

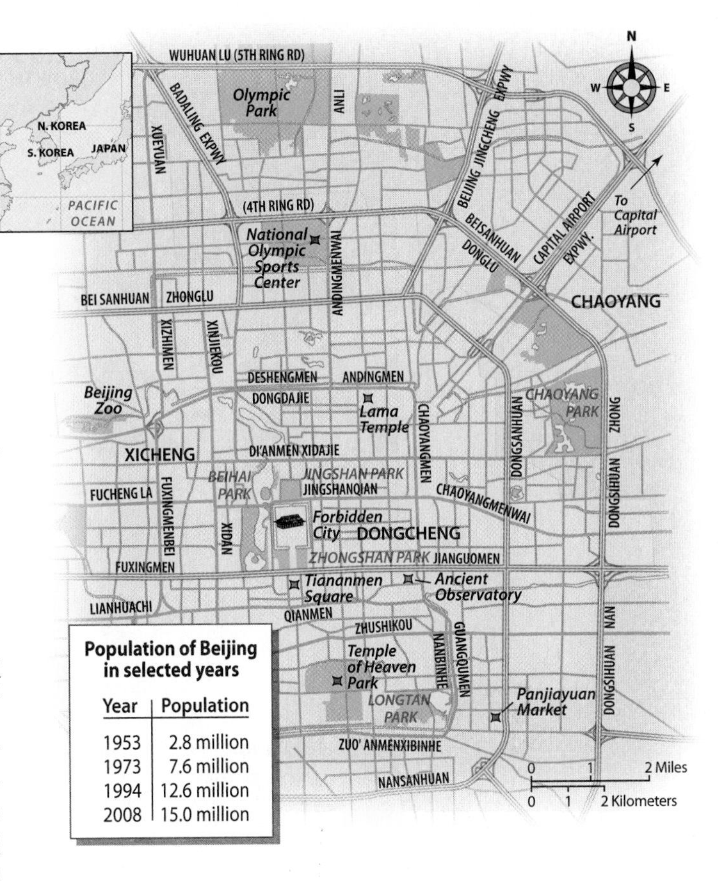

Population of Beijing in selected years

Year	Population
1953	2.8 million
1973	7.6 million
1994	12.6 million
2008	15.0 million

Figure 11.20 Beijing, China The historic capital during the Manchu period (1644–1912), Beijing regained its status as capital city in 1949. Under communist rule, much of Beijing's historical landscape was razed and replaced with large blocks of government offices and massive apartment buildings. Only the historically significant Forbidden City, home to the Manchu rulers, was saved from this transformation of the urban landscape. Tiananmen Square, reputed to be the largest open square in any city, was created by clearing buildings away from the area in front of the Palace Museum.

West), is certainly one of China's premier urban areas. In addition to these four primary Chinese cities are several dozen large and growing urban centers. By 2010, more than 40 of them had more than 1 million inhabitants.

Urban Patterns in Japan, Korea, and Taiwan Unlike China, South Korea and Taiwan are noted for their urban primacy, with their urban population concentrated in a single city. Japan supports a more extensive urban phenomenon called a superconurbation, or megalopolis, which is a huge zone of coalesced metropolitan areas (Figure 11.21).

Seoul, the capital of South Korea, overwhelms all other cities in the country. Seoul itself is home to more than 10 million people, and its metropolitan area contains between 20 and 24 million, almost half of South Korea's population. All of South Korea's major governmental, economic, and cultural institutions are concentrated there. Seoul's explosive and generally unplanned growth has resulted in severe congestion. The South Korean government is promoting industrial growth in other cities, but none of its actions has yet challenged the primacy of the capital.

Taiwan is similarly characterized by a high degree of urban primacy. The capital city of Taipei, located in the far north, mushroomed from some 300,000 people during the Japanese colonial period (from 1895 to 1945) to more than 6 million in the metropolitan area today.

Japan has traditionally been characterized by urban "bipolarity" rather than urban primacy. Until the 1960s, Tokyo, the capital and main business and educational center, together with the neighboring port of Yokohama, was balanced by the mercantile center of Osaka and its port of Kobe. Kyoto, the former imperial capital and the traditional

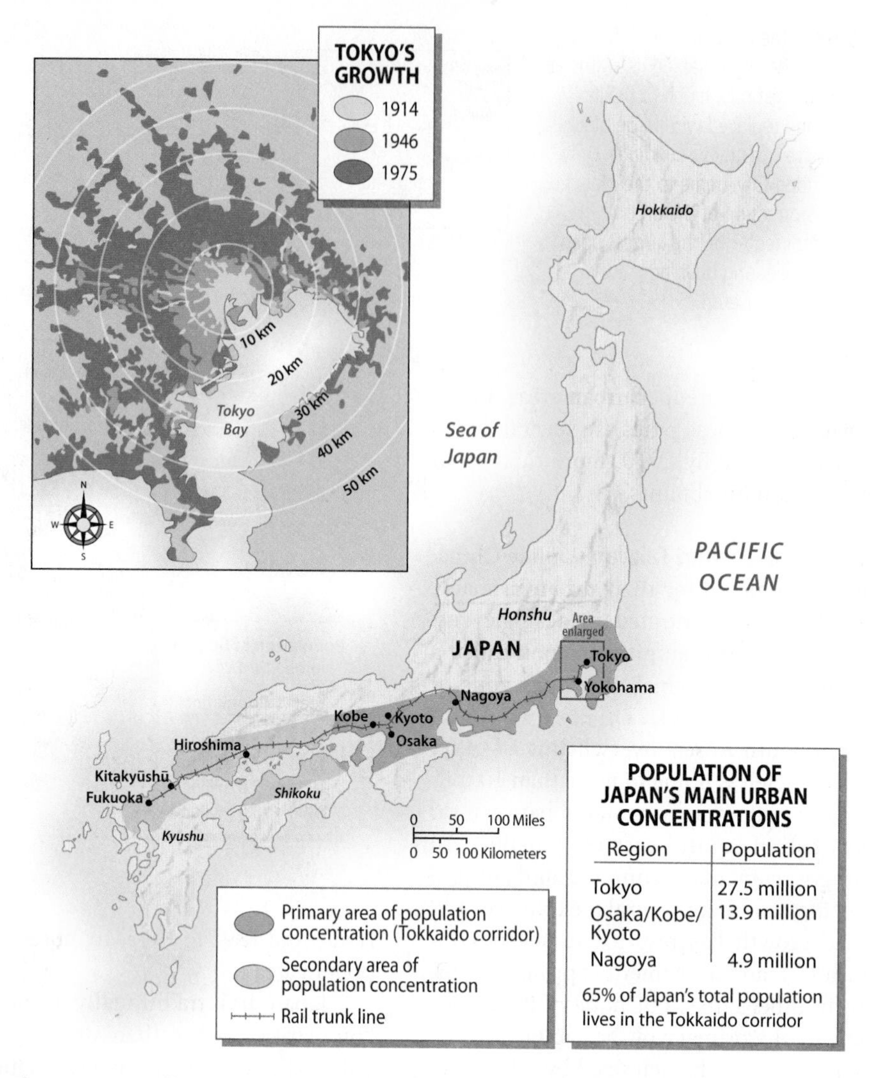

Figure 11.21 Urban Concentration in Japan The inset map shows the rapid expansion of Tokyo in the postwar decades. Today the Greater Tokyo metropolitan area is home to almost 30 million people. The larger map shows the cluster of urban settlements along Japan's southeastern coast. The major area of urban concentration is between Tokyo and Osaka, a distance of some 300 miles (482 kilometers), known as the Tokkaido corridor. By some accounts, 65 percent of Japan's population lives in this area.

cultural center, is also situated in the Osaka region. A host of secondary and tertiary cities balance Japan's urban structure. Nagoya, with a metropolitan area of some 5 million persons, remains the center of the automobile industry and is one of the few large Japanese cities in which it is more efficient to travel by car than by public transportation.

As Japan's economy boomed in the 1960s, 1970s, and 1980s, so did Tokyo (see "Cityscapes: The Varied Neighborhoods of Tokyo"). The capital city then outpaced all other urban areas in almost every urban function. While Tokyo itself has about 13 million inhabitants, the Greater Tokyo area now contains 35 to 39 million people, making it the world's most populous urbanized area. The Osaka–Kobe–Kyoto metropolitan area stands second, with 18 million inhabitants. Concerned about the increasing primacy of Tokyo, the Japanese government has been trying to steer new developments to other parts of the country.

Most of Japan's other main cities have seen modest population gains over the past several decades. Urban growth, in general, has been supported by rural depopulation. Metropolitan expansion has been particularly pronounced in the cities linking Tokyo to Osaka, an area known as the Tokkaido corridor. Transportation connections are superb along this route, and proximity to Tokyo, Osaka, and Nagoya encourages development. The result has been the creation of a superconurbation (see Figure 11.21). One can travel from Tokyo to Osaka on the main rail line, a distance of almost 300 miles (480 kilometers), and never leave the urbanized area. By some accounts, 65 percent of Japan's people are crowded into this narrow supercity.

PEOPLE ON THE MOVE Rural Migrants Flock to China's Cities

China is currently the site of the largest migration stream in human history. Over the last several decades, more than 100 million rural Chinese have moved to the country's burgeoning cities, a number that is expected to increase to nearly 250 million by 2025. According to recent census data, most of these migrants remain in their home provinces. As a result, midsized cities across China are rapidly growing. More than a third of such migrants, however, cross provincial lines, most of them heading from the interior to the booming cities of the coastal zone.

Several factors contribute to this remarkable process of relocation. Agricultural wages have been relatively stagnant for years, whereas those in the urban sector have been rising steadily. China's farms tend to be too small to use labor efficiently; as a result, the country has up to 300 million "surplus" agricultural workers. The rapid growth of urban infrastructure, meanwhile, generates a large demand for unskilled workers. Rural areas also tend to have higher birthrates than cities, as China's one-child policy is not rigorously enforced in the countryside. With few opportunities at home, young people from rural environments often head to the cities when they come of age.

Although the Chinese economy benefits in many ways from such migration, the government is concerned that the process could get out of hand, generating squatter settlements and other forms of urban blight. China also wants to maintain close control over its citizens, and is thus wary of granting the freedom to relocate. Such policies stem both from the authoritarian nature of the Chinese state and from traditional patterns of government. China has long maintained a comprehensive system of household registration, known as the *hukou* system. Registration papers include information about a person's age, family background, marital status, and legal place of residence. Rural migrants in urban areas often have no legal standing in the city; as far as the government is concerned, they are mere visitors (Figure 11.2.1). To obtain governmental services, such people must either return to their home villages or pay hefty bribes.

The hukou system is especially burdensome in regard to education. The estimated 200 million Chinese citizens who live outside of their areas of registration usually have difficulty enrolling their children in schools. If they are to be educated, such children must either move back to the countryside and live with relatives or their parents must find a away around the registration system. But rural schools are often poor in quality, and many migrant parents are too poor to get their children enrolled in urban schools. And even if they are able to attend school in the city, students from unregistered rural backgrounds must return to their home villages to take college entrance exams.

The hukou system has been the subject of intense discussion in China in recent years. Critics contend that it creates a two-tier system of the "haves" (registered urban dwellers) and "have nots" (those registered in the countryside). Some fear that vast numbers of migrant children are not receiving adequate education, thus imperilling the future of the country. As a result, China has recently eased education restrictions on rural migrants. Some voices are calling for the registration system to be abolished altogether, but thus far the government shows no signs of contemplating such a radical reform.

Figure 11.2.1 Rural Migrants in a Chinese City Rural migrants are often unable to acquire official residency in China's cities. As a result, they are often forced to perform difficult and dangerous work for low wages.

Japanese cities sometimes strike foreign visitors as rather gray and monotonous places, lacking historical interest (Figure 11.22). Little of the country's premodern architecture remains intact. Traditional Japanese buildings were made of wood, which survives earthquakes much better than stone or brick. Fires have therefore been a long-standing hazard, and in World War II the U.S. Air Force firebombed most Japanese cities, virtually obliterating them. In addition, Hiroshima and Nagasaki were completely destroyed by atomic bombs. The one exception was Kyoto, the old imperial capital, which was spared devastation. As a result, Kyoto is famous for its beautiful (wooden) Buddhist monasteries and Shinto temples, which ring the basin in which central Kyoto lies.

Other Japanese cities were largely reconstructed in the 1950s and 1960s, a period when Japan was still relatively poor and could afford only inexpensive concrete buildings. In the boom years of the 1980s, however, modern skyscrapers rose in many of the larger cities, and postmodernist architecture began to lend variety, especially to the wealthier neighborhoods. Critics sometimes contend that Japanese cities still have relatively few features that show a distinctive cultural flavor. Others disagree, noting that if one looks closely, one may see aspects of Japanese and, more generally, East Asian culture everywhere.

Figure 11.22 Tokyo Apartments The extremely high population density of Tokyo and other large Japanese cities forces most people to live in crowded apartment blocks. The photograph depicts Tokyo's Danchi high-rise apartment complex.

CULTURAL COHERENCE AND DIVERSITY: A Confucian Realm?

East Asia is in some respects one of the world's most unified cultural regions. Although different parts of East Asia have their own unique cultural features, the entire region shares certain historically rooted ways of life and systems of ideas.

Most of these East Asian commonalties can be traced back to ancient Chinese civilization. Chinese civilization emerged roughly 4,000 years ago, largely in isolation from the Eastern Hemisphere's other early centers of civilization in the valleys of the Indus, Tigris-Euphrates, and Nile rivers. As a result, East Asian civilization developed along several unique lines.

Unifying Cultural Characteristics

The most important unifying cultural characteristics of East Asia are related to religious and philosophical beliefs. Throughout the region, Buddhism and Confucianism have shaped not only individual beliefs but also social and political structures. Although the role of traditional belief systems has been seriously challenged over the past 50 years, especially in China, historically grounded cultural patterns have not disappeared.

Writing Systems The clearest distinction between East Asia and the world's other cultural regions is found in written language. Existing writing systems elsewhere in the world are based on the alphabetic principle, in which each symbol represents a distinct sound. East Asia, in contrast, evolved an entirely different system, **ideographic writing**. In ideographic writing, each symbol (commonly called a character) primarily represents an idea rather than a sound (although the symbols can denote sounds in certain circumstances). As a result, ideographic writing requires the use of a large number of distinct symbols.

The East Asian writing system can be traced to the dawn of Chinese civilization, some 3,200 years ago. As the Chinese Empire expanded and the prestige of Chinese civilization carried its culture to other lands, the Chinese writing system spread. Japan, Korea, and Vietnam all came to use the same system, although in Japan it was substantially modified, while in Korea and Vietnam it was much later largely replaced by alphabetic systems.

The Chinese ideographic writing system has one major disadvantage and one major advantage when compared with alphabetic systems, both of which stem from the fact that it is largely divorced from spoken language. The disadvantage

CITYSCAPES The Varied Neighborhoods of Tokyo

Although Tokyo is a huge city in a vast metropolitan area, it can feel like a small, intimate place. This is partly because it is divided into 23 districts (or *ku*, in Japanese), each of which has its own distinctive character and specializes in specific goods and services (Figure 11.3.1). Fortunately, Tokyo has a superb system of public transportation, making it relatively easy to get from one neighborhood to another (Figure 11.3.2).

Several Tokyo neighborhoods are noted as entertainment districts. For people looking for exciting nightlife and a cosmopolitan atmosphere, Roppongi is the place to go. Emerging around a now-defunct U.S. military base, Roppongi is the most Westernized part of Japan, and thus appeals to many tourists and expatriates. Younger Japanese looking for fun and fashion more often head for Harajuku, famous for its boutiques and sidewalk cafes. Well-heeled business executives, on the other hand, tend to favor the Akasaka nightlife district, celebrated for its expensive hotels and "hostess bars."

Other Tokyo neighborhoods are better known for their shopping potential. Some are highly specialized. Akihabara, for example, has more than 600 stores focused on electronics. Tsukiji, on the other hand, boasts the largest fish market in Japan and perhaps the world. The most expensive shops—and some of the world's most expensive real estate—are found in the Ginza, a swanky neighborhood also noted for its art galleries and exclusive clubs. Tokyo's department store center is found in Shibuya, which also has an active and rather hip nightlife scene. One also can shop, and do much more, in Shinjuku, Tokyo's main business district. Reminiscent of midtown Manhattan, Shinjuku boasts a collection of impressive skyscrapers.

Other Tokyo neighborhoods are better known for their historical significance. Although Hibiya is an important financial district, it is more famous as the site of the Imperial Palace and the center of the old city. Ueno, also part of the old downtown, supports the National Museum, which has the world's largest collection of Japanese art.

Tourists less often venture into Tokyo's other districts, but many are worth the effort. If one wants to see the rougher side of the city—which, by U.S. standards, is not very rough at all—head for Ikebukuro, which has aptly been described as "working man's Tokyo." For the geographically inclined, one can do no better than simply walk the narrow back streets and encounter the unexpected. Turn one corner, and you might find an entire street of small-scale machine shops, each of which is relatively open to visual inspection; turn another, and you might come across a large assemblage of tiny printing establishments. For a pedestrian's encounter of diversity amid globalization, few places are more interesting and accessible than Tokyo.

Figure 11.3.2 Tokyo Train Station Tokyo boasts an extremely efficient public transportation system, based largely on subways and commuter rail lines. Trains and train stations, however, can be very crowded during rush hours.

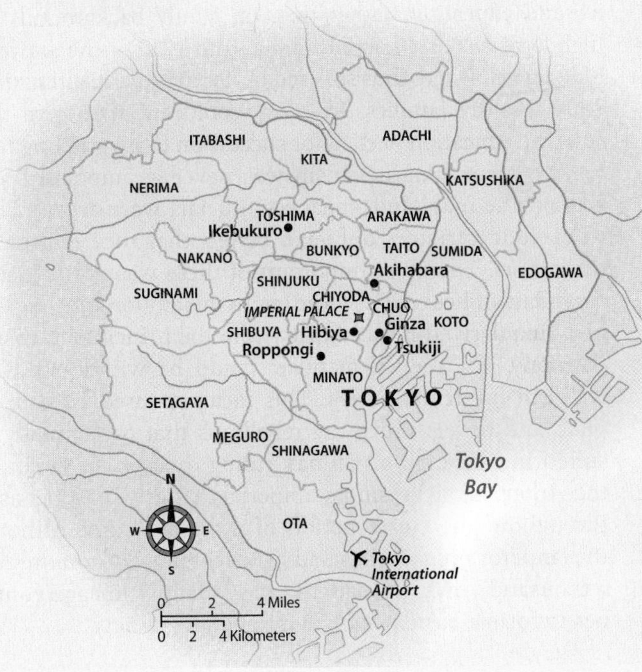

Figure 11.3.1 Neighborhoods and Districts of Tokyo Tokyo is divided into numerous administrative districts, each of which has an individual personality. Each district, in turn, contains a number of distinctive neighborhoods.

is that it is difficult to learn; to be literate, a person must memorize thousands of characters. The advantage is that two literate people do not have to speak the same language to be able to communicate, because the written symbols that they use to express their ideas are the same.

In Korea, Chinese characters were adopted at an early date and were used exclusively for hundreds of years. In the 1400s, however, Korean officials decided that Korea needed its own alphabet. They wanted to allow more widespread literacy, hoping also to more clearly differentiate Korean culture from that of China. The use of the new script spread quickly through the country. Korean scholars and officials, however, continued to use Chinese characters, regarding their own script as suitable only for popular writings. Today the Korean script is used for almost all purposes, but scholarly works still contain scattered Chinese characters.

The writing system of Japan is even more complex (Figure 11.23). Initially the Japanese simply borrowed Chinese characters, referred to in Japanese as kanji. Owing to the grammatical differences between Japanese and Chinese, the exclusive use of kanji resulted in awkward sentence construction. The Japanese solved this quandary by developing a quasi-alphabet known as hiragana that allowed the expression of words and parts of speech not easily represented by Chinese characters. In hiragana, each symbol represents a distinct syllable, or combination of a consonant and a vowel sound. (A different but essentially parallel system, called katakana, is used in Japan for spelling words of foreign origin.) Eventually the two styles of writing merged, and written Japanese came to employ a complex mixture of symbols. Japanese kanji today differ slightly from Chinese characters.

Figure 11.23 Japanese Writing The writing system of Japan was originally based on Chinese characters, known in Japan as kanji. Because of grammatical differences, however, the Japanese developed two unique "alphabets" of syllables, known as katakana and hiragana. Here kanji and katakana symbols are visible.

The Confucian Legacy Just as the use of a common writing system helped forge cultural linkages throughout East Asia, so too **Confucianism** (the philosophy developed by Confucius) came to occupy a significant position in all of the societies of the region. So strong is the heritage of Confucius that some writers refer to East Asia as the "Confucian world."

The premier philosopher of Chinese history, Confucius (or *Kongfuzi*, in Mandarin Chinese) lived during the 6th century BCE, a period of marked political instability. Confucius's goal was to create a philosophy that could generate stability. Although Confucianism is often considered to be a religion, Confucius himself was mostly interested in the "here and now," focusing his attention on how to lead a correct life and organize a proper society. Confucian thought does not deny the existence of a deity or an afterlife, but neither does it give them much consideration.

Confucius stressed deference to the proper authority figures, but he also thought that authority has a responsibility to act in a benevolent manner. The most basic level of the traditional Confucian moral order is the family unit, considered the bedrock of society. The ideal family structure is patriarchal, and children are told to obey and respect their parents—especially their fathers—as well as their elder brothers.

Confucian philosophy also stresses the need for a well-rounded and broadly humanistic education. To a certain extent, Confucianism advocates a kind of meritocracy, holding that an individual should be judged on the basis of behavior and education, rather than on family background. The high officials of Imperial China (pre-1912)—the powerful **Mandarins**—were thus selected by competitive examinations. Only wealthy families, however, could afford to give their sons the education needed for success on those grueling tests.

In Japan, Confucianism was never as important as it was on the mainland. Japanese officials were actually able to exclude certain Confucian beliefs that they considered dangerous. The most important of these was the revocable "mandate of heaven." According to this notion, the emperor of China derived his authority from the principle of cosmic harmony, but such a mandate could be withdrawn if he failed to fulfill his duties. This idea was used both to explain and to legitimize the rebellions that occasionally resulted in a change of China's ruling dynasty. In Japan, on the other hand, a single imperial dynasty has persisted throughout the entire period of written history. Although the emperor of Japan has had little real power for more than a thousand years, the sanctity of his family lineage continues to form a basic principle of Japanese society.

The Modern Role of Confucian Ideology The significance of Confucianism in East Asian development has been hotly debated for the past hundred years. In the early 1900s, many observers believed that the conservatism of the philosophy, derived from its respect for tradition and authority, was responsible for the economically backward position of China and Korea. But because East Asia has enjoyed the world's fastest rates of economic growth over the past several decades, such a position is no longer supportable. Some scholars now argue that Confucianism's respect for education and the social stability that it generates give East Asia an advantage in international competition.

Over the past 100 years, Confucianism has lost much of the hold that it once had on public morality throughout East Asia, especially in China. After the communist revolution, China's rulers sought for decades to discourage if not eliminate Confucian thought. Currently, however, Chinese officials are pushing for a revival of the philosophy, hoping that it will lead to enhanced social stability. Confucian schools are multiplying, and recent books, films, and television shows about Confucius have been popular. The Chinese government also maintains 145 Confucius Institutes in more than 50 countries, designed to promote Chinese culture and language.

Figure 11.24 The Buddhist Landscape Mahayana Buddhism has been traditionally practiced throughout East Asia. This Golden Buddha statue is located in Baomo Park in Chi Lei Village, Guangdong Province, China.

Religious Unity and Diversity

Certain religious beliefs have worked alongside Confucianism to cement together cultures of the East Asian region. The most important culturally unifying beliefs are associated with Mahayana Buddhism. Other religious practices, however, have had a more divisive role.

Mahayana Buddhism Originating in India in the 6th century BCE, Buddhism stresses escape from an endless cycle of rebirths to reach union with the divine cosmic principle (or nirvana). By the 2nd century CE, Buddhism was spreading through China, and within a few hundred years it had expanded throughout East Asia. Today Buddhism remains widespread everywhere in the region, although it is far less significant here than it is in mainland Southeast Asia and Sri Lanka.

The variety of Buddhism practiced in East Asia—Mahayana, or Greater Vehicle—is distinct from the Theravada Buddhism of Sri Lanka and Southeast Asia (Figure 11.24). Most important, Mahayana Buddhism simplifies the quest for nirvana, in part by suggesting that entities (boddhisatvas) exist who refuse divine union for themselves in order to help others spiritually. Mahayana Buddhism is also nonexclusive; in other words, one may follow it while simultaneously practicing other faiths. Thus, many Chinese consider themselves to be both Buddhists and Taoists, whereas most Japanese are at some level both Buddhists and followers of Shinto.

As Mahayana Buddhism spread through East Asia in the medieval period, many different sects emerged. Probably the best known is Zen, which demands that its followers engage in the rigorous practice of "mind emptying." At one time, Buddhist monasteries associated with Zen and other sects were rich and powerful. In all East Asian countries, however, periodic reactions against Buddhism resulted in the persecution of monks and the suppression of monasteries. One reason for this opposition was the fact that government officials often viewed Buddhism as a foreign religion that placed India—rather than China—at the center of the world.

Despite such hardships, East Asian Buddhism was never extinguished. But it also never became the focal point of society that it did in mainland Southeast Asia. In Japan, that position was partially captured by a different religion, Shinto.

Shinto The religious practice of Shinto is so closely bound to the idea of Japanese nationality that it is questionable whether a non-Japanese person can follow it. Shinto began as the animistic worship of nature spirits, but it was gradually refined into a subtle set of beliefs about the harmony of nature and its connections with human existence. Until the late 1800s, Buddhism and Shinto were complexly intertwined. Subsequently, the Japanese government began to disentangle the two faiths while elevating Shinto into a nationalistic cult focused on the divinity of the Japanese imperial family. After World War II, the more excessive aspects of nationalism were removed from the religion.

Shinto is still a place- and nature-centered religion. Certain mountains, particularly the volcanic Mount Fuji, are considered sacred and are thus climbed by large numbers of people. Major Shinto shrines, often located in scenic places, attract numerous pilgrims; most notable is the Ise Shrine south of Nagoya, site of the cult of the emperor. Local Shinto shrines, as well as Buddhist temples, offer leafy oases in otherwise largely treeless Japanese urban neighborhoods.

Taoism and Other Chinese Belief Systems The Chinese religion of Taoism (or Daoism) is similarly rooted in nature worship. Like Shinto, it stresses the acquisition of spiritual harmony and the pursuit of a balanced life. Taoism is indi-

Figure 11.25 The Language Geography of East Asia The linguistic geography of Korea and Japan is very straightforward, as the vast majority of people in those countries speak Korean and Japanese, respectively. In China, the dominant Han Chinese speak a variety of closely related Sinitic languages, the most important of which is Mandarin Chinese. In the peripheral regions of China, a large number of languages—belonging to several different linguistic families—can be found.

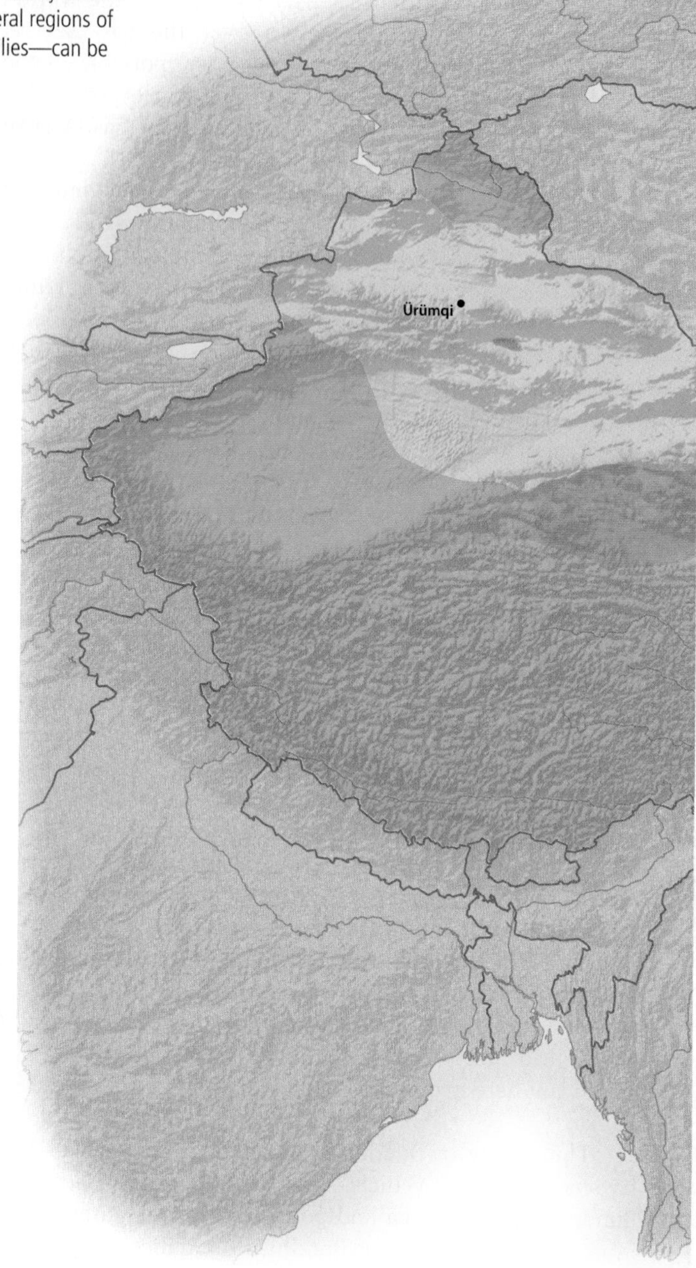

rectly associated with feng shui, also called geomancy, the Chinese and Korean practice of designing buildings in accordance with the spiritual powers that supposedly course through the local topography. Even in hypermodern Hong Kong, skyscrapers worth millions of dollars have occasionally gone unoccupied because their construction failed to accord with geomantic principles. The government of China officially recognizes Taoism as a religion, and seeks to regulate it through the state-run China Taoist Association.

Despite the fact that both Taoism and Buddhism were historically followed throughout the country, traditional religious practice in China has always embraced local particularism; in other words, it has focused on the unique attributes of particular places. Many minor gods were traditionally associated with single cities or other specific areas. In modern-day rural China, village gods are often still honored.

Minority Religions Followers of virtually all world religions can be found in the increasingly cosmopolitan cities of East Asia. Christianity is well represented. Over a million Japanese belong to Christian churches, whereas South Korea is roughly 30 percent Christian (mostly Protestant). South Korea reportedly sends more missionaries abroad than any country except the United States. Some reports indicate that Christianity is growing rapidly in China—despite state discouragement—but reliable information is scarce. In 2008, official statistics put the number of Chinese Protestants at 20 million and Chinese Catholics at 10 million; some experts think that an additional 50 million or so Chinese belong to unofficial "house churches."

China also has a long-established Muslim community, estimated as having between 20 and 60 million adherents. Most Chinese Muslims are members of minority ethnic groups living in the far west. The roughly 10 million Chinese-speaking Muslims, called *Hui,* are concentrated in Gansu and Ningxia in the northwest and in Yunnan province in the south. Smaller clusters of Hui, often segregated in their own villages, live in almost every province of China. The only Muslim congregations in Japan and South Korea, on the other hand, are associated with recent and often temporary immigrants from South, Southeast, and Southwest Asia.

Secularism in East Asia For all of these varied forms of religious expression, East Asia is one of the most secular regions of the world. In Japan, most people occasionally observe Shinto or Buddhist rituals and maintain a small shrine for their ancestors, but only a small segment of the population is deeply religious. As a result, statistics on religious affiliation in Japan vary tremendously, with estimates of the followers of Shinto ranging between 4 million and 119 million! Japan also has a number of "new religions," sometimes called cults, a few of which are noted for their fanaticism. But for Japanese society as a whole, religion is not tremendously important.

After the communist regime took power over China in 1949, all forms of religion and traditional philosophy were discouraged and sometimes severely repressed. Under the new regime, atheistic Marxist philosophy—the communistic belief system developed by Karl Marx—became the official ideology. In the 1960s, many observers thought that traditional forms of Chinese religion would survive only in overseas Chinese communities. With the easing of Marxist orthodoxy during the

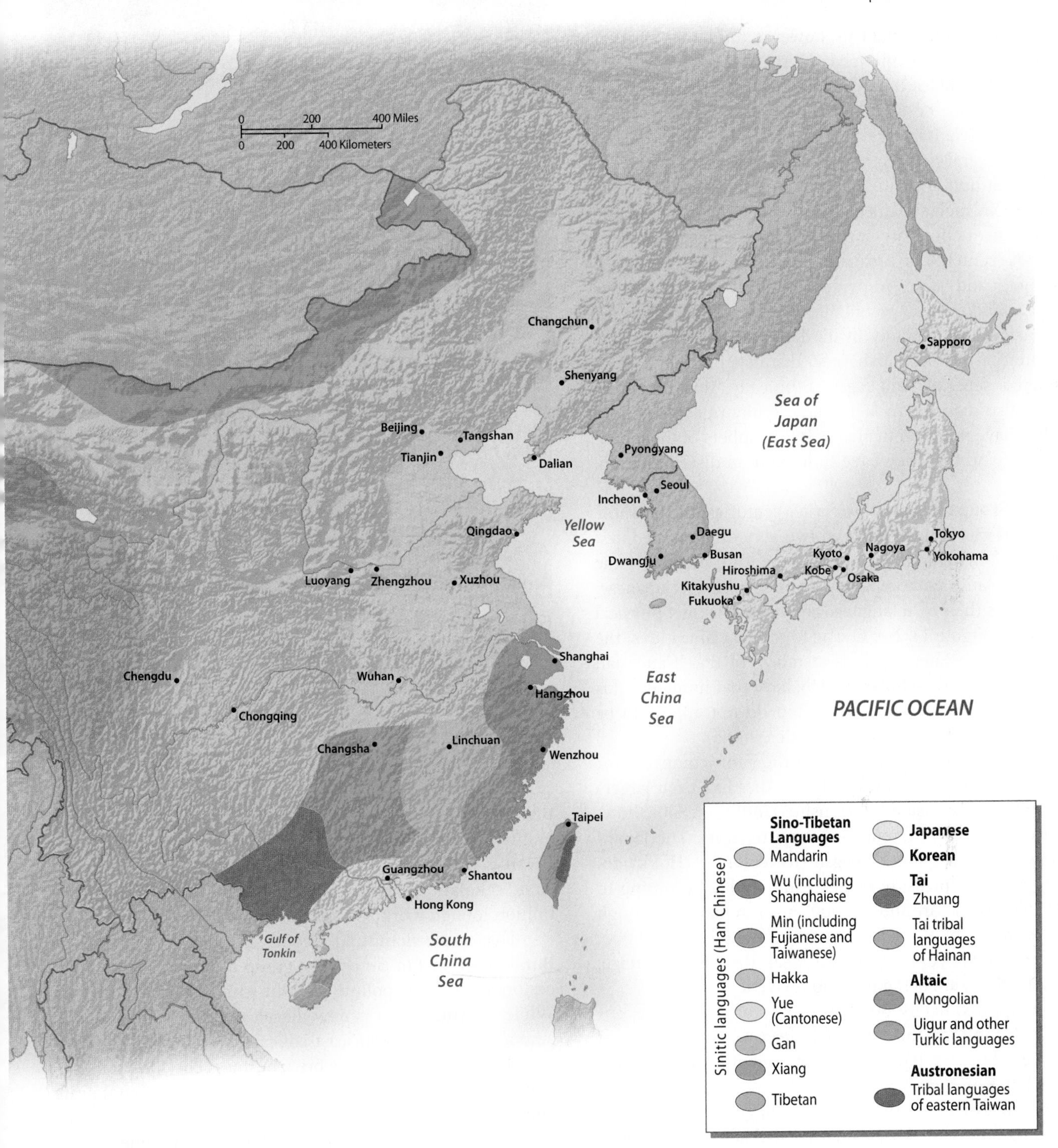

1980s and 1990s, however, many forms of spiritual expression began to return.

Like China, North Korea adopted communist ideology when it became independent after World War II. As a result, religious beliefs were severely repressed. North Korean leaders also supplemented Marxism with a unique official ideology called *juche*, or "self-reliance." Ironically, juche demands absolute loyalty to North Korea's political leaders. It also involves intense nationalism based on the idea of Korean racial purity. Over time, Marxist beliefs have gradually been supplanted by those of juche; in North Korea's revised 2009 constitution, all references to communism were removed.

Linguistic and Ethnic Diversity

Japanese and Mandarin Chinese may partially share a system of writing, but the two languages bear no direct relationship (Figure 11.25). In their grammatical structures, Chinese and Japanese are more different from each other than are Chinese and English. Japanese, however, has adopted many words of Chinese origin.

Language and National Identity in Japan According to most scholars, Japanese is not related to any other language. Korean is also usually classified as the only member of its language family. Many linguists, however, think that Japanese and Korean should be grouped together because they share many grammatical features. Only the mutual distrust between the Japanese and the Koreans, some suggest, has prevented this linguistic relationship from being acknowledged.

From several perspectives, the Japanese form one of the world's most homogeneous peoples, and they tend to regard themselves in such a manner. To be sure, minor cultural and linguistic distinctions are noted between the people of western Japan (centered on Osaka) and eastern Japan (centered on Tokyo), and many individual regions have distinctive customs and lifeways. Overall, however, such differences are of little significance. Only in the Ryukyu Islands does one encounter variants of Japanese so distinct that they can be considered separate languages. Ethnically, this region of Japan is also distinct. In fact, many Ryukyu people believe that they have not been treated as full members of the Japanese nation, and that they have suffered from discrimination.

Figure 11.26 Ainu Men The indigenous Ainu people of northern Japan are much reduced in population, but they still maintain a number of their cultural traditions. In this photograph, Ainu men participate in the Marimo Festival on the northern Japanese island of Hokkaido.

Minority Groups in Japan In earlier centuries, the Japanese archipelago was divided between two very different peoples: the Japanese living to the south, and the Ainu inhabiting the north. The Ainu are completely distinct from the Japanese. They possess their own language and have a distinct physical appearance (Figure 11.26). Owing to their facial features, the Ainu were once categorized as members of the "Caucasian race." However, few scholars now believe that humankind is divided into separate races, and Ainu do not appear to be closely related to Europeans by genetic criteria.

For centuries the Japanese and the Ainu competed for land, and by the 10th century CE the Ainu had been largely driven off the main island of Honshu. Until the 1800s, however, Hokkaido largely remained Ainu territory. The Japanese people subsequently began to colonize Hokkaido, putting renewed pressure on the Ainu. Today, according to official statistics, around 25,000 Ainu remain, although some sources put the actual number as high as 200,000. Fewer than 100 people, however, speak the Ainu language, although efforts are being made to revive it.

The approximately 800,000 people of Korean descent living in Japan today also have felt discrimination. Many of them were born in Japan (their parents and grandparents having left Korea early in the 1900s) and speak Japanese rather than Korean as their primary language. But despite their deep bonds to Japan, such individuals are not easily able to obtain Japanese citizenship. Perhaps as a result of such treatment, many Japanese Koreans hold radical political views and support North Korea.

Starting in the 1980s, other immigrants began to arrive in Japan, mostly from the poorer countries of Asia. Most do not have legal status. Men from China and southern Asia typically work in the construction industry and in other dirty and dangerous jobs; women from Thailand and the Philippines often work as entertainers and sometimes as prostitutes. Roughly 200,000 Brazilians of Japanese ancestry have returned to Japan for the relatively high wages they can earn. However, immigration is less pronounced in Japan than in most other wealthy countries, and relatively few migrants acquire permanent residency, let alone citizenship. Because of Japan's impending population loss, however, some scholars estimate that it may need to import several million foreign workers over the next decade.

The most victimized people in Japan could be the **Burakumin**, or *Eta*, an outcast group of Japanese whose ancestors worked in "polluting" industries such as leathercraft. While discrimination is now illegal, the Burakumin are still among the poorest and least-educated people in Japan. Private detective agencies do a brisk business checking prospective marriage partners and employees for possible Burakumin ancestries. The Burakumin, however, have banded together to demand their rights; the Buraku Liberation League is politically powerful and is reputed to have close connections with the Yakuza (the Japanese mafia). The Burakumin usually live in separate neighborhoods and are concentrated in the Osaka region of western Japan.

Language and Identity in Korea The Koreans, like the Japanese, are a relatively homogeneous people. The vast majority of people in both North and South Korea speak Korean and unquestioningly consider themselves to be members of the Korean nation.

South Korea does, however, have a strong sense of regional identity, which can be traced back to the medieval

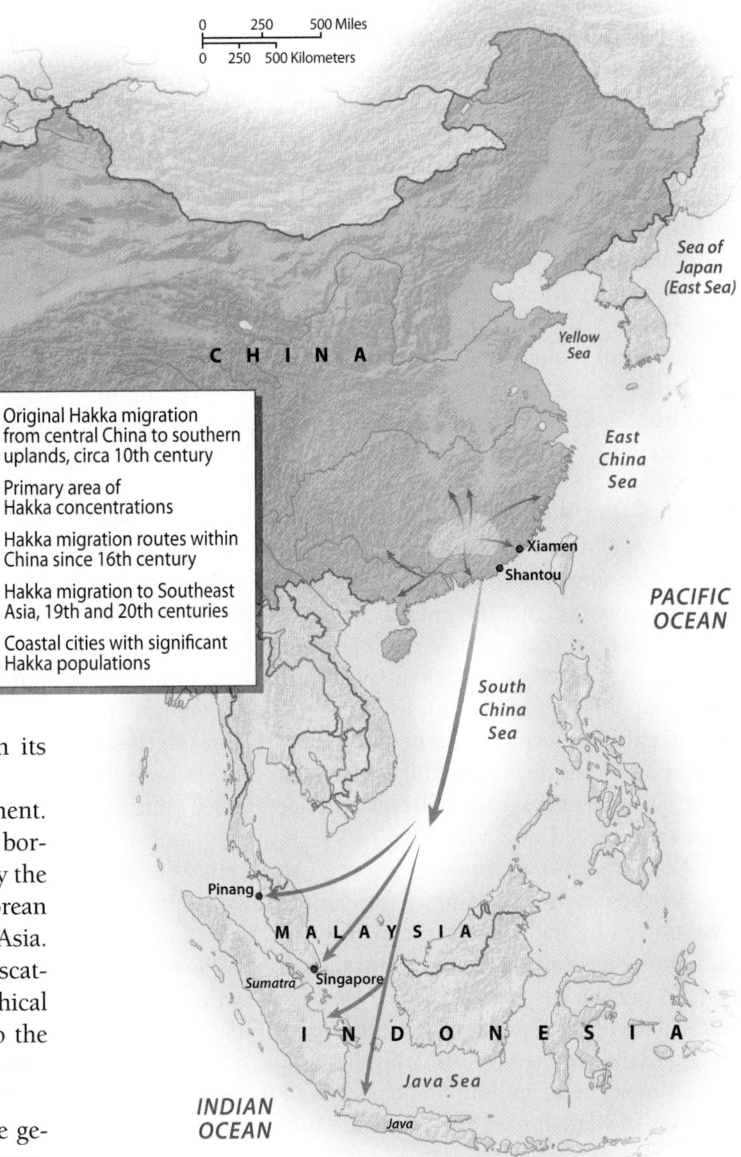

Figure 11.27 The Hakka Diaspora The Hakka form an important and distinctive subgroup of the Han Chinese. They seem to have originated in north-central China but long ago migrated to the area where Fujian, Guangdong, and Jiangxi provinces converge. Here they developed distinctive agricultural patterns well suited to the rough uplands of the region. Later movements took Hakka communities to upland areas throughout much of southern China. In the 1800s and early 1900s, many Hakka moved to Southeast Asia, where they still form important communities.

period when the peninsula was divided into three separate kingdoms. The people of southwestern South Korea, centered on the city of Kwangju, tend to be viewed as distinctive, and many southwesterners believe that they have suffered periodic discrimination. In contract, the Kyongsang region of southeastern South Korea has supplied most of the country's political leaders and has received more than its share of development funds.

Korean identity also has an international component. Several million Korean speakers reside directly across the border in northern China. Because of deportations ordered by the Soviet Union in the mid-20th century, substantial Korean communities also can be found in Kazakhstan in Central Asia. Over the past several decades, a Korean **diaspora**—the scattering of a particular group of people over a vast geographical area—has brought hundreds of thousands of Koreans to the United States, Canada, Australia, and New Zealand.

Language and Ethnicity Among the Han Chinese The geography of language and ethnicity in China is far more complex than that of Korea or Japan. This is true even if one considers only the eastern half of the country, so-called China proper. The most important distinction is that of separating the Han Chinese from the non-Han peoples. The Han, who form the vast majority, are those people who have long been incorporated into the Chinese cultural and political systems and whose languages are expressed through Chinese writing. The Han do not, however, all speak the same language.

Northern, central, and southwestern China—a vast area extending from Manchuria through the middle and upper Yangtze Valley to the valleys and plateaus of Yunnan in the far south—constitute a single linguistic zone. The spoken language here is generally called *Mandarin Chinese*. Mandarin is divided into a number of dialects, but the standard dialect of the Beijing area is gradually spreading. Standard Mandarin—called *Pǔtōnghuà* locally—is China's official national language.

In southeastern China, from the Yangtze Delta to China's border with Vietnam, a number of separate languages are spoken. Peoples speaking these languages are Han Chinese, but they are not native Mandarin speakers. Traveling from south to north, one encounters Cantonese (or Yue) spoken in Guangdong, Fujianese (alternatively Hokkienese, or Min, locally) spoken in Fujian, and Shanghaiese (or Wu), spoken in and around the city of Shanghai and in Zhejiang province. These are true languages, not dialects, since they are not mutually intelligible. They are usually called dialects, however, because they have no distinctive written form.

The Hakka are one group of people speaking a southern Chinese language that are occasionally not considered to be Han Chinese. Evidently, their ancestors fled northern China roughly a thousand years ago to settle in the rough upland area where Guangdong, Fujian, and Jiangxi provinces meet. Later migrations took them throughout southern China, where they typically settled in hilly wastelands (Figure 11.27). The Hakka traditionally made their living by growing upland crops such as sweet potatoes and by working as loggers, stonecutters, and metalworkers. Today they form one of the poorest communities of southern China.

Despite their many differences, all of the languages of the Han Chinese (including Hakka) are closely related to each other and belong to the same Sinitic language subfamily. Since their basic grammars and sound systems are similar, a person speaking one of these languages can learn another relatively easily. It is usually difficult, however, for speakers of European languages—or of Japanese or Korean—to gain fluency in these tongues. All Sinitic languages are **tonal** and monosyllabic; their words are all composed of a single syllable (although compound words can be formed from several syllables), and the meaning of each basic syllable changes according to the pitch in which it is uttered.

The Non-Han Peoples Many of the more remote upland districts of China proper are inhabited by groups of non-Han peoples speaking non-Sinitic languages. According to the Chinese government, the country contains 55 such ethnic groups. Most of these peoples are classified as tribal, implying that they have a traditional social order based on self-governing village communities (Figure 11.28). Such a view is not entirely accurate, however, because some of these groups once had their own kingdoms. What they do have in common is a heritage of cultural and sometimes political tension with the Han Chinese.

Over the course of many centuries, the territory occupied by these non-Han communities has steadily declined in size, caused by the continued expansion of the Han as well as by non-Han emigration. Acculturation into Chinese society and intermarriage with the Han also have reduced many non-Han groups. Their main concentrations today are in the rougher lands of the far north and the far south.

As many as 11 million Manchus live in the more remote portions of Manchuria. The Manchu language is related to those of the tribal peoples of central and southeastern Siberia. Fewer than 100 Manchus, however, still speak their own language, the rest having abandoned it for Mandarin Chinese. This is an ironic situation, because the Manchus ruled the Chinese Empire from 1644 to 1912. Until the end of this period, the Manchus prevented the Han from settling in central and northern Manchuria, which they hoped to preserve as their homeland. Once Chinese were allowed to settle in Manchuria in the 1800s—in part to prevent Russian expansion—the Manchus soon found themselves vastly outnumbered. As they began to intermarry with and adopt the language and customs of the newcomers, their own culture began to disappear.

Larger and more secure communities of non-Han peoples are found in the far south, especially in Guangxi. Most of the inhabitants of Guangxi's more remote areas speak languages of the Tai family, closely related to those of Thailand. Because as many as 18 million non-Han people live in Guangxi, it has been designated an **autonomous region**. Such autonomy was designed to allow non-Han peoples to experience "socialist modernization" at a different pace from that expected of the rest of the country. Critics contend that very little real autonomy has ever existed. (In addition to Guangxi, there are four other autonomous regions in China. Three of these, Xizang [Tibet], Nei Monggol [Inner Mongolia], and Xinjiang, are located in Central Asia and are discussed in Chapter 10. The final autonomous region, Ningxia, located in northwestern China, is distinguished by its large concentration of Hui [Mandarin-speaking Muslims].)

Other areas with sizable numbers of non-Han peoples are Yunnan and Guizhou, in southwestern China, and western Sichuan. Most tribal peoples here practice swidden agriculture (also called "slash and burn"; see Chapters 6 and 13) on rough slopes; flatter lands are generally occupied by rice-growing Han Chinese. A wide variety of separate languages, falling into several linguistic families, are found among the ethnic groups

Figure 11.28 Tribal Villages in South China Non-Han people are usually classified as "tribal" in China, which assumes they have a traditional social order based upon autonomous village communities. The photograph shows a Miao village in the Xiangxi Tujia and Miao Autonomous Prefecture, Hunan province.

living in the uplands. Figure 11.29 shows that in Yunnan, the resulting ethnic mosaic is staggeringly complex.

Language and Ethnicity in Taiwan Taiwan is also noted for its linguistic and ethnic complexity. In the island's mountainous eastern region, several small groups of "tribal" peoples speak languages related to those of Indonesia (belonging to the Austronesian language family). These peoples resided throughout Taiwan before the 16th century. At that time, however, Han migrants began to arrive in large numbers. Most of the newcomers spoke the Fujianese dialect, which evolved into the distinctive language of Taiwanese.

Taiwan was transformed almost overnight in 1949, when China's nationalist forces, defeated by the communists, sought refuge on the island. Most of the nationalist leaders spoke Mandarin, which they made the official language. Taiwan's new leadership discouraged Taiwanese, viewing it as a mere local dialect. As a result, tension developed between the Taiwanese and the Mandarin communities. Only in the 1990s did Taiwanese speakers begin to reassert their language rights, a movement that accelerated after 2000. Interestingly, as Taiwan has become more fully global, its own cultural framework has grown more locally specific.

East Asian Cultures in Global Context

East Asia has long exhibited tensions between an internal orientation and tendencies toward cosmopolitanism. This dichotomy has both a cultural and an economic dimension. Until the mid-1800s, East Asian countries attempted to insulate themselves from Western cultural influences. Japan subsequently opened its doors but remained ambivalent about foreign ideas. Only after its defeat in 1945 did Japan really opt for a globalist orientation. It was followed in this regard by South Korea, Taiwan, and Hong Kong (then a British colony). The Chinese and North Korean governments, to the contrary, decided during the early Cold War decades of the 1950s and 1960s to isolate themselves as much as possible from Western and global culture.

The Cosmopolitan Fringe Japan, South Korea, Taiwan, and Hong Kong are characterized by a vibrant internationalism, which coexists with strong national and local cultural identity. Virtually all Japanese, for example, study English for 6 to 10 years, and, although relatively few learn to speak it fluently, most can read and understand a good deal. Business meetings among Japanese, Chinese, and Korean firms are

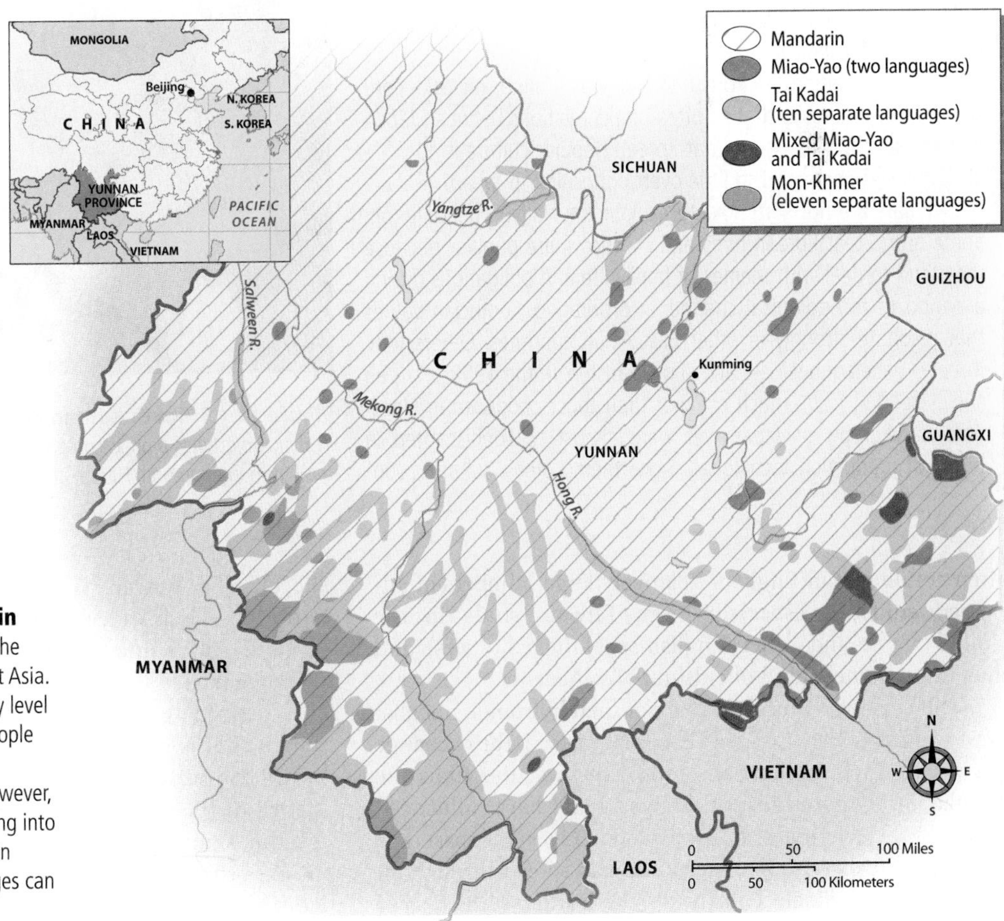

Figure 11.29 Language Groups in Yunnan China's Yunnan Province is the most linguistically complex area in East Asia. In Yunnan's broad valleys and relatively level plateau areas and in its cities, most people speak Mandarin Chinese. In the hills, mountains, and steep-sided valleys, however, a wide variety of tribal languages, falling into several linguistic families, are spoken. In certain areas, several different languages can be found in very close proximity.

GLOBAL TO LOCAL The Korean Wave

A century ago, Korea was commonly referred to as a "hermit kingdom," in reference to its relative isolation from global cultural and political currents. Today, North Korea retains the tradition, largely cutting itself off from the rest of the world. South Korea has taken the opposite approach, and is today one of Asia's most globalized countries. Cultural products as well as technological devices from all over the world are readily available in the streets of Seoul and other large South Korean cities.

In recent years, however, South Korea's most notable contribution to cultural globalization has been as an exporter rather than importer. In the 1990s, South Korean music, movies, television shows and even cuisine began to follow the country's electronic exports into markets throughout East and Southeast Asia and beyond. The Chinese media was so taken by the phenomenon that they dubbed it the *Hallyu,* or "Korean Wave." By 2004, some two-thirds of all television viewers in Japan were tuning into the popular Korean television drama, *Winter Sonata.* By 2006, Chinese television stations devoted more time to South Korean shows than to those of all other foreign countries put together. The Hallyu movement has in turn prompted massive tourism into South Korea, with the number of tourist arrivals going from 2.8 million in 2003 to 3.7 million in 2004. By 2010, Korean hair salons were spreading across the major cities of China, both due to the skill of Korean hairdressers and to the fact that many young Chinese want to look like Korean pop stars.

Not surprisingly, the Korean Wave has generated resentment in other Asian countries. Both Vietnam and Taiwan have threatened to ban or at least limit Korean broadcasting, while in Japan a comic book called "Halting the Korean Wave" was a minor hit. Still, considering the animosity that so often marks relations between Japan and Korea, it is remarkable that Korean popular culture is so well-liked and has generated so little overt opposition in Japan.

Scholars are not sure how to explain the Hallyu phenomenon. Some argue that Korean films and TV shows are popular because they tend to deal with themes of family life and obligations during a period of technological change, incorporating Confucian values that resonate strongly in other East Asian countries. Others look more to issues of business and showmanship. In the 1900s, Korean shows tended to be much cheaper than those produced in North America, Europe, or Japan, yet were almost as sophisticated in terms of production techniques. Since then, Korean cultural products have grown more technologically advanced, enhancing their appeal.

But whatever the explanation, it is undeniable that South Korean popular culture is big business. South Korean stars are now among the highest paid entertainers in the world, with at least 10 actors earning more than $10 million a year (Figure 11.4.1).

Figure 11.4.1 South Korean Film Star South Korea's noted film star Yi-Hyun So attends a press conference in regard to her new film "Jung" at the Pusan International Film Festival.

often conducted in English. Relatively large numbers of advanced students, especially from Taiwan, study in the United States and other English-speaking countries, and thus bring home a kind of cultural bilingualism.

The current cultural flow is not merely from a globalist West to a previously isolated East Asia. Instead, the exchange is growing more reciprocal. Hong Kong's action films are popular throughout most of the world and have influenced filmmaking techniques in Hollywood. Over the past few years, South Korean popular culture has spread throughout East Asia and in much of the rest of the world (see "Global to Local: The Korean Wave"). Japan almost dominates the world market in video games, and its ubiquitous comic-book culture and animation techniques are now following its karaoke bars in their overseas march.

Cultural globalization is, of course, as controversial in Japan as it is elsewhere. Japanese ultranationalists are few but vocal, calling their fellow citizens to resist the decadence of the West and to return to the military traditions of the **samurai**, the warrior class of premodern Japan. Many other Japanese people, however, contend that their country is too insular, worrying that they do not possess

Figure 11.30 Chinese Theme Park As China's economy grows, its people are spending increasing amounts of money on entertainment. Theme parks, which now number over 2,000, are particularly popular. Happy Valley Theme Park in Beijing, shown here, was China's largest when it opened in 2006.

the English-language and global cultural skills necessary to operate effectively in the world economy.

The Chinese Heartland In one sense, Japan is more culturally predisposed to cosmopolitanism than is China. The Japanese have always borrowed heavily from other cultures (particularly from China itself), whereas the Chinese have historically been more self-sufficient. However, the southern coastal Chinese have long tended to be more international in their orientation, with close linkages to the Chinese diaspora communities of Southeast Asia and ultimately to maritime trading circuits extending over much of the globe.

In most periods of Chinese history, the interior orientation of the center prevailed over the external orientation of the southern coast. After the communist victory of 1949, only the small British enclave of Hong Kong was able to pursue international cultural connections. In the rest of the country, a dour and puritanical cultural order was rigidly enforced. While this culture was largely founded on the norms of Chinese peasant society, it was also influenced by the communist-oriented cultural system that had emerged in the Soviet Union.

After China began to liberalize its economy and open its doors to foreign influences in the late 1970s and early 1980s, the southern coastal region suddenly assumed a new prominence. Through this gateway, global cultural patterns began to penetrate the rest of the country. The result has been the emergence of a vibrant but somewhat showy urban popular culture throughout China that is replete with such global features as nightclubs, karaoke bars, fast-food franchises, and theme parks (Figure 11.30).

GEOPOLITICAL FRAMEWORK: Enduring Cold War Tensions

Much of the political history of East Asia revolves around the centrality of China and the ability of Japan to remain outside China's grasp. The traditional Chinese conception of geopolitics was based on the idea of a universal empire: All territories were either supposed to be a part of the Chinese Empire, pay tribute to it and acknowledge its supremacy, or stand outside the system altogether. Until the 1800s, the Chinese government would not recognize any other as its diplomatic equal. When China could no longer maintain its power in the face of European aggression, the East Asian political system fell into disarray. As European power declined in the 1900s, China and Japan contended for regional leadership. After World War II, East Asia was split by larger Cold War rivalries (Figure 11.31).

The Evolution of China

The original core of Chinese civilization was the North China Plain and the Loess Plateau. For many centuries, periods of unification alternated with times of division into competing states. The most important episode of unification occurred in the 3rd century BCE. Once political unity was achieved, the Chinese Empire began to expand vigorously to the south of the Yangtze Valley. Subsequently, the ideal of the imperial unity of China triumphed, with periods of division seen as indicating disorder. This ideology helped cement the Han Chinese into a single people.

Several Chinese dynasties rose and fell between 219 BCE and 1912, most of them controlling roughly the same territory (Figure 11.32). The core of the Chinese Empire remained China proper, excluding Manchuria. Other lands, however, were sometimes ruled as well. The most important of these was a western projection extending north of the Tibetan Plateau into the desert basins of Central Asia (modern Xinjiang). China valued this area primarily because the vital trading route to western Eurasia (the "Silk Road") passed through it.

Various Chinese dynasties attempted to conquer Korea, but the Koreans resisted. Eventually China and Korea worked out an arrangement whereby Korea paid token tribute and acknowledged the supremacy of the Chinese Empire, and in return Korea received trading privileges and retained independence. When foreign armies invaded Korea—as did those of Japan in the late 1500s—China sent troops to support its "vassal kingdom."

For most of the past 2,000 years, the Chinese Empire was Earth's wealthiest and most powerful state. Its only real threat came from the pastoral peoples of Mongolia and Manchuria. Although vastly outnumbered by China, these societies were organized on a highly effective military basis. Usually the Chinese and the Mongols enjoyed a mutually beneficial trading relationship. Periodically, however, they waged war, and on several occasions the northern nomads conquered China (see Chapter 10). The Great Wall along

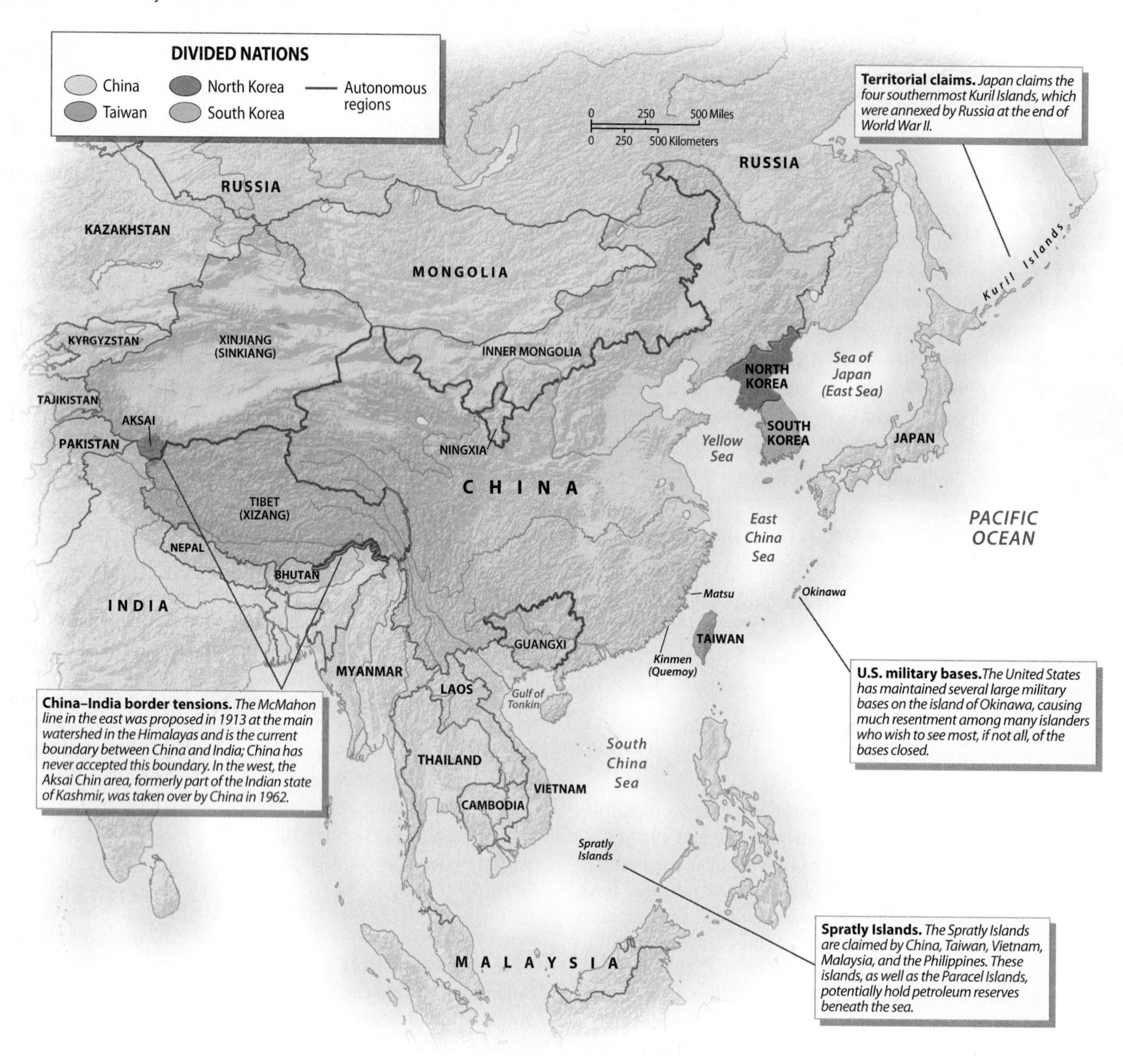

Figure 11.31 Geopolitical Issues in East Asia East Asia remains one of the world's geopolitical hot spots. Tensions are particularly severe between capitalist, democratic South Korea and the isolated communist regime of North Korea, and between China and Taiwan. China has had several border disputes, whereas Japan and Russia have not been able to resolve their quarrel over the southern Kuril Islands.

China's border did not, in other words, provide adequate defense. In time, the conquering armies adopted Chinese customs in order to govern the far more numerous Han people.

The Manchu Qing Dynasty The final and most significant conquest of China occurred in 1644, when the Manchus toppled the Ming Dynasty and replaced it with the Qing (also spelled Ch'ing) Dynasty. Like earlier conquerors, the Manchus retained the Chinese bureaucracy and made few institutional changes. Their strategy was to adapt themselves to Chinese culture, yet at the same time to preserve their own identity as an elite military group. Their system functioned well until the mid-19th century, when the empire began to crumble before the onslaught of European and, later, Japanese power.

China's most significant legacy from the Manchu Qing Dynasty was the extension of its territory to include much of Central Asia. The Manchus subdued the Mongols and eventually established control over eastern Central Asia, including Tibet. Even the states of mainland Southeast Asia sent tribute and acknowledged Chinese supremacy. Never before had the Chinese Empire been so extensive or so powerful (Figure 11.33).

The Modern Era From its height of power and extent in the 1700s, the Chinese Empire descended rapidly in the 1800s as it failed to keep pace with the technological progress of Europe. Threats to the empire had always come from the north, and Chinese and Manchu officials saw little peril from European merchants operating along their coastline. The Europeans were distressed by the amount of silver needed to obtain Chinese silk, tea, and other products, and by the fact that the Chinese disdained their manufactured goods. In response, the British began to sell opium, which Chinese authorities viewed as a threat. When the imperial government tried to suppress the opium trade in the 1839, the British attacked and quickly prevailed.

This first "opium war" ushered in a century of political and economic chaos in China. The British demanded free trade in selected Chinese ports, and in the process overturned the traditional policy of managed trade based on the acknowledgment of Chinese supremacy. As European enterprises penetrated China and undermined local economic interests, anti-Manchu rebellions began to break out. In the 1800s, all such uprisings were crushed, but not before causing tremendous destruction. Meanwhile, European power continued to advance. In 1858, Russia annexed the northernmost reaches of Manchuria, and by 1900 China had been divided into separate **"spheres of influence"** where the in-

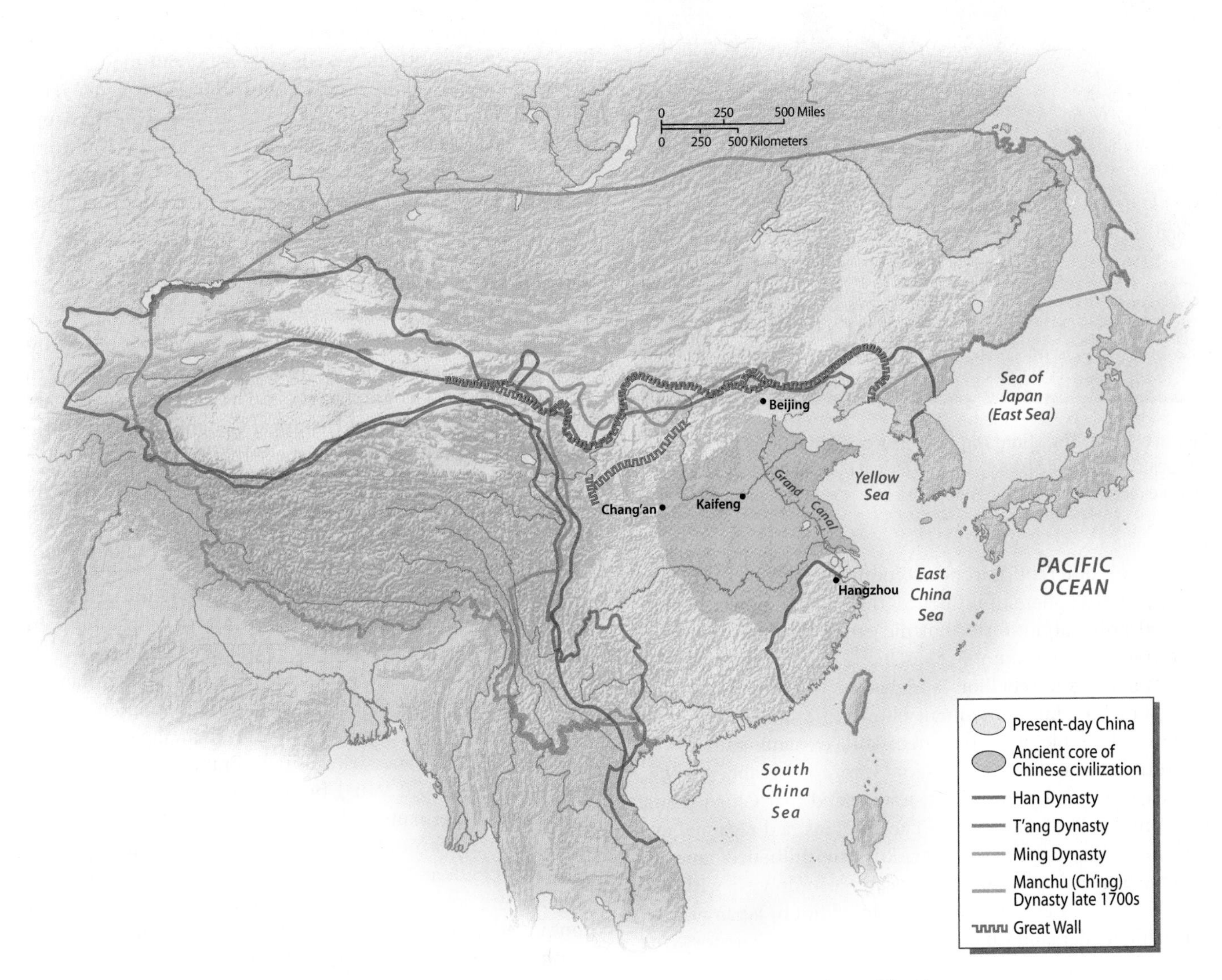

Figure 11.32 The Historical Extent of China China is usually regarded as the world's oldest country, but the territorial extent of the Chinese state has varied greatly over the centuries. The earliest states were limited to the Loess Plateau and North China Plain, but most historical Chinese dynasties controlled the entire core area of modern China as well as the Tarim Basin in Xinjiang. Before the 1600s, however, China seldom held control of Tibet, Inner Mongolia, or central and northern Manchuria.

Figure 11.33 The Qianlong Emperor of China Under the Qianlong Emperor, who reigned from 1736 to 1795, the Manchu-ruled Chinese empire reached its greatest territorial extent.

terests of different European countries prevailed (Figure 11.34). (In a sphere of influence, the colonial power had no formal political authority but did enjoy informal influence and tremendous economic clout.)

A successful rebellion launched in 1911 finally toppled the Manchus, but subsequent efforts to establish a unified Chinese Republic were not successful. In many parts of the country, local military leaders, or "warlords," grabbed power for themselves. By the 1920s, it appeared that China might be completely dismembered. The Tibetans had regained autonomy; Xinjiang was under Russian influence; and in China proper Europeans and local warlords vied with the weak Chinese Republic for power. In addition, Japan was increasing its demands and seeking to expand its territory.

The Rise of Japan

Japan did not emerge as a strong, unified state until the 7th century. From its earliest days, Japan looked to China (and, at first, to Korea as well) for intellectual and political models. Its offshore location insulated Japan from the threat of rule by the Chinese Empire. At the same time, the Japanese conceptualized their islands as a separate empire, equivalent in certain respects to China. Between 1000 and 1580, however, Japan was usually divided into a number of mutually hostile feudal realms.

The Closing and Opening of Japan Around 1600, Japan was reunited by the armies of the Tokugawa Shogunate (a shogun was a supreme military leader who theoretically remained under the emperor). Shortly afterward, Japan attempted to isolate itself from the rest of the world. Until the 1850s, Japan traded with China mostly through the Ryukyu islanders and with Russia through Ainu intermediaries. The only Westerners allowed to trade in Japan were the Dutch, and their activities were strictly limited.

Japan remained largely closed to foreign commerce and influence until U.S. gunboats sailed into Tokyo Bay to demand access in 1853. Aware that they could no longer keep the Westerners out, Japanese leaders set about modernizing their economic, administrative, and military systems. This effort accelerated when the Tokugawa Shogunate was toppled in 1868 by the Meiji Restoration. (It is called a restoration because it was carried out in the emperor's name, but it did not give the emperor any real power.) Unlike China, Japan successfully accomplished most of its reform efforts.

The Japanese Empire Japan's new rulers realized that their country remained threatened by European imperial power. They therefore nurtured the development of a silk export industry, which gave them the funds needed to buy modern equipment (see "Geographic Tools: The East Asian Gazetteer and the Reconstruction of Historical Geography"). They also decided that the only way to meet the European challenge was to become expansionistic themselves. Japan therefore took control over Hokkaido and began to move farther north into the Kuril Islands and Sakhalin.

In 1895, the Japanese government tested its newly modernized army against China, winning a quick and profitable victory that gave it authority over Taiwan. Tensions then mounted with Russia as the two countries vied for power in Manchuria and Korea. The Japanese defeated the Russians in 1905, giving them considerable influence in northern China. With no strong rival in the area, Japan annexed Korea in 1910. Alliance with Britain, France, and the United States during World War I brought further gains, as Japan was awarded Germany's island colonies in Micronesia.

The 1930s brought a global depression, putting a resource-dependent Japan in a difficult situation. The country's leaders sought a military solution, and in 1931 Japan conquered Manchuria. In 1937, Japanese armies moved south, occupying the North China Plain and the coastal cities of southern China. The Chinese government withdrew to the relatively inaccessible Sichuan Basin to continue the struggle. During this period, Japan's relations with the United States steadily deteriorated. When the United States cut off the export of scrap iron, Japan began to experience a resource crunch.

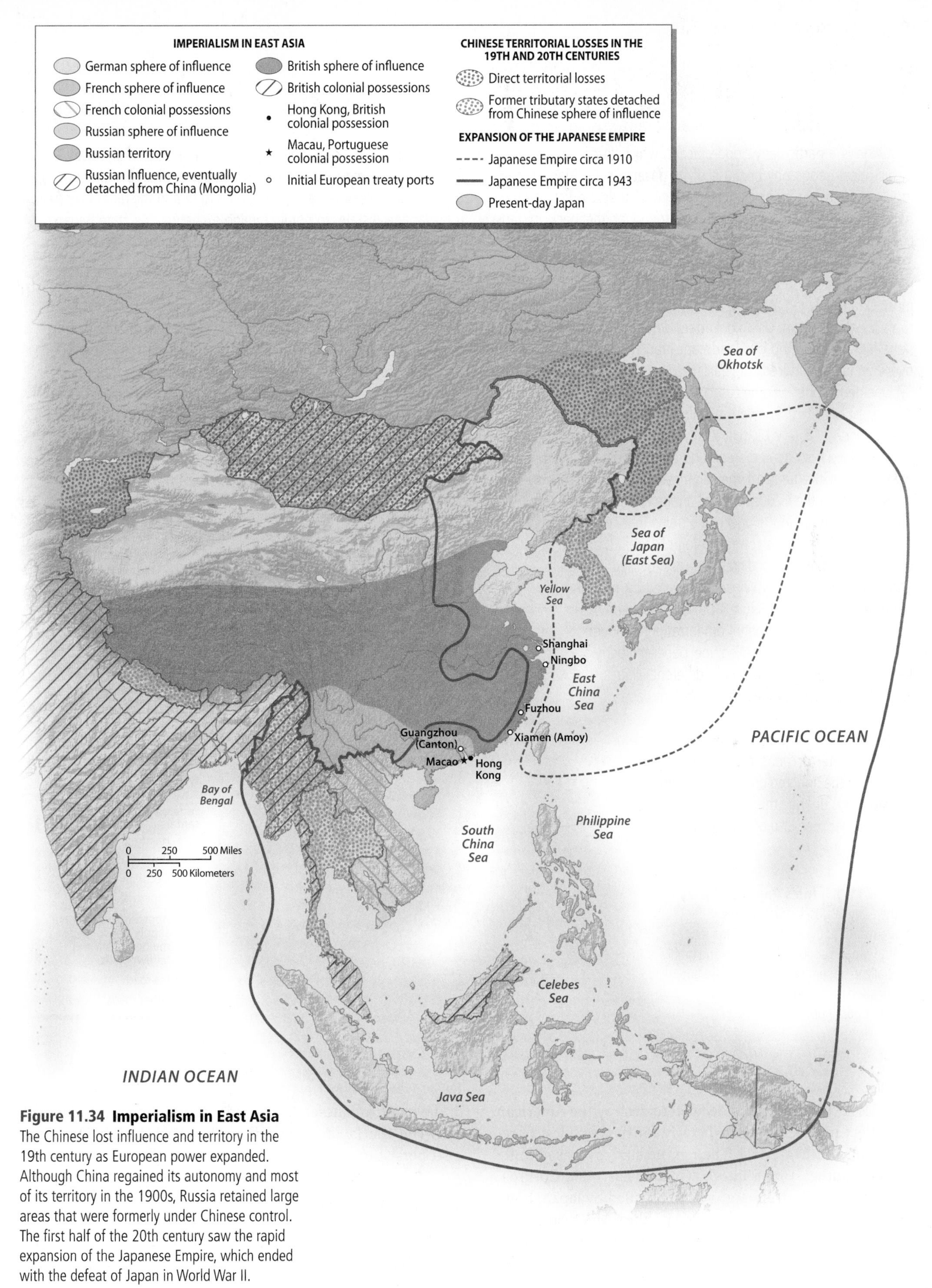

Figure 11.34 Imperialism in East Asia
The Chinese lost influence and territory in the 19th century as European power expanded. Although China regained its autonomy and most of its territory in the 1900s, Russia retained large areas that were formerly under Chinese control. The first half of the 20th century saw the rapid expansion of the Japanese Empire, which ended with the defeat of Japan in World War II.

GEOGRAPHIC TOOLS The East Asian Gazetteer and the Reconstruction of Historical Geography

East Asia is a particularly good place in which to conduct research in historical geography because of its gazetteer tradition. In common usage, a gazetteer is simply a kind of index: a list of place-names and their spatial coordinates found at the back of most comprehensive atlases. But a gazetteer can also be a geographical encyclopedia, providing detailed information about a large array of particular places. Gazetteers of this sort have historically been used most extensively in East Asia, as they were believed to be essential tools for government. Using gazetteer data, geographers and historians literally can map out landscape transformations in East Asia over periods of hundreds of years.

One geographer who has used gazetteer data extensively is Kären Wigen, especially in her award-winning book *The Making of a Japanese Periphery, 1750–1920* (University of California Press, 1995). Here she examined how the region around the city of Ina, located in a sizable valley in the heart of the Japanese Alps, has repeatedly been transformed. In the 1700s, Ina was a minor feudal domain that formed a key node in the packhorse trains that carried goods across the mountainous central region of Japan. It also supported its own major handicraft industry focused on high-quality paper goods. As Wigen discovered by mapping out gazetteer data (see Figure 11.5.1), the paper industry entailed a complex spatial organization; although the entire Ina region supplied the paper-mulberry bark from which paper was made, different villages specialized in the manufacture of different paper-craft items.

As Wigen discovered by examining gazetteers from later decades, the economic geography of the Ina region was completely transformed during the mid- and late 1800s. In part because of changing fashions, decorative paper goods no longer commanded high prices, and as a result the industry underwent a rapid decline. Simultaneously, the regional economy was subordinated to that of the rapidly centralizing Japanese state. To obtain foreign exchange, the state began to export silk thread. As a result, Ina's paper mulberries were ripped out, to be replaced by a different species of mulberry used to feed the voracious silkworms.

The prosperity silk brought to Ina did not prove sustainable. The depression of the 1930s, followed by World War II, combined to destroy the industry. In the postwar period, mulberries were again uprooted, this time to be replaced by apple and other fruit trees. By the turn of the century, however, fruit growing was no longer very profitable, and the farmers of Ina were again looking for alternative crops.

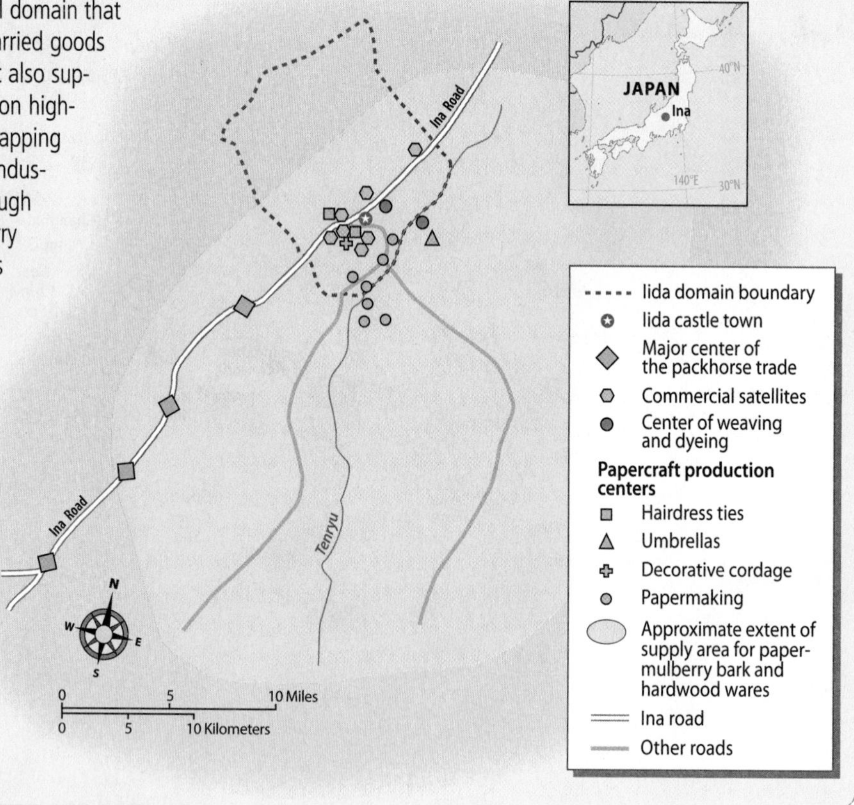

Figure 11.5.1 Ina's Paper Industry In the early 1800s, before Japan's modernization, the area around Ina in central Japan supported a sizable and highly specialized paper-craft industry. Different villages focused on the production of different paper objects.

In 1941, Japan's leaders decided to destroy the American Pacific fleet in order to clear the way for the conquest of resource-rich Southeast Asia. Their grand strategy was to unite East and Southeast Asia into a "Greater East Asia Co-Prosperity Sphere." This "sphere" was to be ruled by Japan, however, and was designed to keep the Americans and Europeans out. Japanese forces elsewhere in East Asia sometimes engaged in brutal acts. In Korea, the colonial government evidently planned to extinguish the Korean language in favor of Japanese. As a result, serious tensions emerged between the Japanese and the other East Asians—tensions that persist to this day.

Postwar Geopolitics

With the defeat of Japan at the end of World War II, East Asia became an arena of rivalry between the United States and the Soviet Union. Initially, American interests prevailed in the maritime fringe, while Soviet influence advanced on the mainland. Soon, however, East Asia began to experience its own revival.

Japan's Revival Japan lost its colonial empire when it lost World War II. Its territory was reduced to the four main islands plus the Ryukyu Archipelago and a few minor outliers. In general, the Japanese government accepted this loss of land. The only remaining territorial conflict concerns the four southernmost islands of the Kuril chain, which were taken by the Soviet Union in 1945. Although Japan still claims these islands, Russia refuses even to discuss relinquishing control, straining Russo-Japanese relations.

After losing its overseas possessions, Japan had to turn to trade to obtain the resources needed for its economy. Japan's military power was strictly limited by the constitution imposed on it by the United States, forcing Japan to rely on the U.S. military for much of its defense needs. The U.S. Navy continues to patrol many of Japan's vital sea-lanes, and U.S. armed forces maintain several bases within the country. This U.S. military presence, however, is controversial. Particularly contentious are the U.S. bases on Okinawa, which take up much of the island's territory. In 2009, the new Japanese prime minister Yukio Hatoyama pledged to force the U.S. military off of Okinawa, but in May 2010, as tensions mounted between Japan and North Korea, he decided to let the bases remain. The resulting controversy, however, was so intense that Hatoyama resigned his position.

Slowly but steadily, Japan's own military has emerged as a strong regional force despite the constitutional limits imposed on it. In recent years, fears about North Korea's nuclear program and China's military growth have led the Japanese government to reconsider its constitutionally imposed limitations on military spending. In 2007, Japan announced that it would make international peacekeeping a military priority, strengthen its missile defenses, and enhance its coordination with U.S. troops.

Other East Asian countries are concerned about the potential threat posed by a remilitarized Japan. Such a perception was strengthened in the early years of the new millennium when Japan's former prime minister visited the Yasukuni Shrine, which contains a military cemetery in which several war criminals from World War II are buried. Anti-Japanese sentiments in China—which have occasionally boiled over into huge public protests—have also been encouraged by the publication of Japanese textbooks that minimize the country's atrocities during the war.

The Division of Korea The aftermath of World War II brought much greater changes to Korea than to Japan. As the end of the war approached, the Soviet Union and the United States agreed to divide Korea at the 38th parallel. Soon two separate regimes were established, the northern one allied with the Soviet Union, the southern one with the United States. In 1950, North Korea invaded South Korea, seeking to reunify the country. The United States, with support from the United Nations, supported the south, while China aided the north. The war ended in a stalemate, and Korea has remained a divided country, its two governments locked in a continuing Cold War (Figure 11.35).

Large numbers of U.S. troops remained in the south after the war. South Korea in the 1960s was a poor, agrarian country that could not defend itself. Over the past 30 years, however, the south has emerged as a wealthy trading nation while the fortunes of the north have plummeted. The southerners' fear of the north gradually lessened, especially among the members of the younger generation. Many South Korean students have resented the presence of U.S. forces, seeking instead to build friendly relations with North Korea. Changing military priorities, meanwhile, led the United States to reduce its presence in Korea, dropping its troop strength in 2004 from roughly 37,000 to 28,500.

In the late 1990s, the South Korean government began to pursue better relations with North Korea, giving it large amounts of food aid and industrial investment in exchange for reduced hostility. In 2008, however, a new South Korean government concluded that this so-called Sunshine Policy had not restrained North Korean belligerency. North Korean nuclear testing further strained relations, as did the

Figure 11.35 The Demilitarized Zone in Korea North and South Korea were divided along the 38th parallel after World War II. Today, even after the conflict of the early 1950s, the demilitarized zone, or DMZ (which runs near the parallel), separates these two states. U.S. armed forces are active in patrolling the DMZ.

Figure 11.36 Chinese Soldiers in Tibet Following a full-scale invasion of Tibet in 1959, China continues to increase its presence through its military forces, the relocation of migrants into the area from other parts of China, and rebuilding programs that mask the traditional Tibetan landscape. Here a Chinese soldier watches Tibetan Buddist monks in robes at a horse festival.

March 2010 sinking of a South Korean naval vessel. Although North Korea denied attacking the ship and killing 46 of its crew members, South Korea presented strong evidence of its guilt. As a result of the attack, tensions in the Korean peninsula reached their highest levels in decades, with the North temporarily cutting off most relations with the South.

The Division of China World War II brought tremendous destruction and loss of life to China. Even before the war began, China had been engaged in a civil conflict between nationalists (who favored an authoritarian capitalist economy) and communists. The communists had originally been based in the middle Yangtze region, but in 1934 nationalist pressure forced them out. Under the leadership of Mao Zedong, they retreated in the "Long March," which took them to the Loess Plateau, an area close to both the traditional power center of northern China and the industrialized zones of Manchuria. After the Japanese invaded China proper in 1937, the two camps cooperated, but as soon as Japan was defeated, China was again embroiled in civil war. In 1949, the communists proved victorious, forcing the nationalists to retreat to Taiwan.

A latent state of war has persisted ever since between China and Taiwan. Although no battles have been fought, gunfire has periodically been exchanged over the Matsu Islands, several small Taiwanese islands just off the mainland. The Beijing government still claims Taiwan as an integral part of China and vows eventually to reclaim it. The nationalists who gained power in Taiwan long maintained that they represented the true government of China. It was actually made a crime in Taiwan to advocate formal Taiwanese independence, because the fiction had to be maintained that Taiwan was merely one province of a temporarily divided China.

The idea of the intrinsic unity of China continues to be influential. In the 1950s and 1960s, the United States recognized Taiwan as the only legitimate government of China, but its policy changed in the 1970s after U.S. leaders decided that it would be more useful and realistic to recognize Mainland China. Soon China entered the United Nations, and Taiwan found itself diplomatically isolated. As of 2010, only 23 countries, mostly small African, Pacific, and Caribbean states, continued to recognize Taiwan and, in return, to receive Taiwanese aid (see Exploring Global Connections: Taiwan Seeks Diplomatic Recognition). In reality, however, Taiwan is a fully separate country. In 2000, after Taiwan elected a former advocate of Taiwanese independence, China threatened to invade if Taiwan were to declare itself a separate country. In 2001, both China and Taiwan entered the World Trade Organization (WTO), but Taiwan was forced to join under the awkward name of "Chinese Taipei," to avoid the suggestion that it is a sovereign country.

In the early years of the new century, Taiwan seemed to be moving toward formal separation from China. The year 2007 also saw the systematic purging of words and phrases suggesting unity with China in Taiwanese textbooks. Thus the common term "cross-strait ties" was replaced with "China–Taiwan ties." In the 2008 Taiwanese election, however, the anti-independence Kuomingtang party returned to power. The new president Ma Ying-Jeou made it clear that he regarded Taiwan and China as two separately governed areas of a single country.

The election of Ma greatly reduced tensions between China and Taiwan. Trade between the island and the mainland is booming, as is tourism. But the de facto independence of Taiwan continues to irritate Beijing. In 2010, China harshly criticized the United States for agreeing to sell $6.4 billion worth of military equipment to the Taiwanese government.

Chinese Territorial Issues Despite the fact that it has been unable to regain Taiwan, China has successfully retained the Manchu territorial legacy. In the case of Tibet, this has required considerable force; resistance by the Tibetans compelled China to launch a full-scale invasion in 1959. The Tibetans, however, have continued to struggle for real autonomy if not actual independence, as they fear that the Han Chinese now moving to Tibet will eventually outnumber them (Figure 11.36; also see Chapter 10).

The postwar Chinese government also retained control over Xinjiang in the northwest, as well as Inner Mongolia (or Nei Monggol), a vast territory stretching along the Mongolian border. Like Tibet, Nei Monggol and Xinjiang

EXPLORING GLOBAL CONNECTIONS Taiwan Seeks Diplomatic Recognition

On March 15, 2010, newspapers across the world announced that Taiwanese President Ma Ying-Jeou would visit his country's allies in the South Pacific: Nauru, Kiribati, Marshall Islands, Palau, Tuvalu, and Solomon Islands. Such headlines were somewhat misleading, as the countries mentioned are not exactly allies of Taiwan.

An ally, according to the common definition, is a "state formally cooperating with another for military or other purposes." It is difficult to imagine Nauru, a mined-out semiwasteland of eight square miles and 14 thousand people, coming to the aid of Taiwan for military or any other purposes. The relationship between these countries and Taiwan is actually one of clientage rather than alliance. In essence, Nauru, Kiribati, Marshall Islands, Tuvalu, Palau, and Solomon Islands sell their diplomatic recognition to Taiwan in exchange for aid. Taiwan thereby gains a small measure of international legitimacy, while these small Pacific countries gain much needed financial resources. They are not alone in the practice, as Taiwan maintains diplomatic recognition with 23 countries (Figure 11.6.1).

Most of Taiwan's diplomatic partners are small countries in the Pacific, the Caribbean, and Central America. Taiwan has had no success in securing recognition by other Asian countries. Most Asian states are too large to be swayed by aid incentives—and too close to China to deny Beijing's power. But Taiwan is also disadvantaged in its quest for recognition by the fact that it claims not just the whole of China but also parts of Russia, Tajikistan, Pakistan, India, Afghanistan, Bhutan, and Burma—as well as Mongolia in its entirety. Officially, Taiwan maintains that all territories controlled by China at the time of the 1911 revolution are rightfully its own.

Despite its formal claims, Taiwan has bent to the demands of reality to informally recognize Mongolia's independence. In 2002, it opened an unofficial embassy in Mongolia called the "Taipei Trade and Economic Representative Office in Ulaanbataar." Taiwan's government simultaneously excluded Mongolia from the oversight of its Mainland Affairs Council, in effect recognizing Mongolia's sovereignty. As a result, Mongolians wanting to visit Taiwan now have to obtain visas, which were not necessary so long as Taipei regarded Mongolia as one of its (temporarily) lost provinces. Still, Taiwan has never formally dropped its constitutional claims to Mongolia. The situation remains ambiguous to say the least.

Nonetheless, Taiwan and Mongolia have developed reasonably close relations. As a sign of friendship, Taiwan recently gave Mongolia a three-story high portrait of Genghis Khan, made out of 437,000 mosaic tiles, based on a rare portrait of the world conqueror held in Taipei's National Palace Museum. But the potential for discord has not vanished. In 2008, a Hong Kong-based company posted a map on its Website showing Mongolia as part of China; when Mongolia protested, China's embarrassed government responded by claiming that the original map had been made in Taiwan.

Figure 11.6.1 Warner Park Sporting Complex, St. Kitts and Nevis Taiwan often rewards its diplomatic partners with lavish subsidies. The Warner Park Sporting Complex, site of the 2007 Cricket World Cup, was mostly financed by Taiwan.

are classified as autonomous regions. The peoples of Xinjiang are asserting their religious and ethnic identities, and separatist sentiments are widespread. However, most Han Chinese regard Nei Monggol and Xinjiang as integral parts of their country, and they regard any talk of secession as treasonous.

China claims several other areas that it does not control. It contends that former Chinese territories in the Himalayas were illegally annexed by Britain when it controlled South Asia, resulting in a border dispute with India. The two countries went to war in 1962 when China occupied an uninhabited highland district in northeastern Kashmir. Tensions between India and China have eased over the past decade, but their territorial disagreement remains unresolved. China also asserts its rights over a number of tiny islands in the East and South China seas. In September 2010, relations between China and Japan reached a crisis point after Japan seized a Chinese ship near the Japanese-controlled Senkaku Islands, which China claims under the name Diaoyu Islands. Japan eventually released the ship's captain, but refused to apologize, angering the Chinese government.

One territorial issue was finally resolved in 1997 when China reclaimed Hong Kong. In the 1950s, 1960s, and 1970s, Hong Kong acted as China's window on the outside world, and it grew prosperous as a capitalist enclave. As Chinese relations with the outer world opened in the 1980s, Britain decided to honor its treaty provisions and return Hong Kong to China. China in turn promised that Hong Kong would retain its fully capitalist economic system and its partially democratic political system for at least 50 years. Civil liberties not enjoyed in China itself were to remain protected in Hong Kong (Figure 11.37).

Hong Kong's autonomy, however, remains insecure. Over the past several years, China has thwarted efforts by its people to install a more representative government. As a result of such incidents, hundreds of thousands of people have taken to the streets of Hong Kong in protest. In 2010, several members of the Legislative Council of Hong Kong resigned in protest against the limitations on democratic governance. Their resignation forced a new election, in which they won back their seats. The Beijing government was angered by this maneuver, which it saw as a waste of taxpayer money.

In 1999 Macao, the last colonial territory in East Asia, was returned to China (Figure 11.38). Like Hong Kong, it is classified as Special Administrative Region, under its own autonomous government and subject to its own laws. This small former Portuguese enclave, located across the estuary from Hong Kong, has functioned largely as a gambling refuge. Now the largest betting center in the world, Macao derives 40 percent of its gross domestic product from gambling.

Global Dimensions of East Asian Geopolitics

In the early 1950s, East Asia was divided into two hostile Cold War camps: China and North Korea were allied with the Soviet Union, while Japan, Taiwan, and South Korea were linked to the United States. The Chinese–Soviet alliance soon deteriorated into mutual hostility, and in the 1970s China and the United States found that they could accommodate each other, sharing an enemy in the Soviet Union. In contrast, North Korea's relations with the United States and many other countries have only grown more heated during the same period.

The North Korean Crisis In 1994, North Korea refused to allow international inspections of its nuclear power facilities—which many countries believed were being used for weapons production—provoking an international crisis. It eventually relented, however, in exchange for energy assistance from the United States. But by 2002, the agreement had fallen apart, as evidence mounted that North Korea was continuing to pursue nuclear weapons. Other complaints by the international community focused on North Korea's traf-

Figure 11.37 Distinctive Hong Kong Although part of China since 1997, Hong Kong remains both politically and culturally distinctive from the rest of China. In this Hong Kong scene, a bride is posing for photographs in front of a sexually suggestive advertisement.

Figure 11.38 Macao Casino Macao, reclaimed by China from Portugal in 1999, retains a unique mixture of Chinese and Portuguese cultural influences. Its economic mainstay remains gambling, which is prohibited in the rest of China.

ficking in illegal drugs and the fact that it reportedly holds more than 200,000 of its own citizens in brutal labor-camp prisons.

In response to North Korea's nuclear ambitions, the United States has advocated multilateral negotiations involving South Korea, China, Japan, and Russia. North Korea, however, insisted for years that it would negotiate only with the United States. Owing in part to pressure from China, its main trading partner, North Korea finally agreed to the broader framework, participating in five rounds of talks between 2003 and 2007. Little progress was made, however, and in 2009 North Korea pulled out of the negotiations, expelled all nuclear inspectors, and resumed nuclear enrichment. In the same year, it detonated a nuclear weapon, launched a partially successful space booster rocket, and tested a series of ballistic missiles. Such actions have greatly increased tensions in the region and have put China, North Korea's main supporter, in a difficult diplomatic situation.

China on the Global Stage The end of the Cold War, coupled with the rapid economic growth of China, reconfigured the balance of power in East Asia. The United States no longer needed China to offset the Soviet Union, and the U.S. military became increasingly worried about the growing power of the Chinese armed forces. China's neighbors also have become more concerned. China now has the largest army in the world, as well as nuclear capability, and sophisticated missile technology. Through the early years of the new century, China's military budget grew at an average annual rate of about 10 percent, creating one of the world's most powerful armed forces. China is also rapidly building up its navy, buying advanced ships from Russia and beginning to construct its own. Like Japan, China is concerned about the vulnerability generated by its reliance on foreign oil, and as a result it is enhancing its naval capacity.

China is thus coming of age as a major force in global politics. Whether it is a force to be feared by other countries is a matter of considerable debate. Chinese leaders insist that they have no expansionistic designs and no intention of interfering in the internal affairs of other countries. They regard concerns expressed by the United States and other countries about their human rights record, as well as their activities in Tibet, as undue meddling in their own internal affairs.

Current opinions on China in the United States vary tremendously. Many U.S. leaders, particularly those in the business community, contend that the two countries should ignore their political differences and develop closer economic and cultural ties. Some East Asian experts similarly argue that the United States must respect China's sovereignty more carefully—or risk a future conflict in the region. Critics, on the other hand, think that China's trade practices are unfair; its labor, human-rights, and environmental records appalling; and its actions in Tibet unsupportable.

Regardless of what one thinks about East Asia's international relations, its economic ascent is undeniable. Yet East Asia does have a number of serious economic problems.

ECONOMIC AND SOCIAL DEVELOPMENT: An Emerging Core of the Global Economy

East Asia exhibits extreme differences in economic and social development. Japan's urban belt contains one of the world's greatest concentrations of wealth, whereas North Korea is one of the world's most impoverished societies. Overall, however, East Asia has experienced rapid economic growth since the 1970s, with two of its economies, those of Taiwan and South Korea, jumping from the ranks of underdeveloped to developed (Table 11.2). Over the past two decades, China has experienced extraordinarily rapid economic growth, and now has the world's second largest economy (Figure 11.39).

TABLE 11.2 Development Indicators

Country	GNI per capita, PPP 2008	GDP Average Annual % Growth 2000–08	Human Development Index (2007)[a]	Percent Population Living Below $2 a Day	Life Expectancy (2010)[b]	Under Age 5 Mortality Rate (1990)	Under Age 5 Mortality Rate (2008)	Gender Equity (2008)[c]
China	6,010	10.4	0.772	36	74	46	21	103
Hong Kong	43,960	5.2	0.944		83			100
Japan	35,190	1.6	0.960		83	6	4	100
North Korea					63	55	55	
South Korea	27,840	4.5	0.937		80	9	5	97
Taiwan					79			

[a] *United Nations, Human Development Report, 2009.*

[b] *Population Reference Bureau, World Population Data Sheet, 2010.*

[c] *Gender Equity—Ratio of female to male enrollments in primary and secondary school. Numbers below 100 have more males in primary/secondary school, numbers above 100 have more females in primary/secondary schools.*

Source: World Bank, World Development Indicators, 2010.

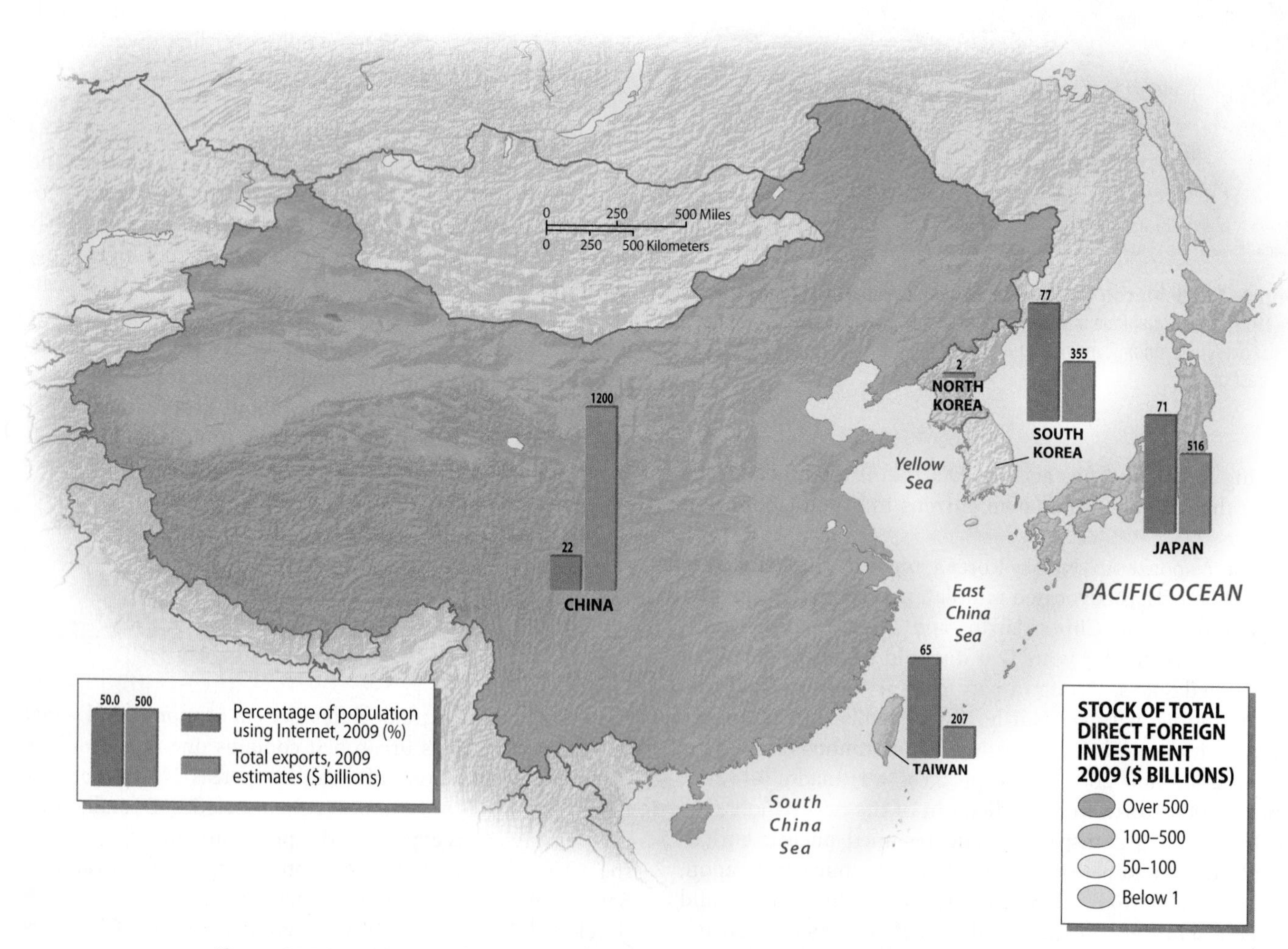

Figure 11.39 East Asia's Global Ties As this map shows, much of East Asia is highly integrated into global economic networks, receiving large amounts of direct foreign investment from other countries and exporting massive quantities of goods. North Korea, however, is a major exception to this pattern.

Figure 11.40 Automated Japanese Auto Factory Part of Japan's economic success has resulted from automation of its factory assembly lines. Here Mazda automobiles are assembled in a Hiroshima plant. These cars are destined for the east coast of the United States.

Japan's Economy and Society

Japan was the pacesetter of the world economy in the 1960s, 1970s, and 1980s. In the early 1990s, however, the Japanese economy experienced a major setback, and growth has been slow ever since. But despite its recent problems, Japan is still the world's third largest economic power.

Japan's Boom and Bust Although Japan's heavy industrialization began in the late 1800s, most of its people remained poor. The 1950s, however, saw the beginnings of the Japanese "economic miracle." Shorn of its empire, Japan was forced to export manufactured products. Beginning with inexpensive consumer goods, Japanese industry moved to more sophisticated materials, including automobiles, cameras, electronics, machine tools, and computer equipment (Figure 11.40). By the 1980s, it was the leader in many segments of the global high-tech economy.

In the early 1990s, Japan's inflated real estate market collapsed, leading to a banking crisis. At the same time, many Japanese companies discovered that producing labor-intensive goods at home had become too expensive. They therefore began to relocate factories to Southeast Asia and China. Because of these and related difficulties, Japan's economy slumped through the 1990s and into the new millennium. The Japanese government made several attempts to revitalize the economy through state spending, resulting in a huge government debt, totaling 192 percent of the country's gross domestic product in 2009. In 2007 it looked like the Japanese economy had turned a corner, expanding as a result of large industrial exports to China. The global economic crisis of 2008–2009, however, hit Japan hard, resulting in an economic decline of 5 percent in 2009.

Despite such problems, Japan remains a core country of the global economic system. Its economic system spans the globe, as Japanese multinational firms invest heavily in production facilities in North America and Europe, as well as in poorer countries. Japan is a world leader in a large array of high-tech fields, including robotics, optics, and machine tools for the semiconductor industry. It is also one of the world's largest creditor nations, despite its own debts, owning a large percentage of U.S. government bonds.

The Japanese Economic System The business environments in Japan and the United States are quite different, reflecting distinct versions of the capitalist economic system. In Japan, the bureaucracy maintains far greater control over the economy than it does in the United States. Japanese corporations are also structured differently from those of the United States. Large groups of companies (called keiretsu) are complexly intertwined, owning each other's stock and buying products and services from each other. Because of these interconnections, Japanese firms are less influenced by investors and stockbrokers than are those of the United States. Two other differences are that Japanese workers seldom switch companies, and the core workforce of each corporation is rarely subjected to layoffs.

Proponents of the Japanese system argue that it creates business and social stability and encourages long-term planning. Opponents argue that it reduces flexibility, results in high prices and low profits, and ultimately will prove too costly to maintain. They also point out that Japanese agriculture, wholesaling, and distribution remain relatively inefficient. Japan has only recently begun to see the rise of large-scale discount stores, which offer lower prices but also threaten the viability of small-scale merchants as well as downtown shopping districts in small cities.

Living Standards and Social Conditions in Japan Despite its affluence, living standards in Japan remain somewhat lower than those of the United States. Housing, food, transportation, and services are particularly expensive in Japan. But while the Japanese live in cramped quarters and pay high prices for basic products, they also enjoy many benefits unknown in the United States. Unemployment remains lower than in the United States; health care is provided by the government; and crime rates are extremely low. By such social measures as literacy, infant mortality, and average longevity, Japan surpasses the United States. Japan also lacks the extreme poverty found in certain pockets of American society.

Japan, of course, has its share of social problems. Koreans and alien residents from other Asian countries suffer discrimination, as do members of the indigenous Japanese underclass, the Burakumin. Japan's more remote rural areas have few jobs, and many have seen population decreases. In many small villages, most of the remaining people are elderly. Farming is an increasingly marginal occupation, and many farm families survive only because one family member works in a factory or office. Professional and managerial occupations in Japan's cities are noted for their long hours and high levels of stress.

Women in Japanese Society Critics often contend that Japanese women have not shared the benefits of their country's success. Advanced career opportunities remain limited, especially for those who marry and have children. Mothers are still expected to devote themselves to their families and to their children's education. Japanese business executives often work, or socialize in bars with their coworkers, late every evening, and thus contribute little to child care. Japan's recent economic problems seem to have further reduced career opportunities for women.

One response to the conditions faced by Japanese women has been a drop in the marriage rate. Many Japanese women are delaying marriage, and a sizable number may be abandoning it altogether. Japan has seen an even more dramatic decline in its fertility rate; beginning in 2005, Japan has seen more deaths than births. Whether this is due to the domestic difficulties of Japanese women or merely the result of the pressures of a postindustrial society is an open question. Fertility rates have, after all, dropped even lower in many parts of Europe. In 2009, the population of Japan did grow slightly, but only because more Japanese returned home than left for other countries. Japanese economic planners are concerned about the increasing dependency burden associated with a shrinking population. As Japan's population ages, increasing numbers of retirees will have to be supported by declining numbers of workers.

The Newly Industrialized Countries

In the 1960s, 1970s, and 1980s, the Japanese path to development was successfully followed by its former colonies, South Korea and Taiwan. Hong Kong also emerged as a newly industrialized economy in this period, although its economic and political systems remained distinctive.

The Rise of South Korea The postwar rise of South Korea was even more remarkable than that of Japan. During the period of Japanese occupation, Korean industrial development was concentrated in the north, which is rich in natural resources. The south, in contrast, remained a densely populated, poor, agrarian region. South Korea emerged from the bloody Korean War as one of the world's least-developed countries.

In the 1960s, the South Korean government initiated a program of export-led economic growth. It guided the economy with a heavy hand and denied basic political freedom to the Korean people. By the 1970s, such policies had proved highly successful in the economic realm. Huge Korean industrial conglomerates, known as chaebol, moved from manufacturing inexpensive consumer goods to heavy industrial products and then to high-tech equipment.

At first South Korean firms remained dependent on the United States and Japan for basic technology. By the 1990s, however, South Korea emerged as one of the world's main producers of semiconductors as well as the world's largest shipbuilder. South Korean wages also rose at a rapid clip. The country has invested heavily in education (by some measures it has the world's most intensive educational system), which has served it well in the global high-tech economy. Large South Korean companies are themselves now strongly multinational, building new factories in the low-wage countries of Southeast Asia and Latin America, as well as in the United States and Europe.

Contemporary Korea The political and social development of South Korea has not been nearly as smooth as its economic progress. Throughout the 1960s and 1970s, student-led protests against the dictatorial government were brutally repressed. As the South Korean middle class expanded and prospered, pressure for democratization grew, and by the late 1980s it could no longer be denied. But even though democratization has been successful, political tension has not disappeared (Figure 11.41).

South Korea's periodic political crises have been accompanied by economic difficulties. In 1997, for example, with the country's banking system in chaos, its economy entered a deep recession. Although the South Korean economy soon recovered, it remained unstable for several years, with periods of fast growth alternating with slowdowns. Critics contend that the South Korean economy needs substantial

Figure 11.41 Protests in South Korea Massive political protests are common in South Korea. In May 2008, some 10,000 cattle breeders took to the streets of Seoul to burn cow effigies, protesting a new policy that would allow increased beef exports from the United States to South Korea.

reforms, in particular the breaking up of the large conglomerates (chaebol). In recent years, however, the South Korean economy has been characterized by steady growth, moderate inflation, low unemployment, and large export surpluses. Even in the global recession year of 2009, the South Korean economy managed a slight expansion.

Under the Sunshine Policy between 1998 and 2008, South Korea attempted to use its own economic success to help North Korea, encouraging Korean companies to invest in jointly operated factories in the North. During the same period, North Korea allowed the limited emergence of public markets. Joint ventures with South Korea, however, were never particularly successful, and during the Korean crisis of 2010 their future was thrown temporarily into doubt. North Korean industry, suffering from parts shortages and decaying infrastructure, continues to decline. The Pyongyang government also moved to restrict its small private markets. In 2009, it ordered a currency reevaluation, forcing citizens to exchange old bank bills for new, less valuable ones. In the process, most of the meager savings of the North Korean people were wiped out, further undermining the economy. In early 2010, the North Korean government admitted that the currency change had been a mistake by executing the top-level bureaucrat who had been in charge of its implementation.

Taiwan and Hong Kong Like South Korea, Taiwan and Hong Kong also have experienced rapid economic growth since the 1960s. The Taiwanese government, like that of South Korea and Japan, has guided the economic development of the country. Taiwan's economy, however, is based more extensively on small to midsized family firms. This characteristically Chinese form of business organization is sometimes said to give Taiwan greater economic flexibility than its northern neighbors, but it has prevented it from entering certain industries that require huge concentrations of capital.

Hong Kong, unlike its neighbors, has been characterized by one of the most laissez-faire economic systems in the world, one with a great amount of market freedom from government interference (laissez-faire is a French term meaning "let it be"). Hong Kong traditionally functioned as a trading center, but in the 1960s and 1970s it emerged as a major producer of textiles, toys, and other consumer goods. By the 1980s, however, such cheap products could no longer be made in such an expensive city. Hong Kong's industrialists subsequently began to move their plants to nearby areas in southern China, while Hong Kong itself increasingly specialized in business services, banking, telecommunications, and entertainment. Fears that its economy would falter after the Chinese takeover in 1997 were not realized. Since 2005, Hong Kong's economy has been strengthened by a number of Chinese companies listing their initial public offerings on the booming Hong Kong stock exchange.

Both Taiwan and Hong Kong have close overseas economic connections. Linkages are particularly tight with Chinese-owned firms located in Southeast Asia. Taiwan's high-technology businesses are also intertwined with those of the United States; there is a constant back-and-forth flow of talent, technology, and money between Taipei and Silicon Valley. Hong Kong's economy also is closely bound with that of the United States (as well as those of Canada and Britain), but its closest connections are with the rest of China. Taiwan is moving in the same direction; China is now its largest export market and its second-largest source of imports after Japan.

Chinese Development

China dwarfs all of the rest of East Asia in both physical size and population. Its economic takeoff is thus reconfiguring the economy of the entire region and, to some extent, that of the world as a whole. But despite its recent growth, China's economy has a number of weaknesses. The future of the Chinese economy is thus one of the biggest uncertainties facing both the East Asian and the global economies.

China Under Communism More than a century of war, invasion, and near-chaos in China ended in 1949 when the communist forces led by Mao Zedong seized power. The new government, inheriting a weak economy, set about nationalizing private firms and building heavy industries. Certain successes were realized, especially in Manchuria, where a large amount of heavy industrial equipment was inherited from the Japanese colonial regime.

In the late 1950s, however, China experienced an economic disaster ironically called the "Great Leap Forward." One of the main ideas behind this scheme was that small-scale village workshops could produce the large quantities of iron needed for sustained industrial growth. Communist Party officials demanded that these inefficient workshops meet unreasonably high production quotas. In some cases the only way they could do so was to melt peasants' agricultural tools. Peasants also were forced to contribute such a large percentage of their crops to the state that many went hungry. The result was a horrific famine that may have killed 20 million people.

The early 1960s saw a return to more pragmatic policies, but toward the end of the decade a new wave of radicalism swept through China. This "Cultural Revolution" was aimed at mobilizing young people to stamp out the remaining vestiges of capitalism. Thousands of experienced industrial managers and college professors were expelled from their positions. Many were sent to villages to be "reeducated" through hard physical labor; others were simply killed. The economic consequences of such policies were devastating.

Toward a Postcommunist Economy When Mao Zedong, revered as an almost superhuman being, died in 1976, China faced a crucial turning point. Its economy was nearly stagnant and its people were desperately poor. A political struggle ensued between pragmatists hoping for change and dedicated communists. The pragmatists emerged victorious, and by the late 1970s it was clear that China would embark on a different economic path. The new China would seek closer connections with the world economy and take a modified capitalist road to development.

China did not, however, transform itself into a fully capitalist country. The state continued to run most heavy industries, and the Communist Party retained a monopoly on political power. Instead of suddenly abandoning the communist model, as the former Soviet Union later did, China allowed cracks to appear in which capitalist ventures could take root and thrive. As recently as 1995, almost 80 percent of the Chinese economy was in government hands, but by 2009 over 70 percent was controlled by the private sector.

One of China's first capitalist openings was in agriculture, which had previously been dominated by large-scale communal farms. Individuals were suddenly allowed to act as agricultural entrepreneurs, selling produce in the open market. Owing to this change, the income of many farmers rose dramatically. By the late 1980s, however, the focus of growth had shifted to the urban-industrial sector. As the government became concerned about inflation, it placed price caps on agricultural products and increased taxes on farmers. By the early years of the new millennium, many rural areas in China's interior provinces were experiencing economic distress, prompting many people to relocate to China's coastal cities or to foreign countries.

Figure 11.42 Shenzhen The city of Shenzhen, adjacent to Hong Kong, was one of China's first Special Economic Zones. It has more recently emerged as a major city in its own right.

Industrial Reform One of China's most important early industrial reforms involved opening Special Economic Zones (SEZs) in which foreign investment was welcome and state interference minimal. The Shenzhen SEZ, adjacent to Hong Kong, proved particularly successful after Hong Kong manufacturers found it a convenient source of cheap land and labor (Figure 11.42). Additional SEZs soon were opened, mostly in the coastal region. The basic strategy was to attract foreign investment that could generate exports, the income from which would supply China with the capital it needed to build its infrastructure and thus achieve conditions for sustained economic growth.

Other market-oriented reforms followed. Former agricultural cooperatives were allowed to transform themselves into capitalist entities. Many of these "township and village enterprises" proved highly successful. From 1980 to 2010, the Chinese economy grew at an average rate of roughly 10 percent a year, perhaps the fastest rate of expansion the world has ever seen. China emerged as a major trading nation, amassing large trade surpluses and foreign reserves. In 2010 China's foreign exchange reserves stood at $2,447 billion, dwarfing second-place Japan's figure of $990 billion.

China's economy has grown so quickly in recent years that many economists have feared that it would overheat, leading to unsustainably high rates of inflation. Thus far, however, inflation has remained moderate. Other critics contend that China must fully abandon centralized planning if its expansion is to be sustainable over the long run. China's leadership, however, has made it clear that reform will be a gradual process, maintaining that it has proved quite successful at managing its economic expansion.

China's economic growth has resulted in increased tensions with the United States. China exports far more to the United States than it imports, leading some U.S. politicians to demand that China allow its currency, the yuan, to rise against the dollar. In 2005, China officially unhooked its currency from the U.S. dollar; by 2008 it had appreciated 21 percent. Since then, the Chinese government has prevented the yuan from rising further. U.S. critics accuse China of unfairly keeping its currency undervalued in order to enhance its exports. China's large and growing holdings of U.S. treasury bonds, however, make it tricky for the United States to exert pressure on this issue.

Social and Regional Differentiation The Chinese economic surge unleashed by the reforms of the late 1970s and 1980s resulted in growing social and regional differentiation. In other words, certain groups of people—and certain portions of the country—prospered far more than others. Despite its official socialism, the Chinese state encouraged the formation of an economic elite, concluding that wealthy individuals are necessary to transform the economy. The least-fortunate Chinese citizens were sometimes left without work, and many millions have migrated to the booming coastal cities. The government has tried to control the transfer of population, but with only partial success. As China's economic boom accelerated, economic disparities mounted. The rapid growth of the elite population made China the world's fastest-growing market for luxury automobiles and even golf-course developments. At the same time, vulnerable state-owned enterprises have increasingly abandoned their provision of housing, medical care, and other social services, leaving many people destitute—particularly the elderly.

Economic disparities in China, as in other countries, are geographically structured. Before the reform period, the communist government attempted to equalize the fortunes of the different regions, giving special privileges to individuals from poor places. Such efforts were not wholly successful, and some provinces continued to be deprived. Since the coming of market reforms, moreover, the process of regional economic differentiation has accelerated (Figure 11.43).

The Booming Coastal Region Most of the benefits from China's economic transformation have flowed to the coastal region and to the capital city of Beijing. The first beneficiaries were the southern provinces of Guangdong and Fujian. This region was perhaps predisposed to the new economy, because the southern Chinese have long been noted for engaging in overseas trade. Guangdong and Fujian have also benefited from their close connections with the overseas Chinese communities of Southeast Asia and North America. Proximity to Taiwan and especially Hong Kong also proved helpful. Vast amounts of capital have flowed to the south coastal region from foreign (and Hong Kong-based) Chinese business networks.

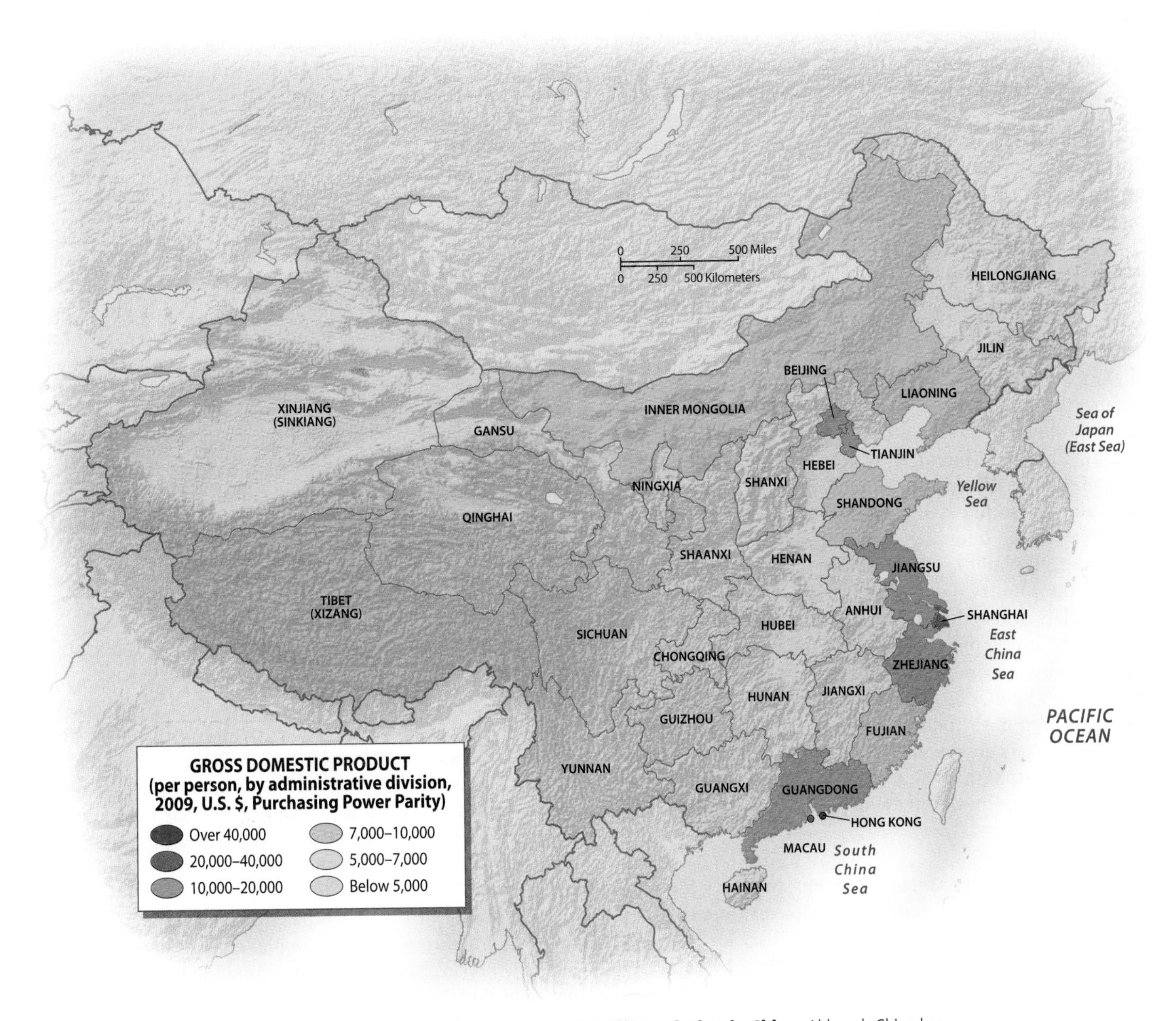

Figure 11.43 Development Issues: Economic Differentiation in China Although China has seen rapid economic expansion since the late 1970s, the benefits of growth have not been evenly distributed throughout the country. Economic prosperity and social development are concentrated on and near the coast. Most of the interior remains relatively poor. The poorest part of China is the upland region of Guizhou in the south-central part of the country.

By the early 2000s, the Yangtze Delta, centered on the city of Shanghai, reemerged as the most dynamic region of China. The delta was the traditional economic (and intellectual) core of China, and before the communist takeover Shanghai was its premier industrial and financial center. The Chinese government, moreover, has encouraged the development of huge industrial, commercial, and residential complexes, hoping to take advantage of the region's dynamism. In 2010, Shanghai was the site of the world's largest and most expensive world's fair, Expo 2010, designed to showcase the "next great world city."

Interior and Northern China The interior regions of China have seen much less economic expansion. Central and northern Manchuria were formerly quite prosperous, owing to fertile soils and early industrialization, but have not participated much in the recent boom. Many of the state-owned heavy industries of the Manchurian "rust belt," or zone of decaying factories, are not efficient. Manchuria's once-productive oil wells, moreover, are largely exhausted. China and Russia have, however, recently constructed an oil pipeline from Siberia to Daqing in Manchuria, hoping to revive the economy of this decaying oil town.

Most of the interior provinces of China likewise largely missed the initial wave of growth in the 1980s and '90s. In many areas, rural populations continued to grow while the natural environment deteriorated. One consequence has been high levels of underemployment and out-migration. By most measures, poverty increases with distance from the coast. As a result of such discrepancies, China is now encouraging development in the west (including Tibet and Xinjiang as well as the western provinces of China proper), focusing on transportation improvement and natural resource extraction. As labor shortages begin to appear in the booming coastal areas, industrialists are responding by building new factories in the interior portions of the country.

Scholars debate the conditions found in the poorer parts of interior China. Some believe that China's official statistics are too positive, hiding significant poverty. Others think that the country's economic boom has substantially raised living standards even in the poorest districts. According to official statistics, China's poverty rate fell from 53 percent in 1981 to less than 3 percent in 2009. Critics note, however, that more than 30 percent of the Chinese population still lives on less than $2 a day.

Rising Tensions China's explosive but uneven economic growth has generated a number of problems. Inflation has made planning difficult and created hardship for those on fixed incomes. Corruption by state officials is by some accounts rampant; success often seems to depend on knowing the right people and having the right connections. In 2009 alone, more than 100,000 public officials were convicted on corruption charges. China's crime rate, moreover, which was extremely low, is rising rapidly. Organized crime in particular is emerging as a major problem in many areas.

An equally significant issue has been the struggle for free expression and political openness. The desire for reform has been enhanced by rising incomes and the development of a sizable middle class. In 1989, however, the state crushed a movement for government accountability and democracy, and forced the opposition to go underground. Whether most Chinese people really want democracy is a controversial issue. If they do, tensions will probably mount as long as China's economy prospers while its rulers deny basic freedoms. An estimated 3 million Chinese citizens took part in some 58,000 separate public protests in 2004, mostly focused on wage disputes, evictions, pension issues, and environmental degradation.

China has responded to such protest movements in part by increasing its restrictions on free speech. Internet censorship has become a particularly contentious issue. The Chinese government is currently estimated to employ an "Internet police force" roughly 30,000 strong. In 2010, in response to such interference, Google temporarily pulled out of China proper, redirecting its Chinese traffic to its servers in Hong Kong. Many Chinese Internet users, however, have learned to use sophisticated techniques to avoid official censorship.

China's political and human-rights policies have complicated its international relations. Tension with the United States and other wealthy countries also stems from economic issues. China's large and growing trade surplus and its reluctance to enforce copyright and patent law irritate many of its trading partners. Numerous U.S. firms have accused Chinese businesses of pirating music, software, and brand names. In 2010, the Unites States accused China of doing to little to control copyright violations; China responded by calling such accusations "groundless." As the global market grows and as popular culture becomes globalized, copyright and trademark infringement are becoming increasingly profitable and hence increasingly difficult to control.

Social Conditions in China

Despite its pockets of persistent poverty, China has achieved significant progress in social development. Since coming to power in 1949, the communist government has made large investments in medical care and education, and today China boasts fairly impressive health and longevity figures. But as China has moved to a market-based economy, its formerly extensive system of rural medical clinics has deteriorated. Some observes fear that levels of both health and education may be declining for the poorest segments of China's population.

Human well-being in China is also geographically structured. The literacy rate, for example, remains lower in many of the poorer parts of China, including the uplands of Yunnan and Guizhou and the interior portions of the North China Plain, than in the most prosperous coastal regions of the country. Increasingly, the largest gap is that which separates the booming cities from the languishing rural areas.

Population policy also remains an unsettling issue for China. With 1.3 billion people highly concentrated in less than half of its territory, China has one of the world's highest effective population densities. By the 1980s, its government had become so concerned that it instituted the famous "one-child policy." Under this plan, couples in normal circumstances are expected to have only a single offspring and can suffer financial and other penalties if they do not comply (Figure 11.44). Although the one-child policy was never fully implemented in many rural areas, China's demographic strategy has been generally successful; the average fertility level is now only 1.6, and the population is growing at the relatively slow rate of 0.6 percent a year. China will likely reach roughly 1.5 billion persons before stabilization occurs.

The decline in Chinese fertility, however, has brought about problems of its own. China is now worried about impending labor shortages as well as its aging population. Thus far, China has not provided adequate facilities for caring for its elderly population. As a result, the government is increasingly allowing couples to have more than one child, hoping that the additional children will care for their parents as they grow old.

China's population policy has also has generated social tensions and human-rights abuses. Particularly troubling is the growing gender imbalance in the Chinese population. In 2009, 119 boys were born for every 100 girls. This asymmetry reflects the practice of honoring one's ancestors; because family lines are traced through male offspring, one must produce a male heir to maintain one's lineage. Many couples are therefore desperate to produce a son. Some opt to bear more than one child regardless of penalties. Another option is gender-selective abortion; if ultrasound reveals a female fetus, the pregnancy is sometimes terminated. Baby girls also are commonly abandoned, and well-substantiated rumors of female infanticide circulate. International women's organizations, as well as antiabortion groups, are concerned about these effects of China's population policy.

Figure 11.44 China's Population Policies One aspect of China's population policy is the expansion of child-care facilities so that mothers can be near their children while at work. This enables women to resume participating in the workforce soon after giving birth. This photo shows a typical day-care center attached to an industrial plant in Guangdong province in coastal China.

The Position of Women Women have historically had a relatively low position in Chinese society, as is true in most other civilizations. One particularly blatant traditional expression of this was the practice of foot binding: the feet of elite girls were usually deformed by breaking and binding them in order to produce a dainty appearance. This crippling and painful practice was eliminated only in the 20th century. In certain areas of southern China it was also common in earlier times for girls to be married, hence to leave their own families, when they were mere toddlers (such marriages, of course, would not be consummated for many years).

Not all women suffered such disabilities in premodern China. Some individuals achieved fame and fortune—a few even through military service. Among the Hakka, women enjoyed a relatively high social position and were seldom subjected to foot binding. Both the nationalist and communist governments have, moreover, sought to begin equalizing the relations between the sexes. Many of their measures have been successful, and women now have a relatively high level of participation in the Chinese workforce. But it is still true that throughout East Asia—in Japan no less than in China—few women have achieved positions of power in either business or government.

Summary

- The economic success of East Asia has been accompanied by severe environmental degradation. Japan, South Korea, and Taiwan have responded by enacting strict environmental laws and by moving many of their most polluting industries overseas. The major environmental issue in the region today concerns the rapid growth of the Chinese economy. Pollution in Chinese cities is so serious that it has major negative effects on human health, while much of the Chinese countryside suffers from such problems as soil erosion and desertification. China is responding, however, with a major program for renewable energy.

- East Asia is a densely populated region, but its birthrates have plummeted in recent decades. Japan is now experiencing population decline, which will put pressure on the Japanese economy. In China, the biggest demographic challenge results from the massive movement of people from the interior to the coast and from rural villages to the rapidly expanding cities. China has been trying to redirect development toward the interior, but thus far has had little success.
- East Asia is united by deep cultural and historical bonds. China has the largest influence because, at one time or another, it ruled most of the region. Although Japan has never been under Chinese rule, it still has profound historical connections to Chinese civilization. The more prosperous parts of East Asia have welcomed cultural globalization over the several past decades.
- Geopolitically, East Asia has been characterized by much strife since the end of World War II. China and Korea are still suspicious of Japan, and they worry that it might rebuild a strong military force. Japan is concerned about the growing military power of China and especially about the nuclear arms and missiles of North Korea. Relations between North and South Korea improved between 2000 and 2008, but subsequently deteriorated. North Korean weapons development is also a major concern in Japan and the United States.
- With the notable exception of North Korea, all East Asian countries have experienced rapid economic growth since the end of World War II. In the 2000s, the most important story has been the rise of China. China's economic expansion has reduced poverty nationwide, but has also generated serious tensions between the wealthier, more globally oriented coastal regions and the less-prosperous interior provinces. The rise of China also has global implications, as many countries in all regions of the world have profited by exporting the raw materials needed by China's booming industries.

Key Terms

anthropogenic landscape *(page 497)*
autonomous region *(page 512)*
Burakumin *(page 510)*
central place theory *(page 500)*
China proper *(page 511)*
Cold War *(page 513)*
Confucianism *(page 506)*
diaspora *(page 511)*
geomancy *(page 508)*
ideographic writing *(page 504)*
laissez-faire *(page 529)*
loess *(page 486)*
Mandarin *(page 506)*
Marxism *(page 509)*
pollution exporting *(page 489)*
regulatory lakes *(page 485)*
rust belt *(page 532)*
samurai *(page 514)*
sediment load *(page 486)*
Shogunate *(page 518)*
social and regional differentiation *(page 530)*
Special Economic Zones (SEZs) *(page 530)*
spheres of influence *(page 517)*
superconurbation *(page 501)*
tonal language *(page 512)*
urban primacy *(page 501)*

Review Questions

1. How does the physiography of the Huang He River valley differ from that of the valley of the Yangtze?
2. Why is Japan so much more heavily forested than China?
3. Why has Shanghai emerged as the most populous urban area in China?
4. What are the major ways in which Japanese cities differ from those of the United States?
5. Where are the non-Han peoples of China concentrated? Why are they concentrated in these areas?
6. What have been the main consequences of the geographical division of Korea into two states?
7. Historically speaking, how did China and Japan act differently as imperial powers?
8. How have the different countries of East Asia followed different paths to economic development?
9. Where in China would one find the most rapid economic development, and why would one find it there?
10. How does the position of women in Japan compare to the position of women in other wealthy, industrialized countries?

Thinking Geographically

1. Discuss the advantages and disadvantages of China's building of major dams.
2. Discuss the ramifications, both positive and negative, of Japan's allowing the importation of rice and opening its agricultural lands to urban development.
3. What might be the consequences of China's granting true autonomy to the Tibetans and other non-Han peoples? Independence?
4. What are the potential implications of Taiwan's declaring itself an independent country?

5. Discuss the potential ramifications of the United States' restricting the importation of Chinese goods in order to put pressure on the Chinese government for human-rights reforms.
6. Discuss the advantages and disadvantages of China's current population policies.
7. Do you think that East Asia will emerge as the center of the world economy in the next century?

Regional Novels and Films

Novels

Sawako Ariyoshi, *The River Ki* (1982, Kodansha)

Xingjian Gao, *Soul Mountain* (2000, Flamingo)

Kazuo Ishiguro, *An Artist of the Floating World* (1989, Vintage Books)

Ho Soo Kim, *Den Haag* (2008, Philmac)

Peter H. Lee, *Flowers of Fire* (1986, University of Hawaii Press)

Wang Shuo, *Playing for Thrills* (1998, Penguin)

Hsueh-Chin Tsao, *The Dream of the Red Chamber* (1958, Doubleday)

Films

The Banquet (2006, China)

Farewell, My Concubine (1993, China)

The Gate of Heavenly Peace (1995, China)

The Host (2006, South Korea)

Little Red Flowers (2006, China)

Nomugi Pass (1979, Japan)

The Old Garden (2006, South Korea)

Princess Mononoke (1997, Japan)

Pulgasari (1985, North Korea)

Seven Samurai (1954, Japan)

A Taxing Woman (1987, Japan)

301, 302 (1995, South Korea)

Twilight Samurai (2002, Japan)

Why Has Bodi-Darma Left for the East? (1989, Korea)

Bibliography

Brown, Melissa. 2004. *Is Taiwan Chinese?: The Impact of Culture, Power, and Migration on Changing Identities*. Berkeley: University of California Press.

The Contemporary Atlas of China. 1988. Boston: Houghton Mifflin.

Cybriwsky, Roman A. 1998. *Tokyo: The Shogun's City and the Twenty-First Century*. New York: Wiley.

Elvin, Mark. 2004. *The Retreat of the Elephant: An Environmental History of China*. New Haven, CT: Yale University Press.

Harris, Mark Edward. 2007. *Inside North Korea*. San Francisco: Chronicle Books.

Knapp, Ronald G., ed. 1992. *Chinese Landscapes: The Village as Place*. Honolulu: University of Hawaii Press.

Kynge, James. 2006. *China Shakes the World: A Titan's Rise and Troubled Future—and the Challenge for America*. Boston: Houghton Mifflin.

Myers, B.R. 2010. *The Cleanest Race: How North Koreans See Themselves and Why It Matters*. New York: Melville House.

Ryu, Je-Hun. 2004. *Reading the Korean Cultural Landscape*. Elizabeth, NJ: Hollym International.

Smil, Vaclav. 2004. *China's Past, China's Future: Energy, Food, Environment*. London: Curzon Press.

Totman, Conrad. 1989. *The Green Archipelago: Forestry in Preindustrial Japan*. Berkeley: University of California Press.

MG™ Mastering Geography

Log in to www.masteringgeography.com for MapMaster™ interactive maps, geography videos, RSS feeds, flashcards, web links, an eText version of *Diversity Amid Globalization,* and self-study quizzes to enhance your study of East Asia.

MapMaster™ presents 13 Place Name and 13 Layered Thematic interactive maps to help students practice and master their geographic literacy, spatial reasoning, and critical thinking skills.

12 South Asia

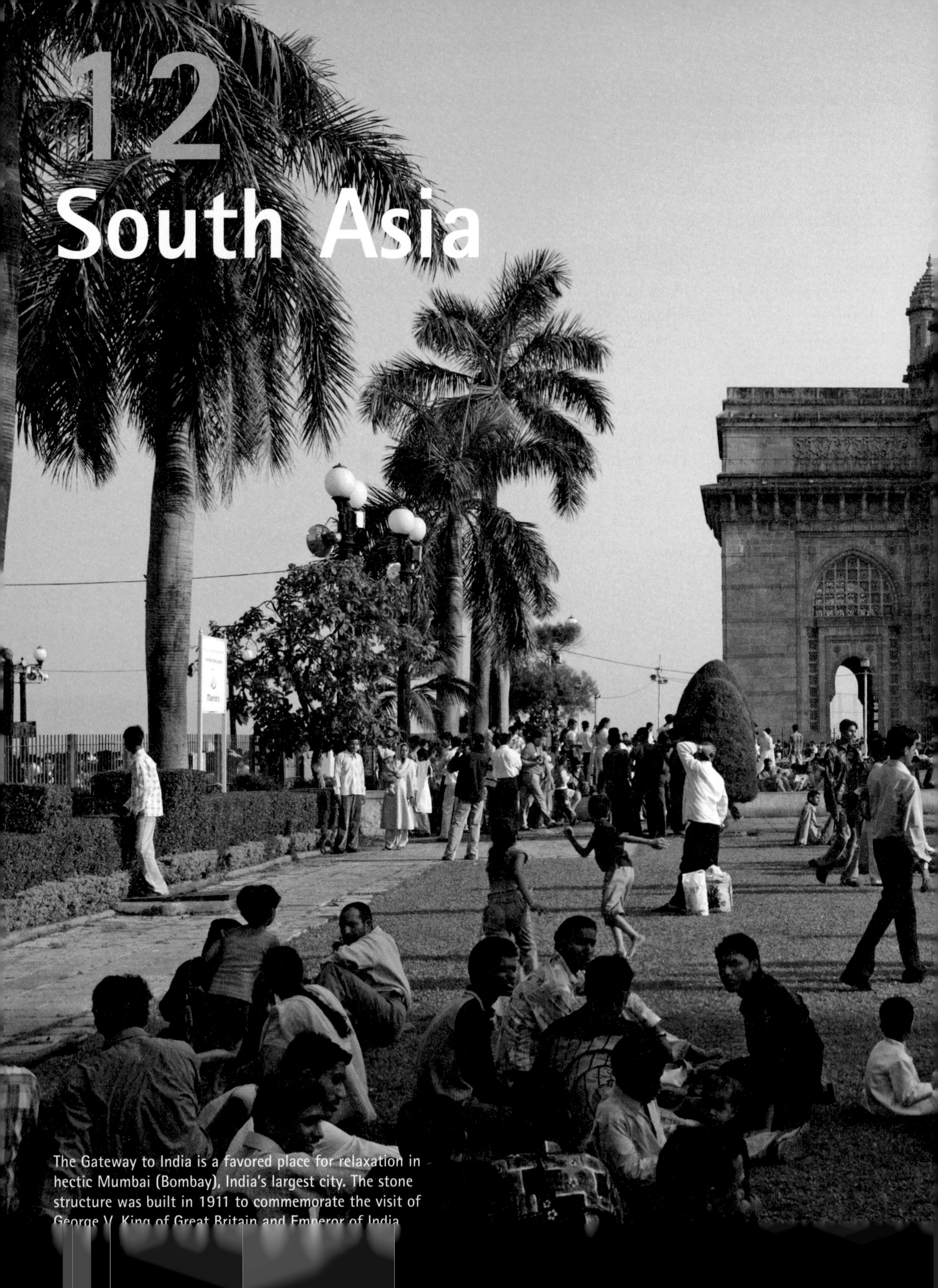

The Gateway to India is a favored place for relaxation in hectic Mumbai (Bombay), India's largest city. The stone structure was built in 1911 to commemorate the visit of George V, King of Great Britain and Emperor of India.

DIVERSITY AMID GLOBALIZATION

From the late 1940s through the 1980s, South Asia was one of the world's least globalized regions. Over the last two decades, however, a number of South Asian cities have emerged as crucial players in worldwide trade, exporting vast quantities of textiles, information technology products, and cultural goods such as films and television shows.

ENVIRONMENTAL GEOGRAPHY

While the arid parts of South Asia suffer from water shortages and the salinization of the soil, the humid areas often experience devastating floods.

POPULATION AND SETTLEMENT

South Asia will soon become the most populous region in the world; although birthrates have come down substantially in India and Bangladesh, they remain high in Pakistan.

CULTURAL COHERENCE AND DIVERSITY

South Asia is one of the most culturally diverse regions of the world, with India alone having more than a dozen official languages as well as numerous followers of most major religions.

GEOPOLITICAL FRAMEWORK

South Asia is burdened not only by a large number of violent secession movements, but also by the struggle between the nuclear-armed countries of India and Pakistan, which threatens both regional and global stability.

ECONOMIC AND SOCIAL DEVELOPMENT

Although South Asia as a whole is one of the poorest regions of the world, parts of India are experiencing rapid economic development based on the high-tech skills of the educated segment of its population.

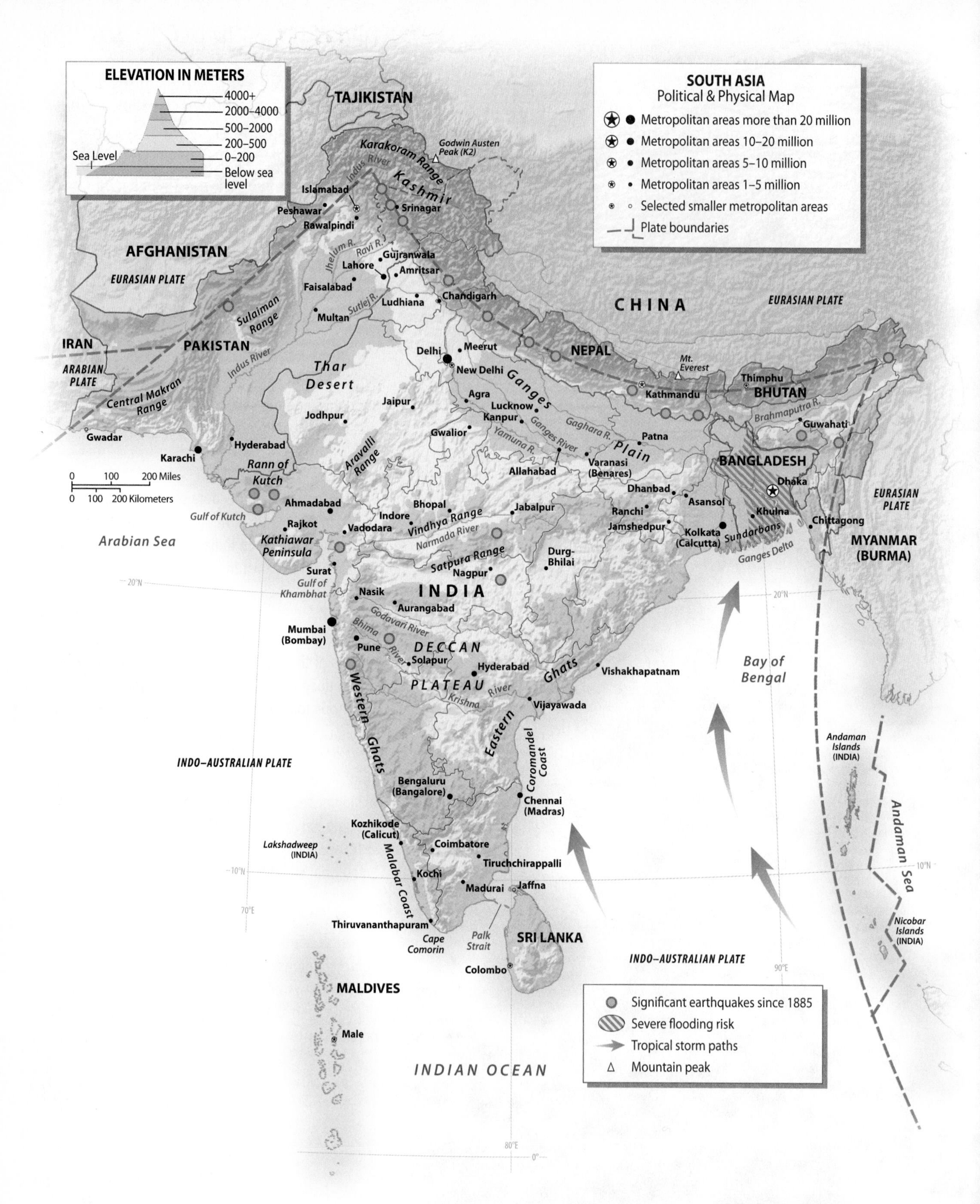

Figure 12.1 South Asia The second most populous region in the world, South Asia is dominated by India, Pakistan, and Bangladesh. The two Himalayan countries of Nepal and Bhutan, along with the island nations of Sri Lanka and the Maldives, round out the region. Four extensive physical subregions cover most of South Asia: the high Himalayan mountains in the north; the expansive Indus–Ganges lowland that reaches from Pakistan to the delta lands of Bangladesh; peninsular India, dominated by the Deccan Plateau; and the island realm that includes Sri Lanka and the Maldives. Many of this seismically active region's landscapes are products of the slow northward movement of the Indo-Australian tectonic plate against the Eurasian plate. Typhoons from the Bay of Bengal pose a significant risk to parts of eastern South Asia.

South Asia, a land of deep historical and cultural interconnections, has experienced intense political conflict for decades. Since independence from Britain in 1947, India and Pakistan, South Asia's two largest countries, have fought several wars and remain locked in a bitter conflict concerning possession of the disputed territory of Kashmir in the northern reaches of the region. Religious divisions are linked to this geopolitical turmoil, for India is primarily a Hindu country with a large Muslim minority, while Pakistan is almost entirely Muslim. Religious and ethnic tensions also abound within both countries, and in the neighboring countries of Bangladesh, Nepal, and Sri Lanka as well.

Parallel to these geopolitical tensions are demographic and economic concerns. Given its current rate of growth, South Asia soon will surpass East Asia as the world's most populous region. Although fertility levels have dropped dramatically in recent years across much of the region, they remain elevated in other areas, leading to concerns about how South Asia will be able to sustain its population. Although agricultural production has kept pace with population growth over the past four decades, many South Asian environments are experiencing pronounced stress. Poverty compounds such problems. Along with Sub-Saharan Africa, South Asia is the poorest part of the world. Roughly one-third of India's people subsist on less than $1 a day.

Despite its poverty, political tensions, and environmental problems, South Asia has also exhibited a significant degree of vitality in recent decades. India's economy in particular has been expanding at a rapid pace, lifting tens of millions of people into the middle class. Since the early 1990s, most South Asian countries have integrated their economies into the global system. Although such globalization has generated a number of controversies, it has brought major benefits to certain areas and groups of people. Parts of India, for example, have recently become major players in information technology (IT), tightly linked to California's Silicon Valley and other centers of the global IT industry.

The bulk of South Asia forms a subcontinent–often called the **Indian Subcontinent**–separated from the rest of Asia by formidable mountain ranges. Located here are the region's major countries, India, Pakistan, and Bangladesh, as well as the smaller mountainous states of Nepal and Bhutan. Also placed within South Asia are a number of islands in the Indian Ocean, including the countries of Sri Lanka and the Maldives, as well as the Indian territories of the Lakshadweep, and the Andaman and Nicobar islands.

India is by far the largest South Asian country, both in size and in population. Covering more than 1 million square miles (2,590,000 square kilometers), India is the world's seventh largest country in area and, with more than 1.1 billion inhabitants, second only to China in population. Pakistan and Bangladesh are next largest in size and population, each with more than 160 million inhabitants. A compact country about the size of Wisconsin, Bangladesh is one of the most densely populated places in the world. Bangladesh has a short border with Burma (Myanmar), but it is otherwise virtually engulfed by India, which wraps around the country to the north and the northeast.

ENVIRONMENTAL GEOGRAPHY: Diverse Landscapes, from Tropical Islands to Mountain Rim

South Asia's environmental geography covers a wide spectrum that ranges from the highest mountains in the world to densely populated delta islands barely above sea level; from one of the wettest places on Earth to dry, scorching deserts; from tropical rain forests to degraded scrublands to coral reefs (Figure 12.1). All of these ecological zones have their own distinct and complex environmental problems. To illustrate the complexity of South Asian environmental issues, let us begin by looking at the building of India's newly constructed "Golden Quadrilateral" highway system.

Building the Quadrilateral Highway

India's need for improved transportation is difficult to deny. As of 2006, the average speed for a trucker traveling between Kolkata (Calcutta) and Mumbai (Bombay) could be as low as 6.8 miles (11 kilometers) per hour. To address the transport needs generated by its booming economy, India has undertaken a massive multi-billion dollar road project designed to connect its four largest cities, New Delhi, Kolkata (Calcutta), Chennai (Madras), and Mumbai (Bombay), with a modern highway (Figure 12.2). The basic four-lane highway was completed in 2010, but a year earlier the Indian government announced that it would begin converting much of it into a

Figure 12.2 The Golden Quadrilateral Highway India's infrastructure is notoriously poor, but the government is now responding with a massive highway construction program. The Golden Quadrilateral Highway, shown here, has generated numerous protests, as villagers object to the destruction of houses, temples, and trees that lie in its path.

six-lane roadway. Although it only makes up about two percent of India's total road network, the Golden Quadrilateral Highway carries roughly 40 percent of India's vehicular traffic.

Building and expanding the new highway system has not proved easy. Truckers have protested the higher taxes and tolls that are needed to finance it, while citizen groups have stopped construction on several occasions with large protests demanding more underpasses, overpasses, and cattle crossings. Some concessions have been made, but many rural Indians are infuriated that their homes have been destroyed and their farms bisected by the massive project. Environmentalists worry that an expanded road system will lead to an increased reliance on trucks and cars, rather than the more environmentally responsible railroad system.

Highway construction has also generated both religious and environmental conflicts. Several Hindu temples have had to be relocated, prompting local worshippers to try to stop the project. Thousands of trees have been destroyed, generating further opposition. Not only are large trees rare in many parts of India, but they are also often regarded as sacred by devout Hindus. As a result, contractors have sometimes been forced to hire Muslims to cut down trees under the cover of night. Such actions, not surprisingly, anger local villagers. Villagers must also worry about a number of other environmental problems, a few of which are examined in the following section.

Environmental Issues in South Asia

As is true in other poor and densely settled regions of the world, a raft of serious ecological issues plague South Asia. The region also suffers from the usual environmental problems of water and air pollution that accompany early industrialization and manufacturing (Figure 12.3). Compounding all of these problems are the immense numbers of new people added each year through natural population growth.

Natural Hazards in Bangladesh The link between population pressure and environmental problems is nowhere clearer than in the delta area of Bangladesh, where the

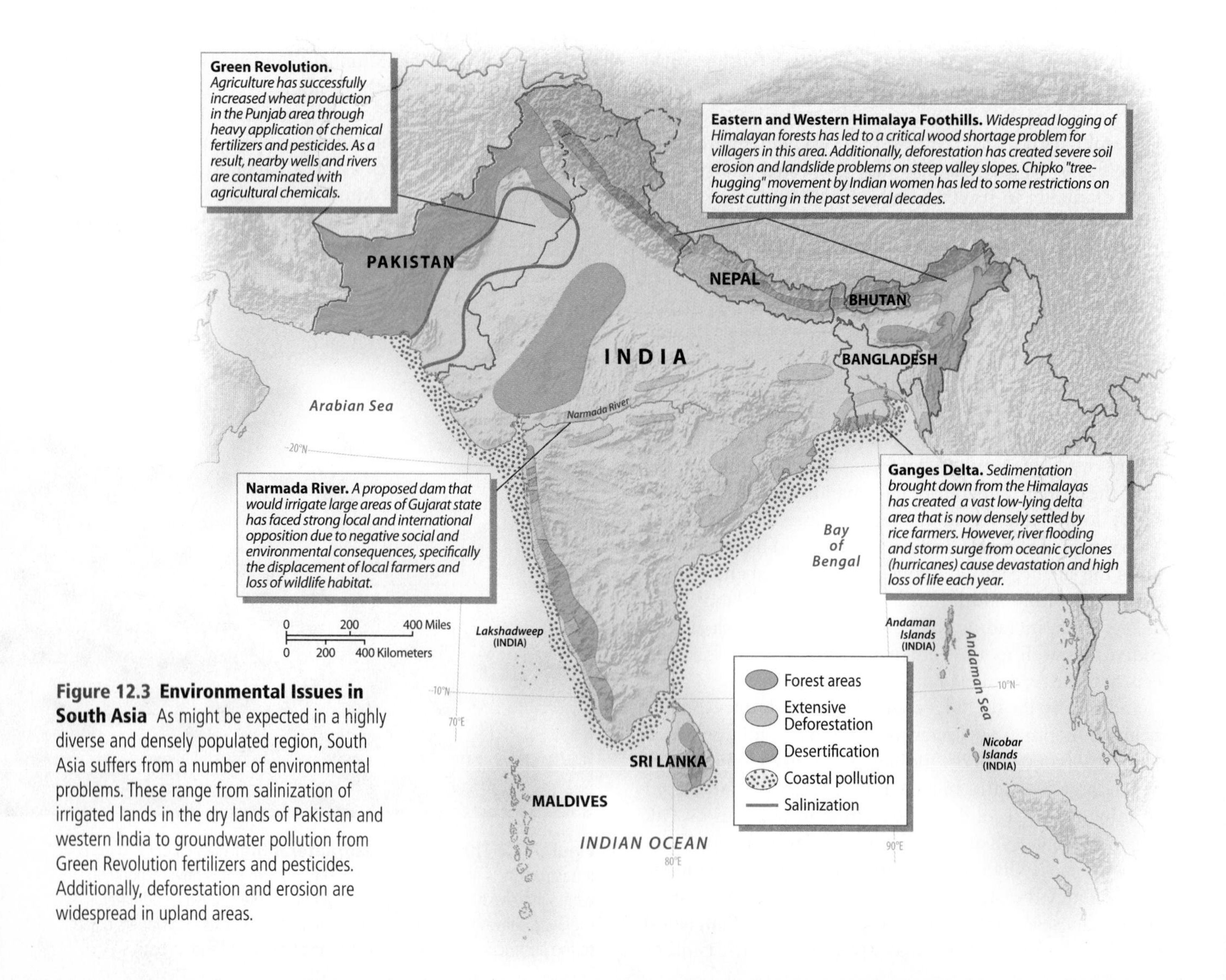

Figure 12.3 Environmental Issues in South Asia As might be expected in a highly diverse and densely populated region, South Asia suffers from a number of environmental problems. These range from salinization of irrigated lands in the dry lands of Pakistan and western India to groundwater pollution from Green Revolution fertilizers and pesticides. Additionally, deforestation and erosion are widespread in upland areas.

Figure 12.4 Flooding in Bangladesh Devastating floods are common in the low-lying delta lands of Bangladesh. Heavy rains come with the southwest monsoon, especially to the Himalayas, and powerful cyclones often develop over the Bay of Bengal.

search for fertile land has driven people into hazardous areas, putting millions at risk from seasonal flooding as well as from the powerful cyclones that form over the Bay of Bengal. For millennia, drenching monsoon rains have eroded and transported huge quantities of sediment from the Himalayan slopes to the Bay of Bengal by the Ganges and Brahmaputra rivers, gradually building this low-lying and fertile deltaic environment. With continual population growth, people have gradually moved into the swamps to transform them into highly productive rice fields. While this agricultural activity has supported Bangladesh's vast population, it has intensified the effects of the area's natural hazards.

Although periodic floods are a natural, even beneficial, phenomenon that enlarges deltas by depositing fertile riverborne sediment, flooding has become a serious problem for people inhabiting these low-lying areas. In September 1998, for example, more than 22 million Bangladeshis were made homeless when water covered two-thirds of the country (Figure 12.4). In June 2010, fairly typical monsoon downpours killed dozens of people and marooned more than 150,000.

With the populations of both Bangladesh and northern India continuing to grow, there is a strong possibility that flooding will take even higher tolls in the coming decade as desperate farmers relocate into the hazardous lower floodplains. Deforestation of the Ganges and Brahmaputra headwaters magnifies the problem. When forest cover and ground vegetation are removed, rainfall does not soak into the ground to slow runoff and replenish groundwater supplies. Thus deforestation in river headwaters results in increased flooding in the wet season, as well as a lower water level during the dry season when river flows are supplemented by groundwater. Bangladesh's water problems are further magnified by the fact that many of its aquifers are contaminated with arsenic, threatening the health of as many as 80 million people.

Forests and Deforestation Forests and woodlands once covered most of South Asia, except for the desert areas in the northeast, but in most places tree cover has vanished as a result of human activities. The Ganges Valley and coastal plains of India, for example, were largely deforested thousands of years ago to make room for agriculture. Elsewhere, forests were cleared more gradually for agricultural, urban, and industrial expansion. Railroad construction in the 19th century was especially destructive. More recently, hill slopes in the Himalayas and in the remote lands of eastern India have been logged for commercial purposes.

Beginning in the 1970s, India has embarked on a number of reforestation projects. Some of these have efforts been successful; according to some sources, Indian forest coverage actually increased by almost 6 percent between 1990 and 2005. As is true in many parts of the world, however, reforested areas are often covered by non-native trees, such as eucalyptus, that support little wildlife (Figure 12.5). According to a 2010 report, India's single-species tree plantations are

Figure 12.5 Logging in India A even-aged stand of small trees is being logged off in the southern Indian state of Kerala. Such forestry practices leave little habitat for wildlife.

expanding by up to 18,000 square kilometers (7,000 square miles) every year. And in India's standing natural forests, wood removal is estimated to exceed growth by a substantial margin.

As a result of deforestation, many villages of South Asia suffer from a shortage of fuelwood for household cooking, forcing people to burn dung cakes from cattle. While this low-grade fuel provides adequate heat, it diverts nutrients that could be used as fertilizers to household fires. It also results in high levels of air pollution, both indoors and outside. Where wood is available, collecting it may involve many hours of female labor because the remaining sources of wood are often far from the villages.

Villagers living in India's forest areas have suffered extensively from deforestation. As early as the 1970s, social movements designed to protect remaining groves gained prominence in some areas. The most noted of these movements, Chipko, mobilized women in northern India to engage in "tree-hugging" campaigns to stop logging. As a result of such social pressure, India began to involve local residents in forestry and conservation projects. Increasingly, villagers have been demanding their own rights to land and resources in wooded areas. In the landmark 2006 Forest Rights Act, the Indian government granted extensive rights to members of forest-dwelling communities. Many urban environmentalists, however, argue that this legislation could actually increase deforestation, as expanding villages often seek to convert wooded lands to agricultural fields. Some also fear that it could harm the efforts to preserve tigers and other large animal species.

Wildlife: Extinction and Protection Although the overall environmental situation in South Asia is rather bleak, wildlife protection inspires some optimism. The region has managed to retain a diverse assemblage of wildlife despite population pressure and intense poverty. The only remaining Asiatic lions live in India's Gujarat state, and even Bangladesh retains a viable population of tigers in the Sundarbans, the mangrove forests of the Ganges Delta. Wild elephants still roam several large reserves in India, Sri Lanka, and Nepal.

The protection of wildlife in India far exceeds that in most other parts of Asia. The official Project Tiger, which currently operates more than 40 preserves, is credited with increasing the country's tiger population from 1,200 in the 1970s to roughly 3,500 by the 1990s. A 2008 wildlife census, however, revealed a sharp drop in numbers, prompting the Indian government to pledge $153 million in further funding, and to consider relocating up to 200,000 villagers to make more room for tiger populations.

As the situation with tigers demonstrates, wildlife conservation in a country as poor and crowded as India is not easy. In many areas, pressure is mounting to convert remaining wildlands to farmlands. The best remaining zone of extensive wildlife habitat is in India's far northeast, an area subjected to rapid immigration and political unrest (see "Environment and Politics: Insurgency and National Parks in Northeastern India"). Moreover, wild animals, particularly tigers and elephants, threaten crops, livestock, and even people living near the reserves. When a rogue elephant herd ruins a crop or a tiger kills livestock, government agents are usually forced to destroy the animal.

The Four Subregions of South Asia

South Asia is separated from the rest of the Eurasian continent by a series of sweeping mountain ranges, including the Himalayas—the highest in the world. To better understand environmental conditions in this vast region, the Indian subcontinent can be broken down into four physical subregions, starting with the high mountain ranges of its northern fringe and extending to the tropical islands of the far south.

Mountains of the North South Asia's northern rim of mountains is dominated by the great Himalayan Range, forming the northern borders of India, Nepal, and Bhutan. These mountains are linked to the equally high Karakoram Range to the west, extending through northern Pakistan. More than two dozen peaks exceed 25,000 feet (7,620 meters), including the world's highest mountain, Everest, on the Nepal–China (Tibet) border at 29,028 feet (8,848 meters). To the east are the lower Arakan Yoma Mountains, forming the border between India and Myanmar (Burma) and separating South Asia from Southeast Asia.

These formidable mountain ranges were produced by tectonic activity caused by peninsular India pushing northward into the larger Eurasia Continental Plate; as a result of the collision between these two tectonic plates, great mountain ranges were folded and upthrust. The entire region is seismically active, putting all of northern South Asia in serious earthquake danger (see Figure 12.1). A massive earthquake in the Pakistani-controlled section of Kashmir on October 8, 2005, for example, resulted in roughly 80,000 deaths and left more than 3 million people homeless.

Indus-Ganges-Brahmaputra Lowlands South of the northern highlands lie vast lowlands created by three major river systems that have carried sediments eroded off the mountains through millions of years, building vast alluvial plains of fertile and easily farmed soils. These lowlands are densely settled and constitute the population core areas of Pakistan, India, and Bangladesh.

Of these three rivers, the Indus is the longest, covering more than 1,800 miles (2,880 kilometers) as it flows southward from the Himalayas through Pakistan to the Arabian Sea, providing much-needed irrigation waters to Pakistan's desert areas. The broad band of cultivated land in the desert zone of central and southern Pakistan that is

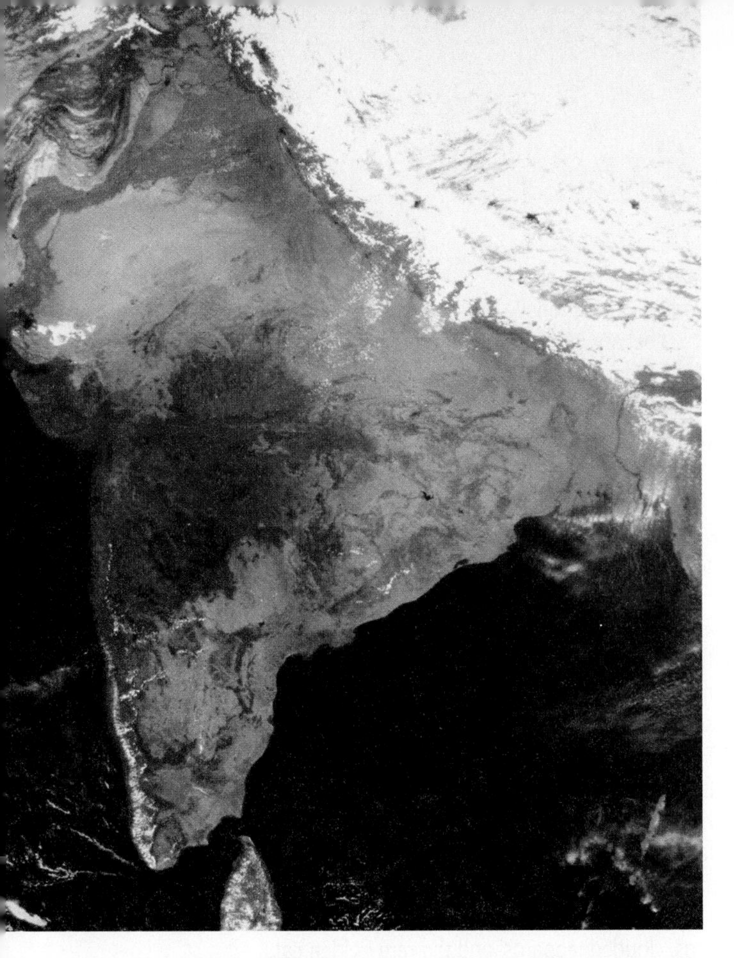

Figure 12.6 South Asia from Space The four physical subregions of South Asia are clearly seen in this satellite photograph, from the snow-clad Himalayas in the north to the islands of the south. The Deccan Plateau is dark, fringed by white clouds as moist air is lifted over the uplands of the Western Ghats.

made possible by the Indus is clearly visible on satellite images (Figure 12.6). Pakistan is currently concerned about India's plans to build dams on several tributaries of the Indus, which could reduce the flow of this all-important river.

Even more densely settled is the vast lowland of the Ganges, which, after flowing out of the Himalayas, travels southeasterly some 1,500 miles (2,400 kilometers) to empty into the Bay of Bengal. The Ganges not only provides the fertile alluvial soil that has made northern India a densely settled area for thousands of years, but has also long served as a vital transportation corridor. Given the central role of this river in South Asia's past and present, it is understandable that Hindus consider the Ganges sacred. But while the waters of the Ganges are reputed to have healing powers, the river is considered to be one of the world's most polluted watercourses.

Although this large South Asian lowland often is referred to as the Indus–Ganges Plain, this term neglects the Brahmaputra River. This river rises on the Tibetan Plateau and flows easterly, then southward and westerly over 1,700 miles (2,720 kilometers), joining the Ganges in central Bangladesh and spreading out over the vast delta, the largest in the world. Unlike the sparsely populated Indus Delta in Pakistan, the Ganges–Brahmaputra Delta is very densely settled, containing more than 3,000 people per square mile (1,200 per square kilometer).

Peninsular India Jutting southward from the Indus-Ganges Plain is the familiar shape of peninsular India, made up primarily of the Deccan Plateau. This plateau zone is bordered on each coast by narrow coastal plains backed by elongated north–south mountain ranges. On the west are the higher Western Ghats, which are generally about 5,000 feet (1,524 meters) but reach more than 8,000 feet (2,438 meters) near the peninsula's southern tip; to the east, the Eastern Ghats are lower and discontinuous, thus forming less of a transportation barrier to the broader eastern coastal plains. On both coastal plains, fertile soils and an adequate water supply support population densities comparable to those of the Ganges lowlands.

Soils are poor or average over much of the Deccan, but in Maharashtra state basaltic lava flows have produced fertile black soils. A reliable water supply for agriculture is a major problem in most areas. Much of the western plateau lies in the rain shadow of the Western Ghats, giving it a semiarid climate. Small reservoirs or tanks have been the traditional method for collecting monsoon rainfall for use during the dry season. More recently, deep wells and powerful pumps have mined groundwater to support irrigated crops and village water needs.

Partly because of the overuse of these aquifers, the Indian government is building a series of large dams to allow more extensive irrigation. These projects are highly controversial, largely because the reservoirs will displace hundreds of thousands of rural residents. A case in point is the Narmada Project, which involves the construction of 30 major dams along the Narmada River. Proponents claims that the scheme will allow India to feed an additional 20 to 30 million people, while opponents claim that the human costs are too high, pointing out that as of 2010 some 200,000 displaced people had not yet received compensation by the Indian government.

The Southern Islands At the southern tip of peninsular India lies the island country of Sri Lanka, which is almost linked to India by a series of small islands called Adam's Bridge. Sri Lanka is ringed by extensive coastal plains and low hills, but mountains reaching more than 8,000 feet (2,438 meters) cover the southern interior, providing a cool, moist climate. Because the main winds arrive from the southwest, that portion of the island is much wetter than the rain-shadow area of the north and east.

Forming a separate country are the Maldives, a chain of more than 1,200 islands stretching south to the equator some 400 miles (640 kilometers) off the southwestern tip of India. The combined land area of these islands is only about 116 square miles (290 square kilometers), and only a quarter of the islands are actually inhabited. The islands of the Maldives are flat, low coral atolls, with a maximum elevation of just over 6 feet (2 meters) above sea level.

ENVIRONMENT AND POLITICS Insurgency and National Parks in Northeastern India

Many of India's best-preserved natural environments are located in its northeastern region, a land with many rugged landscapes and relatively low population density. Unfortunately, it is also a politically troubled region. Northeastern India stands out for its sheer diversity of insurgent groups. The state of Manipur alone has 15 active or proscribed "terrorist/insurgent groups," while nearby Assam has 11, Meghalaya 4, Nagaland and Tripura 3 each, and Mizoram 2. No such groups are listed for Arunachal Pradesh, but it too has seen insurgent violence in recent years—and it is claimed in its entirety by China, greatly complicating Indo-Chinese relations. Insurgent groups in northeastern India have a strong tendency to divide and proliferate. The Kuki people of Manipur, for example, are "represented" by the Kuki Liberation Army, the Kuki National Army, the Kuki Liberation Front, and the United Kuki Liberation Front.

Insurgent activity certainly complicates the preservation of the natural ecosystem, but it does not always make it impossible. Consider, for example, Kaziranga National Park in Assam (Figure 12.1.1).

Figure 12.1.1 Kaziranga National Park. Kazinaga has some of South Asia's best-preserved natural habitats. Two-thirds of the world's Indian rhinoceroses live in this park.

Kaziranga, one of India's largest parks, boasts two-thirds of the world's Indian rhinoceroses as well as the highest density of tigers. The rebel army called the United Liberation Front of Asom has devastated the local economy, attacking oil pipelines, freight trains, and governmental buildings. Its fighters, however, not only respect the park but actually seek to protect it, killing poachers on more than one occasion. Despite rebel activity in the area, tourists have continued to be able to visit Kaziranga.

Through most of northeastern India, insurgency long made tourism impossible. Recently, however, violence has receded. Whereas Nagaland saw 154 insurgency-related deaths in 2007, the 2009 total was only 17; in Meghalaya, the death count dropped from 79 in 2003 to just 4 in 2009. Only in Manipur and Assam have body counts remained high (369 and 371, respectively, in 2009). Due to improved security, India has been opening parts of the northeast to travelers. For those interested in visiting the area, Northeast India Diary (www.northeastindiadiary.com/meghalaya-travel/wildlife-in-meghalaya.html) provides information on local attractions. On a trip to Meghalaya's Balpakram National Park, it claims, one might see "elephants, wild buffaloes, gaur (Indian bison), sambar, barking deer, wild boar, slow loris, capped langur as well as predators such as tigers, leopards, clouded leopards, and the rare golden cat."

In Nagaland and Mizoram, some observers attribute the recent decline in fighting to peacemaking efforts by local church organizations. Owing to successful missionary activities during the colonial period, both states are now strongly Christian: more than 75 percent of the population of Nagaland is Baptist, whereas Mizoram is more than 90 percent Christian (mostly Presbyterian). Missionary schooling has led to high levels of education. Mizoram boasts India's second highest literacy rate (91 percent), trailing only Kerala. Education, however, has not led to economic prosperity. Lack of infrastructure and insecurity are the major problems, but so too are the famines that occur every few decades after the synchronous flowering and then death of the state's massive bamboo groves. When the bamboo flowers and seeds, rodent and insect populations explode; when the plants subsequently perish, rats and bugs invade fields and granaries. The most recent such famines occurred in 2006–2007.

South Asia's Monsoon Climates

The dominant climatic factor for most of South Asia is the **monsoon,** the distinct seasonal change of wind direction that corresponds to wet and dry periods. Most of South Asia has three distinct seasons. First is the warm and rainy season of the southwest monsoon from June through October. This is followed by a relatively cool and dry season, extending from November until February, when the dominant winds are from the northeast. Only a few areas in far northwestern and southeastern South Asia get substantial rainfall during this period. Next comes the hot season from March to early June, which builds up with great heat and humidity until the monsoon's much anticipated and rather sudden "burst."

This monsoon pattern is caused by large-scale meteorological processes that affect much of Asia (Figure 12.7). During the Northern Hemisphere's winter, a large high-pressure system forms over the cold Asian landmass. Cold, dry winds flow outward from the interior of this high-pressure cell, over the Himalayas and down across South Asia. As winter turns to spring, these winds diminish, resulting in the hot, dry season. Eventually this buildup of heat over South and Southwest Asia produces a large thermal low-pressure cell. By early June this low-pressure cell is strong enough to draw in warm, moist air from the Indian Ocean.

Heavy orographic rainfall results from the uplifting and cooling of moist monsoon winds over the Western Ghats. As a result, some stations receive more than 200 inches (508 centimeters) of rain during the 4-month wet season. On the climate map (Figure 12.8), these are the areas of Am, or tropical monsoon, climate. Inland, however, a strong rain-shadow effect dramatically reduces rainfall on the Deccan Plateau. Farther north, as the monsoon winds are forced up and over the mountains, copious amounts of rainfall are characteristic. Cherrapunji, India, at 4,000 feet (1,220 meters), is a strong contender for the title of world's wettest place, with an average rainfall of 450 inches (1,143 centimeters).

Not all of South Asia receives substantial rainfall from the southwest monsoon. In much of Pakistan and the Indian state of Rajasthan, precipitation is low and variable, resulting in steppe and desert climates. In Karachi, the annual total is less than 10 inches (25 centimeters). But even in the drier parts of South Asia, heavy monsoonal rain sometimes hits. In the summer of 2010, for example, prolonged downpours resulted in devastating flooding across much of Pakistan.

Regardless of whether rainfall is heavy or light, the monsoon rhythm affects all of South Asia in many different ways, from the delivery of much-needed water for crops and villages, to the mood of millions of people as they eagerly await relief from oppressive heat (Figure 12.9). Some years the monsoon delivers its promise with abundant moisture; in other years, though, it brings scant rainfall, resulting in crop failure and hardship. A severe drought in northern and central India in 2009, for example, led to a 7 percent decrease in the country's grain harvest. In the near future, the monsoonal patterns may change due to global warming.

Global Warming in South Asia

Many areas of South Asia are particularly vulnerable to the effects of global warming. Even a minor rise in sea level will inundate large areas of the Ganges–Brahmaputra Delta in Bangladesh. Already, over 18,500 acres (7,500 hectares) of swampland in the Sunderbans have been submerged. Several small islands in this area have recently disappeared, due to a combination of sea level rise and subsidence. A 2007 report from the India government suggests that up to 7 million people could be displaced by the end of the century due to a predicted 3.3-foot (1-meter) rise in sea level. If the most severe sea level forecasts come to pass, the atoll nation of the

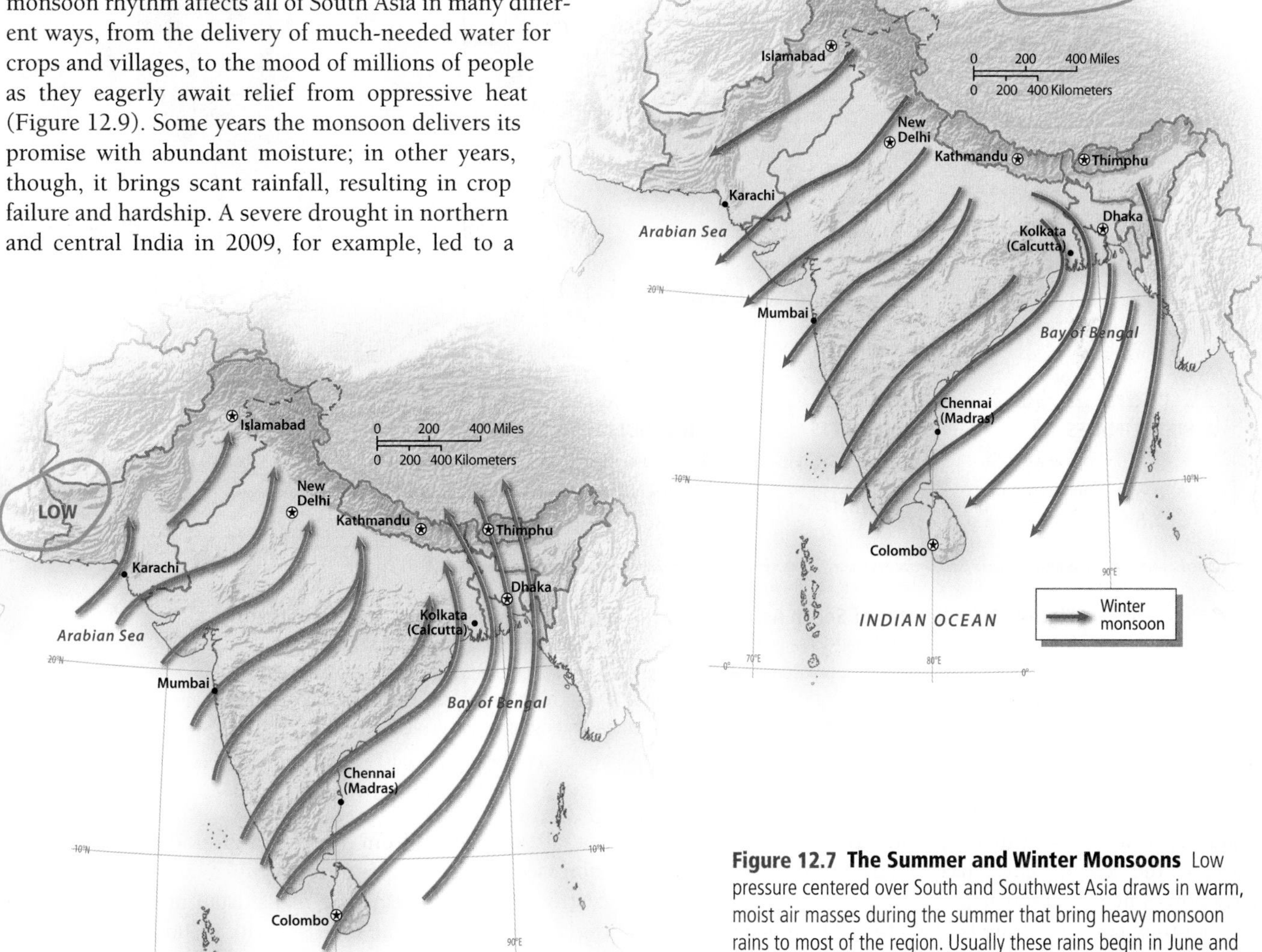

Figure 12.7 The Summer and Winter Monsoons Low pressure centered over South and Southwest Asia draws in warm, moist air masses during the summer that bring heavy monsoon rains to most of the region. Usually these rains begin in June and last for several months. During the winter, high pressure forms over northern Asia. As a result, winds are reversed from those of the summer. During this season, only a few coastal locations along India's east coast and in eastern Sri Lanka receive substantial rain.

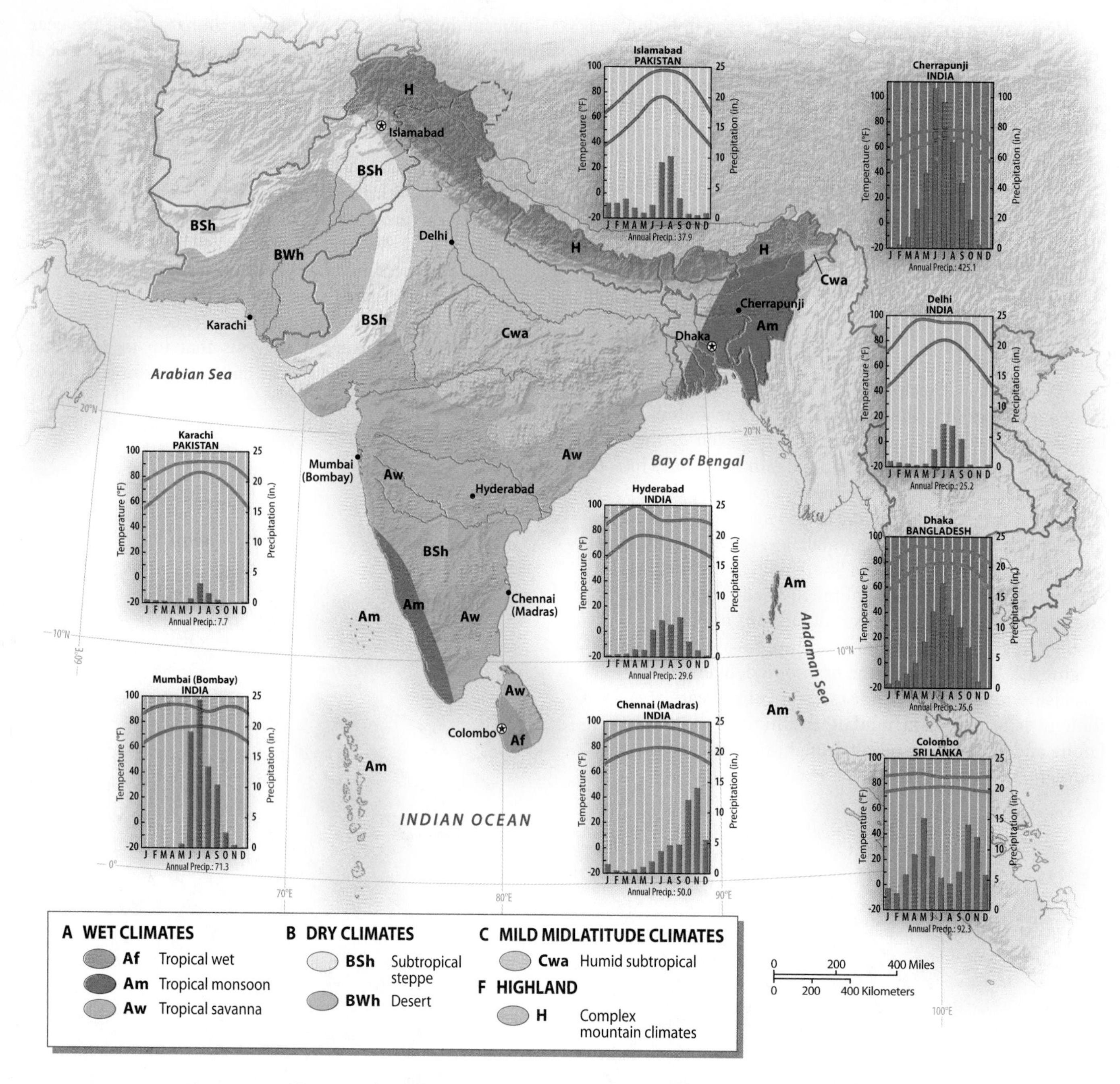

Figure 12.8 Climates of South Asia Except for the extensive Himalayas, South Asia is dominated by tropical and subtropical climates. Many of these climates show a distinct summer rainfall season that is associated with the southwest monsoon. The climographs for Mumbai (Bombay) and Delhi are excellent illustrations. However, the climographs for east-coast locations such as Madras, India, and Colombo, Sri Lanka, show how some locations also receive rains from the northeast monsoon of the winter.

Maldives will simply vanish beneath the waves. Warmer water in the Bay of Bengal has already resulted in the destruction of many coral reefs.

South Asian agriculture is likely to suffer from a number of problems linked to global climate change. Most Himalayan glaciers are rapidly retreating, threatening the dry-season water supplies of the Indus–Ganges Plain, an area that already suffers from overuse of groundwater resources. Increased winter temperatures of up to 6.4°F (3°C) could destroy the vital wheat crop of Pakistan and northwestern India, undermining the food security of both countries. In parts of South Asia, however, global warming could result in increased rainfall due to an intensification of the summer monsoon. Unfortunately, such a new precipitation regime

Figure 12.9 Monsoon Rain During the summer monsoon, some Indian cities such as Mumbai (Bombay) receive more than 70 inches (178 centimeters) of rain in just 3 months. These daily torrents cause floods, power outages, and daily inconvenience. However, the monsoon rains are crucial to India's agriculture. If the rains are late or abnormally weak, crop failure often results.

will likely be characterized by more intense cloudbursts coupled with fewer episodes of gentle, prolonged rain. As a result, more flooding—as well as more soil erosion—may be expected.

India signed the Kyoto Protocol in 2002, but as a developing country has been exempt from the main provisions of the treaty. With its poor and still largely nonindustrial economies, South Asia still has a low per capita output of greenhouse gases. But India's economy in particular is not only growing rapidly, but is heavily dependent on burning coal to generate electricity. According to a 2010 report, India's annual greenhouse gas emissions increased 58 percent between 1994 and 2007. Further increases seem likely; according to official estimates, if India is to maintain an economic growth rate of 8 percent over the next quarter century, it will need to triple or even quadruple its primary energy supply.

South Asia's vulnerability to global warming coupled with its rapidly growing emissions of greenhouse gases has prompted numerous calls to action. Experts were heartened in early 2007 when India's government signed an agreement with Japan aimed at encouraging Japanese companies to invest in India's energy sector and to transfer energy-saving technology. Many South Asian officials, however, argue that the responsibility for addressing global climate change lies with the already industrialized countries, and that local attempts to reduce emissions could undermine economic growth. Some Indian scientists are also skeptical about climate change reports originating in the developed world. In 2010, the Indian government announced that it would create its own organization to monitor global warming, arguing that the United Nations' Intergovernmental Panel on Climate Change could not be fully trusted.

POPULATION AND SETTLEMENT: Continuing Growth in a Crowded Land

South Asia soon will surpass East Asia as the world's most populous region (Table 12.1). Overall, fertility levels have dropped markedly in recent years, but population continues to grow rapidly over much of the region. And although South Asia has made remarkable agricultural gains since the 1960s, there is still widespread concern about the region's ability to feed itself. The threat of crop failure remains, in part because much South Asian farming is vulnerable to the unpredictable monsoon rains.

The Geography of Population Expansion

South Asia's recent decline in human fertility shows distinct geographical patterns. Most of southern and western India, along with Sri Lanka should soon see population stabilization. Over most of northern South Asia, however, the birthrate remains elevated. While all South Asian countries have established family planning programs, the commitment to these policies varies widely from place to place.

India Widespread concern over India's population growth began in the 1960s. To a significant extent, family planning measures, along with general economic and social development, have been successful; the total fertility rate (TFR) dropped from 6 in the 1950s to the current rate of 2.6. Fertility rates vary widely within India, from below 2.0 in the states of Goa, Kerala, Tamil Nadu, Andhra Pradesh, and Himachal Pradesh, to a problematic high of 3.8 in Uttar Pradesh and 4.0 Bihar. (Uttar Pradesh, with more than 183

TABLE 12.1 Population Indicators

Country	Population (millions) 2010	Population Density (per square kilometer)	Rate of Natural Increase (RNI)	Total Fertility Rate	Percent Urban	Percent <15	Percent >65	Net Migration (Rate per 1000) 2005–10[a]
Bangladesh	164.4	1,142	1.5	2.4	25	32	4	–0.7
Bhutan	0.7	15	1.7	3.1	32	31	5	2.9
India	1,188.8	362	1.5	2.6	29	32	5	–0.2
Maldives	0.3	1,070	1.9	2.5	35	30	5	0.0
Nepal	28.0	191	1.9	3.0	17	37	4	–0.7
Pakistan	184.8	232	2.3	4.0	35	38	4	–1.6
Sri Lanka	20.7	315	1.2	2.4	15	26	6	–3.0

[a]*Net Migration Rate from the United Nations, Population Division, World Population Prospects: The 2008 Revision Population Database.*

Source: Population Reference Bureau, World Population Data Sheet, 2010.

million people, would be the sixth largest country in the world if it were independent.) A strong relationship is evident between women's education and family planning; where female literacy has increased most dramatically, fertility levels have rapidly declined (see "Geographic Tools: Survey Work on Contraceptive Use in West Bengal").

A distinct cultural preference for male children is found in most of South Asia, a tradition that complicates family planning. Abortion rates of female fetuses are much higher than those of males, even though the practice of sex-selective abortion is illegal. In northern India, a lower fertility rate seems to be accompanied by an even lower ratio of female-to-male infants. India's prosperous northwestern state of Punjab has recently seen its TFR drop to just below the replacement level, but at the same time its population has grown increasingly male dominated. In 2006, Punjab's sex ratio at birth was only 776 girls for every 1,000 boys.

Pakistan and Bangladesh Pakistan, with a population of more than 184 million, has long had a somewhat ambivalent attitude toward family planning. Although the government's official position is that the birthrate is excessive, the country still lacks an effective, coordinated family planning program. As a result, the TFR remains just under 4.0. At its current birthrate, Pakistan's population would expand to 450 million by 2050. Its concerned government recently launched the National Population Policy 2010 program, which aims to reduce the fertility rate to 3.0 by 2015. The program aims to make family planning more accessible and effective and seeks to recruit religious leaders for support.

Densely settled Bangladesh, with a population about half that of the United States packed into an area smaller than the state of Wisconsin, has made significant strides in population stabilization. As recently as 1975, its TFR was 6.3, but it had dropped to 2.4 by 2010. Its family planning success can be attributed to strong support from the Bangladesh government for advertising through radio and billboards. Also important are more than 35,000 women fieldworkers who take information about family planning into every village in the country (Figure 12.10). In 2010, the president of Bangladesh urged the entire country to adopt the slogan, "Not more than two children, one is better."

Figure 12.10 Family Planning in Bangladesh Bangladesh has been one of the most successful nations in South Asia in reducing its fertility rate through family planning. Many women in Bangladesh use oral contraceptives. This photo shows community health worker Hashina Akhtar giving child nutrition and family planning advice to local villagers.

GEOGRAPHIC TOOLS Survey Work on Contraceptive Use in West Bengal

Geographers, like other social scientists, often make extensive use of surveys in order to acquire raw data. Standardized surveys, in which sizable numbers of local people are asked identical and relatively easily answered questions, allow scholars to build quantitative databases. Statistical tests can then be conducted to determine correlations and to find relationships. Geographers often use such surveys to identify spatial patterns.

Geographers also sometimes follow anthropologists in using informal, open-ended interviews with select people who live in the area under investigation. Such a technique can give a deeper understanding of local attitudes and cultural perceptions and practices. They do not allow, however, for the kinds of statistical testing obtainable with standardized survey methods. As a result, geographers sometimes use both methods to obtain both quantitative and qualitative information.

One geographer who has effectively used both methods is Elizabeth Chacko, of George Washington University, in her study of women's use of contraception in four villages in West Bengal, India. Determining why women in rural South Asia chose to use, or not to use, contraception is particularly important considering the demographic strains faced by the region. But it is a difficult subject to investigate, as women are often reluctant to discuss such issues, particularly with outsiders. Dr. Chacko therefore decided to use both the quantitative method of the formal survey and the qualitative method of in-depth personal conversations and interviews.

In the end, Dr. Chacko discovered that the most important determinants of a woman's contraceptive decisions were age, number of living sons, and religious affiliation. She also found that availability of permanent, village-based government health-care clinics increased the use of contraceptives. Observations and interviews also pointed to the need to provide low-cost, culturally acceptable methods for achieving greater contraceptive usage in the community (Figure 12.2.1). These findings should be of great help to local health-care workers seeking both to empower rural women and to address India's demographic problems.

Figure 12.2.1 Village-Based Health-Care Clinics A rural health-care worker poses behind her medical kit in rural West Bengal. Such local outreach efforts have resulted in a reduced birthrate in much of this area.

Migration and the Settlement Landscape

The most densely settled areas of South Asia still coincide with zones of fertile soils and dependable water supplies (Figure 12.11). The largest rural populations are found in the core area of the Ganges and Indus river valleys and on the coastal plains of India. Settlement is less dense on the Deccan Plateau and is relatively sparse in the arid lands of the northwest. While most of South Asia's northern mountains are too rugged and high to support dense human settlement, there are major population clusters in the Katmandu Valley of Nepal, situated at 4,400 feet (1,341 meters), and the Valley, or Vale, of Kashmir in northern India, at 5,200 feet (1,585 meters).

As is true in most other parts of the world, South Asians have migrated for hundreds of years from poor and densely populated areas to places that are either less densely populated or wealthier. This process is ubiquitous in South Asia today, but several areas stand out as zones of intensive out-migration: Bangladesh, the northern Indian states of Bihar, Punjab, and Rajasthan, and the northern portion of India's Andhra Pradesh (Figure 12.12). Migrants from Bangladesh are settling in large numbers in rural portions of adjacent Indian states, exacerbating ethnic and religious tensions. In Nepal, migrants have been moving over the past 50 years from crowded mountain valleys to formerly malaria-infested lowlands along the Indian border, again generating ethnic strife. Sometimes migrants are forced out by war; sizable streams of people from northern Sri Lanka and from Kashmir have sought security away from their battle-scarred homelands over the past 20 years.

South Asia is one of the least urbanized regions in the world, with only about 35 percent of its population living in

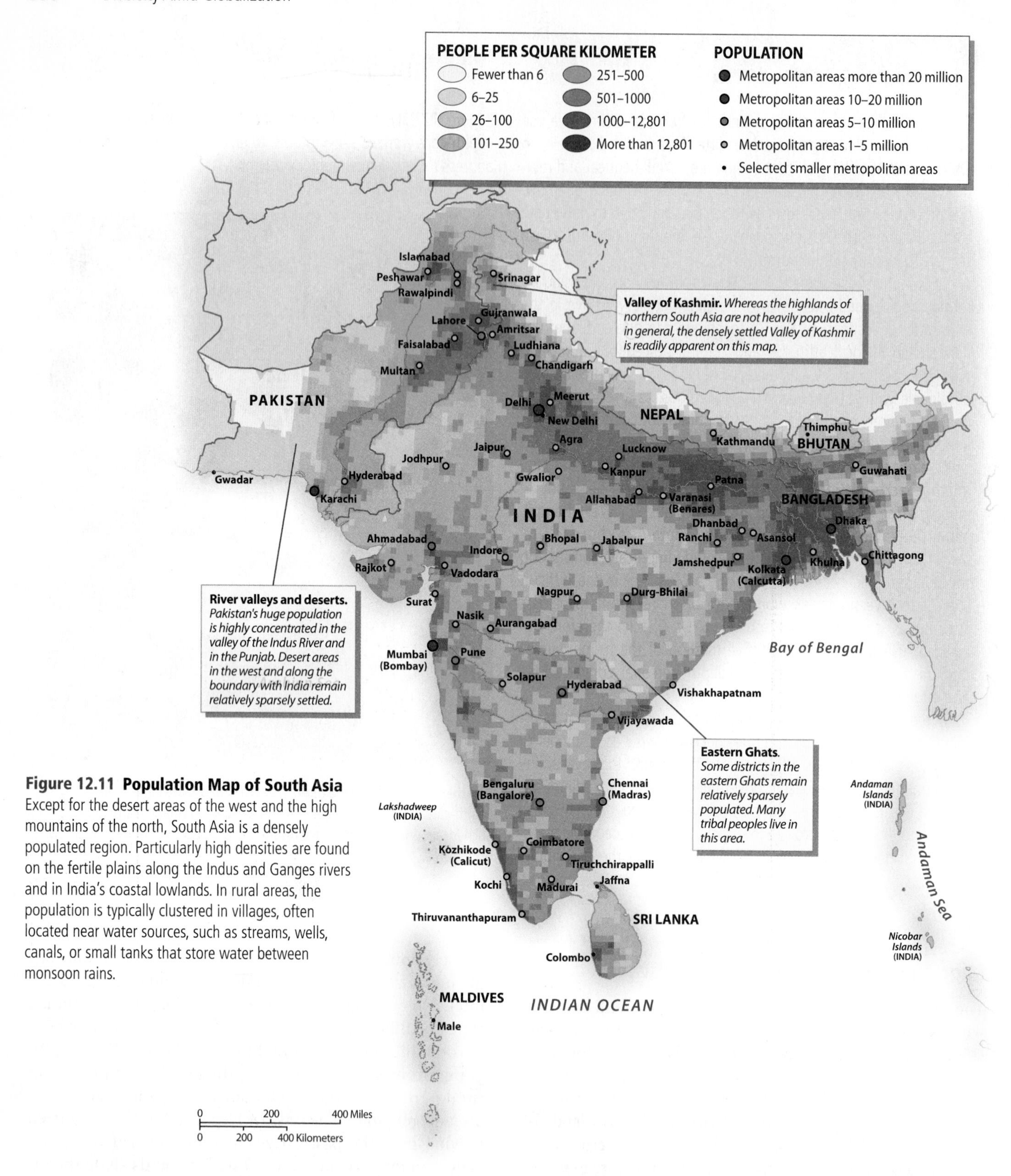

Figure 12.11 Population Map of South Asia Except for the desert areas of the west and the high mountains of the north, South Asia is a densely populated region. Particularly high densities are found on the fertile plains along the Indus and Ganges rivers and in India's coastal lowlands. In rural areas, the population is typically clustered in villages, often located near water sources, such as streams, wells, canals, or small tanks that store water between monsoon rains.

cities. Most South Asians still reside in compact rural villages, but large numbers are streaming into the region's rapidly growing cities. Such urbanization often stems more from desperate conditions in the countryside than from the attractions of city life. As farms begin to mechanize, farm laborers often have no choice but to migrate to urban areas. Many small farmers, moreover, have difficulty competing with their wealthier neighbors, who can more easily afford modern fertilizers and other agricultural inputs. The agricultural situation over much of India has become so desperate that the country is experiencing an epidemic of farmer suicides. In the single month of April 2009, the Indian state

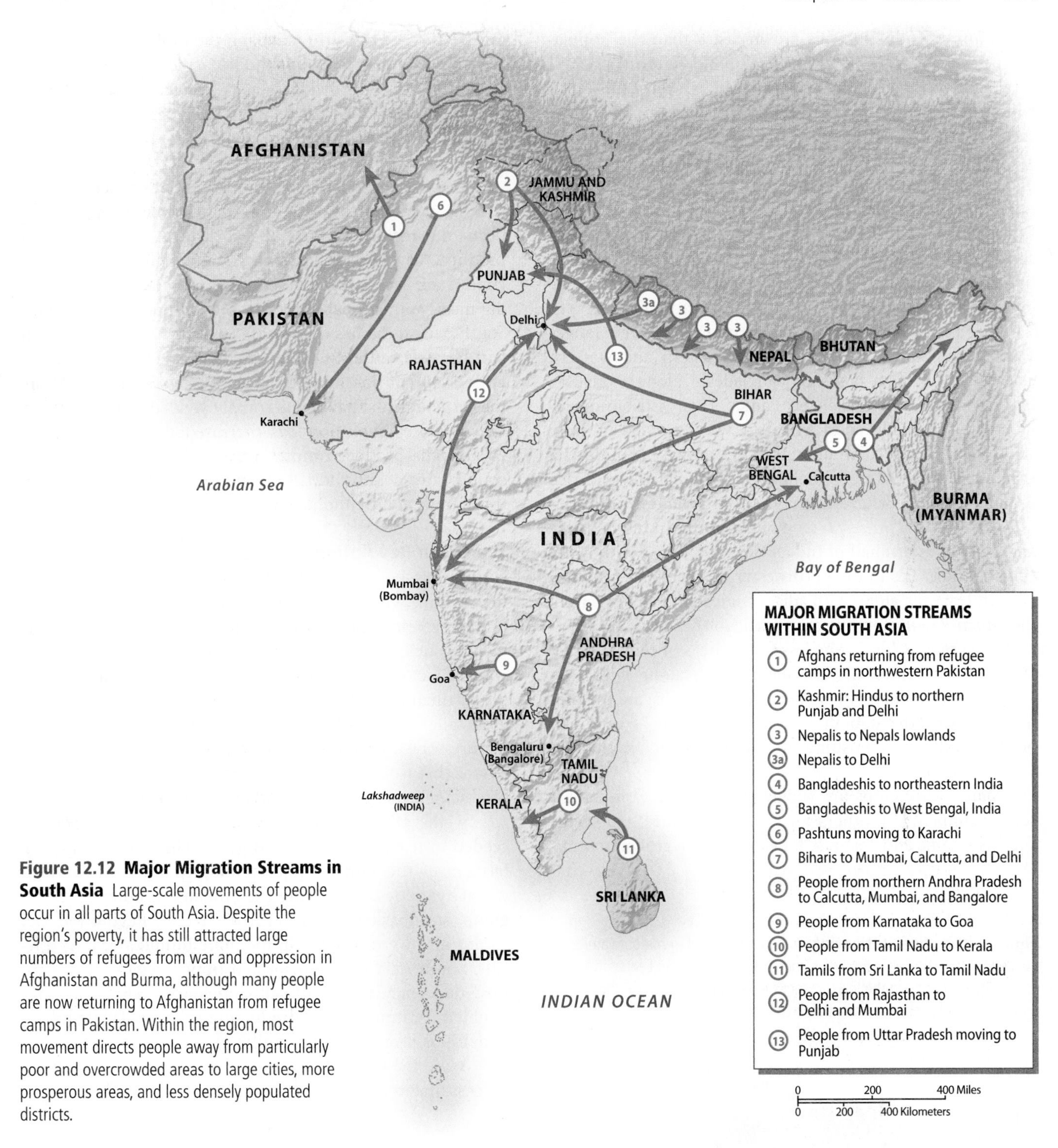

Figure 12.12 Major Migration Streams in South Asia Large-scale movements of people occur in all parts of South Asia. Despite the region's poverty, it has still attracted large numbers of refugees from war and oppression in Afghanistan and Burma, although many people are now returning to Afghanistan from refugee camps in Pakistan. Within the region, most movement directs people away from particularly poor and overcrowded areas to large cities, more prosperous areas, and less densely populated districts.

of Chattisgarh reported some 1,500 farmer suicides, due largely to debt and crop failures.

Agricultural Regions and Activities

South Asian agriculture has historically been much less productive than that of East Asia. Although the reasons behind such poor yields are complex, many experts cite the relatively low social status of most cultivators and the fact that much farmland has long been controlled by wealthy landlords who hire others to work their plots. Some scholars also blame British colonialism, which emphasized export crops for European markets.

Regardless of the causes, low agricultural yields in the context of a huge, hungry, and rapidly growing population constitute a pressing problem. Since the 1970s, however, agricultural production has grown faster than the population, primarily because of the Green Revolution, agricultural cultivation techniques based on hybrid crop strains and the heavy use of industrial fertilizers and pesticides. Because the Green Revolution also carries significant social and environmental costs, it remains controversial, as will be discussed shortly.

Crop Zones South Asia can be divided into several distinct agricultural regions, all with different problems and potentials. The most fundamental division is between the three primary subsistence crops of rice, wheat, and millet.

Rice is the main crop and foodstuff in the lower Ganges Valley, along the lowlands of India's eastern and western coasts, in the delta lands of Bangladesh, along Pakistan's lower Indus Valley, and in Sri Lanka. This distribution reflects the large volume of irrigation water needed to grow rice (Figure 12.13). The amount of rice grown in South Asia is impressive: India ranks behind only China in world rice production, and Bangladesh is the fourth largest producer.

Wheat is the principal crop of the northern Indus Valley and in the western half of India's Ganges Valley. South Asia's "breadbasket" lies in the northwestern Indian states of Punjab and Haryana along with adjacent areas in Pakistan. Here the Green Revolution has been particularly successful in increasing grain yields. In the less-fertile areas of central India, millet and sorghum are the main crops, along with root crops such as manioc. In general, wheat and rice are the preferred staples throughout South Asia, and it is generally poorer people who consume "rough" grains such as the various millets.

Many other crops are widely cultivated in South Asia, some commercially, others for local subsistence. Oil seeds, such as sesame and peanuts, for example, are grown in semi-arid districts, while Sri Lanka and the Indian state of Kerala are noted for their coconut groves, spice gardens, and tea plantations. In both Pakistan and west central India, cotton is widely cultivated, while Bangladesh has long supplied most of the world's jute, a tough fiber used in the manufacture of rope.

Many if not most South Asians receive inadequate protein, as meat consumption is extremely low. This partly reflects the region's poverty, because meat is expensive to produce. In India, religion is equally important, as most Hindus are vegetarians. Despite this prohibition against eating meat, animal husbandry is vitally important throughout South Asia. India has the world's largest cattle population, in part because cattle are sacred in Hinduism but also because milk is one of South Asia's main sources of protein. While the cattle of India have traditionally yielded little milk, a so-called white revolution has increased dairy efficiency. In 2010, India produced more dairy products than any other country by a substantial margin, while Pakistan ranked third, trailing only India and the United States.

The Green Revolution The main reason South Asian agriculture has kept up with population growth is the Green Revolution, which originated during the 1960s in agricultural research stations established by international development agencies. One of the major problems researchers faced was the fact that higher yields could not be attained by simply fertilizing local seed strains, because the plants would grow taller and then fall to the ground before the grain could mature. The solution was to cross-breed new "dwarf" crop strains that would respond to heavy chemical fertilization by producing extra grain rather than longer stems.

By the 1970s, it was clear that these efforts had succeeded in reaching their initial goals. The more prosperous farmers of the Punjab quickly adopted the new "miracle wheat" varieties, solidifying the Punjab's position as the region's breadbasket (Figure 12.14). Green Revolution rice strains also were adopted in the more humid areas. As a result, South Asia was transformed from a region of chronic food deficiency to one of self-sufficiency. India more than doubled its annual grain production between 1970 and the mid-1990s, from 80 to 191 million tons.

While the Green Revolution was clearly an agricultural success, many argue that it has been an ecological and social

Figure 12.13 Rice Cultivation A large amount of irrigation water is needed to grow rice, as is apparent from this photo from Sri Lanka. Rice also is the main crop in the lower Ganges Valley and Delta, along the lower Indus River of Pakistan, and on India's coastal plains.

Figure 12.14 Green Revolution Farming Because of "miracle" wheat strains that have increased yields in the Punjab area, this region has become the breadbasket of South Asia. India more than doubled its wheat production in the last 25 years and has moved from chronic food shortages to self-sufficiency. Increased production, however, has led to both social and environmental problems.

disaster. Serious environmental problems result from the chemical dependency of the new crop strains. Not only do they typically need large quantities of industrial fertilizer, which is both expensive and polluting, but they also require frequent pesticide applications because they often lack natural resistance to local plant diseases and insects. A 2009 study showed that 20 percent of the wells in India's Punjab are seriously contaminated by nitrates, derived largely from artificial fertilizers.

Social problems have also followed the Green Revolution. In many areas, only the more prosperous farmers are able to afford the new seed strains, irrigation equipment, farm machinery, fertilizers, and pesticides necessary to support this new high-technology agriculture. As a result, poorer farmers often go deeply into debt and are then forced off the land when they fail to repay their loans.

Future Food Supply The Green Revolution has fed South Asia's expanding population over the past several decades, but whether it will be able to continue doing so remains unclear. Many of the crop improvements have seemingly exhausted their potential, and much of South Asia's agricultural economy is currently in a state of crisis as farmers fail to make ends meet.

Optimists believe, however, that South Asia's food production could be substantially increased. Further improvements in highways and railroads, for example, would result in less wastage as well as higher profit margins for struggling farmers. Green Revolution techniques, moreover, might be profitably applied to secondary grain crops, and genetic engineering might provide another breakthrough, ushering in a second wave of increased crop yields. Most environmentalists, however, see serious dangers in this new technology.

Another option is expanded water delivery, because many fields remain unirrigated in South Asia's semiarid areas. Even in the humid zone, dry-season fallow is usually the norm. Irrigation, however, brings its own problems. In much of Pakistan and northwestern India, where irrigation has been practiced for generations, soil salinization (see Figure 12.3), or the buildup of salt in agricultural fields, is already a major constraint. An estimated 27 million acres (11 million hectares) of land in Pakistan is now too salty for normal farming. Additionally, groundwater levels are falling in the Punjab, India's breadbasket, because double-cropping has pushed water use beyond the sustainable yield of the underlying aquifers.

Urban South Asia

Although South Asia is one of the least urbanized regions of the world (see Table 12.1), it still has some of the largest urban agglomerations in the world. India alone lists more than 43 cities with populations greater than a million, most of which are growing rapidly. Mumbai's (Bombay) population is now roughly 14 million, with up to 22 million living in the city's metropolitan area (see "Cityscapes: Life on the Streets of Mumbai [Bombay]").

Because of this rapid growth, most South Asian cities have staggering problems with homelessness, poverty, congestion, water shortages, air pollution, and sewage disposal. Kolkata's (Calcutta) homeless are legendary, with perhaps half a million sleeping on the streets each night. In that city and others, sprawling squatter settlements, or bustees, mushroom in and around urban areas, providing temporary shelter for many urban migrants (Figure 12.15). A brief survey of the region's major cities gives clues to the problems and prospects of the region's urban areas.

Figure 12.15 Mumbai (Bombay) Hutments Hundreds of thousands of people in Mumbai (Bombay) live in crude hutments, with no sanitary facilities, built on formerly busy sidewalks. Hutment construction is forbidden in many areas, but wherever it is allowed, sidewalks quickly disappear.

CITYSCAPES Life on the Streets of Mumbai (Bombay)

The largest city in South Asia, Mumbai (Bombay) is India's financial, industrial, and commercial center. The major port on the Arabian Sea, Mumbai is responsible for a disproportionate share of the country's foreign trade. Long noted as the center of India's textile manufacturing, the city also is the hub of its film industry, the largest in the world.

Mumbai occupies a peninsular site originally composed of seven small islands (Figure 12.3.1). Since the 17th century, drainage and reclamation projects have joined the islands into a larger body known as Bombay Island, which basically formed the city area through the colonial period. Two parallel ridges of 100 feet (30 meters) form the spines of the inner city, one occupied by the historic colonial fort, now the commercial center, the other, Malabar Hill, an exclusive residential area. India's financial center is found near the southern end of the island, around Nariman Point, as are many of the city's most exclusive hotels. Another nearby expensive area is Kemps Corner, noted for its boutiques and restaurants.

Bombay Island, only 12 miles long, is notoriously congested. As a result, most growth has taken place to the north and east of the historic peninsula so that now the metropolitan region occupies an area 10 times the size of the original city. Spatial restrictions in downtown

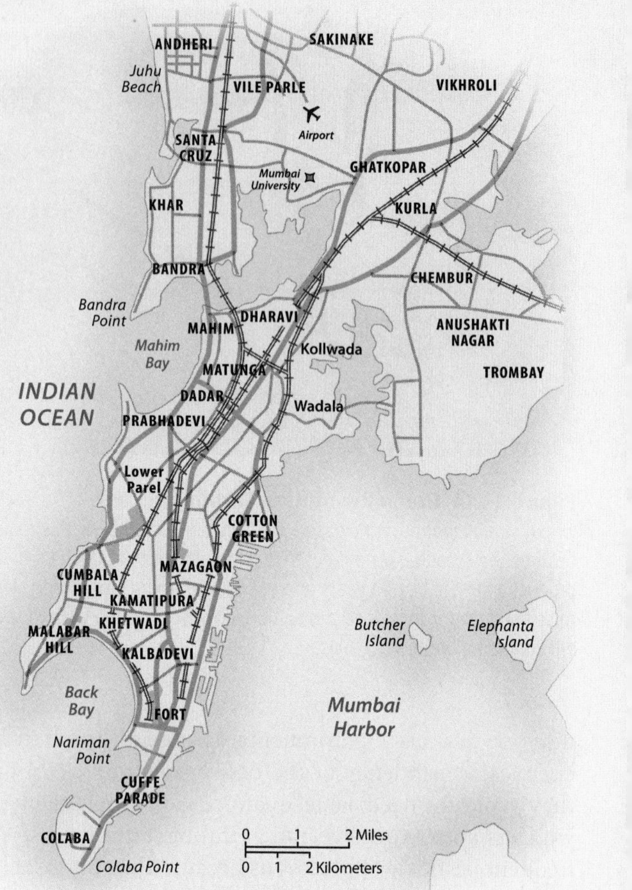

Figure 12.3.1 Map of Mumbai (Bombay) Originally composed of seven separate islands, the core of Mumbai (called Bombay Island) now essentially forms a narrow peninsula some 12 miles (19.3 kilometers) long. Recent urban growth is concentrated in the suburbs to the north and east.

Delhi Delhi, the sprawling capital of India, has roughly 19 million people in its metropolitan area. It consists of two contrasting landscapes expressing its past: Delhi (or old Delhi), a former Muslim capital, is a congested town of tight neighborhoods; New Delhi, in contrast, is a city of wide boulevards, monuments, parks, and expansive residential areas. It was born as a planned city when the British decided to move the colonial capital from Calcutta in 1911. Located here are the embassies, luxury hotels, government office buildings, and airline offices necessary for a vibrant political capital. South of the government area are more expensive residential areas, some of which are focused on expansive parks and ornamental gardens. Rapid growth, along with the government's inability to control auto and industrial emissions, made Delhi one of the world's 10 most polluted cities in the late 1990. As a result, the government began to shut down many of the city's small-scale industries, although local residents have banded together to protect their livelihoods. Overall, significant environmental progress has been made.

Kolkata (Calcutta) To many, Kolkata is emblematic of the problems faced by rapidly growing cities in developing countries. Not only is homelessness widespread, but this metropolitan area of some 15 million falls far short of supplying its residents with water, power, or sewage treatment. Electrical power is so inadequate that every hotel, restaurant, shop, and small business has to have some sort of standby power system. During the wet season, many streets are routinely flooded.

With continued rapid growth as migrants drain in from the countryside, a mixed Hindu-Muslim population that generates ethnic rivalry, a troubled economic base, and an overloaded infrastructure, Kolkata faces a problematic future. Yet it remains a culturally vibrant city noted for its fine educational institutions, theaters, and publishing firms.

Dhaka As the capital and major city of Bangladesh, Dhaka (also spelled *Dacca*) has experienced rapid growth due to migration from the surrounding countryside. In 1971, when the country gained independence from Pakistan, Dhaka had

Mumbai also have resulted in skyrocketing commercial and residential rents, which now are some of the highest in the world (Figure 12.3.2). Even members of the city's thriving middle class find it difficult to secure adequate housing. Hundreds of thousands of less-fortunate immigrants, eager for work in central Mumbai, live in "hutments," crude shelters built on formerly busy sidewalks. The least fortunate sleep on the street or in simple plastic tents, often placed along busy roadways. Meanwhile, slums expand around the city's outskirts, providing crude homes for an estimated 60 percent of Mumbai's population.

Relatively few tourists venture out of the historic and relatively wealthy areas of southern Bombay Island. Thus, they miss much of what Mumbai has to offer. As is true in most other major cities, specific areas have distinctive personalities that must be seen—and heard and smelled—to be appreciated. Turn one corner and you might find yourself in a Muslim neighborhood, with all signs suddenly appearing in Urdu, hence looking much like Arabic. Turn another and you might encounter a several-mile-long stretch of stalled trucks, each waiting to cross a bridge or unload its cargo. The mushrooming suburbs offer their own discoveries. Juhu Beach, for example, offers quality hotels and restaurants that are significantly less expensive than those of the city center, as well as a Sunday night carnival. The ocean here, however, is much too polluted for safe swimming.

Despite Mumbai's extraordinary contrasts of wealth and poverty, it remains in many ways an orderly and relatively crime-free city. In 2007, its crime rate was roughly half of that of other Indian cities with more than one million inhabitants. It is less dangerous to walk the streets of central Mumbai than those of many major North American cities. Organized crime is a problem, especially in the massive film industry, but it has few effects on the lives of the average people.

Figure 12.3.2 Mumbai (Bombay) Central City The heart of this metropolitan area of 16 million is highly congested. Some of the world's highest office and apartment rents are found here, despite the overall poverty of the city.

about 1 million inhabitants; today its metropolitan area numbers almost 13 million (Figure 12.16). Dhaka's economic vitality has increased since independence because it combines the administrative functions of government with the largest industrial concentration in Bangladesh. Cheap and abundant labor in Dhaka has made the city a global center for clothing, shoe, and sports equipment manufacturing. North American shoppers are readily reminded of this new role when looking through clothing tags at any department store.

Karachi Karachi, a rapidly growing port city of approximately 18 million people, is Pakistan's largest urban area and commercial core (Figure 12.17). It also served as the country's capital until 1963, when the new city of Islamabad was created in the northeast. Karachi, however, has suffered little from the exodus of government functions; it is the most cosmopolitan city in Pakistan, with its checkerboard street pattern lined with high-rise buildings, hotels, banks, and travel agencies. In many ways, its landscape conveys the sense that Karachi is a model metropolis for a developing country.

Figure 12.16 Dhaka City View Dhaka, the capital of Bangladesh, has emerged as a vibrant metropolis in recent decades. Although its slums are extensive, its central commercial district is relatively orderly and prosperous.

Figure 12.17 Karachi Landscape Karachi, Pakistan's largest city and main port, is noted for both its economic power and its ethnic violence. Congestion, as seen in this photo, can be severe.

Karachi, however, suffers from serious political and ethnic tensions that have turned parts of the city into armed camps. During the worst of the violence in 1995, more than 200 people were killed on the city's streets each month, and army bunkers were lodged at major urban crossroads. Ethnic conflicts have generally pitted Sindis, the region's indigenous inhabitants, against the Muhajirs, Muslim refugees from India who settled in and around the city after independence in 1947. More recently, the migration of Pashtuns from the tribal areas of northwest Pakistan has further destabilized the situation, as have clashes between Sunni and Shiite Muslims. A number of radical Islamist groups, including the Taliban and Al Qaeda, are reputed to have important bases in the city.

Islamabad Upon independence, Pakistan's leaders determined that Karachi was too far from the center of the country and that an entirely new capital was necessary. This planned city would make a statement through its name—Islamabad—about the religious foundation of Pakistan. Located close to the contested region of Kashmir, it would also make a geopolitical statement. In geographic terms, such a city is referred to as a **forward capital** because it signals—both symbolically and geographically—the intentions of the country. By building its new capital in the north, Pakistan sent a clear message that it would not abandon its claims to the portion of Kashmir controlled by India. Islamabad is closely linked to the historic city of Rawalpindi, once a major British encampment, a few miles away. While these two cities form a single metropolitan region of about 4.5 million people, they are completely different in appearance and character. To avoid congestion, planners designed Islamabad around self-sufficient sectors, each with its own government buildings, residences, and shops.

The closest parallel in India to Islamabad is Chandigarh, the modern, planned city that serves as the capital of two Indian states: Punjab and Haryana. All told, the cities of India and Pakistan have a similar feel, as would be expected considering the two countries' common historical and cultural backgrounds.

CULTURAL COHERENCE AND DIVERSITY: A Common Heritage Undermined by Religious Rivalries

Historically, South Asia forms a well-defined cultural region. A thousand years ago, virtually the entire area was united by ideas and social institutions associated with Hinduism. The subsequent arrival of Islam added a new religious dimension without undercutting the region's cultural unity. British imperialism subsequently imposed a number of cultural features over the entire region, from the widespread use of English to a common passion for cricket. Since the mid-20th century, however, religious and political strife have intensified, leading some to question whether South Asia can still be conceptualized as a culturally coherent world region.

India has been a secular country since its inception, with the Congress Party, its early guiding political organization, struggling to keep politics and religion separate. In doing so, it relied heavily on support from Muslims as well as the lower castes. Since the 1980s, this secular political tradition has come under increasing pressure from Hindu "fundamentalism," which is probably better referred to as **Hindu nationalism**. Hindu nationalists promote the religious values of Hinduism as the essential and fabric of Indian society.

Hindu nationalists gained considerable political power both at the federal level and in many Indian states through the Bharatiya Janata Party (BJP), leading to widespread agitation against the country's Muslim minority in the mid-1990s. In several high-profile instances, Hindu mobs demolished Muslim mosques that were allegedly built on the sites of ancient Hindu temples. The destruction of a mosque in Ayodhya in the Ganges Valley galvanized the nationalistic BJP membership in 1992 (Figure 12.18). In 2002, more than 2,000 people—mostly Muslims—were killed during religious riots in India's state of Gujarat. Two years later, however, India's Hindu nationalist movement suffered a huge electoral defeat, perhaps indicating that the wave of religious strife had peaked.

In predominantly Muslim Pakistan, rising Islamic fundamentalism has generated severe conflict. Radical fundamentalist leaders want to make Pakistan a fully religious state under Islamic law, a plan rejected by the country's secular intellectuals and international businesspeople. The government has attempted to intercede between the two groups, but is often viewed as biased toward the Islamists. Anti-blasphemy laws, for example, have been used to persecute

Figure 12.18 Destruction of the Ayodhya Mosque A group of Hindu nationalists are seen listening to speeches urging them to demolish the mosque at Ayodhya, allegedly built on the site of a former Hindu temple. In 2010, an Indian court decided that two-thirds of the contested site should be given to two Hindu groups, and one-third should be awarded to a Muslim foundation.

members of the country's small Hindu and Christian communities, as well as liberal Muslim writers. Large amounts of money from Saudi Arabia, moreover, have gone into the development of fundamentalist religious schools that are reputed to encourage extremism.

Religious and political extremists in South Asia argue that the current struggles reflect deeply rooted historical divisions. Most scholars disagree, regarding religious conflict as a recent development. To weigh these contrasting claims, it is necessary to examine the cultural history of South Asian civilization.

Origins of South Asian Civilizations

Many scholars think that the roots of South Asian culture extend back to the Indus Valley civilization, which flourished more than 4,000 years ago in what is now Pakistan. This remarkable urban-oriented society vanished almost entirely in the second millennium BCE, after which the record grows dim. By 800 BCE, however, a new urban focus had emerged in the middle Ganges Valley (Figure 12.19). The social, religious, and intellectual customs associated with this civilization eventually spread throughout the lowlands of South Asia.

Hindu Civilization The religious complex of this early South Asian civilization was an early form of Hinduism, a complicated faith that incorporates diverse forms of worship and that lacks any standard system of beliefs (Figure 12.20). Certain deities are recognized, however, by all believers, as is the notion that these various gods are all manifestations of a single divine entity. All Hindus, moreover, share a common set of epic stories, usually written in the sacred language of Sanskrit. Hinduism is noted for its mystical tendencies, which have long inspired many men (and a few women) to seek an ascetic lifestyle, renouncing property and sometimes all regular human relations. One of its hallmarks is a belief in the transmigration of souls from being to being through reincarnation, wherein the nature of one's acts in the physical world influences the courses of these future lives.

Scholars once confidently argued that Hinduism originated from the fusion of two distinct religious traditions: the mystical beliefs of the subcontinent's indigenous inhabitants (including the people of the ill-fated Indus Valley civilization) and the sky-god religion of the Indo-European invaders who swept into the region sometime in the second millen-

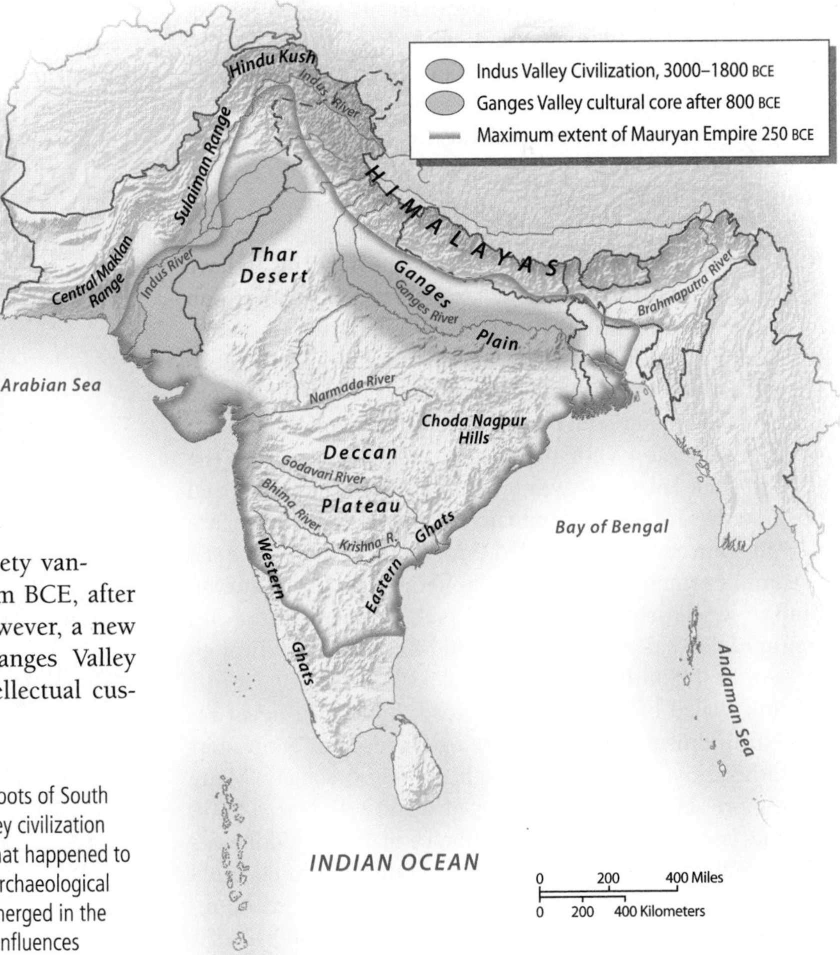

Figure 12.19 Early South Asian Civilizations The roots of South Asian culture may extend back 5,000 years to an Indus Valley civilization based on irrigated agriculture and vibrant urban centers. What happened to that civilization remains a topic of conjecture, because the archaeological record grows dim by 1800 BCE. Later, a new urban focus emerged in the Ganges Valley, from which social, religious, and intellectual influences spread throughout lowland South Asia.

Figure 12.20 Hindu Temple Although India has long been noted for its ancient temples, lavish new Hindu religious complexes continue to be constructed. Visible in this photo is the Akshardham Temple, inaugurated in New Delhi in 2005. Expected to be a major tourist attraction, Akshardham also serves as an educational center focused on the culture of India.

nium BCE from Central Asia. Such a scenario also proved convenient for explaining India's **caste system**, the strict division of society into different hierarchically ranked hereditary groups. The elite invaders, according to this theory, wished to remain separate from the people they had defeated, resulting in an elaborate system of social division. Recent research, however, indicates that the caste system, like Hinduism itself, emerged through more gradual processes of social and cultural evolution.

Buddhism While a caste system of some sort seems to have existed in the early Ganges Valley civilization, it was soon challenged from within by Buddhism. Siddhartha Gautama, the Buddha, was born in 563 BCE in an elite caste. He rejected the life of wealth and power that was laid out before him, however, and sought instead to attain enlightenment, or mystical union with the cosmos. He preached that the path to such "nirvana" was open to all, regardless of social position. His followers eventually established Buddhism as a new religion. Buddhism spread through most of South Asia, becoming something of an official faith under the Mauryan Empire, which ruled much of the subcontinent in the 3rd century BCE. Later centuries saw Buddhism expand through most of East, Southeast, and Central Asia.

But for all of its successes abroad, Buddhism never replaced Hinduism in India. It remained focused on monasteries rather than spreading through the wider society. Many Hindu priests, moreover, struggled against the new faith. One of their techniques was to embrace many of Buddhism's philosophical ideas and enfold them within the intellectual system of Hinduism. By 500 CE, Buddhism was on the retreat throughout South Asia, and within another 500 years it had virtually disappeared from the region. The only major exceptions were the island of Sri Lanka and the high Himalayas, both of which remain mostly Buddhist to this day.

Arrival of Islam The next major challenge to Hindu society—Islam—came from the outside. Arab armies conquered the lower Indus Valley around 700 CE but advanced no farther. Then around the year 1000, Turkish-speaking Muslims began to invade from Central Asia. At first they merely raided, but eventually they began to settle and rule on a permanent basis. By the 1300s, most of South Asia lay under Muslim power, although Hindu kingdoms persisted in southern India and in the arid lands of Rajasthan. Later, during the 16th and 17th centuries, the **Mughal Empire** (also spelled *Mogul*) dominated much of the region from its power center in the upper Indus–Ganges Basin (Figure 12.21).

At first Muslims formed a small ruling elite, but over time increasing numbers of Hindus converted to the new religion, particularly those from lower castes who sought freedom from the rigid social order. Conversions were most pronounced in the northwest and northeast, and eventually the areas now known as Pakistan and Bangladesh became predominantly Muslim.

At first glance, Islam and Hinduism are strikingly divergent faiths. Islam is resolutely monotheistic, austere in its ceremonies, and spiritually egalitarian (all believers stand in the same relationship to God). Hinduism, by contrast, is polytheistic (at least on the surface), lavish in its rituals, and caste structured. Because of these profound differences, Hindu and Muslim communities in South Asia are sometimes viewed as utterly distinct, living in the same region, but not sharing the same culture or civilization. Increasingly, such a view is expressed in South Asia itself. Many residents of modern Pakistan stress the Islamic nature of their country and its complete separation from India.

Figure 12.21 The Red Fort The Red Fort of Delhi, completed in 1648, was the military center of the Moghul Empire. Today this massive fortification, one of the largest in the world, is a major tourist destination.

Yet by overemphasizing the separation of Hindu and Muslim communities, one risks missing much of what is historically distinctive about South Asia. Until the 20th century, Hindus and Muslims usually coexisted on friendly terms; the two faiths stood side by side for hundreds of years, during which time they came to influence each other in many ways. Moreover, aspects of caste organization have persisted among South Asian Muslims, just as they have among India's Christians.

The Caste System

Although caste is one of the historically unifying features of South Asia, the system is not uniformly distributed across the subcontinent. It has never been significant in India's tribal areas; in modern Pakistan and Bangladesh its role is fading; and in the Buddhist society of Sri Lanka its influence has long been somewhat marginal. Even in India, caste is now deemphasized, especially among the more urban and educated segments of society. But all said, caste continues to structure day-to-day social existence for hundreds of millions of Indians. It is estimated that roughly 1,000 young Indians are murdered every year by relatives who will not tolerate their marriages across caste lines.

Caste is actually a rather clumsy term for denoting the complex social order of the Hindu world. The word itself, of Portuguese origin, combines two distinct local concepts: *varna* and *jati*. *Varna* refers to the ancient fourfold social hierarchy of the Hindu world, whereas *jati* refers to the hundreds of local endogamous ("marrying within") groups that exist at each varna level (different jati groups are thus usually called *subcastes*). Jati, like varna, are hierarchically arranged, although the exact order of precedence is not so clear-cut.

It often has been argued that the essence of the caste system is the notion of social pollution. The lower one's position in the hierarchy, the more potentially polluting one's body supposedly is. Members of higher castes were traditionally not supposed to eat or drink with, or even use the same utensils as, members of lower castes.

The Main Caste Groups Three varna groups constitute the traditional elite of Hindu society. At the apex sit the Brahmins, members of the traditional priestly caste. Brahmins perform the high rituals of Hinduism, and they form the traditional intellectual elite of India. Most Brahmins value education highly, and today they are disproportionately represented among India's professional classes. Below the Brahmins in the traditional hierarchy are the Kshatriyas, members of the warrior or princely caste. In premodern India, this group actually had more power and wealth than did the Brahmins; it was they who ruled the old Hindu kingdoms. Next stand the Vaishyas, members of the traditional merchant caste. In earlier centuries, a near monopolization of long-distance trade and money lending in northern India gave many Vaishyas ample opportunities to accumulate wealth. The precepts of vegetarianism and nonviolence are particularly strong among certain merchant subcastes of western India. One prominent representative of this tradition was Mohandas Gandhi, the founder of modern India and one of the 20th century's greatest leaders.

The majority of India's population fits into the fourth varna category, that of the Sudras. The Sudra caste is composed of an especially large array of subcastes (jati), most of which originally reflected occupational groupings. Most Sudra subcastes were traditionally associated with peasant farming, but others were based on craft occupations, including those of barbers, smiths, and potters.

While the Brahmins, Kshatriyas, Vaishyas, and Sudras form the basic fourfold scheme of caste society, another sizable group stands outside the varna system altogether. These are the so-called *untouchables*, or **dalits**, as they are now preferably called. Dalits were not traditionally allowed to enter Hindu temples. Such low-status positions were derived historically from "unclean" occupations, such as those of leather workers (who dispose of dead animals), scavengers, latrine cleaners, and swine herders. In the southern Indian state of Kerala until the 19th century, some groups were actually considered "unseeable"; these unfortunate persons had to hide in bushes and avoid walking on the main roads to protect members of higher castes from "visual pollution." Not surprisingly, many Indian dalits have converted to Islam, Christianity, and Buddhism in an attempt to escape from the caste system. Even so, they continue to suffer discrimination.

The Changing Caste System The caste system is clearly in a state of flux in India today. Its original occupational structure has long been undermined by the necessities of a modern economy, and various social reforms have chipped away at the discrimination that it embodies. The dalit community itself has produced several notable national leaders who have waged partially successful political struggles. Owing to such efforts, the very concept of "untouchability" is now technically illegal in India. The Indian central government also insists that 27 percent of university seats be reserved for students from low caste backgrounds, while a number of Indian states have set higher quotas, both for education and governmental employment. Such "reservations," as they are called, are controversial, as many people think that they unfairly penalize people of higher caste background. In May 2006, some 100,000 people took to the streets of New Delhi to protest a proposal by India's government to increase the number of reserved positions in several of the country's most prestigious universities.

Contemporary Geographies of Religion

South Asia, as we have seen, has a predominantly Hindu heritage overlain by a substantial Muslim imprint. Such a picture fails, however, to capture the enormous diversity of modern religious expression in contemporary South Asia. The following discussion looks specifically at the geographical patterns of the region's main faiths (Figure 12.22).

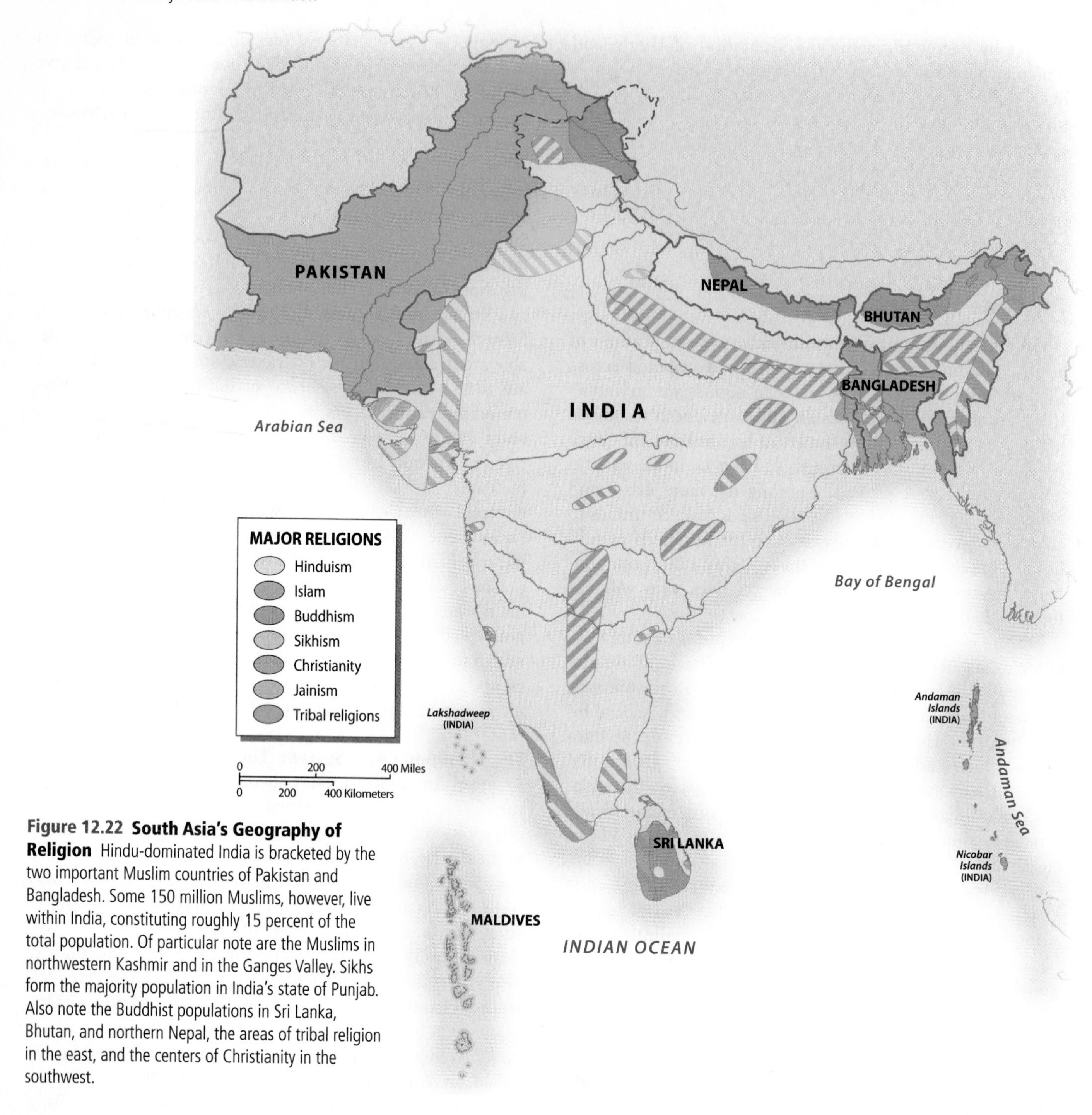

Figure 12.22 South Asia's Geography of Religion Hindu-dominated India is bracketed by the two important Muslim countries of Pakistan and Bangladesh. Some 150 million Muslims, however, live within India, constituting roughly 15 percent of the total population. Of particular note are the Muslims in northwestern Kashmir and in the Ganges Valley. Sikhs form the majority population in India's state of Punjab. Also note the Buddhist populations in Sri Lanka, Bhutan, and northern Nepal, the areas of tribal religion in the east, and the centers of Christianity in the southwest.

Hinduism Less than 1 percent of the people of Pakistan are Hindu, and in Bangladesh and Sri Lanka Hinduism is a distinctly minority religion. Almost everywhere in India, however—and in Nepal, as well—Hinduism is very much the faith of the majority. In east-central India, more than 95 percent of the population is Hindu. Hinduism is itself a geographically complicated religion, with aspects of faith varying in different parts of India. Even within a given Indian state, forms of worship often differ from region to region and from caste to caste.

Islam Islam may be considered a "minority" religion for the region as a whole, but such a designation obscures the tremendous importance of this religion in South Asia. With more than 400 million members, the South Asian Muslim community is the largest in the world. Bangladesh and especially Pakistan are overwhelmingly Muslim. India's Islamic community, although constituting only some 13 to 15 percent of the country's population, is still roughly 160 million strong, making it the world's third largest Muslim group.

Although Muslims live in almost every part of India, they are concentrated in four main areas: in most of India's cities; in Kashmir, particularly in the densely populated Vale of Kashmir; in the upper and central Ganges plain; and in the southwestern state of Kerala. In northern India especially, Muslims tend to be poorer and less educated than their Hindu neighbors, leading some observers to call them the country's "new dalits."

Interestingly, the state of Kerala is about 25 percent Muslim even though it was one of the few parts of India that never experienced prolonged Muslim rule. Islam in Kerala is historically connected not to Central Asia, but rather to trade across the Arabian Sea. Kerala's Malabar Coast long supplied spices and other luxury products to Southwest Asia, enticing many Arabian traders to settle in this part of India. Gradually many of Kerala's native inhabitants converted to the new religion as well. Owing to a similar process, Sri Lanka is approximately 9 percent Muslim, and the Maldives is almost entirely Muslim.

As an overwhelmingly Muslim country, Pakistan officially calls itself an Islamic Republic. As such, Islamic law is technically supposed to override the country's secular laws, although in actuality the situation remains ambiguous. In Pakistan's two most conservative provinces, Balochistan and Khyber Pakhtunkhwa (previously known as the North-West Frontier Province), Islamic law tends to be much more strictly enforced.

Sikhism The tension between Hinduism and Islam in medieval northern South Asia helped give rise to a new religion called **Sikhism**. Sikhism originated in the late 1400s in Punjab, near the modern boundary between India and Pakistan. The Punjab was the site of religious fervor at the time; Islam was gaining converts and Hinduism was increasingly on the defensive. The new faith combined elements of both religions and thus appealed to many who felt trapped between their competing claims. Many orthodox Muslims, however, viewed Sikhism as a heresy precisely because it incorporated elements of their own religion. Periodic bouts of persecution led the Sikhs to adopt a militantly defensive stance, and in the political chaos of the early 1800s they were able to carve out a powerful kingdom for themselves. Even today, Sikh men are noted for their work as soldiers and bodyguards.

At present, the Indian state of Punjab is approximately 60 percent Sikh. Small but often influential groups of Sikhs are scattered across the rest of India. Devout Sikh men are immediately visible, because they do not cut their hair or their beards. Instead, they wear their hair wrapped in a turban and often tie their beards close to their faces.

Buddhism and Jainism Although Buddhism virtually disappeared from India in medieval times, it persisted in Sri Lanka. Among the island's dominant Sinhalese people, Theravada Buddhism developed into a virtual national religion, fostering close connections between Sri Lanka and mainland Southeast Asia. In the high valleys of the Himalayas, Buddhism also survived as the majority religion. Tibetan Buddhism, with its esoteric beliefs and huge monasteries, has been better preserved in Bhutan and in the Ladakh region of northeastern Kashmir than in Chinese-controlled Tibet itself. The small city of Dharmsala in the northern Indian state of Himachal Pradesh is the seat of Tibet's government-in-exile and of its spiritual leader, the Dalai Lama, who fled Tibet in 1959 (Figure 12.23). Another group of Buddhists is found in central India, where 20th-century efforts to convert Hindu dalits to the faith were particularly successful.

Figure 12.23 Dharmasala, India The small highland city of Dharmasala in northern India is noted as the base of the Tibetan government-in-exile. It also boasts a growing tourist industry.

At roughly the same time as the birth of Buddhism (circa 500 BCE), another religion emerged in northern India as a protest against orthodox Hinduism: **Jainism**. This religion took nonviolence to its ultimate extreme. Jains are forbidden to kill any living creatures, and as a result the most devout adherents wear gauze masks to prevent the inhalation of small insects. Agriculture is forbidden to Jains because plowing can kill small creatures. As a result, most members of the faith have looked to trade for their livelihoods. Many have prospered, aided no doubt by the frugal lifestyles required by their religion. Today Jains are concentrated in northwestern India, particularly Gujarat.

Other Religious Groups Even more prosperous than the Jains are the Parsis, or Zoroastrians, concentrated in the Mumbai area. The Parsis arrived as refugees, fleeing from Iran after the arrival of Islam in the 7th century. Zoroastrianism is an ancient religion that focuses on the cosmic struggle between good and evil. Although numbering only a few hundred thousand, the Parsi community has had a major impact on the Indian economy. Several of the country's largest industrial firms were founded by Parsi families, including the giant

EXPLORING GLOBAL CONNECTIONS The Tata Group's Global Network

Most large multinational corporations, which produce goods and services in many different countries, are based in the core areas of the world economy. Such companies generally rely on the financial and technological resources of wealthy cities yet often seek out production sites in poor countries with low wages and few regulations. As a result, multinational corporations have been criticized for their role in maintaining a stark global division between the developed and the underdeveloped worlds. As globalization proceeds, however, such a view is becoming outmoded. Increasing numbers of powerful multinational corporations are based in relatively poor areas, and several are now investing heavily in wealthy countries.

No company illustrates these trends as well as India's Tata Group. Founded in 1868 by Jamshedji Tata, the son of a Parsi Zoroastrian priest, the firm began as an opium-trading firm in Bombay (Mumbai) India. The Tata Group now includes 114 companies operating in 85 companies. Despite its global reach, the company is still headquartered in Mumbai; its current chairman, Ratan Tata, is the fifth generation of the Tata family to exercise control. According to a 2009 survey, international business leaders ranked the Tata Group as the word's 11th most reputable company.

The Tata Group is a classical conglomerate, operating businesses in the steel, motor, consultancy, electrical, chemical, medical, retail, hotel, and information technology industries. Over the past 15 years, it has aggressively expanded its global reach by purchasing foreign companies. In 2008 alone, it acquired firms based in Britain, Spain, Italy, China, South Africa, Norway, and Morocco. Such well-known brands as Tetley Tea and the Taj Hotels now belong to the Tata Group.

One of the most global members of the group is Tata Motors. Although only the world's 11th largest automaker, Tata is the second largest producer of commercial vehicles. It has reached that position in large part through joint ventures and foreign acquisitions. In 2005, for example, it purchased Daewoo Commercial Vehicle, South Korea's second largest bus manufacturer. Tata Motors' most daring international move was probably its 2008 acquisition of British Jaguar Land Rover (JLR) from Ford Motor Corporation for $2 billion. With this purchase, Tata Motors gained control of three manufacturing plants and two design centers in the United Kingdom, as well as two of the best-known luxury car brands.

While targeting high-end automobile consumers with Jaguar, Tata Motors has also taken a greater interest in the opposite end of the market. Its signal product here is the Tata Nano, a "supermini" car that get 73 miles to the gallon and sells for as little as $2,160 (Figure 12.4.1). The two-seater Nano is marketed in a number of African and Asian countries. In 2009, Tata unveiled an upgraded version for the European market that sells for around $6,000, and in 2010 it announced the development of an electric version, the Nano EV, which will be made jointly with a Norwegian company.

Figure 12.4.1 Tata Nano The Tata Nano is the world's least expensive car, selling in India for a little over $2,000.

Tata Group (see "Exploring Global Connections: The Tata Group's Global Network"). Intermarriage and low fertility, however, threaten the survival of the community.

Indian Christians are more numerous than either Parsis or Jains. Their religion arrived some 1,800 years ago; early contact between the Malabar Coast and Southwest Asia brought Christian as well as Muslim traders. A Jewish population also established itself but later declined; today it numbers only a few hundred. Kerala's Christians, by contrast, are counted in the millions, constituting some 20 percent of the state's population. Several Christian sects are represented, but the largest are historically affiliated with the Syrian Church of Southwest Asia. Another stronghold of Christianity is the small Indian state of Goa, a former Portuguese colony, where Roman Catholics make up a little more than a quarter of the population.

During the colonial period, British missionaries went to great efforts to convert South Asians to Christianity. They had very little success, however, in Hindu, Muslim, and Buddhist communities. The remote tribal districts of British India, on the other hand, proved to be more receptive to missionary activity. In the uplands of India's extreme northeast, entire communities abandoned their traditional animist faith in favor of Protestant Christianity. Christian missionaries are still active in many parts of India, but during the 1990s they began to experience severe pressure in many areas from Hindu nationalists. Pakistan's Christian community, some 2.8 million strong, has also been under pressure, in this case from Muslim radicals. In 2009, Pakistani Muslim extremists attacked the Christian community of the town of Gogra, burning 40 houses and a church and killing eight people.

Geographies of Language

South Asia's linguistic diversity matches its religious diversity. In fact, one of the world's most important linguistic boundaries runs directly across India (Figure 12.24). North of the line, languages belong to the Indo-European group, the world's largest linguistic family. The languages of southern India belong to the **Dravidian** family, a linguistic group unique to South Asia. Along the mountainous northern rim of the region, a third linguistic family, Tibeto-Burman, prevails, but this area is marginal to the South Asian cultural sphere. Scattered tribal groups in eastern India speak Austro-Asiatic languages related to those of mainland Southeast Asia. South Asia can thus be divided into two major linguistic zones, the Indo-European north and the Dravidian south. But within these broad divisions there are many different languages, each associated with a distinct culture. In most of South Asia, several languages are spoken within the same region or even city, and multilingualism is common everywhere.

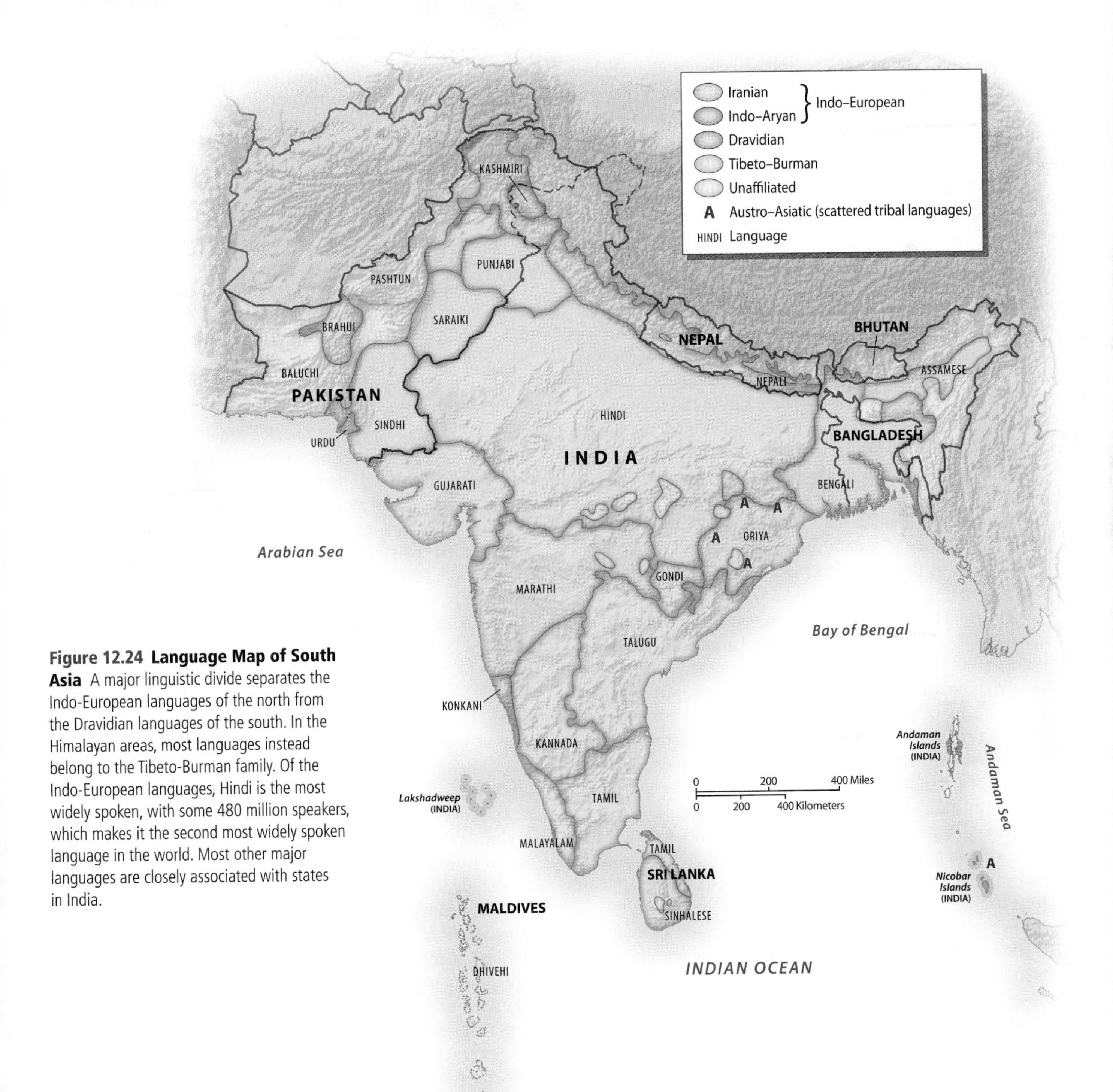

Figure 12.24 Language Map of South Asia A major linguistic divide separates the Indo-European languages of the north from the Dravidian languages of the south. In the Himalayan areas, most languages instead belong to the Tibeto-Burman family. Of the Indo-European languages, Hindi is the most widely spoken, with some 480 million speakers, which makes it the second most widely spoken language in the world. Most other major languages are closely associated with states in India.

How or when Indo-European languages came to South Asia is uncertain, but scholars have traditionally argued that they arrived when pastoral peoples from Central Asia invaded the **subcontinent** in the second millennium BCE, largely supplanting the indigenous Dravidian peoples. According to this hypothesis, offshoots of the same original cattle-herding people also swept across both Iran and Europe, bringing their language to all three places. Supposedly, the ancestral Indo-European tongue introduced to India was similar to Sanskrit. This rather simplistic scenario, however, is now regarded with some suspicion, and many scholars argue for a more gradual infiltration of Indo-European speakers from the northwest.

Any modern Indo-European language of India, such as Hindi or Bengali, is more closely related to English than it is to any **Dravidian language** of southern India, such as Tamil. But South Asian languages on both sides of this linguistic divide do share a number of superficial features. Dravidian languages, for example, have borrowed many words from Sanskrit, particularly those associated with religion and scholarship.

The Indo-European North South Asia's Indo-European languages are themselves divided into two subfamilies: Iranian and Indo-Aryan. Iranian languages, such as Baluchi and Pashto, are found in western Pakistan, near the border with Iran and Afghanistan. Languages of the strictly South Asian Indo-Aryan groups are closely related to each other, yet are still quite distinctive, often written in different scripts. Each of the major languages of India is associated with one or more Indian states. One thus finds Gujarati in Gujarat, Marathi in Maharashtra, Oriya in Orissa, and so on. Two of these languages, Punjabi and Bengali, span India's international boundaries to extend into Pakistan and Bangladesh, respectively, as these borders were established on religious rather than linguistic lines.

The most widely spoken language of South Asia is Hindi—not to be confused with the Hindu religion. With almost 500 million speakers, Hindi is the second most widely spoken language in the world. It occupies a prominent role in contemporary India, both because so many people speak it and because it is the main language of the Ganges Valley, the historical and demographic core of India. Hindi is the dominant tongue of a number of Indian states, including Uttar Pradesh, Madhya Pradesh, and Haryana. In addition, the main forms of speech found in Rajasthan and Bihar are often considered to be dialects of Hindi. Most Indian students learn some Hindi, often as their second or third language.

Bengali is the second most widely spoken language in South Asia. It is the national language of Bangladesh and the main language of the Indian state of West Bengal. Spoken by more than 200 million persons, Bengali is the world's eighth or ninth most widely spoken language. Its significance extends beyond its official status in Bangladesh and its total numerical strength. Equally important is its extensive literature. West Bengal, particularly its capital of Kolkata (Calcutta), has long been one of South Asia's main literary and intellectual centers. Kolkata may be noted for its appalling poverty, but it also has one of the highest levels of cultural production in the world, as measured by the output of drama, poetry, novels, and film (Figure 12.25).

Figure 12.25 Kolkata (Calcutta) Bookstore Although Kolkata (Calcutta) is noted in the West mostly for its abject poverty, the city is also known in India for its vibrant cultural and intellectual life, illustrated by its large number of bookstores, theaters, and publishing firms.

The Punjabi-speaking zone in the west was similarly split at the time of independence, in this case between Pakistan and the Indian state of Punjab. While an estimated 90 million persons speak Punjabi, it does not have the significance of Bengali. Although Punjabi is the main vehicle of Sikh religious writings, it lacks an extensive literary tradition. In recent years, moreover, the people of the southern half of Pakistan's province of Punjab have successfully insisted that their dialect forms a separate language, called Saraiki. And even though Punjabi is the most widely spoken language in Pakistan, did not become the country's national language. Instead, that position was given to Urdu.

Urdu, like Hindi, originated on the plains of northern India. The difference between the two was largely one of religion: Hindi was the language of the Hindu majority, Urdu that of the Muslim minority, including the former ruling class. Owing to this distinction, Hindi and Urdu are written differently—the former in the Devanagari script (derived from Sanskrit) and the latter in a modified version of the Arabic script. Although Urdu contains many words borrowed from Persian, its basic grammar and vocabulary are almost identical to those of Hindi.

With independence in 1947, millions of Urdu-speaking Muslims from the Ganges Valley fled to the new state of Pakistan. Since Urdu had a higher status than Pakistan's indigenous tongues, it was quickly established as the new country's official language. Karachi, Pakistan's largest city, is now largely Urdu-speaking, but elsewhere other languages, such as Punjabi and Sindhi, remain primary. Most Pakistanis, however, do speak Urdu as their common language.

Languages of the South Four thousand years ago, Dravidian languages were probably spoken across most of South Asia. As Figure 12.24 indicates, a Dravidian tongue called *Brahui* is still found in the uplands of western Pakistan. The four main Dravidian languages, however, are confined to the south. As in the north, each language is closely associated with an Indian state: Kannada in Karnataka, Malayalam in Kerala, Telugu in Andhra Pradesh, and Tamil in Tamil Nadu. Tamil is usually considered the most important member of the family because it has the longest history and the largest literature. Tamil poetry dates back to the first century CE, making it one of the world's oldest written languages.

Although Tamil is spoken in northern Sri Lanka, the country's majority population, the Sinhalese, speak an Indo-European language. Apparently the Sinhalese migrated from northern South Asia several thousand years ago. Although this movement is lost to history, the migrants evidently settled on the island's fertile and moist southwestern coastal and central highland areas, which formed the core of a succession of Sinhalese kingdoms. These same people also migrated to the Maldives, where the national language, Divehi, is essentially a Sinhalese dialect. The drier north and east of Sri Lanka, on the other hand, were settled hundreds of years ago by Tamils. In the 1800s, British landowners imported Tamil peasants from the mainland to work on their tea plantations in the central highlands, giving rise to a second population of Tamil-speakers in Sri Lanka.

Linguistic Dilemmas The multilingual countries Sri Lanka, Pakistan, and India are all troubled by linguistic conflicts. Such problems are most complex in India, simply because India is so large and has so many different languages. India's linguistic environment is changing in complicated ways, pushed along by modern economic and political forces.

Indian nationalists have long dreamed of a national language that could help forge the different communities of the country into a more unified nation. But this **linguistic nationalism**, or the linking of a specific language with nationalistic goals, meets the stiff resistance of provincial loyalty, which itself is intertwined with local languages. The obvious choice for a national language would be Hindi, and Hindi was indeed declared as such in 1947. Raising Hindi to this position, however, alienated speakers of important northern languages such as Bengali and Marathi, and even more so the speakers of the Dravidian tongues. As a result of this cultural tension, in the 1950 Indian constitution Hindi was demoted to sharing the position of "official language" of India with 14 other languages. Over time, additional languages were added to this list, such that India now has 23 separate languages with an official status.

Regardless of opposition, the role of Hindi is expanding, especially in the Indo-European-speaking north. Here local languages are closely related to Hindi, which can therefore be learned without too much difficulty. Hindi is spreading through education, but even more significantly through popular media, especially television and motion pictures. Films and television programs are made in several Indian languages, but Hindi remains the primary vehicle. The most popular movies coming out of Mumbai's "Bollywood" film industry actually tend to be delivered in a neutral dialect that is as close to Urdu as to Hindi—and with a good deal of English thrown in as well.

The Role of English Even if Hindi is spreading, it still cannot be considered anything like a common Indian language, even in the north. In the Dravidian south, more importantly, its role remains secondary. National-level political, journalistic, and academic communication thus cannot be conducted in Hindi, or in any other indigenous language. Only English, an "associate official language" of contemporary India, serves this function.

Before independence, many educated South Asians learned English for its political and economic benefits under British colonialism. It therefore emerged as the de facto common tongue of the upper and middle classes. Today a few extreme nationalists want to deemphasize English, but most Indians, and particularly those of the south, advocate it as a neutral national language because all parts of the country have an equal stake in it. Furthermore, English confers substantial international benefits.

English is thus the main integrating language of India, and it remains widely used elsewhere in South Asia. Roughly a third of the people of India are able to carry out a conversation in English. English-medium schools abound in all parts of the region, and many children of the elite learn this global language well before they begin their schooling.

The English spoken in South Asia has forms and vocabulary elements that sometimes make cross-cultural communication difficult. As a result, IT companies in India, particularly those that run international call centers, sometimes ask their employees to watch reruns of popular television shows from the United States in order to gain fluency in American pronunciation and slang.

South Asians in a Global Cultural Context

The widespread use of English in South Asia has not only facilitated the spread of global culture into the region, but it has also helped South Asian cultural production to reach a global audience. The global spread of South Asian literature, however, is nothing new. As early as the turn of the 20th century, Rabindranath Tagore gained international acclaim for his poetry and fiction, earning the Nobel Prize for Literature in 1913. In the 1980s and 1990s, such Indian novelists as Salman Rushdie and Vikram Seth became major literary figures in Europe and North America.

The spread of South Asian culture abroad has been accompanied by the spread of South Asians themselves. Migration from South Asia during the time of the British Empire led to the establishment of large communities in such far-flung places as eastern Africa, Fiji, and the southern Caribbean (Figure 12.26). Subsequent migration targeted the developed world; there are now several million people of

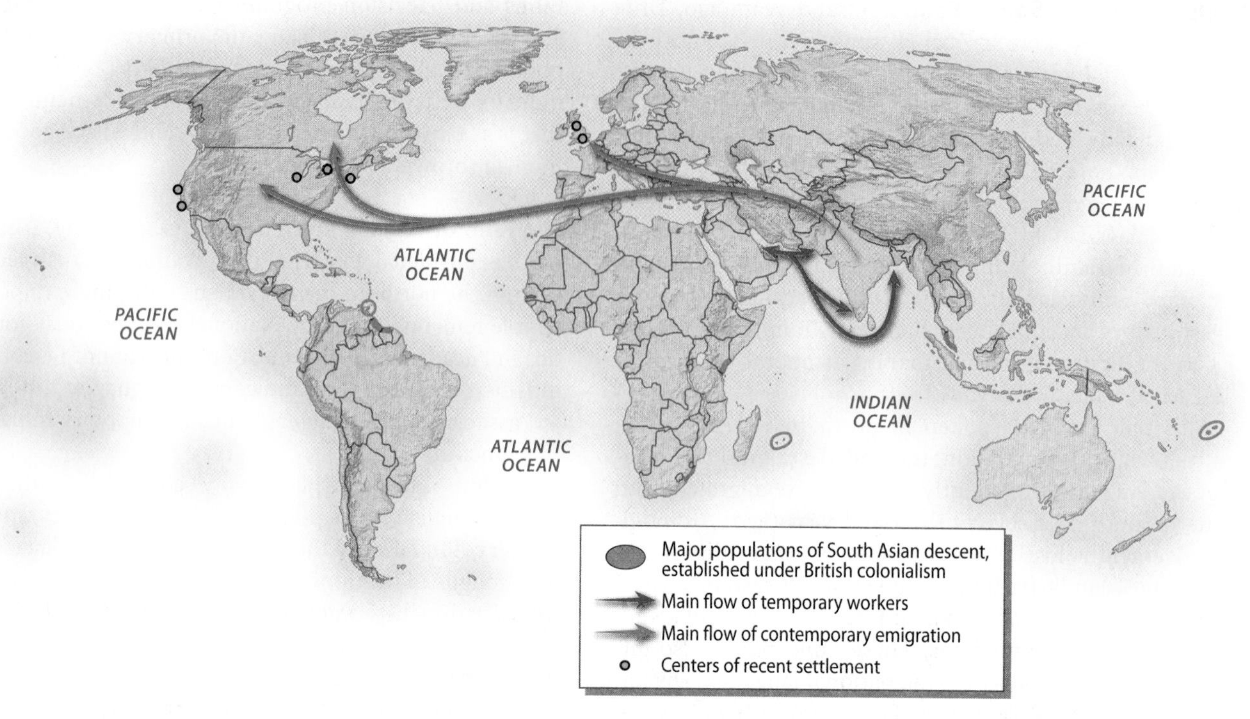

Figure 12.26 The South Asian Global Diaspora During the British imperial period, large numbers of South Asian workers settled in other colonies. Today, roughly 50 percent of the population of such places as Fiji and Mauritius are of South Asian descent. More recently, large numbers have settled, and are still settling, in Europe (particularly Britain) and North America. Large numbers of temporary workers, both laborers and professionals, are employed in the wealthy oil-producing countries of the Persian Gulf.

South Asian descent living in Britain (mostly Pakistani), and a similar number in North America (mostly Indian). Many contemporary migrants to the United States are doctors, software engineers, and other professionals, making Indian Americans one of the country's wealthiest ethnic groups.

In South Asia itself, the globalization of culture has brought tensions as severe as those felt anywhere in the world. Traditional Hindu and Muslim religious customs frown on any overt display of sexuality—a staple feature of global popular culture. While romance is a recurrent theme in the often melodramatic Bollywood films of Mumbai, even kissing is considered risqué. In April 2007, actor Richard Gere sparked massive public protests when he kissed the Indian actress Shilpa Shetty at an AIDS awareness rally in New Delhi (Figure 12.27). Still, the pressures of internationalization are hard to resist. In the tourism-oriented Indian state of Goa, such cultural tensions are on full display. There, German and British sun-worshipers often wear nothing but thong bikini bottoms, whereas Indian women tourists go into the ocean fully clad. Young Indian men, for their part, often simply walk the beach and gawk, good-naturedly, at the outlandish foreigners.

Contemporary popular culture in South Asia thus reveals both global linkages as well as divisions. The same tensions can be seen, and in much stronger form, in the region's geopolitical framework.

Figure 12.27 A Controversial Kiss Actor Richard Gere embraces and then kisses Indian actress Shilpa Shetty during an AIDS awareness program in 2007. Conservative Indian citizens were shocked by this act, which resulted in large-scale protests.

GEOPOLITICAL FRAMEWORK: A Deeply Divided Region

Before the 1800s, South Asia had never been politically united. While a few empires at various times covered most of the subcontinent, none spanned its entire extent. Whatever unity the region had was cultural, not political. The British, however, brought the region into a single political system by the middle of the 19th century. Independence in 1947 witnessed the traumatic separation of Pakistan from India; in 1971, Pakistan itself was divided with the independence of Bangladesh, formerly East Pakistan. Serious internal tensions, moreover, began to rise in several South Asian countries in the late 1900s (Figure 12.28). The most important geopolitical issue is the continuing tension between Pakistan and India, both of which are nuclear powers.

South Asia Before and After Independence in 1947

During the 1500s, when Europeans began to arrive on the coasts of South Asia, most of the northern subcontinent came under the power of the Mughal Empire, a powerful Muslim state ruled by people of Central Asian descent

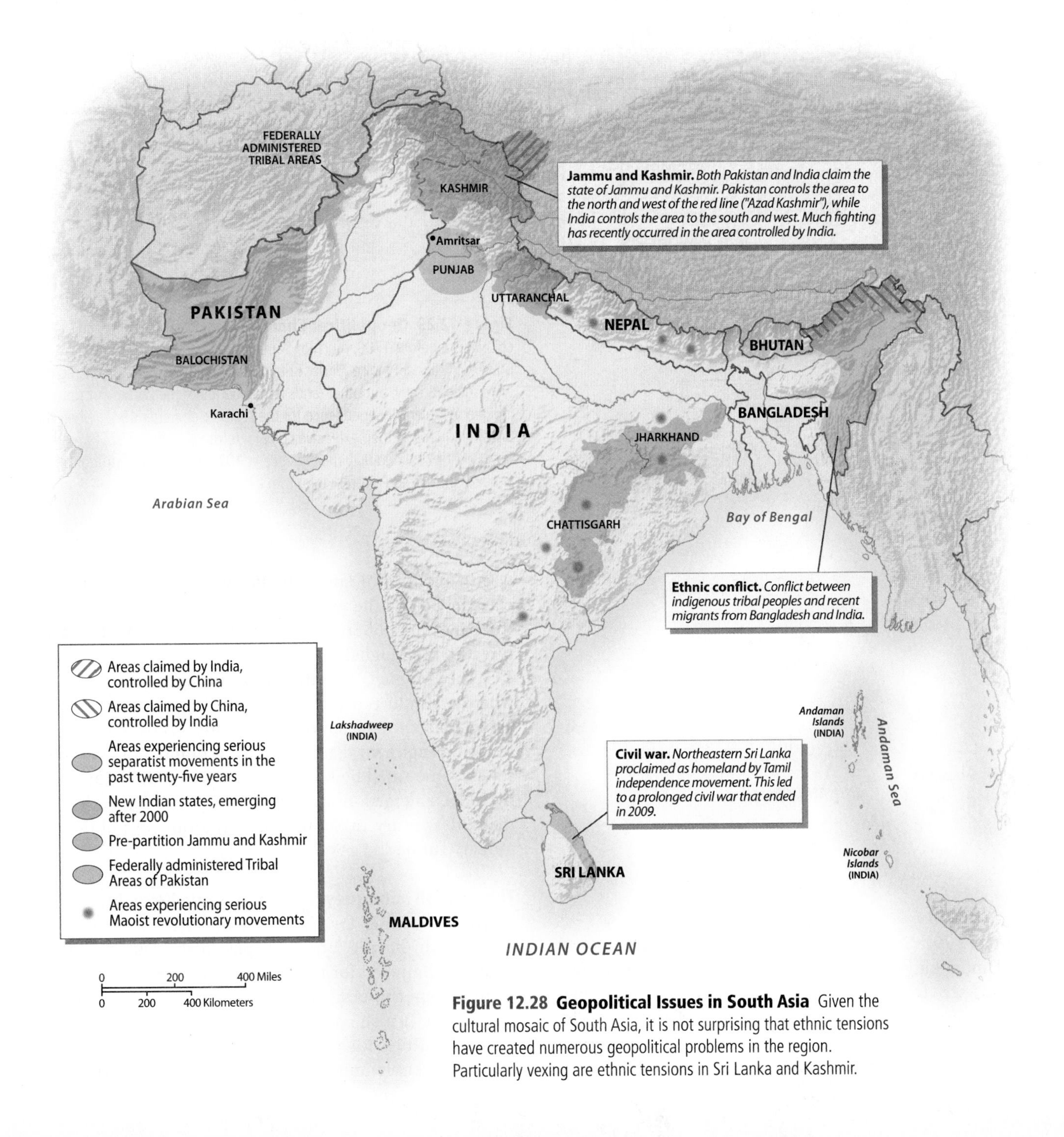

Figure 12.28 Geopolitical Issues in South Asia Given the cultural mosaic of South Asia, it is not surprising that ethnic tensions have created numerous geopolitical problems in the region. Particularly vexing are ethnic tensions in Sri Lanka and Kashmir.

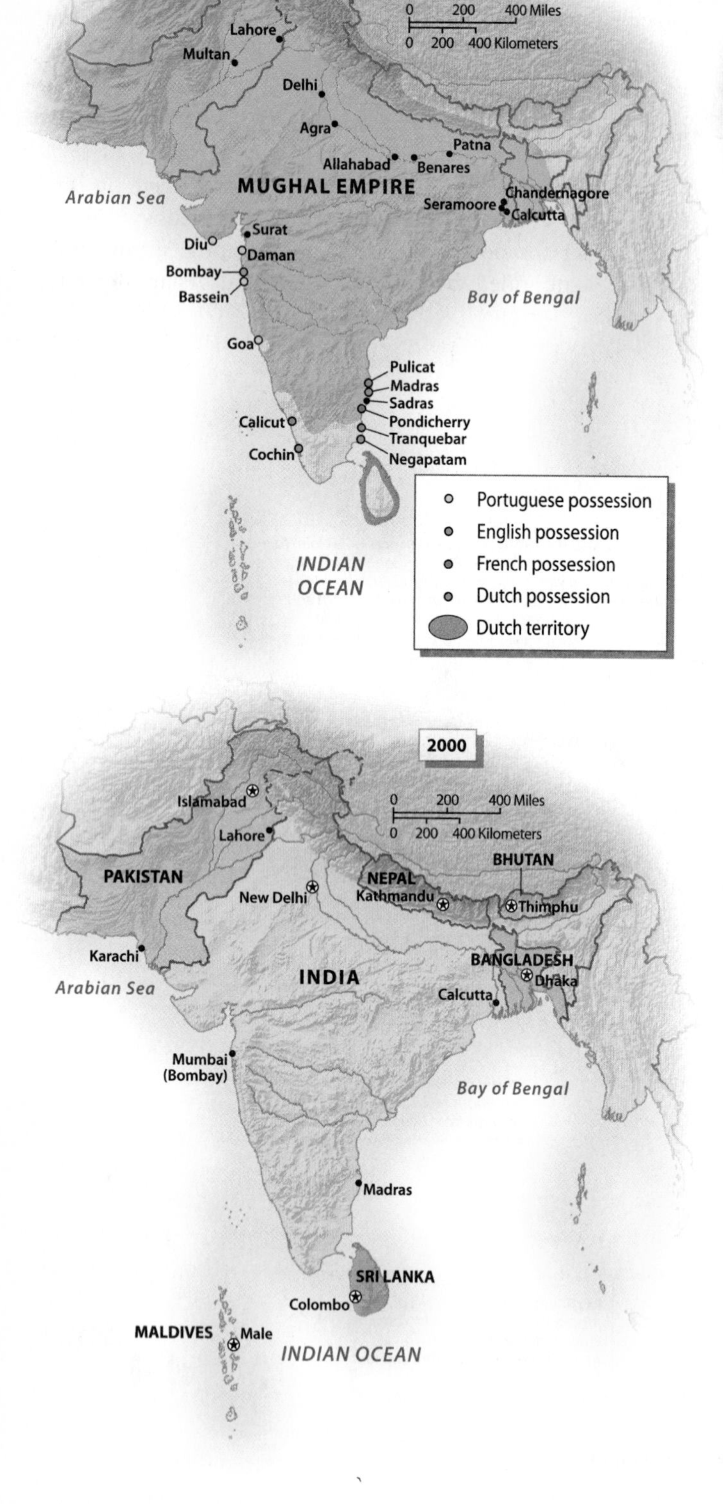

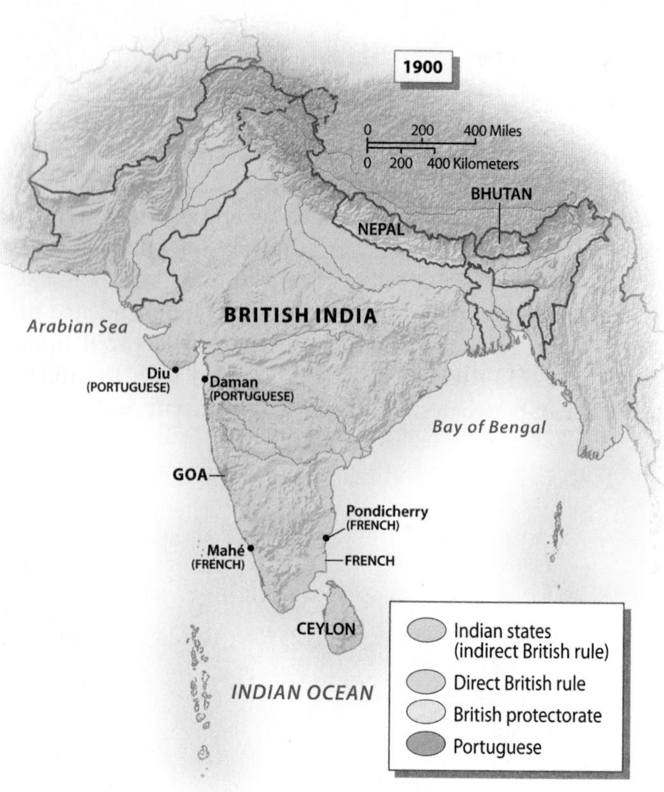

Figure 12.29 Geopolitical Change At the onset of European colonialism before 1700, much of South Asia was dominated by the powerful Mughal Empire. Under Britain, the wealthiest parts of the region were ruled directly, but other lands remained under the partial authority of indigenous rulers. Independence for the region came after 1947, when the British abandoned their extensive colonial territory. Bangladesh, formerly East Pakistan, gained its independence in 1971 after a short struggle against centralized Pakistani rule from the west.

(Figure 12.29). Southern India remained under the control of a Hindu kingdom called Vijayanagara. European merchants, keen to obtain spices, textiles, and other Indian products, established a series of coastal trading posts. The Mughals and other South Asian rulers were little concerned with the growing European naval power, as their own focus was the control of land. The Portuguese carved out an enclave in Goa on the west coast, while the Dutch gained control over much of Sri Lanka in the 1600s, but neither was a threat to the Mughals.

The Mughal Empire grew stronger in the 1600s, whereas Hindu power declined until it was limited to the peninsula's extreme south. In the early 1700s, however, the Mughal Empire weakened rapidly. A number of contending states, some ruled by Muslims, others by Hindus, and a few by Sikhs, emerged in former Mughal territories. As a result, the 18th century in South Asia was a time of political and military turmoil.

The British Conquest These unsettled conditions provided an opening for European imperialism. The British and French, having largely displaced the Dutch and Portuguese, competed for trading posts. Because Indian cotton textiles were the best in the world prior to the Industrial Revolution, British and French merchants needed to obtain huge quantities for their global trading networks. With Britain's overwhelming victory over France in the Seven

Years' War (1756–63), France was reduced to a few marginal coastal possessions in southern India. Britain, or more specifically the **British East India Company**, the private firm that acted as an arm of the British government, now monopolized overseas trade in the area and began to stake out a South Asian empire of its own.

The company's usual method was to make strategic alliances with Indian states to defeat the latter's enemies, most of whose territories it would then grab for itself. As time passed, its army, largely composed of South Asian mercenaries, grew ever more powerful. Several Indian states put up strong resistance, but none could ultimately resist the immense resources of the East India Company. Valuable local allies, as well as a few former enemies, were allowed to remain in power, provided that they no longer threatened British interests. The territories of these indigenous states, however, were gradually whittled back, and British advisors increasingly dictated their policies.

From Company Control to British Colony The continuing reduction in size of the Indian states, coupled with the growing arrogance of British officials, led to a rebellion in 1857 across much of South Asia. When this uprising (called the *Sepoy Mutiny* by the British) was finally crushed, a new political order was implemented. South Asia was now to be ruled by the British government, with the monarch of England serving as its head of state.

Until 1947, the British government maintained direct control over South Asia's most productive and densely populated areas, including virtually the entire Indus-Ganges Valley and most of the coastal plains. The British also ruled Sri Lanka, having supplanted the Dutch in the 1700s. The major areas of indirect rule, where Indian rulers retained their princely states under British advisors, were in Rajasthan, the uplands of central India, southern Kerala, and along the frontiers. The British administered this vast empire through three coastal cities that were largely their own creation: Bombay (Mumbai), Madras (Chennai), and, above all, Calcutta (Kolkata). In 1911, they began building a new capital in New Delhi, near the strategic divide between the Indus and Ganges drainage systems.

While the political geography of British India stabilized after 1857, the empire's frontiers remained unsettled. British officials worried about threats to their immensely profitable colony, particularly from the Russians advancing across Central Asia. In response, they attempted to expand as far to the north as possible. In some cases this merely entailed making alliances with local rulers. In such a manner Nepal and Bhutan retained their independence. In the extreme northeast, a number of small states and tribal territories, most of which had never been part of the South Asian cultural sphere, were more directly brought into the British Empire, hence into India. A similar policy was conducted on the vulnerable northwestern frontier. Here, however, local resistance was much more effective, and the British-Indian army suffered defeat at the hands of the Afghans. Afghanistan thus retained its independence, forming an effective buffer between the British and Russian empires. The British also allowed the tribal Pashto-speaking areas of what is now northwestern Pakistan to retain almost complete autonomy, thus forming a secondary buffer.

Independence and Partition The framework of British India began to unravel in the early 20th century as the people of South Asia increasingly demanded independence. The British, however, were equally determined to stay, and by the 1920s the region was embroiled in massive political protests.

The rising nationalist movement's leaders faced a major dilemma in attempting to organize a potentially independent regime. Many leaders, including Mohandas Gandhi—the main figure of Indian independence—favored a unified state that would encompass all British territories in mainland South Asia. Most Muslim leaders feared that a unified India would leave their people in a vulnerable position. They therefore argued for the division of British India into two new countries: a Hindu-majority India and a Muslim-majority Pakistan. In several parts of northern South Asia, however, Muslims and Hindus were settled in roughly equal proportions. A more significant obstacle was the fact that the areas of clear Muslim majority were located on opposite sides of the subcontinent, in present-day Pakistan and Bangladesh.

No longer able to maintain their world empire after World War II, the British withdrew from South Asia in 1947. As this occurred, the region was indeed divided into two countries: India and Pakistan. Partition itself was a horrific event, resulting in the death of some 200,000 to 1 million people. Roughly 7 million Hindus and Sikhs fled Pakistan, to be replaced by roughly 7 million Muslims fleeing India (Figure 12.30).

The Pakistan that emerged from partition was for several decades a clumsy two-part country, its western section in the Indus Valley, its eastern portion in the Ganges Delta. The Bengalis, occupying the poorer eastern section, complained that they were treated as second-class citizens. In 1971, they launched a rebellion, and, with the help of India, quickly prevailed. Bangladesh then emerged as a new country.

This second partition did not solve Pakistan's problems, however, as it remained politically unstable and prone to military rule. Pakistan retained the British policy of allowing almost full autonomy to the Pashtun tribes of the Federally Administered Tribal Areas of the northwest, a relatively lawless region marked by clan fighting and vengeance feuds. The large and poor province of Balochistan in southwestern Pakistan has been another problem for the national government, as a long-simmering separatist movement continues to fight against Pakistan's military forces.

Bangladesh has also had a troubled political career since achieving independence in 1971. High levels of corruption, growing Islamic radicalism, and street-level fighting between members its two major political parties have undermined its democratic institutions. As tensions mounted, the Bangladeshi army seized power in January 2007, declared a

Figure 12.30 Partition, 1947 Following Britain's decision to leave South Asia, violence and bloodshed broke out between Hindus and Muslims in much of the region. With the creation of Pakistan (originally in two different sectors, west and east), millions of people relocated both to and from the new states. Many were killed in the process, embittering relations between India and Pakistan.

state of emergency, and detained up to 100,000 individuals. Relatively free elections brought Sheikh Hasina to the office of prime minister in 2009, even though she had been arrested and charged with corruption and extortion 2 years earlier.

Geopolitical Structure of India With independence in 1947, the leaders of India, committed to democracy, faced a major challenge in organizing such a large and culturally diverse country. They decided to chart a middle ground between centralization and local autonomy. India itself was thus organized as a federal state, with a significant amount of power being given to its individual states. The national government, however, retained full control over foreign affairs and a large degree of economic authority.

Following independence, India's constituent states were reorganized to match the country's linguistic geography. The idea was that each major language group should have its own state (with the massive Hindi-speaking population having several), hence a degree of political and cultural autonomy. Yet only the largest groups received their own territories, which has led to recurring demands from smaller groups that have felt politically excluded. Over time, several new states have been added to the map. Goa, after the Portuguese were forced out in 1961, became a separate state in 1987 over the objection of its large northern neighbor of Maharashtra. In 2000, three new states were added: Jharkand, Uttaranchal, and Chhattisgarh.

Ethnic Conflicts in South Asia

The movement for new states in India has been largely rooted in ethnic tensions. Unfortunately, violent ethnic conflicts persist in many parts of South Asia. Of these conflicts, the most complex—and perilous—is that in Kashmir.

Kashmir Relations between India and Pakistan were hostile from the start, and the situation in Kashmir has kept the conflict burning (Figure 12.31). During the British period, Kashmir was a large state with a primarily Muslim core joined to a Hindu district in the south (Jammu) and a Tibetan Buddhist district in the far northeast (Ladakh). Kashmir was then ruled by a Hindu maharaja, or a king subject to British advisors. During partition, Kashmir came under severe pressure from both India and Pakistan. Troops linked to Pakistan gained control of western and much of northern Kashmir, at which point the maharaja opted for union with India. India thus retained the core areas of the state, but neither country would accept the other's control over any portion of Kashmir. As a result, India and Pakistan have fought several inconclusive wars over the issue.

Although the Indo-Pakistani boundary has remained fixed, the struggle in Kashmir intensified, flaming out into an open insurgency in 1989. Although some Kashmiris would like to join their homeland to Pakistan, a 2007 poll indicated that 87 percent of the people of Kashmir Valley preferred independence. Opposition to Kashmir's independence is one of the few things on which India and Pakistan can agree. India accuses Pakistan of supporting training camps for Islamist militants and of helping them sneak across the border. As a result, India began the construction of a fortified fence along the so-called line of control that divides Kashmir. The result of all of this activity has been a low-level but periodically brutal war that has claimed over 40,000 lives and displaced one out of every six inhabitants of Kashmir.

The situation in Kashmir began to improve in 2004 when India and Pakistan initiated a serious round of negotiations. India agreed that Pakistan must play a role in any Kashmir peace settlement, whereas Pakistan's government reduced its support of Muslim militants. As a result, the daily death toll from political violence dropped from 10 in 2002 to 2 in early 2007. But in 2010, tensions again increased as massive protests followed allegations that the Indian army had killed 15 Kashmiri civilians, many of them teenagers. The Vale of Kashmir, once one of South Asia's premier tourist destinations with its lush gardens and orchards nestled among some of the world's most spectacular mountains, is now a battle-scarred war zone.

The Punjab Religious conflict has also caused political tensions in India's Punjab. The original Punjab, an area of intermixed Hindu, Muslim, and Sikh communities, was divided between India and Pakistan in 1947. During partition, virtually all Hindus and Sikhs fled from Pakistan's allotted portion, just as Muslims were forced out of the zone awarded to India.

Partition did not solve the problems of India's portion of the Punjab. Most Sikhs resented the fact that the Indian government refused to recognize Sikhism as a separate religion, instead classifying it as a sect of Hinduism. In the 1970s and 1980s, as the area grew increasingly prosperous due to the Green Revolution, Sikh leaders began to strive for autonomy and Sikh radicals began to press for outright independence. Since many Sikh men had maintained the military traditions of their ancestors, this secession movement soon became a formidable fighting force. The Indian government reacted firmly, and tensions mounted. An open rupture occurred in 1984 when the Indian army raided the main Sikh temple at Amritsar, in which a group of militants had barricaded themselves (Figure 12.32). As hostility escalated, the Indian government placed the Punjab under martial law arresting many militants and driving others out of the country. This policy eventually proved successful in quelling political violence.

Figure 12.31 Conflict in Kashmir Unrest in Kashmir inflames the continually hostile relationships between the two nuclear powers of India and Pakistan. Today many Kashmiris wish to join Pakistan, while many others argue for an independent state. Note also Pakistan's Federally Administered Tribal Areas (including the Islamic Emirate of Waziristan) and its North-West Frontier Province, areas that have experienced much recent fighting.

Current concerns in the Indian state of Punjab focus on migration. Many Punjabis have been moving out of the state

Figure 12.32 Sikh Temple at Amritsar Separatism in Indian's Punjab region emerged because of hostility between the Sikh majority and the Indian government. These tensions were magnified in 1984 when the Indian army raided the Golden Temple in Amritsar, shown in this photograph, to dislodge Sikh militants.

PEOPLE ON THE MOVE Punjabis Leaving Agriculture and Leaving India

Punjab is not only India's breadbasket but also one of its most prosperous states. Many landless people from the poorer states of India, such as Bihar, continue to move to Punjab to find work as farm laborers. But despite their relative prosperity, Punjabis have proved eager to migrate, either to India's growing cities or to foreign countries (Figure 12.5.1). Entire villages in central Punjab's "migration belt" are now almost empty, while illegal migration scandals continue to provide stories for local newspapers. Such scandals reached a peak in April 2007, when a member of the Indian Parliament, Babubhai Katara, was arrested for his involvement in a scheme to smuggle Punjabis to other countries in cooperation with a well-known human-trafficking mafia.

India's recent agricultural distress has hit Punjab hard, leading to massive migration of the state's less successful farmers. As the Green Revolution gained momentum, many farmers went deeply into debt to buy tractors, water pumps, and large quantities of agricultural chemicals. As a result, an estimated 120,000 Punjabi farm families were forced off the land between 1991 and 2001. Since that time, the exodus has apparently accelerated.

Caste issues have also contributed to migration out of Punjab. Dalits often have few, if any, other options locally than to work as farm laborers, a marginal and low-paying occupation. Abroad, however, caste barriers disappear, allowing many low-caste families to prosper. The money that they send home—an estimated $3 billion annually for Punjab as a whole—seems to be changing local caste dynamics. The fact that some members of Punjab's artisan castes have taken greater advantage of technical education—highly useful for finding work abroad—than have the higher-ranked landowning castes has generated upheavals in the traditional social structure.

Migration patterns from the agricultural areas of Punjab have a pronounced gender dimension. Punjab has a highly male-biased population, with only 874 females for every 1,000 males. Sex-selective abortion is generally considered to be the primary cause, but differential migration also plays a role. On one hand, men from the poorest parts of India continue to move into Punjab's farming districts. On other hand, rural Punjabi men tend to move to the region's cities. These patterns combine to further skew local sex ratios. In April 2007, local leaders were shocked to discover that Chandigarh, a relatively prosperous and sophisticated city that serves as the capital of both Punjab and neighboring Haryana, had a sex ratio of only 777 females to 1,000 males, one of the lowest figures in the world. A 2009 survey, however, discovered that the city's sex ratio was improving, with 882 girls born for every 1000 boys.

Figure 12.5.1 Punjabi Emigration Young Punjabi children in Paris congregate every Sunday to study the Punjabi language and the Sikh religion. Although one of India's most prosperous states, Punjab has recently experienced pronounced rural economic distress. Partly as a result, it sends many immigrants to other parts of India and to foreign countries.

(see "People on the Move: Punjabis Leaving Agriculture and Leaving India"), but many people from other parts of India have been moving in, attracted by relatively high wages. If presents trends continue, Sikhs could become a minority in Punjab.

The Northeastern Fringe A more complicated ethnic conflict emerged in the late 1900s in the uplands of India's extreme northeast, particularly in the states of Arunachal Pradesh, Nagaland, Manipur, and portions of Assam. One underlying problem stems from demographic change and cultural collision. Much of this area is still relatively lightly populated, and as a result has attracted many migrants from Bangladesh and adjacent provinces of India. Many local inhabitants consider this movement a threat to both their lands and their cultural integrity.

Through much of the northeast, insurgent groups continue to seek autonomy if not statehood. After 2000, the Indian government began to invest more money in this troubled region, hoping to reduce popular support for separatist movements. India is also eager to expand trade with Burma, and has been working with the Burmese government to secure the border zone. As a result of these and other initiatives, a number of local rebel movements have signed cease-fires with the Indian government. Several insurgent groups have even agreed to cooperate with the Indian army against those groups that continue to fight for independence.

Sri Lanka Interethnic violence in Sri Lanka has also been severe. Here the conflict has roots in both religious and linguistic differences. Northern Sri Lanka and parts of its eastern coast are dominated by Hindu Tamils, while the island's majority group is Buddhist in religion and Sinhalese in language. Relations between the two communities have historically been fairly good, but tensions mounted soon after independence.

The basic problem is that Sinhalese nationalists favor a unitary government, some of them going so far as to argue that Sri Lanka ought to be defined as a Buddhist state. Most Tamils, on the other hand, support political and cultural autonomy, and they have accused the government of discriminating against them. Overall, levels of education are higher among the Tamils, but the government has favored the Sinhalese majority. In 1983, war erupted when the rebel force known as the **Tamil Tigers** (or the *Liberation Tigers of Tamil Eelam*) attacked the Sri Lankan army. Both extreme Tamil and extreme Sinhalese nationalists remained unwilling to compromise, prolonging the war (Figure 12.33).

In the early years of the new millennium, a Norwegian-backed peace process achieved some progress in the area, culminating in a 2003 cease-fire. Within a few years, however, fighting resumed and soon intensified. In March 2007, the Tamil Tigers managed to cobble together a rudimentary air force and bomb Sri Lanka's main airbase, adjacent to the country's international airport. The Sri Lankan government subsequently abandoned negotiations, launching instead an all-out offensive. In May 2009, government forces crushingly defeated the Tamil Tiger army, killing the organization's leaders. The defeat of the Tamil Tigers also brought about a humanitarian disaster, as many civilians were killed and over 300,000 were displaced. Sri Lanka is now finally at peace, but ethnic tensions are still pronounced.

Figure 12.33 Civil War in Sri Lanka The majority of Sri Lankans are Sinhalese Buddhist, many of whom maintain that their country should be a Buddhist state. A Tamil-speaking Hindu minority in the northeast strenuously resists this idea. Tamil militants, who have waged war against the Sri Lankan government for several decades until defeated in 2009, hoped to create an independent country in their northern homeland.

The Maoist Challenge

Not all of South Asia's current conflicts are rooted in ethnic or religious differences. Poverty and inequality in east-central India, for example, have generated a persistent revolutionary movement that finds inspiration in the former Chinese communist leader Mao Zedong. Violence associated with this movement resulted in over 1,100 deaths in 2009 alone. In April 2010, Maoist rebels killed 76 Indian troops in the state of Chattisgarh, the largest single death toll of the insurgency. Such persistent violence has prevented investment in some of India's least-developed areas, intensifying the underlying economic and social problems that gave rise to the insurgency in the first place.

While India's Maoist rebellion is too small to effectively challenge the state, the same cannot be said in regard to Nepal. Nepalese Maoists, frustrated by the lack of development in rural areas, emerged as a significant force in the 1990s. In 2002, Nepal's king, citing the communist threat, dissolved parliament and took over total control of the country's government. This move only intensified the struggle, however, and within a few years the rebels had gained control over 70 percent of the country. Both sides committed atrocities, resulting in more than 13,000 deaths. By 2005, Nepal's urban population also turned against the monarchy, launching massive protests in Katmandu. In 2008, the king stepped down, and Nepal became a republic, with the leader of the former Maoist rebels serving as prime minister. But a year later, he resigned after quarrelling with army leaders about the still-armed Maoist militias. By mid 2010, Nepal's parliament was still unable to form a new government, putting the country under the threat of renewed fighting.

Hopes that including the Maoists in Nepal's political mainstream would bring stability have thus been frustrated. Ethnic conflicts also plague other parts of Nepal. In 2007, a new round of political violence broke out in the Terai, a lowland region along the border with India, associated with a group called the Madhesi Peoples' Right Forum. This group claims that the Nepalese lowlanders have been exploited by both Nepal's political elite and its Maoist rebels, and it deeply resents the migration of people from Nepal's mountains and hills into the narrow but fertile lowland belt. Although many Nepalis are working hard to reestablish peace and democratic rule, success will obviously not come easily.

International Geopolitics

South Asia's major international geopolitical problem continues to be the struggle between India and Pakistan (Figure 12.34). Since independence, these two countries have regarded each other as enemies. The stakes now are particu-

Figure 12.34 Border Tensions An Indian officer looks through binoculars in war-torn Kashmir. Relationships between India and Pakistan have remained extremely tense since independence in 1947. Moreover, with both countries now nuclear powers, the fear that border hostilities will escalate into wider warfare has become a nightmarish possibility.

larly high as both India and Pakistan have nuclear capabilities. Tensions were reduced in 2003 and 2004, however, after India initiated a series of talks with Pakistan. Concrete measures have been slow to emerge, but "confidence-building measures"— including cricket matches and the establishment of bus service between the two countries—have led to hopes that an understanding between these rival states might eventually be reached.

During the global Cold War, Pakistan allied itself with the United States and India remained neutral while leaning slightly toward the Soviet Union. Such entanglements fell apart with the end of the superpower conflict in the early 1990s. Subsequently, Pakistan forged an informal alliance with China, from which Pakistan obtained sophisticated military equipment.

China's military connection with Pakistan is rooted in its own animosity toward India. In 1962, China defeated India in a brief war, gaining control over the virtually uninhabited territory of Aksai Chin in northern Kashmir. Growing trade, however, has resulted in a gradual easing of hostilities between India and China. A declaration that 2006 would be an official "year of friendship," however, was undermined when China announced that it still claimed India's northeastern state of Arunachal Pradesh. Most observers think that an eventual settlement will entail India recognizing Chinese sovereignty over Aksai Chin and China giving up its claims in the northeast. But as long as China maintains close military ties with Pakistan, its relationship with India will remain tense.

Pakistan Pakistan's geopolitical situation became more complex in the aftermath of the attacks in the United States on September 11, 2001. Until that time, Pakistan had strongly supported Afghanistan's Taliban regime (see Chapter 10). After the attack on the World Trade Center and Pentagon (orchestrated from Afghanistan by Osama bin Laden), the United States gave Pakistan a stark choice: either it would assist the United States in its fight against the Taliban and receive in return debt reductions and other forms of aid, or it would lose favor. Pakistan's president, Pervez Musharraf, quickly agreed to help, and Pakistan offered both military bases and valuable intelligence to the U.S. military.

Pakistan's decision to help the United States came with large risks. Both Osama bin Laden and the Taliban enjoy substantial support among the Pashtun people of northwestern Pakistan. After suffering several military reversals, Pakistan decided to negotiate with these radical Islamists, and on several occasions it gave them virtual control over sizable areas. From these bases, militants have launched numerous attacks on U.S. forces in Afghanistan and have attempted to gain control over broader swaths of Pakistan's territory. The United States has responded mainly by using unstaffed drones to attack insurgent leaders, a tactic that has resulted in large numbers of civilian casualties and much anti-American sentiment throughout Pakistan.

The security crisis in Pakistan intensified from 2007 to 2010. In November 2008, Pakistan's president declared a state of emergency, and in the following month, Benazir Bhutto, Pakistan's former prime minister, was assassinated as she campaigned for the 2008 election. Although the elections went on relatively smoothly, the new government faced severe challenges as hard-line Islamists expanded their control over remote areas. Islamic militants increasingly gained power not only in the peripheral tribal areas, but also in South Punjab. In April 2009, Pakistan's army launched a major invasion of Swat, a former vacation zone 100 miles (160 kilometers) northwest of the capital city of Islamabad that had been turned over to Islamic militants a few months earlier. As the military advanced, hundreds of thousands of civilians fled, generating a humanitarian catastrophe. The government soon reclaimed Swat as militants retreated, but the entire area remains highly unstable.

Pakistan is a troubled country, facing not only a deep Islamist insurgency and a seemingly unsolvable geopolitical conflict with India but also a low-level ethnic rebellion in its southwestern province (Balochistan). Its relations with India, moreover, deteriorated in November 2008, when terrorists operating from Pakistan launched a series of coordinated attacks on tourist facilities and public places in Mumbai, killing 173 people. Pakistan responded by investigating the event and arresting a number of alleged plotters, but many Indians suspect that the attack had the support of certain elements within Pakistan's government and military. Both Indian and Pakistani leaders, however, decided that it would be in their own best interests to reduce tensions, and in July 2009, they agreed to renew their stalled peace negotiations. In 2010, Pakistan announced that it would crack down on Islamist militants in its core province of Punjab.

Figure 12.35 India-Bangladesh Border Fence India began building a border fence between its territory and that of Bangladesh in 2003 in order to reduce illegal immigration and to stop the influx of militants. Members of the Indian Border Security Force visible in this photograph are patrolling a segment of the fence.

Sceptics, however, fear that Pakistan's security forces have been deeply infiltrated by Islamist radicals, undermining any such moves.

India's Changing Geopolitical Situation The tension between India and Pakistan, and the complex relationships that this entails with China and the United States, overshadows all other international geopolitical issues in South Asia. Elsewhere in the region, the power of India is often overwhelming, if resented. Bangladesh, which owes its very existence to Indian support, long enjoyed relatively cordial relations with India. Their relations began to deteriorate in the late 1990s, however, owing to India's concerns about illegal Bangladeshi immigration and to its fears that Bangladesh was providing refuge to separatist fighters from India's northeastern periphery. As a result, India is building a $1.2 billion security fence along the border between the two countries (Figure 12.35). In other parts of South Asia, India has used its power even more forcefully. For example, in 1975, it annexed the formerly semi-independent country of Sikkim in the Himalayas.

ECONOMIC AND SOCIAL DEVELOPMENT: Burdened by Poverty

South Asia is a land of developmental paradoxes. It is, along with Sub-Saharan Africa, the poorest world region, yet it is also the site of some immense fortunes. Many of South Asia's scientific and technological accomplishments are world-class, but it also has some of the world's highest illiteracy rates. While South Asia's high-tech businesses are closely integrated with centers of the global information economy, the South Asian economy as a whole was long one of the world's most self-contained and inward looking.

It is difficult to exaggerate South Asia's poverty. Roughly 300 million Indians live below their country's official poverty line, which is set at a very meager level (Figure 12.36). Approximately 20 percent of India's citizens are seriously undernourished, as are 30 percent of the people of Bangladesh. By measures such as infant mortality and average longevity, Nepal is in even worse condition. In urban slums throughout South Asia, rapidly growing populations have little chance of finding housing or basic social services. Estimates indicate that up to half a million South Asian children work as virtual slaves in carpet-weaving workshops and other small-scale factories.

Despite such deep and widespread poverty, South Asia should not be regarded as a zone of uniform misery. India especially has a large and growing middle class, as well as a small but wealthy upper class. More than 250 million Indians are able to purchase such modern consumer items as televisions, motor scooters, and washing machines. India's economy grew from the 1950s to the 1990s at a moderate but accelerating pace, and by the new millennium it was booming. Even in the global recession years of 2008 and 2009, the Indian economy expanded at an annual rate of roughly 7 percent. But if a number of Indian states have shown marked economic vitality (Figure 12.37), others have stagnated.

Figure 12.36 Poverty in India India's rampant poverty results in a significant amount of child labor. In this photo, a 10-year-old boy is moving a large burden of plastic waste by bicycle.

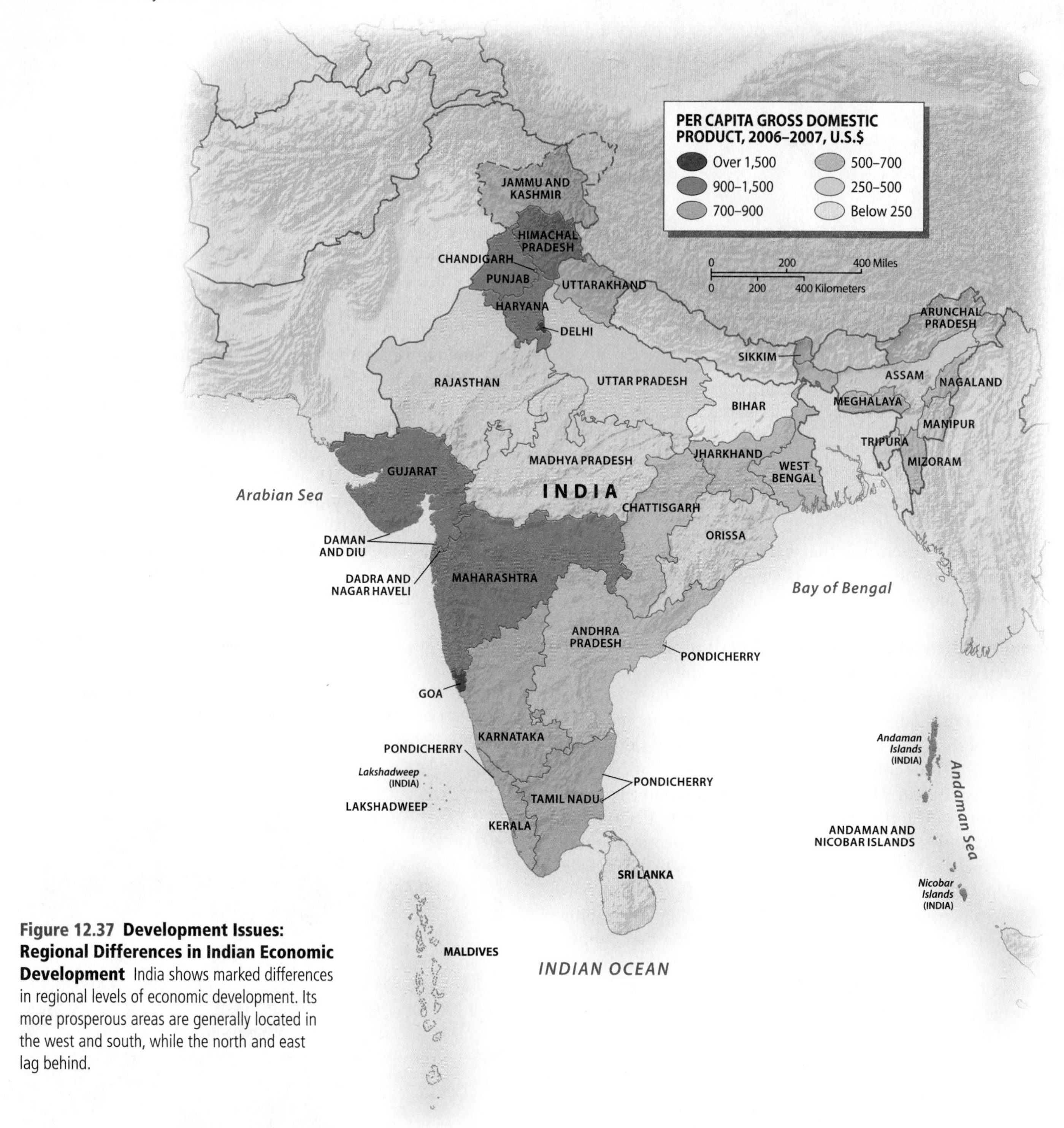

Figure 12.37 Development Issues: Regional Differences in Indian Economic Development India shows marked differences in regional levels of economic development. Its more prosperous areas are generally located in the west and south, while the north and east lag behind.

Geographies of Economic Development

After independence, the governments of South Asia attempted to create new economic systems that would benefit their own people rather than foreign countries or corporations. As in most other parts of the world, planners stressed heavy industry and economic autonomy. While some major gains were realized, the overall pace of development remained slow. Since the 1990s, governments in the region, and especially that of India, have gradually opened their economies to the global economic system. In the process, core areas of economic development have emerged, surrounded by peripheral areas that have lagged behind, creating landscapes of striking economic disparity (Table 12.2).

The Himalayan Countries Both Nepal and Bhutan are disadvantaged by their rugged terrain and remote locations, as well as by the fact that they remain relatively isolated from modern technology and infrastructure. But such measure-

TABLE 12.2 Development Indicators

Country	GNI per capita, PPP 2008	GDP Average Annual % Growth 2000-08	Human Development Index (2007)[a]	Percent Population Living Below $2 a Day	Life Expectancy (2010)[b]	Under Age 5 Mortality Rate (1990)	Under Age 5 Mortality Rate (2008)	Gender Equity (2008)[c]
Bangladesh	1,450	5.8	0.543	81.3	66	149	54	106
Bhutan	4,820		0.619		68			
India	2,930	7.9	0.612	75.6	64	116	69	92
Maldives	5,290		0.771		73			
Nepal	1,110	3.5	0.553	77.6	64	142	51	93
Pakistan	2,590	5.4	0.572	60.3	66	130	89	80
Sri Lanka	4,460	5.5	0.759	39.7	74	29	15	

[a] *United Nations, Human Development Report, 2009.*

[b] *Population Reference Bureau, World Population Data Sheet, 2010.*

[c] *Gender Equity—Ratio of female to male enrollments in primary and secondary school. Numbers below 100 have more males in primary/secondary school, numbers above 100 have more females in primary/secondary schools.*

Source: World Bank, World Development Indicators, 2010.

ments are somewhat misleading, especially for Bhutan, because they fail to take into account the fact that many areas in the Himalayas are still subsistence-oriented.

Bhutan has purposely remained somewhat disconnected from the modern world economy, its small population living in a relatively pristine natural environment. Bhutan is so isolationist that it has only recently allowed tourists to enter—provided that they agree to spend substantial amounts of money while in the country. Its government has made the unusual move of downplaying conventional measures of economic development, attempting to substitute "gross national happiness" for "gross national product."

Bhutan is not, however, completely cut off from the rest of the world. It exports substantial amounts of hydroelectric power to India, helping its economy grow by 19 percent in 2008 and 5 percent in 2009. Such growth has brought immigrants; approximately 100,000 Indian laborers are now working on roads and other infrastructural projects in Bhutan. Local resentment against these and other newcomers from India and especially Nepal has resulted in ethnic tensions, forcing up to 60,000 people to flee from Bhutan to refugee camps in Nepal.

Nepal is more heavily populated and suffers much more severe environmental degradation than Bhutan. It also is more closely integrated with the Indian, and ultimately the world, economy, although three-quarters of its people still depend on small-scale agriculture for their livelihoods. Nepal's economy relies heavily on international tourism (Figure 12.38). Tourism has brought some prosperity to a few favored locations, but often at the cost of heightened ecological damage. Tourism in Nepal began to suffer, moreover, after the country entered a period of political turmoil in 2002. Remittances from Nepalese workers living abroad currently help sustain the country's fragile economy.

Bangladesh The economic figures for Bangladesh are not as low as those of the Himalayan countries, but they are more indicative of widespread hardship because most people there require cash to meet their basic needs. Partly because of the country's massive population, poverty is extreme and widespread.

Environmental degradation has contributed to Bangladesh's impoverishment, as did the partition of 1947. Most of prepartition Bengal's businesses were located in the western area, which went to India. The division of Bengal tore apart an integrated economic region, much to the detriment of the poorer and mainly rural eastern section. Bangladesh also has suffered because of its agricultural emphasis on jute, a plant that yields tough fibers useful for making ropes and burlap bags. Bangladesh failed to

Figure 12.38 Tourism in Nepal. Nepal has long been one of the world's main destinations for adventure tourism, although business has suffered in recent years due to the county's political instability. Many tourists in Nepal stay in rustic lodges, such as the one shown in this photograph.

discover any major alternative export crops as synthetic materials undercut the global jute market.

Not all of the economic news coming from Bangladesh is negative. The country is internationally competitive in textile and clothing manufacture, in part because its wage rate is so low. Low-interest credit provided by the internationally noted Grameen Bank has given hope to many poor women in Bangladesh, allowing the emergence of a number of vibrant small-scale enterprises. By 2000, with the country's birthrate steadily falling, the Bangladeshi economy was finally beginning to grow substantially faster than its population. Even through the global recession of 2008–2009, it continued to grow by 5 to 6 percent a year.

Pakistan Pakistan also suffered the effects of partition in 1947. But unlike Bangladesh, Pakistan at least inherited a reasonably well-developed urban infrastructure. It has a productive agricultural sector, as it shares the fertile Punjab with India. Pakistan also boasts a large textile industry, based in part on its huge cotton crop. Several Pakistani cities, moreover, have developed important niches in the global economy (see "Global to Local: The Sialkot Industrial Complex"). Pakistan's economic growth accelerated after 2001, due in part to the concessions given to it by United States and international economic institutions.

By 2008–2009, however, Pakistan's economy was again faltering, hampered by an inflation rate of over 20 percent. The Pakistani economy is less dynamic than that of India, with a lower potential for growth. Part of the problem is that Pakistan is burdened by extremely high levels of military spending, yet at the same time it has experienced growing internal strife. Additionally, a small but powerful landlord class that pays virtually no taxes to the central government controls many of its best agricultural lands. Unlike India, moreover, Pakistan has not been able to develop a successful IT industry.

Like India, Pakistan has inadequate energy supplies, which has led it to look to both Central Asia and Southwest Asia for fossil fuels. One consequence of this policy has been the construction of a huge new deep-water port at Gwadar near the border with Iran, adjacent to some of the world's most important oil-tanker routes. Built with massive Chinese engineering and financial assistance, the port of Gwadar was turned over to be managed by the Port of Singapore when completed in 2007, showing the increasingly globalized nature of Pakistan's economic planning.

Sri Lanka and the Maldives As can be seen in Table 12.2, Sri Lanka's economy is the second most highly developed in South Asia by conventional criteria. Its exports are concentrated in textiles and agricultural products such as rubber and tea. By global standards, however, it is still a very poor country. Its progress, moreover, was long undercut by its civil war. In 2009, when the war finally ended, the Sri Lankan stock market posted gains of over 100 percent, reflecting renewed confidence in the country's economy. Sri Lanka hopes to benefit more from the prime location of its port of Colombo, from its high levels of education, and from its great tourism potential. To do so, however, it will need to attract large quantities of foreign investment. Sri Lanka currently depends heavily on the remittances sent home by its workers living abroad, estimated to total some $3 billion a year.

The Maldives is the most prosperous South Asian country based on per capita GNI, but its total economy, like its population, is tiny. Most of its revenues are gained from fishing and international tourism. Eighty-eight of the Maldives' otherwise uninhabited islands have been turned over entirely to resorts, which are visited by some 600,000 tourists a year. The benefits from the tourist economy, however, flow mainly to the country's small elite population, resulting in large-scale public discontent and political repression. Tourism is also highly vulnerable to international recessions; in 2009, the economy of the Maldives declined by 4 percent.

India's Less-Developed Areas India's economy, like its population, dwarfs those of the other South Asian countries. While India's per capita GNI is roughly comparable to that of Pakistan, its total economy is many times larger. As the region's largest country, India also exhibits far more internal variation in economic development than its neighbors. The most basic economic division is that between India's more prosperous west and south and its poorer districts in the north and east.

The so-called tribal states of India's northeastern fringe rank low on the economic ladder, as measured by per capita GNI, but the prevalence of subsistence economies makes such statistics misleading. More extreme deprivation is found in the lower Ganges Valley, where cash economies generally prevail. Bihar is India's poorest state by virtually all economic indicators, as well as its most politically corrupt. Neighboring Uttar Pradesh, India's most populous state, also is poverty-stricken. Like Bihar, it is densely populated and has experienced little industrial development. Although both Bihar and Uttar Pradesh have fertile soils, their agricultural systems have not profited as much from the Green Revolution as have those of Punjab. In both states the **caste system** is deeply entrenched, tensions between Hindus and Muslims are bitter, and opportunities for most peasants are limited. Ironically, South Asia's wealth was historically concentrated in the fertile lowlands of the Ganges Valley, yet today the area ranks among the poorest parts of an impoverished world region.

Other eastern India states such as Orissa and Assam are also quite poor, but the large and important state of West Bengal ranks about average for India as a whole. Some of the worst slums in the world are located in West Bengal's Kolkata (Calcutta). But Kolkata also supports a substantial and well-educated middle class, and it is the site of a sizable industrial complex. For most of the period of Indian independence, West Bengal has been governed by a leftist party that has fostered, with little success, heavy, state-led industry. In a dramatic turn-around in the 1990s, West Bengal's Marxist leaders began to advocate internationalization, encouraging large multinational firms to build new factories in

GLOBAL TO LOCAL The Sialkot Industrial Complex

Pakistan does not have a highly globalized economy. Its exports are dominated by bulk cotton goods, mostly yarn, cloth, sheets, and towels rather than clothing, with rice, leather goods, and low-value chemicals rounding out the list. But one Pakistani city, Sialkot, stands out from all the rest, having carved a unique niche in the global economic system though its production of surgical steel implements and sporting goods (Figure 12.6.1). As a result, Sialkot is noted as Pakistan's most entrepreneurial city, enjoying a per capita income roughly double the national average.

During the Mughal period, Sialkot's weapon-makers were famous for their fine steel swords, but changes in warfare put them out of business by the late 1800s. When an American mission hospital was opened in the city, however, local artisans saw an opportunity to supply it with scalpels and other specialized implements. Before long, Sialkot was exporting such goods throughout British India, and then to the rest of the world.

The rise of the city's sporting goods industry is more difficult to explain. Legend says that it began in the late 1800s when an English tennis player broke his racket and asked a local artisan to repair it. He was evidently so impressed with the product that he commissioned additional replacements, and by 1895 a thriving new industry had emerged.

Today Sialkot's sporting goods industry is most famous for its hand-stitched soccer balls, exporting roughly 60 million each year. The city's highly competitive businesses, however, produce a huge array of other athletic goods, including equipment for horseback riding, motorcycle racing, boxing, and martial arts. The global nature of the industry is fully apparent in the production line of Saeed Classical Works (http://sportzbiz.pk/), which advertises such products as "Brazilian Jiu-Jitsu Kimonos" sold under the "Bararian" brand name.

Despite its general success, Sialkot's sporting-goods business has faced a number of serious problems. In the 1990s, international criticism mounted over the use of child labor, particularly for hand-sewing soccer balls, an activity that rarely paid more than $1.50 a day. Since then, increased monitoring of factories has reduced labor violations, but abuses remain widespread. Another threat comes from China, which produces machine-assembled balls that dominate the lower end of the market. In response, Sialkot's producers have been striving to widen their product lines and to move into higher-end merchandise. The government of Pakistan offered help by opening a technical university in 2007. As befits the globalized nature of Sialkot's industrial complex, this new University of Engineering Science and Technology has been established in collaboration with Sweden's Royal Institute of Technology. In 2010, Sialkot firms announced that they would soon begin producing machine-made soccer balls in order to compete with Chinese firms in the global market.

Figure 12.6.1 Sialkot Sporting Goods Firm Young employees work at a soccer-ball factory in Sialkot, Pakistan. Sialkot is noted for its exports of sporting goods and medical implements, but it has also been the site of much exploitative child labor.

"special economic zones" subjected to low levels of taxation. Such programs have generated substantial opposition that occasionally become violent.

Although western India is in general much more prosperous than eastern India, the large western state of Rajasthan ranks well below average. Rajasthan suffers from an arid climate; nowhere else in the world are deserts and semi-deserts so densely populated. It also is noted for its social conservatism. During the British period, almost all of this large state remained outside the sphere of direct imperial power. Here, in the courts of maharajas, the military and political traditions of Hindu India persisted up until recent times. Rajasthan's rulers not only maintained elaborate courts and fortifications, but also supported many traditional Indian arts. Because of this political and cultural legacy, Rajasthan is one of India's most important destinations for international tourists.

India's Centers of Economic Growth North of Rajasthan lie the Indian states of Punjab and Haryana, showcases of the Green Revolution. Their economies have relied largely on agriculture, but recent investments in food processing and other industries have been substantial. Punjab has the lowest levels of malnutrition in India and the most highly developed infrastructure. Despite its relative prosperity, this region has seen rural unrest in recent years. On Haryana's eastern border lies the capital district of New Delhi, where India's political power and much of its wealth are concentrated.

India's west-central states of Gujarat and Maharashtra are noted for their industrial and financial clout, as well as for their agricultural productivity. Gujarat was one of the first parts of South Asia to experience substantial industrialization, and its textile mills are still among the most productive in the region. Gujaratis have long been famed as merchants and overseas traders, and they are disproportionately represented in the **Indian diaspora**, the migration of large numbers of Indians to foreign countries. As a result, cash remittances from these emigrants help to bolster the state's economy.

The state of Maharashtra is usually viewed as India's economic pacesetter. The huge city of Mumbai (Bombay) has long been the financial center, media capital, and manufacturing powerhouse of India. According to official figures, the Mumbai metropolitan area accounts for 25 percent of India's industrial output, 40 percent of its maritime trade, and 70 percent of its major financial transactions. Large industrial zones are located in several other cities of Maharashtra, especially Pune and Nagpur. In recent years, Maharashtra's economy has grown more quickly than those of most other Indian states, reinforcing its primacy.

The center of India's fast-growing high-technology sector lies farther to the south, especially in Karnataka's capital of Bangalore, which in 2007 changed its name to Bengaluru. The Indian government selected the upland Bengaluru area for its fledgling aviation industry in the 1950s. Other technologically sophisticated ventures soon followed. In the 1980s and 1990s, a quickly growing computer software industry emerged, earning Bengaluru the label of "Silicon Plateau." In the 1980s, growth was spurred by the investments of U.S. and other foreign corporations eager to hire relatively inexpensive Indian technical talent. Since the 1990s, these multinational companies have been joined by a rapidly expanding group of locally owned firms. Biotechnology is also thriving. Unfortunately, Bengaluru's rapid growth has stretched the city's infrastructure to the breaking point. Roads are commonly jammed, electricity supplies are inadequate, and many parts of the city can only count on three hours of water a day.

Partly because of Bengaluru's problems, other cities in southern India have recently emerged as rival high-tech centers. Hyderabad in Andhra Pradesh, often called "Cyberabad," is well known for its IT and pharmaceutical firms, as well as its film industry, India's second largest. Chennai (Madras) in Tamil Nadu, recently voted as having the highest quality of life among India's major cities, is noted for its software production as well as its financial services and automobile industry.

India has proved especially competitive in software because software development does not require a sophisticated infrastructure; computer code can be exported via wireless telecommunication systems without the use of modern roads or port facilities. What is necessary, of course, is technical talent, and this India has in abundance. Many Indian social groups are highly committed to education, and India has been a scientific power for decades. With the growth of the software industry, India's brainpower has finally begun to translate into economic gains (Figure 12.39). Whether such developments can spread benefits throughout the country remains to be seen. Most of India's rural areas are not prospering, and malnutrition remains widespread. What is certain, however, is that information technology has tightly linked certain parts of India to the global economy.

Globalization and India's Economic Future

South Asia is not one of the world's most globalized regions by conventional economic criteria. The volume of foreign trade is not huge; foreign direct investment is still modest; and (with the exception of the Maldives) international tourists are few (Figure 12.40). But globalization in South Asia is advancing rapidly.

To understand the low globalization indicators for the region, it is necessary to look at its recent economic history. India's postindependence economic policy, like those of other South Asian countries, was based on widespread private ownership combined with high-tariff barriers and governmental control of planning, resource allocation, and certain heavy industrial sectors. This mixed socialist–capitalist system initially brought a fairly rapid development of heavy industry and allowed India to become virtually self-sufficient.

By the 1980s, however, problems with India's economic model were becoming apparent, and frustration was mounting among the business and political elite. While growth was persistent, it remained in most years only a percentage point or two above the rate of population expansion. The percentage of Indians living below the poverty line, moreover, remained virtually constant. At the same time, countries such as China and Thailand were experiencing rapid development after opening their economies to global forces. Many Indian businesspeople were irritated by the governmental regulations that undermined their ability to expand. In the 1980s, foreign indebtedness began to mushroom, putting further pressure on the economy.

In response to these difficulties, the Indian government began to liberalize its economy in 1991. Many regulations

Figure 12.39 India's Silicon Plateau The Bangalore suburb of Whitefield is usually considered to be the heartland of India's high tech sector. This photograph shows Whitefield's International Technical Park.

were modified and some were eliminated, and the economy was gradually opened to imports and multinational businesses. Other South Asian countries have followed a somewhat similar path. Pakistan, for example, began to privatize many of its state-owned industries in 1994, and in 2000 began to turn over the banking sector to private enterprise.

This gradual internationalization and deregulation of the Indian economy has generated substantial opposition. Foreign competitors are now seriously challenging domestic firms. Cheap manufactured goods from China are seen as an especially serious threat. India has a strong heritage of economic nationalism stemming from the colonial exploitation

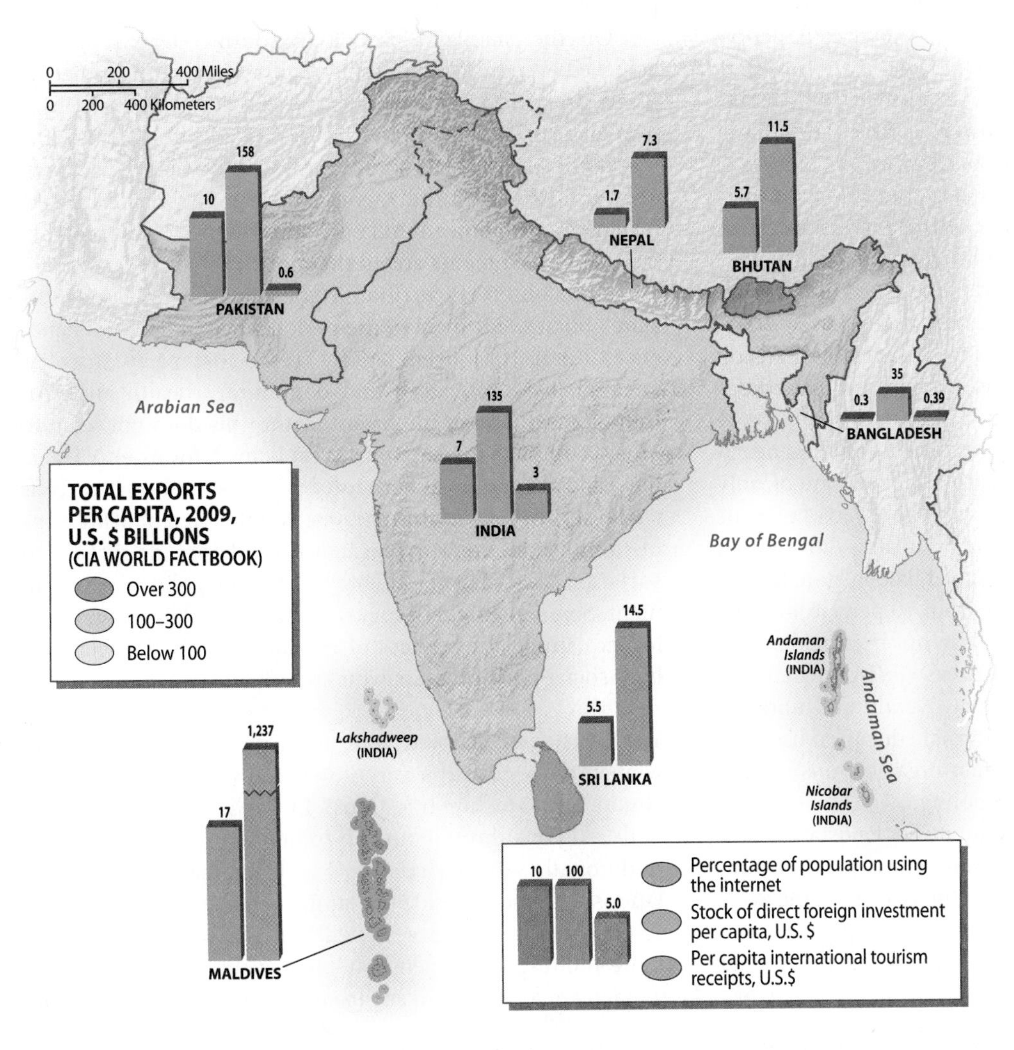

Figure 12.40 South Asia's Global Linkages Despite South Asia's growing global connections, the region as a whole is still relatively self-contained, especially in regard to finance. Internet use remains low, especially in Bangladesh and Nepal.

it long suffered. Agricultural liberalization, a central agenda of the World Trade Organization, became an especially contentious issue by the early 2000s because most of India's huge farming sector is not globally competitive. In 2006, mounting opposition led India's government to stall further plans for the privatization of state-controlled economic activities.

Although economic globalization is rapidly advancing in India, the country has not seen the levels of direct foreign investment that have recently transformed the Chinese economy. India also devotes a far smaller percentage of its national budget to infrastructural improvements than does China. Because of such investment shortages, India, like the rest of South Asia, does not have an adequate transportation system to meet its economic needs, and its supply of electricity is woefully short. As a result, many observers doubt that it will be able to match the growth rates that China has achieved. Other, however, are more optimistic, noting that Indian industrial production rose by over 11 percent between mid-2009 and mid-2010, when much of the world was still in an economic crisis.

Social Development

South Asia's social indices show relatively low levels of health and education, which is hardly surprising considering the region's poverty. Levels of social well-being, not surprisingly, vary greatly across the region. As might be expected, people in the more prosperous areas of western India are healthier, live longer, and are better educated, on average, than people in the poorer areas, such as the lower Ganges Valley. Bihar thus stands at the bottom of most social as well as economic measurements, while Punjab, Gujarat, and Maharashtra stand near the top. In much of Bihar, where almost half of the population lives below the poverty line, the situation is dire; considering the fact that over a third of the state's teachers are absent on any given day, progress has not been easy. By 2008, however, Bihar was at long last making considerable gains on both the social and economic fronts, prompting guarded optimism among informed observers.

Several key measurements of social welfare are higher in India than in Pakistan. Pakistan, with a literacy rate of only about 56 percent, has done a particularly poor job of educating its people. Still, it is important to realize how much progress has been made. The province of Balochistan, for example, saw its literacy rate increase from 10 percent in 1981 to 49 percent in 2008. Such gains have not been realized in the Federally Administered Tribal Areas (FATA), where the literacy rate is still below 20 percent overall, and under 5 percent for women. FATA's lack of formal education has provided an opening for radical Islamist organizations, which often offer the only options for schooling.

In several areas of social development, Pakistan is still ahead of India. Pakistan probably suffers less malnutrition, and the average Pakistani lives a few more years than the average Indian. Pakistan also has fewer beggars and people living on the streets than does India, partly due to its widespread Islamic charity system.

Several discrepancies stand out when one compares South Asia's map of economic development with its map of social well-being. Portions of India's extreme northeast, for example, have high literacy rates despite their poverty, owing to the educational efforts of Christian missionaries. In Mizoram, for example, roughly 90 percent of people over the age of 7 can read and write. Kolkata (Calcutta) and its immediate environs also stand out as a relatively well-educated area, despite the general distress of the lower Ganges Basin. The most pronounced discrepancies, however, are found in the southern reaches of South Asia. In health, longevity, and education, the extreme south far outpaces the rest of the region.

The Educated South Southern South Asia's relatively high levels of social welfare are clearly visible when one examines Sri Lanka. Considering its meager economic resources and long-lasting civil war, Sri Lanka must be considered one of the world's great success stories of social development. It demonstrates that a country can achieve significant health and educational gains even in the context of an "undeveloped" economy. Sri Lanka's average longevity of 74 years stands in favorable comparison with some of the world's industrialized countries, as does its literacy rate of more than 90 percent. The Sri Lankan government has achieved these results by funding universal primary education and inexpensive medical clinics.

On the mainland, Kerala in southwestern India has achieved even more impressive results. Kerala is not a particularly prosperous state. It is extremely crowded, has a high rate of unemployment, and has long had difficulty feeding its population. Kerala's indices of social development, however, are the best in India, comparable to those of Sri Lanka (Figure 12.41). Since Kerala is poorer than Sri Lanka, its social accomplishments are all the more impressive.

Some observers attribute Kerala's social successes to its state policies. For most of the period since Indian independence, Kerala has been led by a socialist party that has stressed mass education and community health care. Although no doubt an important factor, this does not seem to offer a complete explanation. West Bengal, for example, also has a socialist political heritage, but it has not been nearly as successful in its social programs. Kerala's neighboring state of Tamil Nadu, on the other hand, also has made significant social progress despite having a different political environment. Some researchers suggest that one of the key variables for explaining the success of the far south is the relatively high social position of its women.

The Status of Women It is often said that South Asian women are accorded a very low social position in both the Hindu and Muslim traditions. In higher-class families of both religions throughout the Indus–Ganges Basin, women traditionally were secluded to a large degree, their social relations with men outside the family being severely restricted.

Some scholars have argued that the Hindu tradition is more limiting to women than the Muslim tradition. Hindu women are forbidden to engage in certain economic activities

Figure 12.41 Education in Kerala India's southwestern state of Kerala, which has virtually eliminated illiteracy, is South Asia's most highly educated region. It also has the lowest fertility rate in South Asia. Because of this, many argue that women's education and empowerment is the best and most enduring form of contraception.

(such as plowing), and in some areas they are excluded from inheriting land. Throughout northern India, women traditionally leave their own families shortly after puberty to join those of their husbands. As outsiders, often in distant villages, young brides have few opportunities, and it is not uncommon for them to be bullied by their parents-in-law. Widows in the higher castes, moreover, are not supposed to remarry, and instead are encouraged to go into permanent mourning.

Several social indices show that women in the Indus–Ganges Basin still suffer pronounced discrimination. In Pakistan, Bangladesh, and such Indian states as Rajasthan, Bihar, and Uttar Pradesh, female levels of literacy are much lower than those of males. An even more telling statistic is that of gender ratios, the relative proportion of males and females in the population. All things being equal, there should be slightly more women than men in any population because women generally have a longer life expectancy. Northern South Asia, however, contains many more men than women. Nepal is one of the few countries in which men on average live longer than women.

Economics play a major role in India's biased sex ratios. In rural households, boys are usually viewed as a blessing because they typically remain with their families and work for their well-being. In the poorest groups, elderly people (especially widows) subsist largely on what their sons can provide. Girls, on the other hand, marry out of their families at an early age and must be provided with a dowry. They are thus seen as a net economic liability. Considering the economic desperation of the South Asian poor, it is perhaps not surprising to see such shortages of girls and young women.

Evidence suggests that the social position of women is improving, especially in the more prosperous parts of western India where employment opportunities outside the family context are emerging. But even in many of the region's middle-class households, women still suffer major disabilities. Dowry demands seem to be increasing in many areas, and there have been a number of well-publicized murders of young brides whose families failed to provide enough goods. In some parts of South Asia, gender ratios are growing even more male-biased now that technology allows gender-selective abortion. Laws against this practice have generally been ignored.

While the social bias against women across northern South Asia is striking, it is much less evident in southern India and Sri Lanka. In Kerala especially, women have relatively high status, regardless of whether they are Hindus, Muslims, or Christians. Here the overall gender ratio shows the normal pattern, with a slight predominance of females over males. Female literacy is very high in Kerala, which is one reason the state's overall illiteracy rate is so low. Not surprisingly, the high social position of women in southwestern India has deep historical roots. Among the Nairs—Kerala's traditional military and land-holding caste—all inheritance up to the 1920s had to pass through the female line. By 2000, however, census data showed that even in Kerala the number of girls being born relative to the number of boys was declining, showing some evidence of sex-selective abortion.

Summary

- South Asia, a large and complex area, has in many ways been overshadowed by neighboring world regions: by the uneven globalization of Southeast Asia, by the size and political weight of East Asia, and by the geopolitical tensions of Southwest Asia. Much of that is changing, however, as South Asia now figures prominently in discussions of world problems and issues.
- Environmental degradation and instability pose particular problems for South Asia. The region's monsoon climate causes both floods and droughts to be more problematic here than in most other world regions. Rising sea level associated with global climate change directly threatens the low-lying Maldives, and changes in rainfall may play havoc with the monsoon-dependent agricultural systems of India, Pakistan, and Bangladesh.

- Continuing population growth in this already densely populated region demands attention. Although fertility rates have declined rapidly in recent years, Pakistan and northern India cannot easily meet the demands imposed by their expanding populations.
- South Asia's diverse cultural heritage, shaped by peoples speaking several dozen languages and following several major religions, makes for a particularly rich social environment. Unfortunately, cultural differences have often translated into political conflicts. Ethnically or religiously based separatist movements have severely challenged the governments of Pakistan, India, and Sri Lanka. In India, moreover, religious strife between Hindus and Muslims persists, whereas in Pakistan and Bangladesh Islamic radicals clash with the state.
- Geopolitical tensions within South Asia are particularly severe, again demanding global attention. The long-standing feud between Pakistan and India escalated dangerously in the late 1990s, leading many observers to conclude that this was the most likely part of the world to experience a nuclear war. Although tensions between the two countries have lessened, the underlying sources of conflict—particularly the struggle in Kashmir—remain unresolved.
- Although South Asia remains one of the poorest parts of the world, much of the region has seen rapid economic expansion in recent years. Many argue that India in particular is well positioned to take advantage of economic globalization. Large segments of its huge labor force are well educated and speak excellent English, the major language of global commerce. But will these global connections help the vast numbers of India's poor or merely the small number of its economic elite? Advocates of free markets and globalization tend to see a bright future, while skeptics more often see growing problems.

Key Terms

British East India Company *(page 569)*
caste system *(page 558)*
cyclone *(page 541)*
dalit *(page 559)*
Dravidian language *(page 564)*
federal state *(page 570)*
forward capital *(page 556)*
Green Revolution *(page 552)*
Hindu nationalism *(page 556)*
Indian diaspora *(page 580)*
Jainism *(page 561)*
linguistic nationalism *(page 565)*
maharaja *(page 570)*
monsoon *(page 544)*
Mughal Empire (also spelled *Mogul*) *(page 558)*
orographic rainfall *(page 545)*
salinization *(page 553)*
Sikhism *(page 561)*
subcontinent *(page 564)*
Tamil Tigers *(page 573)*

Review Questions

1. What are the four subregions of South Asia? Describe their similarities and differences.
2. What causes the South Asian monsoon? How does it affect different parts of South Asia?
3. What is orographic rainfall? Where is it important for South Asian agriculture?
4. How and why does the birthrate differ geographically within South Asia?
5. What are some of the infrastructural problems faced by South Asian cities? Give some specific examples.
6. Describe the geography of Islam within India. That is, where are the significant Muslim minorities located in India?
7. What are the major Indo-European languages in South Asia? Where are they located? Where are the non-Indo-European languages located?
8. Describe three different regions of geopolitical and ethnic tension within South Asia.
9. Where are the centers or core areas of economic development within the different South Asian countries?
10. What kinds of relationships are seen between women's literacy and different aspects of economic and social development?

Thinking Geographically

1. As a geographer, suggest different strategies for solving (or at least lessening) the serious flooding problems in Bangladesh. Consider the fact that this crowded country must maximize most of its area for agricultural production.
2. What are the advantages and disadvantages of expanding irrigated agriculture in India and Pakistan? How can the disadvantages be reduced to acceptable levels?
3. What are the pros and cons of the Green Revolution as a means of increasing South Asia's food supplies? What is the outlook for the next decade?
4. Discuss the conflict between wildlife protection and rural villages in India by evaluating the tension between preserving habitat and the needs of the rural poor.
5. What are the drawbacks and benefits of using English as a national language in India? Might it help or hinder unity? Would this increase or decrease India's links to the contemporary world?
6. Choose one of the areas of geopolitical tension (Kashmir, Punjab, Sri Lanka, etc.) and, after becoming better acquainted with the complex issues that underlie this conflict, evaluate

the different proposals currently offered for solving (or at least ameliorating) the problem.

7. As a geographer, you work for an international arms reduction agency that is working to reduce tensions that could lead to nuclear war between Pakistan and India. What would you suggest?
8. From a geographical point of view, what is the best course of action for near-term future economic development in Pakistan, India, and Bangladesh?
9. Acquaint yourself with the environmental, social, and economic implications of Nepal's open-door policy toward trekking and other forms of tourism. Do the benefits seem to outweigh the costs?
10. Is the state of Kerala a good model for economic and social development in other parts of India? Why or why not?

Regional Novels and Films

Novels

Vikram Chandra, *Sacred Games* (2006, HarperCollins)

Anita Desai, *Fasting, Feasting* (1999, Houghton Mifflin)

E. M. Forster, *A Passage to India* (1924, Harcourt, Brace and Company)

Bharti Kirchner, *Darjeeling: A Novel* (2002, St. Martin's Press)

Arundhati Roy, *The God of Small Things* (1997, Random House)

Salman Rushdie, *Midnight's Children* (1981, Knopf)

Paul Scott, *The Raj Quartet* (1980, Morrow)

Vikram Seth, *A Suitable Boy* (1993, HarperCollins)

Bapsi Sidhwa, *Cracking India: A Novel* (1991, Milkweed)

Manil Suri, *The Death of Vishnu* (2001, Norton)

Thirty Umrigar, *Bombay Time* (2001, Picador)

Films

August Sun (Sri Lanka, 2003)

Bandit Queen (1994, India)

City of Joy (1992, UK)

Dasavathaaram (2008, India)

Earth (1998, India)

East Is East (1999, UK)

Matir Moina (2002, Bangladesh)

Monsoon Wedding (2001, India)

Saalam Bombay (1988, India)

A Thousand Dreams Such As These (2003, India)

Travellers and Magicians (2003, Australia/Bhutan)

Bibliography

Bose, Sugata, and Jalal, Ayesha. 2002. *Modern South Asia: History, Culture, and Political Economy.* Delhi: Oxford University Press.

Eaton, Richard M. 1993. *The Rise of Islam and the Bengal Frontier, 1204–1760.* Berkeley: University of California Press.

Fox, Richard G., ed. 1977. *Realm and Region in Traditional India.* Durham, NC: Duke University Program in Comparative Studies on Southern Asia.

Hussain, Zahid. 2007. *Frontline Pakistan: The Struggle with Militant Islam.* New York: Columbia University Press.

Guha, Ramachandra. 2008. *India After Gandhi: The History of the World's Largest Democracy.* New York: Harper Perennial.

Schwartzberg, Joseph E. 1992. *A Historical Atlas of South Asia.* Oxford, UK: Oxford University Press.

Spate, O. H. K., and Learmonth, A. T. A. 1967. *India and Pakistan: A General and Regional Geography.* London: Methuen.

Steven, Stanley F. 1993. *Claiming the High Ground: Sherpas, Subsistence, and Environmental Change in the Highest Himalaya.* Berkeley: University of California Press.

Wolpert, Stanley. 1991. *India.* Berkeley: University of California Press.

Zurick, David, and Karan, P. P. 1999. *Himalaya: Life on the Edge of the World.* Baltimore: Johns Hopkins University Press.

Log in to www.masteringgeography.com for MapMaster™ interactive maps, geography videos, RSS feeds, flashcards, web links, an eText version of *Diversity Amid Globalization,* and self-study quizzes to enhance your study of South Asia.

MapMaster™ presents 13 Place Name and 13 Layered Thematic interactive maps to help students practice and master their geographic literacy, spatial reasoning, and critical thinking skills.

13

Southeast Asia

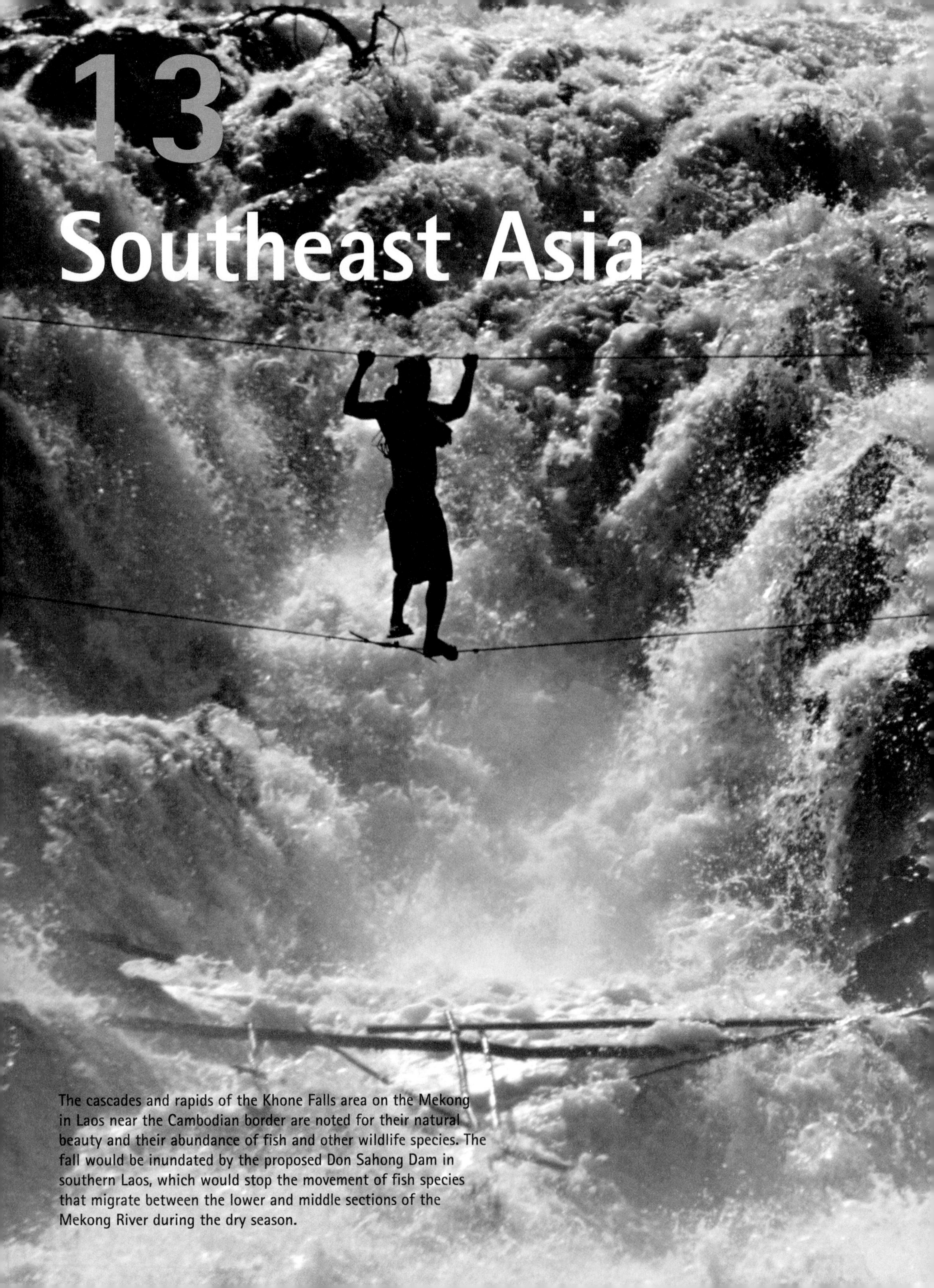

The cascades and rapids of the Khone Falls area on the Mekong in Laos near the Cambodian border are noted for their natural beauty and their abundance of fish and other wildlife species. The fall would be inundated by the proposed Don Sahong Dam in southern Laos, which would stop the movement of fish species that migrate between the lower and middle sections of the Mekong River during the dry season.

DIVERSITY AMID GLOBALIZATION

Southeast Asia has been a key player in global commerce for hundreds of years. It has long attracted migrants from other parts of the world, and has been receptive to foreign cultural influences. In the late 20th century, the Southeast Asia was transformed from a region of Cold War geopolitical struggle to a center of economic globalization. Today Southeast Asian leaders are trying to cope with globalization on their own terms, working to create a more economically and politically united world region.

ENVIRONMENTAL GEOGRAPHY

Southeast Asia's rain forests are vital centers of biological diversity, but they are rapidly diminishing due to commercial logging and agricultural expansion.

POPULATION AND SETTLEMENT

Southeast Asia has a particularly uneven pattern of population distribution, with some areas experiencing serious crowding while others are noted for their sparse settlement.

CULTURAL COHERENCE AND DIVERSITY

Culturally, Southeast Asia is characterized by much more diversity than coherence, hosting significant areas of Muslim, Buddhist, and Christian religions, as well as a vast number of languages.

GEOPOLITICAL FRAMEWORK

Southeast Asia is one of the most geopolitically unified regions of the world, with all but one of its countries belonging to the Association of Southeast Asian Nations (ASEAN).

ECONOMIC AND SOCIAL DEVELOPMENT

Southeast Asia as a whole contains some of the world's most globalized and advanced economies and some of the most isolated and impoverished; it has also experienced marked periods of boom and bust in recent decades.

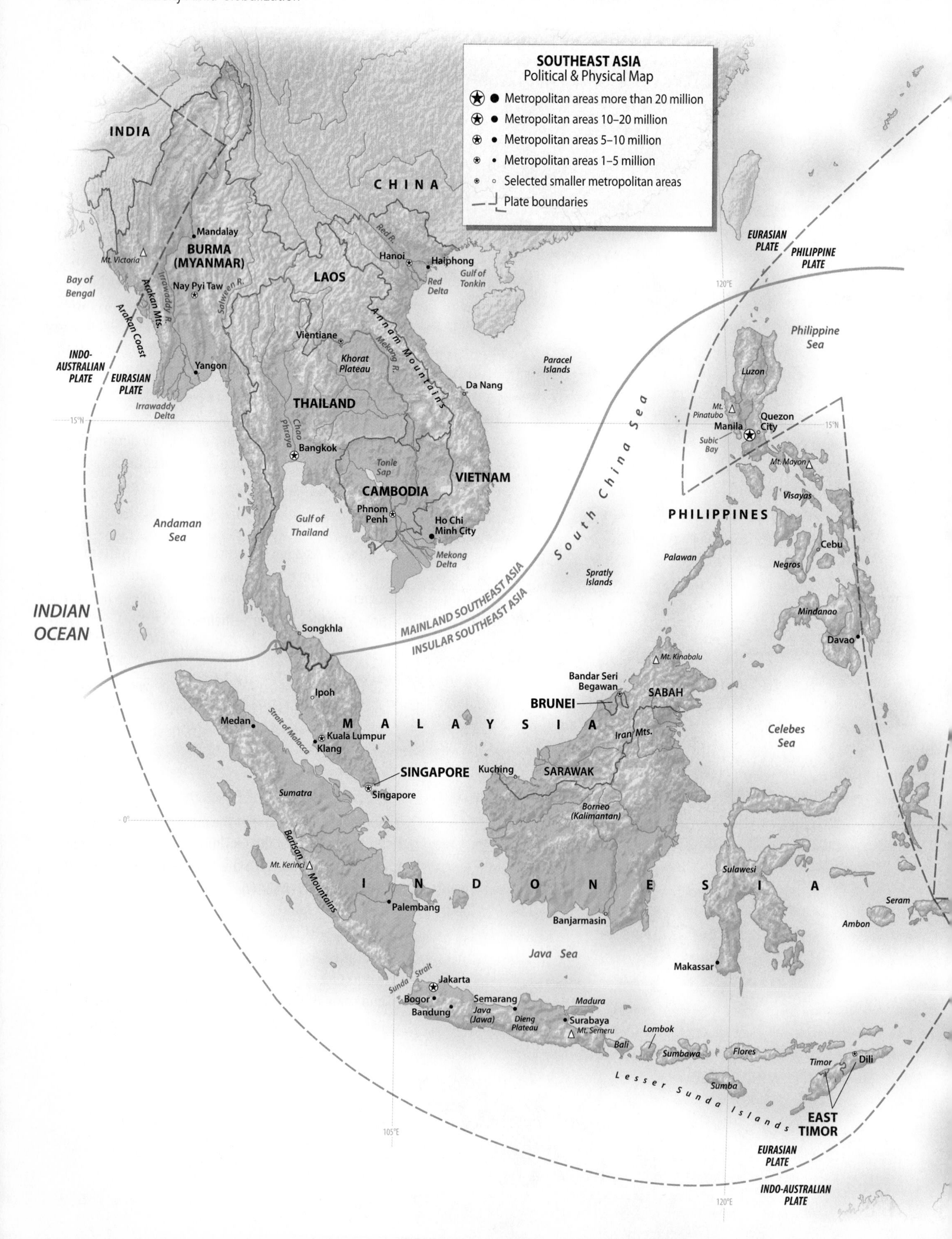
SOUTHEAST ASIA
Political & Physical Map
Metropolitan areas more than 20 million
Metropolitan areas 10–20 million
Metropolitan areas 5–10 million
Metropolitan areas 1–5 million
Selected smaller metropolitan areas
Plate boundaries
INDIA
CHINA
BURMA (MYANMAR)
LAOS
THAILAND
CAMBODIA
VIETNAM
PHILIPPINES
MALAYSIA
BRUNEI
SINGAPORE
INDONESIA
EAST TIMOR
SABAH
SARAWAK
Mandalay
Nay Pyi Taw
Yangon
Hanoi
Haiphong
Vientiane
Da Nang
Bangkok
Phnom Penh
Ho Chi Minh City
Manila
Quezon City
Cebu
Davao
Songkhla
Ipoh
Medan
Kuala Lumpur
Klang
Singapore
Kuching
Bandar Seri Begawan
Palembang
Banjarmasin
Makassar
Jakarta
Bogor
Bandung
Semarang
Surabaya
Dili
Mt. Victoria
Arakan Mts.
Arakan Coast
Irrawaddy R.
Salween R.
Irrawaddy Delta
Bay of Bengal
Red R.
Red Delta
Gulf of Tonkin
Annam Mountains
Mekong R.
Khorat Plateau
Chao Phraya
Tonle Sap
Mekong Delta
Gulf of Thailand
Andaman Sea
INDIAN OCEAN
Paracel Islands
Spratly Islands
South China Sea
Philippine Sea
Luzon
Mt. Pinatubo
Subic Bay
Mt. Mayon
Visayas
Palawan
Negros
Mindanao
Celebes Sea
Mt. Kinabalu
Iran Mts.
Borneo (Kalimantan)
Strait of Malacca
Sumatra
Barisan Mountains
Mt. Kerinci
Sulawesi
Seram
Ambon
Java Sea
Sunda Strait
Java (Jawa)
Dieng Plateau
Madura
Mt. Semeru
Bali
Lombok
Sumbawa
Flores
Sumba
Timor
Lesser Sunda Islands
MAINLAND SOUTHEAST ASIA
INSULAR SOUTHEAST ASIA
INDO-AUSTRALIAN PLATE
EURASIAN PLATE
PHILIPPINE PLATE
15°N
0°
105°E
120°E

Southeast Asia (Figure 13.1) consists of 11 countries that vary widely in spatial extent, population, cultural attributes, and levels of economic and social development. Geographically, the region is commonly divided into an Asian mainland zone and the islands, or the insular realm. The mainland includes Burma (called *Myanmar* by the current government), Thailand, Cambodia, Laos, and Vietnam. Insular Southeast Asia includes the sizable countries of Indonesia, the Philippines, and Malaysia, as well as the small countries of Singapore, Brunei, and East Timor. Although classified as part of the insular realm because of its cultural and historical background, Malaysia actually splits the difference between mainland and islands. Part of its national territory is on the mainland's Malay Peninsula and part is on the large island of Borneo (also known as Kalimantan), some 300 miles (480 kilometers) distant. Borneo also includes the Muslim sultanate of Brunei, a small but oil-rich island country of 400,000 people covering an area slightly larger than Rhode Island. Singapore is essentially a city-state, occupying a small island just to the south of the Malay Peninsula.

Southeast Asia occupies an important place in discussions of globalization. It includes some of the world's most globally networked countries, such as Singapore, as well as some of the countries most resistant to worldwide economic and cultural forces, most notably Burma (Myanmar). Debates about the benefits and drawbacks of economic globalization have often focused on Southeast Asia. Although human well-being has increased across most of the region, the sweatshops associated with low-cost global production are criticized for their low wages and harsh working conditions. Owing to its high levels of economic integration, Southeast Asia is also highly vulnerable to fluctuations in the global economy.

Southeast Asia's involvement with the larger world is not new. Chinese and especially Indian influences date back many centuries. Later, commercial ties with the Middle East opened the doors to Islam, and today Indonesia is the world's most populous country with a Muslim majority. More recently came the heavy imprint of Western colonialism, as Britain, France, the Netherlands, and the United States administered large Southeast Asian colonies. During this period, national territories were rearranged, populations relocated, and new cities built to serve trade and military needs.

Southeast Asia's resources and its strategic location made it a major battlefield during World War II. Yet long after the end of this global conflict in 1945, war continued to be waged in the region. Resistance to imperialism mounted, and after the colonial regimes were replaced by newly independent countries, Southeast Asia became a zone of contention between the world powers and their different philosophical systems. In Vietnam, Laos, and Cambodia, communist forces, supported by China and the Soviet Union, struggled for control of local territory and people. Opposing this movement were the United States and its allies, concerned that communism would rapidly spread throughout Southeast Asia.

Ironically, while communist forces did prevail in Vietnam, Laos, and Cambodia, these countries subsequently opened their economies to global capitalism. With the end of the Cold War, competing political philosophies have taken a back seat to issues of economic integration. Geopolitically, the

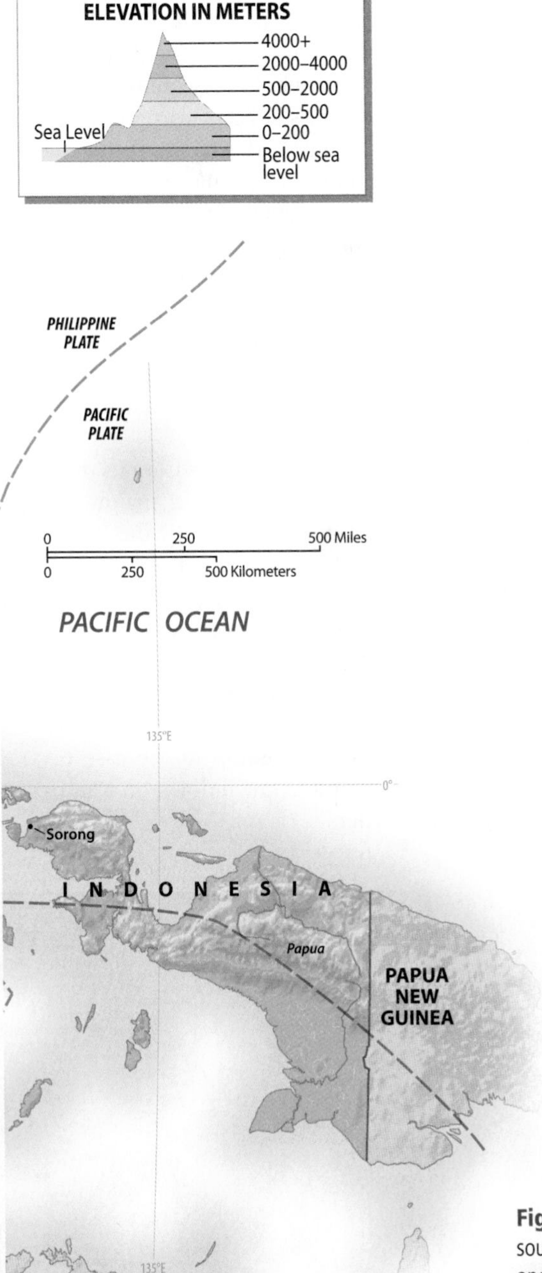

Figure 13.1 Southeast Asia This region includes the large peninsula in the southeastern corner of Asia as well as a vast number of islands scattered to the south and east. It is conventionally divided into two subregions: mainland Southeast Asia, which includes Burma (Myanmar), Thailand, Laos, Cambodia, and Vietnam, and insular (or island) Southeast Asia, which includes Indonesia, the Philippines, Malaysia, Brunei, Singapore, and East Timor. Malaysia includes the tip of the mainland peninsula and most of the northern part of the island of Borneo (Kalimantan).

Association of Southeast Asian Nations (ASEAN), which includes 10 of the region's 11 countries, has brought a new level of regional cooperation.

ENVIRONMENTAL GEOGRAPHY: A Once-Forested Region

The mountainous area along the border between northern Thailand and Burma (Myanmar) is a rugged place. Slopes are steep and thickly wooded; rainfall is often torrential; leeches abound; and strains of medicine-resistant malaria are prevalent. These uplands are home to one of the most distinctive Southeast Asian ethnic groups: the Karen, a people numbering roughly 7 million. Unfortunately, the story of the Karen and their homeland in eastern Burma (Myanmar) is not a happy one. The Burmese army has overrun much of their territory, forcing many of the Karen into squalid refugee camps in Thailand. Furthermore, the Karen's loss of land has been accompanied by the logging of the teak forests of upland Burma (Myanmar). In the saga of the Karen, we can see the tragic meshing of cultural, political, economic, and environmental forces.

The Tragedy of the Karen

The Karen, like other tribal peoples of the upland areas of Burma (Myanmar), were never fully incorporated into the Burmese kingdom, which long ruled the lowlands of the country. With the imposition of British colonial rule in the 1800s, however, the Karen territory became part of Burma. Protestant missionaries educated many Karen and converted over 30 percent of them to Christianity. A number of Karen Christians obtained positions in Burma's colonial government. The Burmans of the lowlands, a strongly Buddhist people, resented this deeply, for they had long viewed the Karen as culturally inferior (according to conventional but confusing terminology, the term *Burmese* refers to all of the inhabitants of Burma [Myanmar], whereas *Burmans* refers only to the country's dominant, Burmese-speaking ethnic group). After independence, the Karen lost their favored position and began to resent Burman cultural and economic domination.

The Karen were soon in open rebellion, building a semi-independent state of their own. They supported their struggle by smuggling goods between Thailand and Burma (Myanmar) and by mining the gemstones of their territory. Because Burma (Myanmar) had one of the world's most protected economies, with high tariffs and governmental controls inhibiting commerce, smuggling was an especially profitable occupation. Some estimates in the 1980s ranked the Karen economy as almost as large as the official Burmese economy.

The Burmese army, however, began to make headway against the rebels in the early 1990s, and by the end of the decade had overrun most of the Karen territory. Although some Karen continue to fight, as a whole they are demoralized and divided. Crucial to Burma's success was an agreement made with Thailand in which the Thai government agreed to prevent Karen soldiers from finding sanctuary on its side of the border, reportedly in exchange for access by Thai timber interests to Burma's teak forests. More than 200,000 Karen people were driven from their homes by the Burmese military between 1994 and 2004 (Figure 13.2). In early 2010, as the Burmese government geared up for national elections, Karen activists living abroad began to press the United Nations to conduct an inquiry on human rights abuses by Burma's armed forces.

Thailand's interests in Burma's forests stems from its own environmental problems. Thailand was once a major exporter of teak and other tropical hardwoods, but by the 1990s the country had been largely deforested. The Thai government responded by banning commercial logging. The only way Thai logging firms could stay in business was to move into the extensive forests of Burma (Myanmar), Laos, and Cambodia, countries that had previously been relatively isolated from the world economy, and therefore little touched by commercial logging. The price of admission in the case of Burma (Myanmar) was for Thailand to crack down on Karen guerrillas on its side of the border.

Figure 13.2 Karen Refugee Camp Many Karen refugees from Burma (Myanmar) still languish in refugee camps in Thailand. Shown here is Tham Hin camp, located 81 miles (130 km) west of Bangkok. The camp is very crowded, but it does have adequate health and education facilities.

The Deforestation of Southeast Asia

While the story of the Karen and Burma's teak forests is unique, deforestation and related environmental problems are major issues throughout most of Southeast Asia (Figure 13.3). Globalization has had a particularly profound effect on the Southeast Asian environment. Export-oriented logging companies have reached deep into the region's forests, cutting trees, damaging watersheds, and reducing biodiversity.

Although colonial powers cut Southeast Asian forests for tropical hardwoods and naval supplies, and indigenous peoples have long cleared small areas of forest for agricultural

Figure 13.3 Environmental Issues in Southeast Asia Southeast Asia was once one of the most heavily forested regions of the world. Most of the tropical forests of Thailand, the Philippines, peninsular Malaysia, Sumatra, and Java, however, have been destroyed by a combination of commercial logging and agricultural settlement. The forests of Borneo (Kalimantan), Burma (Myanmar), Laos, and Vietnam, moreover, are now being rapidly cleared. Water and urban air pollution, as well as soil erosion, are also widespread in Southeast Asia.

use, rampant deforestation came only in the late 1900s with large-scale international commercial logging (Figure 13.4). This activity has largely been driven by the developed world's appetite for wood products such as plywood and paper pulp. Increasingly, China is playing a major role as well. In 2008, it was estimated that over half of forestry exports from Burma and Indonesia went to China.

Although several Southeast Asian countries have been transforming forests into farmlands in order to increase food production and relieve population pressure in their more densely settled areas, agriculture and population growth are not the only cause of deforestation. Most forests are cut so that the wood products can be exported to other parts of the world. Subsequently, some of the logged-off lands are replanted (often with fast-growing, "weedy" tree species), while others are opened for agricultural settlement. In Malaysia and Indonesia, vast areas of former rain forest have been planted in African oil palms, which yield large quantities of edible oil.

Figure 13.4 Commercial Logging Southeast Asia has long been the world's most important supplier of tropical hardwoods. The logging process has destroyed most of the tropical forests of the Philippines and Thailand, as well as on the Indonesian islands of Java and Sumatra.

Local Patterns of Deforestation Malaysia has long been a leading exporter of tropical hardwoods. Most of the primary forests in western (or peninsular) have already been cut, which has resulted in an acceleration of logging in the states of Sarawak and Sabah on the island of Borneo. In these areas, the granting of forestry concessions to Malaysian and foreign firms has harmed local tribal people by disrupting their traditional resource base. Although Malaysia's conservation policies look good on paper, they are often not enforced. As a result, the World Bank estimates that logging is occurring at several times the sustainable rate.

Indonesia, the largest country in Southeast Asia, has fully two-thirds of the region's forest area, including about 10 percent of the world's true tropical rain forests. Its forest coverage, however, is deeply threatened; between 1990 and 2005, Indonesia lost an estimated 69 million acres (28 million hectares) of forest. Most of Sumatra's primary forests are gone, and those of Borneo (Kalimantan) are rapidly diminishing. Indonesia's last forestry frontier is on the island of New Guinea. But Indonesia, which is still covered in 119 million acres (48 million hectares) of primary forests, is fortunate in comparison to the Philippines, which has lost most of its original forests (see "Geographic Tools: Statistical Comparisons and the Assessment of Philippine Deforestation").

Thailand cut more than 50 percent of its forests between 1960 and 1980. Damage to the landscape was severe; flooding increased in lowland areas, and erosion on hillslopes led to the accumulation of silt in irrigation works and hydroelectric facilities As a result, a series of logging bans in the 1990s severely restricted commercial forestry. Many of Thailand's cutover lands are being reforested with fast-growing Australian eucalyptus trees, a nonnative species that cannot support local wildlife. As the forests of Thailand disappeared, mainland Southeast Asia's logging frontier is moved into Burma (Myanmar), Vietnam, Laos, and Cambodia. A 2005 report by the United Nations Food and Agriculture Organization found that Vietnam and Cambodia had the world's second and third fastest rates of primary deforestation, trailing only Nigeria.

In many parts of Southeast Asia, logged-over lands and other degraded areas are gradually returning to forest coverage. As a result, some experts think that Vietnam's total forest area is actually increasing. Indonesia and other Southeast Asian countries also support reforestation projects, which often try to recruit local people to nurture tree growth. The resulting secondary forests are not nearly as biologically rich as primary forests, but over time their biodiversity does increase.

Throughout the coastal areas of Southeast Asia, a more specific problem is the destruction of the mangrove forests that thrive in shallow and silty marine areas. Often mangrove forests are burned for charcoal, but many are converted to fish and shrimp ponds as well as rice fields and oil palm plantations. As mangrove forests serve as nurseries for many fish species, their destruction threatens to undercut a number of important Southeast Asian fisheries. Mangrove

GEOGRAPHIC TOOLS Statistical Comparisons and the Assessment of Philippine Deforestation

The Philippines is one of the most thoroughly deforested countries of Southeast Asia. Forest coverage, which was roughly 90 percent at the time of the Spanish conquest in the sixteenth century, is less than 25 percent today (Figure 13.1.1). The remaining forests, most experts agree, are rapidly retreating and may disappear entirely.

Discussions of Philippine deforestation often are supported by precise statistics showing the steady decline of forest area. Whether such statistics are accurate, however, is a different matter. Geographer David Kummer has conducted research on national-level deforestation data in the Philippines by examining how the data were derived, scrutinizing the methodologies used, and comparing the results. Kummer also traveled widely in the Philippines to confer with local experts and to inspect ongoing forestry practices. Through such work he demonstrates that we actually know surprisingly little about Philippine forest coverage. The numbers that are so confidently used by scholars, government agencies, and environmental activists generally turn out to be little better than educated guesswork.

Figure 13.1.1 Luzon, Philippines Part of northeastern Luzon in the Philippines, which appears dark green on this satellite image, is still thickly forested, but most of the rest of the island has experienced extensive deforestation.

There are several reasons for the poor quality of information on Philippine deforestation. Although we often assume that satellite images give accurate representations of land-use coverage and change, Kummer shows that this simply is not the case. Satellite images are expensive and often hard to obtain, but more importantly they can be difficult to interpret. Cloud coverage, image overlap, shadows in steep areas, and unclear classification categories all create uncertainty. As a result, ground surveys by foresters often result in more accurate data. In the Philippines, much of the survey work done on forests over the past half century has been lost or destroyed. Compounding the problem is the fact that the Philippine government seems to have misstated the extent of deforestation between 1950 and 1987 in order to encourage unregulated logging by politically well-connected firms.

A 2004 government press release indicates that "forest cover" has actually increased since the 1980s. However, the data on which this statement is based may not be reliable, and satellite images remain unverified by on-the-ground investigations. More troubling is the fact that this new survey employs land-cover categories different from those in previous work, making it impossible to evaluate the changes in the Philippines' primary forests—some of the most biologically diverse ecosystems in the world.

Taken as a whole, Kummer's available evidence indicates that primary tropical forests continue to decline as a result of both legal and illegal logging. Poor-quality secondary forests, however, seem to be increasing due to the abandonment of marginal farmland. Biologically impoverished "artificial forests" also are increasing as a result of both governmental reforestation efforts and the planting of fuelwood groves by small-scale farmers. Reforestation with a handful of nonnative, fast-growing tree species has now become a potential threat to biodiversity.

Kummer's work shows that, in poorer parts of the world and especially in countries with high levels of political corruption, official statistics may be questionable. The methods used for this type of statistical evaluation have potential application to other world regions, and to other environmental practices with global implications.

forests also protect coastal areas from storm surges associated with the tropical cyclones that often strike the region.

Protected Areas Despite rampant deforestation, Indonesia, like most other Southeast Asian countries, has created a number of large national parks and other protected areas. Kutai National Park in the province of East Kalimantan, for example, covers more than 741,000 acres (300,000 hectares). Southeast Asian rain forests are among the most biologically diverse areas on the planet, containing a large number of species that are found nowhere else. Conservation officials hope that protected areas will allow animals such as the orangutan, which now lives only in a small portion of northern Sumatra and a somewhat larger area of Borneo (Kalimantan), a chance to survive in the wild. Others are less optimistic, noting that Southeast Asian national parks often exist only on paper, receiving little real protection from loggers or immigrant farmers. Much of Kutai National Park, for example, has been logged and burned. According to some sources, over 80 percent of logging activities in Indonesia are illegal in some manner.

Fires, Smoke, and Air Pollution

Logging operations typically leave large quantities of wood (small trees, broken logs, roots, branches, and so on) on the ground. Exposed to the sun, this remaining "slash" becomes highly flammable. Indeed, it is often burned on purpose in order to clear the ground for agriculture or tree replanting. Commercial forest cutting is responsible for most burning, even though the large logging firms commonly blame small-scale farmers. Wildfires also are common in other Southeast Asian habitats. The most thoroughly deforested areas are often covered by rough grasses. Such grasslands often are purposely burned every year so that cattle can graze on the tender shoots that emerge after a fire passes through. Grassland fires, in turn, prevent forest regeneration, further undercutting the region's biodiversity. In drained wetland areas, organic peat soils often burn, causing extensive air pollution (Figure 13.5).

In the late 1990s, wildfires associated with both logging practices and a severe drought raged so intensely across much of insular Southeast Asia that the region suffered from disastrous air pollution. During that period, a commercial airliner crashed because of poor visibility; countless road accidents resulted; two ferries collided in smoke-laden conditions; and hundreds of thousands of people were admitted to hospitals with life-threatening respiratory problems. The smoke crisis of the late 1990s led several Southeast Asian countries to devote more attention to air quality. Deforestation continues, however, and the fire threat remains. In March 2010, NASA's remote-sensing Aqua satellite noted extremely smoky conditions over northern Thailand, Burma (Myanmar), Laos, and Vietnam.

Efforts to protect Southeast Asia's air quality, moreover, are also hampered by continuing industrial development along with a major increase in vehicular traffic. Southeast Asia's large metropolitan areas have extremely unhealthy levels of pollution, both of air and water. Several cities, including Bangkok and Manila, have built rail-based public transportation systems in order to reduce traffic and vehicular emissions. Bangkok in particular has substantially reduced its levels of ozone and sulfur dioxide, but particulate matter remains at unhealthy levels.

Patterns of Physical Geography

To understand why forestry issues are so important in Southeast Asia, it is necessary to examine its physical geography. The southern portion of the region—insular Southeast Asia—is one of the world's three main zones of tropical rain forest. The northern part of the region—mainland Southeast Asia—is located in the tropical wet-and-dry zone that is noted for particularly valuable timber species, such as teak. Distinguishing between the mainland and the islands is thus one of the keys for understanding the geography of Southeast Asia.

Mainland Environments Mainland Southeast Asia is an area of rugged uplands interspersed with broad lowlands and deltas associated with large rivers. The region's northern boundary lies in a cluster of mountains connected to the highlands of eastern Tibet and south-central China. In the far north of Burma (Myanmar), peaks reach 18,000 feet (5,500 meters). From this point, a series of mountain ranges

Figure 13.5 Burning Peatlands The soil of most wetland areas in Southeast Asia is composed largely of peat, an organic substance that can burn when dry. Draining for agricultural expansion, as well as drought, often results in extensive peat fires. In this photograph, a C130 airplane is dropping water on burning peat in Lampung, Sumatra.

Figure 13.6 Delta Landscape The Mekong Delta in southern Vietnam encompasses a complex maze of waterways. Extensive tracts of fertile farmland can be found between the canals and river channels. The waterways are used for transportation, and provide large quantities of fish and other aquatic resources. This photo shows typical Vietnamese boats in a delta backwater.

radiates out, extending through western Burma (Myanmar), along the Burma–Thailand border, and through Laos into southern Vietnam.

Several large rivers flow southward out of Tibet into mainland Southeast Asia. The longest is the Mekong, which flows through Laos and Thailand, then across Cambodia before entering the South China Sea through an extensive delta in southern Vietnam (Figure 13.6). Second longest is the Irrawaddy, which flows through Burma's central plain before reaching the Bay of Bengal. This river also has a large delta. Two smaller rivers are equally significant: the Red River, which forms a sizable and heavily settled delta in northern Vietnam, and the Chao Phraya, which has created the fertile alluvial plain of central Thailand.

The centermost area of mainland Southeast Asia is Thailand's Khorat Plateau, which is neither a rugged upland nor a fertile river valley. This low, sandstone plateau averages about 500 feet (175 meters) in height and is noted for its thin soils. Water shortages and periodic droughts also plague the area.

Monsoon Climates Mainland Southeast Asia is affected by the seasonally shifting winds known as the monsoon. The climate is characterized by a distinct hot and rainy season from May to October (Figure 13.7). This is followed by dry but still generally warm conditions from November to April. Only the central highlands of Vietnam and a few coastal areas receive significant rainfall during this period. In the far north, the winter months bring mild and sometimes rather cool weather.

Two tropical climate regions are found in mainland Southeast Asia. While both are affected by the monsoon, they differ in the total amount of precipitation. Along the coasts and in the highlands, the tropical monsoon (Am) climate dominates. Rainfall totals for this climate usually register more than 100 inches (254 centimeters) each year. The greater portion of mainland Southeast Asia falls into the tropical savanna (Aw) climate type. Here annual rainfall totals are about half as much. In most cases, this can be explained by interior locations sheltered from the oceanic source of moisture by mountain ranges. A good portion of Thailand receives less than 50 inches (127 centimeters) of annual rainfall. Much of Burma's central Irrawaddy Valley is almost semiarid, with rainfall totals below 30 inches (76 centimeters). Forests in these areas are especially vulnerable, being easily converted by fire and agriculture into rough landscapes of brush and grass.

Insular Environments The signal feature of insular Southeast Asia is its island environment. Indonesia alone is said to contain more than 13,000 islands, while the Philippines supposedly contains 7,000. Borneo (Kalimantan) and Sumatra are the third and sixth largest islands in the world, respectively, while many thousands of others are little more than specks of land rising at low tide from a shallow sea.

Indonesia is dominated by the four great landmasses of Sumatra, Borneo, Java, and the oddly shaped Sulawesi. This island nation also includes the western half of New Guinea and the Lesser Sunda Islands, which extend to the east of Java. A prominent mountain spine runs through these islands as a result of tectonic forces. The two largest and most important islands of the Philippines are Luzon (about the size of Ohio) in the north and Mindanao (the size of South Carolina) in the south. Sandwiched between them are the Visayan Islands, which number roughly a dozen.

Closely related to this impressive collection of islands is the world's largest expanse of shallow seas. These waters cover the **Sunda Shelf**, which is an extension of the continental shelf stretching from the mainland through the Java Sea between Java and Borneo. Here waters are generally less than 200 feet (70 meters) deep. Some local peoples have adopted lifestyles that rely on the rich marine life of this region, essentially living on their boats and setting foot on land only as necessary.

Insular Southeast Asia is less geologically stable than the mainland. Four of Earth's tectonic plates converge here: the Pacific, the Philippine, the Indo-Australian, and the Eurasian. As a result of this tectonic structure, earthquakes occur frequently. Large, often explosive volcanoes are another consequent feature of the insular Southeast Asian landscape. A string of active volcanoes extends the length of eastern Sumatra across Java and into the Lesser Sunda Islands. Volcanic eruptions and earthquakes occasionally result in **tsunamis**, which can devastate coastal regions. A massive earthquake in northern Sumatra on December 26, 2004, for example, caused roughly 100,000 deaths in Indonesia alone. Other geological hazards also confront the region. In late 2006,

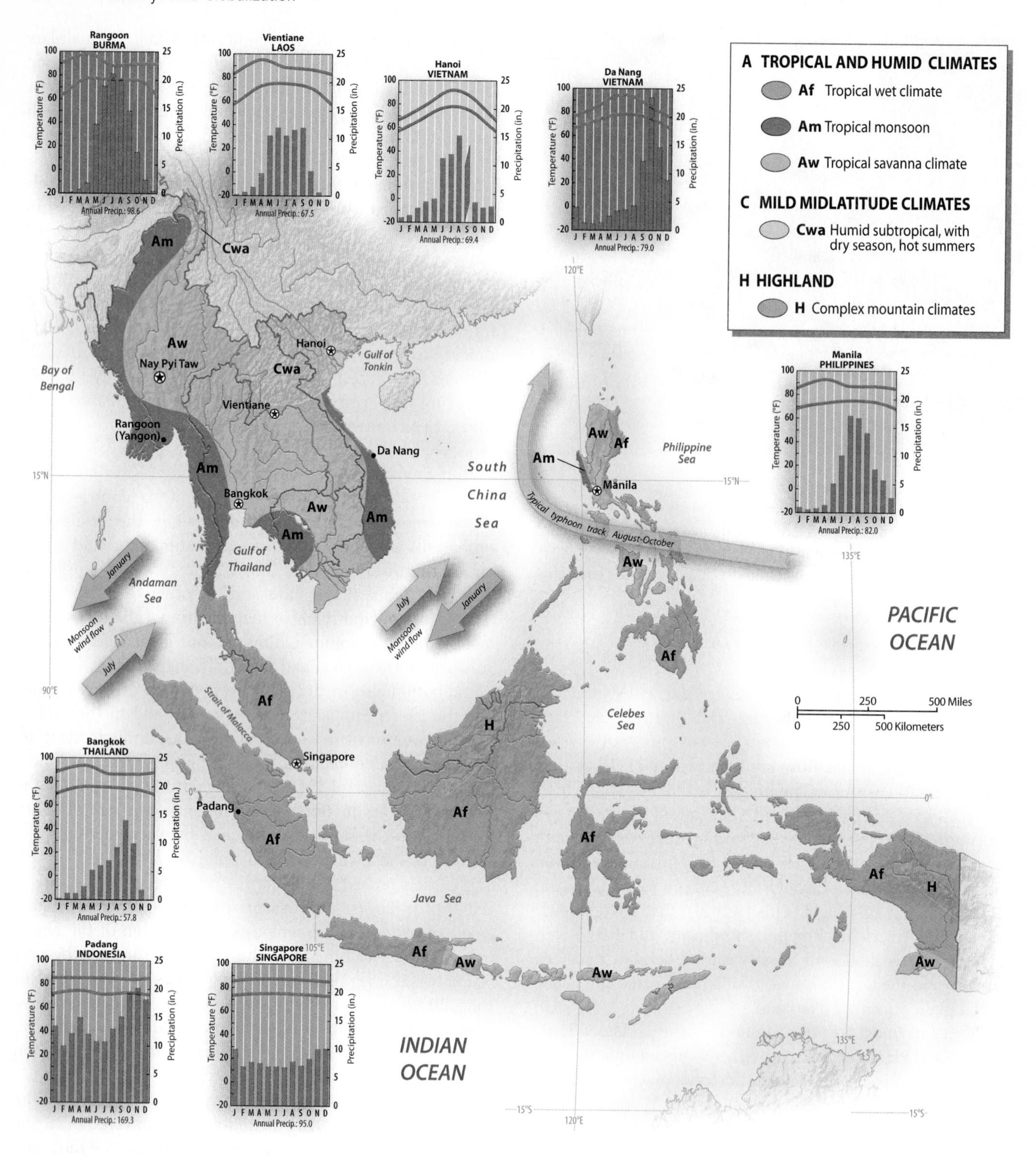

Figure 13.7 Climate Map of Southeast Asia Most of insular Southeast Asia is characterized by the constantly hot and humid climates of the equatorial zone. Mainland Southeast Asia, on the other hand, has the seasonally wet and dry climates of the tropical monsoon and tropical savanna types. Only in the far north are subtropical climates, with relatively cool winters, encountered. The northern half of the region is strongly influenced by the seasonally shifting monsoon winds. Northeastern Southeast Asia—and especially the Philippines—often experiences typhoons from August to October.

13,000 Indonesians had to be evacuated away from a new mud volcano near the city of Surabaya. Roughly 2,500 cubic meters of hot and toxic mud oozes out of the ground everyday, and it is expected to continue flowing for the next 30 years (Figure 13.8).

Island Climates The climates of insular Southeast Asia are more varied than those of the mainland. Most of insular Southeast Asia, unlike the mainland, receives rain during the Northern Hemisphere's winter because the monsoon winds of this season cross large areas of warm equatorial ocean, absorbing moisture. On Sumatra and Java, where north winds blow between November and March, heavy rains occur on the northern side of these east–west-running islands. But during the May–September period, the heaviest rains are found on the southern flanks of these same islands because of the southwesterly winds associated with the Asian summer monsoon.

The climates of Indonesia, Singapore, Malaysia, and Brunei are heavily influenced by their equatorial location, which reduces seasonality. Temperatures remain elevated throughout the year, with little variation. The average high temperature in Singapore, for example, varies between 85°F (29.5°C) in December to 89°F (31.7°C) in April. The equatorial influence also brings high and evenly distributed rainfall to much of the insular realm. As a result, large areas of island Southeast Asia are placed into the tropical rain forest (Af) climate category (see Figure 13.7). The southeastern islands of Indonesia, however, experience a distinct dry season from May to October.

The southern part of the Philippines also has an equatorial climate, but most the country experiences a distinct dry period from December to April. The northern and central Philippines are frequently hit by tropical cyclones, or **typhoons**, especially from August to October. Each year several typhoons strike the Philippines with heavy damage through flooding and landslides. In September 2009, most of central Luzon was declared a "calamity area" after being battered by Typhoon Ketsana, which took almost 500 lives and caused damages of more than $1 billion.

Although typhoons are most pronounced in the Philippines, they do hit mainland Southeast Asia as well. In early May 2008, southern Burma (Myanmar) was slammed by Cyclone Nargis, the worst storm in recent Southeast Asian history. Nargis caused over 150,000 deaths, left as many as 1 million people homeless, and resulted in roughly $10 billion in damages (Figure 13.9). Casualty rates were particularly high because Burma's military government reacted slowly to the crisis, restricting the activities of international aid agencies.

Global Warming in Southeast Asia

Most of Southeast Asia's people live in coastal and delta environments, making the region particularly vulnerable to the rise in sea level associated with global warming. Periodic flooding is already a major problem in many of the regions' low-lying cities. Southeast Asian farmland is also concentrated in delta environments, and thus could suffer from saltwater intrusion and heightened storm surges. A 2009 study by the Asia Development Bank found that Indonesia, Thailand, Vietnam, and the Philippines could suffer climate-change damages of more than twice the global average by the end of the century. Some studies claim that the region's rice harvest could decline by as much as 50 percent by 2100. Sea level rise, moreover, could inundate as many as 2,000 islands in Indonesia alone by the middle of the century.

Figure 13.8 Mud Volcano in Indonesia Villages and farmland in east Java have been devastated by a mud volcano that continues to pour out hot, toxic sludge. The volcano itself was set off when a gas exploration drilling project tapped into a vent of extremely hot mud.

Figure 13.9 Cyclone Nargis Cyclone Nargis slammed into southern Burma (Myanmar) in early May 2008, resulting in the country's worst natural disaster in recorded history. This photograph shows devastated families waiting for relief near their destroyed homes on Haing Guy Island in southwestern Burma.

Changes in precipitation across Southeast Asia brought about by global warming remain highly uncertain. Many experts foresee an intensification of the monsoon pattern, which could bring increased rainfall to much of the mainland. While enhanced precipitation would likely result in more destructive floods, it could bring some agricultural benefits to dry areas such as Burma's central Irrawaddy Valley. Complicating this scenario, however, is the prediction that climate change could intensify the El Niño effect, which would result in more extreme droughts. Some observers think that the prolonged drought that Indonesia experienced in the late 1990s was a harbinger of future conditions.

All Southeast Asian countries have ratified the 1997 Kyoto Accord. But since they are officially classified as developing countries—even wealthy Singapore—none are obligated to reduce their emissions of greenhouse gases. Still, Southeast Asia's overall emissions from conventional sources remain low by global standards. Indonesia, Malaysia, and Vietnam, however, are planning to increase their reliance on coal-fired electricity generation, which would greatly increase the region's emissions. In Thailand, local and international opposition has already led the government to reconsider its coal-based expansion plans. One possible alternative is hydropower, which has significant potential especially in mountainous Laos. But dam building results in its own ecological problems, and is opposed by most environmental groups.

When greenhouse gas emissions associated with deforestation are factored in, Southeast Asia's role in global climate change becomes much larger than it appears at first glance. By some estimates, Indonesia is the world's third largest contributor to the problem, following only China and the United States. Vast quantities of carbon dioxide are released into the atmosphere when slash is burned, a routine practice in many areas. A greater problem, however, is the release of carbon from peat, the partially decayed organic matter that accumulates in perennially saturated soils. The wetlands of coastal Indonesia hold 60 percent of the world's tropical peat, containing roughly 50 billion tons of carbon. During drought periods, peat soils sometimes burn, releasing the stored carbon. More worrisome is the fact that both Indonesia and Malaysia are actively draining coastal wetlands to make room for agricultural expansion, a process that results in the gradual oxidation of the peat. Already some 40 percent of Southeast Asia's peatlands have been lost.

POPULATION AND SETTLEMENT: Densely Settled Lowlands amid Sparsely Settled Uplands

With fewer than 600 million inhabitants, Southeast Asia is not heavily populated by Asian standards. One of the reasons for this relatively low density is the extensive tracts of rugged mountains, which generally remain thinly inhabited. In contrast, relatively dense populations are found in the region's deltas, coastal areas, and zones of fertile volcanic soil (Figure 13.10). Many of the favored lowlands of Southeast Asia have experienced striking population growth over the past half century. Demographic growth and family planning have thus become increasingly important concerns through much of the region. Several countries, especially Singapore and Thailand, have seen rapid reductions in birthrates, but large families are still common in Cambodia and Laos.

Settlement and Agriculture

Much of Insular Southeast Asia has infertile soil, which is unable to support intensive agriculture and high rural population densities. The island rain forests, though lush and biologically rich, typically grow on a poor base. Plant nutrients are locked up in the vegetation itself, rather than being stored in the soil where they would easily benefit agricul-

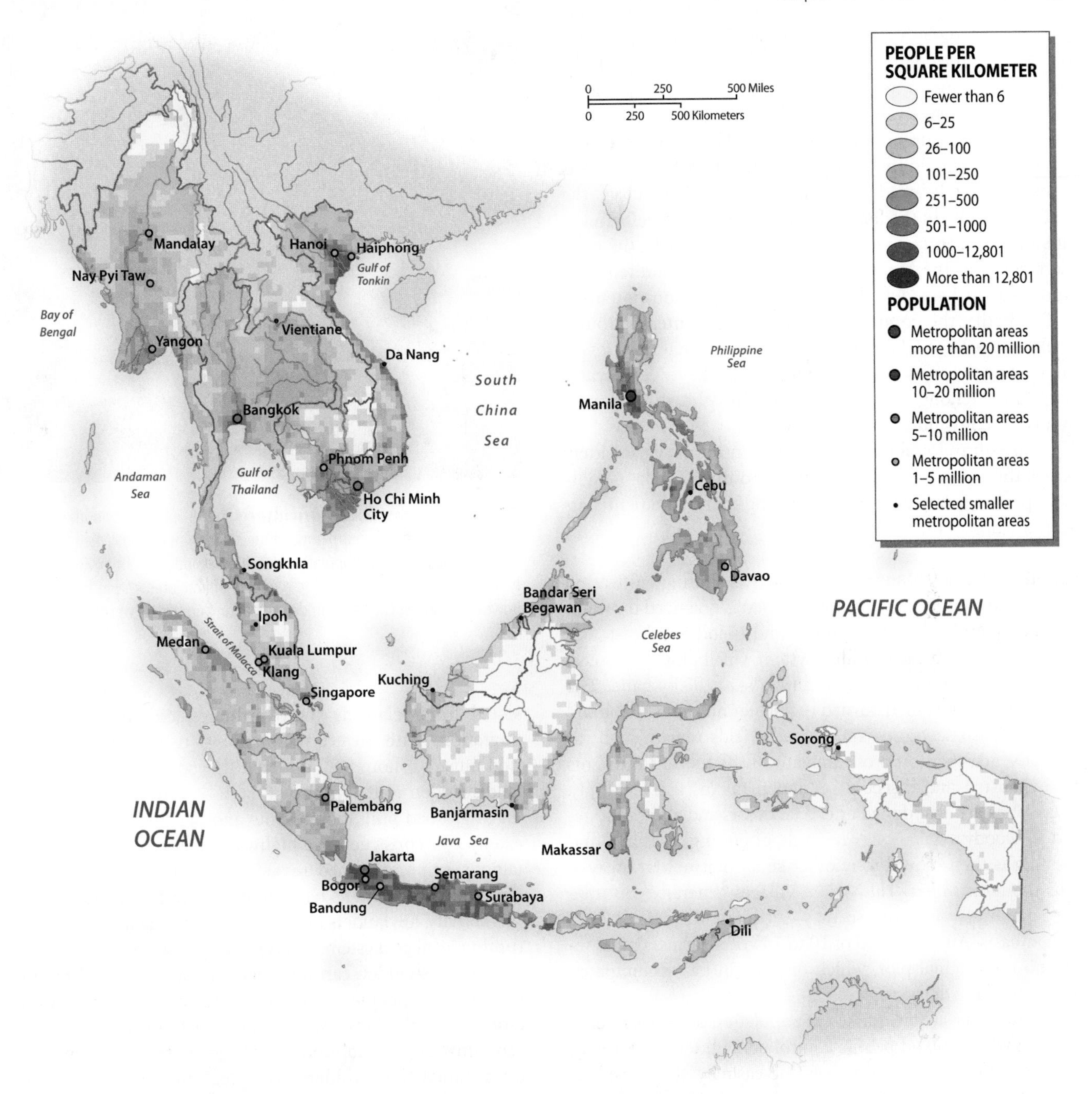

Figure 13.10 Population Map of Southeast Asia In mainland Southeast Asia, population is concentrated in the valleys and deltas of the region's large rivers. In the intervening uplands, population density remains relatively low. In Indonesia, density is extremely high on Java, an island noted for its fertile soil and large cities. Some of Indonesia's outer islands, especially those of the east, remain lightly settled. Overall, population density is high in the Philippines, especially in central Luzon.

ture. The incessant rain of the equatorial zone also tends to wash nutrients out of the soil. Agriculture must be carefully adapted to this limited fertility by constant field rotation or the application of heavy amounts of fertilizer.

Some notable exceptions can be found to this generalization about soil fertility and settlement density in equatorial Southeast Asia. Unusually rich soils connected to volcanic activity are scattered through much of the region, particularly on the island of Java. Java, with more than 50 volcanoes, is blessed with highly productive agriculture that supports a large array of tropical crops and a very high population density. More than 124 million people live on the island—roughly half the total population of Indonesia—in an area smaller than the state of Iowa. Dense populations also are found in pockets of fertile alluvial soils along the coasts of insular Southeast Asia, where people have traditionally supplemented land-based farming with fishing and other commercial activities.

The demographic patterns in mainland Southeast Asia are less complicated than those of the island realm. In all the mainland countries, population is concentrated in the agriculturally intensive valleys and deltas of the large rivers (see Figure 13.10). The population core of Thailand, for example, clusters around the Chao Phraya River, just as Burma's is focused on the Irrawaddy. Vietnam has two distinct foci: the Red River Delta in the far north and the Mekong Delta in the far south. In contrast to these densely settled areas, the middle reaches of the Mekong River provide only limited lowland areas in Laos, which reduces its population potential. In Cambodia, the population historically is centered around Tonle Sap, a large lake with an unusual seasonal flow reversal. During the rainy summer months the lake receives water from the Mekong drainage, but during the drier winter months it contributes to the river's flow.

Agricultural practices and settlement forms vary widely across the complex environments of Southeast Asia. Generally speaking, however, three farming and settlement patterns are apparent.

Figure 13.11 Swidden Agriculture In the uplands of Southeast Asia, swidden (or slash-and-burn) agriculture is widely practiced. When done by tribal peoples with low population densities, swidden is not environmentally harmful. When practiced by large numbers of immigrants from the lowlands, however, swidden can result in deforestation and extensive soil erosion.

Swidden in the Uplands Also known as shifting cultivation or "slash-and-burn" agriculture, swidden is practiced throughout the uplands of both mainland and island Southeast Asia (Figure 13.11). In the **swidden** system, small plots of tropical forest or brush are periodically cut or "slashed" by hand. The fallen vegetation is then burned to transfer nutrients to the soil before subsistence crops are planted. Yields remain high for several years, then drop off dramatically as the soil nutrients are exhausted and insect pests and plant diseases multiply. These plots are abandoned after a few years and allowed to revert to woody vegetation. The cycle of cutting, burning, and planting is then moved to another small plot not far away—thus the term *shifting cultivation*. Villages generally control a large amount of territory so they can rotate their fields on a regular basis. After a period of 10 to 75 years, farmers must return to the original plot, which once again has accumulated nutrients in the dense vegetation.

Swidden is sustainable when population densities remain low and when upland people control enough territory. Today, however, the swidden system is increasingly threatened. It cannot easily support the increasing population that has resulted from relatively high birthrates and, in some cases, migration. With greater population density, the rotation period must be shortened, undercutting soil resources. The upland swidden system is also often undermined by commercial logging. Road building presents another threat. Vietnam, for example, is constructing a highway system through its mountainous spine, designed to aid economic development and more fully integrate its national economy. Partly as a result of this activity, lowlanders are streaming into the mountains, disrupting local ecosystems and indigenous societies.

When swidden can no longer support the local population, upland people sometimes adapt by switching to cash crops that will allow them to participate in the commercial economy. In the mountains of northern Southeast Asia, often called the "**Golden Triangle**," one of the main cash crops is opium, grown by local farmers for the global drug

trade. Burma (Myanmar) is the world's second largest opium producer, after Afghanistan. Recent drug eradication programs proved relatively successful, reducing the area of land devoted to opium production by 85 percent between 1998 and 2006. Some experts think, however, that opium growing began expanding again in 2007. Methamphetamine manufacturing has also emerged as a major problem in much of the Golden Triangle.

Plantation Agriculture With European colonization, Southeast Asia became a center of plantation agriculture, producing high-value specialty crops ranging from rice to rubber. Even in the 19th century, Southeast Asia was closely linked to a globalized economy through the plantation system. Forests were cleared and swamps drained to make room for these commercial farms; labor was supplied, often unwillingly, by indigenous people or by contract laborers brought in from India or China.

Plantations are still an important part of Southeast Asia's geography and economy. Most of the world's natural rubber, for example, is grown in Malaysia, Indonesia, and Thailand, while sugarcane has long been a major plantation crop in the Philippines and Indonesia. More recently, pineapple plantations have spread in both the Philippines and Thailand, which are now the world's leading exporters. Indonesia is the region's leading producer of tea (Figure 13.12). Malaysia has traditionally dominated the production of palm oil, but it has recently been surpassed by Indonesia. Coconut oil and **copra** (dried coconut meat) are widely produced in the Philippines, Indonesia, and elsewhere. In the early 2000s, Vietnam rapidly emerged as the world's second-largest coffee producer, its large harvests reducing global coffee prices.

Rice in the Lowlands The lowland basins and deltas of mainland Southeast Asia are largely devoted to intensive rice cultivation. Through most of Southeast Asia, rice is the preferred staple food. Traditionally, rice was mainly cultivated on a subsistence basis by rural farmers. But as the number of wage laborers in Southeast Asia has grown as a result of economic development, so has the demand for commercial rice cultivation. Thailand is currently the world's largest rice exporter, with Vietnam occupying the number two position. In 2008, Thailand spearheaded the creation of OREC, the Organization of Rice Exporting Nations, which also includes Cambodia, Laos, Burma (Myanmar), and Vietnam. Across most of Southeast Asia, the use of agricultural chemicals and high-yield crop varieties, along with improved water control, have allowed rice production to keep pace with population growth. In 2010, however, an El Niño-linked drought reduced rice yields across most of the region.

In those areas without irrigation, yields remain relatively low. Rice growing on Thailand's Khorat Plateau, for example, depends largely on the uncertain rainfall, without the benefits of the more sophisticated water control methods. In some lowland districts lacking irrigation, dry-field crops, especially sweet potatoes and manioc, form the staple foods of people too poor to buy market rice on a regular basis. In some of the poorest parts of Southeast Asia, population growth has combined with economic stagnation to force larger numbers of people into such meager diets. Elsewhere, economic growth and declining birthrates have significantly reduced the burden of poverty.

Recent Demographic Change

Most Southeast Asian countries have seen a sharp decrease in birthrates over the past several decades. Because the region is not facing the same kind of population pressure as East or South Asia, a wide range of government population policies can be found. In countries with highly uneven population distribution, internal relocation away from densely populated areas to outlying districts is a common response.

Population Contrasts The Philippines, the second most populous country in Southeast Asia, has a relatively high growth rate (Table 13.1). When a popular democratic

Figure 13.12 Tea Harvesting in Indonesia Plantation crops, such as tea, are major sources of exports for several Southeast Asian countries. Coconut, rubber, oil palms, and coffee are other major cash crops. Many of these crops require large amounts of labor, particularly at harvest time.

TABLE 13.1 Population Indicators

Country	Population (millions) 2010	Population Density (per square kilometer)	Rate of Natural Increase (RNI)	Total Fertility Rate	Percent Urban	Percent <15	Percent <65	Net Migration (Rate per 1,000) 2005–10)[a]
Burma (Myanmar)	53.4	79	0.9	2.4	31	27	3	−2.0
Brunei	0.4	66	1.3	1.7	72	27	3	1.8
Cambodia	15.1	83	1.6	3.3	20	35	3	−0.1
East Timor	1.2	77	3.1	5.7	22	45	3	1.8
Indonesia	235.5	124	1.4	2.4	43	28	6	−0.6
Laos	6.4	27	2.1	3.5	27	39	4	−2.4
Malaysia	28.9	87	1.6	2.6	63	32	5	−1.0
Philippines	94.0	313	2.1	3.2	63	33	4	−2.0
Singapore	5.1	7,526	0.6	1.2	100	18	9	22.0
Thailand	68.1	133	0.6	1.8	31	22	7	0.9
Vietnam	88.9	268	1.2	2.1	28	25	8	−0.5

[a]*Net Migration Rate from the United Nations, Population Division, World Population Prospects: The 2008 Revision Population Database.*
Sources: Population Reference Bureau, World Population Data Sheet, 2010.

government replaced a dictatorship in the 1980s, the Philippine Roman Catholic Church pressured the new government to cut funding for family planning programs. As a result, many clinics that had dispensed family planning information were closed. Although high birthrates are not generally associated with Catholicism, the Church's outspoken stand on birth control seems to inhibit the dispersal of family planning information. Still, fertility rates have been declining. In 2010, moreover, the Philippine Secretary of Education came out in favor of teaching family planning in public schools.

Laos and Cambodia, countries of Buddhist religious tradition, also have relatively high fertility rates. Here the elevated birthrate is best explained by the countries' low level of economic and social development. Thailand, which shares cultural traditions with Laos yet is considerably more developed, demonstrates the other end of the spectrum. Here the TFR has dropped from 5.4 in 1970 to 1.8 currently, a figure that will soon bring population stability and then decline. But while economic growth may explain some of this decrease, it is also important to note that the Thai government has promoted family planning for both population and health reasons, including the relatively high incidence of AIDS in the country.

The city-state of Singapore stands out on the demographic charts with its particularly low fertility rate. Unless the deficit is offset by immigration or a dramatic turnabout in the birthrate, Singapore's population will soon begin to decline. The government is concerned about this situation and is actively promoting marriage and childbearing, particularly among the most highly educated segment of its population. As a result, Singaporeans can receive more than $10,000 in direct government subsidies for having a third or fourth child (Figure 13.13). Singapore also encouraged the immigration of highly skilled workers.

Figure 13.13 Pro-Reproduction Advertisement in Singapore
Due to its extremely low birthrate, the population of Singapore will probably begin to decline soon. Shown here are perfumes created by local students and launched in a government-backed "romance" campaign.

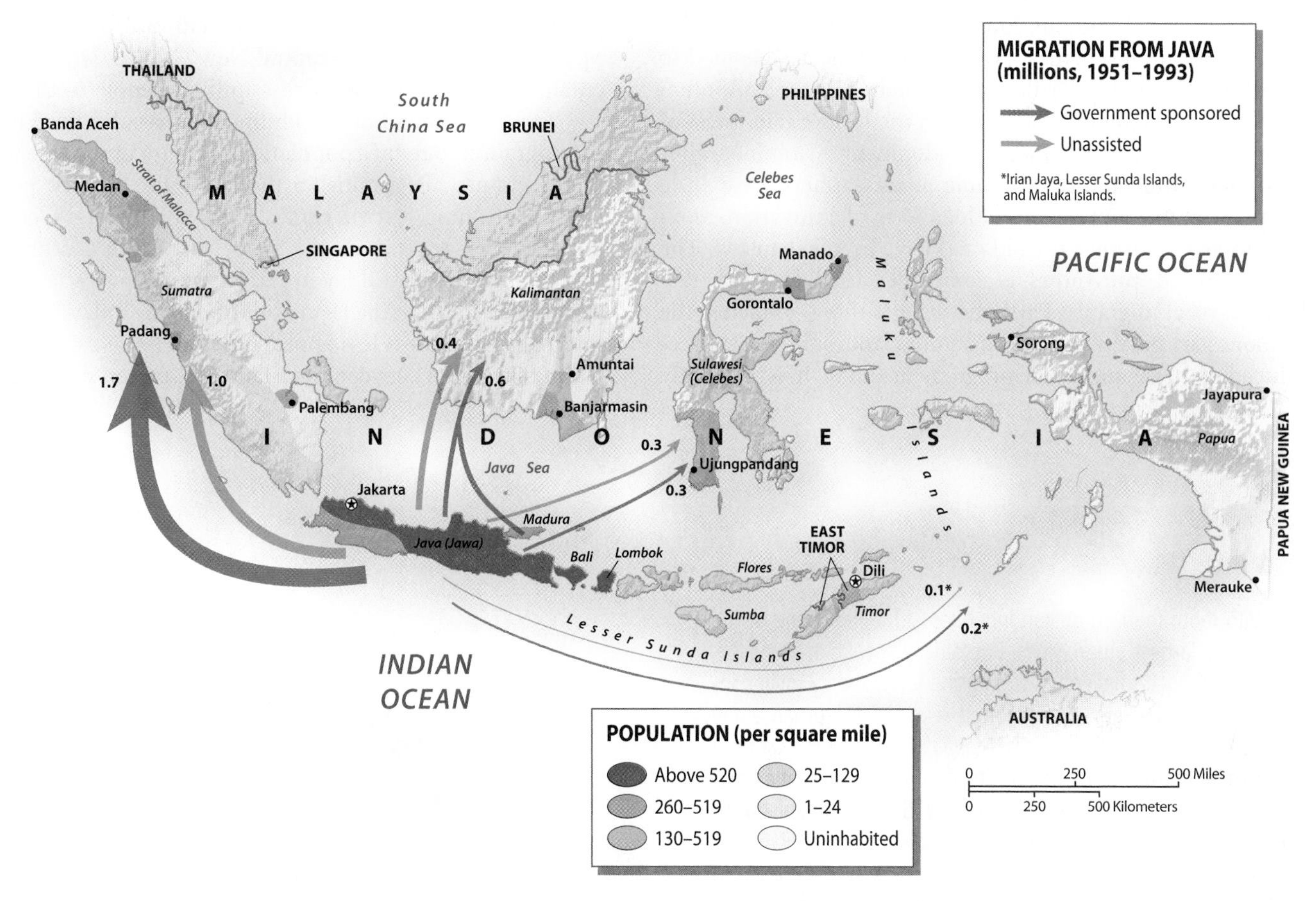

Figure 13.14 Indonesian Transmigration The distribution of population in Indonesia shows a marked imbalance: Java, along with the neighboring island of Madura, is one of the world's most densely settled places, whereas most of the country's other islands remain rather lightly populated. As a result, the Indonesian government has encouraged the resettlement of Javanese and Madurese people to the outer islands, often paying the costs of relocation. This transmigration scheme has resulted in a somewhat more balanced population distribution pattern, but has also caused substantial environmental degradation and intensified several ethnic conflicts.

Indonesia, with the region's largest population at 235 million, also has seen a dramatic decline in fertility in recent decades, with its fertility rate approaching the replacement level. As with Thailand, this drop in fertility seems to have resulted from a strong government family planning effort, coupled with improvements in education.

Growth and Migration Until recently, Indonesia had an official policy of **transmigration**, with the government helping people move from densely populated to lightly populated parts of the country (Figure 13.14). Primarily because of migration from Java and Madura, the population of the outer islands of Indonesia has grown rapidly since the 1970s. The province of East Kalimantan, for example, experienced an astronomical growth rate of 30 percent per year during the last two decades of the 1900s. As a result of this shift in population, many parts of Indonesia outside Java now have moderately high population densities, although many of the more remote districts remain lightly settled (Figure 13.15).

Figure 13.15 Migrant Settlement in Indonesia This sprawling transmigration site in western New Guinea is populated mostly by migrants from Java. Created in the early 1990s, it covers a large area that formerly supported tropical rain forest vegetation. The indigenous inhabitants of the region worry that such schemes will undermine their land rights and traditional subsistence patterns.

High social and environmental costs often accompany these relocation schemes. Javanese peasants, accustomed to working the highly fertile soils of their home island, often fail in their attempts to grow rice in the former rain forest of Borneo (Kalimantan). Field abandonment is common after repeated crop failures. In some areas, farmers have little choice but to adopt a semiswidden form of cultivation, moving to new sites once the old ones have been exhausted. The term **shifted cultivators** is sometimes used for such displaced rural migrants. Partly because of these problems, the Indonesian government restructured and substantially reduced its transmigration program in 2000. It is currently planning, however, a huge new agricultural project in the Merauke region of south-central New Guinea, which could entail the transfer of over half a million people to the area.

The government of the Philippines has also used internal migration to reduce population pressure in the core areas of the country. Beginning in the late 19th century, people began streaming out of central and northwestern central Luzon for the frontier zones. In the early 20th century, this settlement frontier still lay in Luzon; by the postwar years it had moved south to the island of Mindanao. Today, however, internal migration is less appealing, both because the population disparities between the islands have been reduced and

CITYSCAPES Metro Manila, Primate City of the Philippines

With more than 10 million inhabitants, Manila is one of the world's largest urban areas. But, strictly speaking, Manila itself is a much smaller city, supporting fewer than 2 million inhabitants. The term *Manila* often refers to the urban entity more properly called *Metro Manila* (Figure 13.2.1). Metro Manila, consisting of Manila itself plus 16 other cities and towns, was formally constituted in 1975 as an administrative area, the Philippines' National Capital Region (NCR). Most forms of political power, however, remain decentralized among Metro Manila's constituent units, complicating policy initiatives and planning efforts. When a new mayor of Manila decided to crack down on prostitution in the 1990s, for example, most of the city's "go-go bars" simply relocated to Pasay City.

Metro Manila is economically as well as politically decentralized. Manila itself includes the old central business district, now somewhat decayed, as well as sizable slums, a congested "Chinatown" (Binondo) famous for its discount shopping, and several older, wealthy neighborhoods. It also contains the historic zone of Intramuros, the original fortified city of the Spanish colonial period (Figure 13.2.2). Although Manila was almost completely destroyed during World War II, many of Intramuros's buildings have been rebuilt, lending the area charm and historical significance.

Metro Manila's—and the Philippines'—commercial capital now is located in Makati, southeast of the old urban core. From a rural fringe zone in the 1940s, Makati has grown into a mature city of high-rise office buildings and hotels, shopping plazas, and gated communities that

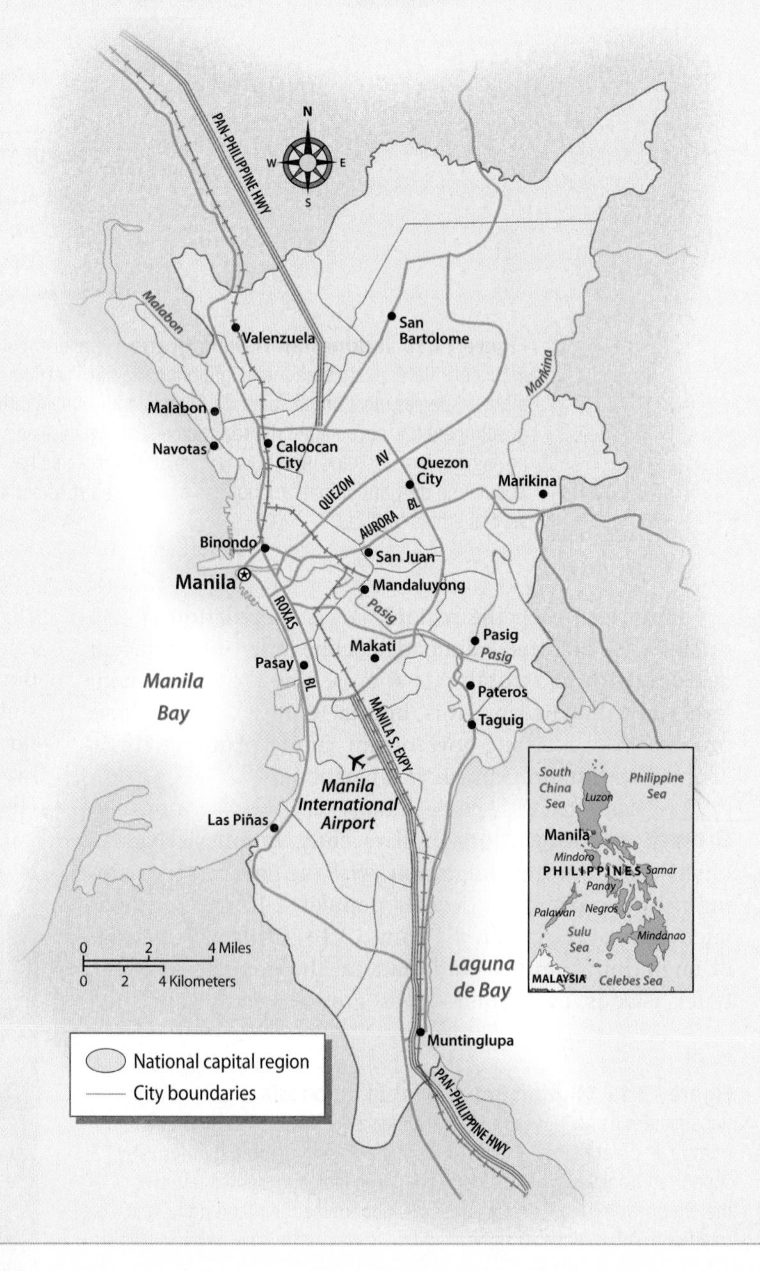

Figure 13.2.1 Metro Manila The Philippines' National Capital Region (Metro Manila) consists of the city of Manila and a number of neighboring cities and suburbs, which together form a single metropolitan area. Such municipalities as Makati and Quezon City are of significance equal to Manila proper.

because of political instability in the south. The emphasis is now on international migration, resulting in a Filipino overseas population of roughly 8 million.

Population issues and policies thus vary greatly across Southeast Asia. Virtually every part of the region, however, has seen a rapid expansion of its urban population in recent decades.

Urban Settlement

Southeast Asia is a not heavily urbanized region. Even Thailand retains a rural flavor, which is somewhat unusual for a country that has experienced so much industrialization. Currently, however, the region's urban population is growing at 1.75 times the world average. As a result, Southeast Asia's urbanization rate is expected to increase from 45 percent in 2008 to 56.6 percent in 2030.

Several Southeast Asian countries have **primate cities**, single, large urban settlements that overshadow all others. Thailand's urban system, for example, is dominated by Bangkok, just as Manila far surpasses all other cities in the Philippines (see "Cityscapes: Metro Manila, Primate City of the Philippines"). Both have recently grown into megacities of more than 10 million residents. More than half of all city-dwellers in

Figure 13.2.2 Intramuros Intramuros, the "city between the walls," was the original, fortified capital of the Spanish Philippines. It now is an important historic district.

house the country's elite. More recently, Metro Manila has seen the rise of other areas that serve the wealthy, most notably Ortigas Center in neighboring Mandaluyong City. Particularly successful here is the SM Megamall, which opened in 1991. This gargantuan six-story enclosed mall contains 12 cinemas and an ice-skating rink, in addition to hundreds of stores and restaurants. SM Megamall attracts tens of thousands of visitors daily, some of whom come simply to experience the modern ambiance and enjoy the air conditioning.

Official government buildings of the Philippines are concentrated in Quezon City in northeast Metro Manila. Quezon City replaced Manila as the national capital between 1948 and 1976. This sprawling city of widely separated office buildings is also an educational center, supporting the country's two most highly regarded universities, the publicly run University of the Philippines (UP) and the Jesuit-run Ateneo de Manila University.

One of Metro Manila's main problems is an extreme shortage of decent housing. The rapid growth of the city has led to skyrocketing land values and rents, forcing roughly half of the city's residents to live in squatters' shacks. Some squatter settlements emerge almost overnight on vacant lots; others sprawl on for miles and have existed for decades. The inadequate infrastructure found in these communities is also characteristic of the rest of the city. Only an estimated 11 percent of Metro Manila's residences, for example, are connected to fully functioning sewer systems.

Figure 13.16 Bangkok Bangkok saw the development of an impressive skyline during its boom years from the late 1970s through the late 1990s. Unfortunately, infrastructure did not keep pace with population and commercial growth, resulting in one of the most congested and polluted urban landscapes in the world.

Thailand live in the Bangkok metropolitan area (Figure 13.16). In both Manila and Bangkok, as with other large Southeast Asian cities, explosive growth has led to housing shortages, congestion, and pollution. In Manila, more than half of the city's population lives in squatter settlements, usually without basic water and electricity service. Both cities also suffer from a lack of parks and other public spaces, enhancing the appeal of massive shopping malls. Bangkok's Paragon Mall has recently emerged as a major urban focus, complete with a conference center and a concert hall.

Thailand, the Philippines, and Indonesia are all making efforts to encourage growth of secondary cities by decentralizing economic functions. The goal is to stabilize the population of the primate cities. In the case of the Philippines, the city of Cebu has emerged in recent years as a relatively dynamic economic center, leading to hopes that a more balanced urban system may be emerging.

Urban primacy is less pronounced in other Southeast Asian countries. Vietnam, for example, has two main cities, Ho Chi Minh City (formerly Saigon) in the south, with 6 million people, and the capital city of Hanoi in the north, with more than 3 million residents. Jakarta is the largest city in Indonesia, but the country has a host of other large and growing metropolitan areas, including Bandung and Surabaya. Rangoon (Yangon), until recently the capital city of Burma (Myanmar), has doubled its population in the last two decades to more than 5 million residents. In Cambodia, the capital of Phnom Penh is still a relatively small city of around 2 million. Vientiane, the capital city of Laos, has only about 200,000 inhabitants.

Kuala Lumpur, the largest city in Malaysia with some 2 million inhabitants, has received heavy investments from

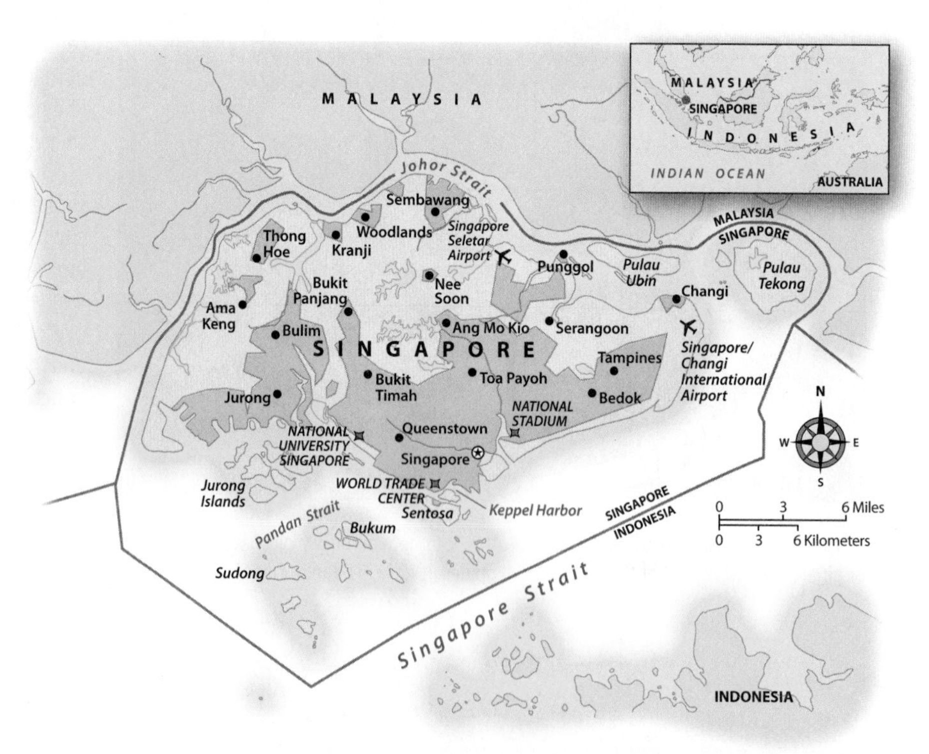

Figure 13.17 Singapore Singapore remains the economic and technological hub of Southeast Asia. It is a very small and extremely crowded country, forming one of the world's few city-states. Despite its high population density, Singapore has devoted almost half of its area to open spaces. The country consists of one main island and some 60 smaller islands.

both the national government and the global business community. This has produced a modern city of grand ambitions that is free of most infrastructural problems plaguing other Southeast Asian cities. The Petronas Towers, owned by the country's national oil company, were the world's tallest buildings at almost 1,500 feet (450 meters) when completed in 1996.

The independent republic of Singapore is essentially a city-state of 5.1 million people on an island of 240 square miles (600 square kilometers) (Figure 13.17). While space is at a premium, Singapore has successfully developed high-tech industries that have brought it great prosperity. Unlike other Southeast Asian cities, Singapore has no squatter settlements or slums. Only in the much diminished Chinatown and historic colonial district does one find older buildings. Otherwise, Singapore is an extremely clean but rather sterile city of modern high-rise skyscrapers, apartment complexes, and space-intensive industry.

Singapore is unique in Southeast Asia not only because it is a city-state, but also because of the Chinese cultural background of most of its people. But as we shall see in the following section, strong Chinese influences also are found in most of the other large cities of the region.

CULTURAL COHERENCE AND DIVERSITY: A Meeting Ground of World Cultures

Unlike many other world regions, Southeast Asia lacks the historical dominance of a single civilization. Instead, the region has been a meeting ground for cultural traditions from South Asia, China, the Middle East, Europe, and even North America. Abundant natural resources, along with the region's strategic location on oceanic trade routes connecting major continents, have long made Southeast Asia attractive to outsiders. As a result, the modern cultural geography of this diverse region reflects a long history of borrowing and combining external influences.

The Introduction and Spread of Major Cultural Traditions

In Southeast Asia, contemporary cultural diversity is embedded in the historical influences connected to the major religions of Hinduism, Buddhism, Islam, and Christianity (Figure 13.18).

South Asian Influences The first major external influence arrived from South Asia roughly 2,000 years ago when small numbers of migrants from South Asia helped local rulers establish Hindu kingdoms in coastal locations in Burma (Myanmar), Thailand, Cambodia, southern Vietnam, Malaysia, and western Indonesia. Although Hinduism faded away in most locations, it persists on the Indonesian islands of Bali and Lombok, and vestiges of the faith remain in many other areas. The ancient Indian script forms the basis for many Southeast Asian writing systems, and in even in Muslim Java the Hindu epic called the **Ramayana** remains a central cultural feature to this day.

A second wave of South Asian religious influence reached mainland Southeast Asia in the 13th century in the form of Theravada Buddhism, which is closely associated with Sri Lanka. Virtually all of the people in lowland Burma (Myanmar), Thailand, Laos, and Cambodia converted to Buddhism at that time. Today this faith forms the foundation for many of the social institutions of Mainland Southeast Asia. Saffron-robed monks, for example, are a common sight in Thailand and Burma (Myanmar), where Buddhist temples abound. In Thailand, the country's revered constitutional monarchy remains closely connected with Buddhism, and in Burma monks have occasionally been politically active (Figure 13.19). While Southeast Asian Theravada Buddhism shares many traits with the Mahayana Buddhism of East Asia, there are enough differences that the two are mapped separately in Figure 13.18.

Chinese Influences Unlike most other mainland peoples, the Vietnamese were not heavily influenced by South Asian civilization. Instead, their early connections were with East Asia. Vietnam was actually a province of China until about a thousand years ago, when the Vietnamese established a kingdom of their own. But while the Vietnamese rejected China's rule, they retained many attributes of Chinese culture. The traditional religious and philosophical beliefs of Vietnam centered around Mahayana Buddhism and Confucianism.

East Asian cultural influences elsewhere in Southeast Asia are linked more directly to immigration from southern China. Although people have been moving to the region from China for hundreds of years, migration reached a peak in the 19th and early 20th centuries. At first, most migrants were single men. Many returned to China after accumulating money, but others married local women and established mixed communities. In the Philippines, the elite population today is often described as "Chinese Mestizo," being of mixed Chinese and Filipino descent. In the 19th century Chinese women begin to migrate in large numbers, allowing the creation of ethnically distinct Chinese settlements. Urban areas throughout much of Southeast Asia still have large and cohesive Chinese communities. In Malaysia, the Chinese minority constitutes a little less than one-third of the population, whereas in the city-state of Singapore three-quarters of the people are of Chinese ancestry.

In much of the region, relationships between the Chinese minority and the indigenous majority are strained. One source of tension is the fact that overseas Chinese communities tend to be relatively prosperous. Chinese emigrants often prospered by becoming merchants, a job that was often ignored by the local people. As a result, they exert a major economic influence on local affairs, which is often resented. In 2000 and 2001, especially severe anti-Chinese rioting broke out in several Indonesian cities.

Figure 13.18 Religion in Southeast Asia Southeast Asia is one of the world's most religiously diverse regions. Most of the mainland is predominantly Buddhist, with Theravada Buddhism prevailing in Burma (Myanmar), Thailand, Laos, and Cambodia, and Mahayana Buddhism (combined with other elements of the so-called Chinese religious complex) prevailing in Vietnam. The Philippines is primarily Christian (Roman Catholic), but the rest of insular Southeast Asia is primarily Muslim. Substantial Muslim minorities are found in the Philippines, Thailand, and Burma (Myanmar). Animist and Christian minorities can be found in remote areas throughout Southeast Asia, especially in Indonesia's portion of the large island of New Guinea.

The Arrival of Islam Muslim merchants from India and Southwest Asia arrived in Southeast Asia more than a thousand years ago, and by the 1200s their religion began to spread. From an initial focus in northern Sumatra, Islam diffused into the Malay Peninsula, through the main population centers in the Indonesian islands, and east to the southern Philippines. By 1650, it had largely replaced Hinduism and Buddhism in Malaysia and Indonesia. The only significant holdout was the small but fertile island of Bali. As Islam spread across Java, thousands of Hindu musicians and artists fled to Bali, giving the island an especially strong tradition of arts and crafts. Partly because of this artistic legacy, Bali is now one of the premier destinations of international tourism.

Figure 13.19 **Protesting Burmese Monks** In September 2007, more than 100,000 people took to the streets to protest Burma's repressive military government. The protests were led by robed monks chanting prayers for peace.

Today the world's most populous Muslim country is Indonesia, where some 88 percent of the people follow Islam. This figure, however, hides a significant amount of internal diversity. In some parts of Indonesia, especially northern Sumatra (Aceh), orthodox forms of Islam took root (Figure 13.20). In others, such as central and eastern Java, a more lax form of worship emerged that retained certain Hindu and even animistic beliefs. Islamic reformers, however, have long been striving to instill more orthodox forms of worship among the Javanese. Today the young people of the island are increasingly turning to mainstream Islam.

In Malaysia and parts of Indonesia, Islamic fundamentalism has recently gained ground. Whereas the current Malaysian government has generally supported the revitalization of Islam, it is wary of the growing power of the fundamentalist movement. Some Malaysian Muslims, moreover, accuse the fundamentalists of advocating social practices derived from the Arabian Peninsula that are not necessarily religious in origin. Adding to Malaysia's religious tension is the fact that almost all ethnic Malays, who narrowly form the country's majority population, follow Islam, whereas most members of the minority communities follow different religions. Religious tensions mounted in 2009, when many Malaysia Muslims were offended by the fact that local Christians were using the term "Allah" to refer to God. In the first three months of 2010, Malaysia witnessed the firebombing or vandalizing of 11 churches, two Muslim prayer halls, a mosque, and a Sikh temple.

Islam was still spreading eastward through insular Southeast Asia when the Europeans arrived in the sixteenth century. When Spain claimed the Philippine Islands in the 1570s, it found the southwestern portion of the archipelago to be thoroughly Islamic. The Muslims resisted the Roman Catholicism introduced by the Spaniards, and much of the southwestern Philippines remains an Islamic stronghold.

Figure 13.20 **Indonesian Mosque** Indonesia is often said to be the world's largest Muslim country, as more Muslims reside here than in any other country in the world. Islam was first established in northern Sumatra, which is still the most devoutly Islamic part of the Indonesia. In this photo, Acehnese people are praying in front of the Baiturrahman Grand Mosque in the city of Banda Aceh.

The Philippines as a whole is currently about 85 percent Roman Catholic, making it, along with East Timor, the only predominantly Christian country in Asia. Several Protestant sects have been spreading rapidly in the Philippines over the past several decades, however, creating a more complex religious environment.

Christianity and Indigenous Cultures Christian missions spread through much of Southeast Asia in the late 19th and early 20th century when European colonial powers controlled the region. While French priests converted many people in Vietnam to Catholicism, they had little influence elsewhere. Beyond Vietnam, missions failed to make headway in areas of Hindu, Buddhist, or Islamic heritage.

Missionaries were far more successful in Southeast Asia's highland areas, where they found a wide array of tribal societies that had never accepted the major lowland religions. These peoples retained their indigenous belief systems, which generally focus on the worship of nature spirits and ancestors. Although some modern tribal groups retain such **animist** beliefs, others converted to Christianity. As a result, notable Christian concentrations are found in the Lake Batak area of north-central Sumatra, the mountainous borderlands between southern Burma (Myanmar) and Thailand, the northern peninsula of Sulawesi, and the highlands of southern Vietnam. Animism is still widespread in the mountains of northern Southeast Asia, central Borneo, far eastern Indonesia, and the highlands of northern Luzon in the Philippines. In Indonesia, however, animist practices are technically illegal, as monotheism is one of the country's founding principles. As a result, many Indonesian tribal groups have been gradually converting to either Islam or Christianity.

Indonesia has experienced pronounced religious strife over the past ten years, especially between its Muslim majority and its Christian minority. Inter-faith relations had been relatively good until the late 1990s. Indonesia, despite its Muslim majority, is a secular state that has always emphasized tolerance among its officially recognized religions (Islam, Christianity, Buddhism, Hinduism, and Confucianism). With the Indonesian economic disaster of the late 1990s, however, relations deteriorated. Religious conflicts broke out in many areas, especially in the Maluku Islands of eastern Indonesia and in central Sulawesi. Between 1999 and 2003, an estimated 2,000 people were killed in Christian–Muslim clashes on the island of Sulawesi alone. Transmigration is also implicated in Indonesia's religious conflicts, as in many instances indigenous Christian and animist groups are now competing against Muslim immigrants from Java and Madura. By 2008, however, religious tensions were decreasing in most parts of the country.

Religion and Communism By 1975, communism had triumphed in Vietnam, Cambodia, and Laos. In all three countries, religious practices were strongly discouraged. At present, Vietnam's officially communist government is struggling against a revival of faith among the country's Buddhist majority and its 8 million Christians. Buddhist monks are frequently detained, and the government reserves for itself the right to appoint all religious leaders. Also expanding is Vietnam's indigenous religion of Cao Dai, a syncretic (or mixed) faith that venerates not only the Buddha, Confucius, and Jesus, but also the French novelist Victor Hugo. Between 3 and 8 million Vietnamese currently belong to this religious community.

Geography of Language and Ethnicity

As with religion, language in Southeast Asia expresses the long history of human movement. The linguistic geography of the region is complicated—too complicated, in fact, to be adequately conveyed in a single map. The several hundred distinct languages of the region can, however, be placed into five major linguistic families. These are *Austronesian,* which covers most of the insular realm; *Tibeto-Burman,* which includes most of the languages of Burma (Myanmar); *Tai-Kadai,* centered on Thailand and Laos; *Mon-Khmer* encompassing most of the languages of Vietnam and Cambodia; and *Papuan,* a Pacific language family found in eastern Indonesia (Figure 13.21).

The Austronesian Languages Austronesian is one of the world's most widespread language families, extending from Madagascar to Easter Island in the eastern Pacific. Linguists believe that this family originated in Taiwan and adjacent areas of East Asia, and then spread widely across the Indian and Pacific oceans by seafaring people who migrated from island to island.

Today almost all insular Southeast Asian languages fall within the Austronesian family. This means that elements of grammar and vocabulary are widely shared across the islands. As a result, it is relatively easy for a person who speaks one of these languages to learn any other. But despite this linguistic commonality, more than 50 distinct Austronesian languages are spoken in Indonesia alone. And in far eastern Indonesia, a variety of languages fall into the completely separate family of Papuan, closely associated with New Guinea.

One language, however, overshadows all others in insular Southeast Asia: Malay. Malay is indigenous to the Malay Peninsula, eastern Sumatra, and coastal Borneo, and was spread historically by merchants and seafarers. As a result, it became a common trade language, or **lingua franca**, understood and used by peoples speaking different languages throughout much of the insular realm. Dutch colonists in Indonesia eventually employed Malay as an administrative language, although they wrote it with the Roman alphabet, rather than in the Arabic-derived script used by native speakers. When Indonesia became an independent country in 1949, its leaders decided to use the lingua franca version of Malay as the basis for a new national language called *Bahasa Indonesia* (or more simply, *Indonesian*). Although Indonesian is slightly different from Malaysian, they essentially form a single, mutually intelligible language.

The goal of the new Indonesian government was to find a common language that would overcome ethnic differences throughout the far-flung state. In general, this policy has been successful. Indonesian is now used in schools, the government, and the media, and most of the country's people understand it. But regional languages, such as Javanese, Balinese, and Sundanese, continue to be primary languages in most homes. More than 75 million people speak Javanese, which makes it one of the world's major tongues.

The eight major languages of the Philippines also belong in the Austronesian family. Despite more than 300 years of colonization by Spain, Spanish never became a unifying force for the islands. During the American period (1898–1946), English served as the language of administration and education. After independence following World War II, Philippine nationalists decided to create a national language that could replace English and help unify the new country. They selected Tagalog, the language spoken in the Manila region. The first task was to standardize and modernize Tagalog, which had many dialects. After this was accomplished, it was renamed *Filipino* (alternatively, *Pilipino*). Today, mainly because of its use in education, television, and movies, Filipino is emerging as a unifying national language.

Tibeto-Burman Languages

Each country of mainland Southeast Asia is closely identified with the national language spoken in its core territory. This does not mean, however, that all the inhabitants of these countries speak these official languages on a daily basis. In

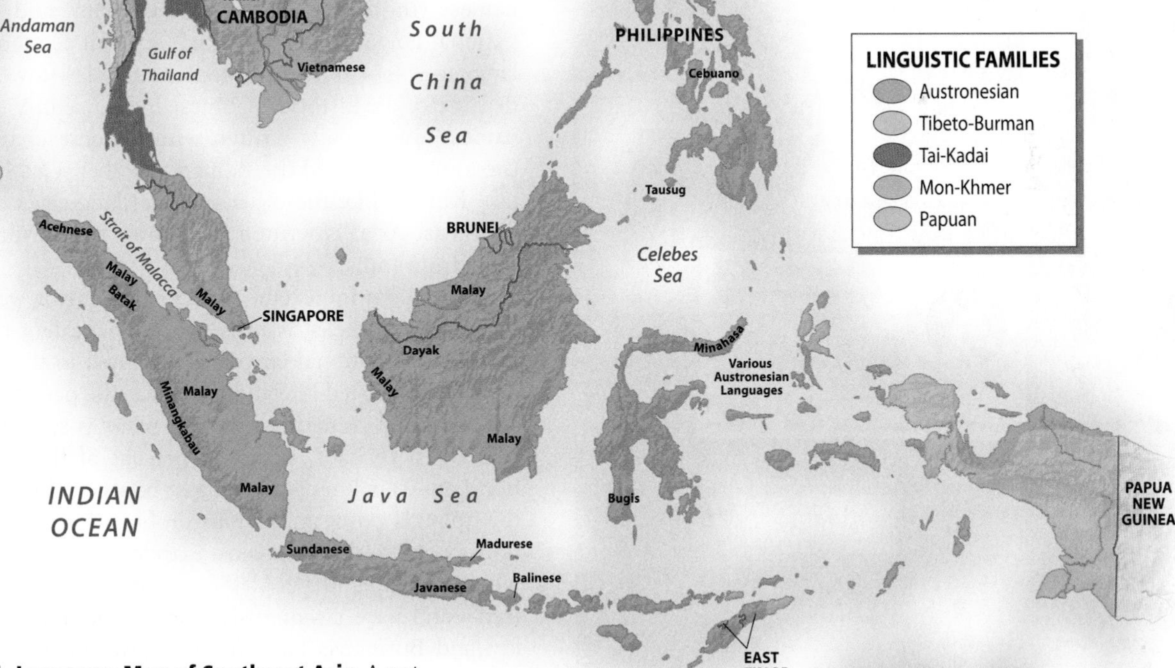

Figure 13.21 Language Map of Southeast Asia A vast number of languages are found in Southeast Asia, but most are tribal tongues spoken by only a few thousand people. In mainland Southeast Asia—the site of three major language families—the central lowlands of each country are dominated by people speaking the national languages: Burmese in Burma (Myanmar), Thai in Thailand, Lao in Laos, and Vietnamese in Vietnam. Almost all languages in insular Southeast Asia belong to the Austronesian linguistic family. There were no dominant languages here before the creation of such national tongues as Filipino and Bahasa Indonesia in the mid-20th century.

the mountains and other remote districts, other languages remain primary. This linguistic diversity reinforces ethnic divisions despite national educational programs designed to foster unity.

Burma (Myanmar) provides good example of ethnic and linguistic diversity. Its national language is Burmese, which is closely related to Tibetan and perhaps, much more distantly, to Chinese. Today some 32 million people speak Burmese as their first language (Figure 13.22). Although the nationalistic military government of Burma (Myanmar) has attempted to force its version of unity on the entire population, a major schism developed with several non-Burman tribal groups inhabiting the rough uplands that flank the Burmese-speaking Irrawaddy Valley. Although most of these peoples speak languages in the Tibeto-Burman family, they are quite distinct from Burmese.

Tai-Kadai Languages The Tai-Kadai linguistic family probably originated in southern China and then spread into Southeast Asia starting around 1200 CE. Today, closely related languages within the Tai subfamily are found through most of Thailand and Laos, in the uplands of northern Vietnam, in Burma's Shan Plateau, and in parts of southern China. Most Tai languages are quite localized, spoken by small ethnic groups. But two of them, Thai and Lao, are national languages, spoken in Thailand and Laos respectively.

Figure 13.22 **Burmese Script** The Burmese script, seen here on posters in Burma's primate city of Rangoon (Yangon), has characters with round shapes, supposedly because the palm leaves that were originally used as writing surfaces would have been split by straight lines. Burma's script, like those of Thailand, Laos, and Cambodia, can be traced back to a writing system from ancient India.

Linguistic terminology here is complicated. Historically, the main language of the Kingdom of Thailand, called *Siamese* (just as the kingdom was called *Siam*), was restricted to the lower Chao Phraya valley, which formed the national core. In the 1930s, however, the country changed its name to Thailand to emphasize the unity of all the peoples speaking closely related Tai languages within its territory. Siamese was similarly renamed *Thai*, and it has gradually become the unifying language for the country. There are still substantial variations in dialect, however, with those of the north often considered separate languages.

Lao, another Tai language, is the national tongue of Laos. More Lao speakers reside in Thailand than in Laos, however, where they form the majority population of the relatively poor Khorat Plateau of the northeast. The dialect of Lao spoken in northeastern Thailand, called Isan locally, seems to be declining in favor of standard Thai. Roughly a third of Thailand's population, however, still speaks Lao (Isan) rather than Thai as their first language.

Mon-Khmer Languages The Mon-Khmer language family probably covered almost all of mainland Southeast Asia 1,500 years ago. It contains two major languages, Vietnamese (the national tongue of Vietnam) and Khmer (the national language of Cambodia), as well as a host of minor languages spoken by hill peoples and a few lowland groups. Because of the historical Chinese influence in Vietnam, Vietnamese was usually written with Chinese characters until the French imposed the Roman alphabet, which remains in use. Khmer, like the other national languages of mainland Southeast Asia, is written in its own script, ultimately derived from India.

The most important aspect of linguistic geography in mainland Southeast Asia is probably the fact that in each country the national language is limited to the core area of densely populated lowlands, whereas the peripheral uplands are inhabited primarily by tribal peoples speaking separate languages. In Laos, up to 40 percent of the population is non-Lao. This linguistic contrast between the lowlands and the uplands poses an obvious problem for national integration. Such problems are most extreme in Burma (Myanmar), where the upland peoples are numerous and well organized—and have strenuously resisted the domination of the lowland Burmans. The Burmese government has, in turn, devoted much of its energies to resisting the influences of global culture.

Southeast Asian Culture in Global Context

The imposition of European colonial rule ushered in a new era of globalization in Southeast Asia, bringing with it European languages, Christianity, and new governmental, economic, and educational systems. This period also deprived most Southeast Asians of their cultural autonomy. As a result, with decolonialization after World War II, several countries attempted to isolate themselves from the cultural and economic influences of the emerging global system. Burma (Myanmar) retreated into its own form of Buddhist socialism, placing strict limits on foreign tourism, which the government viewed as a source of cultural contamination. Although the door has opened substantially since the 1980s, the government of Burma remains wary of foreign practices and influences, an attitude that has become a major source of tension within the country. Burma's government is relatively tolerant of music, however, which has resulted in a surge of spontaneous concerts and underground music festivals. At these events, hip-hop and techobeat styles are often mixed with more traditional Burmese musical forms.

Other Southeast Asian countries have been more receptive to foreign cultural influences. This is particularly true in the case of the Philippines, where American colonialism may have predisposed the country to many of the more popular forms of Western culture. As a result, Filipino musicians and other entertainers are in demand throughout East and Southeast Asia (Figure 13.23). Thailand, which was never subjected to colonial rule, is the most receptive country of the mainland to global culture, with its open policy toward tourism, mass media, and economic interdependence.

Cultural globalization has been recently challenged in some Southeast Asian countries. The Malaysian government, for example, is highly critical of many American films and television programs. Islamic revivalism, in both Malaysia and Indonesia, also presents a challenge to global culture. A recent poll, for example, found that that 62 percent of Indonesians believe that U.S. culture is "disruptive" to Indonesian society. Singapore's leaders have also criticized cultural influences from the West, but the city–state remains committed to globalization. Singapore is also seeking to enhance tourism and bolster its economy by building Las Vegas–style casinos. A huge gambling complex called Resorts World Sentosa opened its doors in February 2010, welcoming 130,000 visitors in its first week of operation.

Figure 13.23 Filipino Entertainers The people of the Philippines have adopted popular forms of Western culture more than most other Southeast Asians, in part because of their long experience with American colonialism. As a result, Filipino performers are often in demand in other Asian countries. This photograph shows a Filipina musician singing in Hong Kong.

As the global language, English causes ambivalence in much of Southeast Asia. On one hand, it is the language of questionable popular culture; yet on the other hand, citizens need proficiency in English if they are to participate in international business and politics. In Malaysia, widespread proficiency in English was challenged in the 1980s as nationalists stressed the importance of their native tongue. This movement distressed the business community, which considers English vital to Malaysia's competitive position, as well as the influential Chinese communities, for which Malaysian is not a native language.

In Singapore, the situation is more complex. Mandarin Chinese, English, Malay, and Tamil (from southeastern India) are all official languages. The local languages of southern China are also common in home environments, as 75 percent of Singapore's population is of southern Chinese ancestry. The Singapore government now encourages Mandarin Chinese and discourages southern Chinese dialects. It also supports English while discouraging *Singlish,* an English-based dialect containing many southern Chinese words and grammatical forms. Singaporean officials have occasionally slapped restrictive ratings on local films for "bad grammar" if they contain too much Singlish dialogue.

In the Philippines, nationalists have long criticized the common use of English, even though widespread fluency has proved beneficial to the millions of Filipinos who have emigrated for better economic conditions or who work as crew members in the globalized shipping industry. The Philippine government has been gradually deemphasizing English in favor of Filipino in public schools and, as a result, some fear that competence in English is slowly declining. Others counter that the spread of cable television, with its many U.S. shows, has enhanced the country's English-language skills.

While the diffusion of global cultural forms is beginning to merge cultures across Southeast Asia, religious revivalism operates in the opposite direction. This same combination of centripetal (merging together) and centrifugal (spreading apart) forces also is found in the geopolitical realm.

GEOPOLITICAL FRAMEWORK: War, Ethnic Strife, and Regional Cooperation

Southeast Asia is sometimes defined as a geopolitical entity composed of the 10 different countries that form the Association of Southeast Asian Nations, or ASEAN (Figure

13.24). Today ASEAN, more than anything else, gives Southeast Asia regional coherence. East Timor, or Timor-Leste, as the country is officially called, gained independence only in 2002, and is not yet a member of ASEAN. It is expected to gain admission, however, in 2012. Despite the general success of ASEAN in reducing international tensions, several Southeast Asian counties are still struggling with serious ethnic and regional conflicts.

Before European Colonialism

The modern countries of mainland Southeast Asia were all indigenous kingdoms at one time or another before the onset of European colonialism. Cambodia emerged earliest, reaching its height in the twelfth century, when it controlled much of what is now Thailand and southern Vietnam. By the 1300s, independent kingdoms had been established by the Burman, Siamese (Thai), Lao, and Vietnamese people in the major river valleys and deltas. These kingdoms were in a nearly constant state of war with one another, often fighting more for labor than for territory. Victors would typically take home thousands of prisoners to settle their lands, leading to considerable ethnic mixing.

The situation in insular Southeast Asia was different, with the premodern map bearing no resemblance to that of the modern nation-states. Many kingdoms existed on the Malay Peninsula and on the islands of Sumatra and Java, but few were territorially stable. In the

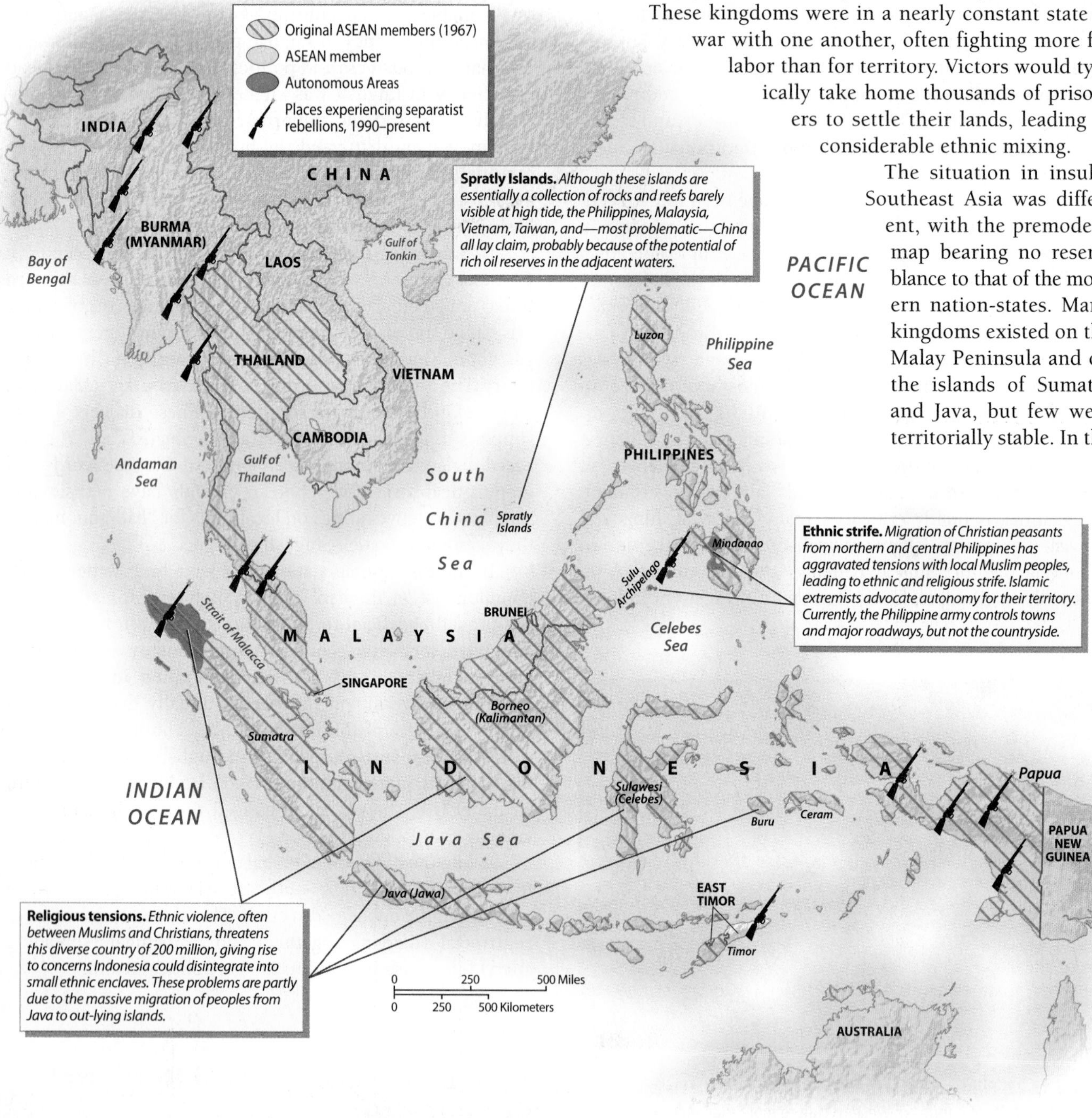

Figure 13.24 Geopolitical Issues in Southeast Asia The countries of Southeast Asia have managed to solve most of their border disputes and other sources of potential conflicts through ASEAN (the Association of Southeast Asian Nations). Internal disputes, however, mostly focused on issues of religious and ethnic diversity, continue to plague several of the region's states, particularly Indonesia and Burma (Myanmar). ASEAN also experiences tension with China over the Spratly Islands of the South China Sea.

Philippines, eastern Indonesia, and central Borneo, most societies were organized at the local level. Indonesia, the Philippines, and Malaysia thus owe their modern territorial configuration almost wholly to the European colonial influence (Figure 13.25).

The Colonial Era

The Portuguese were the first Europeans to arrive (around 1500), lured by the cloves and nutmeg of the Maluku region of eastern Indonesia. In the late 1500s, the Spanish conquered most of the Philippines, which they used as a base for their silver and silk trade between China and the Americas. By the 1600s, the Dutch had started staking out Southeast Asian territory, followed by the British. With superior naval weaponry, the Europeans were quickly able to conquer key ports and strategic trading locales. Yet for the first 200 years of colonialism, except in the Philippines, the Europeans made no major geopolitical gains.

Dutch Power By the 1700s, the Netherlands had become the most powerful force in the region. As a result, a Dutch Empire in the "East Indies" (or Indonesia) began appearing on world maps. Form its original base in northwestern Java, this empire continued to grow into the early 20th century, when the Dutch defeated their last main adversary, the

Figure 13.25 Colonial Southeast Asia With the exception of Thailand, all of Southeast Asia was under Western colonial rule by the early 1900s. The Netherlands had the largest empire in the region, covering the territory that was later to become Indonesia. France maintained a substantial imperial realm in Vietnam, Laos, and Cambodia, as did Britain in Burma (Myanmar) and Malaysia (including Singapore and Brunei). The Philippines was colonized by Spain, but passed to the control of the United States in 1898.

powerful Islamic sultanate of Aceh in northern Sumatra. Later, the Dutch invaded the western portion of New Guinea in response to German and British advances in the eastern half. In a subsequent treaty, these imperial powers sliced New Guinea down the middle, with the Netherlands taking the west.

British, French, and U.S. Expansion The British, preoccupied with their empire in India, concentrated their attention on the sea-lanes linking South Asia to China. As a result, they established several fortified trading outposts along the vital Strait of Malacca, the most notable being on the island of Singapore, founded in 1819. To avoid conflict, the British and Dutch agreed that the British would limit their attention to the Malay Peninsula and northern Borneo. Britain allowed Muslim sultans to retain limited powers, and their descendants still enjoy token authority in several Malaysian states. When Britain left this area in the early 1960s, Malaysia emerged as an independent state. Two small portions of the former British sphere did not join the new country. In northern Borneo, the Sultanate of Brunei gained independence, backed by its large oil reserves. Singapore briefly joined Malaysia, but then withdrew and became fully independent in 1965. This divorce was carried out partly for ethnic reasons. Malaysia was to be a primarily Malay state, but with Singapore included its population would have been almost half Chinese.

In the 1800s, European colonial power spread through most of mainland Southeast Asia. British forces in India fought several wars against the kingdom of Burma before annexing the entire area in 1885, including considerable upland territories that had never been under Burmese rule. During the same period, the French moved into Vietnam's Mekong Delta, gradually expanding their territorial control to the west into Cambodia and north to China's border. Thailand was the only country to avoid colonial rule, although it did lose substantial territories to the British in Malaysia and the French in Laos. Thai independence was maintained partly because it served British and French interests to have a buffer state between their colonial realms.

The final colonial power to enter the area was the United States, which took the Philippines first from Spain and then, after a bitter war, from Filipino nationalists between 1898 and 1900. The U.S. army subsequently conquered the Muslim areas of the southwest, most of which had never been fully under Spanish authority.

Growing Nationalism Organized resistance to European rule began in the 1920s in mainland countries, but it took the Japanese occupation during World War II to shatter the myth of European invincibility. After Japan's surrender in 1945, agitation for independence was renewed throughout Southeast Asia. As Britain realized that it could no longer control its South Asian empire, it also withdrew from adjacent Burma, which achieved independence in 1948. The Netherlands failed to reconquer Indonesia after WWII, and was forced to acknowledge Indonesian independence in 1949. The United States granted long-promised independence to the Philippines on July 4, 1946, although it retained key military bases, as well as considerable economic influence, for several decades.

The Vietnam War and Its Aftermath

Unlike the United States, France was determined to maintain its Southeast Asian empire. Resistance to French rule was organized primarily by communist groups that were deeply rooted in northern Vietnam. As French forces returned to Southeast Asian in 1946, the leader of this resistance movement, Ho Chi Minh, organized a separatist government in the north. Open warfare between French soldiers and the communist forces went on for almost a decade. After a decisive defeat in 1954, France agreed to withdraw. An international peace council in Geneva decided that Vietnam would be divided into two countries. As a result, communist leaders came to power in North Vietnam, allied with the Soviet Union and China. South Vietnam simultaneously emerged as an independent, capitalist-oriented state with close political ties to the United States.

The Geneva peace accord did not end the war. Communist guerrillas in South Vietnam fought to overthrow the new government and unite it with the north. North Vietnam sent troops and war materials across the border to aid the rebels. Most of these supplies reached the south over the Ho Chi Minh Trail, an ill-defined network of forest passages through Laos and Cambodia that steadily drew these two countries into the conflict. In Laos the communist Pathet Lao forces challenged the government, while in Cambodia the **Khmer Rouge** guerrillas gained considerable power. In South Vietnam, the government gradually lost control of key areas, including much of the Mekong Delta, a region perilously close to the capital city of Saigon.

U.S. Intervention By 1962 the United States was sending large numbers of military advisors to South Vietnam. In Washington, DC, the **domino theory** became accepted foreign policy. According to this notion, if Vietnam fell to the communists, then so would Laos and Cambodia; once those countries were lost, Burma (Myanmar) and Thailand, and perhaps even Malaysia and Indonesia, would join the Soviet-dominated communist bloc. By 1968, over half a million U.S. troops were fighting a ferocious land war against the communist guerrillas (Figure 13.26). But despite superiority in arms and troops—and domination of the air—U.S. forces failed to gain control over much of the countryside. As casualties mounted and the antiwar movement back home strengthened, the United States began secret talks with North Vietnam in search of a negotiated settlement. U.S. troop withdrawals began in earnest by the early 1970s.

Communist Victory With the withdrawal of U.S. forces and financial support, the noncommunist governments of the former French zone began to collapse. Saigon fell in 1975, and in the following year Vietnam was officially reunited

Figure 13.26 U.S. Soldier and Vietcong Prisoners The United States maintained a substantial military presence in Vietnam in the 1960s and early 1970s. Although U.S. forces claimed many victories, they were ultimately forced to withdraw, leading to the victory of North Vietnam and the reunification of the country.

under the government of the north. Reunification was a traumatic event in southern Vietnam. Hundreds of thousands of people fled to other countries, especially the United States. The first wave of refugees consisted primarily of wealthy professionals and businesspeople, but later migrants included many relatively poor ethnic Chinese. Most of these refugees fled on small, rickety boats; large numbers suffered shipwreck or pirate attack.

Vietnam proved fortunate compared with Cambodia. There the Khmer Rouge installed one of the most brutal regimes the world has ever seen. City-dwellers were forced into the countryside to become peasants, and most wealthy and educated persons were summarily executed. The Khmer Rouge's goal was to create a wholly new agrarian society by returning to what they called "year zero." After several years of bloodshed that took an estimated 1.5 million lives, neighboring Vietnam invaded Cambodia and installed a far less brutal, but still repressive, regime. Fighting between different factions continued for more than a decade, but by the late 1980s the United Nations was able to broker a peace settlement. A parliamentary system of government subsequently brought peace to the shattered country. Although Cambodia is officially a constitutional monarchy with an elected government, corruption is widespread and democratic institutions remain weak and unstable (see "Global to Local: Poipet and Southeast Asia's Other 'Sin Cities.'")

Vietnam stationed significant numbers of troops in Laos after 1975. Large numbers of Hmong and other tribal peoples—many of whom had fought on behalf of the United States—fled to Thailand and the United States. The Communist Party still maintains a monopoly over political power in Laos, although much of the economy has been opened to private firms. Many Laotian Hmong refugees who fled to the United States after the war, however, continue to seek political change at home. In 2007, the American Hmong community was shocked when U.S. federal agents arrested 11 of their leaders, including the famed general Vang Pao, for plotting to overthrow the government of Laos. In 2009, however, all charges against Vang Pao were dropped.

Geopolitical Tensions in Contemporary Southeast Asia

The most serious geopolitical problems in Southeast Asia today occur within countries rather than between then. In several areas, locally based ethnic groups struggle against centralized national governments that inherited territory from the former colonial powers. Tension also results when tribal groups attempt to preserve their homeland from logging, mining, or interregional migration. Such conflicts are especially pronounced in the large, multiethnic country of Indonesia.

Conflicts in Indonesia When Indonesia gained independence in 1949, it included all the former Dutch possessions in the region except western New Guinea. The Netherlands retained this territory, arguing that its cultural background distinguished it from Indonesia. Although Dutch authorities were preparing western New Guinea for independence, Indonesia demanded the entire territory. Bowing to pressure from the United States, the Netherlands relented and allowed Indonesia to take control in 1963. The Indonesian government formally annexed western New Guinea in 1969.

Tensions in western New Guinea increased in the following decades as Javanese immigrants, along with mining and lumber firms, arrived in the area. Faced with the loss of their land and the degradation of their environment, indigenous residents formed the separatist organization OPM (*Organisesi Papua Merdeka*) and launched a rebellion. Rebel leaders demanded independence, or at least autonomy, but they faced a far stronger force in the Indonesian army. The struggle between the Indonesian government and the OPM is currently a sporadic but occasionally bloody guerrilla affair. Indonesia is determined to maintain control of the region in part because it is home to one of the country's largest taxpayers, the highly polluting Grasberg copper and gold mine run by the Phoenix–based Freeport-McMoRan Corporation (Figure 13.27).

A more intensive war erupted in 1975 on the island of Timor. The eastern half of this poor island had been a Portuguese colony (the only survivor of Portugal's 16th-century foray into the region) and had evolved into a mostly Christian society. The East Timorese expected independence when the Portuguese finally withdrew. Indonesia, however, viewed the area as its own territory and immediately invaded. A ferocious struggle ensued, which the Indonesian army won in part by starving the people of East Timor into submission.

After the economic crisis of 1997, Indonesia's power in the region slipped. A new Indonesian government promised an election in 1999 to see if the East Timorese wanted independence. At the same time, however, Indonesia's army began to organize loyalist militias to intimidate the people of East

Figure 13.27 Grasberg Mine Grasberg, located in the Indonesian province of Papua, is the world's largest gold mine and third largest copper mine. Employing more than 19,000 people, Grasberg is of great economic importance to Indonesia. It is also highly polluting, generating 253,000 tons (230,000 metric tons) of tailings per day and is thus opposed by many local inhabitants.

Timor into voting to remain within the country. When it was clear that the vote would be for independence, the militias began rioting, looting, and slaughtering civilians. Under international pressure, Indonesia finally withdrew, United Nations forces arrived, and the East Timorese began to build a new country.

Considering the devastation caused by the militias, the reconstruction of East Timor was not easy. Even selecting an official language proved difficult; after some deliberation, both Tetum (one of 16 indigenous languages) and Portuguese were selected. In 2006, ethnic rioting further damaged the country, requiring Australian intervention to reestablish peace. A relatively successful election in 2007 brought hope that East Timor could overcome its divisions, but in 2008 a well-coordinated attack by former soldiers and police officers came close to killing the country's president and prime minister.

Secession struggles have occurred elsewhere in Indonesia. In the late 1950s, central Sumatra and northern Sulawesi rebelled, but they were quickly defeated and eventually reconciled to Indonesian rule. Another small-scale war flared up in the late 1990s in western and southern Borneo (Kalimantan). Here the indigenous Dayaks, a tribal people partly converted to Christianity, began to clash with Muslim migrants from Madura, a densely inhabited island north of Java. The Indonesian military has restored order in Kalimantan's cities, but the countryside remains troubled. The southern Maluku Islands, especially Ambon and Seram, have also seen fighting between Muslims and Christians. Even the area's cities are now divided into Muslim and Christian sectors, and crossing to the wrong side can prove dangerous.

Indonesia's most serious regional conflict has long been that of Aceh in northern Sumatra. Many Acehnese, in general Indonesia's most devout Muslim group, have demanded the creation of an independent Islamic state. While the government has allowed Aceh a high degree of autonomy as a "special territory of Indonesia," it has been determined to prevent actual independence. Ironically, the devastation caused by the December 2004 tsunami, which left over 500,000 Acehnese homeless, allowed a peace settlement to be finally reached, as the needs of the area were so great that the separatist fighters agreed to lay down their weapons. A 2006 election brought former rebel leaders into the heart of Aceh's government. In early 2010, however, violence again broke out after the government discovered a militant training camp in the region.

Indonesia has obviously had difficulties creating a unified nation over its vast, sprawling extent and across its numerous cultural and religious communities. As a creation of the colonial period, Indonesia has a weak historical foundation. Some critics have viewed it as something of a Javanese empire, even though many non-Javanese individuals have risen to high governmental positions. While Indonesia now has the highest degree of political freedom in Southeast Asia, its politically influential military remains hostile to movements for regional autonomy.

Regional Tensions in the Philippines The Philippines has long been struggling against a regional secession movement in its Islamic southwest. Although the government made headway in talks with the main rebel group by creating the Autonomous Region in Muslim Mindanao (ARMM), the more extreme Muslim factions continue to demand greater autonomy over a larger area—if not outright independence. The migration of Christian peasants from northern and central Philippines into the Islamic southwest has exacerbated local tensions. The Philippine army generally maintains control over the area's main cities and roadways, but it has little power in the more remote villages. As radical Islamist groups reportedly have close ties with the Al Qaeda network, the Philippines became a key site in the U.S.-led struggle against global terrorism. Several hundred U.S. military advisors are currently working with Philippine military in the Mindanao region. As a result, some Filipino nationalists argue that their country is again falling under U.S. domination.

GLOBAL TO LOCAL Poipet and Southeast Asia's Other "Sin Cities"

A number of Southeast Asian cities have prospered by providing socially objectionable services to an international clientele. Foremost among these is the resort city of Pattaya, located on the Gulf of Thailand 102 miles (165 kilometers) southeast of Bangkok. Although many of the approximately 5 million tourists who annually visit this small city (population 100,000) come primarily for the beaches and the shopping, a bigger draw is the hundreds of "go-go bars" that serve as fronts for prostitution. Although technically illegal, Thai prostitution is a big business, netting some $4.3 billion annually.

Pattaya is not the only small Southeast Asian city to find a niche in the global prostitution business. Before 1992, the economy of Angeles City in the Philippines was based on the nearby U.S. Air Force base of Clark Field. When the American military forces pulled out, Angeles City faced an uncertain economic future. Although the Philippine government created a special economic zone to allow businesses to take advantage of the base's extensive facilities, development has been slow. Angeles City's nightclub district, however, has experienced a successful expansion. Built initially to cater to American servicemen, the prostitution-oriented bars (numbering well over 100) now focus on an international clientele.

The exploitation of prostitutes in Thailand and the Philippines may be severe, but much worse conditions are encountered in the comparatively unregulated brothels of Laos, Cambodia, and Burma (Myanmar). In 2004, *New York Times* journalist Nicholas Kristof brought international attention to the horrific situation faced by child prostitutes in the Cambodian city of Poipet. Kristof went so far as to purchase (for $150 and $203, respectively) and then free two prostitute slave-girls. The sale of illegal drugs, especially methamphetamine, is also widespread in this tawdry Cambodian border town (Figure 13.3.1).

The booming city of Poipet, however, owes its most recent growth to casino gambling. Propelling the rise of this city as the "Las Vegas of Southeast Asia" is the fact that casino gambling is extremely popular, although illegal, in neighboring Thailand, where Thais reportedly spend up to $11 billion annually on illegal gambling activities. In response, Cambodian authorities allowed the creation of a gambling enclave in Poipet directly on the Thai border, the first casino opening in 1999. Tourists from Thailand can enter the casino district without passing through Cambodian immigration, and the gambling halls prefer Thai rather than Cambodian currency. Although 90 percent of Poipet's tourists are from Thailand, increasing numbers are arriving from Malaysia, China, and Vietnam. A number of Poipet's casino managers, moreover, hail from North America, and many casino workers are from the Philippines. Overall, the venture is such an economic success that similar gambling reserves are being planned for several Laotian border towns.

Although Poipet has earned huge sums of money for investors and managers, its broader social effects are disastrous. Poipet is a dangerous city with inadequate infrastructure that lures in desperately poor people, many of whom are forced into work as prostitutes or drug dealers. But considering the poverty and corruption found in Cambodia, it is not surprising that the rise of Poipet has met little national resistance.

Figure 13.3.1 Poipet, Cambodia The Holiday Palace casino is one of seven high-end gambling facilities recently constructed in the Cambodian border city of Poipet.

To maintain a degree of authority in the Muslim areas of the south, the Philippine government has allied itself with powerful local clans and their private armies. The Maguindanao Massacre of November 2009, however, showed the limitations of this policy. Fighters associated with the government-allied Ampatuan family attacked a convoy of a rival clan that was heading to register one of its members as a candidate for an upcoming gubernatorial election, killing 51 people, including as many as 32 journalists (Figure 13.28). The Committee to Protect Journalists describes the incident as the largest single attack on news reporters in world history. The Philippine government subsequently arrested the head of the Ampatuan clan, but it was not able to erase the hostility generated by the incident.

The Philippines' political problems are not limited to the south. A long-simmering communist rebellion has affected almost the entire country. In the mid-1980s, the NPA (New People's Army) controlled one-quarter of the country's villages scattered across all the major islands. Although the NPA's strength has declined since then, it remains a potent force in many areas. The Philippine national government, although democratic, is itself far from stable, suffering from continual coup threats, corruption scandals, and mass protests.

The Quagmire of Burma Burma (Myanmar) has also been a war-ravaged country. Its simultaneous wars of the 1970s, 1980s, and 1990s pitted the central government, dominated by the Burmans, against the country's varied non-Burman societies. Fighting intensified gradually after independence in 1948, until almost half of the country's territory had become a combat zone (Figure 13.29). Burma's troubles, moreover, are by no means limited to the country's ethnic minorities. Since 1988, it has been ruled by a repressive military regime that is bitterly resented by much of the population. Democratic opposition to the government, however, has been vigorously suppressed, with its leader, Nobel Peace Prize winner Aung San Suu Kyi, remaining under house arrest (Figure 13.30).

The rebelling ethnic groups of Burma (Myanmar) have sought to maintain authority over their own cultural traditions, lands, and resources. They generally see the national government as something of a Burman empire seeking to impose its language and, in some cases, religion upon them. Most of the insurgent ethnic groups live in rugged and inaccessible terrain, but even some lowland groups have rebelled. The Muslim peoples of the Arakan coast (the Rohingyas) in western Burma have long faced severe discrimination (see "People on the Move: The Rohingyas Seek Sanctuary"). In the 1980s, many were forced to flee to neighboring Bangladesh, while more recently Burmese Muslims have been fleeing to Yunnan Province in China.

Several of Burma's ethnic insurgencies have been financed largely by opium-growing and heroin manufacture. The Shan, a Tai-speaking people inhabiting a plateau area within the Golden Triangle of drug production, formed a breakaway state in the 1980s and 1990s based largely on the narcotics trade. Their efforts faltered in the late 1990s after Burmese agreements with Thailand reduced the Shan heroin trade just as military operations cut off the supply of raw opium reaching the Shan factories. The government's triumph over the Shan was also made possible by alliances with rival ethnic groups that subsequently moved into the drug trade. The most powerful of these groups is the United Wa State Army (UWSA), which by 2007 controlled an estimated 50 heroin refineries and methamphetamine factories and maintained over 20,000 heavily armed soldiers. In 2009, the Burmese government felt strong enough to order the UWSA and its other allied ethnic militias to fold their command structure into that of the Burmese Army. The militias

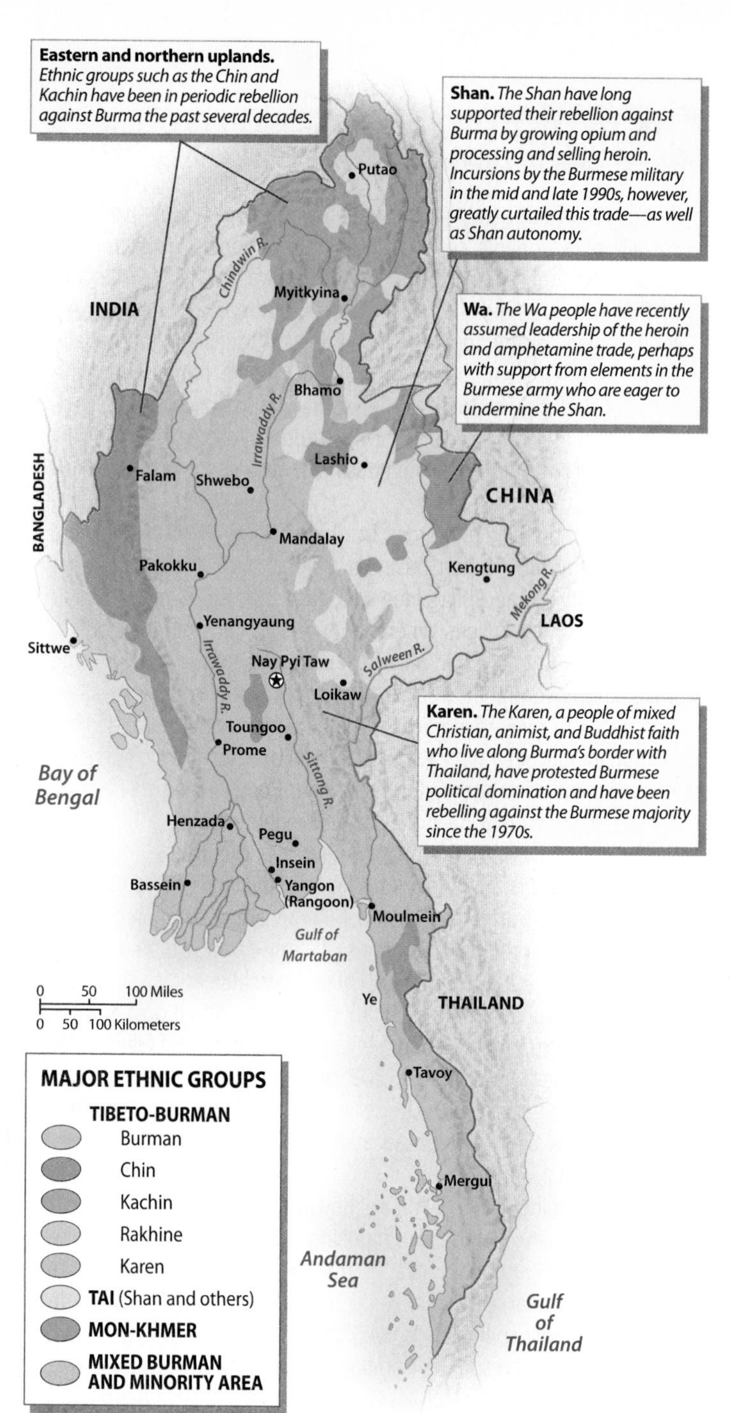

Figure 13.29 Ethnic Conflict in Burma (Myanmar) Although the central lowlands of Burma (Myanmar) are primarily populated by people of the dominant Burman ethnic group, the peripheral highlands and the southeastern lowlands are home to numerous non-Burman peoples. Most of these peoples have been in periodic rebellion against Burma (Myanmar) since the 1970s, owing to their perception that the national government was attempting to impose Burman cultural norms. Economic stagnation and political repression by the central government have intensified these conflicts.

Figure 13.28 Massacre in the Southern Philippines In November 2009, a convoy of political activists and journalists heading to register one of its allies for as a candidate for the governorship of Maguindanao province was attacked by gunmen associated with a rival clan. 57 people were killed, including 34 journalists. The photograph shows a backhoe excavating bodies and vehicles from a shallow grave, where they had been deposited by the perpetrators of the massacre.

Figure 13.30 Aung San Suu Kyi The noted Burmese democratic opposition leader Aung San Suu Kyi received the Nobel Peace Prize in 1991. As of 2010 she remained under house arrest.

have resisted such orders. In early 2010, local reports claimed the UWSA was frantically selling heroin and amphetamine stocks in order to purchase weapons as well as gear to protect its fighters against chemical attack.

Burma's repressive government has provoked international opposition. Both the United States and the European Union maintain trade sanctions. Burma's military rulers do not seem very concerned with global opinion, however, and by joining ASEAN in 1997 they gained international credibility and regional support. Growing economic ties with India and especially China have also helped strengthen Burma's government. The other ASEAN countries have expressed frustration with Burma's military government, but they are wary of provoking it for fear of pushing it closer to China.

Overall, Burma's government remains extremely suspicious of the outside world. In 2005 it mandated the creation of a new capital city at Nay Pyi Taw, located in a remote, forested area 200 miles north of the old capital of Rangoon (Yangon). The new capital, which was supposedly picked on the basis of astrological calculations and defense considerations, is largely closed to the outside world. In 2010, however, Burma agreed to hold multiparty elections, largely as a means of gaining international legitimacy. Most experts, however, viewed the elections as a sham, as neither the major opposition groups nor the Buddhist clergy were allowed to participate, and as free media coverage was prohibited.

Trouble in Thailand Compared to that of Burma, Thailand's recent history appears peaceful and stable. Thailand enjoys basic human freedoms and a thriving free press, although it does have a legacy of military takeovers followed by periods of authoritarian rule. In 2006, mass protests in Bangkok led to the resignation of the corruption-plagued prime minister, Thaksin Shinawatra, who had remained very popular among poor Thai voters in rural areas. Several months later, the Thai army seized power in a bloodless coup, apparently with the blessing of King Bhumibol, Thailand's revered constitutional monarch. Although the new military leaders allowed a quick restoration of democracy, they also banned a number of political movements, substantially reducing the freedom of the Thai people. Through early 2010, Thaksin's supporters, known as the "red-shirts" for their distinctive clothing, were organizing massive demonstrations where they called for fresh elections and the resignation of the prime minister. In May 2010, the Thai government began cracking down on the protestors, resulting in several dozen deaths.

The recent political turmoil in Bangkok is connected with the ethnic tensions that have long plagued Thailand's far south, a primarily Malay-speaking, Muslim region. While minor rebellions have flared up in southern Thailand for decades, the violence sharply escalated in 2004. Prime Minister Thaksin Shinawatra responded to the renewed fighting with harsh military measures, a strategy that many Thai military leaders viewed as counterproductive. But the more conciliatory stance taken by the Thai government after the 2006 coup has been no more effective, as a shadowy Islamist group continues to attack Thai-speaking, Buddhist government officials and as well as ordinary civilians. Between 2004 and March 2010, over 4,000 Thai citizens lost their lives in this conflict. The war has also raised tensions between Thailand and Malaysia, as Thailand accuses its southern neighbor of not doing enough to prevent militants from slipping across the border.

International Dimensions of Southeast Asian Geopolitics

As southern Thailand shows, geopolitical conflicts in Southeast Asia can be complex affairs, involving several different countries as well as non-state organizations. In earlier years, some of the most serious tensions emerged when two countries claimed the same territory. More recently, radical Islamist groups have posed the biggest challenge.

Territorial Conflicts In previous decades, several Southeast Asian countries quarreled over their boundaries. The Philippines, for example, still maintains a "dormant" claim to the Malaysian state of Sabah in northeastern Borneo, based on the fact that in the 1800s the Islamic Sultanate of the Sulu Archipelago in the southern Philippines controlled much of its territory. The Philippines suspended this claim as part of a general movement of Southeast Asian countries to reduce regional tensions. With the rise of ASEAN, national leaders concluded that friendly relations with neighbors are more important than the possible gain of additional territory.

More difficult has been the dispute over the Spratly Islands in the South China Sea, a group of rocks, reefs, and tiny islands that may well lie over substantial undersea oil reserves (Figure 13.31). The Philippines, Malaysia, Vietnam,

PEOPLE ON THE MOVE The Rohingyas Seek Sanctuary

Most contemporary international migrants leave their homeland to seek enhanced opportunity and better economic conditions. Some people, however, move because they are forced out of their own countries by political repression. Although little noted in the global media, the Rohingyas of Burma (Myanmar) have been on the move for some time. A distinct people some 700,000 strong, the Rohingyas appear on few maps of Burma. This erasure of their identity is in keeping with official Burmese policy, which has denied almost all of them Burmese citizenship. Unfortunately, the Rohingyas have been unable to find ready sanctuary elsewhere.

The Rohingyas speak an Indo-European language closely related to Bengali, spoken in Bangladesh and parts of India. Like most eastern Bengali speakers, they follow Islam—the main reason for their persecution by the strongly Buddhist Burmese state. According to Rohingya history, their ancestors began moving to their current homeland as early as the 7th century; Burmese historians contend that they did not arrive until after Burma was conquered by the British, and that they came largely because the British encouraged their migration. As a result, hard-core Burmese nationalists insist that the Rohingyas be regarded as citizens of Bangladesh, not Burma. Bangladesh, not surprisingly, rejects this interpretation.

The persecution of the Rohingyas has been going on for some time. In 1942, after Japanese forces expelled the British from Burma, mob violence took an estimated 100,000 Rohingyas lives. In the late 1970s, renewed harassment sent 250,000 to Bangladesh, where many have continued to languish in wretched refugee camps (Figure 13.4.1). Rohingyas have also fled to refugee camps in Thailand, where an estimated 110,000 currently live. Rohingyas continue to leave Burma, even though they have no place to go. Many have tried to sail to Thailand, but the Thai government tries to prevent them from landing. In 2009, Thai naval authorities reportedly towed five boats loaded with Rohingya refugees out to sea; four of the boats subsequently sank during a storm, and the fifth wrecked on the shore. In March 2010, Malaysian authorities picked up 93 Rohingya refugees in an overcrowded boat who claimed that they had been chased out of Thai waters by the Thai navy.

Bangladeshi authorities also reject Rohingya migrants, arguing that all Rohingyas who entered their country after 1991 are simply illegal immigrants. Tensions between Burma and Bangladesh mounted in the fall of 2009, focused both on the Rohingya issue and on the maritime border between the two countries. The offshore area contains significant energy resources that both countries wish to exploit.

In late December 2009, Burma and Bangladesh finally reached a provisional agreement that involved the "repatriation" of Rohingya refugees. Burma agreed to accept 9,000 out of the estimated 28,000 residing in refugee camps (an additional 300,000 to 400,000 Rohingyas currently live in Bangladesh outside of the camps). But few of the refugees are eager to return. The journalist Nurul Islam quoted one Rohingya man as saying, "We don't have any rights in Myanmar. . . . If we go back, the armed forces will use us as bonded labor. Many will be sent to jail. There are still curbs on practicing our religion or movement from one place to another without the army's permission."

Figure 13.4.1 Rohingya Refugees Thousands of Rohingya refugees live in squalid camps in Bangladesh and Thailand. Many have been expelled from such camps, and must therefore try to seek sanctuary elsewhere. This photo shows the "unregistered" Kutupalong camp near Cox's Bazar in southeastern Bangladesh.

Figure 13.31 The Spratly Islands The Spratly Islands are small and barely above water at high tide, but they are geopolitically important. Oil may exist in large quantities in the vicinity, heightening the competition over the islands. Southeast Asian countries have been especially concerned about China's military activities in the Spratlys.

and Brunei have all advanced territorial claims over the Spratly Islands, as have China and Taiwan. In 2002, all parties pledged to seek a peaceful solution, but in 2010, military analysts highlighted a worrisome naval buildup in the area, particularly on the part of China, Indonesia, Malaysia, and Singapore.

ASEAN and Global Geopolitics The development and enlargement of ASEAN reduced geopolitical tensions across Southeast Asia. Initially, ASEAN was an alliance of nonsocialist countries fearful of communist insurgency. Through the 1980s, the United States maintained naval and air bases in the Philippines, and U.S. military force bolstered the anticommunist coalition. In the early 1990s, however, the United States, under mounting pressure from Filipino nationalists, withdrew from the Philippines. By that time, the struggle between communism and capitalism was no longer an international issue. In 1995, communist Vietnam itself gained membership in ASEAN, followed by Laos and Cambodia.

While ASEAN is on friendly terms with the United States, one purpose of the organization is to prevent the U.S.—or any other country—from gaining undue influence in the region. ASEAN leaders are thus keen to include all Southeast Asian countries within the association. ASEAN's ultimate international policy is to encourage conversation and negotiation over confrontation, and to enhance trade. As a result, the ASEAN Regional Forum (ARF) was established in 1994 as an annual conference in which Southeast Asian leaders could meet with the leaders of both the East Asian and Western powers to attempt to ease tensions. Seeking to further improve relations with East Asia, ASEAN leaders have also established an annual ASEAN+3 meeting where its foreign ministers confer with those of China, Japan, and South Korea.

Global Terrorism and International Relations ASEAN has not defused all of the international political tensions in Southeast Asia. Most worrisome has been rise of radical Islamic fundamentalism in the Muslim parts of the region. One group, Jemaah Islamiya (JI), which calls for the creation of a single Islamic state across Indonesia, Malaysia, southern Thailand, and the southern Philippines, has cooperated extensively with Al Qaeda. In 2002, JI agents blew up a tourist-oriented nightclub in Bali, killing 202 people. The group followed this by bombing an American-run hotel in Jakarta in 2003, setting off explosions in Australia's Indonesian embassy in 2004, and attacking three tourist restaurants in Bali in 2005.

By 2006, however, Indonesian authorities began to make headway against JI. A number of its leaders were arrested, and in 2009 and 2010 top-level operatives were killed in security raids. As a result, some analysts believe that JI is no longer an effective fighting force. Most evidence shows that few Southeast Asian Muslims support radical groups such as JI. Since 2004, even mainstream Muslim political parties have not done particularly well in national elections in Indonesia and Malaysia. Radical Islamist groups in the southern Philippines and southern Thailand, on the other hand, remain violent and militarily capable.

Although radical Islam has relatively few supporters in Southeast Asia, the people of the region have generally been wary of U.S. foreign policy initiatives designed to combat global terrorism. Resentment against the United States is especially pronounced in Indonesia and Malaysia, where most people see the war in Iraq as an assault on Islam. Perhaps as a result, Indonesia's courts have given lenient prison sentences to several JI operatives. The group's spiritual leader, Abu Bakar Basyir, for example, served only 26 months of a 30-month sentence for his alleged involvement in the 2002 Bali bombing.

ECONOMIC AND SOCIAL DEVELOPMENT: The Roller-Coaster Ride of Globalized Development

Until the Asian economic crisis of the late 1990s, Southeast Asia was often held up as a model for a new globalized capitalism. With investment capital flowing first from Japan and the United States, then from international investment portfolios, Thailand, Malaysia, and Indonesia experienced impressive economic booms. In the summer of 1997, however, regional economies suffered a profound crisis, with the currencies of both Thailand and Indonesia being devalued almost 50 percent. Subsequent years have seen economic ups and downs, with expansion between 2002 and 2007 turning again to recession with the global economic crisis of 2008–2009.

Recent efforts to enhance growth by fostering the integration of the region's economies have met with mixed success. All 10 ASEAN countries are now members of the AFT, or the ASEAN Free Trade Area, and they have agreed to reduce tariffs against products originating in other member states. In 2010, the much larger ACFTA, or ASEAN-China Free Trade Area, came into being. ACFTA has proved highly controversial, however, as many Southeast Asian firms are having a difficult time competing against Chinese companies (see "Exploring Global Connections: The ASEAN-China Free Trade Area.")

Uneven Economic Development

Southeast Asia today is a region of strikingly uneven economic and social development. While some countries, such as Indonesia, have experienced both booms and busts, others, such as Burma (Myanmar), Laos, and East Timor, missed

EXPLORING GLOBAL CONNECTIONS ACFTA: The ASEAN–China Free Trade Area

January 1, 2010, saw the emergence of the world's largest free trade area in terms of population, linking China with the 10 members of the Association of Southeast Asian Nations (ASEAN). Disagreements remain as to what to call the new organization. In the English-language press, the favored term is ACFTA, the ASEAN–China Free Trade Area; Chinese newspapers more often call it CAFTA, the China–ASEAN Free Trade Area. "CAFTA" is a potentially misleading term, as the same acronym was used for the Central American Free Trade Agreement. Officially, however, that CAFTA became CAFTA-DR in 2004, when the Dominican Republic joined the club.

Controversies much deeper than those of naming plague the new free trade pact (Figure 13.5.1). On January 7, 2010, thousands of workers took to the streets of the Indonesian city of Bandung to demand a delay in implementation of the agreement. The protestors, expressed fears about mass layoffs, worrying that Indonesian firms would not be able to compete with Chinese businesses. Similar concerns have been expressed in Thailand and other Southeast Asian countries worried about competing with the Chinese manufacturing juggernaut. In the Philippines, highland vegetable farmers expressed fear that cheap Chinese carrots and cabbages would displace their own produce in Philippine markets. In response to such concerns, China announced on January 22, 2010 that it would work with ASEAN countries to make adjustments to the agreement.

Enthusiasm for ACFTA, on the other hand, runs high in the relatively poor regions along the border between southern China and mainland Southeast Asia. The governments of Laos and the Guangxi Zhuang Autonomous Region of China quickly initiated high-level talks to figure out how to take advantage of the free trade area. On January 7, 2010, direct flights began between Laos and Guangxi's capital, Nanning. Officials in China's Yunnan province are equally excited about the new economic possibilities.

As the Website GoKunming reported, the region will soon see "a vast network of highways and rail which will provide cities in Yunnan with cheap overland access to markets in Myanmar, Laos, Vietnam, Cambodia, Thailand, Malaysia and Singapore." The article goes on to exclaim that, "difficult as it may be to imagine, Yunnan's days as an economic and political backwater are officially over."

Economic ties between southern China and the rugged lands of northern Southeast Asia have already been surging in recent years. Such developments can have both positive and negative consequences, but they almost always generate local controversies. To reduce such tensions, China has promised to cooperate more extensively on such matters as tourism, infrastructural development, business partnerships, and cultural interaction.

Figure 13.5.1 Chinese Exports in Indonesia A mother strolls through a toy shop in Jakarta dominated by goods imported from China. Indonesian manufacturers are concerned that the new China-ASEAN Free Trade Agreement will further undermine their ability to compete with Chinese firms.

TABLE 13.2 Development Indicators

Country	GNI per Capita, PPP 2008	GDP Average Annual % Growth (2000–08)	Human Development Index (2007)[a]	Percent of Population Living Below $2 a Day	Life Expectancy (2010)[b]	Under Age 5 Mortality Rate (1990)	Under Age 5 Mortality Rate (2008)	Gender Equity 2008[c]
Burma (Myanmar)			0.586		58	120	98	99
Brunei	50,770		0.920		77			
Cambodia	1,860	9.8	0.593	57.8	61	117	90	90
East Timor	4,690	1.9	0.489	72.8	61	184	93	
Indonesia	3,590	5.2	0.734	60.0	71	86	41	98
Laos	2,050	6.9	0.619	76.8	65	157	61	87
Malaysia	13,730	5.5	0.829	7.8	74	18	6	
Philippines	3,900	5.1	0.751	45.0	72	61	32	102
Singapore	47,940	5.8	0.944		81	7	3	
Thailand	7,760	5.2	0.783	11.5	69	32	14	
Vietnam	2,690	7.7	0.725	48.4	74	56	14	

[a]*United Nations, Human Development Report, 2009.*

[b]*Population Reference Bureau, World Population Data Sheet, 2010.*

[c]*Gender Equity—Ration of female to male enrollments in primary and secondary school. Numbers below 100 have more males in primary/secondary school; numbers above 100 have more females in primary/secondary schools.*

Source: World Bank, World Development Indicators, 2010.

the expansion of the 1980s and 1990s and remain deeply impoverished. Oil-rich Brunei and technologically sophisticated Singapore, on the other hand, remain two of the world's more prosperous countries (Table 13.2). In the Philippines the situation is somewhat more complicated.

The Philippine Decline Fifty years ago, the Philippines was the most highly developed Southeast Asian country and was considered by many to have the brightest prospects in all of Asia. It boasted the best-educated populace in the region, and it seemed to be on the verge of sustained economic development. Per capita GNI in 1960 was higher in the Philippines than in South Korea. By the late 1960s, however, Philippine development had been derailed. Through the 1980s and early 1990s, the country's economy failed to outpace its population growth, resulting in declining living standards. The Philippine people are still well educated and reasonably healthy by world standards, but even the country's educational and health systems declined during the dismal decades of the 1980s and 1990s.

Why did the Philippines fail so spectacularly despite its earlier promise? Although there are no simple answers, it is clear that dictator Ferdinand Marcos (who ruled from 1968 to 1986) squandered billions of dollars while failing to enact programs conducive to genuine development. The Marcos regime instituted a kind of **crony capitalism** in which the president's many friends were granted huge sectors of the economy, while those perceived to be enemies had their properties confiscated. After Marcos declared martial law in 1972 and suspended Philippine democracy, revolutionary activity intensified and the country began to fall into a downward spiral.

While it is tempting to blame the failure of the Philippines on the Marcos regime, such an explanation is only partially adequate. Indonesia and Thailand also saw the development of crony capitalism, yet their economies have proved more competitive. And when the Marcos dictatorship was finally replaced by an elected democratic government in 1986, the Philippine economy continued to languish. Remittances by Filipinos working abroad have kept the economy afloat, but this exodus of labor represents in many respects a tragedy for the country as a whole.

By the beginning of the new millennium, the Philippine economy was showing some signs of revival. The government turned its attention to infrastructural problems, such as electricity generation, and foreign investments began to flow into the country. A vibrant local economy emerged in the former U.S. naval base of Subic Bay. Boasting both world-class facilities and a highly competent local government, Subic Bay has become a major export-processing center. Cebu City, on the central Visayan island of Cebu, also has expanded quickly, giving rise to its local nickname of "Ceboom" (Figure 13.32). Unlike most of its neighbors, the Philippines managed a slight uptick in economic activity in the crisis year of 2009, largely because it had not participated in the financial boom that had preceded the global downturn.

Despite its recent stability, the Philippine economy continues to be undermined by political and social problems. The Philippine political system, modeled on that of the United States, entails elaborate checks and balances between the different branches of government. Critics contend that such "checks" are so effective that little is accomplished. The Philippine government habitually spends far more money than it takes in taxes, leading to a huge public debt

Figure 13.32 Cebu City, Philippines Cebu City in Central Philippines has experienced strong population and economic growth in recent years. The photograph shows a portion of the city's modern central business district.

that consumes a large portion of the national budget. Another obstacle is the fact that the Philippines has the least equitable distribution of wealth in the region. While many members of the elite are fantastically wealthy by any measure, roughly half of the country's people subsist on less than $2 a day. As a result, many Filipinos must seek employment abroad. Women often work as nurses, maids, or nannies, whereas men from the Philippines often work in construction or on ships (Figure 13.33).

The Regional Hub: Singapore Over the past half century, Singapore has been Southeast Asia's greatest developmental success. It has transformed itself from an **entrepôt** port city, a place where goods are imported, stored, and then transshipped, to one of the world's most prosperous and modern states. Singapore, a thriving high-tech manufacturing center, also stands as the communications and financial hub of Southeast Asia. The Singaporean government has played an active role in the development process but has also allowed market forces freedom to operate. Singapore has encouraged investment by multinational companies (especially those involved in technology) and has invested heavily in housing, education, and some social services (Figure 13.34). Although Singapore's economy contracted sharply in 2009, experts see renewed growth in the coming years.

Singapore's political system has nurtured economic development, but it is somewhat repressive and by no means fully democratic. Although elections are held, the government manipulates the process, ensuring that the opposition never gains control of parliament. While many Singaporeans object to such policies, others counter that they have brought fast growth as well as a clean, safe, and remarkably corruption-free society.

Figure 13.33 Filipino Seamen Hundreds of thousands of Filipino men work at sea. Many work in the merchant marines, employed by companies all over the world. Others, like those pictured here, serve as crew members of cruise ships.

Although the Singaporean government has thus far been able to repress dissent, its authoritarian form of capitalism confronts a new challenge in the Internet. National leaders want the communication services that the Net provides, but they are worried about the free expression that it allows, fearing that it will lead to excessive individualism. It will be interesting to see how Singapore responds to this challenge, in part because the governments of China and Vietnam seem to be following the technocratic, authoritarian capitalism pioneered by Singapore.

Malaysia's Insecure Boom Although not nearly as prosperous as Singapore, Malaysia has experienced rapid economic growth over the past several decades and is now generally classified as an upper-middle-income country. Development was initially concentrated in agriculture and natural resource extraction, focused on tropical hardwoods, plantation products (mainly palm oil and rubber), and tin. More recently, manufacturing, especially in labor-intensive high-tech sectors, has become the main engine of growth. As Singapore prospered, many of its enterprises moved their manufacturing operations into neighboring Malaysia. Increasingly, Malaysia's economy is multinational; many Western high-tech firms operate in the country, and several Malaysian companies have themselves established subsidiaries in foreign lands. As a result of its prosperity, Malaysia has attracted hun-

Figure 13.34 Housing in Singapore Despite its free-market approach to economics, the government of Singapore has invested heavily in public housing. Many Singaporeans live in buildings similar to the one depicted in this photograph.

dreds of hundreds of thousands of illegal immigrants, mostly from Indonesia and the Philippines.

Malaysia's economy began to grow rapidly in the 1970s, but it was hard hit by the Asian economic crisis of the late 1990s. The Malaysian government's response to the crisis was different from that of its neighbors. Spurning the advice of the International Monetary Fund, Malaysia instituted temporary currency controls, regulating the flow of international funds in and out of the country. Some observers believe that these policies helped Malaysia recover more rapidly than Thailand and Indonesia, but others are more inclined to credit Malaysia's tight fiscal policies, which have kept its external debt small. Highly dependent on the export of electronic goods, the Malaysian economy remains vulnerable to fluctuations in the North American, European, and Japanese markets. Like Singapore, Malaysia suffered a serious recession in 2009.

The economic geography of modern Malaysia shows large regional variations. Most industrial development has occurred on the west side of peninsular Malaysia, with most of the rest of the country remaining dependent on agriculture and resource extraction. More important, however, are disparities based on ethnicity. The industrial wealth generated in Malaysia has been concentrated in the Chinese community. Ethnic Malays remain less prosperous than Chinese Malaysians, and those of South Asian origin tend to be poorer still. Many of the country's tribal peoples have suffered as development proceeds and their lands are taken away.

The disproportionate prosperity of the local Chinese community is a feature of most Southeast Asian countries. The problem is particularly acute in Malaysia, however, simply because its Chinese minority is so large (some 30 percent of the country's total population). The government's response has been one of aggressive "affirmative action," by which economic clout is transferred to the numerically dominant Malay, or **Bumiputra** ("sons of the soil"), community. This policy has been reasonably successful, although it has not yet reached its main goal of placing 30 percent of the nation's wealth in the hands of the Bumiputras. Because the Malaysian economy as a whole has grown rapidly since the 1970s, the Chinese community has thrived even as its relative share of the country's wealth declined. Considerable resentment, however, is still felt by the Chinese, many of whom believe that both Malaysia's economic and educational systems are biased against them. More recently, several Malay politicians have been criticizing the system, arguing that it undermines economic incentives for the majority population. In 2009, a number of minor economic sectors were removed from the Bumiputra policy regulations.

Thailand's Ups and Downs Thailand also climbed rapidly during the 1980s and 1990s into the ranks of the world's newly industrialized countries, although it remained much less prosperous than Malaysia. Thailand also experienced a major downturn in the late 1990s that undermined much of this development. Recovery began in earnest after 2000, but the military coup of 2006 coupled with the intensifying insurgency in the far south resulted in another slowdown. Export growth of both agricultural goods and machinery, however, remains strong.

Japanese companies were leading players in the Thai boom of the 1980s and early 1990s. As Japan itself became too expensive for assembly and other manufacturing processes, Japanese firms began to relocate factories to such places as Thailand. They were particularly attracted by the country's low-wage, yet reasonably well-educated workforce. Thailand was also seen as politically stable, lacking the severe ethnic tensions found in many other parts of Asia. Although Thailand's Chinese population is large and economically powerful, relations between the Thai and the local Chinese have generally been good.

Thailand's economic expansion has by no means benefited the entire country to an equal extent. Most industrial development has occurred in the historical core, especially in the greater Bangkok metropolitan area. Yet even in Bangkok, the blessings of progress have been mixed. As the city begins to choke on its own growth, industry has begun to spread outward. The entire Chao Phraya lowland area shares to some extent in the general prosperity due both to its proximity to Bangkok and its rich agricultural resources. In northern Thailand, the Chiang Mai area has profited from its ability to attract large numbers of international tourists.

Other parts of the country have not been so fortunate. Thailand's Lao-speaking northeast (the Khorat Plateau) is the country's poorest regions. Soils here are too thin to support highly productive agriculture, yet the population is sizable. Because of the poverty of their homeland, northeasterners often are forced to seek employment in Bangkok. As Lao-speakers, they often experience ethnic discrimination. Men typically find work in the construction industry; northeastern women more often make their living as prostitutes (Figure 13.35). Not coincidentally, northeastern Thailand has strongly supported the ousted populist prime minister, Thaksin Shinawatra. The Muslim area of southern Thailand also remains very poor, contributing to the region's insurgency.

Indonesian Economic Development At the time of independence (1949), Indonesia was one of the poorest countries in the world. The Dutch had used their colony largely as an extraction zone for tropical crops and mineral resources and had invested little in infrastructure or education. The population of Java mushroomed in the 19th and early 20th centuries, leading to serious land shortages.

The Indonesian economy finally began to expand in the 1970s. Oil exports fueled the early growth, as did the logging of tropical forests. But unlike most other petroleum exporters, Indonesia continued to grow even after prices plummeted in the 1980s. Production subsequently declined, and in 2004 Indonesia had to begin importing oil. But like Thailand and Malaysia, Indonesia proved attractive to multinational companies eager to export from a low-wage economy. Large Indonesian firms, some three-quarters of them owned by local Chinese families, also have capitalized on the country's low wages and abundant resources.

But despite rapid growth in the 1980s and 1990s, Indonesia remains a poor country. Until recently, its pace of economic expansion seldom matched that of Thailand, Singapore, and Malaysia, and it has remained more dependent on the unsustainable exploitation of natural resources. The financial crisis of the late 1990s, moreover, hurt Indonesia more severely than any other country, and political instability continues to hamper economic recovery. Yet Indonesia was one of the few large countries to avoid the recent global economic crisis, with its economy growing by more than 4 percent in 2009.

Figure 13.35 Sex Tourism in Thailand Thailand has one of the highest rates of prostitution in the world. While most prostitutes cater to a local clientele, those working in Bangkok's infamous Patpong district are usually hired by foreign men. Prostitution in Thailand is associated with high rates of HIV infection and with the brutal exploitation of women and girls.

As in Thailand, development in Indonesia exhibits pronounced geographical disparities. Northwest Java, close to the capital city of Jakarta, has boomed, and much of the resource-rich and moderately populated island of Sumatra has long been relatively prosperous. Eastern Borneo (Kalimantan), another resource area, also is relatively well off. In the overcrowded rural districts of central and eastern Java, however, many peasants have inadequate land and remain near the margins of subsistence. Far eastern Indonesia has experienced little economic or social development, and throughout the remote areas of the "outer islands" tribal peoples have suffered as their resources have been taken and their lands lost to outsiders.

Divergent Economic Paths: Vietnam, Laos, and Cambodia The three countries of former French Indochina—Vietnam, Cambodia, and Laos—experienced only modest economic expansion during Southeast Asia's boom years of the 1980s and 1990s. This area endured almost continual warfare between 1941 and 1975, which persisted until the mid-1990s in Cambodia. Critics contend that the socialist economic system adopted by these countries prevented sustained economic growth. This debate is now moot, however, since a globalized capitalist model of development has largely been adopted in all three countries.

Vietnam's economy is much stronger than those of Cambodia and Laos. The country's per capita GNI of $2,700, however, is still low by global standards. Postwar reunification in 1975 did not initially bring the anticipated growth, and economic stagnation ensued. Conditions declined in the early 1990s after the fall of the Soviet Union, Vietnam's main supporter and trading partner. Until the mid-1990s, Vietnam remained under embargo by the United States. Frustrated with their country's economic performance, Vietnam's leaders began to embrace the market while retaining the political forms of a communist state. They have, in other words, followed the Chinese model. Vietnam now welcomes multinational corporations, which are attracted by the low wages received by its relatively well-educated workforce (Figure 13.36).

Such efforts seemingly began to pay off after 2000. By 2007, the Vietnamese economy was expanding at roughly 8 percent a year, one of the fastest rates of growth in the world. Foreign investment began to pour into the country, and exports—especially of textiles—began to surge. In January 2007, Vietnam joined the World Trade Organization, which encouraged exports while ensuring the continuation of market-oriented reforms. Like Indonesia, Vietnam continued to grow economically through the global crisis of 2008–2009.

Vietnam's recent economic rise, however, has not been untroubled. Many lowland peasants and upland tribal peoples have been excluded from the boom and are thus growing increasingly discontented. Tensions between the north, center of political authority, and the south, center of economic power, seem to be increasing. Trade disputes have arisen with the United States, which has accused Vietnamese

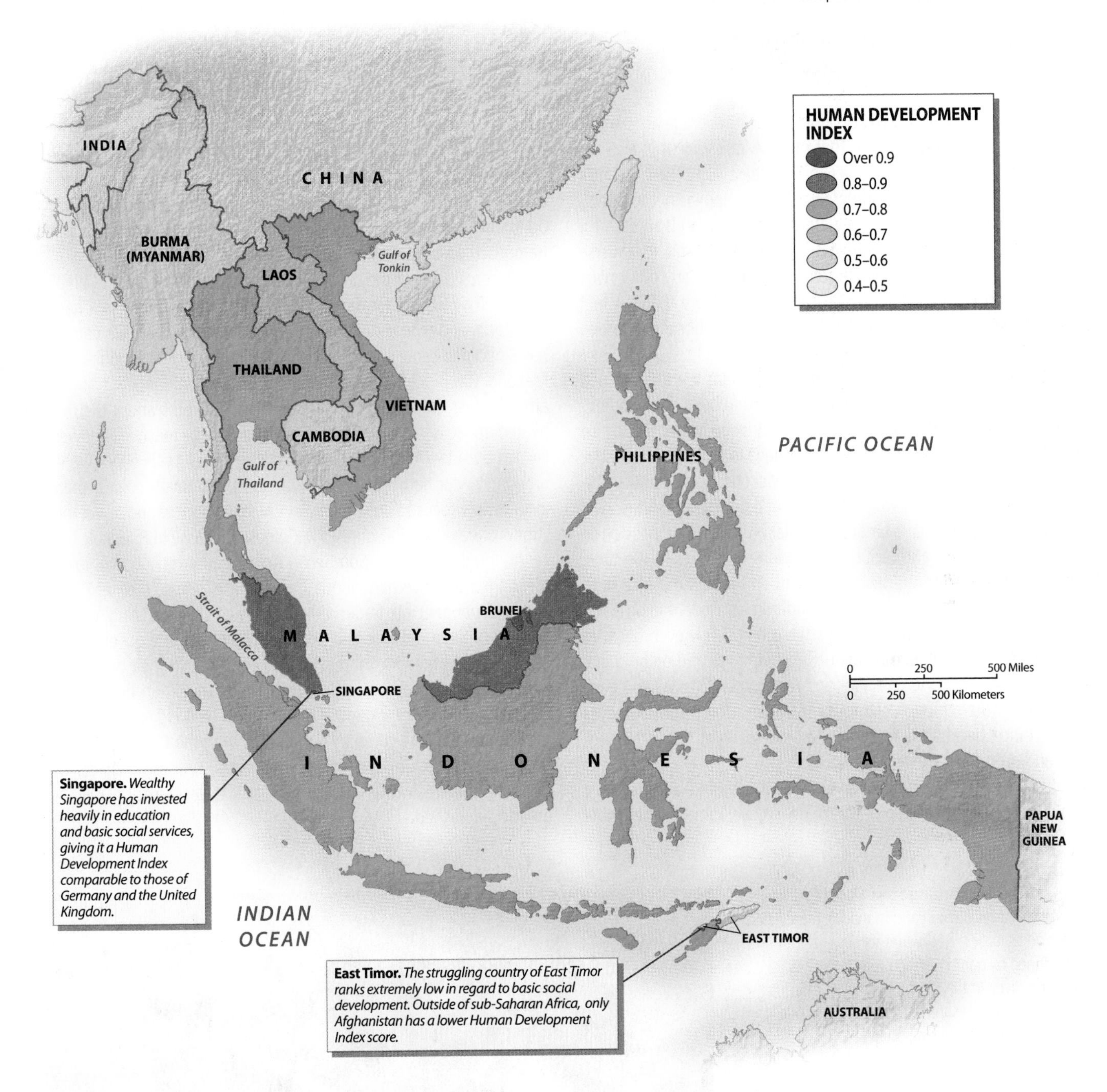

Figure 13.36 Development Issues: Human Development in Southeast Asia Levels of health, education, and other aspects of human development vary widely across Southeast Asia. As the human development figures shown on the map indicate, Singapore, Brunei, and Malaysia have reached high levels of development, whereas Burma, Cambodia, and East Timor lag well behind.

exporters of dumping such products as farm-raised catfish and shrimp on American markets.

Laos and Cambodia face some of the most serious economic problems in the region. In Cambodia, the ravages of war followed by continuing political instability long undermined economic development. Laos faces special difficulties owing to its rough terrain and relative isolation. Both countries also are hampered by a lack of infrastructure; outside the few cities, paved roads and reliable electricity are rare. As a result, the economies of both Cambodia and Laos remain largely agricultural in orientation, relying heavily as well on environmentally destructive logging and mining operations. In Laos, subsistence farming still accounts for more than three-quarters of total employment.

The Laotian government is pinning its economic hopes on dam and road building (see "Environment and Politics:

ENVIRONMENT AND POLITICS Dam Building on the Mekong River

The Mekong is the world's 12th largest river, its average annual discharge being roughly equal to that of the Mississippi. It is also the largest and most important river in Southeast Asia. The Mekong's flow is vital for irrigation and domestic water supplies in Laos, Thailand, Cambodia, and southern Vietnam. The river is also noted for its productive fisheries, worth $2 billion a year, and its diverse fauna. The Mekong has more large fish species that any other river. The Mekong Giant Catfish, which can grow to more than 11 feet (3.3 meters) and 660 pounds (300 kilograms), is usually regarded as the world's largest freshwater fish (Figure 13.6.1).

During the 1950s and 1960s, ambitious plans were developed for "taming" this giant river. Planners wanted to build dams to generate electricity and supply dry-season irrigation water. They also wanted to remove the large rapids that have long impeded navigation. The national governments of the Mekong Basin enthusiastically supported such schemes at first, as did the United States. As a result, local governments created the Mekong River Commission to ensure cooperation and facilitate development. Critics at the time, however, objected to the fact that such dams would displace hundreds of thousands of people in addition to undermining local freshwater fisheries. Their concerns were moot, however, because by the late 1960s all dam-building plans had to be put on hold due to the chaos of the Vietnam War and related conflicts. When peace returned to the region, Mekong River Commission reconsidered its early stance, deciding to focus not on building dams but on managing and preserving existing resources.

Although the Mekong River Commission turned away from a construction-oriented approach to river development, the same cannot be said of China, which is not a member of the Commission even through it controls the river's headwaters and upper reaches. Over the past 30 years, China has constructed three large hydroelectric dams on the river, and is currently building or planning an additional four. China has also been blasting away the rapids that hamper navigation on the upper Mekong.

Most residents of the lower Mekong are not pleased with China's actions. They worry that dam building and rapids blasting will harm fish stocks and reduce the downstream flow of the river. Fish catches have declined, and several endemic animal species, including the Mekong River dolphin and manatee, became endangered shortly after the first Chinese dams were finished. Concerns about the river heated up rapidly in early 2010 when its discharge reached its lowest level in decades. Although scientists claimed that the El Niño-related drought in northern Southeast Asia was the primary culprit, many local people focused their anger on China. As a result, noisy demonstrations were held in front of the Chinese Embassy in Bangkok.

Although the four Southeast Asian countries of the lower Mekong are concerned about the Chinese dams, they are also highly dependent on trade with China, and they certainly do not want to antagonize their giant northern neighbor. The governments of Laos and Cambodia, moreover, are keen to build their own dams so that they can capture hydroelectric and navigation benefits. As they do not have the expertise or the funds to build such dams by themselves, they are currently negotiating with the Chinese government as well as Chinese engineering firms. If these projects go through, the freshwater fisheries of the lower Mekong will be greatly diminished.

Figure 13.6.1 Endangered Species The Mekong Giant Catfish (Pangasianodon gigas), the world's largest freshwater fish, is currently threatened by dam-building projects on the Mekong River. In this photo, taken in 2005, Thai fishermen are measuring the largest fish caught in over 40 years.

Dam Building on the Mekong River"). The country is mountainous and has many large rivers, and could therefore generate large quantities of electricity, in high demand in neighboring Thailand and Vietnam. Up to 100 new dam projects have been proposed for the Mekong River system, most of them to be located in Laos. The Asian Development Bank also is supporting an ambitious road-building program that would pass through Laos in order to link China to Thailand. The Laotian government, however, remains repressive, discouraging foreign investment.

Although still deeply impoverished, Cambodia has experienced a recent economic boom of surprising proportions. Foreign investment has generated an expanding textile sector, tourism is thriving, mining is taking off, and in 2005 important oil and natural gas fields were discovered in Cambodian territorial waters. From 2004 to 2007, the Cambodian grew at an annual rate of over 10 percent. China has recently emerged as Cambodia's closest economic partner. Between 2006 and early 2010, China invested more than $10 billion in Cambodia, a huge figure considering the fact that the country's entire annual economic output is only about $11 billion.

Cambodia's extremely rapid economic expansion was possible only because the country's economic starting point was at such a low level. A lack of both skills and basic infrastructure, as well as problems of political instability and corruption, could easily result in renewed economic difficulties. The recent economic boom has also led to an epidemic of "land-grabbing," in which politically connected elites take over the properties of impoverished peasants in order to develop them (Figure 13.37).

Despite their basic lack of development, Cambodia and Laos are not as poor as one might expect from the official economic statistics. Both countries have relatively low population densities and abundant resources. Their low per capita GNI figures, more importantly, partially reflect the fact that many of their people remain in a subsistence economy. While the Laotian highlanders make few contributions to GNI, most of them at least have adequate shelter and food.

Burma's Troubled Economy Burma (Myanmar) stands near the bottom of the scale of Southeast Asian economic development. For all of its many problems, however, Burma is a land of great potential. It has abundant resources—including natural gas, oil and other minerals, water, and timber—as well as a large expanse of fertile farmland. Its population density is moderate, and its people are reasonably well educated. The country, however, has seen little economic or social development.

Burma's economic woes can be traced in part to the continual warfare that it has experienced. Most observers also blame economic policy. Beginning in earnest in 1962, Burma (Myanmar) attempted to isolate its economic system from global forces in order to achieve self-sufficiency under a system of Buddhist socialism. While its intentions were admirable, the experiment was not successful. Instead of creating a self-contained economy, Burma (Myanmar) found itself burdened by smuggling and black-market activities.

In the early 1990s, Burma began to open its economy to market forces and increase its involvement in foreign trade. By 2000, however, the reform measures had stalled out, and in 2003 a banking crisis resulted in another round of economic havoc. High rates of inflation and worrisome fiscal deficits undermine business confidence, but probably the most damaging economic impediment is Burma's exchange-rate policy. In 2009, when the official exchange rate was 6.4 Burmese kyat to the U.S. dollar, one could get up to 1,090 kyat to the dollar in street markets. The Burmese military and its political allies continue to run much of the economy in a highly inefficient manner. Although Burma's government is now hoping that natural gas exports and increased trade with China and India will help solve its economic problems, its economic policies will probably continue to undermine genuine development.

Figure 13.37 Phnom Penh Severe conflicts have emerged in Cambodia's booming capital city, Phnom Penh, as slum areas are cleared to make way for development projects. In this photograph, a woman wearing a helmet tries to stop a bulldozer from demolishing her home while her daughter tries to pull her away from the dangerous machine.

Globalization and the Southeast Asian Economy

As the previous discussion shows, Southeast Asia has undergone rapid but uneven integration into the global economy (Figure 13.38). Singapore has thoroughly staked its future to the success of multinational capitalism, and several neighboring countries are following suit. According to one measurement, Malaysia and Singapore have, respectively, the fourth and fifth most trade-dependent economies in the world. Even communist Vietnam and once-isolationist Burma (Myanmar) have opened their doors to international commerce, although with far more success in the former case than the latter.

Much debate has arisen among scholars over the roots of Southeast Asia's economic gains, as well as its more recent economic problems. Those who credit primarily the diligence, discipline, and entrepreneurial skills of the Southeast Asian peoples are optimistic about future economic expansion. Skeptics argue, however, that most of the region's growth has come from the application of large quantities of labor and capital unsupported by real advances in productivity. Regardless of how this debate turns out, it is clear that Southeast Asia's globalized economies are heavily dependent upon exports to the international market. In the 1990s, many observers thought that the booming economies of Southeast

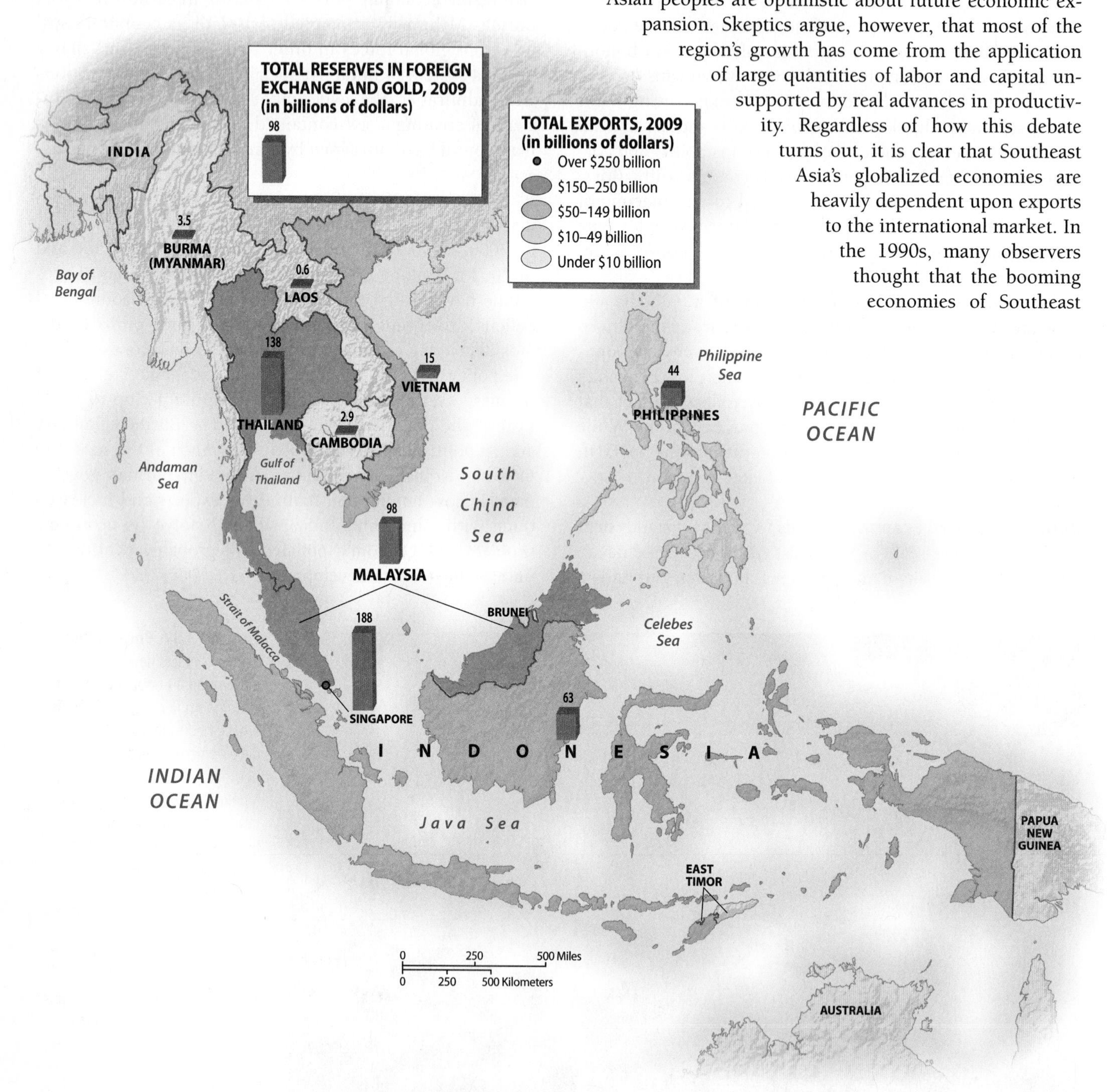

Figure 13.38 Southeast Asia's Global Linkages Much of Southeast Asia is tightly integrated into global trade networks, exporting large amounts of both primary products and manufactured goods, and accumulating large reserves of foreign currency and gold. This is particularly true of Malaysia and Singapore. In contrast, Laos, Cambodia, and especially Burma (Myanmar) remain less connected with the global economic system, although their ties are growing.

Asia had become overly dependent on exports to the United States, but in recent years the rise of China has resulted in a more balanced trading regime. In 2009, for example, 10.9 percent of Thailand's exports went to the United States, while 16.8 percent went to China (including Hong Kong) and another 10.3 went to Japan. Currently, the main concern in Southeast Asia is competition from China in exporting inexpensive consumer goods to the rest of the world. As is true in the United States, many people in the region believe that the Chinese currency is undervalued, giving China an unfair advantage in export markets.

Globalized industrial production in Southeast Asia remains controversial. Consumers in the world's wealthy countries worry that many of their basic purchases, such as sneakers and clothing, are produced under exploitative conditions. Movements have thus emerged to pressure both multinational corporations and Southeast Asian governments to improve the working conditions of laborers in the export industries. The multinational firms in question counter that such exploitative conditions occur not in their own factories, but rather in those of local companies that subcontract for them; activists, however, point out that multinationals have tremendous influence over their local subcontractors. Some Southeast Asian leaders object to the entire debate, accusing Westerners of wanting to prevent Southeast Asian development and protect their own home markets, under the pretext of concern over worker rights.

Issues of Social Development

As might be expected, several key indicators of social development in Southeast Asia correlate well with levels of economic development. Singapore thus ranks among the world leaders in regard to health and education, as does the small, yet oil-rich country of Brunei. East Timor, Laos, and Cambodia, not surprisingly, come out near the bottom for these measures. The people of Vietnam, however, are healthier and better educated than might be expected on the basis of their country's overall economic performance. Burma (Myanmar), on the other hand, has seen a decline in some social indicators. Its per capita health budget is one of the lowest in the world, while its military budget, as a percentage of its overall economic output, is one of the highest.

With the exception of Laos, Cambodia, Burma, and East Timor, Southeast Asia has achieved relatively high levels of social welfare. In Burma, however, life expectancy at birth hovers around a miserable 58 years (as compared to Vietnam's 74 years), and in Laos, female illiteracy rates remain near 40 percent. But even the poorest countries of the region have made improvements in mortality under the age of 5, as is evident from the figures in Table 13.2. Progress has been more pronounced in prosperous countries such as Malaysia and Singapore, whereas war-torn Cambodia has made smaller gains. Thailand has recently begun to provide a social safety net for its citizens, hoping to raise its level of social development to that found in the world's wealthier countries.

Most of the governments of Southeast Asia have placed a high priority on basic education. Literacy rates are relatively high in most countries of the region. Much less success, however, has been realized in university and technical education. As Southeast Asian economies continue to grow, this educational gap is beginning to have negative consequences, forcing many students to study abroad. High levels of basic education, along with general economic and social development, also have led to reduced birthrates through much of Southeast Asia. With population growing much more slowly now than before, economic gains are more easily translated into improved living standards.

Summary

- Some of the most serious problems created by globalization in Southeast Asia are environmental. Given the emphasis placed by global trade on wood products, it is perhaps understandable that Southeast Asia has sacrificed so many of its forests to support economic development. But in most of the region, forests are now seriously depleted. As the Chinese economy expands, market demand for Southeast Asian forest products is increasing rapidly.

- Deforestation in Southeast Asia is also linked to domestic population growth and changes in settlement patterns. As people move from densely populated, fertile lowland areas into remote uplands, both environmental damage and cultural conflicts often follow. Population movements in Southeast Asia also have a global dimension. This is particularly true in regard to the Philippines, which has sent some 8 million workers to more prosperous parts of the world.

- Southeast Asia is characterized today by tremendous cultural diversity. In recent years, conflicts over language and religion have seriously weakened several Southeast Asian countries, including Indonesia, East Timor, and Burma (Myanmar). However, the region has found a new sense of regional identity as expressed through ASEAN, the Association of Southeast Asian Nations.

- The relative success of ASEAN has not solved all of Southeast Asia's political tensions. Many of its countries still argue about geographical, political, and economic issues, while insurgencies remain strong in the Philippines and Thailand. Cambodia, Laos, and especially Burma (Myanmar) have also been held back by repressive and corrupt governments.

- Although ASEAN has played an economic as well as political role, its economic successes have been more limited. Most of the region's trade is still directed outward toward the traditional centers of the global economy: North America, Europe, and East Asia. This orientation is not surprising, considering the export-focused policies of most Southeast Asian countries. A significant question for Southeast Asia's future is whether the region will develop an integrated regional economy. A more important issue is whether social and economic development will be able to lift the entire region out of poverty instead of benefiting just the more fortunate areas.

Key Terms

animism *(page 610)*
Association of Southeast Asian Nations (ASEAN) *(page 590)*
Bumiputra *(page 627)*
copra *(page 601)*
crony capitalism *(page 625)*
domino theory *(page 616)*
entrepôt *(page 626)*
Golden Triangle *(page 600)*
Khmer Rouge *(page 616)*
lingua franca *(page 610)*
primate cities *(page 605)*
Ramayana *(page 607)*
shifted cultivators *(page 604)*
Sunda Shelf *(page 595)*
swidden *(page 600)*
transmigration *(page 603)*
tsunamis *(page 595)*
typhoons *(page 597)*

Review Questions

1. Why are river deltas so important in the settlement pattern of Southeast Asia?
2. Explain how and why the monsoon climates of Southeast Asia differ between mainland and islands.
3. What do we mean when we refer to the "globalization of world forestry"?
4. Explain why smoke and air pollution are so pronounced in Southeast Asia.
5. Compare and contrast the three major types of agriculture in the region—swidden, plantation, and rice cultivation.
6. How might "transmigration" solve and also aggravate a country's population problems?
7. What major religions are found in Southeast Asia? Describe the historical and contemporary patterns for each religion.
8. Why is there ambivalence about the English language in some Southeast Asian countries?
9. What are the ethnic tensions facing Indonesia? Locate the different problem areas on a map.
10. What are the goals of ASEAN, and how have those goals changed in the last several decades?

Thinking Geographically

1. Discuss the ramifications of state-sponsored migration in Indonesia from areas of high population density to areas of low population density in both the "sending" and the "receiving" areas.
2. Why should—or should not—citizens of the United States be concerned about deforestation in Southeast Asia? If they should be, what would be the proper ways they might show such concern?
3. What might the fate of animism be in the new millennium? Consider whether it might be doomed to extinction before the forces of modern economics and national integration, or whether it may persist as tribal peoples struggle to retain their cultural identities.
4. What should be the position of the English language in the educational systems of Southeast Asia? What should be the position of each country's national language? What about local languages?
5. How might ethnic tensions in countries such as Burma (Myanmar) and Indonesia be reduced? Does Indonesia have a reasonable claim to such areas as Irian Jaya?
6. What roles might ASEAN play in the coming years? Should it concentrate on economic or political issues?
7. How could Malaysia successfully integrate its economy into the global system and at the same time regulate the flow of global (or Western) culture? Evaluate whether Singapore can continue to experience economic growth while severely limiting basic freedoms.
8. Is the Southeast Asian economic path of integration into the global economy, which is marked by an openness to multinational corporations, going to prove wise in the long run, or do its potential hazards outweigh its benefits?

Regional Novels and Films

Novels

Anthony Burgess, *The Long Day Wanes: A Malaysian Trilogy* (1965, Norton)

Joseph Conrad, *Victory* (1936, Doubleday)

Graham Greene, *The Quiet American* (1956, Viking Press)

Ninfong Ho, *Rice without Rain* (1990, HarperCollins)

F. Sionil Jose, *The Pretenders* (1962, Philippines)

Mochtar Lubis, *A Road with No End* (1982, Graham Bush)

W. Somerset Maugham, *Borneo Stories* (1976, Heinemann)

Paul Theroux, *The Consul's Fire* (1977, Houghton Mifflin)

Films

Anak-Anak Borobudur (2007, Indonesia)

The Bet Collector (2006, Philippines)

The Big Durian (2003, Malaysia)

The Blossoming of Maximo Oliveros (2005, Philippines)

Eliana, Eliana (2002, Indonesia)

Indochine (1992, France)

Jan Dara (2001, Thailand)

King of the Garbage Dump (2002, Vietnam)

Midnight, My Love (2005, Thailand)

Money No Enough 2 (2008, Singapore)

Platoon (1987, U.S.)

The Scent of Green Papaya (1993, Vietnam)

Sunset at Chaopraya (1996, Thailand)

Bibliography

Acharya, Amitav. 2001. *Constructing a Security Community of Southeast Asia: ASEAN and the Problem of Regional Order*. New York: Routledge.

Broad, Robin, and Cavanagh, John. 1993. *Plundering Paradise: The Struggle for the Environment in the Philippines*. Berkeley: University of California Press.

Broek, Jan O. 1944. "Diversity and Unity in Southeast Asia." *Geographical Review 34*, 175–195.

Dixon, Chris. 1991. *South East Asia in the World Economy: A Regional Geography*. Cambridge, UK: Cambridge University Press.

Gupta, Avijit, (ed.). 2005. *The Physical Geography of Southeast Asia*. New York: Oxford University Press.

Hefner, Robert W. 2001. *The Politics of Multiculturalism: Pluralism and Citizenship in Malaysia, Singapore, and Indonesia*. Honolulu: University of Hawaii Press.

Herring, George. 1995. *America's Longest War: The U.S. and Vietnam 1950–1975*. New York: McGraw-Hill.

Peleggi, Maurizio. 2007. *Thailand: The Worldly Kingdom*. London: Reaktion Books.

Rumney, Thomas A. 2010. *The Geography of Southeast Asia: A Scholarly Bibliography and Guide*. Lanham, MD: University Press of America.

Steinberg, David Joel. 2001. *Burma: The State of Myanmar*. Washington, DC: Georgetown University Press.

Taylor, Jean G. 2004. *Indonesia: Peoples and Histories*. New Haven, CT: Yale University Press.

Ulack, Richard, and Pauer, Gyula. 1989. *Atlas of Southeast Asia*. New York: Macmillan.

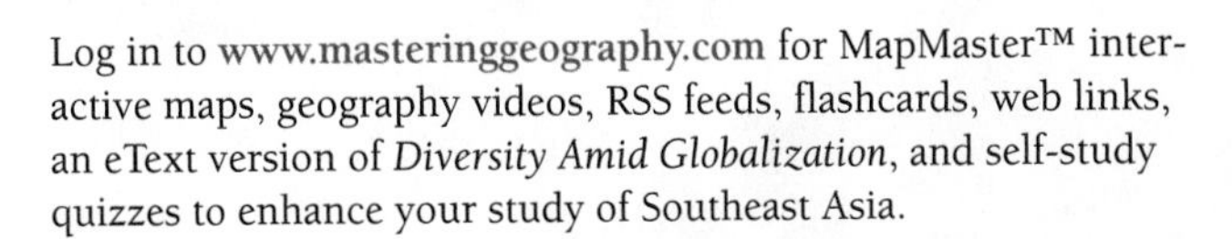

Log in to www.masteringgeography.com for MapMaster™ interactive maps, geography videos, RSS feeds, flashcards, web links, an eText version of *Diversity Amid Globalization*, and self-study quizzes to enhance your study of Southeast Asia.

MapMaster™ presents 13 Place Name and 13 Layered Thematic interactive maps to help students practice and master their geographic literacy, spatial reasoning, and critical thinking skills.

14 Australia and Oceania

Taro root, a native of southeast Asia, has long been an important food source throughout Oceania as well as in Asia and Africa. Besides being a staple food crop, taro is also a prestige and symbolic crop associated with royalty, gift-giving, traditional feasting, and the fulfillment of social obligations. These taro fields are in the Hanalei Valley, Kauai, Hawaii.

DIVERSITY AMID GLOBALIZATION

Australia and Oceania were once isolated from the world; however, globalization has now integrated them into the global community through a complex history of colonization, geopolitical arrangements, and economic ties. Traditionally oriented toward Europe because of shared colonial histories, today the region is becoming increasingly linked to Asia.

ENVIRONMENTAL GEOGRAPHY

Global warming may result in Australia's semiarid interior agricultural lands suffering from warmer temperatures and drought, while sea level rise threatens low-lying Pacific islands. Other environmental issues result from the impact nonnative plants and animals, which include snakes, rabbits, and feral pigs, are having on the region's unique biodiversity.

POPULATION AND SETTLEMENT

Australia and New Zealand, because of their higher living standards, continue to be migration targets for immigration from both within and outside of Oceania. Conversely, higher rates of natural growth on the Pacific islands, coupled with high unemployment, encourage high out-migration.

CULTURAL COHERENCE AND DIVERSITY

Until 1973, Australia protected its European ethnic roots with the White Australia Policy; however, in the last decades immigration from Africa and Asia has created a new multicultural society.

GEOPOLITICAL FRAMEWORK

From Hawaii to Australia, native peoples are demanding ownership or, minimally, access to their ancestral lands. More often than not, however, these land claims are fraught with controversy and tension.

ECONOMIC AND SOCIAL DEVELOPMENT

Increased trade linkages with China have brought economic benefits to many countries in Oceania, particularly Australia, even though those benefits also involve liabilities because of China's political agenda and the fickleness of its economy.

This vast world region, dominated mostly by water, includes the island continent of Australia as well as **Oceania**, a collection of islands that reach from New Guinea and New Zealand to the U.S. state of Hawaii in the mid-Pacific (Figure 14.1). Although native peoples settled the area long ago, more recent European and North American colonization began the process of globalization. Today, China and other Asian countries play significant roles in creating new and sometimes unsettled environmental, cultural, and political geographies.

Ongoing political and ethnic unrest in Fiji illustrates how the heat of 21st-century globalization has fired the cauldron of change (Figure 14.2). Currently, the country struggles to deal with new political and economic relationships between the two dominant ethnic groups, indigenous Fijians and the descendants of South Asian sugarcane workers (commonly called Indo-Fijians) who were brought to the islands in the 19th century to resolve labor shortages in the cane fields. Strong cultural barriers exist between the two groups, resulting in extremely low rates of intermarriage. Generally speaking, the Indo-Fijians dominate the country's commercial life

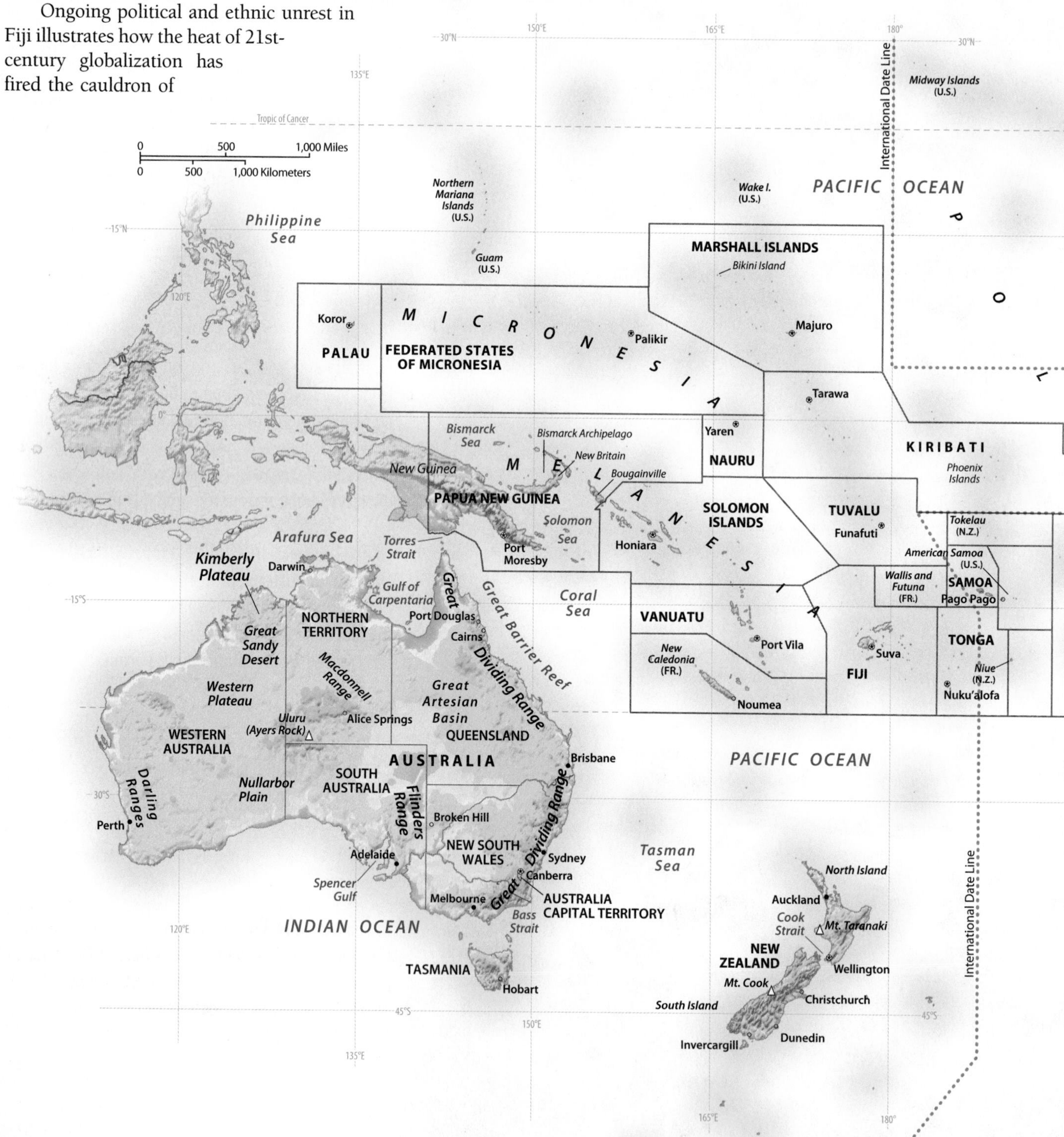

Figure 14.1 Australia and Oceania More water than land, the Australia and Oceania region sprawls across the vast reaches of the western Pacific Ocean. Australia dominates the region, both in its physical size and in its economic and political clout. Along with New Zealand, Australia represents largely Europeanized settlement in the South Pacific. Elsewhere, however, the island subregions of Melanesia, Micronesia, and Polynesia contain large native populations that have mixed in varied ways with later European, Asian, and American arrivals.

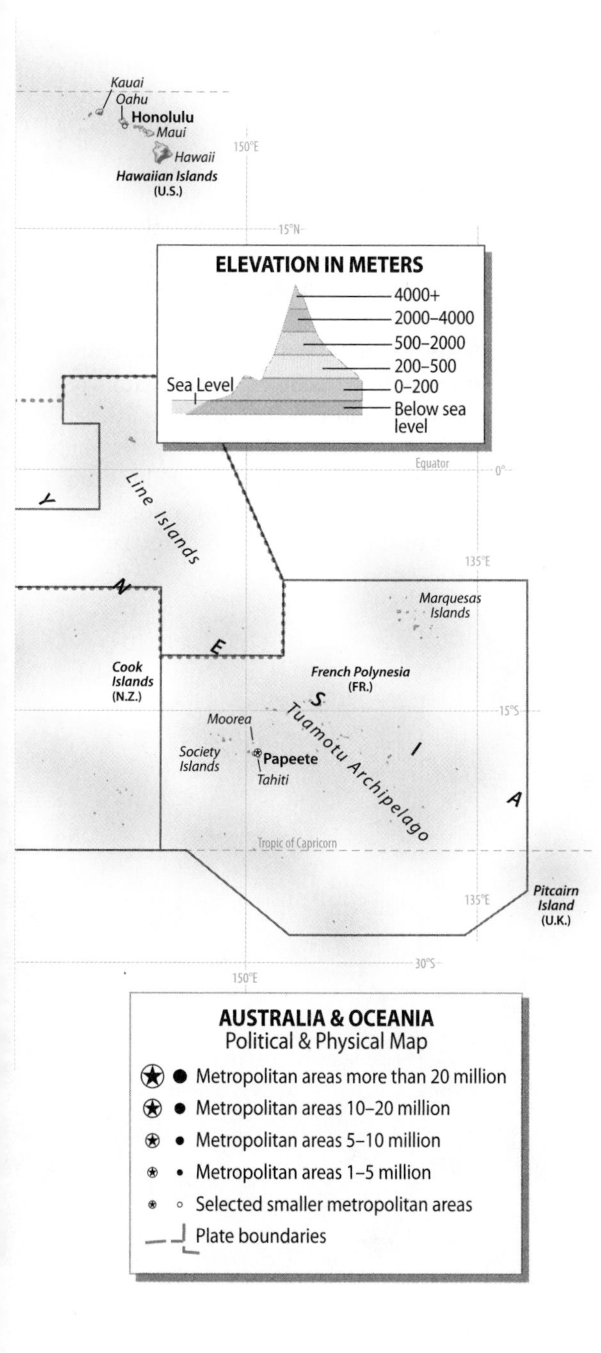

Figure 14.2 Unrest in Fiji South Asians were brought to Fiji as workers in the sugarcane fields. Today, the Indo-Fijians generally control the sugar industry, which is a source of tension with indigenous Fijians (such as this cane cutter), many of whom support politicians who advocate Fijian control of the industry.

and are better off economically than the Fijians, even though the indigenous tribal communities own the land that produces much of the country's wealth. These fields are leased from Fijian tribes, and now, as the leases expire, the Fijian tribes are less likely to renew them in hopes of improving their own economic position by farming the fields themselves.

While tensions have long caused conflict between these two groups, this unrest has been heightened by modern-day globalization as foreign investors push for economic restructuring and market development of the sugar industry. As a result, island governance has been unsettled by a series of coups and counter-coups over the past two decades. In May 2000, armed Fijians took hostage the nation's first Indo-Fijian prime minister, creating both an internal and international crisis as other countries boycotted Fiji until democracy was restored. Elections in 2001 and 2006 led to a Fijian-dominated government hostile to Indo-Fijians, and, as a result, large numbers of that group fled the islands, leaving the commercial and economic life of the country in shambles.

In December 2006, yet another Fijian military coup installed the head of the Fijian military forces, Frank Bainimarama, as interim prime minister. When this coup was declared illegal by the Fijian appeals court in 2009, the figurehead president abolished the country's constitution and gave all political power to Prime Minister Bainimarama, who, in turn, then ousted the country's judges. After New Zealand and Australia protested this consolidation of power in a military dictatorship with a trade ban, the diplomatic corps were ousted from Fiji.

But attempts to isolate Fiji and force a democratic election have largely failed because China, which has increasingly looked to Oceania for resources to stoke its economic boom, complicated matters by intervening with economic aid and trade agreements with Fiji.

In many ways, the tensions found in Fiji between indigenous peoples and "outsiders" (new and old) are also found in many other island countries of Oceania and, to a lesser degree, even within Australia and New Zealand. These tensions serve to remind us of the way environmental, settlement, cultural, geopolitical, and economic activities are inseparably linked in the contemporary, globalized world.

Australia and New Zealand dominate the Pacific World economically and politically and, in fact, share many geographic characteristics. Major population clusters in both countries are located in the middle latitudes rather than the tropics. Australia's 22.4 million residents occupy a vast land area of 2.97 million square miles (7.69 million square kilometers), while New Zealand's combined North and South Islands (104,000 square miles, or 269,000 square kilometers) are home to 4.4 million people. Most residents of both countries live in urban settlements near the coasts (Figure 14.3). Australia's huge and dry interior, commonly called the **Outback,** is as thinly settled as North Africa's Sahara Desert, and much of the New Zealand countryside is a visually spectacular but sparsely occupied collection of volcanic peaks and rugged, glaciated mountain ranges (Figure 14.4). Taken together, the land areas of these two South Pacific nations almost equals that of the United States, but their populations total less than 10 percent of the U. S. All three countries, however, share a European cultural heritage, the product of common global processes that sent Europeans far from their homelands over the past several centuries. But unlike the United States, both Australia and New Zealand still retain particularly close cultural links to Britain.

Punctuated with isolated chains of sand-fringed and sometimes mountainous islands, the blue waters of the tropical Pacific dominate much of the rest of the region. Three major subregions of Oceania each contain a surprising variety of human settlements and political units. Farthest west, **Melanesia** (meaning "dark islands") contains the culturally complex, generally darker-skinned peoples of New Guinea, the Solomon Islands, Vanuatu, and Fiji. The largest of these countries, Papua New Guinea (179,000 square miles, or 463,000 square kilometers), includes the eastern half of the island of New Guinea (the western half is part of Indonesia), as well as nearby portions of the northern Solomon Islands.

Figure 14.3 Sydney, Australia Most Australians live in cities, and the country's urban landscapes often resemble their North American counterparts. This view of Sydney features its world-famous harbor and displays the dramatic interplay of land and water.

Figure 14.4 The Australian Outback Arid and generally treeless, the vast lands of the Australian Outback resemble some of the dry landscapes of the U.S. West. In this photo, wildflowers blossom along a dirt road near Tom Price, in the Pilbara region of Western Australia.

Its population of 6.8 million people is larger than that of New Zealand.

To the east, the small island groups, or **archipelagos**, of the central South Pacific are called **Polynesia** (meaning "many islands"); this linguistically unified subregion includes French-controlled Tahiti in the Society Islands, the Hawaiian Islands, and smaller political states such as Tonga, Tuvalu, and Samoa. New Zealand is also often considered a part of Polynesia because its native peoples, known collectively as the **Maori**, share many cultural and physical characteristics with the somewhat lighter-skinned peoples of the mid-Pacific region. Finally, the more culturally diverse region of **Micronesia** (meaning "small islands") is north of Melanesia and west of Polynesia and includes microstates such as Nauru and the Marshall Islands, as well as the U.S. territory of Guam.

ENVIRONMENTAL GEOGRAPHY: A Varied Natural and Human Habitat

The region's physical setting speaks to the power of space: The geology and climate of the seemingly limitless Pacific Ocean define much of the physical geography of Oceania, while the expansive interior of Australia shapes the basic physical geography of that island continent.

Environments at Risk

Despite their relatively small populations, many areas in Australia and Oceania face significant human-induced environmental problems. Some environmental challenges are caused by natural events that increasingly affect larger and more widely distributed human populations. For instance, Pacific Rim earthquakes, periodic Australian droughts, and tropical cyclones now pose greater threats than they once did, as new settlements have made increasing populations vulnerable to these problems. Other environmental issues, however, are even more directly related to human causes (Figure 14.5).

Global Resource Pressures Globalization has exacted an environmental toll on Australia and Oceania. Specifically, the region's considerable base of natural resources has been opened to development, much of it by outside interests. While

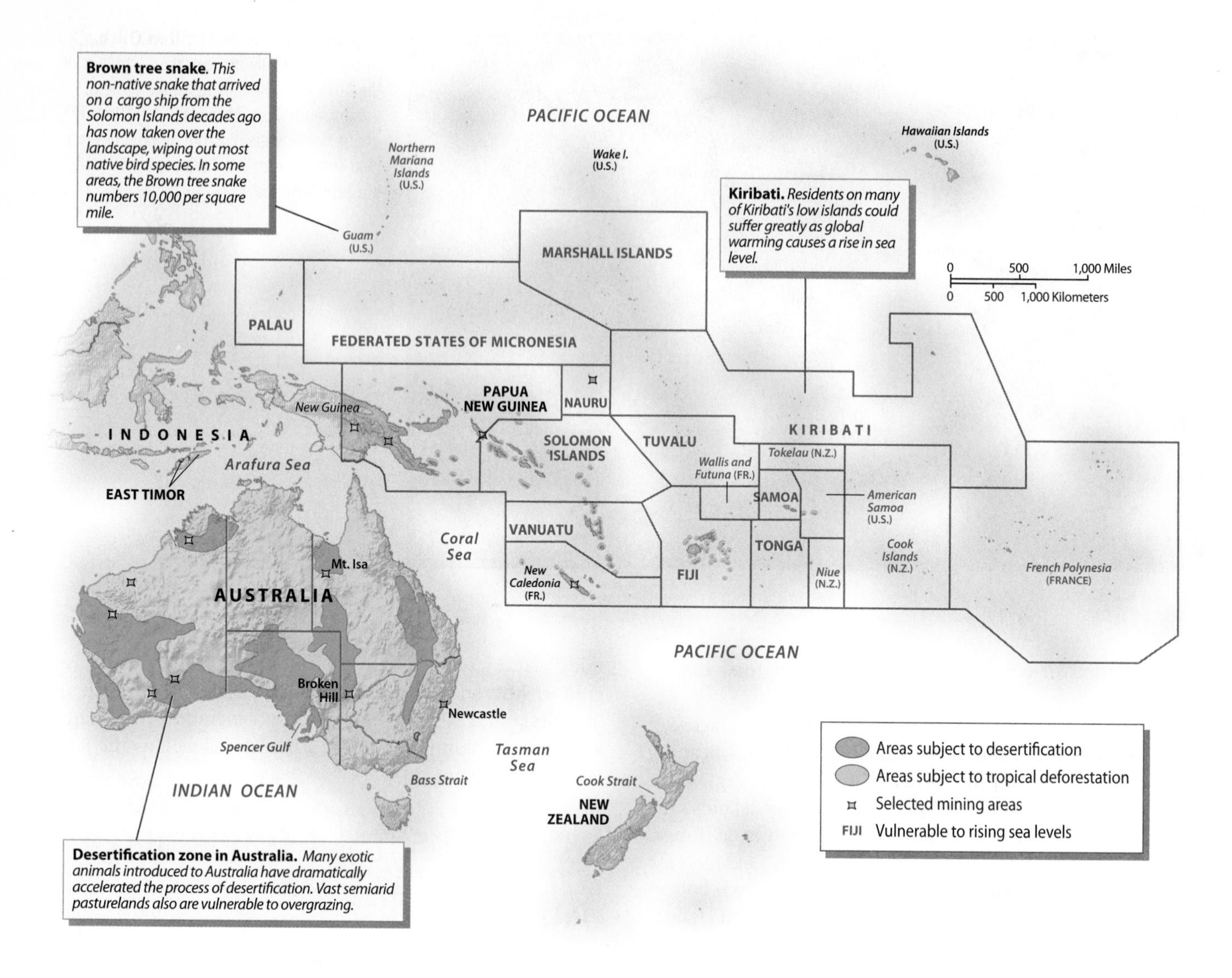

Figure 14.5 Environmental Issues in Australia and Oceania Modern environmental problems belie the region's myth that it is an earthly paradise. Tropical deforestation, extensive mining, and a long record of nuclear testing by colonial powers have brought varied challenges to the region. Human settlements have also extensively modified the pattern of natural vegetation. Future environmental threats loom for low-lying Pacific islands as sea levels rise from global warming.

gaining from the benefits of global investment, the region has also paid a considerable price for encouraging development, and the result is an increasingly threatened environment.

Major mining operations have greatly affected Australia, Papua New Guinea, New Caledonia, and Nauru. Some of Australia's largest gold, silver, copper, and lead mines are located in sparsely settled portions of Queensland and New South Wales, putting watersheds in these semiarid regions at risk for metals pollution. In Western Australia, huge open-pit iron mines dot the landscape, unearthing ore that is usually bound for global markets, particularly China and Japan. To the north, Papua New Guinea's Bougainville copper mine has transformed the Solomon Islands, while even larger gold-mining ventures have raised increasing environmental concerns on the island of New Guinea (Figure 14.6). Elsewhere, Micronesia's tiny Nauru has been virtually turned inside out as much of the island's jungle cover has been removed to get at some of the world's richest phosphate deposits (see "Global to Local: Nauru and the Mixed Benefits of Globalization").

Deforestation is another major environmental threat across the region. Vast stretches of Australia's eucalyptus woodlands, for example, have been destroyed to create better pastures. In addition, coastal rain forests in Queensland are only a fraction of their original area, although a growing environmental movement in the region is fighting to save the remaining forest tracts. Tasmania has also been an environmental battleground, particularly given the biodiversity

Figure 14.6 Mining in Papua New Guinea Open-pit mining for gold, silver, copper, and lead mark the landscapes of Papua New Guinea, New Caledonia, and Nauru. Although bringing some economic benefit to local peoples, these activities also cause immense environmental damage to the region. In New Guinea, for example, sediments from upland mines have severely damaged the Fly River ecosystem.

of its midlatitude forest landscapes. While the island's earlier European and Australian development featured many logging and pulp mill operations, more than 20 percent of the island is now protected by national parks.

Many islands in Oceania are also threatened by deforestation. With limited land areas, islands are subject to rapid tree loss, which in turn often leads to soil erosion. Although rain forests still cover 70 percent of Papua New Guinea, more than 37 million acres (15 million hectares) have been identified as suitable for logging (Figure 14.7). Some of the world's most biologically diverse environments are being threatened in these operations, but landowners see the quick cash sales to loggers as attractive, even though this nonsustainable practice is contrary to their traditional lifestyles.

Exotic Plants and Animals The introduction of exotic (nonnative) plants and animals has caused problems for endemic (native) species throughout the Pacific region. In Australia, for example, nonnative rabbits successfully multiplied in an environment that lacked the diseases and predators that kept their numbers in check in Europe. Before long, rabbit populations had reached plague-like proportions, stripping large sections of land of vegetation. The animals were brought under control only through the purposeful introduction of the rabbit disease myxomatosis. Introduced sheep and cattle populations have also stressed the region's environment by increasing soil erosion and contributing to desertification.

The introduction of exotic plants and animals to island environments has had similar effects. For example, many small islands possessed no native land mammals, and their native bird and plant species proved vulnerable to the ravages of introduced rats, pigs, and other animals. The larger islands of the region, such as those of New Zealand, originally supported several species of large, flightless birds that filled some of the ecological niches held by mammals on the continents. The largest of these, the moas, were substantially larger than ostriches. During the first wave of human settlement in New Zealand some 1,500 years ago, moa numbers fell rapidly as they were hunted by humans and their eggs consumed by invading rats. By 1800, the moas had been completely exterminated.

Figure 14.7 Logging in Oceania Foreign logging companies have made large investments in tropical Pacific countries such as Papua New Guinea and Samoa. While bringing new jobs, these ventures dramatically alter local environments, as precious hardwood forests are harvested for export.

GLOBAL TO LOCAL Nauru and the Mixed Benefits of Globalization

The microstate of Nauru, a tiny country of 14,000 people clustered on a small island of 8 square miles (21 square kilometers), offers insight into the mixed consequences of globalization as Nauru travels the bumpy road from self-sufficient fishing society to bankrupt ward of the world community.

Nauru is one of several small islands in the Pacific where, over centuries, seabird droppings have collected to form rich deposits of phosphate, a coveted mineral with many uses, including agricultural fertilizer and gunpowder. Although Nauru first allowed foreign mining companies access to its phosphate deposits as far back as the 1920s, it was only in the 1960s when Nauru became an independent country that it granted full rights to an Australian mining company. As a result of the mining royalties, Nauru became known as the Kuwait of the Pacific, with its small populations enjoying one of the world's highest per capita incomes.

With handsome payouts in their pockets, Naurans spent lavishly, building mansions, buying luxury cars, and traveling widely to faraway destinations for golf holidays. But with their new riches, Naurans also turned away from their traditional lifeways and substituted imported food (mainly junk food, some would say) for local fish, shellfish, and fruits and vegetables. Unfortunately, this change of diet took an immediate toll; in 1975, close to half of the adult population was diagnosed with diabetes, a direct result of dietary change and obesity. In fact, Nauru has one of the highest rates of diabetes anywhere in the world still today, with its population suffering high rates of blindness and limb amputations.

Furthermore, the phosphate resources are now exhausted (Figure 14.1.1). Although Nauru's government had planned for a day when the mining royalties would end, its strategies for keeping the cash flowing through global connections have gone awry. Bad investments in Southeast Asian real estate and failed London musicals stand out, sapping the gains made from successful ventures in Oregon subdivisions and high-end Washington, DC, condominiums.

To compensate, Nauru developed an offshore Internet banking facility, and while this activity appeared to be successful at first, the international financial community shut it down when it was discovered that the major activity was laundering money for the Russian Mafia.

Nauru then tried selling citizenship in its small nation, complete with Nauruan passports to those who bought into the scheme. Not surprisingly, global antiterrorist organizations took a very dim view of this activity, and once again the international community applied pressure to cease operations.

More recently, Nauru tried to capitalize on global mobility by building a detention center (some would call it a prison) for people who entered Australia illegally and then asked for political asylum. This plan was supported by the former conservative Australian prime minister, who thought it best to detain these refugees elsewhere while their applications for asylum were slowly processed. However, the current Australian government considered this a very bad idea and recently withdrew support for the Nauru detention center.

Today, Nauru's future is grim, a poster child for the complexities of globalization.

Figure 14.1.1 The Globalization of Nauru A Nauruan local points out the scarred landscape left from decades of intensive phosphate mining by Australian mining companies. With no mineral riches left, the islanders have tried several different strategies to find a place in the globalized economy.

Figure 14.8 Island Pest The brown tree snake, which arrived in Guam accidentally in the 1950s, has now taken over large parts of the island's forestlands and killed off most native bird species. Because these snakes, which reach 10 feet in length, climb along electrical wires, they frequently cause power outages throughout Guam.

The spread of nonnative species continues today, perhaps at even greater pace. In Guam, the brown tree snake, which arrived accidentally by cargo ship from the Solomon Islands in the 1950s, has taken over the landscape (Figure 14.8). In forest areas, with more than 10,000 snakes per square mile, they have wiped out nearly all the native bird species. In addition, the snakes cause frequent power outages as they crawl along electrical wires. While the brown tree snake has already done its damage to Guam, it threatens other islands as well because it readily hides in cargo containers shipped to other island destinations.

Global Warming in Oceania

Even though Oceania contributes relatively little atmospheric pollution to the global atmosphere, the harbingers of climate change are already widespread and problematic. In New Zealand, mountain glaciers are melting away while Australia suffers from frequent droughts and devastating wildfires. Warmer ocean waters have caused widespread bleaching of the Great Barrier Reef off Australia's coast, and rising sea levels are flooding several low-lying island nations, forcing residents to migrate to higher land (Figure 14.9). United Nations projections for the future are also highly disturbing: Stronger tropical cyclones could devastate Pacific islands, with widespread damage to land and life; island inhabitants will suffer from reduced coastal resources as ocean waters warm; and increased wildfire will threaten population centers in southeast Australia, while agriculture will suffer from droughts and severe water shortages.

In response to these threats, the actions and policies taken by Oceania's countries vary considerably. Until a 2007 change of government, Australia was the only industrial country besides the United States to not ratify the Kyoto Protocol. Perhaps the fact that Australia is the world's largest exporter of coal influenced that decision; not to be overlooked is that most of that country's emissions are produced by coal-fired power plants. While the former government rejected the Kyoto Protocol and downplayed the threat of global warming, the current government has proposed a 5 percent reduction in greenhouse gas emissions (GHG) by year 2020 based upon an emissions trading program. Opposition to this plan, however, comes from three powerful components of the Australia economy: the coal industry, aluminum manufacturers (which use coal-based power plants to generate the energy needed to refine aluminum), and agriculture. As is the case in the United States, Australian politics will determine whether

Figure 14.9 Sea Level Rise and Pacific Islands The world's oceans are forecast to rise at least 3.3 feet (1 meter) by the end of the century, flooding low-lying islands not just in Oceania but throughout the globe. This photo is of Funafuti Atoll, home to half of Tuvalu's population of just over 12 thousand people. With its highest point 16 feet (5 meters) above sea level, Tuvalu is already flooded regularly by storm surf and seasonal high tides, events that will occur more often in the future because of global warming. Not surprisingly, Tuvalu is one of the most outspoken countries on the topic of international measures to alleviate global warming.

Figure 14.10 New Zealand's Greenhouse Gas Problem Worldwide, cattle and sheep emit the greenhouse gas methane, but only in New Zealand are these livestock emissions greater than those produced by human activities. While New Zealand has discussed a "flatulence tax" on sheep farms, the country's scientists are also working on an anti-flatulence inoculation that would reduce sheep emissions.

or not any sort of emissions reduction plan becomes a reality.

When New Zealand ratified the Kyoto agreement, it committed itself to a 5 percent reduction over its 1990 baseline; however, a booming economy resulted in a 50 percent increase of emissions over the past several decades. In response to this increase the government is now proposing a 10 to 20 percent reduction in GHG to be achieved by 2020.

A major component of New Zealand's greenhouse gas pollution is methane emission from the country's large livestock population. These emissions, in fact, account for over half of the global warming pollution coming from that small country and, as a result, New Zealand is discussing a much-publicized "flatulence tax" that would be levied on livestock operations (Figure 14.10).

Many Pacific nations, most notably Tuvalu, Kiribati, and the Marshall Islands, maintain they are already experiencing problems from sea level rise, coral bleaching, and degraded fishery resources because of warming ocean waters, and have banded together into a strident political union lobbying for a global solution to climate change. At the December 2009 Copenhagen meetings, these small island nations were outspoken in their demands that developed nations such as the United States, Japan, and Europe provide the island nations with financial aid to mitigate damage from global warming.

Related, Hawaii was one of the first U.S. states to formally address global warming with, first, an analysis of the sources for its GHG emissions, and, second, setting a target for reduction of those emissions. Since most of Hawaii's emissions result (as is true in many states and countries) from oil and coal-fired powered plants, the state, like Australia and other countries, is actively promoting sustainable energy generation through wind, tidal, and solar power (Figure 14.11).

Australian and New Zealand Environments

Curiously, Australia is one of the world's most urbanized societies, yet most people associate the country with its vast and arid Outback, a sparsely settled land of sweeping distances, scrubby vegetation, and unusual animals. In contrast, the two small islands that make up New Zealand are

Figure 14.11 Wind Power in Australia Currently Australia produces most of its energy from coal-fired power plants, which is not surprising given the wealth of its coal resources. However, like many other countries attempting to cut down on their carbon emissions, Australia is building an increasing number of wind farms so that today about 2 percent of their energy needs are met with wind farm generation. This array of turbines is on the coast of Western Australia.

Figure 14.12 The Great Barrier Reef Stretching along the eastern Queensland coast, the famed Great Barrier Reef is one of the world's most spectacular examples of coral reef-building. Threatened by varied forms of coastal pollution, much of the reef is now protected in a national marine park.

known for their humid landscapes of rolling foothills and rugged mountains.

Landform Regions Three major landform regions dominate Australia's physical geography. The Western Plateau occupies more than half of the continent. Most of the region is a vast, irregular plateau that averages only 1,000 to 1,800 feet in height (305 to 550 meters). Further east, the Interior Lowland Basins stretch north to south for more than 1,000 miles (1,609 kilometers) from the swampy coastlands of the Gulf of Carpentaria to the Murray and Darling valleys, Australia's largest river system. Finally, more forested and mountainous country exists near Australia's east coast. The Great Dividing Range extends from the Cape York Peninsula in northern Queensland to southern Victoria. Nearby, off the eastern coast of Queensland, the Great Barrier Reef offers a final dramatic subsurface feature: Over the past 10,000 years, one of the world's most spectacular examples of coral reef-building has produced a living legacy now protected by the Great Barrier Reef Marine Park (Figure 14.12).

Part of the Pacific Rim of Fire, New Zealand owes its geologic origins to volcanic mountain-building that produced two rugged and spectacular islands in the South Pacific. The North Island's active volcanic peaks, reaching heights of more than 9,100 feet (2,775 meters), and geothermal features reveal the country's fiery origins (Figure 14.13). Even higher and more rugged mountains comprise the western spine of the South Island. Mantled by high mountain glaciers and surrounded by steeply sloping valleys, the Southern Alps are some of the world's most visually spectacular mountains, complete with narrow, fjordlike valleys that indent much of the South Island's isolated western coast.

Climate Generally, zones of somewhat higher precipitation encircle Australia's arid center (Figure 14.14). In the tropical low-latitude north, seasonal changes are dramatic and unpredictable. For example, Darwin can experience drenching monsoonal rains in the summer (December to March), followed by bone-dry winters (June to September). Indeed, life across the region is shaped by this annual rhythm of what is called locally "the wet" and "the dry." By the end of the dry

Figure 14.13 Mt. Taranaki New Zealand's North Island contains several volcanic peaks, including Mt. Taranaki. The 8,000-foot (2,440-meter) peak offers everything from subtropical forests to challenging ski slopes and attracts both local and international tourists.

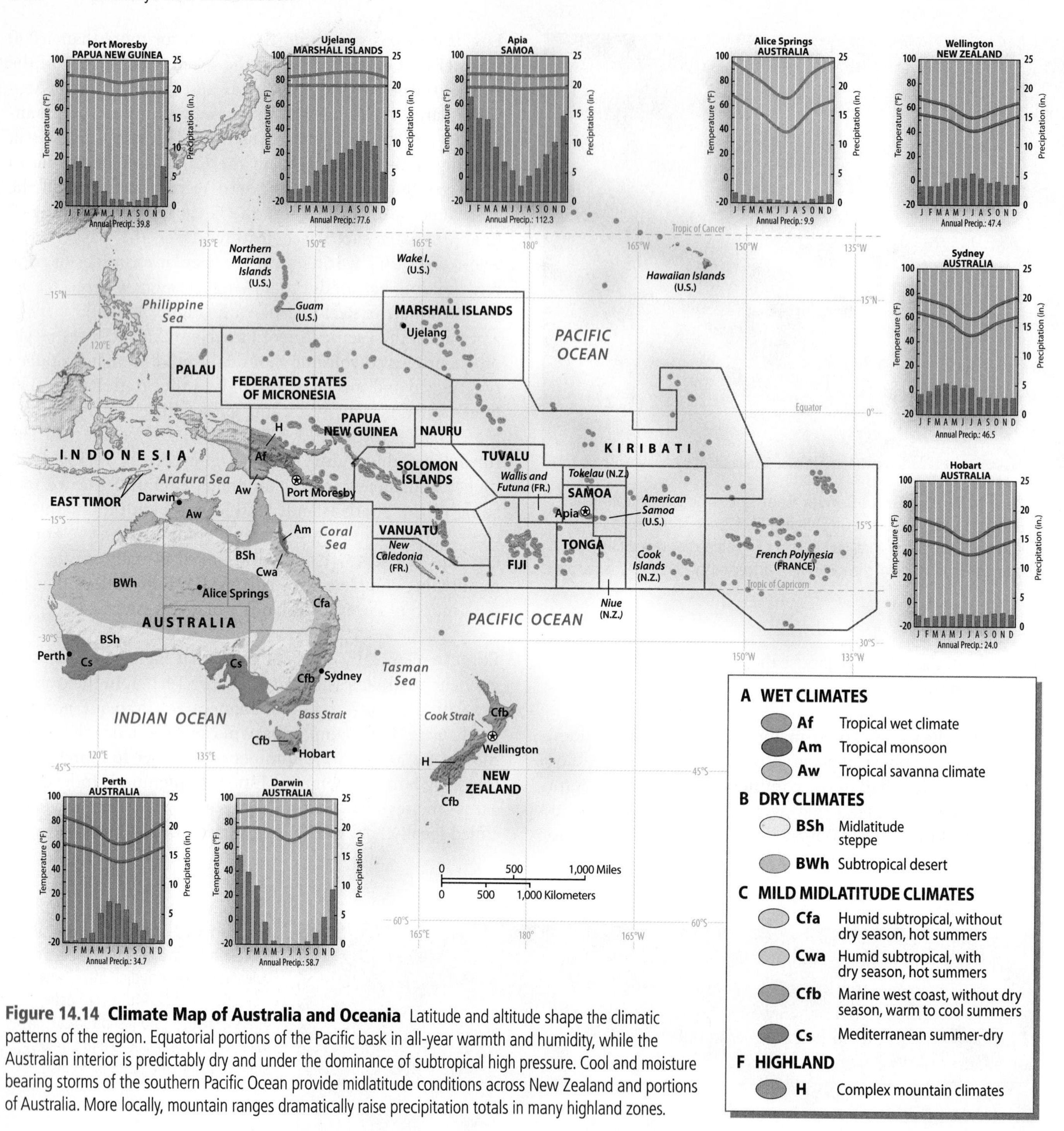

Figure 14.14 Climate Map of Australia and Oceania Latitude and altitude shape the climatic patterns of the region. Equatorial portions of the Pacific bask in all-year warmth and humidity, while the Australian interior is predictably dry and under the dominance of subtropical high pressure. Cool and moisture bearing storms of the southern Pacific Ocean provide midlatitude conditions across New Zealand and portions of Australia. More locally, mountain ranges dramatically raise precipitation totals in many highland zones.

season, wildfires usually dot the landscape of northern Australia (Figure 14.15) (see "Geographic Tools: Wildfire Ecology and Management in Australia").

Along the east coast of Queensland, precipitation remains high (60 to 100 inches, or 153 to 254 centimeters), but it diminishes rapidly as one moves into the interior. Rainfall at interior locations such as the Northern Territory's Alice Springs averages less than 10 inches (25 centimeters) annually. South of Brisbane, more midlatitude influences dominate eastern Australia's climate. Coastal New South Wales, southeastern Victoria, and Tasmania experience the country's most dependable year-round rainfall, which averages 40 to 60 inches (102 to 152 centimeters) of precipitation per year. Nearby mountains see frequent winter snows. Farther west, summers are hot and dry in much of South Australia and in the southwest corner of Western Australia. These zones of Mediterranean climate produce the **mallee** vegetation, a scrubby eucalyptus woodland.

Climates in New Zealand are influenced by a combination of latitude, the moderating effects of the Pacific Ocean,

Figure 14.15 Australian Wildfires Huge and savage dry season wildfires (known as bushfires in Australia) threaten both rural settlement and sprawling city suburbs in the southeast. This fire in February 2009, just 70 miles from the heart of Melbourne, was the worst fire disaster in 25 years and may be a harbinger of even more damaging fires accompanying global warming.

and proximity to local mountains or mountain ranges. Most of the North Island is distinctly subtropical; the coastal lowlands near Auckland, for example. are mild and wet year-round. Still, local variations can be striking as the area's volcanic peaks create their own microclimates. On the South Island, conditions become distinctly cooler as one moves poleward. Indeed, the island's southern edge feels the seasonal breath of Antarctic chill as it lies more than 46° south of the equator. Mountain ranges on New Zealand's South Island also display incredible local variations in precipitation: West-facing slopes are drenched with more than 100 inches (254 centimeters) of precipitation annually, while lowlands to the east average only 25 inches (64 centimeters) per year. The Otago region, inland from Dunedin, sits partially in the rain shadow of the Southern Alps, and its rolling, open landscapes resemble the semiarid expanses of North America's Intermountain West (Figure 14.16).

Island Climates Many Pacific islands receive abundant precipitation, and high islands in particular are often noted for their heavy rainfall and dense tropical forests. In American Samoa, this environment has been protected in one of the nation's newest national parks. Much of the island zone is located in the rainy tropics or in a tropical wet-dry climate region where abundant summer rains and even tropical cyclones can bring heavy seasonal precipitation. In contrast to the high islands, low-lying atolls usually receive less precipitation than high islands and very often experience water shortages. During dry periods, the limited stores of water on these islands are quickly depleted.

The Oceanic Realm

The vast expanse of Pacific waters reveals another set of environmental settings that are as rich and complex as they are fragile. Oceanic currents and wind patterns have historically served as natural highways of movement between these island worlds. Those same forces define the rhythm of weather patterns across the region in a broad band that extends more than 20° north (Hawaiian Islands) and south (New Caledonia) of the equator.

Creating Island Landforms Much of Melanesia and Polynesia is part of the seismically active Pacific Basin. As a result, volcanic eruptions, major earthquakes, and **tsunamis,** or earthquake-induced sea waves, are not uncommon across the region, and they impose major environmental hazards on the population. For example, volcanic eruptions and earthquakes on the island of New Britain (Papua New Guinea) forced more than 100,000 people from their homes in 1994. Only 4 years later, a massive tsunami triggered by an offshore earthquake swept across the north coast of New Guinea, killing 3,000 residents and destroying numerous villages. Such events are unfortunately a part of life in this geologically active part of the world.

Most of the islands of Polynesia and Micronesia are truly oceanic, having originated from volcanic activity on the ocean floor. The larger active and recently active volcanoes form **high islands** that often rise to a considerable elevation and cover a large area. The island of Hawaii, the largest and youngest of the Pacific's high islands, is more than 80 miles (128 kilometers) across and rises to a height of more than 13,000 feet (3,980 meters). Indeed, the entire Hawaiian archipelago is a geological **hot spot** where slowly moving oceanic

Figure 14.16 Central Otago, South Island On New Zealand's South Island, the Southern Alps capture rainfall on the west coast but leave areas to the east in a drier rain shadow. As a result, the Central Otago region has a semiarid landscape resembling portions of the U.S. West.

GEOGRAPHIC TOOLS Fire Ecology and Management in Australia

Wildfires are common everywhere in Australia, from the northern "Top End" to the island of Tasmania off the country's southeastern coast. In Queensland and the Northern Territory, tropical savanna fires scorch the countryside during the winter dry season, fueled by grass and shrub growth from the summer monsoon, whereas in southeast Australia, forest and brush fires singe the suburbs of major cities during the dry summer season. Each year, hundreds of thousands of acres burn across the country.

Fire, however, is not just a hazard that threatens lives and property. It is also a resource management tool for achieving environmental—and even social—goals. Since fire is a natural part of Australia's ecology, certain kinds of fire are necessary to enhance biodiversity and protect endangered plant and animal species. Fire also serves Aboriginal people in many different ways. To illustrate, for Aboriginal tribes, fire is part of a ritual to "clean up the countryside" by cleansing it of evil spirits. Fire also is important for manipulating vegetation to ensure that certain plant and animal species are available for these hunting and gathering peoples.

As a result, many different geographic tools are needed to understand fire as both a resource and a hazard. Lesley Head, a geographer at the University of Wollongong, south of Sydney, is one of the leading researchers on Australia's fire ecology. To determine how often fires have occurred in prehistory, Professor Head and her students dig soil trenches several feet deep to find buried charcoal layers from past fires that can then be dated by laboratory analysis. For a more recent picture of burning, she and other fire ecologists commonly use remote sensing images to map and measure burned areas. Because satellites pass over Australia daily, they provide a continuing record of fires (Figure 14.2.1).

These satellite pictures are used not only by researchers but also by fire managers who decide which fires to let burn and which to extinguish. By tracking fires on a daily basis, they decide which fires can provide resource benefits by being allowed to burn and which fires pose a hazard to Outback ranches and settlements and thus should be extinguished.

Even before fires start, various tools, from satellite images to ground measurements, are used to produce maps of potential fire risk. Satellite photos show the different vegetation types—grass, brush, trees—for an area that constitutes potential fuel for fires. In the field, fire managers collect information on the dryness of these different vegetation types. All of this information is then put into a geographic information system (GIS) that combines the different factors influencing fire behavior—vegetation types, fuel load and condition, and topography—with maps of settlement and structures that need to be protected. The resulting map provides an invaluable tool for everything from predicting where fires might occur to determining how to best manage them when they do start.

These same geographic tools are used in North America, and students who master them can usually find jobs with the U.S. Forest Service and other agencies responsible for fire management and ecology.

Figure 14.2.1 Australian Wildfires This satellite photo taken in early October 2004 shows numerous fires burning in Australia's Northern Territory. Wildfires are common at the end of the dry season (October–November) before the monsoon rains begin in December. Using satellite images like this one, fire managers can decide which fires to let burn and which to suppress. (NASA)

crust passes over a vast supply of magma from Earth's interior, thus creating a chain of volcanic islands.

Many of the islands of French Polynesia, including Bora Bora, are smaller examples of high islands (Figure 14.17) and are widely scattered throughout

Figure 14.17 Bora Bora The jewel of French Polynesia, Bora Bora displays many of the classic features of Pacific high islands. As the island's central volcanic core retreats, surrounding coral reefs produce a mix of wave-washed sandy shores and shallow lagoons.

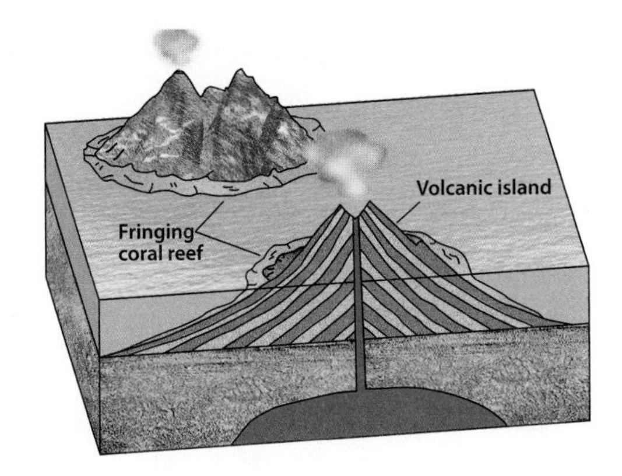

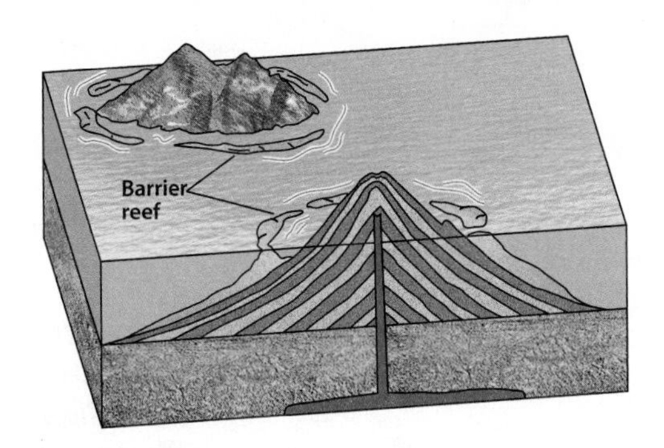

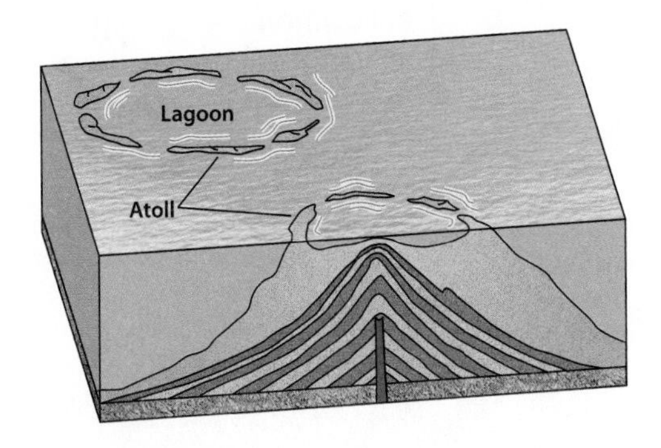

Figure 14.18 Evolution of an Atoll Many Pacific islands begin as rugged volcanoes (a) with fringing coral reefs. However, as the extinct volcano subsides and erodes away, the coral reef expands, becoming a larger barrier reef (b). The term *barrier reef* comes from the hazards these features pose to navigation when approaching the island from sea. Finally, all that remains (c) is a coral atoll surrounding a shallow lagoon.

Micronesia and Polynesia. In tropical latitudes, most high islands are ringed by coral reefs that quickly grow in the shallow waters near the shore.

The combination of narrow sandy islands, barrier coral reefs, and shallow central lagoons is known as an **atoll.** The islands and reefs of an atoll characteristically form a circular or oval shape, although some are quite irregular (Figure 14.18). The world's largest atoll, Kwajalein in Micronesia's Marshall Islands, is 75 miles (120 kilometers) long and 15 miles (24 kilometers) wide. Polynesia and Micronesia are dotted with extensive atoll systems, as is Melanesia as well.

POPULATION AND SETTLEMENT: A Diverse Cultural Landscape

Modern population patterns across the region reflect the combined influences of indigenous and European settlement. In countries such as New Zealand, Australia, and the Hawaiian Islands, Anglo-European migration has structured the distribution and concentration of contemporary populations. In contrast, on smaller islands elsewhere in Oceania, population geographies are determined by the needs of native peoples (Figure 14.19). More recently, however, migration has taken place from outlying islands to Australia and New Zealand because of a combination of push forces, including unemployment, resource depletion, and the threat of flooding associated with global warming (Table 14.1).

Contemporary Population Patterns

Despite the stereotypes of life in the Outback, modern Australia has one of the most highly urbanized populations in the world. Indeed, 91 percent of the country's residents live within either the Sydney or Melbourne metropolitan areas. Australia's eastern and southern areas are home to the majority of its 22.4 million people. Inland, population densities decline as rapidly as the rainfall: Semiarid hills west of the Great Dividing Range still contain significant rural settlement, but the state's southwestern periphery remains sparsely populated. New South Wales is the country's most heavily populated state, and its sprawling capital city of Sydney (over 4 million), focused around one of the world's most magnificent natural harbors, is the largest metropolitan area in the entire South Pacific. In the nearby state of Victoria, Melbourne's 3.8 million residents have long competed with Sydney for status as Australia's premiere city, claiming cultural and architectural supremacy over their slightly larger neighbor (Figure 14.20). In between these two metropolitan giants, the location of the much smaller federal capital of Canberra (population 325,000) represents a classic geopolitical compromise in the same spirit that created Washington, DC, located midway between the populous southern and northern portions of the eastern United States.

Smaller clusters of population are found inland from Australia's eastern coast. More fertile farmlands in the interior of New South Wales and Victoria feature higher population densities than are found in the nation's arid heartland (see "People on the Move: Does Australia Have a Population Problem?"). Inland Aboriginal populations are widely but thinly scattered across Western Australia and South Australia, as well as in the Northern Territory, accounting for smaller but regionally important centers of settlement.

The population geography of the rest of Oceania shows a broad distribution of peoples, both native and European, who have clustered near favorable resource opportunities. In New Zealand, more than 70 percent of the country's 4.4 million residents live on the North Island, with the Auckland region (over 1.1 million) dominating the metropolitan scene in the north and the capital city of Wellington (165,000) an-

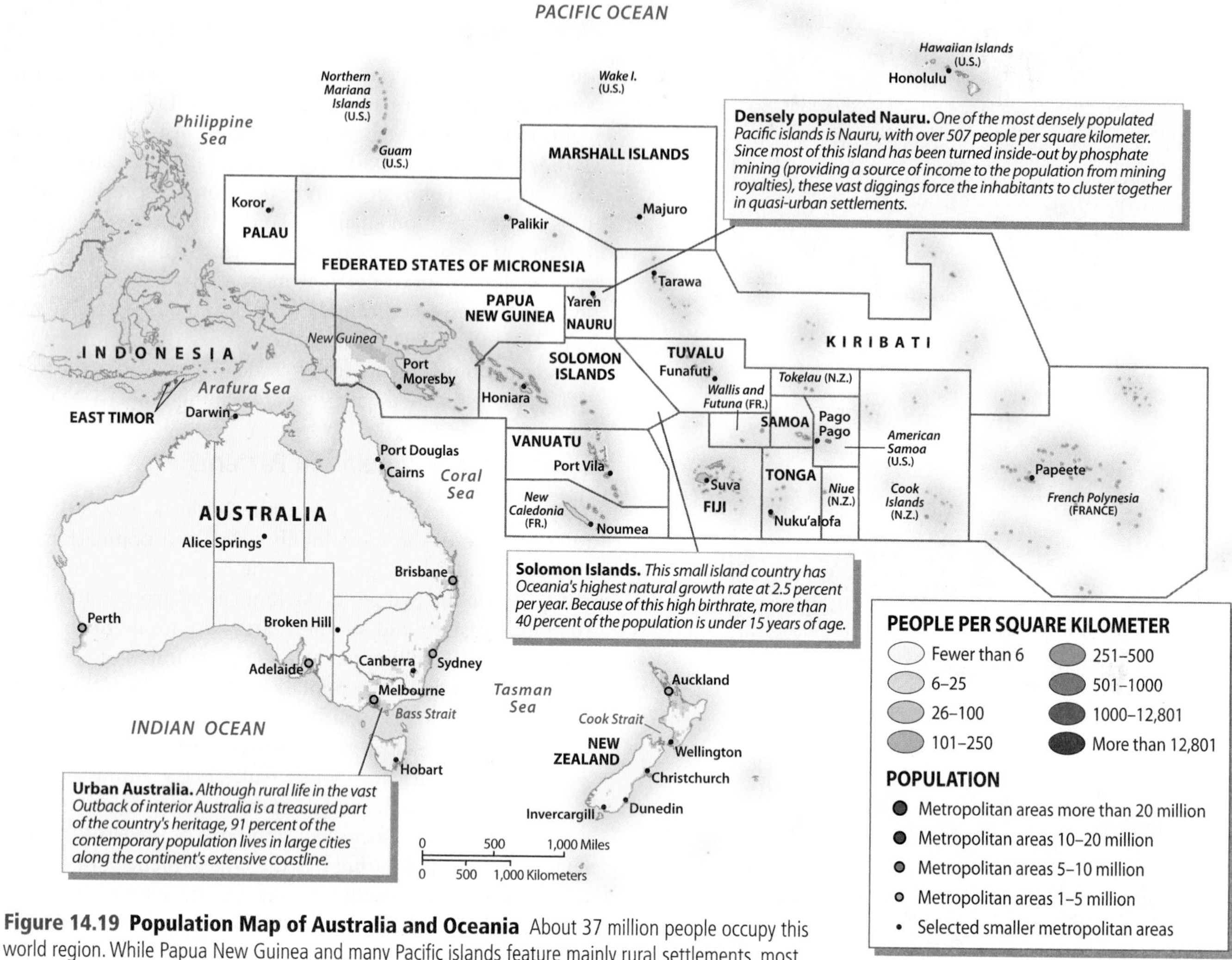

Figure 14.19 Population Map of Australia and Oceania About 37 million people occupy this world region. While Papua New Guinea and many Pacific islands feature mainly rural settlements, most regional residents live in the large urban areas of Australia and New Zealand. Sydney and Melbourne account for almost half of Australia's population, and most New Zealand residents live on the North Island, home to Auckland and Wellington.

choring settlement along the Cook Strait in the south. Settlement on the South Island is mostly located in the somewhat drier lowlands and coastal districts east of the mountains, with Christchurch (340,000) serving as the largest urban center. Elsewhere, rugged and mountainous terrain on both the North and South Islands produce much lower densities.

In Papua New Guinea only 13 percent of the country's population is urban, with most people living in the isolated, interior highlands. The nation's largest city is the capital, Port Moresby (around 200,000), located along the narrow coastal lowland in the far southeastern corner of the country.

Figure 14.20 Downtown Melbourne Metropolitan Melbourne lies along the Yarra River. Capital of the Australian state of Victoria, Melbourne resembles many growing North American cities, with its high-rise office buildings, entertainment districts, and downtown urban redevelopment.

TABLE 14.1 Population Indicators

Country	Population (millions) 2010	Population Density (per square kilometer)	Rate of Natural Increase (RNI)	Total Fertility Rate	Percent Urban	Percent <15	Percent >65	Net Migration (Rate per 1000) 2005–10[a]
Australia	22.4	3	0.7	1.9	82	19	13	4.8
Fed. States of Micronesia	0.1	158	1.9	3.9	22	37	4	−16.3
Fiji	0.9	47	1.7	2.6	51	29	5	−8.3
French Polynesia	0.3	68	1.3	2.2	53	26	6	0.0
Guam	0.2	344	1.5	2.7	93	28	7	0.0
Kiribati	0.1	139	1.8	3.5	44	36	4	
Marshall Islands	0.1	298	2.8	4.3	68	41	2	
Nauru	0.01	507	1.9	3.2	100	39	1	
New Caledonia	0.3	14	1.2	2.1	58	28	6	4.5
New Zealand	4.4	16	0.8	2.1	86	21	13	2.4
Palau	0.02	45	0.6	2.0	78	24	6	
Papua New Guinea	6.8	15	2.2	4.1	13	40	2	0.0
Samoa	0.2	68	2.0	4.2	22	40	5	−18.4
Solomon Islands	0.5	19	2.5	4.4	17	41	3	0.0
Tonga	0.1	139	2.2	4.2	23	38	6	−17.5
Tuvalu	0.01	376	1.4	3.7	47	32	6	
Vanuatu	0.2	20	2.5	4.0	24	40	3	0.0

[a]*Net Migration Rate from the United Nations, Population Division, World Population Prospects: The 2008 Revision Population Database.*
Source: Population Reference Bureau, World Population Data Sheet, 2010.

In stark contrast to New Guinea, the largest urban area on the northern margin of Oceania is Honolulu (1 million), on the island of Oahu, where rapid metropolitan growth since World War II has occurred because of U.S. statehood and the scenic attractions of its mid-Pacific setting.

Historical Settlement

The region's remoteness from the world's early population centers meant that it lay beyond the dominant migratory paths of earlier peoples. Even so, prehistoric settlers eventually found their way to the isolated Australian interior and even the far reaches of the Pacific. Later, the pace of new in-migrations increased once Europeans explored the region and identified its resource potential.

Peopling the Pacific The large islands of New Guinea and Australia, given their nearness to the Asian landmass, were settled much earlier than the more distant islands of the Pacific. By 60,000 years ago, the ancestors of today's native Australian, or **Aborigine**, populations were making their way out of Southeast Asia and into Australia (Figure 14.21). The first Australians most likely arrived using some kind of watercraft, but because such boats were probably not capable of more lengthy voyages, the more distant islands remained inaccessible to humankind for tens of thousands of years. During the last glacial period, however, sea levels were much lower than they are now, which would have allowed easier movement to Australia across relatively narrow spans of water from what is now called Southeast Asia. It is not known whether the original Australians arrived in one wave of people or in many, but the available evidence suggests that they soon occupied large portions of the continent, including Tasmania, which was at that time connected to the mainland by a land bridge because of the lower sea level.

Eastern Melanesia was settled much later than Australia and New Guinea. By 3,500 years ago, certain Pacific peoples had mastered long-distance sailing and navigation, which eventually opened the entire oceanic realm to human habitation. In that era, people gradually moved east to occupy New Caledonia, the Fiji Islands, and Samoa. From there, later movements took seafaring folk north into Micronesia, with the Marshall Islands occupied around 2,000 years ago.

Continuing movements from Asia further complicated the story of these migrating Melanesians. Some of the migrants mixed culturally and eventually reached western Polynesia, where they formed the core population of the Polynesian people. By 800 CE, they had reached such distant places as New Zealand, Hawaii, and Easter Island. Prehistorians hypothesize that population pressures may have quickly reached crisis stage on the relatively small islands,

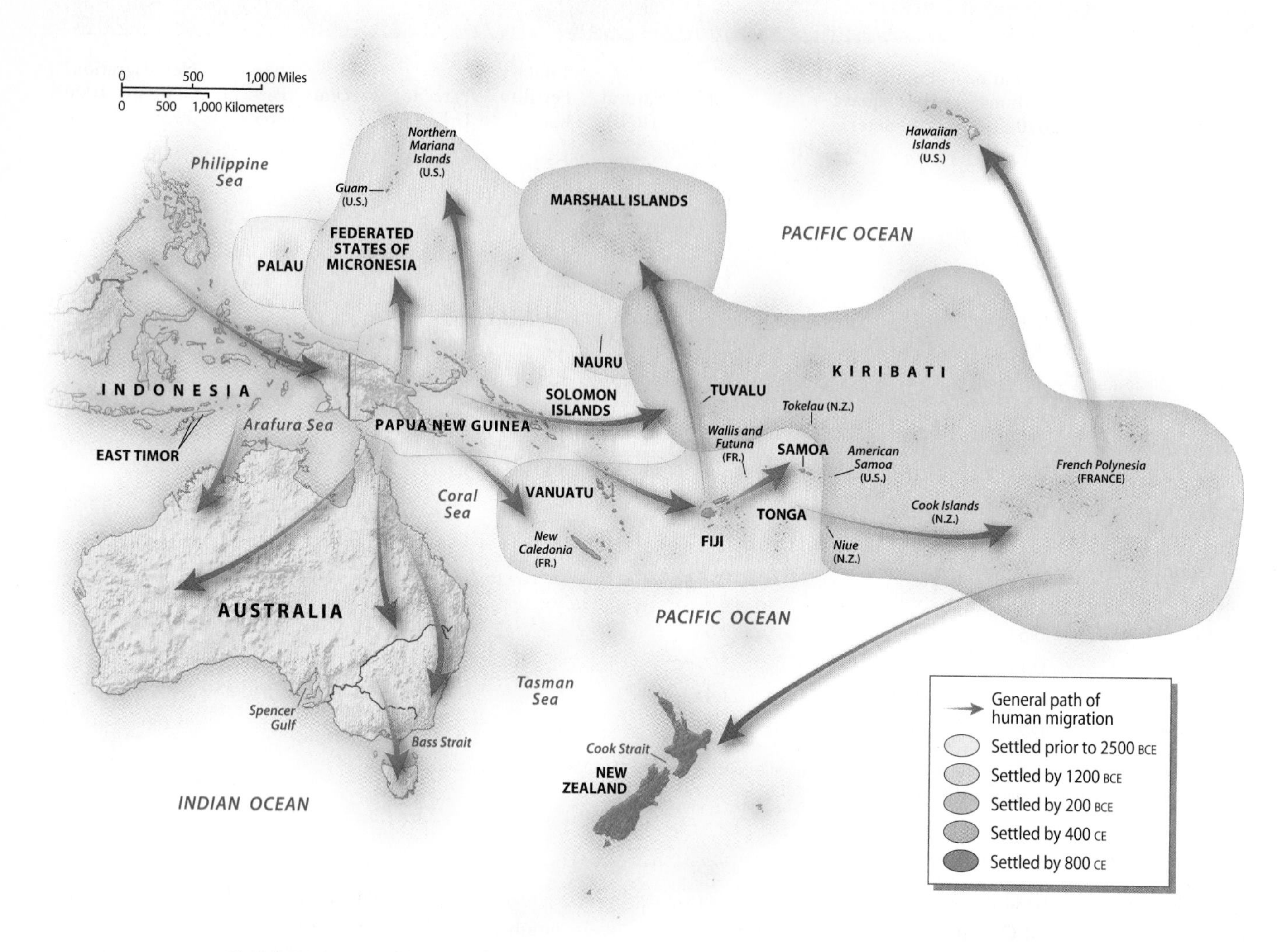

Figure 14.21 Peopling the Pacific Ancestors of Australia's aboriginal population may have made their way into the island continent more than 60,000 years ago. Much more recent settlement of Pacific islands by Austronesian peoples from Southeast Asia shaped cultural patterns across the oceanic portions of the realm. Eastward migrations through the Solomon Islands, Fiji, and the Cook Islands were followed by late movements to the north and south.

leading people to attempt dangerous voyages to colonize other Pacific islands. Equipped with sturdy outrigger sailing vessels and ample supplies of food, the Polynesians were quickly able to colonize most of the islands they discovered.

European Colonization About six centuries after the Maori brought New Zealand into the Polynesian realm, Dutch navigator Abel Tasman spotted the islands on his global exploration of 1642. Tasman's initial sighting marked the beginning of a new chapter in the human occupation of the South Pacific. Late in the following century, more lasting European contacts were made. British sea captain James Cook surveyed the shorelines of both New Zealand and Australia between 1768 and 1780. Cook and others believed that these distant lands might be worthy of European development. In addition, other European expeditions were exploring the Pacific, placing most of Oceania's major island groups on colonial maps by the end of the 18th century.

European colonization of the region began in Australia when the British needed a remote penal colony to which convicts could be exiled. The southeastern coast of Australia was selected as an appropriate site, and in 1788 the First Fleet arrived with 750 prisoners in Botany Bay, near what is now Sydney. Other fleets and more convicts soon followed, as did boatloads of free settlers. Before long, free settlers outnumbered the convicts, who were themselves gradually being released after serving their sentences. The growing population of English-speaking people soon moved inland and also settled other favorable coastal areas. British and Irish settlers were attracted by the agricultural and stock-raising potential of the distant colony as well as by the lure of gold and other minerals (a major gold rush

PEOPLE ON THE MOVE Does Australia Have a Population Problem?

According to politicians, talk radio pundits, and activists of different stripes, Australia is facing a severe population crisis because either the country (a) already has too many people, or, alternatively, (b) is not growing fast enough and its population is too small.

Here, simplified, are the contrasting positions to this heated population controversy.

Groups like Sustainable Population Australia (SPA) call for an environmentally responsible population policy that restricts population size because so much of the country is either arid or semiarid, and thus has a limited rural carrying capacity that demands vast quantities of water to survive and prosper. Global warming will make matters even worse, they argue, as Australia's dry interior expands, forcing even more people into the crowded coastal urban strips where seasonal brush fires are already a serious problem. Tim Flannery, scientist, author, activist, and Australian of the Year in 2007, argues that the country is already overpopulated by some 8 million people, consequently Australia should formulate a population policy that limits immigration and is sensitive to the country's ecological constraints.

Although sometimes uneasy with their company, environmental groups advocating population limits often find their bandwagon crowded with an array of anti-immigration groups and their baggage of ethnic discrimination and racism (Figure 14.3.1).

"While we would not suggest that a call for reduced immigration is necessarily racist, we would argue that such a call is not acceptable," says *Friends of the Earth Australia.* "There is no doubt that humans are having a dramatic and devastating effect on the natural systems of Australia, We would argue that this is largely a result of how we live here, not how many people live here."

On the other side of the controversy are those who say Australia's low birthrate will lead to a shrinking population unless immigration is increased. For example, the Australian Labor Party's minister for family and community services has said that current birthrates could halve the population by the end of the century. "What we face is not just the fade-out of the Australian family but also the fade-out of our nation," he told the *Daily Telegraph* last year, noting that most of the baby boomers will retire in the next 20 years, placing a massive burden on the current workforce that must pay taxes for social services such as pensions and health care.

To combat this so-called "baby bust," the Labor Party and business coalitions such as the Australian Population Institute are calling for maternity leave, tax breaks and, most of all, increased immigration to shore up Australia's population. Enhanced population growth, they argue, is absolutely necessary for the national economy, with a larger market, labor force, and broader base of taxpayers.

In a country basically the size of the United States, yet with a total population only slightly larger than the greater Los Angeles metropolitan area, many find it difficult to believe Australia is already overpopulated, particularly those living on densely populated Pacific islands threatened by rising sea levels who cast a longing eye at Australia's empty spaces.

Figure 14.3.1 Immigrants in Australia Each year Australia allows around 100,000 skilled immigrants into the country, most of whom come from Asia. As is the case in many other countries, immigration remains a controversial topic in Australia, particularly by those advocating a population policy of slow or even no growth.

occurred in Australia during the 1850s). The British government also encouraged the emigration of its own citizens, often paying the transportation fare of those too poor to afford it themselves.

The new settlers came into conflict with the Aborigines almost immediately after arriving. No treaties were signed, however, and in most cases, Aborigines were simply displaced from their lands. In some places, most notably Tasmania, they were hunted down and killed. In mainland Australia, the Aborigines were greatly reduced in numbers by disease, removal from their lands, and pure economic hardship. By the mid-19th century, Australia was primarily an English-speaking land as the native peoples were driven into submission.

The lush and fertile lands of New Zealand also attracted British settlers. European whalers and sealers arrived shortly before 1800, with more permanent agricultural settlement taking shape after 1840 as the British formally declared sovereignty over the region. As new arrivals grew in number and the scope of planned settlement colonies on the North and South islands expanded, tensions with the native Maori population increased. Organized in small kingdoms, or chiefdoms, the Maori were formidable fighters. In 1845 widespread Maori wars began that engulfed New Zealand until 1870. The British eventually prevailed, however, and as in Australia, the native Maori lost most of their land.

The native Hawaiians also lost control of their lands to immigrants. Hawaii emerged as a united and powerful kingdom in the early 1800s, and for many years its native rulers limited U.S. and European claims to their islands. Increasing numbers of missionaries and settlers from the United States were allowed in, however, and by the late 19th century, control of the Hawaiian economy had largely passed to foreign plantation owners. In 1893, United States interests were strong enough to overthrow the Hawaiian monarchy, resulting in formal political annexation in 1898.

Figure 14.22 Port Moresby, Papua New Guinea Urban poverty and high crime haunt the city of Port Moresby, the capital of Papua New Guinea. The city's slums, many built out on the water, reflect stresses of recent urban growth as rural residents emigrate from nearby highlands.

Settlement Landscapes

The settlement geography of Australia and Oceania offers an interesting mixture of local and global influences. While the contemporary cultural landscape still reflects the imprint of indigenous peoples in those areas where native populations remain numerically dominant, elsewhere, patterns of recent colonization have produced a modern scene mainly shaped by Europeans. The result includes everything from German-owned vineyards in South Australia to houses on New Zealand's South Island that appear to be plucked directly from the British Isles. In addition, processes of economic and cultural globalization have resulted in urban forms that make cities such as Perth or Auckland look strikingly similar to such places as San Diego or Seattle.

The Urban Transformation Both Australia and New Zealand are highly urbanized, Westernized societies, and thus the vast majority of their populations live in city and suburban environments. As in Europe and North America, much of this urban transformation came during the 20th century as the rural economy became less labor intensive and as opportunities for urban manufacturing and service employment grew. As urban landscapes evolved, they took on many of the characteristics of their largely European populations, yet blended with a strong dose of North American influences. The result is an urban landscape in which many North Americans are quite comfortable even though the varied local accents heard on the street and many features of the metropolitan scene are reminders of the strong and lasting attachments to British traditions.

The affluent Western-style urban environments in Australia and New Zealand offer a dramatic contrast with the urban landscapes found in less-developed cities in the region. Walk the streets of Port Moresby in Papua New Guinea, and a very different urban landscape is evidence of the large gap between rich and poor within Oceania (Figure 14.22). Rapid growth in Port Moresby, the country's political capital and largest commercial center, has produced many of the classic problems of urban underdevelopment: There is a shortage of adequate housing, the building of roads and schools lags far behind the need while street crime and alcoholism rise. Elsewhere, urban centers such as Suva (Fiji), Noumea (New Caledonia), and Apia (Samoa) also reflect the economic and cultural tensions generated as local populations are exposed to Western influences (see "Cityscapes: Apia, Capital of Samoa"). Rapid growth is a common problem in the smaller cities of Oceania because native people from rural areas and nearby islands gravitate toward the job opportunities available. In the past 50 years, the huge global growth of tourism in places such as Fiji and Samoa has also transformed the urban scene replacing traditional village life by a landscape of souvenir shops, honking taxicabs, and crowded seaside resorts (Figure 14.23).

The Rural Scene Rural landscapes across Australia and the Pacific region express a complex mosaic of cultural and economic influences. In some settings, Australian Aborigines or native Papua New Guinea Highlanders can still be found in their familiar homelands, their traditional lifeways and settlements barely changed from pre-European times. Yet such settlement landscapes are becoming increasingly rare. Global influences penetrate the scene as the cash economy, foreign

Figure 14.23 Tourism in Oceania On many islands, traditional ways of life are being replaced by tourist activities, both in the city (such as this scene in Fiji) as well as the countryside. While the economic benefits are welcome, the accompanying social, cultural, and political side effects of this form of globalization are not always beneficial.

tourism and investment, and the currents of popular culture work their way from city to countryside.

Much of rural Australia is too dry for farming or serves as only marginally valuable agricultural land. Although modest in size, the area in crops has doubled since 1960, as increased use of fertilizers, more widespread irrigation, and more aggressive rabbit eradication efforts have opened up new areas for development. Much of the remainder of the interior, however, features range-fed livestock, areas beyond the pale of any agricultural potential, and isolated areas where Aboriginal peoples still pursue their traditional forms of hunting and gathering.

Sheep and cattle dominate rural Australia's livestock economy. Many rural landscapes in the interior of New South Wales, Western Australia, and Victoria, for example, are oriented around isolated sheep stations, ranch operations that move the flocks from one large pasture to the next. Cattle can sometimes be found in these same areas, although many of the more extensive, range-fed cattle operations are concentrated farther north, in Queensland. Croplands also vary across the region. Sometimes mingling with the sheep country, a band of commercial wheat farming includes southern Queensland; the moister interiors of New South Wales, Victoria, and South Australia; and a swath of favorable land east and north of Perth. Elsewhere, specialized sugarcane operations thrive along the narrow, warm, and humid coastal strip of Queensland. To the south and west, productive irrigated agriculture has developed in places such as the Murray River Basin, allowing for the production of orchard crops and vegetables. **Viticulture**, or grape cultivation, increasingly shapes the rural scene in places such as South Australia's Barossa Valley, the Riverina district in New South Wales, and Western Australia's Swan Valley. Indeed, the area under grape cultivation grew by 50 percent between 1991 and 1998, as the popular Chardonnay, Cabernet Sauvignon, and Shiraz varieties boosted wine production to revenues of more than $540 million per year.

New Zealand's Landscapes Although much smaller in area than Australia, New Zealand's rural settlement landscape includes a variety of agricultural activities. Pastoral activities clearly dominate the New Zealand scene, with the vast majority of agricultural land devoted to livestock production, particularly sheep grazing and dairying. Commercial livestock outnumber people in New Zealand by a ratio of more than 20 to 1, and this is apparent everywhere in the countryside. Dairy operations are present mostly in the lowlands of the north, where they sometimes mingle with suburban landscapes in the vicinity of Auckland. One of the largest zones of more specialized cropping spreads across the fertile Canterbury Plain near Christchurch (Figure 14.24).

Figure 14.24 Canterbury Plain The varied agricultural landscape of South Island's Canterbury Plain offers a mix of grain fields, livestock, orchard crops, and vegetable gardens. The rugged Southern Alps are a dramatic backdrop to this productive region.

CITYSCAPES Apia, Capital of Samoa

The great cities of Australia and Oceania—Sydney, Auckland, and Honolulu—although exciting and attractive in their own ways, are essentially Western places, familiar to the North American eye with their array of skyscrapers, banks, and look-alike hotels connected by expressways to suburban shopping malls and neighborhoods of single-family homes.

But for a real Pacific Island experience, journey to the town of Apia, the capital of Samoa and home to 35,000 people. Here, in what is called the "heart of Polynesia"—about 2,500 (4,000 kilometers) miles southwest of Hawaii—the pulse beats more slowly. Or at least it did until lately.

With globalization, the pace is faster now than it was just 20 years ago, and outside influences are changing the town's traditional landscape. On Beach Road, overlooking Apia harbor, the new eight-story government building and the neighboring seven-story Central Bank of Samoa seem out of place in a town dominated by colonial-era structures and thatch-roofed, open-sided Polynesian *fala*, or longhouses (Figures 14.4.1 and 14.4.2).

Figure 14.4.2 Apia Governmental Center The Samoan parliament building gives a hint of the traditional architecture of that island nation with its thatched roof and longhouse appearance.

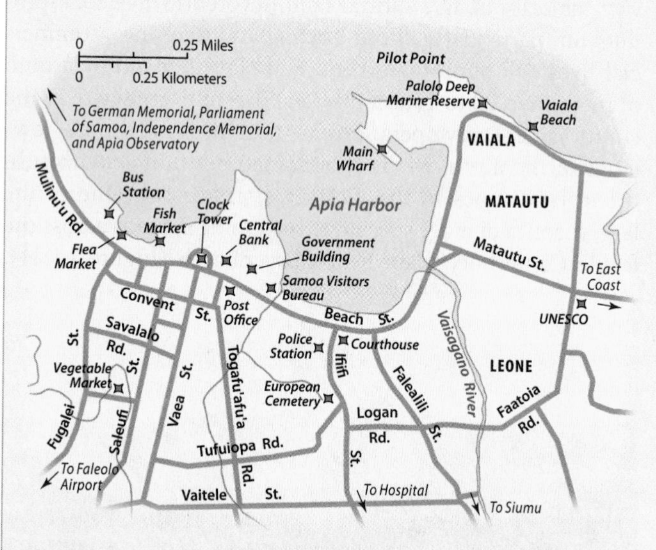

Figure 14.4.1 Map of Apia This map shows some of the major features of Apia, capital city of Samoa. The new auto parts factory and other industry are located along Mulinu'u Road, leading away from the town in the northwest.

The colonial past is apparent everywhere. Until moving into their new high-rise home in 1994, the Samoan government was housed in the wooden courthouse on Ififi Street built in the 1880s by German interests. Closer to the harbor, at the intersection of Beach Road and Vaea Street, is the World War I memorial, constructed by New Zealand and British colonials not only to honor their fallen compatriots but also to mark the end of the German period.

West of the harbor, along the shoreline road leading to the airport and overlooking the harbor full of inter-island ferries, transient yachts, and long-line tuna fishing boats, are several new industrial parks giving clue to Samoa's search for new economic activities to supplement its long reliance on copra (coconut) exports. Indicative of the country's emerging linkages to the globalized world is a Japanese auto parts factory employing 2,000 women who assemble electronic car parts. Farther down the road is the Vailima brewery, a local beer maker that now serves foreign markets, especially those in New Zealand and Australia.

This spectacular South Island setting proved fertile ground for English settlement and continues to feature a varied landscape of pastures, grain fields, orchards, and vegetable gardens, all spread beneath the towering peaks of the Southern Alps.

Rural Oceania Elsewhere in Oceania, varied influences shape the rural landscape. On high islands with more water, denser populations take advantage of more diverse agricultural opportunities than are usually found on the more barren low islands where fishing is often more important. Several types of rural settlement can be identified across the island realm. In rural New Guinea, village-centered shifting cultivation dominates: Farmers clear a patch of forest and then, after a few years, shift to another patch, thus practicing a form of land rotation. Subsistence foods such as sweet potatoes, taro (another starchy root crop), coconut palms, bananas, and other garden crops are often found in the same field, and growing numbers of planters also include commercial crops such as coffee. In other parts of Oceania, traditional agricul-

Figure 14.25 Yam Harvest These farmers on the Melanesian island of Vakuta (Papua New Guinea) are harvesting yams. Traditional tropical agriculture features a mix of crops, often grown in the same field. Where possible, fields are periodically rotated to maintain productivity.

tural patterns are similar (Figure 14.25). Commercial plantation agriculture has also made its mark in many more accessible rural settings. In these places, settlements consist of worker housing near crops that are typically controlled by absentee landowners. For example, copra (coconut), cocoa, and coffee operations have transformed many agricultural settings in places such as the Solomon Islands and Vanuatu. Sugarcane plantations have reshaped other island settings, particularly in Fiji and Hawaii.

Diverse Demographic Paths

A variety of population-related issues face residents of the region today. In Australia and New Zealand, while populations grew rapidly (mostly from natural increases) in the 20th century, today's low birthrates parallel the pattern in North America. Just as in the United States and Canada, however, significant population shifts within these countries continue to create challenges. To illustrate, the departure of farmers from Australia's wheat-growing and sheep-raising interior mirrors similar processes at work in the rural Midwest of the United States and in the Canadian prairies. Communities see many of their productive young people and professionals leave for the better employment opportunities of the city.

Different demographic challenges grip many less-developed island nations of Oceania. High population growth rates over 2 percent per year are not uncommon, as in Vanuatu and the Marshall Islands, and while the larger islands of Melanesia contain some room for settlement expansion, competitive pressures from commercial mining and logging operations limit the amount of new agricultural land available. On some of the smaller island groups in Micronesia and Polynesia, population growth is a more pressing problem. Tuvalu (north of Fiji), for example, has just over 12,000 inhabitants, but they are crowded onto a land area of about 10 square miles (26 square kilometers), making it one of the world's most densely populated countries.

Out-migration from several island nations is very high, such as Tonga and Samoa, where the lack of employment is a considerable push force. As mentioned earlier in this chapter, ethnic tensions in Fiji explain its high rate of emigration. In contrast, Australia and New Zealand remain attractive to migrants, as does New Caledonia because of its recent mining boom.

CULTURAL COHERENCE AND DIVERSITY: A Global Crossroads

The Pacific world offers excellent examples of how culture is transformed as different groups migrate to a region, interact with one another, and evolve over time. As Europeans and other outsiders arrived in Oceania, colonization forced native peoples to resist or adjust. Worldwide processes of globalization have also redefined the region's cultural geography, provoking fears of homogenization in some places while at the same time promoting cultural preservation efforts in others as native groups attempt to protect their heritage.

Multicultural Australia

Australia's cultural patterns illustrate many of the fundamental processes of globalization at work. Today, while still dominated by its colonial European roots, the country's multicultural character is becoming increasingly visible as native peoples assert their cultural identity and as varied immigrant populations play larger roles in society, particularly in major metropolitan areas.

Aboriginal Imprints For thousands of years, Australia's Aborigines dominated the cultural geography of the continent. They never practiced agriculture, opting instead for a hunting-and-gathering way of life that persisted up to the time of the European conquest. As the population consisted of foragers and hunters living in a relatively dry land, settlement densities remained low; tribal groups were often isolated from one another, and overall populations probably never numbered more than 300,000 inhabitants. To survive, Aborigines developed great adaptive skills, subsisting in harsh environments that Europeans avoided. Because these people were clustered in many different areas, their language became fragmented. Although precise counts vary, there were probably 250 languages spoken at the time of European contact, and almost 50 indigenous languages can still be found today.

Radical cultural and geographic changes accompanied the arrival of Europeans, and Aboriginal populations were decimated in the process. The geographic results of colonization were striking, as Aboriginal settlements were relocated to the sparsely settled interior, particularly in northern and central Australia, where fewer Europeans competed for land. In most cases, the European attitude toward the Aboriginal population was even more discriminatory than it was toward native peoples of the Americas.

Today, Aboriginal cultures persevere in Australia, and a growing native peoples movement is similar to activities in the Americas. Indigenous peoples account for approximately 2 percent (or 430,000) of Australia's population, but their geographic distribution changed dramatically over the past century. Aborigines account for almost 30 percent of the Northern Territory's population (many of these near Darwin), and other large native reserves are located in northern Queensland and Western Australia. Most native peoples, however, live in the same urban areas that dominate the country's overall population geography. Indeed, more than 70 percent of Aborigines live in cities, and very few of them still practice traditional hunting-and-gathering lifestyles. Processes of cultural assimilation are clearly at work: Urban Aborigines are frequently employed in service occupations, Christianity has often replaced traditional animist religions, and only 13 percent of the native population still speaks a native language.

Still, forces of diversity are at work, suggesting a growing Aboriginal interest in preserving traditional cultural values. Particularly in the Outback, a handful of Aboriginal languages remain strong and have growing numbers of speakers. In addition, cultural leaders are preserving some aspects of Aboriginal spiritualism, and these religious practices often link local populations to surrounding places and natural features that are considered sacred. In fact, a growing number of these sacred locations are at the center of land-use controversies (such as mining on sacred lands) between Aboriginal populations and Australia's European majority. The future for Aboriginal cultures remains unclear: Pressures for cultural assimilation will be intense as many native peoples move to more Western-oriented urban settlements and lifestyles. At the same time, rapid rates of natural increase (almost twice the national average) and a growing cultural awareness of Aboriginal traditions will work to preserve elements of the country's indigenous cultures.

A Land of Immigrants Most Australians reflect the continent's more recent European-dominated migration history, but even these patterns have become more complex as a rising tide of Asian cultures becomes increasingly important. Overall, more than 70 percent of Australia's population continues to reflect a British or Irish cultural heritage. These groups dominated many of the 19th- and early 20th-century migrations into the country, and, as a result, the close cultural ties to the British Isles remain strong.

A need for laborers along the fertile Queensland coast caused European plantation owners to import inexpensive workers from the Solomons and New Hebrides in the late nineteenth century. These Pacific Island laborers, known as **kanakas**, were spatially and socially segregated from their Anglo employers but further diversified the cultural mix of Queensland's "sugar coast." Historically, however, nonwhite migrations to the country were strictly limited by what is often termed the **White Australia Policy**, in which governmental guidelines promoted European and North American immigration at the expense of other groups. This remained national policy until 1973.

Recent migration trends have reversed this historical bias, and more diverse inflows of new workers and residents are adding to the country's multicultural character. Since the 1970s, the government's Migration Program has been dominated by a variety of people chosen on the basis of their educational background and potential for succeeding economically in Australian society. For example, a growing number of families have come from places such as China, India, Malaysia, and the Philippines. Smaller numbers have arrived as refugees from troubled parts of the world, such as Southeast Asia and the Balkans. The result is a more diverse foreign-born population. Indeed, today 25 percent of Australia's people are foreign born, reflecting the country's global popularity as a migration destination. In the period 2000–2010, almost 40 percent of the settlers arriving in the country were from Asia. Major cities offer particularly attractive possibilities: Sydney's Asian population already exceeds 10 percent and is growing rapidly, while Perth's culture and economy are increasingly linked to Asian countries.

Cultural Patterns in New Zealand

New Zealand's cultural geography broadly reflects the patterns seen in Australia, although the precise cultural mix differs slightly. Native Maori populations are more numerically

Figure 14.26 Maori Artisans New Zealand's native Maori population actively preserves its cultural traditions and has recently increased its political role in national affairs. These artisans are carving decorations for a traditional Maori canoe.

important and culturally visible in New Zealand than their Aboriginal counterparts in Australia. While British colonization clearly mandated the dominance of Anglo cultural traditions by the late 19th century, Maori populations survived, although they lost most of their land in the process. After the initial decline, native populations began rebounding in the 20th century, and today the Maori account for more than 8 percent of the country's 4 million residents. Geographically, the Maori remain most numerous on the North Island, including a sizable concentration in metropolitan Auckland. While urban living is on the rise, many Maori, like their Aboriginal counterparts, are also committed to preserving their religion, traditional arts, and Polynesian lifeways (Figure 14.26). In addition, Maori is now an official language in the country (along with English).

While many New Zealanders still identify with their largely British heritage, the country's cultural identity has increasingly separated from its British roots. Several processes have forged New Zealand's special cultural character. As Britain tightened its own links with the European continent after World War II, New Zealanders increasingly formed a more independent and diverse identity. In many ways, popular culture ties the country ever more closely to Australia, the United States, and continental Europe, a function of increasingly global mass media. A number of major movies, for example, have been filmed in New Zealand, including *Avatar, The Lord of the Rings, Whale Rider,* and *The Piano.*

The Mosaic of Pacific Cultures

Native and colonial influences produce a variety of cultures across the islands of the South Pacific. In more isolated places, traditional cultures are largely insulated from outside influences. In most cases, however, modern life in the islands revolves around an intricate cultural and economic interplay of local and Western influences, illustrating that the relative cultural insularity of the past is gone forever, and in its place is a Pacific realm rapidly adjusting to powerful forces of globalization.

Language Geography A modern language map reveals some significant cultural patterns that both unite and divide the region (Figure 14.27). Most of the native languages of Oceania belong to the Austronesian language family, which encompasses wide expanses of the Pacific, much of insular Southeast Asia, and Madagascar. Linguists hypothesize that the first prehistoric wave of oceanic mariners spoke Austronesian languages and thus spread them throughout this vast realm of islands and oceans. Within this broad language family, the Malayo-Polynesian subfamily includes most of the related languages of Micronesia and Polynesia, suggesting a common cultural and migratory history for these widespread peoples.

Melanesia's language geography is more complex and still incompletely understood by outside experts: While coastal peoples often speak languages brought to the region by the seafaring Austronesians, more isolated highland cultures, particularly on the island of New Guinea, speak varied Papuan languages. Indeed, the linguistic complexity of that island is so complex—more than 1,000 languages have been identified—that many experts question whether they even constitute a unified "Papuan family" of related languages. Some scholars estimate that half of New Guinea's languages are spoken by fewer than 500 persons, suggesting that the region's rugged topography plays a strong role in isolating cultural groups. These New Guinea highlands may hold some of the

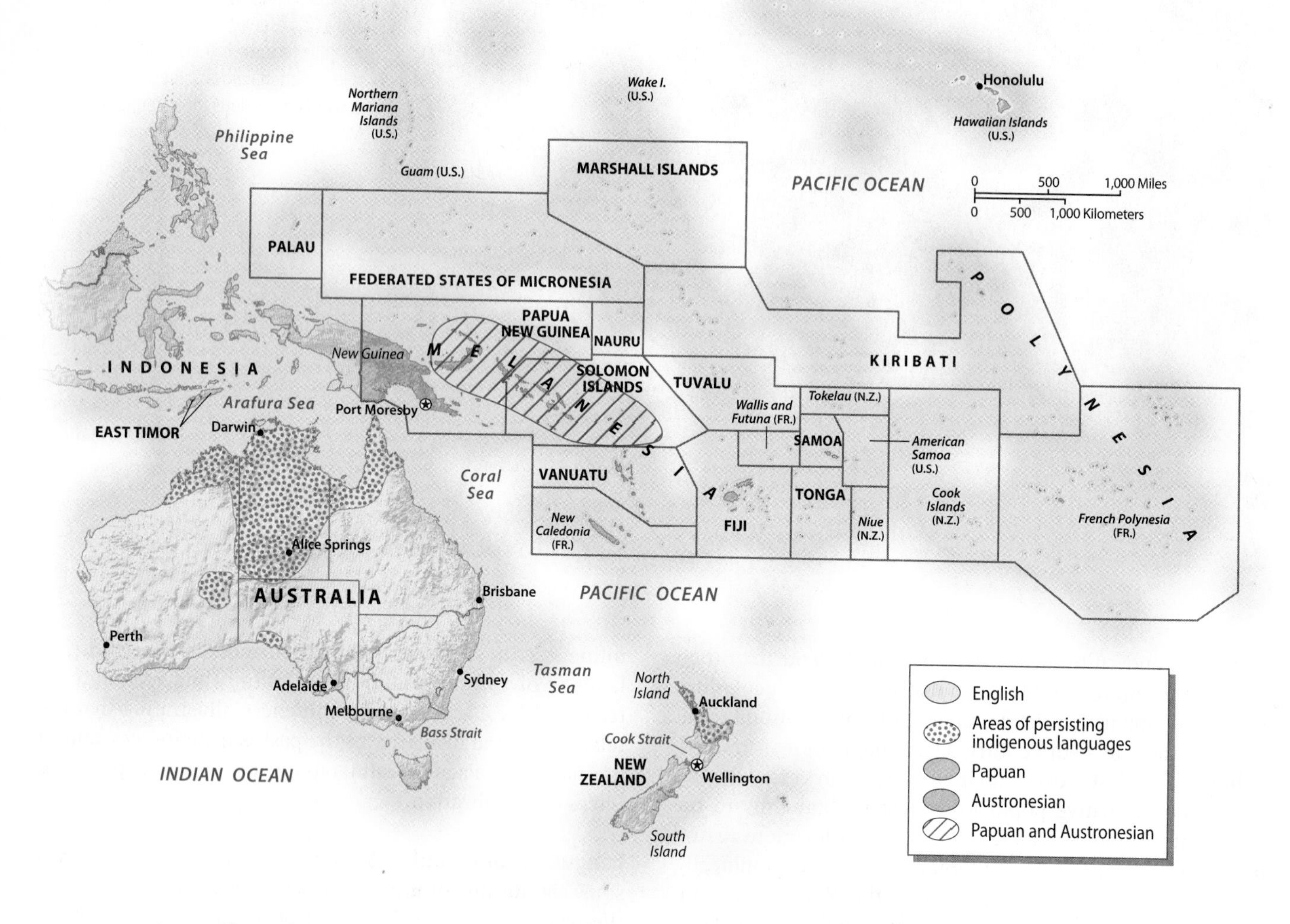

Figure 14.27 Language Map of Australia and Oceania While English is spoken by most residents, native peoples and their linguistic traditions remain an important cultural and political force in both Australia and New Zealand. Elsewhere, traditional Papuan and Austronesian languages dominate Oceania. The French colonial legacy also persists in select Pacific locations. Tremendous linguistic diversity has shaped the cultural geography of Melanesia, and more than 1,000 languages have been identified in Papua New Guinea.

world's few remaining **uncontacted peoples**, cultural groups that have yet to be "discovered" by the Western world.

Village Life Traditional patterns of social life are as complex and varied as the language map. In many cases life revolves around predictable settings. For example, across much of Melanesia, including Papua New Guinea, most people live in small villages, often occupied by a single clan or family group. Many of these traditional villages contain fewer than 500 residents, although some larger communities may house more than 1,000 people. Life often revolves around the gathering and growing of food, annual rituals and festivals, and complex networks of kin-based social interactions.

Traditional Polynesian culture also focuses on village life (Figure 14.28), although there are often strong class-based relationships between local elites who are often religious leaders and ordinary residents. Polynesian villages are also more likely linked to other islands by wider cultural and political ties. Despite the Western stereotype of depicting Polynesian communities in idyllic and peaceful terms, violent warfare was actually quite common across much of the region prior to European contact.

External Cultural Influences While traditional culture persists in some areas, most Pacific islands have witnessed tremendous cultural transformations in the past 150 years. Outsiders from Europe, the United States, and Asia brought new settlers, values, and technological innovations that have forever changed Oceania's cultural geography and its place in the larger world. The result is a modern setting

Figure 14.28 Tonga Village Although the economic and technological effects of globalization have arrived in Tonga, village life remains important in many Polynesian settings. Most village housing in these tropical environments reflects the use of locally available construction materials.

where Pidgin English has mostly replaced native languages, Hinduism is practiced on remote Pacific Islands, and traditional fishing peoples now work at resort hotels and golf course complexes.

European colonialism transformed the cultural geography of the Pacific world by introducing new political and economic systems. In addition, the region's cultural makeup was changed by new people migrating into the Pacific islands. Hawaii illustrates the pattern. By the mid-19th century, Hawaii's King Kamehameha was already entertaining a varied assortment of whalers, Christian missionaries, traders, and navy officers from Europe and the United States. A small elite group of **haoles**, or light-skinned European and American foreigners, were successfully profiting from commercial sugarcane plantations and Pacific shipping contracts. Labor shortages on the islands led to the importation of Chinese, Portuguese, and Japanese workers who further complicated the region's cultural geography. By 1900, the Japanese had become a dominant part of the island workforce.

After the United States formally annexed the islands in 1898 the cultural mix revealed in the Hawaiian census of 1910 suggests the magnitude of culture change: More than 55 percent of the population was Asian (mostly Japanese and Chinese), native peoples made up another 20 percent, and about 15 percent (mostly imported European workers) were white. By the end of the 20th century, however, the Asian population was less dominant as about 40 percent of Hawaii's residents were white; additionally the small number of remaining native Hawaiians had been joined by an increasingly diverse group of other Pacific Islanders. In addition, ethnic mixing between these groups has produced a rich mosaic of Hawaiian cultures that offer a unique blend of North American, Asian, Pacific Island, and European influences (Figure 14.29).

Hawaii's story has also played out in many other Pacific Island locations. In the Mariana Islands, Guam was absorbed into America's Pacific empire as part of the Spanish-American War in 1898. Thereafter, not only did native peoples feel the effects of Americanization (the island remains a self-governing U.S. territory today), but thousands of Filipinos were moved there to supplement its modest labor force. To the southeast, the British-controlled Fiji Islands offered similar opportunities for redefining Oceania's cultural mix. The same sugar plantation economy that spurred changes in Hawaii prompted the British to import thousands of South Asian laborers to Fiji. The descendants of these Indians (most practicing Hinduism) now constitute almost half the island

Figure 14.29 Multicultural Hawaiians Many residents of the Hawaiian Islands represent a blend of Pacific Island, Asian, and European influences. These young women express a mixture of Polynesian and Asian ancestors.

Figure 14.30 South Asians in Fiji British sugar plantation owners imported thousands of South Asian workers to Fiji during the colonial era. Today, almost half of Fiji's population is South Asian and is often in conflict with indigenous Fijians.

country's population and, as noted earlier, often come into sharp conflict with the native Fijians (Figure 14.30). In French-controlled portions of the realm, small groups of traders and plantation owners filtered into the Society Islands (Tahiti), but a larger group of French colonial settlers (many originally a part of a penal colony) had a major impact on the cultural makeup of New Caledonia. Still a French colony, New Caledonia's population is more than one-third French, and its capital city of Noumea reveals a cultural setting forged from French and Melanesian traditions.

Given the frequency of contact between different island cultures, it is no surprise that people have generated new forms of intercultural communication. For example, several forms of **Pidgin English** (also known simply as *Pijin*) are found in the Solomons, Vanuatu, and New Guinea, where it is the major language used between ethnic groups. In Pijin, a largely English vocabulary is reworked and blended with Melanesian grammar. Pijin's origin is commonly traced to 19th-century Chinese sandalwood traders ("pijin" is the Chinese pronunciation of the word for "business"). While of historical origin, Pijin is now becoming a globalized language of sorts in Oceania as trade and political ties develop between different native island groups.

A tidal wave of outside influences since World War II has produced cultural changes as well as growing indigenous responses designed to preserve traditional values. Some groups, particularly in Melanesia, remain more isolated from the outside world, although even there growing demands for natural resources offer an avenue for increasing Western or Asian contacts. As well the global growth of tourism has brought Oceania into the relatively easy reach of wealthier Europeans, North Americans, Asians, and Australians. The Hawaiian Islands, Fiji, French Polynesia, and American Samoa are being joined by an increasing number of other island tourist destinations. While offering tremendous economic benefits to certain places, the onrush of tourists and their consumer-driven values has often come into sharp conflict with native cultures.

GEOPOLITICAL FRAMEWORK: A Region of Dynamic Polities

Pacific geopolitics reflect a complex interplay of local, colonial-era, and global-scale forces (Figure 14.31). These complexities become apparent in the story of Micronesia's Marshall Islands. This sprinkling of islands and atolls (covering 70 square miles, or 180 square kilometers, of land) historically consisted of many ethnic groups that made up small political units. In 1914, the Japanese moved into the islands, and the area remained under their control until 1944 when U.S. troops occupied the region. Following World War II, a United Nations trust territory (administered by the United States) was created across a wide swath of Micronesia, including the Marshall group. Demands for local self-government grew during the 1960s and 1970s, resulting in a new constitution and independence for the Marshall Islanders by the early 1990s. Today, still benefiting from U.S. aid, government officials in the modest capital city on Majuro Atoll struggle to unite island populations, protect large maritime sea claims, and resolve a generation of legal and medical problems that grew from U.S. nuclear bomb testing in the region. Similar stories are typical across the realm, suggesting a 21st-century political geography that is still very much in the making.

Roads to Independence

The newness and fluidity of the region's political boundaries are remarkable: The region's oldest independent states are Australia and New Zealand, and both were 20th-century creations that are still considering whether they want to complete their formal political separation from the British Crown. Elsewhere, political ties between colony and mother country are closer and, perhaps, more enduring. Even many of the newly independent Pacific **microstates**, with their tiny overall land areas, keep special political and economic ties to countries such as the United States.

Independent Australia (1901) and New Zealand (1907) gradually created their own political identities, yet both still struggle with the final shape of these identities. Although Australia became a commonwealth in 1901, it still acknowledges the British Crown as the symbolic head of its government. A national referendum in 1999 asked Australians to decide whether they would like to drop this remaining tie to Britain and instead become a genuine republic with its own president replacing the British queen as head of state. Despite a strong movement towards complete independence from Britain, a slight majority (55 percent) voted to retain Australia's ties to the Crown, thus they remain still today. In New Zealand, formal legislative links with Great Britain were not

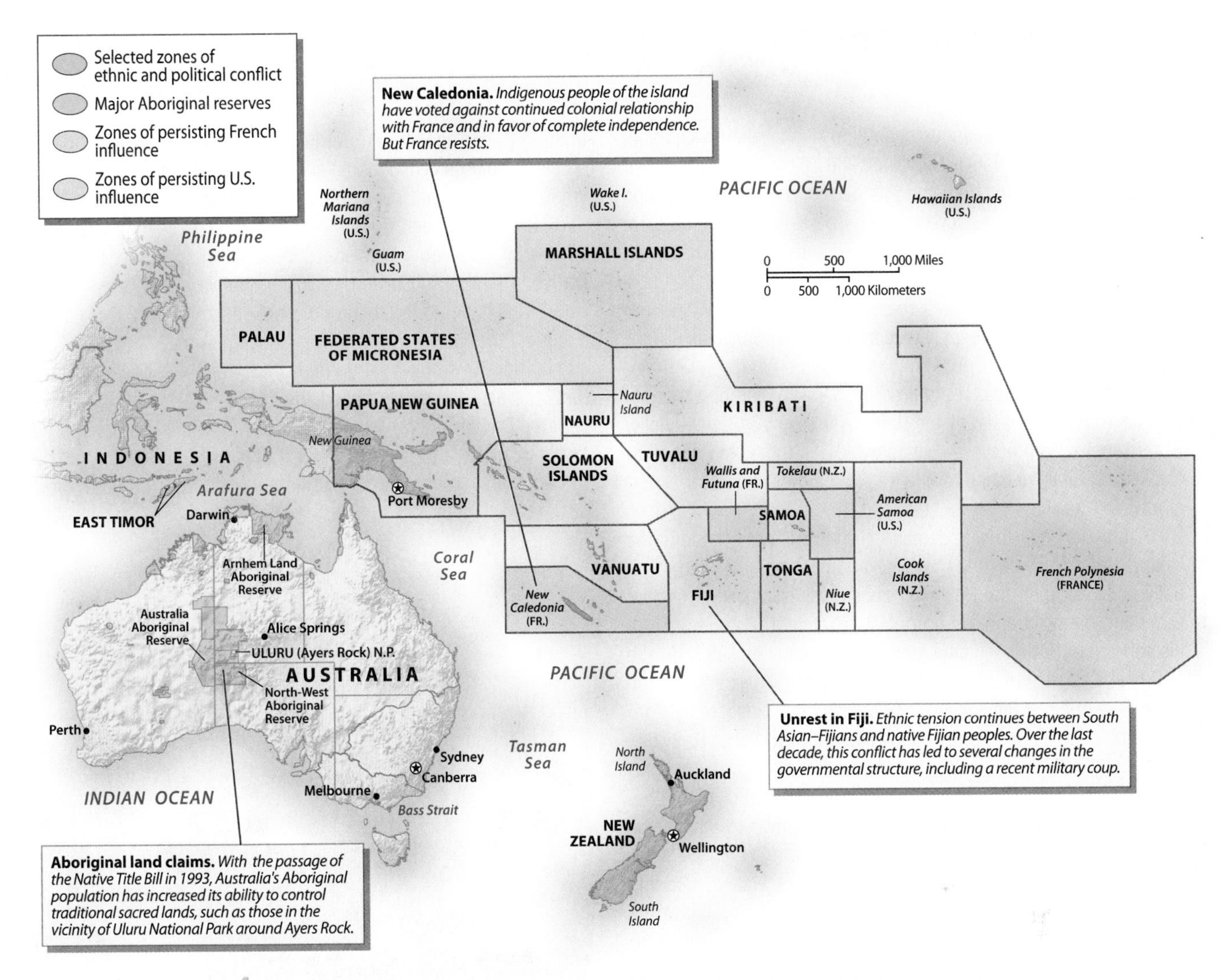

Figure 14.31 Geopolitical Issues in Australia and Oceania Native land claim issues increasingly shape domestic politics in Australia and New Zealand. Elsewhere, ethnic conflicts have raised political tensions in countries such as Fiji and Papua New Guinea. Colonialism's impact endures as well: American and French interests remain particularly visible in the region, including a legacy of nuclear testing that continues to affect certain Pacific Island populations.

broken until 1947 and, today, New Zealand is discussing the same formal break with the British Crown being debated by the Australians.

Elsewhere in the Pacific, colonial ties were cut even more slowly, and the process has not yet been completed. In the 1970s, Britain and Australia began giving up their colonial empires in the Pacific. Fiji (Great Britain) gained independence in 1970, followed by Papua New Guinea (Australia) in 1975 and the Solomon Islands (Great Britain) in 1978. The small island nations of Kiribati and Tuvalu (Great Britain) also became independent in the late 1970s.

The United States has recently turned over most of its Micronesian territories to local governments, while still holding a large influence in the area. After gaining these islands from Japan in the 1940s, the U.S. government provided large monetary subsidies to islanders and also utilized a number of islands for military purposes. Bikini Atoll was destroyed by nuclear tests, and the large lagoon of Kwajalein Atoll was used as a giant missile target. Moreover, a major naval base was established in Palau, the westernmost archipelago of Oceania. By the early 1990s, both the Marshall Islands and the Federated States of Micronesia (including the Caroline Islands) had gained independence. Their ties to the United States, however, remain close. A number of other Pacific islands remain under U.S. administration. Palau is a U.S. "trust territory," which gives Palauans some local autonomy. The people of the Northern Marianas chose to become a "self-governing commonwealth in association with the United States," a rather vague political position that allows them to become U.S. citizens. The residents of self-governing Guam and American Samoa are also U.S. citizens. Hawaii became a full-fledged U.S. state in 1959, yet debate

ENVIRONMENT AND POLITICS Land, Native Rights, and Sovereignty in Hawaii

As is the case with indigenous peoples in Australia, New Zealand, and North America, Native Hawaiians also have issues with their government concerning human rights, access to ancestral land, and the political standing of native peoples. And although attempts are being made to resolve these contentious issues in Hawaii, the path forward is uncertain.

Native Hawaiians, who call themselves *Kanaka maoli,* are descendants of Polynesian people who arrived in the Hawaiian Islands a thousand years or so ago, and who lived in an independent and sovereign state that was recognized by major foreign powers until the islands were annexed by the United States in 1898. The legality of this annexation, however, is contested still today and underlies Native Hawaiian demands for a return of their historical sovereignty.

Giving credence to this demand is that on two occasions the United States has admitted the annexation process was illegal; initially in 1893 when the Hawaiian queen was displaced (President Cleveland opposed annexation of Hawaii at that time because of his concern about its legality), and more recently in 1993, on the 100th anniversary of that earlier event, when President Clinton signed the Apology Bill, which was passed without debate by Congress and once again referred to the illegal overthrow of Hawaiian sovereignty in 1893.

Today, many Hawaiian nationalists support a return to Polynesian sovereignty, and one pathway might be similar to that of the Navajo nation in Arizona, Utah, and New Mexico. Extremists, however, reject the idea of Native Hawaiians having the same standing as U.S. Indian tribes and, instead, demand the UN invalidate the 1898 annexation, thereby granting Native Hawaiians a return to complete international sovereignty (Figure 14.5.1). Were that to happen Native Hawaiian ancestral lands would be a separate country within the U.S. state of Hawaii.

Understandably, this extreme solution is resisted by private and corporate land owners in Hawaii since it could call into question the legality of their ownership, be it residential properties in the Honolulu suburbs or hotel parcels along Waikiki Beach.

Another land issue is the disposition of those lands historically administered by the Hawaiian queen, land that was ceded to the U.S. government in the 1898 annexation. These lands, which constitute almost half of the land area of the Hawaiian Islands, were transferred to the local government in 1959 upon statehood with the stipulation that they be used for the benefit of Native Hawaiians. While many argue these lands should be turned over to the Native Hawaiians to use as they please, this notion is resisted by the state government since these lands generate revenue by renting them to public agencies (a portion of the Honolulu airport, for example, is built on ceded lands) and the U.S. military.

To resolve these issues, the U.S. Congress is pondering a "Native Hawaiian Government Reorganization Act." If passed this bill would give Native Hawaiians legal standing similar to that of Native Americans. Currently, unlike Native Americans, Native Hawaiians have no legal standing nor recognition in U.S. courts, a legality that prevents them from seeking redress for acts of discrimination or settling land claims. As well, and equally important, the act (commonly called the "Akaka Bill" after its author, Senator Daniel Akaka) would, for the first time, provide a legal mechanism for interaction between Native Hawaiians and the United States government, a legal necessity that currently does not exist.

Figure 14.5.1 Native Hawaiian Nationalism Thousands of red-shirted supporters of Native Hawaiian sovereignty gather to march along Waikiki Beach in Honolulu in the annual "Justice for Hawaiians" rally.

continues regarding native land claims and sovereignty (see "Environment and Politics: Land, Native Rights, and Sovereignty in Hawaii").

Other colonial powers appear less inclined to give up their oceanic possessions. New Zealand still controls substantial territories in Polynesia, including the Cook Islands, Tokelau, and the island of Niue. France has even more extensive holdings in the region. Its largest maritime possession is French Polynesia, which includes a large expanse of mid-Pacific territory. To the west, France still controls the much smaller territory of Wallis and Futuna in Polynesia and the larger island of New Caledonia in Melanesia.

Figure 14.32 Aboriginals at Uluru National Park Australian Aborigines gathered recently at Uluru National Park to celebrate their increased political control over the region. Since then, the Native Title Bill has promoted numerous land cessions and further legal settlements.

Persisting Geopolitical Tensions

Cultural diversity, colonial legacy, youthful states, and a rapidly changing political map contribute to ongoing geopolitical tensions in the Pacific world. Indeed, some of these conflicts have consequences that extend far beyond the boundaries of the region. Others are more locally based but are still reminders of the difficulties that occur as political space is redefined in varied natural and cultural settings.

Native Rights in Australia and New Zealand Indigenous peoples in both Australia and New Zealand have used the political process to gain more control over land and resources in their two countries, and the strategies these native groups have used parallel efforts in North America and elsewhere. In Australia, Aboriginal groups are discovering newfound political power from both more effective lobbying efforts by native groups and a more sympathetic federal government interested in rectifying historical discrimination that left native peoples with no legal land rights. More recently, the Australian government established a number of Aboriginal reserves, particularly in the Northern Territory, and expanded Aboriginal control over sacred national parklands such as Uluru (Ayers Rock) (Figure 14.32). Further concessions to indigenous groups were made in 1993, as the government passed the **Native Title Bill**, which compensated Aborigines for lands already given up, and gave them the right to gain title to unclaimed lands they still occupied, as well as providing them with legal standing to deal with mining companies in native-settled areas.

Efforts to expand Aboriginal land rights, however, have met strong opposition. In 1996, an Australian court ruled that pastoral leases (the form of land tenure held by the cattle and sheep ranchers who control most of the Outback) do not necessarily negate or replace Aboriginal land rights. Grazing interests were infuriated, which led the government to respond that Aboriginal claims allow the visiting of sacred sites and some hunting and gathering but do not give native peoples complete economic control over the land (Figure 14.33).

In New Zealand, Maori land claims have generated similar controversies in recent years. Because the Maori constitute a far larger proportion of the overall population, coupled with the fact the lands they claim tend to be more valuable

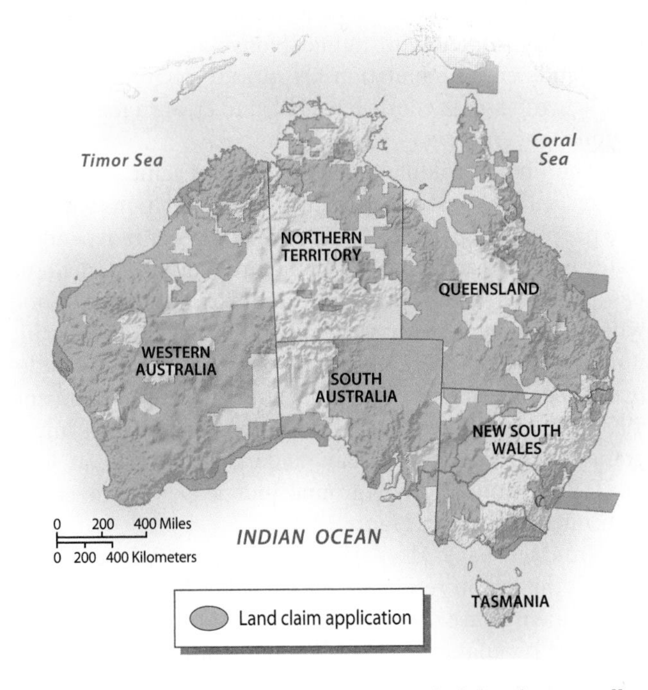

Figure 14.33 Applications for Native Land Claims in Australia This map shows the applications for native land claims in Australia filed by different Aboriginal groups as of 2004. Important to note is that these are applications only—not government-approved claims. Nevertheless, the widespread extent of the claims shows why the topic is so contentious and controversial.

than those rural Aborigine lands in Australia, the issues are even more complicated. Recent protests include civil disobedience, ever-increasing Maori land claims over much of North and South islands, and a call to return the country's name to the indigenous **Aotearoa**, "Land of the Long White Cloud." The government response, including a 1995 visit from Queen Elizabeth, has been to increasingly recognize Maori land and fishing rights. In 2009, for example, the Maori political party's support was crucial to formalizing New Zealand's global warming emissions trading scheme.

Ongoing Conflicts in Oceania Geopolitical issues simmer elsewhere in the Pacific, periodically threatening to further redefine the region's territorial boundaries. As discussed earlier, ethnic differences in Fiji have threatened to tear apart that small island nation.

Papua New Guinea also contends with ethnic tensions. The country is composed of different cultural groups, many of which have a long history of mutual hostility. Most of these peoples now get along with each other reasonably well, although tribal fighting occasionally breaks out in highland market towns. A much bigger problem for the national government has been the rebellion on Bougainville. This sizable island, which has large reserves of copper and other minerals, is located in the Solomon archipelago but belongs to Papua New Guinea because Germany colonized it in the late 1800s, thus it became politically attached to the eastern portion of New Guinea. Many of Bougainville's native residents believe that their resources are being exploited by foreign interests and by an unsympathetic national government, therefore they demand local control. Papua New Guinea has reacted with military force; about 5 percent of the island's entire population has already been killed in the conflict, and recent efforts have failed to create a more stable regional government.

The French colonial presence in the Pacific region has also created political uncertainties, both in relations with native peoples and between the French and other independent states in the area. Continued French rule in New Caledonia has provoked much local opposition on this large island with rich mineral reserves (especially nickel) and sizable numbers of French colonists. By the 1980s, a local independence movement was gaining strength, but in 1987 and 1998, the island's residents (indigenous and French immigrants alike) voted to remain under French rule, at least until 2018. Still, an underground independence movement continues to operate (Figure 14.34).

Troubles have also arisen in French Polynesia. Although the region receives large subsidies from the French, and residents have voted to remain a colony, a large minority of the population opposes French control and demands independence.

Influences from the Two Chinas Some political analysts foresee Asian influences in Oceania replacing the historic linkages to the United States, Australia, and New Zealand. Taiwan, also known as the Republic of China, or ROC, has long provided economic aid to the island nations of Nauru, Tuvalu, the Solomon Islands, Marshall Islands, and Palau, in return for their political allegiance. As a result it is thought by some experts that ROC political influence has been behind recent unrest in the Solomon Islands and Fiji. But now, with the People's Republic of China's emergence as a regional player, matters could become even more complicated.

China appears to have two objectives as it expands its influence into Oceania: to draw upon the region's natural resources to fuel its domestic economy, and, second, to neutralize—and then replace—Taiwan's political and economic influence in the region. Since China's economic and political resources are far greater than Taiwan's, it seems inevitable that they will achieve those economic and political objectives sometime soon. Recent interaction between the current military leader of Fiji and China suggest that Fiji's former allegiance to ROC is negotiable.

But Australia and New Zealand still play key political roles in the South Pacific. Although these two countries sometimes disagree on strategic and military matters, their size, wealth, and collective political influence in the region make them important forces for political stability. Addition-

Figure 14.34 Unrest in New Caledonia The indigenous Melanesians of New Caledonia, known as *kanaks,* recently voted overwhelmingly in favor of ending French colonial power and creating an independent government. France, however, is resisting this change. This photo is of protesters waving the flag of Kanaky independence.

ally, special colonial relationships still connect these two nations with many Pacific islands. Australia maintains close political ties to its former colony of Papua New Guinea, and New Zealand's continuing control over Niue, Tokelau, and the Cook Islands in Polynesia confirm that its political influence extends well beyond its borders. How these two countries will respond to China's growing influence, however, is unclear.

ECONOMIC AND SOCIAL DEVELOPMENT: A Difficult Path to Paradise

As with all other world regions, the Pacific realm contains a diversity of economic situations resulting in both wealth and poverty. Even within affluent Australia and New Zealand there are pockets of pronounced poverty, and large economic disparities exist as well between those Pacific countries with global trade ties and small island nations lacking resources and external trade. While tourism offers some relief from abject poverty, the whims and economics of foreign tourists can be fickle, creating a Pacific version of boom and bust economies. Tourism in Hawaii, for example, suffered deeply during the global recession of 2008–10. Overall, the economic future of the Pacific realm remains highly variable because of its small domestic markets, its peripheral position in the global economy, and a diminishing resource base.

The Australian and New Zealand Economies

Much of Australia's past economic wealth has been built on the cheap extraction and export of abundant raw materials. Export-oriented agriculture, for example, has long been one of the key supports of Australia's economy. Australian agriculture is highly productive in terms of labor input, and it produces a wide variety of both temperate and tropical crops, as well as huge quantities of beef and wool for world markets. While farm exports are still important to the economy, the mining sector has grown more rapidly since 1970 making Australia one of the world's mining superpowers.

This industry has grown in the last two decades primarily due to increased trade with China, an activity that has made Australia the world's largest exporter of iron and coal. Besides coal and iron ore, Australia has an assortment of other metals, namely bauxite (for aluminum), copper, gold, nickel, lead, and zinc. As a result, the New South Wales–based Broken Hill Proprietary Company (BHP) is one of the world's largest mining corporations.

Growing numbers of Asian immigrants and economic links with potential Asian markets also offer promise for Australia's economic future. In addition, an expanding tourism industry is helping to diversify the economy. More than 7 percent of the nation's workforce is now devoted to serving the needs of more than 4 million visitors annually.

Figure 14.35 Queensland's Gold Coast Many of these luxury hotels in the Surfer's Paradise section of the Gold Coast are owned by Japanese firms specializing in accommodations for Asian tourists.

Popular destinations include Melbourne and Sydney, as well as Queensland's resort-filled Gold Coast, the Great Barrier Reef, and the vast, arid Outback. Along the Gold Coast, most luxury hotels are owned by Japanese firms and provide a bilingual resort experience for their Asian clientele (Figure 14.35).

New Zealand is also a relatively wealthy country, although somewhat less well off than Australia. Before the 1970s, New Zealand relied heavily on exports to Great Britain, especially exports of agricultural products such as wool and butter. Problems developed, however, with these colonial trade linkages in 1973 when Britain joined the European Union with its strict agricultural protection policies. Unlike Australia, New Zealand lacked a rich base of mineral resources to export to global markets. As a result the country had slipped into a serious recession by the 1980s. Eventually, the New Zealand government enacted drastic reforms and the country that had previously been noted for its lofty taxes, high levels of social welfare, and state ownership of large companies changed dramatically. Privatization became the watchword, and most state industries were sold off to private parties. As a result, New Zealand has been transformed into one of the most market-oriented countries of the world and has largely recovered from its earlier recession.

Oceania's Economic Diversity

Varied economic activities shape the Pacific Island nations. While one way of life is oriented around subsistence-based economies, such as shifting cultivation or fishing, in other places, a commercial extractive economy dominates, with large-scale plantations, mines, and timber activities competing for land and labor with the traditional subsistence sector. Elsewhere, the huge growth in global tourism has transformed the economic geographies of many island settings, forever changing the way that people make a living. In addition, many island nations still benefit from direct subsidies and economic assistance that come from present and former colonial powers and is designed to promote development and stimulate employment.

Melanesia is the least-developed and poorest part of Oceania because these countries have benefited less from tourism and from subsidies from wealthy colonial and ex-colonial powers. Still today most Melanesians live in remote villages that remain somewhat isolated from the modern economy. The Solomon Islands, for example, with few industries other than fish canning and coconut processing, has a per capita gross national income (GNI) of only $2,130 per year. Similarly, Papua New Guinea's economy produces a per capita GNI of $2,030. Although traditional exports such as coconut products and coffee have increasingly been supplemented with the rapid development of tropical hardwoods, the economic returns remain low. Further, gold and copper mining have dramatically transformed the landscape, although political instability has often interfered with mineral production in places such as Bougainville. In New Guinea's interior highlands, much village life is focused on subsistence activities. In contrast, Samoa is the most prosperous Melanesian country, with a per capita GNI of $4,410, largely because of its tourist economy, popular with North Americans and Japanese.

The Economic Impact of Mining Among the smaller islands of Melanesia and Micronesia, mining economies dominate New Caledonia and Nauru. New Caledonia's nickel reserves, the world's second largest, are both a blessing and a curse (Figure 14.36). While they currently sustain much of the island's export economy, income from nickel mining will lessen in the near future as the reserves dwindle. Dramatic price fluctuations for the industrial economy also hamper economic planning for the French colony. Other activities include coffee growing, cattle grazing, and tourism. To the north, the tiny, phosphate-rich island of Nauru also depended on mining; however, that day is now past with deposits exhausted and the future uncertain

Micro- and Polynesian Economies Throughout Micronesia and Polynesia, economic conditions depend on both local subsistence economies and economic linkages to the wider world beyond. Many archipelagos export a few food products, but native populations survive mainly on fish, coconuts, bananas, and yams. Some island groups, though, enjoy large subsidies from either France or the United States, even though such support often comes with a political price. Change is also occurring in some of these islands: Japan agreed to build a spaceport for its future shuttlecraft on Mi-

Figure 14.36 Nickel Mine in New Caledonia The North Star nickel mine is the second largest in the world and largely responsible for New Caledonia having a quarter of the world's nickel resources. While mining is a mainstay of the New Caledonia economy, fluctuations in the demand and price for nickel cause sporadic problems for the country. Further, the fiscal future is not particularly bright because of dwindling reserves in the North Star mine.

Figure 14.37 Tahitian Resort Luxury resort settings in Tahiti (near Papeete) exemplify a growth industry in the South Pacific. While bringing important investment capital, such ventures reorder the region's social and economic structure, as well as refashion its cultural landscape.

cronesia's Christmas Island (Kiribati), and the Marshall Islands are the site of a planned industrial park financed by mainland Chinese. Palau, however, is the Micronesian country with the highest estimated GNI per capita, a result of tourism, fishing, and—not unimportant—government assistance from the United States.

Other Polynesian island groups have been completely transformed by tourism. In Hawaii, more than one-third of the state's economy flows directly from tourist dollars. With almost 7 million visitors annually (including more than 1.5 million from Japan), Hawaii represents all the classic benefits and risks of the tourist economy. While job creation and economic growth have reshaped the island realm, congested highways, high prices, and the unpredictable spending habits of tourists have put the region at risk for future problems. Elsewhere, French Polynesia has long been a favored destination of the international jet set. More than 20 percent of French Polynesia's GNI is derived from tourism, making it one of the wealthiest areas of the Pacific (Figure 14.37). More recently, Guam has emerged as a favorite destination of Japanese and Korean tourists, especially those on honeymoons.

The Global Economic Setting

Many international trade flows link the area to the far reaches of the Pacific and beyond. Australia and New Zealand dominate global trade patterns in the region. In the past 30 years, ties to Great Britain, the British Commonwealth, and Europe have weakened in comparison with growing trade links to Japan, East Asia, the Middle East, and the United States. Australia, for example, now imports more manufactured goods from China, Japan, and the United States than it does from Britain and Europe (Figure 14.38; also, see "Exploring Global Connections: China Buys Australia"). Other global economic ties have come in the form of capital investment in the region. U.S. and Japanese banks and other financial institutions now dot the South Pacific landscape from Sydney to Suva. Both Australia and New Zealand also participate in the **Asia-Pacific**

Figure 14.38 Australia's Trade with China Containers from Asia testify to the recent explosion of two-way trade between Australia and China. While raw materials, mainly iron ore, are exported to China, consumer goods flow from China into Australia, making that country a key beneficiary of China's recent economic growth.

EXPLORING GLOBAL CONNECTIONS China Buys Australia

A bit more than 600 miles (1020 km) northeast of Perth, Australia, in the Pilbara region of Western Australia, in a land so vast that road signs warn that the nearest gas station is hundreds of miles away, lies Mt. Whaleback, the world's largest iron ore mine. Mt. Whaleback, once 1,500 feet high (457 meters), however, is no longer a mountain (Figure 14.6.1). Instead, it's a very large hole in the ground, for the mountain, rich in iron ore, was recently shipped piecemeal to China.

Ton by ton, Australia's coal, iron ore, and natural gas is being bought up by China, resulting in one of the world's largest transfers of natural resources. The result has both enriched and alarmed many Australians.

Hardly slowed by the recent global recession, China's appetite for natural resources is huge . . . equalled, apparently, only by Australia's willingness to sell its resources. Even during the global recession of 2008–09, China pumped $40 billion in the Australian economy. Besides natural resources, half a million Chinese tourists visited Australia and added to the local economy by keeping busy lifeguards, blackjack dealers, and real estate brokers. As well there are 70,000 Chinese students currently attending Australian universities.

"China is remaking the social and political fabric of this country," said Chen Jie, a senior lecturer in international relations at the University of Western Australia who immigrated to Australia from China 20 years ago. "China is intruding into society itself." Australia

Figure 14.6.1 Mt Whaleback Mine This iron ore mine in Western Australia is the largest in the world and has been a major supplier of iron ore for China's economic boom. Mt Whaleback itself was once 1500 feet (457 meters) high, but as can be seen in this photo, is now mostly a deep pit.

Economic Cooperation Group (APEC), an organization designed to encourage economic development in Southeast Asia and the Pacific Basin. The region's economic ties to Asia also carry risks, however; the Asian downturn in the late 1990s slowed the popular Korean tourist trade to New Zealand and also lowered China's demand for a variety of Australian raw material exports.

To promote more economic integration within the region Australia and New Zealand signed the **Closer Economic Relations (CER) Agreement** in 1982 that successfully slashed trade barriers between the two countries. As a result New Zealand benefited from the opening of larger Australian markets to New Zealand exports, and Australian corporate and financial interests gained new access to New Zealand business opportunities. Since the CER Agreement's signing, trade between the two countries has expanded almost 10 percent per year. Today, more than 20 percent of New Zealand's imports and exports come from Australia, and this pattern of regional free trade is likely to strengthen in the future.

Smaller nations of Oceania, while often closely tied to countries such as Japan, the United States, and France, also benefit from their proximity to Australia and New Zealand. More than half of Fiji's imports come from those two nearby nations, and other countries, such as Papua New Guinea, Vanuatu, and the Solomon Islands, enjoy a simiis even speaking Chinese: former Prime Minister Kevin Rudd was the first Western leader to speak fluent Mandarin.

No city better illustrates the China boom than Perth, the capital of Western Australia, a state five times as big as Texas that holds the bulk of Australia's mineral wealth. The state's average income has jumped $10,000 in 5 years to more than $70,000, thanks to China's purchases of iron ore, natural gas, and other resources, and in Perth, the median price for a home just broke $500,000. The city's unemployment rate is a measly 2.6 percent while the national rate is 5.5 percent however, that figure is predicted to drop in the near future thanks to even more trade with China.

Beyond the numbers, Perth even looks Chinese, with a skyline filled with skyscrapers and cranes. The city is bristling with more than $1 billion in new construction, including a hospital, museum, highways, offices, and a vast indoor entertainment and sports complex (Figure 14.6.2). More than 110 people move into Western Australia each day, making it the fastest growing state in Australia, yet there is still a labor shortage. Statewide, projects worth more than $95 billion are transforming the economy of the region and contributing to a historic shift in power from "old Australia" in Sydney and Melbourne, to the resource-rich Australia of the west.

Source: Adapted from "Australian welcomes China's investment, if not its influence," *Washington Post,* February 14, 2010.

Figure 14.6.2 Booming Perth in Western Australia Largely because of the rich iron and coal resources in the Western Australia region, commodities sold to China and Japan, Perth has turned into a boom town of low unemployment and ubiquitous construction projects. Because of this economic vitality, some opine that the political and economic geography of Australia is shifting to the west and away from the traditional centers of Sydney, Melbourne, and Canberra in the Southeast.

larly close trading relationship with their more developed Pacific neighbors.

Continuing Social Challenges

Australians and New Zealanders enjoy high levels of social welfare but face some of the same challenges evident elsewhere in the developed world (Table 14.2). Life spans average about 80 years in both countries, and rates of child mortality have fallen greatly since 1960. But echoing patterns in North America and Europe, cancer and heart disease are leading causes of death, and alcoholism is a continuing social problem, particularly in Australia. Furthermore, Australia's rate of skin cancer is among the world's highest, the result of having a largely fair-skinned, outdoors-oriented population from northwest Europe in a sunny, low-latitude setting. Overall, Australia's Medicare program (initiated in 1984) and New Zealand's system of social services provide high-quality health care to their populations.

Not surprisingly, the social conditions of the Aborigines and Maoris are much less favorable than those of the population overall. Schooling is irregular for many native peoples, and levels of postsecondary education for Aborigines (12 percent) and Maoris (14 percent) remain far below the national averages (32 to 34 percent). Many other social

TABLE 14.2 Development Indicators

Country	GNI per capita, PPP 2008	GDP Average Annual % Growth 2000–08	Human Development Index (2007)[a]	Percent Population Living Below $2 a Day	Life Expectancy (2010)[b]	Under Age 5 Mortality Rate (1990)	Under Age 5 Mortality Rate (2008)	Gender Equity (2008)[c]
Australia	37,250	3.3	0.970		81	9	6	98
Fed. States of Micronesia	3,270				68	58	40	102
Fiji	4,320		0.741		68	22	18	104
French Polynesia					74			
Guam					79			
Kiribati	3,610				61	88	63	107
Marshall Islands					66	92	54	100
Nauru					56			
New Caledonia					76			
New Zealand	25,200	3.1	0.950		80	11	6	102
Palau					69	21	10	102
Papua New Guinea	2,030	2.9	0.541	58	59	91	69	
Samoa	4,410		0.771		73	50	27	105
Solomon Islands	2,130		0.610		62	121	70	93
Tonga	3,980		0.768		70	32	23	99
Tuvalu					64			
Vanuatu	3,480		0.693		67	62	34	

[a] *United Nations, Human Development Report, 2009.*

[b] *Population Reference Bureau, World Population Data Sheet, 2010.*

[c] *Gender Equity—Ratio of female to male enrollments in primary and secondary school. Numbers below 100 have more males in primary/secondary school, numbers above 100 have more females in primary/secondary schools.*

Source: World Bank, World Development Indicators, 2010.

measures reflect the pattern, as well. Fewer than one-third of Aboriginal households own their own homes, while more than 70 percent of white Australian households are homeowners. Furthermore, considerable discrimination against native peoples continues in both countries, a situation that has been aggravated and publicized with the recent assertion of indigenous political rights and land claims. As with North American African-American, Hispanic, and Native American populations, simple social policies do not yet exist as solutions to these lasting problems.

Even in Hawaii the social welfare of native peoples is problematic as the proportion of Native Hawaiians living below the poverty level is much higher than any other ethnic group (Figure 14.39). Related, this group also has the

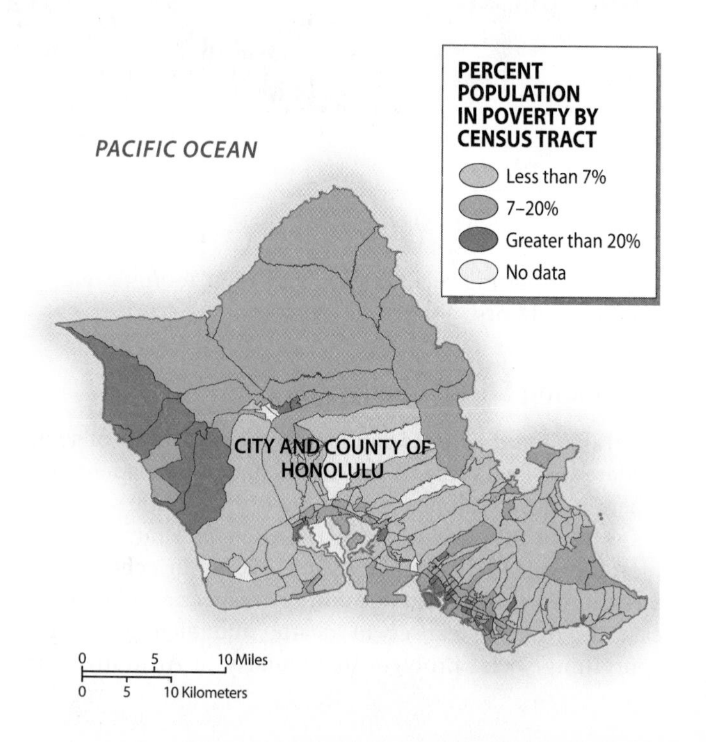

Figure 14.39 Development Issues: Poverty in Hawaii This map of the distribution of poverty on the Hawaiian island of Oahu shows the highest levels in urban pockets around Honolulu in the southeast as well as in the rural western census tracts with a high number of Native Hawaiians. Another area of poverty surrounds military bases at Pearl Harbor.

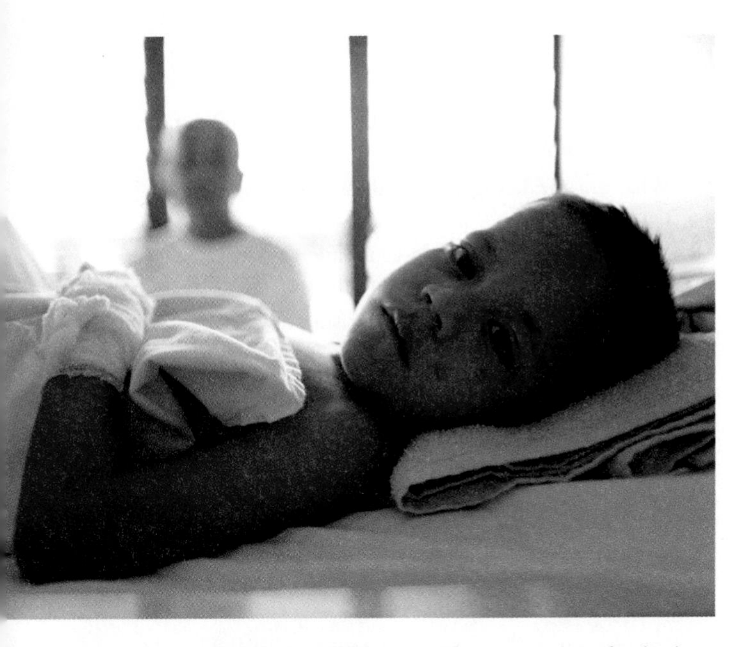

Figure 14.40 Health Services in Oceania In general, Pacific island governments have invested heavily in health and social services despite their modest budgets. As a result life expectancy is higher and infant mortality rates lower than in other countreis with similar modest national budgets. In this photo a 4-year old casualty of a recent tsunami recovers in a Samoan hospital.

shortest life expectancy and the highest infant mortality rate; death from cancer and heart disease is almost 50 percent higher than other groups in the United States, and Native Hawaiian women have the highest rate of breast cancer in the world. Further, 55 percent of Native Hawaiians do not complete high school, and only 7 percent have college degrees.

In other parts of Oceania levels of social welfare are higher than one might expect based on the region's economic situation. Many of its countries and colonies have invested heavily in health and education services and have achieved considerable success (Figure 14.40). For example, the average life expectancy in the Solomon Islands, one of the world's poorer countries as measured by per capita GNI figures, is a respectable 62 years. By other social measures as well, the Solomon Islands and a number of other Oceania states have reached higher levels of human well-being than exist in most Asian and African countries with similar levels of economic output. This is partly a result of successful policies; but it also reflects the relatively healthy natural environment of Oceania, for many of the tropical diseases that are so troublesome in Africa simply do not exist in the Pacific islands.

Summary

- Globalization has brought fresh interconnections to Oceania as Japanese-financed golf courses pop up along tropical shores, Canadian and Chinese mining companies invest in Australian iron ore and coal, and Korean newlyweds honeymoon beneath coconut palms on tropical islands. How native cultures will respond and what new cultural hybrids will emerge as Pacific, European, North American, and Asian peoples mingle and interact is uncertain. Nonetheless, whatever the expression, these issues will give Oceania a new globalized face.
- The natural environment, which has been transformed over millennia by indigenous and colonial settlement, has witnessed accelerating change in the past 50 years as urbanization, tourism, extractive economic activity, exotic species, and climate change from global warming have reconfigured the landscape and increased the vulnerability of island environments.
- The region's contemporary political geography reveals a fluid and changing character as countries struggle to disentangle themselves from colonial ties by asserting their own political identities. Globalization complicates the process, with new economic and political linkages replacing historic colonial ties.
- If Australia sits at the edge of Asia, then New Zealand is on the edge of Polynesia with its Maori population growing rapidly as native peoples immigrate from other parts of Polynesia. Further, New Zealand is taking an active role in the affairs of the entire Pacific Basin, thus becoming Oceania's leading state, economically and politically, as Australia becomes increasingly oriented to Asia.
- The global recession of 2008–2010 took a heavy toll on the Pacific world between China reducing its consumption of coal and iron and the tourist economies of Tahiti and Hawaii suffering heavily. While China's upturn in late 2010 has helped economic matters in Australia, tourist destinations in other parts of Oceania still languish.

Key Terms

Aborigine *(page 653)*
Aotearoa *(page 668)*
archipelago *(page 641)*
Asia-Pacific Economic Cooperation Group (APEC) *(page 671)*
atoll *(page 651)*
Closer Economic Relations (CER) Agreement *(page 672)*
haoles *(page 663)*
high island *(page 649)*
hot spot *(page 649))*
kanakas *(page 660)*
mallee *(page 648)*
Maori *(page 641)*
Melanesia *(page 640)*
Micronesia *(page 641)*
microstate *(page 664)*
Native Title Bill *(page 667)*
Oceania *(page 638)*
Outback *(page 640)*
Pidgin English (Pijin) *(page 664)*
Polynesia *(page 641)*
tsunami *(page 649)*
uncontacted people *(page 662)*
viticulture *(page 657)*
White Australia Policy *(page 660)*

Review Questions

1. Why does it make sense to group together Australia, New Zealand, and the other Pacific islands as a region? What complexities and contradictions are involved in doing so?
2. What effects have nonnative plants and animals had on the environment of Australia and the Pacific islands?
3. How might global warming effect Australia and Oceania? What measures are different countries taking in the region to address these problems?
4. Describe how a Pacific high island is created and how it is transformed into a low island. How do these environments produce distinctive settings for human settlement?
5. What are some notable similarities between Australian and North American cities?
6. Where—and where not—in Oceania is natural population growth a problem? Why?
7. What are the key geopolitical issues facing Oceania in this decade?
8. Why is there such a wide economic gap separating affluent Australia and New Zealand from the rest of Oceania?
9. Do the economic benefits of tourism outweigh the social and environment costs in Tahiti and other Pacific islands?
10. What are the benefits and liabilities of Australia's economic linkages with China?

Thinking Geographically

1. Identify what you see as the three principal economic challenges facing Papua New Guinea. Discuss how the country's physical and human geography may affect its future prospects.
2. Should Australia welcome or be wary of increased economic and political ties with China? Include in your response perspectives from different segments of Australia's population, such as miners, ranchers, Aborigines, farmers, and middle-class urbanites.
3. As a Maori activist in touch with your North American counterparts, what are some of the similarities and differences in the plight of your people?
4. As a maker of educational films, you have been hired to film a three-part series titled *The Essential South Pacific.* Episodes are to be set in (a) Australia, (b) New Zealand, and (c) Polynesia. Because the budget is limited, you can film in only a single location in each region. Identify three specific locations (one in each region) and then describe what you would film to capture the essence of those countries.
5. You are part of a team responsible for a comprehensive economic and social development plan for a small South Pacific island. Choose a specific island, become conversant with its needs and resources, define your goals, and describe a plan for achieving them in three or four pages, complete with multimedia support.

Regional Novels and Films

Novels

Murray Ball, *Eucalyptus* (1998, Harvest)

Derek Hansen, *Sole Survivor* (1999, Simon and Schuster)

Keri Hulme, *The Bone People* (1985, Hodder and Stoughton)

Joan Lindsay, *Picnic at Hanging Rock* (1967, Viking)

James Michener, *Hawaii* (1959, Random House)

Judy Nunn, *Floodtide* (2007, Random House)

Elliot Perlman, *Three Dollars* (1999, Macmillan)

Silvia Watanabe, *Talking to the Dead* (1992, Doubleday)

Albert Wendt, *Leaves of the Banyan Tree* (1979, University of Hawaii)

Tim Winton, *Dirt Music* (2002, Scribner's)

Films

The Adventures of Priscilla, Queen of the Desert (1994, Australia)

Australia (2008, Australia)

Australian Rules (2002, Australia)

Cedar Boys (2009, Australia)

Charlie and Boots (2009, Australia)

Hawaii: A Voice for Sovereignty (2009, Hawaii)

The Land Has Eyes (2004, Fiji Islands)

Mauri (1988, New Zealand)

Noho Hewa, The Wrongful Occupation of Hawaii (2008, Hawaii)

Once Were Warriors (1994, New Zealand)

Oscar and Lucinda (1997, U.S.)

The Price of Milk (2000, New Zealand)

Rabbit-Proof Fence (2002, Australia)

Samson and Delilah (2009, Australia)

Walkabout (1971, Australia)

Whaledreamers (2008, Australia and New Zealand)

Whale Rider (2003, New Zealand)

Bibliography

Bambrick, Susan, ed. 1994. *The Cambridge Encyclopedia of Australia*. New York: Cambridge University Press.

Coffman, Tom. 2009. *Nation Within: The History of the American Occupation of Hawaii*. Kihei, HI: Koa Books.

Forster, Clive A. 1995. *Australian Cities: Continuity and Change*. Melbourne: Oxford University Press.

Hammerton, A. James, and Alistair Thomson. 2005. *Ten Pound Poms: A Life History of British Postwar Emigration to Australia*. Manchester: Manchester University Press.

Head, Lesley. 2000. *Second Nature: The History and Implications of Australia as Aboriginal Landscape*. Syracuse, NY: Syracuse University Press.

Heathcote, R. L. 1994. *Australia*. New York: Wiley.

Kirch, Patrick V., and Terry Hunt. 1997. *Historical Ecology in the Pacific Islands: Prehistoric Environmental and Landscape Change*. New Haven, CT: Yale University Press.

McKnight, Tom L. 1995. *Oceania: The Geography of Australia, New Zealand, and the Pacific Islands*. Upper Saddle River, NJ: Prentice Hall.

Pilger, John. 1991. *A Secret Country: The Hidden Australia*. New York: Knopf.

Rudiak-Gould, Peter. 2009 *Surviving Paradise: One Year on a Disappearing Island*. New York: Union Square Press.

Stanley, David. 2000. *South Pacific Handbook*. 7th ed. Emeryville, CA: Avalon Travel Publishing.

Wesley-Smith, Terrance, and Edgar Porter, eds. 2007. *China in Oceania. Towards a New Regional Order.* Honolulu: East West Center.

Log in to www.masteringgeography.com for MapMaster™ interactive maps, geography videos, RSS feeds, flashcards, web links, an eText version of *Diversity Amid Globalization*, and self-study quizzes to enhance your study of Australia and Oceania.

MapMaster™ presents 13 Place Name and 13 Layered Thematic interactive maps to help students practice and master their geographic literacy, spatial reasoning, and critical thinking skills.

Glossary

Aborigine An indigenous inhabitant of Australia.

acid rain Harmful form of precipitation high in sulfur and nitrogen oxides. Caused by industrial and auto emissions, acid rain damages aquatic and forest ecosystems in regions such as eastern North America and Europe.

African Union (AU) Founded in 1963, the organization grew to include all the states of the continent except South Africa, which finally was asked to join in 1994. In 2004 the body changed its name from the Organization of African Unity (OAU) to the African Union. It is mostly a political body that has tried to resolve regional conflicts.

agrarian reform A popular but controversial strategy to redistribute land to peasant farmers. Throughout the 20th century, various states redistributed land from large estates or granted land titles from vast public lands in order to reallocate resources to the poor and stimulate development. Agrarian reform occurred in various forms, from awarding individual plots or communally held land to creating state-run collective farms.

agribusiness The practice of large-scale, often corporate farming in which business enterprises control closely integrated segments of food production, from farm to grocery store.

agricultural density The number of farmers per unit of arable land. This figure indicates the number of people who directly depend upon agriculture, and it is an important indicator of population pressure in places where rural subsistence dominates.

alluvial fan A fan-shaped deposit of sediments dropped by a river or stream flowing out of a mountain range.

Altiplano The largest intermontane plateau in the Andes, which straddles Peru and Bolivia and ranges in elevation from 10,000 to 13,000 feet (3,000 to 4,000 meters).

altitudinal zonation The relationship between higher elevations, cooler temperatures, and changes in vegetation that result from the environmental lapse rate (averaging 3.5°F for every 1,000 feet [6.5°C for every 1,000 meters]). In Latin America, four general altitudinal zones exist: tierra caliente, tierra templada, tierra fria, and tierra helada.

animism A wide variety of tribal religions based on the worship of nature spirits and human ancestors.

animist One who follows an animist religion.

anthropocentric Human-centered.

anthropogenic Caused by human activities.

anthropogenic landscapes A landscape heavily transformed by humans.

Aotearoa The ancestral Maori word, meaning "Land of the Long White Cloud," for New Zealand.

apartheid The policy of racial separateness that directed separate residential and work spaces for white, blacks, coloreds, and Indians in South Africa for nearly 50 years. It was abolished when the African National Congress came to power in 1994.

archipelagos Island groups, often oriented in an elongated pattern.

areal differentiation The geographic description and explanation of spatial differences on Earth's surface, including both physical as well as human patterns.

areal integration The geographic description and explanation of how places, landscapes, and regions are connected, interact, and are integrated with each other.

Asia-Pacific Economic Cooperation Group (APEC) An international group of Asian and Pacific Basin nations that fosters coordinated economic development within the region.

Association of Southeast Asian Nations (ASEAN) An international organization linking together the 10 most important countries of Southeast Asia.

atoll Low, sandy islands made from coral, often oriented around a central lagoon.

autonomous areas Minor political subunits created in the former Soviet Union and designed to recognize the special status of minority groups within existing republics.

autonomous region In the context of China, provinces that have been granted a certain degree of political and cultural autonomy, or freedom from centralized authority, owing to the fact that they contain large numbers of non-Han Chinese people. Critics contend that they have little true autonomy.

Baikal-Amur Mainline (BAM) Railroad Key central Siberian railroad connection completed in the Soviet era (1984), which links the Yenisey and Amur rivers and parallels the Trans-Siberian Railroad.

balkanization Geopolitical process of fragmentation of larger states into smaller ones through independence of smaller regions and ethnic groups. The term takes its name from the geopolitical fabric of the Balkan region.

Berlin Conference The 1884 conference that divided Africa into European colonial territories. The boundaries created in Berlin satisfied European ambition but ignored indigenous cultural affiliations. Many of Africa's civil conflicts can be traced to ill-conceived territorial divisions crafted in 1884.

biofuels Energy sources derived from plants or animals. Throughout the developing world, wood, charcoal, and dung are primary energy sources for cooking and heating.

biome Ecologically interactive flora and fauna adapted to a specific environment. Examples are deserts or tropical rain forests.

bioregion A spatial unit or region of local plants and animals adapted to a specific environment such as a tropical savanna.

Bodhisattva In the religion of Mahayana Buddhism, a spiritual being that helps others attain enlightenment.

Bolsheviks A faction within the Russian communist movement led by Lenin that successfully took control of the country in 1917.

boreal forest Coniferous forest found in high-latitude or mountainous environments of the Northern Hemisphere.

brain drain Migration of the best-educated people from developing countries to developed nations where economic opportunities are greater.

brain gain The potential of return migrants to contribute to the social and economic development of their country of origin with the experiences they have gained abroad.

British East India Company Private trade organization with its own army that acted as an arm of Britain in monopolizing trade in South Asia until 1857, when it was abolished and replaced by full governmental control.

bubble economy A highly inflated economy that cannot be sustained. Bubble economies usually result from the rapid influx of international capital into a developing country.

buffer zone An array of nonaligned or friendly states that "buffer" a larger country from invasion. In Eurasia, maintaining a buffer zone has been a long-term policy of Russia (and also of the former Soviet Union) to protect its western borders from European invasion.

Bumiputra The name given to native Malays (literally, "sons of the soil"), who are given preference for jobs and schooling by the Malaysian government.

Burakumin The indigenous outcast group of Japan, a people whose ancestors reputedly worked in leathercraft and other "polluting" industries.

cap-and-trade energy policy An energy policy in which overall carbon emissions are limited or *capped* by a central authority within a system that allows polluters to buy and sell carbon credits (or *allowances*) in an emissions-trading marketplace.

capital leakage The gap between the gross receipts an industry (such as tourism) brings into a developing area and the amount of capital retained.

Caribbean Community and Common Market (CARICOM) A regional trade organization established in 1972 that includes mostly former English Caribbean colonies as its members.

caste system Complex division of South Asian society into different hierarchically ranked hereditary groups. Most explicit in Hindu society, the caste system is also found in other South Asian cultures to a lesser degree.

Central American Free Trade Association (CAFTA) A 2004 free trade agreement between the United States and five Central American countries—Guatemala, Nicaragua, El Salvador, Honduras, and Costa Rica—plus the Dominican Republic.

central place theory A theory used to explain the distribution of cities, and the relationships between different cities, based on retail marketing.

centralized economic planning An economic system in which the state sets production targets and controls the means of production.

chain migration A pattern of migration in which a sending area becomes linked to a particular destination, such as Dominicans with New York City.

China proper The eastern half of the country of China where the Han Chinese form the dominant ethnic group. The vast majority of China's population is located in China proper.

circular migration Temporary labor migration, in which an individual seeks short-term employment overseas, earns money, and then returns home.

clan A social unit that is typically smaller than a tribe or ethnic group but larger than a family, based on supposed descent from a common ancestor.

climate region A region of similar climatic conditions. An example would be the marine west-coast climate regions found on the west coasts of North America and Europe.

climographs Graph of average annual temperature and precipitation data by month and season.

Closer Economic Relations (CER) Agreement An agreement signed in 1982 between Australia and New Zealand designed to eliminate all economic and trade barriers between the two countries.

Cold War The ideological struggle between the United States and the Soviet Union that was conducted between 1946 and 1990.

colonialism The formal and established rule over local peoples by a larger imperialist government.

coloured A racial category used throughout South Africa to define people of mixed European and African ancestry.

Columbian exchange An exchange of people, diseases, plants, and animals between the Americas (New World) and Europe/Africa (Old World) initiated by the arrival of Christopher Columbus in 1492.

command economy Centrally planned and controlled economies, generally associated with socialist or communist countries, in which all goods and services, along with agricultural and industrial products, are strictly regulated. This was done during the Soviet era in both the Soviet Union and its eastern European satellites.

Commonwealth of Independent States (CIS) A loose political union of former Soviet republics (without the Baltic states) established in 1992 after the dissolution of the Soviet Union.

concentric zone model A simplified description of urban land use: a well-defined central business district (CBD) is surrounded by concentric zones of residential activity, with higher-income groups living on the urban periphery.

Confucianism The philosophical system developed by Confucius in the 6th century BCE.

connectivity The degree to which different locations are linked with one another through transportation and communication infrastructure.

continentality Inland climates, removed from the ocean, with hot summers and cold winters such as those found in interior North America and Russia.

convection cells Large areas of slow-moving molten rock in Earth's interior that are responsible for moving tectonic plates.

convergent plate boundary Where two tectonic plates converge or are being forced together by convection cells. The Andes Mountains, for example, are formed by the converging Nazca and South America plates.

copra Dried coconut meat.

core-periphery model According to this scheme, the United States, Canada, western Europe, and Japan constitute the global economic core, while other regions make up a less-developed economic periphery from which the core extracts resources.

Cossacks Highly mobile Slavic-speaking Christians of the southern Russian steppe who were pivotal in expanding Russian influence in 16th- and 17th-century Siberia.

creolization The blending of African, European, and even some Amerindian cultural elements into the unique socio-cultural systems found in the Caribbean.

crony capitalism A system in which close friends of a political leader are either legally or illegally given business advantages in return for their political support.

cultural assimilation The process in which immigrants are culturally absorbed into the larger host society.

cultural imperialism The active promotion of one cultural system over another, such as the implantation of a new language, school system, or bureaucracy. Historically, this has been primarily associated with European colonialism.

cultural landscape Primarily the visible and tangible expression of human settlement (house architecture, street patterns, field form, etc.), but also includes the intangible, value-laden aspects of a particular place and its association with a group of people.

cultural nationalism A process of protecting, either formally (with laws) or informally (with social values), the primacy of a specific cultural system against influences from another culture.

cultural syncretism or hybridization The blending of two or more cultures, which then produces a synergistic third culture or specific behavior that exhibits traits from all cultural parents.

culture Learned and shared behavior by a group of people empowering them with a distinct "way of life"; it includes both material (technology, tools, etc.) and non-material (speech, religion, values, etc.) components.

culture hearth An area of historical cultural innovation.

cyclones Large storms, marked by well-defined air circulation around a low-pressure center. Tropical cyclones are typically called hurricanes in the Atlantic Ocean and typhoons in the western Pacific.

Cyrillic alphabet Based on the Greek alphabet and used by Slavic languages heavily influenced by the Eastern Orthodox Church. Attributed to the missionary work of St. Cyril in the 9th century.

dacha A Russian country cottage used especially in the summer.

dalits The so-called untouchable population of India; people often considered socially polluting because of their historical connections with occupations classified as unclean, such a leatherworking and latrine-cleaning.

decolonialization The process of a former colony's gaining (or regaining) independence over its territory and establishing (or reestablishing) an independent government.

demographic transition model A four-stage scheme that explains different rates of population growth over time through differing birth and death rates.

denuclearization The process whereby nuclear weapons are removed from an area and dismantled or taken elsewhere.

dependency theory A popular theory to explain patterns of economic development in Latin America. Its central premise is that underdevelopment was created by the expansion of European capitalism into the region that served to develop "core" countries in Europe and to impoverish and make dependent peripheral areas such as Latin America.

desertification The spread of desert conditions into semi-arid areas owing to improper management of the land.

diaspora The scattering of a particular group of people over a vast geographical area. Originally, the term referred to the migration of Jews out of their original homeland, but now it has been generalized to refer to any ethnic dispersion.

divergent plate boundary A place where two tectonic plates move away from each other. Here, magma often flows from Earth's interior, producing volcanoes on the surface such as in Iceland. In other divergent zones large trenches or rift valleys are created.

dollarization An economic strategy in which a country adopts the U.S. dollar as its official currency. A country can be partially dollarized, using U.S. dollars alongside its national currency, or fully dollarized, when the U.S. dollar becomes the only medium of exchange and a country gives up its own national currency. Panama fully dollarized in 1904; more recently, Ecuador became fully dollarized in 2000, followed by El Salvador in 2001.

domestication The purposeful selection and breeding of wild plants and animals for cultural purposes.

domino theory A U.S. geopolitical policy of the 1970s that stemmed from the assumption that if Vietnam fell to the communists, the rest of Southeast Asia would soon follow.

Dravidian language A language that belongs to a large linguistic family that is native to South Asia. Once spoken throughout South Asia, Dravidian languages are now largely confined to southern India and northern Sri Lanka.

Eastern Orthodox Christianity A loose confederation of self-governing churches in eastern Europe and Russia that are historically linked to Byzantine traditions and to the primacy of the patriarch of Constantinople (Istanbul).

economic convergence The notion that globalization will result in the world's poorer countries gradually catching up with more developed economies.

El Niño An abnormally warm current that appears periodically in the eastern Pacific Ocean and usually influences storminess along the western coasts of the Americas. During an El Niño event, which can last several years, torrential rains often bring devastating floods to the Pacific coasts of North, Central, and South America.

entrepôt A city and port that specializes in transshipment of goods.

environmental lapse rate The decline in temperature as one ascends higher in the atmosphere. On average, the temperature declines 3.5°F for every 1,000 feet ascended, or 6.5°C for every 1,000 meters.

ethnicity A shared cultural identity held by a group of people with a common background or history, often as a minority group within a larger society.

ethnic religions A religion closely identified with a specific ethnic or tribal group, often to the point of assuming the role of the major defining characteristic of that group. Normally, ethnic religions do not actively seek new converts.

Euroland The assemblage of European countries belonging to the European Monetary Union (EMU) and using the euro as their currency.

European Union (EU) The current association of 27 European countries that are joined together in an agenda of economic, political, and cultural integration.

exclave A portion of a country's territory that lies outside its contiguous land area.

exotic river A river that issues from a humid area and flows into a dry area otherwise lacking streams.

fair trade An international certification movement to identify primary commodities exported from the developing world in which farmers earn a better price for their product. Commodities such as coffee, tea, and forest products are certified "fair trade" when small-scale producers earn more for their product and production methods are viewed as environmentally and socially sustainable.

federal states Political system in which a significant amount of power is given to individual states. In India, these states were created upon independence in 1947 and were drawn primarily along linguistic lines, so that today state power is often associated with specific ethnic groups within the nation.

Fertile Crescent An ecologically diverse zone of lands in Southwest Asia that extends from Lebanon eastward to Iraq and that is often associated with early forms of agricultural domestication.

free trade zone (FTZ) A duty-free and tax-exempt industrial park created to attract foreign corporations and create industrial jobs.

forward capital A capital city deliberately positioned near a contested territory, signifying the state's interest and presence in this zone of conflict.

fossil water Water supplies that were stored underground during wetter climatic periods.

gender gap A term often used to describe gender differences in salary, working conditions, or political power.

geomancy The traditional Chinese and Korean practice of designing buildings in accordance with the principles of cosmic harmony and discord that supposedly course through the local topography.

genocide The deliberate and systematic killing of a racial, political, or cultural group.

gentrification A process of urban revitalization in which higher income residents displace lower-income residents in central-city neighborhoods.

glasnost A policy of greater political openness initiated during the 1980s by Soviet president Mikhail Gorbachev.

globalization The increasing interconnectedness of people and places throughout the world through converging processes of economic, political, and cultural change.

global warming The increase in global temperatures as a result of the magnification of the natural greenhouse effect due to human-produced pollutants. The resulting climate change could cause profound—and probably damaging—changes to Earth's environments.

Golden Triangle An area of northern Thailand, Burma, and Laos that is known as a major source region for heroin and is plugged into the global drug trade.

Gondwanaland The ancient megacontinent that included Africa, South America, Antarctica, Australia, Madagascar, and Saudi Arabia. Some 250 million years ago, it began to split apart due to plate tectonics.

grassification The conversion of tropical forest into pasture for cattle ranching. Typically, this process involves introducing species of grasses and cattle, mostly from Africa.

Great Escarpment A landform that rims southern Africa from Angola to South Africa. It forms where the narrow coastal plains meet the elevated plateaus in an abrupt break in elevation.

Greater Antilles The four large Caribbean islands of Cuba, Jamaica, Hispaniola, and Puerto Rico.

Green Revolution Term applied to the development of agricultural techniques used in developing countries that involve new, genetically altered seeds that provide higher yields than native seeds when combined with high inputs of chemical fertilizer, irrigation, and pesticides.

greenhouse effect The natural process of lower atmosphere heating that results from the trapping of incoming and reradiated solar energy by water moisture, clouds, and other atmospheric gases.

Group of Eight (G-8) A collection of powerful countries that confer regularly on key global economic and political issues. It includes the United States, Canada, Japan, Great Britain, Germany, France, Italy, and Russia.

gross domestic product (GDP)/gross national product (GNP) GDP is the total value of goods and services produced within a given country (or other geographical unit) in a single year. GNP is a somewhat broader measure that includes the inflow of money from other countries in the form of the repatriation of profits and other returns on investments, as well as the outflow to other countries for the same purposes.

gross national income (GNI) The value of all final goods and services produced within a country's borders (gross domestic product, or GDP), plus the net income from abroad (formerly referred to as gross national product, or GNP).

gross national income (GNI) per capita The figure that results from dividing a country's GNI by the total population.

guest workers Workers from Europe's agricultural periphery—primarily Greece, Turkey, southern Italy, and the former Yugoslavia—solicited to work in Germany, France, Sweden, and Switzerland during chronic labor shortages in Europe's boom years (1950s to 1970s).

Gulag Archipelago A collection of Soviet-era labor camps for political prisoners, made famous by writer Aleksandr Solzhenitsyn.

Hajj An Islamic religious pilgrimage to Makkah. One of the five essential pillars of the Muslim creed to be undertaken once in life, if an individual is physically and financially able to do it.

haoles Light-skinned Europeans or U.S. citizens in the Hawaiian Islands.

high islands Larger, more elevated islands, often focused around recent volcanic activity.

Hindu nationalism A contemporary "fundamental" religious and political movement that promotes Hindu values as the essential—and exclusive—fabric of Indian society. As a political movement, it appears to have less tolerance of India's large Muslim minority than other political movements.

homelands Nominally independent ethnic territories created for blacks under the grand apartheid scheme. Homelands were on marginal land, overcrowded, and poorly serviced. In the post-apartheid era, they were eliminated.

hot spot A supply of magma that produces a chain of midocean volcanoes atop a zone of moving oceanic crust.

Horn of Africa The northeastern corner of Sub-Saharan Africa that includes the states of Somalia, Ethiopia, Eritrea, and Djibouti. Since the 1980s, drought, famine, ethnic conflict, and political turmoil have undermined development efforts in this area.

Human Development Index (HDI) For the past three decades the United Nations has tracked social development in the world's countries through the Human Development Index (HDI), which combines data on life expectancy, literacy, educational attainment, gender equity, and income.

hurricanes Storm systems with an abnormally low-pressure center sustaining winds of 74 mph or higher. Each year during hurricane season (July–October), a half dozen to a dozen hurricanes form in the warm waters of the Atlantic and Caribbean, bringing destructive winds and heavy rain.

ideographic writing A writing system in which each symbol represents not a sound but rather a concept.

indentured labor Foreign workers (usually South Asians) contracted to labor on Caribbean agricultural estates for a set period of time, often several years. Usually the contract stipulated paying off the travel debt incurred by the laborers. Similar indentured labor arrangements have existed in most world regions.

Indian diaspora The historical and contemporary propensity of Indians to migrate to other countries in search of better opportunities. This has led to large Indian populations in South Africa, the Caribbean, and the Pacific islands, along with western Europe and North America.

informal sector A much-debated concept that presupposes a dual economic system consisting of formal and informal sectors. The informal sector includes self-employed, low-wage jobs that are usually unregulated and untaxed. Street vending, shoe shining, artisan manufacturing, and even self-built housing are considered part of the informal sector. Some scholars include illegal activities such as drug smuggling and prostitution in the informal economy.

insolation Incoming solar energy that enters the atmosphere adjacent to Earth.

internally displaced persons Groups and individuals who flee an area due to conflict or famine but still remain in their country of origin. These populations often live in refugee-like conditions but are harder to assist because they technically do not qualify as refugees.

irredentism A state or national policy of reclaiming lost lands or those inhabited by people of the same ethnicity in another nation-state.

Iron Curtain A term coined by British leader Winston Churchill during the Cold War that defined the western border of Soviet power in Europe. The notorious Berlin Wall was a concrete manifestation of the Iron Curtain.

Islamic fundamentalism A movement within both the Shiite and Sunni Muslim traditions to return to a more conservative, religious-based society and state. Often associated with a rejection of Western culture and with a political aim to merge civic and religious authority.

Islamism A political movement within the religion of Islam that challenges the encroachment of global popular culture and blames colonial, imperial, and Western elements for many of the region's problems. Adherents of Islamism advocate merging civil and religious authority.

isolated proximity A concept that explores the contradictory position of the Caribbean states, which are physically close to North America and economically dependent upon that region. At the same time, Caribbean isolation fosters strong loyalties to locality and limited economic opportunity.

Jainism A religious group in South Asia that emerged as a protest against orthodox Hinduism in the 6th century BCE. Jains are noted for their practice of nonviolence, which prohibits them from taking the life of any animal.

kanakas Melanesian workers imported to Australia, historically concentrated along Queensland's "sugar coast."

Khmer Rouge Literally, "Red (or communist) Cambodians." The left-wing insurgent group led by French-educated Marxists that ruled Cambodia from 1975 to 1979, during which time it engaged in genocidal acts against the Cambodian people.

kibbutzim Collective farms in Israel.

kleptocracy A state where corruption is so institutionalized that politicians and bureaucrats siphon off a large percentage of a country's wealth for personal gain.

laissez-faire An economic system in which the state has minimal involvement and in which market forces largely guide economic activity.

latifundia A large estate or landholding in Latin America.

Lesser Antilles The arc of small Caribbean islands from St. Maarten to Trinidad.

Levant The eastern Mediterranean region.

lingua franca An agreed-upon common language to facilitate communication on specific topics such as international business, politics, sports, or entertainment.

linguistic nationalism The promotion of one language over others that is, in turn, linked to shared notions of political identity. In India, some nationalists promote Hindi as the unifying language of the country, yet this is resisted by many non-Hindi-speaking peoples.

location factors The various influences that explain why an economic activity takes place where it does.

loess A fine, wind-deposited sediment that makes fertile soil but is very vulnerable to water erosion.

Maghreb A region in northwestern Africa, including portions of Morocco, Algeria, and Tunisia.

maharaja Historical term for Hindu royalty, usually a king or prince, who ruled specific areas of South Asia before independence, but who was usually subject to overrule by British colonial advisers.

mallee A tough and scrubby eucalyptus woodland of limited economic value that is common across portions of interior Australia.

mandarins A member of the high-level bureaucracy of Imperial China (before 1911). Mandarin Chinese is the official spoken language of the country and is the native tongue of the vast majority of people living in north, central, and southwestern China.

Maori Indigenous Polynesian people of New Zealand.

maquiladoras Assembly plants on the Mexican border built and owned by foreign companies. Most of their products are exported to the United States.

marine west-coast climate Moderate climate with cool summers and mild winters that is heavily influenced by maritime conditions. Such climates are usually found on the west coasts of continents between latitudes of 45 to 50 degrees.

maritime climate Climate moderated by proximity to oceans or large seas. It is usually cool, cloudy, and wet, and lacks the temperature extremes of continental climates.

maroons Runaway slaves who established communities rich in African traditions throughout the Caribbean and Brazil.

Marxism A term referring to the political philosophy developed by Karl Marx in the 1800s and based on the ideas of communism.

medieval landscape Urban landscapes from 900 to 1500 CE characterized by narrow, winding streets, three- or four-story structures (usually in stone, but sometimes wooden), with little open space except for the market square. These landscapes are still found in the centers of many European cities.

medina The original urban core of a traditional Islamic city.

Mediterranean climate A unique climate, found in only five locations in the world, that is characterized by hot, dry summers with very little rainfall. These climates are located on the west side of continents, between 30 and 40 degrees latitude.

megacities Urban conglomerations of more than 10 million people.

megalopolis A large urban region formed as multiple cities grow and merge with one another. The term is often applied to the string of cities in eastern North America that includes Washington, DC; Baltimore; Philadelphia; New York City; and Boston.

Melanesia Pacific Ocean region that includes the culturally complex, generally darker-skinned peoples of New Guinea, the Solomon Islands, Vanuatu, New Caledonia, and Fiji.

Mercosur The Southern Common Market established in 1991 that calls for free trade among member states and common external tariffs for nonmember states. Argentina, Paraguay, Brazil, and Uruguay are full members and Venezuela's full membership is pending; the Andean countries of Colombia, Ecuador, Peru, Bolivia and Chile are associate members.

mestizo A person of mixed European and Indian ancestry.

Micronesia Pacific Ocean region that includes the culturally diverse, generally small islands north of Melanesia. Includes the Mariana Islands, Marshall Islands, and Federated States of Micronesia.

microstates Usually independent states that are small in both area and population.

mikrorayons Large, state-constructed urban housing projects built during the Soviet period in the 1970s and 1980s.

Millennium Development Goals Part of a group of programs implemented since 2000 to foster development in the world's poorest countries. The Millennium Development Goals are part of a global United Nations effort to reduce extreme poverty by 2015.

minifundia A small landholding farmed by peasants or tenants who produce food for subsistence and the market.

monocrop production Agriculture based upon a single crop, often for export.

Monroe Doctrine A proclamation issued by U.S. President James Monroe in 1823 that the United States would not tolerate European military action in the Western Hemisphere. Focused on the Caribbean as a strategic area, the doctrine was repeatedly invoked to justify U.S. political and military intervention in the region.

monotheism A religious belief in a single God.

monsoon The seasonal pattern of changes in winds, heat, and moisture in South Asia and other regions of the world that is a product of larger meteorological forces of land and water heating, the resultant pressure gradients, and jet-stream dynamics. The monsoon produces distinct wet and dry seasons.

Mughal Empire (also spelled Mogul) The Muslim-dominated state that covered most of South Asia from the early 16th to late 17th centuries. The last vestiges of the Mughal dynasty were dissolved by the British following the rebellion of 1857.

nation-state A relatively homogeneous cultural group (a nation) with its own political territory (the state). While useful conceptually, the reality of today's globalized world is that there are very few countries that fit this simplistic definition because of the influx of migrants and/or the presence of minority ethnic groups.

Native Title Bill Australian legislation signed in 1993 that provides Aborigines with enhanced legal rights over land and resources within the country.

neocolonialism Economic and political strategies by which powerful states indirectly (and sometimes directly) extend their influence over other, weaker states.

neoliberal policies Economic policies widely adopted in the 1990s that stress privatization of services, export production, and few restrictions on imports.

net migration rate A statistic that assesses whether more people are entering or leaving a country by measuring the amount of immigration (in-migration) and emigration (out-migration). A positive figure means the population is growing because of in-migration, whereas a negative number means a population is shrinking because of out-migration.

new urbanism An urban design movement stressing higher density, mixed-use, pedestrian-scaled neighborhoods where residents might be able to walk to work, school, and local entertainment.

nonmetropolitan growth A pattern of migration in which people leave large cities and suburbs and move to smaller towns and rural areas.

North American Free Trade Agreement (NAFTA) An agreement made in 1994 among Canada, the United States, and Mexico that established a 15-year plan for reducing all barriers to trade among the three countries.

Oceania A major world subregion that includes New Zealand and the major islands of Melanesia, Micronesia, and Polynesia.

offshore banking Islands or microstates that offer financial services that are typically confidential and tax-exempt. As part of a global financial system, offshore banks have developed a unique niche, offering their services to individual and corporate clients for set fees. The Bahamas and Cayman Islands are leaders in this sector.

Organization of American States (OAS) Founded in 1948 and headquartered in Washington, DC, the organization advocates hemispheric cooperation and dialog. Most states in the Americas belong except Cuba.

Organization of the Petroleum Exporting Countries (OPEC) An international organization of 12 oil-producing nations (formed in 1960) that attempts to influence global prices and supplies of oil. Algeria, Gabon, Indonesia, Iran, Iraq, Kuwait, Libya, Nigeria, Qatar, Saudi Arabia, UAE, and Venezuela are members.

orographic rainfall Enhanced precipitation over uplands that results from lifting (and cooling) of air masses as they are forced over mountains.

Ottoman Empire A large, Turkish-based empire (named for Osman, one of its founders) that dominated large portions of southeastern Europe, North Africa, and Southwest Asia between the 16th and 19th centuries.

Outback Australia's large, generally dry, and thinly settled interior.

outsourcing A business practice that transfers portions of a company's production and service activities to lower-cost settings, often located overseas.

over-urbanization A process in which the rapid growth of a city, most often because of in-migration, exceeds the city's ability to provide jobs, housing, water, sewers, and transportation.

Palestinian Authority (PA) A quasi-governmental body that represents Palestinian interests in the West Bank and Gaza.

Pan-African Movement Founded in 1900 by U.S. intellectuals W. E. B. Du Bois and Marcus Garvey, this movement's slogan was "Africa for Africans," and its influence extended across the Atlantic.

pastoralists Nomadic and sedentary peoples who rely upon livestock (especially cattle, camels, sheep, and goats) for their sustenance and livelihood.

pastoral nomadism A traditional subsistence agricultural system in which practitioners depend on the seasonal movements of livestock within marginal natural environments.

perestroika A program of partially implemented, planned economic reforms (or restructuring) undertaken during the Gorbachev years in the Soviet Union designed to make the Soviet economy more efficient and responsive to consumer needs.

permafrost A cold-climate condition in which the ground remains permanently frozen.

physiological density A population statistic that relates the number of people in a country to the amount of arable land.

Pidgin English (Pijin) A version of English that also incorporates elements of other local languages, often utilized to foster trade and basic communication between different culture groups.

plantation America A cultural region that extends from midway up the coast of Brazil, through the Guianas and the Caribbean, and into the southeastern United States. In this coastal zone, European-owned plantations, worked by African laborers, produced agricultural products for export.

plate tectonics The theory that explains the gradual movement of large geological platforms (or plates) along Earth's surface.

podzol soils A Russian term for an acidic soil of limited fertility, typically found in northern forest environments.

pollution exporting The process of exporting industrial pollution and other waste material to other countries. Pollution exporting can be direct, as when waste is simply shipped abroad for disposal, or indirect, as when highly polluting factories are constructed abroad.

Polynesia Pacific Ocean region, broadly unified by language and cultural traditions, that includes the Hawaiian Islands, Marquesas Islands, Society Islands, Tuamotu Archipelago, Cook Islands, American Samoa, Samoa, Tonga, and Kiribati.

population density The number of people per areal unit, usually measured in people per square kilometer or per square mile.

population pyramid A graph representing the structure of a population, including the percentage of young and old. The percentages of all different age groups are plotted along a vertical axis that divides the population into male and female. In a fast growing population with a large percentage of young people and a small percentage of elderly, the graph will have a wide base and a narrow tip, giving it a pyramidal shape.

postindustrial economy An economy in which the tertiary and quaternary sectors dominate employment and expansion.

prairie An extensive area of grassland in North America. In the more humid eastern portions, grasses are usually longer than in the drier western areas, which are in the rain shadow of the Rocky Mountain range.

primate cities Massive urban settlements that dominate all other cities in a given country. Often—yet not always—the primate city is also the country's capital.

privatization The process of moving formerly state-owned firms into the contemporary capitalist private sector.

protectorates During the period of global Western imperialism, a state or other political entity that remained autonomous, yet sacrificed its foreign affairs to an imperial power in exchange for "protection" from other imperial powers.

purchasing power parity (PPP) A method of reducing the influence of inflated currency rates by adjusting a local currency to a composite baseline of one U.S. dollar based upon its ability to purchase a standardized "market basket" of goods.

qanat system A traditional system of gravity-fed irrigation that uses gently sloping tunnels to capture groundwater and direct it to needed fields.

Quran (also spelled Koran) A book of divine revelations received by the prophet Muhammad that serves as a holy text in the religion of Islam.

Ramayana One of the two main epic poems of the Hindu religion, the Ramayana is also commonly performed in the shadow puppet theaters of the predominately Muslim island of Java.

rate of natural increase (RNI) The standard statistic used to express natural population growth per year for a country, region, or the world based upon the difference between birth and death rates. RNI does not consider population change from migration. Though most often a positive figure (such as 1.7 percent), RNI can also be expressed either as zero or even as a negative number for no-growth countries.

refugee A person who flees his or her country because of a well-founded fear of persecution based on race, ethnicity, religion, ideology, or political affiliation.

regulatory lakes A term applied to a series of lakes in the middle Yangtze Valley of China. Regulatory lakes take excess water from the river during flood periods, and supply water to the river during dry periods.

remittances Money sent by immigrants to their country of origin to support family members left behind. For many countries, remittances are a principal source of foreign exchange.

Renaissance-Baroque landscape An urban landscape generally constructed during the period from 1500 to 1800 characterized by wide ceremonial boulevards, large monumental structures, and ostentatious houses for the urban elite.

renewable energy sources Energy sources such as hydroelectric, solar, wind, and geothermal that are known for their enduring availability and their potentially lower environmental costs

rift valleys A surface landscape feature formed where two tectonic plates are diverging or moving apart. Usually, this forms a depression or large valley.

rimland The mainland coastal zone of the Caribbean, beginning with Belize and extending along the coast of Central America to northern South America.

rural-to-urban migration The flow of internal migrants from rural areas to cities that began in the 1950s and intensified in the 1960s and 1970s.

Russification A policy of the Soviet Union designed to spread Russian settlers and influences to non-Russian areas of the country.

rust belt Regions of heavy industry that experience marked economic decline after their factories cease to be competitive.

Sahel The semidesert region at the southern fringe of the Sahara, and the countries that fall within this region, which extends from Senegal to Sudan. Droughts in the 1970s and early 1980s caused widespread famine and dislocation of population.

salinization The accumulation of salts in the upper layers of soil, often causing a reduction in crop yields, resulting from irrigation with water of high natural salt content and/or irrigation of soils that contain a high level of mineral salts.

samurai The warrior class of traditional Japan. After 1600, the military role of the samurai declined as they assumed administrative positions, but their military ethos remained alive until the class was abolished in 1868.

Schengen Agreement The 1985 agreement between some—but not all—European Union member countries to reduce border formalities in order to facilitate free movement of citizens between member countries of this new "Schengenland." For example, today there are no border controls between France and Germany, or between France and Italy.

sectoral transformation The evolution of a labor force from being highly dependent on the primary sector to being oriented around more employment in the secondary, tertiary, and quaternary sectors.

secularization The widespread movement in western Europe away from regular participation and engagement with traditional organized religions such as Protestantism or Catholicism.

sediment load The amount of sand, silt, and clay carried by a river.

Shanghai Cooperation Organization (SCO) An international organization composed of China, Russia, Kazakhstan, Kyrgyzstan, Tajikistan, and Uzbekistan that aims to enhance security and economic cooperation in Central Asia.

shields Large upland areas of very old exposed rocks that range in elevation from 600 to 5,000 feet (200 to 1,500 meters). The three major shields in South America are the Guiana, Brazilian, and Patagonian.

shifted cultivators Migrants farmers who are either transplanted by government relocation schemes or forced to move on their own when their lands are expropriated.

Shiites Muslims who practice one of the two main branches of Islam; especially dominant in Iran and nearby southern Iraq.

shogunate The political order of Japan before 1868, in which power was held by the military leader known as the shogun, rather than by the emperor, whose authority was merely symbolic.

Sikhism An Indian religion combining Islamic and Hindu elements, founded in the Punjab region in the late 15th century.

Slavic peoples A group of peoples in eastern Europe and Russia who speak Slavic languages, a distinctive branch of the Indo-European language family.

social and regional differentiation "Social differentiation" refers to a process by which certain classes of people grow richer when others grow poorer; "regional differentiation" refers to a process by which certain places grow more prosperous while others become less prosperous.

socialist realism An artistic style once popular in the Soviet Union that was associated with realistic depictions of workers in their patriotic struggles against capitalism.

Spanglish A hybrid combination of English and Spanish spoken by Hispanic Americans.

Special Economic Zones (SEZs) Relatively small districts in China were fully opened to global capitalism after China began to reform its economy in the 1980s.

spheres of influence In countries not formally colonized in the 19th and early 20th centuries (particularly China and Iran), limited areas called "spheres of influence" were gained by particular European countries for trade purposes and more generally for economic exploitation and political manipulation.

squatter settlements Makeshift housing on land not legally owned or rented by urban migrants, usually in unoccupied open spaces within or on the outskirts of a rapidly growing city.

steppe Semiarid grasslands found in many parts of the world. Grasses are usually shorter and less dense than in prairies.

structural adjustment programs Controversial yet widely implemented programs used to reduce government spending, encourage the private sector, and refinance foreign debt. Typically, these IMF and World Bank policies trigger drastic cutbacks in government-supported services and food subsidies, which disproportionately affect the poor.

subcontinent A large segment of land separated from the main land mass on which it sits by lofty mountains or other geographical barriers. South Asia, separated from the rest of Eurasia by the Himalayas, is often called the "Indian subcontinent."

subduction zone Areas where two tectonic plates are converging or colliding. In these areas, one plate usually sinks below another. They are characterized by earthquakes, volcanoes, and deep oceanic trenches.

subnational organizations Groups that form along ethnic, ideological, or regional lines that can induce serious internal divisions within a state.

Suez Canal Pivotal waterway connecting the Red Sea and the Mediterranean opened by the French in 1869.

Sunda Shelf An extension of the continental shelf from the Southeast Asia mainland to the large islands of Indonesia. Because of the shelf, the overlying sea is generally shallow (less than 200 feet [61 meters] deep).

Sunnis Muslims who practice the dominant branch of Islam.

superconurbation A massive urban agglomeration that results from the coalescing of two or more formerly separate metropolitan areas.

supranational organizations Governing bodies that include several states, such as trade organizations, and often involve a loss of some state powers to achieve the organization's goals.

sustainable agriculture A system of agriculture where organic farming principles, a limited use of chemicals, and an integrated plan of crop and livestock management combine to offer both producers and consumers environmentally friendly alternatives.

sweatshops Small manufacturing units in developing countries where menial tasks are performed by largely unskilled and poorly paid workers.

swidden A form of tropical cultivation in which forested or brushy plots are cleared of vegetation, burned, and then planted in crops, only to be abandoned a few years later as soil fertility declines. Also called slash-and-burn agriculture or shifting cultivation.

syncretic religions The blending of different belief systems. In Latin America, many animist practices were folded into Christian worship.

taiga The vast coniferous forest of Russia that stretches from the Urals to the Pacific Ocean. The main forest species are fir, spruce, and larch.

Taliban A harsh, Islamic fundamentalist political group that ruled most of Afghanistan in the late 1990s. In 2001, the Taliban lost power, but it later regrouped in Pakistan and continues to fight against U.S.-led forces in southern and eastern Afghanistan.

Tamil Tigers The common name of the rebel forces in Sri Lanka (officially known as the Liberation Tigers of Tamil Eelam, or LTTE) that fought the Sri Lankan army from 1983 until their defeat in 2009.

tectonic plates The basic building blocks of Earth's crust; large blocks of solid rock that move very slowly over the underlying semimolten material.

theocracy A government run by religious leaders.

theocratic state A political state led by religious authorities.

tonal language A language in which the meaning of a syllable varies in accordance with the tone in which it is uttered.

total fertility rate (TFR) The average number of children who will be borne by women of a hypothetical, yet statistically valid, population, such as that of a specific cultural group or within a particular country. Demographers consider TFR a more reliable indicator of population change than the crude birthrate.

townships Racially segregated neighborhoods created for nonwhite groups under apartheid in South Africa. They are usually found on the outskirts of cities and classified as black, coloured, or South Asian.

Trans-Siberian Railroad Key southern Siberian railroad connection completed during the Russian empire (1904) that links European Russia with the Russian Far East terminus of Vladivostok.

transhumance A form of pastoralism in which animals are taken to high-altitude pastures during the summer months and returned to low-altitude pastures during the winter.

transmigration The planned, government-sponsored relocation of people from one area to another within a state territory.

transnational firms A firm or corporation that does international business through an array of global subsidiaries.

transnational migration Complex social and economic linkages that form between home and host countries through international migration. Unlike earlier generations of migrants, 21st century immigrants can maintain more enduring and complex ties to their home countries as a result of technological advances.

Treaty of Tordesillas A treaty signed in 1494 between Spain and Portugal that drew a north–south line some 300 leagues west of the Azores and Cape Verde islands. Spain received the land to the west of the line and Portugal the land to the east.

tribalism Allegiance to a particular tribe or ethnic group rather than to the nation-state. Tribalism is often blamed for internal conflict within Sub-Saharan states.

tribes A group of families or clans with a common kinship, language, and definable territory but not an organized state.

tsars A Russian term (also spelled czar) for "Caesar," or ruler; the authoritarian rulers of the Russian empire before its collapse in the 1917 revolution.

tsetse flies A fly that is a vector for a parasite that causes sleeping sickness (typanosomiasis), a disease that especially affects humans and livestock. Livestock is rarely found in those areas of Sub-Saharan Africa where the tsetse fly is common.

tsunamis Very large sea waves induced by earthquakes.

tundra Arctic region with a short growing season in which vegetation is limited to low shrubs, grasses, and flowering herbs.

typhoons Large tropical storms, synonymous to hurricanes, that form in the western Pacific Ocean in tropical latitudes and cause widespread damage to the Philippines and coastal Southeast and East Asia.

uncontacted peoples Cultures that have yet to be contacted and influenced by the Western world.

unitary states A political system in which power is centralized at the national level.

universalizing religions A religion, usually with an active missionary program, that appeals to a large group of people regardless of local culture and conditions. Christianity and Islam both have strong universalizing components.

urban decentralization The process in which cities spread out over a larger geographical area.

urban form The physical arrangement or landscape of the city, made up of building architecture and style, street patterns, open spaces, housing types, and so on.

urbanized population That percentage of a country's population living in settlements characterized as cities. Usually, high rates of urbanization are associated with higher levels of industrialization and economic development because these activities are usually found in and around cities. Conversely, lower urbanized populations (less than 50 percent) are characteristic of developing countries.

urban primacy The situation found in a country in which a disproportionately large city, such as London, New York, or Bangkok, dominates the urban system and is the center of economic, political, and cultural life.

urban realms model A simplified description of urban land use, especially descriptive of the modern North American city. It features a number of dispersed, peripheral centers of dynamic commercial and industrial activity linked by sophisticated urban transportation networks.

urban structure The distribution and pattern of land use, such as commercial, residential, or manufacturing, within the city. Often, commonalities give rise to models of urban structure characteristic of the cities of a certain region or of a shared history, such as cities shaped by European colonialism.

viticulture Grape cultivation.

water stress An environmental planning tool used to predict areas that have—or will have—serious water problems based upon the per capita demand and supply of freshwater.

White Australia Policy Before 1975, a set of stringent Australian limitations on nonwhite immigration to the country. Largely replaced by a more flexible policy today.

World Trade Organization (WTO) Formed as an outgrowth of the General Agreement on Tariffs and Trade (GATT) in 1995, the WTO is a large collection of member states dedicated to reducing global barriers to trade.

Photo Credits

Chapter 1 opening photo Image Source/Photolibrary **1.1** Rob Crandall **1.2** Sherwin Crasto/Reuters/Corbis **1.3** Sean Sprague/The Image Works **1.4** Rob Crandall **1.5** Stephanie Kuykendal/Corbis **1.7** Rob Crandall **1.8** Tim Boyle/Getty Images **1.9** AP Images **1.11** Bobby Yip/Reuters/Landov **1.12** Thorvaldur Kristmundsson/AP Images **1.13** Caroline Penn/Panos Pictures **1.14** Ramzi Haidar/Newscom **1.1.1** Les Rowntree **1.15** Brynjar Gauti/AP Images **1.16** Photo Researchers, Inc. **1.18** Maurice Joseph/Robert Harding **1.2.2** Bill Bachmann/Photo Edit **1.19** Philippe Michel/Robert Harding **1.22** Chris Stowers/Panos Pictures **1.3.1** Desire Martin/AFP/Getty Images **1.23** Trygve Bolstad/Panos Pictures **1.26** Liz Gilbert/India Government Tourist Office **1.27a** Craig Aurness/Corbis **1.27b** Tony Waltham/Robert Harding **1.29** Rob Crandall **1.30** Jay Directo/AFP/Getty Images **1.4.1** Leonid Serebrennikov/AGE Fotostock **1.31** Ajit Solanki/AP Images **1.32a** Rob Crandall **1.32b** Rob Crandall **1.32c** Rob Crandall **1.32d** Rob Crandall **1.5.1** Alfredo Estrella/AFP/Getty Images **1.33** Arko Datta/Reuters/Corbis **1.35** Rob Crandall **1.37** Rob Crandall **1.38** Photodisc/Getty Images **1.39** GAIZKA IROZ/AFP/Getty Images **1.42** Catherine Karnow/Woodfin Camp& Associates, Inc. **1.43** Hubert Boesl/dpa/Corbis **1.45** Ajay Verma/Reuters/Corbis **1.6.1** Rafael Ramirez Lee/Shutterstock **1.47** Rob Crandall **1.50** SUPRI/Reuters/Corbis **1.51** Ron Giling/Panos Pictures

Chapter 2 opening photo James Kay/DanitaDelimont.com **2.1** Chris James/Peter Arnold, Inc./Photolibrary **2.4** aerialarchives.com/Alamy **2.5** Paul A. Souders/Corbis **2.6** Jorge Silva/Reuters **2.8** Alberto Garcia/Corbis **2.9** iStockphoto **2.11** Jim Edds/Corbis **2.15** Steven J. Kazlowski/Alamy **2.16** Reuters/Corbis **2.1.1** imagebroker/Alamy **2.17** Liu Liqun/Corbis **2.19** Ajay Verma/Reuters **2.21** Gary Braasch/Woodfin Camp & Associates, Inc. **2.22** James P. Blair/Getty Images **2.2.1** Nathalie van Vliet/Center for International Forestry Research/Handout/Reuters **2.23** Wofgang Kaehler/Alamy **2.24** Calvin Larsen/Photo Researchers, Inc. **2.25** EPA/Chun Lei/Corbis **2.26** Mark Edwards/Photolibrary/Peter Arnold, Inc. **2.27** Martin Wendler/Photolibrary/Peter Arnold, Inc.

Chapter 3 opening photo Martin Thomas Photography/Alamy **3.2** Bill Wyckoff **3.1.1** Kevin Schafer/CORBIS **3.5** Danny E. Hooks/Shutterstock **3.6** Bill Wyckoff **3.8** Earth Satellite Corporation/Photo Researchers, Inc. **3.9** Joe Sohm/The Image Works **3.2.1** Gabriel Bouys/AFP/Getty Images **3.11** Courtesy of GNP Archives and Karen Holzer, USGS **3.15** Jeremy Woodhouse/Getty Images **3.17** Rob Crandall/The Image Works **3.18** Soffer Organization **3.19** Bill Wyckoff **3.3.1** Mansur Real Estate Services **3.20** Craig Aurness/Corbis **3.26** Jim Noelker/The Image Works **3.4.2** Rob Crandall **3.4.3** Rob Crandall **3.28** Bill Wyckoff **3.30** Worldsat International Inc./Photo Researchers, Inc. **3.32** Alison Wright/Corbis **3.33** Bill Wyckoff **3.35** Rob Crandall **3.5.2** Richard Hamilton Smith/CORBIS **3.36** Getty Images **3.37** David R. Frazier/DanitaDelimont/Newscom **3.38** George Hall/Woodfin Camp & Associates, Inc. **3.39** Dennis MacDonald/Photo Edit **3.41** Bill Wyckoff **3.42** Comstock/Photolibrary **3.6.1** Michael Newman/PhotoEdit

Chapter 4 opening photo Pablo Caridad/Alamy **4.2** Rob Crandall **4.4** Rob Crandall **4.5a** Biological Resources Division, U.S. Geological Survey **4.5b** Biological Resources Division, U.S. Geological Survey **4.1.1** Rob Crandall **4.6** Rob Crandall **4.7** Jorge Uzon/AFP/Getty Images **4.8** Hubert Stadler/Corbis **4.9** Stephaine Maze/National Geographic Image Collection **4.10** Rob Crandall **4.12** Sue Cunningham/Worldwide Picture Library/Alamy **4.15** Rob Crandall **4.17a** Rob Crandall **4.17b** Rob Crandall **4.17c** Rob Crandall **4.17d** Rob Crandall **4.18** Rob Crandall **4.2.1** Bruno Domingos/Reuters **4.2.2** VANDERLEI ALMEIDA/AFP/Getty Images **4.20** Holger Leue/age fotostock **4.21** ORLANDO KISSNER/AFP/Getty Images **4.22** Eric Gay/AP Images **4.23** Rob Crandall **4.24** Rob Crandall **4.25** Rob Crandall **4.27** Rob Crandall **4.28** Fernando Soutello/Reuters/Corbis **4.3.1** Bruno Domingos/Reuters **4.29a** Rob Crandall **4.29b** Christian Hebb/Aurora Photos, Inc. **4.31** Rob Crandall **4.4.1** Susana Gonzalez/Bloomberg/Getty Images **4.34** Eltana Aponte/Reuters/Bettmann/Corbis **4.35** Bob Daemmrich/Corbis **4.36** Rob Crandall **4.37** Paulo Fridman/Corbis **4.38** Rob Crandall **4.6.1** Rob Crandall **4.6.2** Rob Crandall **4.41** Alejandro Pagni/AFP/Getty Images

Chapter 5 opening photo Cameron Davidson/Corbis **5.2** Rob Crandall **5.3** Gideon Mendel/Corbis **5.5** Rob Crandall **5.6** Rob Crandall **5.7** European Space Agency **5.1.2** Hal Pierce/Laboratory of Atmospheres/NASA/Goddard Space Flight Center **5.9** Lucas Oleniuk/Zuma Press **5.11** Steve Bly/Alamy **5.13** Rob Crandall **5.14** Rob Crandall **5.2.1** AGE Fotostock/Robert Harding **5.15** Rob Crandall **5.2.2** Philip Wolmuth/Alamy **5.16** Michael Holford/Michael Holford Photographs **5.17** Steve Raymer/Corbis **5.19** Robert Caputo/Getty Images **5.20b** AFP/Getty Images **5.3.1** Rob Crandall **5.3.2** Paivi Varisanen/Association of Samba Schools Findland **5.22** Rob Crandall **5.23** Rob Crandall **5.4.2** Stock Connection Blue/Alamy **5.26** David Frazier/The Image Works **5.5.1** David Mercado/Reuters/Newscom **5.27** AP Images **5.29** Frank Heuer/Redux Pictures **5.31** Duane StevensAlamy **5.32** Rob Crandall **5.6.1** Angel Valentin/The Guardian

Chapter 6 opening photo Eric Nathan/Alamy **6.2** Dung Vo Trung/Sygma/Corbis **6.3** Rob Crandall **6.4** Buddy Mays/Corbis **6.5** Images of Africa Photobank/Alamy **6.7** Rob Crandall **6.8** Louis Gubb/The Image Works **6.10** Daniel Berehulak/Getty Images **6.1.1a** USGS Center for Earth Resources Observation and Science **6.1.1b** USGS Center for Earth Resources Observation and Science **6.11** Louise Gubb/SABA/Corbis **6.12** Roberto Schmidt/AFP/Getty Images **6.13** Rob Crandall **6.15** Nomi Baumgart/Photolibrary/Peter Arnold, Inc. **6.17** Jorgen Schytte/Photolibrary **6.19** Hereward Holland/Reuters **6.20** Ton Koene/Picture Contact BV/Alamy **6.21** Jeremy Horner/Corbis **6.22** James Marshall/Corbis **6.24** Rob Crandall **6.2.2** Alexander Joe/AFP/Getty Images **6.2.3** Hoberman Collection/Photolibrary **6.27** Rob Crandall **6.28** Ed Kashi/Corbis **6.30** Greenshoots Communication/Alamy **6.3.1** Rob Crandall **6.32** Frans Lemmens/Alamy **6.33** David Eulitt/Newscom **6.37** Englebert Photography Inc. **6.40** David Wall/Alamy **6.4.1** Des Willie/Alamy **6.42** David Lewis/Reuters **6.43** Marc & Evelyn Bernheim/Woodfin Camp & Associates **6.44** Tom Kirkwood/Reuters **6.5.1** Shao Jie/AP Images **6.46** PRNewsFoto/Marathon Oil Corporation/Newscom **6.6.1** John Hrusa/epa/Corbis **6.6.2** Anup Shah/Nature Picture Library **6.47** Washington Post Writers Group

Chapter 7 opening photo Motty Levy/Photolibrary **7.2** Rob Crandall **7.3** Minosa/Scorpio/Sygma Collection **7.4** Chris Hellier/Corbis **7.7** NASA **7.8** Ccharleson/Dreamstime.com **7.9** NASA **7.10** Alan Carey/Corbis **7.1.1** Patrick Baz/Pool/Getty Images **7.12** Buiten-Beeld/Alamy **7.13** Dott Giorgio Gualco/Photo Shot **7.14** Patrick Ben Luke/Lonely Planet Images **7.16** imagebroker/Alamy **7.18** Glen Allison/Getty Images **7.19** NASA **7.2.1** Imagestate Media Partners Limited-Impact Photos/Alamy **7.20** Monique Jacot/Woodfin Camp & Associates, Inc. **7.21** Stuart Franklin/Magnum

Photos 7.22 Dave Bartruff/Corbis 7.24 Michel Setboun/Corbis 7.25 Rob Crandall 7.28 Awad Awad/AFP/Getty Images 7.3.1 Rob Crandall 7.3.2 Rob Crandall 7.4.1 Walter Bibikow/Robert Harding 7.32 Desert Heat Management 7.34 Frans Lemmens/Robert Harding 7.37 Nir Elias/Reuters/Corbis 7.38 Kevin Unger-KPA/Zuma Press 7.42 Earth Satellite Corporation/Science Photo Library/Photo Researchers, Inc 7.43 Ricki Rosen/Corbis Saba Press Photos 7.5.1 Design Pics Inc./Alamy 7.44 Shawn Baldwin/New York Times/CORBIS 7.6.1 Gavin Hellier/Robert Harding 7.6.2 EPA/JORGE FERRARI/CORBIS 7.6.3 Rob Crandall 7.45 Francois Perri/Woodfin Camp & Associates, Inc.

Chapter 8 opening photo Pegaz/Alamy 8.2 Jerome Delay/AP Images 8.4 P.Vauthey/Corbis 8.1.1 Allison Kidder 8.5 Christope Boisvieux 8.7 Pascal Perret/Getty Images 8.9 Karol Agentur/Agentur Bilderburg 8.10 Filip Horvat 8.11 Martin Zwick/Photoshot 8.12 National Geographic Image Collection 8.4.1 BERTRAND GUAY/AFP/Getty Images 8.2.1 David Brauchili/AP Images 8.15 Les Rowntree 8.16 Les Rowntree 8.3.2 Les Rowntree 8.17 Les Rowntree 8.18 Les Rowntree 8.19 Les Rowntree 8.21 AP Images 8.23 Les Rowntree 8.24 Cathal McNaughton/Getty Images 8.25 Embassy of the Federal Republic of Germany 8.4.1 BERTRAND GUAY/AFP/Getty Images 8.26 Joanna B. Pinneo/Aurora 8.29 Jean Gaumy/Magnum Photos **8.30a** Corbis/Bettman **8.30b** Anthony Suau 8.5.2 Alfred Buellesbach/VISUM/Still Pictures/Specialist Stock 8.32 ZUMAPress-Gamma 8.33 James Mashall/Corbis 8.35 Agentur Bilderburg GmbH 8.37 Rob Crandall 8.6.1 Newscom 8.38 Ton Koene/Photolibrary/Peter Arnold, Inc.

Chapter 9 opening photo Ivan Vdovin/Alamy 9.2 Sisse Brimberg/National Geographic Image Collection 9.3 Scanpix/Photolibrary 9.4a Wolfgang Kaehler/Corbis 9.4b Dmitry Beliakov/Redux Pictures 9.6 Bettman/Corbis 9.7 Bodryaskin Valery/Landov Media 9.1.1 Konrad Wothe/Photolibrary 9.10 Art Directors & Trip Library 9.11 NASA 9.2.1 Bo Zaunders/CORBIS 9.2.2 Lilyana Vynogradova/Alamy 9.12 Mikhail Metzel/AP Images 9.15 Photo Researchers, Inc. 9.16 Jon Arnold ImagesDanitaDelimont.com 9.18 Ria Novosti/The Image Works 9.4.1 Iain Masterton/Alamy 9.19 Grigory Dukor/Reuters/Corbis 9.20 Ivan Vdovin/Alamy 9.22 Belinsky Yuri/ Landov Media 9.24 Hulton Archieve Photos/Getty Images 9.27 Dean Conger/Corbis 9.29 Iain Masterton/Alamy 9.30 AP Images 9.31 Sergei Ilnitsky/CORBIS 9.5.1 Artyom Korotayev/AP Images 9.5.2 Jim Wark/Photolibrary 9.34 Kurt Scholz/SuperStock 9.35 Gleb Garanich/Reuters 9.36 Sergei Chirikov/Corbis 9.37 Tom Schulze/Sovfoto Eastfoto 9.39 Alexander Chelmodeev/Shutterstock 9.42 Ivan Sekretarev/AP Images 9.43 Sovfoto/Eastfoto 9.6.1 Heidi Bradner/The Image Works 9.6.2 Sovfoto/Eastfoto

Chapter 10 opening photo Mikhail Metzel/AP Images 10.1.1a NASA 10.1.1b NASA 10.2.1 EPA/BENEDIKT VON IMHOF/CORBIS 10.2 Ogoniok Louchine/Corbis 10.4 Kazuyoshi Nomachi/Corbis 10.6 Jerry Kobalenko/Getty Images 10.9 David Turnley/Corbis 10.10 Aurora Photos/Alamy 10.3.1 Buddy Mays/Corbis 10.4.1 Gonzalo Azumendi/Robert Harding 10.4.2 GM Photo Images/Alamy 10.15 iStockphoto 10.16 Lemoyne/Getty Images 10.17 Tian Zhan/Shutterstock 10.18 Mathew Ashton/Alamy 10.21 Dimitri Borko/Getty Images 10.22 Ng Han Guan/AP Images 10.23 AP Images 10.5.1 Corbis 10.24 AP Images 10.26 John Spaull/Panos Pictures 10.6.1 Gavriel Vecan/OnAsia 10.27 Paula Bronstein/Getty 10.29 Dean Conger/Corbis

Chapter 11 opening photo Hu guolin/AP Images 11.3 Panarama Images/The Image Works 11.4 JTB Photo/Photolibrary 11.5 Bob Sacha/Corbis 11.6 CNS/Photolibrary 11.1.1 John Novis/OnAsia.com 11.7 Natalie Behring/OnAsia.com 11.10 George Mobley/NGS Image Collection 11.11 Robert Holmes/Corbis/Bettmann 11.13 Corbis 11.16 Corbis 11.17 Christopher Liu/China Stock Photo Library 11.18 Itsuo Inouye/AP Images 11.19 JPImages/Alamy 11.2.1 Yu Haibo/Photolibrary 11.22 Corbis 11.3.2 Hiroshi Harada/Dumo/Photo Researchers, Inc. 11.23 Hiroshi Harada/DUNQ/Photo Researchers, Inc. 11.24 Rob Crandall 11.26 Masa Uemura/Alamy 11.28 imagebroker/Alamy 11.4.1 Yuan Hongyan/ChinaFotoPress/Zuma 11.30 China Photos/Getty Images 11.33 Gordon Sinclair/Alamy 11.35 Yonhap/AP Image 11.36 Justin Guariglia/National Geographic/Getty 11.6.1 Julian Herbert/Getty Images 11.37 Rob Crandall 11.38 Rob Crandall 11.40 Jodi Cobb/NGS Image Collection 11.41 Jeon Heon-Kyun/epa/Corbis 11.42 Panorama Media (Beijing) Ltd/Alamy 11.44 China National Tourist Office

Chapter 12 opening photo Frans Lemmens/SuperStock 12.2 Tyler Hicks/The New York Times/Redux Pictures 12.4 Baldev/Corbis 12.5 Vivek Menon/Nature Picture Library 12.6 Earth Satellite Corporation/Science Photo Library/Photo Researchers,Inc. 12.1.1 Ann and Steve Toon/Alamy 12.9 Joerg Boethling/Still Pictures/Specialist Stock 12.10 Mark Edwards/Still Pictures/Specialist Stock 12.2.1 Elizabeth Chacko 12.13 Dereck Hall/Corbis 12.14 Earl Kowall/Corbis 12.15 Rob Crandall 12.3.2 Rob Crandall 12.16 dbtravels/Alamy 12.17 Peter/Georgina Bowater/Creative Eye/MIRA.com 12.18 Sunil Malhotra 12.20 Harish Tyagi/epa/Corbis 12.21 Lindsay Hebberd/Corbis 12.23 Chris Fredriksson/Alamy 12.4.1 WENN/Newscom 12.25 Earl & Nazima Kowall/Corbis 12.27 Tanushree Punwani/Landov Media 12.30 Bettmenn/Corbis 12.32 Giovanni Mereghetti/Makal/Alamy 12.5.1 Noshir Desai/Corbis 12.34 Danish Ismail/Reuters/Corbis/Bettmann 12.35 Jayanta Shaw/Reuters/CORBIS 12.36 Deshakalyan Chowdhury/AFP/Getty Images 12.38 MarionTipple/Amati Images 12.6.1 Ed Kashi/Corbis 12.39 Dan Vincent/Alamy 12.41 Rob Crandall

Chapter 13 opening photo Suthep Kritsanavarin/Onasia.com 13.2 Sukree Sukplang/Reuters 13.4 Barry Calhoun 13.1.1 The Image Works 13.5 Michael S. Yamashita/Corbis 13.6 Julia Rogers/Alamy 13.8 Vinai Dithajohn/OnAsia 13.9 Getty Images 13.11 Paula Bronstein/Getty Images 13.12 Peter Bowater/Alamy 13.13 Ed Wray/AP Images 13.15 George Steinmetz/Corbis 13.2.2 Jose Fuste Raga/Age Fotostock 13.16 Saeed Khan/AFP/Getty Images 13.19 Getty Images 13.20 Corbis 13.22 Topham/The Image Works 13.23 Andy Maluche Photography 13.26 Getty Images 13.27 Tetra Images/Alamy 13.3.1 Nacolas Lainez/OnAsia 13.28 ERIK DE CASTRO/Reuters/CORBIS 13.30 David VanderVeen/EPA/Corbis 13.4.1 Thierry Falise/OnAsia 13.31 China National Tourist Office 13.5.1 ADI WEDA/epa/Corbis 13.32 Richard Handley/Alamy 13.33 Jeffery Greenburg/The Image Works 13.34 Dr. Pradeep Kumar/ProPhotoz 13.35 Stephen Shaver/AFP/Getty Images 13.6.1 Suthep Kritsanavarin/OnAsia 13.37 Reuters/Corbis

Chapter 14 opening photo Andre Seale/Specialist Stock 14.2 Will Burgess/Reuters/Corbis 14.3 Rob Crandall 14.4 Rob Crandall 14.6 Fredrich Stark/Peter Arnold, Inc./Photolibrary 14.7 David Austen/Woodfin Camp & Associates, Inc. 14.1.1 Torsten Blackwood/Newscom 14.8 John Mitchell/Photo Researchers, Inc. 14.9 TORSTEN BLACKWOOD/AFP/Getty Images 14.10 Mula Eshet/Robert Harding 14.11 Alex Craig/Photolibrary 14.12 Hilarie Kavanagh/Getty Images 14.13 David Wall/Lonely Planet Images/Getty Images 14.15 AP Images 14.16 John Lamb/Getty Images 14.2.1 NASA 14.17 Chad Ehlers/Getty Images 14.20 Fritz Prenzel/Photolibrary/Peter Arnold, Inc. 4.3.1 Tim Wimbourne/Reuters/Corbis 14.22 Chris Rainier/Corbis 14.23 David Crausby/Alamy 14.24 Robert Frerck/Woodfin Camp & Associates, Inc. 14.4.2 Topham/The Image Works 14.25 Peter Essick/Aurora Photos 14.26 Arno Gasteiger/Bilderberg/Aurora 14.28 Ted Streshinsky/Corbis 14.29 Porterfield/Chickering/Photo Researchers 14.30 Frank Frames/Woodfin Camp & Associates, Inc. 14.5.1 Lucy Pemoni/Reuters 14.32 Michael Jensen/Auscape International Proprietary Ltd. 14.34 WITT/SIPA 14.35 Peter Hendrie/Getty Images 14.36 Remi Benali/CORBIS 14.37 Shutterstock 14.38 Torsten Blackwood/Getty Images 14.6.1 Roel Loopers/Photolibrary 14.6.2 Roel Loopers/Photolibrary 14.40 Tim Wimborne/Reuters

Index

Page numbers in italics indicate figures and tables.